Springer Monographs in Mathematics

Editors-in-Chief

Minhyong Kim, School of Mathematics, Korea Institute for Advanced Study, Seoul, South Korea
International Centre for Mathematical Sciences, Edinburgh, UK

Katrin Wendland, School of Mathematics, Trinity College Dublin, Dublin, Ireland

Series Editors

Sheldon Axler, Department of Mathematics, San Francisco State University, San Francisco, CA, USA

Maria Chudnovsky, Department of Mathematics, Princeton University, Princeton, NJ, USA

Tadahisa Funaki, Department of Mathematics, University of Tokyo, Tokyo, Japan

Isabelle Gallagher, Département de Mathématiques et Applications, Ecole Normale Supérieure, Paris, France

Sinan Güntürk, Courant Institute of Mathematical Sciences, New York University, New York, NY, USA

Claude Le Bris, CERMICS, Ecole des Ponts ParisTech, Marne la Vallée, France

Pascal Massart, Département de Mathématiques, Université de Paris-Sud, Orsay, France

Alberto A. Pinto, Department of Mathematics, University of Porto, Porto, Portugal

Gabriella Pinzari, Department of Mathematics, University of Padova, Padova, Italy

Ken Ribet, Department of Mathematics, University of California, Berkeley, CA, USA

René Schilling, Institute for Mathematical Stochastics, Technical University Dresden, Dresden, Germany

Panagiotis Souganidis, Department of Mathematics, University of Chicago, Chicago, IL, USA

Endre Süli, Mathematical Institute, University of Oxford, Oxford, UK

Shmuel Weinberger, Department of Mathematics, University of Chicago, Chicago, IL, USA

Boris Zilber, Mathematical Institute, University of Oxford, Oxford, UK

This series publishes advanced monographs giving well-written presentations of the "state-of-the-art" in fields of mathematical research that have acquired the maturity needed for such a treatment. They are sufficiently self-contained to be accessible to more than just the intimate specialists of the subject, and sufficiently comprehensive to remain valuable references for many years. Besides the current state of knowledge in its field, an SMM volume should ideally describe its relevance to and interaction with neighbouring fields of mathematics, and give pointers to future directions of research.

Piernicola Bettiol • Richard B. Vinter

Principles of Dynamic Optimization

 Springer

Piernicola Bettiol
Department of Mathematics
University of Brest
Brest, France

Richard B. Vinter
Electrical & Electronic Engineering
Imperial College London
London, UK

ISSN 1439-7382   ISSN 2196-9922 (electronic)
Springer Monographs in Mathematics
ISBN 978-3-031-50088-6   ISBN 978-3-031-50089-3 (eBook)
https://doi.org/10.1007/978-3-031-50089-3

Mathematics Subject Classification: 49-02, 49K15, 49K21

© The Editor(s) (if applicable) and The Author(s), under exclusive license to Springer Nature Switzerland AG 2024

This work is subject to copyright. All rights are solely and exclusively licensed by the Publisher, whether the whole or part of the material is concerned, specifically the rights of translation, reprinting, reuse of illustrations, recitation, broadcasting, reproduction on microfilms or in any other physical way, and transmission or information storage and retrieval, electronic adaptation, computer software, or by similar or dissimilar methodology now known or hereafter developed.

The use of general descriptive names, registered names, trademarks, service marks, etc. in this publication does not imply, even in the absence of a specific statement, that such names are exempt from the relevant protective laws and regulations and therefore free for general use.

The publisher, the authors, and the editors are safe to assume that the advice and information in this book are believed to be true and accurate at the date of publication. Neither the publisher nor the authors or the editors give a warranty, expressed or implied, with respect to the material contained herein or for any errors or omissions that may have been made. The publisher remains neutral with regard to jurisdictional claims in published maps and institutional affiliations.

This Springer imprint is published by the registered company Springer Nature Switzerland AG
The registered company address is: Gewerbestrasse 11, 6330 Cham, Switzerland

If disposing of this product, please recycle the paper.

To Francis Clarke, il miglior fabbro

Preface

What control strategy will transfer a space vehicle from one circular orbit to another in least time or, alternatively, with minimum fuel consumption? What should be the strategy for harvesting a renewable resource (a fish population, say) to maximize financial returns while satisfying sustainability constraints? In chemo-immunotherapy for cancer, what treatment regime (concentration and frequency of cytotoxic drug doses) will minimize the tumour cell population while maintaining the blood cell population above a critical level? Mitigation strategies are available to counter an epidemic, including vaccination, livestock culling and host removal; how should we deploy these strategies while minimizing the social and economic costs involved? How should a batch distillation column be operated to maximize the yield, subject to specified constraints on product purity?

There are a number of common features in these questions. First, they all concern phenomena where the relevant 'state of nature' (relating, for example, to the position of a space vehicle or the size of a diseased population) is dynamic, in the sense that it evolves with time. Second, the evolution of the state of nature, or state as we shall simply call it, is affected by the choice of a control strategy. Third, we can attach a cost to a control strategy and the evolving state to which it gives rise. The underlying problem is to choose a control strategy that minimizes the cost.

In certain cases, problems in the classical calculus of variations ('minimization of an integral functional over arcs and their derivatives') match this description. Here, the independent variable is interpreted as time, the 'state' is the value of the arc at the current time and the control its rate of change. But techniques for their solution provided by this earlier theory fail to take account of the dynamic constraints that are so often encountered today, in engineering, applied science and economics. Here, by 'dynamic constraints' we mean the mathematical relations governing future evolution of the state, which will depend on the control strategy.

Dynamic optimization is the name given to the systematic study of optimization problems with dynamic constraints. General study of optimization problems with dynamic constraints dates from the late 1950s, which saw several crucial advances, one conceptual and two technical. The conceptual advance, due by L. S. Pontryagin et al., was the realization that optimization problems where the dynamic constraint took the form of a controlled differential equation covered a wide range of engineering control problems involving mechanical systems such as space vehicles and, furthermore, was amenable to analysis. As for the two technical advances, one was Pontryagin's maximum principle, a set of necessary conditions for a control strategy to be optimal. The other was dynamic programming, a procedure initiated by R. Bellman, which reduces the search for an optimal strategy to finding the solution to a partial differential equation (the Hamilton Jacobi equation).

'Dynamic optimization' is synonymous with 'optimal control'. We have chosen the nomenclature dynamic optimization in this book, to convey the idea that optimization problems with general dynamic constraints merit study in their own right and that the field has widespread application, within and beyond engineering. We seek then to avoid the specificity of 'optimal control', a name introduced to describe a branch of control engineering, in which the control design objectives are expressed in terms minimizing a cost, rather than, say, in terms of stability and robustness requirements.

From the mid-1970s, it became apparent that progress in the study of dynamic optimization problems was being impeded by a lack of suitable analytic tools for investigating local properties of functions which are nonsmooth, i.e. not differentiable in the traditional sense. Nonsmooth functions were encountered at first attempts to put Dynamic Programming on a rigorous footing, specifically attempts to relate value functions and solutions to the Hamilton Jacobi equation. It was found that, for many dynamic optimization problems of interest, the only 'solutions' to the Hamilton Jacobi equation have discontinuous derivatives. How should we interpret these solutions? New ideas were required to answer this question since the Hamilton Jacobi equation of dynamic optimization is a nonlinear partial differential equation for which traditional interpretations of generalized solutions, based on the distributions they define, are inadequate.

Nonsmooth functions surfaced once again when efforts were made to extend the applicability of necessary conditions such as the maximum principle. A notable feature of the maximum principle (and one which distinguishes it from necessary conditions derivable using classical techniques) is that it can take account of pathwise constraints on values of the control functions. For some practical problems, the constraints on values of the control depend on the vector state variable. In flight mechanics, for example, the maximum and minimum thrust of a jet engine (a control variable) will depend on the altitude (a component of the state vector). The maximum principle in its original form is not, in general, valid for problems involving state-dependent control constraints. One way to derive necessary conditions for these problems, and others not covered by the maximum principle, is to reformulate them as generalized problems in the calculus of variations, the cost integrands for which include penalty terms to take account of the constraints. The reformulation comes at a price, however. To ensure equivalence with the original problems, it is necessary to employ penalty terms with discontinuous derivatives. So the route to necessary conditions via generalized problems in the calculus of variations can be followed only if we know how to adapt traditional necessary conditions to allow for nonsmooth cost integrands.

Two important breakthroughs occurred in the 1970s. One was the end product of a long quest for effective, local descriptions of 'non-smooth' functions, based on generalizations of the concept of 'subdifferentials' of convex functions, to larger function classes. F. H. Clarke's theory of generalized gradients, by achieving this goal, launched the field of nonsmooth analysis and provided a bridge to necessary conditions of optimality for nonsmooth variational problems (and in particular dynamic optimization problems reformulated as generalized problems in the calculus of variations). The other breakthrough, a somewhat later development, was the concept of viscosity solutions, due to M. G. Crandall and P.-L. Lions, which provides a framework for proving existence and uniqueness of generalized solutions to Hamilton Jacobi equations arising in dynamic optimization.

Nonsmooth analysis and viscosity methods were introduced to overcome obstacles in dynamic optimization. But they have come to have a significant impact on nonlinear analysis as a whole. Nonsmooth analysis provides an important new perspective: useful properties of functions, even differentiable functions, can be proved by examining related nondifferentiable functions, in the same way that trigonometric identities relating to real numbers can sometimes simply be derived by a temporary excursion into the field of complex numbers. Viscosity methods, on the other hand, provide a fruitful approach to studying generalized solutions to broad classes of nonlinear partial differential equations which extend beyond Hamilton Jacobi equations of dynamic optimization and their approximation for computational purposes. The calculus of variations (in its modern guise as dynamic optimization) continues to uphold a long tradition then, as a stimulus to research in other areas of mathematics.

The main purpose of this book is to bring together as a single comprehensive, up-to-date publication major advances in the theory dynamic optimization, with emphasis on those accomplished through the use of nonsmooth analytical techniques. Necessary conditions receive special attention. But other topics are covered as well. Material on the important topic of minimizer regularity provides a showcase for the application of nonsmooth necessary conditions to derive qualitative information about solutions to variational problems. The chapter on dynamic programming stands a little apart from other sections of the book, as it is complementary to mainstream research in the area based on viscosity methods (and which in any case is the subject matter of a number of substantial expository texts). Instead we concentrate on aspects of dynamic programming well matched to the analytic techniques of this book, notably the characterization (in terms of the Hamilton Jacobi equation) of extended-valued value functions associated with problems having endpoint and state constraints, inverse verification theorems, sensitivity relationships and links with the maximum principle.

A subsidiary purpose is to meet the needs of readers with little prior exposure to modern dynamic optimization who seek quick answers to the questions: what are the main results, what were the deficiencies of the 'classical' theory and to what extent have they been overcome? Chapter 1 provides, for their benefit, a lengthy overview, in which analytical details are suppressed and the emphasis is placed instead on communicating the underlying ideas.

To render this book self-contained, preparatory chapters are included on nonsmooth analysis, measurable multifunctions and differential inclusions. Much of this material is implicit in the books of R. T. Rockafellar and J. B. Wets [177] and Clarke et al. [85], and of J.-P. Aubin and H. Frankowska [14]. It is expected, however, that readers, whose main interest is in optimization rather than in broader application areas of nonsmooth analysis which require additional techniques, will find these chapters helpful, because of the strong focus on topics relevant to optimization.

Dynamic optimization is a large field and the choice of material for this is necessary selective. The techniques used here to derive necessary conditions of optimality are, for the most part, within a tradition of research pioneered and developed by Clarke, Ioffe, Loewen, Mordukhovich, Rockafellar, Vinter and others, based on perturbation, elimination of constraints and passage to the limit. The necessary conditions are 'state of the art', as far as this tradition is concerned. Alternative approaches, based on set separation ideas, also make an appearance, but principally for comparison purposes and historical perspective. We do not enter into the topic of higher order necessary conditions nor computational aspects of dynamic optimization.

This book is similar in structure and content to the 2000 book *Optimal Control* [194]. It brings up to date this earlier publication by, in many instances, providing new, simpler proofs of key theorems, where these have become available, and by broadening the applicability of the theory. We provide, for the first time in book form, recent improvements to necessary conditions of optimality for problems for dynamic optimization problems involving a differential inclusion constraint, referred to as the Ioffe refinement. It draws on recent research developments, unavailable at the time of the earlier publication, to provide a thorough discussion and analysis of necessary conditions in the form of Clarke's Hamiltonian inclusion. The book includes new material on necessary conditions for problems with mixed state/control constraints and on problems with free end-times, drawing on latest research in these areas. Also included is a new framework for dynamic programming treating dynamic constraints with discontinuous time dependence.

Brest, France   Piernicola Bettiol
London, UK   Richard B. Vinter

Acknowledgements

We wish to thank our colleagues at Imperial College, including Martin Clarke, David Angeli and Alessandro Astolfi, and at University of Brest, including Rainer Buckdahn, Chloé Jimenez, Vuk Milisic, Marc Quincampoix and Miloud Sadkane for creating working environments in which writing this book has been possible. Many people have influenced our thinking on the contents and presentation of this book, which is indeed the product of many years of collaboration with colleagues and graduate students in the control systems and applied analysis communities. We make special mention of the insights we have gained from Francis Clarke, whose pioneering work in nonsmooth analysis, perturbation methods and applications in nonsmooth optimization laid the foundation for the work reported here. An incomplete list of our collaborators is Zvi Artstein, Aram Arutyunov, Julien Bernis, Andrea Boccia, Frédéric Bonnans, Bernard Bonnard, Alberto Bressan, Arrigo Cellina, Giovanni Colombo, Maria do Rosário de Pinho, Asen Dontchev, Paola Falugi, Margarida Ferreira, Fernando Fontes, Helene Frankowska, Alexander Ioffe, Nathalie Khalil, Yuri Ledyaev, Philip Loewen, Carlo Mariconda, Helmut Maurer, Monica Motta, Michele Palladino, Fernando Lobo Pereiera, Franco Rampazzo, Alain Rapaport, Ralph Tyrrell Rockafellar, Javier Rosenblueth, Jérémy Rouot, Geraldo Silva, Peter Wolenski, and Harry Zheng.

Finally, and most importantly,

'I express my profound gratitude to my wife Giorgia who has provided unwavering and irreplaceable support in my projects, including this book. She and my wonderful children (Eloïse, Gabriele and Leonardo) with their presence, enthusiasm and understanding made it possible for me to dedicate time and effort to bring this book to fruition. I would like to extend my heartfelt appreciation and recognition to Richard; working alongside him has been an enlightening and inspiring experience.' (Piernicola)

'My wife, Donna, and children, Magdalena, Becky and Hannah, have given me unconditional support and encouragement in everything I have wanted to do. This book was no exception. Writing a book like this takes time; I wish to express my deepest thanks to them, especially to Donna for giving me that time, generously and with such good grace. I add my thanks to Piernicola, both for his friendship and for his indispensable part in our fruitful mathematical collaboration.' (Richard)

Contents

1	**Overview**		1
	1.1	Dynamic Optimization	2
	1.2	The Calculus of Variations	9
	1.3	Existence of Minimizers and Tonelli's Direct Method	22
	1.4	Sufficient Conditions and the Hamilton Jacobi Equation	25
	1.5	The Maximum Principle	30
	1.6	Dynamic Programming	36
	1.7	Nonsmoothness	41
	1.8	Nonsmooth Analysis	45
	1.9	Nonsmooth Dynamic Optimization	59
	1.10	Epilogue	62
	1.11	Appendix: Proof of the Classical Maximum Principle	66
	1.12	Exercises	86
	1.13	Notes for Chapter 1	89
2	**Set Convergence, Measurability and Existence of Minimizers**		91
	2.1	Introduction	91
	2.2	Convergence of Sets and Continuity of Multifunctions	92
	2.3	Measurable Multifunctions	95
	2.4	The Generalized Bolza Problem	107
	2.5	Exercises	115
	2.6	Notes for Chapter 2	116
3	**Variational Principles**		119
	3.1	Introduction	120
	3.2	Exact Penalization	121
	3.3	Ekeland's Theorem	123
	3.4	Quadratic Inf Convolution	127
	3.5	Variational Principles with Smooth Perturbation Terms	132
	3.6	Mini-Max Theorems	134
	3.7	Exercises	145
	3.8	Notes for Chapter 3	146

4 Nonsmooth Analysis ... 149
- 4.1 Introduction ... 150
- 4.2 Normal Cones ... 151
- 4.3 Subdifferentials ... 156
- 4.4 Difference Quotient Representations ... 162
- 4.5 Nonsmooth Mean Value Inequalities ... 167
- 4.6 Characterization of Limiting Subgradients ... 173
- 4.7 Subgradients of Lipschitz Continuous Functions ... 177
- 4.8 The Distance Function ... 184
- 4.9 Criteria for Lipschitz Continuity ... 190
- 4.10 Relations Between Normal and Tangent Cones ... 194
- 4.11 Interior of Clarke's Tangent Cone ... 201
- 4.12 Appendix: Proximal Analysis in Hilbert Space ... 203
- 4.13 Exercises ... 213
- 4.14 Notes for Chapters 4 and 5 ... 214

5 Subdifferential Calculus ... 217
- 5.1 Introduction ... 217
- 5.2 A Marginal Function Principle ... 220
- 5.3 Partial Limiting Subgradients ... 224
- 5.4 A Sum Rule ... 226
- 5.5 A Nonsmooth Chain Rule ... 230
- 5.6 Lagrange Multiplier Rules ... 232
- 5.7 Max Rule for an Infinite Family of Functions ... 236
- 5.8 Exercises ... 238
- 5.9 Notes for Chapter 5 ... 240

6 Differential Inclusions ... 241
- 6.1 Introduction ... 242
- 6.2 Existence and Estimation of F Trajectories ... 243
- 6.3 Perturbed Differential Inclusions ... 256
- 6.4 Existence of Minimizing F Trajectories ... 261
- 6.5 Relaxation ... 264
- 6.6 Estimates on Trajectories Confined to a Closed Subset ... 271
- 6.7 Exercises ... 292
- 6.8 Notes for Chapter 6 ... 293

7 The Maximum Principle ... 295
- 7.1 Introduction ... 295
- 7.2 Clarke's Nonsmooth Maximum Principle ... 297
- 7.3 A Preliminary Maximum Principle, for Dynamic Optimization Problems with no End-Point Constraints ... 303
- 7.4 Proof of Theorem 7.2.1 ... 319
- 7.5 Exercises ... 323
- 7.6 Notes for Chapter 7 ... 327

8 The Generalized Euler-Lagrange and Hamiltonian Inclusion Conditions ... 333
- 8.1 Introduction ... 334
- 8.2 Pseudo Lipschitz Continuity ... 338
- 8.3 Unbounded Differential Inclusions ... 341
- 8.4 The Generalized Euler Lagrange Condition ... 342
- 8.5 Special Cases ... 347
- 8.6 Proof of Theorem 8.4.3 ... 350
- 8.7 The Hamiltonian Inclusion for Convex Velocity Sets ... 377
- 8.8 The Hamiltonian Inclusion for Non-convex Velocity Sets ... 381
- 8.9 Discussion and a Counter-Example ... 386
- 8.10 Appendix: Dualization of the Euler Lagrange Inclusion ... 391
- 8.11 Exercises ... 404
- 8.12 Notes for Chapter 8 ... 406

9 Free End-Time Problems ... 411
- 9.1 Introduction ... 412
- 9.2 Lipschitz Time Dependence ... 415
- 9.3 Essential Values ... 424
- 9.4 Measurable Time Dependence ... 427
- 9.5 Proof of Theorem 9.4.1 ... 430
- 9.6 A Free End-Time Maximum Principle ... 448
- 9.7 Appendix: Metrics on the Space of Free End-Time Trajectories ... 463
- 9.8 Exercises ... 466
- 9.9 Notes for Chapter 9 ... 468

10 The Maximum Principle for Problems with Pathwise Constraints ... 471
- 10.1 Introduction ... 472
- 10.2 Problems with Pure State Constraints: Preliminary Discussion ... 474
- 10.3 Convergence of Measures ... 477
- 10.4 The Maximum Principle (Pure State Constraints) ... 482
- 10.5 Proof of Theorem 10.4.1 ... 485
- 10.6 Maximum Principles for Free End-Time Problems with State Constraints ... 498
- 10.7 Non-degenerate Conditions ... 503
- 10.8 Mixed Constraints ... 521
- 10.9 Exercises ... 537
- 10.10 Notes for Chapter 10 ... 543

11 The Euler-Lagrange and Hamiltonian Inclusion Conditions in the Presence of State Constraints ... 547
- 11.1 Introduction ... 548
- 11.2 The Euler Lagrange Inclusion ... 549
- 11.3 Proof of Theorem 11.2.1 ... 552
- 11.4 Free End-Time Problems with State Constraints ... 567
- 11.5 Non-degenerate Necessary Conditions ... 574

	11.6	Exercises	588
	11.7	Notes for Chapter 11	589
12	**Regularity of Minimizers**		591
	12.1	Introduction	592
	12.2	Tonelli Regularity	599
	12.3	Proof of The Generalized Tonelli Regularity Theorem	604
	12.4	Lipschitz Continuous Minimizers	613
	12.5	Nonautonomous Variational Problems with State Constraints	619
	12.6	Bounded Controls	630
	12.7	Lipschitz Continuous Controls	633
	12.8	Exercises	638
	12.9	Notes for Chapter 12	640
13	**Dynamic Programming**		643
	13.1	Introduction	644
	13.2	Invariance Theorems	654
	13.3	The Value Function and Generalized Solutions of the Hamilton Jacobi Equation	667
	13.4	Local Verification Theorems	685
	13.5	Costate Trajectories and Gradients of the Value Function	697
	13.6	State Constrained Problems	708
	13.7	Proofs of Theorems 13.6.1 and 13.6.2	711
	13.8	Costate Trajectories and Gradients of the Value Functions for State-Constrained Problems	719
	13.9	Semiconcavity and the Value Function	723
	13.10	The Infinite Horizon Problem	728
	13.11	The Minimum Time Problem	735
	13.12	Viscosity Solutions of the Hamilton Jacobi Equation	741
	13.13	A Comparison Theorem for Viscosity Solutions	754
	13.14	Exercises	759
	13.15	Notes for Chapter 13	762
References			769
Index			779

Notation

$\mathbb{B}$	Closed unit ball in Euclidean space
$\lvert x \rvert$	Eulidean norm of x
$a \wedge b$	$\min\{a, b\}$
$a \vee b$	$\max\{a, b\}$
$\mathbb{R}_+$	Non-negative real numbers
$d_C(x)$	Euclidean distance of x from C
$\overset{\circ}{C}$, int C	Interior of C
∂C, bdy C	Boundary of C
$\bar{C}$	Closure of C
co D	Convex hull of D
$\overline{\text{co}}\, D$	Closure of the convex hull of D
$N_C^P(x)$	Proximal normal cone to C at x
$\hat{N}_C(x)$	Strict normal cone to C at x
$N_C(x)$	Limiting normal cone to C at x
$T_C(x)$	Bouligand tangent cone to C at x
$\bar{T}_C(x)$	Clarke tangent cone to C at x
$\partial^P f(x)$	Proximal subdifferential of f at x
$\hat{\partial} f(x)$	Strict subdifferential of f at x
$\partial f(x)$	Limiting subdifferential of f at x
$\partial_P^\infty f(x)$	Asymptotic proximal subdifferential of f at x
$\hat{\partial}^\infty f(x)$	Asymptotic strict subdifferential of f at x
$\partial^\infty f(x)$	Asymptotic limiting subdifferential of f at x
dom f	(Effective) domain of f
Gr F	Graph of F
epi f	Epigraph of f

$f^0(x; v)$	Generalized directional derivative of f at x in the direction v
$\Psi_C(x)$	Indicator function of the set C at the point x
$\nabla f(x)$	Gradient vector of f at x
$x_i \xrightarrow{C} x$	$x_i \to x$ and $x_i \in C \quad \forall i$
$x_i \xrightarrow{f} x$	$x_i \to x$ and $f(x_i) \to f(x)$
supp μ	Support of the measure μ
$\mathcal{L}$	Lebesgue subsets of $I \subset \mathbb{R}$
$\mathcal{B}^k$	Borel subsets of $\mathbb{R}^k$
$L^1(I; \mathbb{R}^n)$	Integrable functions $f : I \to \mathbb{R}^n$
$L^1(S, T)$	Integrable functions $f : [S, T] \to \mathbb{R}$
$W^{1,1}(I; \mathbb{R}^n)$	Absolutely continuous functions $f : I \to \mathbb{R}^n$
$NBV([S, T]; \mathbb{R}^n)$	Functions of bounded variation $f : [S, T] \to \mathbb{R}^n$, right continuous on (S, T)
$H, \mathcal{H}$	Hamiltonian, un-maximized Hamiltonian

Chapter 1
Overview

Abstract Dynamic optimization emerged as a distinct field of research in the late 1950's, to address new kinds of optimization problems, in aerospace, economics and other areas. The distinctive feature of these problems was an underlying dynamic constraint, typically in the form of a controlled differential equation, which placed these problems beyond the scope of earlier variational techniques. Rapid advances were made in the 1970's and 80's, with the discovery of the maximum principle and methodologies (dynamic programming) that linked optimal strategies and the Hamilton Jacobi equation. These were the main elements in what, today, is known as the classical theory of dynamic optimization. While classical dynamic optimization was adequate for many applications, deficiencies became apparent, leading to a new body of theory in the 1980's, including Clarke's nonsmooth maximum principle and generalized solutions of Hamilton-Jacobi equations, based on techniques of nonsmooth analysis.

The purpose of this overview chapter is twofold. First, it provides a self-contained exposition of the classical theory suitable for a first course in dynamic optimization (at undergraduate or graduate level). It includes motivating examples, a derivation of the classical maximum principle, optimality conditions of dynamic programming type expressed in terms of solutions to the Hamilton Jacobi equation, and extensive discussion. Second, it gives answers to the questions: what are the shortcomings of the classical theory and how are they surmounted by more recent developments? We argue that many of the deficiencies of the earlier theory arise from the lack of appropriate analytic techniques for constructing useful local approximations of non-differentiable functions and closed sets with irregular boundaries. We then cover rudiments of nonsmooth analysis, which was developed precisely for this purpose, and show how we can use it to derive new, improved optimality conditions, unshackled by the restrictive hypotheses of classical dynamic optimization. We thereby offer readers the 'big picture' in preparation for later chapters, and also to equip them better to understand the contemporary literature.

1.1 Dynamic Optimization

Dynamic optimization emerged as a distinct field of research in the 1950's, to address in a unified fashion optimization problems arising in scheduling and the control of engineering devices, beyond the reach of earlier analytical and computational techniques. This field was initially called optimal control, but this earlier name is increasingly giving way to dynamic optimization, to convey a wider range of potential application, beyond control engineering. Aerospace engineering is an important source of such problems, and the relevance of dynamic optimization to the American and Russian space programmes gave powerful initial impetus to research in this area. A simple example is:

The Maximum Orbit Transfer Problem A rocket vehicle is in a circular orbit. What is the radius of the largest possible co-planar orbit to which it can be transferred over a fixed period of time? The motion of the vehicle during the manoeuvre is governed by the rocket thrust and by the rocket thrust orientation, both of which can vary with time. See the Fig. 1.1. The variables involved are

$$
\begin{aligned}
r &= \text{radial distance of vehicle from attracting centre,} \\
u &= \text{radial component of velocity,} \\
v &= \text{tangential component of velocity,} \\
m &= \text{mass of vehicle,} \\
T_r &= \text{radial component of thrust,} \\
T_t &= \text{tangential component of thrust.}
\end{aligned}
$$

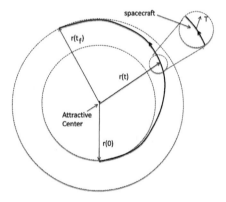

Fig. 1.1 The maximum orbit transfer problem

1.1 Dynamic Optimization

The constants are

$$
\begin{aligned}
r_0 &= \text{initial radial distance,} \\
m_0 &= \text{initial mass of vehicle,} \\
\gamma_{\max} &= \text{maximum fuel consumption rate,} \\
T_{\max} &= \text{maximum thrust,} \\
\mu &= \text{gravitational constant of attracting centre,} \\
t_f &= \text{duration of manoeuvre.}
\end{aligned}
$$

A precise formulation of the problem, based on an idealized point mass model of the space vehicle, is as follows:

$$
\begin{cases}
\text{Minimize } -r(t_f) \\
\text{over radial and tangential components of the thrust history,} \\
\quad (T_r(t), T_t(t)),\ 0 \le t \le t_f,\ \text{satisfying} \\
\dot{r}(t) = u, \\
\dot{u}(t) = v^2(t)/r(t) - \mu/r^2(t) + T_r(t)/m(t), \\
\dot{v}(t) = -u(t)v(t)/r(t) + T_t(t)/m(t), \\
\dot{m}(t) = -(\gamma_{\max}/T_{\max})(T_r^2(t) + T_t^2(t))^{1/2}, \\
(T_r^2(t) + T_t^2(t))^{1/2} \le T_{\max}, \\
m(0) = m_0,\ r(0) = r_0,\ u(0) = 0,\ v(0) = \sqrt{\mu/r_0}, \\
u(t_f) = 0,\ v(t_f) = \sqrt{\mu/r(t_f)}.
\end{cases}
$$

Here $\dot{r}(t)$ denotes $dr(t)/dt$, etc. It is standard practice in dynamic optimization to formulate optimization problems as minimization problems. Accordingly, the problem of maximizing the radius of the terminal orbit $r(t_f)$ is replaced by the equivalent problem of minimizing the 'cost' $-r(t_f)$. Notice that knowledge of the *control function* or *strategy* $(T_r(t), T_t(t))$, $0 \le t \le t_f$ permits us to calculate the cost $-r(t_f)$: we solve the differential equations, for the specified boundary conditions at time $t = 0$, to obtain the corresponding *state trajectory* $(r(t), u(t), v(t), m(t))$, $0 \le t \le t_f$, and thence determine $-r(t_f)$. The control strategy therefore has the role of choice variable in the optimization problem. We seek a control strategy which minimizes the cost, from among the control strategies whose associated state trajectories satisfy the specified boundary conditions at time $t = t_f$.

For the following values of relevant dimensionless parameters:

$$\frac{T_{\max}/m_0}{\mu/r_0^2} = 0.1405,\quad \frac{\gamma_{\max}}{T_{\max}/\sqrt{\mu/r_0}} = 0.07487,\quad \frac{t_f}{\sqrt{r^3/\mu}} = 3.32,$$

the radius of the terminal circular orbit is

$$r(t_f) = 1.5\, r_0.$$

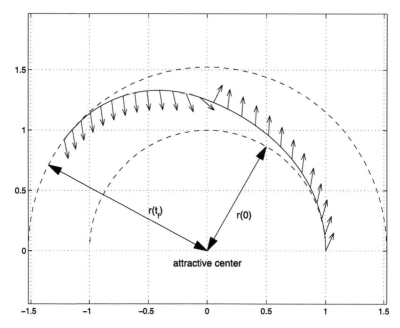

Fig. 1.2 A control strategy for the maximum orbit transfer problem

In Fig. 1.2, the arrows indicate the magnitude and orientation of the thrust at times $t = 0, 0.1t_f, 0.2t_f, \ldots, t_f$. As indicated, full thrust is maintained. The thrust is outward for (approximately) the first half of the manoeuvre and inward for the second.

Suppose, for example, that the attracting centre is the Sun, the space vehicle weighs 10,000 lb, the initial radius is 1.50 million miles (the radius of a circle approximating the Earth's orbit), the maximum thrust is 0.85 lb (i.e. a force equivalent to the gravitational force on a 0.85 lb mass on the surface of the earth, which corresponds to $T_{\max} = 3.778$ N, the maximum rate of fuel consumption is 1.81 lb/day and the transit time is 193 days. Corresponding values of the constants are

$$T_{\max} = 3.778 \text{ N}, \quad m_0 = 4.536 \times 10^3 \text{ kg},$$
$$r_0 = 1.496 \times 10^{11} \text{ m}, \quad \gamma_{\max} = 0.9496 \times 10^{-5} \text{ kg s}^{-1},$$
$$t_f = 1.6675 \times 10^7 \text{ s}, \quad \mu = 1.32733 \times 10^{20} \text{ m}^3 \text{ s}^{-2}.$$

Then the terminal radius of the orbit is 2.44 million miles. (This is the radius of a circle approximating the orbit of the planet Mars.)

Numerical methods, inspired by necessary conditions of optimality akin to the maximum principle of Chap. 7, were used to generate the above control strategy.

1.1 Dynamic Optimization

Optimal Control of a Growth/Consumption Model Dynamic optimization problems are encountered also in the field of economics. One example is the 'growth versus consumption' problem of neoclassical macro-economics, based on the Ramsey model of economic growth. The question here is, what balance should be struck between investment and consumption to maximize overall spending on social programmes over a fixed period time? A simple formulation of the problem is as follows.

$$\begin{cases} \text{Minimize } -\int_0^T (1-u(t))x^\alpha(t)dt \\ \text{subject to} \\ \dot{x}(t) = -ax(t) + bx^\alpha(t)u(t) \quad \text{for a.e. } t \in [0,T], \\ u(t) \in [0,1] \quad \text{for a.e. } t \in [0,T], \\ x(t) \geq 0 \quad \text{for all } t \in [0,T], \\ x(0) = x_0 \,. \end{cases}$$

Here, $a > 0$, $b > 0$, $x_0 \geq 0$ and $\alpha \in (0,1)$ are given constants and $[0,T]$ is a given interval.

It has the following interpretation: x denotes global economic output. The rate of financial return $r(x)$ from economic output x is modelled as

$$r(x) = bx^\alpha \,.$$

The term $-ax$ takes account of fixed costs reducing growth (wages, etc.).

To describe the solution to this problem, we introduce the constants

$$\hat{x} := \left(\frac{\alpha b}{a}\right)^{\frac{1}{1-\alpha}} \quad \text{and } \Delta := \frac{1}{a\alpha} \ln\left(\frac{1}{1-\alpha}\right)$$

and also the state feedback function $\chi : [0,T] \times (0,\infty) \to [0,1]$:

$$\chi(t,x) := \begin{cases} 0 & \text{if } x > \bar{y}(t) \\ 1 & \text{if } x < \bar{y}(t) \\ \alpha & \text{if } x = \bar{y}(t) \text{ and } t \leq T - \Delta \\ 0 & \text{if } x = \bar{y}(t) \text{ and } t > T - \Delta \,, \end{cases}$$

in which $\bar{y} : (-\infty, T] \to (0, \infty)$ is the function

$$\bar{y}(t) := \begin{cases} \hat{x} & \text{if } t \leq T - \Delta \\ \left[\frac{b}{a}(1 - e^{-a\alpha(T-t)})\right]^{\frac{1}{1-\alpha}} & \text{if } t > T - \Delta \,. \end{cases}$$

Techniques of dynamic programming covered in Chap. 13 provide the following solution to this problem:

Given arbitrary initial data $(t_0, x_0) \in [0,T] \times (0, \infty)$, the optimal output x^* is the unique solution in the space of Lipschitz continuous functions on $[t_0, T]$ of the differential equation

$$\begin{cases} \dot{x}^*(t) = -ax^*(t) + bx^{*\alpha}(t)\chi(t, x^*(t)) & \text{a.e. } t \in [t_0, T], \\ x(t_0) = x_0. \end{cases} \quad (1.1.1)$$

The optimal proportion of financial return for investment u^* is unique (w.r.t. the equivalence class of almost everywhere equal functions) and is given by

$$u^*(t) = \chi(t, x^*(t)), \text{ for a.e. } t \in [t_0, T].$$

Notice that the solution above is expressed in *state feedback* form; that is, the optimal control u^* is expressed as a function of the current state. For any given initial state and time t_0, the optimal state expressed as a function of time, i.e. in *open loop form*, is the solution to the 'closed loop' state equation (1.1.1) for the given initial state and time t_0. We then obtain the optimal control as a function of time (open loop form) by plugging the optimal state trajectory into the state feedback function. Notice that the feedback form captures, within a single relation, the optimal strategies for every initial state x_0 and time t_0.

Intuition would suggest that if, at the start of the time interval, economic output is low, the optimal control should have a first phase of maximum investment during which economic output builds up to some critical value, followed by a second phase of intermediate investment over which economic output is maintained and, finally, a third phase over which there is no investment because the remaining time is too small for the benefits of investment to show through. This is indeed the optimal control, with the qualification that, if the initial output is high, the optimal control is pure consumption in the first phase. There are also values of the initial investment and T such that there is no first phase or no first and second phase. Analysis is required, of course, to determine precisely the times separating the phases, the critical value of output and the proportion of financial return for investment required to maintain it; also to identify the situations when there are fewer than three phases. Optimal state trajectories, for various choices of initial data, are illustrated in Fig. 1.3.

Optimal Control in Anti-Cancer Treatment

We illustrate applications of dynamic optimization in medicine. Chemotherapy is a treatment aimed at destroying cancer cells by means of a cocktail of drugs,

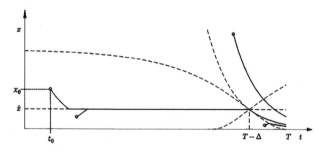

Fig. 1.3 Optimal trajectories for the consumption/growth problem

1.1 Dynamic Optimization

administered either at specific times or continuously. It is typically part of a complex overarching treatment plan, in which chemotherapy is following up by procedures, surgical or drug-based, for inhibiting renewed tumour growth.

Traditionally, chemotherapy treatments have been based on the maximum tolerated dose paradigm. But a side effect of chemotherapy, a 'two-edged sword', is damage to normal cells. Modern day treatments aim to improve outcomes by balancing destruction of cancer cells and suppression of side effects. Empirical design of treatment plans based on clinical trials is time consuming and extremely expensive. Mathematical models of the underlying pharmaco-dynamic processes involved have an important role, because they can be used to simulate on the computer the effects of different treatment strategies, simply and at low cost. Dynamic optimization is the appropriate tool for designing optimal treatment strategies based on these models.

The following formulation of treatment planning as a dynamic optimization problem is taken from [188]. The underlying dynamic model involves the time-varying state variable components c_1, c_2, n and w and the control variable u:

c_1 = concentration of administered anti-cancer drug in plasma,
c_2 = active drug concentration at the tumour cellular level,
n = number of tumour cells,
w = number of white blood cells (WBCs),
u = drug dosage.

The evolution of the state variable components for some control strategy $u(t)$, $0 \leq t \leq T$, is governed by the differential equations over the fixed time interval $[0, T]$

$$\left.\begin{aligned} \dot{c}_1(t) &= -(k_1 + k_2)c_1(t) + \left(\frac{1}{V_1}\right)u(t) \\ \dot{c}_2(t) &= k_{12}\left(\frac{V_1}{V_2}\right)c_1(t) - k_2 c_2(t) \\ \dot{n}(t) &= \Lambda\psi(n(t)) - K\max\{c_2(t) - C_{\min}, 0\} \\ \dot{w}(t) &= r_c - Vw(t) - \mu w(t)c_1(t) \end{aligned}\right\} \quad (1.1.2)$$

in which ψ is the function $\psi(n) := n\log_e\left(\frac{\theta}{n}\right)$.

The initial conditions on state variable components are

$$c_1(0) = 0, c_2(0) = 0, n(0) = n_0 \text{ and } w(0) = w_0.$$

The first differential equation relates the administered drug concentration to the drug dosage. The second relates the active drug concentration to the administered drug concentration. The third is a Gompertz-type differential equation governing tumour growth with an exogenous term to account for the suppressive effects of the active drug concentration. The fourth determines how the WBC population, whose decrease reflects chemotherapy toxicity, responds to the administered drug concentration.

The chosen values of parameters in the model are as in Table 1.1.

Table 1.1 Values of the parameters in the model

Par.	Description	Value
V_1	Volume of distribution in first compartment	25 litres
V_2	Volume of distribution in second compartment	15 litres
k_1	Process of drug elimination from plasma compartment	1.6 day^{-1}
k_{12}	Link process between two compartments	0.4 day^{-1}
θ	Largest tumour	10^{12}
n_0	Initial size of tumour at $t=0$	30×10^9 cells
Λ	Gompertz growth parameter for tumour	3×10^{-3} day^{-1}
$C_{\min}$	Threshold below which no tumour cells are killed	0.0001 gml^{-1}
K	Rate of cell killing	30 g^{-1} litres day^{-1}
μ	Delayed toxicity of drug concentration on WBCs	80 g^{-1} litres day^{-1}
w_0	Initial physiology level of WBCs at $t=0$	8×10^9 litres^{-1}
V	Nominal turnover constant	0.15 day^{-1}
r_c	Rate of WBCs production	0.2×10^9 litre^{-1} day^{-1}
$C_{\max}$	Maximum allowable drug concentration	0.01 gml^{-1}
W_D	Absolute leukopenia level	2×10^9 litres^{-1}
T	Terminal time	40 days

The control problem is to minimize a weighted sum of the tumour volume and the total amount of drug, subject to upper and lower bounds at each time on drug toxicity and white blood cell population respectively, both of which affect the patient's health.

$$\begin{cases} \text{Minimize } \int_0^T (\alpha n(t) + u(t))dt, \\ \text{over control strategies } u : [0, T] \to \mathbb{R} \\ \qquad\qquad\qquad\qquad \text{ and state trajectories } (c_1, c_2, n, w) \\ \text{satisfying} \\ u(t) \in [0, 1] \text{ for } t \in [0, T], \\ c_1(t) \leq C_{\max} \text{ for } t \in [0, T], \\ w(t) \geq W_D \text{ for } t \in [0, T]. \end{cases}$$

The upper and lower bounds, $C_{\max}$ and W_D, are as given in Table 1.1.

In [188] a combination of analytical and computational techniques are employed to determine a control strategy $\bar{u}$ which satisfies necessary conditions of optimality, when the weighting factor is chosen to be $\alpha = (3/5) \times 10^{-10}$. The control $\bar{u}$ gives rise to a quite complicated, 5-subarc state trajectory structure, involving two short bang-bang pulses and three subarcs, in each of which $u(t)$ takes a constant value.

By neglecting the bang-bang pulses, we arrive at a simpler, and therefore more practical, drug treatment strategy, with only slightly increased cost. See Fig. 1.4. According to this strategy the dosage is held constant at a higher level over an initial period, reduced to a lower level for a subsequent period and finally reduced to 0 for the final period:

1.2 The Calculus of Variations

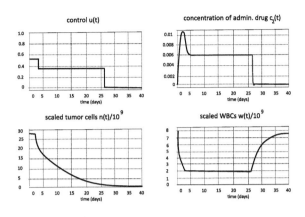

Fig. 1.4 Optimized drug strategy

$$\bar{u}(t) = \begin{cases} u_1 & \text{for } 0 \le t < t_1 \\ u_2 & \text{for } t_1 \le t < t_2 \\ 0 & \text{for } t_2 \le t < 40. \end{cases}$$

Here, $t_1 = 2.2786$ days, $t_2 = 25.855$ days, $u_1 = 0.60254$ and $u_2 = 0.33737$.

1.2 The Calculus of Variations

From a mathematical perspective, dynamic optimization is an outgrowth of the calculus of variations (in one independent variable) that takes account of new kinds of constraints (differential equation constraints, pathwise constraints on control functions 'parameterizing' the differential equations, etc.) encountered in advanced engineering design and dynamic decision making. A number of key developments in dynamic optimization have resulted from marrying old ideas from the calculus of variations and modern analytical techniques. For purposes both of setting dynamic optimization in its historical context and of illuminating later developments in dynamic optimization, we pause to review relevant material from the classical calculus of variations.

The basic problem in the calculus of variations is that of finding an arc $\bar{x}$ which minimizes the value of an integral functional

$$J(x) = \int_S^T L(t, x(t), \dot{x}(t)) dt$$

over some class of arcs satisfying the boundary condition

$$x(S) = x_0 \text{ and } x(T) = x_1.$$

Here $[S, T]$ is a given interval, $L : [S, T] \times \mathbb{R}^n \times \mathbb{R}^n \to \mathbb{R}$ is a given function, and x_0 and x_1 are given points in $\mathbb{R}^n$.

Fig. 1.5 The Brachistochrone problem

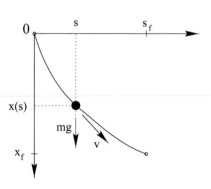

The Brachistochrone Problem An early example of such a problem was the *brachistochrone problem* circulated by Johann Bernoulli in the late seventeenth century. Positive numbers s_f and x_f are given. A frictionless bead, initially located at the point $(0, 0)$, slides along a wire under the force of gravity. The wire, which is located in a fixed vertical plane, joins the points $(0, 0)$ and (s_f, x_f). What should the shape of the wire be, in order that the bead arrives at its destination, the point (s_f, x_f), in minimum time? See Fig. 1.5.

There are a number of possible formulations of this problem. We now describe one of them. Denote by s and x the horizontal and vertical distances of a point on the path of the bead (vertical distances are measured downward). We restrict attention to wires describable as the graph of a suitably regular function $x(s)$, $0 \le s \le s_f$. For any such function x, the speed $v(s)$ is related to the downward displacement $x(s)$, when the horizontal displacement is s, according to

$$mgx(s) = \frac{1}{2}mv^2(s) \qquad (1.2.1)$$

('loss of potential energy equals gain of kinetic energy'). For any $s \in [0, s_f]$, we denote by $t(s)$ the time elapsed when the position of the bead is $(s, x(s))$. If it is assumed that speed v is positive valued, the functions t and v are related by the equation

$$v(s)\frac{dt}{ds}(s) = \sqrt{1 + |\frac{dx}{ds}(s)|^2}, \qquad \text{for } t \in [0, s_f].$$

Denote by t_f the transit time: $t_f = t(s_f)$. The change of independent variable $t(s) = \int_0^s v^{-1}(s') \sqrt{1 + |dx(s')/ds|^2}\, ds'$ now gives the following formula for t_f:

$$t_f = \int_0^{t_f} dt = \int_0^{s_f} \frac{\sqrt{1 + |dx(s)/ds|^2}}{v(s)} ds.$$

1.2 The Calculus of Variations

Using (1.2.1) to eliminate $v(s)$, we arrive at a formula for the transit time:

$$J(x) = \int_0^{s_f} L(s, x(s), \dot{x}(s)) ds,$$

in which

$$L(s, x, w) := \frac{\sqrt{1 + |w|^2}}{\sqrt{2gx}}.$$

The problem is to minimize $J(x)$ over some class of arcs x satisfying

$$x(0) = 0 \text{ and } x(s_f) = x_f.$$

This is an example of the basic problem of the calculus of variations, in which $(S, x_0) = (0, 0)$ and $(T, x_1) = (s_f, x_f)$. Suppose that we seek a minimizer in the class of absolutely continuous arcs. It can be shown that the minimum time t^* and the minimizing arc $(x(t), s(t))$, $0 \leq t \leq t^*$ (expressed in parametric form with independent variable time t) are given by the formulae

$$x(t) = a\left(1 - \cos\sqrt{\frac{g}{a}}t\right) \quad \text{and} \quad s(t) = a\left(\sqrt{\frac{g}{a}}t - \sin\sqrt{\frac{g}{a}}t\right).$$

Here, a and t^* are constants which uniquely satisfy the conditions

$$x(t^*) = x_f,$$
$$s(t^*) = t_f,$$
$$0 \leq \sqrt{\frac{g}{a}}t^* \leq 2\pi.$$

The minimizing curve is a cycloid, with infinite slope at the point of departure: it coincides with the locus of a point on the circumference of a disc of radius a, which rolls without slipping along a line of length t_f.

Problems of this kind, the minimization of integral functionals, may perhaps have initially attracted attention as individual curiosities. But throughout the eighteenth and nineteenth centuries their significance became increasingly evident, as the list lengthened of laws of physics which identified states of nature with minimizing curves and surfaces. Some examples of rules of the minimum are as follows:

Fermat's Principle in Optics The path of a light ray achieves a local minimum of the transit times over paths between specified end-points which visit the relevant reflecting and refracting boundaries. The principle predicts Snell's Laws of Reflection and Refraction, and the curved paths of light rays in inhomogeneous media. See Fig. 1.6.

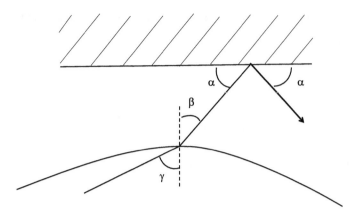

Fig. 1.6 Fermat's principle predicts Snell's laws

Dirichlet's Principle Take a bounded, open set $\Omega \subset \mathbb{R}^2$ with boundary $\partial\Omega$, in which a static two-dimensional electric field is distributed. Denote by $V(x)$ the voltage at point $x \in \Omega$. Then $V(x)$ satisfies Poisson's equation

$$\Delta V(x) = 0 \text{ for } x \in \Omega$$
$$V(x) = \bar{V}(x) \text{ for } x \in \partial\Omega.$$

Here, $\bar{V} : \partial\Omega \to \mathbb{R}$ is a given function, which supplies the boundary data.

Dirichlet's principle characterizes the solution to this partial differential equation as the solution of a minimization problem

$$\begin{cases} \text{Minimize } \int_\Omega \nabla V(x) \cdot \nabla V(x) dx \\ \text{over surfaces } V \text{ satisfying } V(x) = \bar{V}(x) \text{ on } \partial\Omega. \end{cases}$$

This optimization problem involves finding a *surface* which minimizes a given integral functional. See Fig. 1.7.

Dirichlet's principle and its generalizations are important in many respects. They are powerful tools for the study of existence and regularity of solutions to boundary value problems. Furthermore, they point the way to Galerkin methods for computing solutions to partial differential equations, such as Poisson's equation: the solution is approximated by the minimizer of the Dirichlet integral above over some finite dimensional subspace $\mathcal{S}_N$ of the domain of the original optimization problem, spanned by a finite collection of 'basis' functions $\{\phi_i\}_{i=1}^N$,

$$\mathcal{S}_N = \{\sum_{i=1}^N \alpha_i \phi_i(x) : \alpha \in \mathbb{R}^N\}.$$

The widely used finite element methods are modern implementations of Galerkin's method.

1.2 The Calculus of Variations

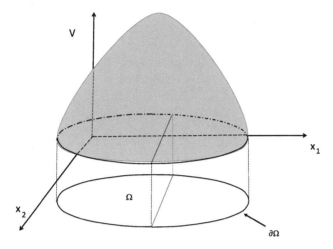

Fig. 1.7 A minimizer for the Dirichlet integral

The Action Principle Let $x(t)$ be the vector of generalized coordinates of a conservative mechanical system. The action principle asserts that $x(t)$ evolves in a manner to minimize (strictly speaking, to render stationary) the 'action', namely

$$\int [T(x(t), \dot{x}(t)) - V(x(t))]dt.$$

Here $T(x, \dot{x})$ is the kinetic energy and $V(x)$ is the potential energy. Suppose, for example, $x = (r, \theta)$, the polar coordinates of an object of mass m moving in a plane under the influence of a radial field (the origin is the centre of gravity of a body, massive in relation to the object). Then

$$T(x, \dot{x}) = \frac{1}{2}m(\dot{r}^2 + r^2\dot{\theta}^2)$$

and

$$V(r) = K/r,$$

for some constant K. The action in this case is

$$\int \left(\frac{1}{2}m[\dot{r}^2(t) + r^2(t)\dot{\theta}^2(t)] - K/r(t) \right) dt.$$

In this case, the premise 'the action is minimized' predicts the motion of a single planet about a much larger sun in a Newtonian universe. See Fig. 1.8. The action principle has proved a fruitful starting point for deriving the dynamical equations of complex interacting systems, and for studying their qualitative properties (existence of periodic orbits with prescribed energy, etc.).

Fig. 1.8 The action principle predicts planetary motion

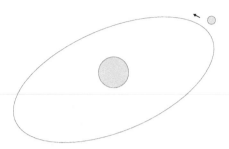

Necessary Conditions

Consider the optimization problem

$$\text{(CV)} \begin{cases} \text{Minimize } J(x) := \int_S^T L(t, x(t), \dot{x}(t)) dt \\ \text{over arcs } x \text{ satisfying} \\ x(S) = x_0 \text{ and } x(T) = x_1, \end{cases}$$

in which $[S, T]$ is a fixed interval, $L : \mathbb{R} \times \mathbb{R}^n \times \mathbb{R}^n \to \mathbb{R}$ is a given C^2 function and x_0 and x_1 are given points in $\mathbb{R}^n$. Precise formulation of problem (CV) requires us to specify the domain of the optimization problem. We take this to be $W^{1,1}([S, T]; \mathbb{R}^n)$, the class of absolutely continuous $\mathbb{R}^n$ valued arcs on $[S, T]$, for reasons which will be discussed presently.

The systematic study of minimizers $\bar{x}$ for this problem was initiated by Euler, whose seminal paper of 1744 provided the link with the equation:

$$\frac{d}{dt} L_v(t, \bar{x}(t), \dot{\bar{x}}(t)) = L_x(t, \bar{x}(t), \dot{\bar{x}}(t)). \tag{1.2.2}$$

(In this equation, L_x and L_v are the gradients of $L(t, x, v)$ with respect to the second and third arguments respectively.)

The Euler equation (1.2.2) is, under appropriate hypotheses, a necessary condition for an arc $\bar{x}$ to be a minimizer. Notice that, if the minimizer $\bar{x}$ is a C^2 function, then the Euler equation is a second order, n-vector differential equation:

$$L_{vt}(t, \bar{x}(t), \dot{\bar{x}}(t)) + L_{vx}(t, \bar{x}(t), \dot{\bar{x}}(t)) \cdot \dot{\bar{x}} + L_{vv}(t, \bar{x}(t), \dot{\bar{x}}(t)) \cdot \ddot{\bar{x}}(t) = L_x(t, \bar{x}(t), \dot{\bar{x}}(t)).$$

A standard technique for deriving the Euler equation is to reduce the problem to a scalar optimization problem, by consideration of a one-parameter family of *variations*. The calculus of variations, incidentally, owes its name to these ideas. (*Variations* of the minimizing arc cannot reduce the cost; conditions on minimizers are then derived by processing this information, with the help of a suitable *calculus* to derive necessary conditions of optimality). Because of its historical importance and its continuing influence on the derivation of necessary conditions, we now describe the technique in detail.

1.2 The Calculus of Variations

Fix attention on a minimizer $\bar{x}$. Further hypotheses are required to derive the Euler equation. We assume that there exists some number K such that

$$|L(t, x, v) - L(t, y, w)| \leq K(|x - y| + |v - w|) \tag{1.2.3}$$

for all $x, y \in \mathbb{R}^n$ and all $v, w \in \mathbb{R}^n$.

Take an arbitrary C^1 arc y, which satisfies the homogeneous boundary conditions

$$y(S) = y(T) = 0.$$

Then, for any $\epsilon > 0$, the 'variation' $x + \epsilon y$, which satisfies the end-point constraints, must have cost not less than that of $\bar{x}$. It follows that

$$\epsilon^{-1}[J(\bar{x} + \epsilon y) - J(\bar{x})] \geq 0.$$

Otherwise expressed

$$\int_S^T \epsilon^{-1}[L(t, \bar{x}(t) + \epsilon y(t), \dot{\bar{x}}(t) + \epsilon \dot{y}(t)) - L(t, \bar{x}(t), \dot{\bar{x}}(t))]dt \geq 0.$$

Under hypothesis (1.2.3), the dominated convergence theorem permits us to pass to the limit under the integral sign. We thereby obtain the inequality

$$\int_S^T [L_x(t, \bar{x}(t), \dot{\bar{x}}(t)) \cdot y(t) + L_v(t, \bar{x}(t), \dot{\bar{x}}(t)) \cdot \dot{y}(t)]dt \geq 0.$$

This relationship holds, we note, for all continuously differentiable functions y satisfying the boundary conditions $y(S) = 0$ and $y(T) = 0$. By homogeneity, the inequality can be replaced by equality.

Now apply integration by parts to the first term on the left. This gives

$$\int_S^T [-\int_S^t L_x(s, \bar{x}(s), \dot{\bar{x}}(s))ds + L_v(t, \bar{x}(t), \dot{\bar{x}}(t))] \cdot \dot{y}(t)dt = 0.$$

Take any continuous function $w : [S, T] \to \mathbb{R}^n$ which satisfies

$$\int_S^T w(t)dt = 0. \tag{1.2.4}$$

Then the continuously differentiable arc $y(t) \equiv \int_S^t w(s)ds$ vanishes at the endtimes. Consequently

$$\int_S^T [-\int_S^t L_x(s, \bar{x}(s), \dot{\bar{x}}(s))ds + L_v(t, \bar{x}(t), \dot{\bar{x}}(t))] \cdot w(t)dt = 0, \tag{1.2.5}$$

a relationship which holds for all continuous arcs w satisfying (1.2.4). To advance the analysis, we require

Lemma (Du Bois-Reymond) *Take a function $a \in L^2([S, T]; \mathbb{R}^n)$. Suppose that*

$$\int_S^T a(t) \cdot w(t) \, dt = 0 \tag{1.2.6}$$

for every continuous function w which satisfies

$$\int_S^T w(t) \, dt = 0. \tag{1.2.7}$$

Then there exists some vector $d \in \mathbb{R}^n$ such that

$$a(t) = d \quad \text{for a.e. } t \in [S, T].$$

Proof Write A for the subset of constant functions in the Hilbert space $L^2([S, T]; \mathbb{R}^n)$, endowed with the standard inner product $\langle w, v \rangle_{L^2} = \int_S^T w(t) \cdot v(t) \, dt$. We shall prove that A is the orthogonal complement, denoted by $W^\perp$, of the subspace

$$W := \left\{ w \in L^2([S, T]; \mathbb{R}^n) \; : \; \int_S^T w(t) \, dt = 0 \right\}.$$

Clearly, if $a \in A$ then, for every $w \in W$, we have $\langle a, w \rangle_{L^2} = a \cdot \int_S^T w(t) \, dt = 0$. Therefore $A \subset W^\perp$.

Now, take any $f \in W^\perp$. Set $\bar{f} := \int_S^T f(t) \, dt$ and consider $\bar{w} := f - \bar{f}$. Observe that $\bar{f} \in A$ and $\bar{w} \in W$. But, we know that $A \subset W^\perp$ and, so, it follows that $\langle \bar{f}, \bar{w} \rangle_{L^2} = 0$. We deduce that

$$0 = \langle f, \bar{w} \rangle_{L^2} - \langle \bar{f}, \bar{w} \rangle_{L^2} = \langle f - \bar{f}, \bar{w} \rangle_{L^2} = \|f - \bar{f}\|_{L^2}^2.$$

Thus, $f = \bar{f}$ a.e. on $[S, T]$, which implies $f \in A$. We conclude that $W^\perp \subset A$. The lemma is proved. □

Return now to the derivation of the Euler equation. We identify the function a of the lemma with

$$t \to -\int_S^t L_x(s, \bar{x}(s), \dot{\bar{x}}(s)) \, ds + L_v(t, \bar{x}(t), \dot{\bar{x}}(t)).$$

Assume that $t \to L_x(t, \bar{x}(t), \dot{\bar{x}}(t))$ and $t \to L_v(t, \bar{x}(t), \dot{\bar{x}}(t))$ are integrable and square integrable functions respectively. Taking note of (1.2.5), we deduce from the lemma informs that there exists a vector d such that

$$-\int_S^t L_x(s, \bar{x}(s), \dot{\bar{x}}(s)) \, ds + L_v(t, \bar{x}(t), \dot{\bar{x}}(t)) = d, \text{ a.e. } t \in [S, T]. \tag{1.2.8}$$

1.2 The Calculus of Variations

Since $L_x(t, \bar{x}(t), \dot{\bar{x}}(t))$ is integrable, it follows that $t \to L_v(t, \bar{x}(t), \dot{\bar{x}}(t))$ is almost everywhere equal to an absolutely continuous function and

$$\frac{d}{dt} L_v(t, \bar{x}(t), \dot{\bar{x}}(t)) = L_x(t, \bar{x}(t), \dot{\bar{x}}(t)), \text{ a.e. } t \in [S, T].$$

We have verified the Euler equation and given it a precise interpretation, when the domain of the optimization problem is the class of absolutely continuous arcs.

The above analysis conflates arguments assembled over several centuries. The first step was to show that smooth minimizers $\bar{x}$ satisfy the pointwise Euler equation. Euler's original derivation made use of discrete approximation techniques. Lagrange's alternative derivation introduced variational methods similar to those outlined above (though differing in the precise nature of the 'integration by parts' step). Erdmann subsequently discovered that, if the domain of the optimization problem is taken to be the class of piecewise C^1 functions (i.e. absolutely continuous functions with piecewise continuous derivatives) then

'the function $t \to L_v(t, \bar{x}(t), \dot{\bar{x}}(t))$ has removable discontinuities'.

This condition is referred to as the *first Erdmann condition*.

For piecewise C^1 minimizers, the integral version of the Euler equation (1.2.8), was first regarded as a convenient way of combining the pointwise Euler equation and the first Erdmann condition. We shall refer to it as the *Euler Lagrange condition*. An analysis in which absolutely continuous minimizers substitute for piecewise C^1 minimizers is an early twentieth century development, due to Tonelli.

Another important property of minimizers can be derived in situations when L is independent of the t (write $L(x, v)$ in place of $L(t, x, v)$). In this 'autonomous' case the second order n vector differential equation of Euler can be integrated. There results a first order differential equation involving a 'constant of integration' c:

$$L_v(\bar{x}(t), \dot{\bar{x}}(t)) \cdot \dot{\bar{x}}(t) - L(\bar{x}(t), \dot{\bar{x}}(t)) = c . \qquad (1.2.9)$$

This condition is referred to as the *Second Erdmann condition* or *constancy of the Hamiltonian condition*. It is easily deduced from the Euler Lagrange condition when $\bar{x}$ is a C^2 function. Fix t. We calculate in this case

$$\frac{d}{dt}(L_v \cdot \dot{\bar{x}}(t) - L) = \frac{d}{dt} L_v \cdot \dot{\bar{x}}(t) + L_v \cdot \ddot{\bar{x}}(t) - L_x \cdot \dot{\bar{x}}(t) - L_v \cdot \ddot{\bar{x}}(t)$$

$$= (\frac{d}{dt} L_v - L_x) \cdot \dot{\bar{x}}(t)$$

$$= 0.$$

(In the above relationships, L, L_v, etc., are evaluated at $(\bar{x}(t), \dot{\bar{x}}(t))$). Equation (1.2.9) follows.

A more sophisticated analysis leads to an 'almost everywhere' version of this condition for autonomous problems, when the minimizer $\bar{x}$ in question is assumed to be merely absolutely continuous.

'Variations' of the type $\bar{x}(t) + \epsilon y(t)$ lead to the Euler Lagrange condition. Necessary conditions supplying additional information about minimizers have been derived by considering other kinds of variations. We note in particular the *Weierstrass Condition* or the *maximization of the Hamiltonian condition*:

$$L_v(t, \bar{x}(t), \dot{\bar{x}}(t)) \cdot \dot{\bar{x}}(t) - L(t, \bar{x}(t), \dot{\bar{x}}(t))$$
$$= \max_{v \in \mathbb{R}^n} \{L_v(t, \bar{x}(t), \dot{\bar{x}}(t)) \cdot v - L(t, \bar{x}(t), v)\}.$$

Suppressing the $(t, \bar{x}(t))$ argument in the notation and expressing the Weierstrass condition as

$$L(v) - L(\dot{\bar{x}}(t)) \geq L_v(\dot{\bar{x}}) \cdot (v - \dot{\bar{x}}(t)) \text{ for all } v \in \mathbb{R}^n,$$

we see that it conveys no useful information when $L(t, x, v)$ is convex with respect to the v variable: in this case it simply interprets L_v as a subgradient of L in the sense of convex analysis. In general however, it tells us that L coincides with its 'convexification' (with respect to the velocity variable) along the optimal trajectory, i.e.

$$L(t, \bar{x}(t), \dot{\bar{x}}(t)) = \tilde{L}(t, \bar{x}(t), \dot{\bar{x}}(t)).$$

Here $\tilde{L}(t, x, .)$ is the function with epigraph set co $\{$epi $L(t, x, .)\}$. See Fig. 1.9.

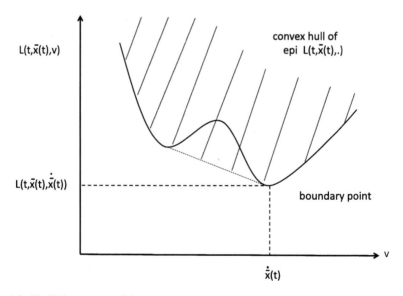

Fig. 1.9 The Weierstrass condition

1.2 The Calculus of Variations

The above necessary conditions (the Euler Lagrange condition, the Weierstrass condition and the second Erdmann condition) have convenient formulations in terms of the 'costate arc'

$$p(t) = L_v(t, \bar{x}(t), \dot{\bar{x}}(t)).$$

They are

$$(\dot{p}(t), p(t)) = L_{x,v}(t, \bar{x}(t), \dot{\bar{x}}(t)), \tag{1.2.10}$$

$$p(t) \cdot \dot{\bar{x}}(t) - L(t, \bar{x}(t), \dot{\bar{x}}(t)) = \max_{v \in \mathbb{R}^n} [p(t) \cdot v - L(t, \bar{x}(t), v)]$$

and, in the case when L does not depend on t,

$$p(t) \cdot \dot{\bar{x}}(t) - L(\bar{x}(t), \dot{\bar{x}}(t)) = c$$

for some constant c.

To explore the qualitative properties of minimizers it is often helpful to reduce the Euler Lagrange condition to a system of specially structured first order differential equations. This was first achieved by Hamilton for Lagrangians L arising in mechanics. In this analysis the *Hamiltonian*,

$$H(t, x, p) := \max_{v \in \mathbb{R}^n} \{p \cdot v - L(t, x, v)\},$$

has an important role.

Suppose that the right side has a unique maximizer v_{max}. This will depend on (t, x, p) and so we can use it to define a function $\chi : [S, T] \times \mathbb{R}^n \times \mathbb{R}^n \to \mathbb{R}^n$:

$$\chi(t, x, p) := v_{max}.$$

Then

$$H(t, x, p) = p \cdot \chi(t, x, p) - L(t, x, \chi(t, x, p))$$

and

$$\nabla_v (p \cdot v - L(t, x, v))|_{v=\chi(t,x,p)} = 0.$$

The last relationship implies

$$p = L_v(t, x, \chi(t, x, p)). \tag{1.2.11}$$

(The mapping from x vectors to p vectors implicit in this relationship, for fixed t, is referred to as the *Legendre transformation*.)

Now consider an arc $\bar{x}$ and associated costate arc p which satisfy the Euler Lagrange and Weierstrass conditions, namely

$$(\dot{p}(t), p(t)) = \nabla_{x,v} L(t, \bar{x}(t), \dot{\bar{x}}(t)) \tag{1.2.12}$$

and

$$p(t) \cdot \dot{\bar{x}}(t) - L(t, \bar{x}(t), \dot{\bar{x}}(t)) = \max{}_{v \in \mathbb{R}^n} \{p(t) \cdot v - L(t, \bar{x}(t), v)\}.$$

Since it is assumed that the 'maximizer' in the definition of the Hamiltonian is unique, it follows from the Weierstrass condition that

$$\dot{\bar{x}}(t) = \chi(t, \bar{x}(t), p(t)).$$

Fix t. Let us assume that $\chi(t, ., .,)$ is differentiable. Then we can calculate the gradients of $H(t, ., .)$:

$$H_x(t, x, p)|_{x=\bar{x}(t), p=p(t)}$$
$$= \nabla_x(p \cdot \chi(t, x, p) - L(t, x, \chi(t, x, p)))|_{x=\bar{x}(t), p=p(t)}$$
$$= p \cdot \chi_x(t, x, p) - L_x(t, x, \chi(t, x, p))$$
$$\quad - L_v(t, x, \chi(t, x, p)) \cdot \chi_x(t, x, p)|_{x=\bar{x}(t), p=p(t)}$$
$$= (p - L_v(t, x, \chi(t, x, p))) \cdot \chi_x(t, x, p) - L_x(t, x, \chi(t, x, p))|_{x=\bar{x}(t), p=p(t)}.$$
$$= 0 - \dot{p}(t).$$

(The last step in the derivation of these relations makes use of (1.2.11) and (1.2.12).) We have evaluated the x-derivative of H:

$$H_x(t, \bar{x}(t), p(t)) = -\dot{p}(t).$$

As for the p-derivative, we have

$$H_p(t, x, p)|_{x=\bar{x}(t), p=p(t)}$$
$$= \nabla_p(p \cdot \chi(t, x, p) - L(t, x, \chi(t, x, p)))|_{x=\bar{x}(t), p=p(t)}$$
$$= \chi(t, x, p) + p \cdot \chi_p(t, x, p)$$
$$\quad - L_v(t, x, \chi(t, x, p)) \cdot \chi_p(t, x, p)|_{x=\bar{x}(t), p=p(t)}$$
$$= \dot{\bar{x}}(t) + (p(t) - L_v(t, \bar{x}(t), \dot{\bar{x}}(t))) \cdot \chi_p(t, \bar{x}(t), p(t))$$
$$= \dot{\bar{x}}(t) + 0.$$

Combining these relations, we arrive at the system of first order differential equations of interest, namely the *Hamilton condition*

$$(-\dot{p}(t), \dot{\bar{x}}(t)) = H_{x,p}(t, \bar{x}(t), p(t)). \tag{1.2.13}$$

So far, we have limited attention to the problem of minimizing an integral functional of arcs with fixed end-points. A more general problem is that in which the arcs are constrained to satisfy the boundary condition

$$x(S) = x_0 \text{ and } x(T) \in C, \tag{1.2.14}$$

1.2 The Calculus of Variations

for some specified point $x_0 \in \mathbb{R}^n$ and subset $C \subset \mathbb{R}^n$. (The fixed end-point problem is a special case, in which C is chosen to be $\{x_1\}$.) The above necessary conditions remain valid when we pass to this more general end-point constraint, since a minimizer $\bar{x}$ over arcs satisfying (1.2.14) is also a minimizer for the fixed end-point problem in which $C = \{\bar{x}(T)\}$. But something more is required: replacing the fixed end-point constraint by other types of end-point constraint which allow end-point variation introduces extra degrees of freedom into the optimization problem, which should be reflected in supplementary necessary conditions. These conditions are conveniently expressed in terms of a boundary condition on the costate arc p:

$$-p(T) \text{ is an outward normal to } C \text{ at } \bar{x}(T).$$

They are collectively referred to as the *transversality condition*, because they assert that a vector composed from end-points of the costate arc is orthogonal or 'transverse' to the tangent hyperplane to C at $\bar{x}(T)$.

The transversality condition too can be derived by considering suitable 'variations' (variations which, on this occasion, allow perturbations of the end-point values of the minimizer $\bar{x}$). In favourable circumstances the transversality condition combines with the boundary condition (1.2.14) to supply the appropriate number of $2n$ boundary conditions to accompany the system of $2n$ Eq. (1.2.13).

The following theorem brings together these classical conditions and gives precise hypotheses under which they are satisfied.

Theorem 1.2.1 *Let $\bar{x}$ be a minimizer for*

$$\begin{cases} \text{Minimize } J(x) := \int_S^T L(t, x(t), \dot{x}(t)) dt \\ \text{over absolutely continuous arcs } x \text{ satisfying} \\ x(S) = x_0 \text{ and } x(T) \in C, \end{cases}$$

for some given interval $[S, T]$, function $L : [S, T] \times \mathbb{R}^n \times \mathbb{R}^n \to \mathbb{R}$, point $x_0 \in \mathbb{R}^n$ and closed set $C \subset \mathbb{R}^n$. We list the hypotheses

(i) $L(., x, v)$ *is measurable for each (x, v) and $L(t, ., .)$ is continuously differentiable for each $t \in [S, T]$,*
(ii) *there exists $k \in L^1$ and $\beta > 0$, $\epsilon > 0$ such that*

$$|L(t, x, v) - L(t, x', v)| \leq (k(t) + \beta |v|)|x - x'|$$

$$\text{for all } x, x' \in \bar{x}(t) + \epsilon \mathbb{B}, \ v \in \mathbb{R}^n,$$

(iii) $L(t,x,.)$ *is convex for all $(t, x) \in [S, T] \times \mathbb{R}^n$,*
(iv) $H(t,.,.)$ *is continuously differentiable for each $t \in [S, T]$ where*

$$H(t, x, p) := \sup_{v \in \mathbb{R}^n} \{p \cdot v - L(t, x, v)\}.$$

Suppose that hypotheses (i) and (ii) are satisfied. Then there exists an absolutely continuous arc $p : [S, T] \to \mathbb{R}^n$ satisfying the following conditions:

The Euler Lagrange Condition

$$(\dot{p}(t), p(t)) = L_{x,v}(t, \bar{x}(t), \dot{\bar{x}}(t)) \quad a.e.,$$

The Weierstrass Condition

$$p(t) \cdot \dot{\bar{x}}(t) - L(t, \bar{x}(t), \dot{\bar{x}}(t)) = \max_{v \in \mathbb{R}^n} \{p(t) \cdot v - L(t, \bar{x}(t), v)\} \quad a.e.,$$

The Transversality Condition

$$-p(T) \in N_C(\bar{x}(T)).$$

If additionally hypotheses (iii) and (iv) are satisfied, then p can be chosen also to satisfy

Hamilton's Condition

$$(-\dot{p}(t), \dot{\bar{x}}(t)) = H_{x,p}(t, \bar{x}(t), p(t)) \quad a.e. \ .$$

Finally, if the additional hypothesis

(v) $L(t, x, v)$ is independent of t,

is imposed, then the above assertions can be strengthened to require that also that p satisfies

Constancy of the Hamiltonian Condition

$$p(t) \cdot \dot{\bar{x}}(t) - L(t, \bar{x}(t), \dot{\bar{x}}(t)) = c \quad a.e.$$

for some constant c.

The theorem is a special case of general necessary conditions which will be proved in Chap. 8.

In this theorem, the transversality condition is expressed in a way that makes sense when C is assumed to be a closed set. It makes reference to the 'limiting normal cone' $N_C(\bar{x}(T))$. This will be defined in the overview of nonsmooth analysis provided in Sect. 1.8.

1.3 Existence of Minimizers and Tonelli's Direct Method

Deriving the Euler Lagrange condition and related conditions would appear to reduce the problem of finding a minimizer to one merely of solving a differential equation. But this is to overlook the fact that satisfaction of the Euler Lagrange condition is only a *necessary* condition for optimality; we cannot be certain, without further analysis, that an arc satisfying these conditions is truly a minimizer.

1.3 Existence of Minimizers and Tonelli's Direct Method

Early in the twentieth century, Tonelli recognized the significance of establishing the *existence* of minimizers before using necessary conditions of optimality to try to identify them. Tonelli's *direct method* for obtaining a minimizer consists of the following steps:

Step 1: Show that a minimizer exists.

Step 2: Search among arcs satisfying the necessary conditions for an arc with lowest cost.

These steps, when successfully carried out, are guaranteed to yield a minimizer. Notice however that, if we neglect Step 1, then Step 2 alone can be positively misleading. The point is illustrated by the following example:

$$\begin{cases} \text{Minimize } \int_0^1 x(t)\dot{x}^2(t)dt \\ \text{over absolutely continuous arcs } x \text{ satisfying} \\ x(0) = 0 \text{ and } x(1) = 0. \end{cases}$$

This example is not precisely matched to the necessary conditions of Theorem 1.2.1 because the integrand L fails to satisfy hypothesis (ii). It can nevertheless be shown that any *Lipschitz continuous* minimizer $\bar{x}$ satisfies the Euler Lagrange, Weierstrass and Transversality conditions. In this case the Euler Lagrange condition takes the form:

$$2\frac{d}{dt}\left\{\bar{x}(t)\dot{\bar{x}}(t)\right\} = \dot{\bar{x}}^2(t) \quad \text{a.e.} \,.$$

Obviously the arc

$$\bar{x} \equiv 0$$

satisfies the Euler equation and the specified end-point conditions. It is in fact the *unique* Lipschitz continuous function so doing. To see this note that, for any Lipschitz continuous arc y which vanishes at the end-times and satisfies the Euler Lagrange Condition, we have

$$\frac{d^2}{dt^2} y^2(t) = \dot{y}^2(t) \quad \text{a.e.} \,.$$

It follows that, for some constant c,

$$\frac{d}{dt} y^2(t) = c + \int_0^t \dot{y}^2(s)ds \quad \text{for all } t.$$

Since y vanishes at the left end-point and is assumed to have bounded slope, we deduce from this last condition that $c = 0$. If $y \not\equiv 0$ then $\dot{y}$ must be nonzero on a set of positive measure, in order to satisfy the end-point constraints. But then

$$y(1)^2 = 0 + \int_0^1 \int_0^s \dot{y}^2(t)\,ds\,dt > 0,$$

in violation of the end-point constraints. We have confirmed that $\bar{x} \equiv 0$ is the unique Lipschitz continuous arc satisfying the Euler Lagrange condition.

Notice, however, that $\bar{x}$ is not a minimizer, not even in a local sense. Indeed by choosing the parameter $\epsilon > 0$ sufficiently small we can arrange that the arc $t \to -\epsilon t(1-t)$, which has cost strictly less than 0, is arbitrarily close to $\bar{x}$ with respect to the supremum norm.

To summarize, in this example there is a unique Lipschitz continuous arc satisfying the constraints of the problem and also the Euler Lagrange, Weierstrass and transversality conditions. Yet it is not a minimizer. This is a case in which a naive belief in the power of necessary conditions to identify minimizers leads us astray. The pathological feature of this example is, of course, that there are no minimizers in the class of Lipschitz continuous functions. The collection of arcs satisfying necessary conditions for a Lipschitz continuous arc to be a minimizer will not therefore contain a minimizer, even though it is non-empty.

Clearly, it makes sense to speak of a minimizer only if we specify the class of functions from which minimizers are chosen. Unfortunately, hypotheses under which we can guarantee existence of minimizers in the 'elementary' function spaces (C^1 functions, piecewise C^1 functions, etc.) are, for many purposes, unacceptably restrictive. One of Tonelli's most significant contributions in the calculus of variations was to show that, by enlarging the domain of (CV) to include $W^{1,1}([S, T]; \mathbb{R}^n)$ arcs, the existence of minimizers could be guaranteed under broad, directly verifiable hypotheses on the Lagrangian L. An existence theorem, in the spirit of Tonelli's pioneering work in this field, is:

Theorem 1.3.1 *Consider the minimization problem*

$$\begin{cases} \textit{Minimize } \int_S^T L(t, x(t), \dot{x}(t))dt \\ \textit{over } x \in W^{1,1}([S, T]; \mathbb{R}^n) \textit{ satisfying} \\ x(S) = x_0 \textit{ and } x(T) = x_1, \end{cases}$$

for given $[S, T] \subset \mathbb{R}$, $L : [S, T] \times \mathbb{R}^n \times \mathbb{R}^n \to \mathbb{R}$, $x_0 \in \mathbb{R}^n$ and $x_1 \in \mathbb{R}^n$. Assume that

(i) L is continuous,
(ii) $L(t, x, .)$ is convex for all $t \in [S, T]$ and $x \in \mathbb{R}^n$,
(iii) there exist constants $c > 0$, $d > 0$ and $\alpha > 0$ such that

$$L(t, x, v) \geq c|v|^{1+\alpha} - d \ \textit{ for all } t \in [S, T], \ x \in \mathbb{R}^n \ \textit{ and } \ v \in \mathbb{R}^n.$$

Then there exists a minimizer (in the class $W^{1,1}([S, T]; \mathbb{R}^n)$).

A more general version of this theorem will be proved in Chap. 2.

Nowadays, existence of minimizers is recognized as an important topic in its own right, which gives insights into whether variational problems have been properly formulated and also into the regularity properties of minimizers. But, as we have already observed, existence of minimizers was first studied as an ingredient in the direct method, i.e. to justify seeking a minimizer from the class of arcs satisfying the necessary conditions.

Unfortunately, existence theorems such as Theorem 1.3.1 are not entirely adequate for purposes of applying the direct method. The difficulty is this: implicit in the direct method is the assumption that a single set of hypotheses on the data for the optimization problem at hand, simultaneously guarantees existence of minimizers and also validity of the necessary conditions. Yet for certain variational problems of interest, even though existence of a minimizer $\bar{x}$ is guaranteed by Theorem 1.3.1, it is not clear a priori that the hypotheses are satisfied under which necessary conditions, such as those listed in Theorem 1.2.1, are valid at $\bar{x}$. For these problems we cannot be sure that a search over arcs satisfying the Euler Lagrange condition and related conditions will yield a minimizer.

There are two ways out of this dilemma. One is to refine the existence theory by showing that $W^{1,1}$ minimizers are confined to some subset of $W^{1,1}$ for which the standard necessary conditions are valid. The other is to replace the standard necessary conditions by conditions which are valid under hypotheses similar to those of existence theory. Tonelli's achievements along these lines, together with recent significant developments, provide much of the subject matter for Chap. 12.

1.4 Sufficient Conditions and the Hamilton Jacobi Equation

We have discussed pitfalls in seeking minimizers among arcs satisfying the necessary conditions. The steps initiated by Tonelli for dealing with them, namely investigating hypotheses under which minimizers exist, is a relatively recent development. They were preceded by various procedures, some of a rather *ad hoc* nature, for testing whether an arc satisfying known necessary conditions, which we have somehow managed to find, is truly a minimizer. For certain classes of variational problems, an arc $\bar{x}$ satisfying the necessary conditions of Theorem 1.2.1 is a minimizer (at least in some 'local' sense) if it can be shown to satisfy also certain higher order conditions (such as the strengthened Legendre or Jacobi conditions). In other cases we might hope to carry out calculations, 'completing the square' or construction of a 'verification function' for example, which make it obvious that an arc we have obtained and have reason to believe is a minimizer, is truly a minimizer. This ragbag of 'indirect' procedures, none of which may apply, contrasts with Tonelli's direct method, which is in some sense a more systematic approach to finding minimizers.

However one 'indirect' method for confirming that a given arc is a minimizer figures prominently in current research and we discuss it here. It is based on the relationship between, on the one hand, the variational problem

$$(Q) \begin{cases} \text{Minimize } \int_S^T L(t, x(t), \dot{x}(t))dt + g(x(T)) \\ \text{over Lipschitz continuous arcs } x \text{ satisfying} \\ x(S) = x_0 \text{ and } x(T) \in \mathbb{R}^n \end{cases}$$

(in which $L : [S, T] \times \mathbb{R}^n \times \mathbb{R}^n \to \mathbb{R}$ and $g : \mathbb{R}^n \to \mathbb{R}$ are given continuous functions and x_0 is a given n-vector) and, on the other, the Hamilton Jacobi partial differential equation:

$$\text{(HJE)} \begin{cases} \phi_t(t, x) + \min_{v \in \mathbb{R}^n} \{\phi_x(t, x) \cdot v + L(t, x, v)\} = 0 \\ \qquad\qquad\qquad\qquad\qquad\qquad\qquad \text{for all } t \in (S, T), x \in \mathbb{R}^n \\ \phi(T, x) = g(x) \text{ for all } x \in \mathbb{R}^n. \end{cases}$$

We shall sometimes find it convenient to express the first of these equations in terms of the Hamiltonian H:

$$H(t, x, p) := \sup_{v \in \mathbb{R}^n} \{p \cdot v - L(t, x, v)\}$$

thus

$$-\phi_t(t, x) + H(t, x, -\phi_x(t, x)) = 0 \text{ for all } t \in (S, T), x \in \mathbb{R}^n.$$

(It is helpful in the present context to consider a variational problem in which the right end-point is unconstrained and a function of the terminal value of the arc is added to the cost.)

The following proposition summarizes the approach, which is referred to as Carathéodory's verification technique :

Proposition 1.4.1 *Let $\bar{x}$ be a Lipschitz continuous arc satisfying $\bar{x}(S) = x_0$. Suppose that a continuously differentiable function $\phi : \mathbb{R} \times \mathbb{R}^n \to \mathbb{R}$ can be found satisfying (HJE) and also*

$$\phi_x(t, \bar{x}(t)) \cdot \dot{\bar{x}}(t) + L(t, \bar{x}(t), \dot{\bar{x}}(t))$$
$$= \text{Min}_{v \in \mathbb{R}^n} \{\phi_x(t, \bar{x}(t)) \cdot v + L(t, \bar{x}(t), v)\} \quad a.e. \ t \in [S, T].$$

Then $\bar{x}$ is a minimizer for (Q) and

$$\phi(S, x_0) = \inf(Q).$$

Proof Take any Lipschitz continuous arc $x : [S, T] \to \mathbb{R}^n$ for which $x(S) = x_0$. Then $t \to \phi(t, x(t))$ is a Lipschitz continuous function and

$$\frac{d\phi}{dt}(t, x(t)) = \phi_t(t, x(t)) + \phi_x(t, x(t)) \cdot \dot{x}(t) \quad \text{a.e.} \ .$$

1.4 Sufficient Conditions and the Hamilton Jacobi Equation

Now express $\phi(t, x(t))$ as the integral of its derivative:

$$\phi(S, x_0) = -\int_S^T \frac{d}{dt}\phi(t, x(t))dt + g(x(T))$$

(we have used the facts that $x(S) = x_0$ and $\phi(T, x) = g(x)$)

$$= -\int_S^T [\phi_t(t, x(t)) + \phi_x(t, x(t)) \cdot \dot{x}(t) + L(t, x(t), \dot{x}(t))]dt$$

$$+ \int_S^T L(t, x(t), \dot{x}(t))dt + g(x(T)).$$

But the first term on the right satisfies

$$-\int_S^T [\phi_t(t, x(t)) + \phi_x(t, x(t)) \cdot \dot{x}(t) + L(t, x(t), \dot{x}(t))]dt$$

$$\leq -\int_S^T [\phi_t(t, x(t)) + \inf_v \{\phi_x(t, x(t)) \cdot v + L(t, x(t), v)\}]dt$$

$$= 0. \tag{1.4.1}$$

It follows that

$$\phi(S, x_0) \leq \int_S^T L(t, x(t), \dot{x}(t))dt + g(x(T)). \tag{1.4.2}$$

Now repeat the above arguments with $\bar{x}$ in place of x. Since, by hypothesis,

$$\phi_x(t, \bar{x}(t)) \cdot \dot{\bar{x}}(t) + L(t, \bar{x}(t), \dot{\bar{x}}(t)) = \min_{v \in \mathbb{R}^n} \{\phi_x(t, \bar{x}(t)) \cdot v + L(t, \bar{x}(t), v)\},$$

we can replace inequality (1.4.1) by equality:

$$\int_S^T [-\phi_t(t, \bar{x}(t)) + \phi_x(t, \bar{x}(t)) \cdot \dot{\bar{x}}(t) + L(t, \bar{x}(t), \dot{\bar{x}}(t))]dt = 0.$$

Consequently

$$\phi(S, x_0) = \int_S^T L(t, \bar{x}(t), \dot{\bar{x}}(t))dt + g(\bar{x}(T)). \tag{1.4.3}$$

We see from (1.4.2) and (1.4.3) that $\bar{x}$ has cost $\phi(S, x_0)$ and, on the other hand, any other Lipschitz continuous arc satisfying the constraints of problem (Q) has cost not less than $\phi(S, x_0)$. It follows that $\phi(S, x_0)$ is the minimum cost and $\bar{x}$ is a minimizer. □

In favourable circumstances, it is possible to find a continuously differentiable solution ϕ to the Hamilton Jacobi equation such that the function

$$v \to \phi_x(t, x) \cdot v + L(t, x, v)$$

has a unique minimizer (write it $d(t, x)$ to emphasize that it will depend on (t, x)) and the differential equation

$$\dot{x}(t) = d(t, x(t)) \text{ a.e.}$$
$$x(0) = x_0$$

has a Lipschitz continuous solution $\bar{x}$. Then, Proposition 1.4.1 informs us, $\bar{x}$ must be a minimizer.

Success of this method for finding minimizers hinges of course on finding a solution ϕ to the Hamilton Jacobi equation. Recall that, if a solution ϕ confirms the optimality of some putative minimizer, then $\phi(S, x_0)$ coincides with the minimum cost for (Q). (This is the final assertion of Proposition 1.4.1). It follows that the natural candidate for solution to the Hamilton Jacobi equation is the *value function* $V : [S, T] \times \mathbb{R}^n \to \mathbb{R}$ for (Q):

$$V(t, x) = \inf(Q_{t,x}) .$$

Here, the right side denotes the infimum cost of a variant of problem (Q), in which the initial data (S, x_0) is replaced by (t, x):

$$(Q_{t,x}) \begin{cases} \text{Minimize } \int_t^T L(s, y(s), \dot{y}(s))ds + g(y(T)) \\ \text{over Lipschitz continuous functions } y \text{ satisfying} \\ y(t) = x. \end{cases}$$

The following proposition gives precise conditions under which the value function is a solution to (HJE):

Theorem 1.4.2 *Let V be the value function for (Q). Suppose that*

(i) *V is a continuously differentiable function,*
(ii) *For each $(t, x) \in [S, T] \times \mathbb{R}^n$, the optimization problem $(Q_{t,x})$ has a minimizer which is continuously differentiable.*

Then V is a solution to (HJE).
(Here and elsewhere, we say a vector valued function defined on a closed subset D of a finite dimensional linear space is continuously differentiable if it is the restriction to D of some continuously differentiable function on a neighbourhood of D.)

1.4 Sufficient Conditions and the Hamilton Jacobi Equation

Proof Take $(t, x) \in [S, T] \times \mathbb{R}^n$, $\tau \in [t, T]$, a continuously differentiable function $y : [t, T] \to \mathbb{R}^n$ and a continuously differentiable minimizer $\bar{y}$ for $(Q_{t,x})$. Simple contradiction arguments lead to the following relationships

$$V(t, x) \leq \int_t^\tau L(s, y(s), \dot{y}(s))ds + V(\tau, y(\tau))$$

$$V(t, x) = \int_t^\tau L(s, \bar{y}(s), \dot{\bar{y}}(s))ds + V(\tau, \bar{y}(\tau))$$

Take arbitrary $v \in \mathbb{R}^n$, $\epsilon \in (0, T - t)$ and choose $y(s) = x + (s - t)v$. Then

$$\epsilon^{-1}[V(t + \epsilon, x + \epsilon v) - V(t, x)] + \epsilon^{-1} \int_t^{t+\epsilon} L(s, x + sv, v)ds \geq 0.$$

and

$$\epsilon^{-1}[V(t + \epsilon, x + \int_t^{t+\epsilon} \dot{\bar{y}}(s)ds) - V(t, x)] + \epsilon^{-1} \int_t^{t+\epsilon} L(s, \bar{y}(s), \dot{\bar{y}}(s))ds = 0.$$

Passage to the limit as $\epsilon \downarrow 0$ gives

$$V_t(t, x) + V_x(t, x) \cdot v + L(t, x, v) \geq 0$$

and

$$V_t(t, x) + V_x(t, x) \cdot \dot{\bar{y}}(t) + L(t, x, \dot{\bar{y}}(t)) = 0.$$

Bearing in mind that v is arbitrary, we conclude that

$$V_t(t, x) + \min_{v \in \mathbb{R}^n} \{V_x(t, x) \cdot v + L(t, x, v)\} = 0.$$

We have confirmed that V satisfies (HJE). The fact that V also satisfies the boundary condition $V(T, x) = g(x)$ follows directly from the definition of the value function. □

How might we find solutions to the Hamilton Jacobi equation? One approach, 'construction of a field of extremals', is inspired by the above interpretation of the value function. The idea is to generate a family of continuously differentiable arcs $\{y_{t,x}\}$, parameterized by points $(t, x) \in [S, T] \times \mathbb{R}^n$, satisfying the necessary conditions. We then choose

$$\phi(t, x) = \int_t^T L(t, y_{t,x}(s), \dot{y}_{t,x}(s))ds + g(y_{t,x}(T)).$$

If it turns out that ϕ is a continuously differentiable function and the extremal $y_{t,x}$ really is a minimizer for $(Q_{t,x})$ (a property of which we have no a priori knowledge), then $V \equiv \phi$ and ϕ is a solution to the Hamilton Jacobi equation.

The Hamilton Jacobi equation is a nonlinear partial differential equation of hyperbolic type. From a classical perspective, constructing a field of extremals (a procedure for building up a solution to a partial differential equation from the solutions to a family of ordinary differential equations) amounts to solving the Hamilton Jacobi equation by the method of characteristics.

1.5 The Maximum Principle

A convenient framework for studying minimization problems, encountered in the optimal selection of flight trajectories and other areas of advanced engineering design and dynamic decision making, is to regard them as special cases of the problem:

$$(P) \begin{cases} \text{Minimize } g(x(T)) + \int_S^T L(t, x(t), u(t))dt \\ \text{over measurable functions } u : [S, T] \to \mathbb{R}^m \text{ and} \\ \qquad\qquad\qquad\qquad\qquad \text{arcs } x \in W^{1,1}([S, T]; \mathbb{R}^n) \text{ satisfying} \\ \dot{x}(t) = f(t, x(t), u(t)) \text{ a.e.,} \\ u(t) \in U(t) \text{ a.e.,} \\ x(S) = x_0 \text{ and } x(T) \in C, \end{cases}$$

the data for which comprise an interval $[S, T]$, functions $f : [S, T] \times \mathbb{R}^n \times \mathbb{R}^m \to \mathbb{R}^n$, $L : [S, T] \times \mathbb{R}^n \times \mathbb{R}^m \to \mathbb{R}$ and $g : \mathbb{R}^n \to \mathbb{R}$, a point $x_0 \in \mathbb{R}^n$, a set $C \subset \mathbb{R}^n$, and $\{U(t) \subset \mathbb{R}^m : S \le t \le T\}$ is a given family of sets.

This formulation of the dynamic optimization problem (or variants on it in which, for example, the cost depends also on the left end-point, or the end-points of the time interval $[S, T]$ are included among the choice variables, or pathwise constraints are imposed on values of x) is referred to as the Pontryagin formulation. The importance of this formulation is that it embraces a wide range of significant optimization problems which are beyond the reach of traditional variational techniques and, at the same time, it is very well suited to the derivation of general necessary conditions of optimality.

In (P), the n-vector dependent variable x is called the *state*. The function describing its time evolution, $x(t)$, $S \le t \le T$, is called the *state trajectory*. The state trajectory depends on our choice of control function $u(t)$, $S \le t \le T$. The object is to choose a control function u to minimize the value of the cost $g(x(T)) + \int_S^T L(t, x(t), u(t))dt$ resulting from our choice of u.

There follows a statement of the maximum principle, whose discovery by L. S. Pontryagin *et al.* in the 1950's was an important milestone in the emergence of dynamic optimization as a distinct field of research.

1.5 The Maximum Principle

Theorem 1.5.1 (Maximum Principle) *Let $(\bar{x}, \bar{u})$ be a minimizer for (P). Assume that*

(i): *g is continuously differentiable on a neighbourhood of $\bar{x}(T)$,*
(ii): *C is a closed set,*
(iii): *(f, L) is continuous, $(f, L)(t, ., u)$ is continuously differentiable for each (t, u) and there exist $\epsilon > 0$, $k \in L^1$ and $c \in L^1$ such that*

$$|(f, L)(t, x, u) - (f, L)(t, x', u)| \le k(t)|x-x'| \text{ and } |(f, L)(t, x, u)| \le c(t)$$

for all $x, x' \in \bar{x}(t) + \epsilon \mathbb{B}$ and $u \in U$, a.e.,
(iv): *$U(t) \equiv U$ is a Borel set.*

Then there exist an arc $p \in W^{1,1}([S, T]; \mathbb{R}^n)$ and $\lambda \ge 0$, not both zero, such that the following conditions are satisfied:

The Costate Equation

$$-\dot{p}(t) = p(t) \cdot f_x(t, \bar{x}(t), \bar{u}(t)) - \lambda L_x(t, \bar{x}(t), \bar{u}(t)), \quad \text{a.e.,}$$

The Generalized Weierstrass Condition

$$p(t) \cdot f(t, \bar{x}(t), \bar{u}(t)) - \lambda L(t, \bar{x}(t), \bar{u}(t))$$
$$= \max_{u \in U}\{p(t) \cdot f(t, \bar{x}(t), u) - \lambda L(t, \bar{x}(t), u)\} \quad \text{a.e.,}$$

The Transversality Condition

$$-p(T) = \lambda \nabla g(\bar{x}(T)) + \eta$$

for some $\eta \in N_C(\bar{x}(T))$.

The limiting normal cone N_C of the right end-point constraint set C featuring in the above transversality condition will be defined presently; in the case that C is a smooth manifold it reduces to the set of outward pointing normals.

A Maximum Principle for Problems with Functional Endpoint Constraints In many applications, the endpoint constraint C set in problem (P) takes the form of a collection of functional inequality and equality constraints, thus

$$C = \{x : \phi_j(x) \le 0, j = 1, \ldots, n_\phi \text{ and } \psi_i(x) = 0, i = 1, \ldots, n_\psi\}, \quad (1.5.1)$$

in which $\phi : \mathbb{R}^n \to \mathbb{R}^{n_\phi}$ and $\psi : \mathbb{R}^n \to \mathbb{R}^{n_\psi}$ are given C^1 functions. Can we, in this case, replace the transversality condition by a condition expressed directly in terms the constraint functions ϕ and ψ? We show how this can be done.

Define the index set of active constraints

$$I(\bar{x}) := \{j : \phi_j(\bar{x}(T)) = 0\}$$

and the 'multiplier domain' for the system of functional constraints defining C:

$$\Lambda(\bar{x}(T)) := \{(\lambda^\phi, \lambda^\psi) \in (\mathbb{R}_+)^{n_\phi} \times \mathbb{R}^{n_\psi} : \lambda_j^\phi = 0 \text{ for all } j \notin I(\bar{x}(T))\}.$$

Now assume the non-degeneracy condition

$$\left.\begin{array}{l} \lambda^\phi \cdot \phi_x(\bar{x}(T)) + \lambda^\psi \cdot \nabla\psi(\bar{x}(T)) = 0 \\ (\lambda^\phi, \lambda^\psi) \in \Lambda(\bar{x}(T)) \end{array}\right\} \implies (\lambda^\phi, \lambda^\psi) = 0, \qquad (1.5.2)$$

which requires the gradients of the active inequality constraint function components and all the equality constraint function components to be linearly independent ('positively' linearly independent w.r.t. to the active inequality constraint function components.) Then it can be shown that

$$N_C(\bar{x}(T)) \subset \{\lambda^\phi \cdot \nabla\phi(\bar{x}(T)) + \lambda^\psi \cdot \nabla\psi(\bar{x}(T)) : (\lambda^\phi, \lambda^\psi) \in \Lambda(\bar{x}(T))\}.$$

These observations lead to the following version of the maximum principle:

Theorem 1.5.2 (Maximum Principle for Functional Endpoint Constraints) *Let $(\bar{x}, \bar{u})$ be a minimizer for (P) in the special case when C has the representation (1.5.1) for some C^1 functions $\phi : \mathbb{R}^n \to \mathbb{R}^{n_\phi}$ and $\psi : \mathbb{R}^n \to \mathbb{R}^{n_\psi}$. Assume that*

(i): g is continuously differentiable on a neighbourhood of $\bar{x}(T)$,
(ii): (f, L) is continuous, $(f, L)(t, ., u)$ is continuously differentiable for each (t, u) and there exist $\epsilon > 0$, $k \in L^1$ and $c \in L^1$ such that

$$|(f, L)(t, x, u) - (f, L)(t, x', u)| \leq k(t)|x - x'| \text{ and } |(f, L)(t, x, u)| \leq c(t)$$

for all $x, x' \in \bar{x}(t) + \epsilon\mathbb{B}$ and $u \in U$, a.e. $t \in [S, T]$,
(iii): $U(t) \equiv U$ is a Borel set.

Then there exist an arc $p \in W^{1,1}([S, T]; \mathbb{R}^n)$, $\lambda \geq 0$ and $(\lambda^\phi, \lambda^\psi) \in \Lambda(\bar{x}(T))$ such that

(a): $(p, \lambda, \lambda^\phi, \lambda^\psi) \neq (0, 0, 0, 0)$,
(b): $-\dot{p}(t) = p(t) \cdot f_x(t, \bar{x}(t), \bar{u}(t)) - \lambda L_x(t, \bar{x}(t), \bar{u}(t))$, a.e. $t \in [S, T]$,
(c): $p(t) \cdot f(t, \bar{x}(t), \bar{u}(t)) - \lambda L(t, \bar{x}(t), \bar{u}(t)) =$
$\max_{u \in U}\{p(t) \cdot f(t, \bar{x}(t), u) - \lambda L(t, \bar{x}(t), u)\}$ a.e. $t \in [S, T]$,
(d): $-p(T) = \lambda g_x(\bar{x}(T)) + \lambda^\phi \cdot \phi_x(\bar{x}(T)) + \lambda^\psi \cdot \psi_x(\bar{x}(T))$.

(The validity of maximum principle in this form follows from the earlier stated maximum principle (Theorem 1.5.1) has been established in the preceding analysis, in the case when the non-degeneracy condition (1.5.2) is satisfied, If the data violates

1.5 The Maximum Principle

the non-degeneracy condition, however, there exists $(\bar{\lambda}^\phi, \bar{\lambda}^\psi) \in \Lambda(\bar{x}(T))$ such that $(\bar{\lambda}^\phi, \bar{\lambda}^\psi) \neq (0, 0)$; in this case the assertions of the theorem are still true, in a trivial sense, with $p \equiv 0$, $\lambda = 0$ and $(\lambda^\phi, \lambda^\phi) = (\bar{\lambda}^\phi, \bar{\lambda}^\psi)$.)

A self contained proof of the maximum principle for functional endpoint constraints is given in the Appendix to this chapter.

Alternative statements of the maximum principle (either form) can be given in terms of the *un-maximized Hamiltonian*

$$\mathcal{H}^\lambda(t, x, p, u) := p \cdot f(t, x, u) - \lambda L(t, x, u).$$

The costate equation (augmented by the state equation $\dot{x} = f$) and the generalized Weierstrass condition can be written

$$(-\dot{p}(t), \dot{\bar{x}}(t)) = \mathcal{H}^\lambda_{x,p}(t, \bar{x}(t), p(t), \bar{u}(t)) \quad \text{a.e.}$$

and

$$\mathcal{H}^\lambda(t, \bar{x}(t), p(t), \bar{u}(t)) = \max_{u \in U} \mathcal{H}^\lambda(t, \bar{x}(t), p(t), u) \quad \text{a.e.,}$$

a form of the conditions which emphasizes their affinity with Hamilton's system of equations in the calculus of variations.

In favorable circumstances, we are justified in setting the cost multiplier $\lambda = 1$, and the generalized Weierstrass condition permits us to express u as a function of x and p

$$u = u^*(x, p).$$

The maximum principle then asserts that a minimizing state trajectory $\bar{x}$ is the first component of a pair of absolutely continuous functions $(\bar{x}, p)$ satisfying the differential equation

$$(-\dot{p}(t), \dot{\bar{x}}(t)) = \mathcal{H}^{\lambda=1}_{x,p}(t, \bar{x}(t), p(t), u^*(\bar{x}(t), p(t))) \quad \text{a.e.} \qquad (1.5.3)$$

together with the end-point conditions

$$\bar{x}(S) = x_0, \quad \bar{x}(T) \in C$$

and, for the general maximum principle Theorem 1.5.1,

$$-p(T) \in g_x(\bar{x}(T)) + N_C(\bar{x}(T)).$$

The minimizing control is given by the formula

$$\bar{u}(t) = u^*(\bar{x}(t), p(t)).$$

Notice that the (vector) differential Eq. (1.5.3) is a system of $2n$ scalar, first order differential equations for $(p, \bar{x})$. If C is a $(n-k)$-dimensional manifold, specified by k scalar functional constraints on the end-points of $\bar{x}$, then the endpoint constraints on state trajectories impose $n+k$ boundary conditions. The transversality condition on the right end-point of p imposes a further $n-k$ conditions. Thus the 'two point boundary value problem' which we must solve to obtain $(p, \bar{x})$ and p has the 'right' number of end-point conditions, namely $n + k + (n-k) = 2n$.

To explore the relationship between the maximum principle and classical conditions in the calculus of variations, let us consider the following refinement of the basic problem of the calculus of variations, in which a pathwise constraint on the velocity variable '$\dot{x}(t) \in U$' is imposed. (U is a given Borel subset of $\mathbb{R}^n$.)

$$\begin{cases} \text{Minimize } \int_S^T L(t, x(t), \dot{x}(t)) dt \\ \text{over arcs } x \text{ satisfying} \\ \dot{x}(t) \in U, \\ x(S) = x_0 \text{ and } x(T) \in C. \end{cases} \quad (1.5.4)$$

This problem will be recognized as a special case of (P), in which the dynamic and pathwise control constraints are $\dot{x}(t) = u(t)$ and $u(t) \in U$, respectively.

Let $\bar{x}$ be a minimizer for (1.5.4). Under appropriate hypotheses, the maximum principle supplies $\lambda \geq 0$ and a costate arc q satisfying

$$(q, \lambda) \neq (0, 0), \quad (1.5.5)$$

$$-\dot{q}(t) = -\lambda L_x(t, \bar{x}(t), \dot{\bar{x}}(t)), \quad (1.5.6)$$

$$q(T) = \xi'$$

for some $\xi' \in N_C(\bar{x}(T))$, and

$$q(t) \cdot \dot{\bar{x}}(t) - \lambda L(t, \bar{x}(t), \dot{\bar{x}}(t)) = \max_{v \in U}\{q(t) \cdot v - \lambda L(t, \bar{x}(t), v)\}. \quad (1.5.7)$$

We now impose the 'normality condition':

(CQ): λ can be chosen strictly positive.

Two special cases, in either of which (CQ) is automatically satisfied, are

(i) $\dot{\bar{x}}(t) \in \text{int } U$ on a subset of $[S, T]$ having positive measure,
(ii) $\bar{x}(T) \in \text{int } C$.

(It is left to the reader to check that, if the condition $\lambda > 0$ is violated, then, in either case, (1.5.6) and (1.5.7) imply $(q, \lambda) = (0, 0)$, in contradiction of (1.5.5).)

In terms of $p(t) := \lambda^{-1} q(t)$ and $\xi := \lambda^{-1} \xi'$, these conditions can be expressed

$$\dot{p}(t) = L_x(t, \bar{x}(t), \dot{\bar{x}}(t)) \quad \text{a.e.} \quad (1.5.8)$$

$$-p(T) = \xi, \quad (1.5.9)$$

1.5 The Maximum Principle

for some $\xi \in N_C(\bar{x}(T))$ and

$$p(t) \cdot \dot{\bar{x}}(t) - L(t, \bar{x}(t), \dot{\bar{x}}(t)) = \max_{v \in U}\{p(t) \cdot v - L(t, \bar{x}(t), v)\}. \quad (1.5.10)$$

(We have used the fact that $N_C(\bar{x}(T))$ is a cone). When $U = \mathbb{R}^n$, the Generalized Weierstrass Condition (1.5.10) implies

$$p(t) = L_v(t, \bar{x}(t), \dot{\bar{x}}(t)) \text{ a.e.}.$$

(L_v denotes the derivative of $L(t, x, .)$.) This combines with (1.5.8) to give

$$(\dot{p}(t), p(t)) = L_{x,v}(t, \bar{x}(t), \dot{\bar{x}}(t)) \text{ a.e.}.$$

We have shown that the maximum principle subsumes the Euler Lagrange condition and the Weierstrass condition. But is has much wider implications, because it covers problems with pathwise velocity constraints.

The innovative aspects of the maximum principle, in relation to classical optimality conditions, are most clearly revealed when it is compared with the Hamilton condition. For Problem (1.5.4), the maximum principle associates with a minimizer $\bar{x}$ a costate arc p satisfying relationships (1.5.8)–(1.5.10) above. These can be expressed in terms of the un-maximized Hamiltonian

$$\mathcal{H}(t, x, p, v) = p \cdot v - L(t, x, v)$$

as

$$(-\dot{p}(t), \dot{\bar{x}}(t)) = \mathcal{H}_{x,p}(t, \bar{x}(t), p(t), \dot{\bar{x}}(t))$$

and

$$\mathcal{H}(t, \bar{x}(t), p(t), \dot{\bar{x}}(t)) = \max_{v \in U}\{\mathcal{H}(t, \bar{x}(t), p(t), v)\}.$$

On the other hand, classical conditions in the form of Hamilton's equations, which are applicable in the case $U = \mathbb{R}^n$ are also a set of differential equations satisfied by $\bar{x}$ and a costate arc p

$$(-\dot{p}(t), \dot{\bar{x}}(t)) = H_{x,p}(t, \bar{x}(t), p(t)),$$

where H is the Hamiltonian

$$H(t, x, p) = \max_{v \in \mathbb{R}^n}\{p \cdot v - L(t, x, v)\}.$$

Both sets of necessary conditions are intimately connected with the function $\mathcal{H}$ and its supremum H with respect to the velocity variable. However there is a crucial difference. In Hamilton's classical necessary conditions, the supremum-taking operation is applied *before* the function $\mathcal{H}$ is differentiated and inserted into the vector differential equation for p and $\bar{x}$. In the maximum principle applied to Problem (1.5.4), by contrast, supremum-taking is carried out only *after* the differential equations for p and $\bar{x}$ have been assembled.

One advantage of postponing supremum taking is that we can dispense with differentiability hypotheses on the Hamiltonian $H(t, ., .)$. (If the data $L(t, ., .)$ is differentiable, it follows immediately that $\mathcal{H}(t, ., ., .)$ is differentiable, as required for derivation of the maximum principle, but it does *not* follow in general that the derived function $H(t, ., .)$ is differentiable, as required for derivation of the Hamilton condition.)

Other significant advantages are that conditions (1.5.8)–(1.5.10) are valid in circumstances when the velocity $\bar{x}(t)$ is constrained to lie in some set U, and that they may be generalized to cover problems involving differential equation constraints. These are the features which make the maximum principle so well suited to problems in advanced engineering design, economics and elsewhere.

1.6 Dynamic Programming

We have seen how necessary conditions from the calculus of variations evolved into the maximum principle, to take account of pathwise constraints encountered in advanced engineering design and dynamic decision making. What about optimality conditions related to the Hamilton Jacobi equation? Here, too, long-established techniques in the calculus of variations can be adapted to cover present day applications.

It is convenient to discuss these developments, referred to as dynamic programming, in relation to the special case of the problem studied earlier, in which the right end-point of the state trajectory is unconstrained.

$$(I) \begin{cases} \text{Minimize } J(x) := \int_S^T L(t, x(t), u(t))dt + g(x(T)) \\ \text{over measurable functions } u : [S, T] \to \mathbb{R}^m \\ \qquad \text{and } x \in W^{1,1}([S, T]; \mathbb{R}^n) \text{ satisfying} \\ \dot{x}(t) = f(t, x(t), u(t)) \text{ a.e.,} \\ u(t) \in U \text{ a.e.,} \\ x(S) = x_0, \end{cases}$$

the data for which comprise an interval $[S, T]$, a set $U \subset \mathbb{R}^m$, a point $x_0 \in \mathbb{R}^n$ and functions $f : [S, T] \times \mathbb{R}^n \times \mathbb{R}^m \to \mathbb{R}^n$, $L : [S, T] \times \mathbb{R}^n \times \mathbb{R}^m \to \mathbb{R}$ and $g : \mathbb{R}^n \to \mathbb{R}$.

It is assumed that f, L and g are continuously differentiable functions and that f satisfies additional assumptions ensuring that, in particular, the differential equation $\dot{y}(s) = f(s, y(s), u(s))$, $y(t) = \xi$, has a unique solution y on $[t, T]$ for an arbitrary control function u, initial time $t \in [S, T]$ and initial state $\xi \in \mathbb{R}^n$.

1.6 Dynamic Programming

The Hamilton Jacobi equation for this problem is:

$$(HJE)' \begin{cases} \phi_t(t,x) + \min_{u \in U}\{\phi_x(t,x) \cdot f(t,\xi,u) + L(t,\xi,u)\} = 0 \\ \qquad\qquad\qquad\qquad\qquad\text{for all } (t,x) \in (S,T) \times \mathbb{R}^n, \\ \phi(T,x) = g(x) \text{ for all } x \in \mathbb{R}^n. \end{cases}$$

Finding a suitable solution to this equation is one possible approach to verifying optimality of a putative minimizer. This extension of Carathéodory's verification technique for problem (I) is summarized as the following optimality condition: the derivation is along very similar lines to the proof of Proposition 1.4.1.

Theorem 1.6.1 *Let $(\bar{x}, \bar{u})$ satisfy the constraints of the dynamic optimization problem (I). Suppose that there exists $\phi \in C^1$ satisfying $(HJE)'$ and also*

$$\phi_x(t, \bar{x}(t)) \cdot f(t, \bar{x}(t), \bar{u}(t)) + L(t, \bar{x}(t), \bar{u}(t))$$
$$= \min_{u \in U}\{\phi_x(t, \bar{x}) \cdot f(t, \bar{x}(t), u) + L(t, \bar{x}(t), u)\}. \qquad (1.6.1)$$

Then

(a) $(\bar{x}, \bar{u})$ is a minimizer for (I),
(b) $\phi(S, x_0)$ is the minimum cost for (I).

The natural candidate for 'verification function' ϕ is the value function $V : [S, T] \times \mathbb{R}^n \to \mathbb{R}$ for (I). For $(t, x) \in [S, T] \times \mathbb{R}^n$ we define

$$V(t, x) := \inf(I_{t,x})$$

where the right side indicates the infimum cost for a modified version of (I), in which the initial time and state (S, x_0) is replaced by (t, x):

$$(I_{t,x}) \begin{cases} \text{Minimize } \int_t^T L(s, y(s), u(s))ds + g(x(T)) \\ \text{over measurable functions } u : [t, T] \to \mathbb{R}^m \\ \qquad\qquad \text{and } y \in W^{1,1}([t, T]; \mathbb{R}^n) \text{ satisfying} \\ \dot{y}(s) = f(s, y(s), u(s)) \text{ a.e. } s \in [t, T], \\ u(s) \in U \text{ a.e. } s \in [t, T], \\ x(t) = x. \end{cases}$$

Indeed, we can mimic the proof of Theorem 1.4.2 to show

Theorem 1.6.2 *Let V be the value function for (I). Suppose that*

(i) V is a continuously differentiable function,
(ii) For each $(t, x) \in [S, T] \times \mathbb{R}^n$, the optimization problem $(I_{t,x})$ has a minimizer with continuous control function.

Then V is a solution to (HJE).

One reason for the special significance of the value function in dynamic optimization is its role in the solution of the 'optimal synthesis problem'. A feature of many engineering applications of dynamic optimization is that knowledge of an optimal control strategy is required for a variety of initial states and times. This is because the initial state typically describes the deviation of plant variables from their nominal values, due to disturbances; the requirements of designing a control system to correct this deviation, regardless of the magnitudes and time of occurrence of the disturbances, make necessary consideration of more than one initial state and time. The *optimal synthesis problem* is that of obtaining a solution to $(I_{t,x})$ in feedback form such that the functional dependence involved is independent of the data point $(t, x) \in [S, T] \times \mathbb{R}^n$. To be precise, we seek a function $G : [S, T] \times \mathbb{R}^n \to \mathbb{R}^m$ such that the feedback equation

$$u(s) = G(s, y(s)),$$

together with the dynamic constraints and initial condition,

$$\begin{cases} \dot{y}(s) = f(s, y(s), u(s)) \text{ for a.e. } s \in [\tau, T], \\ y(\tau) = \xi \end{cases}$$

can be solved to yield a minimizer (y, u) for $(I_{\tau,\xi})$. Furthermore, it is required that G is independent of (τ, ξ). We mention that feedback implementation, which is a dominant theme in control engineering, has many significant benefits besides the scope it offers for treatment of arbitrary initial data.

Consider again the dynamic programming sufficient condition of optimality. Let ϕ be a continuously differentiable function that satisfies (HJE)'. For each (t, x) define the set $Q(s, x) \subset \mathbb{R}^m$

$$Q(t, x) := \{v' : \phi_x(t, x) \cdot f(t, x, v') + L(t, x, v') = \min_{v \in U} \{\phi_x(t, x) \cdot f(t, x, v) + L(t, x, v)\}.$$

Let us assume that Q is single valued and that, for each $(t, x) \in [S, T] \times \mathbb{R}^n$, the equations

$$\begin{aligned} \dot{y}(s) &= f(s, y(s), u(s)) \quad \text{a.e. } s \in [t, T], \\ u(s) &= Q(s, y(s)) \quad \text{a.e. } s \in [t, T], \\ y(t) &= x \end{aligned}$$

have a solution y on $[t, T]$. Then, by the sufficient condition, $(y, u(s) = Q(s, x(s))$ is a minimizer for $(I_{t,x})$, whatever the initial data point (t, x) happens to be. Evidently, Q solves the synthesis problem.

We next discuss the relationship between dynamic programming and the maximum principle. The question arises, if ϕ is a solution to (HJE)' and if $(\bar{x}, \bar{u})$ is a minimizer for (I), can we interpret costate arcs for $(\bar{x}, \bar{u})$ in terms of ϕ? The nature of the relationship can be surmised from the following calculations.

1.6 Dynamic Programming

Assume that ϕ is twice continuously differentiable. We deduce from (HJE)' that, for each $t \in [S, T]$,

$$\phi_x(t, \bar{x}(t)) \cdot f(t, \bar{x}(t), \bar{u}(t)) + L(t, \bar{x}(t), \bar{u}(t))$$
$$= \min_{v \in U}\{\phi_x(t, \bar{x}(t)) \cdot f(t, \bar{x}(t), v) + L(t, \bar{x}(t), v)\} \qquad (1.6.2)$$

and

$$\phi_t(t, \bar{x}(t)) + \phi_x(t, \bar{x}(t)) \cdot f(t, \bar{x}(t), \bar{u}(t)) + L(t, \bar{x}(t), \bar{u}(t)) = 0. \qquad (1.6.3)$$

Since ϕ is assumed to be twice continuously differentiable, we deduce from (1.6.3) that

$$\phi_{tx} + \phi_{xx} \cdot f + \phi_x \cdot f_x + L_x = 0. \qquad (1.6.4)$$

(In this equation, all terms are evaluated at $(t, \bar{x}(t), \bar{u}(t))$.)

Now define the continuously differentiable functions p and h as follows:

$$p(t) := -\phi_x(t, \bar{x}(t)), \quad h(t) := \phi_t(t, \bar{x}(t)). \qquad (1.6.5)$$

We have

$$-\dot{p}(t) = \phi_{tx}(t, \bar{x}(t)) + \phi_{xx}(t, \bar{x}(t)) \cdot f(t, \bar{x}(t), \bar{u}(t))$$
$$= -\phi_x(t, \bar{x}(t)) \cdot f_x(t, \bar{x}(t), \bar{u}(t)) - L_x(t, \bar{x}(t), \bar{u}(t)),$$

by (1.6.4). It follows from (1.6.5) that

$$-\dot{p}(t) = p(t) \cdot f_x(t, \bar{x}(t), \bar{u}(t)) - L_x(t, \bar{x}(t), \bar{u}(t)) \text{ for all } t \in [S, T]. \qquad (1.6.6)$$

We deduce from the boundary condition for (HJE)' and the definition of p that

$$-p(T) \ (= \phi_x(T, \bar{x}(T))) \ = \ \nabla g(\bar{x}(T)). \qquad (1.6.7)$$

From (1.6.2) and (1.6.5)

$$p(t) \cdot f(t, \bar{x}(t), \bar{u}(t)) - L(t, \bar{x}(t), \bar{u}(t))$$
$$= \max_{v \in U}\{p(t) \cdot f(t, \bar{x}(t), v) - L(t, \bar{x}(t), v)\}. \qquad (1.6.8)$$

We deduce from (1.6.3) and (1.6.8) that

$$h(t) = H(t, \bar{x}(t), p(t)) \qquad (1.6.9)$$

where, as usual, H is the Hamiltonian

$$H(t, x, p) := \max_{v \in U} \{p(t) \cdot f(t, \bar{x}(t), v) - L(t, \bar{x}(t), v)\}.$$

Conditions (1.6.6)–(1.6.9) tell us that, when a smooth verification function ϕ exists, then a form of the maximum principle is valid, in which the costate arc p is related to the gradient $\nabla \phi$ of ϕ according to

$$(H(t, \bar{x}(t), p(t)), -p(t)) = \nabla \phi(t, \bar{x}(t)) \quad \text{for all } t \in [S, T]. \tag{1.6.10}$$

As a proof of the maximum principle, the above analysis is deficient because the underlying hypothesis that there exists a smooth verification function is difficult to verify outside simple special cases.

However the relationship (1.6.10) is of considerable interest, for the following reasons. It is sometimes useful to know how the minimum cost of a dynamic optimization problem depends on parameter values. In engineering applications for example, such knowledge helps us to assess the extent to which optimal performance is degraded by parameter drift (variation of parameter values due to ageing of components, temperature changes, etc.). If the parameters comprise the initial time and state components (t, x), the relevant information is supplied by the value function which, under favourable circumstances, can be obtained by solving the Hamilton Jacobi equation. However for high dimensional, nonlinear problems it is usually not feasible to solve the Hamilton Jacobi equation. On the other hand, computational schemes are available for calculating pairs of functions $(\bar{x}, \bar{u})$ satisfying the conditions of the maximum principle and accompanying costate arcs p, even for high dimensional problems. Of course the costate arc is calculated for the specified initial data $(t, x) = (S, x_0)$ and does not give a full picture of how the minimum cost depends on the initial data in general. The costate arc does however at least supply gradient (or 'sensitivity') information about this dependence near the nominal initial data $(t, x) = (S, x_0)$: in situations where there is a unique costate arc associated with a minimizer and the value function V is a smooth solution to the Hamilton Jacobi equation, we have from (1.6.10) that

$$\nabla V(S, x_0) = (H(S, x_0, p(S)), -p(S)).$$

A major weakness of classical dynamic programming, as summarized above, is that it is based on the hypothesis that the value function is continuously differentiable. (For some aspects of the classical theory, even stronger regularity properties must be invoked.) This hypothesis is violated in many cases of interest.

The historical development of necessary conditions of optimality and of dynamic programming in dynamic optimization could not be more different. Necessary conditions made a rapid start, with the early conceptual breakthroughs involved in the formulation and rigorous derivation of the maximum principle. The maximum principle remains a key optimality condition to this day. Dynamic programming, by contrast, was a slow beginner. Early developments (in part reviewed above) were fairly obvious extensions of well-known techniques in the calculus of variations. In the 1960's dynamic programming was widely judged to lack a rigorous foundation: it aimed to provide a general approach to the solution of dynamic optimization problems yet dynamic programming, as originally formulated, depended on regularity properties of the value function which frequently it did not possess. The conceptual breakthroughs, required to elevate the status of dynamic programming from that of a useful heuristic tool to a cornerstone of modern dynamic optimization, did not occur until the 1970's. They involved a complete re-appraisal of what we mean by 'solution' to the Hamilton Jacobi equation.

1.7 Nonsmoothness

By the early 1970's, it was widely recognized that attempts to broaden the applicability of available 'first order' necessary conditions and also to put dynamic programming on a rigorous footing were being impeded by a common obstacle: a lack of suitable techniques for analysing local properties of non-differentiable functions and of sets with non-differentiable boundaries.

Consider first dynamic programming, where the need for new analytic techniques is most evident. This focuses attention on the relationship between dynamic optimization problems such as

$$(I) \begin{cases} \text{Minimize } J(x) := \int_S^T L(t, x(t), \dot{x}(t))dt + g(x(T)) \\ \text{over measurable functions } u : [S, T] \to \mathbb{R}^m \\ \qquad \text{and absolutely continuous arcs } x \text{ satisfying} \\ \dot{x}(t) = f(t, x(t), u(t)) \text{ a.e.,} \\ u(t) \in U \text{ a.e.,} \\ x(S) = x_0 \end{cases}$$

and the Hamilton Jacobi equation

$$(\text{HJE})' \begin{cases} \phi_t(t, x) + \min_{u \in U} \{\phi_x(t, x) \cdot f(t, x, u) + L(t, x, u))\} = 0 \\ \qquad\qquad\qquad\qquad\qquad \text{for all } (t, x) \in (S, T) \times \mathbb{R}^n, \\ \phi(T, x) = g(x) \text{ for all } x \in \mathbb{R}^n. \end{cases}$$

The elementary theory tells us that, if the value function V is continuously differentiable (and various other hypotheses are satisfied) then V is a solution to (HJE)'. One shortcoming is that it does not exclude the existence of other solutions to the (HJE)' and, therefore, does not supply a full characterization of the value function in terms of solutions to (HJE)'. Another, more serious, shortcoming is the severity of the hypothesis 'V is continuously differentiable', which is extensively invoked in the elementary analysis. In illustration of problems in which this hypothesis is violated, consider the following example:

$$\begin{cases} \text{Minimize } x(1) \\ \text{over measurable functions } u : [0, 1] \to \mathbb{R} \\ \qquad \text{and } x \in W^{1,1}([0, 1]; \mathbb{R}) \text{ satisfying} \\ \dot{x}(t) = xu \text{ a.e.,} \\ u(t) \in [-1, +1] \text{ a.e.,} \\ x(0) = 0. \end{cases}$$

The Hamilton Jacobi equation in this case takes the form

$$\begin{cases} \phi_t(t, x) - |\phi_x(t, x)x| = 0 & \text{for all } (t, x) \in (0, 1) \times \mathbb{R}, \\ \phi(1, x) = x & \text{for all } x \in \mathbb{R}. \end{cases}$$

The value function is

$$V(t, x) = \begin{cases} xe^{-(1-t)} & \text{if } x \geq 0 \\ xe^{+(1-t)} & \text{if } x < 0. \end{cases}$$

We see that V satisfies the Hamilton Jacobi equation on $\{(t, x) \in (0, 1) \times \mathbb{R} : x \neq 0\}$. However V cannot be said to be a classical solution because V is non-differentiable on the subset $\{(t, x) \in (0, 1) \times \mathbb{R} : x = 0\}$.

What is significant about this example is that the non-differentiability of the value function here encountered is by no means exceptional. It is a simple instance of a kind of dynamic optimization problem frequently encountered in engineering design, in which the dynamic constraint

$$\dot{x} = f(t, x, u)$$

has right side affine in the u variable, the control variable is subject to simple magnitude constraints and the cost depends only on the terminal value of the state variable; for these problems, we *expect* the value function to be non-differentiable.

How can the difficulties associated with non-differentiable value functions be overcome? Of course we can choose to define non-differentiable solutions of a given partial differential equation in a number of different ways. However the particular challenge in dynamic programming is to come up with a definition of solution, according to which the value function is the unique solution.

Unfortunately we must reject traditional interpretations of non-differentiable solution to partial differential equations based on distributional derivatives, since the theory of 'distribution sense solutions' is essentially a *linear* theory which is ill-matched to the *nonlinear* Hamilton Jacobi equation. It might be thought, on the other hand, that a definition based on 'almost everywhere' satisfaction of the Hamilton

1.7 Nonsmoothness

Jacobi equation would meet our requirements. But such a definition is simply too coarse to provide a characterization of the value function under hypotheses of any generality.

Clearing the 'nondifferentiable value functions' bottleneck in dynamic programming, encountered in the late 1970's, was a key advance in dynamic optimization. Success in relating the value function and the Hamilton Jacobi equation was ultimately achieved under hypotheses of considerable generality by introducing new solution concepts, 'viscosity' solutions and related notions of a generalized solution, based on a fresh look at local approximation of nondifferentiable functions.

The need to 'differentiate the un-differentiable' arises also when we attempt to derive necessary conditions for dynamic optimization problems not covered by the maximum principle. Prominent among these are problems

$$\begin{cases} \text{Minimize } g(x(T)) \\ \text{over measurable functions } u : [S, T] \to \mathbb{R}^m \\ \quad \text{and } x \in W^{1,1}([S, T]; \mathbb{R}^n) \text{ satisfying} \\ \dot{x}(t) = f(t, x(t), u(t)) \text{ a.e.,} \\ u(t) \in U(t, x(t)) \text{ a.e.,} \\ x(S) = x_0 \text{ and } x(T) \in C, \end{cases} \qquad (1.7.1)$$

involving a *state dependent* control constraint set $U(t, x) \subset \mathbb{R}^m$. Problems with state dependent control constraint sets are encountered in flight mechanics applications where, for example, the control variable $u(t)$ is the engine thrust. For large excursions of the state variables, it is necessary to take account of the fact that the upper and lower bounds on the thrust depend on atmospheric pressure and therefore on altitude. Since altitude is a state component, the control function is required to satisfy a constraint of the form

$$u(t) \in U(t, x)$$

in which

$$U(t, x) := \{u : a^-(t, x) \le u \le a^+(t, x)]$$

and $a^-(t, x)$ and $a^+(t, x)$ are respectively the state dependent lower and upper bounds on the thrust at time t. In this simple case, it is possible to reduce to the state-independent control constraint case, by re-defining the problem. (The state dependence of the control constraint set can be absorbed into the dynamic constraint.) But in other cases, involving several interacting control variables, each subject to state-dependent constraints, this may no longer be possible or convenient.

A natural framework for studying problems with state dependent control constraints is provided by:

$$\begin{cases} \text{Minimize } g(x(T)) \\ \text{over arcs } x \in W^{1,1} \text{ satisfying} \\ \dot{x}(t) \in F(t, x(t)) \text{ a.e.,} \\ x(S) = x_0 \text{ and } x(T) \in C. \end{cases} \quad (1.7.2)$$

Now, the dynamic constraint takes the form of a *differential inclusion*:

$$\dot{x}(t) \in F(t, x(t)) \quad \text{a.e.}$$

in which, for each (t, x), $F(t, x)$ is a given subset of $\mathbb{R}^n$. The perspective here is that, in dynamic optimization, the fundamental choice variables are state trajectories, not control function/state trajectory pairs, and the essential nature of the dynamic constraint is made explicit by identifying the set $F(t, x(t))$ of allowable values of $\dot{x}(t)$. The earlier problem, involving a state dependent control constraint set can be fitted to this framework by choosing F to be the multifunction

$$F(t, x) := \{v = f(t, x, u) : u \in U(t, x)\}. \quad (1.7.3)$$

Indeed it can be shown, under mild hypotheses on the data for problem (1.7.1), that the two problems (1.7.1) and (1.7.2) are equivalent for this choice of F in the sense that $\bar{x}$ is a minimizer for (1.7.2) if and only if $(\bar{x}, \bar{u})$ is a minimizer for (1.7.1) (for some $\bar{u}$).

By what means can we derive necessary conditions for problem (1.7.2)? One approach is to seek necessary conditions for a related problem, from which the dynamic constraint has been eliminated by means of a penalty function:

$$\begin{cases} \text{Minimize } \int_S^T L(t, x(t), \dot{x}(t))dt + g(x(T)) \\ \text{over } x \in W^{1,1}([S, T]; \mathbb{R}^n) \text{ satisfying} \\ x(S) = x_0 \text{ and } x(T) \in C. \end{cases} \quad (1.7.4)$$

Here

$$L(t, x, v) := \begin{cases} 0 & \text{if } v \in F(t, x) \\ +\infty & \text{if } v \notin F(t, x). \end{cases}$$

(Notice that problems (1.7.2) and (1.7.4) have the same cost at an arc x satisfying the constraints of (1.7.2). On the other hand, if x violates the constraints of (1.7.2), it is excluded from consideration as a minimizer for (1.7.4), since the penalty term in the cost ensures that it has infinite cost.)

We have arrived at a problem in the calculus of variations, albeit one with discontinuous data. It is natural then to seek necessary conditions for a minimizer $\bar{x}$ in the spirit of classical conditions, including the Euler Lagrange and Hamilton conditions:

$$(\dot{p}(t), p(t)) \in \text{`}\partial_{x,v}\text{'} L(t, \bar{x}(t), \dot{\bar{x}}(t))$$

and

$$(-\dot{p}(t), \dot{\bar{x}}(t)) \in \text{`}\partial_{x,p}\text{'} H(t, \bar{x}(t), p(t))$$

for some p. Here

$$H(t, x, p) := \max_{v \in \mathbb{R}^n} \{p \cdot v - L(t, x, v)\}.$$

These conditions involve derivatives of the functions L and H. Yet, in the present context, L is discontinuous and H is, in general, nondifferentiable. How then should these relationships be interpreted?

These difficulties were decisively overcome in the 1970's and a new chapter of dynamic optimization was begun, with the introduction by F. H. Clarke, in his seminal paper [57], of new concepts of local approximation of non-differentiable functions and of a supporting calculus for applications in variational analysis.

1.8 Nonsmooth Analysis

Nonsmooth analysis is a branch of nonlinear analysis concerned with the local approximation of non-differentiable functions and of sets with non-differentiable boundaries. Since its inception in the early 1970's, there has been a sustained and fruitful interplay between nonsmooth analysis and dynamic optimization. A familiarity with nonsmooth analysis is, therefore, essential for an in-depth understanding of present day research in dynamic optimization. This section provides a brief, informal review of material on nonsmooth analysis, which will be covered in far greater detail in Chaps. 4 and 5.

The guiding question is:

How should we adapt classical concepts of outward normals to subsets of vector spaces with smooth boundaries and of gradients of differentiable functions, to cover situations in which the subsets have nonsmooth boundaries and the functions are nondifferentiable?

There is no single answer to this question. The need to study local approximations of nonsmooth functions arises in many branches of analysis and a wide repertoire of techniques has been developed to meet the requirements of different applications.

In this book emphasis is given to proximal normal vectors, proximal subgradients and limits of such vectors originating in Clarke's early work [57]. This reflects their prominence in applications to dynamic optimization.

Consider first generalizations of 'outward normal vector' to general closed sets (in a finite dimensional space).

Fig. 1.10 A proximal normal vector

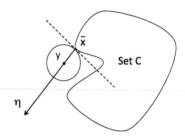

The Proximal Normal Cone *Take a closed set $C \subset \mathbb{R}^k$ and a point $\bar{x} \in C$. A vector $\eta \in \mathbb{R}^k$ is said to be a proximal normal vector to C at $\bar{x}$ if there exists $M \geq 0$ such that*

$$\eta \cdot (x - \bar{x}) \leq M|x - \bar{x}|^2 \quad \text{for all } x \in C. \tag{1.8.1}$$

The cone of all proximal normal vectors to C at $\bar{x}$ is called the proximal normal cone to C at $\bar{x}$ and is denoted by $N_C^P(\bar{x})$:

$$N_C^P(\bar{x}) := \{\eta \in \mathbb{R}^k : \exists M \geq 0 \text{ such that } (1.8.1) \text{ is satisfied}\}.$$

Figure 1.10 provides a geometrical interpretation of proximal normal vectors: η is a proximal normal vector to C at $\bar{x}$ if their exists a point $y \in \mathbb{R}^k$ such that $\bar{x}$ is the closest point to y in C and η is a scaled version of $y - \bar{x}$, i.e. there exists $\alpha \geq 0$ such that

$$\eta = \alpha(y - \bar{x}).$$

The Limiting Normal Cone *Take a closed set $C \subset \mathbb{R}^k$ and a point $\bar{x} \in C$. A vector η is said to be a limiting normal vector to C at $\bar{x} \in C$ if there exist sequences $x_i \xrightarrow{C} \bar{x}$ and $\eta_i \to \eta$ such that*

$$\eta_i \in N_C^P(x_i) \quad \text{for all } i.$$

The cone of limiting normal vectors to C at $\bar{x}$ is denoted $N_C(\bar{x})$:

$$N_C(\bar{x}) := \{\eta \in \mathbb{R}^k : \exists x_i \xrightarrow{C} x \text{ and } \eta_i \to \eta \text{ such that } \eta_i \in N_C^P(x_i) \text{ for all } i\}.$$

(The notation $x_i \xrightarrow{C} x$ indicates that $x_i \to x$ and $x_i \in C$ for all i.)

Figure 1.11 illustrates limiting normal cones at various points in a set with nonsmooth boundary. It can be shown that, if $\bar{x}$ is a boundary point of C, then the limiting normal cone contains non-zero elements. Limiting normal cones are closed cones, but they are not necessarily convex.

We consider next generalizations of the concept of 'gradient' to functions which are not differentiable.

1.8 Nonsmooth Analysis

Fig. 1.11 Limiting normal cones

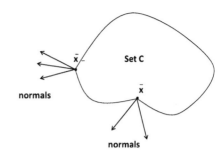

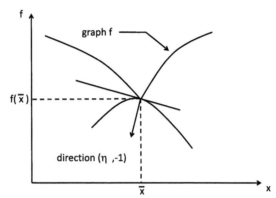

Fig. 1.12 A geometric interpretation of proximal subgradients

The Proximal Subdifferential *Take an extended valued, lower semi-continuous function $f : \mathbb{R}^k \to \mathbb{R} \cup \{+\infty\}$ and a point $\bar{x} \in \text{dom}\{f\}$. A vector $\eta \in \mathbb{R}^k$ is said to be a proximal subgradient of f at $\bar{x}$ if there exist $\epsilon > 0$ and $M \geq 0$ such that*

$$\eta \cdot (x - \bar{x}) \leq f(x) - f(\bar{x}) + M|x - \bar{x}|^2 \tag{1.8.2}$$

for all points x which satisfy $|x - \bar{x}| \leq \epsilon$.

The set of all proximal subgradients of f at $\bar{x}$ is called the proximal subdifferential of f at $\bar{x}$ and is denoted by $\partial^P f(\bar{x})$:

$$\partial^P f(\bar{x}) := \{\text{there exist } \epsilon > 0 \text{ and } M \geq 0 \text{ such that } (1.8.2) \text{ is satisfied }\}.$$

(The notation dom$\{f\}$ denotes the set $\{y : f(y) < +\infty\}$.)

Figure 1.12 provides a geometric interpretation of proximal subgradients: a proximal subgradient to f at $\bar{x}$ is the slope at $x = \bar{x}$ of a paraboloid,

$$y = \eta \cdot (x - \bar{x}) + f(\bar{x}) - M|x - \bar{x}|^2,$$

which coincides with f at $x = \bar{x}$ and which lies on or below the graph of f on a neighbourhood of $\bar{x}$.

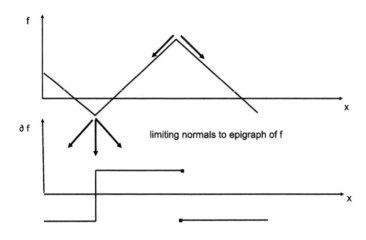

Fig. 1.13 Limiting subdifferentials

The Limiting Subdifferential *Take an extended valued, lower semi-continuous function* $f : \mathbb{R}^k \to \mathbb{R} \cup \{+\infty\}$ *and a point* $\bar{x} \in \text{dom}\{f\}$. *A vector* $\eta \in \mathbb{R}^k$ *is said to be a limiting subgradient of* f *at* $\bar{x}$ *if there exist sequences* $x_i \xrightarrow{f} \bar{x}$ *and* $\eta_i \to \eta$ *such that*

$$\eta_i \in \partial^P f(x_i) \text{ for all } i.$$

The set of all limiting subgradients of f *at* $\bar{x}$ *is called the limiting subdifferential and is denoted by* $\partial f(\bar{x})$:

$$\partial f(\bar{x}) := \{\eta : \exists\, x_i \xrightarrow{f} x \text{ and } \eta_i \to \eta \text{ such that } \eta_i \in \partial^P f(x_i) \text{ for all } i\}.$$

(The notation $x_i \xrightarrow{f} x$ indicates that $x_i \to x$ and $f(x_i) \to f(x)$ as $i \to \infty$.)

Figure 1.13 illustrates how the limiting subdifferential depends on the base point for a nonsmooth function. The limiting subdifferential is a closed set, but it need not be convex. It can happen that

$$\partial f(\bar{x}) \neq -\partial(-f)(\bar{x}).$$

This is because the inequality (1.8.2) in the definition of proximal subgradients does not treat positive and negative values of the function f in a symmetrical fashion.

We mention that there are, in fact, a number of equivalent ways of defining limiting subgradients. As so often in mathematics, it is a matter of expository convenience what we regard as definitions and what we regard as consequences of these definitions. The 'difference quotient' definition above has been chosen for

1.8 Nonsmooth Analysis

this review chapter, because it is often the simplest to use in applications. Another, equivalent, definition based on limiting normals to the epigraph set is chosen in Chap. 4 below, as a more convenient starting point for an in-depth analysis.

We now list a number of important properties of limiting normal cones and limiting subdifferentials.

(A): (Closed Graph)

(i) Take a closed set $C \subset \mathbb{R}^k$ and a point $x \in C$. Then for any convergent sequences $x_i \xrightarrow{C} x$ and $\eta_i \to \eta$ such that

$$\eta_i \in N_C(x_i) \quad \text{for all } i,$$

we have $\eta \in N_C(x)$.

(ii) Take a lower semi-continuous function $f : \mathbb{R}^k \to \mathbb{R} \cup \{+\infty\}$ and a point $x \in \text{dom}\{f\}$. Then for any convergent sequences $x_i \xrightarrow{f} x$ and $\xi_i \to \xi$ such that

$$\xi_i \in \partial f(x_i) \quad \text{for all } i,$$

we have $\xi \in \partial f(x)$.

(B): (Links Between Limiting Normals and Subdifferentials) Take a closed set $C \subset \mathbb{R}^k$ and a point $\bar{x} \in C$. Then

$$N_C(\bar{x}) = \partial \Psi_C(\bar{x})$$

and

$$N_C(\bar{x}) \cap \mathbb{B} = \partial d_C(\bar{x}).$$

Here, $\mathbb{B}$ is the closed unit Euclidean ball, centre the origin, in $\mathbb{R}^k$, $\Psi_C : \mathbb{R}^k \to \mathbb{R} \cup \{+\infty\}$ is the indicator function of the set C

$$\Psi_C(x) := \begin{cases} 0 & \text{if } x \in C \\ +\infty & \text{if } x \notin C \end{cases}$$

and $d_C : \mathbb{R}^k \to \mathbb{R}$ is the distance function of the set C

$$d_C(x) := \inf_{y \in C} |x - y|.$$

It is well known that a Lipschitz continuous function f on $\mathbb{R}^k$ is differentiable almost everywhere with respect to Lebesgue measure. (This is Rademacher's theorem.) It is natural then to ask, what is the relationship between the limiting subdifferential and limits of gradients at neighbouring points? F. H. Clarke provided an answer this question, by proving the following important representation of the limiting subdifferential (or, more precisely, its convex hull):

(C): (Limits of Derivatives) *Consider a lower semi-continuous function $f : \mathbb{R}^k \to \mathbb{R} \cup \{+\infty\}$ and a point $x \in \mathrm{dom}\, f$. Suppose that f is Lipschitz continuous on a neighbourhood of $\bar{x}$. Then, for any subset $S \subset \mathbb{R}^k$ of zero k-dimensional Lebesgue measure, we have*

$$\mathrm{co}\, \partial f(\bar{x}) = \mathrm{co}\, \{\eta : \exists\, x_i \to x \text{ such that } \nabla f(x_i) \text{ exists and}$$
$$x_i \notin S \text{ for all } i \text{ and } \nabla f(x_i) \to \eta\}.$$

Calculations of the limiting subdifferential for a specific function, based on the defining relations, is often a challenge, because of their indirect nature. If the function in question is Lipschitz continuous and we are content to estimate the limiting subdifferential $\partial f(\bar{x})$ by its convex hull $\mathrm{co}\, \partial f(\bar{x})$, then property (C) can simplify the task. A convenient feature of this characterization is that, in examining limits of neighbouring gradients, we are allowed to exclude gradients on an arbitrary null-set S. When f is piecewise smooth, for example, we can exploit this flexibility by hiding in S problematical points (on ridges, at singularities, etc.) and calculating $\mathrm{co}\, \partial f(\bar{x})$ merely in terms of gradients of f on subdomains were it is smooth.

Property (C) is significant also in other respects. It tells us, for example, that, if a lower semi-continuous function $f : \mathbb{R}^n \to \mathbb{R} \cup \{+\infty\}$ is continuously differentiable near a point $\bar{x}$, then $\partial f(\bar{x}) = \{\nabla f(\bar{x})\}$. Since f is differentiable at x if and only if $-f$ is differentiable at x and $\nabla f(x) = -\nabla(-f)(x)$), we also deduce:

(D): (Homogeniety of the Convexified Limiting Subdifferential) *Consider a lower semi-continuous function $f : \mathbb{R}^k \to \mathbb{R} \cup \{+\infty\}$ and a point $\bar{x} \in \mathrm{dom}\, f$. Suppose that f is Lipschitz continuous on a neighbourhood of $\bar{x}$. Then*

$$\mathrm{co}\, \partial f(\bar{x}) = -\mathrm{co}\, \partial(-f)(\bar{x}).$$

Another useful property of Lipschitz continuous functions is

(E): (Bounds on Limiting Subdifferentials) *Consider a lower semi-continuous function $f : \mathbb{R}^k \to \mathbb{R} \cup \{+\infty\}$ and a point $x \in \mathrm{dom}\, f$. Suppose that f is Lipschitz continuous on a neighbourhood of $\bar{x}$ with Lipschitz constant K. Then*

$$\partial f(\bar{x}) \subset K\mathbb{B}.$$

A fundamental property of the gradient of a differentiable function is that it vanishes at a local minimizer of the function. The proximal subdifferential has a similar role in identifying possible minimizers:

(F): (Limiting Subdifferentials at Minima) *Take a lower semi-continuous function $f : \mathbb{R}^k \to \mathbb{R} \cup \{+\infty\}$ and a point $\bar{x} \in \mathrm{dom}\, f$. Assume that, for some $\epsilon > 0$*

$$f(\bar{x}) \leq f(x) \text{ for all } x \in \bar{x} + \epsilon \mathbb{B}.$$

1.8 Nonsmooth Analysis

Then

$$0 \in \partial^P f(\bar{x}).$$

It follows that $0 \in \partial f(\bar{x})$.

In applications, it is often necessary to estimate limiting subdifferentials of composite functions in terms of limiting subgradients of their constituent functions. Fortunately, an extensive calculus is available for this purpose. Some important calculus rules are as follows:

(G): (Positive Homogeneity) *Take a lower semi-continuous function* $f : \mathbb{R}^k \to \mathbb{R} \cup \{+\infty\}$, *a point* $\bar{x} \in \text{dom } f$ *and* $\alpha \geq 0$. *Then*

$$\partial(\alpha f)(\bar{x}) = \alpha \partial f(\bar{x}).$$

(H): (Sum Rule) *Consider a collection* $f_i : \mathbb{R}^k \to \mathbb{R} \cup \{+\infty\}, i = 1, .., n$ *of lower semi-continuous extended valued functions and a point* $\bar{x} \in \cap_{i=1,...,n} \text{dom } f_i$. *Assume that all the* f_i's *except possibly one are Lipschitz continuous on a neighbourhood of* $\bar{x}$. *Then the limiting subdifferential of* $(f_1 + \cdots + f_n)(x) = f_1(x) + \cdots + f_n(x)$ *satisfies*

$$\partial(f_1 + .. + f_n)(\bar{x}) \subset \partial f_1(\bar{x}) + .. + \partial f_n(\bar{x}).$$

(I): (Max Rule) *Consider a collection* $f_i : \mathbb{R}^k \to \mathbb{R} \cup \{+\infty\}, i = 1, .., n$ *of lower semi-continuous extended valued functions and a point* $\bar{x} \in \cap_{i=1,...,n} \text{dom } f_i$. *Assume that all the* f_i's, *except possibly one, are Lipschitz continuous on a neighbourhood of* $\bar{x}$. *Then the limiting subdifferential of* $(\max\{f_1, \ldots, f_n\})(x) = \max\{f_1(x), \ldots, f_n(x)\}$ *satisfies*

$$\partial(\max_{i=1,..,n} f_i)(\bar{x}) \subset \{\sum_{i=1}^n \alpha_i \partial f_i(\bar{x}) : \{\alpha_1, \ldots, \alpha_n\} \in (\mathbb{R}_+)^n, \text{ such that } \sum_{i=1}^n \alpha_i = 1$$

$$\text{and } \alpha_i = 0 \text{ for } \alpha_i \notin I(\bar{x})\},$$

in which

$$I(\bar{x}) := \{i \in \{1, .., n\} : f_i(\bar{x}) = \max_j f_j(\bar{x})\}.$$

(J): (Chain Rule) *Take Lipschitz continuous functions* $G : \mathbb{R}^k \to \mathbb{R}^m$ *and* $g : \mathbb{R}^m \to \mathbb{R}$ *and a point* $\bar{x} \in \mathbb{R}^k$. *Then the limiting subdifferential* $\partial(g \circ G)(\bar{x})$ *of the composite function* $x \to g(G(x))$ *at* $\bar{x}$ *satisfies:*

$$\partial(g \circ G)(\bar{x}) \subset \{\eta : \eta \in \partial(\gamma \cdot G)(\bar{x}) \text{ for some } \gamma \in \partial g(G(\bar{x}))\}.$$

Many of these rules are extensions of familiar principles of differential calculus. (Notice however that, in most cases, equality has been replaced by set inclusion). An exception is the max rule above. The max rule is a truly nonsmooth principle; it has no parallel in traditional analysis because the operation of taking the pointwise maximum of a collection of smooth functions usually generates a nonsmooth function. ('corners' typically occur when there is a change in the function on which the maximum is achieved).

We illustrate the use of the calculus rules by proving a multiplier rule for a mathematical programming problem with functional inequality constraints:

$$\begin{cases} \text{Minimize } f(x) \\ \text{over } x \in \mathbb{R}^k \text{ satisfying} \\ g_i(x) \le 0, \text{ for } i = 1, .., n. \end{cases} \quad (1.8.3)$$

Here $f : \mathbb{R}^k \to \mathbb{R}$ and $g_i : \mathbb{R}^k \to \mathbb{R}$, $i = 1, .., n$, are given functions.

Multiplier Rule I (Inequality Constraints) *Let $\bar{x}$ be a minimizer for problem (1.8.3). Assume that $f, g_i, i = 1, .., n$, are Lipschitz continuous.*

Then there exist non-negative numbers $\lambda_0, \lambda_1, .., \lambda_n$ such that

$$\sum_{i=0}^{n} \lambda_i = 1,$$

$$\lambda_i = 0 \text{ if } g_i(\bar{x}) < 0 \text{ for } i = 1, .., n$$

and

$$0 \in \lambda_0 \partial f(\bar{x}) + \sum_{i=1}^{n} \lambda_i \partial g_i(\bar{x}). \quad (1.8.4)$$

(A related, but sharper, multiplier rule will be proved in Chap. 5.)

Proof Consider the function

$$\phi(x) = \max\{f(x) - f(\bar{x}), g_1(x), .., g_n(x)\}.$$

We claim that $\bar{x}$ minimizes ϕ over $x \in \mathbb{R}^k$. If this were not the case, if would be possible to find x' such that

$$\phi(x') < \phi(\bar{x}) = 0.$$

It would then follow that

$$f(x') - f(\bar{x}) < 0, \ g_1(x') < 0, .., g_n(x') < 0,$$

in contradiction of the minimality of $\bar{x}$. The claim has been confirmed.

1.8 Nonsmooth Analysis

According to Rule (F)

$$0 \in \partial \phi(\bar{x}).$$

We deduce from Rule (I) that there exist non-negative numbers $\lambda_0, .., \lambda_n$ such that

$$\sum_{i=0}^{n} \lambda_i = 1,$$

and

$$0 \in \lambda_0 \partial f(\bar{x}) + \sum_{i=1}^{n} \lambda_i \partial g_i(\bar{x}).$$

Since, for index values i such that $g_i(\bar{x}) < 0$, we have

$$g_i(\bar{x}) < \phi(\bar{x}).$$

Rule (I) supplies the supplementary information that

$$\lambda_i = 0 \quad \text{if} \quad g_i(\bar{x}) < 0.$$

The multiplier rule is proved. □

In the case when $f, g_1, .., g_n$ are continuously differentiable, inclusion (1.8.4) implies

$$0 = \lambda_0 \nabla f(\bar{x}) + \sum_{i=1}^{n} \lambda_i \nabla g_i(\bar{x}),$$

the traditional Fritz-John condition for problems with functional inequality constraints. Nonsmooth analysis supplies a new, simple proof of a classical theorem in the case when the functions involved are continuously differentiable and extends it, by supplying information even in the situations when the functions are not differentiable. Application of nonsmooth calculus rules to some derived, nonsmooth function (in this case ϕ) is a staple of 'nonsmooth' methodology. Notice that, in the above proof, we can expect ϕ to be nonsmooth, even if the functions from which it is constructed are smooth.

Calculus rules are some of the tools of the trade in nonsmooth analysis. In many applications however, these are allied to variational principles. A 'variational principle' is an assertion that some quantity is minimized. Two which are widely used in nonsmooth analysis are the exact penalization theorem and Ekeland's theorem. Suppose that the point $\bar{x} \in \mathbb{R}^k$ is a minimizer for the constrained minimization problem

$$\begin{cases} \text{Minimize } f(x) \text{ over } x \in \mathbb{R}^k \\ \text{satisfying } x \in C, \end{cases}$$

for which the data comprise a function $f : \mathbb{R}^k \to \mathbb{R}$ and a set $C \subset \mathbb{R}^k$. For both analytical and computational reasons, it is often convenient to replace this by an unconstrained problem, in which the new cost

$$f(x) + p(x)$$

is the sum of the former cost and a term $p(x)$ which penalizes violation of the constraint $x \in C$. In the case that f is Lipschitz continuous, the exact penalization theorem says that $\bar{x}$ remains a minimizer when the penalty function p is chosen to be the nonsmooth function

$$p(x) = \hat{K} d_C(x),$$

for some constant $\hat{K}$ sufficiently large. Notice that the penalty term, while nonsmooth, is Lipschitz continuous.

The Exact Penalization Theorem *Take a function $f : \mathbb{R}^k \to \mathbb{R}$ and a closed set $C \subset \mathbb{R}^k$. Assume that f is Lipschitz continuous, with Lipschitz constant K. Let $\bar{x}$ be a minimizer for*

$$\begin{cases} \text{Minimize } f(x) \text{ over } x \in \mathbb{R}^k \\ \text{satisfying } x \in C. \end{cases}$$

Then, for any $\hat{K} \geq K$, $\bar{x}$ is a minimizer also for the unconstrained problem

$$\begin{cases} \text{Minimize } f(x) + \hat{K} d_C(x) \\ \text{over points } x \in \mathbb{R}^k. \end{cases}$$

A more general version of this theorem is proved in Chap. 3.

One important consequence of the exact penalization theorem is that, if $\bar{x}$ is a minimizer for the optimization problem of the theorem statement then, under the hypotheses of the exact penalization theorem,

$$0 \in \partial f(\bar{x}) + N_C(\bar{x}).$$

(It is a straightforward matter to deduce this necessary condition of optimality for an optimization problem with an 'implicit' constraint from the fact that $\bar{x}$ also minimizes $f(x) + K d_C(x)$ over $\mathbb{R}^n$ and Rules (F), (H) and (B) above.)

We turn next to Ekeland's theorem. In plain terms, this states that, if a point x_0 approximately minimizes a function f, then there is a point $\bar{x}$ close to x_0 which is a minimizer for some new function, obtained by adding a small perturbation term to the original function f. See Fig. 3.1 of Chap. 3.

1.8 Nonsmooth Analysis

Ekeland's Theorem Take a complete metric space (X, d), a lower semi-continuous function $f : X \to \mathbb{R} \cup \{+\infty\}$, a point $\bar{x} \in \mathrm{dom}\, f$ and a number $\epsilon > 0$ such that

$$f(\bar{x}) \leq \inf_{x \in X} f(x) + \epsilon.$$

Then there exists $x' \in X$ such that

$$d(x', \bar{x}) \leq \epsilon^{\frac{1}{2}}$$

and

$$f(x) + \epsilon^{\frac{1}{2}} d(x, x')|_{x=x'} = \inf_{x \in X} \{f(x) + \epsilon^{\frac{1}{2}} d(x, x')\}.$$

Ekeland's theorem and the preceding nonsmooth calculus rules can be used, for example, to derive a multiplier rule for a mathematical programming problem involving a functional equality constraint:

$$\begin{cases} \text{Minimize } f(x) \\ \text{over } x \in \mathbb{R}^k \text{ satisfying} \\ G(x) = 0. \end{cases} \qquad (1.8.5)$$

The data for this problem comprise functions $f : \mathbb{R}^k \to \mathbb{R}$ and $G : \mathbb{R}^k \to \mathbb{R}^m$.

Multiplier Rule II (Equality Constraints) *Let $\bar{x}$ be a minimizer for problem (1.8.5). Assume that f and G are Lipschitz continuous.*

Then there exist number $\lambda_0 \geq 0$ and a vector $\lambda_1 \in \mathbb{R}^m$ such that

$$\lambda_0 + |\lambda_1| = 1$$

and

$$0 \in \lambda_0 \partial f(\bar{x}) + \partial(\lambda_1 \cdot G)(\bar{x}). \qquad (1.8.6)$$

Proof The simple proof above of the multiplier rule for inequality constraints suggests that equality constraints can be treated in the same way. We might hope to use the fact that, for the problem with equality constraints, x_0 is an unconstrained minimizer of the function

$$\gamma(x) := \max\{f(x) - f(\bar{x}), |G(x)|\}.$$

By Rule (F), $0 \in \partial \gamma(\bar{x})$. We deduce from the Max Rule (I) that there exist multipliers $\lambda_0 \geq 0$ and $\nu_1 \geq 0$ such that $\lambda_0 + \nu_1 = 1$ and

$$0 \in \lambda_0 \partial f(\bar{x}) + \nu_1 \partial |G(x)||_{x=\bar{x}}. \qquad (1.8.7)$$

We aim then to estimate vectors in $\partial |G(x)| |_{x=\bar{x}}$ using nonsmooth calculus rules, in such a manner as to deduce the asserted multiplier rule. The flaw in this approach is that $0 \in \partial |G(x)| |_{x=\bar{x}}$ with the result that (1.8.7) is trivially satisfied by the multipliers $(\lambda_0, \nu_1) = (0, 1)$. In these circumstances, (1.8.7) cannot provide useful information about $\bar{x}$.

As shown by Clarke [60], Ekeland's Theorem can be used to patch up the argument, however. The difficulty we have encountered is that $0 \in \partial |G(x)| |_{x=\bar{x}}$. The idea is to *perturb* the function γ in such a way that, along some sequence $x_i \to \bar{x}$, (1.8.7) is approximately satisfied at x_i but, significantly, $0 \notin \partial |G(x)| |_{x=x_i}$. The perturbed version of (1.8.7) is no longer degenerate and we can expect to recover the multiplier rule in the limit as $i \to \infty$.

Take a sequence $\epsilon_i \downarrow 0$. For each i define the function

$$\phi_i(x) := \max\{f(x) - f(\bar{x}) + \epsilon_i, |G(x)|\}.$$

Note that $\phi_i(\bar{x}) = \epsilon_i$ and $\phi_i(x) \geq 0$ for all x. It follows that

$$\phi_i(\bar{x}) \leq \inf_{x \in \mathbb{R}^k} \phi(x) + \epsilon_i.$$

In view of the fact that ϕ_i is lower semi-continuous (indeed Lipschitz continuous) with respect to the metric induced by the Eulidean norm, we deduce from Ekeland's theorem that there exists $x_i \in \mathbb{R}^k$ such that

$$|x_i - \bar{x}| \leq \epsilon_i^{\frac{1}{2}}$$

and

$$\tilde{\phi}_i(x_i) \leq \inf_{x \in \mathbb{R}^k} \tilde{\phi}_i(x),$$

where

$$\tilde{\phi}_i(x) := \max\{f(x) - f(\bar{x}) + \epsilon_i, |G(x)|\} + \epsilon_i^{\frac{1}{2}} |x - x_i|.$$

Since the function $x \to |x - x_i|$ is Lipschitz continuous with Lipschitz constant 1, we know that its limiting subgradients are contained in the closed unit ball $\mathbb{B}$. By Rules (E), (F) and (H) then,

$$0 \in \partial \tilde{\phi}_i(x_i) + \epsilon_i^{\frac{1}{2}} \mathbb{B}.$$

Otherwise expressed,

$$0 \in z_i + \epsilon_i^{\frac{1}{2}} \mathbb{B}$$

1.8 Nonsmooth Analysis

for some point

$$z_i \in \partial \max\{f(x) - f(\bar{x}) + \epsilon_i, |G(x)|\}|_{x=x_i}.$$

We now make an important observation: for each i,

$$\max\{f(x_i) - f(\bar{x}) + \epsilon_i, |G(x)|\}|_{x=x_i} > 0.$$

Indeed, if this were not the case, we would have $f(x_i) \le f(\bar{x}) - \epsilon_i$ and $G(x_i) = 0$, in violation of the optimality of $\bar{x}$.

It follows that

$$|G(x_i)| = 0 \quad \text{implies} \quad f(x_i) - f(\bar{x}) + \epsilon_i > |G(x_i)|. \tag{1.8.8}$$

Now apply the max rule (I) to $\tilde{\phi}_i$. We deduce the existence of non-negative numbers λ_0^i and v_1^i such that

$$\lambda_0^i + v_1^i = 1,$$

$$z_i \in \lambda_0^i \partial f(x_i) + v_1^i \partial |G(x_i)|.$$

Furthermore, by (1.8.8),

$$|G(x_i)| = 0 \quad \text{implies} \quad v_1^i = 0. \tag{1.8.9}$$

We note next that, if $y' \ne 0$, then all limiting subgradients of $y \to |y|$ at y' have unit Euclidean norm. (This follows from the fact that $\nabla |y| = y/|y|$ for $y \ne 0$ and rule (C)). We deduce from the chain rule and (1.8.9) that there exists a vector $a_i \in \mathbb{R}^m$ such that $|a_i| = 1$ and

$$z_i \in \lambda_0^i \partial f(x_i) + v_1^i \partial(a_i \cdot G)(x_i).$$

Writing $\lambda_1^i := v_1^i a_i$, we have, by positive homogeniety,

$$\lambda_0^i + |\lambda_1^i| = 1 \tag{1.8.10}$$

and

$$z_i = \lambda_0^i \partial f(x_i) + \partial(\lambda_1^i \cdot G)(x_i).$$

It follows that, for each i, there exist points

$$\gamma_0^i \in \partial f(x_i), \ \gamma_1^i \in \partial(\lambda_1^i \cdot G)(x_i) \text{ and } e_i \in \epsilon_i^{\frac{1}{2}} \mathbb{B}$$

such that
$$0 = \lambda_0^i \gamma_0^i + \gamma_1^i + e_i. \tag{1.8.11}$$

By rule (E), and in view of (1.8.10) and of the Lipschitz continuity of f and G, the sequences

$$\{\gamma_0^i\}, \{\gamma_1^i\}, \{\lambda_0^i\} \text{ and } \{\lambda_1^i\}$$

are bounded. We can therefore arrange, by extracting subsequences, that they have limits γ_0, γ_1, λ_0, λ_1 respectively. Since the λ_0^i's are non-negative, it follows from (1.8.10) that

$$\lambda_0 \geq 0 \quad \text{and} \quad \lambda_0 + |\lambda_1| = 1.$$

But $e_i \to 0$. We deduce from (1.8.11), in the limit as $i \to \infty$, that

$$0 = \lambda_0 \gamma_0 + \gamma_1.$$

It is a straightforward matter to deduce from the fact that the limiting subdifferential has closed graph, the sum rule (H) and Rule (E) that

$$\gamma_0 \in \partial f(\bar{x}) \text{ and } \gamma_1 \in \partial(\lambda_1 \cdot G)(\bar{x}).$$

If follows that

$$0 \in \lambda_0 \partial f(\bar{x}) + \partial(\lambda_1 \cdot G)(\bar{x}).$$

This is what we set out to show. □

In the case when f and g are continuously differentiable, condition (1.8.6) implies

$$0 = \lambda_0 \nabla f(\bar{x}) + \lambda_1 \nabla G(\bar{x}).$$

Once again, the tools of nonsmooth analysis provide both an alternative derivation of a well-known multiplier rule in nonlinear programming for differentiable data and also extensions to nonsmooth cases.

The above proof is somewhat complicated, but it is no more so (if we have the above outlined techniques of nonsmooth analysis at our disposal) than traditional proofs of Lagrange multiplier rules for problems with equality constraints (treating the smooth case), based on the classical inverse function theorem, say. The Lagrange multiplier rule for problems with equality constraints, even for differentiable data, is not an elementary result and we must expect to work for it.

Many known relations in classical analysis involving continuously differentiable functions (calculus rules and first order necessary conditions of optimality in nonlinear programming, for example) can be generalized, with the help of nonsmooth analysis, to allow for functions that are merely Lipschitz continuous or even lower semi-continuous. Variational principles often play a key role in achieving these extensions. The proof of the Lagrange multiplier rule above is included in this review chapter as an illuminating prototype. The underlying ideas will be used repeatedly in future chapters to derive necessary conditions of optimality in dynamic optimization. In many applications, variational principles such as Ekeland's theorem (which concern properties of nondifferentiable functions) carry the burden traditionally borne by the inverse function theorem and its relatives in classical analysis (which concern properties of differentiable functions).

1.9 Nonsmooth Dynamic Optimization

The analytical techniques of the preceding section were developed primarily for the purpose of applying them to dynamic optimization. We now briefly describe how they have been used to overcome fundamental difficulties earlier encountered in this field.

Consider first necessary conditions. As discussed, the following optimization problem, in which the dynamic constraint takes the form of a differential inclusion, serves as a convenient paradigm for a broad class of dynamic optimization problems, including some problems to which the maximum principle is not directly applicable:

$$\begin{cases} \text{Minimize } g(x(T)) \\ \text{over arcs } x \in W^{1,1} \text{ satisfying} \\ \dot{x}(t) \in F(t, x(t)) \text{ a.e.,} \\ x(S) = x_0 \text{ and } x(T) \in C. \end{cases} \quad (1.9.1)$$

Here, $g : \mathbb{R}^n \times \mathbb{R}^n \to \mathbb{R}$ is a given function, $F : [S, T] \times \mathbb{R}^n \rightsquigarrow \mathbb{R}^n$ is a given multifunction and $C \subset \mathbb{R}^n$ is a given subset. We proposed reformulating it as a generalized problem in the calculus of variations:

$$\begin{cases} \text{Minimize } \int_S^T L(t, x(t), \dot{x}(t))dt + g(x(T)) \\ \text{over absolutely continuous arcs } x \text{ satisfying} \\ x(S) = x_0 \text{ and } x(T) \in C, \end{cases}$$

in which L is the extended valued function

$$L(t, x, v) := \begin{cases} 0 & \text{if } v \in F(t, x) \\ +\infty & \text{if } v \notin F(t, x), \end{cases}$$

and deriving generalizations of classical necessary conditions to take account of nonsmoothness of the data. This can be done, with recourse to nonsmooth analysis. The details are given in future chapters.

A decisive advance in this direction was Clarke's Hamiltonian inclusion. This was a generalization, to a nonsmooth setting, of Hamilton's system of equations in the calculus of variations. Clarke's proof of this condition, under an additional convexity hypothesis on the values of the multifunction F, was an early landmark in nonsmooth dynamic optimization. As usual we write

$$H(t, x, p) = \max_{v \in F(t,x)} p \cdot v.$$

Theorem (Clarke's Hamiltonian Inclusion) *Let $\bar{x}$ be a minimizer for problem (1.9.1). Assume that, for some $\epsilon > 0$, the data for the problem satisfy:*

(a) *$F(t, x)$ is nonempty for each $(t, x) \in [S, T] \times \mathbb{R}^n$, $\mathrm{Gr}\, F(t, .)$ is closed for each $t \in [S, T]$ and F is $\mathcal{L} \times \mathcal{B}^n$ measurable,*
(b) *there exists $c \in L^1$ such that*

$$F(t, x) \subset c(t)\mathbb{B} \quad \forall\, x \in \bar{x}(t) + \epsilon\mathbb{B}, \text{ a.e.,}$$

(c) *there exist $k \in L^1$ such that*

$$F(t, x) \subset F(t, x') + k(t)|x - x'|\mathbb{B} \quad \forall\, x, x' \in \bar{x}(t) + \epsilon\mathbb{B}, \text{ a.e.,}$$

(d) *C is closed and g is Lipschitz continuous on $\bar{x}(T) + \epsilon\mathbb{B}$.*

Then there exist $p \in W^{1,1}([S, T]; \mathbb{R}^n)$ and $\lambda \geq 0$ such that

(i) *$(p, \lambda) \neq (0, 0)$,*
(ii) *$(-\dot{p}(t), \dot{\bar{x}}(t)) \in \mathrm{co}\, \partial_{x,p} H(t, \bar{x}(t), p(t))$ a.e. $t \in [S, T]$,*
(iii) *$-p(T) \in \lambda \partial g(\bar{x}(T)) + N_C(\bar{x}(T))$.*

(The sigma algebra $\mathcal{L} \times \mathcal{B}^n$, referred to in hypothesis (a), will be defined in Chap. 2. $\partial_{x,p} H$ denotes the limiting subdifferential of $H(t, ., .)$.)

The essential feature of this extension is that Hamilton's system of equations for $(p, \bar{x})$ has been replaced by a differential inclusion, involving the convex hull of limiting subdifferential of the Hamiltonian H with respect to the x and p variables. The limiting subdifferential and the limiting normal cone are required to give meaning to the transversality condition (iii).

We stress that the Hamiltonian inclusion is an intrinsically nonsmooth condition. Even when the multifunction F is the velocity set of some smooth control system

$$\dot{x}(t) = f(t, x(t), u), \quad u(t) \in U \quad \text{a.e.,}$$

1.9 Nonsmooth Dynamic Optimization

('smooth' in the sense that f is a smooth function and U has a smooth boundary), in which case

$$F(t, x) = f(t, x, U),$$

we can still expect that $H(t, ., .)$ will be nonsmooth and that the apparatus of nonsmooth analysis will be required to make sense of the Hamiltonian inclusion.

Other nonsmooth necessary conditions are derived in future chapters. These include a generalization of the Euler Lagrange condition for problem (1.9.1) and a generalization of the maximum principle ('Clarke's nonsmooth maximum principle') for the dynamic optimization problem of Sect. 1.5.

Consider next dynamic programming. This, we recall, concerns the relationship between the value function V, which describes how the minimum cost of a dynamic optimization problem depends on the initial data, and the Hamilton Jacobi equation.

For now, we focus on the dynamic optimization problem

$$\begin{cases} \text{Minimize } g(x(T)) \\ \text{over measurable functions } u \text{ and arcs } x \in W^{1,1} \text{ satisfying} \\ \dot{x}(t) \in f(t, x(t), u(t)) \text{ a.e.,} \\ u(t) \in U, \text{ a.e.,} \\ x(S) = x_0, \; x(T) \in C, \end{cases}$$

to which corresponds the Hamilton Jacobi equation:

$$\begin{cases} \phi_t(t, x) + \min_{u \in U} \phi_x(t, x) \cdot f(t, \xi, u) = 0 \\ \qquad\qquad\qquad\qquad \text{for all } (t, x) \in (S, T) \times \mathbb{R}^n \\ \phi(T, x) = g(x) + \Psi_C(x) \text{ for all } x \in \mathbb{R}^n. \end{cases}$$

(Ψ_C denotes the indicator function of the set C.)

It has been noted that the value function is often nonsmooth. It cannot therefore be identified with a classical solution to the Hamilton Jacobi equation above, under hypotheses of any generality. However the machinery of nonsmooth analysis provides a characterization of the value function in terms of the Hamilton Jacobi equation, even in cases when V is discontinuous. There are a number of ways to do this. One involves the concept of viscosity solution. Another is based on properties of proximal subgradients. In this book we follow this latter approach because it is better integrated with our treatment of other topics and because it is particularly well suited to problems with end-point constraints.

Specifically, we shall show that, under mild hypotheses, V is the unique lower semi-continuous function which is a generalized solution of the Hamilton Jacobi equation. Here, 'generalized solution' means a function ϕ taking values in the extended real line $\mathbb{R} \cup \{+\infty\}$, such that the following three conditions are satisfied:

(i): For each $(t, x) \in ((S, T) \times \mathbb{R}^n) \cap \text{dom } \phi$

$$\eta^0 + \inf_{v \in F(t,x)} \eta^1 \cdot v = 0 \quad \forall \, (\eta^0, \eta^1) \in \partial^P \phi(t, x),$$

(ii): $\phi(T, x) = g(x) + \Psi_C(x) \quad \forall x \in \mathbb{R}^n,$

(iii): $\phi(S, x) = \liminf_{t' \downarrow S, x' \to x} \phi(t', x') \quad \forall x \in \mathbb{R}^n,$
$\phi(T, x) = \liminf_{t' \uparrow T, x' \to x} \phi(t', x') \quad \forall x \in \mathbb{R}^n.$

Of principle interest here is condition (i). (ii) merely reproduces the earlier boundary condition. (iii) is a regularity condition on ϕ, which is automatically satisfied when ϕ is continuous.

If ϕ is a C^1 function on $(S, T) \times \mathbb{R}^n$, condition (i) implies that ϕ is a classical solution to the Hamilton Jacobi equation. This follows from the fact that, in this case, $\partial^P V(t, x) \subset \{\nabla V(t, x)\}$.

What is remarkable about this characterization is the unrestrictive nature of the conditions it imposes on V, for it to be a value function. Condition (i) merely requires us to test certain relationships on the subset of $(S, T) \times \mathbb{R}^n$:

$$\mathcal{D} = \{(t, x) \in ((S, T) \times \mathbb{R}^n) \cap \text{dom } V : \partial^P V(t, x) \neq \emptyset\},$$

a subset which can be very small indeed (in relation to $(S, T) \times \mathbb{R}^n$); it need not even be dense.

Another deficiency of classical theory of dynamic programming is its failure to provided precise sensitivity information about the value function V, though our earlier formal calculations suggest that, if p is the adjoint arc associated with a minimizer $(\bar{x}, \bar{u})$, then

$$(\max_{u \in U} p(t) \cdot f(t, \bar{x}, u), -p(t)) = \nabla V(t, \bar{x}(t)), \quad \text{a.e. } t \in [S, T]. \qquad (1.9.2)$$

Indeed for many cases of interest V fails to be differentiable and (1.9.2) does not even make sense. Once again, nonsmooth analysis fills the gap. The sensitivity relation (1.9.2) can be given precise meaning as an inclusion in which the gradient ∇V is replaced by a suitable 'nonsmooth' subdifferential, namely the convexified limiting subdifferential co ∂V.

1.10 Epilogue

Our overview of dynamic optimization is concluded. To summarize, dynamic optimization concerns the properties of minimizing arcs. Its roots are in the investigation of minimizers of integral functionals, which has been an active area of research for over 200 years, in the guise of the calculus of variations. The distinguishing feature of dynamic optimization is that it can take account of dynamic and pathwise constraints, of a nature encountered in areas of advance engineering design and operations such as path planning for space vehicles, and in other areas of dynamic decision making that include resource economics, epidemiology and climate change mitigation.

1.10 Epilogue

Key early advances were the maximum principle and an intuitive understanding of the relationship between the value function and the Hamilton Jacobi equation of dynamic programming. The antecedents of dynamic programming in the branch of the calculus of variations known as field theory are apparent. On the other hand the nature of the maximum principle, and the techniques first employed to prove it, based on approximation of reachable sets, suggested that essentially new kinds of necessary conditions and analytical techniques were required to deal with the constraints arising in dynamic optimization problems.

Subsequent developments in dynamic optimization, aimed at extending the range of application of available necessary conditions of optimality, emphasize its similarities rather than its differences with the calculus of variations. Here, the constraints of dynamic control are replaced by extended valued penalty terms in the integral functional to be minimized. Problems in dynamic optimization are thereby reformulated as extended problems in the calculus of variations with nonsmooth data. It is then possible to derive, in the context of dynamic optimization, optimality conditions of remarkable generality, analogous to classical necessary conditions in the calculus of variations, in which classical derivatives are replaced by 'generalized' derivatives of nonsmooth functions. Nonsmooth analysis, which gives meaning to generalized derivatives has, of course, had a key role in these developments. nonsmooth analysis, by supplying the appropriate machinery for interpreting generalized solutions to the Hamilton Jacobi equation, can be used also to clarify the relationship between the value function and the Hamilton Jacobi equation, hinted at in the 1950's.

This overview stresses developments in dynamic optimization which provide the background to material in this book. It omits a number of important topics, such as higher order conditions of optimality, optimality conditions for a wide variety of nonstandard dynamic optimization problems (impulsive control problems, problems involving functional differential equations, etc.), computational dynamic optimization and an analysis of the dependence of minimizers on parameters, such as that provided in [148]. It also emphasizes only one possible route to proving general necessary conditions of optimality, which highlights their affinity with classical conditions, and to providing a rigorous foundation for dynamic programming, based on the pathwise analysis and the application of invariance/viability theorems. A complete picture would include the 'set separation' approaches to deriving necessary conditions, associated with Neustadt [163], Halkin [121], Dubovitskii and Milyutin [96], Warga [204] and Sussmann [185], which aim at generalizations of the maximum principle, and viscosity solution methods to establish the link between the Hamilton Jacobi equation and the value function in dynamic programming.

Dynamic optimization has been an active research field for over 60 years, much longer if you regard it as an outgrowth of the calculus of variations. Newcomers will expect, with reason, to find an armoury of easily applied techniques for computing optimal strategies and an extensive library of solved problems. This is not exactly what they will find. The truth is, solving dynamic optimization problems of scientific or engineering interest is often extremely difficult. Analytical tools of dynamic optimization, such as the maximum principle, provide equations satisfied by minimizers. They do not tell us how to solve them or, if we do succeed in solving them, that the 'solutions' are the right ones. Over two centuries ago Euler wrote of fluid mechanics (he could have commented thus equally about the calculus of variations)

> *If it is not permitted to us to penetrate to a complete knowledge concerning the motion of fluids, it is not to mechanics or to the insufficiency of the known principles of motion, that we must attribute the cause. It is analysis itself which abandons us here.*

As regards dynamic optimization, the picture remains the same today. It is quite clear that general methodologies giving 'complete knowledge' of solutions to large classes of optimal control problems will never be forthcoming, if by 'solution' is meant formulae describing the solution in terms of standard functions.

Even in cases when we can completely solve a variational problem by analysis, using general theory to find a minimizer for a specific problem is often a major undertaking. Often we find that for seemingly simple classical problems of longstanding interest, the hypotheses are violated, under which the general theory guarantees existence of minimizers and supplies information about them. In cases when optimality conditions are valid, the information they supply can be indirect and incomplete.

Some of these points are illustrated by the brachistochrone problem of Sect. 1.2. It is easy to come up with a candidate $\bar{x}$ for minimizing path by solving the Euler Lagrange Condition, namely a cycloid satisfying the boundary conditions. But is it truly a minimizer? If we adopt the formulation of Sect. 1.2, in which we seek a minimizer among curves describable as the graph of an absolutely continuous function in the vertical plane, the answer is yes. But this is by no means easy to establish. The cost integrand violates the Tonelli criteria for existence of minimizers (though other criteria can be invoked in this case to establish existence of a minimum time path). Then it is not even clear at the outset that minimizers will satisfy the Euler Lagrange condition because the standard Lipschitz continuity hypothesis commonly invoked to derive the condition, namely hypothesis (ii) of Theorem 1.2.1, is violated. Various arguments can be used to overcome these difficulties. (We can study initially a simpler problem in which, to eliminate the troublesome infinite slope at the left end-point, the ball is assumed to have a small initial velocity, or a pathwise constraint is imposed on the slope of the arcs considered, for example, and subsequently examine limits). But none of them are straightforward. Solution of the brachistochrone problem is discussed at length in, for example, [187, 191].

The brachistrochrone problem can be solved by analysis, albeit with some hard work. The maximal orbit transfer and cancer treatment problems of Sect. 1.1 are more typical of advanced engineering applications. We made no claims that this problem has been solved. But numerical schemes, inspired by modern day Lagrange multiplier rules and the maximum principle, generate the control strategy of Fig. 1.5, which, from a strictly rigorous viewpoint, is merely a *candidate* for an optimal strategy.

1.10 Epilogue

It is only when we adopt a more realistic view of the scope of dynamic optimization, that we can truly appreciate the benefits of research in this field:

Simple Problems While we cannot expect the theory of dynamic optimization alone to provide complete solutions to complex dynamic optimization problems of advanced engineering design, it does often provide optimal strategies for simplified versions of these problems. Optimal strategies for a simpler problem can serve as adequate close-to-optimal strategies (synonymously 'suboptimal' strategies) for the original problem. Alternatively, they can be used as initial guesses at an optimal strategy for use in numerical schemes to solve the original problem. This philosophy underlies some applications of dynamic optimization to resource economics, where simplifying assumptions about the underlying dynamics lead to problems which can be solved explicitly by analytical means [65]. In a more general context, many dynamic optimization problems of practical importance, in which there are only small excursions of the state variables, can be approximated by the linear quadratic control problem

$$(LQ) \begin{cases} \text{Minimize } \int_0^T (x^T(t)Qx(t) + u^T(t)Ru(t))dt + x^T(T)Gx(T) \\ \text{over } u \in L^2 \text{ and } x \in W^{1,1} \text{ satisfying} \\ \dot{x}(t = Ax(t) + Bu(t) \quad \text{a.e.,} \\ x(0) = x_0, \end{cases}$$

with data the number $T > 0$, the $n \times n$ symmetric matrices $Q \geq 0$ and $G \geq 0$, the $m \times m$ symmetric matrix $R > 0$, the $n \times n$ matrix A, the $n \times m$ matrix B and the n-vector x_0. The linear state equation here corresponds to linearization of nonlinear dynamics about an operating point. Weighting matrices in the cost are chosen to reflect economic criteria and to penalize constraint violations. In engineering applications, solving (LQ) (or related problems involving an approximate, linear, model) is a common ingredient in controller design.

Numerical Methods For many dynamic optimization problems which cannot be solved by the application of analysis alone, we can seek a minimizer with the help of numerical methods. Among the successes of dynamic optimization theory is the impact it has had on computational dynamic optimization, in the areas of both algorithm design and algorithm convergence analysis. Ideas such as exact penalization, for example, Lagrange multiplier rules and techniques for local approximation of nonsmooth functions, so central to the theory, have taken root also in general computational procedures for solving dynamic optimization problems. Early contributors in this area were Polak and Mayne, who exploited nonsmooth constructs and analytic methods from dynamic optimization in a computational context [150]. (See [151, 166, 169] for additional references.)

A widely used procedure for the computation of optimal strategies for a (continuous time) dynamic optimization problem involves several steps. First, we employ discretization methods, such as those discussed in Betts' book [38], to reduce the problem to a non-linear program. This can be configured with the help of the Applied Modelling Programming Language (AMPL) [114] to interface with one of several available optimization solvers, such as the interior point optimization solver IPOPT, developed by Wächler and Biegler [202].

Global convergence analysis, which amounts to establishing that limit points satisfy first order conditions of optimality, is of course intimately connected with dynamic optimization theory. The significance of global convergence analysis is that, on the one hand, it makes explicit classes of problems for which the algorithm is effective and, on the other, it points to possible modifications to existing algorithms to increase the likelihood of convergence.

Qualitative Properties Dynamic optimization can also provide useful qualitative information about minimizers. Often there are various ways of formulating an engineering design problem as an optimization problem. The theory of dynamic optimization provides simple criteria for existence of solutions, which, at the very least, can give useful insights into whether a sensible choice of optimization problem has been made. The theory also supplies a priori information about regularity of minimizers. This is of benefit for computation of optimal controls, since a 'best choice' of optimization algorithm can depend on minimizer regularity, as discussed, for example in [120]. Minimizer regularity is also relevant to the modelling of physical phenomena. In applications to nonlinear elasticity for example, existence of minimizers with infinite gradients has been linked to material failure and so an analysis identifying classes of problems for which minimizers have bounded derivatives gives insights into mechanisms of fracture [16]. Finally, and perhaps most significantly from the point of view of solving engineering problems, necessary conditions can supply structural information which greatly simplifies the computation of optimal controls. In the computation of optimal flight trajectories, for instance, information obtained from known necessary conditions of optimality concerning bang-bang subarcs, relationships governing dependent variables on interior arcs, bounds on the number of switches, etc., is routinely used to reduce the computational burden [50].

1.11 Appendix: Proof of the Classical Maximum Principle

The maximum principle is the cornerstone of dynamic optimization. In later chapters we shall derive general versions of the maximum principle for problems with non-differentiable data as well as extensions to allow for end-point constraint sets arbitrary closed sets, pathwise state constraints and other features. Even understanding the statement of the maximum principle in these general settings requires the apparatus of non-smooth analysis and the proofs are long and complicated.

Yet many dynamic optimization problems encountered in applications are basic ones, in the sense that there are no pathwise state constraints, the data is continuously differentiable w.r.t. the state variable state and the state end-point constraints take the form of smooth functional equality and inequality constraints. For basic

1.11 Appendix: Proof of the Classical Maximum Principle

problems in this sense, the statement of the maximum problem simplifies and its derivation becomes more straightforward.

In this appendix we give two proofs of the maximum principle for these basic dynamic optimization problems. The first is based on Clarke's perturbation methods (supported by a variational principle), similar to those deployed in future chapters to derive necessary conditions of optimality for a broader class of problems but which, here, are employed in a much more direct fashion. The second proof is based on set separation methods.

The proof of the maximum principle by perturbation methods is intended to serve the needs of readers who wish to gain an understanding of this important optimality condition and apply it to basic problems of dynamic optimization, based on a minimal use of non-standard analytical techniques. For other readers intending to enter more deeply into the field, this proof is an introduction to ideas for deriving necessary conditions used throughout this book, but which, here, take a simpler, more transparent form.

The earliest proofs of the maximum principle, including that in L. S. Pontragin et al.'s book [167], are based, not on Clarke's perturbation methods that are centre stage in this book, but on set separation methods. We include in this appendix a second proof of the maximum principle for the basic problem of dynamic optimization, based on these methods. This is partly to acknowledge their historical importance. But we have included it also because set separation ideas have a continuing influence on the modern day dynamic optimization research and a rounded understanding of the current research landscape requires familiarity with both perturbation and set separation approaches.

Consider the following 'basic' dynamic optimization problem in which the endpoint state constraints take the form of functional inequalities and equalities:

$$
(P) \begin{cases} \text{Minimize } g(x(T)) + \int_S^T L(t, x(t), u(t))dt \\ \text{over absolutely cont. functions } x : [S, T] \to \mathbb{R}^n \\ \qquad\qquad \text{and meas. functions } u : [S, T] \to \mathbb{R}^m \text{ such that} \\ \dot{x}(t) = f(t, x(t), u(t)), \text{ a.e. } t \in [S, T], \\ u(t) \in U(t), \text{ a.e. } t \in [S, T], \\ x(S) = x_0 \text{ and } x(T) \in C. \end{cases}
$$

Here, $f : [S, T] \times \mathbb{R}^n \times \mathbb{R}^m \to \mathbb{R}^n$, $L : [S, T] \times \mathbb{R}^n \times \mathbb{R}^m \to \mathbb{R}$ and $g : \mathbb{R}^n \to \mathbb{R}$ are given functions, $\{U(t) \subset \mathbb{R}^m : S \leq t \leq T\}$ is a given family of sets, x_0 is a given point in $\mathbb{R}^n$ and C is the set

$$C = \{x \in \mathbb{R}^n : \phi(x) \leq 0 \text{ and } \psi(x) = 0\}$$

for given functions $\phi : \mathbb{R}^n \to \mathbb{R}^{n_\phi}$ and $\psi : \mathbb{R}^n \to \mathbb{R}^{n_\psi}$.

As usual, we take an admissible process to be a pair of functions (x, u), comprising an absolutely continuous function $x : [S, T] \to \mathbb{R}^n$ (called a 'state trajectory') and a measurable function $u : [S, T] \to \mathbb{R}^m$ (called a 'control'), if they satisfy the conditions listed in problem (P). An admissible process $(\bar{x}, \bar{u})$ is said to be an L^∞ *local minimizer* if there exists a number $\beta > 0$ such that $g(x(T)) \geq g(\bar{x}(T))$ for all admissible processes (x, u) such that $||x - \bar{x}||_{L^\infty} \leq \beta$. Under circumstances when there is a unique state trajectory associated with each control u, we write this state trajectory x^u.

For problem (P), the maximum principle takes the following form:

Theorem 1.11.1 (Maximum Principle) *Take an L^∞ local minimizer $(\bar{x}, \bar{u})$. Assume, for some $k \in L^1$, $c \in L^1$ and $\epsilon > 0$,*

(A): $(f, L)(., x, .)$ *is $\mathcal{L} \times \mathcal{B}^m$ measurable for each $x \in \mathbb{R}^n$ and $\{(t, u) \in [S, T] \times \mathbb{R}^m : u \in U(t)\}$ is $\mathcal{L} \times \mathcal{B}^m$ measurable, where $\mathcal{L} \times \mathcal{B}^m$ denotes the product σ-field generated by the Lebesgue subsets of $[S, T]$ and the Borel subsets of $\mathbb{R}^m$,*

(B): g, ϕ *and ψ are C^1 on $\bar{x}(T) + \epsilon \mathbb{B}$, and $(f, L)(t, ., u)$ is C^1 on $\bar{x}(t) + \epsilon \mathbb{B}$ for all $u \in U(t)$, a.e. $t \in [S, T]$,*

(C): $|(f, L)(t, x, u) - (f, L)(t, x', u)| \leq k(t)|x - x'|$ *for all $x, x' \in \bar{x}(t) + \epsilon \mathbb{B}$ and $u \in U(t)$, a.e. $t \in [S, T]$,*

(D): $|(f, L)(t, x, u)| \leq c(t)$ *for all $x \in \bar{x}(t) + \epsilon \mathbb{B}$ and $u \in U(t)$, a.e. $t \in [S, T]$.*

Then there exist $\lambda^0 \geq 0$, $\lambda^\phi \in (\mathbb{R}_+)^{n_\phi}$, $\lambda^\psi \in \mathbb{R}^{n_\psi}$ and $p \in W^{1,1}([S, T]; \mathbb{R}^n)$ such that $\lambda_k^\phi = 0$ if $\phi_k(\bar{x}(T)) < 0$, $k = 1, \ldots, n_\phi$ and

(i): $(p, \lambda^0, \lambda^\phi, \lambda^\psi) \neq (0, 0, 0, 0)$,

(ii): $-\dot{p}(t) = p(t) \cdot \nabla_x f(t, \bar{x}(t), \bar{u}(t)) - \lambda^0 \nabla_x L(t, \bar{x}(t), \bar{u}(t))$, *a.e. $t \in [S, T]$,*

(iii): $-p(T) = \lambda^0 g_x(\bar{x}(T)) + \lambda^\phi \cdot \nabla \phi^1(\bar{x}(T)) + \lambda^\psi \cdot \nabla \psi(\bar{x}(T))$,

(iv): $p(t) \cdot f(t, \bar{x}(t), \bar{u}(t)) - \lambda^0 L(t, \bar{x}(t), \bar{u}(t)) =$
$$\max_{u \in U(t)} \left(p(t) \cdot f(t, \bar{x}(t), u) - \lambda^0 L(t, \bar{x}(t), u) \right) \text{ a.e. } t \in [S, T].$$

If there is no inequality end-point constraint, i.e. $C = \{x : \psi(x) = 0\}$, then these relations are valid, with $\lambda^\phi = 0$ and '$\lambda^\phi \cdot \nabla \phi(\bar{x}(T))$' deleted from (iii). If there is no equality end-point constraint, i.e. $C = \{x : \phi(x) \leq 0\}$, then these relations are valid with $\lambda^\psi = 0$ and '$\lambda^\psi \cdot \nabla \psi(\bar{x}(T))$' deleted from (iii).

A: Proof of the Maximum Principle by Clarke's Perturbation Methods

Preliminary Analysis The following lemma justifies the imposition of a number of additional, simplifying hypotheses:

Lemma 1.11.2 (Hypothesis Reduction) *Assume that the assertions of the theorem, for some L^∞ local minimizer $(\bar{x}, \bar{u})$, are valid in the special case when*

(a): *(C) and (D) are replaced by: there exist $k \in L^1$ and $c \in L^1$ such that*

(C)': $|(f, L)(t, x, u) - (f, L)(t, x', u)| \leq k(t)$ *for all $x, x' \in \mathbb{R}^n$ and $u \in U(t)$, a.e. $t \in [S, T]$,*

(D)': $|(f, L)(t, x, u))| \leq c(t)$ *for all $x \in \mathbb{R}^n$ and $u \in U(t)$, a.e. $t \in [S, T]$,*

and g, ϕ and ψ are globally Lipschitz continuous,

1.11 Appendix: Proof of the Classical Maximum Principle

and

(b): $L \equiv 0$,

(c): $\phi_j(\bar{x}(T)) = 0$ for $j = 1, \ldots n_\phi$; that is, we can assume that all the end-point inequality constraints are active.

Then the assertions of the theorem are valid without these restrictions.

Proof

(a): Consider a modification of (P), in which $(f, L)(t, x, u)$, $(g, \phi, \psi)(x)$ are replaced by their 'truncations'

$$(f, L)(t, \bar{x}(t) + tr_\epsilon(x - \bar{x}(t)), u), (g, \phi, \psi)(\bar{x}(T) + tr_\epsilon(x - \bar{x}(T)))$$

in which $tr_\epsilon(y) := \begin{cases} y & \text{if } |y| \leq \epsilon \\ \epsilon |y|^{-1} y & \text{if } |y| > \epsilon \end{cases}$. The stronger hypotheses are now satisfied (with a modified Lipschitz bound k and parameter $\epsilon > 0$). The process $(\bar{x}, \bar{u})$ retains its status as an L^∞ local minimizer and the statement of the necessary conditions is unaffected, when we change the data in this way. So applying the necessary conditions, restricted to problems for which the data is 'globally' Lipschitz continuous yields the same necessary conditions for the problem involving the original data.

(b): Assume the assertions of the theorem are valid when $L \equiv 0$. Define $\bar{y}(t) := \int_0^t L(t, \bar{x}(s), \bar{u}(s)) ds$. Then $((\bar{x}, \bar{y}), \bar{u})$ is an L^∞ local minimizer for the problem with 'augmented state':

$$\begin{cases} \text{Minimize } g(x(T)) + y(T) \\ \text{s.t.} \\ (\dot{x}(t), \dot{y}(t)) = (f(t, x(t), u(t)), L(t, x(t), u(t))), \text{ a.e. } t \in [S, T], \\ u(t) \in U(t), \text{ a.e. } t \in [S, T] \\ (x(S), y(S)) = (x_0, 0), \phi(x(T)) \leq 0 \text{ and } \psi(x(T)) = 0. \end{cases}$$

(This follows easily from the facts that the cost of a process $((x, y), u)$ for the augmented state problem is $g(x(T)) + z(T) = g(x(T)) + \int_0^T L(s, x(s), u(s))ds$, which is the cost of the process (x, u) for the original problem, and that $||x||_{L^\infty} \leq ||(x, y)||_{L^\infty}$.)

The augmented state problem has no integral cost term so, under our assumption, we may apply the maximum principle. This yields a costate trajectory (p, q), whose components take values in $\mathbb{R}^n$ and $\mathbb{R}$ respectively, and Lagrange multipliers $\lambda^0, \lambda^\phi, \lambda^\psi$. Because $(f, L)(t, x, u)$ does not depend on the y variable, the costate equation tells us that $\dot{q} \equiv 0$. We can deduce from the transversality condition that $-q(T) = \lambda^0 \partial_y(g(\bar{x}(T)) + \bar{y}(T)) = \lambda^0$. It follows that we can eliminate the q variable from the necessary conditions by the substitution '$q \equiv -\lambda^0$'. The necessary conditions for the state augmented problems, expressed in this way, are precisely the necessary conditions for the original problem.

(c): We show that we can assume that $\phi_j(\bar{x}(T)) = 0$ for all $j = 1,\ldots n_\phi$. If, to the contrary, $\phi_j(\bar{x}(T)) < 0$ for some j then, since ϕ_j is a continuous function, $(\bar{x}, \bar{u})$ continues to be an L^∞ local minimizer when this constraint is removed from the problem formulation. The necessary conditions for the problem without this constraint are the same as the necessary conditions for the problem with the constraint, in which the multiplier λ_j^ϕ is set to 0. Notice that we can thereby arrange that the Lagrange multiplier λ^ϕ automatically satisfies the condition: $\lambda_k^\phi = 0$ if $\phi_k(\bar{x}(T)) < 0, k = 1, \ldots, n_\phi$. □

Henceforth, we assume that the stronger hypotheses (C)′ and (D)′ are imposed on the data, $L \equiv 0$ and all the end-point inequality constraints are satisfied at $\bar{x}(T)$. According to the preceding lemma, no loss of generality is involved. Define

$$\mathcal{U} := \{u : [S, T] \to \mathbb{R}^m : u \text{ is } \mathcal{L}\text{-meas and } u(t) \in U(t) \text{ a.e. } t \in [S, T]\}.$$

Take any $u \in \mathcal{U}$. By standard existence/uniqueness properties of differential equations under these strengthened hypotheses, there is a unique absolutely continuous solution to

$$\begin{cases} \dot{x}(t) = f(t, x(t), u(t)) \text{ a.e. } t \in [S, T] \\ x(S) = x_0, \end{cases}$$

which we can write, unambiguously, as x^u. For $u \in \mathcal{U}$ define

$$J(u) := g(x^u(T)).$$

Lemma 1.11.3 *Take $u, u' \in \mathcal{U}$. Then*

$$\|x^u - x^{u'}\|_{L^\infty} \leq 2K \int_{A(u,u')} c(t) dt, \qquad (1.11.1)$$

in which $K := \exp(\int_S^T k(t) dt)$ and $A(u, u') := \{t \in [S, T] : u(t) \neq u'(t)\}$.

Proof For every $t \in [S, T]$, the continuous function $t \to x^u(t) - x^{u'}(t)$ satisfies

$$|x^u(t) - x^{u'}(t)| \leq \int_{[S,t]} (|f(s, x^u(s), u(s)) - f(s, x^{u'}(s), u(s))|$$

$$+ |f(s, x^{u'}(s), u(s)) - f(s, x^{u'}(s), u'(s))|) ds$$

$$\leq \int_{[S,t]} (k(s)|x^u(s) - x^{u'}(s)| + |f(s, x^{u'}(s), u(s)) - f(s, x^{u'}(s), u'(s))|) ds.$$

1.11 Appendix: Proof of the Classical Maximum Principle

By Gronwall's lemma (Lemma 6.2.4)

$$||x^u - x^{u'}||_{L^\infty} \le K \int_S^T |f(s, x^{u'}(s), u(s)) - f(s, x^{u'}(s), u'(s))| ds$$

$$\le 2K \int_{A(u,u')} c(t) dt,$$

in which $K = \exp(\int_S^T k(t)dt)$. This is the required estimate. □

Lemma 1.11.4 (Needle Variations) *Take any $u, u' \in \mathcal{U}$. Write $x' := x^{u'}$. Take the point $\bar{t} \in (S, T)$ to be such that $\bar{t}$ is a Lebesgue point of $t \to f(t, x'(t), u(t)) - f(t, x'(t), u'(t))$, $t \to k(t)$ and $t \to c(t)$ anad $\bar{u}(\bar{t}) \in U(\bar{t})$. (Such points comprise a set of full measure.)*

Let the absolutely continuous function p be the solution to

$$\begin{cases} -\dot{p}(t) = p(t) \cdot \nabla_x f(t, x'(t), u'(t)) \text{ a.e. } t \in [S, T] \\ -p(T) = \nabla_x g(x'(T)). \end{cases} \quad (1.11.2)$$

Take $\epsilon_i \downarrow 0$ such that $\epsilon_i < T - \bar{t}$ for each i. For each $t \in [S, T]$ and $i = 1, 2, \ldots$ define

$$u_i(t) := \begin{cases} u(t) & \text{if } t \in [\bar{t}, \bar{t} + \epsilon_i] \\ u'(t) & \text{if } t \notin [\bar{t}, \bar{t} + \epsilon_i] \end{cases}.$$

Then

$$\lim_{i \to \infty} \epsilon_i^{-1} (J(u_i) - J(u')) = -p(\bar{t}) \cdot [f(\bar{t}, x'(\bar{t}), u(\bar{t})) - f(\bar{t}, x'(\bar{t}), u'(\bar{t}))]. \quad (1.11.3)$$

Proof Write $x_i := x^{u_i}$. Since x_i and x' satisfy the dynamic constraint, we know that, for each i,

$$J(u_i) - J(u') = g(x_i(T)) - g(x'(T))$$

$$+ \int_{\bar{t}+\epsilon_i}^T p(s) \cdot \Big(\dot{x}_i(s) - \dot{x}'(s) - f(s, x_i(s), u_i(s)) + f(s, x'(s), u'(s))\Big) ds.$$

Performing an integration by parts (on the term $\int_{\bar{t}+\epsilon_i}^T p(s) \cdot (\dot{x}_i(s) - \dot{x}'(s)) ds$), noting the defining relations for p and also that

$$x_i(\bar{t} + \epsilon_i) - x'(\bar{t} + \epsilon_i) = \int_{\bar{t}}^{\bar{t}+\epsilon_i} (f(s, x_i(s), u_i(s)) - f(s, x'(s), u'(s))) ds,$$

we arrive at the following identity for the integral term on the right of the previous equation

$$\int_{\bar{t}+\epsilon_i}^{T} p(s) \cdot \Big(\dot{x}_i(s) - \dot{x}'(s) - f(s, x_i(s), u_i(s)) + f(s, x'(s), u'(s)) \Big) ds$$

$$= p(T) \cdot (x_i(T) - x'(T)) - p(\bar{t}+\epsilon_i) \cdot (x_i(\bar{t}+\epsilon_i) - x'(\bar{t}+\epsilon_i))$$

$$- \int_{\bar{t}+\epsilon_i}^{T} \Big(\dot{p}(s) \cdot (x_i(s) - x'(s))$$

$$+ p(s) \cdot (f(s, x_i(s), u_i(s)) - f(s, x'(s), u'(s))) \Big) ds$$

$$= -\nabla g(x'(T)) \cdot (x_i(T) - x'(T))$$

$$- p(\bar{t}+\epsilon_i) \cdot \int_{\bar{t}}^{\bar{t}+\epsilon_i} (f(s, x_i(s), u_i(s)) - f(s, x'(s), u'(s))) ds$$

$$- \int_{\bar{t}+\epsilon_i}^{T} p(s) \cdot \Big(f(s, x(s), u_i(s)) - f(s, x'(s), u'(s))$$

$$- \nabla_x f(t, x'(s), u'(s)) \cdot (x_i(s) - x'(s)) \Big) ds .$$

Assembling there relations and dividing across by ϵ_i yields:

$$\epsilon_i^{-1}(J(u_i) - J(u')) = a_i^{(1)} + a_i^{(2)} - a_i^{(3)} , \tag{1.11.4}$$

in which
$a_i^{(1)} := -p(\bar{t}+\epsilon_i) \cdot \epsilon_i^{-1} \int_{\bar{t}}^{\bar{t}+\epsilon_i} (f(s, x_i(s), u_i(s)) - f(s, x'(s), u'(s))) ds,$
$a_i^{(2)} := \epsilon_i^{-1} \Big(g(x_i(T)) - g(x'(T)) - \nabla g(x'(T)) \cdot (x_i(T) - x'(T)) \Big)$ and
$a_i^{(3)} := \int_{\bar{t}+\epsilon_i}^{T} \ell_i^{(3)}(s) ds,$
in which
$\ell_i^{(3)}(s) := p(s) \cdot \epsilon_i^{-1} \Big(f(s, x_i(s), u_i(s)) - f(s, x'(s), u'(s)) - \nabla_x f(t, x'(s), u(s')) \cdot (x_i(s) - x'(s)) \Big).$

Our next task is to examine the numbers $a_i^{(1)}, a_i^{(2)}$ and $a_i^{(3)}$, in the limit as $i \to \infty$. Notice first that $\mathcal{L}\text{-meas}\{t \in [S, T] : u_i(t) \neq u'(t)\} \to 0$, as $i \to \infty$. It follows from Lemma 1.11.3 that

$$||x_i - x'||_{L^\infty} \to 0, \text{ as } i \to \infty . \tag{1.11.5}$$

Since $\bar{t}$ is a Lebesgue point of c, we know that $\epsilon_i^{-1} \int_{\bar{t}}^{\bar{t}+\epsilon_i} c(t) dt$, $i = 1, 2, \ldots$, is a bounded sequence. It follows from Lemma 1.11.3 that, for some $L > 0$

$$\epsilon^{-1} ||x_i - x'||_{L^\infty} \leq L, \text{ for } i = 1, 2, \ldots \tag{1.11.6}$$

1.11 Appendix: Proof of the Classical Maximum Principle

For each i,

$$\left| \int_{\bar{t}}^{\bar{t}+\epsilon_i} (f(t, x_i(s), u_i(s)) - f(t, x'(s), u'(s)))ds \right|$$

$$= \left| \int_{\bar{t}}^{\bar{t}+\epsilon_i} (f(t, x'(s), u(s)) - f(t, x'(s), u'(s)))ds \right| \le \gamma_i,$$

where $\gamma_i := \epsilon_i \times \int_{\bar{t}}^{\bar{t}+\epsilon_i} c(s)ds$. Using the facts that $\epsilon_i^{-1}\gamma_i \to 0$ as $i \to 0$, that p is continuous and that $\bar{t}$ is a Lebesgue point of $s \to f(s, x'(s), u(s)) - f(s, x'(s), u'(s))$, we deduce from this estimate that

$$a_i^{(1)} \to -p(\bar{t}) \cdot (f(\bar{t}, x'(\bar{t}), u(\bar{t})) - f(\bar{t}, x'(\bar{t}), u'(\bar{t}))), \text{ as } i \to \infty. \quad (1.11.7)$$

Making use of the fact that g is continuously differentiable on a neighbourhood of $x'(T)$, we deduce from the mean value theorem that

$$a_i^{(2)} \to 0, \text{ as } i \to \infty. \quad (1.11.8)$$

Since, for a.e. $t \in [S, T]$, $x \to p'(t) \cdot f(t, x, u'(t))$ is continuously differentiable on a neighbourhood of $\bar{x}(t)$, we can deduce from (1.11.5) and (1.11.6) and the mean value theorem that

$$\ell_i^{(3)}(s) \to 0, \text{ a.e. } s \in [S, T].$$

The functions $\ell_i^{(3)}$, $i = 1, 2, \ldots$ are majorized by the common integrable function $L(\|p\|_{L^\infty} + 1)k(s)$, in which L is the constant of (1.11.6). By the dominated convergence theorem

$$a_i^{(3)} \to 0, \text{ as } i \to \infty. \quad (1.11.9)$$

It follows from (1.11.4), (1.11.7), (1.11.8) and (1.11.9) that the limit $\lim_{i \to \infty} \epsilon_i^{-1} (J(u_i) - J(u'))$ exists and is given by (1.11.3). □

With these analytical tools, we are now ready to prove the maximum principle.

Step 1: The Maximum Principle for a Problem with Free Right End-Point

Proposition 1.11.5 *Let $(\bar{x}, \bar{u})$ be an L^∞ local minimizer for the special case of (P), in which $C = \mathbb{R}^n$. Assume hypotheses (A), (B), (C) and (D) are satisfied. Then the maximum principle (Theorem 1.11.1) is valid with $\lambda^0 = 1$ and terms in the theorem statement involving λ^ϕ and λ^ψ deleted.*

Proof We can assume (C) and (D) have been strengthened to (C)' and (D)' and $L \equiv 0$. Let p be the solution to the costate Eq. (1.11.2), in which we take $(x', u') =$

$(\bar{x}, \bar{u})$. All the assertions of the proposition follow from the definition of p, with the choice $\lambda^0 = 1$, with the exception of condition (iv). To prove this last condition suppose, to the contrary, that

$$p(t) \cdot f(t, \bar{x}(t), \bar{u}(t)) < \max_{u \in U(t)} p(t) \cdot f(t, \bar{x}(t), u),$$

for all t in a subset of $[S, T]$ of positive measure. A measurable selection theorem (whose application is justified, given the hypothesized measurability properties of f and U) supplies a control function $u \in \mathcal{U}$ such that

$$p(t) \cdot f(t, \bar{x}(t), \bar{u}(t)) < p(t) \cdot f(t, \bar{x}(t), u(t))$$

for all t in a subset of $[S, T]$ of positive measure. This contradicts relation (1.11.3) in Lemma 1.11.4 . So condition (iv) is also satisfied. □

Step 2: Completion of the Proof
Take $\epsilon_i \downarrow 0$. Define, for each i, the function

$$d^i(x) := \sqrt{((g(x) - g(\bar{x}(T)) + \epsilon_i) \vee 0)^2 + \sum_{j=1}^{n_\phi}(\phi_i(x) \vee 0)^2 + |\psi(x)|^2},$$
(1.11.10)

in which $a \vee b := \max\{a, b\}$.

Consider, for each i, the dynamic optimization problem

$$(P)_i \begin{cases} \text{Minimize } d^i(x(T)) \\ \text{s.t.} \\ \dot{x}(t) = (f(t, x(t), u(t)) \text{ and } u(t) \in U(t), \text{ a.e. } t \in [S, T], \\ x(S) = x_0 \text{ and } ||x - \bar{x}||_{L^\infty} \leq \epsilon. \end{cases}$$

Define

$$\mathcal{M} := \{u : [S, T] \to \mathbb{R}^m : u \text{ is } \mathcal{L}\text{-meas.}, u(t) \in U(t), \text{ a.e., and } ||x_u - \bar{x}||_{L^\infty} \leq \epsilon\}.$$

Problem $(P)_i$ with the specified end-point constraint can be posed as an optimization problem with domain control functions $u \in \mathcal{M}$, namely

$$\text{Minimize } \{\Psi_i(u) : u \in \mathcal{M}\},$$

in which $\Psi_i(u) := d^i(x_u(T))$.

Now equip $\mathcal{M}$ with the Ekeland metric

$$d_{\mathcal{E}}(u, u') := \mathcal{L}\text{-meas}\{t \in [S, T] : u(t) \neq u'(t)\}.$$

$\mathcal{M}$ is a closed, non-empty set and Ψ_i is a continuous function $\mathcal{M}$ w.r.t. this metric. Notice that

$$\Psi_i(\bar{u}) = \epsilon_i \text{ and } \Psi_i(u) \geq 0, \text{ for all } u \in \mathcal{M}.$$

1.11 Appendix: Proof of the Classical Maximum Principle

By Ekeland's theorem, there exists $u_i \in \mathcal{M}$ such that

$$\Psi_i(u_i) = \min\{\Psi_i(u) + \epsilon_i^{\frac{1}{2}} d_{\mathcal{E}}(u, u_i), u \in \mathcal{M}\}$$

and

$$d_{\mathcal{E}}(u_i, \bar{u}) \leq \epsilon_i^{\frac{1}{2}}. \tag{1.11.11}$$

Define

$$m_i(t, u) := \begin{cases} 1 & \text{if } u \neq u_i(t) \\ 0 & \text{if } u = u_i(t). \end{cases}$$

Write $x_i := x_{u_i}$. It follows from Lemma 1.11.3 that

$$||x_i - \bar{x}||_{L^\infty} \to 0, \text{ as } i \to \infty.$$

It is important to observe that, for i sufficiently large,

$$d^i(x_i(T)) > 0. \tag{1.11.12}$$

Indeed, the contrary assertion '$d^i(x_i(T)) = 0$' implies that $g(x_i(T)) - g(\bar{x}(T)) = -\epsilon_i$, $\phi(x_i(T)) \leq 0$ and $\psi(x_i(T)) = 0$. Since $||x_i - \bar{x}||_{L^\infty} \to 0$, as $i \to \infty$, this violates the optimality of $\bar{u}$ (relative to controls $u \in \mathcal{U}$ that are admissible and satisfy $||x^u - \bar{x}||_{L^\infty} \leq \beta$, for some arbitrary, pre-assigned $\beta > 0$). Referring back to the definition of d^i, we see that d^i is C^1 near $x_i(T)$.

Since

$$\mathcal{L}\text{-meas}\{t \in [S, T] : u \neq u_i(t)\} = \int_S^T m_i(t, u(t)) dt,$$

the above minimization property implies: for sufficiently large i, (x_i, u_i) is an L^∞ local minimizer for the dynamic optimization problem

$$\begin{cases} \text{Minimize } d^i(x(T)) + \epsilon_i^{\frac{1}{2}} \int_S^T m_i(t, u(t)) dt \\ \text{subject to} \\ \dot{x}(t) = f(t, x(t), u(t)) \text{ and } u(t) \in U(t), \text{ a.e. } t \in [S, T], \\ x(S) = x_0. \end{cases}$$

This is a free right end-point problem to which the special case of the maximum principle, Proposition 1.11.5, is applicable. (Notice that the relevant hypotheses are satisfied, including the continuous differentiability of d^i on a neighbourhood of $x_i(T)$.)

Fix $u \in \mathcal{U}$. We deduce the existence of $p_i \in W^{1,1}$ such that, for a.e. $t \in [S, T]$,

$$-\dot{p}_i(t) = p_i(t) \cdot f_x(t, x_i(t), u_i(t)), \tag{1.11.13}$$

$$p_i(t) \cdot f(t, x_i(t), u_i(t)) \geq p_i(t) \cdot f(t, x_i(t), u(t)) - \epsilon_i^{\frac{1}{2}} \tag{1.11.14}$$

and

$$-p_i(T) = d_x^i(x_i(T)). \tag{1.11.15}$$

We deduce from (1.11.10) and (1.11.12) that

$$d_x^i(x_i(T)) := (\lambda_i^0, \lambda_i^\phi, \lambda_i^\psi) \cdot (g_x(x_i(T)), \phi_x(x_i(T)), \psi_x(x_i(T))),$$

in which

$$(\lambda_i^0, \lambda_i^\phi, \lambda_i^\psi) := \frac{1}{d^i(x_i(T))} \left((g(x_i(T)) - g(\bar{x}(t)) + \epsilon_i) \vee 0 \right),$$

$$\{(\phi_j(x_i(T)) \vee 0)\}_{j=1}^{n_\phi}, \psi(x_i(T)).$$

Observe that, by (1.11.10) and (1.11.12), we have

$$d^i(x_i(T)) = \sqrt{\left\{ ((g(x_i(T)) - g(\bar{x}(T)) + \epsilon_i) \vee 0)^2 + \sum_{j=1}^{n_\phi} (\phi_j(x_i(T)) \vee 0)^2 + |\psi(x_i(T))|^2 \right\}} > 0.$$

We see that

$$\lambda_i^0 \geq 0, \ \lambda_i^\phi \in (\mathbb{R}_+)^{n_\phi}, \ \lambda_i^\psi \in \mathbb{R}^{n_\psi} \text{ and } |(\lambda_i^0, \lambda_i^\phi, \lambda_i^\psi)| = 1, \text{ for each } i. \tag{1.11.16}$$

Our next task is to pass to the limit, as $i \to \infty$, in relations (1.11.13), (1.11.14), (1.11.15) and (1.11.16), which will be recognized as perturbed versions of assertions (i)–(iv) in the proposition statement.

In view of (1.11.16), we can arrange, by subsequence extraction, that

$$(\lambda_i^0, \lambda_i^\phi, \lambda_i^\psi) \to (\lambda^0, \lambda^\phi, \lambda^\psi)$$

for some $(\lambda_i^0, \lambda_i^\phi, \lambda_i^\psi)$ such that $\lambda_i^0 \geq 0$, $\lambda_i^\phi \in (\mathbb{R}_+)^{n_\phi}$, $\lambda_i^\psi \in \mathbb{R}^{n_\psi}$ and $|(\lambda_i^0, \lambda_i^\phi, \lambda_i^\psi)| = 1$.

1.11 Appendix: Proof of the Classical Maximum Principle

We know from (1.11.11) that $\mathcal{L}$-meas$\{t \in [S, T] : u_i(t) \neq \bar{u}(t)\} \to 0$ as $i \to \infty$. We can therefore arrange, by subsequence extraction, that

$$\sum_{i=0}^{\infty} \mathcal{L}\text{-meas}\{t \in [S, T] : u_i(t) \neq \bar{u}(t)\} < \infty. \tag{1.11.17}$$

Define the nested sequence of sets

$$\mathcal{A}_j := \{t \in [S, T] : u_i(t) \neq \bar{u}(t) \text{ for some } i \geq j\}, \quad j = 1, 2, \ldots$$

and write

$$\mathcal{A} := \cap_{j=1}^{\infty} \mathcal{A}_j.$$

In view of (1.11.17), $\mathcal{L}$-meas$\{\mathcal{A}_j\} \to 0$, as $j \to \infty$. It follows that $\mathcal{A}$ has measure 0. Now define

$$\mathcal{T} := [S, T] \setminus \mathcal{A}.$$

$\mathcal{T}$ is a measurable subset of $[S, T]$ having full measure. It has representation

$$\mathcal{T} := \{t \in [S, T] : \text{ there exists an integer } i(t)$$
$$\text{such that } u_j(t) = \bar{u}(t) \text{ for all } j \geq i(t)\}.$$

From (1.11.13) and (1.11.16), the continuous function $q_i(s) := p_i(T - s), 0 \leq s \leq T - S$ satisfies

$$|q_i(s)| \leq |(\lambda_i^0, \lambda_i^\phi, \lambda_i^\psi) \cdot (g_x(x_i(T)), \phi_x(x_i(T)), \psi_x(x_i(T)))| + \int_0^s |q_i(s')| \cdot \alpha_i(s') ds',$$

in which $\alpha_i(s) := |\nabla_x f(T - s, x_i(T - s), u_i(T - s))|$. Since, for each i, $\|\alpha_i\|_{L^\infty} \leq \|k\|_{L^\infty}$, it follows from Gronwall's lemma that q_i are uniformly bounded and therefore $\|p_i\|_{L^\infty} \leq c_1$ for some $c_1 > 0$, independent of i. But then, from (1.11.13), the $\dot{p}_i$'s are dominated by the integrable function $t \to c_1 k(t)$. It can be deduced from Ascoli's theorem that, along some subsequence, $p_i \to p$, uniformly, and $\dot{p}_i \to \dot{p}$, in the weak L^1 topology, for some absolutely continuous function p.

For each $t \in [S, T]$, we deduce from (1.11.13) that

$$p_i(t) = (\lambda_i^0, \lambda_i^\phi, \lambda_i^\psi) \cdot (g_x(x_i(T)), \phi_x(x_i(T)), \psi_x(x_i(T)))|$$
$$- \int_t^T p_i(s) \cdot f_x(s, x_i(s), u_i(s)) ds.$$

The integrand is bounded by the integrable function $t \to c_1 \times k(t)$ and at all points t lying in the set $\mathcal{T}$ (a set of full measure), the integrand converges to $s \to p(s) \cdot f_x(s, \bar{x}(s), \bar{u}(s))$. Applying the dominated convergence theorem and using the convergence properties of $\{p_i\}$ and $\{(\lambda_i^0, \lambda_i^\phi, \lambda_i^\psi)\}$, we can conclude that the limiting integrand is integrable and

$$p(t) = (\lambda^0, \lambda^\phi, \lambda^\psi) - \int_t^T p(s) \cdot f_x(s, \bar{x}(s), \bar{u}(s)) ds.$$

Expressing this relation as a differential equation with boundary conditions we arrive at conditions (ii) and (iii) of the proposition statement. We have also shown that the non-triviality condition (i) is also satisfied.

For every point $t \in \mathcal{T}$, both sides of relation (1.11.14) converge and yield in the limit as $i \to \infty$,

$$p(t) \cdot f(t, \bar{x}(t), \bar{u}(t)) \geq p(t) \cdot f(t, \bar{x}(t), u(t)). \tag{1.11.18}$$

(Recall that, here, u is an arbitrary element in $\mathcal{U}$.) If assertion (iv) were false, then there would exist a subset of points t, of positive measure, on which

$$p(t) \cdot f(t, \bar{x}(t), \bar{u}(t)) < \sup_{u \in U(t)} p(t) \cdot f(t, \bar{x}(t), u).$$

We could then choose the control function u such that

$$p(t) \cdot f(t, \bar{x}(t), \bar{u}(t)) < p(t) \cdot f(t, \bar{x}(t), u(t)),$$

for all t in a set of positive measure. This contradicts (1.11.18). Assertion (iv) is confirmed. □

B: Proof of the Maximum Principle by Set Separation Methods

Take a control system

$$(S) \begin{cases} \dot{x}(t) = f(t, x(t), u(t)), \text{ a.e. } t \in [S, T], \\ u(t) \in U(t), \text{ a.e. } t \in [S, T], \\ x(S) = x_0, \end{cases}$$

with data as in the dynamic optimization problem (P). Assume that f and $U(t)$, $S \leq t \leq T$, satisfy hypotheses (A), (C) and (D) of Theorem 1.11.1. Fix a process $(\bar{x}, \bar{u})$ for (S) and $\beta > 0$.

Take a function $d: \mathbb{R}^n \to \mathbb{R}^k$ and $\beta > 0$. Define the β-local reachable set for (S) at time T, relative d, to be

$$\mathcal{R}_d(T) := \{\xi \in \mathbb{R}^k : \xi = d(x^u(T)) \text{ for some } u \in \mathcal{U} \text{ such that } ||x^u - \bar{x}||_{L^\infty} \leq \beta\}.$$

Here, as before x^u denotes the state trajectory for (S) corresponding to $u \in \mathcal{U}$.

The following theorem provides sufficient conditions for a point to be a boundary point of $\mathcal{R}_d(T)$.

1.11 Appendix: Proof of the Classical Maximum Principle

Theorem 1.11.6 (Maximum Principle for Boundary Processes) *Let $(\bar{x}, \bar{u})$ be a process for (S) such that*

$$d(\bar{x}(T)) \in bdy \, \mathcal{R}_d(T).$$

Assume that f and $U(t)$, $S \le t \le T$, satisfy hypotheses (A), (C) and (D) of Theorem 1.11.1 and there exists $\epsilon > 0$ such that $f(t, ., u)$ is C^1 on $\bar{x}(t) + \epsilon \mathbb{B}$ for all $u \in U(t)$, a.e. $t \in [S, T]$, and d is C^1 on $\bar{x}(T) + \epsilon \mathbb{B}$. Then there exist $\lambda \in \mathbb{R}^k$ and $p \in W^{1,1}([S, T]; \mathbb{R}^n)$ such that $\lambda \ne 0$ and

(i): $-\dot{p}(t) = p(t) \cdot f_x(t, \bar{x}(t), \bar{u}(t))$ *a.e.* $t \in [S, T]$,
(ii): $p(t) \cdot f(t, \bar{x}(t), \bar{u}(t)) = \max_{u \in U(t)} p(t) \cdot f(t, \bar{x}(t), u)$, *a.e.* $t \in [S, T]$,
(iii): $-p(T) = \lambda \cdot d_x(\bar{x}(T))$.

Before giving the proof of this theorem, we show that it leads to the maximum principle for the optimization problem (P), which we reproduce here as a corollary:

Corollary 1.11.7 *Let $(\bar{x}, \bar{u})$ be a an L^∞ local minimizer for problem (P). Assume hypotheses (A)–(D) of Theorem 1.11.1. Then there exist $\lambda^0 \ge 0$ and $\lambda^\phi \in (\mathbb{R}_+)^{n_\phi}$ and $\lambda^\psi \in \mathbb{R}^{n_\psi}$ and $p \in W^{1,1}([S, T]; \mathbb{R}^n)$ such that $\lambda_k^\phi = 0$ if $\phi_k(\bar{x}(T)) < 0$, $k = 1, \ldots, n_\phi$, and*

(i): $(p, \lambda^0, \lambda^\phi, \lambda^\psi) \ne (0, 0, 0, 0)$,
(ii): $-\dot{p}(t) = p(t) \cdot f_x(t, \bar{x}(t), \bar{u}(t)) - \lambda^0 L_x(t, \bar{x}(t), \bar{u}(t))$, *a.e.* $t \in [S, T]$,
(iii): $-p(T) = \lambda^0 g_x(\bar{x}(T)) + \lambda^\phi \cdot \phi_x^1(\bar{x}(T)) + \lambda^\psi \cdot \psi_x(\bar{x}(T))$,
(iv): $p(t) \cdot f(t, \bar{x}(t), \bar{u}(t)) - \lambda^0 L(t, \bar{x}(t), \bar{u}(t))$
$= \max_{u \in U(t)} \left(p(t) \cdot f(t, \bar{x}(t), u) - \lambda^0 L(t, \bar{x}(t), u) \right)$ *a.e.* $t \in [S, T]$.

Proof As we have shown in Section A, which provided the proof of the maximum principle by perturbational methods, it is required to proof the corollary only in the special case when $L = 0$ and the inequality end-point constraints are all active.

Step 1: We prove the assertions of the maximum principle, with the exception of the non-negativity conditions on the cost and inequality constraint Lagrange multipliers.

Choose the function $d : \mathbb{R}^n \to \mathbb{R}^{1+n_\phi+n_\psi}$ in Theorem 1.11.6 to be

$$d(x) := (g(x) - g(\bar{x}(T)), \phi(x), \psi(x)) \quad \text{for } x \in \mathbb{R}^n$$

and let $\beta > 0$ to be a number such that $(\bar{x}, \bar{u})$ is minimizing relative to admissible processes (x, u) such that $\|x - \bar{x}\|_{L^\infty} \le \beta$. Then, for this β,

$$(0, 0, 0) \in bdy \, \mathcal{R}_d(T). \qquad (1.11.19)$$

This is because, otherwise, there would exist $\gamma > 0$ and a process (x, u) such that $g(x(T)) - g(\bar{x}(T)) = -\gamma$, $\phi(x(T)) = 0$, $\psi(x(T)) = 0$ and $\|x - \bar{x}\|_{L^\infty} \le \beta$. This violates the local optimality of $(\bar{x}, \bar{u})$.

Assertions (i)–(iii) of the Corollary 1.11.7, for some $(\lambda^0, \lambda^\phi, \lambda^\psi) \neq (0, 0, 0)$, now follow from Theorem 1.11.6, with the exception of the non-negativity requirements: $\lambda^0 \geq 0$, $\lambda^\phi \in (\mathbb{R}_+)^{n_\phi}$ and $\lambda^\psi \in \mathbb{R}^{n_\psi}$.

Step 2: Consider the problem

$$(\mathcal{Q}) \begin{cases} \text{Minimize } g(x(T)) + z_0(T) \\ \text{s.t.} \\ (\dot{x}(t), \dot{z}(t)) = (f(t, x(t), u(t)), v(t)), \text{ a.e. } t \in [S, T], \\ u(t) \in U(t), v(t) \in [0, 1]^{n_\phi+1} \text{ a.e. } t \in [S, T], \\ (x(S), z(S)) = (0, 0), \phi_1(x(T)) + z_1(T), \ldots, \phi_{n_\phi}(x(T)) \\ \quad + z_{n_\phi}(T) = 0 \text{ and } \psi(x(T)) = 0. \end{cases}$$

Let $\bar{v}, \bar{z} : [S, T] \to \mathbb{R}^{n_\phi+1}$ be the functions $\bar{v} \equiv 0$ and $\bar{z} \equiv 0$.

The process $((\bar{x}, \bar{z}), (\bar{u}, \bar{v}))$ is clearly admissible for (Q). We claim that it is also an L^∞ local minimizer for (Q). To justify this, suppose that this assertion is not true. Then, for any $\beta > 0$, we can find a process $((x, z), (u, v))$ for (Q) such that

$$g(x(T)) < g(\bar{x}(T)) - \int_0^T v_0(t)dt, \; \phi_j(x(T)) = -\int_0^T v_j(t)dt,$$

$$\text{for } j = 1, \ldots, n_\phi,$$

$$\psi(x(T)) = 0 \text{ and } \|(x, y) - (\bar{x}, \bar{z})\|_{L^\infty}^\infty \leq \beta$$

and $\|(x, z) - (\bar{x}, \bar{z})\|_{L^\infty} \leq \beta$.

Taking note of the fact that the v_j's are non-negative valued functions, we conclude that $g(x(T)) < g(\bar{x}(T))$, $\phi_j(x(T)) \leq 0$ for $j = 1, \ldots n_\phi$, $\psi(x(T)) = 0$ and $\|x - \bar{x}\|_{L^\infty} \leq \beta$. Since $\beta > 0$ is arbitrary, this means that $(\bar{x}, \bar{u})$ cannot be an L^∞ local minimizer for the original problem. We have confirmed the claim.

Now apply the necessary conditions of Step 1, with reference to $((\bar{x}, \bar{z}), (\bar{u}, \bar{v}))$. Write p and $(q_0, q = (q_1, \ldots, q_{n_\phi}))$ for costate function components corresponding to the x and z variables. Let λ^0, λ^ϕ and λ^ψ be the Lagrange multipliers associated with the cost and end-point constraints. Then

$(p, \lambda^0, \lambda^\phi, \lambda^\psi) \neq (0, 0, 0, 0)$,
$-\dot{p}(t) = p(t) \cdot f_x(t, \bar{x}(t), \bar{u}(t))$ and $-(\dot{q}_0, \dot{q})(t) = 0$, a.e. $t \in [S, T]$,
$-p(T) = \lambda^0 g_x(\bar{x}(T)) + \lambda^\phi \cdot \phi_x^1(\bar{x}(T)) + \lambda^\psi \cdot \psi_x(\bar{x}(T))$
and $-(q_0, q)(T) = (\lambda^0, \lambda^\phi)$,

and

$$p(t) \cdot f(t, \bar{x}(t), \bar{u}(t)) = \max_{u \in U(t)} p(t) \cdot f(t, \bar{x}(t), u)$$
$$+ \sum_{j=0}^{n_\phi} \max_{v_j \in [0,1]} q_j(t) v_j, \text{ for a.e. } t \in [S, T].$$

1.11 Appendix: Proof of the Classical Maximum Principle

Setting $u = \bar{u}(t)$ in the last condition yields the information that $q_j(t) \le 0$, for $j = 0, \ldots, n_\phi$, a.e. $t \in [S, T]$. Since $-q_0 \equiv \lambda^0$ and $-q \equiv \lambda^\phi$, we can conclude that $\lambda^0 \ge 0$ and $\lambda^\phi \ge 0$. Contained in the preceding conditions are all the assertions of the theorem statement. □

Proof of Theorem 1.11.6 (*Maximum Principle for Boundary Processes*) Take a collection of control functions $u_j \in \mathcal{U}$, $j = 1, \ldots, r$. Define

$$\Lambda^{(r)} := \{\alpha \in (\mathbb{R}_+)^r : \alpha_1 + \ldots + \alpha_r \le 1\}.$$

For any $\alpha \in \Lambda$ define $x^{(\alpha)}$ to be the solution of

$$(S)_{\text{relaxed}} \begin{cases} \dot{x}(t) = f(t, x(t), \bar{u}(t)) \\ \qquad + \sum_{j=1}^{r} \alpha_j \Big(f(t, x(t), u_j(t)) - f(t, x(t), \bar{u}(t)) \Big) \text{ a.e.,} \\ x(S) = x_0. \end{cases}$$

Notice that elements $x^{(\alpha)}$ may fail to be state trajectories, but they are 'relaxed' state trajectories, in the sense that they satisfy the differential inclusion

$$\dot{x}^{(\alpha)}(t) \in \text{co}\{f(t, x^{(\alpha)}(t), u) : u \in U(t)\}, \text{ for a.e. } t \in [S, T]. \quad (1.11.20)$$

Here 'co' denotes 'convex hull'. Define $y^{(\alpha)}$ to be the solution of

$$(S)_{\text{lin}} \begin{cases} \dot{y}(t) = f_x(t, \bar{x}(t), \bar{u}(t)) y(t) \\ \qquad + \sum_{j=1}^{r} \alpha_j \Big(f(t, \bar{x}(t), u_j(t)) - f(t, \bar{x}(t), \bar{u}(t)) \Big) \text{ a.e.,} \\ y(S) = 0. \end{cases}$$

Preliminaries
The following estimate justifies the interpretation $(S)_{\text{lin}}$ as a local linearization of $(S)_{\text{relaxed}}$:

Lemma 1.11.8 (**Basic Estimates**) *There exist $K > 0$ and a function $\theta : \mathbb{R}_+ \to \mathbb{R}_+$ such that $\lim_{s \downarrow 0} \theta(s) = 0$ and, for all $\alpha \in \Lambda^{(r)}$,*

$$\|(x^{(\alpha)} - \bar{x}) - y^{(\alpha)}\|_{L^\infty} \le |\alpha|_1 \times \theta(|\alpha|_1) \quad (1.11.21)$$

and

$$\|y^{(\alpha)}\|_{L^\infty} \le K|\alpha|_1, \quad (1.11.22)$$

in which $|\alpha|_1 := \sum_j \alpha_j$.

The proof of this lemma, which is based on the definitions of $x^{(\alpha)}$ and $y^{(\alpha)}$ as solutions to the differential equations $(S)_{\text{relaxed}}$ and $(S)_{\text{lin}}$, respectively, and Gronwall's lemma, is omitted.

Relaxation theorems in control theory tells us that solutions to the differential inclusion (1.11.20) can be mapped into neighbouring state trajectories for (S). The proof of the maximum principle by set separation methods makes use of a relaxation theorem, in which the mapping, restricted to a finitely generated class of relaxed state trajectories, can be chosen to be continuous.

Lemma 1.11.9 (Relaxation) *For any given $\delta > 0$, there exists a continuous mapping $\chi_\delta : \Lambda^{(r)} \to C([S, T]; \mathbb{R}^n)$ such that, for each $\alpha \in \Lambda^{(r)}$,*

(i): $\chi_\delta(\alpha)$ is a state trajectory for (S),

and

(ii): $\|\chi_\delta(\alpha) - x^{(\alpha)}\|_{L^\infty} \leq \delta$.

The proof of this lemma, which is available in a number of standard texts, e.g. [21], is not reproduced here.

With these preliminaries behind us, we are ready to return to the proof. The following set should be regarded as a first order approximation to the reachable set $\mathcal{R}_d(T)$

$$\mathcal{R}_d^{approx}(T) := \{\nabla d(\bar{x}(T)) \cdot y^u(T) : u \in \mathcal{U}\},$$

in which y^u denotes the solution of the linearized system

$$\begin{cases} \dot{y}(t) = f_x(t, \bar{x}(t), \bar{u}(t))y(t) + f(t, \bar{x}(t), u(t)) - f(t, \bar{x}(t), \bar{u}(t)), \text{ a.e. } t \in [S, T] \\ y(S) = 0. \end{cases}$$

The key step is to establish that

'$0 \in \text{int co } \mathcal{R}_d^{approx}(T)$' implies '$d(\bar{x}(T)) \in \text{int } \mathcal{R}_d(T)$'. (1.11.23)

To prove this relation, assume that $0 \in \text{int co } \mathcal{R}_d^{approx}(T)$. It follows that, for some $r_0 > 0$, $r_0 \mathbb{B} \subset \text{co} \mathcal{R}_d^{approx}(T)$. Take any $\bar{\xi} \in r_0 \mathbb{B}$, such that $\bar{\xi} \neq 0$. Then there exists a simplex S generated by 0 and the linearly independent vectors $\xi_1 \ldots, \xi_r$ in the (unconvexified) set $\mathcal{R}_g^{approx}(T)$ such that $\bar{\xi}$ is interior to S.

The vectors in S are in a one-to-one relation with the elements in $\Lambda^{(r)}$:

$$S = \{\xi \in \mathbb{R}^r : \xi = F\alpha, \alpha \in \Lambda^{(r)}\},$$

in which $F \in \mathbb{R}^{r \times r}$ is the invertible matrix $F := \text{row}\{\xi_1, \ldots, \xi_r\}$.

We can deduce from the property that $\bar{\xi} \in \text{int } S$ that there exists $\rho > 0$ such that $\bar{\xi} + \rho \mathbb{B} \subset S$. Define $\alpha(\xi) := F^{-1}\xi$. Since $\bar{\xi} + \rho \mathbb{B} \subset S$, we know that

$$\alpha(\xi) \in \Lambda^{(r)}, \text{ for all } \xi \in \bar{\xi} + \rho \mathbb{B} \subset S.$$

1.11 Appendix: Proof of the Classical Maximum Principle

For $i = 1, 2, \ldots, r$, we can choose $u_i \in \mathcal{U}$ such that

$$\xi_i = d_x(\bar{x}(T)) y^{u_i}(T), \text{ for } i = 1, \ldots, r.$$

Now let us assume that the arbitrary collection of control functions $\{u_1, \ldots, u_r\}$, introduced at the at the beginning of the proof, have been chosen to be the control functions in the preceding relation.

Take any $\epsilon \in (0, 1)$ and $\delta > 0$. Now consider the mapping $\Gamma_{\epsilon,\delta} : \bar{\xi} + \rho \mathbb{B} \to \mathbb{R}^r$ defined as follows:

$$\Gamma_{\epsilon,\delta}(\xi) := \epsilon^{-1}\Big(-d(\chi_\delta(\epsilon\alpha(\xi))(T)) + d(\bar{x}(T))\Big) + \bar{\xi} + \nabla d(\bar{x}(T)) \cdot y^{(\epsilon\alpha(\xi))}.$$

Using the properties of the mapping χ_δ provided by Lemma 1.11.9 and also the local Lipschitz continuity of d, we can show that

$$|\Gamma_{\epsilon,\delta}(\xi) - \bar{\xi}| \leq \epsilon^{-1}\Big|-d(x^{(\epsilon\alpha(\xi))}(T)) + d(\bar{x}(T)) + \nabla d(\bar{x}(T)) \cdot y^{(\epsilon\alpha(\xi))}(T)\Big| + \epsilon^{-1} e(\delta),$$

in which $e : \mathbb{R}_+ \to \mathbb{R}_+$ is a mapping such that $e(s) \downarrow 0$ as $s \downarrow 0$. But then,

$$|\Gamma_{\epsilon,\delta}(\xi) - \bar{\xi}| \leq \epsilon^{-1}\Big|-d(y^{(\epsilon\alpha(\xi))}(T) + \bar{x}(T))$$
$$+ d(\bar{x}(T)) + \nabla d(\bar{x}(T)) \cdot y^{(\epsilon\alpha(\xi))}(T)\Big|$$
$$+ \epsilon^{-1} e(\delta) + k_d \theta(\epsilon|\alpha(\xi)|_1),$$

in which θ is the modulus of continuity of Lemma 1.11.8 and k_d is a Lipschitz constant for d. Employing the mean value function and Lemma 1.11.8 to estimate the first term on the right of this inequality and noting that $|\alpha|_1 \leq 1$ for $\alpha \in \Lambda^{(r)}$, we deduce that

$$|\Gamma_{\epsilon,\delta}(\xi) - \bar{\xi}| \leq \epsilon^{-1} e(\delta) + k_d \theta(\epsilon) + K\theta_1(K\epsilon).$$

(θ_1 is the modulus of continuity for ∇d near $\bar{x}(T)$ and K is the constant of Lemma 1.11.8.)

It follows that, by choosing $\epsilon > 0$ and $\delta > 0$ such that $\epsilon^{-1} e(\delta) + k_d \theta(\epsilon) + K\theta_1(K\epsilon) \leq \rho$, we can arrange that $\Gamma_{\epsilon,\delta}$ maps the closed ball $\bar{\xi} + \rho \mathbb{B}$ into itself. We may also arrange, by making a further reduction in the size of ϵ, that, $||\chi_\delta(\epsilon\alpha(\xi)) - \bar{x}||_{L^\infty} \leq \beta$ for all $\xi \in \bar{\xi} + \rho \mathbb{B}$. (See Lemma 1.11.8.) The mapping $\Gamma_{\epsilon,\delta}$ is continuous. We conclude from Schauder's fixed point theorem that $\Gamma_{\epsilon,\delta}$ has a fixed point ξ. But, by definition of F, $\xi = F\alpha = \nabla d(\bar{x}(T)) \cdot y^{(\epsilon\alpha(\xi))}$. It follows from the fixed point property that

$$d(\chi_\delta(\epsilon\alpha(\xi)(T))) = d(\bar{x}(T)) + \epsilon\bar{\xi}.$$

We have shown that for every non-zero $\bar{\xi} \in r_0 \mathbb{B}$ there exists $u \in \mathcal{U}$ such that $\|x^u - \bar{x}\|_{L^\infty} \leq \beta$ and $d(x^u(T)) = d(\bar{x}(T)) + \epsilon\bar{\xi}$. This assertion is obviously true, with $u = \bar{u}$, when $\bar{\xi} = 0$. It has been confirmed that $d(\bar{x}(T))(= 0)$ is interior to $\mathcal{R}_d(T)$.

We now carry out the final step of the proof. Notice that the maximum principle can be equivalently expressed as follows: there exists a non-trivial vector λ such that

$$\lambda \cdot \nabla d(\bar{x}(T)) \int_S^T S(T,s)[f(s,\bar{x}(s),u(s)) - f(s,\bar{x}(s),\bar{u}(s))]ds \leq 0, \text{ for all } u \in \mathcal{U}. \quad (1.11.24)$$

Here, $S(t,s)$ is the fundamental matrix for the linear differential equation $\dot{y} = \nabla_x f(t,\bar{x}(t),\bar{u}(t))y(t)$, that is $S(.,s)$ is the $n \times n$ matrix value function that satisfies, for each $s \in [S,T]$, the matrix differential equation $d/dt\, S(t,s) = \nabla_x f(t,\bar{x}(t),\bar{u}(t))S(t,s)$ on $[S,T]$, with boundary condition $S(s,s) = I$.

We can rewrite relation (1.11.24), using our earlier notation, as

$$\lambda \cdot \nabla d(\bar{x}(T))y^u(T) \leq 0, \text{ for all } u \in \mathcal{U}. \quad (1.11.25)$$

We argue by contraposition. Suppose that the maximum principle is not satisfied. Then there exists no non-zero vector λ such that (1.11.25) is satisfied. This means that the point 0 is not separated from the convex set co $\mathcal{R}_d^{\text{approx}}(T)$. It follows from the separating hyperplane theorem that 0 is interior to co $\mathcal{R}_d^{\text{approx}}(T)$. But then, as we have seen, $d(x(T))$ is an interior point of $\mathcal{R}_d(T)$. This means that, if $d(x(T))$ is a boundary point of $\mathcal{R}_d(T)$, then the maximum principle is valid. The proof is complete. $\square$

Discussion

We see in the preceding proofs two very different approaches to proving the maximum principle. To simplify the discussion, let us assume that the L^∞ local minimizer of interest is a minimizer and that there is a unique state trajectory associated with each control, so we can regard the domain of the dynamic optimization problem (P) as the set of controls. The perturbation method can be summarized as follows:

Clarke's Perturbation Approach take a minimizer $\bar{u}$ for the original problem (P).

- Choose $\epsilon > 0$. Find, with the help of Ekeland's theorem, a control u_ϵ that is a minimizer for a perturbed problem $(P)_\epsilon$ and whose proximity to the minimizing control $\bar{u}$ is governed by the parameter ϵ. The perturbed problem is one for which it is easy to derive necessary condition of optimality; to be specific, the perturbed problem has no right end-point constraints.
- Write down necessary conditions for the $(P)_\epsilon$, with reference to the minimizer u_ϵ.
- Obtain necessary conditions for the original problem (P), with reference to $\bar{u}$, by passing to the limit, as $\epsilon \downarrow 0$, in the necessary conditions for the perturbed problem.

1.11 Appendix: Proof of the Classical Maximum Principle

Set Separation Approach Note that a minimizer $\bar{u}$ defines a boundary point of the reachable set that comprises images of the terminal values of the state trajectories under a function having components the cost and equality constraint functions.

- Construct a (convex) approximating reachable set. Use the Schauder fixed point theorem and a relaxation lemma to show that, if 0 is an interior point of the approximating reachable set, then $\bar{u}$ must be associated with an interior point of the reachable set, i.e. $\bar{u}$ does not define a boundary point of the reachable set.
- Since 'there exists a linear hyperplane supporting the approximating reachable set at 0' implies 'the maximum principle is valid' and the approximation cone is convex, we can deduce from the separating hyperplane theorem that 'the maximum principle is not valid' implies '0 is interior to the approximating reachable set'.

These relations combine to give a proof of the maximum principle in contrapositive form: 'the maximum principle is not valid' $\implies$ '0 is an interior point of the approximating reachable set' $\implies$ '$\bar{u}$ does not define an interior point of the reachable set' $\implies$ '$\bar{u}$ is not a minimizer'.

Note that the relative simplicity of the two proofs should not be judged by their length alone. This is because gradient calculations ('needle' variations) and detailed justification of reductions to special cases are included in the first proof, but not in the second.

Both approaches are powerful ones, and extend to provide necessary conditions of optimality for a wide-range of dynamic optimization problems, including those with nonsmooth data and a variety of different pathwise constraints. As regards extensions based on perturbation methods, Clarke's nonsmooth maximum principle of 1976 stands out for the unrestrictive nature of the hypotheses under which it was proved [59]. Other group of contributors to the field who employed such methods included Ioffe, Loewen, Rockafellar and Vinter. The body of work on extensions to the maximum principle, based on set separation methods, is large; a far from complete list of contributors include Pontryagin and his collaborators; also Gamkrelidze, Dubovistkii, Milyutin, Dmitruk, Osmolovski, Mordukhovich, Neustadt, Warga, Sussmann and Hestenes. It has often been, but not always, the case that advances in the field, based on one approach, have been matched by subsequent independent proofs based on the other approach. For example, necessary conditions for general problems with data which are not continuously differentiable w.r.t. the state variable (so-called nonsmooth problems) was introduced independently by Clarke, using perturbation methods, and Warga, who followed a set separation approach. Even when both approaches can be used to provide necessary conditions for a particular class of dynamic optimization problems, they differ often, in terms of transparency of the proofs and the relative strengths of the hypotheses that are invoked. Both approaches then are distinct and valuable weapons in the armoury of modern day research.

1.12 Exercises

1.1 Consider a generalization of the fixed left endpoint problem (P) of Sect. 1.5, in which the initial state is included in the choice variables:

$$(P1) \begin{cases} \text{Minimize } g(x(S), x(T)) + \int_S^T L(t, x(t), u(t))dt \\ \text{over absolutely cont. functions } x : [S, T] \to \mathbb{R}^n \\ \qquad \text{and meas. functions } u : [S, T] \to \mathbb{R}^m \text{ such that} \\ \dot{x}(t) = f(t, x(t), u(t)), \text{ a.e. } t \in [S, T], \\ u(t) \in U(t), \text{ a.e. } t \in [S, T], \\ (x(S), x(T)) \in C. \end{cases}$$

(The data are as for fixed left end-point problem (P) of Sect. 1.5 except, now, g has domain $\mathbb{R}^n \times \mathbb{R}^n$ and C is a subset of $\mathbb{R}^n \times \mathbb{R}^n$.)

Taking as starting point the maximum principle (Theorem 1.5.1) for the fixed left end-point problem (P), derive the following maximum principle for $(P1)$:

Let $(\bar{x}, \bar{u})$ be an L^∞ local minimizer for (P1). Assume f, L and U satisfy the hypotheses of Theorem 1.5.1, C is closed and g is C^1 on a neighbourhood of $(\bar{x}(S), \bar{x}(T))$. Then there exist $\lambda \geq 0$ and $p \in W^{1,1}([S, T]; \mathbb{R}^n)$ such that

(i): $(p, \lambda) \neq (0, 0)$,
(ii): $-\dot{p}(t) = p(t) \cdot \nabla_x f(t, \bar{x}(t), \bar{u}(t)) - \lambda \nabla_x L(t, \bar{x}(t), \bar{u}(t))$, a.e. $t \in [S, T]$,
(iii): $p(t) \cdot f(t, \bar{x}(t), \bar{u}(t)) - \lambda L(t, \bar{x}(t), \bar{u}(t)) =$
$$\max_{u \in U(t)} \Big(p(t) \cdot f(t, \bar{x}(t), u) - \lambda L(t, \bar{x}(t), u) \Big), \text{ a.e. } t \in [S, T],$$
(iv): $(p(S), -p(T)) \in \lambda \nabla g(\bar{x}(S), \bar{x}(T)) + N_C(\bar{x}(S), \bar{x}(T))$.

Hint: Show that the process $((\bar{x}_1, \bar{x}_2), (\bar{u}_1, \bar{u}_2))$ on $[S - 1, T]$, in which

$$(\bar{x}_1, \bar{u}_1) \equiv (\bar{x}(S), 0), \text{ and } (\bar{x}_2, \bar{u}_2) = \begin{cases} (\bar{x}(S), 0) & \text{if } t \in [S - 1, S) \\ (\bar{x}(t), \bar{u}(t)) & \text{if } t \in [S, T] \end{cases}$$

is an L^∞ local minimizer to the fixed left end-point problem

$$(P1)' \begin{cases} \text{Minimize } g(x_1(T), x_2(T)) + \int_{S-1}^T \begin{cases} 0 & \text{if } t \in [S - 1, S) \\ L(t, x_2(t), u_2(t)) & \text{if } t \in [S, T] \end{cases} dt \\ \text{subject to} \\ (\dot{x}_1(t), \dot{x}_2(t)) = \begin{cases} (u_1, u_1) & \text{if } t \in [S - 1, S) \\ (0, f(t, x_2(t), u_2(t))) & \text{if } t \in [S, T] \end{cases}, \\ u_1(t) \in \begin{cases} [-1, +1]^n & \text{if } t \in [S - 1, S) \\ \{0\} & \text{if } t \in [S, T] \end{cases} \\ \text{and } u_2(t) \in \begin{cases} \{0\} & \text{if } t \in [S - 1, S) \\ U(t) & \text{if } t \in [S, T] \end{cases} \text{ a.e.} \\ (x_1(S - 1), x_2(S - 1)) = (\bar{x}(S), \bar{x}(S)) \text{ and } (x_1(T), x_2(T)) \in C. \end{cases}$$

Now apply the fixed left end-point maximum principle.

1.12 Exercises

1.2 (Catalytic Conversion Problem) This is a dynamic optimization problem arising in chemical engineering. Here, the independent variable is not time, but displacement s along a one-dimensional, isothermic catalytic reactor. A mixture of three chemicals A, B and C is propelled through a longitudinal catalytic converter. The objective is to maximize the mole fraction of the chemical C at some point along the reactor. Let x_1, x_2 and x_3 denote the mole fractions of the three chemicals A, B and C, at displacement s along the reactor. Catalyst a promotes a reversible reaction between A and B. Catalyst b promotes an irreversible reaction that converts chemical B into C. Intuition suggests that the mixture should first be exposed to catalyst a (to convert A to B) and, further along the reactor, to expose it to catalyst b (to convert B into C). But this is incorrect. We can deduce from the maximum principle that the maximum yield of C is obtained by mixing the catalysts along the reactor. The problem can be formulated at follows:

$$(P2) \begin{cases} \text{Minimize } -x_3(L) \\ \text{subject to} \\ \dot{x}_1(s) = u(s)(k_2 x_2(s) - k_1 x_1(s)) \\ \dot{x}_2(s) = u(s)(k_1 x_1(s) - k_2 x_2(s)) - (1-u(s))k_3 x_2(t) \quad \text{a.e. } s \in [0, L], \\ \dot{x}_3(s) = (1-u(s))k_3 x_2(t) \\ u(t) \in [0, 1] \text{ a.e. } s \in [0, L], \\ (x_1(0), x_2(0), x_3(0)) = (1, 0, 0), \\ (x_1(L), x_2(L), x_3(L)) \in \mathbb{R}^3. \end{cases}$$

Here, the control $u(s)$ is the proportion of catalyst a at displacement s along the reactor. k_1 and k_2 are the reaction constants of catalyst a and k_3 is the reaction constant for catalyst b.

Show, using the maximum principle, that, for L sufficiently large, the optimal catalyst mixture $\bar{u}(s), 0 \leq s \leq L$ has the following structure: For some intermediate displacements $0 < S_1 < S_2 < L$ and some constant $u^* \in (0, 1)$,

$$\bar{u}(s) = \begin{cases} 1 & \text{if } s \in [0, S_1) \\ u^* & \text{if } s \in [S_1, S_2) \\ 0 & \text{if } s \in [S_2, L]. \end{cases}$$

Remark
This example illustrates that in applied dynamic optimization problems, the independent variable need not have the interpretation 'time'; here is it displacement along a reactor duct.

1.3 [Linear Quadratic Control] This problem arises in control engineering where, typically, x and u describe deviations from a nominal state and control. The first order controlled differential equation is often a local linear approximation to a non-linear dynamic constraint. The cost function penalizes large deviations of state and control.

$$(LQ) \begin{cases} \text{Minimize } \int_0^T (x^T(t)Q(t)x(t) + u^T(t)R(t)u(t))dt + x^T(T)Gx(T) \\ \text{over } u \in L^2 \text{ and } x \in W^{1,1} \text{ satisfying} \\ \dot{x}(t) = A(t)x(t) + B(t)u(t) \quad \text{a.e. } t \in [0,T], \\ u(t) \in \mathbb{R}^m, \text{ a.e. } t \in [0,T], \\ x(0) = x_0, \end{cases}$$

with data the number $T > 0$, symmetric, essentially bounded, measurable matrix valued functions $Q : [0,T] \to \mathbb{R}^{n \times n}$, $R : [0,T] \to \mathbb{R}^{m \times m}$, a symmetric matrix $G \in \mathbb{R}^{n \times n}$, essentially bounded, measurable, matrix valued functions $A : [0,T] \to \mathbb{R}^{n \times n}$ and $B : [0,T] \to \mathbb{R}^{n \times m}$ and $x_0 \in \mathbb{R}^n$. Assume that

$G \geq 0$ and there exists $\epsilon > 0$ such that $Q(t) \geq 0$, $R(t) \geq \epsilon I_{m \times m}$ for a.e. $t \in [0,T]$.

Use dynamic programming to show that the minimizing process $(\bar{x}, \bar{u})$ is generated by the linear feedback law

$$\bar{u}(t) = -R^{-1}(t)B^T(t)P(t)\bar{x}(t),$$

in which $P : [0,T] \to \mathbb{R}^{n \times n}$ is the unique absolutely continuous, symmetric matrix value function satisfying the *Riccati equation*

$$\begin{cases} -\dot{P}(t) = A^T(t)P(t) + P(t)A(t) + Q - P(t)B(t)R^{-1}(t)B^T(t)P(t) \text{ a.e..} \\ P(T) = G. \end{cases}$$

Hint: Seek a solution V to the Hamilton Jacobi equation having the structure $V(t,x) = x^T P(t)x$.

1.4 Adapt the analysis of Exercise 1.3, to determine the minimizing process, when we allow for an inhomogeneous term $r : [0,T] \to \mathbb{R}^n$ in the controlled differential equation, which now becomes

$$\dot{x}(t) = A(t)x(t) + B(t)u(t) + r(t) \quad \text{a.e. } t \in [0,T].$$

Hint: Seek a solution V to the Hamilton Jacobi equation of the form $V(t,x) = x^T P(t)x + 2r^T(t)x$.

1.5 (Infinite Horizon Control) Consider the problem

$$(I) \begin{cases} \text{Minimize } \liminf_{T \to \infty} \int_0^T L(x(t), u(t))dt \\ \text{over locally absolutely cont. } x : [0, \infty) \to \mathbb{R}^n \text{ and meas. } u : [0, \infty) \to \mathbb{R}^m \\ \text{such that} \\ \dot{x}(t) = f(x(t), u(t)), \text{ a.e. } t \in [0, \infty), \\ u(t) \in U, \text{ a.e. } t \in [0, \infty), \\ x(0) = x_0. \end{cases}$$

The data comprise functions $f : \mathbb{R}^n \times \mathbb{R}^m \to \mathbb{R}^n$, $L : \mathbb{R}^n \times \mathbb{R}^m \to \mathbb{R}$, a compact set $U \subset \mathbb{R}^m$ and a point $x_0 \in \mathbb{R}^n$. It is assumed that $f(., u)$ and $L(., u)$ are C^1 for all $u \in U$, and f, f_x, L, L_x are continuous functions. Let $(\bar{x}, \bar{u})$ be a minimizing process.

Use the finite horizon maximum principle to show that there exist $\lambda \geq 0$ and a locally absolutely continuous function $p : [0, \infty) \to \mathbb{R}^n$, not both zero, such that

$$-\dot{p}(t) = p(t) \cdot f_x(\bar{x}(t), \bar{u}(t)) - \lambda L_x(\bar{x}(t), \bar{u}(t))dt, \text{ a.e.} t \in [0, \infty)$$

$$p(t) \cdot f(\bar{x}(t), \bar{u}(t)) - \lambda L(\bar{x}(t), \bar{u}(t))dt = \max_{u \in U} \Big(p(t) \cdot f(\bar{x}(t), u) - \lambda L(\bar{x}(t), u)dt \Big),$$
$$\text{a.e. } t \in [0, \infty).$$

Hint: Take $T_i \to \infty$. For each i, consider the corresponding finite horizon problem with underlying time interval $[0, T_i]$ with cost $\int_0^{T_i} L(x(t), u(t))dt$ and with right endpoint constraint $x(T_i) = \bar{x}(T_i)$. Show the restriction of $(\bar{u}, \bar{x})$ to $[0, T_i]$ is a minimizer for this problem. Apply the (finite horizon) maximum principle, for each i. Prove the infinite horizon necessary conditions by scaling the multipliers, subsequence extraction and passage to the limit as $T_i \to \infty$.

1.6 Since problem (I) of question 5 has no terminal cost and terminal constraints, we might expect that the necessary conditions include a transversality condition of the type

$$\lim_{T \to \infty} p(T) = 0.$$

This conjecture was shown to be false by Halkin, who provided the following counter example. Take the data for problem (I) to be $n = m = 1$, $f(x, u) = (1-x)u$, $L(x, u) = -(1-x)u$, $U = [0, 1]$ and $x_0 = 0$. Show that $(\bar{x} \equiv (1 - \exp(-t)), \bar{u} \equiv 1)$ is a minimizer. Determine Lagrange multipliers (λ, p) for this problem. Show that the necessary conditions cannot be satisfied, when they are supplemented by the above boundary condition on the costate arc.

1.13 Notes for Chapter 1

This chapter maps out major themes of the book and links them to earlier developments in mathematics. It is not intended as a comprehensive historical review of dynamic optimization.

For a more detailed exposition of the material in Sect. 1.2 on optimality conditions in the calculus of variations we recommend Clarke [68] and Troutman [191]. L. C. Young's inspirational, if individualistic, book [209] is notable for tracking the evolution of the classical theory into modern day dynamic optimization. The brachistochrone problem of Sect. 1.2 is analysed in many publications, including [191] and [209].

From numerous older books on classical dynamic optimization, we single out for clarity and accessibility [21, 104, 132] and [210]. Warga [204] and Cesari [54] provide more detailed coverage, centred on their own contributions to the theory of existence of minimizers and necessary conditions. Bryson and Ho's widely read book [49] (and also [48]), which follow a traditional applied mathematics approach to dynamic optimization, includes a wealth of engineering-inspired examples. Betts' monograph [38] also provides many examples, while focusing on computational techniques for their solution. Shättler and Ledzewicz's book [137] emphasizes geometric aspects of dynamic optimization .

The two books [68] and [85] by Clarke et al. are source references on nonsmooth analysis, covering different approaches in both in finite and infinite dimensional vector spaces, including the one based on the proximal methods that feature prominently in this publication. Other authoritative books including material on nonsmooth analysis are those of Rockafellar and Wets [177] and Aubin and Frankowska [14]. As regards nonsmooth dynamic optimization, Clarke's book [65], which had a major role in winning an audience for the field, remains a standard reference. Clarke's more recent book [68] includes later developments in the field. See also [85]. Expository accounts of this material (and extensions) are to be found in [64] and [141].

Two proofs of the classical maximum principle are given. The proof based on older set separation methods is standard. (See, for example, [21] or [132]). The proof based on Clarke's perturbation techniques is essentially that in [65], but exploits the simplifications that are possible when the data is smooth. See also [98].

The maximum orbital transfer problem discussed in Sect. 1.1 is a refinement of a problem first studied by Kopp and McGill, and reported in [49], to allow for thrusts of variable magnitude. See also [48].A second order, feasible directions algorithm due to R. Pytlak, documented in [168, 169] was used in the computations summarized in the diagrams of Sect. 1.1. The solution to the growth versus consumption problem from economics is taken from [152], which contains references to earlier dynamic optimization based on the Ramsay growth model. Other applications of dynamic optimization theory to economics are to be found in, for example, [160]. The application of dynamic optimization to chemotherapy has been an active field since the late 1980's. A notable early contribution was [161]. Ledzewicz and Schättler have been prominent among researchers in this field [138]. The example considered in this chapter is based on [188], following removal of a time delay from the dynamic model. The numerical results and graphs are taken from this paper. Many formulations of the dynamic optimization problem have been proposed, involving different cost functions, constraints and underlying dynamics. Methods of dynamic optimization have been used also to design integrated treatments for tumour angiogenesis and chemotherapy [139].

Examples of applications of the theory of dynamic optimization to other areas referred to in this chapter are to be found in [55] (resource economics) and [180] (epidemiology).

Chapter 2
Set Convergence, Measurability and Existence of Minimizers

Abstract This chapter covers basic concepts and definitions concerning set convergence, measurability and multifunctions. Multifunctions are mappings whose values are not single points in the range space but subsets of such points. They are widely encountered in dynamic optimization; for example, it is helpful to interpret a time dependent control constraint set as a multifunction and admissible control functions as selectors of that multifunction. The chapter provides proofs of key properties, with special emphasis on properties relevant to dynamic optimization. A prior study of measurability is an important step in establishing existence of solutions to dynamic optimization problems. The chapter concludes with a theorem concerning existence of minimizers for the generalized problem of Bolza. The underlying optimization problem resembles the problem of Bolza from the classical calculus of variations, involving integral and endpoint costs. But it is in fact a far broader problem formulation, because we depart from the classical framework by allowing the cost integrand and endpoint constraint functions to take values in the extended real line. This formulation subsumes a wide range of problems involving dynamic constraints (in the form of controlled differential equations or differential inclusions), which can be accommodated by the use of extended valued indicator functions added to the cost integrand.

2.1 Introduction

Early developments in variational analysis, which predated the measurability concepts introduced by Lebesgue, sought to provide information, such the Euler condition, about an arc that minimizes a given integral functional of the arc and its derivative. Here, 'arc' was understood to be an element in the class of continuous functions with piecewise continuous derivatives. When however, in the early twentieth century, the field of study broadened also to address the questions of existence, 'when does a minimizing arc exist?', the need arose to consider larger classes of arcs. In modern day dynamic optimization, it is customary to take the underlying family of arcs to be the space of absolutely continuous functions (or some subset of this space that takes account of endpoint and pathwise constraints) since, in this framework, it is possible to guarantee existence of minimizers under unrestrictive, directly verifiable hypotheses. So we must now consider minimizers that belong to the class of indefinite integrals of measurable/summable functions. The formulation of dynamic constraints as either controlled differential equations or differential inclusions requires us to consider, not just measurable functions but also measurable multifunctions and the interpretation of integrals of set-valued functions.

This chapter supplies basic definitions and concepts from the theory of measurable multifunctions, extensive use of which is made throughout this book. It also provides derivations of multifunction properties having special significance in dynamic optimization. Earlier we referred to the link between existence and measurability. Appropriately then, the chapter concludes with a theorem giving conditions guaranteeing existence of minimizers (in the class of absolutely continuous functions) for the generalized problem of Bolza. The underlying optimization problem resembles the problem of Bolza from the classical calculus of variations, that is the variational problem with both integral and endpoint costs. But it is in fact a far broader problem formulation, because, in contrast to classical studies, we allow the cost integrand and endpoint constraint functions to take values in the *extended* real line $(-\infty, \infty) \cup \{+\infty\}$. (Hence our use of the qualifier 'generalized'.) This formulation subsumes a wide variety of problems involving dynamic constraints (in the form of controlled differential equations or differential inclusions), which can be accommodated by the use of extended valued penalty functions added to the cost integrand.

2.2 Convergence of Sets and Continuity of Multifunctions

Take a sequence of sets $\{A_i\}$ in $\mathbb{R}^n$. There are number of ways of defining limit sets. For our purposes, 'Kuratowski sense' limit operations will be the most useful. The set

$$\liminf_{i \to \infty} A_i$$

(the Kuratowski lim inf) comprises all points $x \in \mathbb{R}^n$ satisfying the condition: there exists a sequence $x_i \to x$ such that $x_i \in A_i$ for all i.

The set

$$\limsup_{i \to \infty} A_i$$

(the Kuratowski lim sup) comprises all points $x \in \mathbb{R}^n$ satisfying the condition: there exist a subsequence $\{A_{i_j}\}$ of $\{A_i\}$ and a sequence $x_j \to x$ such that $x_j \in A_{i_j}$ for all j.

2.2 Convergence of Sets and Continuity of Multifunctions

$\liminf_{i\to\infty} A_i$ and $\limsup_{i\to\infty} A_i$ are (possibly empty) closed sets, related according to

$$\liminf_{i\to\infty} A_i \subset \limsup_{i\to\infty} A_i.$$

In the event $\liminf_{i\to\infty} A_i$ and $\limsup_{i\to\infty} A_i$ coincide, we say that $\{A_i\}$ has a limit (in the Kuratowski sense) and write

$$\lim_{i\to\infty} A_i := \liminf_{i\to\infty} A_i \, (= \limsup_{i\to\infty} A_i).$$

The sets $\liminf_{i\to\infty} A_i$ and $\limsup_{i\to\infty} A_i$ are succinctly expressed in terms of the distance function

$$d_A(x) = \inf_{y\in A} |x - y|,$$

thus

$$\liminf_{i\to\infty} A_i = \{x : \limsup_{i\to\infty} d_{A_i}(x) = 0\},$$

and

$$\limsup_{i\to\infty} A_i = \{x : \liminf_{i\to\infty} d_{A_i}(x) = 0\}.$$

These concepts can be generalized to define limit points of multifunctions. Take a set D. A multifunction $\Gamma : D \rightsquigarrow \mathbb{R}^n$ is a mapping from D into the space of subsets of $\mathbb{R}^n$. For each $x \in D$ then, $\Gamma(x)$ is a subset of $\mathbb{R}^n$. In this way we obtain a family of sets $\{\Gamma(x) \subset \mathbb{R}^n : x \in D\}$, parameterized by points $x \in D$. We shall often refer to a multifunction as *convex*, *closed* or *non-empty* depending on whether $\Gamma(x)$ has the referred-to property for all $x \in D$. The graph of Γ is the set

$$\mathrm{Gr}\,\Gamma := \{(x, \xi) \in D \times \mathbb{R}^n : \xi \in \Gamma(x)\}.$$

Consider now the case in which $D \subset \mathbb{R}^k$. Fix a point $x \in \mathbb{R}^k$. The set

$$\liminf_{y \xrightarrow{D} x} \Gamma(y)$$

(the Kuratowski lim inf) comprises all points ξ satisfying the condition: corresponding to any sequence $y_i \xrightarrow{D} x$, there exists a sequence $\xi_i \to \xi$ such that $\xi_i \in \Gamma(y_i)$ for all i. The set

$$\limsup_{y \xrightarrow{D} x} \Gamma(y)$$

(the Kuratowski lim sup) comprises all points ξ satisfying the condition: there exist sequences $y_i \xrightarrow{D} x$ and $\xi_i \to \xi$ such that $\xi_i \in \Gamma(y_i)$ for all i. (In the above, '$y_i \xrightarrow{D} x$' means '$y_i \to x$ and $y_i \in D$ for all i'.)

If D is a neighbourhood of x we write $\liminf_{y \to x} \Gamma(y)$ in place of $\liminf_{y \xrightarrow{D} x} \Gamma(y)$, etc.

Here, too, we have convenient characterizations of the limit sets in terms of the distance function on $\mathbb{R}^n$:

$$\liminf_{y \xrightarrow{D} x} \Gamma(y) = \{\xi \in \mathbb{R}^n : \limsup_{y \xrightarrow{D} x} d_{\Gamma(y)}(\xi) = 0\}$$

and

$$\limsup_{y \xrightarrow{D} x} \Gamma(y) = \{\xi \in \mathbb{R}^n : \liminf_{y \xrightarrow{D} x} d_{\Gamma(y)}(\xi) = 0\}.$$

We observe that $\liminf_{y \xrightarrow{D} x} \Gamma(y)$ and $\limsup_{y \xrightarrow{D} x} \Gamma(y)$ are closed (possibly empty) sets, related according to

$$\liminf_{y \xrightarrow{D} x} \Gamma(y) \subset \limsup_{y \xrightarrow{D} x} \Gamma(y).$$

When the $\liminf_{y \xrightarrow{D} x} \Gamma(y)$ and $\limsup_{y \xrightarrow{D} x} \Gamma(y)$ are equal, we say that the limit of Γ at x along D exists and we set

$$\lim_{y \xrightarrow{D} x} \Gamma(y) := \liminf_{y \xrightarrow{D} x} \Gamma(y) \, (= \limsup_{y \xrightarrow{D} x} \Gamma(y)).$$

To reconcile these definitions, we note that, give a sequence of sets $\{A_i\}$ in $\mathbb{R}^n$,

$$\limsup A_i = \limsup_{y \xrightarrow{D} x} \Gamma(y) \text{ etc.,}$$

when we identify D with the subset $\{1, 1/2, 1/3, \ldots\}$ of the real line, choose $x = 0$ and define $\Gamma(y) = A_i$ when $y = i^{-1}$, $i = 1, 2, \ldots$.

Closely related concepts concern the (semi-)continuity of multifunctions. $\Gamma : D \rightsquigarrow \mathbb{R}^n$ is said to be *upper semi-continuous* at $x \in D$ if for any $\epsilon > 0$ there exists $r > 0$ such that

$$\Gamma(y) \subset \Gamma(x) + \epsilon \mathbb{B}, \quad \text{for all } y \in (x + r\mathbb{B}) \cap D,$$

and Γ is called *upper semi-continuous* if it is upper semi-continuous at x for all $x \in D$. $\Gamma : D \rightsquigarrow \mathbb{R}^n$ is said to be *lower semi-continuous* at $x \in D$ if for any $\xi \in \Gamma(x)$

and any sequence $x_i \xrightarrow{D} x$, there exists a sequence $\xi_i \in \Gamma(x_i)$ such that $\xi_i \to \xi$. Γ is called *lower semi-continuous* if it is lower semi-continuous at x for all $x \in D$. Clearly Γ is lower semi-continuous at $x \in D$ if and only if $\Gamma(x) \subset \liminf_{y \xrightarrow{D} x} \Gamma(y)$. Observe also that, if Γ is upper semi-continuous at $x \in D$, then

$$\limsup_{y \xrightarrow{D} x} \Gamma(y) \subset \overline{\Gamma(x)}.$$

The multifunction $\Gamma : D \rightsquigarrow \mathbb{R}^n$ is *continuous* at $x \in D$ if it both lower semi-continuous and upper semi-continuous at x, and Γ is continuous if it is continuous at each $x \in D$.

2.3 Measurable Multifunctions

Recall that a measurable space $(\Omega, \mathcal{F})$ comprises a set Ω and a family $\mathcal{F}$ of subsets of Ω which is a σ-field, i.e.

(i) $\emptyset \in \mathcal{F}$,
(ii) $F \in \mathcal{F}$ implies that $\Omega \setminus F \in \mathcal{F}$,
(iii) $F_1, F_2, \cdots \in \mathcal{F}$ implies $\cup_{i=1}^{\infty} F_i \in \mathcal{F}$.

Definition 2.3.1 Let $(\Omega, \mathcal{F})$ be a measurable space. Take a multifunction $\Gamma : \Omega \rightsquigarrow \mathbb{R}^n$. Γ is *measurable* when the set

$$\Gamma^{-1}(W) = \{x \in \Omega : \Gamma(x) \cap W \neq \emptyset\}$$

is $\mathcal{F}$ measurable for every open set $W \subset \mathbb{R}^n$.

Fix a Lebesgue subset $I \subset \mathbb{R}$. Let $\mathcal{L}$ denote the Lebesgue subsets of I. If Ω is the set I then '$\Gamma : I \rightsquigarrow \mathbb{R}^n$ is measurable' is taken to mean that the multifunction is $\mathcal{L}$ measurable.

Denote by $\mathcal{B}^k$ the Borel subsets of $\mathbb{R}^k$. The product σ-algebra $\mathcal{L} \times \mathcal{B}^k$ (that is the smallest σ-algebra of subsets of $I \times \mathbb{R}^k$ that contains all product sets $A \times B$ with $A \in \mathcal{L}$ and $B \in \mathcal{B}^k$) is often encountered in hypotheses invoked to guarantee measurability of multifunctions, validity of certain representations for multifunctions, etc.

A first taste of such results is provided by:

Proposition 2.3.2 *Take an $\mathcal{L} \times \mathcal{B}^m$ measurable multifunction $F : I \times \mathbb{R}^m \rightsquigarrow \mathbb{R}^k$ and a Lebesgue measurable function $u : I \to \mathbb{R}^m$. Then $G : I \rightsquigarrow \mathbb{R}^k$ defined by*

$$G(t) := F(t, u(t))$$

is an $\mathcal{L}$ measurable multifunction.

Proof For an arbitrary choice of set $A \in \mathcal{L}$ and $B \in \mathcal{B}^m$, the set

$$\{t \in I : (t, u(t)) \in A \times B\}$$

is a Lebesgue subset because it is expressible as $A \cap u^{-1}(B)$ and u is Lebesgue measurable. Denote by $\mathcal{D}$ the family of subsets $E \subset I \times \mathbb{R}^m$ for which the set

$$\{t \in I : (t, u(t)) \in E\}$$

is Lebesgue measurable. $\mathcal{D}$ is a σ-field, as is easily checked. We have shown that it contains all product sets $A \times B$ with $A \in \mathcal{L}$, $B \in \mathcal{B}^m$. So $\mathcal{D}$ contains the σ-field $\mathcal{L} \times \mathcal{B}^m$.

Take any open set $W \subset \mathbb{R}^k$. Then, since F is $\mathcal{L} \times \mathcal{B}^m$ measurable, $E := \{(t, u) : F(t, u) \cap W \neq \emptyset\}$ is $\mathcal{L} \times \mathcal{B}^m$ measurable. But then

$$\{t \in I : G(t, u(t)) \cap W \neq \emptyset\} = \{t \in I : (t, u(t)) \in E\}$$

is a Lebesgue measurable set since $E \in \mathcal{D}$. Bearing in mind that W is an arbitrary open set, we conclude that $t \rightsquigarrow G(t) := F(t, u(t))$ is a Lebesgue measurable multifunction. □

Specializing to the point valued case we obtain:

Corollary 2.3.3 *Consider a function $g : I \times \mathbb{R}^m \to \mathbb{R}^k$ and a Lebesgue measurable function $u : I \to \mathbb{R}^m$. Suppose that g is $\mathcal{L} \times \mathcal{B}^m$ measurable. Then the mapping $t \to g(t, u(t))$ is Lebesgue measurable.*

Functions $g(t, u)$ arising in dynamic optimization to which Corollary 2.3.3 are often applied are composite functions of a nature covered by the following proposition.

Proposition 2.3.4 *Consider a function $\phi : I \times \mathbb{R}^n \times \mathbb{R}^m \to \mathbb{R}^k$ satisfying the following hypotheses:*

(a) $\phi(t, \cdot, u)$ is continuous for each $(t, u) \in I \times \mathbb{R}^m$,
(b) $\phi(\cdot, x, \cdot)$ is $\mathcal{L} \times \mathcal{B}^m$ measurable for each $x \in \mathbb{R}^n$.

Then for any Lebesgue measurable function $x : I \to \mathbb{R}^n$, the mapping $(t, u) \to \phi(t, x(t), u)$ is $\mathcal{L} \times \mathcal{B}^m$ measurable.

Proof Let $\{r_j\}$ be an ordering of the set of n-vectors with rational coefficients. For each integer k define

$$\phi_k(t, u) := \phi(t, r_j, u)$$

where j is chosen (j will depend on k and t) such that

$$|x(t) - r_j| \leq 1/k \text{ and } |x(t) - r_i| > 1/k \text{ for all } i \in \{1, 2.., j-1\}.$$

(These conditions uniquely define j.)

2.3 Measurable Multifunctions

Since $\phi(t, \cdot, u)$ is continuous,

$$\phi_k(t, u) \to \phi(t, x(t), u) \text{ as } k \to \infty$$

for every $(t, u) \in I \times \mathbb{R}^m$. It suffices then to show that ϕ_k is $\mathcal{L} \times \mathcal{B}^m$ measurable for an arbitrary choice of k.

For any open set $V \subset \mathbb{R}^k$,

$$\phi_k^{-1}(V) = \{(t, u) \in I \times \mathbb{R}^m : \phi_k(t, u) \in V\}$$
$$= \cup_{j=1}^{\infty} \big(\{(t, u) \in I \times \mathbb{R}^m : \phi(t, r_j, u) \in V\}$$
$$\cap \ \{(t, u) \in I \times \mathbb{R}^m : |x(t) - r_j| \leq 1/k$$
$$\text{and } |x(t) - r_i| > 1/k \text{ for } i = 1, ..., j-1\}\big).$$

Since the set on the right side is a countable union of $\mathcal{L} \times \mathcal{B}^m$ measurable sets, we have established that ϕ_k is a $\mathcal{L} \times \mathcal{B}^m$ measurable function. □

The $\mathcal{L} \times \mathcal{B}^m$ measurability hypothesis of Proposition 2.3.4 is unrestrictive. It is satisfied, for example, by the Carathéodory functions:

Definition 2.3.5 A function $g : I \times \mathbb{R}^m \to \mathbb{R}^k$ is said to be a Carathéodory function if

(a) $g(\cdot, u)$ is Lebesgue measurable for each $u \in \mathbb{R}^m$,
(b) $g(t, \cdot)$ is continuous for each $t \in I$.

Proposition 2.3.6 *Consider a function $g : I \times \mathbb{R}^m \to \mathbb{R}^k$. Assume that g is a Carathéodory function. Then g is $\mathcal{L} \times \mathcal{B}^m$ measurable.*

Proof Let $\{r_1, r_2...\}$ be an ordering of the set of m-vectors with rational components. For every positive integer k, $t \in I$ and $u \in \mathbb{R}^m$, define

$$g_k(t, u) := g(t, r_j),$$

in which the integer j is uniquely defined by the relations:

$$|r_j - u| \leq 1/k \text{ and } |r_i - u| > 1/k \text{ for } i = 1, ..., j-1.$$

Since $g(t, \cdot)$ is assumed to be continuous, we have

$$g_k(t, u) \to g(t, u)$$

as $k \to \infty$, for each fixed $(t, u) \in I \times \mathbb{R}^m$. It suffices then to show that g_k is $\mathcal{L} \times \mathcal{B}^m$ measurable for each k. However this follows from the fact that, for any open set $V \subset \mathbb{R}^k$, we have

$$g_k^{-1}(V) = \{(t,u) \in I \times \mathbb{R}^m : g_k(t,u) \in V\}$$
$$= \cup_{j=1}^\infty \{(t,u) \in I \times \mathbb{R}^m : g(t,r_j) \in V, |u - r_j| \le 1/k$$
$$\text{and } |u - r_i| > 1/k, \ i = 1, .., j-1\}$$
$$= \cup_{j=1}^\infty (\{t \in I : g(t,r_j) \in V\} \times \{u \in \mathbb{R}^m : |u - r_j| \le 1/k,$$
$$\text{and } |u - r_i| > 1/k \text{ for } i = 1,2,..,j-1\})$$

and this last set is $\mathcal{L} \times \mathcal{B}^m$ measurable, since it is expressible as a countable union of sets of the form $A \times B$ with $A \in \mathcal{L}$ and $B \in \mathcal{B}^m$. □

The preceding propositions combine incidentally to provide an answer to the following question concerning an appropriate framework for the formulation of variational problems: under what hypotheses on the function $L : I \times \mathbb{R}^n \times \mathbb{R}^n \to \mathbb{R}$ is the integrand of the Lagrange functional

$$\int L(t, x(t), \dot{x}(t)) dt$$

Lebesgue measurable for an arbitrary absolutely continuous arc $x \in W^{1,1}$? The two preceding propositions guarantee that the integrand is Lebesgue measurable if

(i) $L(\cdot, x, \cdot)$ is $\mathcal{L} \times \mathcal{B}^n$ measurable for each $x \in \mathbb{R}^n$

and

(ii) $L(t, \cdot, u)$ is continuous for each $(t, u) \in I \times \mathbb{R}^n$.

This is because Proposition 2.3.4 tells us that $(t, u) \to L(t, x(t), u)$ is $\mathcal{L} \times \mathcal{B}^n$ measurable, in view of assumptions (i) and (ii), and since $x : I \to \mathbb{R}^n$ is Lebesgue measurable. Corollary 2.3.3 permits to conclude, since $t \to \dot{x}(t)$ is Lebesgue measurable, that $t \to L(t, x(t), \dot{x}(t))$ is indeed Lebesgue measurable.

The following theorem, a proof of which is to be found in [53], lists important characterizations of closed multifunctions which are measurable. (Throughout, I is a Lebesgue subset of $\mathbb{R}$.)

Theorem 2.3.7 *Take a multifunction $\Gamma : I \rightsquigarrow \mathbb{R}^n$ and define $D := \{t \in I : \Gamma(t) \ne \emptyset\}$. Assume that Γ is closed. Then the following statements are equivalent:*

(a) *Γ is an $\mathcal{L}$ measurable multifunction,*
(b) *Gr Γ is an $\mathcal{L} \times \mathcal{B}^n$ measurable set,*
(c) *D is a Lebesgue subset of I and there exists a sequence $\{\gamma_k : D \to \mathbb{R}^n\}$ of Lebesgue measurable functions such that*

$$\Gamma(t) = \overline{\cup_{k=1}^\infty \{\gamma_k(t)\}} \quad \text{for all } t \in D. \tag{2.3.1}$$

2.3 Measurable Multifunctions

The representation of a multifunction in terms of a countable family of Lebesgue measurable functions according to (2.3.1) is called the *Castaing Representation* of Γ.

Our aim now is to establish the measurability of a number of frequently encountered multifunctions derived from other multifunctions.

Proposition 2.3.8 *Take a measurable space $(\Omega, \mathcal{F})$ and a measurable multifunction $\Gamma : \Omega \rightsquigarrow \mathbb{R}^n$. Then the multifunction $\tilde{\Gamma} : \Omega \rightsquigarrow \mathbb{R}^n$ is also measurable in each of the following cases:*

(i) $\tilde{\Gamma}(y) := \overline{\Gamma(y)}$ for all $y \in \Omega$,
(ii) $\tilde{\Gamma}(y) := \operatorname{co} \Gamma(y)$ for all $y \in \Omega$.

Proof

(i): $\tilde{\Gamma}$ is measurable in this case since, for any open set $W \in \mathbb{R}^n$,

$$\{y \in \Omega : \overline{\Gamma(y)} \cap W \neq \emptyset\} = \{y \in \Omega : \Gamma(y) \cap W \neq \emptyset\}.$$

(ii): Define the multifunction $\Gamma^{(n+1)} : \Omega \rightsquigarrow \mathbb{R}^{n \times (n+1)}$ to be

$$\Gamma^{(n+1)}(y) := \Gamma(y) \times \cdots \times \Gamma(y) \quad \text{for all } y \in \Omega.$$

Then $\Gamma^{(n+1)}$ is measurable, since $\{y \in \Omega : \Gamma^{(n+1)}(y) \cap W \neq \emptyset\}$ is obviously measurable for any set W in $\mathbb{R}^{n \times (n+1)}$ which is a product of open sets of $\mathbb{R}^n$, and therefore for any open set W, since an arbitrary open set in the product space can be expressed as a countable union of such sets.

Define also

$$\Lambda := \left\{ (\lambda_0, .., \lambda_n) : \lambda_i \geq 0 \text{ for all } i, \sum_{i=0}^{n} \lambda_i = 1 \right\}.$$

Take any open set W in $\mathbb{R}^n$ and define

$$W^{(n+1)} := \left\{ (w_0, \ldots w_n) : \sum_i \lambda_i w_i \in W, (\lambda_0, \ldots, \lambda_n) \in \Lambda \right\}.$$

Obviously, $W^{(n+1)}$ is an open set.
We must show that the set

$$\{y \in \Omega : \operatorname{co} \Gamma(y) \cap W \neq \emptyset\}$$

is measurable. But this follows immediately from the facts that $\tilde{\Gamma}^{(n+1)}$ is measurable and $W^{(n+1)}$ is open, since

$$\{y : \operatorname{co} \Gamma(y) \cap W \neq \emptyset\} = \{y : \Gamma^{(n+1)}(y) \cap W^{(n+1)} \neq \emptyset\}.$$

□

The next proposition concerns the measurability properties of limits of sequences of multifunctions. We make reference to Kuratowski sense limit operations, defined in Sect. 2.2.

Theorem 2.3.9 *Consider closed multifunctions* $\Gamma_j : I \rightsquigarrow \mathbb{R}^n$, $j = 1, 2, \ldots$ *Assume that Γ_j is $\mathcal{L}$ measurable for each j. Then the closed multifunction $\Gamma : I \rightsquigarrow \mathbb{R}^n$ is also $\mathcal{L}$ measurable when Γ is defined in each of the following ways*

(a) $\Gamma(t) := \overline{\cup_{j \geq 1} \Gamma_j(t)}$,
(b) $\Gamma(t) := \cap_{j \geq 1} \Gamma_j(t)$,
(c) $\Gamma(t) := \limsup_{j \to \infty} \Gamma_j(t)$,
(d) $\Gamma(t) := \liminf_{j \to \infty} \Gamma_j(t)$.

(c) and (d) imply in particular that if $t \to \{\Gamma_j(t)\}$ has a limit as $j \to \infty$ for almost every t, then $t \to \lim_{j \to \infty} \Gamma_j(t)$ is measurable.

Proof
(a) ($\Gamma(t) = \overline{\cup_{j \geq 1} \Gamma_j(t)}$)

Take any open set $W \subset \mathbb{R}^n$. Then, since W is an open set, we have

$$\begin{aligned}\{t \in I : \Gamma(t) \cap W \neq \emptyset\} &= \{t \in I : \overline{\cup_{j \geq 1} \Gamma_j(t)} \cap W \neq \emptyset\} \\ &= \{t \in I : \cup_{j \geq 1} \Gamma_j(t) \cap W \neq \emptyset\} \\ &= \cup_{j \geq 1} \{t \in I : \Gamma_j(t) \cap W \neq \emptyset\}.\end{aligned}$$

This establishes the measurability of $t \rightsquigarrow \Gamma(t)$ since the set on the right side, a countable union of Lebesgue measurable sets, is Lebesgue measurable.

(b) ($\Gamma(t) = \cap_{j \geq 1} \Gamma_j(t)$)

In this case,

$$\operatorname{Gr} \Gamma = \operatorname{Gr} \{t \rightsquigarrow \cap_{j \geq 1} \Gamma_j(t)\} = \cap_{j \geq 1} \operatorname{Gr} \Gamma_j.$$

Gr Γ then is $\mathcal{L} \times \mathcal{B}^n$ measurable since each Gr Γ_j is $\mathcal{L} \times \mathcal{B}^n$ measurable by Theorem 2.3.7. Now apply again Theorem 2.3.7.

(c) ($\Gamma(t) = \limsup_{j \to \infty} \Gamma_j(t)$.)

The measurability of $t \rightsquigarrow \Gamma(t)$ in this case follows from (a) and (b) and the following characterization of $\limsup_{j \to \infty} \Gamma_j(t)$:

$$\limsup_{j \to \infty} \Gamma_j(t) = \cap_{J \geq 1} \overline{\cup_{j \geq J} \Gamma_j(t)}.$$

(d) ($\Gamma(t) = \liminf_{j \to \infty} \Gamma_j(t)$)

2.3 Measurable Multifunctions

Define
$$\Gamma_j^k(t) := \Gamma_j(t) + (1/k)\mathbb{B}.$$

Notice that Γ_j^k is measurable since, for any closed set $W \subset \mathbb{R}^n$,
$$\{t : \Gamma_j^k(t) \cap W \neq \emptyset\} = \{t : \Gamma_j(t) \cap (W + (1/k)\mathbb{B}) \neq \emptyset\}$$
and the set $W + (1/k)\mathbb{B}$ is closed.

The measurability of Γ in this case too follows from (a) and (b) in view of the identity
$$\liminf_{j\to\infty} \Gamma_j(t) = \cap_{k\geq 1}\overline{\cup_{J\geq 1}\cap_{j\geq J} \Gamma_j^k(t)}.$$

$\square$

Proposition 2.3.10 *Take a multifunction* $F : I \times \mathbb{R}^n \rightsquigarrow \mathbb{R}^k$ *and a Lebesgue measurable function* $\bar{x} : I \to \mathbb{R}^n$. *Assume that*

(a) *for each* x, $F(\cdot, x) : I \rightsquigarrow \mathbb{R}^k$ *is a $\mathcal{L}$ measurable, non-empty, closed multifunction,*
(b) *for each t, $F(t, \cdot)$ is continuous at $x = \bar{x}(t)$, in the sense that*
$$y_i \to \bar{x}(t) \text{ implies } F(t, \bar{x}(t)) = \lim_{i\to\infty} F(t, y_i).$$

Then $G : I \rightsquigarrow \mathbb{R}^k$ defined by
$$G(t) := F(t, \bar{x}(t))$$

is a closed $\mathcal{L}$ measurable multifunction.

Proof Let $\{r_i\}$ be an ordering of n-vectors with rational entries. For each integer l and for each $t \in I$, define
$$F_l(t) := F(t, r_j),$$
in which j is chosen according to the rule
$$|\bar{x}(t) - r_j| \leq 1/l \text{ and } |\bar{x}(t) - r_i| > 1/l \text{ for } i = 1, ..., j-1.$$

In view of the continuity properties of $F(t, \cdot)$,
$$F(t, \bar{x}(t)) = \lim_{l\to\infty} F_l(t).$$

By Theorem 2.3.9 then it suffices to show that F_l is measurable for arbitrary l.

We observe however that

$$\text{Gr } F_l = \cup_j (\text{Gr } F(\cdot, r_j) \cap (A_j \times \mathbb{R}^k))$$

where

$$A_j := \{t : |\bar{x}(t) - r_j| \le 1/l, |\bar{x}(t) - r_i| > 1/l \text{ for } i = 1, .., j-1\}.$$

Since $F(\cdot, r_j)$ has $\mathcal{L} \times \mathcal{B}^k$ measurable graph (see Theorem 2.3.7) and $\bar{x}$ is Lebesgue measurable, we see that Gr F_l is $\mathcal{L} \times \mathcal{B}^k$ measurable. Applying Theorem 2.3.7 again, we see that the closed multifunction F_l is measurable. □

The derivation of necessary conditions of optimality for dynamic optimization problems with non-smooth data will require us to establish measurability properties of multifunctions involving the limiting subdifferential $\partial g(\bar{x})$ of a lower semi-continuous function $g : \mathbb{R}^k \to \mathbb{R} \cup \{+\infty\}$, at a point $\bar{x} \in \text{dom } g$. The limiting subdifferential and its properties feature prominently in Chaps. 4 and 5. Here, we merely anticipate its definition. The definition is in two stages: first we define the proximal subdifferential $\partial_g^P(x)$ at a point $x \in \text{dom } g$:

$$\partial^P g(x) := \{\xi : \text{there exists } M \text{ such that}$$
$$\xi \cdot (x' - x) \le g(x') - g(x) + M|x' - x|^2 \text{ for all } x' \in \mathbb{R}^k\}.$$

The limiting subdifferential is then obtained by 'closing the graph' of the multifunction $x \to \partial^P g(x)$, thus

$$\partial g(\bar{x}) := \limsup_{\epsilon \downarrow 0} \{\xi : \xi = \partial^P g(x) \text{ for some } x \in \mathbb{R}^k \text{ such that}$$
$$|(x, g(x)) - (\bar{x}, g(\bar{x}))| \le \epsilon\}.$$

The following proposition concerns the measurability properties of the multifunction $t \to \partial_x f(t, \bar{x}(t))$, for given functions $(t, x) \to f(t, x)$ and $t \to \bar{x}(t)$, where ∂_x denotes the (partial) limiting subdifferential w.r.t. the x variable.

Proposition 2.3.11 *Take a function* $f : I \times \mathbb{R}^n \to \mathbb{R}$ *such that*

(a) $f(., x) : I \to \mathbb{R}$ *is a $\mathcal{L}$ measurable for each $x \in \mathbb{R}^n$,*
(b) $f(t, .) : \mathbb{R}^n \to \mathbb{R}$ *is locally Lipschitz continuous for each $t \in I$.*

Take a Lebesgue measurable function $\bar{x} : I \to \mathbb{R}^n$. *Consider the multifunction* $G : I \rightsquigarrow \mathbb{R}^n$ *defined by*

$$G(t) := \partial_x f(t, \bar{x}(t)).$$

Then G and co G are closed $\mathcal{L}$ measurable multifunctions.

2.3 Measurable Multifunctions

Proof Since the limiting subdifferential is closed it immediately follows that G and co G take closed values. We also observe that, owing to Proposition 2.3.8, it suffices to prove that G is a closed $\mathcal{L}$ measurable multifunction. For each integer $i \geq 1$ we consider the function $\phi_i : I \times \mathbb{R}^n \times \mathbb{R}^n \to \mathbb{R}$ defined by

$$\phi_i(t, x, \xi) := \min_{y \in i^{-1}\mathbb{B}} f(t, x+y) - f(t, x) - \xi \cdot y + i|y|^2.$$

Observe that, since $\phi_i(t, x, \xi) = \inf_{y \in (i^{-1}\mathbb{B}) \cap \mathbb{Q}^n} \{f(t, x+y) - f(t, x) - \xi \cdot y + i|y|^2\}$, it is easy to see that, for each $(x, \xi) \in \mathbb{R}^n \times \mathbb{R}^n$, $\phi_i(., x, \xi)$ is Lebesgue measurable. Moreover, since $f(t, .)$ is locally Lipschitz continuous for each $t \in I$, we can easily deduce that $\phi_i(t, ., .)$ is locally Lipschitz. We claim that $\xi \in \partial_{x,P} f(t, x)$ if and only if $\phi_i(t, x, \xi) = 0$. Indeed, assume that $\xi \in \partial_{x,P} f(t, x)$, then, by the definition of (partial) proximal subdifferential we can find $M > 0$ and $\epsilon > 0$ such that

$$\xi \cdot (x' - x) \leq f(t, x') - f(t, x) + M|x' - x|^2, \quad \text{for all } x' \in x + \epsilon \mathbb{B}. \tag{2.3.2}$$

Then taking an integer i_0 such that $i_0 \geq \max\{1/\epsilon, M\}$ and setting $x' = y - x$ we deduce that

$$0 \leq f(t, x+y) - f(t, x) - \xi \cdot y + i_0|y|^2, \quad \text{for all } y \in i_0^{-1}\mathbb{B}, \tag{2.3.3}$$

and so $\phi_i(t, x, \xi) = 0$. Suppose now that $\phi_{i_0}(t, x, \xi) = 0$ for some $i_0 \geq 1$, which means that (2.3.3) is satisfied. Then, from the characterization of proximal subgradients, taking $\epsilon = i_0^{-1}$ and $M = i_0$, we obtain that $\xi \in \partial_{x,P} f(t, x)$.

Take a closed set $W \subset \mathbb{R}^n$. The proof will be complete if we show that the following set E is measurable:

$$E := \{t \in I \mid \partial_x f(t, \bar{x}(t)) \cap W \neq \emptyset\}.$$

We observe that

$$E = \bigcap_{j \geq 1} \bigcup_{i \geq 1} E_{ij}$$

where, for each $i \geq 1$ and $j \geq 1$, $E_{ij} := \{t \in I \mid F_{ij}(t) \neq \emptyset\}$, i.e. the (effective) domain of the multivalued function $F_{ij} : I \rightsquigarrow \mathbb{R}^n \times \mathbb{R}^n$ defined as

$$F_{ij}(t) := \{(x, \xi) \in (\bar{x}(t) + j^{-1}\mathbb{B}) \times (W + j^{-1}\mathbb{B}) \mid \phi_i(t, x, \xi) = 0\}.$$

Invoking Theorem 2.3.7 we can easily deduce that the multifunctions F_{ij}'s are measurable and therefore E is measurable. □

Take a multifunction $\Gamma : I \rightsquigarrow \mathbb{R}^k$. We say that a function $x : I \to \mathbb{R}^k$ is a *measurable selection* for Γ if

(i) x is Lebesgue measurable,

and

(ii) $x(t) \in \Gamma(t)$ a.e..

We obtain directly from Theorem 2.3.7 the following conditions for Γ to have a measurable selection:

Theorem 2.3.12 *Let $\Gamma : I \rightsquigarrow \mathbb{R}^k$ be a non-empty multifunction. Assume that Γ is closed and measurable. Then Γ has a measurable selection.*

(In fact Theorem 2.3.7 tells us rather more than this: not only does there exist a measurable selection under the stated hypotheses, but the measurable selections are sufficiently numerous to 'fill out' the values of the multifunction.)

The above measurable selection theorem is inadequate for certain applications in which the multifunction is not closed. An important extension (see [203]) is

Theorem 2.3.13 (Aumann's Measurable Selection Theorem) *Let $\Gamma : I \rightsquigarrow \mathbb{R}^k$ be a non-empty multifunction. Assume that*

$$\text{Gr } \Gamma \text{ is } \mathcal{L} \times \mathcal{B}^k \text{ measurable.}$$

Then Γ has a measurable selection.

This can be regarded as a generalization of Theorem 2.3.12 since if Γ is closed and measurable then, by Theorem 2.3.7, Gr Γ is automatically $\mathcal{L} \times \mathcal{B}^k$ measurable.

Of particularly significance in applications to dynamic optimization is the following measurable selection theorem involving the composition of a function and a multifunction.

Theorem 2.3.14 (The Generalized Filippov Selection Theorem) *Consider a non-empty multifunction $U : I \rightsquigarrow \mathbb{R}^m$ and a function $g : I \times \mathbb{R}^m \to \mathbb{R}^n$ satisfying*

(a) The set Gr U is $\mathcal{L} \times \mathcal{B}^m$ measurable,
(b) The function g is $\mathcal{L} \times \mathcal{B}^m$ measurable.

Then for any measurable function $v : I \to \mathbb{R}^n$, the multifunction $U' : I \rightsquigarrow \mathbb{R}^m$ defined by

$$U'(t) := \{u \in U(t) : g(t, u) = v(t)\}$$

has $\mathcal{L} \times \mathcal{B}^m$ measurable graph. Furthermore, if

$$v(t) \in \{g(t, u) : u \in U(t)\} \qquad a.e. \qquad (2.3.4)$$

then there exists a measurable function $u : I \to \mathbb{R}^m$ satisfying

$$u(t) \in U(t) \qquad a.e., \qquad (2.3.5)$$

$$g(t, u(t)) = v(t) \qquad a.e.. \qquad (2.3.6)$$

2.3 Measurable Multifunctions

Notice that condition (2.3.4) is just a rephrasing of the hypothesis

'$U'(t)$ is non-empty for a.e. $t \in I$'

and the final assertion can be expressed in measurable selection terms as

'The multifunction U' has a measurable selection',

a fact which follows directly from Aumann's selection theorem.

We mention that the name Filippov's selection theorem usually attaches to the final assertion of the theorem concerning existence of a measurable function $u : I \to \mathbb{R}^m$ satisfying (2.3.5) and (2.3.6) under (2.3.4) and strengthened forms of hypotheses (a) and (b), namely

(a)' U is a measurable, closed multifunction,
(b)' g is a Carathéodory function.

Proof By redefining $U'(t)$ on a null-set if required, we can arrange that $U'(t)$ is non-empty for all $t \in I$. In view of the preceding discussion it is required to show merely that Gr U' is $\mathcal{L} \times \mathcal{B}^m$ measurable. But this follows directly from the relation

$$\text{Gr } U' = \phi^{-1}(\{0\}) \cap \text{Gr } U$$

in which $\phi(t, u) := g(t, u) - v(t)$, since under the hypotheses both $\phi^{-1}(\{0\})$ and Gr U are $\mathcal{L} \times \mathcal{B}^m$ measurable sets. $\square$

The relevance of Filippov's theorem in a control systems context is illustrated by the following application. Take a function $f : [S, T] \times \mathbb{R}^n \times \mathbb{R}^m \to \mathbb{R}^n$ and a multifunction $U : [S, T] \rightsquigarrow \mathbb{R}^m$. The class of state trajectories for the control system

$$\dot{x}(t) = f(t, x(t), u(t)) \quad \text{a.e. } t \in [S, T]$$
$$u(t) \in U(t) \quad \text{a.e. } t \in [S, T]$$

comprises absolutely continuous functions $x : [S, T] \to \mathbb{R}^n$ such that the above relations are satisfied for some measurable $u : [S, T] \to \mathbb{R}^m$. We often need to view them as solutions of the differential inclusion $\dot{x}(t) \in F(t, x(t))$ with

$$F(t, x) := \{f(t, x, u) : u \in U(t)\}.$$

The question then arises whether the state trajectories for the control system are *precisely* the absolutely continuous functions x satisfying

$$\dot{x}(t) \in F(t, x(t)) \quad \text{a.e..} \qquad (2.3.7)$$

Clearly a necessary condition for an absolutely continuous function to be a state trajectory for the control system is that (2.3.7) is satisfied. Filippov's theorem tells us that (2.3.7) is also a sufficient condition (i.e. the differential inclusion provides an equivalent description of state trajectories) under the hypotheses:

(i) $f(\cdot, x, \cdot)$ is $\mathcal{L} \times \mathcal{B}^m$ measurable and $f(t, \cdot, u)$ is continuous
(ii) Gr U is $\mathcal{L} \times \mathcal{B}^m$ measurable.

To see this, apply the generalized Filippov selection theorem with $g(t, u) = f(t, x(t), u)$ and $v = \dot{x}$. (The relevant hypotheses, (b) and (2.3.4), are satisfied in view of Proposition 2.3.4 and since $\dot{x}(t) \in F(t, x(t))$ a.e..) This yields a measurable function $u : [S, T] \to \mathbb{R}^m$ satisfying $\dot{x}(t) = f(t, x(t), u(t))$ and $u(t) \in U(t)$ a.e. thereby confirming that x is a state trajectory for the control system.

Having once again an eye for future dynamic optimization applications, we now establish measurability of various derived functions and the existence of measurable selections for related multifunctions. The source of a number of useful results is the following theorem concerning the measurability of a 'marginal' function.

Theorem 2.3.15 *Consider a function $g : I \times \mathbb{R}^k \to \mathbb{R}$ and a closed, non-empty multifunction $\Gamma : I \rightsquigarrow \mathbb{R}^k$. Assume that*

(a) g is a Carathéodory function,
(b) Γ is a measurable multifunction.

Define the extended valued function $\eta : I \to \mathbb{R} \cup \{-\infty\}$

$$\eta(t) = \inf_{\gamma \in \Gamma(t)} g(t, \gamma) \qquad \text{for } t \in I.$$

Then η is a Lebesgue measurable function. Furthermore, if we define

$$I' := \{t \in I : \inf_{\gamma' \in \Gamma(t)} g(t, \gamma') = g(t, \gamma) \text{ for some } \gamma \in \Gamma(t)\}$$

(i.e. I' is the set of points t for which the infimum of $g(t, \cdot)$ is achieved over $\Gamma(t)$) then I' is a Lebesgue measurable set and there exists a measurable function $\gamma : I' \to \mathbb{R}^k$ such that

$$\eta(t) = g(t, \gamma(t)) \qquad a.e\ t \in I'. \tag{2.3.8}$$

Proof Since Γ is closed, non-empty and measurable, it has a Castaing representation in terms of some countable family of measurable functions $\{\gamma_i : I \to \mathbb{R}^k\}$. Since $g(t, \cdot)$ is continuous and $\{\gamma_i(t)\}$ is dense in $\Gamma(t)$

$$\eta(t) = \inf\{g(t, \gamma_i(t)) : i \text{ an integer}\}$$

for all $t \in I$.

Now according to Proposition 2.3.6 and Corollary 2.3.3, $t \to g(t, \gamma_i(t))$ is a measurable function. It follows from a well-known property of measurable functions that η, which we have expressed as the pointwise infimum of a countable family of measurable functions, is Lebesgue measurable and $\text{dom}\{\eta\} := \{t \in I : \eta(t) > -\infty\}$ is a Lebesgue measurable set.

Now apply the generalized Filippov selection theorem (identifying η with v and Γ with U, and replacing I by $\text{dom}\{\eta\}$). If $I' = \emptyset$, there is nothing to prove. Otherwise, since

$$I' = \{t \in \text{dom}\{\eta\} : \{\gamma \in \Gamma(t) : g(t, \gamma) = \eta(t)\} \neq \emptyset\},$$

I' is a nonempty, Lebesgue measurable set and there exists a measurable function $\gamma : I' \to \mathbb{R}^k$ such that

$$\eta(t) = g(t, \gamma(t)) \quad \text{a.e. } t \in I'.$$

□

2.4 The Generalized Bolza Problem

In Sect. 6.4 we shall give conditions for the existence of minimizers in the context of minimization problems over some class of arcs satisfying a given differential inclusion $\dot{x}(t) \in F(t, x(t))$. These conditions restricted attention to problems for which the velocity sets $F(t, x)$ are bounded. For traditional variational problems and also for many dynamic optimization problems of interest there are no constraints on permitted velocities. To deal in a unified manner with problems with bounded and unbounded velocity sets, it is convenient to adopt a new framework for the optimization problems involved, namely to regard them as special cases of the generalized Bolza problem:

$$(GBP) \begin{cases} \text{Minimize } \Lambda(x) := l(x(S), x(T)) + \int_S^T L(t, x(t), \dot{x}(t)) dt \\ \text{over arcs } x \in W^{1,1}([S, T]; \mathbb{R}^n), \end{cases}$$

in which $[S, T]$ is a given interval, and $l : \mathbb{R}^n \times \mathbb{R}^n \to \mathbb{R} \cup \{+\infty\}$ and $L : [S, T] \times \mathbb{R}^n \times \mathbb{R}^n \to \mathbb{R} \cup \{+\infty\}$ are given extended valued functions. Provided we arrange that $t \to L(t, x(t), \dot{x}(t))$ is measurable and minorized by an integrable function for every $x \in W^{1,1}$ (our hypotheses take care of this), Λ will be a well-defined $\mathbb{R} \cup \{+\infty\}$ valued functional on $W^{1,1}$.

Notice that the functions l and L are permitted to takes values $+\infty$. So they can be used implicitly to take account of constraints. For example the 'differential inclusion' problem

$$\begin{cases} \text{Minimize } g(x(S), x(T)) \\ \text{over arcs } x \in W^{1,1}([S, T]; \mathbb{R}^n) \text{ satisfying} \\ \dot{x}(t) \in F(t, x(t)), \\ (x(S), x(T)) \in C. \end{cases}$$

is a special case of the generalized Bolza problem in which

$$L(t, x, v) = \begin{cases} 0 & \text{if } v \in F(t, x) \\ +\infty & \text{otherwise} \end{cases}$$

and

$$l(x_0, x_1) = \begin{cases} g(x_0, x_1) & \text{if } (x_0, x_1) \in C \\ +\infty & \text{otherwise.} \end{cases}$$

In existence theorems covering problems with unbounded velocity sets, superlinear growth hypotheses on the cost integrand are typically invoked to compensate for unbounded velocity sets. The key advantage of the generalized Bolza problem as a vehicle for such theorems is that unrestrictive hypotheses ensuring existence of minimizers, which require coercivity of the cost integrand in precisely those 'directions' in which velocities are unconstrained, can be economically expressed as conditions on the extended valued function L.

Theorem 2.4.1 (*Generalized Bolza Problem: Existence of Minimizers*) *Assume that the data for (GBP) satisfy the following hypotheses:*

(H1): l is lower semi-continuous and there exist a lower semi-continuous function $l^0 : \mathbb{R}_+ \to \mathbb{R}$ satisfying

$$\lim_{r \uparrow +\infty} l^0(r) = +\infty$$

and either

$$l^0(|x_0|) \leq l(x_0, x_1) \quad \text{for all } x_0, x_1$$

or

$$l^0(|x_1|) \leq l(x_0, x_1) \quad \text{for all } x_0, x_1,$$

(H2): L is $\mathcal{L} \times \mathcal{B}^{n \times n}$ measurable,
(H3): $L(t, ., .)$ is lower semi-continuous for each $t \in [S, T]$,
(H4): For each $(t, x) \in [S, T] \times \mathbb{R}^n$, $L(t, x, .)$ is convex and dom $L(t, x, .) \neq \emptyset$,
(H5): For all $t \in [S, T]$, $x \in \mathbb{R}^n$ and $v \in \mathbb{R}^n$,

$$L(t, x, v) \geq \theta(|v|) - \alpha |x|,$$

2.4 The Generalized Bolza Problem

for some $\alpha \geq 0$ and and some lower semi-continuous, convex function θ : $\mathbb{R}_+ \to \mathbb{R}_+$ satisfying

$$\lim_{r \uparrow +\infty} \theta(r)/r = +\infty.$$

Then (GBP) has a minimizer. (We allow the possibility that $\Lambda(x) = +\infty$ for all $x \in W^{1,1}$. In this case all arcs x are regarded as minimizers.)

Remark
The proof of this theorem, which follows shortly, exploits properties of the Hamiltonian

$$H(t, x, p) := \sup_{v \in \mathbb{R}^n} \{p \cdot v - L(t, x, v)\}.$$

To a large extent then, the role of the growth condition (H5) is to ensure that the Hamiltonian has the required properties to furnish existence theorems. Loosely speaking, growth conditions on the Lagrangian translate into (one-sided) boundedness conditions on the Hamiltonian. One direction for generalizing this theorem is to replace (H5) by less restrictive conditions imposed directly on the Hamiltonian. Rockafellar has shown that the conclusions of Theorem 2.4.1 remain valid when (H5) is replaced by:

(H5)′ For all $t \in [S, T]$ and $x, p \in \mathbb{R}^n$

$$H(t, x, p) \leq \mu(t, p) + |x|(\sigma(t) + \rho(t)|p|),$$

for some integrable functions σ, ρ and some function μ such that $\mu(., p)$ is integrable for each $p \in \mathbb{R}^n$.

The ensuing analysis calls upon some properties of convex functions, which it is now convenient to summarize. Take any function $f : \mathbb{R}^n \to \mathbb{R} \cup \{+\infty\}$; the *conjugate* of f is the function $f^* : \mathbb{R}^n \to \mathbb{R} \cup \{+\infty\}$, defined by the *Legendre-Fenchel transformation*:

$$f^*(y) := \sup_{x \in \mathbb{R}^n} \{y \cdot x - f(x)\}.$$

We say that a function $f : \mathbb{R}^n \to \mathbb{R} \cup \{+\infty\}$ is *proper* if dom $f \neq \emptyset$. Important facts are that, for any proper lower semi-continuous, convex function $f : \mathbb{R}^n \to \mathbb{R} \cup \{+\infty\}$, its conjugate f^*, too, is proper lower semi-continuous, convex. Furthermore, f can be recovered from f^* by means of a second application of the Legendre-Fenchel transformation:

$$f(x) = \sup_{y \in \mathbb{R}^n} \left(y \cdot x - f^*(y) \right).$$

One consequence of these relations is

Proposition 2.4.2 (Jensen's Inequality) *Take any proper lower semi-continuous, convex function $f : \mathbb{R}^n \to \mathbb{R} \cup \{+\infty\}$. Then, for any set $I \subset \mathbb{R}$ of positive measure and any $v \in L^1(I; \mathbb{R}^n)$, $t \to f(v(t))$ is a measurable function, minorized by an integrable function, and*

$$\int_I f(v(t))dt \geq |I| f\left(|I|^{-1} \int_I v(t)dt\right),$$

where $|I|$ denotes the Lebesgue measure of I.

Proof Take any $y \in \text{dom } f^*$. Then, for each $t \in I$,

$$f(v(t)) \geq v(t) \cdot y - f^*(y).$$

It follows that the (measurable) function $t \to f(v(t))$ is minorized by an integrable function. Also,

$$\int_I f(v(t))dt \geq \left(\int_I v(t)dt\right) \cdot y - f^*(y)|I|$$

$$= |I|\left[\left(|I|^{-1}\int_I v(t)dt\right) \cdot y - f^*(y)\right].$$

This inequality is valid for all $y \in \text{dom } f^*$. Maximizing over $\text{dom } f^*$ and noting that f is obtained from f^* by applying the Legendre-Fenchel transformation, we obtain

$$\int_I f(v(t))dt \geq |I| f\left(|I|^{-1} \int_I v(t)dt\right),$$

as claimed. □

Proof of Theorem 2.4.1 We assume that $l^0(|x_0|) \leq l(x_0, x_1)$ for all (x_0, x_1). (The case $l^0(|x_1|) \leq l(x_0, x_1)$ for all (x_0, x_1) is treated similarly.) Under the hypotheses, l^0 and θ are bounded below. By scaling and adding a constant to $\int L dt + l$ (this does not effect the minimizers) we can arrange that $l^0 \geq 0$ and $\theta \geq 0$. We can also arrange that the constant α is arbitrary small.

Choose α such that $e^{\alpha |T-S|}|T - S|\alpha < 1$.

Since θ has superlinear growth, we can define $k : \mathbb{R}_+ \to \mathbb{R}_+$:

$$k(\beta) := \sup\{r \geq 0 : r = 0 \text{ or } \theta(r) \leq \beta r\}.$$

Step 1: Fix $M \geq 0$. We show that the level set

$$\mathcal{S}_M := \{x \in W^{1,1} : \Lambda(x) \leq M\}$$

2.4 The Generalized Bolza Problem

is weakly sequentially pre-compact, i.e. any sequence $\{x_i\}$ in $\mathcal{S}_M$ has a subsequence which converges, with respect to the weak $W^{1,1}$ topology, to some point in $W^{1,1}$.

Take any $x \in \mathcal{S}_M$ and define the L^1 function

$$q(t) := L(t, x(t), \dot{x}(t)).$$

Then

$$|\dot{x}(t)| \leq k(1) + \theta(|\dot{x}(t)|) \leq k(1) + q(t) + \alpha|x(t)| \quad \text{a.e..} \tag{2.4.1}$$

But, for each $t \in [S, T]$,

$$\int_S^t q(s)ds = \Lambda(x) - \int_t^T q(s)ds \leq M - \int_t^T \theta(|\dot{x}(s)|)ds + \alpha \int_t^T |x(s)|ds$$

$$\leq M + \alpha|T - S||x(t)| + \int_t^T (\alpha|T - S||\dot{x}(s)| - \theta(|\dot{x}(s)|))\, ds$$

$$\leq M + \alpha|T - S||x(t)| + \alpha|T - S|^2 k(\alpha|T - S|).$$

It follows from (2.4.1) and Gronwall's lemma that

$$|x(t)| \leq e^{\alpha(t-S)} \left[|x(S)| + \int_S^t (k(1) + q(s))ds \right]$$

$$\leq e^{\alpha(T-S)} \Big[|x(S)| + k(1)|T - S| + M$$
$$+ \alpha|T - S|^2 k(\alpha|T - S|) + \alpha|T - S||x(t)| \Big].$$

Therefore

$$|x(t)| \leq A|x(S)| + B \tag{2.4.2}$$

where the constants A and B (they do not depend on x) are

$$A := \left(1 - \alpha|T - S|e^{\alpha|T-S|}\right)^{-1} e^{\alpha|T-S|}$$

and

$$B := \left(1 - \alpha|T - S|e^{\alpha|T-S|}\right)^{-1} \left(k(1)|T - S| + M + \alpha|T - S|^2 k(\alpha|T - S|)\right) e^{\alpha(T-S)}.$$

We deduce from (2.4.2) and the fact that $l(x_0, x_1) \geq l^0(|x_0|)$ that

$$|x(S)| \leq K, \tag{2.4.3}$$

where $K > 0$ is any constant (it can be chosen independent of x) such that

$$l^0(r) - \alpha(T - S)[Ar + B] > M \quad \text{for all } r \geq K.$$

Now, for any set $I \subset [S, T]$ of positive measure, Jensen's inequality yields

$$\theta(|I|^{-1} \int_I |\dot{x}(t)|dt) \leq |I|^{-1} \int_I \theta(|\dot{x}(t)|)dt$$

$$\leq |I|^{-1} \int_S^T \theta(|\dot{x}(t)|)dt$$

$$\leq |I|^{-1} \left(\int_S^T L(t, x(t), \dot{x}(t))dt + \alpha \int_S^T |x(t)|dt \right)$$

$$\leq |I|^{-1}(M + \alpha(T - S)(AK + \alpha B)).$$

We conclude that, if $\int_I |\dot{x}(t)|dt > 0$,

$$\frac{\theta(\int_I |\dot{x}(t)|dt/|I|)}{\int_I |\dot{x}(t)|dt/|I|} \leq \frac{(M + \alpha(T - S)(AK + B))}{\int_I |\dot{x}(t)|dt}.$$

Since θ has superlinear growth, it follows from this inequality that there exists a function $\omega : \mathbb{R}_+ \to \mathbb{R}_+$ (which does not depend on x) such that $\lim_{\sigma \downarrow 0} \omega(\sigma) = 0$ and

$$\int_I |\dot{x}(t)|dt \leq \omega(|I|) \quad \text{for all measurable } I \subset [S, T]. \tag{2.4.4}$$

Take any sequence $\{x_i\}$ in $\mathcal{S}_M$. Then, by (2.4.3), $\{x_i(S)\}$ is a bounded sequence. On the other hand, $\{\dot{x}_i\}$ is an equicontinuous sequence, by (2.4.4). Invoking the Dunford Pettis criterion for weak sequential compactness in L^1 (Theorem 6.3.1), we deduce that, along a subsequence,

$$x_i(S) \to x(S) \text{ and } \dot{x}_i \to \dot{x} \text{ weakly in } L^1.$$

Otherwise expressed,

$$x_i \to x \quad \text{weakly in } W^{1,1},$$

for some $x \in W^{1,1}$. This is what we set out to prove.

Step 2: Take an $\mathcal{L} \times \mathcal{B}^n$ function $\phi : [S, T] \times \mathbb{R}^n \to \mathbb{R} \cup \{+\infty\}$ which satisfies the conditions:

(a): For each $t \in [S, T]$, $\phi(t, .)$ is lower semi-continuous and finite at some point,

2.4 The Generalized Bolza Problem

(b): For some $\tilde{p} \in L^\infty$, the function $t \to \phi(t, \tilde{p}(t))$ is minorized by an integrable function.

We shall show that

$$\int_S^T \hat{\phi}(t)dt = \sup_{p \in L^\infty} \int_S^T \phi(t, p(t))dt, \qquad (2.4.5)$$

where

$$\hat{\phi}(t) := \sup_{p \in \mathbb{R}^n} \phi(t, p).$$

(Note that, under the hypotheses, $\hat{\phi}$ is measurable and minorized by an integrable function. So the left side of (2.4.5) is well-defined. The right side is interpreted as the supremum of the specified integral over p's such that the integrand is minorized by some integrable function.)

For any $p \in L^\infty$, $\hat{\phi}(t) \geq \phi(t, p(t))$ for all t. It immediately follows that (2.4.5) holds, when '$\geq$' replaces '='.

It suffices then to validate (2.4.5) when '$\leq$' replaces '='. To this end, choose any $r \in \mathbb{R}^n$ such that

$$\int_S^T \hat{\phi}(t)dt > r.$$

We can also choose $K > 0$ and $\epsilon > 0$ such that, writing

$$\hat{\hat{\phi}}(t) := \min\{\hat{\phi}(t), K\},$$

we have

$$\int_S^T \hat{\hat{\phi}}(t)dt > r$$

and

$$\int_S^T (\hat{\hat{\phi}}(t) - \epsilon)dt > r.$$

Define the multifunction

$$\Gamma(t) := \{p \in \mathbb{R}^n : \phi(t, p) > \hat{\hat{\phi}}(t) - \epsilon\}.$$

Under the hypotheses, Γ takes values non-empty (open) sets and $\text{Gr}\,\Gamma$ is $\mathcal{L} \times \mathcal{B}^n$ measurable. According to Aumann's measurable selection theorem then, Γ has a measurable selection, which we write $\bar{p}$.

However, since $t \to \phi(t, \bar{p}(t))$ and $t \to \phi(t, \tilde{p}(t))$ are minorized by integrable functions, we can find a measurable set E such that $\bar{p}$ restricted to $[S, T] \setminus E$ is essentially bounded and

$$\int_{[S,T] \setminus E} \phi(t, \bar{p}(t)) dt + \int_E \phi(t, \tilde{p}(t)) dt > r.$$

It follows that

$$\int_S^T \phi(t, p(t)) dt > r,$$

in which p is the essentially bounded function

$$p(t) := \begin{cases} \tilde{p}(t) & \text{if } t \in E, \\ \bar{p}(t) & \text{otherwise.} \end{cases}$$

Since r is an arbitrary strict lower bound on $\int_S^T \hat{\phi}(t) dt$, the desired inequality is confirmed.

Step 3: We show that Λ is weakly sequentially lower semi-continuous (w.r.t. the $W^{1,1}$ topology).

Since weak $W^{1,1}$ convergence implies uniform convergence, we deduce from the lower semicontinuity of l that $x \to l(x(S), x(T))$ is weakly sequentially lower semi-continuous. It remains therefore to show that

$$\tilde{\Lambda}(x) := \int_S^T L(t, x(t), \dot{x}(t)) dt$$

is also weakly sequentially lower semi-continuous. Take any $x \in W^{1,1}$. Then, since $L(t, x(t), \cdot)$ is a proper lower semi-continuous, convex function for each t,

$$\tilde{\Lambda}(x) = \int_S^T \sup_{p \in \mathbb{R}^n} [p \cdot \dot{x}(t) - H(t, x(t), p)] dt$$

$$= \sup_{p \in L^\infty} \int_S^T [p(t) \cdot \dot{x}(t) - H(t, x(t), p(t))] dt.$$

(We have used the results of Step 2 to justify the last equality. Note that the function $\phi(t, p) = p \cdot \dot{x}(t) - H(t, x(t), p)$ satisfies the relevant hypotheses. In particular, $\phi(t, \tilde{p}(t))$ is minorized by an integrable function for the choice $\tilde{p} \equiv 0$.)

For fixed $p \in L^\infty$, consider now the integral functional

$$\Lambda_p(x) := \int_S^T [p(t) \cdot \dot{x}(t) - H(t, x(t), p(t))] dt.$$

We claim that Λ_p is weakly sequentially lower semi-continuous.

To verify this assertion, take any weakly convergent sequence $y_i \to y$ in $W^{1,1}$. Then $\dot{y}_i \to \dot{y}$ weakly in L^1 and $y_i \to y$ uniformly. Since $H(t,\cdot,p(t))$ is upper semi-continuous for each t and the functions $t \to -H(t, y_i(t), p(t))$ are minorized by a common integrable function (this last property follows from hypothesis (H5)), we deduce from Fatou's lemma

$$\liminf_{i \to \infty} \Lambda_p(y_i) = \lim\inf_{i \to \infty} \int_S^T [p(t) \cdot \dot{y}_i(t) - H(t, y_i(t), p(t))]\, dt$$
$$\geq \int_S^T [p(t) \cdot \dot{y}(t) - H(t, y(t), p(t))]\, dt$$
$$= \Lambda_p(y).$$

Weak sequential lower semicontinuity of Λ_p is confirmed.

We have shown that, for each $x \in W^{1,1}$,

$$\tilde{\Lambda}(x) = \sup_{p \in L^\infty} \Lambda_p(x).$$

But the upper envelope of a family of weakly sequentially lower semi-continuous functionals on $W^{1,1}$ is also weakly sequentially lower semi-continuous. It follows that $\tilde{\Lambda}$ is weakly sequentially lower semi-continuous.

Conclusion We have shown in Steps 1 and 3 that Λ is sequentially lower semi-continuous and that the level sets of Λ are sequentially compact with respect to the weak $W^{1,1}$ topology. These properties guarantee existence of a minimizer. (We allow the possibility that $\Lambda(x) = +\infty$ for all x. In this case all x's are minimizers.) □

2.5 Exercises

2.1 Let $U : [S, T] \rightsquigarrow \mathbb{R}^m$ be a non-empty multifunction. Write $\mathcal{U} := \{u : [S, T] \to \mathbb{R}^m \ \mathcal{L}-\text{measurable} : u(t) \in U(t), \ \text{a.e.}\ t \in [S, T]\}$. Assume that $\mathcal{U}$ is nonempty. Consider the map $d_\mathcal{E} : \mathcal{U} \times \mathcal{U} \to \mathbb{R}_+$ defined by

$$d_\mathcal{E}(u', u) := \text{meas}\,\{t \in [S, T] : u'(t) \neq u(t)\}.$$

Show that $d_\mathcal{E}$ is a metric on $\mathcal{U}$ and $(\mathcal{U}, d_\mathcal{E})$ is a complete metric space.

2.2 Consider the control system

$$(CS) \begin{cases} \dot{x}(t) = f(t, x(t), u(t)) \ \text{a.e.}\ t \in [S, T], \\ u(t) \in U(t), \ \text{a.e.}\ t \in [S, T], \\ x(S) \in C. \end{cases}$$

Assume that

(a): $f(., x, .)$ is $\mathcal{L} \times \mathcal{B}^m$ measurable for each $x \in \mathbb{R}^n$, and $\mathrm{Gr}\, U = \{(t, u) \in [S, T] \times \mathbb{R}^m : u \in U(t)\}$ is a $\mathcal{L} \times \mathcal{B}^m$ measurable set,
(b): for some $k_f, c_f \in L^1(S, T)$
$|f(t, x, u) - f(t, x', u)| \le k_f(t)|x - x'|$ and $|f(t, x, u)| \le c_f(t)$, for all $x, x' \in \mathbb{R}^n$ and $u \in U(t)$, a.e. $t \in [S, T]$,
(c): $C \subset \mathbb{R}^n$ is a closed set.

Consider the (nonempty) set $\mathcal{X} := \{\text{admissible processes } (x, u) \text{ for } (CS)\}$, and the map $d_\mathcal{X} : \mathcal{X} \times \mathcal{X} \to \mathbb{R}_+$ defined by

$$d_\mathcal{X}((x', u'), (x, u)) := |x'(S) - x(S)| + \operatorname{meas} \{t \in [S, T] : u'(t) \ne u(t)\}.$$

Show that $d_\mathcal{X}$ defines a metric on $\mathcal{X}$ and $(\mathcal{X}, d_\mathcal{X})$ is a complete metric space.

2.3 Let $U : [S, T] \rightsquigarrow \mathbb{R}^m$ be a non-empty multifunction such that $\mathrm{Gr}\, U$ is an $\mathcal{L} \times \mathcal{B}^m$ measurable set. Take an absolutely continuous strictly increasing function $\psi : [S, T] \to [S, T]$. Show that the graph of multifunction $\widehat{U} : [S, T] \rightsquigarrow \mathbb{R}^m$, defined by $\widehat{U}(s) := U(\psi^{-1}(s))$, is an $\mathcal{L} \times \mathcal{B}^m$ measurable set.

Remark
The property expressed in this exercise has a role in some reduction techniques to simplify the analysis, employing transformations of the time variable (cf. the proof of Theorem 9.6.2).

Hint: Show that, if $\psi : [S, T] \to [S, T]$ *is a homeomorphism, then* $\psi(B) \in \mathcal{B}$ *for all* $B \in \mathcal{B}$ *($\mathcal{B}$ is the σ-algebra of Borel sets in $[S, T]$). And prove that, if* $\psi : [S, T] \to \mathbb{R}$ *is an absolutely continuous function, then* $\mathcal{L} - \operatorname{meas}(\psi(N)) = 0$, *for all* $N \subset [S, T]$ *such that* $\mathcal{L} - \operatorname{meas}(N) = 0$ *(i.e. ψ enjoys the 'Lusin property').*

2.6 Notes for Chapter 2

The 'Kuratowski sense' limit operations on sequences of sets of Sect. 2.2 are so called for they were employed in Kuratowski's influential 1933 monograph, reprinted as [135]. But their definitions were anticipated in Painlevé's earlier lecture notes. (See [177, Notes to Chapter 4].) Closely related concepts concern convergence of set valued functions. A detailed discussion of the differing definitions and terminologies that have been proposed are also to be found in [177, Notes to Chapter 5]. We shall introduce various continuity concepts for set valued mappings (in addition to pointwise convergence in the Kuratowski sense) in subsequent chapters, as required.

2.6 Notes for Chapter 2

Measure and integration theory of set valued functions on a measure space arose out of applications to control theory [103] economics ([15, 89]) and other fields. Castaing is credited with initiating a systematic study of the field, establishing different fundamental properties associated with the concept of multifunctions measurability in a very general framework, see [53]. These become equivalent when the multifunctions take values closed sets in $\mathbb{R}^n$ (cf. [53, Chapter III]), which is often, but not always, the case in applications to dynamic optimization. We follow Clarke [65], Ioffe and Tihomirov [132] and Rockafellar and Wets [177] who, to simplify certain constructions, adopted as defining property of a measurable set value function G the requirement that $G^{-1}(\mathcal{O})$ is measurable for all open sets $\mathcal{O} \subset \mathbb{R}^n$. We recommend [177, Chapter 14] for its up-to-date and comprehensive exposition of this material.

For problems in the calculus of variations (one independent variable) Tonelli was the first to identify the key hypotheses for existence of minimizers, namely convexity and uniform, superlinear growth of the Lagrangian w.r.t. the velocity variable. See [189]. Tonelli considered smooth Lagrangians. Rockafellar's formulation and systemic investigation of the generalized problem of Bolza [174], in which the Lagrangian is permitted to be a nonsmooth, extended valued function was an important advance. Its significance is that it brings together, within a single framework a wide range of dynamic optimization problems, including those incorporating controlled differential equations, differential inclusions and mixed pathwise constraints. In particular, Rockafellar identified minimal properties of the Lagrangian, under which the integral of its evaluation along an arbitrary arc is well defined or under which level sets are pre-compact. (These properties are implicit in the hypotheses of Theorem 2.4.1, though they take a more complicated form ('normal integrands') when the Lagrangian is not convex in the velocity variable.) This framework has been successfully employed both as a basis for existence theory and in the derivation of necessary conditions of optimality.

Chapter 3
Variational Principles

Abstract The term 'variational principle', which was formerly attached to laws of nature asserting that some quantity is minimized, is now used to describe any procedure in which a property of interest is shown to imply that some quantity is minimized. Variational principles, taken to mean 'some function is minimized', make sense even if the function is not differentiable, and the part they play in nonsmooth analysis is not then all that surprising.

This chapter brings together a number of variational principles that feature prominently in nonsmooth analysis and its applications. It also introduces a regularization procedure 'quadratic inf convolution' that is frequently used hand-in-hand with variational principles to reproduce, in a nonsmooth setting, formerly known properties of smooth functions.

The first variational principle is the exact penalization theorem. It tells us that if a point minimizes a Lipschitz function over a closed set, then the point remains a minimizer of an *unconstrained* problem, in which the constraint is accommodated by nonsmooth penalty function (the distance function to the set scaled by the Lipschitz constant of the objective function).

A lower semi-continuous function with domain a topological space may fail to have a minimizer if the domain is not compact. It is however possible to assure existence of a minimizer if we add to the function a suitable perturbation term. The remaining variational principles provide different procedures for constructing a perturbation term to ensure this property. The oldest of these, Ekeland's theorem, features a nonsmooth perturbation. It not only ensures existence of a minimizer, but locates that minimizer near an approximate minimizer for the original problem and tells us we can control the size of the perturbation according to the accuracy of the approximation. For some problems the presence of a nonsmooth perturbed term (obtained after applying a variational principle) can be inconvenient. The Borwein/Preiss theorem can be regarded as a variant of Ekeland's theorem, in which restrictions are placed on the class of functions to which it applies, but the perturbation term is smooth. Stegall's theorem gives conditions under which we choose a linear perturbation term, but gives no information about the location of the minimizer for the perturbed problem.

The chapter ends with a section on mini-max theory. This area is of interest in its own right, in large part, because of its relevance to game theory. But it is included in this chapter because it also has important applications to optimization, for example in the derivation of Lagrange multiplier rules for constrained optimization problems, where the Lagrange multipliers are interpreted as secondary player in some two player game.

3.1 Introduction

The name 'variational principle' is traditionally attached to a law of nature asserting that some quantity is minimized. Examples are Dirichlet's principle (the spatial distribution of an electrostatic field minimizes some quadratic functional), Fermat's principle (a light ray follows a shortest path) or Hamilton's principle of least action (the evolution of a dynamical system is an 'extremum' for the action functional). These principles are called 'variational' because working through their detailed implications entails solving problems in the calculus of variations.

Nowadays the term 'variational principle' is used to describe any procedure in which a property of interest is shown to imply that some quantity is minimized. Variational principles in this broader sense are at the heart of nonsmooth analysis and its applications. It is no exaggeration to say that they have as significant a role in a nonsmooth setting as, say, the inverse function theorem and fixed point theorems do in traditional real analysis. Variational principles, 'some function is minimized', make sense even if the function is not differentiable, and the part they play in nonsmooth analysis is not therefore all that surprising.

This chapter brings together a number of variational principles that feature prominently both in the fundamental theory of nonsmooth analysis and its applications. Take an extended valued function $f : X \to \mathbb{R} \cup \{+\infty\}$. A point $x_0 \in \text{dom } f$ is said to be an ϵ-minimizer if

$$f(x_0) \leq \inf_{x \in X} f(x) + \epsilon.$$

The variational principles we consider all address the question of how to construct a modified version of the function f (taking the form of the original function with additive perturbation term), to ensure that the modified problem has a minimizer and, for certain of them, to locate this minimizer in a neighbourhood of a given ϵ-minimizer, whose size is controlled by the parameter ϵ. They differ according to the assumed regularity of f, the nature of the underlying space X and the form of the perturbations.

The first and simplest of these is the exact penalization theorem. Here X is a metric space and f is a Lipschitz continuous function. It asserts that if a point x_0 is a minimizer of f over a subset $C \subset X$, then x_0 is also a minimizer over the whole space of a modified function, which is the extension to all of X of the restriction of f to C. The additive perturbation term is the scaled distance function to the subset C. We also take the opportunity in this chapter to introduce the quadratic inf convolution operation, which permits us to approximate a Lipschitz continuous function by a function having a one-sided differentiability property and which has an important role throughout this book in passing from known optimality conditions for minimizers of continuously differentiable functions to analogous optimality conditions which are valid for functions that are merely Lipschitz continuous.

The other variational principles covered in this chapter, the theorems of Ekeland, Borwein/Preiss and Stegall, all concern extended valued lower semi-continuous functions, but provide differing procedures for perturbing them to guarantee the existence of a minimizer, and for locating this minimizer. In Ekeland's theorem, the domain X is allowed to be an arbitrary complete metric space and the perturbation term is non-differentiable. Non-differentiability can sometimes be an obstacle to analysis and, in such circumstances, the Borwein/Preiss theorem, in which the perturbation term is quadratic, may be preferred, even though the Borwein/Preiss theorem places a restriction on the underlying space X, which must be a real Hilbert space. Stegall's theorem, in which X is also required to be a real Hilbert space, is less precise than the Borwein/Priess theorem because it fails to locate the minimizer of the perturbed problem. But it introduces a *linear* perturbation term that is advantageous in certain applications.

The chapter concludes with a section on the theory of mini-max problems, also referred as 'two person zero-sum games'. This field is of great intrinsic interest, with applications in economics, management science, population dynamics under evolutionary change, etc. But mini-max theory also has important applications to optimization (even thought optimization problems involve only a primary player), notably in the derivation of Lagrange multiplier rules for constrained optimization problems, in which context the Lagrange multipliers are interpreted as secondary player in some two player games. Thus the theory adds to our arsenal of techniques for studying optimization problems.

3.2 Exact Penalization

Take a metric space (X, m), a non-empty subset $C \subset X$ and a function $f : X \to \mathbb{R}$. We can use the metric m to define a 'distance function' d_C on X, with respect to the set C:

$$d_C(x) := \inf_{x' \in C} m(x, x') \text{ for each } x \in X.$$

It is easy to show that, for any $x \in C$, '$x \in C$' implies '$d_C(x) = 0$' and, if C is a closed subset, the converse is also true.

We say that the function f satisfies a Lipschitz condition on X (with Lipschitz constant K) if

$$|f(x) - f(x')| \le K m(x, x') \text{ for all } x, x' \in X.$$

Consider the problem of minimizing the function $f : X \to \mathbb{R}$ over the set $C \subset X$. We would like to replace this problem (or at least approximate it) by a more amenable one involving no constraint; an approach of long standing is to drop the constraint, but to compensate for its absence by adding a 'penalty term' $Kg(x)$ to the cost (K is the penalty parameter). The function g is chosen to be zero on C and positive outside C. The larger K, the more severe the penalty for violating the constraint, so one would expect that solving the 'penalized problem'

$$\text{Minimize } f(x) + Kg(x) \text{ over } x \in X$$

for large K would yield a point x which approximately minimizes the cost for the original problem and approximately satisfies the constraint. Now suppose that C is closed and f satisfies a Lipschitz condition on X. A remarkable feature of the distance function is that, if it is adopted as the penalty function, then penalization is 'exact', in the sense that a minimizer for the original problem is also a minimizer for the penalized problem. Justification of this assertion is provided by the following theorem.

Theorem 3.2.1 (Exact Penalization Theorem) *Let (X, m) be a metric space. Take a set $C \subset X$ and a function $f : X \to \mathbb{R}$. Assume that f satisfies a Lipschitz condition on X with Lipschitz constant K. Let $\bar{x}$ be a minimizer for the constrained minimization problem*

$$\text{Minimize } f(x) \text{ over points } x \in X \text{ satisfying } x \in C. \tag{3.2.1}$$

Choose any $\hat{K} \geq K$. Then $\bar{x}$ is a minimizer also for the unconstrained minimization problem

$$\text{Minimize } f(x) + \hat{K} d_C(x) \text{ over points } x \in X. \tag{3.2.2}$$

If $\hat{K} > K$ and C is a closed set, then the converse assertion is also true: any minimizer $\bar{x}$ for the unconstrained problem (3.2.2) is also a minimizer for the constrained problem (3.2.1) and so, in particular, $\bar{x} \in C$.

Proof Let $\bar{x}$ be a minimizer for (3.2.1) and let $\hat{K} \geq K$. Suppose that, contrary to the claims of the theorem, $\bar{x}$ fails to be a minimizer for (3.2.2). Then there exist a point $y \in X$ and $\epsilon > 0$ such that $f(y) + \hat{K} d_C(y) < f(\bar{x}) - \hat{K}\epsilon$. Choose a point $z \in C$ such that $m(y, z) \leq d_C(y) + \epsilon$. Since $\hat{K}$ is a Lipschitz constant for f on X,

$$f(z) \leq f(y) + \hat{K} m(y, z) \leq f(y) + \hat{K}(d_C(y) + \epsilon) < f(\bar{x}).$$

This is not possible since $\bar{x}$ minimizes f over C.

Suppose next that $\hat{K} > K$ and C is closed. Let $\bar{x}$ be a minimizer for (3.2.2). Choose any $\epsilon > 0$. Then we can find a point $z \in C$ such that

$$d_C(\bar{x}) > m(\bar{x}, z) - \hat{K}^{-1}\epsilon.$$

We have

$$f(z) \leq f(\bar{x}) + Km(\bar{x}, z)$$
$$\leq f(\bar{x}) + Kd_C(\bar{x}) + (K/\hat{K})\epsilon$$
$$< f(\bar{x}) + \hat{K}d_C(\bar{x}) - (\hat{K} - K)d_C(\bar{x}) + \epsilon$$
$$\leq f(z) - (\hat{K} - K)d_C(\bar{x}) + \epsilon.$$

It follows that $(\hat{K} - K)d_C(\bar{x}) < \epsilon$. Since $\epsilon > 0$ is arbitrary, $d_C(\bar{x}) = 0$. But then $\bar{x} \in C$, because C is closed. We deduce that $f(c) \geq f(\bar{x})$ for all $c \in C$. In other words, $\bar{x}$ is a minimizer for (3.2.1). □

3.3 Ekeland's Theorem

Take a complete metric space (X, d), a lower semi-continuous function $f : X \to \mathbb{R} \cup \{+\infty\}$, a point $x_0 \in \text{dom } f$ and some $\epsilon > 0$.

Suppose that x_0 is an ϵ-minimizer for f. This means

$$f(x_0) \leq \inf_{x \in X} f(x) + \epsilon.$$

In these circumstances, as we shall see, there exists some $\bar{x} \in \text{dom } f$ satisfying

$$d(\bar{x}, x_0) \leq \epsilon^{\frac{1}{2}}$$

which is a minimizer for the perturbed function

$$f_\epsilon(x) = f(x) + \epsilon^{\frac{1}{2}} d(x, \bar{x}),$$

i.e.

$$f_\epsilon(\bar{x}) = \inf_{x \in X} f_\epsilon(x).$$

This is a version of Ekeland's variational principle. It tells us that we can perturb f in such a way as to ensure that a minimizer $\bar{x}$ for the perturbed problem exists. Furthermore we can arrange that both the distance of the minimizer $\bar{x}$ for the perturbed function from x_0 and also the perturbation term are small, if ϵ is small.

The essential idea is captured by Fig. 3.1 Here X is $\mathbb{R}$ with the metric induced by the Euclidean norm, and f is a strictly monotone decreasing function. The function f has no minimizer 'close' to the ϵ-minimizer x_0. (In fact f has no minimizers at all!) However an appropriately chosen point $\bar{x}$, close to x_0, is a minimizer for the perturbed function

$$x \to f(x) + \epsilon^{\frac{1}{2}} d(x, \bar{x}).$$

Fig. 3.1 Ekeland's thoerem

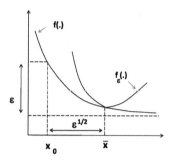

The perturbation term penalizes deviation from $\bar{x}$; by raising the graph of the function away from $\bar{x}$, we force $\bar{x}$ to be a minimizer. In this example it is essential that the perturbation term be a nonsmooth function and also that we allow $\bar{x}$ to be different from x_0.

Widespread use of this variational principle has been made in the field of optimization and, indeed, nonlinear analysis generally. Its main role in optimization has been to justify techniques for deriving necessary conditions for a minimization problem based on applying available necessary conditions to a simpler, perturbed problem and passage to the limit. The fact that it allows (X, d) to be an arbitrary complete metric space adds greatly to it flexibility.

The variational principle summarized above is a special case of a slightly more general theorem in which, by making different choices of parameter $\alpha > 0$ and $\lambda > 0$, we can trade off the size of the perturbation term and the distance of $\bar{x}$ from x_0. (The preceding version involves the choices $\lambda = \epsilon^{\frac{1}{2}}$ and $\alpha = \epsilon^{\frac{1}{2}}$.)

Theorem 3.3.1 (Ekeland's Variational Principle) *Take a complete metric space (X, d), a lower semi-continuous function $f : X \to \mathbb{R} \cup \{+\infty\}$, a point $x_0 \in \operatorname{dom} f$ and numbers $\alpha > 0$ and $\lambda > 0$. Assume that*

$$f(x_0) \leq \inf_{x \in X} f(x) + \lambda \alpha. \qquad (3.3.1)$$

Then there exists $\bar{x} \in X$ such that

(i) $f(\bar{x}) \leq f(x_0)$,
(ii) $d(x_0, \bar{x}) \leq \lambda$,
(iii) $f(\bar{x}) \leq f(x) + \alpha d(x, \bar{x})$ for all $x \in X$.

Proof It suffices to find some $\bar{x}$ such that

(a): $f(\bar{x}) + \alpha d(x_0, \bar{x}) \leq f(x_0)$,
and
(b): $f(\bar{x}) < f(x) + \alpha d(x, \bar{x})$ for all $x \neq \bar{x}$.

Indeed (a) implies (i) and (b) implies (iii). Notice also that (3.3.1) and (a) imply

$$f(\bar{x}) + \alpha d(x_0, \bar{x}) \leq f(x_0) \leq \inf_{x \in X} f(x) + \lambda \alpha \leq f(\bar{x}) + \lambda \alpha.$$

3.3 Ekeland's Theorem

Since $f(\bar{x})$ is finite (from (a)) and $\alpha > 0$ we conclude that $d(x_0, \bar{x}) \leq \lambda$. The remaining assertion (ii) is also confirmed.

The proof of (a) and (b) is as follows. Define a multifunction $T : X \leadsto X$

$$T(x) := \{y \in X : f(y) + \alpha d(x, y) \leq f(x)\}.$$

Notice that for each $x \in X$, $T(x)$ is a closed set. It is clear that

$$x \in T(x) \text{ for all } x \in X. \tag{3.3.2}$$

A further significant property of the multifunction T is

$$y \in T(x) \text{ implies } T(y) \subset T(x). \tag{3.3.3}$$

This is obviously true if $x \notin \operatorname{dom} f$ (for then $T(x) = X$). So assume $x \in \operatorname{dom} f$. Since $y \in T(x)$,

$$f(y) + \alpha d(x, y) \leq f(x).$$

Take any $z \in T(y)$. Then

$$f(z) + \alpha d(z, y) \leq f(y).$$

These inequalities, together with the triangle inequality imply

$$f(y) + f(z) + \alpha d(z, x) \leq f(x) + f(y).$$

However $f(y) < \infty$, since $f(x) < \infty$. We conclude that

$$f(z) + \alpha d(z, x) \leq f(x).$$

This inequality implies $z \in T(x)$, which is the desired condition.

Now define $\xi : \operatorname{dom} f \to \mathbb{R} \cup \{+\infty\}$

$$\xi(x) := \inf_{y \in T(x)} f(y). \tag{3.3.4}$$

We see that

$$y \in T(x) \text{ implies } \xi(x) \leq f(x) - \alpha d(x, y).$$

It follows that

$$\operatorname{diam} T(x) \leq 2\alpha^{-1}(f(x) - \xi(x)) \tag{3.3.5}$$

where

$$\operatorname{diam} T(x) = \sup\{d(y, y') : y \in T(x), y' \in T(x)\}.$$

Now construct a sequence $x_0, x_1, \ldots$, starting with the x_0 of the theorem statement, as follows: for $k = 0, 1, \ldots$ choose $x_{k+1} \in T(x_k)$ to satisfy

$$f(x_{k+1}) \leq \xi(x_k) + 2^{-k}.$$

This is possible since, by definition,

$$\xi(x_k) = \inf_{x \in T(x_k)} f(x).$$

Fix $k \geq 0$. Since $x_{k+1} \in T(x_k)$, we deduce from (3.3.3) that

$$\xi(x_k) \leq \xi(x_{k+1}).$$

On the other hand, property (3.3.2) implies

$$\xi(x) \leq f(x) \text{ for all } x \in X.$$

It follows

$$\xi(x_{k+1}) \leq f(x_{k+1}) \leq \xi(x_k) + 2^{-k} \leq \xi(x_{k+1}) + 2^{-k}.$$

From the preceding inequalities we deduce that

$$0 \leq f(x_{k+1}) - \xi(x_{k+1}) \leq 2^{-k}.$$

Recalling (3.3.5) we see that

$$\operatorname{diam} T(x_{k+1}) \leq 2^{-k} \cdot 2\alpha^{-1}.$$

We have shown that $\{T(x_k)\}$ is a nested sequence of non-empty closed sets in X whose diameters tend to zero as $k \to \infty$. $\{x_k\}$ is a Cauchy sequence then and, since (X, d) is complete, $x_k \to \bar{x}$ for some $\bar{x} \in X$. Furthermore we have

$$\bigcap_{k=0}^{\infty} T(x_k) = \{\bar{x}\}. \tag{3.3.6}$$

It remains to show that, for this choice of $\bar{x}$, properties (a) and (b) are satisfied. But (3.3.6) implies $\bar{x} \in T(x_0)$ which simply means

$$f(\bar{x}) + \epsilon d(x_0, \bar{x}) \leq f(x_0).$$

We have verified (a). On the other hand (3.3.3) and (3.3.6) imply

$$T(\bar{x}) \subset \cap_{k=0}^{\infty} T(x_k) = \{\bar{x}\}.$$

It follows that if $x \neq \bar{x}$ then $x \notin T(\bar{x})$ and so

$$f(\bar{x}) < f(x) + \alpha d(x, \bar{x}).$$

We have verified (b). □

3.4 Quadratic Inf Convolution

Situations are encountered throughout applied analysis, in which we would like to establish properties of some function $g : \mathbb{R}^k \to \mathbb{R} \cup \{+\infty\}$ by means of analytical techniques which would be applicable if the function satisfied certain conditions, but we are prevented from doing so because these conditions are violated. A standard approach to dealing with this difficulty is to construct a family of functions $\{g_\alpha : \alpha > 0\}$, each member of which satisfies the required conditions, such that g can be identified, in some way, with $\lim_{\alpha \to \infty} g_\alpha$. The idea is then to demonstrate the desired properties of g by verifying them for each α (by means of available techniques) and to show that the properties are preserved in the limit as $\alpha \to \infty$.

A common manifestation of this difficulty is that a continuous function $g : \mathbb{R}^k \to \mathbb{R}$ fails to have first (or higher) order derivatives. Here it is often useful to 'mollify' g, to achieve a higher order of differentiability, by convolving it with a smooth mollifier ϕ_α, depending on the parameter $\alpha > 0$, which has the properties that it is non-negative valued, vanishes outside $\alpha^{-1} \mathbb{B}$ and

$$\int \ldots \int \phi_\alpha(x) dx_1 \ldots dx_k = 1.$$

Thus we choose

$$g_\alpha(x) = \int \ldots \int g(y) \phi_\alpha(x - y) dy_1 \ldots dy_k.$$

We refer to the operation of replacing g by g_α as 'integral convolution'.

In later chapters of this book, where we investigate optimization problems involving a Lipschitz continuous function g, it will sometimes be convenient to employ techniques that are available only under the following conditions

(i): for each $x \in \mathbb{R}^k$, there exist $\eta \in \mathbb{R}^k$ and $M > 0$ such that

$$g(y) - g(x) \leq \eta \cdot (x - y) + M|y - x|^2 \text{ for all } y \in \mathbb{R}^k$$

and

(ii): the vector η can be identified as a proximal subgradient of g at some point y near x. (Here, 'proximal subgradient' is understood as in Chap. 1).

A vector η satisfying the inequality in condition (i) (for given x) is called a *proximal supergradient* of g at x. Another way of describing condition (i) is: the proximal subdifferential of $x \to -g(x)$ is non-empty at all points in its domain. A Lipschitz continuous function may fail to have a proximal superdifferential at some points in its domain.

It turns out that the integral convolution procedure described above is not well suited to this purpose. This is because, although the integral mollification of g is everywhere differentiable, it is not easy to link the derivatives of the mollified function with proximal subgradients of the original function g.

It is helpful to consider, instead, a procedure, called *inf convolution*, in which the integral operation in replaced by a minimization operation. Now, the approximating functions $\{g_\alpha : \alpha > 0\}$ are taken to be:

$$g_\alpha(x) := \inf_{y \in \mathbb{R}^k} \{g(y) + \alpha |x - y|^2\}. \tag{3.4.1}$$

We say that the function g_α given by this formula is the quadratic *inf convolution* of g (with parameter $\alpha > 0$). The family of functions $\{g_\alpha : \alpha > 0\}$ is elsewhere referred to as the Moreau-Yosida envelope of g.

Key properties of the quadratic inf convolution operation are brought together in the following theorem.

Theorem 3.4.1 (Quadratic Inf Convolution, I) *Take a function* $g : \mathbb{R}^k \to \mathbb{R}$, *with Lipschitz constant* k_g. *Let* g_α *be its quadratic inf convolution (with parameter* $\alpha > 0$). *Take any* $x \in \mathbb{R}^k$. *Then there exists* $y \in \mathbb{R}^k$ *such that*

(i): g_α is Lipschitz continuous, with Lipschitz constant k_g,
(ii): $g_\alpha(x) \leq g(x) \leq g_\alpha(x) + \frac{1}{4\alpha} k_g^2$,
(iii): $g_\alpha(x') - g_\alpha(x) \leq 2\alpha(x - y) \cdot (x' - x) + \alpha |x' - x|^2$, for all $x' \in \mathbb{R}^k$,
(iv): $2\alpha(x - y) \in \partial^P g(y)$,
(v): $|y - x| \leq \alpha^{-1} k_g$.

Proof For all $x, x' \in \mathbb{R}^k$ we have

$$g_\alpha(x) = \inf_{z \in \mathbb{R}^k} \{g(z) + \alpha |x - z|^2\} = \inf_{w \in \mathbb{R}^k} \{g(w - (x' - x)) + \alpha |w - x'|^2\}$$

$$\leq \inf_{w \in \mathbb{R}^k} \{g(w) + \alpha |x' - w|^2\} + k_g |x - x'| = g_\alpha(x') + k_g |x - x'|.$$

Exchanging the roles of x and x' we obtain that $|g_\alpha(x') - g_\alpha(x)| \leq k_g |x' - x|$ for all $x, x' \in \mathbb{R}^k$ confirming the fact that g_α is Lipschitz continuous of Lipschitz constant k_g. This is (i).

3.4 Quadratic Inf Convolution

Observe that $g_\alpha(x) = \inf\{g(y') + \alpha|y' - x|^2 : y' \in \mathbb{R}^k\} \le g(x)$. It follows that $g_\alpha(x) \le g(x)$. Notice also that, for any y',

$$g(y') + \alpha|y' - x|^2 \ge g(x) - k_g|y' - x| + \alpha|y' - x|^2$$
$$\ge g(x) + \min\{\alpha d^2 - k_g d : d \ge 0\} = g(x) - \frac{1}{4\alpha}k_g^2.$$

Taking the infimum of the left side over y' and recalling the definition of g_α, we conclude that $g_\alpha(x) \ge g(x) - \frac{1}{4\alpha}k_g^2$. We have shown (ii).

Next note that, for any y' such that $|y' - x| > \alpha^{-1}k_g$, we have

$$g(y') + \alpha|y' - x|^2 \ge g(x) - k_g|y' - x| + \alpha|y' - x|^2 > g(x) \ge g_\alpha(x).$$

It follows that there exists $y \in \mathbb{R}^n$ such that

$$|y - x| \le \alpha^{-1}k_g$$

and

$$g_\alpha(x) = \inf_{y' \in \mathbb{R}^k} \left(g(y') + \alpha|y' - x|^2\right)$$
$$= \inf_{\{y': |y' - x| \le \alpha^{-1}k_g\}} \left(g(y') + \alpha|y' - x|^2\right) = g(y) + \alpha|y - x|^2,$$

since $y' \to g(y') + \alpha|y' - x|^2$ is continuous and the second infimum in the preceding relation is taken over a compact set. We have shown that the infimum in the definition of $g_\alpha(x)$ is attained.

Take any $x' \in \mathbb{R}^k$. Then

$$g_\alpha(x') - g_\alpha(x) \le g(y) + \alpha|y - x'|^2 - g(y) - \alpha|y - x|^2$$
$$= 2\alpha(x - y) \cdot (x' - x) + \alpha|x' - x|^2.$$

We have confirmed (iii) and (v).

It follows from the definition of g_α that, for every $y' \in \mathbb{R}^k$,

$$g(y) + \alpha|y - x|^2 \le g(y') + \alpha|y' - x|^2.$$

Rearranging this inequality, we arrive at the relation

$$g(y') - g(y) \ge \alpha(|y - x|^2 - |y' - x|^2) = 2\alpha(x - y) \cdot (y' - y) - \alpha|y' - y|^2,$$

which can be equivalently expressed $2\alpha(x - y) \in \partial^P g(y)$. We have confirmed (iv). $\square$

The main purpose of the following example is to illustrate how quadratic inf convolution can be used in dynamic optimization, specifically to derive necessary conditions for problems with non-smooth data and, by so doing, to provide a foretaste of techniques employed extensively in subsequent chapters. Notice that the role of quadratic inf convolution in this example (and elsewhere in this book) is to construct a smooth optimization problem, approximating the original non-smooth problem of interest. The fact that such a smooth problem can be constructed can be viewed as some kind of variational principle. To this extent, the example links quadratic inf convolution to the main themes of this chapter.

Example

Consider the dynamic optimization problem

$$(P) \begin{cases} \text{Minimize } g(x(T)) \\ \text{over arcs } x \in W^{1,1}([S, T]; \mathbb{R}^n) \\ \qquad \text{and meas. functions } u : [S, T] \to \mathbb{R}^m \text{ such that} \\ \dot{x}(t) = f(t, x(t), u(t)), \text{ a.e. } t \in [S, T], \\ u(t) \in U(t), \text{ a.e. } t \in [S, T], \\ x(S) = x_0. \end{cases}$$

in which $f : [S, T] \times \mathbb{R}^n \times \mathbb{R}^m \to \mathbb{R}^n$ and $g : \mathbb{R}^n \to \mathbb{R}$ are given functions, $\{U(t) \subset \mathbb{R}^m : S \le t \le T\}$ is a given family of sets and x_0 is a given point in $\mathbb{R}^n$. This is a special case of a dynamic optimization addressed in Chap. 1.

Necessary conditions (in the form of a maximum principle), for a given process $(\bar{x}, \bar{u})$ to be an L^∞ local minimizer, were provided by Theorem 1.11.1, under hypotheses that included the requirement that the terminal cost function g be continuously differentiable. We illustrate the application of Theorem 3.4.1 by showing that it can be used to derive a form of "nonsmooth' maximum principle, in which g is permitted to be an arbitrary Lipschitz continuous (with Lipschitz constant k_g) but in which, in all otherwise respects, the hypotheses Theorem 1.11.1 remain in place.

Take $\epsilon_i \downarrow 0$. For each i and let g_i be the quadratic inf convolution of g (with parameter $\alpha = \frac{1}{4\epsilon_i} k_g^2$). Then according to property (ii) of the inf convolution in Theorem 3.4.1, the process $(\bar{x}, \bar{u})$ is an ϵ_i minimizer for a modified version of (P), in which the terminal cost g is replaced by g_i. Because there is a fixed left end-point there is, under the given hypotheses, a unique state trajectory x^u corresponding to a given control function u. It follows from the assumption that $(\bar{x}, \bar{u})$ is an L^∞ local minimizer for (P), that $\bar{u}$ is an ϵ_i-minimizer for the problem

$$\text{Minimize } \{J(u) := g_i(x^u(T)) : u \in \mathcal{M}\},$$

in which, for some $\gamma > 0$,

$$\mathcal{M} := \{u : u \text{ is a control for } (P) \text{ such that } ||x^u - \bar{x}||_{L^\infty} \le \gamma\}.$$

3.4 Quadratic Inf Convolution

The parameter γ takes account of the facts that $(\bar{x}, \bar{u})$ is only an L^∞ local minimizer and that the hypotheses (B), (C) and (D) of Theorem 1.11.1 are local (not global) Lipschitz continuity and boundedness conditions on the velocity function f.

Under the hypotheses of Theorem 1.11.1, $\mathcal{M}$ is a complete metric space for the distance function

$$d_{\mathcal{E}}(u', u) := \mathcal{L}\text{-meas } \{t : u'(t) \neq u(t)\}.$$

Furthermore J is continuous on $\mathcal{M}$, with respect to this metric. It follows from Ekeland's theorem that, for i sufficiently large, there exists $u_i \in \mathcal{M}$ such that u_i is a minimizer for

$$\text{Minimize } \{J_i(u) := g_i(x^u(T)) + \epsilon_i^{\frac{1}{2}} \int_S^T m_i(t, u(t))dt : u \in \mathcal{M},$$

in which $m_i(t, u) = \begin{cases} 0 \text{ if } u = u_i(t) \\ 1 \text{ if } u \neq u_i(t) \end{cases}$. (Notice that $d_{\mathcal{E}}(u, u_i) = \int_S^T m_i(t, u(t))dt$.)

Write $x_i := x^{u_i}$.

We also know that $d_{\mathcal{E}}(u_i, \bar{u}) \leq \epsilon_i^{\frac{1}{2}}$, from which we conclude:

$$x_i \to \bar{x}, \text{ uniformly, and } \mathcal{L}\text{-meas. } \{t : u_i(t) \neq \bar{u}(t)\} \to 0. \tag{3.4.2}$$

Next notice that, by properties (iii), (iv) and (v) of the quadratic inf convolution in Theorem 3.4.1, there exists

$$\xi_i \in \partial^P g(y_i) \text{ for some } y_i \in x_i(T) + \frac{4\epsilon_i}{k_g} \mathbb{B},$$

such that

$$\begin{cases} g_i(x) \leq \tilde{g}_i(x) \text{ for all } x \in \mathbb{R}^k \\ g_i(x) = \tilde{g}_i(x) \text{ if } x = x_i(T). \end{cases}$$

Here $\tilde{g}_i$ is the quadratic function

$$\tilde{g}_i(x) := g_i(x_i(T)) + \xi_i \cdot (x - x_i(T)) + \frac{k_g^2}{4\epsilon_i}|x - x_i(T)|^2.$$

Observe that the function $\tilde{g}_i$ dominates the function g, and the graphs of these two functions touch at the point $(x_i(T), g_i(x_i(T)))$. It follows that u_i remains a minimizer when we replace the terminal cost function g_i by $\tilde{g}_i$. We have shown that (x_i, u_i) is an L^∞ local minimizer for the problem

$$(P_i) \begin{cases} \text{Minimize } \tilde{g}_i(x(T)) + \epsilon_i^{\frac{1}{2}} \int_S^T m_i(t, u(t)) dt \\ \text{subject to} \\ \dot{x}(t) = f(t, x(t), u(t)), \text{ a.e. } t \in [S, T], \\ u(t) \in U(t), \text{ a.e. } t \in [S, T], \\ x(S) = x_0. \end{cases}$$

Now apply the special case of the 'smooth' maximum principle, when there is no right endpoint constraint to (P_i). (See Proposition 1.11.5.) This is permissible because the terminal cost function $\tilde{g}$ is now a continuously differentiable (indeed quadratic) function. We deduce the existence of an absolutely continuous function p_i on $[S, T]$ such that

(i): $-\dot{p}_i(t) = p_i(t) \cdot \nabla_x f(t, x_i(t), u_i(t))$ a.e. $t \in [S, T]$,
(ii): $-p_i(T) = \xi_i$ for some $\xi_i \in \partial^P g(y_i)$,
(iii): $p_i(t) \cdot f(t, x_i(t), u_i(t)) \geq \max_{u \in U(t)} \left(p_i(t) \cdot f(t, x_i(t), u) \right) - \epsilon_i^{\frac{1}{2}}$, a.e. $t \in [S, T]$.

We know that $\|x_i - \bar{x}\|_{L^\infty} \to 0$ and $\mathcal{L}$-meas $\{t : u_i(t) \neq \bar{u}(t)\} \to 0$. A similar convergence analysis to that employed in the proof of the smooth maximum principle (Theorem 1.11.1) permits us to extract subsequences and pass to the limit in the above relations, as $i \to \infty$. We arrange in particular that $p_i \to p$ uniformly, for some absolutely continuous function p and $\xi_i \to \xi$ for some $\xi \in \mathbb{R}^n$. We conclude:

$$-p(T) = \xi = \lim_{i \to \infty} \xi_i \in \partial g(\bar{x}(T))$$

since, by the characterization of the limiting subdifferential (see Theorem 4.6.2)

$$\partial g(\bar{x}(T)) := \{\eta : \text{ there exist } z_i \to \bar{x}(T) \text{ and } \eta_i \to \eta$$

such that $\eta_i \in \partial^P g(z_i)$ for all $i\}$.

3.5 Variational Principles with Smooth Perturbation Terms

Take a lower semi-continuous, extended valued function $g : X \to \mathbb{R} \cup \{+\infty\}$ on a complete metric space (X, d). Take $\epsilon > 0$ and a point $\bar{x} \in X$ which is an an ϵ minimizer of the function g over X, that is

$$g(\bar{x}) \leq \inf_{x \in X} g(x) + \epsilon.$$

Recall that, according to Ekeland's theorem, there exists a point $y \in X$ such that

$$d(y, \bar{x}) \leq \epsilon^{\frac{1}{2}} \tag{3.5.1}$$

3.5 Variational Principles with Smooth Perturbation Terms

and

$$x \to g(x) + \epsilon^{\frac{1}{2}} d(y, x) \text{ is minimized over } X \text{ at } y.$$

A distinguishing feature of Ekeland's theorem is that the additive perturbation term '$+\epsilon^{\frac{1}{2}} d(y, x)$' in the modified function is non-smooth. The lack of smoothness can be troublesome for some applications and, for this reason, other variational principles, involving more amenable perturbation terms, have been investigated.

We now provide two variants on Ekeland's theorem in this spirit. Both of them concern the properties of a lower semi-continuous function $g : X \to \mathbb{R} \cup \{+\infty\}$ but now we restrict attention to the case when X is a real Hilbert space. (Write the inner product $\langle x, y \rangle_X$ and the induced norm $\|x\|_X$.)

Take an ϵ minimizer $\bar{x}$ for the function g. In a Hilbert space setting, Ekeland's theorem supplies a minimizer y (close to $\bar{x}$ with respect to the X norm, if ϵ is small) for a function with additive perturbation term $+\epsilon^{\frac{1}{2}} \|x - y\|_X$. To arrive at a variational principle with a smooth perturbation term, we might consider replacing the norm of x (about the base point y) by the *square* of the norm, thus $+\epsilon^{\frac{1}{2}} \|x - y\|_X^2$. To see that this is not possible, we have only to consider the example of the function $g(x) := e^{-x}$ on $\mathbb{R}$. Indeed,

$$x \to e^{-x} + K|x - y|^2 \text{ is not minimized at } y \text{ for any point } y \text{ or } K > 0.$$

The insight of Borwein and Preiss [43] was to understand that modifying the perturbation term in this way is possible, if we replace the base point y by a new base point z (close to $\bar{x}$ if ϵ is small).

Theorem 3.5.1 (Borwein and Preiss) *Take a real Hilbert space X, a lower semi-continuous function $g : X \to \mathbb{R} \cup \{+\infty\}$, which is bounded below, and $\epsilon > 0$. Suppose that $\bar{x} \in X$ such that $g(\bar{x}) < \inf_{x \in X} g(x) + \epsilon$. Then, for any $\lambda > 0$ there exist points $y, z \in X$ such that*

$$\|z - \bar{x}\|_X < \lambda, \ \|y - z\|_X < \lambda, \ g(y) \le g(\bar{x})$$

and such that the function

$$x \to g(x) + \frac{\epsilon}{\lambda^2} \|x - z\|_X^2$$

attains a unique minimum at $x = y$.

The second variant, like its predecessor, asserts the existence of a minimizer for a perturbed function, in which the perturbation term is a linear function of arbitrarily small slope. Notice however that the simplicity of the perturbation come at a price: it fails to provide any information about the proximity of the minimizer to a given ϵ minimizer for the original, unperturbed, function.

Theorem 3.5.2 (Stegall's Variational Principle) *Consider a nonempty, closed and bounded subset Y of a real Hilbert space X and a lower semi-continuous function $g : X \to \mathbb{R} \cup \{+\infty\}$ which is bounded below on Y. Suppose that $Y \cap \operatorname{dom} g \neq \emptyset$. Then there exists a dense set of points $x \in X$ such that*

$$z \to g(z) - \langle x, z \rangle_X$$

attains a unique minimum over Y.

Proofs of Theorems 3.5.1 and 3.5.2, based on proximal normal analysis in real Hilbert spaces, are given in the Appendix to Chap. 4.

3.6 Mini-Max Theorems

Take non-empty sets X and Y and a function $F : X \times Y \to \mathbb{R}$. The central question addressed in this section is: under what circumstances is the relation

$$\inf_{x \in X} \sup_{y \in Y} F(x, y) = \sup_{y \in Y} \inf_{x \in X} F(x, y) \tag{3.6.1}$$

valid? In other words, when do the operations of taking the supremum over Y and taking the infimum over X commute?

Notice that, without the imposition of any additional hypotheses whatsoever, we are assured that the two sides of (3.6.1) are related by inequality:

Proposition 3.6.1 *Let X, Y and F be as above. Then*

$$\inf_{x \in X} \sup_{y \in Y} F(x, y) \geq \sup_{y \in Y} \inf_{x \in X} F(x, y).$$

Proof For any fixed $x' \in X$ and $y' \in Y$

$$\sup_{y \in Y} F(x', y) \geq F(x', y') \geq \inf_{x \in X} F(x, y').$$

It follows that

$$\inf_{x \in X} \sup_{y \in Y} F(x, y) \geq \inf_{x \in X} F(x, y')$$

and consequently

$$\inf_{x \in X} \sup_{y \in Y} F(x, y) \geq \sup_{y \in Y} \inf_{x \in X} F(x, y).$$

$\square$

3.6 Mini-Max Theorems

The challenge then is to establish when the reverse inequality holds:

$$\inf_{x \in X} \sup_{y \in Y} F(x, y) \leq \sup_{y \in Y} \inf_{x \in X} F(x, y)$$

for this combines with the assertion of Proposition 3.6.1 to give (3.6.1).

A related question, and one of great independent interest, is whether there exists a point $(x^*, y^*) \in X \times Y$ satisfying

$$\sup_{y \in Y} F(x^*, y) = F(x^*, y^*) = \inf_{x \in X} F(x, y^*).$$

A pair (x^*, y^*) having this property is called a *saddlepoint*.

The connection is that existence of a saddlepoint is a sufficient condition for the commutability condition (3.6.1):

Proposition 3.6.2 *Suppose that (x^*, y^*) is a saddlepoint. Then*

$$\inf_{x \in X} \sup_{y \in Y} F(x, y) = F(x^*, y^*) = \sup_{y \in Y} \inf_{x \in X} F(x, y).$$

Furthermore,

$$\sup_{y \in Y} F(x^*, y) = \sup_{y \in Y} \inf_{x \in X} F(x, y) \tag{3.6.2}$$

and

$$\inf_{x \in X} F(x, y^*) = \inf_{x \in X} \sup_{y \in Y} F(x, y). \tag{3.6.3}$$

Proof By definition of a saddlepoint and in view of Proposition 3.6.1 we have

$$F(x^*, y^*) = \sup_{y \in Y} F(x^*, y) \geq \inf_{x \in X} \sup_{y \in Y} F(x, y)$$

$$\geq \sup_{y \in Y} \inf_{x \in X} F(x, y) \geq \inf_{x \in X} F(x, y^*) = F(x^*, y^*).$$

These relations therefore hold with equality. The assertions of the proposition follow immediately. □

Relations (3.6.2) and (3.6.3) do in fact fully characterize a saddlepoint:

Proposition 3.6.3 *There exists a saddlepoint if and only if the following conditions both hold:*

(a) There exists $x^* \in X$ such that

$$\sup_{y \in Y} F(x^*, y) = \sup_{y \in Y} \inf_{x \in X} F(x, y),$$

(b) There exists $y^* \in Y$ such that

$$\inf_{x \in X} F(x, y^*) = \inf_{x \in X} \sup_{y \in Y} F(x, y).$$

Furthermore, if (a) and (b) both hold, then (x^*, y^*) is a saddlepoint.

Proof In view of the preceding proposition all we have to show is that, if (x^*, y^*) satisfies conditions (a) and (b), then (x^*, y^*) is a saddlepoint. However (a) implies

$$\inf_{x \in X} \sup_{y \in Y} F(x, y) \leq \sup_{y \in Y} F(x^*, y) = \sup_{y \in Y} \inf_{x \in X} F(x, y).$$

This combines with the assertions of Proposition 3.6.1 to give

$$\inf_{x \in X} \sup_{y \in Y} F(x, y) = \sup_{y \in Y} \inf_{x \in X} F(x, y).$$

So if (b) also holds, then

$$F(x^*, y^*) \leq \sup_{y \in Y} F(x^*, y) =$$

$$\sup_{y \in Y} \inf_{x \in X} F(x, y) = \inf_{x \in X} \sup_{y \in Y} F(x, y) = \inf_{x \in X} F(x, y^*) \leq F(x^*, y^*).$$

It follows that

$$\sup_{y \in Y} F(x^*, y) = F(x^*, y^*) = \inf_{x \in X} F(x, y^*).$$

This is the saddlepoint condition. □

Establishing existence of a saddlepoint is by no means straightforward and requires the imposition of stringent hypotheses (see Von Neumann's mini-max theorem below). It turns out however that these hypotheses can be relaxed significantly if we are willing to settle for just one of the two 'one-sided' properties (a) and (b) characterizing existence of a saddlepoint. The one-sided mini-max theorem, giving hypotheses under which property (a) of Proposition 3.6.3 holds, has come to be recognized as a powerful analytic tool with important implications for optimization. Numerous applications will be made in future chapters.

We make one final observation. It is that, while property (a) falls somewhat short of guaranteeing existence of a saddlepoint, it nonetheless implies the commutability of the infimum and supremum operations. Validity of this assertion is a by-product of the proof of Proposition 3.6.3. (Replacing F by $-F$ so that 'inf' becomes 'sup' and vice versa, we arrive at an analogous statement in relation to property (b).)

3.6 Mini-Max Theorems

Proposition 3.6.4 *Suppose that either of the following conditions hold*

(a) there exists x^* such that

$$\sup_{y \in Y} F(x^*, y) = \sup_{y \in Y} \inf_{x \in X} F(x, y),$$

(b) there exists y^* such that

$$\inf_{x \in X} F(x, y^*) = \inf_{x \in X} \sup_{y \in Y} F(x, y).$$

Then

$$\inf_{x \in X} \sup_{y \in Y} F(x, y) = \sup_{y \in Y} \inf_{x \in X} F(x, y).$$

The main theorems of the section now follow.

Theorem 3.6.5 (Aubin One-Sided Mini-Max Theorem) *Consider a function $F: X \times Y \to \mathbb{R}$ in which X is a subset of a linear space and Y is a subset of a topological linear space. Assume that*

(i) X and Y are convex sets,
(ii) $F(., y)$ is convex for every $y \in Y$,
(iii) $F(x, .)$ is concave and upper semi-continuous for every $x \in X$,
(iv) Y is compact.

Then there exists $y^* \in Y$ such that

$$\inf_{x \in X} F(x, y^*) = \inf_{x \in X} \sup_{y \in Y} F(x, y).$$

This implies in particular (see Proposition 3.6.4) that

$$\inf_{x \in X} \sup_{y \in Y} F(x, y) = \sup_{y \in Y} \inf_{x \in X} F(x, y).$$

Proof Define

$$\eta^- := \sup_{y \in Y} \inf_{x \in X} F(x, y)$$

and

$$\eta^+ := \inf_{x \in X} \sup_{y \in Y} F(x, y).$$

For any subset $K \subset X$ write

$$\eta_K^- := \sup_{y \in Y} \inf_{x \in K} F(x, y).$$

Now let $\mathcal{S}$ denote the class of subsets of X comprising only a finite number of points and define:

$$\tilde{\eta}^- := \inf_{K \in \mathcal{S}} \sup_{y \in Y} \inf_{x \in K} F(x, y) = \inf_{K \in \mathcal{S}} \eta_K^-.$$

Step 1: We show that

$$\eta^- \leq \tilde{\eta}^- \leq \eta^+.$$

Take any $K \in \mathcal{S}$. Then

$$\sup_{y \in Y} \inf_{x \in K} F(x, y) \geq \sup_{y \in Y} \inf_{x \in X} F(x, y) = \eta^-.$$

Taking the infimum of the left side over $K \in \mathcal{S}$ we obtain

$$\tilde{\eta}^- \geq \eta^+.$$

On the other hand, for any x we have $\{x\} \subset \mathcal{S}$ whence

$$\sup_{y \in Y} F(x, y) \geq \inf_{K \in \mathcal{S}} \sup_{y \in Y} \inf_{x \in K} F(x, y) = \tilde{\eta}^-.$$

Taking the infimum of the left side over x yields

$$\eta^+ \geq \tilde{\eta}^-,$$

as required.

Step 2: We show that there exists $y^* \in Y$ such that

$$\inf_{x \in X} F(x, y^*) \geq \tilde{\eta}^-.$$

For each $x \in X$, define $E_x \subset X$ to be the set

$$E_x := \{y \in Y : F(x, y) \geq \tilde{\eta}^-\}.$$

We must show that

$$\cap_{x \in X} E_x \neq \emptyset. \tag{3.6.4}$$

3.6 Mini-Max Theorems

Notice however that, because $F(x, .)$ is an upper semi-continuous function, the 'level set' E_x is compact for every $x \in X$. Property (3.6.4) will follow then if we can show that, for an arbitrary finite set $\{x_1, .., x_m\}$ of points in X,

$$\cap_{i=1}^m E_{x_i} \neq \emptyset.$$

Take any finite subset $K = \{x_1, \ldots, x_m\} \subset X$. Then

$$\cap_{i=1}^m E_{x_i} = \{y \in Y : \min_{i=1,\ldots,m} F(x_i, y) \geq \tilde{\eta}^-\}.$$

But $\min_{i=1,\ldots,m} F(x_i, .)$ is upper semi-continuous and its maximum value is therefore achieved over the compact set Y at some $y^* \in Y$. We have

$$\inf_{x \in K} F(x, y^*) = \sup_{y \in Y} \inf_{x \in K} F(x, y) \geq \tilde{\eta}^-.$$

It follows that $\cap_{i=1}^m E_{x_i}$ contains the point y^* and is therefore non-empty.

Step 3: For a finite subset $K = \{x_1, ..x_m\} \subset X$, define

$$\zeta_K := \inf_{(\lambda_1,\ldots,\lambda_m) \in \Sigma^m} \sup_{y \in Y} \sum_i^m \lambda_i F(x_i, y),$$

in which

$$\Sigma^m := \{(\lambda_1, .., \lambda_m) : \lambda_i \geq 0 \text{ for each } i \text{ for all } \sum_{i=1}^m \lambda_i = 1\}.$$

We show that

$$\zeta_K \leq \eta_K^-.$$

(Recall that

$$\eta_K^- := \sup_{y \in Y} \inf_{x \in K} F(x, y) \quad .)$$

Indeed suppose, contrary to the claim, that there exists $\alpha > 0$ such that

$$\zeta_K - \alpha > \sup_{y \in Y} \inf_{x \in K} F(x, y).$$

This implies

$$(\zeta_K - \alpha)(1, .., 1) \notin C$$

where $C \subset \mathbb{R}^m$ is the subset

$$C := \{(\xi_1, \ldots, \xi_m) : \text{ there exists } y \in Y$$
$$\text{such that } \xi_i \leq F(x_i, y) \text{ for } i = 1, .., m\}.$$

However it may be deduced from the concavity of the functions $F(x_i, .)$ and the convexity of the set Y, that C is a convex set.

By the separation theorem there exists a non-zero vector $\lambda \in \mathbb{R}^m$ (we may assume that $\sum_{i=1}^m |\lambda_i| = 1$) such that

$$(\zeta_K - \alpha)(1, .., 1) \cdot \lambda \geq c \cdot \lambda \text{ for all } c \in C.$$

This inequality can be satisfied only if $\lambda_i \geq 0$ for all i. It follows that $\lambda \in \sum^m$ and $(1, .., 1) \cdot \lambda = 1$. Inserting

$$c := (F(x_1, y), .., F(x_m, y))$$

into this relation for arbitrary $y \in Y$ gives

$$\zeta_K - \alpha \geq \sum_{i=1}^m \lambda_i F(x, y).$$

Taking the supremum of the right side over $y \in Y$, we arrive at

$$\zeta_K - \alpha \geq \sup_{y \in Y} \sum_{i=1}^m \lambda_i F(x_i, y)$$

$$\geq \inf_{\lambda \in \Sigma^m} \sup_{y \in Y} \sum_{i=1}^m \lambda_i F(x_i, y) = \zeta_K.$$

From this contradiction we conclude that

$$\zeta_K \leq \bar{\eta}_K.$$

Step 4: We show that

$$\tilde{\eta}^- \geq \eta^+.$$

Take any finite set $K = \{x_1, .., x_m\}$ and $\lambda \in \Sigma^m$. Define

$$x_\lambda := \sum_{i=1}^m \lambda_i x_i.$$

3.6 Mini-Max Theorems

Since X is convex and $F(., y)$ is a convex function for each $y \in Y$, we have

$$\sup_{y \in Y} \sum_{i=1}^{m} \lambda_i F(x_i, y) \geq \sup_{y \in Y} F(x_\lambda, y) \geq \inf_{x \in X} \sup_{y \in Y} F(x, y) = \eta^+.$$

Taking the infimum on the left over $\lambda \in \Sigma^m$ we deduce that

$$\zeta_K \geq \eta^+.$$

But then, by Step 3,

$$\eta_K^- \geq \eta^+.$$

Taking the infimum of the left side over $K \subset S$ we obtain

$$\tilde{\eta}^- \geq \eta^+.$$

Conclusion

Steps 1, 2 and 4 give

$$\eta^- \leq \tilde{\eta}^- \leq \eta^+,$$

$$\eta^- \left(= \sup_{y \in Y} \inf_{x \in X} F(x, y) \right) \geq \inf_{x \in X} F(x, y^*) \geq \tilde{\eta}^-$$

for some $y^* \in Y$ and

$$\tilde{\eta}^- \geq \eta^+ = \left(\inf_{x \in X} \sup_{y \in Y} F(x, y) \right).$$

We conclude that

$$\inf_{x \in X} F(x, y^*) = \inf_{x \in X} \sup_{y \in Y} F(x, y).$$

This is what we set out to prove. □

Theorem 3.6.6 (Von Neumann Mini-Max Theorem) *Consider a function* $F : X \times Y \to \mathbb{R}$ *in which X and Y are subsets of topological linear spaces. Assume that*

(i) X and Y are convex, compact sets,
(ii) $F(., y)$ is convex and lower semi-continuous for every $y \in Y$,
(iii) $F(x, .)$ is concave and upper semi-continuous for every $x \in X$.

Then there exists an element $(x^*, y^*) \in X \times Y$ which is a saddlepoint for F, i.e.

$$\sup_{y \in Y} F(x^*, y) = F(x^*, y^*) = \inf_{x \in X} F(x, y^*). \tag{3.6.5}$$

This implies in particular (see Proposition 3.6.2)

$$\inf_{x \in X} \sup_{y \in Y} F(x, y) = \sup_{y \in Y} \inf_{x \in X} F(x, y).$$

Proof Apply Theorem 3.6.5 to $F(x, y)$ and also to $-F(x, y)$ (in the latter case interchanging the roles of X and Y). We deduce existence of some $(x^*, y^*) \in X \times Y$ such that

$$\inf_{x \in X} F(x, y^*) = \inf_{x \in X} \sup_{y \in Y} F(x, y)$$

and

$$\sup_{y \in Y} \inf_{x \in X} F(x, y) = \sup_{y \in Y} F(x^*, y).$$

In view of Proposition 3.6.4 it follows that

$$F(x^*, y^*) \leq \sup_{y \in Y} F(x^*, y) = \inf_{x \in X} F(x, y^*) \leq F(x^*, y^*).$$

This implies the saddlepoint condition (3.6.5). □

Compactness of the underlying sets X and Y (or at least of one of them) plays an essential part in proving the above mini-max theorems. The compactness hypotheses on X and Y can be replaced in certain circumstances by coercivity hypotheses on the 'objective function' F; here compactness of the level sets of certain constructed functions in some sense substitutes for that of X and Y.

We illustrate the point by proving existence of a saddlepoint for an objective function $F(x, y)$ neither of whose variables x or y are confined to compact sets. This particular objective function is of interest because, in Chap. 8, it will be used to establish a link between necessary conditions of optimality for a general class of dynamic optimization problems related, on the one hand, to the classical Euler Lagrange conditions and, on the other, Hamilton's system of equation.

Proposition 3.6.7 *Take a lower semi-continuous, convex function $h : \mathbb{R}^n \to \mathbb{R} \cup \{+\infty\}$ such that $\text{dom } h \neq \emptyset$, a number $\sigma > 0$ and vectors $\bar{x} \in \mathbb{R}^n$ and $\bar{y} \in \mathbb{R}^n$.*

$$F(x, y) := x \cdot (y - \bar{y}) + \sigma |x - \bar{x}|^2 - h(y).$$

3.6 Mini-Max Theorems

Then there exists a point $(x^*, y^*) \in \mathbb{R}^n \times \text{dom } h$ such that

$$F(x^*, y) \leq F(x^*, y^*) \leq F(x, y^*) \text{ for all } x, y \in \mathbb{R}^n.$$

Furthermore

$$x^* = \bar{x} - \frac{1}{2}\sigma^{-1}(y^* - \bar{y}).$$

Proof Translating $(\bar{x}, \bar{y})$ to the origin and replacing the function h by $y \to \bar{x} \cdot y + h(y + \bar{y})$, we reduce consideration to the case when $\bar{x} = \bar{y} = 0$.

Define $D := \text{dom } h$. By a basic property of lower semi-continuous $\mathbb{R} \cup \{+\infty\}$ valued convex functions, h is minorized by an affine function. It follows that there exists $\alpha > 0$ and $\beta > 0$ such that

$$h(y) \geq -\alpha - \beta|y| \text{ for all } y \in \mathbb{R}^n. \tag{3.6.6}$$

Let $\psi : \mathbb{R}^n \to \mathbb{R} \cup \{-\infty\}$ be the function

$$\psi(y) := \inf_{x \in \mathbb{R}^n} F(x, y).$$

We explicity calculate

$$\psi(y) = -\frac{1}{4}\sigma^{-1}|y|^2 - h(y) \text{ for all } y \in \mathbb{R}^n.$$

Notice that $\psi(y') > -\infty$ for some $y' \in \mathbb{R}^n$ (since $D \neq \emptyset$). Also, by (3.6.6),

$$\limsup_{|y| \to \infty} \psi(y) \leq \limsup_{|y| \to \infty} \{-\frac{1}{4}\sigma^{-1}|y|^2 + \alpha + \beta|y|\} = -\infty.$$

It follows that the number d defined by

$$d := \sup_{y \in \mathbb{R}^n} \psi(y)$$

is finite and there exists $K > 0$ such that

$$\sup_{y \in \mathbb{R}^n} \psi(y) = \sup_{y \in K\mathbb{B} \cap D} \psi(y).$$

The right side is expressible as

$$\sup_{y \in K\mathbb{B} \cap D} \inf_{x \in \mathbb{R}^n} F(x, y).$$

Applying the one-sided mini-max theorem (Theorem 3.6.5), we obtain a point $y^* \in D \cap KB$ such that

$$\inf_{x \in \mathbb{R}^n} F(x, y^*) = \sup_{y \in K\mathbb{B} \cap D} \inf_{x \in \mathbb{R}^n} F(x, y) = \sup_{y \in D} \inf_{x \in \mathbb{R}^n} F(x, y). \qquad (3.6.7)$$

Now choose $k > 0$ such that

$$-(k - 2\beta - \frac{\alpha}{\sigma k})\sigma k < d - 1 \text{ and } \beta < k, \qquad (3.6.8)$$

and define, for each $y \in \mathbb{R}^n$,

$$\psi_k(y) := \inf_{x \in k B} F(x, y).$$

We easily calculate

$$\psi_k(y) = \begin{cases} \psi(y) & \text{if } |y| < 2\sigma k \\ -k|y| + \sigma k^2 - h(y) & \text{if } |y| \geq 2\sigma k. \end{cases}$$

Note however that, in view of (3.6.6) and (3.6.8),

$$\psi_k(y) \leq d - 1 = \sup_{y \in D} \psi(y) - 1 \text{ if } |y| \geq 2\sigma k.$$

Since ψ_k majorizes ψ and $\psi_k = \psi$ on $2\sigma k \mathbb{B}$, this last relation can be true only if

$$\sup_{y \in D} \psi(y) = \sup_{y \in D} \inf_{x \in k\mathbb{B}} F(x, y). \qquad (3.6.9)$$

Now apply the one-sided mini-nax theorem (Theorem 3.6.5) to

$$\inf_{x \in k\mathbb{B}} \sup_{y \in D} F(x, y).$$

This gives the existence of x^* such that

$$\sup_{y \in D} F(x^*, y) = \sup_{y \in D} \inf_{x \in k\mathbb{B}} F(x, y) = \sup_{y \in D} \inf_{x \in \mathbb{R}^n} F(x, y) \qquad (3.6.10)$$

by (3.6.9). But then, by Proposition 3.6.4,

$$\sup_{y \in D} \inf_{x \in \mathbb{R}^n} F(x, y) = \inf_{x \in \mathbb{R}^n} \sup_{y \in D} F(x, y).$$

We conclude from (3.6.7) that

$$\inf_{x \in \mathbb{R}^n} F(x, y^*) = \inf_{x \in \mathbb{R}^n} \sup_{y \in D} F(x, y). \qquad (3.6.11)$$

According to Proposition 3.6.3, assertions (3.6.10) and (3.6.11) imply that

$$\sup_{y \in D} F(x^*, y) = F(x^*, y^*) = \inf_{x \in \mathbb{R}^n} F(x, y^*).$$

This is the saddlepoint property.

Notice that, since $y^* \in D$, x^* minimizes the function

$$x \to F(x, y^*) = x \cdot y^* + \sigma |x|^2 - h(y^*)$$

over $\mathbb{R}^n$. It follows that the gradient of $F(., y^*)$ vanishes at $x = x^*$. We conclude that

$$x^* = -\frac{1}{2\sigma} y^*.$$

This is the final property to be verified and the proof is complete. □

3.7 Exercises

3.1 Use Ekeland's theorem to prove Caristi's fixed point theorem:
'Take a complete metric space V and a mapping $f : V \to V$, such that

$$d_V(u, f(u)) \le \phi(u) - \phi(f(u)) \quad \text{for all } u \in V$$

for some given lower semi-continuous, lower bounded function $\phi : V \to \mathbb{R}$. Then there exists $\bar{v} \in V$ such that $f(\bar{v}) = \bar{v}$.'
Hint: Apply Ekeland's theorem (Theorem 3.3.1) to the function ϕ, with $\alpha = \lambda = \frac{1}{2}$, to find a point v such that $\phi(w) \ge \phi(v) - \frac{1}{2} d_V(v, w)$ for all $w \in V$. Use this inequality, when $w = f(u)$, together with $d_V(u, f(u)) \le \phi(u) - \phi(f(u))$, to show $d(v, f(v)) = 0$.

3.2 Use Ekeland's theorem to prove Takahashi's minimization theorem:
'Let (X, d) be a complete metric space. Take a proper bounded below lower semi-continuous function $f : X \to \mathbb{R} \cup \{+\infty\}$, which satisfies the following property: for every $x \in \text{dom } f$ with $f(x) > \inf_{x' \in X} f(x')$ there exists a point $y \in \text{dom } f$, $y \ne x$, such that

$$f(y) + d(x, y) \le f(x).$$

Then f attains its minimum on X: there exists $\bar{x} \in X$ such that $f(\bar{x}) = \inf_{x' \in X} f(x')$.'

3.3 Let (X, d) be a complete metric space. Take a function $\phi : X \to X$. The *lower derivative* of ϕ at $x \in X$ in the direction of $y \in X$, written $\underline{D}\phi(x; y)$, is defined to be

$$\underline{D}\phi(x; y) := \begin{cases} 0 & \text{if } y = x \\ \liminf_{z \to x,\, z \in (x,y)} \frac{d(\phi(z), \phi(x))}{d(z,x)} & \text{otherwise} \end{cases}$$

where (x, y) denotes the *open interval* between x and y:

$$(x, y) := \{z \in X : z \neq x,\ z \neq y \text{ for all } d(x, z) + d(z, y) = d(x, y)\}.$$

The map ϕ is called a *weak directional contraction* on X if ϕ is continuous and there exists a number $\sigma \in [0, 1)$ such that

$$\underline{D}\phi(x; \phi(x)) \leq \sigma \quad \text{for all } x \in X.$$

Show that every weak directional contraction on X has a fixed point. (This fixed point theorem is due to Clarke.)

Hint: Consider the continuous function $f : X \to \mathbb{R}_+$ defined to be $f(x) := d(\phi(x), x)$, for all $x \in X$. Apply Ekeland's theorem (Theorem 3.3.1) to f and construct a sequence $\{x_k\}_{k \geq 1}$ in X such that $\underline{D}\phi(x_k; \phi(x_k)) \geq 1 - \frac{1}{k}$, for all $k \geq 1$.

3.8 Notes for Chapter 3

The important role of exact penalization in the derivation of necessary conditions for constrained optimization problems was early recognized by Clarke [60]. The proof of the exact penalization theorem, Theorem 3.2.1, is taken from [65], adapted to allow for an underlying space which is a (possibly incomplete) metric space in place of a Banach space.

Ekeland's variational principle was initially devised to show that approximate minimizers approximately satisfy necessary conditions of optimality [98]. Our proof is taken from [13]. An early application to dynamic optimization, which provided a pattern for later research into constrained optimization, was Clarke's derivation of necessary conditions for nonsmooth dynamic optimization problems with endpoint constraints [59]. Since then the principle has been put to many and diverse uses in nonlinear analysis, some of which are described in [99]. The proofs of both the Borwein and Preiss and Stegall theorems (deferred to the appendix of Chap. 4), based on properties of quadratic inf convolutions in Hilbert spaces are those given by Clarke et al. [85].

The inf convolution technique, also referred to as Moreau-Yoshida regularization, was introduced in the 1960s, to approximate a convex function by a continuously differentiable function whose derivatives are related to those of the original function.

3.8 Notes for Chapter 3

More recently, the properties of quadratic inf convolutions for non-convex problems have been used broadly in nonlinear analysis (including the proof of variational principles), the derivation of necessary conditions of optimality, Hamilton Jacobi analysis, Lyapunov theory and computational optimization.

The father of mini-max results is von Neumann. Mini-max theorems appear, at first sight, rather specialized affairs, because they concern saddle properties of (from some perspectives) the rather narrow class of convex-concave functions. However they are powerful tools in non-convex optimization, because they can be used to convert the non-negativity property of the first variation (which can be interpreted as a one-sided mini-max property) into a multiplier rule. The centrepiece of Sect. 3.6, from this point of view, is Aubin's one-sided mini-max theorem, which retains part of Von Neumann's mini-max theorem under reduced hypotheses and which is well suited to such applications. We reproduce the proof in [8].

Chapter 4
Nonsmooth Analysis

Abstract Local properties both of closed sets in topological vector spaces with smooth boundary and also of smooth functions on such spaces have traditionally been investigated via normal and tangent spaces (linear subspaces) and function derivatives (affine functions) respectively. This is no longer possible when the sets have nonsmooth boundaries and the functions are not differentiable in a traditional sense. Nonsmooth analysis provides techniques for the local approximation of closed sets with non-smooth boundaries and functions that are not differentiable. The key idea is to use, instead, cones and *families* of affine mappings in place of linear subspaces and affine mappings to achieve such approximations, in this more general setting. This and the following chapter provide a self-contained treatment of those aspects of nonsmooth analysis of special relevance to dynamic optimization. Key concepts are the limiting normal cone to a set (at a given base point) and the limiting subdifferential (at a point in the domain of the function). A number of approaches have been proposed. We follow Clarke in constructing the limiting normal cone as the cone comprising limits of proximal normal vectors at neighbouring points in the set. We work, for the most part, in the framework of real, finite dimensional vector spaces, though briefly describe how some concepts generalize to a Hilbert space setting.

This chapter introduces different kinds of normal cones to a set. Each of these constructs gives rise to a related concept of subdifferential of a function, defined via the normal cone to the epigraph the function. We then provide useful 'finite difference' representations of these subdifferentials and their asymptotic relatives. Properties of the limiting subdifferentials of locally Lipschitz functions are explored. Here, the distance function receives special attention. We also establish relations between the different kinds of normal and tangent cones that feature in the theory.

4.1 Introduction

Let $\bar{x} \in \mathbb{R}^k$ be a point in the manifold

$$C := \{x : g_i(x) = 0 \text{ for } i = 1, .., m\}$$

in which $g_i : \mathbb{R}^k \to \mathbb{R}, i = 1, .., m$ are given continuously differentiable functions such that $\nabla g_1(\bar{x}), \ldots, \nabla g_m(\bar{x})$ are linearly independent. Then the set of normal vectors to C at $\bar{x}$ is

$$\left\{\sum_{i=1}^{m} \lambda_i \nabla g_i(\bar{x}) : \lambda_1, \ldots, \lambda_m \in \mathbb{R}\right\},$$

and its orthogonal complement (translated to $\bar{x}$)

$$\bar{x} + \{y : y \cdot \nabla g_i(\bar{x}) = 0 \text{ for } i = 1, 2, \ldots, m\}$$

is an affine subspace of $\mathbb{R}^k$ which provides a local approximation to C 'near' $\bar{x}$.

If, on the other hand, we are given a continuously differentiable function $f : \mathbb{R}^k \to \mathbb{R}$ and a point $\bar{x}$, then

$$x \to f(\bar{x}) + \nabla f(\bar{x}) \cdot (x - \bar{x})$$

is an affine function which approximates f near $\bar{x}$.

We see here how local approximations to smooth manifolds and functions are traditionally constructed: they take the form of affine subspaces for smooth manifolds and of affine functions for smooth functions. The importance of these approximations is that it is often possible to predict qualitative properties of smooth manifolds and functions from properties of their 'affine' approximations, which are in almost all cases simpler to investigate. A case in point is the inverse function theorem, which tells us that a continuously differentiable function $f : \mathbb{R}^k \to \mathbb{R}^k$ is invertible on a neighbourhood of a point $\bar{x} \in \mathbb{R}^k$ if its affine approximation $x \to f(\bar{x}) + \nabla f(\bar{x}) \cdot (x - \bar{x})$ is invertible.

What general principles and techniques may be developed, governing the approximation of sets and functions in topological vector spaces when the regularity and differentiability hypotheses underpinning traditional methods are violated? These are precisely the questions that nonsmooth analysis aims to answer.

The key idea is to abandon the notion of affine approximation. In nonsmooth analysis, closed sets (and in particular manifolds) are approximated, not by affine subspaces, but by *cones* and functions are approximated, not by a single affine function, but by a *family* of affine functions.

Once this idea is accepted, it is possible to extend to a nonsmooth setting many principles of traditional nonlinear analysis and to assemble a calculus of rules for estimating approximations to specific sets and functions of interest. There has been a great deal of research activity in nonsmooth analysis since 1980s. A sophisticated and far-reaching theory has now been put together, the significance of which is confirmed by a growing body of applications in mathematical programming, variational analysis, dynamic optimization and other fields of applied nonlinear analysis.

Our aim in this and the next chapter is to provide a self-contained treatment of those aspects of nonsmooth analysis of particular relevance to dynamic optimization. In this chapter, basic constructs of nonsmooth analysis are defined and relations between them are explored. In the next, a 'generalized calculus' is developed.

4.2 Normal Cones

Take a closed set $C \subset \mathbb{R}^k$ and a point $x \in C$. We wish to give meaning to 'normals' to C at x, that is vectors which in some sense point out of C from the 'basepoint' x. We introduce three notions of normals: proximal normals, strict normals and limiting normals.

Proximal normals are vectors which satisfy an outward pointing condition with quadratic error term. Proximal normals have a simple geometric interpretation and, for this reason, are often easy to characterize in specific applications. Strict normals are similarly defined, except an 'order' term replaces the quadratic error term in the defining condition. Limiting normals, as their name suggests, are limits of proximal normals at neighbouring basepoints; they can be equivalently defined as limits of strict normals at neighbouring basepoints, as we shall see.

Limiting normals feature most prominently in the applications to optimization explored in this book, owing to their superior analytic properties. Many of these properties are consequences of the fact that the cone of limiting normals has closed graph, regarded as a set valued function of the basepoint. Proximal normals have the role of building blocks of limiting normals. Strict normals provide an alternative route to generating limiting normals via limit taking.

Definition 4.2.1 Given a closed set $C \subset \mathbb{R}^k$ and a point $x \in C$, the *proximal normal cone* to C at x, written $N_C^P(x)$, is the set

$$N_C^P(x) := \{p \in \mathbb{R}^k : \exists\, M > 0 \text{ such that condition (4.2.1) is satisfied}\}$$

$$p \cdot (y - x) \leq M|y - x|^2, \text{ for all } y \in C. \tag{4.2.1}$$

Elements in $N_C^P(x)$ are called *proximal normals* to C at x.

The defining condition (4.2.1) for a proximal normal, which is referred to as the 'proximal normal inequality for cones', can be expressed as follows: there exists $M > 0$ such that

$$|(x + (2M)^{-1}p) - y| \geq (2M)^{-1}|p|, \text{ for all } y \in C. \tag{4.2.2}$$

In terms of $r = 2M$ and $z = x + (2M)^{-1}p$, the condition amounts to:

$$|z - x| = \min\{|z - y| : y \in C\} \text{ and } p = r(z - x).$$

These observations lead to the following geometric interpretation of proximal normals.

Proposition 4.2.2 *Take a closed set $C \subset \mathbb{R}^k$ and points $x \in C$ and $p \in \mathbb{R}^k$. Then p is a proximal normal to C at x if and only if there exist a point $z \in \mathbb{R}^k$ and a scaling factor $r > 0$ such that*

$$|z - x| = \min\{|z - y| : y \in C\} \text{ and } p = r(z - x).$$

This proposition tells us that p is a non-zero proximal normal to C at x precisely when there is some $z \in \mathbb{R}^k$ lying outside C such that x is the closest point in C to z (with respect to the Euclidean distance function) and the vector p points in the direction $(z - x)$. See *Fig. 1.10*.

The derivation of many useful relations involving normal cones to a set C at x involves limit taking with respect to the basepoint x. This is often not possible if the cones in question are proximal normal cones, because membership of proximal normal cones is not in general preserved under such operations. We can construct a normal cone from the proximal normal cone which does have the desired closure properties with respect to changes of the basepoint, by adding extra normals which are limits of proximal normals as nearby basepoints. In this way we arrive at the limiting normal cone. See *Fig. 1.11*.

Definition 4.2.3 Given a closed set $C \subset \mathbb{R}^k$ and a point $x \in C$, the *limiting normal cone* to C at x, written $N_C(x)$, is the set

$$N_C(x) := \left\{ p : \text{ there exists } x_i \xrightarrow{C} x, p_i \to p \text{ such that } p_i \in N_C^P(x_i) \text{ for all } i \right\}$$

Elements in $N_C(x)$ are called *limiting normals* to C at x.

Another normal cone of interest is the strict normal cone:

Definition 4.2.4 Given a closed set $C \subset \mathbb{R}^k$ and a point $x \in C$, the strict normal cone to C at x, written $\hat{N}_C(x)$, is the set

$$\hat{N}_C(x) = \left\{ p : \limsup_{\substack{y \to x \\ y \xrightarrow{C}}} \frac{p \cdot (y - x)}{|y - x|} \leq 0 \right\}.$$

Elements in $\hat{N}_C(x)$ are called *strict normals* to C at x.

Otherwise expressed, a strict normal p to C at x is a vector satisfying

$$p \cdot (y - x) \leq o(|y - x|) \text{ for all } y \in C, \tag{4.2.3}$$

4.2 Normal Cones

for some function $o : \mathbb{R}_+ \to \mathbb{R}_+$ such that $\lim_{\epsilon \downarrow 0} o(\epsilon)/\epsilon = 0$; in other words p satisfies (to 'within first order') conditions asserting that it is an outward normal to a hyperplane supporting C at x. Notice that the defining relation (4.2.3) for strict normals is weaker than the defining relation (4.2.1) for proximal normals: for a strict normal p also to be a proximal normal, we stipulate that a special form of error modulus $o(\cdot)$ can be used, namely a quadratic function. It follows that

$$N_C^P(x) \subset \hat{N}_C(x).$$

Proximal normal cones and strict normal cones are distinct concepts: it is not difficult to supply examples in which

$$N_C^P(x) \overset{\text{strict}}{\subset} \hat{N}_C(x).$$

However limits of strict normals at basepoints converging to x generate the *same* set ($N_C(x)$) as limits of proximal normals. It is for this reason that strict normals, in place of proximal normals, provide an alternative foundation on which to define and establish properties of limiting normal cones.

Proposition 4.2.5 *Take a closed set $C \subset \mathbb{R}^k$ and points $x \in C$ and $p \in \mathbb{R}^k$. Then the following assertions are equivalent:*

(i) $p \in N_C(x)$,
(ii) *there exist sequences $x_i \overset{C}{\to} x$ and $p_i \to p$ such that $p_i \in \hat{N}_C(x_i)$ for all i.*

Proof (i) implies (ii), by the definition of $N_C(x)$ and since $N_C^P(x) \subset \hat{N}_C(x)$.

To establish that (ii) implies (i), we show that strict normals can be suitably approximated by proximal normals. Fix $q \in \hat{N}_C(y)$ for some $y \in C$. It suffices to demonstrate that sequences $y_i \overset{C}{\to} y$ and $q_i \to q$ may be found such that $q_i \in N_C^P(y_i)$ for all i. For then we can show, by constructing a suitable diagonal sequence, that if a vector p is the limit of a sequence of strict normals at neighbouring points, then it is also the limit of proximal normals; that is, it lies in the limiting normal cone.

Take any $\epsilon_i \downarrow 0$. Then for each i, we can find a point y_i which is a closest point to $(y + \epsilon_i q)$ in C. Since $\epsilon_i q \to 0$, we have that $y_i \to y$. For each i define $q_i := \epsilon_i^{-1}(y + \epsilon_i q - y_i)$. Then $q_i \in N_C(y_i)$. We see that $q_i = q + \epsilon_i^{-1}(y - y_i)$, so the proof will be complete if we can show that $\epsilon_i^{-1}(y - y_i) \to 0$. However by the 'closest point' property of y_i,

$$|(y + \epsilon_i q) - y_i|^2 \leq |(y + \epsilon_i q) - z|^2, \text{ for all } z \in C.$$

Choosing $z = y$ we arrive at

$$0 \geq |y_i - (y + \epsilon_i q)|^2 - \epsilon_i^2 |q|^2 = |y_i - y|^2 - 2\epsilon_i (y_i - y) \cdot q.$$

But $q \in \hat{N}_C(y)$. It follows that

$$|y_i - y|^2 \leq 2\epsilon_i(y_i - y) \cdot q \leq 2\epsilon_i o(|y_i - y|)$$

for some function $o : \mathbb{R}_+ \to \mathbb{R}_+$ such that $\lim_{\epsilon \downarrow 0} o(\epsilon)/\epsilon = 0$. We conclude that $\epsilon_i^{-1}|y_i - y| \to 0$. □

Some elementary properties of the cones which have been introduced are now listed without proof.

Proposition 4.2.6 *Take a closed set $C \subset \mathbb{R}^k$ and a point $x \in C$. Then:*

(i) *$N_C^P(x)$, $\hat{N}_C(x)$ and $N_C(x)$ are all cones in $\mathbb{R}^k$, containing $\{0\}$ and*

$$N_C^P(x) \subset \hat{N}_C(x) \subset N_C(x),$$

(ii) *$N_C^P(x)$ is convex (but possibly not closed),*
(iii) *$\hat{N}_C(x)$ is closed and convex,*
(iv) *the set valued mapping $N_C(.) : C \rightsquigarrow \mathbb{R}^k$ has closed graph, in the sense that, for any sequences $y_i \xrightarrow{C} y$ and $p_i \to p$ such that $p_i \in N_C(y_i)$ for all i, we have $p \in N_C(x)$.*

We note also

Proposition 4.2.7 *Take a closed set $C \subset \mathbb{R}^k$ and a point $x \in C$. Then:*

(i) *$x \in \text{int}\{C\}$ implies $N_C(x) = \{0\}$ and hence*

$$N_C^P(x) = \hat{N}_C(x) = \{0\}.$$

(ii) *$x \in \text{bdy}\{C\}$ implies $N_C(x)$ contains non-zero elements.*

Proof (i) Suppose $x \in \text{int}\{C\}$. It follows from Proposition 4.2.2 that all proximal normals at points in C near to x are zero. So $N_C(x) = \{0\}$.

(ii) Assume $x \in \text{bdy}\{C\}$. Then there exists a sequence $x_i \to x$ such that $x_i \notin C$ for all i. For each i select a closest point c_i to x_i in C. Then $x_i - c_i \neq 0$ and $\xi_i := |x_i - c_i|^{-1}(x_i - c_i)$ is a proximal normal to C at c_i, of unit norm. But $|x - c_i| \leq |x - x_i| + |x_i - c_i| \leq 2|x - x_i|$. So $c_i \xrightarrow{C} x$. Along a subsequence then $\xi_i \to \xi$. The vector ξ has unit norm and belongs to $\xi \in N_C(x)$. □

We note also

Proposition 4.2.8 *Take closed subsets $C_1 \subset \mathbb{R}^m$ and $C_2 \subset \mathbb{R}^n$, and a point $(x_1, x_2) \in C_1 \times C_2$. Then*

$$N_{C_1 \times C_2}^P(x_1, x_2) = N_{C_1}^P(x_1) \times N_{C_2}^P(x_2)$$

$$\hat{N}_{C_1 \times C_2}(x_1, x_2) = \hat{N}_{C_1}(x_1) \times \hat{N}_{C_2}(x_2)$$

$$N_{C_1 \times C_2}(x_1, x_2) = N_{C_1}(x_1) \times N_{C_2}(x_2).$$

4.2 Normal Cones

Proof The first two relations are direct consequences of the definitions of proximal normals and strict normals. The last relation is a consequence of the first relation and the manner in which limiting normals are obtained as limits of proximal normals. □

In the convex case, the normal cones above all coincide with the normal cone in the sense of convex analysis.

Proposition 4.2.9 *Take a closed, convex set $C \subset \mathbb{R}^k$ and a point $\bar{x} \in C$. Then*

$$N_C^P(\bar{x}) = \hat{N}_C(\bar{x}) = N_C(\bar{x})$$
$$= \{\xi : \xi \cdot (x - \bar{x}) \leq 0 \text{ for all } x \in C\}.$$

Proof For $y \in C$ write

$$S(y) := \{\xi : \xi \cdot (y' - y) \leq 0 \text{ for all } y' \in C\}.$$

Since the conditions for membership of $S(\bar{x})$ are more severe than those for membership of $N_C^P(\bar{x})$, we know that

$$S(\bar{x}) \subset N_C^P(\bar{x}).$$

Take an arbitrary element $\xi \in N_C^P(\bar{x})$. By definition, there exists $M \geq 0$ such that

$$\xi \cdot (x' - \bar{x}) \leq M|x' - \bar{x}|^2,$$

for all $x' \in C$. Choose any $x \in C$. Since C is convex, $x' = \epsilon x + (1 - \epsilon)\bar{x} \in C$ for any $\epsilon \in (0, 1)$. Inserting this choice of x' into the above inequality, dividing across by ϵ and passing to the limit as $\epsilon \downarrow 0$ gives

$$\xi \cdot (x - \bar{x}) \leq 0.$$

It follows that $\xi \in S(\bar{x})$. This establishes that $S(\bar{x}) \supset N_C^P(\bar{x})$. We conclude that

$$S(\bar{x}) = N_C^P(\bar{x}). \tag{4.2.4}$$

Now, take sequences $x_i \xrightarrow{C} \bar{x}$ and $\xi_i \to \xi$ such that $\xi_i \in S(x_i)$ for all i. It is easy to deduce that $\xi \in S(\bar{x})$. It follows from (4.2.4) and the definition of $N_C(\bar{x})$ that

$$N_C(\bar{x}) \subset S(\bar{x}).$$

The remaining relations in the proposition to be proved follow from Proposition 4.2.6 (i). □

4.3 Subdifferentials

Consider a function $f : \mathbb{R}^k \to \mathbb{R}$ and a point $x \in \mathbb{R}^k$. Suppose that f is C^2 (twice continuously differentiable). Then the gradient $\nabla f(x)$ of f at x is the unique vector $\xi \in \mathbb{R}^k$ such that, for some $M > 0$ and $\epsilon > 0$,

$$|f(y) - f(x) - \xi \cdot (y - x)| \leq M|y - x|^2 \text{ for all } y \in x + \epsilon \mathbb{B}. \tag{4.3.1}$$

This description evokes the familiar 'difference quotient' characterization of the gradient. The fact that we assume f is a C^2 function permits us to use a quadratic error term on the right side.

An alternative description of $\nabla f(x)$ is in terms of the epigraph of f, epi f,

$$\text{epi } f := \{(x, \alpha) \in \mathbb{R}^k \times \mathbb{R} : \alpha \geq f(x)\}.$$

It can be shown that $\nabla f(x)$ is the unique vector $\xi \in \mathbb{R}^k$ such that, for some $M \geq 0$,

$$(\xi, -1) \cdot ((y, \alpha) - (x, f(x))) \leq M(|y - x|^2 + |\alpha - f(x)|^2) \tag{4.3.2}$$

for all $(y, \alpha) \in$ epi f. (By considering a first order Taylor expansion of f around x, with quadratic remainder term, we can show that, for some $M > 0$, this inequality is satisfied 'locally' when $\xi = \nabla f(x)$; the 'extra' error term $|\alpha - f(x)|^2$ ensures that the inequality holds for all y, not just y's near x). To see that this last condition does indeed identify $\nabla f(x)$, we note that $(y, f(y)) \in$ epi f. Hence

$$0 \leq \liminf_{y \to x} \frac{f(y) - f(x) - \xi \cdot (y - x)}{|y - x|}$$
$$= \liminf_{y \to x} \left[\frac{f(y) - f(x) - \nabla f(x) \cdot (y - x)}{|y - x|} + \frac{\nabla f(x) - \xi) \cdot (y - x)}{|y - x|} \right]$$
$$= 0 + \liminf_{y \to x} \frac{(\nabla f(x) - \xi) \cdot (y - x)}{|y - x|}$$
$$= -|\nabla f(x) - \xi|.$$

It follows that $\xi = \nabla f(x)$.

The reasons for introducing relation (4.3.2) may not be immediately apparent. All becomes clear however when it is interpreted in terms of normal cones: (4.3.2) can be equivalently stated

$$(\xi, -1) \in N^P_{\text{epi } f}(x, f(x)), \tag{4.3.3}$$

where, we recall, $N^P_{\text{epi } f}(x, f(x))$ denotes the proximal normal cone.

Up to this point, we have discussed different ways of looking at the gradient of a C^2 function f. Our ulterior motive of course has been to come up with

4.3 Subdifferentials

some description which makes sense when f is no longer 'differentiable' in the conventional sense.

Two routes are open to us, to generalize either (4.3.1) (the analytical approach) or (4.3.3) (the geometric approach). As they stand, conditions (4.3.1) and (4.3.3) are often rather restrictive: even for Lipschitz continuous functions f there will be many points x at which the sets of ξ's satisfying either (4.3.1) or (4.3.3) will be empty. It is therefore often helpful to play variations on these themes, namely to consider limits of ξ_i's satisfying (4.3.1) along a sequence of points $x_i \to x$, or to replace the quadratic error term $M|y - x|^2$ by an error modulus $o(|y - x|)$ (if we are going down the analytical road), or to insert different kinds of normal cones in place of $N^P_{\text{epi } f}(x, f(x))$ (if we want to take a more geometric approach).

Different choices are available regarding what we deem to be definitions and what to be consequences of the definitions. The theory would appear to unfold most simply when the basic definitions are given in terms of normal cones to epigraphs, and this therefore is our chosen approach.

To what class of functions should we aim to generalize the traditional notion of derivatives? The class of lower semi-continuous functions $f : \mathbb{R}^k \to \mathbb{R} \cup \{+\infty\}$ is a natural choice here. This is because generalized derivatives will be defined in terms of normal cones to the epigraph set epi f, so we would like to arrange that the epigraph set is closed. The class of lower semi-continuous functions is precisely the class of functions with closed epigraph sets.

Definition 4.3.1 Take a lower semi-continuous function $f : \mathbb{R}^k \to \mathbb{R} \cup \{+\infty\}$ and a point $x \in \text{dom } f := \{y \in \mathbb{R}^k : f(y) < +\infty\}$.

(i) The *proximal subdifferential* of f at x, written $\partial^P f(x)$, is the set

$$\partial^P f(x) := \{\xi : (\xi, -1) \in N^P_{\text{epi } f}(x, f(x))\}.$$

Elements in $\partial^P f(x)$ are called *proximal subgradients*.

(ii) The *strict subdifferential* of f at x, written $\hat{\partial} f(x)$, is the set

$$\hat{\partial} f(x) := \{\xi : (\xi, -1) \in \hat{N}_{\text{epi } f}(x, f(x))\}.$$

Elements in $\hat{\partial} f(x)$ are called *strict subgradients*.

(iii) The *limiting subdifferential* of f at x, written $\partial f(x)$, is the set

$$\partial f(x) := \{\xi : (\xi, -1) \in N_{\text{epi } f}(x, f(x))\}.$$

Elements in $\partial f(x)$ are called *limiting subgradients*.

Proximal subgradients and limiting subdifferentials are illustrated in *Fig. 1.12* and *Fig. 1.13* respectively.

Notice that, since

$$N_C^P(x) \subset \hat{N}_C(x) \subset N_C(x),$$

we have

$$\partial^P f(x) \subset \hat{\partial} f(x) \subset \partial f(x).$$

These different subdifferentials provide local information about a function f near a point x. The most noteworthy departure from classical real analysis is that, for lower semi-continuous functions, the subdifferentials are in general *set valued*.

Example

In each of the following cases, f is a function from $\mathbb{R}$ to $\mathbb{R}$.

(i) $f(y) = |y|$ and $x = 0$. Then

$$\partial^P f(x) = \hat{\partial} f(x) = \partial f(x) = [-1, +1].$$

(ii) $f(y) = -|y|$ and $x = 0$. Then

$$\partial^P f(x) = \hat{\partial} f(x) = \emptyset \text{ and } \partial f(x) = \{-1\} \cup \{+1\}.$$

(iii) $f(y) = |y|^{1/2}$ and $x = 0$. Then

$$\partial^P f(x) = \hat{\partial} f(x) = \partial f(x) = (-\infty, +\infty).$$

(iv) $f(y) = \text{sgn}\{y\}|y|^{1/2}$ and $x = 0$. Then

$$\partial^P f(x) = \hat{\partial} f(x) = \partial f(x) = \emptyset.$$

Unbounded or empty subdifferentials (as occur in (iii) and (iv)) give warning that the slopes of the function on an arbitrary small ϵ-ball about the base point are unbounded. In the presence of such pathologies, we would like to know the *direction* of these arbitrarily large slopes. Information of this nature is conveyed by some new constructs; they are the 'asymptotic' analogues of the subdifferentials already defined.

Definition 4.3.2 Take a lower semi-continuous function $f : \mathbb{R}^k \to \mathbb{R} \cup \{+\infty\}$ and a point $x \in \text{dom } f$.

(i) The *asymptotic proximal subdifferential* of f at x, written $\partial_P^\infty f(x)$, is

$$\partial_P^\infty f(x) := \{\xi : (\xi, 0) \in N_{\text{epi } f}^P(x, f(x))\}.$$

Elements in $\partial_P^\infty f(x)$ are called *asymptotic proximal subgradients*.

4.3 Subdifferentials

(ii) The *asymptotic strict subdifferential* of f at x, written $\hat{\partial}^\infty f(x)$, is

$$\hat{\partial}^\infty f(x) := \{\xi : (\xi, 0) \in \hat{N}_{\text{epi } f}(x, f(x))\}.$$

Elements in $\hat{\partial}^\infty f(x)$ are called *asymptotic strict subgradients*.

(iii) The *asymptotic limiting subdifferential* of f at x, written $\partial^\infty f(x)$, is

$$\partial^\infty f(x) := \{\xi : (\xi, 0) \in N_{\text{epi } f}(x, f(x))\}.$$

Elements in $\partial^\infty f(x)$ are called *asymptotic limiting subgradients*.

Example
Consider the function $f : \mathbb{R} \to \mathbb{R}$ given by $f(y) = \text{sgn}\{y\}|y|^{1/2}$ and the point $x = 0$. As we have noted $\partial f(x) = \emptyset$. However $\partial^\infty f(x) = [0, \infty)$. The calculation of $\partial^\infty f(x)$ has revealed to us that the 'infinite' slopes near x are positive.

The asymptotic limiting subdifferential $\partial^\infty f(x)$ is of interest also because of the information that non-zero asymptotic limiting subgradients give about 'non-Lipschitz behaviour' of a function f near a point $x \in \text{dom } f$. It is to be expected then that the asymptotic limiting subdifferential of a Lipschitz continuous function is the trivial set $\{0\}$.

Proposition 4.3.3 *Take a lower semi-continuous function $f : \mathbb{R}^k \to \mathbb{R} \cup \{+\infty\}$ and a point $\bar{x} \in \mathbb{R}^k$. Assume that f is Lipschitz continuous on a neighbourhood of $\bar{x}$ with Lipschitz constant K. Then:*

(i) $\partial f(\bar{x}) \subset K\mathbb{B}$,
(ii) $\partial^\infty f(\bar{x}) = \{0\}$.

Proof Take a point

$$(\xi, -\lambda) \in N_{\text{epi } f}(\bar{x}, f(\bar{x})).$$

Then there exist convergent sequences

$$(x_i, \alpha_i) \stackrel{\text{epi } f}{\to} (\bar{x}, f(\bar{x})) \text{ and } (\xi_i, -\lambda_i) \to (\xi, -\lambda)$$

such that

$$(\xi_i, -\lambda_i) \in N^P_{\text{epi } f}(x_i, \alpha_i) \text{ for all } i \text{ sufficiently large,}$$

and such that f is Lipschitz continuous on a neighbourhood of x_i with Lipschitz constant K. □

We know that $\alpha_i = f(x_i) + \beta_i$ for some $\beta_i \geq 0$. It follows now from the definition of proximal subgradients that

$$\xi_i \cdot (x - x_i) - \lambda_i(\beta + f(x) - \beta_i - f(x_i))$$
$$\leq M(|x - x_i|^2 + |\beta - \beta_i + f(x) - f(x_i)|^2)$$

for all $\beta \geq 0$ and $x \in \mathbb{R}^n$.

Setting $x = x_i$ we deduce

$$-\lambda_i(\beta - \beta_i) \leq M|\beta - \beta_i|^2 \text{ for all } \beta \geq 0.$$

This implies $\lambda_i \geq 0$. Passing to the limit as $i \to \infty$, we obtain $\lambda \geq 0$.

Setting $\beta = \beta_i$ we deduce from the Lipschitz continuity of f that

$$\xi_i \cdot (x - x_i) \leq \lambda_i K|x - x_i| + (M + MK^2)|x - x_i|^2$$

for all x in a neighbourhood of x_i. Choosing $x = x_i + r\xi_i$ for $r > 0$ sufficiently small gives

$$r|\xi_i|^2 \leq rK\lambda_i|\xi_i| + M(1 + K^2)r^2|\xi_i|^2.$$

Dividing across by r and passing to the limit as $r \downarrow 0$ gives

$$|\xi_i|^2 \leq K\lambda_i|\xi_i|.$$

We deduce that

$$|\xi_i| \leq K\lambda_i \text{ for each } i.$$

Passing to the limit as $i \to \infty$, we arrive at

$$|\xi| \leq K\lambda.$$

It follows that $\partial^\infty f(\bar{x}) = \{0\}$ and $\partial f(\bar{x}) \subset K\mathbb{B}$. □

A converse of the above proposition is proved later in the chapter (Theorem 4.9.1 and Corollary 4.9.2).

Limiting subgradients and asymptotic limiting subgradients are defined in terms of the limiting normal cone to the epigraph set. Taking a reverse view, we can represent the limiting normal cone to the epigraph of a function by means of the subgradients of the function.

Proposition 4.3.4 *Take a lower semi-continuous function $f : \mathbb{R}^k \to \mathbb{R} \cup \{+\infty\}$ and a point $x \in \text{dom } f$. Then:*

$$N_{\text{epi } f}(x, f(x)) = \{(\lambda\xi, -\lambda) : \lambda > 0, \; \xi \in \partial f(x)\} \cup (\partial^\infty f(x) \times \{0\}).$$

Either $\partial f(x)$ is non-empty or $\partial^\infty f(x)$ contains non-zero elements.

4.3 Subdifferentials

Proof The first part of the proposition merely describes how the limiting subdifferential and asymptotic limiting subdifferential are constructed from the limiting normal cone. Since $(x, f(x))$ is a boundary point of epi f, $N_{\text{epi } f}(x, f(x))$ contains non-zero elements (see Proposition 4.2.7). The fact that either $\partial f(x)$ is non-empty or $\partial^\infty f(x)$ contains non-zero elements is an immediate consequence. □

The following 'closure' properties of the limiting subdifferential and the asymptotic limiting subdifferential follow from their representation in terms of $N_{\text{epi } f}(x, f(x))$ and from the fact that the set valued function $N_{\text{epi } f} : \text{epi } f \rightsquigarrow \mathbb{R}^k \times \mathbb{R}$ has closed graph.

Proposition 4.3.5 *Take a lower semi-continuous function $f : \mathbb{R}^k \to \mathbb{R} \cup \{+\infty\}$ and a point $x \in \text{dom } f$. Then:*

(i) *$\partial f(x)$ is a closed set. Given sequences $x_i \xrightarrow{f} x$ and $\xi_i \to \xi$ such that $\xi_i \in \partial f(x_i)$ for all i, then $\xi \in \partial f(x)$,*

(ii) *$\partial^\infty f(x)$ is a closed cone. Given sequences $x_i \xrightarrow{f} x$ and $\xi_i \to \xi$ such that $\xi_i \in \partial^\infty f(x_i)$ for all i, then $\xi \in \partial^\infty f(x)$.*

Finally we note that, for convex functions, the definitions of proximal, strict and limiting subdifferential all coincide with the subdifferential in the sense of convex analysis.

Proposition 4.3.6 *Take a lower semi-continuous, convex function $f : \mathbb{R}^k \to \mathbb{R} \cup \{+\infty\}$ and a point $\bar{x} \in \text{dom } f$. Then*

$$\partial^P f(\bar{x}) = \hat{\partial} f(\bar{x}) = \partial f(\bar{x})$$
$$= \{\xi : \xi \cdot (x - \bar{x}) \leq f(x) - f(\bar{x}) \text{ for all } x \in \mathbb{R}^k\}.$$

Proof Define

$$D(\bar{x}) = \{\xi : \xi \cdot (x - \bar{x}) \leq f(x) - f(\bar{x}) \text{ for all } x \in \mathbb{R}^k\}.$$

It is easy to see that $D(\bar{x})$ can be alternatively expressed in terms of the normal cone (in the sense of convex analysis) to epi f at $(\bar{x}, f(\bar{x}))$:

$$D(\bar{x}) = \{\xi : (\xi, -1) \cdot ((x, \alpha) - (\bar{x}, f(\bar{x}))) \leq 0 \text{ for all } (x, \alpha) \in \text{epi } f\}.$$

All the assertions of the proposition now follow from Proposition 4.2.9 and the definitions of the proximal, strict and limiting normal cones. □

4.4 Difference Quotient Representations

Subgradients have been defined, somewhat indirectly, via normals to epigraph sets. This approach has certain advantages from the point of view of unifying the treatment of normal vectors and subgradients and of deriving a number of important properties of subgradients. This section provides alternative conditions for vectors to be proximal subgradients (and strict subgradients), in terms of limits of difference quotients. These often simplify the task of investigating the detailed properties of subgradients in specific cases.

We start with a very useful characterization of proximal subgradients.

Proposition 4.4.1 *Take a lower semi-continuous function $f : \mathbb{R}^k \to \mathbb{R} \cup \{+\infty\}$ and points $x \in \text{dom } f$ and $\xi \in \mathbb{R}^k$. Then the following two statements are equivalent:*

(i) $\xi \in \partial^P f(x)$,
(ii) there exist $M > 0$ and $\epsilon > 0$ such that

$$\xi \cdot (y - x) \leq f(y) - f(x) + M|y - x|^2 \qquad (4.4.1)$$

for all $y \in x + \epsilon \mathbb{B}$.

The significance of condition (ii) (which is referred to as the proximal normal inequality for functions) is the absence of the term $M|f(y) - f(x)|^2$ from the right side of inequality (4.4.1), a term which you would expect to be required to provide a 'difference quotient' characterization of proximal subgradient. In situations when f is not continuous this troublesome error term complicates the analysis of limits of proximal normal vectors. Notice that (4.4.1) is required to hold only locally.

Proof Assume (ii). Inequality (4.4.1) implies

$$\xi \cdot (y - x) \leq \alpha - f(x) + M[|y - x|^2 + |\alpha - f(x)|^2]$$

for any $(y, \alpha) \in \text{epi } f \cap [(x, f(x)) + \epsilon \mathbb{B}]$. The inequality can be written

$$(\xi, -1) \cdot ((y, \alpha) - (x, f(x))) \leq M|(y, \alpha) - (x, f(x))|^2. \qquad (4.4.2)$$

The geometric interpretation of this relation is that the open ball, with centre $(x, f(x)) + (2M)^{-1}(\xi, -1)$ and of radius $(2M)^{-1}|(\xi, -1)|$, is disjoint from epi $f \cap [(x, f(x)) + \epsilon \mathbb{B}]$, and the closure of this ball contains $(x, f(x))$. (See Proposition 4.2.2 and preceding comments). So for $\tilde{M} \geq M$ sufficiently large, when we substitute $\tilde{M}$ in place of M, this open ball is contained in $(x, f(x)) + \epsilon \overset{\circ}{\mathbb{B}}$, and is consequently disjoint from epi f. (We denote by $y + r\overset{\circ}{\mathbb{B}}$ the open ball in $\mathbb{R}^k$ with centre y and of radius $r > 0$.) Inequality (4.4.2), which is valid when M is replaced by the larger $\tilde{M}$, then tells that

4.4 Difference Quotient Representations

$$\Big((x, f(x)) + (2\tilde{M})^{-1}(\xi, -1) + (2\tilde{M})^{-1}|(\xi, -1)|\mathbb{B}\Big) \cap \text{epi } f = \emptyset.$$

Again appealing to the geometric interpretation of proximal normals, we deduce that $(\xi, -1) \in N^P_{\text{epi } f}(x, f(x))$. We have shown that (i) is true.

Assume (i). We may suppose that $(x, f(x)) = (0, 0)$ (this amounts simply to translating the origin in $\mathbb{R}^k \times \mathbb{R}$ to $(x, f(x))$).

By assumption there exists $M > 0$ such that

$$(\xi, -1) \cdot (y, \alpha) \leq M(|y|^2 + |\alpha|^2) \text{ for all } (y, \alpha) \in \text{epi } f.$$

Set $\lambda = (2M)^{-1}$.

According to Proposition 4.2.2 and the preceding discussion $(0, 0) + \{s + \rho \overset{\circ}{\mathbb{B}}\}$ is disjoint from epi f for

$$s = (s_1, s_2) := (\lambda\xi, -\lambda) \text{ and } \rho := \lambda\sqrt{1 + |\xi|^2}.$$

For each $x \in s_1 + \rho\overset{\circ}{\mathbb{B}}$ consider the vertical line $\{x\} \times \mathbb{R}$. This penetrates the ball $s + \rho\overset{\circ}{\mathbb{B}}$. Define the functionals $\phi^+(x)$ and $\phi^-(x)$ to take values the vertical coordinates of the points at which this line intercepts the upper and lower hemispheres respectively.

For each $x \in s_1 + \rho\overset{\circ}{\mathbb{B}}$, $\phi^+(x) > \phi^-(x)$ and consequently

$$\{x\} \times (\phi^-(x), \phi^+(x))$$

is a non-empty subset of $s + \rho\overset{\circ}{\mathbb{B}}$. But $\{x\} \times [f(x), \infty) \subset \text{epi } f$. Since epi f and $s + \rho\overset{\circ}{\mathbb{B}}$ are disjoint, we conclude that

$$f(x) \geq \phi^+(x) \text{ for all } x \in s_1 + \rho\overset{\circ}{\mathbb{B}}. \tag{4.4.3}$$

In particular this inequality holds for all x in the neighbourhood $\epsilon'\overset{\circ}{\mathbb{B}}$ of the origin, in which ϵ' is the positive number

$$\epsilon' := \frac{1}{2}(\lambda\sqrt{1 + |\xi|^2} - \lambda|\xi|).$$

A simple calculation yields the following formula for $\phi^+(x)$:

$$\phi^+(x) = \sqrt{\lambda^2 + \lambda^2|\xi|^2 - |x - \lambda\xi|^2} - \lambda$$

for $x \in \epsilon'\overset{\circ}{\mathbb{B}}$. $\phi^+(x)$ is analytic on $\epsilon'\overset{\circ}{\mathbb{B}}$. The gradient $\nabla \phi^+(0)$ and Hessian $\nabla^2 \phi^+(0)$ at the origin are

$$\nabla \phi^+(0) = \xi$$
$$\nabla^2 \phi^+(0) = -\lambda^{-1}(I + \xi\xi^T).$$

Since $\lambda = (2M)^{-1}$ and $\phi^+(0) = 0$, we have

$$\phi^+(x) = \phi^+(0) + \nabla \phi^+(0) \cdot x + \frac{1}{2} x \cdot \left(\nabla^2 \phi^+(0)\right)(x) + o(x)$$
$$= \xi \cdot x - M|x|^2 - M|\xi \cdot x|^2 + o(x). \tag{4.4.4}$$

Here, o is some function which satisfies $\lim_{x \to 0} |o(x)|/|x|^2 = 0$. Now take any $\epsilon \in (0, \epsilon')$ such that

$$|o(x)| < M(1 + |\xi|^2)|x|^2, \quad \text{for all } x \in \epsilon\overset{\circ}{\mathbb{B}}.$$

It follows then from (4.4.3) and (4.4.4) that

$$\xi \cdot x \leq f(x) + M|x|^2 + M|\xi \cdot x|^2 - o(x)$$
$$\leq f(x) + M(1 + |\xi|^2)|x|^2 + M(1 + |\xi|^2)|x|^2$$
$$= f(x) + \tilde{M}|x|^2, \quad \text{for all } x \in \epsilon\overset{\circ}{\mathbb{B}},$$

where $\tilde{M} := 2M(1 + |\xi|^2)$. This is what we set out to prove. □

We note for future use a variant on the preceding characterization of proximal normals, valid in the Lipschitz case.

Proposition 4.4.2 *Take a Lipschitz continuous function* $f : \mathbb{R}^k \to \mathbb{R}$ *and points* $x \in \mathbb{R}^k$ *and* $\xi \in \mathbb{R}^k$. *Then the following two statements are equivalent:*

(i) $\xi \in \partial^P f(x)$,
(ii) *there exists* $M > 0$ *such that*

$$\xi \cdot (y - x) \leq f(y) - f(x) + M|y - x|^2 \tag{4.4.5}$$

for all $y \in \mathbb{R}^k$.

The difference, of course, is that, now, the 'proximal inequality' (4.4.5) must be satisfied for *all* y's in $\mathbb{R}^k$, not just y's in a neighbourhood of x.

Proof Assume (ii). Then (i) follows from Proposition 4.4.1.

Assume (i). Then we know from Proposition 4.4.1 that there exist $M > 0$ and $\epsilon > 0$ such that (4.4.5) is valid for all $y \in x + \epsilon\mathbb{B}$. Let K be a Lipschitz constant for f. Define $N := \epsilon^{-1}(|\xi| + K)$. Then

4.4 Difference Quotient Representations

$$\xi \cdot (y - x) - (f(y) - f(x)) \leq |\xi| |y - x| + K |y - x| \leq N |y - x|^2 \text{ if } y \notin x + \epsilon \mathbb{B}.$$

It follows that (4.4.5) is satisfied for all y when M is replaced by $\tilde{M} := \max\{M, N\}$. □

The next result provides a 'one-sided difference quotient' characterization of the strict subdifferential.

Proposition 4.4.3 *Take a lower semi-continuous functions function $f : \mathbb{R}^k \to \mathbb{R} \cup \{+\infty\}$ and points $x \in \text{dom } f$ and $\xi \in \mathbb{R}^k$. Then the following two statements are equivalent:*

(i) $\xi \in \hat{\partial} f(x)$,
(ii)

$$\limsup_{y \to x} \frac{\xi \cdot (y - x) - (f(y) - f(x))}{|y - x|} \leq 0. \tag{4.4.6}$$

Reference is made in the proof to the Bouligand tangent cone $T_C(x)$ to a closed set $C \subset \mathbb{R}^k$ at the point $x \in C$, defined as

$$T_C(x) := \{\eta : \text{ there exists } y_i \xrightarrow{C} y \text{ and } t_i \downarrow 0 \text{ such that } t_i^{-1}(y_i - x) \to \eta\},$$

whose properties will be explored later in the chapter.

Proof Assume (ii). Condition (4.4.6) can be expressed:

$$\xi \cdot (y - x) - (f(y) - f(x)) \leq o(|y - x|) \text{ for all } y \in \mathbb{R}^k,$$

in which $o(\cdot) : \mathbb{R}_+ \to \mathbb{R}_+$ is a function satisfying $o(\epsilon)/\epsilon \to 0$ as $\epsilon \downarrow 0$. It follows that

$$(\xi, -1) \cdot (z - (x, f(x))) \leq o(|z - (x, f(x))|) \text{ for all } z \in \text{epi } f.$$

Otherwise expressed

$$(\xi, -1) \in \hat{N}_{\text{epi } f}(x, f(x)),$$

which implies (i).

Assume now (i). Suppose that condition (4.4.6) is not satisfied. This means that there exists a number $\alpha > 0$ and a sequence $y_i \to x$ such that $y_i \neq x_i$ and

$$\xi \cdot (y_i - x) - (f(y_i) - f(x)) \geq \alpha |y_i - x|, \text{ for all } i. \tag{4.4.7}$$

This inequality can be rearranged to give

$$f(y_i) \leq f(x) + \xi \cdot (y_i - x) - \alpha |y_i - x|, \text{ for all } i. \tag{4.4.8}$$

We see that $\limsup_{i\to\infty} f(y_i) \leq f(x)$. Since f is lower semi-continuous, it follows that $f(y_i) \to f(x)$ as $i \to \infty$.

Define $z_i := (y_i, f(y_i))$, $t_i := |z_i - (x, f(x))|$ and $\eta_i := t_i^{-1}(z_i - (x, f(x)))$. Observe that, for each i, $|\eta_i| = 1$. Following extraction of subsequences we have that $\eta_i \to \eta$ for some non-zero vector η. Notice that $z_i \stackrel{\text{epi } f}{\to} (x, f(x))$ and $t_i \to 0$. By definition of the Bouligand tangent cone then $\eta \in T_{\text{epi } f}(x, f(x))$.

We claim that

$$(\xi, -1) \cdot \eta > 0.$$

If this is true the proof will be complete because, according to Theorem 4.10.4 below, the strict normal cone $\hat{N}_{\text{epi } f}(x, f(x))$ is contained in the polar set of the tangent cone $T_{\text{epi } f}(x, f(x))$. This implies $(\xi, -1) \cdot \eta \leq 0$, a contradiction.

To verify the claim, we make use of the following identity, which follows from the definition of η_i:

$$(\xi, -1) \cdot \eta_i = t_i^{-1}(\xi \cdot (y_i - x) - (f(y_i) - f(x))). \tag{4.4.9}$$

Now (4.4.8) implies that

$$\limsup_{i\to\infty} |y_i - x|^{-1}(f(y_i) - f(x)) < \infty.$$

There are therefore two possibilities to be considered:

(a) there exists $K > 0$ such that

$$f(y_i) - f(x) \leq K|y_i - x| \text{ for all } i, \tag{4.4.10}$$

(b) along a subsequence

$$\lim_{i\to\infty} |y_i - x|^{-1}(f(y_i) - f(x)) = -\infty.$$

Assume (a) is true. By (4.4.8), (4.4.9) and (4.4.10), we have:

$$(\xi, -1) \cdot \eta_i \geq t_i^{-1}\alpha|y_i - x| = \frac{\alpha|y_i - x|}{|(y_i, f(y_i)) - (x, f(x))|} \geq (1 + K^2)^{-1/2}\alpha > 0.$$

Since $\eta_i \to \eta$, we must have $(\xi, -1) \cdot \eta > 0$.

Finally assume (b) is true. (Along the subsequence) $t_i^{-1}|y_i - x| \to 0$ and $t_i^{-1}(f(y_i) - f(x)) \to -1$. But then by (4.4.9), $(\xi, -1) \cdot \eta = \lim_i (\xi, -1) \cdot \eta_i = +1$. This confirms the claim in case (b) also. $\square$

4.5 Nonsmooth Mean Value Inequalities

We pause to establish some consequences of the preceding theory, namely generalizations of the classical mean value theorem. The results of this section will be used, both to carry out a deeper investigation of subdifferentials, normal cones, etc. and also to provide information about solutions to optimization problems.

It will be convenient to use the following notation. Given a point $\bar{x} \in \mathbb{R}^k$ and a subset $Y \subset \mathbb{R}^k$, we denote by $[\bar{x}, Y]$ the set

$$[\bar{x}, Y] := \{x : x = \lambda \bar{x} + (1 - \lambda)y \text{ for some } y \in Y \text{ and } \lambda \in [0, 1]\}.$$

$(\bar{x}, Y)$ denotes the related set, in which the qualifier '$\lambda \in [0, 1]$' is replaced by '$\lambda \in (0, 1)$'.

If $Y = \{\bar{y}\}$ we write $[\bar{x}, \bar{y}]$ in place of $[\bar{x}, Y]$.

The classical mean value theorem tells us that if $f : \mathbb{R}^n \to \mathbb{R}$ is a C^1 function on an open set containing the line segment $[\bar{x}, \bar{y}]$ then, for some $\bar{z} \in [\bar{x}, \bar{y}]$,

$$\nabla f(\bar{z}) \cdot (\bar{y} - \bar{x}) = f(\bar{y}) - f(\bar{x}).$$

We might ask whether there is an analogue of this result covering situations where, say, f is a Lipschitz continuous function, expressed in terms of a limiting subgradient. The answer to this question is no, as is evident from the following example.

Example
Define the function $f : \mathbb{R} \to \mathbb{R}$ to be

$$f(x) = -|x|$$

and set $\bar{x} = -1$ and $\bar{y} = +1$. A simple calculation shows that

$$\partial f(x) \subset \{-1\} \cup \{+1\} \text{ for all } x \in \mathbb{R}.$$

It follows that, for any $z \in \mathbb{R}$ and any $\xi \in \partial f(z)$,

$$0 = f(\bar{y}) - f(\bar{x}) \notin \xi \cdot (\bar{y} - \bar{x}).$$

The above example rules out a direct generalization of the mean value theorem involving limiting subgradients, to allow for the function f to be nonsmooth. We do notice however in this example that the inequality

$$f(\bar{y}) - f(\bar{x}) \leq \xi \cdot (\bar{y} - \bar{x})$$

is satisfied for some $\xi \in \partial f(\bar{z})$ with $\bar{z} \in [\bar{x}, \bar{y}]$. Possible choices are $\bar{z} = -0.5$ and $\xi = +1$.

It is the 'inequality' version of the mean value theorem which may be extended to a nonsmooth setting. Nonsmooth mean value inequalities have many uses, as an aid to proving fundamental relations in nonsmooth analysis and in the analysis of solutions to optimization problems. The role of these theorems is usually to provide some upper bound and the fact that they take the form only of an inequality is seldom a significant shortcoming.

The mean value inequality proved here is due to Clarke and Ledyaev [71]. Besides allowing the function concerned to be nonsmooth, it departs from the classical mean value theorem in one other respect. It is to assert that the inequality involved holds in some *uniform* sense.

In the setup for this mean value inequality, the line segment $[\bar{x}, \bar{y}]$ is replaced by the 'generalized' line segment joining a point $\bar{x}$ to a compact, convex set $Y \subset \mathbb{R}^k$, namely $[\bar{x}, Y]$. A natural extension of the smooth mean value inequality would be: for each $y \in Y$ there exists $\bar{z} \in [\bar{x}, Y]$, where

$$[\bar{x}, Y] := \{\epsilon x + (1-\epsilon) y : \epsilon \in [0, 1] \text{ and } y \in Y\},$$

and $\xi \in \partial f(\bar{z})$ such that

$$f(y) - f(\bar{x}) \leq \xi \cdot (y - \bar{x}).$$

What comes as a surprise is that, if we replace the left side by

$$\min_{y \in Y} f(y) - f(\bar{x}),$$

then the inequality is valid whatever choice is made of y in Y for the *same* ξ (a limiting subgradient of f evaluated at some point in $[\bar{x}, Y]$). This result, when first proved, was new even for smooth functions.

The main step is to prove the following 'approximate' mean value inequality involving proximal subgradients.

Theorem 4.5.1 (The Proximal Mean Value Inequality) *Take a lower semicontinuous function* $f : \mathbb{R}^n \to \mathbb{R} \cup \{+\infty\}$, *a point* $\bar{x} \in \text{dom } f$ *and a compact, convex set* $Y \subset \mathbb{R}^n$. *Define* $\hat{r} \in \mathbb{R} \cup \{+\infty\}$ *to be*

$$\hat{r} := \inf_{y \in Y} f(y) - f(\bar{x}).$$

Then for any finite $r < \hat{r}$ *and* $\epsilon > 0$, *we can choose* $\bar{z} \in [\bar{x}, Y] + \epsilon \mathbb{B}$ *and* $\xi \in \partial^P f(\bar{z})$ *such that*

$$r < \xi \cdot (y - \bar{x}) \text{ for all } y \in Y$$

and

$$f(\bar{z}) - f(\bar{x}) \leq \max\{r, 0\}.$$

4.5 Nonsmooth Mean Value Inequalities

Proof By translation of the origin in $\mathbb{R}^n \times \mathbb{R}$ we can arrange that $\bar{x} = 0$ and $f(\bar{x}) = 0$. Fix $r < \hat{r}$ and $\epsilon > 0$. We deduce from the compactness of Y and the lower semicontinuity of f that positive numbers δ, M and k may be chosen such that $0 < \delta < \epsilon$ and

$$f(z) \geq -M \text{ for all } z \in [0, Y] + \delta \mathbb{B},$$

$$\min_{y \in Y + \delta \mathbb{B}} f(y) \geq (r + \hat{r})/2$$

and

$$k > (M + |r| + 1)/\delta^2.$$

The proof hinges on examining the properties of the function $g : [0, 1] \times \mathbb{R}^n \times \mathbb{R}^n \to \mathbb{R} \cup \{+\infty\}$:

$$g(t, y, z) := f(z) + k|ty - z|^2 - tr.$$

Write

$$S := Y \times ([0, Y] + \delta \mathbb{B}).$$

Notice that, for any $y \in Y$,

$$(0, y, 0) \in [0, 1] \times S \text{ and } g(0, y, 0) = 0.$$

Since S is compact and g is lower semi-continuous, the minimization problem

$$\text{Minimize } g(t, y, z) \text{ over } (t, y, z) \in [0, 1] \times S$$

has a solution $(\bar{t}, \bar{y}, \bar{z})$ and

$$\min_{(t,y,z) \in [0,1] \times S} g(t, y, z) \leq 0.$$

This relation implies that

$$g(\bar{t}, \bar{y}, \bar{z}) \leq 0, \qquad (4.5.1)$$

and so

$$f(\bar{z}) - \max\{0, r\} \leq 0.$$

In particular, $f(\bar{z}) < +\infty$. Let us show that

$$\bar{z} \in [0, Y] + \delta \text{ int } \mathbb{B}.$$

If $\bar{z}$ does not satisfy this condition then $|\bar{z} - \bar{t}\bar{y}| \geq \delta$, because $\bar{t}\bar{y} \in [0, Y]$. But then

$$g(\bar{t}, \bar{y}, \bar{z}) = f(\bar{z}) + k|\bar{t}\bar{y} - \bar{z}|^2 - \bar{t}r > -M + (M + |r| + 1) - |r| > 1.$$

We have contradicted (4.5.1).

We note next that

$$\bar{t} < 1. \tag{4.5.2}$$

This is because if $\bar{t} = 1$, it would follow that

$$g(\bar{t}, \bar{y}, \bar{z}) = f(\bar{z}) + k|\bar{y} - \bar{z}|^2 - r$$
$$\geq \begin{cases} f(\bar{z}) + k|\bar{y} - \bar{z}|^2 - r & \text{if } |\bar{y} - \bar{z}| \leq \delta \\ f(\bar{z}) + k\delta^2 - r & \text{if } |\bar{y} - \bar{z}| > \delta \end{cases}$$
$$\geq \min\{(r + \hat{r})/2 - r, -M + M + |r| + 1 - r\} > 0.$$

But then, in view of (4.5.1), $(\bar{t}, \bar{y}, \bar{z})$ cannot be a minimizer. We deduce from this contradiction the validity of (4.5.2).

We know that, for any z in the open set $[0, Y] + \delta$ int $\mathbb{B}$,

$$g(\bar{t}, \bar{y}, z) \geq g(\bar{t}, \bar{y}, \bar{z}).$$

Using the identity

$$|\bar{t}\bar{y} - z|^2 - |\bar{t}\bar{y} - \bar{z}|^2 = -2(\bar{t}\bar{y} - \bar{z}) \cdot (z - \bar{z}) + |z - \bar{z}|^2,$$

we can write the preceding inequality as

$$2k(\bar{t}\bar{y} - \bar{z}) \cdot (z - \bar{z}) \leq f(z) - f(\bar{z}) + k|z - \bar{z}|^2.$$

This holds, in particular, for all points z in some neighbourhood of $\bar{z}$. Define

$$\xi = 2k(\bar{t}\bar{y} - \bar{z}).$$

According to Proposition 4.4.1,

$$\xi \in \partial^P f(\bar{z}).$$

Two cases now need to be considered

Case (a): $\bar{t} = 0$. Here $\xi = -2k\bar{z}$. Choose $t \in (0, 1]$. For any $y \in Y$ we have

$$0 \leq g(t, y, \bar{z}) - g(0, \bar{y}, \bar{z})$$
$$= f(\bar{z}) + k|ty - \bar{z}|^2 - tr - f(\bar{z}) - k|\bar{z}|^2$$
$$= t^2 k|y|^2 - 2kt\bar{z} \cdot y - tr.$$

4.5 Nonsmooth Mean Value Inequalities

Dividing across the inequality by t and passing to the limit as $t \downarrow 0$ we obtain

$$\xi \cdot y = -2k\bar{z} \cdot y \geq r.$$

This is the required relation.

Case (b): $\bar{t} \in (0, 1)$. The fact that $\bar{y}$ minimizes the quadratic, convex function $y \to g(\bar{t}, y, \bar{z})$ over the convex set Y implies

$$2\bar{t}k(\bar{t}\bar{y} - \bar{z}) \cdot (y - \bar{y}) \geq 0, \quad \text{for all } y \in Y.$$

Since $\bar{t} > 0$, it follows that

$$\xi \cdot (y - \bar{y}) \geq 0.$$

But $\bar{t}$ minimizes the quadratic function $t \to g(t, \bar{y}, \bar{z})$ over $(0, 1)$ so

$$2k(\bar{t}\bar{y} - \bar{z}) \cdot \bar{y} - r = 0.$$

It follows that $\xi \cdot \bar{y} = r$. We conclude that

$$\xi \cdot y \geq r.$$

The theorem is proved. □

Performing a simple exercise in limit taking now yields an 'exact' mean value inequality for Lipschitz functions in terms of limiting subdifferentials.

Theorem 4.5.2 (The Lipschitz Case) *Take a lower semi-continuous function* $f : \mathbb{R}^n \to \mathbb{R} \cup \{+\infty\}$, *a point* $\bar{x} \in \mathbb{R}^n$ *and a compact, convex set* $Y \subset \mathbb{R}^n$. *Assume that, for some* $\delta > 0$, f *is Lipschitz continuous on* $[\bar{x}, Y] + \delta \mathbb{B}$.
Then there exists $\bar{z} \in [\bar{x}, Y]$ *and* $\xi \in \partial f(\bar{z})$ *such that*

$$\min_{y \in Y} f(y) - f(\bar{x}) \leq \xi \cdot (y - \bar{x}) \quad \text{for all } y \in Y$$

and

$$f(\bar{z}) - f(\bar{x}) \leq \max\{\min_{y \in Y} f(y) - f(\bar{x}), 0\}.$$

Proof Choose sequences $\epsilon_i \downarrow 0$ and $r_i \uparrow \hat{r}$, where

$$\hat{r} := \min_{y \in Y} f(y) - f(\bar{x}).$$

Now apply Theorem 4.5.1 with ϵ and r taken to be ϵ_i and r_i respectively for $i = 1, 2, \ldots$. For each i, we deduce the existence of some $z_i \in [\bar{x}, Y] + \epsilon_i \mathbb{B}$ and $\xi_i \in \partial^P f(z_i)$ such that

$$r_i < \xi_i \cdot (y - \bar{x}) \text{ for all } y \in Y \tag{4.5.3}$$

and

$$f(z_i) - f(\bar{x}) \leq \max\{r_i, 0\}. \tag{4.5.4}$$

$\{z_i\}$ and $\{\xi_i\}$ are bounded sequences because Y is compact and f is Lipschitz continuous on $[\bar{x}, Y] + \delta \mathbb{B}$ (see Proposition 4.3.3). We may arrange then by extracting subsequences that $z_i \to \bar{z}$ and $\xi_i \to \xi$ for some $\bar{z} \in [\bar{x}, Y]$ and some $\xi \in \mathbb{R}^n$. It follows now from Proposition 4.3.5 that $\xi \in \partial f(\bar{z})$. The property that ξ and $\bar{z}$ satisfy the desired inequalities is now recovered from (4.5.3) and (4.5.4), in the limit as $i \to \infty$. □

If we specialize to the case $Y = \{\bar{y}\}$ and substitute the convexified limiting subdifferential $\mathrm{co}\,\partial f$ in place of ∂f, then a full (two sided) mean value theorem is valid. This result was earlier proved by Lebourg. To prove it, we make use of properties of limiting subgradients, derived in the next chapter.

Theorem 4.5.3 (A Two Sided Mean Value Theorem) *Take a locally Lipschitz continuous function* $f : \mathbb{R}^n \to \mathbb{R}$ *and a line segment* $[\bar{x}, \bar{y}] \subset \mathbb{R}^n$. *Then there exists* $z \in \{\epsilon \bar{x} + (1-\epsilon)\bar{y} : 0 < \epsilon < 1\}$ *and* $\xi \in \mathrm{co}\,\partial f(z)$ *such that*

$$f(\bar{y}) - f(\bar{x}) = \xi \cdot (\bar{y} - \bar{x}).$$

Proof Consider the function $g : \mathbb{R} \to \mathbb{R} \cup \{+\infty\}$ defined by

$$g(t) := f(\bar{x} + t(\bar{y} - \bar{x})) - t(f(\bar{y}) - f(\bar{x})) + \Psi_{[0,1]}(t),$$

in which $\Psi_{[0,1]}$ denotes the indicator function of the set $[0, 1]$. ($\Psi(t)_{[0,1]}$ takes the value 0 or ∞ depending on whether, or not, $t \in [0, 1]$.) Notice that $g(0) = g(1) = f(\bar{x})$. So either of the following conditions are satisfied:

(a): g has a minimizer at some 'interior' point $\tau \in (\bar{x}, \bar{y})$,

(b): $-g$ has a minimizer at some 'interior' point $\sigma \in (\bar{x}, \bar{y})$.

Here $(\bar{x}, \bar{y})$ denotes the relative interior of $[\bar{x}, \bar{y}]$.

Consider case (a). We deduce from the proximal normal inequality that $0 \in \partial^P g(\tau)$. It follows from the chain rule (see Chap. 5) that

$$0 = \xi \cdot (\bar{y} - \bar{x}) - (f(\bar{y}) - f(\bar{x}))$$

for some $\xi \in \partial f(\bar{x} + \tau(\bar{y} - \bar{x}))$ as required.

Consider case (b). The chain rule applied to $-g$ now gives

$$0 = \xi \cdot (\bar{y} - \bar{x}) + (f(\bar{y}) - f(\bar{x}))$$

for some $\xi \in \partial(-f)(\sigma)$. Since, however,

$$\text{co}\partial f(x) = -\text{co}\partial(-f(x))$$

(see Theorem 4.7.5) we conclude that

$$0 = \xi' \cdot (\bar{y} - \bar{x}) - (f(\bar{y}) - f(\bar{x}))$$

for some $\xi' \in \partial f(\sigma)$ in this case also. □

4.6 Characterization of Limiting Subgradients

We now seek an analytic description of limiting subgradients (and their asymptotic relatives). Our goal is to describe these objects in terms of limits of proximal subgradients, for which an analytic characterization is already available.

The fact that limiting subgradients can be expressed as limits of proximal subgradients at neighbouring points is a straightforward consequence of the definition of limiting subgradients in terms of limiting normal vectors to the epigraph set and the property that limiting normal vectors are limits of proximal normal vectors. This is part (a) of Theorem 4.6.2 below.

It is easy also to show that asymptotic limiting subgradients are expressible as limits of vectors which (apart from a positive scale factor) are either proximal subgradients or asymptotic proximal subgradients. *There is, however, a more convenient description of asymptotic limiting subgradients which involves sequences of proximal subgradients alone.* This is part (b) of Theorem 4.6.2. It has important implications because the validity of various subdifferential calculus rules centres on the properties of asymptotic subdifferentials of the functions involved, and any results which restrict the possible ways in which asymptotic limiting subdifferentials can arise simplify the analysis of pathological phenomena.

As a first step we prove a lemma, whose conclusions may be summarized 'proximal normals to an epigraph set can be approximated by non-horizontal proximal normals'.

Lemma 4.6.1 *Take a lower semi-continuous function $f : \mathbb{R}^k \to \mathbb{R} \cup \{+\infty\}$, a point $\bar{x} \in \text{dom } f$ and a non-zero point $\xi \in \mathbb{R}^k$ such that*

$$(\xi, 0) \in N^P_{\text{epi } f}(\bar{x}, \bar{\alpha}),$$

in which $(\bar{x}, \bar{\alpha}) \in \text{epi } f$. Then there exist convergent sequences $x_i \xrightarrow{f} \bar{x}$, $\xi_i \to \xi$ and a sequence of positive numbers $\lambda_i \downarrow 0$ such that

$$(\xi_i, -\lambda_i) \in N^P_{\text{epi } f}(x_i, f(x_i)) \text{ for all } i.$$

Use is made in the proof of properties of the distance function derived later in the chapter. (See Lemmas 4.8.3 and 4.8.4 and Theorem 4.8.5 below.)

Proof To begin the proof, we show that we can restrict attention to the case when $\bar{\alpha} = f(\bar{x})$. Indeed, since $(\xi, 0)$ is a non-zero 'horizontal' proximal normal, there exist a point $(x, \bar{\alpha}) \notin \text{epi } f$ and a number $\sigma > 0$ such that $(\bar{x}, \bar{\alpha})$ is a closest point in epi f to $(x, \bar{\alpha})$ and $\xi = \sigma(x - \bar{x})$. We can arrange that $(\bar{x}, \bar{\alpha})$ is the *unique* closest point, and also that there exists some $\beta > 0$ such that $(\bar{x}, \bar{\alpha})$ is the closest point in epi f to the point

$$(\bar{x} + (1+\beta)(x - \bar{x}), \bar{\alpha}) = (\bar{x} + \sigma^{-1}(1+\beta)\xi, \bar{\alpha}).$$

This can be achieved by replacing x by a point on the 'open' line segment joining x and $\bar{x}$ and by increasing σ appropriately. Therefore, for all $y \in \text{dom } f$ and $a \geq 0$ we have

$$|(y, f(y)+a) - (\bar{x}+\sigma^{-1}(1+\beta)\xi, \bar{\alpha})| \geq |(\bar{x}, \bar{\alpha}) - (\bar{x}+\sigma^{-1}(1+\beta)\xi, \bar{\alpha})| \geq \sigma^{-1}(1+\beta)|\xi|.$$

Taking $a = (\bar{\alpha} - f(\bar{x})) + a'$ in the above inequality, we obtain

$$|(y, f(y)+a') - (\bar{x}+\sigma^{-1}(1+\beta)\xi, f(\bar{x}))| \geq \sigma^{-1}(1+\beta)|\xi|$$

for all $y \in \text{dom } f$, $a' \geq 0$.

In view of the proximal normal characterization provided by Proposition 4.2.2, it follows that $(\xi, 0) \in N^P_{\text{epi } f}(\bar{x}, f(\bar{x}))$. We have confirmed our claim.

We may now deduce from Lemmas 4.8.3, 4.8.4 and Theorem 4.8.5 the following properties.

First,

$$(|\xi|^{-1}\xi, 0) \in \partial^P d_{\text{epi } f}(x, f(\bar{x})).$$

Second, if for some sequences $(y_i, r_i) \to (x, f(\bar{x}))$ and $\{(\xi'_i, -\lambda'_i)\}$ in $\mathbb{R}^n \times \mathbb{R}$

$$(\xi'_i, -\lambda'_i) \in \partial^P d_{\text{epi } f}(y_i, r_i) \text{ for all } i, \tag{4.6.1}$$

then there exists (x_i, s_i), a unique closest point to (y_i, r_i) in epi f, such that

$$(\xi'_i, -\lambda'_i) \in \partial^P d_{\text{epi } f}(x_i, s_i) \text{ for all } i \tag{4.6.2}$$

$$(x_i, s_i) \to (\bar{x}, f(\bar{x})) \tag{4.6.3}$$

4.6 Characterization of Limiting Subgradients

and
$$(\xi'_i, -\lambda'_i) \to (|\xi|^{-1}\xi, 0). \qquad (4.6.4)$$

We shall show presently that

$$d_{\text{epi } f}(x, f(\bar{x}) - t) > d_{\text{epi } f}(x, f(\bar{x})) \text{ for all } t > 0. \qquad (4.6.5)$$

Assuming this to be the case, we take a sequence $t_i \downarrow 0$ and, for each i, apply the proximal mean value inequality (Theorem 4.5.1) to the function $d_{\text{epi } f}$ with base point $\{(x, f(\bar{x}))\}$ and 'Y set' $\{(x, f(\bar{x}) - t_i)\}$. The theorem tells us that there exist sequences $(y_i, r_i) \to (x, f(\bar{x}))$ and $\{(\xi'_i, -\lambda'_i)\}$ in $\mathbb{R}^n \times \mathbb{R}$ satisfying (4.6.1) and, for all i,

$$0 < (\xi'_i, -\lambda'_i) \cdot ((x, f(\bar{x}) - t_i) - (x, f(\bar{x}))).$$

This inequality implies that $\lambda'_i > 0$ for all i. It follows from (4.6.2) that in fact $s_i = f(x_i)$. Equation (4.6.3) may therefore be replaced by $x_i \xrightarrow{f} \bar{x}$.

Now define

$$(\xi_i, -\lambda_i) := (|\xi|\xi'_i, -|\xi|\lambda'_i) \text{ for each } i.$$

Note that, since $\xi \neq 0$, $\lambda_i > 0$ for each i. We have from (4.6.2) and Proposition 4.8.2 that

$$(\xi_i, -\lambda_i) \in N^P_{\text{epi } f}(x_i, f(x_i)) \text{ for each } i$$

and, in view of (4.6.4)

$$(\xi_i, -\lambda_i) \to (\xi, 0) \text{ as } i \to \infty.$$

The sequences $\{x_i\}, \{\xi_i\}$ and $\{\lambda_i\}$ therefore have the properties asserted in the lemma. It remains to confirm (4.6.5). If it is untrue, there exists $t > 0$ such that

$$d_{\text{epi } f}(x, f(\bar{x}) - t) \leq d_{\text{epi } f}(x, f(\bar{x})).$$

We may find $(y, r) \in \mathbb{R}^n \times \mathbb{R}$ with $r \geq f(y)$ such that

$$d_{\text{epi } f}(x, f(\bar{x}) - t) = |(x, f(\bar{x}) - t) - (y, r)|.$$

Notice that $y \neq \bar{x}$. Indeed if this were not the case then

$$|(x, f(\bar{x}) - t) - (\bar{x}, r)| = |(x - \bar{x}, (f(\bar{x}) - t - r))| > |x - \bar{x}| = d_{\text{epi } f}(x, f(\bar{x})),$$

since $t > 0$ and $r \geq f(\bar{x})$, in contradiction of our premise. It follows that

$$d_{\text{epi } f}(x, f(\bar{x})) \geq d_{\text{epi } f}(x, f(\bar{x}) - t)$$
$$= |(x, f(\bar{x}) - t) - (y, r)| = |(x, f(\bar{x})) - (y, r + t)|.$$

But $(y, r+t) \in \text{epi } f$ since $t \geq 0$. It follows that $(y, r+t)$ is a closest point in epi f to $(x, f(\bar{x}))$ with $y \neq \bar{x}$. This is impossible because $(\bar{x}, f(\bar{x}))$ is the unique closest point. □

It is now a simple matter to prove:

Theorem 4.6.2 (Asymptotic Description of Limiting Subgradients) *Take a lower semi-continuous function $f : \mathbb{R}^k \to \mathbb{R} \cup \{+\infty\}$ and points $x \in \text{dom } f$ and $\xi \in \mathbb{R}^k$.*

(a) The following conditions are equivalent:

(i) $\xi \in \partial f(x)$,

(ii) there exist $x_i \xrightarrow{f} x$ and $\xi_i \to \xi$ such that $\xi_i \in \partial^P f(x_i)$ for all i.

(b) The following conditions are equivalent:

(iii) $\xi \in \partial^\infty f(x)$,

(iv) there exist $x_i \xrightarrow{f} x$, $t_i \downarrow 0$ and $\xi_i \to \xi$ such that $t_i^{-1} \xi_i \in \partial^P f(x_i)$ for all i.

Proof (a) Assume (i). Then there exist sequences $(\xi_i', -t_i) \to (\xi, -1)$ and $z_i \xrightarrow{\text{epi } f} (x, f(x))$ such that $(\xi_i', -t_i) \in N_{\text{epi } f}^P(z_i)$ for all i. We may assume $t_i > 0$ for all i. We deduce from the proximal inequality that $z_i = (x_i, f(x_i))$ for all i and hence $x_i \xrightarrow{f} x$. Define $\xi_i := t_i^{-1} \xi_i'$. Then $\xi_i \in \partial^P f(x_i)$ for each i. Since $\xi_i \to \xi$ as $i \to \infty$, (ii) follows.

Assume (ii). For the given sequences $\{x_i\}$ and $\{\xi_i\}$, we have that $(\xi_i, -1) \in N_{\text{epi } f}^P(x_i, f(x_i))$ for each i and $(x_i, f(x_i)) \to (x, f(x))$ as $i \to \infty$. It follows from the closure properties of the limiting normal cone that $(\xi, -1) \in N_{\text{epi } f}(x, f(x))$. But then $\xi \in \partial f(x)$.

(b) Assume (iii). Suppose to begin with that $\xi \neq 0$. By Lemma 4.6.1, there exist sequences $(\xi_i', -t_i) \to (\xi, 0)$ and $x_i \xrightarrow{f} x$ such that $t_i > 0$ and $(\xi_i', -t_i) \in N_{\text{epi } f}^P(x_i, f(x_i))$ for each i. Define $\xi_i := t_i^{-1} \xi_i'$ for each i. Then $(\xi_i, -1) \in N_{\text{epi } f}^P(x_i, f(x_i))$, whence $\xi_i \in \partial f(x_i)$ for each i. We have shown (iv).

Suppose on the other hand that $\xi = 0$. Since $x \in \text{dom } f$, it follows from Proposition 4.2.7 that there exists $(\xi', -\beta) \neq (0, 0)$ such that $(\xi', -\beta) \in N_{\text{epi } f}(x, f(x))$. Since limiting normals are limits of proximal normals and proximal normals to epigraph sets have second coordinate a non-negative number, $\beta \geq 0$. Whether or not $\beta = 0$, we deduce from part (a) of Theorem 4.6.2 or part (b) (in the case $\xi' \neq 0$) that there exist sequences $x_i \xrightarrow{f} x$, $\xi_i' \to \xi'$ and $\beta_i \to \beta$ such that

$\beta_i > 0$ and $\beta_i^{-1}\xi_i' \in \partial^P f(x_i)$ for all i.

Choose a sequence $\alpha_i \downarrow 0$ such that

$$\alpha_i \beta_i \to 0 \text{ as } i \to \infty.$$

Now set $\xi_i = \alpha_i \xi_i'$, $t_i = \alpha_i \beta_i$ for each i.
We see that

$$\xi_i \to \xi (= 0), \ t_i \downarrow 0 \text{ and } t_i^{-1}\xi_i \in \partial^P f(x_i) \text{ for all } i.$$

We have shown (iv) in this case also.

Suppose (iv). For the given sequences $\{x_i\}$, $\{t_i\}$ and $\{\xi_i\}$, we have $(\xi_i, -t_i) \in N_{\text{epi } f}^P(x_i, f(x_i))$ for all i. Since $(\xi_i, -t_i) \to (\xi, 0)$ and $(x_i, f(x_i)) \stackrel{\text{epi } f}{\to} (x, f(x))$, it follows $(\xi, 0) \in N_{\text{epi } f}^P x, f(x)$. This is (iii). $\square$

A companion piece to the above theorem is a characterization of limiting subgradients and asymptotic limiting subgradients in terms of limits of strict (rather than proximal) subgradients.

Theorem 4.6.3 *Take a lower semi-continuous functions function $f : \mathbb{R}^k \to \mathbb{R} \cup \{+\infty\}$ and points $x \in \text{dom } f$ and $\xi \in \mathbb{R}^k$.*

(a) The following conditions are equivalent:

  (i) $\xi \in \partial f(x)$,
  (ii) there exist $x_i \stackrel{f}{\to} x$ and $\xi_i \to \xi$ such that $\xi_i \in \hat{\partial} f(x_i)$, for all i.

(b) The following conditions are equivalent:

  (iii) $\xi \in \partial^\infty f(x)$,
  (iv) there exists $x_i \stackrel{f}{\to} x$, $t_i \downarrow 0$ and $\xi_i \to \xi$ such that $t_i^{-1}\xi_i \in \hat{\partial} f(x_i)$ for all i.

The most significant assertions in this theorem are the implications (i) $\Rightarrow$ (ii) and (iii) $\Rightarrow$ (iv); but these follow directly from Theorem 4.6.2 since proximal normals are certainly strict normals. The reverse implications are routine consequences of the fact that limiting normal vectors to epi f are limits of neighbouring strict normal vectors (see Proposition 4.2.5).

4.7 Subgradients of Lipschitz Continuous Functions

The techniques so far assembled provide local descriptions of Lipschitz continuous functions, since they are special cases of lower semi-continuous functions. But Lipschitz continuous functions are encountered so frequently in applications of nonsmooth analysis that it is important to exploit the extra structure they introduce and alternative approaches to local approximation.

The Lipschitz continuity hypothesis is essentially a hypothesis that the function in question has bounded slope. It is hardly surprising then that the subdifferentials of Lipschitz continuous functions are bounded sets, a fact which follows easily from the preceding analysis.

Proposition 4.7.1 *Take a lower semi-continuous function $f : \mathbb{R}^k \to \mathbb{R} \cup \{+\infty\}$ and a point $x \in \mathbb{R}^k$. Assume that f is Lipschitz continuous on a neighbourhood of x with Lipschitz constant K. Then:*

(i) *$\partial f(x)$ is non-empty and $\partial f(x) \subset K\mathbb{B}$,*
(ii) *$\partial^\infty f(x) = \{0\}$.*

Proof Take ξ and y such that $\xi \in \partial^P f(y)$ and f is Lipschitz continuous on a neighbourhood of y. Then there exist $M > 0$ and $\epsilon > 0$ such that

$$\xi \cdot (z - y) \leq f(z) - f(y) + M|z - y|^2 \quad \text{for all } z \in y + \epsilon \mathbb{B}.$$

It follows that, for all $\lambda > 0$, sufficiently small

$$\lambda |\xi|^2 \leq f(y + \lambda \xi) - f(y) + M\lambda^2 |\xi|^2 \leq \lambda K |\xi| + M\lambda^2 |\xi|^2.$$

Dividing across by λ and passing to the limit gives

$$|\xi|^2 \leq K|\xi|.$$

We conclude that $|\xi| \leq K$. This shows that for all y sufficiently close to x,

$$\xi \in \partial^P f(y) \text{ implies } |\xi| \leq K.$$

The assertions of the proposition now follow from Theorem 4.6.2. □

The above proposition is a special case of more far-reaching results, covered by Corollary 4.9.2 below.

We now examine an alternative approach to defining subdifferentials of Lipschitz continuous functions, based on convex approximations and duality ideas. It is important partly because of the insights it gives into the relation between subdifferentials of nonsmooth functions and their counterparts in convex analysis, but also because the approach will provide new representations of subdifferentials which are extremely useful in applications.

The starting point is the definition of the (Clarke) generalized directional derivative of a locally Lipschitz continuous function.

Definition 4.7.2 *Take a function $f : \mathbb{R}^k \to \mathbb{R}$ and points $x \in \mathbb{R}^k$ and $v \in \mathbb{R}^k$. Assume that f is Lipschitz continuous on a neighbourhood of x. The generalized directional derivative of f at x in the direction v, written $f^0(x, v)$, is the number*

4.7 Subgradients of Lipschitz Continuous Functions

$$f^0(x, v) := \limsup_{y \to x, t \downarrow 0} t^{-1}[f(y + tv) - f(y)].$$

Notice that, because f is assumed merely to be Lipschitz continuous, the right side would not make sense if 'lim' replaced 'limsup'.

We list some salient properties of $f^0(x, v)$:

Proposition 4.7.3 *Take a function $f : \mathbb{R}^k \to \mathbb{R}$ and a point $x \in \mathbb{R}^k$. Assume that f is Lipschitz continuous on a neighbourhood of x with Lipschitz constant K. Then the function $v \to f^0(x, v)$ with domain $\mathbb{R}^k$ has the following properties:*

(i) it is finite valued, Lipschitz continuous with Lipschitz constant K, and positively homogeneous, in the sense that

$$f^0(x, \alpha v) = \alpha f^0(x, v) \text{ for all } v \in \mathbb{R}^k \text{ and } \alpha \geq 0,$$

(ii) it is convex.

Proof (i) These properties are straightforward consequences of the definition of f^0 and the assumed Lipschitz continuity of f (near x).

(ii) To show convexity of $f^0(x, \cdot)$, take any $v, w \in \mathbb{R}^k$. Since $f^0(x, \cdot)$ is positively homogeneous, it suffices to show 'sub-additivity', namely $f^0(x, v+w) \leq f^0(x, v) + f^0(x, w)$. We know that there exist sequences $y_i \to x$ and $t_i \downarrow 0$ such that

$$f^0(x, v + w) = \lim_i \{t_i^{-1}(f(y_i + t_i(v + w)) - f(y_i))\}.$$

But the term between braces on the right can be expressed

$$t_i^{-1}(f(z_i + t_i v) - f(z_i)) + t_i^{-1}(f(y_i + t_i w) - f(y_i)),$$

in which $z_i = y_i + t_i w$. Since $z_i \to x$, we conclude

$$f^0(x, v + w) \leq \limsup_{z \to x, t \downarrow 0} \{t^{-1}(f(z + tv) - f(z))\}$$

$$+ \limsup_{y \to x, t \downarrow 0} \{t^{-1}(f(y + tw) - f(y))\}$$

$$= f^0(x, v) + f^0(x, w),$$

as claimed. □

We have constructed a convex function $f^0(x, \cdot)$ which approximates f 'near' x. What would be more natural then than to introduce a 'subdifferential', we write it $\bar{\partial} f(x)$, which is the subdifferential in the sense of convex analysis of the convex function $v \to f^0(x, v)$ at $v = 0$? Since $f^0(x, 0) = 0$, this approach gives

$$\bar{\partial} f(x) := \{\xi : f^0(x, v) \geq \xi \cdot v \ \text{ for all } v \in \mathbb{R}^k\}.$$

$\bar{\partial} f(x)$ is called the *Clarke subdifferential* (alternatively referred to as the *Clarke generalized gradient*) of f at x. $\bar{\partial} f(x)$ is a non-empty, compact, convex set (for fixed x). Elements in $\bar{\partial} f(x)$ are uniformly bounded in Euclidean norm by the Lipschitz constant of f on a neighbourhood of x.

The generalized directional derivative can be interpreted as the support function of $\bar{\partial} f(x)$:

Proposition 4.7.4 *Take a function $f : \mathbb{R}^k \to \mathbb{R}$ which is Lipschitz continuous on a neighbourhood of a point $x \in \mathbb{R}^k$. Then*

$$f^0(x, v) = \max\{v \cdot \xi : \xi \in \bar{\partial} f(x)\} \ \text{for all } v \in \mathbb{R}^k.$$

Proof We have $f^0(x, v) \geq \max\{v \cdot \xi : \xi \in \bar{\partial} f(x)\}$ for all v, by definition of $\bar{\partial} f(x)$. Fix any v. It remains to show equality for this v. Choose ξ to be a subgradient (in the sense of convex analysis) to $f^0(x, \cdot)$ at v. Then

$$f^0(x, w) - f^0(x, v) \geq \xi \cdot (w - v) \ \text{ for all } w \in \mathbb{R}^k.$$

Take $w = \alpha \bar{w}$ for arbitrary $\bar{w} \in \mathbb{R}^k$ and $\alpha > 0$. Since $f^0(x, w) = \alpha f^0(x, \bar{w})$, we may divide across the inequality by α and pass to the limit as $\alpha \to \infty$. This gives $f^0(x, \bar{w}) \geq \xi \cdot \bar{w}$. We conclude that $\xi \in \bar{\partial} f(x)$. On the other hand, setting $w = 0$, we obtain $\xi \cdot v \geq f^0(x, v)$. It follows $f^0(x, v) = \max\{v \cdot \xi : \xi \in \bar{\partial} f(x)\}$. □

The Clarke subdifferential commutes with -1.

Proposition 4.7.5 *Take a function $f : \mathbb{R}^k \to \mathbb{R}$ which is Lipschitz continuous on a neighbourhood of a point $x \in \mathbb{R}^k$. Then*

$$\bar{\partial}(-f)(x) = -\bar{\partial} f(x).$$

Proof It suffices to show that the two sets have the same support function, which (as we now know) are $(-f)^0(x, v)$ and $f^0(x, -v)$ for any v. But

$$(-f)^0(x, v) = \limsup_{x' \to x, t \downarrow 0} \frac{(-f)(x' + tv) - (-f)(x')}{t},$$

which, following the substitution $y' = x' + tv$, becomes

$$(-f)^0(x, v) = \limsup_{y' \to x, t \downarrow 0} \frac{f(y' - tv) - f(y')}{t} = f^0(x, -v).$$

This is the required relation. □

4.7 Subgradients of Lipschitz Continuous Functions

Note that assertions of Proposition 4.7.5 are in general false if the limiting subdifferential is substituted for the Clarke subdifferential. This point is illustrated by functions (i) and (ii) in Example 4.3.

How does the Clarke subdifferential $\bar{\partial} f(x)$ fit in with our other nonsmooth constructs? It relates very neatly to the limiting subdifferential $\partial f(x)$.

Proposition 4.7.6 *Take a lower semi-continuous functions function $f : \mathbb{R}^k \to \mathbb{R}$ which is Lipschitz continuous on a neighbourhood of some point $x \in \mathbb{R}^k$. Then*

$$\bar{\partial} f(x) = \operatorname{co} \partial f(x).$$

Proof Since the two sets are closed convex sets, it suffices to show that they have the same support function, i.e. to show $f^0(x, v) = \max\{v \cdot \xi : \xi \in \partial f(x)\}$ for all $v \in \mathbb{R}^k$.

(a): We show $f^0(x, v) \geq \max\{v \cdot \xi : \xi \in \partial f(x)\}$ for all $v \in \mathbb{R}^k$.

Choose any $\xi \in \partial f(x)$ and any $v \in \mathbb{R}^k$. Then there exist sequences $x_i \xrightarrow{f} x$ and $\xi_i \to \xi$ such that $\xi_i \in \partial^P f(x_i)$ for each i, by Theorem 4.6.2. We know that for each i there exist $M_i > 0$ and $\delta_i > 0$ such that

$$f(x) - f(x_i) \geq \xi_i \cdot (x - x_i) - M_i |x - x_i|^2 \quad \text{for all } x \in x_i + \delta_i \mathbb{B}.$$

It follows that we can find a sequence $t_i \downarrow 0$ such that

$$f(x_i + t_i v) - f(x_i) \geq t_i \xi_i \cdot v - i^{-1} t_i \quad \text{for all } i.$$

Then

$$f^0(x, v) \geq \limsup_{i \to \infty} t_i^{-1} (f(x_i + t_i v) - f(x_i)) \geq \xi \cdot v.$$

Since ξ and v were arbitrary, we have

$$f^0(x, v) \geq \max\{v \cdot \xi : \xi \in \partial f(x)\} \quad \text{for all } v \in \mathbb{R}^k.$$

(b): We show $f^0(x, v) \leq \max\{v \cdot \xi : \xi \in \partial f(x)\}$ for all $v \in \mathbb{R}^k$.

Pick an arbitrary $v \in \mathbb{R}^k$. Then sequences $x_i \to x$ and $t_i \downarrow 0$ can be found such that

$$f^0(x, v) = \lim_{i \to \infty} t_i^{-1} (f(x_i + t_i v) - f(x_i)).$$

The mean value inequality (Theorem 4.5.2) applied to the function f with basepoint x_i and 'Y set' $\{x_i + t_i v\}$ for each i yields the following information: for each i there exists $z_i \in x_i + t_i |v| \mathbb{B}$ and $\xi_i \in \partial f(z_i)$ such that

$$f(x_i + t_i v) - f(x_i) \leq t_i \xi_i \cdot v. \tag{4.7.1}$$

Clearly $z_i \to x$. Because f is Lipschitz continuous on a neighbourhood of x, the ξ_i's are uniformly bounded. Along a subsequence then $\xi_i \to \xi$, for some $\xi \in \mathbb{R}^k$. By Theorem 4.6.2, $\xi \in \partial f(x)$. Dividing across (4.7.1) by t_i and passing to the limit gives

$$f^0(x, v) \leq \xi \cdot v \text{ for some } \xi \in \partial f(x).$$

We conclude that

$$f^0(x, v) \leq \max\{v \cdot \xi : \xi \in \partial f(x)\} \text{ for all } v.$$

$\square$

Consider now a function $f : \mathbb{R}^k \to \mathbb{R}$ and a point $x \in \mathbb{R}^k$ such that f is Lipschitz continuous on a neighbourhood of x. According to Rademacher's theorem, f is differentiable almost everywhere on this neighbourhood (with respect to k-dimensional Lebesgue measure). Can we construct limiting subgradients of f at x as limits of neighbouring derivatives? The answer to this question, supplied by the following theorem, is a qualified yes: the convex hull of the set of limits of neighbouring derivatives coincides with the convex hull of the limiting subdifferential at x. The same is true, even if we exclude neighbouring derivatives on some specified subset of measure zero. This important theorem is of intrinsic interest (it relates classical and modern concepts of derivative in a very concrete way), but also provides a computational tool for the convex hull of the limiting subdifferential of great power. Often the simplest approach to calculating co $\partial f(x)$, by far, is to look at limits of neighbouring derivatives.

When a function $f : \mathbb{R}^k \to \mathbb{R}$ is Lipschitz continuous on a neighbourhood of x, the concepts of Gâteaux, Hadamard and Fréchet differentiability all coincide. For concreteness, when we say that a Lipschitz continuous function f is differentiable at x, we shall mean Gâteaux differentiable, i.e. there is a unique vector $\xi \in \mathbb{R}^k$, the Gâteaux derivative $\nabla f(x)$ of f at x, such that

$$\lim_{h \downarrow 0} h^{-1}(f(x + hv) - f(x) - h\xi \cdot v) = 0 \text{ for all } v \in \mathbb{R}^k. \tag{4.7.2}$$

Notice that if f is differentiable at x (and Lipschitz continuous on a neighbourhood of x) then

$$\nabla f(x) \in \text{co } \partial f(x).$$

This is true since (4.7.2) (with $\nabla f(x)$ substituted in place of ξ) implies $f^0(x, v) \geq \xi \cdot v$ for every v, whence (by Proposition 4.7.6) $\nabla f(x) \in \text{co } \partial f(x)$.

Theorem 4.7.7 *Take a function $f : \mathbb{R}^k \to \mathbb{R}$, a point $x \in \mathbb{R}^k$ and any subset $\Omega \subset \mathbb{R}^k$ having Lebesgue measure zero. Assume that f is Lipschitz continuous on a neighbourhood of x. Then*

4.7 Subgradients of Lipschitz Continuous Functions

$$\text{co}\, \partial f(x) = \text{co}\{\xi \, : \, \exists x_i \to x, x_i \notin \Omega, \nabla f(x_i) \text{ exists and } \nabla f(x_i) \to \xi\}.$$

Proof We know that $y \rightsquigarrow \partial f(y)$ has closed graph on a neighbourhood of x (see Proposition 4.3.5). However the vectors in $\partial f(y)$ are uniformly bounded as y ranges over this neighbourhood, as may be deduced from Theorem 4.7.1. It follows from Carathéodory's theorem that the set valued function $y \rightsquigarrow \text{co}\, \partial f(y)$ has closed graph on a neighbourhood of x. But, we have observed, if f is differentiable at a point y near x, that $\nabla f(y) \in \text{co}\, \partial f(y)$. It follows that $S \subset \text{co}\, \partial f(x)$, where S is the set on the right side in the theorem statement.

Both the set S and $\text{co}\, \partial f(x)$ are compact, convex sets and, as we have shown, $S \subset \text{co}\, \partial f(x)$. So it remains to establish that the support functions of the two sets satisfy the inequality: for any $v \neq 0$

$$\sup\{\xi \cdot v : \xi \in \partial f(x)\} \leq \sup\{\xi \cdot v : \xi \in S\}.$$

Choose an arbitrary vector $v \neq 0$. In view of Proposition 4.7.4, it suffices to show that

$$f^0(x, v) \leq \sup\{\xi \cdot v : \xi \in S\}.$$

In fact we need only demonstrate that

$$f^0(x, v) \leq \alpha, \tag{4.7.3}$$

where the number α is defined to be

$$\alpha := \limsup\{\nabla f(y) \cdot v : f \text{ is differentiable at } y, y \notin \Omega, y \to x\}.$$

Choose any ϵ. Denote by $\mathcal{S}$ the subset of $\mathbb{R}^k$ on which f fails to be differentiable. By definition of α, there exists $\delta > 0$ such that

$$y \in x + \delta \mathbb{B} \text{ and } y \notin \Omega \cup \mathcal{S} \text{ implies } \nabla f(y) \cdot v \leq \alpha + \epsilon. \tag{4.7.4}$$

We suppose that $\delta > 0$ has been chosen small enough that $x + \delta \mathbb{B}$ lies in the neighbourhood on which f is Lipschitz continuous. By Rademacher's theorem then, $\Omega \cup \mathcal{S}$ has k dimensional Lebesgue measure zero in $x + \delta \mathbb{B}$. For each $y \in \mathbb{R}^k$, consider the line segments $L_y := \{y + tv : 0 < t < \delta/(2|v|)\}$. Since $\Omega \cup \mathcal{S}$ has k dimensional Lebesgue measure 0 in $x + \delta \mathbb{B}$, it follows from Fubini's theorem that, for almost every y in $x + (\delta/2)\mathbb{B}$, the line segment L_y intersects $\Omega \cup \mathcal{S}$ on a set of zero one-dimensional Lebesgue measure. Take y to be any point in $x + (\delta/2)\mathbb{B}$ having this property, and take any $t \in (0, \delta/(2|v|))$. Then

$$f(y + tv) - f(y) = \int_0^t \nabla f(y + sv) \cdot v \, ds.$$

This formula is valid because f is Lipschitz continuous on

$$\{y + sv : 0 \leq s \leq t\}.$$

Since $y + sv \in x + \delta \mathbb{B}$ and $y + sv \notin \Omega \cup \mathcal{S}$ for almost all $s \in (0, t)$, we conclude from (4.7.4) that

$$f(y + tv) - f(y) \leq t(\alpha + \epsilon).$$

This inequality holds for all $y \in x + (\delta/2)\mathbb{B}$, excluding a subset of measure zero and for all $t \in (0, \delta/(2|v|))$. But then, since f is continuous, it is true for all $y \in x + (\delta/2)\mathbb{B}, t \in (0, \delta/(2|v|))$. This implies that

$$f^0(x, v) = \limsup_{y \to x, t \downarrow 0} t^{-1}[f(y + tv) - f(y)] \leq \alpha + \epsilon.$$

Since $\epsilon > 0$ was arbitrary, we conclude (4.7.3). The proof is complete. □

4.8 The Distance Function

Take a set $C \subset \mathbb{R}^k$. The distance function $d_C : \mathbb{R}^k \to \mathbb{R}$ is

$$d_C(x) := \inf\{|x - y| : y \in C\}.$$

The distance function features prominently in nonsmooth analysis. This is not surprising since a key building block is the proximal normal, whose very definition revolves around the concept of 'closest point' (a point which achieves the infimum in the definition of the distance function).

Chapter 3 has already provided a foretaste of the distance function's significance. Take a set A in $\mathbb{R}^n$ and a Lipschitz continuous function $f : A \to \mathbb{R}$. Take also a subset $C \subset A$. Then, for the 'exact penalty' parameter K chosen sufficiently large, a minimizer for the constrained optimization problem

$$\text{Minimize } f(x) \text{ over points } x \in A \text{ satisfying } x \in C$$

is a minimizer also for the unconstrained problem

$$\text{Minimize } f(x) + K d_C(x) \text{ over points } x \in A.$$

This is a consequence of the exact penalization theorem (Theorem 3.2.1), which permits us to trade off the complications associated with a constraint against an extra, nonsmooth, term in the cost.

4.8 The Distance Function

Another important property of d_C is that it provides a description of a set C in terms of a Lipschitz continuous function.

Proposition 4.8.1 *For any set $C \subset \mathbb{R}^k$, the distance function $d_C : \mathbb{R}^k \to \mathbb{R}$ is Lipschitz continuous with Lipschitz constant 1.*

Proof Take any $x, y \in \mathbb{R}^k$. Choose $\epsilon > 0$. By definition of d_C, there exists $z \in C$ such that $d_C(y) \geq |y - z| - \epsilon$. Then

$$d_C(x) \leq |x - z| \leq |x - y| + |y - z| \leq |x - y| + d_C(y) + \epsilon.$$

But $\epsilon > 0$ was arbitrary. So $d_C(x) - d_C(y) \leq |x - y|$. Since the roles of x and y are interchangeable, we conclude $|d_C(x) - d_C(y)| \leq |x - y|$, which implies d_C is Lipschitz continuous with Lipschitz constant 1. □

Our goal now is to describe the limiting subgradients of d_C at a point in C. As usual, we start by studying proximal normals.

Proposition 4.8.2 *Take a closed set $C \subset \mathbb{R}^k$ and a point $x \in C$. Then*

$$\partial^P d_C(x) = N_C^P(x) \cap \mathbb{B}.$$

Proof
(a): We show $\partial^P d_C(x) \subset N_C^P(x) \cap \mathbb{B}$.

Suppose $\xi \in \partial^P d_C(x)$. Then, since d_C is Lipschitz continuous with Lipschitz constant 1, by Proposition 4.4.2, there exists $M > 0$ such that

$$\xi \cdot (y - x) \leq d_C(y) - d_C(x) + M|y - x|^2 \quad \text{for all } y \in \mathbb{R}^k.$$

Since d_C vanishes on C, we have

$$\xi \cdot (y - x) \leq M|y - x|^2 \quad \text{for all } y \in C.$$

But this implies $\xi \in N_C^P(x)$. Hence $|\xi| \leq 1$, by Proposition 4.3.3, so $\xi \in N_C^P(x) \cap \mathbb{B}$.
(b): We show $\partial^P d_C(x) \supset N_C^P(x) \cap \mathbb{B}$.

Take any $\xi \in N_C^P(x) \cap \mathbb{B}$. We need to show that $\xi \in \partial^P d_C(x)$. This is certainly true for $\xi = 0$ so we may assume $\xi \neq 0$. We know that there exists $M > 0$ such that

$$\xi \cdot (y - x) \leq M|y - x|^2 \quad \text{for all } y \in C.$$

This fact can be expressed:

$$C \cap Q = \emptyset$$

where Q is the open ball

$$Q := \{y \in \mathbb{R}^k : |(x + (2M)^{-1}\xi - y| < (2M)^{-1}|\xi|\}.$$

It follows that, for any $y \in Q$,

$$d_C(y) \geq g(y) \qquad (4.8.1)$$

in which

$$g(v) := (2M)^{-1}|\xi| - |v - x - (2M)^{-1}\xi|.$$

($g(v)$ will be recognized as the distance of v from the surface of the ball.) Since d_C is nonnegative, however, and g is nonpositive on the complement of Q, (4.8.1) is actually valid for all y.

The function g is analytic on a neighbourhood of x (since $\xi \neq 0$). We calculate $\nabla g(x) = |\xi|^{-1}\xi$. It follows that there exist some $\alpha > 0$ and $\epsilon > 0$ such that

$$g(y) - g(x) \geq |\xi|^{-1}\xi \cdot (y - x) - \alpha|y - x|^2 \text{ for all } y \in x + \epsilon \mathbb{B}.$$

Since $g(x) = d_C(x) = 0$, we deduce from (4.8.1) that

$$d_C(y) - d_C(x) \geq g(y) - g(x) \geq |\xi|^{-1}\xi \cdot (y - x) - \alpha|y - x|^2$$

for all $y \in x + \epsilon \mathbb{B}$. Noting however that $|\xi| \leq 1$ and $d_C(x) = 0$, we deduce that

$$d_C(y) - d_C(x) \geq |\xi|(d_C(y) - d_C(x)) \geq \xi \cdot (y - x) - \alpha|\xi||y - x|^2$$

for all $y \in x + \epsilon \mathbb{B}$. From this we conclude (via Proposition 4.4.1) that $\xi \in \partial^P d_C(x)$. □

Our next objective is to derive a similar description of limiting normal cones in terms of limiting subdifferentials of the distance function. First of all however we need to assemble more information about proximal subgradients $\xi \in \partial^P d_C(x)$ at basepoints $x \notin C$. The following lemma tells us that $\xi \in \partial^P d_C(\bar{x})$ for some $\bar{x} \in C$, and a little bit more.

Lemma 4.8.3 *Take a closed set $C \subset \mathbb{R}^k$, a point $x \notin C$.*

(i): For any vector $\xi \in \partial^P d_C(x)$, x has a unique closest point $\bar{x}$ in C such that

$$\xi = |x - \bar{x}|^{-1}(x - \bar{x})$$

and therefore $\xi \in \partial^P d_C(\bar{x})$.
(ii): $\partial d_C(x) \subset \{\xi : |\xi| = 1\}$.

Proof
(i): Take any $\xi \in \partial^P d_C(x)$. By Proposition 4.4.2, there exists $M > 0$ such that

$$\xi \cdot (y - x) \leq d_C(y) - d_C(x) + M|y - x|^2 \text{ for all } y \in \mathbb{R}^k. \qquad (4.8.2)$$

4.8 The Distance Function

Let $\bar{x}$ be a closest point to x in C. Then

$$|x - \bar{x}| = d_C(x) \, (\neq 0).$$

Evidently then $\bar{x}$ is the closest point to $\bar{x}+(1-\alpha)(x-\bar{x})$ for all $\alpha \in (0, 1)$, a property which can be expressed analytically as

$$d_C(\bar{x} + (1-\alpha)(x-\bar{x})) = (1-\alpha)|x-\bar{x}| = (1-\alpha)d_C(x).$$

Combining this relation with the proximal inequality (4.8.2), we deduce

$$-\alpha|x-\bar{x}| = d_C(\bar{x} + (1-\alpha)(x-\bar{x})) - d_C(x)$$
$$\geq -\alpha\xi \cdot (x-\bar{x}) - \alpha^2 M |x-\bar{x}|^2 \text{ for all } \alpha \in (0,1).$$

Divide across this inequality by α and pass to the limit as $\alpha \downarrow 0$. This gives

$$|x - \bar{x}| \leq \xi \cdot (x - \bar{x}).$$

But, by Proposition 4.8.2, $|\xi| \leq 1$. Since $x - \bar{x} \neq 0$, we deduce that

$$\xi = |x - \bar{x}|^{-1}(x - \bar{x}).$$

If $\bar{y}$ is any other closest point to x in C, the above reasoning gives $d_C(x) = |x - \bar{y}| = |x - \bar{x}| \neq 0$ and $\xi = |x - \bar{y}|^{-1}(x - \bar{y}) = |x - \bar{x}|^{-1}(x - \bar{x})$, from which we conclude $\bar{x} = \bar{y}$.

Of course, the fact that $\xi \in \partial^P d_C(\bar{x})$ follows from Proposition 4.8.1.

(ii): This assertion follows from (i), since C is closed and any limiting subgradient is the limit of a convergent sequence of proximal subgradients at neighbouring points to x in $\mathbb{R}^n \backslash C$. □

We note also the following partial converse of part (i) of the preceding proposition.

Lemma 4.8.4 *Take a closed set $C \subset \mathbb{R}^k$ and points $x \notin C$ and $\bar{x} \in C$. Assume that, for some $\alpha > 0$, $y = \bar{x}$ minimizes the function*

$$y \to |\bar{x} + (1+\alpha)(x-\bar{x}) - y|$$

over C. Define

$$\xi := |x - \bar{x}|^{-1}(x - \bar{x}).$$

We have:

(i) $\xi \in \partial^P d_C(x)$,

(ii) *if $\{x_i\}$ and $\{\xi_i\}$ are sequences such that $x_i \to x$ as $i \to \infty$ and $\xi_i \in \partial^P d_C(x_i)$ for each i, then there exists a sequence of points $\{y_i\}$ in C such that $\xi_i = |x_i - y_i|^{-1}(x_i - y_i)$ for all i, and $y_i \to \bar{x}$ and $\xi_i \to \xi$ as $i \to \infty$.*

Proof We arrange, by translation of the origin, that $\bar{x} = 0$.
(i): The minimizing property of $\bar{x}$ can be expressed geometrically: the open ball $(1+\alpha)x + (1+\alpha)|x|\mathbb{B}$ is disjoint from C. It follows that for any point $v \in \mathbb{R}^k$ with $|v - \alpha x| \leq |x|(1+\alpha)$

$$d_C(x+v) \geq \psi(v)$$

where

$$\psi(v) := (1+\alpha)|x| - |x+v-(1+\alpha)x| = (1+\alpha)|x| - |v-\alpha x|.$$

(the function ψ measures the distance of the point $x+v$ in the ball $(1+\alpha)x + (1+\alpha)|x|\mathbb{B}$ to its boundary.)

Notice that $d_C(x) = |x| = \psi(0)$, so

$$d_C(x+v) - d_C(x) \geq \psi(v) - \psi(0).$$

But ψ is analytic on $\alpha|x|\overset{\circ}{\mathbb{B}}$. (Notice $x \neq 0$ since $x \notin C$.) We calculate its gradient at $v = 0$ to be

$$\nabla\psi(0) = |x|^{-1}x.$$

It follows that, for some $M > 0$ and $\epsilon > 0$

$$d_C(x+v) - d_C(x) \geq |x|^{-1}v \cdot x - M|v|^2, \quad \text{for all } v \in \epsilon\mathbb{B}.$$

By Proposition 4.4.1, then

$$|x|^{-1}x \in \partial^P d_C(x).$$

(ii): We begin by showing that $\bar{x}(=0)$ is the unique minimizer of $y \to |x-y|$ over C. Let y' be an arbitrary such minimizer. Since $0 \in C$, $|x - y'|^2 \leq |x|^2$. Hence

$$-2x \cdot y' + |y'|^2 \leq 0.$$

But the minimizing property of $\bar{x}(=0)$ tells us that

$$|(1+\alpha)x - y'|^2 \geq (1+\alpha)^2|x|^2,$$

and hence that

$$-2(1+\alpha)x \cdot y' + |y'|^2 \geq 0.$$

4.8 The Distance Function

These relations combine to give

$$|y'|^2 - (1+\alpha)|y'|^2 \geq 0.$$

This is possible only if $y' = 0$. So $\bar{x}(= 0)$ is indeed the unique minimizer.

Now take sequences $\{x_i\}$ and $\{\xi_i\}$ with the stated properties. Since $x \notin C$ and C is closed, we may assume $x_i \notin C$ for all i. According to Lemma 4.8.3 then there exists a sequence $\{y_i\}$ in C such that

$$d_C(x_i) = |x_i - y_i| \tag{4.8.3}$$

and

$$\xi_i = |x_i - y_i|^{-1}(x_i - y_i). \tag{4.8.4}$$

Extract an arbitrary subsequence. $\{y_i\}$ is bounded. Along a further subsequence then we have $y_i \to \bar{y}$ for some $\bar{y} \in C$. By the continuity of the distance function, we deduce from (4.8.3) that $d_C(x) = |x - \bar{y}|$. In view of our earlier observations then, $\bar{y} = 0$. Noting (4.8.4), we see that

$$y_i \to 0 \text{ and } \xi_i \to |x|^{-1}x = \xi. \tag{4.8.5}$$

Since the limits here are independent of the subsequence initially selected, (4.8.5) is in fact true for the original sequence. □

Simple limit taking procedures will now permit to relate the limiting normal cone to a set and the limiting subdifferential of the distance function.

Theorem 4.8.5 *Take a closed set $C \subset \mathbb{R}^k$ and a point $x \in C$. Then*

$$\partial d_C(x) = N_C(x) \cap \mathbb{B}.$$

Proof
(a): We show $\partial d_C(x) \subset N_C(x) \cap \mathbb{B}$.
Take a vector $\xi \in \partial d_C(x)$. Then there exist $x_i \to x$ and $\xi_i \to \xi$ such that $\xi_i \in \partial^P d_C(x_i)$ for all i (by Theorem 4.6.2). In consequence of Proposition 4.8.2 and Lemma 4.8.3, there exists a sequence of points $\{z_i\}$ in C such that $z_i \xrightarrow{C} x$, and $\xi_i \in N_C^P(z_i)$ and $|\xi_i| \leq 1$ for each i. But then by definition of N, $\xi \in N_C(x) \cap \mathbb{B}$.

(b): We show $\partial d_C(x) \supset N_C(x) \cap \mathbb{B}$.
Since $N_C(x)$ is a cone, generated by taking limits of proximal normals and $x \rightsquigarrow \partial d_C(x)$ has closed graph, it suffices to show that $N_C^P(x) \cap \mathbb{B} \subset \partial^P d_C(x)$. But this property is supplied by Proposition 4.8.2. □

4.9 Criteria for Lipschitz Continuity

On occasions, we wish to establish Lipschitz continuity of some specific function. In the case when an explicit formula is available for evaluation of the function, we can expect to be able to assess whether this property holds directly. There are, too, certain elementary criteria for Lipschitz continuity which are sometimes of use: 'the pointwise supremum of a uniformly bounded family of Lipschitz continuous functions with common Lipschitz constant is Lipschitz continuous', for example. But for certain implicitly defined functions, most notably for many of the 'value' functions considered in optimization, the task of establishing Lipschitz continuity on a neighbourhood of a basepoint can be a challenging one. Here conditions involving boundedness of subdifferentials are often useful. The following theorem makes precise the intuitive notion that lack of Lipschitz continuity should be reflected in the presence of 'unbounded' neighbouring derivatives.

Theorem 4.9.1 *Take a lower semi-continuous function* $f : \mathbb{R}^k \to \mathbb{R} \cup \{+\infty\}$, *a point* $x \in \text{dom } f$ *and a constant* $K > 0$. *Then the following conditions are equivalent:*

(i) *f is Lipschitz continuous with Lipschitz constant K on some neighbourhood of x,*
(ii) *there exists $\epsilon > 0$ such that for every $x \in \mathbb{R}^k$ and $\xi \in \partial^P f(x)$ satisfying*

$$|x - \bar{x}| \leq \epsilon \text{ and } |f(x) - f(\bar{x})| \leq \epsilon,$$

we have

$$|\xi| \leq K.$$

The theorem tells us that, in order to check Lipschitz continuity, we have only to test boundedness of neighbouring proximal subgradients, at points where they are defined.

This theorem and earlier derived properties of subdifferentials lead to an alternative criterion for Lipschitz continuity involving the asymptotic subdifferential.

Corollary 4.9.2 *Take a lower semi-continuous function* $f : \mathbb{R}^k \to \mathbb{R} \cup \{+\infty\}$ *and a point $\bar{x} \in \text{dom } f$. The following conditions are equivalent:*

(i) *f is Lipschitz continuous on a neighbourhood of $\bar{x}$,*
(ii) *$\partial^\infty f(\bar{x}) = \{0\}$.*

Proof Condition (i) implies condition (ii), by Proposition 4.7.1. To show the reverse implication, assume that (i) is not true. Then, by Theorem 4.9.1, there exist sequences $x_i \xrightarrow{f} \bar{x}$ and $\{\xi_i\}$ such that

$$\xi_i \in \partial^P f(x_i) \text{ for each } i$$

4.9 Criteria for Lipschitz Continuity

and
$$|\xi_i| \to \infty \text{ as } i \to \infty.$$

For each i, we have
$$(|\xi_i|^{-1}\xi_i, -|\xi_i|^{-1}) \in N^P_{\text{epi}f}(x_i, f(x_i)).$$

Along a subsequence however
$$(|\xi_i|^{-1}\xi_i, -|\xi_i|^{-1}) \to (\bar{\xi}, 0)$$

for some $\bar{\xi}$ with $|\bar{\xi}| = 1$. In view of the closure properties of the limiting normal cone
$$(\bar{\xi}, 0) \in N_{\text{epi}f}(\bar{x}, f(\bar{x})).$$

This means that $\bar{\xi}$ is a non-zero vector in $\partial^\infty f(\bar{x}) = \{0\}$. We have shown that (ii) is not true. $\square$

As a preliminary step in the proof of Theorem 4.9.1, we establish some properties of proximal subgradients of a special 'min' function.

Lemma 4.9.3 *Take a lower semi-continuous function* $f : \mathbb{R}^k \to \mathbb{R} \cup \{+\infty\}$, $x \in \mathbb{R}^k$ *and* $\beta \in \mathbb{R}$. *Define*
$$g(y) := \min\{f(y), \beta\}.$$

Suppose that ξ is a non-zero vector such that
$$\xi \in \partial^P g(x).$$

Then
$$f(x) \leq \beta \text{ and } \xi \in \partial^P f(x).$$

Proof Since $\xi \in \partial^P g(x)$, there exists $\alpha > 0$ such that
$$\min\{f(x'), \beta\} - \min\{f(x), \beta\} \geq \xi \cdot (x' - x) - M|x' - x|^2$$

for all $x' \in x + \alpha \mathbb{B}$.

Suppose that $f(x) > \beta$. Then
$$0 = \beta - \beta \geq \min\{f(x'), \beta\} - \min\{f(x), \beta\} \geq \xi \cdot (x' - x) - M|x' - x|^2$$

for all $x' \in x + \alpha \mathbb{B}$. It follows that $\xi = 0$. But this is impossible since we have assumed $\xi \neq 0$. We have shown that

$$f(x) \leq \beta.$$

Notice also that

$$f(x') - f(x) \geq \min\{f(x'), \beta\} - \min\{f(x), \beta\}$$
$$\geq \xi \cdot (x' - x) - M|x' - x|^2$$

for all $x' \in x + \alpha \mathbb{B}$. It follows that $\xi \in \partial^P f(x)$. □

Proof of Theorem 4.9.1 That condition (i) implies condition (ii) follows from Theorem 4.7.1. To prove the reverse implication, assume that f is not Lipschitz continuous with Lipschitz constant K on any neighbourhood of $\bar{x}$. We shall show that condition (ii) cannot be satisfied.

Choose any $\epsilon > 0$. Since f is lower semi-continuous, we can find $\delta > 0$, $0 < \delta < \epsilon$, such that

$$z \in \bar{x} + \delta \mathbb{B} \text{ implies } f(z) > f(\bar{x}) - \epsilon. \quad (4.9.1)$$

Under the hypothesis there exist sequences

$$x_i \to \bar{x}, \ y_i \to \bar{y}$$

such that

$$x_i \neq y_i$$

and

$$f(y_i) - f(x_i) > K|y_i - x_i| \quad (4.9.2)$$

for all i. (The case $f(y_i) = +\infty$ is permitted.) We can assume that for some subsequence (we do not relabel)

$$f(x_i) \to f(\bar{x}) \text{ as } i \to \infty. \quad (4.9.3)$$

This is because if this property fails to hold then, in view of the lower semicontinuity of f, we have

$$\liminf_{i \to \infty} f(x_i) > f(\bar{x}).$$

But $|x_i - \bar{x}| \to 0$. It follows that (4.9.2) remains valid (for all i sufficiently large) when we redefine $x_i := \bar{x}$. Of course (4.9.3) is automatically satisfied.

4.9 Criteria for Lipschitz Continuity

In view of these properties, we can choose a sequence $\epsilon_i \downarrow 0$ such that

$$f(x_i) - f(\bar{x}) \leq \epsilon_i \tag{4.9.4}$$

and

$$K|y_i - x_i| < f(\bar{x}) - f(x_i) + \epsilon_i. \tag{4.9.5}$$

Fix a value of the index i such that

$$|x_i - \bar{x}| < \tfrac{1}{2}\delta, \ |y_i - \bar{x}| < \tfrac{1}{2}\delta \text{ and } \epsilon_i < \epsilon. \tag{4.9.6}$$

Define the function

$$g_i(y) := \min\{f(\bar{x}) + \epsilon_i, f(y)\}.$$

It follows from (4.9.2) and (4.9.5) that

$$g_i(y_i) - g_i(x_i)$$
$$= \min\{f(\bar{x}) - f(x_i) + \epsilon_i, f(y_i) - f(x_i)\} > K|y_i - x_i|.$$

Since g_i is a lower semi-continuous function which is finite at x_i and y_i, the following conclusions can be drawn from this last relation and the mean value inequality (Theorem 4.5.1): there exist

$$z \in \bar{x} + \delta\mathbb{B} \text{ and } \xi \in \partial^P g_i(z)$$

such that

$$K|y_i - x_i| < \xi \cdot (y_i - x_i).$$

But $y_i \neq x_i$. It follows that

$$|\xi| > K.$$

Recall however that $\delta < \epsilon$. So $|z - \bar{x}| < \epsilon$. From (4.9.1) we deduce

$$f(z) - f(\bar{x}) > -\epsilon.$$

Since the proximal subgradient ξ is non-zero, we deduce, from Lemma 4.9.3 and the preceding relations, that

$$|z - \bar{x}| < \epsilon, \ |f(z) - f(\bar{x})| < \epsilon \text{ and } \xi \in \partial^P f(z).$$

But ϵ is an arbitrary positive number. It follows that condition (ii) is not satisfied. $\square$

4.10 Relations Between Normal and Tangent Cones

Up to this point, tangent vectors have put in only an occasional appearance. In this optimization oriented treatment of the theory, it is natural to emphasize the significance of normal vectors, since it is in terms of these objects (and their close relatives, subgradients) that Lagrange multiplier rules and other important principles of nonsmooth optimization are expressed. There is however a rich web of relations involving tangent vectors and normal vectors, as we now reveal.

Take a closed set $C \subset \mathbb{R}^k$ and a point $x \in C$. With the help of the set limit notation of Sect. 2.1, we now review the various concepts of normal cone and their interrelation.

The *proximal normal cone* to C at x, $N_C^P(x)$, is

$$N_C^P(x) := \{\xi : \exists\, M > 0 \text{ such that } \xi \cdot (y - x) \leq M|y - x|^2 \text{ for all } y \in C\}.$$

The *strict normal cone* to C at x, $\hat{N}_C(x)$, is

$$\hat{N}_C(x) := \{\xi : \limsup_{y \xrightarrow{C} x} |y - x|^{-1} \xi \cdot (y - x) \leq 0\}.$$

The *limiting normal cone* is expressible in terms of our new notation as

$$N_C(x) = \limsup_{y \xrightarrow{C} x} N_C^P(y).$$

Proposition 4.2.5 provides us also with the following representation of $N_C(x)$:

$$N_C(x) = \limsup_{y \xrightarrow{C} x} \hat{N}_C(y).$$

We recall that $N_C^P(x)$, $\hat{N}_C(x)$ and $N_C(x)$ are all cones containing $\{0\}$. $N_C^P(x)$ is convex, $\hat{N}_C(x)$ is closed and convex, $N_C(x)$ is closed and

$$N_C^P(x) \subset \hat{N}_C(x) \subset N_C(x).$$

The tangent cones of primary interest here are the *Bouligand* and the *Clarke* tangent cones defined as follows

Definition 4.10.1 Take a closed set $C \subset \mathbb{R}^k$ and a point $x \in C$.
The *Bouligand tangent cone* to C at $x \in C$, written $T_C(x)$, is the set

$$T_C(x) := \limsup_{t \downarrow 0} t^{-1}(C - x).$$

4.10 Relations Between Normal and Tangent Cones

The *Clarke tangent cone* to C at x, written $\bar{T}_C(x)$, is the set

$$\bar{T}_C(x) := \liminf_{t\downarrow 0,\, y\xrightarrow{C} x} t^{-1}(C - y).$$

Equivalent 'sequential' definitions are as follows:

$T_C(x)$ comprises vectors ξ corresponding to which there exist some sequence $\{c_i\}$ in C and some sequence $t_i \downarrow 0$ such that $t_i^{-1}(c_i - x) \to \xi$.

$\bar{T}_C(x)$ comprises vectors ξ such that for any sequences $x_i \xrightarrow{C} x$ and $t_i \downarrow 0$ there exists a sequence $\{c_i\}$ in C such that $t_i^{-1}(c_i - x_i) \to \xi$.

We state without proof the following elementary properties.

Proposition 4.10.2 *Take a closed set $C \subset \mathbb{R}^k$ and a point $x \in C$. Then $T_C(x)$ and $\bar{T}_C(x)$ are closed cones containing the origin and $\bar{T}_C(x) \subset T_C(x)$.*

A less obvious property is:

Proposition 4.10.3 *Take a closed set $C \subset \mathbb{R}^k$ and a point $x \in C$. Then $\bar{T}_C(x)$ is a convex set.*

Proof Take any u and v in $\bar{T}_C(x)$. Since $\bar{T}_C(x)$ is a cone, to establish convexity we must show that $u + v \in \bar{T}_C(x)$. Take any sequence $x_i \xrightarrow{C} x$ and $t_i \downarrow 0$. Because $u \in \bar{T}_C(x)$, there exists a sequence of points $u_i \to u$ such that $x_i + t_i u_i \in C$ for each i. But $x_i + t_i u_i \to x$, so, since $v \in \bar{T}_C(x)$ also, there exists a sequence of points $v_i \to v$ such that $(x_i + t_i u_i) + t_i v_i \in C$ for each i. A rearrangement of this inclusion gives $x_i + t_i(u_i + v_i) \in C$. But this implies $u + v \in \bar{T}_C(x)$. □

Denote by S^* the polar cone of a set $S \subset \mathbb{R}^k$, namely

$$S^* := \{\xi : \xi \cdot x \le 0 \text{ for all } x \in S\}.$$

Theorem 4.10.4 *Take a closed set $C \subset \mathbb{R}^k$ and a point $x \in C$. Then the strict normal cone $\hat{N}_C(x)$ and the Bouligand tangent cone $T_C(x)$ are related according to*

$$\hat{N}_C(x) = T_C(x)^*.$$

Proof
(a): We show that $\hat{N}_C(x) \subset T_C(x)^*$.
Take any $\xi \in \hat{N}_C(x)$. Then

$$\xi \cdot (y - x) \le o(|y - x|) \quad \text{for all } y \in C \tag{4.10.1}$$

for some $o(\cdot) : \mathbb{R}_+ \to \mathbb{R}_+$ which satisfies $o(s)/s \to 0$ as $s \downarrow 0$. Choose any $v \in T_C(x)$. Then there exists $x_i \xrightarrow{C} x$ and $t_i \downarrow 0$ such that, if we define $v_i := t_i^{-1}(x_i - x)$ for each i, then $v_i \to v$. It is claimed that $\xi \cdot v \le 0$; it will follow that $\xi \in T_C(x)^*$.

The claim is certainly valid when $v = 0$. So assume $v \neq 0$. For each i then

$$\xi \cdot v_i = t_i^{-1} \xi \cdot (x_i - x) \leq |v_i| \|x_i - x\|^{-1} o(|x_i - x|)$$

by (4.10.1). Passing to the limit as $i \to \infty$, we obtain $\xi \cdot v \leq 0$.

(b): We show that $\hat{N}_C(x) \supset T_C(x)^*$.

Suppose that $\xi \notin \hat{N}_C(x)$. Then there exist $\epsilon > 0$ and $x_i \xrightarrow{C} x$ such that

$$\xi \cdot (x_i - x) > \epsilon |x_i - x| \text{ for all } i.$$

Note that $x_i \neq x$ for each i. Set $t_i = |x_i - x|$ and $v_i := t_i^{-1}(x_i - x)$. The v_i's all have unit length. Along a subsequence then, $v_i \to v$ for some $v \in T_C(x)$ with $|v| = 1$. Along the subsequence we have

$$\xi \cdot v_i = t_i^{-1} \xi \cdot (x_i - x) > \epsilon t_i^{-1} |x_i - x| = \epsilon |v_i| = \epsilon.$$

In the limit, we obtain $\xi \cdot v \geq \epsilon$. It follows $\xi \notin T_C(x)^*$. □

The next theorem tells us (among other things) that we get the Clarke tangent cone by applying the lim inf operation (with respect to the base point) to the Bouligand tangent cone.

Theorem 4.10.5 *Take a closed set $C \subset \mathbb{R}^k$ and a point $x \in C$. Then the Bouligand tangent cone $T_C(x)$, its closed convex hull and the Clarke tangent cone $\bar{T}_C(x)$ are related as follows:*

$$\liminf_{y \xrightarrow{C} x} \overline{\mathrm{co}}\, T_C(y) = \liminf_{y \xrightarrow{C} x} T_C(y) = \bar{T}_C(x).$$

Proof Since $T_C(y) \subset \overline{\mathrm{co}}\, T_C(y)$ for each y, it suffices to show

(i) $\liminf_{y \xrightarrow{C} x} \overline{\mathrm{co}}\, T_C(y) \subset \bar{T}_C(x)$,

(ii) $\bar{T}_C(x) \subset \liminf_{y \xrightarrow{C} x} T_C(y)$,

since these inclusions imply

$$\liminf_{y \xrightarrow{C} x} \overline{\mathrm{co}} T_C(y) \subset \bar{T}_C(x) \subset \liminf_{y \xrightarrow{C} x} T_C(y) \subset \liminf_{y \xrightarrow{C} x} \overline{\mathrm{co}} T_C(y)$$

from which the required relations follow.

We show

$$\liminf_{y \xrightarrow{C} x} \overline{\mathrm{co}} T_C(y) \subset \bar{T}_C(x).$$

4.10 Relations Between Normal and Tangent Cones

Take any $v \in \liminf_{y \xrightarrow{C} x} \overline{co} T_C(y)$. Let $x_i \xrightarrow{C} x$ and $t_i \downarrow 0$ be arbitrary sequences. Suppose we are able to find a sequence $\epsilon_i \downarrow 0$ such that

$$t_i^{-1} d_C(x_i + t_i v) \leq \epsilon_i. \tag{4.10.2}$$

Then we have

$$\limsup_i t_i^{-1} d_C(x_i + t_i v) = 0.$$

This implies $v \in \bar{T}_C(x)$ which is the required relation.

It remains to show (4.10.2). For each i, we consider the function

$$g_i(t) := d_C(x_i + tv), \ 0 \leq t \leq t_i.$$

Choose $z_i(t) \in \arg\min\{|x_i + tv - z| : z \in C\}$. Notice that, since $x_i \in C$,

$$|x - z_i(t)| \leq |x_i + tv - z_i(t)| + |x_i - x + tv| \leq 2t_i|v| + |x_i - x|$$

for $0 \leq t \leq t_i$ and $i = 1, 2, \ldots$
Evidently then

$$\sup_{0 \leq t \leq t_i} |x - z_i(t)| \to 0 \text{ as } i \to \infty.$$

Since $v \in \liminf_{y \xrightarrow{C} x} \overline{co} T_C(y)$, there exist $\epsilon_i \downarrow 0$ and functions $w_i(\cdot) : [0, t_i] \to \mathbb{R}^n$ such that

$$w_i(t) \in \overline{co} T_C(z_i(t)) \text{ and } |w_i(t) - v| < \epsilon_i \text{ for all } t \in [0, t_i].$$

Fix i. We verify (4.10.2) for the above choice of ϵ_i.

Let us first investigate the properties of a point $t \in [0, t_i)$ at which $g_i > 0$ and where g_i is differentiable. To simplify the notation, write z for $z_i(t)$ and set $p = x_i + tv$. We have for sufficiently small $h > 0$

$$g_i(t + h) - g_i(t) \leq |x_i + tv + hv - z| - |x_i + tv - z|$$
$$= |p - z + hv| - |p - z|.$$

Since $p \neq z$ ($g_i(t) > 0$, remember) the function $u \to |p - z + u|$ is differentiable at any point sufficiently close to the origin, in particular at hv (for $h > 0$ sufficiently small). In view of the convexity of the norm then the 'subgradient inequality' for convex functions gives

$$g_i(t + h) - g_i(t) \leq h|p - z + hv|^{-1}(p - z + hv) \cdot v.$$

We conclude that

$$h^{-1}(g_i(t+h) - g_i(t)) \le \left(\frac{p-z+hv}{|p-z+hv|} - \frac{p-z}{|p-z|}\right) \cdot v + \frac{p-z}{|p-z|} \cdot v$$

$$\le \left(\frac{p-z+hv}{|p-z+hv|} - \frac{p-z}{|p-z|}\right) \cdot v$$

$$+ \frac{p-z}{|p-z|} \cdot w_i + \epsilon_i. \qquad (4.10.3)$$

Since $p - z \in N_C^P(z)$ and $w_i(t) \in \bar{co}T_C(z)$ and in view of the fact that

$$\hat{N}_C(z) = T_C^*(z) = (\bar{co}T_C(z))^*,$$

it follows that

$$(p-z) \cdot w_i(t) \le 0.$$

We therefore retain inequality (4.10.3) when we drop the second term on the right. Passing to the limit as $h \downarrow 0$, we obtain

$$\frac{d}{dt} g_i(t) \le \epsilon_i.$$

Since g_i is Lipschitz continuous and therefore almost everywhere differentiable, this inequality is valid for all points $t \in \{s \in [0, t_i) : g_i(s) > 0\}$ excluding a nullset.

Condition (4.10.2) is automatically satisfied if $g_i(t_i) = 0$. Suppose then that $g_i(t_i) > 0$. Define

$$\tau := \inf\{t \in [0, t_i] : g_i(s) > 0 \text{ for all } s \in [t, t_i]\}.$$

Since g_i is continuous and $g_i(0) = 0$, we conclude $g_i(\tau) = 0$. It follows that

$$g_i(t_i) = g_i(\tau) + \int_\tau^{t_i} \dot{g}_i(\sigma) d\sigma \le 0 + t_i \epsilon_i.$$

But then

$$t_i^{-1} d_C(x_i + t_i v)(= t_i^{-1} g_i(t_i)) \le \epsilon_i.$$

Condition (4.10.2) is therefore satisfied in this case also. The proof is complete.

We show that

$$\bar{T}_C(x) \subset \liminf_{y \xrightarrow{C} x} T_C(y).$$

4.10 Relations Between Normal and Tangent Cones

Take $v \in \bar{T}_C(x)$. Consider an arbitrary sequence $x_i \xrightarrow{C} x$. It follows from the definition of $\bar{T}_C(x)$ that there exists $\epsilon_i \downarrow 0$, $t_i \downarrow 0$ and $N_i \uparrow \infty$ such that

$$t^{-1} d_C(x_k + tv) \leq \epsilon_i \quad \text{for all } 0 \leq t \leq t_i, \text{ and } k \geq N_i.$$

Relabel $\{x_{N_i}\}_{i=1}^\infty$ as $\{x_i\}$ and $\{t_{N_i}\}_{i=1}^\infty$ as $\{t_i\}$. Then for each $i = 1, 2, \ldots$ and for all $t \in [0, t_i]$, we have in particular

$$t^{-1} d_C(x_i + tv) \leq \epsilon_i.$$

Fix i and $t \in [0, t_i]$, choose $z_i(t) \in C$ such that

$$d_C(x_i + tv) = |x_i + tv - z_i(t)|.$$

Define

$$v_i(t) := t^{-1}(z_i(t) - x_i), \quad \text{for all } t \in (0, t_i], \, i = 1, 2, \ldots \quad (4.10.4)$$

Then

$$|v_i(t) - v| \leq \epsilon_i \quad \text{for all } t \in (0, t_i], \text{ for } i = 1, 2, \ldots$$

Fix i, choose $\bar{v}_i$ to be a cluster point of $\{v_i(t)\}_{t>0}$ at $t = 0$ (such $\bar{v}_i$ exists since $v_i(\cdot)$ is bounded on $(0, t_i]$). We have $|\bar{v}_i - v| \leq \epsilon_i$. By (4.10.4) and in view of the definition of $T_C(x_i)$, we have that

$$\bar{v}_i \in T_C(x_i) \text{ and } \bar{v}_i \to v.$$

Starting with an arbitrary sequence $\{x_i\}$ in C converging to x, we have found a subsequence $\{x_{i_n}\}_{n=1}^\infty$ and a sequence $\{\bar{v}_n\}$ such that

$$\bar{v}_n \in T_C(x_{i_n}) \text{ for all } n \text{ and } \bar{v}_n \to v \text{ as } n \to \infty.$$

It follows that $v \in \liminf_{y \xrightarrow{C} x} T_C(y)$. □

Our next objective is to relate the Clarke tangent cone and the limiting normal cone. First a lemma is required on the interaction between limit taking and the construction of polar sets.

Lemma 4.10.6 *Take a set valued function $S : \mathbb{R}^k \rightsquigarrow \mathbb{R}^n$, a set $C \subset \mathbb{R}^k$ and a point $x \in \mathbb{R}^k$. Assume that $S(y)$ is a closed convex cone for all y. Then*

$$\liminf_{y \xrightarrow{C} x} S(y) = [\limsup_{y \xrightarrow{C} x} S(y)^*]^*.$$

Proof

(a): We show that

$$\liminf_{y \xrightarrow{C} x} S(y) \subset [\limsup_{y \xrightarrow{C} x} S(y)^*]^*.$$

Take $v \in \liminf_{y \xrightarrow{C} x} S(y)$. Choose any $\xi \in \limsup_{y \xrightarrow{C} x} S(y)^*$. Then there exist sequences $x_i \xrightarrow{C} x$ and $\{\xi_i\}$ in $\mathbb{R}^n$ such that

$$\xi_i \in S(x_i)^* \text{ for all } i \text{ and } \xi_i \to \xi.$$

We may also choose a sequence $v_i \to v$ such that $v_i \in S(x_i)$ for each i. For each i then we have $\xi_i \cdot v_i \le 0$. In the limit we get $\xi \cdot v \le 0$. It follows that

$$v \in [\limsup_{y \xrightarrow{C} x} S(y)^*]^*.$$

(b): We show that

$$\liminf_{y \xrightarrow{C} x} S(y) \supset [\limsup_{y \xrightarrow{C} x} S(y)^*]^*.$$

Take $v \in [\limsup_{y \xrightarrow{C} x} S(y)^*]^*$. Assume, contrary to the assertions of the lemma that $v \notin \liminf_{y \xrightarrow{C} x} S(y)$. Then there is some $\epsilon > 0$ and a sequence $x_i \xrightarrow{C} x$ such that

$$(v + \epsilon \mathbb{B}) \cap S(x_i) = \emptyset, \text{ for each } i.$$

Applying the separation theorem we obtain, for each i, a vector ξ_i such that $|\xi_i| = 1$ and

$$\sup_{w \in S(x_i)} w \cdot \xi_i \le \inf_{w \in v + \epsilon \mathbb{B}} w \cdot \xi_i = v \cdot \xi_i - \epsilon .$$

Since $S(x_i)$ is a cone, we deduce that $\sup_{w \in S(x_i)} w \cdot \xi_i = 0$ and so $\xi_i \in S(x_i)^*$. Along a subsequence $\xi_i \to \xi$ for some ξ with $|\xi| = 1$. We have that $\xi \in \limsup_{y \xrightarrow{C} x} S(y)^*$. We also have $0 \le v \cdot \xi_i - \epsilon$. In the limit we arrive at $v \cdot \xi \ge \epsilon$, from which we conclude that $v \notin [\limsup_{y \xrightarrow{C} x} S(y)^*]^*$. This contradiction concludes the proof. □

Theorem 4.10.7 *Take a closed set $C \subset \mathbb{R}^k$ and $x \in C$. The Clarke tangent cone $\bar{T}_C(x)$ and the limiting normal cone $N_C(x)$ are related according to*

$$\bar{T}_C(x) = N_C(x)^*.$$

4.11 Interior of Clarke's Tangent Cone

Proof Apply the preceding lemma to $S(y) := \overline{\text{co}} T_C(y)$. This gives

$$\liminf_{y \xrightarrow{C} x} \overline{\text{co}}\, T_C(y) = (\limsup_{y \xrightarrow{C} x} \overline{\text{co}}\, T_C(y)^*)^*.$$

But $\liminf_{y \xrightarrow{C} x} \overline{\text{co}} T_C(y) = \bar{T}_C(x)$ by Theorem 4.10.5. Also, for each y we have $(\overline{\text{co}} T_C(y))^* = T_C(y)^* = \hat{N}_C(y)$ by Theorem 4.10.4. But then, by Proposition 4.2.5, $N_C(x) = \limsup_{y \xrightarrow{C} x} (\overline{\text{co}} T_C(y))^*$. Assembling those relations, we get $\bar{T}_C(x) = N_C(x)^*$. □

The relations we have established between $N_C^P(x)$, $\hat{N}_C(x)$, $N_C(x)$, $\bar{T}_C(x)$ and $T_C(x)$ associated with a closed set $C \in \mathbb{R}^k$ and point $x \in C$, are summarized in the following diagram, in which '$*$' denotes 'take the polar cone':

$$\begin{array}{ccc} T_C(x) & \xrightarrow{\liminf} & \bar{T}_C(x) \\ *\downarrow & & \uparrow * \\ \hat{N}_C(x) & \xrightarrow{\limsup} N_C(x) \xleftarrow{\limsup} & N_C^P(x) \end{array}$$

4.11 Interior of Clarke's Tangent Cone

A useful characterization of the interior of Clarke's tangent cone is provided by the following theorem.

Theorem 4.11.1 *Take a closed set $C \subset \mathbb{R}^k$ and $x \in C$. the following statements are equivalent:*

(i) $v \in \text{int}\, \bar{T}_C(x)$,
(ii) *there exists $\epsilon > 0$ such that*

$$d_C(y + tw) \leq d_C(y), \quad \text{for all } y \in x + \epsilon \mathbb{B},\ w \in v + \epsilon \mathbb{B} \text{ and } t \in [0, \epsilon],$$
(4.11.1)

(iii) *there exists $\epsilon > 0$ such that*

$$y + [0, \epsilon](v + \epsilon \mathbb{B}) \subset C, \quad \text{for all } y \in (x + \epsilon \mathbb{B}) \cap C.$$
(4.11.2)

Proof Since (iii) is an immediate consequence of (ii), we shall show that '(i) $\Rightarrow$ (ii)' and '(iii) $\Rightarrow$ (i)'.

'(i) $\Rightarrow$ (ii)'.
Assume first that $v \in \text{int}\, \bar{T}_C(x)$. If $v = 0$, then it means that $\bar{T}_C(x) = \mathbb{R}^n$. Using well-known polarity properties we deduce that $\{0\} = (\bar{T}_C(x))^* = \overline{\text{co}} N_C(x)$, and so

$N_C(x) = \{0\}$. Then, from Proposition 4.2.7 it follows that $x \in \overset{\circ}{C}$. Then taking a suitably small $\epsilon > 0$ we obtain (4.11.2).

So, we can reduce attention to the case when $0 \neq v \in \operatorname{int} \bar{T}_C(x)$. From Theorem 4.10.7 we know that $\bar{T}_C(x) = N_C(x)^*$, and so there exists $\rho > 0$ such that

$$\xi \cdot v \leq -\rho, \quad \text{for all } \xi \in N_C(x).$$

Suppose now that (4.11.1) is not satisfied. Then, there would exist sequences $y_i \to x$, $v_i \to v$ and $t_i \downarrow 0$ such that

$$d_C(y_i + t_i v_i) - d_C(y_i) > 0, \quad \text{for all } i \geq 1.$$

Take any sequence $\epsilon_i \downarrow 0$. Then, applying the proximal mean value inequality (Theorem 4.5.1), we can find, for each $i \geq 1$, a point $x_i \in [y_i, y_i + t_i v_i] + \epsilon_i \mathbb{B}$ and a vector $\xi_i \in \partial^P d_C(x_i)$ such that

$$\xi_i \cdot v_i > 0.$$

Using Proposition 4.8.2 (when $x_i \in C$) and Lemma 4.8.3 (when $x_i \notin C$), we also know that $\xi_i/|\xi_i| \in \partial^P d_C(x_i) \cap \partial \mathbb{B}$. Therefore, bearing in mind Theorem 4.6.2 (which provides a characterization of the limiting subgradients) and Theorem 4.8.5, extracting a subsequence if necessary, we see that ξ_i converges to a vector $\xi \in \partial d_C(x) = N_C(x) \cap \mathbb{B}$ such that $|\xi| = 1$. As a consequence we would obtain that

$$-\rho = -\rho|\xi| \geq v \cdot \xi = \lim_{i \to \infty} |\xi_i|^{-1} \xi_i \cdot v_i \geq 0,$$

and thereby arrive at a contradiction. We deduce the validity of (4.11.1).

'(iii) $\Rightarrow$ (i)'.

Suppose now that (4.11.2) is in force. Then it follows that the generalized directional derivative $d_C^0(x, w) \leq 0$ for all $w \in v + \epsilon \mathbb{B}$. In view of Propositions 4.7.4 and 4.7.6 we have that, for all $w \in v + \epsilon \mathbb{B}$,

$$w \cdot \xi \leq 0, \quad \text{for all } \xi \in \operatorname{co} \partial d_C(x),$$

and, recalling that $\partial d_C(x) = N_C(x) \cap \mathbb{B}$ (from Theorem 4.8.5), we deduce that

$$w \in N_C(x)^*.$$

Since $\bar{T}_C(x) = N_C(x)^*$ (from Theorem 4.10.7), we conclude that $w \in \bar{T}_C(x)$ for all $w \in v + \epsilon \mathbb{B}$ and, therefore, $v \in \operatorname{int} \bar{T}_C(x)$. □

4.12 Appendix: Proximal Analysis in Hilbert Space

Nonsmooth analysis has a pivotal role in this book, both in the derivation of first order necessary conditions of optimality and also in establishing the link between the value function of dynamic optimization and the Hamilton Jacobi equation. Despite the fact that the dynamic optimization problem is an optimization problem over function spaces (and is therefore infinite dimensional), the application of nonsmooth analysis for these purposes is based, almost exclusively, on properties of nonsmooth functions and sets with nonsmooth boundaries in the setting of a finite dimensional space. Many aspects of nonsmooth analysis, in which the underlying spaces are finite dimensional linear spaces, have analogues when the function domains and also the sets involved are Hilbert spaces.

Nonsmooth analysis in a real Hilbert space setting does find useful applications to some areas of dynamic optimization. These include derivation of necessary conditions of optimality for distributed parameter control systems and sensitivity analysis relating to perturbations of parameters in a function space. While such topics are not covered in this book, it is of interest to know that many of the ideas in the chapter generalize to an infinite dimensional setting. We show, in this appendix, how proximal subgradients of functions on real Hilbert spaces can be defined and prove an important density theorem. This broader theory is then used to prove two variational principles stated in Chap. 3.

Henceforth we take X to be a real Hilbert space endowed with a scalar product written $\langle .,.\rangle$. We write $||.||$ the associated norm and $\mathbb{B}_X$ the closed unit ball in X; $\overset{\circ}{\mathbb{B}}_X$ is the open unit ball in X.

Definition 4.12.1 Take a lower semi-continuous function $f : X \to \mathbb{R} \cup \{+\infty\}$ and a point $x \in \text{dom } f := \{y \in X : f(y) < +\infty\}$. We say that $\xi \in X$ is a *proximal subgradient* if there exist $M > 0$ and $\epsilon > 0$ such that

$$\langle \xi, y - x \rangle \leq f(y) - f(x) + M||y - x||^2, \quad \text{for all } y \in x + \epsilon \mathbb{B}_X. \tag{4.12.1}$$

The *proximal subdifferential* of f at x, written $\partial^P f(x)$, is the set

$$\partial^P f(x) := \{\xi : \text{there exist } M > 0 \text{ and } \epsilon > 0 \text{ such that condition (4.12.1) is satisfied }\}.$$

Theorem 4.12.2 (Proximal Density) *Take a real Hilbert space X and a lower semi-continuous function $f : X \to \mathbb{R} \cup \{+\infty\}$. Let $x_0 \in \text{dom } f$ and $\epsilon > 0$ be given. Then there exists a point $y \in x_0 + \epsilon \overset{\circ}{\mathbb{B}}_X$ satisfying $\partial^P f(y) \neq \emptyset$ and $|f(y) - f(x_0)| \leq \epsilon$.*

Proof Since f is lower semi-continuous there exists $\delta \in (0, \epsilon)$ such that

$$f(x) \geq f(x_0) - \epsilon, \quad \text{for all } x \in x_0 + \delta \mathbb{B}_X. \tag{4.12.2}$$

In the particular case when X is finite dimensional, say $X = \mathbb{R}^n$, the proof becomes very simple. Indeed we define the function

$$\phi(x) := \begin{cases} \frac{1}{\delta^2 - |x-x_0|^2} & \text{if } \in x_0 + \delta \overset{\circ}{\mathbb{B}} \\ +\infty & \text{otherwise.} \end{cases}$$

Observe that $\phi(x) \to +\infty$ as $|x - x_0| \uparrow \delta$, ϕ is lower semi-continuous, but ϕ is of class C^2 on $x_0 + \delta \overset{\circ}{\mathbb{B}}$. Then, if we consider the function $f + \phi$ we obtain a lower semi-continuous function which is bounded below on $X = \mathbb{R}^n$. As a consequence $f + \phi$ achieves the minimum value at a point $y \in x_0 + \delta \mathbb{B}$, and from the definition of ϕ we can also deduce that necessarily $y \in x_0 + \delta \overset{\circ}{\mathbb{B}}$. From Proposition 5.1.1 and Lemma 5.1.2 it follows that $0 \in \partial^P(f+\phi)(y) = \partial^P f(y) + \{\nabla \phi(y)\}$. Therefore $-\nabla \phi(y) \in \partial^P f(y)$ and so, in particular, we have $\partial^P f(y) \neq \emptyset$. Since y is a minimum for $f + \phi$ and clearly $\phi(x_0) \leq \phi(y)$ we also deduce that $f(y) \leq f(x_0) + (\phi(x_0) - \phi(y)) \leq f(x_0)$. In view of this inequality and (4.12.2) we obtain that $|f(y) - f(x_0)| \leq \epsilon$. The proof is then complete when $X = \mathbb{R}^n$.

We proceed now with the general case. Take $\alpha > 2\epsilon/\delta^2$. We claim that there exists $y_0 \in x_0 + \delta \mathbb{B}_X$ such that the function $x \to f(x) + \alpha \|x - y_0\|^2$ attains a minimum on $x_0 + \delta \overset{\circ}{B}_X$ at some point $y \in x_0 + \delta \overset{\circ}{\mathbb{B}}_X$, and $f(y) \leq f(x_0)$.

Write $X_0 := x_0 + \delta \mathbb{B}_X$ and

$$X_1 := \{x \in X_0 : f(x) + \frac{\alpha}{2}\|x - x_0\|^2 \leq f(x_0)\}.$$

Observe that $x_0 \in X_1$, and since $x \to f(x) + \frac{\alpha}{2}\|x - x_0\|^2$ is lower semi-continuous, X_1 is closed. Moreover, if $x \in X_1$, then from (4.12.2) and the choice of α we deduce that

$$\|x - x_0\|^2 \leq \frac{2}{\alpha}(f(x_0) - f(x)) < \frac{\delta^2}{\epsilon} \times \epsilon = \delta^2,$$

and, so,

$$X_1 \subset x_0 + \delta \overset{\circ}{B}_X \subset X_0. \tag{4.12.3}$$

We can find $x_1 \in X_1$ such that

$$f(x_1) + \frac{\alpha}{2}\|x_1 - x_0\|^2 \leq \inf_{x \in X_1}\{f(x) + \frac{\alpha}{2}\|x - x_0\|^2\} + \frac{\alpha}{4},$$

and we define the closed set

$$X_2 := \left\{x \in X_1 : f(x) + \frac{\alpha}{2}\left(\|x - x_0\|^2 + \frac{\|x - x_1\|^2}{2}\right) \leq f(x_1) + \frac{\alpha}{2}\|x_1 - x_0\|^2\right\},$$

4.12 Appendix: Proximal Analysis in Hilbert Space

which is nonempty since $x_1 \in X_2$. Using an induction argument we obtain a sequence of points $\{x_k\}$ and a sequence of sets $\{X_k\}$ such that $x_k \in X_k$ for all $k \geq 0$, and for each $k \geq 1$ we have

$$f(x_k) + \frac{\alpha}{2}\sum_{i=0}^{k-1}\frac{||x_k - x_i||^2}{2^i} \leq \inf_{x \in X_k}\{f(x) + \frac{\alpha}{2}\sum_{i=0}^{k-1}\frac{||x - x_i||^2}{2^i}\} + \frac{\alpha}{4^k} \qquad (4.12.4)$$

and

$$X_k := \left\{ x \in X_{k-1} : f(x) + \frac{\alpha}{2}\sum_{i=0}^{k-1}\frac{||x - x_i||^2}{2^i} \leq f(x_{k-1}) + \frac{\alpha}{2}\sum_{i=0}^{k-1}\frac{||x_{k-1} - x_i||^2}{2^i} \right\}. \qquad (4.12.5)$$

Clearly $x_{k-1} \in X_k$ for all $k \geq 1$, and so $\{X_k\}$ is a nested sequence of nonempty closed sets. Moreover, for each $k \geq 1$ given, from (4.12.4) and (4.12.5) we deduce that, for all $x \in X_{k+1}$,

$$\frac{\alpha}{2}\frac{||x - x_k||^2}{2^k} \leq f(x_k) + \frac{\alpha}{2}\sum_{i=0}^{k-1}\frac{||x_k - x_i||^2}{2^i} - \left(f(x) + \frac{\alpha}{2}\sum_{i=0}^{k-1}\frac{||x - x_i||^2}{2^i}\right)$$

$$\leq \frac{\alpha}{4^k}$$

and, so,

$$\sup_{x \in X_{k+1}} ||x - x_k|| \leq 2^{1-k}.$$

It follows that diam (X_k) $(:= \sup\{||x - x'|| : x, x' \in X_k\}) \to 0$ as $k \to \infty$. Bearing in mind that X is an Hilbert space (and therefore it is complete), we deduce that there exists a point $y \in X$ such that $\cap_{k=0}^{\infty} X_k = \{y\}$. Observe that in particular we have $y \in X_1 \subset x_0 + \delta \overset{\circ}{\mathbb{B}}_X$ (recall that we have the inclusion (4.12.3)) and so $||y - x_0|| < \delta < \epsilon$, and (invoking again the definition of X_1) $f(y) \leq f(y) + \frac{\alpha}{2}||y - x_0||^2 \leq f(x_0)$. This inequality combined with (4.12.2) yields $|f(y) - f(x_0)| \leq \epsilon$. Set

$$y_0 := \frac{1}{2}\sum_{i=0}^{\infty}\frac{x_i}{2^i} \quad \text{and} \quad c := \frac{1}{2}\sum_{i=0}^{\infty}\frac{||x_i||^2}{2^i} - ||y_0||^2.$$

Observe that, for each $x \in X$, we have

$$||x - y_0||^2 = \frac{1}{2}\sum_{i=0}^{\infty}\frac{||x - x_i||^2}{2^i} - c. \qquad (4.12.6)$$

Now we show that the function $x \to f(x) + \alpha \|x - y_0\|^2$ achieves its minimum value on $x_0 + \delta \, \mathring{\mathbb{B}}_X$ at y. Indeed, consider the sequence of real numbers $\{\beta_k\}$ where, for each $k \geq 1$,

$$\beta_k := f(x_k) + \frac{\alpha}{2} \sum_{i=0}^{k} \frac{\|x_k - x_i\|^2}{2^i}.$$

We know that $x_k \in X_k$ for all k, so from (4.12.5) we deduce that the sequence $\{\beta_k\}$ is nonincreasing. Now fix any $x \in x_0 + \delta \mathring{\mathbb{B}}_X$ with $x \neq y$, and write $\hat{k}$ the integer such that $x \in X_{\hat{k}} \setminus X_{\hat{k}+1}$. Take any $k \geq \hat{k}+1$. Since $y \in X_{k+1}$ and $\{\beta_k\}$ is nonincreasing, from (4.12.5) it follows that

$$f(y) + \frac{\alpha}{2} \sum_{i=0}^{k} \frac{\|y - x_i\|^2}{2^i} \leq f(x_k) + \frac{\alpha}{2} \sum_{i=0}^{k} \frac{\|x_k - x_i\|^2}{2^i}$$

$$\leq f(x_{k-1}) + \frac{\alpha}{2} \sum_{i=0}^{k-1} \frac{\|x_{k-1} - x_i\|^2}{2^i}$$

$$\leq \cdots$$

$$\leq f(x_{\hat{k}}) + \frac{\alpha}{2} \sum_{i=0}^{\hat{k}} \frac{\|x_{\hat{k}} - x_i\|^2}{2^i}. \qquad (4.12.7)$$

Recalling that $x \in X_{\hat{k}} \setminus X_{\hat{k}+1}$, from (4.12.5) we also have that

$$f(x_{\hat{k}}) + \frac{\alpha}{2} \sum_{i=0}^{\hat{k}} \frac{\|x_{\hat{k}} - x_i\|^2}{2^i} < f(x) + \frac{\alpha}{2} \sum_{i=0}^{\hat{k}} \frac{\|x - x_i\|^2}{2^i}$$

$$\leq f(x) + \frac{\alpha}{2} \sum_{i=0}^{\infty} \frac{\|x - x_i\|^2}{2^i}. \qquad (4.12.8)$$

Using the inequalities (4.12.7) and (4.12.8), we obtain that, for all $k \geq \hat{k} + 1$,

$$f(y) + \frac{\alpha}{2} \sum_{i=0}^{k} \frac{\|y - x_i\|^2}{2^i} \leq f(x) + \frac{\alpha}{2} \sum_{i=0}^{\infty} \frac{\|x - x_i\|^2}{2^i}$$

and, letting $k \to \infty$, we arrive at

$$f(y) + \frac{\alpha}{2} \sum_{i=0}^{\infty} \frac{\|y - x_i\|^2}{2^i} \leq f(x) + \frac{\alpha}{2} \sum_{i=0}^{\infty} \frac{\|x - x_i\|^2}{2^i}.$$

4.12 Appendix: Proximal Analysis in Hilbert Space

Adding to both sides the quantity '$-\alpha c$' and invoking the relation (4.12.6) we deduce that

$$f(y) + \alpha ||y - y_0||^2 \leq f(x) + \alpha ||x - y_0||^2,$$

which, since $x \in x_0 + \delta \overset{\circ}{\mathbb{B}}_X$ was an arbitrary point such that $x \neq y$, confirms our claim.

To conclude the proof, we observe that $g(x) := \alpha ||x - y_0||^2$ is of class C^2 and that $Dg(x) = 2\alpha(x - y_0)$. Therefore, since y is a minimum for $f + g$, it follows that $0 \in \partial^P(f + g)(y) = \partial^P f(y) + \{Dg(y)\}$, where $Dg(y)$ is the Fréchet derivative of g at y (this can easily be shown by adapting the analysis in the proofs of Proposition 5.1.1 and Lemma 5.1.2 to the case when X is an Hilbert space, in place of $\mathbb{R}^n$). Therefore $-2\alpha(y - y_0) \in \partial^P f(y)$ and $\partial^P f(y) \neq \emptyset$. □

In Sect. 3.4 we introduced the quadratic inf convolution of a lower semicontinuous function on a finite dimensional linear space. We may carry out a similar construction when the finite dimensional linear space is replaced by a real Hilbert space X. Now we define the quadratic inf convolution of a lower semi-continuous function $g : X \to \mathbb{R} \cup \{+\infty\}$ to be

$$g_\alpha(x) := \inf_{y \in X} \{g(y) + \alpha ||x - y||^2\}. \tag{4.12.9}$$

We shall make use of the properties collected together in the following theorem:

Theorem 4.12.3 *(Inf-Convolution, II) Take a real Hilbert space X and a proper lower semi-continuous function $g : X \to \mathbb{R} \cup \{+\infty\}$ which is bounded below by a constant c. Then, for each $\alpha > 0$, $g_\alpha : X \to \mathbb{R}$ is bounded below by c, and is Lipschitz on each bounded set of X (and in particular is finite valued). Now take $\alpha > 0$ and a point $x \in X$ is such that*

$$\partial^P g_\alpha(x) \neq \emptyset.$$

Then there exists a point $y \in X$ satisfying the following conditions,

(i): *if $\{y_i\}$ is a minimizing sequence for the infimum in (4.12.9), then $\lim_{i \to \infty} y_i = y$,*
(ii): *the infimum in (4.12.9) is attained uniquely at y,*
(iii): *the Fréchet derivative $Dg_\alpha(x)$ exists and $Dg_\alpha(x) = 2\alpha(x - y)$; furthermore, $\partial^P g(x) = \{2\alpha(x - y)\}$ and*

$$g_\alpha(x') - g_\alpha(x) \leq \langle 2\alpha(x - y), x' - x \rangle + \alpha ||x' - x||^2, \quad \text{for all } x' \in X,$$

(iv): $2\alpha(x - y) \in \partial^P g(y)$,
(v): *if, in addition, g is Lipschitz continuous with Lipschitz constant k_g, then g_α is Lipschitz continuous with the same Lipschitz constant k_g.*

Proof Consider a proper lower semi-continuous function $g : X \to \mathbb{R} \cup \{+\infty\}$ which is bounded below by a constant c. Fix $\alpha > 0$. From the definition of g_α it immediately follows that $g_\alpha(x) \geq c$, for all $x \in X$. Now, let $B \subset X$ be a bounded set, and take $R > 0$ such that $B \subset R\mathbb{B}_X$. Fix any $x_0 \in \operatorname{dom} g \ (\neq \emptyset)$. Observe that

$$g_\alpha(x) \leq g(x_0) + \alpha \|x_0 - x\|^2, \quad \text{for all } x \in X,$$

and therefore, restricting attention to the bounded set B, we have

$$g_\alpha(x) \leq g(x_0) + \alpha(\|x_0\| + R)^2, \quad \text{for all } x \in B.$$

Write $N := g(x_0) + \alpha(\|x_0\| + R)^2$ and, for any $\delta > 0$ define the (nonempty) set

$$C_\delta := \{z \in X \ : \ \text{there exists } u \in B \text{ s.t. } g(z) + \alpha\|u - z\|^2 \leq N + \delta\}.$$

Since B is bounded and g is bounded below, it follows that C_δ is bounded in X. Fix an arbitrary $\delta > 0$. Take any $x, x' \in B$. From the definition of g_α we know that we can find $z \in C_\delta$ such that

$$g_\alpha(x') \geq g(z) + \alpha\|x' - z\|^2 - \delta.$$

Therefore we obtain:

$$\begin{aligned}
g_\alpha(x) - g_\alpha(x') &\leq g_\alpha(x) - g(z) - \alpha\|x' - z\|^2 + \delta \\
&\leq g(z) + \alpha\|x - z\|^2 - g(z) - \alpha\|x' - z\|^2 + \delta \\
&= \alpha\|x - x'\|^2 + 2\alpha\langle x - x', x' - z\rangle + \delta \\
&\leq \alpha\|x - x'\|(\|x - x'\| + 2\|x' - z\|) + \delta \\
&\leq k_B\|x - x'\| + \delta,
\end{aligned}$$

in which $k_B := 6\alpha R + 2(N + \delta + |c|) \ (\geq \alpha \sup\{\|y - y'\| + 2\|y' - z\| \ : \ y, y' \in B \text{ and } z \in C_\delta\})$. Exchanging the roles of x and x' and letting $\delta \downarrow 0$, we obtain that g_α is Lipschitz continuous of Lipschitz constant k_B on B.

Suppose now that $x \in X$ is such that $\partial^P g(x) \neq \emptyset$. Take a vector $\xi \in \partial^P g(x)$. From the definition of proximal subgradient, there exist $M > 0$ and $\epsilon > 0$ such that

$$\langle \xi, x' - x\rangle \leq g_\alpha(x') - g_\alpha(x) + M\|x' - x\|^2, \quad \text{for all } x' \in x + \epsilon \mathbb{B}_X. \tag{4.12.10}$$

Consider a minimizing sequence $\{y_i\}$ for the infimum in (4.12.9). Then we can find a sequence $\delta_i \downarrow 0$, as $i \to \infty$ such that

$$g_\alpha(x) \leq g(y_i) + \alpha\|y_i - x\|^2 = g_\alpha(x) + \delta_i^2. \tag{4.12.11}$$

4.12 Appendix: Proximal Analysis in Hilbert Space

Recalling the definition of g_α we also know that

$$g_\alpha(x') \leq g(y_i) + \alpha ||y_i - x'||^2, \quad \text{for all } x' \in x + \epsilon \mathbb{B}_X. \tag{4.12.12}$$

Using (4.12.10), (4.12.11) and (4.12.12), it follows that, for all $x' \in x + \epsilon \mathbb{B}_X$,

$$\begin{aligned}\langle \xi, x' - x \rangle &\leq \alpha ||y_i - x'||^2 - \alpha ||y_i - x||^2 + \delta_i^2 + M||x' - x||^2 \\ &= -2\alpha \langle y_i, x' - x \rangle + \alpha ||x'||^2 - \alpha ||x||^2 + \delta_i^2 + M||x' - x||^2 \\ &= 2\alpha \langle x - y_i, x' - x \rangle + (M + \alpha) ||x' - x||^2 + \delta_i^2 ,\end{aligned}$$

and so

$$\langle \xi - 2\alpha(x - y_i), x' - x \rangle \leq (M + \alpha) ||x' - x||^2 + \delta_i^2, \quad \text{for all } x' \in x + \epsilon \mathbb{B}_X. \tag{4.12.13}$$

We claim that for all i large enough we have

$$||\xi - 2\alpha(x - y_i)|| \leq \delta_i (M + \alpha + 1) . \tag{4.12.14}$$

Indeed, for each i, either $\xi = 2\alpha(x - y_i)$ (and then (4.12.14) is trivially valid) or $\xi \neq 2\alpha(x - y_i)$; in the latter case we take the point $x' = x + \delta_i \frac{\xi - 2\alpha(x - y_i)}{||\xi - 2\alpha(x - y_i)||}$, which for i large enough belongs to $x + \epsilon \mathbb{B}_X$, and so we can easily deduce (4.12.14) from (4.12.13). Now we define $y := x - \frac{\xi}{2\alpha}$. Therefore, letting $i \to \infty$ in (4.12.14), we obtain (i).

Notice also that invoking (4.12.9), property (i), (4.12.11) and the lower semicontinuity of g we see that

$$g_\alpha(x) \leq g(y) + \alpha ||y - x||^2 \leq \liminf_{i \to \infty}(g(y_i) + \alpha ||y_i - x||^2) = g_\alpha(x) ,$$

and so $g_\alpha(x) = g(y) + \alpha ||y - x||^2$ which means that y achieves the minimum in (4.12.9). Moreover, y is necessarily unique, since, if $w \in X$ is another minimizer for (4.12.9), then we can consider the minimizing sequence $\{w_i \equiv w\}$ and, from (i) we know that $\lim_{i \to \infty} w_i = y$, which yields $w = y$. This confirms (ii).

To see (iii) we first observe that for all $x' \in X$ we have $g_\alpha(x') \leq g(y) + \alpha ||x' - y||^2$, and taking $x' = x$, from the analysis above we also know that $g_\alpha(x) = g(y) + \alpha ||x - y||^2$. Then we obtain

$$g_\alpha(x) - g_\alpha(x') \geq 2\alpha \langle x' - x, x - y \rangle - \alpha ||x' - x||^2$$

and so

$$g_\alpha(x') - g_\alpha(x) - 2\alpha \langle x' - x, x - y \rangle \leq \alpha ||x' - x||^2. \tag{4.12.15}$$

Then, from (4.12.10) (with $\xi = 2\alpha(x-y)$) and (4.12.15) we deduce

$$\frac{|g_\alpha(x') - g_\alpha(x) - \langle 2\alpha(x-y), x'-x\rangle|}{||x'-x||} \leq \max\{M, \alpha\}||x'-x||, \text{ for all } x' \in x+\epsilon\mathbb{B}_X,$$

which means that g_α is Fréchet differentiable at x and $Dg_\alpha(x) = 2\alpha(x-y)$. We claim that $\partial^P g(x) = \{2\alpha(x-y)\} = \{Dg_\alpha(x)\}$. Indeed, take any other vector $\hat{\xi} \in \partial^P g(x)$; from the definition of proximal subgradient we have, for some $\hat{M} > 0$ and $\hat{\epsilon} > 0$,

$$\langle \hat{\xi}, x'-x \rangle \leq g_\alpha(x') - g_\alpha(x) + \hat{M}||x'-x||^2, \text{ for all } x' \in x + \hat{\epsilon}\mathbb{B}_X.$$

Then we obtain

$$0 \leq \liminf_{x' \to x} \frac{g_\alpha(x') - g_\alpha(x) - \langle \hat{\xi}, x'-x \rangle}{||x'-x||}$$

$$\leq \liminf_{x' \to x} \left[\frac{g_\alpha(x') - g_\alpha(x) - \langle \xi, x'-x \rangle}{||x'-x||} + \frac{\langle \xi - \hat{\xi}, x'-x \rangle}{||x'-x||} \right]$$

$$= 0 + \liminf_{x' \to x} \frac{\langle \xi - \hat{\xi}, x'-x \rangle}{||x'-x||}$$

$$\leq -||\xi - \hat{\xi}||,$$

that implies $\hat{\xi} = \xi$, confirming the claim. To conclude with the proof of (iii), we observe that (4.12.15) yields also

$$g_\alpha(x') - g_\alpha(x) \leq \langle 2\alpha(x-y), x'-x\rangle + \alpha||x'-x||^2, \quad \text{for all } x' \in X.$$

We show now (iv). Observe that $x' \to ||x'-x||^2$ is of class C^2, and $D[x' \to ||x'-x||^2](y) = 2(y-x)$ and the function $x' \to \phi(x') := g(x') + \alpha||x'-x||^2$ has its minimum value at $x' = y$. Therefore, $0 \in \partial^P \phi(y) = \partial^P g(y) + 2\alpha(y-x)$ (recall that the same arguments employed to derive Proposition 5.1.1 and Lemma 5.1.2 in the finite dimensional case can be easily extended to the case when X is a general Hilbert space). Then $-2\alpha(y-x) \in \partial^P g(y)$.

Assume now that $g : X \to \mathbb{R}$ is Lipschitz continuous with Lipschitz constant k_g. For all x, $x' \in X$ we have

$$g_\alpha(x) = \inf_{z \in X}\{g(z) + \alpha||x-z||^2\} = \inf_{w \in X}\{g(w-(x'-x)) + \alpha||w-x'||^2\}$$

$$\leq \inf_{w \in X}\{g(w) + \alpha||w-x'||^2\} + k_g||x-x'||$$

$$= g_\alpha(x') + k_g||x-x'||.$$

4.12 Appendix: Proximal Analysis in Hilbert Space

Exchanging the roles of x and x' we obtain that $|g_\alpha(x) - g_\alpha(x')| \leq k_g ||x - x'||$ for all $x, x' \in X$ confirming the fact that g_α is Lipschitz continuous of Lipschitz constant k_g. This concludes the proof. □

These analytical tools can be used to prove the variational principle of Borwein and Preiss and also that of Stegall, stated in Chap. 3.

Proof of Theorem 3.5.1 (Borwein and Preiss) Take a lower semi-continuous function $f : X \to \mathbb{R} \cup \{+\infty\}$ which is bounded below (on the Hilbert space X), a point $x_0 \in \text{dom } f$ and a number $\epsilon > 0$. Assuming that

$$f(x_0) < \inf_{x \in X} f(x) + \epsilon. \tag{4.12.16}$$

we want to show that, for any $\lambda > 0$, there exist $\bar{x}$ and z in X such that

(i) $f(\bar{x}) \leq f(x_0)$,
(ii) $||z - x_0|| < \lambda$ and $||\bar{x} - z|| < \lambda$,
(iii) the function $x \to f(x) + \frac{\epsilon}{\lambda^2}||x - z||^2$ has a unique minimizer on X at $x = \bar{x}$.

Fix any $\lambda > 0$. Consider the inf-convolution of f with $\alpha := \frac{\epsilon}{\lambda^2}$:

$$f_\alpha(x) := \inf_{y \in X}\{f(y) + \frac{\epsilon}{\lambda^2}||y - x||^2\} \quad \text{for all } x \in X. \tag{4.12.17}$$

From Theorem 4.12.3 (on the inf-convolution properties) we know that f_α is finite valued on X (so $x_0 \in \text{dom } f_\alpha = X$). Owing to the proximal density theorem (Theorem 4.12.2) we can find a point $z \in x_0 + \lambda \overset{\circ}{\mathbb{B}}_X$ such that $\partial^P f_\alpha(z) \neq \emptyset$ and

$$f_\alpha(z) \leq f_\alpha(x_0). \tag{4.12.18}$$

Now, property (ii) of Theorem 4.12.3 (on inf-convolution) guarantees the existence of a unique point $\bar{x} \in X$ such that the infimum which provides the definition of $f_\alpha(z)$ (cf. (4.12.17)) is attained:

$$f_\alpha(z) = \inf_{y \in X}\{f(y) + \frac{\epsilon}{\lambda^2}||y - z||^2\} = f(\bar{x}) + \frac{\epsilon}{\lambda^2}||\bar{x} - z||^2. \tag{4.12.19}$$

This yields (iii). Moreover, since clearly $f_\alpha \leq f$, from (4.12.18) and (4.12.19) we deduce also that

$$f(\bar{x}) \leq f(\bar{x}) + \frac{\epsilon}{\lambda^2}||\bar{x} - z||^2 = f_\alpha(z) \leq f_\alpha(x_0) \leq f(x_0), \tag{4.12.20}$$

confirming property (i). Concerning (ii), since from the analysis above we already have $z \in x_0 + \lambda \overset{\circ}{\mathbb{B}}_X$, it remains to show only that $||\bar{x} - z|| < \lambda$. In view of (4.12.20) and (4.12.16) it follows that

$$f(\bar{x}) + \frac{\epsilon}{\lambda^2}||\bar{x} - z||^2 = f_\alpha(z) \leq f(x_0) < \inf_{x \in X} f(x) + \epsilon,$$

which, subtracting the term $f(\bar{x})$, implies that

$$\frac{\epsilon}{\lambda^2}||\bar{x} - z||^2 < \inf_{x \in X} f(x) - f(\bar{x}) + \epsilon \leq \epsilon,$$

and, so, $||\bar{x} - z|| < \lambda$. This concludes the proof. □

Proof of Theorem 3.5.2 (Stegall) Take a lower semi-continuous function $f : X \to \mathbb{R} \cup \{+\infty\}$ which is bounded below on a bounded subset Y of a real Hilbert space X. We assume that $Y \cap \text{dom } f \neq \emptyset$. Our aim is to show that there exists a dense set of points $x \in X$ such that

$$z \to f(z) - \langle x, z \rangle$$

attains a unique minimum over Y.

Define the function

$$\varphi(x) := \inf_{y \in X} \{f(y) + \Psi_Y(y) - \frac{1}{2}||y||^2 + \frac{1}{2}||x - y||^2\} \quad (4.12.21)$$

where Ψ_Y denotes the indicator function of the subset $Y \subset X$ (recall that $\Psi_Y(z)$ takes the value 0 if $z \in Y$, otherwise it takes the value $+\infty$). Observe that φ is the inf-convolution of the function $g(x) := f(x) + \Psi_Y(x) - \frac{1}{2}||x||^2$, $x \in X$, with $\alpha = \frac{1}{2}$. Notice that g is bounded below since f is bounded below and Y is bounded. Moreover the term which defines φ (4.12.21) can be easily simplified to

$$\varphi(x) = \inf_{y \in Y}\{f(y) - \langle x, y \rangle\} + \frac{1}{2}||x||^2. \quad (4.12.22)$$

Fix $x \in X$. Observe that the set of points $y \in X$ achieving an infimum in (4.12.21) or in (4.12.22) or also in the following expression

$$\inf_{y \in Y}\{f(y) - \langle x, y \rangle\} \quad (4.12.23)$$

is the same and is contained in Y. Theorem 4.12.3 (on the inf-convolution properties) guarantees that dom $\varphi = X$. The proximal density theorem (Theorem 4.12.2) tells us that there exists a dense set $D \subset X$ such that $\partial^P \varphi(x) \neq \emptyset$ for all $x \in D$. Invoking again Theorem 4.12.3 we have that for each $x \in D$ the infimum in (4.12.21) is uniquely attained at a point $y_x \in Y$. Therefore, for each $x \in D$, also the infimum in (4.12.23) is uniquely achieved at the (same) point $y_x \in Y$. □

4.13 Exercises

4.1 (*This Exercise Provides Some Details of the Proofs of Theorems 4.7.7 and 5.7.1.*) Take a non-zero vector $v \in \mathbb{R}^n$. Let $\Omega \subset \mathbb{R}^n$ be a null-set w.r.t. n-dimensional Lebesgue measure. For $x \in \mathbb{R}$, denote by L_x the line $L_x := \{x + tv : t \in \mathbb{R}\}$. Show that there exists a subset $\mathcal{S} \subset \mathbb{R}^n$, whose complement in $\mathbb{R}^n$ has zero n-dimensional Lebesgue measure, with the following property: for any $x \in \mathcal{S}$, $L_x \cap \Omega$ has zero one-dimensional Lebesgue measure, i.e. the line L_x touches the set Ω on a set of zero one-dimensional Lebesgue measure.
Hint: We can assume $|v| = 1$, since L_y is unaffected by scaling v. By considering the orthogonal (Lebesgue measure preserving) transformation $x \to Tx$, in which $T \in \mathbb{R}^{n \times n}$ is an orthonormal matrix whose first column is the vector v, we can justify restricting attention to the case when $v = (1, 0, \ldots, 0)$. Define $m : \mathbb{R}^n \to \mathbb{R}$ to be $m(x) = \begin{cases} 1 & \text{if } x \in \Omega \\ 0 & \text{otherwise} \end{cases}$. Since m is the indicator function of the nullset Ω, m is integrable and

$$\int_{-\infty}^{\infty} \ldots \int_{-\infty}^{\infty} m(x_1, \ldots, x_n) dx_1 \ldots dx_n = 0.$$

Define $m_1 : \mathbb{R}^{n-1} \to \mathbb{R}$ to be $m_1(x_2, \ldots, x_n) := \int_{-\infty}^{\infty} m(x_1, x_2, \ldots, x_n) dx_1$. By Fubini's theorem,

$$\int_{-\infty}^{\infty} \ldots \int_{-\infty}^{\infty} m_1(x_2, \ldots, x_n) dx_2 \ldots dx_n = 0.$$

But, then, there exists a set $\mathcal{M}_{n-1}$, whose complement in $\mathbb{R}^{n-1}$ has zero $(n-1)$-dimensional Lebesgue measure, such that $m_1(x_2, \ldots, x_n) = 0$ for all $(x_2, \ldots, x_n) \in \mathcal{M}_{n-1}$. Now define $\mathcal{S} \subset \mathbb{R}^n$ to be

$$\mathcal{S} := \mathbb{R} \times \mathcal{M}_{n-1}.$$

Show that $\mathcal{S}$ is the complement of a set of n-dimensional Lebesgue measure in $\mathbb{R}^n$ and has the stated intersection properties.

4.2 Take a multifunction $F : [S, T] \times \mathbb{R}^n \rightsquigarrow \mathbb{R}^n$. Assume that F takes values compact sets. Consider the Hamiltonian

$$H(t, x, p) := \max_{v \in F(t,x)} p \cdot v.$$

Show that

(i): if $F(t, .)$ is $k(t)$-Lipschitz, then $x \to H(t, x, p)$ is $|p|k(t)$-Lipschitz continuous, (ii): if $(\xi, \eta) \in \mathrm{co}\, \partial_{x,p} H(t, x, p)$, then $\eta \in \partial_p H(t, x, p)$, and $H(t, x, p) = \eta \cdot p$; moreover,

(iii): $\partial_p H(t, x, p) \subset \mathrm{co}\, F(t, x)$.

Remark

The properties expressed in this exercise are used several times in the book (cf. Chaps. 8 and 11). Observe, in particular, that relations (ii) and (iii) can be used to derive the Weierstrass condition from the Hamiltonian inclusion or even the partially convexified Hamiltonian inclusion.

Hint: For (ii) and (iii) use Theorem 4.7.7 and the fact that the map $p \to H(t, x, p)$ is convex.

4.3 Let $C \subset \mathbb{R}^k$ be a closed nonempty set. Take a point $x \in C$. Show that $\xi \in N_C^P(x)$ if and only if there exist $M > 0$ and $\epsilon > 0$ such that

$$\xi \cdot (y - x) \leq M|y - x|^2, \quad \text{for all } y \in (x + \epsilon \mathbb{B}) \cap C.$$

Remark

This exercise highlights a useful (cf. the proof of Lemma 8.6.3) local property of proximal normal vectors; the property stated in this exercise tells us also that

$$\xi \in N_C^P(x) \quad \text{if and only if } \xi \in \partial^P \Psi_C(x).$$

(Ψ_C is the indicator function of the set C.)

4.14 Notes for Chapters 4 and 5

Developments in convex analysis in 1960s, centred substantially on Rockafellar's contributions, revealed that, for purposes of characterizing minimizers to convex optimization problems with possibly non-differentiable data, tangent cones, normal cones and set-valued subdifferentials can in many ways do the work of tangent spaces, co-tangent spaces and derivatives respectively in traditional analysis.

A desire to reproduce some of these successes in a nonconvex setting was the impetus behind the field of nonsmooth analysis, initiated in the following decade. The breakthrough was the introduction by Clarke in his 1973 thesis [56] of various 'robust' nonsmooth constructs, including the generalized gradient of a Lipschitz continuous function and what are now widely referred to as the Clarke normal cone and Clarke tangent cone to a closed set [56, 57, 65]. Numerous earlier definitions of 'derivative' of a nondifferentiable function had previously been proposed, but the generalized gradient was the first to stand out for its generality, its extensive calculus, geometric interpretations and the breadth of its applications, notably in the derivation of necessary conditions of optimality.

4.14 Notes for Chapters 4 and 5

Once the idea of local approximation of nonsmooth functions and sets with nonsmooth boundaries by means of cones and set valued 'gradients' was out the bag, there was an explosion of definitions. Aubin and Frankowska's book [14], which includes a systematic study of the 'menagerie' of tangent cones, obtained by attaching different qualifiers to the limit operations in their definitions, and derivatives which they induce by consideration of tangent cones to graphs of functions, is evidence of this.

Chapters 4 and 5 concern those aspects of nonsmooth analysis required to support future chapters on necessary conditions and dynamic programming in dynamic optimization. Here, fortunately, we need to consider only relatively few constructs. Much of the material in these chapters focuses on the limiting normal cone (and the related limiting subdifferential of a lower semi-continuous function, defined via limiting normals to the epigraph set). But these chapters also feature a limited repertoire of tangent cones of relevance to our analysis. All material is more or less standard. Proofs of basic properties of limiting normals and subdifferentials follow, in many respects, those in [175] and [141] and ideas implicit in [65]. Proofs of the relations between the various normal and tangent cones considered here in many respects follow those in [14]. Material on generalized gradients in Sect. 4.7 is taken from [65]. Proofs of the results contained in Sect. 4.11 and in the Appendix this chapter follow those of Clarke et al. in [85].

The limiting normal cone was introduced by Mordukhovich in his 1976 paper [154], where it was used to formulate generalized transversality conditions in dynamic optimization. The limiting normal cone is also referred to as the approximate normal cone by Ioffe [123] and simply as the normal cone (by Mordukhovich [157] and, recently, by Rockafellar and Wets [177]). The names 'co-derivative' and 'subdifferential' are also used for the limiting subdifferential, in [157] and [177] respectively. These constructs were studied in detail in papers of Mordukhovich, Kruger and Ioffe, including [134, 156] and [123]. More recent expository treatments are provided in [157, 177] and [141]. Proximal normal vectors, the generators of the limiting normal cone, were used earlier by Clarke to prove properties of (Clarke) normal cones and generalized gradients [57]. But focusing attention on limiting normal cones as interesting objects in their own right was a significant departure.

The proximal normal cone, limiting normal cone and its convex hull the Clarke normal cone, together with their associated subdifferentials (the proximal subdifferential, limiting subdifferential and the generalized gradient respectively), all figure prominently in nonsmooth dynamic optimization. Of these constructs, the generalized gradient is well suited to the formulation of adjoint inclusions for nonsmooth dynamics, by virtue of its convexity properties. On the other hand, the limiting normal cone is a natural choice of normal cone to express transversality conditions for general endpoint constraint sets. The proximal subdifferential is a convenient vehicle for the interpretation of generalized solutions to the Hamilton Jacobi equation of dynamic programming.

Lebourg [136] proved the first nonsmooth mean value theorem (the two sided mean value theorem for Lipschitz functions). Clarke and Ledyaev's approximate mean value inequality [70], the uniform nature of which was novel even for smooth functions, has found numerous applications—in fixed point theorems, the interpretation of generalized solutions to partial differential inequalities and other areas. (See also [71] and [85].) Approximate generalizations of the mean value theorem were earlier investigated by Zagrodny [211] and Loewen [142].

We stick for the most part to nonsmooth analysis in finite dimensional spaces. This suffices for most topics in dynamic optimization in this book. However dynamic optimization problems are inherently infinite dimensional. The investigation of certain issues in dynamic optimization, such as sensitivity of the minimum cost to 'infinite dimensional perturbations' of the dynamics [63], as well as some alternative proofs of results in this book, involve the application of nonsmooth analysis in infinite dimensional spaces. Clarke has shown that his theory of generalized gradients can be developed, with the accompanying calculus largely intact, in the context of Lipschitz continuous functions on Banach spaces [65]. A number of researchers have been involved in building alternative frameworks for nonsmooth analysis in infinite dimensions, including Clarke, Borwein, Ioffe, Loewen and Mordukhovich. Borwein and Zhu has provided a helpful, detailed review [44], see also [45]. An appealing approach for its simplicity and broad applicability, followed by Clarke, Ledyaev, Stern and Wolenski [85], centres on proximal normals in Hilbert space. Proximal normal cones and proximal subdifferentials do not have a satisfactory *exact* calculus. They do however admit a rich *fuzzy* calculus. Fuzzy calculus, initiated by Ioffe (see [125]), provides rules, involving ϵ error terms which can be made arbitrarily small, for the estimation of proximal subdifferentials of composite functions. The idea is to retain ϵ terms in the general theory and to attempt to dispose of them only at the applications stage, using special features of the problem at hand. The Appendix (to Chap. 4) contains some material on proximal analysis in Hilbert spaces, as required to prove the Borwein and Preiss and Stegall variational principles of Chap. 3.

Chapter 5
Subdifferential Calculus

Abstract The subject matter of this chapter is a calculus governing the limiting subdifferentials of composite functions, that is functions that are composed from lower semi-continuous extended valued functions and indicator functions of closed sets. It is remarkable the extent to which a subdifferential calculus can be developed, reproducing aspects of classical calculus governing differential properties of smooth functions, when the functions involved are no longer smooth. A distinctive feature of subdifferential calculus is that, typically, it does not provide precise representations of the limiting subdifferentials of composite functions, but only estimates sets for these limiting subdifferentials. We have already encountered a nonsmooth mean value theorem in the previous chapter. The subdifferential calculus presented in this chapter provides versions of the sum rule, and chain rule. It also provides a widely used 'max rule' for composite functions, which has no parallel classical calculus. To give a foretaste of techniques employed in later chapters, we derive a very general Lagrange multiplier in nonlinear programming, using subdifferential calculus in harness with a variational principle.

5.1 Introduction

In this chapter, we assemble a number of useful rules for calculating and estimating the limiting subdifferentials of composite functions in terms of their constituent mappings. A typical rule is the 'sum rule'

$$\partial(f+g)(x) \subset \partial f(x) + \partial g(x). \qquad (5.1.1)$$

(The right side denotes, of course, the set of all vectors ξ expressible as $\xi = \eta_1 + \eta_2$ for some $\eta_1 \in \partial f(x)$ and $\eta_2 \in \partial g(x)$.)

We notice at once that the rule is 'one sided': it takes the form of a set inclusion not a set equivalence. Fortunately the inclusion that comes naturally is in the helpful direction: we want an estimate of $\partial(f+g)$ in terms of the subgradients of the (usually simpler) functions f and g of which $f+g$ is constituted. It is inevitable that inclusions feature in these rules if they are to handle functions which are locally

Lipschitz continuous (or even less regular). This point is illustrated in the following example.

Example
Take $f : \mathbb{R} \to \mathbb{R}$ and $g : \mathbb{R} \to \mathbb{R}$ to be the Lipschitz continuous functions

$$f(x) = \min\{0, x\} \text{ and } g(x) = -\min\{0, x\}.$$

Then

$$\partial(f + g)(0) = \{0\}.$$

Yet

$$\partial f(0) = \{0\} \cup \{1\} \text{ and } \partial g(0) = [-1, 0].$$

We see that

$$\partial(f + g)(0) \stackrel{\text{strict}}{\subset} \partial f(0) + \partial g(0).$$

It is desirable to have calculus rules which apply to general lower semi-continuous functions. But even a 'one sided' rule such as (5.1.1) may fail unless some non-degeneracy hypothesis is imposed. The precise nature of the hypotheses required will differ from rule to rule, but in each case they will eliminate certain kinds of interaction of the relevant asymptotic limiting subdifferentials. These nondegeneracy hypotheses are automatically satisfied when the functions involved are Lipschitz continuous.

Example
Take the function $s : \mathbb{R} \to \mathbb{R}$ to be

$$s(x) := \text{sgn}\{x\}|x|^{\frac{1}{2}}$$

and define the functions $f : \mathbb{R} \to \mathbb{R}$ and $g : \mathbb{R} \to \mathbb{R}$ according to

$$f(x) = s(x) \text{ and } g(x) = -s(x).$$

We find that $\partial f(0) = \partial g(0) = \emptyset$ and $\partial(f + g)(0) = \{0\}$. So

$$\{0\} = \partial(f + g)(0) \not\subset \partial f(0) + \partial g(0) = \emptyset.$$

Notice that $\partial^\infty f(0) = [0, \infty)$ and $\partial^\infty g(0) = (-\infty, 0]$. The pathological aspect of this example is that there exist nonzero numbers a and b, such that $a \in \partial^\infty f(0)$, $b \in \partial^\infty g(0)$ and $a + b = 0$. (Take $a = 1$ and $b = -1$ for example.) The data then fail to satisfy the condition:

5.1 Introduction

'$a \in \partial^\infty f(0)$, $b \in \partial^\infty g(0)$ and $a + b = 0$' implies '$a = b = 0$',

which will turn out to be precisely the nondegeneracy hypothesis appropriate to the sum rule.

For a concept of subdifferential to be useful in the field of optimization, we require it, at the very least, to have the property: 'If f achieves its minimum value at $\bar{x} \in \text{dom } f$, then $\{0\}$ is contained in the subdifferential of f at $\bar{x}$'. Fortunately this is true even of the 'tightest' subdifferential we have introduced, the proximal subdifferential.

Proposition 5.1.1 *Take a lower semi-continuous function $f : \mathbb{R}^k \to \mathbb{R} \cup \{+\infty\}$ and a point $x \in \text{dom } f$. Assume that x achieves the minimum value of f over a neighbourhood of x, then*

$$0 \in \partial^P f(x).$$

Proof The fact that x is a local minimum means that there exists $\epsilon > 0$ such that

$$\xi \cdot (y - x) \leq f(y) - f(x) + M|y - x|^2 \text{ for all } y \in x + \epsilon \mathbb{B}$$

when we take $\xi = 0$ and any $M \geq 0$. We conclude from Proposition 4.4.1 that $0 \in \partial^P f(x)$. □

To make a start, we shall need the following rudimentary sum rule:

Lemma 5.1.2 *Take functions $f : \mathbb{R}^k \to \mathbb{R} \cup \{+\infty\}$ and $g : \mathbb{R}^k \to \mathbb{R}$, a closed set $C \subset \mathbb{R}^k$ and a point $x \in (\text{int } C) \cap (\text{dom } f)$. Assume that f is lower semi-continuous and g is of class C^2 on a neighbourhood of x. Then*

$$\partial^P(f + g + \Psi_C)(x) = \partial^P f(x) + \{\nabla g(x)\},$$
$$\partial(f + g + \Psi_C)(x) = \partial f(x) + \{\nabla g(x)\},$$
$$\partial^\infty(f + g + \Psi_C)(x) = \partial^\infty f(x).$$

(As usual, Ψ_C denotes the indicator function of the set C.)

Proof For some $\epsilon > 0$ such that $x + \epsilon \mathbb{B} \subset \text{int } C$, there exists $m > 0$ such that

$$|g(x') - g(x) - \nabla g(x) \cdot (x' - x)| \leq m|x' - x|^2$$

for all $x' \in x + \epsilon \mathbb{B}$, since g is assumed C^2 on a neighbourhood of x. Take any $y \in x + \epsilon \mathbb{B}$, $\xi \in \mathbb{R}^k$. Then, in consequence of this observation, there exists $M > 0$ and $\delta > 0$ such that

$$(\xi + \nabla g(x)) \cdot (x' - x) \leq (f + g + \Psi_C)(x') - (f + g + \Psi_C)(x) + M|x' - x|^2$$

for all $x' \in x + \delta\mathbb{B}$ if and only if there exists $M' > 0$ and $\delta' > 0$ such that

$$\xi \cdot (x' - x) \leq f(x') - f(x) + M'|x' - x|^2$$

for all $x' \in x + \delta'\mathbb{B}$.

It follows that $\xi + \nabla g(x) \in \partial^P(f + g + \Psi_C)(x)$ if and only if $\xi \in \partial^P f(x)$. We have confirmed the first assertion of the lemma.

The remaining assertions follow from Theorem 4.6.2, upon limit taking, when we note that $x' \xrightarrow{f} x$ if and only if $x' \xrightarrow{f+g+\Psi_C} x$. □

5.2 A Marginal Function Principle

Take a function $F : \mathbb{R}^k \times \mathbb{R}^l \to \mathbb{R} \cup \{+\infty\}$. Let $f : \mathbb{R}^k \to \mathbb{R} \cup \{+\infty\}$ be the marginal function:

$$f(x) := F(x, 0) \text{ for all } x \in \mathbb{R}^k$$

and let $\bar{x}$ be a minimizer for f:

$$f(\bar{x}) = \min\{f(x) : x \in \mathbb{R}^k\}.$$

In convex optimization there is a standard procedure for constructing a function $g : \mathbb{R}^l \to \mathbb{R} \cup \{-\infty\}$ such that

$$f(\bar{x}) \geq \sup\{g(v) : v \in \mathbb{R}^l\}. \tag{5.2.1}$$

It is to introduce the Fenchel conjugate functional $F^* : \mathbb{R}^k \times \mathbb{R}^l \to \mathbb{R} \cup \{+\infty\}$:

$$F^*(y, v) := \sup\{x \cdot y + u \cdot v - F(x, u) : (x, u) \in \mathbb{R}^k \times \mathbb{R}^l\}$$

and set

$$g(v) := -F^*(0, v).$$

The validity of the 'weak duality' condition (5.2.1) follows directly from the definition of F^*. If F is jointly convex in its arguments and satisfies an appropriate nondegeneracy hypothesis then condition (5.2.1) can be replaced by the stronger condition: there exists $\bar{v} \in \mathbb{R}^l$ such that

$$f(\bar{x}) = \sup\{g(v) : v \in \mathbb{R}^l\} = g(\bar{v}). \tag{5.2.2}$$

5.2 A Marginal Function Principle

Assertions of this nature ('duality principles') have an important role in the derivation of optimality conditions in convex optimization and can be used as a starting point for developing a subdifferential calculus for convex functions.

Now suppose that F is no longer convex. Then it will be possible to find $\bar{v}$ such that (5.2.2) is true only under very special circumstances. Surprisingly however we can still salvage some ideas from the above constructions for analysing nonconvex functions.

The key observation is that condition (5.2.2) can be expressed

$$F(\bar{x}, 0) + F^*(0, \bar{v}) = \bar{x} \cdot 0 + 0 \cdot \bar{v}$$

or, alternatively (in terms of the subdifferential of F in the sense of convex analysis),

$$(0, \bar{v}) \in \partial F(\bar{x}, 0). \tag{5.2.3}$$

The above duality principle can therefore be formulated as: there exists $\bar{v} \in \mathbb{R}^l$ such that (5.2.3) is satisfied. The advantage of writing the condition in this way is that it makes no reference to conjugate functionals and is suitable for generalization to situations where F is no longer convex.

We now prove a theorem, a corollary of which will assert the following: Suppose $\bar{x}$ is a minimizer for $x \to F(x, 0)$. Then, under a mild nondegeneracy hypothesis on F, there exists $\bar{v}$ such that (5.2.3) is true, when ∂F is interpreted as a limiting subdifferential. As we have discussed, this 'marginal function' principle has a similar role in the derivation of nonsmooth calculus rules as that of duality principles in convex analysis.

Theorem 5.2.1 (Marginal Function Principle) *Take a lower semi-continuous function* $F : \mathbb{R}^n \times \mathbb{R}^m \to \mathbb{R} \cup \{+\infty\}$ *and a point* $(\bar{x}, \bar{u}) \in \text{dom } F$. *Assume that*

(i) There exists a bounded set $K \subset \mathbb{R}^n$ *such that*

$$\text{dom } F(., u) \subset K \text{ for all } u \in \mathbb{R}^m,$$

(ii) $\bar{x}$ is the unique minimizer of $x \to F(x, \bar{u})$.

Define $V : \mathbb{R}^m \to \mathbb{R} \cup \{+\infty\}$ *to be*

$$V(u) := \min_{x \in \mathbb{R}^n} F(x, u).$$

Then V is a lower semi-continuous function and

$$\partial V(\bar{u}) \subset \{\xi : (0, \xi) \in \partial F(\bar{x}, \bar{u})\},$$
$$\partial^\infty V(\bar{u}) \subset \{\xi : (0, \xi) \in \partial^\infty F(\bar{x}, \bar{u})\}.$$

Proof We begin by establishing the following properties of F and V:

(a): For every $u \in \mathbb{R}^m$, $x \to F(x, u)$ has a minimizer (with infinite cost if dom $F(., u) = \emptyset$),

(b): V is lower semi-continuous,

(c): Given any sequences $u_i \xrightarrow{V} \bar{u}$ and $\{x_i\}$ such that $V(u_i) = F(x_i, u_i)$ for each i, then $(x_i, u_i) \xrightarrow{F} (\bar{x}, \bar{u})$.

Take any $u \in \mathbb{R}^m$. If dom $F(., u) = \emptyset$, then any x is a minimizer. If dom $F(., u) \neq \emptyset$ then the search for the minimizer of $F(., u)$ must be carried out over the non-empty set dom $F(., u)$. But this set is compact, since F is lower semi-continuous and dom $F(., u)$ is assumed bounded. A minimizer exists then in this case too. We have shown (a).

Take any sequence $u_i \to u$. We wish to show $\liminf_i V(u_i) \geq V(u)$. If all but a finite number of the $V(u_i)$'s are infinite there is nothing to prove. Otherwise we can replace the sequence by a subsequence along which $V(u_i)$ is finite. For each i choose a minimizer x_i of the function $x \to F(x, u_i)$. By (i), we can arrange by extracting a further subsequence that $x_i \to x$ for some $x \in \mathbb{R}^m$. We know then that $(x_i, u_i) \to (x, u)$. It follows now from the lower semicontinuity of F that

$$\liminf_i V(u_i) = \liminf_i F(x_i, u_i) \geq F(x, u) \geq V(u).$$

We have confirmed property (b).

Take any sequence $u_i \xrightarrow{V} \bar{u}$ and $\{x_i\}$ such that $V(u_i) = F(x_i, u_i)$ for each i. Then $V(u_i) < +\infty$ (for i sufficiently large), and the x_i's are confined to a compact set. Along a subsequence then, $x_i \to x$ for some $x \in \mathbb{R}^n$. By lower semicontinuity of F,

$$V(\bar{u}) = \lim_i V(u_i) = \lim F(x_i, u_i) \geq F(x, \bar{u}) \geq V(\bar{u}).$$

It follows that

$$F(x_i, u_i) \to F(x, \bar{u}).$$

We see also that $x = \bar{x}$ since $F(., \bar{u})$ has a unique minimizer. From the fact that the limits are independent of the subsequence extracted, we conclude that, for the original sequence,

$$(x_i, u_i) \xrightarrow{F} (\bar{x}, \bar{u}).$$

Property (c) is confirmed.

We are now ready to estimate subgradients. Let F, V and $(\bar{x}, \bar{u})$ be as in the theorem. Take any $\xi \in \partial V(\bar{u})$. According to Theorem 4.6.2 there exist sequences

5.2 A Marginal Function Principle

$u_i \xrightarrow{V} \bar{u}$ and $\xi_i \to \xi$ such that $\xi_i \in \partial^P V(u_i)$ for all i. For each i, there exist $\epsilon_i > 0$ and $M_i > 0$ such that

$$\xi_i \cdot (u - u_i) \le V(u) - V(u_i) + M_i |u - u_i|^2 \text{ for all } u \in u_i + \epsilon_i \mathbb{B}. \qquad (5.2.4)$$

Let x_i be a minimizer for $x \to F(x, u_i)$. Then, since $V(u) \le F(x, u)$ for all x and u, we deduce from (5.2.4) that

$$(0, \xi_i) \cdot (x - x_i, u - u_i) \le F(x, u) - F(x_i, u_i) + M_i |(x, u) - (x_i, u_i)|^2$$

for all $(x, u) \in (x_i, u_i) + \epsilon_i \mathbb{B}$. It follows that

$$(0, \xi_i) \in \partial^P F(x_i, u_i).$$

In view of property (c) above, $(x_i, u_i) \xrightarrow{F} (\bar{x}, \bar{u})$. By Theorem 4.6.2 then

$$(0, \xi) \in \partial F(\bar{x}, \bar{u}).$$

Next take $\xi \in \partial^\infty V(\bar{u})$. We know that there exist sequences $u_i \xrightarrow{V} \bar{u}$, $\xi_i \to \xi$ and $t_i \downarrow 0$ such that $t_i^{-1} \xi_i \in \partial^P V(u_i)$ for all i. The preceding arguments yield

$$t_i^{-1}(0, \xi_i) \in \partial^P F(x_i, u_i).$$

Here x_i is a minimizer for $x \to F(x, u_i)$. Recalling that $(x_i, u_i) \xrightarrow{F} (\bar{x}, \bar{u})$, we conclude that $(0, \xi) \in \partial^\infty F(\bar{x}, \bar{u})$. $\square$

Frequently the theorem is used in the form of the following corollary:

Corollary 5.2.2 *Take a lower semi-continuous function $F : \mathbb{R}^n \times \mathbb{R}^m \to \mathbb{R} \cup \{+\infty\}$ and a point $(\bar{x}, \bar{u}) \in \text{dom } F$. Assume that $\bar{x}$ minimizes $x \to F(x, \bar{u})$ over some neighbourhood of $\bar{x}$. Suppose that*

$$\{\eta : (0, \eta) \in \partial^\infty F(\bar{x}, \bar{u})\} = \{0\}.$$

Then there exists a point $\xi \in \mathbb{R}^m$ such that

$$(0, \xi) \in \partial F(\bar{x}, \bar{u}).$$

Proof Since $\bar{x}$ is a local minimizer of $F(., \bar{u})$, there exists $\epsilon > 0$ such that $F(\bar{x}, \bar{u}) = \min\{F(x, \bar{u}) : x \in \bar{x} + \epsilon \mathbb{B}\}$. Set $C := \bar{x} + \epsilon \mathbb{B}$. Choose any $r > 0$. Define $\tilde{F} : \mathbb{R}^n \times \mathbb{R}^m \to \mathbb{R} \cup \{+\infty\}$:

$$\tilde{F}(x, u) := F(x, u) + r|x - \bar{x}|^2 + \Psi_C(x)$$

Define also

$$V(u) := \inf_x \tilde{F}(x, u).$$

The function $\tilde{F}$ satisfies the hypotheses of Theorem 5.2.1. Notice in particular that the 'penalty term' $r|x - \bar{x}|^2$ ensures that $\tilde{F}(., \bar{u})$ has a unique minimizer. Bearing in mind that the limiting subgradients and asymptotic limiting subgradients of F and $\tilde{F}$ at $(\bar{x}, \bar{u})$ coincide (Lemma 5.1.2), we deduce that

$$\partial V(\bar{u}) \subset \{\xi : (0, \xi) \in \partial F(\bar{x}, \bar{u})\} \text{ and } \partial^\infty V(\bar{u}) \subset \{\xi : (0, \xi) \in \partial^\infty F(\bar{x}, \bar{u})\}.$$

But under the non-degeneracy hypothesis, $\partial^\infty V(\bar{u}) = \{0\}$. We know from Corollary 4.9.2 and Proposition 4.7.1 that $\partial V(\bar{u})$ is non-empty. It follows that there exists some $\xi \in \mathbb{R}^m$ such that $(0, \xi) \in \partial F(\bar{x}, \bar{u})$. □

5.3 Partial Limiting Subgradients

The first calculus rule relates the partial limiting subdifferential of a function of two variables and the projection of the 'total' limiting subdifferential on to the relevant coordinate.

Theorem 5.3.1 (Partial Limiting Subgradients) *Take a lower semi-continuous function $f : \mathbb{R}^n \times \mathbb{R}^m \to \mathbb{R} \cup \{+\infty\}$ and a point $(\bar{x}, \bar{y}) \in \mathrm{dom}\, f$. Assume that*

$$(0, \eta) \in \partial^\infty f(\bar{x}, \bar{y}) \text{ implies } \eta = 0.$$

Then:

$$\partial_x f(\bar{x}, \bar{y}) \subset \{\xi : \text{there exists } \eta \text{ such that } (\xi, \eta) \in \partial f(\bar{x}, \bar{y})\}$$
$$\partial_x^\infty f(\bar{x}, \bar{y}) \subset \{\xi : \text{there exists } \eta \text{ such that } (\xi, \eta) \in \partial^\infty f(\bar{x}, \bar{y})\}.$$

Proof Take any $\xi \in \partial_x f(\bar{x}, \bar{y})$. We know then that there exist $\xi_i \to \xi$ and $x_i \xrightarrow{f(., \bar{y})} \bar{x}$ such that $\xi_i \in \partial_x^P f(x_i, \bar{y})$ for all i. For each i then we may choose $\epsilon_i > 0$ and $M_i > 0$ such that x_i is the unique minimizer of $F_i(., \bar{y})$, where

$$F_i(x, y) := f(x, y) - \xi_i \cdot (x - x_i) + M_i |x - x_i|^2 + \Psi_{x_i + \epsilon_i \mathbb{B}}(x).$$

Corollary 5.2.2 and Lemma 5.1.2 applied to F_i yield: either there exists $\eta' \in \mathbb{R}^m$ such that $(\xi_i, \eta') \in \partial f(x_i, \bar{y})$ or there exists a non-zero $\eta'' \in \mathbb{R}^m$ such that $(0, \eta'') \in \partial^\infty f(x_i, \bar{y})$. There are two possible situations:

5.3 Partial Limiting Subgradients

(a): for an infinite number of i's there exists $\eta_i \in \mathbb{R}^m$ such that

$$(\xi_i, \eta_i) \in \partial f(x_i, \bar{y})$$

(b): for an infinite number of i's there exists $\eta_i \in \mathbb{R}^m$ such that $|\eta_i| = 1$ and

$$(0, \eta_i) \in \partial^\infty f(x_i, \bar{y})$$

(we can arrange by scaling that $|\eta_i| = 1$ since $\partial^\infty f(x_i, \bar{y})$ is a cone).

Case (b) however is never encountered. This is because, in case (b), the η_i's which satisfy $(0, \eta_i) \in \partial^\infty f(x_i, \bar{y})$ have an accumulation point η satisfying $(0, \eta) \in \partial^\infty f(\bar{x}, \bar{y})$ and $|\eta| = 1$, in view of the fact that $(x_i, \bar{y}) \xrightarrow{f} (\bar{x}, \bar{y})$. This is in violation of the non-degeneracy hypothesis.

It remains then to attend to (a). We restrict attention to a subsequence, thereby ensuring that η_i, with the stated properties, exists for each i.

It may be assumed that $\{\eta_i\}$ is a bounded sequence for, otherwise, we are in a situation where, following a further subsequence extraction, $|\eta_i| \to \infty$ and $|\eta_i|^{-1}\eta_i \to v$ for some v with $|v| = 1$. But then, for $t_i := |\eta_i|^{-1}$, we have

$$t_i \downarrow 0 \text{ and } t_i(\xi_i, \eta_i) \to (0, v).$$

Since $(x_i, \bar{y}) \xrightarrow{f} (\bar{x}, \bar{y})$, we conclude

$$(0, v) \in \partial^\infty f(\bar{x}, \bar{y}).$$

This is ruled out by the non-degeneracy hypothesis. We have confirmed that $\{\eta_i\}$ is a bounded sequence.

By extracting a subsequence we can arrange that $\eta_i \to \eta$ for some $\eta \in \mathbb{R}^m$. Along the subsequence $(\xi_i, \eta_i) \in \partial f(x_i, \bar{y})$ and $(x_i, \bar{y}) \xrightarrow{f} (\bar{x}, \bar{y})$. It follows that $(\xi, \eta) \in \partial f(\bar{x}, \bar{y})$. We have shown that

$$\partial_x f(\bar{x}, \bar{y}) \subset \{\xi : \text{there exists } \eta \text{ such that } (\xi, \eta) \in \partial f(\bar{x}, \bar{y})\}.$$

Verification of the estimate governing the asymptotic partial limiting subdifferential

$$\partial_x^\infty f(\bar{x}, \bar{y}) \subset \{\xi : \text{there exists } \eta \text{ such that } (\xi, \eta) \in \partial^\infty f(\bar{x}, \bar{y})\}$$

is along similar lines. A sketch of the arguments involved is as follows. Take any $\xi \in \partial_x^\infty f(\bar{x}, \bar{y})$. We deduce from Theorem 4.6.2 that there exist $\xi_i \to \xi$, $x_i \xrightarrow{f(.,\bar{y})} \bar{x}$ and $t_i \downarrow 0$ such that

$$\xi_i \in t_i \partial_x^P f(x_i, \bar{y}).$$

Once again there are two possibilities to be considered:

(a)': for an infinite number of i's there exists $\eta_i \in \mathbb{R}^m$ such that $(\xi_i, \eta_i) \in t_i \partial f(x_i, \bar{y})$,
(b)': for an infinite number of i's there exists $\eta_i \in \mathbb{R}^m$ such that $|\eta_i| = 1$ and

$$(0, \eta_i) \in t_i \partial^\infty f(x_i, \bar{y}).$$

(b)' is analogous to (b), a possibility which, as we have noted, cannot occur. So we may assume (a)'. We deduce from the non-degeneracy hypothesis that the η_i's are bounded and so have an accumulation point η. We deduce (as before) the required relation:

$$(\xi, \eta) \in \partial^\infty f(\bar{x}, \bar{y}).$$

$\square$

5.4 A Sum Rule

A general sum rule is another consequence of the marginal function principle.

Theorem 5.4.1 (Sum Rule) *Take lower semi-continuous functions $f_i : \mathbb{R}^n \to \mathbb{R} \cup \{+\infty\}$, $i = 1, \ldots, m$, and a point $\bar{x} \in \cap_i \mathrm{dom}\, f_i$. Define $f = f_1 + \ldots + f_m$. Assume that*

$$v_i \in \partial^\infty f_i(\bar{x})\ \text{for}\ i = 1, \ldots, m\ \text{and}\ \sum_i v_i = 0\ \text{imply}\ v_i = 0\ \text{for all}\ i.$$

Then

$$\partial f(\bar{x}) \subset \partial f_1(\bar{x}) + \ldots + \partial f_m(\bar{x})$$

and

$$\partial^\infty f(\bar{x}) \subset \partial^\infty f_1(\bar{x}) + \ldots + \partial^\infty f_m(\bar{x}).$$

Proof Consider the function $F : \mathbb{R}^n \times (\mathbb{R}^n)^m \to \mathbb{R} \cup \{+\infty\}$ defined by

$$F(x, y) := \sum_{i=1}^m f_i(x + y_i)\ \text{for}\ y = (y_1, \ldots, y_m).$$

Evidently $\partial f(\bar{x}) = \partial_x F(\bar{x}, 0)$. This formula relating limiting subgradients of f and partial limiting subgradients of F provides the link with the marginal function principle.

5.4 A Sum Rule

Take $(\xi, \eta) \in \partial^P F(x, y)$ at some point $(x, y) \in \text{dom } F$. Then there exists $M > 0$ such that

$$F(x', y') - F(x, y) \geq (\xi, \eta) \cdot ((x', y') - (x, y)) - M|(x', y') - (x, y)|^2$$

for all (x', y') in some neighbourhood of (x, y). Expressed in terms of the f_i's this inequality informs us that

$$\sum_i [f_i(x' + y_i') - f_i(x + y_i)] \geq \xi \cdot (x' - x) + \sum_i \eta_i \cdot (y_i' - y_i) - M|(x', y') - (x, y)|^2.$$

Fix an index value $j \in \{1, \ldots, m\}$. Set $x' = x$ and $y_i' = y_i$ for $i \neq j$ in the above inequality. There results

$$f_j(x + y_j') - f_j(x + y_j) \geq \eta_j \cdot (y_j' - y_j) - M|y_j' - y_j|^2$$

for all y_j''s close to y_j. It follows that $\eta_j \in \partial^P f_j(x + y_j)$.

Next, for any x' close to x, set $y_i' = x + y_i - x'$ for $i = 1, \ldots, m$. This gives

$$0 \geq \xi \cdot (x' - x) - \left(\sum_i \eta_i\right) \cdot (x' - x) - (m+1)M|x' - x|^2.$$

We conclude that $\xi = \sum_i \eta_i$. To summarize:

$$\partial^P F(x, y) \subset \{(\eta_1 + \ldots + \eta_m, \eta) : \eta = (\eta_1, \ldots, \eta_m) \text{ and}$$
$$\eta_i \in \partial^P f_i(x + y_i) \text{ for } i \in \{1, \ldots, m\}\}.$$

Now take any $(\xi, \eta) \in \partial F(\bar{x}, 0)$. Then $(\xi, \eta) = \lim_k (\xi_k, \eta^k)$ for some sequences $\{(\xi_k, \eta^k)\}$ and $\{(x_k, y^k)\}$ such that $(\xi_k, \eta^k) \in \partial^P F(x_k, y^k)$ for each k, and $(x_k, y^k) \xrightarrow{F} (\bar{x}, 0)$. As we have shown

$$\eta_i^k \in \partial^P f_i(x_k + y_i^k) \text{ for } i = 1, \ldots, m$$

and $\xi_k = \sum_{i=1}^m \eta_i^k$. We claim also that

$$f_i(x_k + y_i^k) \to f_i(\bar{x}) \text{ for each } i.$$

To confirm this relation, consider

$$\gamma_i := \limsup_{k \to \infty} (f_i(x_k + y_i^k) - f_i(\bar{x})) \text{ for } i = 1, \ldots, m.$$

Since the f_i's are lower semi-continuous, we must show that $\gamma_i = 0$ for each i. By extracting subsequences if necessary, we can arrange that

$$\gamma_i = \lim_{k\to\infty} (f_i(x_k + y_i^k) - f_i(\bar{x})) \text{ for } i = 1, \ldots, m.$$

('lim' has replaced 'lim sup' here). But since $(x_k, y^k) \xrightarrow{F} (\bar{x}, 0)$,

$$\lim_k \sum_{i=1}^m f_i(x_k + y_i^k) = \sum_{i=1}^m f_i(\bar{x}),$$

whence

$$\sum_i \liminf_{k\to\infty} (f_i(x_k + y_i^k) - f_i(\bar{x})) \leq 0.$$

By lower semicontinuity, each term in the summation is non-negative. It follows that, for each i,

$$0 = \liminf_{k\to\infty} (f_i(x_k + y_i^k) - f_i(\bar{x})) = \lim_{k\to\infty} (f_i(x_k + y_i^k) - f_i(\bar{x})) = \gamma_i.$$

Our claim then is justified.

Since

$$x_k + y_i^k \xrightarrow{f} \bar{x}, \ \eta_i^k \to \eta_i \quad \text{as } k \to \infty,$$

$$\eta_i^k \in \partial^P f_i(x_k + y_i^k) \quad \text{for all } k,$$

we have

$$\eta_i \in \partial f_i(\bar{x}) \text{ for each } i.$$

We have shown that

$$\partial F(\bar{x}, 0) \subset \{(\eta_1 + \ldots + \eta_m, \eta) :$$
$$\eta = (\eta_i, \ldots, \eta_m) \text{ and } \eta_i \in \partial f_i(\bar{x}) \text{ for } i \in \{1, \ldots, m\}\}.$$

Similar arguments based on the representation of $\partial^\infty F$ in terms of scaled proximal subgradients at neighbouring points yield

$$\partial^\infty F(\bar{x}, 0) \subset \{(\eta_1 + \ldots + \eta_m, \eta) :$$
$$\eta = (\eta_i, \ldots, \eta_m) \text{ and } \eta_i \in \partial^\infty f_i(\bar{x}) \text{ for each } i\}.$$

5.4 A Sum Rule

Now apply Theorem 5.3.1 to F. Note that the non-degeneracy hypothesis of Theorem 5.3.1 is satisfied since, if $(0, \eta) \in \partial^\infty F(\bar{x}, 0)$ and $\eta \neq 0$, then $\sum_i \eta_i = 0$ and $\eta_i \in \partial^\infty f_i(\bar{x})$ for each i, a possibility excluded by our hypotheses. Recalling that $f(x) = F(x, 0)$ for all $x \in \mathbb{R}^n$, we deduce that

$$\partial f(\bar{x}) = \partial_x F(\bar{x}, 0) \subset \partial f_1(\bar{x}) + \ldots + \partial f_m(\bar{x})$$

and

$$\partial^\infty f(\bar{x}) = \partial_x^\infty F(\bar{x}, 0) \subset \partial^\infty f_1(\bar{x}) + \ldots + \partial^\infty f_m(\bar{x}).$$

□

As a first application of the sum rule, we estimate the normal cone to the graph of a Lipschitz continuous map.

Proposition 5.4.2 *Take a lower semi-continuous map $G : \mathbb{R}^n \to \mathbb{R}^m$ and a point $u \in \mathbb{R}^n$. Assume that G is Lipschitz continuous on a neighbourhood of u. Then*

$$N_{\mathrm{Gr}\, G}(u, G(u)) \subset \{(\xi, -\eta) : \xi \in \partial(\eta \cdot G)(u), \eta \in \mathbb{R}^m\}.$$

Proof Take any $(\xi, -\eta) \in N_{\mathrm{Gr}\, G}(u, G(u))$. Then $(\xi, -\eta) = \lim_i (\xi_i, -\eta_i)$ for sequences $\{(\xi_i, -\eta_i)\}$ and $u_i \xrightarrow{G} u$ such that

$$(\xi_i, -\eta_i) \in N^P_{\mathrm{Gr}\, G}(u_i, G(u_i)) \quad \text{for all } i.$$

For each i, there exists $M_i > 0$ such that

$$(\xi_i, -\eta_i) \cdot (u' - u_i, G(u') - G(u_i)) \leq M_i |u' - u_i|^2$$

for all u' close to u_i. This inequality can be re-arranged to give

$$\eta_i \cdot G(u') \geq \eta_i \cdot G(u_i) + \xi_i \cdot (u' - u_i) - M_i |u' - u_i|^2,$$

from which we deduce that

$$\xi_i \in \partial^P (\eta_i \cdot G)(u_i) .$$

If follows from Proposition 4.7.1 and the sum rule (Theorem 5.4.1) that

$$\xi_i \in \partial(\eta \cdot G)(u_i) + \partial((\eta_i - \eta) \cdot G)(u_i)$$

for each i. Also by Proposition 4.7.1. then, for sufficient large i,

$$\xi_i \in \partial(\eta \cdot G)(u_i) + K|\eta_i - \eta|\mathbb{B}.$$

Here K is a Lipschitz constant for G on a neighbourhood of u. Since the limiting normal cone is 'robust' under limit taking, we obtain $\xi \in \partial(\eta \cdot G)(u)$ in the limit as $i \to \infty$. □

5.5 A Nonsmooth Chain Rule

We now derive a far-reaching chain rule.

Theorem 5.5.1 (A Chain Rule) *Take a locally Lipschitz continuous function $G : \mathbb{R}^n \to \mathbb{R}^m$, a lower semi-continuous function $g : \mathbb{R}^m \to \mathbb{R} \cup \{+\infty\}$ and a point $\bar{u} \in \mathbb{R}^n$ such that $G(\bar{u}) \in \text{dom } g$. Define the lower semi-continuous function $f(u) := g \circ G(u)$. Assume that:*

The only vector $\eta \in \partial^\infty g(G(\bar{u}))$ such that $0 \in \partial(\eta \cdot G)(\bar{u})$ is $\eta = 0$.

Then

$$\partial f(\bar{u}) \subset \{\xi : \text{there exists } \eta \in \partial g(G(\bar{u})) \text{ such that } \xi \in \partial(\eta \cdot G)(\bar{u})\} \quad (5.5.1)$$

and

$$\partial^\infty f(\bar{u}) \subset \{\xi : \text{there exists } \eta \in \partial^\infty g(G(\bar{u})) \text{ such that } \xi \in \partial(\eta \cdot G)(\bar{u})\}. \quad (5.5.2)$$

Otherwise expressed, inclusion (5.5.1) tells us that, given some $\xi \in \partial f(\bar{u})$, we may find some $\eta \in \partial g$, evaluated at $G(\bar{u})$, such that $\xi \in \partial d(\bar{u})$, where d is the scalar-valued function $d(u) := (\eta \cdot G)(u)$. ((5.5.2) can be likewise interpreted.)

Proof Define the lower semi-continuous function

$$F(x, u) := g(x) + \Psi_{\text{Gr}G}(u, x) + \Psi_{\{\bar{u}, G(\bar{u})\} + \mathbb{B}}(u, x). \quad (5.5.3)$$

We see that $f(u) = \min_x F(x, u)$. Notice that $(G(\bar{u}), \bar{u}) \in \text{dom } F$, $\text{dom } F$ is bounded and $G(\bar{u})$ is the unique minimizer of $F(., \bar{u})$. The scene is set then for applying Theorem 5.2.1. This tells us that

$$\partial f(\bar{u}) \subset \{\xi : (0, \xi) \in \partial F(G(\bar{u}), \bar{u})\}$$

and

$$\partial^\infty f(\bar{u}) \subset \{\xi : (0, \xi) \in \partial^\infty F(G(\bar{u}), \bar{u})\}.$$

Now the sum rule (Theorem 5.4.1) is applied to estimate the limiting subdifferential and asymptotic limiting subdifferential of F given by (5.5.3). The last term in (5.5.3) makes no contribution to the subdifferentials and can be ignored.

5.5 A Nonsmooth Chain Rule

It is necessary to check the sum rule non-degeneracy condition. Now, as is easily shown, a general element v_1 in the asymptotic limiting subdifferential of $(x, u) \to g(x)$ at $(G(\bar{u}), \bar{u})$ is expressible as $v_1 = (\gamma, 0)$ where $\gamma \in \partial^\infty g(G(\bar{u}))$. On the other hand, by Proposition 5.4.2,

$$\partial\{(x', u') \to \Psi_{\mathrm{Gr}\, G}(u', x')\}(x, u)$$
$$= \{(-\eta, \xi) : (\xi, -\eta) \in \partial \Psi_{\mathrm{Gr}\, G}(u, x)\}$$
$$\subset \{(-\eta, \xi) : \xi \in \partial(\eta \cdot G)(u)\}.$$

So an element in $\partial\{(x, u) \to \Psi_{\mathrm{Gr}\, G}(u, x)\}(G(\bar{u}), \bar{u})$ is expressible as $v_2 = (-\eta, \xi)$ for some ξ and η which satisfy $\xi \in \partial(\eta \cdot G)(\bar{u})$. We must examine the consequences of $v_1 + v_2 = 0$. They are that $\xi = 0$ and $0 \in \partial(\eta \cdot G)(\bar{u})$ for some $\eta \in \partial^\infty g(G(\bar{u}))$. But then $\eta = 0$ under the present hypotheses. So $v_1 + v_2 = 0$ implies $v_1 = v_2 = 0$. The non-degeneracy hypothesis is therefore satisfied. The sum rule (Theorem 5.4.1) and previous considerations now give

$$\partial f(\bar{u}) \subset \{\xi : (0, \xi) = (v, 0) + (-\eta, \xi), v \in \partial g(G(\bar{u})), \xi \in \partial(\eta \cdot G)(\bar{u})\}$$
$$= \{\xi : \text{there exists } \eta \in \partial g(G(\bar{u})) \text{ such that } \xi \in \partial(\eta \cdot G)(\bar{u})\}.$$

Similarly

$$\partial^\infty f(\bar{u}) \subset \{\xi : \text{there exists } \eta \in \partial^\infty g(G(\bar{u})) \text{ such that } \xi \in \partial(\eta \cdot G)(\bar{u})\}.$$

These are the relations we set out to prove. □

A number of important calculus rules are now obtainable as corollaries of the nonsmooth chain rule.

Theorem 5.5.2 (Max Rule) *Take locally Lipschitz continuous functions* $f_i : \mathbb{R}^n \to \mathbb{R}$, $i = 1, \ldots, m$, *and a point* $\bar{x} \in \mathbb{R}^n$. *Define* $f(x) = \max_i f_i(x)$ *and* $\Lambda := \{\lambda = (\lambda_1, \ldots, \lambda_m) \in \mathbb{R}^m : \lambda_i \geq 0, \sum_i \lambda_i = 1\}$. *Then*

$$\partial f(\bar{x}) \subset \{\partial(\sum_{i=1}^m \lambda_i f_i)(\bar{x}) : \lambda \in \Lambda, \text{ and } \lambda_i = 0 \text{ if } f_i(\bar{x}) < f(\bar{x})\}.$$

Let us be clear about what this theorem tells us: given $\xi \in \partial f(\bar{x})$, there exists a 'convex combination' $\{\lambda_i\}$ of 'active' f_i's such that

$$\xi \in \partial(\sum_i \lambda_i f_i)(\bar{x}).$$

This implies (via the sum rule and in view of the positive homogeneity properties of the subdifferential)

$$\xi \in \sum_i \lambda_i \partial f_i(\bar{x}),$$

another, slightly weaker, version of the max rule.

Proof We readily calculate the limiting subgradient of $g(y) := \max\{y_1, \ldots, y_m\}$ for $y = (y_1, \ldots, y_m)$. It is

$$\partial g(y) = \{\lambda = (\lambda_1, \ldots, \lambda_m) \in \Lambda : \lambda_i = 0 \text{ if } y_i < \max_j y_j\}.$$

Now we apply the chain rule (Theorem 5.5.1) with this g and $G(x) := (f_1, \ldots, f_m)(x)$, noting that the non-degeneracy hypothesis is satisfied in view of Proposition 4.7.1. □

Theorem 5.5.3 (Product Rule) *Take locally Lipschitz continuous functions $f_i : \mathbb{R}^n \to \mathbb{R}$, $i = 1, \ldots, m$, and a point $\bar{x} \in \mathbb{R}^n$. Define $f(x) = f_1(x) f_2(x) \ldots f_m(x)$. Then*

$$\partial f(\bar{x}) \subset \partial (\sum_i^m \Pi_{j \neq i} f_j(\bar{x}) f_i)(\bar{x}).$$

This theorem tells us for example that, in the case $m = 2$,

$$\partial f(\bar{x}) \subset \partial (f_1(\bar{x}) f_2 + f_2(\bar{x}) f_1)(\bar{x}).$$

In the event that $f_1(\bar{x}) \geq 0$, $f_2(\bar{x}) \geq 0$, this condition implies

$$\partial f(\bar{x}) \subset f_1(\bar{x}) \partial f_2(\bar{x}) + f_2(\bar{x}) \partial f_1(\bar{x})$$

Proof Apply the chain rule with $g(y) := \Pi_i y_i$ for $y = (y_1, \ldots, y_m)$ and $G(x) := (f_1, \ldots, f_m)(x)$ for $x \in \mathbb{R}^n$. □

5.6 Lagrange Multiplier Rules

Finally we make contact with optimization. A general theorem is proved, providing necessary conditions for a point to be a minimizer of a composite function. Choosing different ingredients for this function supplies a variety of Lagrange multiplier rules, one example of which we investigate in detail.

Theorem 5.6.1 (Generalized Multiplier Rule) *Take Lipschitz continuous functions $f : \mathbb{R}^n \to \mathbb{R}$ and $F : \mathbb{R}^n \to \mathbb{R}^m$, a lower semi-continuous function $h : \mathbb{R}^m \to \mathbb{R} \cup \{+\infty\}$, a closed set $C \subset \mathbb{R}^n$ and a point $\bar{x} \in C$ such that*

5.6 Lagrange Multiplier Rules

$h(F(\bar{x})) < \infty$. Define the lower semi-continuous function $l : \mathbb{R}^n \to \mathbb{R} \cup \{+\infty\}$ to be

$$l(x) := f(x) + h(F(x)) + \Psi_C(x).$$

Let $\bar{x}$ attain the minimum of l. Assume that

$$0 \in \partial(\eta \cdot F)(\bar{x}) + N_C(\bar{x}) \text{ for some } \eta \in \partial^\infty h(F(\bar{x})) \text{ implies } \eta = 0.$$

Then

$$0 \in \partial f(\bar{x}) + \partial(\eta \cdot F)(\bar{x}) + N_C(\bar{x}) \text{ for some } \eta \in \partial h(F(\bar{x})).$$

Proof Choose $g(x, y) := f(x) + h(y) + \Psi_C(x)$ and $G(x) = (x, F(x))$. Then $\bar{x}$ minimizes $l(x) := g(G(x))$, and so

$$0 \in \partial^P l(\bar{x}) \subset \partial(g \circ G)(\bar{x}).$$

Set $\bar{y} = F(\bar{x})$. Now apply the chain rule (Theorem 5.5.1). To begin with, we need to check the non-degeneracy hypothesis. Take any

$$\eta = (\eta_1, \eta_2) \in \partial^\infty g(\bar{x}, \bar{y}) = \partial^\infty(f(x) + h(y) + \Psi_C(x))|_{(x,y)=(\bar{x},\bar{y})}.$$

It is a straightforward exercise to check that the conditions under which the sum rule (Theorem 5.4.1) supplies an estimate for (η_1, η_2) are satisfied. Consequently there exist $\eta_1 \in N_C(\bar{x})$ and $\eta_2 \in \partial^\infty h(F(\bar{x}))$. We see that if $0 \in \partial(\eta \cdot G)(\bar{x})$ then $0 \in \eta_1 + \partial(\eta_2 \cdot F)(\bar{x})$. Under current hypotheses then η_2 (and therefore also η_1) is zero, so the hypotheses under which we may apply the chain rule are satisfied.

A further application of the sum rule tells us that if $(\xi, \eta) \in \partial g(G(\bar{x}))$ then $\xi \in \partial f(\bar{x}) + N_C(\bar{x})$ and $\eta \in \partial h(\bar{y})$. From the chain rule then

$$0 \in \partial(g \circ G)(\bar{x}) \subset \partial f(\bar{x}) + \partial(\eta \cdot F)(\bar{x}) + N_C(\bar{x})$$

for some $\eta \in \partial h(F(\bar{x}))$. This is what we set out to prove. □

In applications the function h in the generalized multiplier theorem is usually taken to be the indicator function of some closed set E (comprising allowable values for $F(x)$). In this case a minimizer $\bar{x}$ for l is a minimizer for the constrained optimization problem

$$\text{Minimize}\{f(x) : F(x) \in E \text{ and } x \in C\}.$$

Now

$$\partial \psi_E(F(\bar{x})) = \partial^\infty \psi_E(F(\bar{x})) = N_E(F(\bar{x})).$$

It follows from Theorem 5.6.1 that, if

$$0 \in \partial(\eta \cdot F)(\bar{x}) + N_C(\bar{x}) \text{ and } \eta \in N_E(F(\bar{x})) \text{ implies } \eta = 0, \qquad (5.6.1)$$

then there exists

$$\eta \in N_E(F(\bar{x}))$$

such that

$$0 \in \partial f(\bar{x}) + \partial(\eta \cdot F)(\bar{x}) + N_C(\bar{x}).$$

The vector η is a Lagrange multiplier, which must be directed into the normal cone $N_E(F(\bar{x}))$ of the constraint set E at $F(\bar{x})$. The non-degeneracy condition (5.6.1) under which this Lagrange multiplier rule is valid requires, when F is a smooth function, that there does not exist a non-zero vector

$$\eta = \{\eta_1, \ldots, \eta_n\}$$

such that $-\eta$ is a limiting normal to E at $F(\bar{x})$ and the linear combination $\sum_i \eta_i \nabla F_i(\bar{x})$ of gradients of the components of F at $\bar{x}$ is a limiting normal to C at $\bar{x}$. This condition is a generalization of Mangasarian Fromovitz type constraint qualifications invoked in the mathematical programming literature.

A widely studied optimization problem to which the above theorem is applicable is

(NLP)   Minimize $f_0(x)$ over $x \in X \cap C$

where

$$X := \{x \in \mathbb{R}^n : f_1(x) \leq 0, \ldots, f_p(x) \leq 0, g_1(x) = 0, \ldots, g_q(x) = 0\}.$$

Here $f_0, f_1, \ldots, f_p$ and $g_1, \ldots, g_q$ are all $\mathbb{R}$-valued functions with domain $\mathbb{R}^n$, and C is a closed subset of $\mathbb{R}^n$.

Theorem 5.6.2 (Lagrange Multiplier Rule) *Let $\bar{x}$ be a local minimizer for (NLP). Assume that $f_0, \ldots, f_p$ and $g_1, \ldots, g_q$ are locally Lipschitz continuous functions. Then there exist $\lambda_0 \geq 0, \lambda_1 \geq 0, \ldots, \lambda_p \geq 0$ (with $\lambda_0 = 0$ or 1) and real numbers $\gamma_1, \ldots, \gamma_q$ such that*

$$\lambda_i = 0 \text{ if } f_i(\bar{x}) < 0 \text{ for } i = 1, \ldots, p, \qquad (5.6.2)$$

$$\sum_{i=0}^{p} \lambda_i + \sum_{i=1}^{q} |\gamma_i| \neq 0$$

5.6 Lagrange Multiplier Rules

and

$$0 \in \partial[\sum_{i=0}^{p} \lambda_i f_i + \sum_{i=1}^{q} \gamma_i g_i](\bar{x}) + N_C(\bar{x}).$$

Proof By replacing C by $C \cap (\bar{x} + \epsilon \mathbb{B})$, for $\epsilon > 0$ sufficiently small, we can arrange that $\bar{x}$ is a minimizer of f_0 over $X \cap C$ ($N_C(\bar{x})$ is unaffected). We can arrange also that, if $f_i(\bar{x}) < 0$, for some $i \in \{1, \ldots, p\}$ then $f_i(x) < 0$ for all $x \in \bar{x} + \epsilon \mathbb{B}$. The constraint $f_i(x) < 0$ is thereby rendered irrelevant and we can ignore it. Let us assume that index values corresponding to inequality constraints inactive at $\bar{x}$ have been removed and we have labelled the remaining inequality constraint functions by indices in some new (possibly reduced) index set. The assertions of the theorem for the new index set (excluding the complementary slackness condition (5.6.2)) imply those for the original index set (including this last condition). Complementary slackness therefore takes care of itself, if we attend to the other assertions of the theorem.

It is clear that $(\bar{\alpha}, \bar{x})$ (where $\bar{\alpha} = f_0(\bar{x})$) is a minimizer for the optimization problem:

$$\text{Minimize } \{\alpha : f_0(x) - \alpha \leq 0 \text{ and } x \in X \cap C\}.$$

Set

$$f(\alpha, x) = \alpha, \; F(\alpha, x) = (f_0(x) - \alpha, f_1(x), \ldots, f_p(x), g_1(x), \ldots, g_q(x))$$

and

$$E = (-\infty, 0]^{p+1} \times \{0\}^q.$$

We find $N_E(0) = M$ where

$$M = \{(\lambda_0, \lambda_1, \ldots, \lambda_p, \gamma_1, \ldots, \gamma_q) : \lambda_i \geq 0 \text{ for } i \in \{0, \ldots, p\}\}.$$

Let us now examine the non-degeneracy hypothesis of Theorem 5.6.1 for the above identifications of f and F and for $h = \Psi_E$. If $\eta \in \partial^\infty h(0)$, then

$$\eta = (\lambda_0, \lambda_1, \ldots, \lambda_p, \gamma_1, \ldots, \gamma_q) \in M.$$

It follows that the relation

$$0 \in \partial(\eta \cdot F)(\bar{\alpha}, \bar{x}) + \{0\} \times N_C(\bar{x})$$

can be written

$$0 \in \{-\lambda_0\} \times \partial(\sum_{i=0}^{p} \lambda_i f_i + \sum_{i=1}^{q} \gamma_i g_i)(\bar{x}) + \{0\} \times N_C(\bar{x}).$$

which implies

$$0 \in \{0\} \times \partial(\sum_{i=1}^{p} \lambda_i f_i + \sum_{i=1}^{q} \gamma_i g_i)(\bar{x}) + \{0\} \times N_C(\bar{x}).$$

It follows that if the non-degeneracy hypothesis is violated, then

$$0 \in \partial(\sum_{i=1}^{p} \lambda_i f_i + \sum_{i=1}^{q} \gamma_i g_i)(\bar{x}) + N_C(\bar{x})$$

for some nonzero $(\lambda_0, \lambda_1, \ldots, \lambda_p, \gamma_1, \ldots, \gamma_q) \in M$ and $\lambda_0 = 0$. The assertions of the theorem are valid then (albeit only in a degenerate sense).

On the other hand, if the non-degeneracy hypothesis is satisfied then there exists $\eta = (\lambda_0, \ldots, \lambda_p, \gamma_1, \ldots, \gamma_q) \in M$ such that

$$0 \in \partial f(\bar{\alpha}, \bar{x}) + \partial(\eta \cdot F)(\bar{\alpha}, \bar{x}) + \{0\} \times N_C(\bar{x}).$$

Since $\partial f(\bar{\alpha}, \bar{x}) = \{(1, 0, \ldots, 0)\}$, we conclude

$$0 \in \partial(\sum_{i=0}^{p} \lambda_i f_i + \sum_{i=1}^{q} \gamma_i g_i)(\bar{x}) + N_C(\bar{x}),$$

for some $(\lambda_0, \lambda_1, \ldots, \lambda_p, \gamma_1, \ldots, \gamma_q) \in M$ with $\lambda_0 = 1$. □

5.7 Max Rule for an Infinite Family of Functions

We have proved a max rule estimating the subdifferential of the pointwise supremum of a finite family number of lower semi-continuous functions (see Theorem 5.5.2). What if the pointwise supremum is taken over a family of functions $x \to f(t, x)$ indexed by points $t \in T$, where T is a set containing possibly an infinite number of points? The following theorem, due to Clarke, provides an answer to this question. It characterizes the convexified limiting subdifferential of $x \to \sup_{t \in T}\{f(t, x)\}$, i.e. the Clarke generalized gradient, in circumstances where no restrictions are place on the index set, under a finiteness hypothesis and when we require the functions to be uniformly Lipschitz continuous, regarding their x dependence.

Theorem 5.7.1 *Take a non-empty abstract set T, a function $f : T \times \mathbb{R}^n \to \mathbb{R}$ and a point $x \in \mathbb{R}^n$. Define*

$$f_{\sup}(x) := \sup_{t \in T} f(t, x), \text{ for } x \in \mathbb{R}^n.$$

5.7 Max Rule for an Infinite Family of Functions

Assume that, for some open neighbourhood U of x, there exists $K > 0$ such that

(H1): $|f(t, x'') - f(t, x')| \leq K|x'' - x'|$ for all $t \in T$ and $x'', x' \in U$,
(H2): $f_{\sup}(x') < \infty$ for some $x' \in U$.

Then, for any subset $S \subset U$ of zero n-dimensional Lebesgue measure,

$$\text{co } \partial f_{\sup}(x) \subset C,$$

where $C := \text{co}\{\lim_{i \to \infty} \nabla_x f(t_i, x_i) : t_i \in T, x_i \notin S \text{ for all } i \text{ and } f(t_i, x_i) \to f_{\sup}(x)\}$.

Proof Notice that, in consequence of (H1), (H2) implies $f_{\sup}(x') < \infty$ for all $x' \in U$. It is easily checked that $f_{\sup}$ is Lipschitz continuous on U (with Lipschitz constant K). Observe that the set C is nonempty and compact.

Take any $v \in \mathbb{R}^n$ and $\epsilon > 0$. Write

$$m := \max_{\xi \in C} \xi \cdot v.$$

Recalling that the Clarke directional derivative $f_{\sup}^0(x; \cdot)$ is the support function of the closed convex set $\text{co } \partial f_{\sup}(x)$ (Propositions 4.7.4 and 4.7.6), we see that the assertions of the theorem will have been confirmed if we can show

$$f_{\sup}^0(x; v) \leq m + \epsilon.$$

Since the function dependencies concerned are homogeneous, we can assume, without loss of generality, that $|v| = 1$.

It follows from the definition of C that there exists $\delta > 0$ such that $x + 2\delta \mathbb{B} \subset U$ and, for any $y \in x + 2\delta \mathbb{B}$ and $t \in T$ satisfying

$$y \notin S, \nabla_x f(t, y) \text{ exists and } f(t, x) \geq f_{\sup}(x) - \delta,$$

we have

$$\nabla_x f(t, y) \cdot v \leq m + \epsilon. \tag{5.7.1}$$

Take any $t \in T$ such that $f(t, x) \geq f_{\sup}(t, x) - \delta$. Write $\Omega := \{x' \in U : \nabla_x f(t, x') \text{ does not exist}\}$. Now take y to be any point in the ball $x + \delta \mathbb{B}$ such that the intersection of the line segment $[y, y + \delta v]$ with $S \cup \Omega$ has zero one-dimensional Lebesgue measure. (It can be deduced from Fubini's theorem that the set of such y's is a subset of $x + \delta \mathbb{B}$ with full n-dimensional Lebesgue measure.)

Now take any $\lambda \in (0, \delta)$. It follows from (5.7.1) and the Lipschitz continuity of $f(t, \cdot)$ that

$$f(t, y + \lambda v) - f(t, y) = \int_0^\lambda \nabla_x f(t, y + \sigma v) \cdot v \, d\sigma \leq \lambda(m + \epsilon). \tag{5.7.2}$$

Since this relation is valid for y's in a dense set, it follows from the continuity of $f(t,.)$ that it is also valid for all $y \in x + \delta \mathbb{B}$.

Now take $r \in (0, \delta)$ such that $r^2 + 4rK < \delta$.

Claim: for every $y \in x + r\mathbb{B}$ and $\lambda \in (0, r)$

$$f_{\sup}(y + \lambda v) - f_{\sup}(y) \leq \lambda(m + \epsilon) + \lambda^2.$$

This will mean that

$$f^0_{\sup}(x; v) = \{\limsup_{i \to \infty} \lambda_i^{-1}(f_{\sup}(y_i + \lambda_i v) - f_{\sup}(y_i)) : \lambda_i \downarrow 0, y_i \to y\} \leq m + \epsilon$$

as required for completion of the proof.

To validate the claim, take any $y \in x + r\mathbb{B}$ and $\lambda \in (0, r)$. Let t be any point in T such that $f(t, y + \lambda v) \geq f_{\sup}(y + \lambda v) - \lambda^2$. Then

$$\begin{aligned}
f(t, x) &\geq f(t, y + \lambda v) - K|x - y - \lambda v| \\
&\geq f_{\sup}(y + \lambda v) - \lambda^2 - K|x - y - \lambda v| \\
&\geq f_{\sup}(x) - \lambda^2 - 2K|x - y - \lambda v| \\
&\geq f_{\sup}(x) - r^2 - 4Kr, \text{ since } |v| \leq 1, \\
&\geq f_{\sup}(x) - \delta.
\end{aligned}$$

We have shown that t and λ chosen as above meet the criteria for validity of (5.7.2). It follows that

$$f_{\sup}(y + \lambda v) - f_{\sup}(y) \leq f(t, y + \lambda v) - f(t, y) + \lambda^2 \leq \lambda(m + \epsilon) + \lambda^2.$$

□

5.8 Exercises

5.1 Let $A \subset \mathbb{R}^n$ be a closed set with representation

$$A = \{x : \phi(x) \leq 0, \ \psi(x) = 0 \text{ and } x \in \Omega\},$$

for some locally Lipschitz continuous functions $\phi : \mathbb{R}^n \to \mathbb{R}^{k_1}$ and $\psi : \mathbb{R}^n \to \mathbb{R}^{k_2}$ and some closed set $\Omega \subset \mathbb{R}^n$.

5.8 Exercises

Take $\bar{x} \in \mathbb{R}^n$ and assume that

$$\{(\mu, \nu) \in (\mathbb{R}_+)^{k_1} \times \mathbb{R}^{k_2} : \partial\big(\mu \cdot \phi + \nu \cdot \psi\big)(\bar{x}) \cap \big(-N_\Omega(\bar{x})\big)$$

and $\mu_i = 0$ if $\phi_i(\bar{x}) < 0\} = \{(0, 0)\}$.

Take $\eta \in N_A(\bar{x})$. Show that there exists $(\mu, \nu) \in (\mathbb{R}_+)^{k_1} \times \mathbb{R}^{k_2}$ such that $\mu_i = 0$ if $\phi_i(\bar{x}) < 0$ and

$$\eta \in \partial\big(\mu \cdot \phi + \nu \cdot \psi\big)(\bar{x}) + N_\Omega(\bar{x}).$$

Remark
The relations derived in this exercise enable us to deduce first order necessary conditions of optimality for nonlinear programming problems involving inequality and equality state constraints, expressed in terms of Lagrange multipliers, from general necessary conditions, such as those provided by Theorem 5.6.2, expressed in terms of the limiting normal cone of the constraint set.

5.2 (Nonsmooth Derivatives of Vector Valued Functions) Take a vector valued function $G : \mathbb{R}^n \to \mathbb{R}^k$ and $\bar{x} \in \mathbb{R}^n$. Write $G(x) = [g_1(x), \ldots, g_k(x)]$. Assume that G is Lipschitz continuous on a neighbourhood of $\bar{x}$. The Clarke generalized Jacobian of G at $\bar{x}$ is the subset of the space of $k \times n$ matrices defined by

$$J^C G(\bar{x}) := \mathrm{co}\{M \in \mathbb{R}^{k \times n} : \exists\, x_i \to \bar{x} \text{ s.t. } G \text{ is differentiable at } x_i$$
$$\text{for each } i \text{ and } \nabla G(x_i) \to M\}$$

(i): Show that, for each $p \in \mathbb{R}^k$,

$$p \cdot J^C G(\bar{x}) = \mathrm{co}\, \partial(p \cdot G)(\bar{x}).$$

Hint: For each p, use the characterization of $\mathrm{co}\, \partial(p \cdot G)(\bar{x})$ provided by Theorem 4.7.7, taking as 'bad' set points in $\mathbb{R}^n$ at which G fails to be differentiable.

(ii): Show that

$$J^C G(\bar{x}) \subset \mathrm{co}\, \partial g_1(\bar{x}) \times \ldots \times \mathrm{co}\, \partial g_k(\bar{x}).$$

Give an example where this inclusion relation is strict.

Remark
The multifunction $D^* G(\bar{x}) : \mathbb{R}^k \to \mathbb{R}^n$ defined by

$$D^* G(\bar{x})(p) := \partial(p \cdot G)(\bar{x}),$$

the convex hull of whose values feature in the above representation of Clarke's generalized Jacobian, is referred to as the coderivative of G at $\bar{x}$ in [177]. It properties were systematically investigated in [123] and it features in the early nonsmooth constructions of Mordukhovich (See references in [159] and [160].)

5.9 Notes for Chapter 5

(See previous chapter)

Chapter 6
Differential Inclusions

Abstract Differential inclusions are generalizations of first order, vector differential equations, in which set inclusion replaces equality and the right side is no longer a point valued, but a set valued function ('multifunction') of the current time and state. Differential inclusions feature prominently in dynamic optimization. They are used to formulate a dynamic constraint, thereby serving as an alternative to the controlled differential equation framework of the early literature. Even when the original dynamic constraint is of a different nature, it is often helpful (regarding the formulation of hypotheses for existence of minimizers, or under which nonsmooth optimality conditions can be derived) to consider the associated differential inclusion.

Many properties of first order differential equations have analogues in the theory of differential inclusions. These include existence of global solutions with a specified initial value, when a Lipschitz continuity hypothesis is imposed on the differential inclusion (strictly speaking, on the associated multifunction), and sensitivity properties of solutions under data perturbations. Differential inclusions fail to have unique solutions even in the simplest cases, however. It is therefore of interest to consider the whole set of solutions for a given initial state and ask: is this set non-empty, is it compact and, if so, w.r.t. what topology, or is it stable under perturbations of the differential inclusion? This chapter provides answers to questions of this nature.

The starting point is the generalized Filippov existence theorem, which asserts existence of solutions to Lipschitz continuous differential inclusions. But it supplies additional useful information, in the form of estimates of the distance of a solution from a given arc that is an approximate solution of the differential inclusion. This is followed by the compactness of trajectories theorem, which concerns properties of sequences of arcs with uniformly integrably bounded velocities, approximately satisfying the given differential inclusion with an increasing degree of accuracy. We find that accumulation points of such sequences (in the uniform topology) satisfy, not the nominal differential inclusion but its convexification, in which the values of the associated multifunction are replaced by their convex hulls. We deduce as a straightforward consequence that, under a Lipschitz continuity hypothesis, the set of solutions to a given differential inclusion is closed, provided the differential inclusion is convex (that is, has values convex sets). If the differential inclusion fails to be convex, it can be shown that, nonetheless, an arbitrary solution to the convexified differential inclusion can be approximated uniformly closely (in the uniform norm) by solutions to the original, unconvexified, differential inclusion; this is the relaxation theorem. In the final section of the chapter, we introduce a state constraint. Conditions are given for existence of solutions that satisfy the state constraint. We also provide estimates of the distance of this solution from a nominal solution that violates the state constraint. Distance estimates of this nature are used extensively in the theory of state constrained dynamic optimization, to establish non-degeneracy of optimality conditions and stability of the infimum cost under data perturbations.

6.1 Introduction

A differential inclusion is a relation of the form

$$\dot{x}(t) \in F(t, x(t)) \text{ a.e. } t \in I, \tag{6.1.1}$$

in which I is a given interval and $F : I \times \mathbb{R}^n \rightsquigarrow \mathbb{R}^n$ is a given multifunction. Absolutely continuous functions that satisfy this relation are called F trajectories. Differential inclusions feature prominently in modern treatments of dynamic optimization. This has come about for several reasons. One is that condition (6.1.1), summarizing constraints on allowable velocities, provides a convenient framework for stating hypotheses under which dynamic optimization problems have solutions and optimality conditions may be derived. Another is that, even when we choose not to formulate a dynamic optimization problem in terms of a differential inclusion, in cases when the data is nonsmooth, often the very statement of optimality conditions makes reference to differential inclusions. It is convenient then at this stage to assemble important properties of multifunctions and differential inclusions of particular relevance in dynamic optimization.

Differential inclusions are generalizations of (first order) differential equations. Many aspects of the theory of first order differential equations carry across to this more general setting. These include existence of solutions with a specified initial value and sensitivity properties of solutions under data perturbation. An exception of course is 'uniqueness of solutions'. Indeed we can expect there to be a multiplicity of F trajectories having a given initial state since the differential inclusion merely imposes a constraint on possible values of the velocity (at each time and state), but may fail to specify that velocity. For this reason, important questions underlying the theory of differential inclusions concern properties of the whole family of F trajectories, not just individual members. Is it non-empty, is it closed and, if so, w.r.t. which topology on the underlying function space? How does the family behave under perturbations? Is it non-empty if we require the F trajectories to satisfy a path-wise constraint? All of these questions are addressed in this chapter.

6.2 Existence and Estimation of F Trajectories

Lipschitz continuity of the right side of a differential equation w.r.t. the state guarantees existence of global solutions and stability properties under perturbations. Likewise, Lipschitz continuity features prominently in the theory of differential inclusions as a hypothesis, to ensure non-emptiness of the class of F trajectories and serve as a starting point for a perturbational analysis. (Here, of course, 'Lipschitz continuity' of F w.r.t. to the state variable requires interpretation, since F is set valued.)

The starting point of this chapter is the generalized Filippov existence theorem. This fundamental theorem tells us, as its name suggests, that an F trajectory exists (under a Lipschitz continuity hypothesis). But it is important also for the additional information it supplies, namely estimates on the distance of this F trajectory from a given absolutely continuous arc that may fail to satisfy the differential inclusion.

At the heart of our investigation of the closure properties of F trajectories families is the compactness of trajectories theorem. This reveals limit behaviour of sequences of absolutely continuous functions, approximately satisfying the given differential inclusion $\dot{x} \in F$ with an increasing degree of accuracy. A key feature here is convexity: under a uniform integrability hypothesis, bounded sequences of F trajectories have accumulation points (in the uniform topology) that satisfy, not the given differential inclusion $\dot{x} \in F(t, x)$, but its *convexification*, namely $\dot{x} \in \operatorname{co} F(t, x)$.

The compactness of trajectories theorem points directly to simple criteria for existence of minimizers, for dynamic optimization problems in which the dynamic constraint takes the form of the differential inclusion (6.1.1), when F takes values convex sets. If F fails to be convex valued, there may be no minimizers in a traditional sense. But existence of minimizers can still be guaranteed, if the domain of the dynamic optimization problem is enlarged to include also convexified F trajectories. This is the theme of relaxation.

In the final section of the chapter, we undertake a more refined study of the existence of F trajectories, now to take account of a (path-wise) state constraint. We give conditions which ensure the existence of an F trajectory satisfying the state constraint. We also provide estimates concerning the distance of this F trajectory from some nominal F trajectory that violates the state constraint. Distance estimates of this nature are used extensively in the theory of state constrained dynamic optimization.

6.2 Existence and Estimation of F Trajectories

Fix an interval $[S, T]$ and a relatively open set $\Omega \subset [S, T] \times \mathbb{R}^n$. For $t \in [S, T]$, define

$$\Omega_t := \{x : (t, x) \in \Omega\}.$$

Take a continuous function $y : [S, T] \to \mathbb{R}^n$ and $\epsilon > 0$. Then the ϵ tube about y is the set

$$T(y, \epsilon) := \{(t, x) \in [S, T] \times \mathbb{R}^n : t \in [S, T], |x - y(t)| \leq \epsilon\}.$$

Consider a multifunction $F : \Omega \rightsquigarrow \mathbb{R}^n$. We say that an arc $x \in W^{1,1}([S, T]; \mathbb{R}^n)$ is an F trajectory if $\operatorname{Gr} x \subset \Omega$ and

$$\dot{x}(t) \in F(t, x(t)) \text{ a.e. } t \in [S, T].$$

Naturally we would like to know when F trajectories exist. We shall make extensive use of a local existence theorem which gives conditions under which an F trajectory exists near a nominal arc $y \in W^{1,1}([S, T]; \mathbb{R}^n)$. This theorem will also provide important supplementary information about how 'close' to y the F trajectory x may be chosen. Just how close will depend on the extent to which the nominal arc y fails to satisfy the differential inclusion, as measured by the function Λ_F,

$$\Lambda_F(y) := \int_S^T \rho_F(t, y(t), \dot{y}(t)) dt.$$

Here

$$\rho_F(t, x, v) := \inf\{|\eta - v| : \eta \in F(t, x)\}.$$

(We shall make use of function $\Lambda_F(y)$ only when the integrand above is Lebesgue measurable and F is a closed multifunction; then $\Lambda_F(y)$ is a non-negative number which is zero if and only if y is an F trajectory.)

We pause for a moment however to list some relevant properties of ρ_F which, among other things, give conditions under which the integral $\Lambda_F(y)$ is well-defined.

Proposition 6.2.1 *Take a multifunction* $F : [S, T] \times \mathbb{R}^n \rightsquigarrow \mathbb{R}^n$.

(a) Fix $(t, x) \in \operatorname{dom} F$. *Then*

$$|\rho_F(t, x, v) - \rho_F(t, x, v')| \leq |v - v'| \text{ for all } v, v' \in \mathbb{R}^n.$$

(b) Fix $t \in [S, T]$. *Suppose that, for some* $\epsilon > 0$ *and* $k > 0$, $\bar{x} + \epsilon \mathbb{B} \subset \operatorname{dom} F$ *and*

$$F(t, x) \subset F(t, x') + k|x - x'|\mathbb{B} \text{ for all } x, x' \in \bar{x} + \epsilon \mathbb{B}.$$

Then

$$|\rho_F(t, x, v) - \rho_F(t, x', v')| \leq |v - v'| + k|x - x'|,$$

for all $v, v' \in \mathbb{R}^n$ *and* $x, x' \in \bar{x} + \epsilon \mathbb{B}$.

6.2 Existence and Estimation of F Trajectories

(c) *Assume that F is $\mathcal{L} \times \mathcal{B}^n$ measurable. Then for any Lebesgue measurable functions $y : [S, T] \to \mathbb{R}^n$ and $v : [S, T] \to \mathbb{R}^n$ such that $\mathrm{Gr}\, y \subset \mathrm{dom}\, F$, we have that $t \to \rho_F(t, y(t), v(t))$ is a Lebesgue measurable function on $[S, T]$.*

Proof
(a) Choose any $\epsilon > 0$ and $v, v' \in \mathbb{R}^n$. Since $F(t, x) \neq \emptyset$, there exists $\eta \in F(t, x)$ such that

$$\rho_F(t, x, v) \geq |v - \eta| - \epsilon$$
$$\geq |v' - \eta| - |v - v'| - \epsilon$$
$$\geq \rho_F(t, x, v') - |v - v'| - \epsilon.$$

(The second line follows from the triangle inequality.) Since $\epsilon > 0$ is arbitrary, and v and v' are interchangeable, it follows that

$$|\rho_F(t, x, v') - \rho_F(t, x, v')| \leq |v - v'|.$$

(b) Choose any $x, x' \in \bar{x} + \epsilon \mathbb{B}$ and $v, v' \in \mathbb{R}^n$. Take any δ. Since $F(t, x') \neq \emptyset$, there exists $\eta' \in F(t, x')$ such that

$$\rho_F(t, x', v') > |v' - \eta'| - \delta.$$

Under the hypotheses, there exists $\eta \in F(t, x)$ such that $|\eta - \eta'| \leq k|x - x'|$. Of course, $\rho_F(t, x, v) \leq |v - \eta|$. It follows from these relations and the triangle inequality that

$$\rho_F(t, x, v) - \rho_F(t, x', v')$$
$$\leq |v - \eta| - |v' - \eta'| + \delta \leq |v - v'| + |v' - \eta| - |v' - \eta'| + \delta$$
$$\leq |v - v'| + |\eta - \eta'| + \delta \leq |v - v'| + k|x - x'| + \delta.$$

Since the roles of (x, v) and (x', v') are interchangeable and $\delta > 0$ is arbitrary, we conclude that

$$|\rho_F(t, x, v) - \rho_F(t, x', v')| \leq |v - v'| + k|x - x'|.$$

(c) For each v, the function $t \to \rho_F(t, y(t), v)$ is measurable. This follows from the identity

$$\{t \in [S, T] : \rho_F(t, y(t), v) < \alpha\} = \{t : F(t, y(t)) \cap (v + \alpha \,\mathrm{int}\, \mathbb{B}) \neq \emptyset\},$$

valid for any $\alpha \in \mathbb{R}$, and the fact that $t \rightsquigarrow F(t, y(t))$ is measurable (see Proposition 2.3.2).

By part (a) of the lemma, the function $\rho_F(t, y(t), .)$ is continuous, for each $t \in [S, T]$. It follows that $(t, v) \to \rho_F(t, y(t), v)$ is a Carathéodory function. But then $t \to \rho_F(t, y(t), v(t))$ is (Lebesgue) measurable on $[S, T]$, by Proposition 2.3.2 and Corollary 2.3.3. □

The following Proposition, concerning regularity properties of projections onto continuous, convex multifunctions, will also be required.

Proposition 6.2.2 *Take a continuous multifunction* $\Gamma : [S, T] \rightsquigarrow \mathbb{R}^n$ *and a function* $u : [S, T] \to \mathbb{R}^n$. *Assume that*

(i) u is continuous,
(ii) $\Gamma(t)$ is non-empty, compact and convex for each $t \in [S, T]$ and Γ is continuous, i.e. there exists $o : \mathbb{R}_+ \to \mathbb{R}_+$ with $\lim_{\alpha \downarrow 0} o(\alpha) = 0$ such that

$$\Gamma(s) \subset \Gamma(t) + o(|t - s|)\mathbb{B} \text{ for all } t, s \in [S, T].$$

Let $\hat{u} : [S, T] \to \mathbb{R}^n$ be the function defined according to

$$|u(t) - \hat{u}(t)| = \min_{u' \in \Gamma(t)} |u(t) - u'| \text{ for all } t.$$

(There is a unique minimizer for each t since $\Gamma(t)$ is non-empty, closed and convex.) Then $\hat{u}$ is a continuous function.

Proof Suppose that $\hat{u}$ is not continuous. Then there exist $\epsilon > 0$ and sequences $\{s_i\}$ and $\{t_i\}$ in $[S, T]$ such that $|t_i - s_i| \to 0$ and

$$|\hat{u}(t_i) - \hat{u}(s_i)| > \epsilon, \text{ for all } i. \tag{6.2.1}$$

Since the multifunction Γ is compact valued and continuous and has bounded domain, $\operatorname{Gr}\Gamma$ is bounded. It follows that, for some $K > 0$,

$$u(s_i), u(t_i), \hat{u}(s_i), \hat{u}(t_i) \in K\mathbb{B}.$$

Since Γ is continuous, there exist $\{y_i\}$ and $\{z_i\}$ such that

$$y_i \in \Gamma(t_i), z_i \in \Gamma(s_i), \text{ for all } i \tag{6.2.2}$$

and

$$|y_i - \hat{u}(s_i)| \to 0, |z_i - \hat{u}(t_i)| \to 0 \text{ as } i \to \infty. \tag{6.2.3}$$

Since Γ is a convex multifunction, $u(s_i) - \hat{u}(s_i)$ and $u(t_i) - \hat{u}(t_i)$ are normal vectors to $\Gamma(s_i)$ and $\Gamma(t_i)$ respectively. Noting (6.2.2), we deduce from the normal inequality for convex sets that

6.2 Existence and Estimation of F Trajectories

$$(u(s_i) - \hat{u}(s_i)) \cdot (z_i - \hat{u}(s_i)) \leq 0$$
$$(u(t_i) - \hat{u}(t_i)) \cdot (y_i - \hat{u}(t_i)) \leq 0.$$

It follows that

$$(u(s_i) - \hat{u}(s_i)) \cdot (\hat{u}(t_i) - \hat{u}(s_i)) \leq (u(s_i) - \hat{u}(s_i)) \cdot (\hat{u}(t_i) - z_i)$$
$$(u(t_i) - \hat{u}(t_i)) \cdot (\hat{u}(s_i) - \hat{u}(t_i)) \leq (u(t_i) - \hat{u}(t_i)) \cdot (\hat{u}(s_i) - y_i).$$

Adding these inequalities gives

$$(u(s_i) - u(t_i)) \cdot (\hat{u}(t_i) - \hat{u}(s_i)) + |\hat{u}(t_i) - \hat{u}(s_i)|^2$$
$$\leq (u(s_i) - \hat{u}(s_i)) \cdot (\hat{u}(t_i) - z_i) + (u(t_i) - \hat{u}(t_i)) \cdot (\hat{u}(s_i) - y_i).$$

We deduce that

$$|\hat{u}(t_i) - \hat{u}(s_i)|^2 \leq |u(s_i) - u(t_i)| \cdot |\hat{u}(t_i) - \hat{u}(s_i)|$$
$$+ |u(s_i) - \hat{u}(s_i)| \cdot |\hat{u}(t_i) - z_i| + |u(t_i) - \hat{u}(t_i)| \cdot |\hat{u}(s_i) - y_i|.$$

It follows that

$$|\hat{u}(t_i) - \hat{u}(s_i)|^2 \leq 2K(|u(s_i) - u(t_i)| + |\hat{u}(t_i) - z_i| + |\hat{u}(s_i) - y_i|).$$

But the right side has limit 0 as $i \to \infty$, since u is continuous, and by (6.2.3). We conclude that $|\hat{u}(t_i) - \hat{u}(s_i)| \to 0$ as $i \to \infty$. We have arrived at a contradiction of (6.2.1). $\hat{u}$ must therefore be continuous. $\square$

We shall refer to the 'truncation' function $tr_\epsilon : \mathbb{R}^n \to \mathbb{R}^n$, defined to be

$$tr_\epsilon(\xi) := \begin{cases} \xi & \text{if } |\xi| \leq \epsilon \\ \epsilon |\xi|^{-1}\xi & \text{if } |\xi| > \epsilon \end{cases}. \tag{6.2.4}$$

Recall that $T(y, \epsilon)$ denotes the ϵ tube about the arc y:

$$T(y, \epsilon) := \{(t, x) \in [S, T] \times \mathbb{R}^n : t \in [S, T], |x - y(t)| \leq \epsilon\}.$$

Theorem 6.2.3 (Generalized Filippov Existence Theorem) *Let Ω be a relatively open set in $[S, T] \times \mathbb{R}^n$. Take a multifunction $F : \Omega \rightsquigarrow \mathbb{R}^n$, an arc $y \in W^{1,1}([S, T]; \mathbb{R}^n)$, a point $\xi \in \mathbb{R}^n$ and $\epsilon \in (0, +\infty) \cup \{+\infty\}$ such that $T(y, \epsilon) \subset \Omega$. Assume that*

(i) *$F(t, x')$ is a closed, non-empty set for all $(t, x') \in T(y, \epsilon)$, and F is $\mathcal{L} \times \mathcal{B}^n$ measurable,*

(ii) there exists $k \in L^1([S, T]; \mathbb{R})$ such that

$$F(t, x') \subset F(t, x'') + k(t)|x' - x''|\mathbb{B} \qquad (6.2.5)$$

for all $x', x'' \in y(t) + \epsilon\mathbb{B}$, a.e. $t \in [S, T]$.

Assume further that

$$\exp\left(\int_S^T k(t)dt\right)\left(|\xi - y(S)| + \int_S^T \rho_F(t, y(t), \dot{y}(t))dt\right) \leq \epsilon. \qquad (6.2.6)$$

Then there exists an F trajectory x satisfying $x(S) = \xi$ such that for all $t \in (S, T]$

$$\|x - y\|_{L^\infty(S,t)} \leq |x(S) - y(S)| + \int_S^t |\dot{x}(s) - \dot{y}(s)|ds$$

$$\leq \exp\left(\int_S^t k(s)ds\right)\left(|\xi - y(S)| + \int_S^t \rho_F(s, y(s), \dot{y}(s))ds\right) (6.2.7)$$

Now suppose that (i) and (ii) are replaced by the stronger hypotheses

(i)' $F(t, x')$ is a non-empty, compact, convex set for all $(t, x') \in T(y, \epsilon)$,
(ii)' there exists a function $o : \mathbb{R}_+ \to \mathbb{R}_+$ and $k_\infty > 0$ such that $\lim_{\alpha \downarrow 0} o(\alpha) = 0$ and

$$F(s', x') \subset F(s'', x'') + k_\infty |x' - x''|\mathbb{B} + o(|s' - s''|)\mathbb{B} \qquad (6.2.8)$$

for all $(s', x'), (s'', x'') \in T(y, \epsilon)$.

Then, if y is continuously differentiable, x can be chosen also to be continuously differentiable.
(If $\epsilon = +\infty$ then in the above hypotheses $T(y, \epsilon)$ and $\epsilon\mathbb{B}$ are interpreted as $[S, T] \times \mathbb{R}^n$ and $\mathbb{R}^n$ respectively, and the left side of condition (6.2.6) is required to be finite.)

Remarks

(i) The hypotheses of Theorem 6.2.3 do not require F to be convex valued. For many developments in dynamic optimization the requirement that F is convex is crucial; fortunately, proving this basic existence theorem is not one of them.
(ii) The proof of Theorem 6.2.3 is by construction. The iterative procedure used is a generalization to differential inclusions of the well-known Picard iteration scheme for obtaining a solution to the differential equation

$$\dot{x}(t) = f(t, x(t)), \quad x(S) = \xi.$$

6.2 Existence and Estimation of F Trajectories

An initial guess y at a solution is made. It is then improved by 'successive approximations' $x_0, x_1, x_2, \ldots$. These arcs are generated by the recursive equations

$$x_{i+1}(t) = \xi + \int_S^t f(s, x_i(s))ds$$

with starting condition $x_0(t) = y(t) + (\xi - y(S))$.

Proof We may assume without loss of generality that $\epsilon = \infty$. Indeed if $\epsilon = \bar{\epsilon}$ for some finite $\bar{\epsilon}$ then we consider $\tilde{F}$ in place of F, where

$$\tilde{F}(t, x) := F(t, y(t) + tr_{\bar{\epsilon}}(x - y(t))).$$

(See (6.2.4) for definition of $tr_{\bar{\epsilon}}$.)
$\tilde{F}$ satisfies the hypotheses (in relation to y) with $\epsilon = \infty$ and with the same $k \in L^1$ as before. Of course

$$\tilde{F}(t, x) = F(t, x) \text{ for } x \in y(t) + \bar{\epsilon}\mathbb{B}.$$

Now apply the '$\epsilon = +\infty$' case of the theorem to $\tilde{F}$. This gives an $\tilde{F}$ trajectory x such that $x(S) = \xi$ and (6.2.7) is satisfied (when $\tilde{F}$ replaces F). If, however,

$$K\left(|\xi - y(S)| + \int_S^T \rho_F(t, y(t), \dot{y}(t))dt\right) \leq \bar{\epsilon},$$

where $K := \exp\left(\int_S^T k(t)dt\right)$, then the theorem tell us that

$$||x - y||_{L^\infty} \leq \bar{\epsilon},$$

and therefore x is an F trajectory because $F(t, .)$ and $\tilde{F}(t, .)$ coincide on $y(t) + \bar{\epsilon}\mathbb{B}$. This justifies setting $\epsilon = +\infty$.

It suffices to consider only the case $\xi = y(S)$. To show this, suppose that $\xi \neq y(S)$. Replace the underlying time interval $[S, T]$ by $[S - 1, T]$ and ξ by $\tilde{\xi} = y(S)$. Replace also F by $\tilde{F} : [S - 1, T] \times \mathbb{R}^n \rightsquigarrow \mathbb{R}^n$ and y by $\tilde{y} : [S - 1, T] \to \mathbb{R}^n$, defined as follows:

$$\tilde{F}(t, x) := \begin{cases} F(t, x) & \text{for } (t, x) \in [S, T] \times \mathbb{R}^n \\ \{\xi - y(S)\} & \text{for } (t, x) \in [S - 1, S) \times \mathbb{R}^n \end{cases}$$

and

$$\tilde{y}(t) := \begin{cases} y(t) & \text{for } t \geq S \\ y(S) & \text{for } t < S. \end{cases}$$

Now apply the special case of the theorem to find $\tilde{x} \in W^{1,1}([S-1, T]; \mathbb{R}^n)$ such that $\tilde{x}(S-1) = \tilde{y}(S-1) = \tilde{\xi} \equiv y(S)$. Take x to be the restriction of $\tilde{x}$ to $[S, T]$. We readily deduce that for all $t \in (S, T]$

$$||\dot{\tilde{x}} - \dot{\tilde{y}}||_{L^1([S-1, t]; \mathbb{R}^n)} \le \exp\left(\int_S^t k(s)ds\right) \int_{S-1}^t \rho_{\tilde{F}}(s, \tilde{y}(s), \dot{\tilde{y}}(s))ds.$$

Now, to derive the desired estimate, we have merely to note that

$$||\dot{\tilde{x}} - \dot{\tilde{y}}||_{L^1([S-1, t]; \mathbb{R}^n)} = |\xi - y(S)| + ||\dot{x} - \dot{y}||_{L^1([S, t]; \mathbb{R}^n)}$$

and

$$\int_{S-1}^t \rho_{\tilde{F}}(s, \tilde{y}(s), \dot{\tilde{y}}(s))ds = |\xi - y(S)| + \int_S^t \rho_F(s, y(s), \dot{y}(s))ds.$$

Henceforth, then, we assume that $\epsilon = +\infty$ and $y(S) = \xi$; we must find an F trajectory x such that $x(S) = \xi$ and, for all $t \in (S, T]$,

$$||\dot{x} - \dot{y}||_{L^1([S, t]; \mathbb{R}^n)} \le \exp\left(\int_S^t k(s)ds\right) \int_S^t \rho_F(s, y(s), \dot{y}(s))ds.$$

Write $x_0(t) = y(t)$. According to Theorem 2.3.12, we may choose a measurable function v_1 satisfying

$$v_1(t) \in F(t, x_0(t)) \text{ a.e. } t \in [S, T]$$

and

$$\rho_F(t, x_0(t), \dot{x}_0(t)) = |v_1(t) - \dot{x}_0(t)| \text{ a.e. } t \in [S, T].$$

This is because

$$t \rightsquigarrow G(t) := \{v \in F(t, x_0(t)) : \rho_F(t, x_0(t), \dot{x}_0(t)) = |v - \dot{x}_0(t)|\}$$

is a closed nonempty measurable multifunction. (We use Proposition 2.3.2 and the fact that

$$(t, v) \to g(t, v) := \rho_F(t, x_0(t), \dot{x}_0(t)) - |v - \dot{x}_0(t)|$$

is a Carathéodory function.) Under the hypotheses, $t \to \rho_F(t, x_0(t), \dot{x}_0(t))$ is integrable. Since $\dot{x}_0$ is integrable, v_1 is integrable too. We may therefore define x_1 according to

6.2 Existence and Estimation of F Trajectories

$$x_1(t) := y(S) + \int_S^t v_1(s)ds.$$

Note that

$$\rho_F(t, x_0(t), \dot{x}_1(t)) = 0 \quad \text{a.e..}$$

Again appealing to Theorem 2.3.12, we choose a measurable function v_2 satisfying

$$v_2(t) \in F(t, x_1(t)) \quad \text{a.e.}$$

and

$$|v_2(t) - \dot{x}_1(t)| = \rho_F(t, x_1(t), \dot{x}_1(t)) \quad \text{a.e..} \quad (6.2.9)$$

In view of the Lipschitz continuity properties of $\rho_F(t, ., v)$ and since $\dot{x}_1$ is integrable, we readily deduce from the integrability of $t \to \rho_F(t, x_0(t), \dot{x}_0(t))$ that $t \to \rho_F(t, x_1(t), \dot{x}_1(t))$ is also integrable. It then follows from (6.2.9) that v_2 is integrable and we may define

$$x_2(t) = y(S) + \int_S^t v_2(s)ds.$$

We proceed in this way to construct a sequence of absolutely continuous arcs $\{x_m\}$ satisfying

$$\rho_F(t, x_m(t), \dot{x}_{m+1}(t)) = 0 \quad \text{a.e.}$$

$$|\dot{x}_{m+1}(t) - \dot{x}_m(t)| = \rho_F(t, x_m(t), \dot{x}_m(t)) \quad \text{a.e.}$$

for $m = 0, 1, 2\ldots$ and

$$\rho_F(t, x_0(t), \dot{x}_0(t)) = \rho_F(t, y(t), \dot{y}(t)) \quad \text{a.e..}$$

Notice that

$$\|x_1 - x_0\|_{L^\infty} \leq \int_S^T |\dot{x}_1(t) - \dot{x}_0(t)|dt$$

$$= \int_S^T \rho_F(t, x_0(t), \dot{x}_0(t))dt = \Lambda_F(y). \quad (6.2.10)$$

Applying Proposition 6.2.1, we deduce that, for $m \geq 1$ and a.e. t,

$$|\dot{x}_{m+1}(t) - \dot{x}_m(t)| \le \rho_F(t, x_{m-1}(t), \dot{x}_m(t)) + k(t)|x_m(t) - x_{m-1}(t)|$$
$$= k(t)|x_m(t) - x_{m-1}(t)|. \quad (6.2.11)$$

Since $x_{m+1}(S) = x_m(S)$, it follows that, for all $m \ge 1$ and a.e. $t \in [S, T]$,

$$|x_{m+1}(t) - x_m(t)|$$
$$\le \int_S^t k(t_1)|x_m(t_1) - x_{m-1}(t_1)|dt_1$$
$$\le \int_S^t k(t_1)\int_S^{t_1} k(t_2)\ldots \int_S^{t_{m-1}} k(t_m)|x_1(t_m) - x_0(t_m)|dt_m\ldots dt_2 dt_1$$
$$\le S_m(t)\Lambda_F(y), \quad (6.2.12)$$

in view of (6.2.10). Here

$$S_m(t) := \int_S^t k(t_1)\int_S^{t_1} k(t_2)\ldots \int_S^{t_{m-1}} k(t_m)dt_m..dt_2 dt_1.$$

The right side can be reduced, one indefinite integral at a time with the help of the integration by parts formula. There results:

$$S_m(t) = \frac{\left(\int_S^t k(t)dt\right)^m}{m!}.$$

It follows from (6.2.10)–(6.2.12) that, for any integers $M > N \ge 0$, and for all $t \in (S, T]$, we have

$$\|\dot{x}_M - \dot{x}_N\|_{L^1([S,t];\mathbb{R}^n)}$$
$$\le \|\dot{x}_M - \dot{x}_{M-1}\|_{L^1([S,t];\mathbb{R}^n)} + \ldots + \|\dot{x}_{N+1} - \dot{x}_N\|_{L^1([S,t];\mathbb{R}^n)}$$
$$\le \left[\frac{\left(\int_S^t k(s)ds\right)^{M-1}}{(M-1)!} + .. + \frac{\left(\int_S^t k(s)ds\right)^N}{N!}\right]\int_S^t \rho_F(s, y(s), \dot{y}(s))ds$$
$$\le \left[\frac{\left(\int_S^T k(s)ds\right)^{M-1}}{(M-1)!} + .. + \frac{\left(\int_S^T k(s)ds\right)^N}{N!}\right]\Lambda_F(y). \quad (6.2.13)$$

(Here $\left(\int_S^t k(t)dt\right)^m/m! := 1$ when $m = 0$.) It is clear from this inequality that $\{\dot{x}_m\}$ is a Cauchy sequence in L^1. It follows that

$$\dot{x}_m \to v \text{ in } L^1,$$

6.2 Existence and Estimation of F Trajectories

for some $v \in L^1$. Define $x \in W^{1,1}$ according to

$$x(t) := \xi + \int_S^t v(s)ds.$$

Since

$$\|x - x_m\|_{L^\infty} \leq \int_S^T |v(s) - \dot{x}_m(s)|ds$$

and $\dot{x}_m \to v$ in L^1, we know that

$$x_m \to x \text{ uniformly.}$$

By extracting a subsequence (we do not relabel), we can arrange that

$$\dot{x}_m \to \dot{x} \text{ a.e..}$$

Define $\mathcal{O}$ to be the subset of points $t \in [S, T]$ such that $\dot{x}_m(t) \in F(t, x_{m-1}(t))$ for all index values $m = 1, 2, ..$ and such that $\dot{x}_m(t) \to \dot{x}(t)$. Take any $t \in \mathcal{O}$. Then

$$\dot{x}_m(t) \in F(t, x_{m-1}(t)).$$

Since $F(t, .)$ has closed graph, we obtain in the limit

$$\dot{x}(t) \in F(t, x(t)).$$

But $\mathcal{O}$ has full measure. It follows that x is an F trajectory.

Next observe that by setting $N = 0$ in inequality (6.2.13), for all $t \in (S, T]$, we arrive at

$$\|\dot{x}_M - \dot{y}\|_{L^1([S,t];\mathbb{R}^n)} \leq \exp\left(\int_S^t k(s)ds\right) \int_S^t \rho_F(s, y(s), \dot{y}(s))ds$$

for $M = 1, 2, \ldots$. Since $\dot{x}_M \to \dot{x}$ in L^1 as $M \to \infty$ we deduce that, for all $t \in (S, T]$,

$$\|\dot{x} - \dot{y}\|_{L^1([S,t];\mathbb{R}^n)} \leq \exp\left(\int_S^T k(t)dt\right) \int_S^t \rho_F(s, y(s), \dot{y}(s))ds.$$

This is the required estimate.

It remains to prove that x can be chosen continuously differentiable when the 'comparison function' y is continuously differentiable, under the additional hypotheses.

Construct a sequence $\{x_i\}$ as above. Under the additional hypotheses, $t \to F(t, x_0(t))$ is a continuous multifunction. In view of Proposition 6.2.2, $\dot{x}_1$ is a continuous function. ($\dot{x}_1(t)$ is the projection of $\dot{y}(t) = \dot{x}_0(t)$ onto $F(t, x_0(t))$). Arguing inductively, we conclude that

$$t \to \dot{x}_i(t) \text{ is continuous for all } i \geq 0.$$

Fix any $i \geq 0$. For arbitrary t, $\dot{x}_{i+1}(t) \in F(t, x_i(t))$ and $\dot{x}_{i+2}(t)$ minimizes $v \to |\dot{x}_{i+1}(t) - v|$ over $v \in F(t, x_{i+1}(t))$, by construction. Under the hypotheses, we can find $w \in F(t, x_{i+1}(t))$ such that

$$|w - \dot{x}_{i+1}(t)| \leq k_\infty |x_{i+1}(t) - x_i(t)| \text{ for all } t.$$

Since $x_{i+2}(S) = x_{i+1}(S)$ and i was chosen arbitrarily, we conclude that

$$|\dot{x}_{i+2}(t) - \dot{x}_{i+1}(t)| \leq k_\infty |x_{i+1}(t) - x_i(t)| \text{ for all } t.$$

Now hypothesis (ii)$'$ implies hypothesis (ii) with $k(t) = k_\infty$ for all t. By (6.2.12) then, for any integers $M > N \geq 2$, we have

$$\|\dot{x}_M - \dot{x}_N\|_C \leq k_\infty \left[\frac{\left(\int_S^T k_\infty dt\right)^{M-2}}{(M-2)!} + \ldots + \frac{\left(\int_S^T k_\infty dt\right)^{N-1}}{(N-1)!} \right] \Lambda_F(y).$$

It follows that $\dot{x}_i$ is a Cauchy sequence in C. But C is complete, so the sequence has a strong C limit, some continuous function v. But v must coincide with $\dot{x}$, the strong L^1 limit of $\{\dot{x}_i\}$, following adjustment on a null-set. It follows that x is a continuously differentiable function. □

Naturally, if we specialize to the point valued case, we recover an existence theorem for differential equations. In this important special case the solution is unique. We shall require

Lemma 6.2.4 (Gronwall's Lemma) *Take an absolutely continuous function $z : [S, T] \to \mathbb{R}^n$. Assume that there exist non-negative integrable functions k and v such that*

$$\left|\frac{d}{dt} z(t)\right| \leq k(t)|z(t)| + v(t) \text{ a.e. } t \in [S, T].$$

Then

$$|z(t)| \leq \exp\left(\int_S^t k(\sigma) d\sigma\right) \left[|z(S)| + \int_S^t \exp\left(-\int_S^\tau k(\sigma) d\sigma\right) v(\tau) d\tau\right]$$

for all $t \in [S, T]$.

6.2 Existence and Estimation of F Trajectories

Proof Since z is absolutely continuous so too is $t \to |z(t)|$. Let $\mathcal{O} \subset [S, T]$ be the subset of points t such that z and $|z|$ are both differentiable at t. $\mathcal{O}$ has full measure and, it is straightforward to show,

$$\frac{d}{dt}|z(t)| \leq |\dot{z}(t)| \quad \text{for all } t \in \mathcal{O}.$$

Now define the absolutely continuous function

$$\eta(t) := \exp\left(-\int_S^t k(\sigma)d\sigma\right)|z(t)| \quad \text{for all } t \in [S, T].$$

Then for every $t \in \mathcal{O}$ we have

$$\dot{\eta}(t) = \exp\left(-\int_S^t k(\sigma)d\sigma\right)\left[\frac{d}{dt}|z(t)| - k(t)|z(t)|\right]$$

$$\leq \exp\left(-\int_S^t k(\sigma)d\sigma\right)[|\dot{z}(t)| - k(t)|z(t)|]$$

$$\leq \exp\left(-\int_S^t k(\sigma)d\sigma\right) v(t).$$

It follows that for each $t \in [S, T]$,

$$\eta(t) \leq \eta(S) + \int_S^t \exp\left(-\int_S^\tau k(\sigma)d\sigma\right) v(\tau)d\tau.$$

Since $\eta(S) = |z(S)|$ and

$$|z(t)| = \exp\left(\int_S^t k(\sigma)d\sigma\right)\eta(t),$$

we deduce

$$|z(t)| \leq \exp\left(\int_S^t k(\sigma)d\sigma\right)\left[|z(S)| + \int_S^t \exp\left(-\int_S^\tau k(\sigma)d\sigma\right) v(\tau)d\tau\right].$$

□

Corollary 6.2.5 (ODE's: Existence and Uniqueness of Solutions) *Take a function $f : [S, T] \times \mathbb{R}^n \to \mathbb{R}^n$, an arc $y \in W^{1,1}([S, T]; \mathbb{R}^n)$, $\epsilon \in (0, \infty) \cup \{+\infty\}$ and a point $\xi \in \mathbb{R}^n$. (The case $\epsilon = +\infty$ is interpreted as in the statement of Theorem 6.2.2.) Assume that*

(i) *$f(., x)$ is measurable for each $x \in \mathbb{R}^n$,*
(ii) *there exists $k \in L^1$ such that*

$$|f(t, x') - f(t, x'')| \le k(t)|x' - x''|$$

for all $x', x'' \in y(t) + \epsilon \mathbb{B}$, a.e. $t \in [S, T]$.

Assume further that

$$K\left(|\xi - y(S)| + \int_S^T |\dot{y}(t) - f(t, y(t))|dt\right) \le \epsilon,$$

where $K := \exp\left(\int_S^T k(t)dt\right)$.

Then there exists a unique solution to the differential equation

$$\dot{x}(t) = f(t, x(t)), \quad a.e. \ t \in [S, T]$$
$$x(S) = \xi$$

which satisfies

$$||x - y||_{L^\infty} \le |x(S) - y(S)| + \int_S^T |\dot{x}(t) - \dot{y}(t)| \, dt$$
$$\le K\left(|\xi - y(S)| + \int_S^T |\dot{y}(t) - f(t, y(t))|dt\right).$$

Proof All the assertions of the corollary follow immediately from the generalized Filippov existence theorem (Theorem 6.2.3), with the exception of 'uniqueness'. Suppose however that there are two solutions x' and x'' to the differential equation which satisfy $||x' - y||_{L^\infty} \le \epsilon$ and $||x'' - y||_{L^\infty} \le \epsilon$. Define $z(t) := x'(t) - x''(t)$. Then under the hypotheses, for almost every $t \in [S, T]$,

$$|\dot{z}(t)| = |f(t, x'(t)) - f(t, x''(t))|$$
$$\le k(t)|x'(t) - x''(t)| = k(t)|z(t)|.$$

Since $z(S) = 0$, it follows from Gronwall's lemma (Lemma 6.2.4) that $z \equiv 0$. We conclude that $x' = x''$. Uniqueness is proved. □

6.3 Perturbed Differential Inclusions

Consider a sequence of arcs whose elements satisfy perturbed versions of a 'nominal' differential inclusion such that the perturbation terms in some sense tend to zero as we proceed along the sequence. When can we extract a subsequence with limit a solution to the nominal differential inclusion? As we shall see this is possible under unrestrictive hypotheses on the differential inclusion and on the nature of the

6.3 Perturbed Differential Inclusions

perturbations. Necessary conditions in nonsmooth dynamic optimization are usually obtained by deriving necessary conditions for simpler, perturbed versions of the dynamic optimization problem of interest and passing to the limit. The significance of the results of this section is that they will justify the limit taking procedures.

Use will be made of a characterization of subsets of L^1 which are relatively, sequentially compact.

Theorem 6.3.1 (Dunford Pettis Theorem) *Let S be a bounded subset of $L^1([S, T]; \mathbb{R}^n)$. Then the following conditions are equivalent:*

(i) *every sequence in S has a subsequence converging to some L^1 function, with respect to the weak L^1 topology,*
(ii) *for every $\epsilon > 0$ there exists $\delta > 0$ such that for every measurable set $D \subset [S, T]$ and $x \in S$ satisfying meas$\{D\} < \delta$, we have $\int_D x(t)dt < \epsilon$.*

Proof See [97]. □

When condition (ii) above is satisfied, we shall say 'the family of functions S is equi-integrable'. A simple criterion for equi-integrability (a sufficient condition to be precise) is that the family of functions is 'uniformly integrably bounded' in the sense that there exists an integrable function $\alpha \in L^1$ such that

$$|x(t)| \leq \alpha(t) \text{ a.e. } t \in [S, T]$$

for all $x \in S$.

We shall require also certain properties of the Hamiltonian $H(t, x, p)$ associated with a given multifunction $F : [S, T] \times \mathbb{R}^n \rightsquigarrow \mathbb{R}^n$, defined at points $(t, x, p) \in \text{dom } F \times \mathbb{R}^n$.

$$H(t, x, p) := \sup_{v \in F(t,x)} p \cdot v. \quad (6.3.1)$$

Proposition 6.3.2 *Consider a multifunction $F : [S, T] \times \mathbb{R}^n \rightsquigarrow \mathbb{R}^n$, which has values closed sets.*

(a) *Fix $(t, x) \in \text{dom } F$. Assume that there exists $c \geq 0$ such that $F(t, x) \subset c\mathbb{B}$. Then*

$$|H(t, x, p)| \leq c|p| \text{ for every } p \in \mathbb{R}^n,$$

and $H(t, x, .)$ is Lipschitz continuous with Lipschitz constant c.
(b) *Fix $t \in [S, T]$. Take convergent sequences $x_i \to x$ and $p_i \to p$ in $\mathbb{R}^n$. Assume that there exists $c \geq 0$ such that*

$$(t, x_i) \in \text{dom } F \text{ and } F(t, x_i) \subset c\mathbb{B} \text{ for all } i,$$

and that $\text{Gr } F(t, .)$ is closed. Then

$$\limsup_{i\to\infty} H(t, x_i, p_i) \le H(t, x, p).$$

(c) *Take measurable functions* $x : [S, T] \to \mathbb{R}^n$, $p : [S, T] \to \mathbb{R}^n$ *such that* $\operatorname{Gr} x \subset \operatorname{dom} F$. *Assume that*

$$F \text{ is } \mathcal{L} \times \mathcal{B}^n \text{ measurable.}$$

Then $t \to H(t, x(t), p(t))$ *is a measurable function.*

Proof

(a): Take $(t, x) \in \operatorname{dom} F$. Choose any $p, p' \in \mathbb{R}^n$. Since $F(t, x)$ is a compact, non-empty set, there exists $v \in F(t, x)$ such that

$$H(t, x, p) = p \cdot v.$$

Of course, $H(t, x, p') \ge p' \cdot v$. But then

$$H(t, x, p) - H(t, x, p') \le (p - p') \cdot v \le |v||p - p'|.$$

Since $|v| \le c$, and the roles of p and p' can be interchanged, we deduce that

$$H(t, x, p) - H(t, x, p') \le c|p - p'|.$$

Notice, in particular, that $|H(t, x, p)| \le c|p|$, since $H(t, x, 0) = 0$.

(b): Fix $t \in [S, T]$ and take any sequences $x_i \to x$ and $p_i \to p$ in $\mathbb{R}^n$. $\limsup_i H(t, x_i, p_i)$ can be replaced by $\lim_i H(t, x_i, p_i)$, following extraction of a suitable subsequence (we do not relabel). For each i, $F(t, x_i)$ is a nonempty compact set (since $(t, x_i) \in \operatorname{dom} F$); consequently, there exists $v_i \in F(t, x_i)$ such that $H(t, x_i, p_i) = p_i \cdot v_i$. But $|v_i| \le c$ for $i = 1, 2, \ldots$. We can therefore arrange, by extracting another subsequence, that $v_i \to v$ for some $v \in \mathbb{R}^n$. Since $\operatorname{Gr} F(t, .)$ is closed, $v \in F(t, x)$ and so

$$\lim_i H(t, x_i, p_i) = \lim_i p_i \cdot v_i = p \cdot v \le H(t, x, p).$$

(c): According to Proposition 2.3.2 and Theorem 2.3.12, we can find a measurable selection v of $t \rightsquigarrow F(t, x(t))$. Fix $k > 0$ and define for all $(t, x) \in [S, T] \times \mathbb{R}^n$,

$$F'_k(t, x) := F(t, x) \cap (v(t) + k\mathbb{B})$$

and

$$F_k(t, x) = \begin{cases} F'_k(t, x) & \text{if } (t, x) \in \operatorname{dom} F'_k \\ v(t) + k\mathbb{B} & \text{otherwise} \end{cases}.$$

6.3 Perturbed Differential Inclusions

It is a straightforward exercise to show that F_k is $\mathcal{L} \times \mathcal{B}^n$ measurable. Since dom $F_k = [S, T] \times \mathbb{R}^n$, we can define

$$H_k(t, x, p) := \max\{p \cdot v : v \in F_k(t, x)\}$$

for all $(t, x, p) \in [S, T] \times \mathbb{R}^n \times \mathbb{R}^n$. Clearly

$$H_k(t, x(t), p(t)) \to H(t, x(t, p(t))) \text{ as } k \to \infty,$$

for all t. It therefore suffices to show that $t \to H_k(t, x(t), p(t))$ is measurable.

However, by Corollary 2.3.3, this will be true if we can show that $(t, (x, p)) \to H_k(t, (x, p))$ is $\mathcal{L} \times \mathcal{B}^{2n}$ measurable.

Fix $r \in \mathbb{R}$. We shall complete the proof by showing that

$$D := \{(t, x, p) \in [S, T] \times \mathbb{R}^n \times \mathbb{R}^n : H_k(t, x, p) \geq r\}$$

is $\mathcal{L} \times \mathcal{B}^{2n}$ measurable. Let $\{v_i\}$ be a dense subset of $\mathbb{R}^n$. Take a sequence $\epsilon_j \downarrow 0$. It is easy to verify that $D = \mathcal{D}$, where

$$\mathcal{D} := \cap_{j=1}^{\infty} \cup_{i=1}^{\infty} (A_{i,j} \times B_{i,j}),$$

in which

$$A_{i,j} := \{(t, x) \in [S, T] \times \mathbb{R}^n : (v_i + \epsilon_j \mathbb{B}) \cap F_k(t, x) \neq \emptyset\}$$

and

$$B_{i,j} := \{p \in \mathbb{R}^n : p \cdot v_i > r - \epsilon_j\}.$$

(That $D \subset \mathcal{D}$ is obvious; to show that $\mathcal{D} \subset D$, we exploit the fact that Gr $F_k(t, .)$ is a compact set.)

But D is obtainable from $\mathcal{L} \times \mathcal{B}^{2n}$ measurable sets, by means of a countable number of union and intersection operations. It follows that D is $\mathcal{L} \times \mathcal{B}^{2n}$ is measurable, as claimed. □

With these preliminaries behind us we are ready to answer the question posed at the beginning of the section.

Theorem 6.3.3 (Compactness of Trajectories) *Take a relatively open subset $\Omega \subset [S, T] \times \mathbb{R}^n$ and a multifunction $F : \Omega \rightsquigarrow \mathbb{R}^n$.*

Assume that, for some closed multifunction $X : [S, T] \rightsquigarrow \mathbb{R}^n$ such that Gr $X \subset \Omega$, the following hypotheses are satisfied:

(i) *F is a closed, convex, non-empty multifunction,*
(ii) *F is $\mathcal{L} \times \mathcal{B}^n$ measurable,*
(iii) *for each $t \in [S, T]$, the graph of $F(t, .)$ restricted to $X(t)$ is closed.*

Consider a sequence $\{x_i\}$ of $W^{1,1}([S, T]; \mathbb{R}^n)$ functions, a sequence $\{r_i\}$ in $L^1([S, T]; \mathbb{R})$ such that $||r_i||_{L^1} \to 0$ as $i \to \infty$ and a sequence $\{A_i\}$ of measurable subsets of $[S, T]$ such that meas $A_i \to |T - S|$ as $i \to \infty$.

Suppose that:

(iv) $\operatorname{Gr} x_i \subset \operatorname{Gr} X$ for all i,
(v) $\{\dot{x}_i\}$ is a sequence of uniformly integrably bounded functions on $[S, T]$ and $\{x_i(S)\}$ is a bounded sequence,
(vi) There exists $c \in L^1$ such that

$$F(t, x_i(t)) \subset c(t)\mathbb{B}$$

for a.e. $t \in A_i$ and for $i = 1, 2, \ldots$.

Suppose further that

$$\dot{x}_i(t) \in F(t, x_i(t)) + r_i(t)\mathbb{B} \quad a.e. \ t \in A_i.$$

Then along some subsequence (we do not relabel)

$$x_i \to x \text{ uniformly} \quad \text{and} \quad \dot{x}_i \to \dot{x} \text{ weakly in } L^1$$

for some $x \in W^{1,1}([S, T]; \mathbb{R}^n)$ satisfying

$$\dot{x}(t) \in F(t, x(t)) \quad a.e. \ t \in [S, T].$$

Proof The $\dot{x}_i$'s are uniformly integrably bounded on $[S, T]$. According to the Dunford Pettis theorem, we can arrange, by extracting a subsequence (we do not relabel), that $\dot{x}_i \to v$ weakly in L^1 for some L^1 function v. Since $\{x_i(S)\}$ is a bounded sequence we may arrange by further subsequence extraction that $x_i(S) \to \xi$ for some $\xi \in \mathbb{R}^n$. Now define

$$x(t) := \xi + \int_S^t v(s)ds.$$

By weak convergence, $x_i(t) \to x(t)$ for every $t \in [S, T]$. Clearly too $\dot{x}_i \to \dot{x}$ weakly in L^1.

Now consider the Hamiltonian $H(t, \xi, p)$ defined by (6.3.1). Choose any p and any Lebesgue measurable subset $V \subset [S, T]$. For almost every $t \in V \cap A_i$, we have

$$H(t, x_i(t), p) \geq p \cdot \dot{x}_i(t) - r_i(t)|p|.$$

Since all terms in this inequality are integrable, we deduce

$$\int_{V \cap A_i} p \cdot \dot{x}_i(t)dt - \int_{V \cap A_i} r_i(t)|p|dt \leq \int_{V \cap A_i} H(t, x_i(t), p)dt.$$

Because the $\dot{x}_i$'s are uniformly integrably bounded, meas$[A_i] \to |T - S|$, $\dot{x}_i \to \dot{x}$ weakly in L^1 and $||r_i||_{L^1} \to 0$, we see that the left side of this relation has limit $\int_V p \cdot \dot{x}(t)$. It follows that

$$\int_V p \cdot \dot{x}(t) dt \leq \limsup_{i \to \infty} \int_V \chi_i(t) H(t, x_i(t), p) dt.$$

Here χ_i denotes the characteristic function of the set A_i.

Since $\chi_i(t) H(t, x_i(t), p)$ is bounded above by $c(t)|p|$, we deduce from Fatou's Lemma that

$$\int_V p \cdot \dot{x}(t) \leq \int_V \limsup_i \chi_i(t) H(t, x_i(t), p) dt.$$

From the upper semicontinuity properties of H then (see Proposition 6.3.2)

$$\int_V (H(t, x(t), p) - p \cdot \dot{x}(t)) \, dt \geq 0.$$

Let $\{p_i\}$ be an ordering of the set of n-vectors having rational coefficients. Define $D \subset [S, T]$ to be the subset of points $t \in [S, T]$ such that t is a Lebesgue point of $t \to H(t, x(t), p_i) - p_i \cdot \dot{x}(t)$ for all i. D is a set of full measure. For any $t \in D \cap [S, T)$ and any i

$$H(t, x(t), p_i) - p_i \cdot \dot{x}(t) =$$
$$\lim_{\delta \downarrow 0} \frac{1}{\delta} \int_t^{t+\delta} [H(\sigma, z(\sigma, x(\sigma), p_i) - p_i \cdot \dot{x}(\sigma)] d\sigma \geq 0.$$

Since $H(t, x(t), .)$ is continuous for each t, it follows that

$\sup\{p \cdot e - p \cdot \dot{x}(t) : e \in F(t, x(t))\} \geq 0$ for all $p \in \mathbb{R}^n$ a.e. $t \in [S, T]$.

But $F(t, x(t))$ is closed and convex. We deduce with the help of the separation theorem that

$$\dot{x}(t) \in F(t, x(t)) \text{ a.e. } t \in [S, T].$$

We have confirmed that x is an F trajectory. □

6.4 Existence of Minimizing F Trajectories

Take a relatively open subset $\Omega \subset [S, T] \times \mathbb{R}^n$, a multifunction $F : \Omega \rightsquigarrow \mathbb{R}^n$, a closed multifunction $X : [S, T] \rightsquigarrow \mathbb{R}^n$ with the property that $\text{Gr } X \subset \Omega$ and a closed set $C \subset \mathbb{R}^n \times \mathbb{R}^n$. We define the set of admissible F trajectories (associated with the constraint sets $X(t)$, $S \le t \le T$ and C) to be

$$\mathcal{R}_F(X, C) := \{x \in C([S, T]; \mathbb{R}^n) : x \text{ is an } F \text{ trajectory},$$
$$x(t) \in X(t) \text{ for all } t \in [S, T] \text{ and } (x(S), x(T)) \in C\}.$$

We shall deduce from the results of the previous section the following criteria for compactness of the set of admissible F trajectories.

Proposition 6.4.1 *Take Ω, F, X and C as above. Assume that*

(i) *F is a closed, $\mathcal{L} \times \mathcal{B}^n$ measurable multifunction,*
(ii) *for each $t \in [S, T]$, the graph of $F(t, .)$ restricted to $X(t)$ is closed,*
(iii) *there exist $\alpha \in L^1$ and $\beta \in L^1$ such that*

$$F(t, x) \subset (\alpha(t)|x| + \beta(t))\mathbb{B} \text{ for all } (t, x) \in \text{Gr } X,$$

(iv) *either $X(s)$ is bounded for some $s \in [S, T]$ or one of the following two sets*

$$C_0 := \{x_0 \in \mathbb{R}^n : (x_0, x_1) \in C \text{ for some } x_1 \in \mathbb{R}^n\}$$

$$C_1 := \{x_1 \in \mathbb{R}^n : (x_0, x_1) \in C \text{ for some } x_0 \in \mathbb{R}^n\}$$

is bounded.

Assume further that

(v) *$F(t, x)$ is convex for all $(t, x) \in \text{Gr } X$.*

Then $\mathcal{R}_F(X, C)$ is compact with respect to the supremum norm topology.

Proof Since the supremum norm topology is a metric topology, it suffices to prove sequential compactness. Accordingly, take any sequence of admissible F trajectories $\{x_i\}$. We must show that there exists an F trajectory x satisfying the constraints $x(t) \in X(t)$ for all $t \in [S, T]$ and $(x(S), x(T)) \in C$ such that

$$x_i \to x \text{ uniformly}$$

along some subsequence. But these conclusions can be drawn from Theorem 6.3.3 provided we can show that the set $\mathcal{R}_F(X, C)$ is bounded with respect to the supremum norm. By hypothesis (iv) however there exists $k > 0$ and $\bar{s} \in [S, T]$ such that for any admissible F trajectory y we have

$$|y(\bar{s})| \le k.$$

By hypothesis (iii)

$$|\dot{y}(t)| \le \alpha(t)|y(t)| + \beta(t) \text{ a.e..}$$

6.4 Existence of Minimizing F Trajectories

It follows from Gronwall's lemma (applied 'backwards' in time on the interval $[S, \bar{s}]$ and 'forwards' on $[\bar{s}, T]$) that

$$|y(t)| \leq K \text{ for all } t \in [S, T]$$

where

$$K = e^{\|\alpha\|_{L^1}} \left(k + \|\beta\|_{L^1} \right).$$

We have confirmed that $\mathcal{R}(X, C)$ is bounded with respect to the supremum norm. □

It is a simple step now to supply conditions for existence of solutions to the dynamic optimization problem

$$(P) \begin{cases} \text{Minimize } g(x(S), x(T)) \text{ over } x \in W^{1,1}([S, T]; \mathbb{R}^n) \\ \text{which satisfy} \\ \dot{x}(t) \in F(t, x(t)) \text{ a.e. } t \in [S, T], \\ x(t) \in X(t) \text{ for all } t \in [S, T], \\ \text{and} \\ (x(S), x(T)) \in C, \end{cases}$$

in which $g : \mathbb{R}^n \times \mathbb{R}^n \to \mathbb{R}$ is a given lower semi-continuous function.

Indeed (P) can be equivalently formulated as a problem of seeking a minimizer of a lower semi-continuous function over a compact subset of $C([S, T]; \mathbb{R}^n)$ equipped with the supremum norm, namely

$$\text{Minimize } \psi(y) \text{ over } y \in \mathcal{R}_F(X, C)$$

where

$$\psi(y) := g(y(S), y(T)).$$

(P) therefore has a minimizer provided $\mathcal{R}_F(X, C)$ is non-empty. We have proved:

Proposition 6.4.2 *Take Ω, F, X, C and g as above. Assume that*

(i) *F is a closed, $\mathcal{L} \times \mathcal{B}^n$ measurable multifunction,*
(ii) *For each $t \in [S, T]$, the graph of $F(t, .)$ restricted to $X(t)$ is closed,*
(iii) *There exist $\alpha \in L^1$ and $\beta \in L^1$ such that*

$$F(t, x) \subset (\alpha(t)|x| + \beta(t)) \mathbb{B} \text{ for all } (t, x) \in \text{Gr } X,$$

(iv) *Either $X(s)$ is bounded for some $s \in [S, T]$ or one of the following two sets*

$$C_0 := \left\{ x_0 \in \mathbb{R}^n : (x_0, x_1) \in C \text{ for some } x_1 \in \mathbb{R}^n \right\}$$

$$C_1 := \{x_1 \in \mathbb{R}^n : (x_0, x_1) \in C \text{ for some } x_0 \in \mathbb{R}^n\}$$

is bounded.

Assume further that

(a) The set of admissible F trajectories $\mathcal{R}_F(X, C)$ is non-empty,
(b) $F(t, x)$ is convex for each $(t, x) \in \text{Gr } X$.

Then (P) has a minimizer.

6.5 Relaxation

Suppose that an optimization problem of interest fails to have a minimizer. 'Relaxation' is the procedure of adding extra elements to the domain of the optimization problem to ensure existence of minimizers.

For a relaxation scheme to be of interest it usually needs to be accompanied by the information that an element $\bar{x}$ in the extended domain can be approximated by an element y in the original domain of the optimization problem (to the extent that we can arrange that the cost of y is arbitrarily close to that of $\bar{x}$). In these circumstances, we can find a sub-optimal element for the original problem (i.e. one whose cost is arbitrarily close to the infimum cost) by finding a minimizer in the extended domain and approximating it.

Relaxation will now be examined in connection with the optimization problem (P) of the preceding section, which for convenience we reproduce

$$(P) \begin{cases} \text{Minimize } g(x(S), x(T)) \text{ over } x \in W^{1,1}([S, T]; \mathbb{R}^n) \\ \text{which satisfy} \\ \dot{x}(t) \in F(t, x(t)) \quad \text{a.e. } t \in [S, T], \\ x(t) \in X(t) \quad \text{for all } t \in [S, T], \\ (x(S), x(T)) \in C. \end{cases}$$

We impose the hypotheses of Proposition 6.4.2 with the exception of the convexity hypothesis

'$F(t, x)$ is convex for all $(t, x) \in \text{Gr } X$'.

In these circumstances (P) may fail to have a minimizer. The point is illustrated by the following example.

Example
Consider problem (P) when the state vector x is 2 dimensional. We write the components of the state vector $x = (y, z)$. Set

6.5 Relaxation

$$[S, T] = [0, 1], \quad \Omega = [0, 1] \times \mathbb{R}^2, \quad X(t) = \mathbb{R}^2,$$
$$g((y_0, z_0), (y_1, z_1)) = z_1,$$

$$C = \{((y_0, z_0), (y_1, z_1)) : z_0 = 0\},$$

$$F(y, z) := (\{-1\} \cup \{+1\}) \times \{|y|\}.$$

Notice that, if an arc (y, z) satisfies the constraints, then

$$z(t) = \int_0^t |y(s)| ds. \tag{6.5.1}$$

Evidently then this special case of (P) is a disguised version, arrived at through state augmentation, of the optimization problem

$$(P)' \begin{cases} \text{Minimize } J(y) = \int_0^1 |y(s)| ds \\ \text{over } y \in W^{1,1}([0, 1]; \mathbb{R}) \text{ satisfying} \\ \dot{y}(t) \in \{-1\} \cup \{+1\} \text{ a.e..} \end{cases}$$

(If (y, z) is admissible for (P) then y and z are related by (6.5.1), y is admissible for $(P)'$ and the costs are the same; also, if y is admissible for $(P)'$ then (y, z), with z given by (6.5.1), is admissible for (P) and again the costs are the same.)

However, as we shall see, $(P)'$ has no minimizers. It follows that, in this case, (P) fails to have a solution. The fact that all the hypotheses of Proposition 6.4.2 are satisfied with the exception of (b) confirms that the convexity hypothesis cannot be dispensed with.

To show that $(P)'$ does not have a solution, notice first of all that

$$J(y) \geq 0 \tag{6.5.2}$$

for all arcs y satisfying the constraint of $(P)'$. Consider next the sequence of admissible arcs $\{y_i\}$,

$$y_i(t) = \int_0^t v_i(s) ds$$

where

$$v_i(s) = \begin{cases} +1 & \text{for } s \in A_i \cap [0, 1] \\ -1 & \text{for } s \notin A_i \cap [0, 1] \end{cases}$$

and

$$A_i = \cup_{j=0}^\infty [(2i)^{-1} 2j, (2i)^{-1}(2j + 1)].$$

An easy calculation yields:

$$J(y_i) = 2^{-(i+1)} \text{ for } i = 1, 2, \ldots.$$

Since $J(y_i) \to 0$ as $i \to \infty$ we conclude from (6.5.2) that the infimum cost is 0. If there exists a minimizer $\bar{y}$ then

$$J(\bar{y}) = \int_0^1 |\dot{\bar{y}}(s)| ds = 0.$$

This implies that $\bar{y} \equiv 0$. It follows that

$$\dot{\bar{y}}(t) = 0 \text{ a.e..}$$

But then $\bar{y}$ fails to satisfy the differential inclusion

$$\dot{\bar{y}}(t) \in (\{-1\} \cup \{+1\}) \text{ . a.e..}$$

It follows that no minimizer exists.

The pathological feature of the above problem is that the limit point $\bar{y}$ of any minimizing sequence satisfies only the convexified differential inclusion

$$\dot{\bar{y}}(t) \in \text{co} (\{-1\} \cup \{+1\}) \text{ a.e..}$$

In light of this example, a natural relaxation procedure for us to adopt is to allow arcs which satisfy the convexified differential inclusion

$$\dot{x}(t) \in \text{co } F(t, x(t)) \text{ a.e. } t \in [S, T]. \tag{6.5.3}$$

Accordingly, an arc $x \in W^{1,1}$ satisfying this dynamic constraint is called a 'relaxed' F trajectory. When it is necessary to emphasize the distinction with relaxed F trajectories, we sometimes call F trajectories 'ordinary' F trajectories.

As earlier discussed, for this concept of relaxed trajectory to be useful we need to know that arcs satisfying the 'relaxed' dynamic constraint (6.5.3) can be adequately approximated by arcs satisfying the original, unconvexified, constraint. That this can be done is a consequence of the generalized Filippov existence theorem and the following theorem of R.J. Aumann on integrals of multifunctions.

Theorem 6.5.1 (Aumann's Theorem) *Take a Lebesgue measurable multifunction $\Gamma : [S, T] \rightsquigarrow \mathbb{R}^n$ which is closed and non-empty. Assume that there exists $c \in L^1$ such that*

$$\Gamma(t) \in c(t)\mathbb{B} \text{ for all } t \in [S, T].$$

6.5 Relaxation

Then

$$\int_S^T \Gamma(s)ds = \int_S^T co\,\Gamma(s)ds,$$

where

$$\int_S^T A(s)ds := \left\{ \int_S^T \gamma(s)ds : \gamma \text{ is a Lebesgue measurable selection of } A \right\}$$

for $A = \Gamma$ and $A = co\,\Gamma$.

The ground has now been prepared for:

Theorem 6.5.2 (Relaxation Theorem) *Take a relatively open set $\Omega \subset [S, T] \times \mathbb{R}^n$ and a $\mathcal{L} \times \mathcal{B}^n$ measurable multifunction $F : \Omega \rightsquigarrow \mathbb{R}^n$ which is closed and non-empty. Assume that there exist $k \in L^1$ and $c \in L^1$ such that*

$$F(t, x') \subset F(t, x'') + k(t)\mathbb{B} \text{ for all } (t, x'), (t, x'') \in \Omega$$

and

$$F(t, x) \subset c(t)\mathbb{B} \text{ for all } (t, x) \in \Omega.$$

Take any relaxed F trajectory x with $\text{Gr } x \subset \Omega$ and any $\delta > 0$. Then there exists an ordinary F trajectory y which satisfies $y(S) = x(S)$ and

$$\max_{t \in [S,T]} |y(t) - x(t)| < \delta.$$

Proof Choose $\epsilon > 0$ such that $T(x, 2\epsilon) \subset \Omega$ and let α be such that

$$0 < \alpha < \min \left\{ \frac{\epsilon}{K \ln K}, \epsilon, \frac{\delta}{1 + K \ln K} \right\},$$

where $K := \exp(\|k\|_{L^1})$. Let $h > 0$ be such that

$$\int_I c(t)dt < \alpha/2$$

for any subinterval $I \subset [S, T]$ of length no greater than h.

Let $\{S = t_0, t_1, .., t_k = T\}$ be a partition of $[S, T]$ such that meas $I_i < h$ for $i = 1, 2, .., k$ where $I_i := [t_{i-1}, t_i)$ for $i = 1, .., k-1$ and $I_k = [t_{k-1}, t_k]$. The multifunction $t \rightsquigarrow F(t, x(t))$ is Lebesgue measurable (see Proposition 2.3.2) and satisfies

$$F(t, x(t)) \subset c(t)\mathbb{B} \text{ for all } t \in [S, T].$$

Recalling that x is a co F trajectory, we deduce from Aumann's theorem (Theorem 6.5.1) that there exist measurable functions $f_i : [S, T] \to \mathbb{R}^n$ such that $f_i(t) \in F(t, x(t))$ a.e. $t \in I_i$ and

$$\int_{I_i} f_i(t)dt = \int_{I_i} \dot{x}(t)dt$$

for $i = 1, .., k$. Define

$$f(t) := \sum_{i=1}^{k} f_i(t)\chi_{I_i}(t)$$

where χ_{I_i} denotes the characteristic function of I_i and set

$$z(t) = x(S) + \int_S^t f(s)ds \text{ for } t \in [S, T].$$

Fix $t \in [S, T]$. Then for some $j \in \{1, .., k\}$,

$$|z(t) - x(t)| = |\sum_{i=1}^{j-1} \int_{I_i} (f_i(\sigma) - \dot{x}(\sigma))d\sigma + \int_{I_j \cap [S,t]} (f_j(\sigma) - \dot{x}(\sigma))d\sigma|$$

$$\leq 0 + 2\int_{I_j} c(t)dt < \alpha.$$

It follows that

$$||z - x||_{L^\infty} < \alpha. \tag{6.5.4}$$

Since $\alpha < \epsilon$, we have

$$T(z, \epsilon) \subset \Omega.$$

Notice that, since $\dot{z}(t) \in F(t, x(t))$ a.e. and in view of Proposition 6.2.1,

$$\rho_F(t, z(t), \dot{z}(t)) \leq \rho_F(t, x(t), \dot{z}(t)) + k(t)|x(t) - z(t)|$$
$$< k(t)\alpha \text{ a.e..}$$

We have

$$\Lambda_F(z) := \int_S^T \rho_F(t, z(t), \dot{z}(t))dt < \alpha \ln K.$$

6.5 Relaxation

Since $\alpha K \ln K < \epsilon$, we deduce from the generalized Filippov existence theorem that there exists an F trajectory y such that $y(S) = x(S)$ and

$$||y - z||_{L^\infty} \le K \Lambda_F(z) < \alpha K \ln K.$$

By (6.5.4) then

$$||y - x||_{L^\infty} \le ||y - z||_{L^\infty} + ||z - x||_{L^\infty} \le \alpha(K \ln K + 1).$$

Since however

$$\alpha(K \ln K + 1) < \delta,$$

we conclude that the F trajectory y satisfies

$$||y - x||_{L^\infty} < \delta.$$

$\square$

The optimization problem obtained when we replace the dynamic constraint $\dot{x} \in F$ in (P) by $\dot{x} \in \mathrm{co}\, F$ will be denoted $(P)_{\text{relaxed}}$. Minimizers for $(P)_{\text{relaxed}}$ will be called relaxed minimizers for (P).

The following proposition provides information clarifying the relation between (P) and $(P)_{\text{relaxed}}$.

Proposition 6.5.3 *Take Ω, F, X, C and g as above. Assume that*

(i) *F is a compact, $\mathcal{L} \times \mathcal{B}^n$ measurable multifunction,*
(ii) *For each $t \in [S, T]$, the graph of $F(t, .)$ restricted to $X(t)$ is closed,*
(iii) *There exist $\alpha \in L^1$ and $\beta \in L^1$ such that*

$$F(t, x) \subset (\alpha(t)|x| + \beta(t))\mathbb{B} \quad \text{for all } (t, x) \in \mathrm{Gr}\, X.$$

(iv) *Either $X(s)$ is bounded for some $s \in [S, T]$ or one of the following two sets*

$$C_0 := \{x_0 \in \mathbb{R}^n : (x_0, x_1) \in C \text{ for some } x_1 \in \mathbb{R}^n\}$$

$$C_1 := \{x_1 \in \mathbb{R}^n : (x_0, x_1) \in C \text{ for some } x_0 \in \mathbb{R}^n\}$$

is bounded.

Assume further that the set of admissible F trajectories $\mathcal{R}_F(X, C)$ is non-empty. Then $(P)_{\text{relaxed}}$ has a minimizer.

If, in addition, we assume that

(a) *there exists $k \in L^1$ such that*

$$F(t, x) \subset F(t, x') + k(t)|x - x'|\mathbb{B} \text{ for all } (t, x), (t, x') \in \Omega,$$

(b) g is continuous,

and for some relaxed minimizer $\bar{x}$ and $\epsilon > 0$

(c)

$$\bar{x}(t) + \epsilon\mathbb{B} \subset X(t) \text{ for all } t \in [S, T], \tag{6.5.5}$$

(d)

$$\text{either } (\bar{x}(S) + \epsilon\mathbb{B}) \times \{\bar{x}(T)\} \subset C \text{ or } \{\bar{x}(S)\} \times (\bar{x}(T) + \epsilon\mathbb{B}) \subset C. \tag{6.5.6}$$

Then

$$\inf(P)_{relaxed} = \inf(P).$$

The right and left sides of the last relation denote the infimum cost for (P) and $(P)_{relaxed}$ respectively.

Proof Existence of a minimizer for $(P)_{relaxed}$ follows immediately from Proposition 6.4.2 applied to the modified version of (P) in which co F replaces F. (Notice that co F inherits the measurability properties of F according to Proposition 2.3.8, as well as the linear growth properties.)

Suppose that there exists a relaxed minimizer $\bar{x}$ and $\epsilon > 0$ such that

$$\bar{x}(t) + \epsilon\mathbb{B} \subset X(t) \text{ for all } t \in [S, T]$$

and

$$\{\bar{x}(S)\} \times (\bar{x}(T) + \epsilon\mathbb{B}) \subset C.$$

(The case $(\bar{x}(S) + \epsilon\mathbb{B}) \times \{\bar{x}(T)\} \subset C$ is treated by 'reversing time'.) Take any $\alpha > 0$. Then noting the continuity of the function g and applying the relaxation theorem, Theorem 6.5.2, we can find an ordinary F trajectory $\bar{y}$ such that

$$\bar{y}(S) = \bar{x}(S), ||\bar{y} - \bar{x}||_{L^\infty} < \epsilon$$

and

$$g(\bar{y}(S), \bar{y}(T)) < g(\bar{x}(S), \bar{x}(T)) + \alpha.$$

Clearly $\bar{y}$ satisfies the constraints

$$(\bar{y}(S), \bar{y}(T)) \in C \text{ and } \bar{y}(t) \in X(t) \text{ for all } t \in [S, T].$$

6.6 Estimates on Trajectories Confined to a Closed Subset

In other words, $\bar{y}$ is an admissible (ordinary) F trajectory. It follows that

$$\inf(P)_{\text{relaxed}} = g(\bar{x}(S), \bar{x}(T)) > g(\bar{y}(S), \bar{y}(T)) - \alpha \geq \inf(P) - \alpha.$$

Since $\alpha > 0$ is arbitrary and

$$\inf(P)_{\text{relaxed}} \leq \inf(P),$$

we conclude that

$$\inf(P)_{\text{relaxed}} = \inf(P).$$

□

Notice the crucial role of the 'interiority' hypotheses (6.5.5) and (6.5.6). If all relaxed minimizers violate these hypotheses then the relaxation theorem does not automatically imply that the infimum costs of (P) and $(P)_{\text{relaxed}}$ coincide; while it is true that any relaxed minimize $\bar{x}$ can be uniformly approximated by an F trajectory y, we cannot in general guarantee that y will satisfy the constraints for it to qualify as an admissible F trajectory.

6.6 Estimates on Trajectories Confined to a Closed Subset

In the final section of this chapter, we provide information about F trajectories x that satisfy a pathwise constraint $x(t) \in A$, for a given subset A of the state space. Conditions for the existence of such F trajectories is the subject matter of viability/invariance theory. The focus of this section is on providing estimates of the distance of a given F trajectory from the family of F trajectories that satisfy the state constraint. Such estimates, known as 'distance estimates', are important tools in the study of constrained dynamic optimization problems, where they are used both in dynamic programming and in the derivation of necessary conditions of optimality.

Consider the state-constrained differential inclusion:

$$\begin{cases} \dot{x}(t) \in F(t, x(t)) & \text{for a.e. } t \in [S, T] \\ x(t) \in A & \text{for all } t \in [S, T], \end{cases} \quad (6.6.1)$$

in which $[S, T]$ is a given interval $(T > S)$, $F : [S, T] \times \mathbb{R}^n \rightsquigarrow \mathbb{R}^n$ is a given multifunction with closed, non-empty values, and $A \subset \mathbb{R}^n$ is a given closed set. We recall that, given a subinterval (possibly closed or left open) $I \subset [S, T]$, we refer to an absolutely continuous function $x : I \to \mathbb{R}^n$ which satisfies $\dot{x}(t) \in F(t, x(t))$ a.e. as an F trajectory (on I). An F trajectory x on I is said to be 'admissible' (on I) if $x(t) \in A$ for all $t \in I$, and 'strictly admissible' (on I) if $x(t) \in \text{int } A$ for all $t \in I$.

Theorem 6.6.1 (Distance Estimate) *Take a multifunction $F : [S, T] \times \mathbb{R}^n \rightsquigarrow \mathbb{R}^n$ and a closed set $A \subset \mathbb{R}^n$. Fix $r_0 > 0$. Assume that, for some constant $c_0 > 0$, some functions c, $k_F \in L^1(S, T)$ and for $R_0 := e^{\int_S^T c(t)\, dt}(r_0 + 1)$, the following hypotheses (H1), (H2), (H3), (IPC) and (BVL) are satisfied:*

(H1): $F : [S, T] \times \mathbb{R}^n \rightsquigarrow \mathbb{R}^n$ *takes closed, non-empty values, $F(., x)$ is $\mathcal{L}$-measurable for all $x \in \mathbb{R}^n$,*

(H2):   (i) $F(t, x) \subset c(t)(1 + |x|)\,\mathbb{B}$ *for all $x \in \mathbb{R}^n$ and for a.e. $t \in [S, T]$,*
         (ii) $F(t, x) \subset c_0\,\mathbb{B}$ *for all $(t, x) \in [S, T] \times R_0\mathbb{B}$,*

(H3): $F(t, x') \subset F(t, x) + k_F(t)|x - x'|\,\mathbb{B}$
                     for all $x, x' \in R_0\mathbb{B}$ and a.e. $t \in [S, T]$,

(IPC):   *for each $(t, x) \in [S, T] \times (R_0\mathbb{B} \cap \partial A)$,*

$$\left(\liminf_{(t',x') \xrightarrow{D} (t,x)} \mathrm{co}\, F(t', x') \right) \cap \mathrm{int}\, \bar{T}_A(x) \neq \emptyset ,$$

where $D := [S, T] \times A$,

(BVL):   *$F(., x)$ has bounded variation from the left uniformly over $x \in (\partial A + \eta_0\mathbb{B}) \cap R_0\mathbb{B}$ for some $\eta_0 > 0$, in the following sense: there exists a non-decreasing bounded variation function $\eta : [S, T] \to \mathbb{R}$ (called a 'modulus of variation of $F(., x)$') such that, for every $[s, t] \subset [S, T]$ and $x \in (\partial A + \eta_0\mathbb{B}) \cap R_0\mathbb{B}$,*

$$F(s, x) \subset F(t, x) + (\eta(t) - \eta(s)) \times \mathbb{B} .$$

Then there exists a constant $K > 0$ with the following property: Given any interval $[t_0, t_1] \subset [S, T]$, any F trajectory $\hat{x}$ on $[t_0, t_1]$ with $\hat{x}(t_0) \in A \cap \left(e^{\int_S^{t_0} c(t)\, dt}(r_0 + 1) - 1\right)\mathbb{B}$, and any $\rho > 0$ such that

$$\rho \geq \max\{d_A(\hat{x}(t)) : t \in [t_0, t_1]\} ,$$

we can find an F trajectory x on $[t_0, t_1]$ such that $x(t_0) = \hat{x}(t_0)$,

$$x(t) \in \mathrm{int}\, A \quad \text{for all } t \in (t_0, t_1]$$

and

$$\|\hat{x} - x\|_{L^\infty(t_0, t_1)} \leq K\,\rho . \tag{6.6.2}$$

(Recall that $\bar{T}_A(x)$ denotes the Clarke tangent cone to A at x.)

Such estimates are referred to as linear L^∞ *distance estimates* for F trajectories confined to a closed set. They provide an upper bound on the distance of a given F trajectory $\hat{x}$ from the set of F trajectories satisfying a state constraint, in terms of the expression $\max_{t \in [t_0, t_1]} d_A(\hat{x}(t))$, which can be interpreted as a measure of the

6.6 Estimates on Trajectories Confined to a Closed Subset

state constraint violation by $\hat{x}$. The assertions of the theorem cover two cases, each of independent interest:

Case A: $\max\{d_A(\hat{x}(t)) : t \in [t_0, t_1]\} > 0$ ($\hat{x}$ is *not* admissible).

In this case, an F trajectory x with initial value $\hat{x}(t_0)$ and strictly admissible on $(t_0, t_1]$ exists, which satisfies the linear distance estimate

$$\|\hat{x} - x\|_{L^\infty(t_0, t_1)} \leq K \max\{d_A(\hat{x}(t)) : t \in [t_0, t_1]\} .$$

(This follows from the theorem statement, after setting $\rho := \max\{d_A(\hat{x}(t)) : t \in [t_0, t_1]\}$.)

Case B: $\max\{d_A(\hat{x}(t)) : t \in [t_0, t_1]\} = 0$ ($\hat{x}$ is admissible).

In this case, for arbitrary $\epsilon > 0$, there exists an F trajectory x, with initial value $\hat{x}(t_0)$ and strictly admissible on $(t_0, t_1]$ such that

$$\|\hat{x} - x\|_{L^\infty(t_0, t_1)} \leq \epsilon .$$

(This follows from the theorem statement, after setting $\rho := \epsilon/K$.)

The next theorem is a variant on the preceding distance estimate, in which it is asserted that, if $F(t, x)$ has a two-sided bounded variation property w.r.t. the time variable (in place of the one-sided property required in Theorem 6.6.1), a linear L^∞ estimate is still valid, under a less restrictive version of the inward pointing condition.

Theorem 6.6.2 *The assertions of Theorem 6.6.1 remain valid when assumptions (IPC) and (BVL) are replaced by the following hypotheses (IPC)$'$ and (BV)$_A$ (for $R_0 := e^{\int_S^T c(t)\, dt}(r_0 + 1)$):*

(IPC)$'$: *for each $t \in [S, T)$, $s \in (S, T]$ and $x \in R_0 \mathbb{B} \cap \partial A$,*

$$\left(\liminf_{x' \xrightarrow{A} x} \limsup_{t' \downarrow t} \operatorname{co} F(t', x') \right) \cap \operatorname{int} \bar{T}_A(x) \neq \emptyset ,$$

$$\left(\liminf_{x' \xrightarrow{A} x} \limsup_{s' \uparrow s} \operatorname{co} F(s', x') \right) \cap \operatorname{int} \bar{T}_A(x) \neq \emptyset ,$$

(BV)$_A$: $F(., x)$ *has bounded variation uniformly over $x \in (\partial A + \eta_0 \mathbb{B}) \cap R_0 \mathbb{B}$ for some $\eta_0 > 0$, in the following sense: there exists a non-decreasing bounded variation function $\eta : [S, T] \to \mathbb{R}$ such that, for every $[s, t] \subset [S, T]$ and $x \in (\partial A + \eta_0 \mathbb{B}) \cap R_0 \mathbb{B}$,*

$$d_H(F(s, x), F(t, x)) \leq \eta(t) - \eta(s) .$$

Here, $d_H(A, B)$ is the Hausdorff distance function between two arbitrary non-empty closed sets in $\mathbb{R}^n$ A and B:

$$d_H(A, B) := \max \left\{ \sup_{a \in A} d_B(a), \sup_{b \in B} d_A(b) \right\}. \tag{6.6.3}$$

As a preliminary step in the proof of Theorem 6.6.1, we establish some technical lemmas.

The first lemma, whose proof exploits properties of the interior of Clarke's tangent cone, allows us to select suitable inward pointing vectors which belong to the convex hull of the velocity set F or are limits of such vectors.

Lemma 6.6.3 *Suppose the multifunction $F : [S, T] \times \mathbb{R}^n \rightsquigarrow \mathbb{R}^n$ and the closed set A satisfy hypothesis (IPC) (for some $R_0 > 0$). Then there exist $M > 0$, $\epsilon > 0$ and $\bar{\eta} > 0$ with the following property: for any $(t, x) \in [S, T] \times \left((\partial A + \bar{\eta}\mathbb{B}) \cap R_0 \mathbb{B} \cap A\right)$, there exists $v \in \text{co}\, F(t, x) \cap M\mathbb{B}$ such that*

$$y + [0, \epsilon](v + \epsilon\mathbb{B}) \subset A, \quad \text{for all } y \in (x + \epsilon\mathbb{B}) \cap A. \tag{6.6.4}$$

Proof

Step 1: We claim that for each $(t, x) \in [S, T] \times \left(R_0\mathbb{B} \cap \partial A\right)$ there exist $M_{t,x} > 0$, $\epsilon_{t,x} \in (0, 1)$ and $\delta_{t,x} \in (0, \epsilon_{t,x}]$ such that, given any $(t', x') \in ((t, x) + \delta_{t,x}\mathbb{B}) \cap ([S, T] \times A)$, a vector $v' \in \text{co}\, F(t', x')$ can be found such that $|v'| \leq M_{t,x}$ and

$$y' + [0, \epsilon_{t,x}](v' + \epsilon_{t,x}\mathbb{B}) \subset A, \quad \text{for all } y' \in (x' + \epsilon_{t,x}\mathbb{B}) \cap A.$$

Indeed, take any $(t, x) \in [S, T] \times (R_0\mathbb{B} \cap \partial A)$ and chose any vector

$$v \in \left(\liminf_{(t', x') \xrightarrow{[S,T] \times A} (t,x)} \text{co}\, F(t', x')\right) \cap \text{int}\, \bar{T}_A(x).$$

By the characterization of the interior of the Clarke tangent cone (see Theorem 4.11.1), there exists $\epsilon \in (0, 1)$ such that

$$y + [0, \epsilon](v + 2\epsilon\mathbb{B}) \subset A, \quad \text{for all } y \in (x + 2\epsilon\mathbb{B}) \cap A. \tag{6.6.5}$$

On the other hand, by definition of the limit inferior operation, there exists $\delta \in (0, \epsilon]$ such that, given any $(t', x') \in ((t, x) + \delta\mathbb{B}) \cap ([S, T] \times A)$, there exists $v' \in \text{co}\, F(t', x')$ satisfying $|v - v'| \leq \epsilon$. Then, $|v'| \leq |v| + 1 (=: M_{t,x})$. Now take any $y' \in (x' + \epsilon\mathbb{B}) \cap A$. Then, since $x' + \epsilon\mathbb{B} \subset x + 2\epsilon\mathbb{B}$ and $v' \in v + \epsilon\mathbb{B}$, we may conclude from (6.6.5) that

$$y' + [0, \epsilon](v' + \epsilon\mathbb{B}) \subset A, \quad \text{for all } y' \in (x' + \epsilon\mathbb{B}) \cap A.$$

Step 2: By a standard compactness argument, we can find a finite number of points $(t_i, x_i) \in [S, T] \times \left(R_0\mathbb{B} \cap \partial A\right)$ and numbers $M_i > 0$, $\epsilon_i > \delta_i > 0$, for $i = 1, \ldots, m$, such that

6.6 Estimates on Trajectories Confined to a Closed Subset

$$\bigcup_{i=1,\ldots,m} \left((t_i, x_i) + \delta_i \, \overset{\circ}{\mathbb{B}}\right) \supset [S, T] \times (R_0 \mathbb{B} \cap \partial A), \tag{6.6.6}$$

and for each $(t', x') \in ((t_i, x_i) + \delta_i \mathbb{B}) \cap ([S, T] \times A)$, there exists a vector $v' \in \text{co } F(t', x')$ such that $|v'| \leq M_i$ and

$$y' + [0, \epsilon_i](v' + \epsilon_i \mathbb{B}) \subset A, \qquad \text{for all } y' \in (x' + \epsilon_i \mathbb{B}) \cap A.$$

Notice also that there exists $\bar{\eta} \in (0, \min_{i=1,\ldots,m} \delta_i)$ such that

$$\bigcup_{i=1,\ldots,m} \left((t_i, x_i) + \delta_i \, \overset{\circ}{\mathbb{B}}\right) \supset [S, T] \times ((\partial A + \bar{\eta} \mathbb{B}) \cap R_0 \mathbb{B}),$$

otherwise we could find a sequence of points $(s_j, y_j) \in ([S, T] \times R_0 \mathbb{B}) \setminus \bigcup_{i=1,\ldots,m} \left((t_i, x_i) + \delta_i \, \overset{\circ}{\mathbb{B}}\right)$ such that $(s_j, y_j) \to (s, y) \in [S, T] \times (R_0 \mathbb{B} \cap \partial A)$, which would contradict (6.6.6).

To conclude we just take $\epsilon := \min_{i=1,\ldots,m} \epsilon_i$, $M := \max_{i=1,\ldots,m} M_i$ and the assertions of the lemma immediately follow. □

The following lemma summarizes some implications of the 'bounded variation' hypotheses (BVL) and (BV)$_A$:

Lemma 6.6.4 *Take a multifunction* $F : [S, T] \times \mathbb{R}^n \rightsquigarrow \mathbb{R}^n$ *that satisfies hypotheses (H1), (H2), (H3) and (BVL) for some* $R_0 > 0$. *Take also any interval* $[a, b] \subset [S, T]$. *Define* $\widetilde{F} : [a, b] \times \mathbb{R}^n \rightsquigarrow \mathbb{R}^n$ *to be*

$$\widetilde{F}(t, x) := \begin{cases} \limsup_{t \downarrow a} F(t, x) & \text{if } t = a \\ F(t, x) & \text{if } t \in (a, b) \\ \limsup_{t \uparrow b} F(t, x) & \text{if } t = b, \end{cases} \tag{6.6.7}$$

for any $x \in \mathbb{R}^n$. *Then* $\widetilde{F}$ *takes values, closed non-empty sets,*

$$\widetilde{F}(t, x) \subset c_0 \mathbb{B}, \quad \text{for all } (t, x) \in [a, b] \times R_0 \mathbb{B}$$

and

$$\widetilde{F}(t, x) \subset \widetilde{F}(t, x') + k_F(t)|x - x'| \mathbb{B} \quad \text{for all } x, x' \in R_0 \mathbb{B}, \text{ and a.e. } t \in [a, b].$$

Furthermore

$$\widetilde{F}(s, x) \subset \widetilde{F}(t, x) + (\tilde{\eta}(t) - \tilde{\eta}(s)) \times \mathbb{B} \quad \text{for all } x \in (\partial A + \eta_0 \mathbb{B}) \cap R_0 \mathbb{B} \tag{6.6.8}$$

for any $[s, t] \subset [a, b]$, *where*

$$\tilde{\eta}(t) := \begin{cases} 0 & \text{for } t = a \\ \eta(t) - \eta(a^+) & \text{for } t \in (a,b) \\ \eta(b^-) - \eta(a^+) & \text{for } t = b, \end{cases} \quad (6.6.9)$$

in which η is the 'modulus of variation' in hypothesis (BVL).

Assume, in addition, that also $(BV)_A$ is satisfied. Then, we can supplement the assertions above with the following extra information

$$\sup_{x \in (\partial A + \eta_0 \mathbb{B}) \cap R_0 \mathbb{B}} d_H(\widetilde{F}(s,x), \widetilde{F}(t,x)) \leq \tilde{\eta}(t) - \tilde{\eta}(s). \quad (6.6.10)$$

Proof Suppose first that assumptions (H1), (H2), (H3) and (BVL) are satisfied. We omit the proof of the assertions that the constructed $\widetilde{F}$ takes values non-empty closed sets, and inherits the Lipschitz continuity and boundedness properties of the original multifunction F (with the same constants), since this is a straightforward exercise. Consider the final assertion. Take any $[s, t] \subset [a, b]$. We must show (6.6.8). This relation is obviously true for $a < s < t < b$. Consider the case $[s, t] = [a, b]$. (The remaining cases $[s, t] = [a, t']$ or $[t', b]$, for some $t' \in (a, b)$ are similar, but simpler, to deal with.) Fix any $x \in (\partial A + \eta_0 \mathbb{B}) \cap R_0 \mathbb{B}$. Take any $v \in \widetilde{F}(a, x)$ and we must find $w \in \widetilde{F}(b, x)$ such that

$$|v - w| \leq \eta(b^-) - \eta(a^+). \quad (6.6.11)$$

By definition of the 'lim sup' operation, there exist $s_i \downarrow a$ and $v_i \to v$ such that $v_i \in F(s_i, x)$ for each i. Now take any sequence $t_i \uparrow b$. Then by the properties of $F(t, x)$, which coincides with $\widetilde{F}(t, x)$ for $t \in (a, b)$ we have: for each x, there exists $w_i \in F(t_i, x)$ such that

$$|v_i - w_i| \leq \eta(t_i) - \eta(s_i) \quad (6.6.12)$$

The sequence $\{w_i\}$ is bounded, in view of hypothesis (H2). So, by restricting attention to a subsequence, we can arrange that $w_i \to w$ for some $w \in \mathbb{R}^n$. But then, again by definition of the 'lim sup' operation, $w \in \widetilde{F}(b, x)$. Passing to the limit in (6.6.12), and noting that η, as a monotone function, has left and right limits everywhere, we arrive at (6.6.11).

If the stronger assumption $(BV)_A$ is also satisfied, then (6.6.10) follows from the argument above in view of the symmetric property of the Hausdorff distance. The proof is complete. □

Proof of Theorem 6.6.1 We take F to be the multifunction in the statement of Theorem 6.6.1, satisfying (H1), (H2), (H3), (IPC) and (BVL) for some positive numbers r_0, c_0, η_0 (and $R_0 := \exp\{\int_S^T c(t)\,dt\}(r_0 + 1)$), some integrable functions c and k_F, and some non-decreasing function η.

6.6 Estimates on Trajectories Confined to a Closed Subset

Remark
We take note, for future use, of an implication of Lemmas 6.6.3 and 6.6.4. Take any $[t_0, t_1] \subset [S, T]$ and define $\widetilde{F} : [t_0, t_1] \times \mathbb{R}^n \rightsquigarrow \mathbb{R}^n$ according to (6.6.7). Take an arbitrary point $(t, x) \in [t_0, t_1] \times \big((\partial A + \eta_0 \mathbb{B}) \cap R_0 \mathbb{B} \cap A\big)$. Then there exists a vector $v \in \operatorname{co} \widetilde{F}(t, x)$ such that (6.6.4) is true for all $y \in (x + \epsilon \mathbb{B}) \cap A$. Here ϵ is the positive constant as in the lemma statement. This follows from the way (IPC) is formulated and the fact that, for any $(t, x) \in [t_0, t_1] \times \mathbb{R}^n$:

$$\liminf_{(t',x') \xrightarrow{D} (t,x)} \operatorname{co} F(t', x') \subset \liminf_{(t',x') \xrightarrow{D} (t,x)} \operatorname{co} \widetilde{F}(t', x'),$$

where $D = [S, T] \times A$.

Henceforth, we take $\epsilon > 0$, $\bar{\eta} \in (0, \eta_0)$ and $M = c_0$ to be the constants referred to in Lemma 6.6.3: we can arrange (by reducing the size of $\bar{\eta}$, if necessary) that $\bar{\eta} \in (0, \eta_0)$ and $M = c_0$ are also the constants appearing in Lemma 6.6.3. (This is true since, if the statement of Lemma 6.6.3 is valid for some $\bar{\eta}$ and M, then it remains valid for any lower positive value of the parameter $\bar{\eta}$ and bigger value of the constant M.)

The Lemma 6.6.5 below are aimed at simplifying the proof of Theorem 6.6.1, by showing that attention can be restricted to a special case. □

Lemma 6.6.5 (Hypothesis Reduction) *Fix $r_0 > 0$. Assume that, for $\delta > 0$, $\bar{\rho} > 0$ and $\gamma > 0$ sufficiently small, that the assertions of Theorem 6.6.1 are valid when conditions (H1), (H2), (H3), (IPC) and (BVL) are satisfied (for some constants r_0, c_0, η_0, $R_0 := \exp\{\int_S^T c(t)\, dt\}(r_0 + 1)$, some functions c, $k_F \in L^1(S, T)$ and some non-decreasing bounded variation function η), and under the additional hypotheses:*

(H2)': $F(t, x) \subset c_0 \mathbb{B}$ *for all* $(t, x) \in [S, T] \times \mathbb{R}^n$,
(H3)': $F(t, x') \subset F(t, x) + k_F(t)|x - x'|\mathbb{B}$
 for all $x, x' \in \mathbb{R}^n$ *and a.e.* $t \in [S, T]$,
(H4): $F(t, x)$ *is convex for all* $(t, x) \in [S, T] \times \mathbb{R}^n$,

and when the following conditions are imposed on the reference F trajectory $\hat{x} : [t_0, t_1] \to \mathbb{R}^n$, with $\hat{x}(t_0) \in A \cap \big(e^{\int_S^{t_0} c(t)\, dt}(r_0 + 1) - 1\big)\mathbb{B}$, and the positive number $\rho \geq \max\{d_A(\hat{x}(t)) : t \in [t_0, t_1]\}$:

(i): $t_1 - t_0 \leq \delta$,
(ii): $\rho \leq \bar{\rho}$,
(iii): $\eta(t_1) - \eta(t_0) \leq \gamma$ *and* $\eta(t_0) = 0$.

Then the assertions are valid under (H1), (H2), (H3), (IPC) and (BVL) alone, without conditions (i)–(iii) on $\hat{x}$ and ρ, and when the extra hypotheses (H2)', (H3)' and (H4) are no longer required to be satisfied.

Proof In what follows we shall make use of the following fact: given any $r_0 > 0$ and any interval $[t_0, t_1] \subset [S, T]$, for any F trajectory x on $[t_0, t_1]$ with $x(t_0) \in$

$\left(e^{\int_S^{t_0} c(t)\,dt}(r_0+1)-1\right)\mathbb{B}$ (c is the summable function of assumption (H2)), standard a-priori estimates yield: $x(t) \in \left(e^{\int_S^t c(s)\,ds}(r_0+1)-1\right)\mathbb{B}$ ($\subset R_0\mathbb{B}$) for all $t \in [t_0, t_1]$.

Step 1: Assume that the assertions of the lemma are valid (with constant K) when conditions (H1), (H2), (H3), (IPC) and (BVL) are satisfied together with the additional hypotheses (H2)', (H3)' and (H4), and when it is assumed that $\hat{x}$ on $[t_0, t_1]$ satisfies conditions (i)–(iii). We show that they remain valid (with a modified K) even if condition (iii) is no longer required to be satisfied.

Choose any $\gamma > 0$, $\delta > 0$, $\bar{\rho} > 0$ and $K > 0$, such that the assertions of the lemma are valid.

The function η is increasing, and can therefore be decomposed into the sum of a continuous function η^c and a countable family of functions $\{s_i\}$ satisfying, for each i,

$$s_i(t) = a_i \times \begin{cases} 1 \text{ if } t > \sigma_i \\ 0 \text{ if } t < \sigma_i, \end{cases}$$

in which $\{\sigma_i\}$ is a sequence of distinct points in $[S, T]$ (the 'jump times') and $\{a_i\}$ is a sequence of non-negative numbers (the 'jumps'). (In the analysis to follow, we do not have to take account of the value of s_i at its jump time σ_i.) The collection of jumps $\{a_i\}$ satisfies

$$\sum_i a_i < \infty.$$

In view of this last relation, there exists a finite index set $J \subset \{1, 2, \ldots\}$ such that

$$\sum_{i \notin J} a_i < \gamma/2.$$

Since the jump times $\{\sigma_i\}$ are distinct, there exists $\bar{\alpha} > 0$ such that

$$|\sigma_i - \sigma_j| > \bar{\alpha} \text{ for } i, j \in J, i \neq j.$$

By reducing the size of $\delta > 0$, if necessary, we can also ensure that

$$\eta^c(t') - \eta^c(s') < \gamma/2$$

for all subinterval $[s', t'] \subset [S, T]$ such that $t' - s' \leq \delta$. Now further reduce the size of δ, if necessary, to ensure also that $\delta < \bar{\alpha}$. □

Take any interval $[t_0, t_1] \subset [S, T]$ and F trajectory $\hat{x}$ on $[t_0, t_1]$ with $\hat{x}(t_0) \in A \cap \left(e^{\int_S^{t_0} c(t)\,dt}(r_0+1)-1\right)\mathbb{B}$ such that conditions (i) and (ii) are satisfied, i.e.

$$t_1 - t_0 \leq \delta \text{ and } \max\{d_A(\hat{x}(t)) : t \in [s, t]\} \leq \bar{\rho}.$$

6.6 Estimates on Trajectories Confined to a Closed Subset

Since we have arranged that $\delta < \bar{\alpha}$, there is at most one jump time $\bar{t} = \sigma_j$, with $j \in J$, located in the interval $[t_0, t_1]$. It follows that

$$\eta(\bar{t}^-) - \eta(t_0^+) \leq \gamma \quad \text{and} \quad \eta(t_1^-) - \eta(\bar{t}^+) \leq \gamma. \tag{6.6.13}$$

Case 1: $\bar{t} \in (t_0, t_1)$.

Define $\tilde{F}_1 : [t_0, \bar{t}] \times \mathbb{R}^n \rightsquigarrow \mathbb{R}^n$ according to (6.6.7) and $\tilde{\eta}_1$ according to (6.6.9) when $[a, b] = [t_0, \bar{t}]$. Similarly, set $\tilde{F}_2 : [\bar{t}, t_1] \times \mathbb{R}^n \rightsquigarrow \mathbb{R}^n$ according to (6.6.7) and $\tilde{\eta}_2$ according to (6.6.9) when $[a, b] = [\bar{t}, t_1]$.

$\tilde{F}_1$ satisfies assumptions (H1), (H2), (H3), (IPC) and (BVL) together with the additional hypotheses (H2)′, (H3)′ and (H4) of F, restricted to $[t_0, \bar{t}] \times \mathbb{R}^n$, with the same constants c_0, R_0 and $\bar{\eta}$, and functions c, k_F and η. Also, the assertions of Lemma 6.6.4 continue to be true, with the same constants $\epsilon > 0$ and $\bar{\eta}$, when $\tilde{F}_1$ replaces F. By Lemma 6.6.4 and (6.6.13) however, the variation modulus $\tilde{\eta}_1$ of $\tilde{F}_1$ is such that

$$\tilde{\eta}_1(\bar{t}) - \tilde{\eta}_1(t_0) = \eta(\bar{t}^-) - \eta(t_0^+) < \gamma \quad \text{and} \quad \tilde{\eta}_1(t_0) = 0.$$

We have shown that $\hat{x}$ restricted to $[t_0, \bar{t}]$ satisfies not just conditions (i) and (ii), but also condition (iii), in the lemma statement, when $\hat{x}$ is interpreted as a $\tilde{F}_1$ trajectory. We have confirmed that $\tilde{F}_1$ satisfies all the conditions required in the lemma statement. This means that, if we take any $\rho > 0$ such that

$$\rho \geq \max\{d_A(\hat{x}(t)) : t \in [t_0, t_1]\}$$

and $\rho \leq \bar{\rho}$, there exists an $\tilde{F}_1$ trajectory x_1 (which is also an F trajectory) satisfying $x_1(t_0) = \hat{x}(t_0)$ and

$$\|\hat{x} - x_1\|_{L^\infty(t_0, \bar{t})} \leq K\rho.$$

Also, $x_1(t) \in \text{int } A$ for $t \in (t_0, \bar{t}]$. By Filippov's existence theorem (Theorem 6.2.3), there exists an F trajectory y on $[\bar{t}, t_1]$ such that $y(\bar{t}) = x_1(\bar{t})$ and

$$\|y - \hat{x}\|_{L^\infty(\bar{t}, t_1)} \leq K_1 K\rho,$$

where $K_1 := e^{\int_S^T k_F(t) \, dt}$. We see that

$$\max\{d_A(y(t)) : t \in [t_0, \bar{t}]\} \leq (1 + K_1 K)\rho.$$

Define $\rho_1 > 0$ to be

$$\rho_1 := (1 + K_1 K)\rho.$$

Arguing as above, now in relation to $\widetilde{F}_2$, we conclude that there exists an $\widetilde{F}_2$ trajectory x_2 on $[\bar{t}, t_1]$ (which is also an F trajectory) such that

$$\|y - x_2\|_{L^\infty(\bar{t},t_1)} \leq K\rho_1 = (K_1 K + 1)K\rho,$$

such that $x_2(t) \in \text{int } A$ for all $t \in [\bar{t}, t_1]$. (We have used the fact that $y(\bar{t}) = x_1(\bar{t}) \in \text{int } A$.) But then, by the triangle inequality,

$$\|\hat{x} - x_2\|_{L^\infty(\bar{t},t_1)} \leq (1 + K_1 + K_1 K)K\rho.$$

Now define x to be the F trajectory on $[t_0, t_1]$ obtained by concatenating x_1 (on $[t_0, \bar{t}]$) and x_2 (on $[\bar{t}, t_1]$). Then $x(t_0) = \hat{x}(t_0)$ and $x(t) \in \text{int } A$ for $t \in (t_0, \bar{t}]$. Furthermore

$$\|\hat{x} - x\|_{L^\infty(t_0,t_1)} \leq \max\{K_1 K\rho, (1 + K_1 + K_1 K)K\rho\} = K'\rho$$

where

$$K' := (1 + K_1 + K_1 K)K.$$

Case 2: $\bar{t} \notin (t_0, t_1)$.

In this case, $\bar{t}$ coincides with either t_0 or t_1, so either $[t_0, \bar{t}]$ or $[\bar{t}, t_1]$ degenerate to a single point. Invoking the hypothesis of the lemma just once yields an F trajectory x on $[t_0, t_1]$ such that $x(t_0) = \hat{x}(t_0)$, $x(t) \in \text{int } A$ for $t \in (t_0, t_1]$ and satisfying

$$\|\hat{x} - x\|_{L^\infty(t_0,t_1)} \leq K\rho \leq \max\{K_1 K\rho, (1 + K_1 + K_1 K)K\rho\} = K'\rho.$$

In either case, we have exhibited a new K, written K', with the desired properties.

Step 2: Suppose that the assertions of the lemma are valid (with constant K) when conditions (H1), (H2), (H3), (IPC) and (BVL) are satisfied together with the additional hypotheses (H2)', (H3)' and (H4), and when it is assumed that $\hat{x}$ on $[t_0, t_1]$ satisfies conditions (i) and (ii). We show that they remain valid (with a modified K) even if condition (ii) is no longer required to be satisfied.

By assumption, the assertions are valid (with constant K) if $\rho \leq \bar{\rho}$. Suppose that $\rho > \bar{\rho}$. By the weak invariance theorem for time-varying systems (Theorem 13.2.2), there exists some admissible F trajectory x on $[t_0, t_1]$, with $x(t_0) = \hat{x}(t_0)$. Now apply the special case of the theorem we assume to be valid, treating x as the reference trajectory, to justify replacing x by an F trajectory (we do not re-label) that is *strictly* admissible on $(t_0, t_1]$. Then by (H2)'

$$\|x - \hat{x}\|_{L^\infty} \leq 2c_0(T - S) \leq 2\bar{\rho}^{-1}c_0(T - S) \times \rho.$$

So the assertions of the theorem are valid, in absence of the condition (ii), with the larger constant K

$$\max\{K, 2\bar{\rho}^{-1}c_0(T-S)\}.$$

Step 3: Suppose that the assertions of the lemma are valid (with constant K) when conditions (H1), (H2), (H3), (IPC) and (BVL) are satisfied together with the additional hypotheses (H2)′, (H3)′ and (H4), and when it is assumed that $\hat{x}$ on $[t_0, t_1]$ satisfies condition (i). We show that they remain valid (with a modified K) even if condition (i) is violated.

Observe that the constant c_0 will be the same upper bound for the velocities of all the F trajectories considered in this step. Choose N_0 to be the smallest integer such that $N_0^{-1}(T-S) \leq \delta$. Let N be the smallest integer such that $N^{-1}(t_1 - t_0) \leq \delta$. Observe that $N \leq N_0$. Write $x_0 = \hat{x}$. Partition $[t_0, t_1]$ as a family of N contiguous intervals $\{[t_0^i, t_1^i]\}_{i=1}^N$ with $t_0^1 = t_0$ and $t_1^N = t_1$, each of length at most δ ($< 1/c_0$). Now apply the special case of the theorem (in which condition (i) is assumed to hold) with $\hat{x}|_{[t_0^1, t_1^1]}$ as reference trajectory, to yield an F trajectory x_1 on $[t_0^1, t_1^1]$ such that x_1 is strictly admissible on $(t_0^1, t_1^1]$, $x_1(t_0^1) = \hat{x}(t_0^1)$ and

$$\|x_1 - x_0\|_{L^\infty(t_0^1, t_1^1)} \leq K\rho.$$

Invoking Filippov's existence theorem (Theorem 6.2.3), we can extend x_1 as an F trajectory to $[t_0^1, t_1^N]$ (we do not re-label) such that

$$\|x_1 - x_0\|_{L^\infty(t_0^1, t_1^N)} \leq K_1 K\rho = K_1 K \left((\max\{d_A(x_0(t)) : t \in [t_0^1, t_1^N]\}) \vee \rho\right),$$

in which $K_1 := \exp\{\int_S^T k_F(t)\, dt\}$. Now apply the special case of the theorem (in which condition (i) is satisfied), taking as reference trajectory x_1 restricted to $[t_0^2, t_1^2]$, to yield an F trajectory x_2 on $[t_0^1, t_1^2]$ that is strictly admissible on $(t_0^1, t_1^2]$, which we extend to $[t_0^1, t_1^N]$, and so on. We thereby generate a sequence of F trajectories x_i on $[t_0^1, t_1^N]$, $i = 1, \ldots, N$, such that for each $i = 1, \ldots, N$, x_i is strictly admissible on $(t_0^1, t_1^i]$ and

$$\|x_i - x_{i-1}\|_{L^\infty(t_0^1, t_1^N)} \leq K_1 K \left((\max\{d_A(x_{i-1}(t)) : t \in [t_0^1, t_1^N]\}) \vee \rho\right).$$

We also have

$$(\max\{d_A(x_i(t)) : t \in [t_0^1, t_1^N]\}) \vee \rho \leq \left((\max\{d_A(x_{i-1}(t)) : t \in [t_0^1, t_1^N]\}) \vee \rho\right)$$
$$+ \|x_i - x_{i-1}\|_{L^\infty(t_0^1, t_1^N)}.$$

Write $x = x_N$. Then x is strictly admissible on $(t_0^1, t_0^N] = (t_0, t_1]$. Now, for $i = 1, \ldots, N$, define $a_i := \|x_i - x_{i-1}\|_{L^\infty(t_0^1, t_1^N)}$ and $b_i := (\max\{d_A(x_i(t)) : t \in [t_0^1, t_1^N]\}) \vee \rho$. Then, from the preceding relations, we have, for $i = 2, \ldots, N$

$$a_i \leq K_1 K b_{i-1}, \qquad b_i \leq b_{i-1} + a_i.$$

Eliminating a_i from these inequalities gives

$$b_i \leq (1 + K_1 K) b_{i-1}.$$

Hence

$$b_i \leq (1 + K_1 K)^i b_0$$

and

$$a_i \leq K_1 K (1 + K_1 K)^{i-1} b_0 \leq K_1 K (1 + K_1 K)^{i-1} \rho.$$

But

$$\|x - \hat{x}\|_{L^\infty(t_0,t_1)} = \|x_N - x_0\|_{L^\infty(t_0,t_1)} \leq \sum_{i=1}^{N} \|x_i - x_{i-1}\|_{L^\infty(t_0,t_1)}$$

$$\leq \left(\sum_{i=1}^{N} (1 + K_1 K)^{i-1} \right) K_1 K \rho$$

$$= \left((1 + K_1 K)^N - 1 \right) \rho$$

$$\leq \left((1 + K_1 K)^{N_0} - 1 \right) \rho.$$

It follows that

$$\|x - \hat{x}\|_{L^\infty(t_0,t_1)} \leq \bar{K} \rho,$$

in which $\bar{K} = (1 + K_1 K)^{N_0} - 1$. The assertions are therefore valid even if (i) is not satisfied, when we replace K by $\bar{K}$.

Step 4: Assume that the assertions of the lemma are valid (with constant K) when conditions (H1), (H2), (H3), (IPC) and (BVL) are satisfied together with the additional hypotheses (H2)', (H3)' and (H4). We show that they remain valid even if (H4) is violated, i.e. F is not convex valued.

Assume that the above hypotheses are satisfied, with the exception of (H4). Replace F by co F. Then the above hypotheses, including (H4), are satisfied. The special case of the theorem yields a constant K (independent of the choice of reference trajectory $\hat{x}$ on $[t_0, t_1]$) and a co F trajectory $x' : [t_0, t_1] \to \mathbb{R}^n$, which is strictly admissible on $(t_0, t_1]$, such that

$$\|x' - \hat{x}\|_{L^\infty(t_0,t_1)} \leq K \rho.$$

6.6 Estimates on Trajectories Confined to a Closed Subset

Choose a decreasing sequence $\{s_i\}$ in $(t_0, t_1]$, with $s_1 = t_1$, such that $s_i \downarrow t_0$. Since x' is strictly admissible on $(t_0, t_1]$ we can find a sequence of positive numbers $\epsilon_i \in (0, \rho)$ such that $\epsilon_i \downarrow 0$ and, for $i = 1, 2, \ldots$

$$x'(\sigma) + \epsilon_i \mathbb{B} \subset \text{int } A \quad \text{for all } \sigma \in [s_i, t_1] . \tag{6.6.14}$$

Take a sequence of positive numbers $\{\alpha_i\}$. (We shall place restrictions on the α_i's presently.) By the relaxation theorem Theorem 6.5.2 (which asserts the density, with respect to the L^∞ norm, of the set of F trajectories with a fixed initial state in the set of co F trajectories, with the same initial state), there exists a sequence of F trajectories $x_i : [s_i, t_1] \to \mathbb{R}^n$ such that, for all integer $i \geq 2$, we have $x_i(s_i) = x'(s_i)$ and

$$\|x_i - x'\|_{L^\infty(s_i, t_1)} \leq \alpha_i . \tag{6.6.15}$$

For each integer $j \geq 2$, we construct an F trajectory $y_j : [s_j, t_1] \to \mathbb{R}^n$ as follows: y_j restricted to $[s_j, s_{j-1}]$ coincides with x_j. y_j restricted to $[s_{j-1}, s_{j-2}]$ is an F trajectory with initial state $y_j(s_{j-1})$, obtained by applying the Filippov existence theorem (Theorem 6.2.3) with reference trajectory x_{j-1}, and so on, until y_j has been constructed on the whole interval $[s_j, s_1 = t_1]$.

Now fix an integer $j > 2$. We deduce from Theorem 6.2.3 that, for each $2 \leq i < j$,

$$\|y_j - x_i\|_{L^\infty(s_i, s_{i-1})} \leq K_1 |y_j(s_i) - x'(s_i)|$$

$$\|\dot{y}_j - \dot{x}_i\|_{L^1(s_i, s_{i-1})} \leq K_1 |y_j(s_i) - x'(s_i)| .$$

(Here $K_1 = \exp\{\int_S^T k_F(t) dt\}$. We have also used the fact that $x_i(s_i) = x'(s_i)$.) From these relations and (6.6.15) it follows that for each $2 \leq i < j$ and any integer m, we have

$$\|y_j - x'\|_{L^\infty(s_i, s_{i-1})} \leq \sum_{k=i}^{j} K_1^{k-i} \alpha_k \tag{6.6.16}$$

$$\|\dot{y}_{j+m} - \dot{y}_j\|_{L^1(s_i, s_{i-1})} \leq 2 \sum_{k=i+1}^{j+m} K_1^{k-i} \alpha_k . \tag{6.6.17}$$

Notice that for each $j \geq 2$, $y_j(s_j) = x'(s_j)$. So we can extend each F trajectory y_j as a co F trajectory to all of $[t_0, t_1]$, setting $y_j(\sigma) = x'(\sigma)$ for $\sigma \in [t_0, s_j]$. (We do not re-label.)

Now choose the sequence $\{\alpha_k\}$ to satisfy

$$\sum_{k=i}^{\infty} K_1^k \alpha_k < \epsilon_i/2, \quad \text{for all } i \geq 2 . \tag{6.6.18}$$

This condition is satisfied, in particular, if we assume that $\epsilon_i < 1/3$, for all $i \geq 2$, and we chose $\alpha_k = (\epsilon_k/K_1)^k$.

Since the y_i's have initial value $\hat{x}(t_0)$ and in view of hypotheses (H1)-(H2), we can extract a subsequence (we do not re-label) converging uniformly to a co F trajectory x on $[t_0, t_1]$, with initial value $\hat{x}(t_0)$. We conclude from (6.6.14), (6.6.16) and (6.6.18) that x is strictly admissible on $(t_0, t_1]$. To see this, take any $\sigma \in (t_0, t_1]$ and note that $\sigma \in (s_i, s_{i-1}]$ for some $i \geq 2$. But then from (6.6.16) and (6.6.18) we have

$$y_j(\sigma) \in x'(\sigma) + \frac{\epsilon_i}{2}\mathbb{B} \subset \text{int } A, \quad \text{for all } j > i.$$

Since the y_j's converge uniformly to x,

$$x(\sigma) \in x'(\sigma) + \frac{\epsilon_i}{2}\mathbb{B} \subset \text{int } A.$$

On the other hand, for each $k \geq 2$ and for every $i \geq k$, the y_i's, restricted to $[s_k, t_1]$, are F trajectories, which, owing to (6.6.17), define a Cauchy sequence in $W^{1,1}(s_k, s_{k-1})$. It may be deduced that the limiting co F trajectory x is actually an F trajectory. Finally we note that, since each $\epsilon_i \leq \rho$,

$$\|\hat{x} - x\|_{L^\infty(t_0, t_1)} \leq \|\hat{x} - x'\|_{L^\infty(t_0, t_1)} + \|x - x'\|_{L^\infty(t_0, t_1)} \leq \bar{K}\rho,$$

where $\bar{K} = K + 1$. This is the desired distance estimate, with the modified constant $\bar{K}$.

Step 5: Suppose that the assertions of the lemma are valid (with constant K) when conditions (H1), (H2), (H3), (IPC) and (BVL) are satisfied together with the additional hypotheses (H2)', (H3)'. We show that they remain valid (with the same K) under (H1), (H2), (H3) alone, and when (IPC) and (BVL) are satisfied (for $R_0 := \exp\{\int_S^T c(t)\,dt\}(r_0 + 1)$).

Assume that (H1), (H2), (H3), and that (IPC) and (BVL) are satisfied with $R_0 = e^{\int_S^T c(t)\,dt}(r_0 + 1)$. We observe that, given any interval $[t_0, t_1] \subset [S, T]$, for any F trajectory x on $[t_0, t_1]$ with $x(t_0) \in A \cap \left(e^{\int_S^{t_0} c(t)\,dt}(r_0 + 1) - 1\right)\mathbb{B}$, we have $x(t) \in R_0\mathbb{B}$ for all $t \in [t_0, t_1]$. As a consequence, if (H2)' or (H3)' are violated we can redefine $F(t, x)$ for $|x| > R_0$ (employing for instance the multifunction $F_{R_0}(t, x) := F(t, tr_{R_0}(x))$, where tr_{R_0} is the truncation function defined in (6.2.4)) so that (H2)' and (H3)' are satisfied together with condition (H1). □

End of the Proof of Theorem 6.6.1 Fix $r_0 > 0$. In view of the preliminary analysis we may assume that the multifunction F and set A in the theorem statement satisfy conditions (H1), (H2), (H3), (IPC) and (BVL) together with the additional hypotheses (H2)', (H3)' and (H4) with constant $c_0 > 0$ and function $k_F \in L^1(S, T)$. (The constant c_0 bounds the velocities of F trajectories x on subintervals of $[S, T]$ originating in $\left(e^{\int_S^{t_0} c(t)\,dt}(r_0+1)-1\right)\mathbb{B}$.) Let η be the modulus of variation appearing in (BVL) (and also in Lemma 6.6.3), and let $\epsilon > 0$ and $\bar{\eta} > 0$ be the constants from Lemma 6.6.3.

6.6 Estimates on Trajectories Confined to a Closed Subset

We know (see Lemma 6.6.3) that, given any $(t, x) \in [S, T] \times \big((\partial A + \bar{\eta}\mathbb{B}) \cap R_0\mathbb{B} \cap A\big)$, $v \in \operatorname{co} F(t, x)$ can be found such that

$$x' + [0, \epsilon](v + \epsilon\mathbb{B}) \subset A, \quad \text{for all } x' \in (x + \epsilon\mathbb{B}) \cap A. \tag{6.6.19}$$

By hypothesis (BVL)

$$F(s, x) \subset F(t, x) + (\eta(t) - \eta(s))\,\mathbb{B} \tag{6.6.20}$$

for all points $x \in (\partial A + \bar{\eta}\mathbb{B}) \cap R_0\mathbb{B}$ and subintervals $[s, t] \subset [S, T]$.

Let $\omega : [0, T - S] \to [0, \infty)$ be the function

$$\omega(\alpha) := \sup\{\int_I k_F(s)\,ds\}$$

where the supremum is taken over sub-intervals $I \subset [S, T]$ of length not greater than α. Since $k_F \in L^1(S, T)$, by properties of integrable functions, ω is well-defined on $[0, T - S]$, and $\omega(\alpha) \to 0$, as $\alpha \downarrow 0$.

Fix $k > 0$ such that $k > \epsilon^{-1}$ and take constants $\delta > 0$, $\bar{\rho} > 0$ and $\gamma > 0$ in such a manner that

$$\delta \leq \epsilon, \quad \bar{\rho} + c_0\delta < \epsilon, \quad k\bar{\rho} < \delta, \quad \bar{\rho} \leq \bar{\eta}, \quad 4\delta c_0 \leq \bar{\eta}, \tag{6.6.21}$$

and

$$e^{\omega(\delta)}(\gamma + \omega(\delta)c_0)(T - S) < \epsilon, \quad 2e^{\omega(\delta)}(\gamma + \omega(\delta)c_0)\,k < (k\epsilon - 1) \tag{6.6.22}$$

The assertions of the theorem will be confirmed provided that we establish the existence a constant $K > 0$ such that, for any interval $[t_0, t_1] \subset [S, T]$, given any F trajectory $\hat{x}$ on $[t_0, t_1]$ with $\hat{x}(t_0) \in A \cap \big(e^{\int_S^{t_0} c(t)\,dt}(r_0 + 1) - 1\big)\mathbb{B}$, and any $\rho > 0$ satisfying $\rho \geq \max\{d_A(\hat{x}(t)) : t \in [t_0, t_1]\}$, we can find an admissible F trajectory x on $[t_0, t_1]$ with $x(t_0) = \hat{x}(t_0)$ and such that

$$\|\hat{x} - x\|_{L^\infty(t_0, t_1)} \leq K\rho$$

and

$$x(t) \in \operatorname{int} A \quad \text{for } t \in (t_0, t_1].$$

Owing to the reduction Lemma 6.6.5, we can restrict attention, without loss of generality, to the case in which F is convex valued,

(i): $t_1 - t_0 \leq \delta$,
(ii): $\rho \leq \bar{\rho}$ and
(iii): $\eta(t_1) - \eta(t_0) \leq \gamma$ and $\eta(t_0) = 0$.

Observe that, if $\hat{x}(t_0) \in \left(A \cap \left(e^{\int_s^{t_0} c(t)\, dt}(r_0 + 1) - 1\right)\mathbb{B}\right) \setminus (\partial A + \frac{\bar{\eta}}{2}\mathbb{B})$, then the fifth condition in (6.6.21) implies that $x = \hat{x}$ is an admissible F trajectory having the required properties. Therefore, it is not restrictive to assume also that $\hat{x}(t_0) \in (\partial A + \frac{\bar{\eta}}{2}\mathbb{B}) \cap A \cap \left(e^{\int_s^{t_0} c(t)\, dt}(r_0 + 1) - 1\right)\mathbb{B}$. An immediate consequence (from the definition of R_0) is that

$$\hat{x}(t_0) \in (\partial A + \frac{\bar{\eta}}{2}\mathbb{B}) \cap R_0\mathbb{B} \cap A.$$

Then, recalling the fact that the multifunction F can be considered convex valued, there exists a vector $v \in F(t_0, \hat{x}(t_0))$ satisfying property (6.6.19) for $(t, x) = (t_0, \hat{x}(t_0))$. Now, consider the arc $y : [t_0, t_1] \to \mathbb{R}^n$ such that $y(t_0) = \hat{x}(t_0)$ and

$$\dot{y}(t) = \begin{cases} v & \text{if } t \in [t_0, (t_0 + k\rho) \wedge t_1] \\ \dot{\hat{x}}(t - k\rho) & \text{if } t \in (t_0 + k\rho, t_1] \text{ and if } \dot{\hat{x}}(t - k\rho) \text{ exists}. \end{cases}$$

If $t_0 + k\rho < t_1$, then it immediately follows that, for all $t \geq t_0 + k\rho$,

$$y(t) = \hat{x}(t - k\rho) + k\rho v. \tag{6.6.23}$$

Recalling that c_0 constitutes an upper bound for the magnitude for both v and $\|\dot{\hat{x}}\|_{L^\infty}$, we can deduce that

$$\|\hat{x} - y\|_{L^\infty(t_0, t_1)} \leq 2c_0 k\rho. \tag{6.6.24}$$

In addition, from (6.6.20) we can also deduce that, for all $s \in [t_0, (t_0 + k\rho) \wedge t_1]$,

$$d_{F(s,y(s))}(\dot{y}(s)) \leq d_{F(s,y(t_0))}(v) + k_F(s)|y(s) - y(t_0)|$$
$$\leq (\eta(s) - \eta(t_0)) + k_F(s)|y(s) - y(t_0)|$$
$$\leq \gamma + k_F(s)c_0(s - t_0). \tag{6.6.25}$$

Invoking Filippov's existence theorem (Theorem 6.2.3) and taking into account condition (6.6.25), we can find an F trajectory x on $[t_0, (t_0 + k\rho) \wedge t_1]$ with $x(t_0) = y(t_0)$ and such that, for any $t \in [t_0, (t_0 + k\rho) \wedge t_1]$

$$\|x - y\|_{L^\infty(t_0, t)} \leq e^{\omega(\delta)}(\gamma + \omega(\delta)c_0)(t - t_0). \tag{6.6.26}$$

Consider now the case in which $t_0 + k\rho < t_1$. Conditions (6.6.20), (6.6.21) and (6.6.23) imply that, for a.e. $s \in [t_0 + k\rho, t_1]$,

$$d_{F(s,y(s))}(\dot{y}(s)) = d_{F(s,k\rho v + \hat{x}(s - k\rho))}(\dot{\hat{x}}(s - k\rho))$$
$$\leq k_F(s)c_0 k\rho + d_{F(s,\hat{x}(s - k\rho))}(\dot{\hat{x}}(s - k\rho))$$
$$\leq k_F(s)c_0 k\rho + (\eta(s) - \eta(s - k\rho)) + d_{F(s - k\rho, \hat{x}(s - k\rho))}(\dot{\hat{x}}(s - k\rho))$$
$$= k_F(s)c_0 k\rho + (\eta(s) - \eta(s - k\rho)) + 0.$$

6.6 Estimates on Trajectories Confined to a Closed Subset

Then it follows that, for any $t \in [t_0 + k\rho, t_1]$,

$$\int_{t_0+k\rho}^t d_{F(s,y(s))}(\dot{y}(s))\, ds \leq (\omega(\delta)c_0 + \eta(t_1))k\rho \leq (\omega(\delta)c_0 + \gamma)k\rho.$$

Thus, the F trajectory x can be extended from $[t_0, (t_0 + k\rho) \wedge t_1]$ to $[t_0, t_1]$ by applying again Filippov's theorem, in such a manner that

$$\|x - y\|_{L^\infty(t_0, t_1)} \leq 2e^{\omega(\delta)}(\gamma + \omega(\delta)c_0)k\rho. \tag{6.6.27}$$

From (6.6.24) and (6.6.27) we deduce the required estimate:

$$\|\hat{x} - x\|_{L^\infty(t_0, t_1)} \leq K\rho$$

in which

$$K := 2\left(c_0 + e^{\omega(\delta)}(\gamma + \omega(\delta)c_0)\right)k.$$

To complete the proof we must show that

$$x(t) \in \text{int } A \quad \text{for } t \in (t_0, t_1].$$

There are two cases to consider (we are assuming that $t_0 + k\rho < t_1$, otherwise only the first case occurs):

Case 1: $t \in (t_0, t_0 + k\rho]$. Since $y(t) = \hat{x}(t_0) + (t - t_0)v$ and $t - t_0 \leq \epsilon$, it follows from (6.6.19) that

$$y(t) + (t - t_0)\epsilon\mathbb{B} = \hat{x}(t_0) + (t - t_0)(v + \epsilon\mathbb{B}) \subset A.$$

Then, conditions (6.6.26) and (6.6.22) immediately yield $x(t) \in \text{int } A$, for all $t \in (t_0, t_0 + k\rho]$.

Case 2: $t \in (t_0 + k\rho, t_1]$. Write $z(t)$ a projection on A of the arc $t \to \hat{x}(t - k\rho)$. It means that, for each $t \in (t_0 + k\rho, t_1]$, we select $z(t) \in A$ such that

$$|\hat{x}(t - k\rho) - z(t)| = d_A(\hat{x}(t - k\rho)) \ (\leq \rho).$$

As a consequence, invoking (6.6.23), we obtain

$$y(t) \in z(t) + k\rho v + \rho\mathbb{B}, \tag{6.6.28}$$

and, since $|\hat{x}(t - k\rho) - \hat{x}(t_0)| \leq c_0(t_1 - t_0)$ for all $t \in (t_0 + k\rho, t_1]$, appealing once again to (6.6.21), we also have

$$|z(t) - \hat{x}(t_0)| \leq \bar{\rho} + c_0\delta < \epsilon.$$

Thus bearing in mind (6.6.19) and (6.6.21), we see that

$$z(t) + k\rho v + k\rho \epsilon \mathbb{B} \subset A,$$

and, owing to (6.6.28),

$$y(t) + (k\epsilon - 1)\rho \mathbb{B} \subset A.$$

Taking into account (6.6.22) and (6.6.27), we deduce that $x(t) \in \text{int } A$ in this case as well, confirming all the assertions of the theorem. □

If the data of the state-constrained differential inclusion (6.6.1) satisfy condition (IPC), then, in consequence of 'stability' properties of the interior of Clarke tangent cone, Lemma 6.6.3 guarantees the existence of vectors in co $F(t, x)$ pointing uniformly inward A, whenever x is close to the boundary of A. The crucial step in the proof of Theorem 6.6.2 is to show that, if we impose just condition (IPC)', which is weaker than condition (IPC), still we can obtain implications similar to those ones stated in Lemma 6.6.3. The next lemma ensures that the required properties are valid if, in addition, we are allowed to make use of condition $(BV)_A$.

Lemma 6.6.6 *Suppose that the multifunction $F : [S, T] \times \mathbb{R}^n \rightsquigarrow \mathbb{R}^n$ and the closed set A satisfy hypotheses (H1), (H2), (H3), (IPC)' and $(BV)_A$ for some $R_0 > 0$. Then there exist $\epsilon > 0$, $\bar{\eta} \in (0, \eta_0)$ and a finite time set $\{\tau_j\}_{j \in J} \subset [S, T]$ with the following property: for any $(t, x) \in [S, T] \times \left((\partial A + \bar{\eta} \mathbb{B}) \cap R_0 \mathbb{B} \cap A\right)$, we can find*

$$v \in \begin{cases} \operatorname{co} F(t, x) & \text{if } t \notin \{\tau_j\}_{j \in J} \\ \left(\limsup_{\tau \downarrow t} \operatorname{co} F(\tau, x)\right) & \text{if } t \in \{\tau_j\}_{j \in J}, \end{cases}$$

such that

$$y + [0, \epsilon](v + \epsilon \mathbb{B}) \subset A, \quad \text{for all } y \in (x + \epsilon \mathbb{B}) \cap A. \tag{6.6.29}$$

Proof Fix any $(t, x) \in [S, T] \times (R_0 \mathbb{B} \cap \partial A)$. From (IPC)' we can find vectors

$$v_1 \in \left(\liminf_{x' \xrightarrow{A} x} \limsup_{t' \downarrow t} \operatorname{co} F(t', x')\right) \cap \text{int } \bar{T}_A(x),$$

and

$$v_2 \in \left(\liminf_{x' \xrightarrow{A} x} \limsup_{t' \uparrow t} \operatorname{co} F(t', x')\right) \cap \text{int } \bar{T}_A(x).$$

In view of the characterization of the interior of the Clarke tangent cone (see Theorem 4.11.1), there exist $\gamma \in (0, 1/2)$ and $r \in (0, \eta_0/4)$ such that

6.6 Estimates on Trajectories Confined to a Closed Subset

$$y + [0, \gamma](v_1 + 2\gamma \mathbb{B}) \subset A, \quad \text{for all } y \in (x + 2r\mathbb{B}) \cap A \tag{6.6.30}$$

and

$$y + [0, \gamma](v_2 + 2\gamma \mathbb{B}) \subset A, \quad \text{for all } y \in (x + 2r\mathbb{B}) \cap A. \tag{6.6.31}$$

From the definition of the lim inf operator (in the sense of Kuratowski), we can deduce the existence of $\bar{r} \in (0, r)$ such that

$$\sup_{x' \in (x+\bar{r}\mathbb{B}) \cap A} \liminf_{t' \downarrow t} d_{\text{co } F(t',x')}(v_1) < \frac{\gamma}{2} \tag{6.6.32}$$

and

$$\sup_{x' \in (x+\bar{r}\mathbb{B}) \cap A} \liminf_{t' \uparrow t} d_{\text{co } F(t',x')}(v_2) < \frac{\gamma}{2}. \tag{6.6.33}$$

Taking into account assumption $(BV)_A$ and the same argument employed in the proof of Lemma 6.6.3, we denote η the modulus of variation of F, which can be decomposed into a sum of a continuous function η^c and a singular (discontinuous) component. Write $\{\sigma_i\}$ and $\{a_i\}$ respectively the sequence of the countable distinct jump times and the sequence of the (countable) non-negative jumps of η in $[S, T]$. We can find numbers $\delta > 0$ and $\bar{\alpha} > 0$ such that

$$\eta^c(t') - \eta^c(s') < \gamma/2 \tag{6.6.34}$$

for each subinterval $[s', t'] \subset [S, T]$ with $t' - s' \leq \delta$, and

$$\sum_{i \notin J} a_i < \gamma/2, \tag{6.6.35}$$

where $J \subset \{1, 2, \ldots\}$ is a finite index set for which

$$|\sigma_i - \sigma_j| > \bar{\alpha} \text{ for } i, j \in J, \ i \neq j.$$

Take $\delta_t \in (0, \delta)$ such that $([t - \delta_t, t + \delta_t] \setminus \{t\}) \cap \{\sigma_i : i \in J\} = \emptyset$. Define $\tilde{F}_1 : [t - \delta_t, t] \times \mathbb{R}^n \rightsquigarrow \mathbb{R}^n$ according to (6.6.7) and $\tilde{\eta}_1$ according to (6.6.9) when $[a, b] = [t - \delta_t, t]$. Similarly, define $\tilde{F}_2 : [t, t + \delta_t] \times \mathbb{R}^n \rightsquigarrow \mathbb{R}^n$ according to (6.6.7) and $\tilde{\eta}_2$ according to (6.6.9) when $[a, b] = [t, t + \delta_t]$. $\tilde{F}_1$ and $\tilde{F}_2$ satisfy the hypotheses (H1), (H2)', (H3)', $(BV)_A$ of F, restricted respectively to $[t - \delta_t, t] \times \mathbb{R}^n$ and $[t, t + \delta_t] \times \mathbb{R}^n$, with the same constants $c_0 > 0$ and $\eta_0 > 0$, and function k_F. Moreover, the assertions of Lemma 6.6.4 hold true when $\tilde{F}_1$ and $\tilde{F}_2$ replace $\tilde{F}$. Then, from (6.6.10), for all $s' \in [t - \delta_t, t]$, we have

$$\sup_{x \in (\partial A + \eta_0 \mathbb{B}) \cap R_0 \mathbb{B}} d_H(\tilde{F}_1(s', x), \tilde{F}_1(t, x)) \leq \tilde{\eta}_1(t) - \tilde{\eta}_1(s') \tag{6.6.36}$$

and for all $t' \in [t, t + \delta_t]$, we have

$$\sup_{x \in (\partial A + \eta_0 \mathbb{B}) \cap R_0 \mathbb{B}} d_H(\tilde{F}_2(t, x), \tilde{F}_2(t', x)) \leq \tilde{\eta}_2(t') - \tilde{\eta}_2(t) . \tag{6.6.37}$$

Also conditions (6.6.32) and (6.6.33) can be rephrased in terms of $\tilde{F}_1$ and $\tilde{F}_2$: for all $x' \in (x + \bar{r}\mathbb{B}) \cap A$

$$(v_1 + \frac{\gamma}{2}\mathbb{B}) \cap \mathrm{co}\, \tilde{F}_1(t, x') \neq \emptyset \quad \text{and} \quad (v_2 + \frac{\gamma}{2}\mathbb{B}) \cap \mathrm{co}\, \tilde{F}_2(t, x') \neq \emptyset .$$

Thus, for all $x' \in (x + \bar{r}\mathbb{B}) \cap A$ and for all $s' \in [t - \delta_t, t)$, there exists $v' \in \mathrm{co}\, \tilde{F}_1(s', x')$ such that

$$|v_1 - v'| \leq \frac{\gamma}{2} + \tilde{\eta}_1(t) - \tilde{\eta}_1(s') \leq \frac{3}{2}\gamma ,$$

where, taking account of (6.6.34) and (6.6.35), we have used the following fact:

$$\tilde{\eta}_1(t) - \tilde{\eta}_1(s') \leq \eta(t^-) - \eta(s'^+) < \gamma .$$

Similarly, for all $x' \in (x + \bar{r}\mathbb{B}) \cap A$ and for all $t' \in (t, t + \delta_t]$, there exists $w' \in \mathrm{co}\, \tilde{F}_2(t', x')$ such that

$$|v_2 - w'| \leq \frac{\gamma}{2} + \tilde{\eta}_2(t) - \tilde{\eta}_2(t') \leq \frac{3}{2}\gamma .$$

(Observe that in the relations above, $\tilde{F}_1(s', x') = F(s', x')$ for all $s' \in (t - \delta_t, t)$, and $\tilde{F}_2(t', x') = F(t', x')$ for all $t' \in (t, t + \delta_t)$.)
Therefore, for all $t' \in (t - \delta_t, t + \delta_t)$, $t' \neq t$, and for all $x' \in (x + \bar{r}\mathbb{B}) \cap A$, we can find $v' \in \mathrm{co}\, F(t', x')$ such that

$$y + [0, \gamma](v' + \frac{\gamma}{2}\mathbb{B}) \subset A , \quad \text{for all } y \in (x' + \bar{r}\mathbb{B}) \cap A .$$

And, if $t' = t$, for all $x' \in (x + \bar{r}\mathbb{B}) \cap A$, there exists $\tilde{v} \in \mathrm{co}\, \tilde{F}_2(t, x')$ satisfying

$$y + [0, \gamma](\tilde{v} + \frac{\gamma}{2}\mathbb{B}) \subset A , \quad \text{for all } y \in (x' + \bar{r}\mathbb{B}) \cap A .$$

Set $\varepsilon_{t,x} := \min\{\delta_t, \bar{r}, \gamma/2\}$. Then we have proved the following property: for any $(t, x) \in [S, T] \times (R_0 \mathbb{B} \cap \partial A)$ there exists $\varepsilon_{t,x} > 0$ such that for all $(t', x') \in [S, T] \times A$ with $|x - x'| < \varepsilon_{t,x}$ and $|t - t'| < \varepsilon_{t,x}$, we can find

$$v' \in \begin{cases} \mathrm{co}\, F(t', x') & \text{if } t' \neq t \\ \mathrm{co}\, \tilde{F}_2(t', x') & \text{if } t' = t , \end{cases}$$

6.6 Estimates on Trajectories Confined to a Closed Subset

satisfying

$$y + [0, \varepsilon_{t,x}](v' + \varepsilon_{t,x}\mathbb{B}) \subset A, \quad \text{for all } y \in (x' + \varepsilon_{t,x}\mathbb{B}) \cap A.$$

By a standard compactness argument, we can find a finite number of points $(t_i, x_i) \in [S, T] \times (R_0\mathbb{B} \cap \partial A)$ and numbers $\epsilon_i > 0$, for $i = 1, \ldots, m$, such that

$$\bigcup_{i=1,\ldots,m} \left((t_i, x_i) + \epsilon_i \overset{\circ}{\mathbb{B}}\right) \supset [S, T] \times (R_0\mathbb{B} \cap \partial A) \tag{6.6.38}$$

and for each $(t', x') \in ((t_i, x_i) + \epsilon_i \mathbb{B}) \cap ([S, T] \times A)$, there exists a vector

$$v' \in \begin{cases} \operatorname{co} F(t', x') & \text{if } t' \neq t \\ \operatorname{co} \tilde{F}_2(t', x') & \text{if } t' = t, \end{cases}$$

such that

$$y' + [0, \epsilon_i](v' + \epsilon_i \mathbb{B}) \subset A, \quad \text{for all } y' \in (x' + \epsilon_i \mathbb{B}) \cap A.$$

Notice also that there exists $\bar{\eta} \in (0, \min_{i=1,\ldots,m} \epsilon_i)$ (observe that $\bar{\eta} < \eta_0$) such that

$$\bigcup_{i=1,\ldots,m} \left((t_i, x_i) + \epsilon_i \overset{\circ}{\mathbb{B}}\right) \supset [S, T] \times \left((\partial A + \bar{\eta}\mathbb{B}) \cap R_0\mathbb{B}\right),$$

otherwise we could find a sequence of points $(s_j, y_j) \in ([S, T] \times R_0\mathbb{B}) \setminus \bigcup_{i=1,\ldots,m} \left((t_i, x_i) + \epsilon_i \overset{\circ}{\mathbb{B}}\right)$ such that $(s_j, y_j) \to (s, y) \in [S, T] \times (R_0\mathbb{B} \cap \partial A)$, which would contradict (6.6.38).

To conclude we just take $\epsilon := \min_{i=1,\ldots,m} \epsilon_i$ and the assertions of the lemma immediately follow. □

Proof of Theorem 6.6.2 In view of Lemma 6.6.6, we can define the following multifunction

$$\widehat{F}(t, x) := \begin{cases} F(t, x) & \text{if } t \notin \{\tau_j\}_{j \in J} \\ \limsup_{\tau \downarrow t} F(\tau, x) & \text{if } t \in \{\tau_j\}_{j \in J}. \end{cases}$$

Then, the approach used in the proof of Theorem 6.6.1 is applicable to the state-constrained differential inclusion (6.6.1) when $\widehat{F}$ replaces F. Since an $\widehat{F}$ trajectory is also an F trajectory and vice-versa, we obtain the validity of the required properties. □

6.7 Exercises

6.1 (Intermediate Minimizers) Take $1 \le s \le \infty$. An admissible F trajectory $\bar{x}$ for a dynamic optimization problem (with cost J, underlying time domain $[S, T]$ and differential inclusion constraint $\dot{x} \in F(t, x)$) is said to be a $W^{1,s}$ local minimizer if there exists $\epsilon > 0$ such that, for all admissible F trajectories such that $\|x - \bar{x}\|_{W^{1,s}} \le \epsilon$, we have $J(\bar{x}) \le J(x)$.

For $r \in (1, \infty)$ set

$$\phi_r(t) := t^{1-1/r}.$$

Consider

$$(P)_r \begin{cases} \text{Minimize } J(x) := x_1(1) \\ \text{subject to} \\ (\dot{x}_1, \dot{x}_2, \dot{x}_3)(t) \in F(t, x(t)) := \{(0, |x_1(t)\phi_r(t) + x_3(t)|, e) \\ \quad : e \in [-\dot{\phi}_r(t), \dot{\phi}_r(t)]\}, \text{ a.e. } t \in [0, 1], \\ (x(0), x(1)) \in (\mathbb{R} \times \{0\} \times \mathbb{R})^2. \end{cases}$$

Notice that, for any $r \in (1, \infty)$, the data for $(P)_r$ satisfy the standard convexity, integrable boundedness and integrable Lipschitz continuity hypotheses invoked in Chap. 6 and subsequent chapters.

Consider the admissible F trajectory $\bar{x} \equiv 0$. Take any $r \in (1, \infty)$ and s and s' such that

$$1 \le s < r \le s' \le \infty.$$

Show that $\bar{x}$ is a $W^{1,s'}$ local minimizer but not a $W^{1,s}$ local minimizer.

6.2 (Infinite Horizon Problems) Consider the following dynamic optimization problem with discounted running cost over an infinite horizon:

$$(P_\infty) \begin{cases} \text{Minimize } \int_0^\infty e^{-\lambda t} L(x(t), u(t)) dt \\ \text{over } x \in W^{1,1}_{\text{loc}}([0, \infty); \mathbb{R}^n) \\ \text{and measurable functions } u : [0, \infty) \to \mathbb{R}^m \text{ satisfying} \\ \dot{y}(t) = f(y(t), u(t)), \text{ a.e. } t \in [0, \infty), \\ u(t) \in U, \text{ a.e. } t \in [0, \infty), \\ y(0) = x_0, \end{cases}$$

in which $f : \mathbb{R}^n \times \mathbb{R}^m \to \mathbb{R}^n$ and $L : \mathbb{R}^n \times \mathbb{R}^m \to \mathbb{R}$ are given functions, $U \subset \mathbb{R}^m$ is a given set, $\lambda > 0$ and $x_0 \in \mathbb{R}^n$. Assume

(H1): $\lambda > 0$, f is continuous, U is compact and

$$\{(v, \alpha) \in \mathbb{R}^n \times \mathbb{R} : v \in f(x, u), \alpha \geq L(x, u)\} \text{ is convex for each } x \in \mathbb{R}^n,$$

(H2): there exist positive constants k_f and c_f such that

$$|f(x, u) - f(y, u)| \leq k_f |x - y| \text{ and } |f(x, u)| \leq c_f(1 + |x|)$$

$$\text{for all } x, y \in \mathbb{R}^n \text{ and } u \in \mathbb{R}^m,$$

(H3): there exist positive constants k_L and c_L such that

$$|L(x, u) - L(y, u)| \leq k_L |x - y| \text{ and } |L(x, u)| \leq c_L$$

$$\text{for all } x, y \in \mathbb{R}^n \text{ and } u \in \mathbb{R}^m.$$

Show that (P_∞) has a minimizer (in $W^{1,1}_{\text{loc}}$).

Remark
The object of this exercise is to complete some of the details of the proof of Proposition 13.10.1 part (i).

6.3 (Proximal Gronwall's Inequality) Let $f : [S, T] \to \mathbb{R}$ be lower semi-continuous function. Assume that there exists $\beta \geq 0$ such that

$$|\xi| \leq \beta f(s) \quad \text{for all } \xi \in \partial^P f(s) \text{ and } s \in [S, T].$$

(Here, the definition of $\partial^P f(s)$ when $s = S$ or T can be provided extending f to $\mathbb{R}$ by setting values to $+\infty$ outside $[S, T]$.) Show that

(i): f is Lipschitz continuous on $[S, T]$ and, if $\beta > 0$, then $f \geq 0$,
(ii): $f(t) \leq f(S)e^{\beta(t-S)}$, for all $t \in [S, T]$,
(iii): $|f(t') - f(t)| \leq f(t)\left(e^{\beta|t'-t|} - 1\right)$, for all $t, t' \in [S, T]$,
(iv): there exits a constant $\alpha \leq (e^{\beta(T-S)} + 1)\dfrac{e^{\beta(T-S)} - 1}{T - S}$ such that

$$|f(t_2) - f(t_1)| \leq \alpha f(t)|t_2 - t_1| \quad \forall t, t_1, t_2 \in [S, T].$$

6.8 Notes for Chapter 6

There is a substantial literature on differential inclusions. Material in Sects. 6.2–6.4, which restricts attention largely to differential inclusions involving multifunctions which are Lipschitz continuous with respect to the state, only scratches the surface of this extensive field. (See [11] and [93].) The important existence theorem Theorem 6.2.3 and accompanying estimates are essentially due to Filippov [102]. The proof given here is an adaptation of that in [11], to allow for measurable time dependence. The compactness of trajectories theorem Theorem 6.3.3 is a refinement of that in [65]. Implicit in this theorem are early ideas for establishing existence of optimal controls under a convexity hypothesis on the 'velocity set' associated with Tonelli, L. C. Young, Filippov, Gamkrelidze, Roxin and others.

The concept of relaxation has its origins in the work of L. C. Young (see references in [209]), where it provided a framework for expanding the domain of problems in the calculus of variations to ensure 'closure' properties of families of trajectories, and thereby guarantee existence of minimizers. Relaxation for dynamic optimization associated with controlled differential equations was introduced by Warga [204] and by Gamkrelidze [118]. The related closure properties of families of solutions to a differential inclusion, which are the subject matter of Sect. 6.5 are a by-product of the theory of set valued integrals [15] and, specifically, relations between the integral of a set valued functions and the integral of its convex hull. More recently, the study of relaxation schemes has broadened out to apply to other classes of optimization problems, for example variational problems in several independent variables [7] and dynamic optimization problems with time delay [178].

Distance estimates for state constrained dynamic systems, of the kind covered in Sect. 6.6 originate in work of Soner [182], who proved an L^∞ linear estimate for autonomous controlled differential equations and a state constraint with smooth boundary, under an 'inward pointing' condition. Subsequent research has provided extensions to allow for control systems described by time-varying differential inclusions and state constraint set take to be arbitrary closed sets. Theorems 6.6.1 and 6.6.2 are a refinement of [30, Thms. 1 and 2] (and of [33]). There has been interest, too, in $W^{1,1}$ distance estimates and applications. When we replace the L^∞ norm by the stronger $W^{1,1}$ norm, it is still possible to derive linear distance estimates under inward pointing conditions, but only if we impose additional hypotheses, concerning the time dependence of the differential inclusion $F(t, x)$, the regularity of the boundary of the state constraint set or the precise nature of the inward pointing condition (cf. [32, 109]). It is possible, however, to derive super-linear $W^{1,1}$ distance estimates under weaker hypotheses (see [47] and [23]).

Distance estimates have been widely employed in the analysis of state constrained dynamic optimization problems: to ensure the validity of necessary optimality conditions in normal or nondegenerate form (cf. [27, 36, 170]), to establish regularity properties of the value function as well as to justify interpreting the value function as a unique (generalized) solution of the Hamilton-Jacobi equation (cf. [23, 28, 108, 110, 111]), to derive sensitivity relations (cf. [35]), and as a key element in other regularity investigations (see, e.g., [86]). Some constructions (employed in the derivation of distance estimates) can be refined in such a manner they are also 'nonanticipative', a property that is crucial in the differential games context (cf. [37]).

Chapter 7
The Maximum Principle

Abstract The classical maximum principle was introduced in Chap. 1. This fundamental theorem is a set of necessary conditions of optimality for dynamic optimization problems with endpoint cost and endpoint constraints, in which the dynamic constraint takes the form of a controlled differential equation. It unifies earlier optimality conditions in the calculus of variations and extends them to take account of the dynamic constraint.

This chapter provides a generalization of these conditions, known as the Clarke nonsmooth maximum principle, that covers problems in which the endpoint constraint and cost are expressed in terms of nonsmooth functions and general closed sets, and the right side of the controlled differential equation is nonsmooth w.r.t. the state variable. Special cases of interest and simple extensions are discussed in detail.

The derivation of Clarke's nonsmooth principle appearing in this chapter is based on his perturbation technique. The necessary conditions in their full generality are arrived at in stages, starting with necessary conditions for a simple problem with a smooth endpoint cost function and no endpoint constraints. Subsequent stages introduce refinements (additional constraints, nonsmoothness, removal of temporary simplifying hypotheses). The key idea is that, at each stage, when we seek necessary conditions for a newly refined version of the problem, we construct a minimizer to a perturbed problem, which is simpler and for which necessary conditions are available from the previous stage. Necessary conditions for the new problem are then obtained in the limit, from the necessary conditions for the perturbed problems. Construction of the perturbed problems is accomplished by various techniques, which include application of Ekeland's theorem and the use of quadratic inf convolutions. Limit taking of the necessary conditions is carried out with the help of subdifferential calculus.

7.1 Introduction

In this chapter we derive a set of optimality conditions for dynamic optimization problems, known as the maximum principle. Many competing sets of optimality conditions are now available, but the maximum principle retains a special significance. An early version of the maximum principle due to Pontryagin et al. was after all the breakthrough marking the emergence of dynamic optimization as a distinct field of research. Also, whatever additional information about minimizers is provided by dynamic programming, higher order conditions and the analysis of the geometry of state trajectories, first order necessary conditions akin to the maximum principle remain the principal vehicles for the solution of specific dynamic optimization problems (either directly or indirectly via the computational procedures they inspire), or at least for generating 'suspects' for their solution.

A number of first order necessary conditions will be derived in this book. We attend to the maximum principle at the outset, because it was the first set of necessary conditions to handle in a satisfactory way constraints typically encountered in practical dynamic optimization problems.

The dynamic optimization problem studied here is

$$(P) \begin{cases} \text{Minimize } g(x(S), x(T)) \\ \text{over } x \in W^{1,1}([S, T]; \mathbb{R}^n) \text{ and measurable } u \text{ satisfying} \\ \dot{x}(t) = f(t, x(t), u(t)) \text{ and } u(t) \in U(t) \text{ a.e. } t \in [S, T], \\ (x(S), x(T)) \in C, \end{cases}$$

the data for which comprise an interval $[S, T]$, functions $g : \mathbb{R}^n \times \mathbb{R}^n \to \mathbb{R}$, $f : [S, T] \times \mathbb{R}^n \times \mathbb{R}^m \to \mathbb{R}^n$, a non-empty multifunction $U : [S, T] \rightsquigarrow \mathbb{R}^m$ and a closed set $C \subset \mathbb{R}^n \times \mathbb{R}^n$.

A measurable function $u : [S, T] \to \mathbb{R}^m$ which satisfies

$$u(t) \in U(t) \quad \text{a.e.} \quad t \in [S, T]$$

is called a *control function*. The set of all control functions is written $\mathcal{U}$.

A *process* (x, u) comprises a function $x \in W^{1,1}([S, T]; \mathbb{R}^n)$ and a control function u such that x is a solution to the differential equation

$$\dot{x}(t) = f(t, x(t), u(t)) \quad \text{a.e.} \quad t \in [S, T].$$

A *state trajectory* x is the first component of some process (x, u). A process (x, u) is said to be *admissible* for (P) if the state trajectory x satisfies the end-point constraint

$$(x(S), x(T)) \in C.$$

The maximum principle and related necessary conditions of optimality are satisfied by all minimizers. But, as we shall see, these conditions are satisfied also by processes which are merely 'local' minimizers for (P). Different choices of topology on the set of processes give rise to different notions of local minimizer. The one adopted in this chapter is that of $W^{1,1}$ local minimizer, though we make reference in the proofs also to L^∞ local minimizers.

A admissible process $(\bar{x}, \bar{u})$ is

(a): a $W^{1,1}$ *local minimizer* if there exists $\epsilon > 0$ such that

$$g(x(S)), x(T)) \geq g(\bar{x}(S), \bar{x}(T)),$$

for all admissible processes (x, u) which satisfy $\|x - \bar{x}\|_{W^{1,1}} \leq \epsilon$.

(b): an L^∞ *local minimizer* if there exists $\epsilon > 0$ such that

$$g(x(S)), x(T)) \geq g(\bar{x}(S), \bar{x}(T)),$$

for all admissible processes (x, u) which satisfy $\|x - \bar{x}\|_{L^\infty} \leq \epsilon$.

The point in showing that the optimality conditions are valid for local minimizers is to focus attention on the limitations of these conditions: they may lead us to a local minimizer in place of a global minimizer of primary interest. The $W^{1,1}$ norm is stronger than the L^∞ norm and therefore the class of $W^{1,1}$ local minimizers is larger than the class of L^∞ local minimizers. It follows that, by choosing to work with $W^{1,1}$ local minimizers, we are carrying out a more precise analysis of the local nature of the maximum principle than would be the case if we chose to derive conditions satisfied by L^∞ local minimizers.

7.2 Clarke's Nonsmooth Maximum Principle

We denote by $\mathcal{H} : [S, T] \times \mathbb{R}^n \times \mathbb{R}^n \times \mathbb{R}^m \to \mathbb{R}$ the un-maximized Hamiltonian function, namely

$$\mathcal{H}(t, x, p, u) := p \cdot f(t, x, u), \quad \text{for } (t, x, p, u) \in [S, T] \times \mathbb{R}^n \times \mathbb{R}^n \times \mathbb{R}^m.$$

Theorem 7.2.1 (Clarke, 1976) *Let $(\bar{x}, \bar{u})$ be a $W^{1,1}$ local minimizer for (P). Assume that, for some $\bar{\epsilon} > 0$, the following hypotheses are satisfied:*

(H1) *for fixed x, $f(., x, .)$ is $\mathcal{L} \times \mathcal{B}^m$ measurable; there exists an $\mathcal{L} \times \mathcal{B}^m$ measurable function $k : [S, T] \times \mathbb{R}^m \to \mathbb{R}_+$ such that $t \to k(t, \bar{u}(t))$ is integrable and, for a.e. $t \in [S, T]$,*

$$|f(t, x, u) - f(t, x', u)| \leq k(t, u)|x - x'|$$

for all $x, x' \in \bar{x}(t) + \bar{\epsilon}\mathbb{B}$ and $u \in U(t)$,

(H2) *Gr U is an $\mathcal{L} \times \mathcal{B}^m$ measurable set,*

(H3) *g is Lipschitz continuous on $(\bar{x}(S), \bar{x}(T)) + \bar{\epsilon}\mathbb{B}$.*

Then there exist $p \in W^{1,1}([S, T]; \mathbb{R}^n)$ and $\lambda \geq 0$ such that

(a): (non-triviality condition)
$$(p, \lambda) \neq (0, 0),$$
(b): (co-state inclusion)
$$-\dot{p}(t) \in \operatorname{co} \partial_x \mathcal{H}(t, \bar{x}(t), p(t), \bar{u}(t)) \quad \text{a.e. } t \subset [S, T],$$
(c): (Weierstrass condition)
$$\mathcal{H}(t, \bar{x}(t), p(t), \bar{u}(t)) = \max_{u \in U(t)} \mathcal{H}(t, \bar{x}(t), p(t), u) \quad \text{a.e. } t \in [S, T],$$
(d): (transversality condition)
$$(p(S), -p(T)) \in \lambda \partial g(\bar{x}(S), \bar{x}(T)) + N_C(\bar{x}(S), \bar{x}(T)).$$

Now assume, also, that

$$f(t, x, u) \text{ and } U(t) \text{ are independent of } t.$$

Then, in addition to the above conditions, there exists a constant r such that

(e): $\mathcal{H}(t, \bar{x}(t), p(t), \bar{u}(t)) = r \quad \text{a.e. } t \in [S, T].$

($\partial_x \mathcal{H}$ denotes the limiting subdifferential of $\mathcal{H}(t, \cdot, p, u)$ for fixed (t, p, u).)

The proof of the theorem is deferred to a later section.

Elements (p, λ) whose existence is asserted in the maximum principle are called *Lagrange multipliers* for (P). The components p and λ are referred to as the *co-state trajectory* and *cost multiplier* respectively.

Remarks

(a): The co-state inclusion (condition (ii) in the theorem statement) can be stated in terms of the Clarke *generalized Jacobian* introduced in Exercise 5.2, that we recall here:
take a point $y \in \mathbb{R}^n$ and a function $\phi : \mathbb{R}^n \to \mathbb{R}^m$ which is Lipschitz continuous on a neighbourhood of y. Then the *Clarke generalized Jacobian* $J^C \phi(y)$ of ϕ at y is the set of $m \times n$ matrices:

$$J^C \phi(y) := \operatorname{co}\{M \in \mathbb{R}^{m \times n} : \exists\, y_i \to y \text{ such that}$$
$$\nabla \phi(y_i) \text{ exists } \forall i \text{ and } \nabla \phi(y_i) \to M\}.$$

A noteworthy property of the generalized Jacobian $J^C \phi(y)$ of function $\phi : \mathbb{R}^n \to \mathbb{R}^m$, which is Lipschitz continuous on a neighbourhood of a point y, is that, for any vector $r \in \mathbb{R}^m$,

$$r \cdot J^C \phi(y) = \operatorname{co} \partial(r \cdot \phi)(y)$$

(see Exercise 5.2). Here, $\partial(r \cdot \phi)(y)$ is the limiting subdifferential of the function $y \to r \cdot \phi(y)$. It follows immediately that the adjoint inclusion can be equivalently written

7.2 Clarke's Nonsmooth Maximum Principle

$$-\dot{p}(t) \in p(t) \cdot J_x^C f(t, \bar{x}(t), \bar{u}(t)),$$

in which $J_x^C f(t, x, u)$ denotes the generalized Jacobian with respect to the x variable. This alternative, but equivalent, statement of the costate inclusion is sometimes encountered in the dynamic optimization literature.

(b): We observe finally that the necessary conditions of Theorem 7.2.1 are homogeneous with respect to the Lagrange multipliers (p, λ). This means that if (p, λ) serves as a set of Lagrange multipliers then, for any $\alpha > 0$, $(\alpha p, \alpha \lambda)$ also serves. Since $(p, \lambda) \neq 0$, we can always arrange by choosing an appropriate α that

$$\|p\|_{L^\infty} + \lambda = 1.$$

Scaling Lagrange multipliers in this way is often carried out to assist convergence analysis.

The above maximum principle is easily adapted to cover problems with the more general formulation:

$$(PI) \begin{cases} \text{Minimize } g(x(S), x(T)) + \int_S^T L(t, x(t), u(t))dt \\ \text{over } x \in W^{1,1}([S, T]; \mathbb{R}^n) \text{ and} \\ \qquad \text{measurable functions } u : [S, T] \to \mathbb{R}^m \text{ satisfying} \\ \dot{x}(t) = f(t, x(t), u(t)) \quad \text{a.e.,} \\ u(t) \in U(t) \quad \text{a.e.,} \\ (x(S), x(T)) \in C, \end{cases}$$

in which an integral term $+ \int_S^T L(t, x(t), u(t))dt$ has been added to the cost. This is accomplished by introducing a new state variable z which satisfies the relations

$$\begin{cases} \dot{z}(t) = L(t, x(t), u(t)) \\ z(S) = 0. \end{cases}$$

The 'mixed' cost can then be replaced by the pure end-point cost

$$z(T) + g(x(S), x(T)).$$

In this way we arrive at problem to which Theorem 7.2.1 is applicable. (Of course we have to make suitable additional assumptions about the function L to ensure the hypotheses of Theorem 7.2.1 are satisfied.) Applying the theorem and interpreting the conclusions in terms of the data for the original problem gives the following maximum principle for problem (PI):

Corollary 7.2.2 (The Maximum Principle for End-Point and Integral Cost)
Let $(\bar{x}, \bar{u})$ be a $W^{1,1}$ local minimizer for (PI). Assume that (H1) and (H2) of Theorem 7.2.1 and, in place of (H3), the hypothesis

(H3̄): hypothesis (H3) of Theorem 7.2.1 is satisfied when the $\mathbb{R}^{1+n}$-valued function (L, f) is inserted in place of f.

Then the assertions of Theorem 7.2.1 are valid, when the function $\mathcal{H}$ is replaced by

$$\mathcal{H}_\lambda(t, x, p, u) := p \cdot f(t, x, u) - \lambda L(t, x, u).$$

The procedure employed above, for reducing a problem with integral and endpoint cost terms to one with an endpoint cost term alone, is widely used. It is called *state augmentation*.

Examples of the following dynamic optimization problem, in which the endpoint constraints are specified as functional constraints, are often encountered.

$$(FEC) \begin{cases} \text{Minimize } g(x(S), x(T)) \\ \text{over } x \in W^{1,1}([S, T]; \mathbb{R}^n) \text{ and} \\ \qquad \text{measurable functions } u : [S, T] \to \mathbb{R}^m \text{ satisfying} \\ \dot{x}(t) = f(t, x(t), u(t)) \quad \text{a.e.,} \\ u(t) \in U(t) \quad \text{a.e.,} \\ \phi_i(x(S), x(T)) \le 0 \text{ for } i = 1, 2, .., k_1, \\ \psi_i(x(S), x(T)) = 0 \text{ for } i = 1, 2, .., k_2. \end{cases}$$

The new ingredients here are the constraint functions $\phi_i : \mathbb{R}^n \times \mathbb{R}^n \to \mathbb{R}$, $i = 1, .., k_1$ and $\psi_i : \mathbb{R}^n \times \mathbb{R}^n \to \mathbb{R}$, $i = 1, ..., k_2$. We permit the cases $k_1 = 0$ (no inequality constraints) and $k_2 = 0$ (no equality constraints).

Of course the necessary conditions of Theorem 7.2.1 apply to the special case of (P), in which the endpoint constraint set is chosen to be

$$C = \{(x_0, x_1) : \phi_i(x_0, x_1) \le 0 \text{ for } i = 1, \ldots, k_1$$
$$\text{and } \psi_i(x_0, x_1) = 0 \text{ for } i = 1 \ldots, k_2\}. \tag{7.2.1}$$

But they are rather cumbersome to apply, since this involves construction of a normal cone to an end-point constraint set defined implicitly as an admissible region for a collection of functional inequality and equality constraints. It is convenient then to have at hand necessary conditions expressed directly in terms of the constraint functionals themselves, by means of additional Lagrange multipliers. Such conditions are provided by the next theorem.

Theorem 7.2.3 (The Maximum Principle for Functional End-Point Constraints) *Suppose that $(\bar{x}, \bar{u})$ is a $W^{1,1}$ local minimizer for (FEC). Assume that hypotheses (H1)–(H3) of Theorem 7.2.1 (for the data of the special case of (P) in which the endpoint constraint set C takes the form (7.2.1)) are satisfied for some $\bar{\epsilon} > 0$. Assume also that the functions ϕ_i, $i = 1, k_1$, and ψ_i, $i = 1, k_2$, defining C are Lipschitz continuous on a neighbourhood of $(\bar{x}(S), \bar{x}(T))$.*

Then there exist $p \in W^{1,1}$, $\lambda \ge 0$, and sets of numbers $\{\alpha_i \ge 0\}_{i=1}^{k_1}$ and $\{\beta_i\}_{i=1}^{k_2}$, such that

7.2 Clarke's Nonsmooth Maximum Principle

(a): $\|p\|_{L^\infty} + \lambda + \sum_{i=1}^{k_1} \alpha_i + \sum_{i=1}^{k_2} |\beta_i| \neq 0$,

(b): $-\dot{p}(t) \in \mathrm{co}\, \partial_x \mathcal{H}(t, \bar{x}(t), p(t), \bar{u}(t))$ a.e. $t \in [S, T]$,

(c): $\mathcal{H}(t, \bar{x}(t), p(t), \bar{u}(t)) = \max_{u \in U(t)} \mathcal{H}(t, \bar{x}(t), p(t), u)$ a.e. $t \in [S, T]$,

(d): $(p(S), -p(T)) \in \partial\Big(\lambda g(\bar{x}(S), \bar{x}(T)) + \sum_{i=1}^{k_1} \alpha_i \phi_i(\bar{x}(S), \bar{x}(T))$
$+ \sum_{i=1}^{k_2} \beta_i \partial \psi_i(\bar{x}(S), \bar{x}(T))\Big)$,

and

$\alpha_i = 0$ for all $i \in \{1, .., k_1\}$ such that $\phi_i(\bar{x}(S), \bar{x}(T)) < 0$.

(In the preceding relations, $\sum_{i=1}^k$ is interpreted as 0 if $k = 0$.)
If, furthermore,

$$f(t, x, u) \text{ and } U(t) \text{ are independent of } t,$$

then there exists a constant r such that

(e) $\mathcal{H}(t, \bar{x}(t), p(t), \bar{u}(t)) = r$ a.e..

Note that if the functions

$$g, \phi_1, \ldots, \phi_{k_1}, \psi_1, \ldots, \psi_{k_2}$$

are continuously differentiable and if $f(t, x, u)$ is continuously differentiable with respect to the x variable, then the 'nonsmooth' conditions (b) and (d) can be replaced by relations involving classical (Fréchet) derivatives

$$-\dot{p}(t) = \nabla_x \mathcal{H}(t, \bar{x}(t), p(t), \bar{u}(t)) \quad \text{a.e.}$$

and

$$(p(S), -p(T)) = \lambda \nabla g(\bar{x}(S), \bar{x}(T)) + \sum_{i=1}^{k_1} \alpha_i \nabla \phi_i(\bar{x}(S), \bar{x}(T))$$
$$+ \sum_{i=1}^{k_2} \beta_i \nabla \psi_i(\bar{x}(S), \bar{x}(T)).$$

This version of the maximum principle (addressing problems with functional end-point constraints and smooth data) does not require, for its formulation, the modern apparatus of subdifferentials etc. Many of the more recent alternative necessary conditions of optimality, such as the generalized Euler Lagrange and generalized Hamilton conditions which are the subject matter of Chap. 8, are, by contrast, inherently nonsmooth: even if the dynamic constraint originates as a differential equation parameterized by a control variable, with smooth right side, the statement of these necessary conditions involves constructs of nonsmooth analysis (outside rather special cases).

Proof of Theorem 7.2.3 We assume that $k_1 \geq 1$ and $k_2 \geq 1$ and that all the end-point inequality constraints are active. There is no loss of generality in so doing. To see this, note that $(\bar{x}, \bar{z}_0 \equiv 0, \bar{z}_1 \equiv 0, \bar{u})$ is a $W^{1,1}$ local minimizer for a modified version of problem (FEC) obtained by deleting the inactive end-point inequality constraints, replacing the state vector x by $(x, z_0, z_1) \in \mathbb{R}^n \times \mathbb{R} \times \mathbb{R}$ and by requiring the new state trajectory components z_0 and z_1 to satisfy the differential equations

$$\dot{z}_0(t) = 0, \dot{z}_1(t) = 0$$

and the end-point constraints

$$z_0(T) \leq 0, z_1(T) = 0.$$

The additional hypotheses (non-emptiness of the sets of equality and inequality constraints and all inequality constraints are active) are met. If the assertions of the theorem were true in these circumstances, we would be able to find a co-state trajectory which we write (p, q_0, q_1) satisfying the conditions of the maximum principle with functional end-point constraints. But because z_0 and z_1 are unconstrained at $t = S$, we conclude from the transversality conditions that $q_0 \equiv 0$ and $q_1 \equiv 0$. The resulting conditions may be interpreted as the assertions of the theorem for the original problem when we associate with each inactive inequality end-point constraint a zero Lagrange multiplier.

Define $G : \mathbb{R}^n \times \mathbb{R}^n \to \mathbb{R} \times \mathbb{R}^{k_1} \times \mathbb{R}^{k_2}$ according to

$$G(x_0, x_1) := \Big(g(x_0, x_1) - g(\bar{x}(S), \bar{x}(T)),$$

$$\phi_1(x_0, x_1), \ldots, \phi_{k_1}(x_0, x_1), \psi_1(x_0, x_1) \ldots, \psi_{k_2}(x_0, x_1) \Big).$$

Now observe that $(\bar{x}, \bar{y} \equiv 0, \bar{u})$ is a $W^{1,1}$ local minimizer for the dynamic optimization problem with state vector $(x, y = (y_0, .., y_{k_1+k_2})) \in \mathbb{R}^n \times \mathbb{R}^{1+k_1+k_2}$:

$$\begin{cases} \text{Minimize } y_0(T) \\ \text{over } x \in W^{1,1}, (y_0, .., y_{k_1+k_2}) \in W^{1,1} \\ \quad \text{and measurable functions } u \text{ satisfying} \\ \dot{x} = f, \dot{y}_0 = 0, .., \dot{y}_{k_1+k_2} = 0, \quad \text{a.e.,} \\ u(t) \in U(t) \quad \text{a.e.,} \\ (x(S), x(T), y(S)) \in \text{Gr } G, \\ y_i(T) \leq 0 \text{ for } i = 1, .., k_1, \\ y_i(T) = 0 \text{ for } i = k_1 + 1, .., k_1 + k_2. \end{cases}$$

The assertions of Theorem 7.2.1, which are valid for this problem, are expressed in terms of a co-state trajectory, which we write

$$(p, q \equiv (-\alpha_0, .., -\alpha_{k_1}, -\beta_1, .., -\beta_{k_2})),$$

and a cost multiplier $\lambda \geq 0$, not both zero. (Here, we have made use of the fact that, since $f(t, x, u)$ does not depend on $(y_0, \ldots y_{k_1+k_2})$ and from the co-state inclusion, q does not depend on t.) From the transversality condition of Theorem 7.2.1, we know

$$(p(S), -p(T), -(\alpha_0, .., \alpha_{k_1}, \beta_1, .., \beta_{k_2})) \in N_{\mathrm{Gr}\,G}(\bar{x}(S), \bar{x}(T), 0, \ldots, 0)$$

$$\alpha_0 = \lambda, \ \alpha_1, .., \alpha_{k_1} \geq 0.$$

With the help of Proposition 5.4.2, however, we deduce that

$$(p(S), -p(T)) \in \partial \left(\lambda g + \sum_{i=1}^{k_1} \alpha_i \phi_i + \sum_{i=1}^{k_2} \beta_i \psi_i \right) (\bar{x}(S), \bar{x}(T)).$$

This is the transversality condition as it appears in the theorem statement. All the other assertions of the theorem, for 'Lagrange multipliers' $p, \alpha_1, .., \alpha_{k_1}, \beta_1, .., \beta_{k_2}$ and λ as above, follow from Theorem 7.2.1. □

7.3 A Preliminary Maximum Principle, for Dynamic Optimization Problems with no End-Point Constraints

The proof of the maximum principle, as stated in Theorem 7.2.1, is built up in several stages. The first stage, the conclusions of which are summarised as Proposition 7.3.1 below, is accomplished in this section. We prove a version of the maximum principle for a variant of the dynamic optimization problem (P), the most significant feature of which is the absence of end-point constraints. The ultimate goal will be to prove the full maximum principle, for problems with end-point constraints; to this end, we shall, in the next section, apply the proposition to each of a sequence of endpoint-free problems, constructed with the help of Ekeland's theorem, to generate a sequence of sets of Lagrange multipliers which, in the limit, satisfy the conditions of the maximum principle for problems with end-point constraints. The dynamic optimization problem considered in this section, which, in anticipation of these applications, also includes an integral cost term to take account of perturbations terms introduced by the application of Ekeland's theorem, is as follows:

$$(F) \begin{cases} \text{Minimize } J(x, u) := g(x(S), x(T)) + \int_{[S,T]} L(t, u(t)) \, dt, \\ \text{subject to} \\ \dot{x}(t) = f(t, x(t), u(t)) \text{ a.e. } t \in [S, T], \\ u(t) \in U(t) \text{ a.e. } t \in [S, T], \end{cases}$$

the data for which comprise an interval $[S, T]$, functions $g : \mathbb{R}^n \times \mathbb{R}^n \to \mathbb{R}$, $f : [S, T] \times \mathbb{R}^n \times \mathbb{R}^m \to \mathbb{R}^n$, $L : [S, T] \times \mathbb{R}^m \to \mathbb{R}$, a non-empty multifunction $U : [S, T] \rightsquigarrow \mathbb{R}^m$.

A admissible process (x, u) for (F) will be interpreted now as a pair of functions in which $x \in W^{1,1}([S, T]; \mathbb{R}^n)$ and u is a measurable selector of U, such that $t \to L(t, u(t))$ is integrable. The concepts of $W^{1,1}$ local minimizers and L^∞ local minimizers, in this context, have their obvious meanings.

The following proposition will be recognized as a special case of Theorem 7.2.1, extended, by state augmentation, to allow for the integral cost term, and in which the cost multiplier $\lambda = 1$. Notice that the admissible process $(\bar{x}, \bar{u})$ under consideration is assumed to be an L^∞ local minimizer, not merely a $W^{1,1}$ minimizer.

For given $\lambda \geq 0$, we define the integral cost-related un-maximized Hamiltonian, $\mathcal{H}_\lambda : [S, T] \times \mathbb{R}^n \times \mathbb{R}^n \times \mathbb{R}^m \to \mathbb{R}$, to be

$$\mathcal{H}_\lambda(t, x, p, u) := p \cdot f(t, x, u) - \lambda L(t, u).$$

Proposition 7.3.1 *Let $(\bar{x}, \bar{u})$ be an L^∞ local minimizer for (F). Assume that, for some $\bar{\epsilon} > 0$, the following hypotheses are satisfied:*

(F1): Gr U is an $\mathcal{L} \times \mathcal{B}^m$ measurable set, f is an $\mathcal{L} \times \mathcal{B}^{n \times m}$ measurable function and there exist integrable functions $c_0 : [S, T] \to \mathbb{R}$ and $k_0 : [S, T] \to \mathbb{R}$ such that

$$|f(t, x, u) - f(t, x', u)| \leq k_0(t)|x - x'| \quad \text{and} \quad |f(t, x, u)| \leq c_0(t)$$
for all $x, x' \in \bar{x}(t) + \bar{\epsilon} \mathbb{B}$, $u \in U(t)$, a.e. $t \in [S, T]$,

(F2): g is Lipschitz continuous on $(\bar{x}(S), \bar{x}(T)) + \bar{\epsilon} \mathbb{B} \times \bar{\epsilon} \mathbb{B}$,
(F3): L is a bounded, $\mathcal{L} \times \mathcal{B}^m$ measurable function.

Then there exist $p \in W^{1,1}([S, T]; \mathbb{R}^n)$ such that, for $\lambda = 1$,

(b): $-\dot{p}(t) \in \mathrm{co}\, \partial_x \mathcal{H}_\lambda(t, \bar{x}(t), p(t), \bar{u}(t))$ a.e. $t \in [S, T]$,
(c): $\mathcal{H}_\lambda(t, \bar{x}(t), p(t), \bar{u}(t)) = \max_{u \in U(t)} \mathcal{H}_\lambda(t, \bar{x}(t), p(t), u)$ a.e. $t \in [S, T]$,
(d): $(p(S), -p(T)) \in \lambda \partial g(\bar{x}(S), \bar{x}(T))$.

(These conditions omit stating explicitly that $(p, \lambda) \neq (0, 0)$, i.e. the 'non-triviality of the Lagrange multipliers' condition, because, in view of the fact that $\lambda = 1$, this condition is automatically satisfied.)

Proof of Proposition 7.3.1 We note at the outset that, without loss of generality, we can assume that the (F1) and (F2) (which are of a local nature w.r.t. the x variable) have been replaced by the stronger, global, hypotheses

(F1)′: Gr U is an $\mathcal{L} \times \mathcal{B}^m$ measurable set, f is an $\mathcal{L} \times \mathcal{B}^{n \times m}$ measurable function and there exist integrable functions $k_0 : [S, T] \to \mathbb{R}$ and $c_0 : [S, T] \to \mathbb{R}$ such that

$$|f(t, x, u) - f(t, x', u)| \leq k_0(t)|x - x'| \quad \text{and} \quad |f(t, x, u)| \leq c_0(t)$$
for all $x, x' \in \mathbb{R}^n$, $u \in U(t)$, a.e. $t \in [S, T]$,

(F2)′: g is Lipschitz continuous (with Lipschitz constant k_g) on $\mathbb{R}^n \times \mathbb{R}^n$.

7.3 A Preliminary Maximum Principle, for Dynamic Optimization Problems...

This is because, if the stronger hypotheses are violated, we can replace $f(t, x, u)$ and $g(x_0, x_1)$ by the functions $\tilde{f}(t, x, u) := f(t, \bar{x}(t) + \text{tr}_{\bar{\epsilon}}(x - \bar{x}(t)), u)$ and $\tilde{g}(x_0, x_1) := g(\bar{x}(S) + \text{tr}_{\bar{\epsilon}}(x - \bar{x}(S)), \bar{x}(T) + \text{tr}_{\bar{\epsilon}}(x - \bar{x}(T)))$, in which $\text{tr}_{\bar{\epsilon}} : \mathbb{R}^n \to \mathbb{R}^n$ (the 'truncation function') is

$$\text{tr}_{\bar{\epsilon}}(x) := \begin{cases} x & \text{if } |x| \leq \bar{\epsilon} \\ \bar{\epsilon} \frac{1}{|x|} x & \text{otherwise.} \end{cases}$$

Note that the truncation function $\text{tr}_{\bar{\epsilon}}$ has the following properties: for any function $d : \mathbb{R}^r \to \mathbb{R}$ that is Lipschitz continuous on $\bar{\epsilon}\mathbb{B}$ (with Lipschitz constant k_d)

- $(d \circ \text{tr}_{\bar{\epsilon}})(x) = d(x)$ for all $x \in \bar{\epsilon}\mathbb{B}$
- $(d \circ \text{tr}_{\bar{\epsilon}})(x) \leq \sup_{x \in \bar{\epsilon}\mathbb{B}} d(x)$ for all $x \in \mathbb{R}^r$
- $(d \circ \text{tr}_{\bar{\epsilon}})(x)$ is Lipschitz continuous on $\mathbb{R}^r$ with Lipschitz constant k_d.

It will be clear from these properties that the modified functions $\tilde{f}$ and $\tilde{g}$ satisfy the stronger hypotheses, with the original integrable bounds and Lipschitz constants. Furthermore, $(\bar{x}, \bar{u})$ remains an L^∞ local minimizer for the modified data and the assertions of the proposition, in terms of the original and modified functions, are equivalent.

We can also arrange, without loss of generality, that the integrable function k_0 in hypothesis (F1) satisfies $k_0 \geq 1$ by adding, if necessary, a positive constant.

Step 1: We prove the proposition under the additional hypothesis that

$$\text{(S):} \begin{cases} g(x_0, x_1) = \ell(x_0, x_1) + \alpha|x_0 - \zeta|, \\ \text{for some continuously differentiable function } \ell : \mathbb{R}^n \times \mathbb{R}^n \to \mathbb{R}, \, \zeta \in \mathbb{R}^n \text{ and} \\ \alpha \geq 0. \end{cases}$$

Since $(\bar{x}, \bar{u})$ is an L^∞ local minimizer for problem (F), there exists $\epsilon > 0$, such that $(\bar{x}, \bar{u})$ is a minimizer for

$$(F^\epsilon) \begin{cases} \text{Minimize } J(x, u) := \ell(x(S), x(T)) + \alpha|x(S) - \bar{x}(S)| + \int_{[S,T]} L(t, u(t))\, dt, \\ \text{subject to} \\ \dot{x}(t) = f(t, x(t), u(t)) \text{ a.e. } t \in [S, T], \\ u(t) \in U(t) \text{ a.e. } t \in [S, T], \\ \|x - \bar{x}\|_{L^\infty} \leq \epsilon. \end{cases}$$

For each positive integer i, consider the related problem:

$$(F_i) \begin{cases} \text{Minimize } J_i(x, y, u) := \ell(x(S), x(T)) + \alpha|x(S) - \bar{x}(S)| \\ \qquad + \int_{[S,T]} L(t, u(t))\, dt + i \times \int_{[S,T]} k_0(t)|y(t) - x(t)|^2 dt \\ \text{over } y \in L^1_{k_0}([S, T]; \mathbb{R}^n) \text{ and meas. functions } u \text{ s.t.} \\ \dot{x}(t) = f(t, y(t), u(t)), \text{ a.e. } t \in [S, T], \\ u(t) \in U(t), \text{ a.e. } t \in [S, T], \\ \|x - \bar{x}\|_{L^\infty} \leq \epsilon. \end{cases}$$

Notice that, to construct each (F_i) problem, we have changed the dynamic constraint from the controlled differential equation $\dot{x} = f(t, x, u)$ to a new controlled differential equation $\dot{x} = f(t, y, u)$, in which y in interpreted as a new, control-like, variable; then, to take account of the possible discrepancy between x and y, we have introduced a quadratic penalty term, with penalty parameter i.

In (F_i), k_0 is the integrable bound appearing in the hypotheses of the proposition statement, and $L^1_{k_0}([a, b]; \mathbb{R}^n)$ denotes the 'weighted L^1 space' of measurable functions $\phi : [a, b] \to \mathbb{R}^n$ such that $t \to k_0(t)\phi(t)$ is an L^1 function. Observe that the cost in problem (F_i) can be infinite, because y is allowed to be a weighted L^1 function and the cost involves a weighted L^2 norm. Write the infimum costs of (F_i) and (F^ϵ) as $\inf(F_i)$ and $\inf(F^\epsilon)$, respectively.

Lemma 7.3.2

$$\lim_{i \to \infty} \inf(F_i) = \inf(F^\epsilon).$$

Proof It is obvious from the special structure of the (F_i)'s and (F^ϵ) that

$$-\infty < \inf(F_i) \leq \inf(F_j) \leq \inf(F^\epsilon) \quad \text{for any index values } i < j. \quad (7.3.1)$$

Fix $i > 0$ and take any admissible process (x, y, u) for (F_i) such that $J_i(x, y, u) < \infty$. (Such a admissible process, namely $(\bar{x}, \bar{y} \equiv \bar{x}, \bar{u})$, exists.)

By Filippov's existence theorem, applied to the reference trajectory x and initial state $\xi = x(S)$, there exists a admissible process (x_i, u) for (F^ϵ) (with the same u) such that $x_i(S) = x(S)$ and

$$||x_i - x||_{L^\infty} \leq K \times \left(\int_{[S,T]} k_0(t)|y(t) - x(t)|dt \right).$$

(K is a number that does not depend on our choice of (x, y, u).) Notice that, from Hölder's inequality, we know that $(\int_{[S,T]} k_0(t)|y(t) - x(t)|dt)^2 \leq (\int_{[S,T]} k_0(t)dt) \times \int_{[S,T]} k_0(t)|y(t) - x(t)|^2 dt$. Observe also that $|g(x(S), x(T)) - g(x_i(S), x_i(T))| \leq k_g ||x_i - x||_{L^\infty}$ in which k_g is a Lipschitz constant for g. It follows that

$$J_i(x, y, u) - J_i(x_i, x_i, u) = J_i(x, y, u) - J(x_i, u)$$

$$\geq \int_{[S,T]} k_0(t) \left(i \times |y(t) - x(t)|^2 - Kk_g|y(t) - x(t)| \right) dt,$$

$$\geq \left(\left(\int_{[S,T]} k_0(t)dt \right)^{-1} \times i \times z^2 - Kk_g z \right), \quad \left(z := \int_{[S,T]} k_0(t)|y(t) - x(t)|dt \right)$$

$$\geq \min_{z \in \mathbb{R}} \left(\left(\int_{[S,T]} k_0(t)dt \right)^{-1} \times i \times z^2 - Kk_g z \right) \geq -\gamma^2 \times i^{-1},$$

7.3 A Preliminary Maximum Principle, for Dynamic Optimization Problems...

in which $\gamma^2 := \frac{1}{4} \times (Kk_g)^2 \times ||k_0||_{L^1}$. Since (x, y, u) was chosen arbitrarily, we deduce

$$\inf(F_i) \geq \inf(F^\epsilon) - \gamma^2 \times i^{-1}.$$

This inequality, combined with (7.3.1) tell us that $\lim_{i \to \infty} \inf(F_i) = \inf(F^\epsilon)$. □

Now write problem (F_i) as

$$\text{Minimize } \{J_i(x, y, u) : (x, y, u) \in \mathcal{A}_\epsilon\},$$

in which

$$\mathcal{A}_\epsilon := \{(x, y, u) : x \in W^{1,1}([S, T]; \mathbb{R}^n), \ y \in L^1_{k_0}([S, T]; \mathbb{R}^n),$$
$$u \text{ is a measurable selector of } U, \ \dot{x}(t) = f(t, y(t), u(t)) \text{ a.e.},$$
$$||x - \bar{x}||_{L^\infty} \leq \epsilon\}.$$

Equip $\mathcal{A}_\epsilon$ with the metric

$$d_\mathcal{E}((x', y', u'), (x, y, u)) :=$$
$$|x'(S) - x(S)| + \int_{[S,T]} k_0(t)|y'(t) - y(t)| \, dt$$
$$+ \text{meas } \{t \in [S, T] : u'(t) \neq u(t)\}.$$

We shall use the following properties of $\mathcal{A}_\epsilon$ and $d_\mathcal{E}$, which are consequences of the hypotheses of the proposition statement:

(i): $d_\mathcal{E}$ defines a metric on $\mathcal{A}_\epsilon$ and $(\mathcal{A}_\epsilon, d_\mathcal{E})$ is a complete metric space,
(ii): If $(x_j, y_j, u_j) \to (x, y, u)$ in $(\mathcal{A}_\epsilon, d_\mathcal{E})$, then $||x_j - x||_{L^\infty} \to 0$ as $j \to \infty$,
(iii): The function J_i is lower semi-continuous on $(\mathcal{A}_\epsilon, d_\mathcal{E})$.

We comment only on the proof of (iii). Note that the function J_i is a sum of terms. Taking note of property (ii) above, we see that the first three terms are continuous. So we need to attend only to the last term. Take $(x_j, y_j, u_j) \to (x, y, u)$ w.r.t. $d_\mathcal{E}$. Since $k_0 \geq 1$, we know $y_j \to y$ in L^1 and therefore, for a subsequence, $y_j \to y$, a.e.. Using these facts and Fatou's lemma, which may be invoked because the integrands are non-negative, we have

$$\liminf_{j \to \infty} \int_{[S,T]} k_0(t)|y_j(t) - x_j(t)|^2 dt \geq \int_{[S,T]} \liminf_{j \to \infty} (k_0(t)|y_j(t) - x_j(t)|^2) dt$$
$$= \int_{[S,T]} k_0(t)|y(t) - x(t)|^2 dt.$$

(These relations are valid, even if some of the integrals involved are infinite.) Since the lower limit is independent of the subsequence that is chosen, the above relation is true also for the original sequence. We have confirmed the lower semi-continuity of the third term and therefore of J_i.

By Lemma 7.3.2, there exists $\gamma_i \downarrow 0$ such that, for each i, $(\bar{x}, \bar{y} \equiv \bar{x}, \bar{u})$ is a γ_i^2 minimizer for (F_i). According to Ekeland's theorem, there exists, for each i, an element $(x_i, y_i, u_i) \in \mathcal{A}_\epsilon$, which is a minimizer for the optimization problem:

$(\tilde{F}_i)$ Minimize $\{J_i(x, y, u) + \gamma_i d_{\mathcal{E}}((x, y, u), (x_i, y_i, u_i)) : (x, y, u) \in \mathcal{A}_\epsilon\}$,

and

$$d_{\mathcal{E}}((x_i, y_i, u_i), (\bar{x}, \bar{x}, \bar{u})) \leq \gamma_i . \tag{7.3.2}$$

We know from the properties of the metric $d_{\mathcal{E}}$ that $||x_i - \bar{x}||_{L^\infty} \to 0$, as $i \to \infty$. Consequently, for i sufficiently large,

$$||x_i - \bar{x}||_{L^\infty} \leq \epsilon/2.$$

Define $m_i : [S, T] \times \mathbb{R}^m \to \mathbb{R}$ as

$$m_i(t, u) := \begin{cases} 0 \text{ if } u = u_i(t) \\ 1 \text{ if } u \neq u_i(t) . \end{cases} \tag{7.3.3}$$

The significance of this definition is that it permits us to write the perturbation term $\text{meas}\{t : u(t) \neq u_i(t)\}$ as an integral term, since, for any control function u,

$$\text{meas}\{t : u(t) \neq u_i(t)\} = \int_{[S,T]} m_i(t, u(t))dt.$$

The cost function for $(\tilde{F}_i)$ can be written

$$\tilde{J}_i(x, y, u) := i \times \left(\int_{[S,T]} k_0(t)|y(t) - x(t)|^2 dt \right)$$
$$+ \gamma_i \int_{[S,T]} k_0(t)|y(t) - y_i(t)|dt + \int_{[S,T]} (L(t, u(t)) + \gamma_i m_i(t, u(t)))dt$$
$$+ \ell(x(S), x(T)) + \alpha|x(S) - \bar{x}(S)| + \gamma_i |x(S) - x_i(S)|.$$

Fix i. By choosing i sufficiently large, we can arrange that $||x_i - \bar{x}||_{L^\infty} \leq \epsilon/2$. We now assemble, in the following lemma, a set of conditions, in the form of a maximum principle, consequent on the fact that (x_i, y_i, u_i) is an L^∞ local minimizer for a modification of $(\tilde{F})_i$, in which we drop the constraint $||x - \bar{x}||_{L^\infty} \leq \epsilon$.

7.3 A Preliminary Maximum Principle, for Dynamic Optimization Problems...

Lemma 7.3.3 *Define the function* $p : [S, T] \to \mathbb{R}^n$ *according to*

$$\begin{cases} -\dot{p}(t) = 2 \times i \times k_0(t)(y_i(t) - x_i(t)) \text{ a.e. } t \in [S, T] \\ -p(T) = \nabla_{x_1}\ell(x_i(S), x_i(T)). \end{cases}$$

Then

(b)': $-\dot{p}(t) \in \text{co } \partial_x (p(t) \cdot f(t, y_i(t), u_i(t))) + \gamma_i k_0(t)\mathbb{B}$, *a.e.* $t \in [S, T]$,

(c)': $\mathcal{H}_{\lambda=1}(t, y_i(t), p(t), u_i(t)) \geq \sup_{u \in U(t)} \mathcal{H}_{\lambda=1}(t, y_i(t), p(t), u) - \gamma_i$,

a.e. $t \in [S, T]$,

(d)': $p(S) \in \nabla_{x_0}\ell(x_i(S), x_i(T)) + (\gamma_i + \alpha)\mathbb{B}, \quad -p(T) = \nabla_{x_1}\ell(x_i(S), x_i(T))$.

Proof Since $\tilde{J}_i(x_i, y_i, u_i) < +\infty$, we deduce from the presence of the penalty quadratic term in J_i that $y_i \in L^2_{k_0}([S, T]; \mathbb{R}^n)$. Take $\delta > 0, \xi \in \mathbb{R}^n$ such that $|\xi| \leq 1$, a function $y \in L^2_{k_0}([S, T]; \mathbb{R}^n)$ and a selector u of U. Let x be the corresponding state trajectory, with initial condition $x(S) = x^i(S) + \delta\xi$. Assume that $x(t) \in \bar{x}(t) + \epsilon/2\mathbb{B}$ for all $t \in [S, T]$.

By 'optimality' of (x_i, y_i, u_i) and since $\dot{x} = f(t, y, u)$ and $\dot{x}_i = f(t, x_i, u_i)$, we can show that

$$\delta^{-1} \int_{[S,T]} \left(i \times k_0(t) \left(|y(t) - x(t)|^2 - |y_i(t) - x_i(t)|^2 \right) \right.$$

$$+ \gamma_i \times k_0(t)|y(t) - y_i(t)| \Big) dt + \delta^{-1} \Big(\ell(x_i(S) + \delta\xi, x(T)) - \ell(x_i(S), x_i(T))$$

$$+ \gamma_i \delta|\xi| + \alpha(|x_i(S) + \delta\xi - \bar{x}(S)| - |x_i(S) - \bar{x}(S)|) \Big)$$

$$+ \delta^{-1} \int_{[S,T]} \Big((L(t, u(t)) - L(t, u_i(t))) + \gamma_i(m_i(t, u(t)) - m_i(t, u_i(t))) \Big) dt$$

$$+ \delta^{-1} \int_{[S,T]} p(t) \cdot \Big(\dot{x}(t) - \dot{x}_i(t) - [f(t, y(t), u(t)) - f(t, y_i(t), u_i(t))] \Big) dt \geq 0.$$

(7.3.4)

An integration by parts yields the identity

$$\int_{[S,T]} p(t) \cdot (\dot{x}(t) - \dot{x}_i(t))dt = -\int_{[S,T]} \dot{p}(t) \cdot (x(t) - x_i(t))dt$$

$$+ p(T) \cdot (x(T) - x_i(T)) - p(S) \cdot (x(S) - x_i(S)).$$

Substituting this expression into (7.3.4), employing the expansion

$$|y(t) - x(t)|^2 = |y(t) - x_i(t)|^2 + |x(t) - x_i(t)|^2$$

$$- 2(y_i(t) - x_i(t)) \cdot (x(t) - x_i(t)) - 2(x(t) - x_i(t)) \cdot (y(t) - y_i(t))$$

and using the estimates

$$\ell(x(S), x(T)) - \ell(x_i(S), x_i(T)) - \nabla_{x_0}\ell(x_i(S), x_i(T)) \cdot (x(S) - x_i(S))$$
$$-\nabla_{x_1}\ell(x_i(S), x_i(T)) \cdot (x(T) - x_i(T)) \leq \theta\Big(\delta|\xi| + |x(T) - x_i(T)|\Big)$$

and

$$|x(S) - \bar{x}(S)| - |x_i(S) - \bar{x}(S)| \leq \delta|\xi|,$$

(for some function $\theta : [0, \infty) \to [0, \infty)$ such that $\lim_{\alpha \downarrow 0} s^{-1}\theta(s) = 0$), we arrive at

$$\delta^{-1} \int_{[S,T]} \Big(i \times k_0(t) \Big(|y(t) - x_i(t)|^2 - |y_i(t) - x_i(t)|^2\Big)$$
$$+ \gamma_i \times k_0(t)|y(t) - y_i(t)|\Big) dt$$
$$+ (\nabla_{x_0} l(x_i(S), x_i(T)) - p(S)) \cdot \xi + (\gamma_i + \alpha)|\xi|$$
$$+ \delta^{-1} \left(\int_{[S,T]} (L(t, u(t)) - L(t, u_i(t))) + \gamma_i (m_i(t, u(t)) - m_i(t, u_i(t))) \, dt \right)$$
$$+ \delta^{-1} \int_{[S,T]} -p(t) \cdot [f(t, y(t), u(t)) - f(t, y_i(t), u_i(t)))] \, dt$$
$$+ E_1(y, x, y_i, x_i) + E_2(y, x, y_i, x_i) \geq 0. \qquad (7.3.5)$$

in which the first 'error' term $E_1(y, x, y_i, x_i)$ is

$$E_1(y, x, y_i, x_i) :=$$
$$\delta^{-1} \int_{[S,T]} \Big(- 2ik_0(t)(y_i(t) - x_i(t)) - \dot{p}(t) \Big) \cdot \Big(x(t) - x_i(t)\Big) dt$$
$$+ \delta^{-1} \Big(\nabla_{x_1}\ell(x_i(S), x_i(T)) + p(T)\Big) \cdot (x(T) - x_i(T)),$$

and the second 'error' term $E_2(y, x, y_i, x_i)$ is

$$E_2(y, x, y_i, x_i) :=$$
$$\delta^{-1} \int_{[S,T]} i \times k_0(t) \Big(|x(t) - x_i(t)|^2 - 2(y(t) - y_i(t)) \cdot (x(t) - x_i(t))\Big) dt$$
$$+ \delta^{-1}\Big(\ell(x_i(S) + \delta\xi, x(T)) - \ell(x_i(S), x_i(T))$$
$$- \nabla \ell(x_i(S), x_i(T)) \cdot (\delta\xi, x(T) - x_i(T))\Big).$$

7.3 A Preliminary Maximum Principle, for Dynamic Optimization Problems... 311

We can estimate $E_2(y, x, y_i, x_i)$ as follows:

$$|E_2(y, x, y_i, x_i)| \leq \delta^{-1}\theta\Big(\delta|\xi| + |x(T) - x_i(T)|\Big)$$
$$+ \delta^{-1} \times i \times \|k_0\|_{L^1} \times \|x - x_i\|_{L^\infty}^2 + \gamma_i \int_{[S,T]} k_0(t)|y(t) - y_i(t)|dt$$
$$+ \delta^{-1} \times i \times \int_{[S,T]} 2k_0(t)|y(t) - y_i(t)|dt \times \|x - x_i\|_{L^\infty}. \quad (7.3.6)$$

Note that $E_1(y, x, y_i, x_i) \equiv 0$, because of the defining relations for p.

We now confirm the assertions of the lemma by examining inequality (7.3.5), for various choices of $\delta > 0$, ξ, y and u.

Confirmation of (d)': Notice that $-p(T) = \nabla_{x_1}\ell(x_i(S), x_i(T))$, by definition of p. To verify the full transversality condition, take any $\xi \in \mathbb{R}^n$ such that $|\xi| \leq 1$ and sequence $\delta_j \downarrow 0$. For each j let x^j be the state trajectory corresponding to y_i and u_i, and with initial value $x^j(S) = x_i(S) + \delta_j \xi$. We can deduce from Filippov's existence theorem that

$$\|x^j - x_i\|_{L^\infty} \to 0, \text{ as } j \to \infty \quad (7.3.7)$$

and there exists a number B such that, for $j = 1, 2, \ldots,$

$$\delta_j^{-1}\|x^j - x_i\|_{L^\infty} \leq B. \quad (7.3.8)$$

Now consider (7.3.5) when $\delta = \delta_j$, $y = y_i$, $u = u_i$ and $x = x^j$. From (7.3.7) and (7.3.8) we see that

$$E_2(y_i, x^j, y_i, x_i) \to 0 \text{ as } j \to \infty.$$

We may pass to the limit as $j \to \infty$, to obtain

$$\big(\nabla_{x_0}\ell(x_i(S), x_i(T)) - p(S)\big) \cdot \xi + (\gamma_i + \alpha)|\xi| \geq 0.$$

Since this inequality is valid for every $\xi \in \mathbb{R}^n$ such that $|\xi| \leq 1$, We can conclude that $p(S) \in \nabla_{x_0}\ell(x_i(S), x_i(T)) + (\gamma_i + \alpha)\mathbb{B}$.

Confirmation of (b)': Choose any measurable function $y : [S, T] \to \mathbb{R}^n$ such that

$$y(t) \in y_i(t) + (1 + k_0(t))^{-1}\mathbb{B} \text{ a.e. } t. \quad (7.3.9)$$

Observe that $y \in L^2_{k_0}([S, T]; \mathbb{R}^n)$. Define

$$\mathcal{S} := \{\bar{t} \in (S, T) : \bar{t} \text{ is a Lebesgue point of}$$
$$t \to k_0(t)(|y(t) - x_i(t)|^2 - |y_i(t) - x_i(t)|^2 + \gamma_i|y(t) - y_i(t)|)$$
$$\text{and } t \to f(t, y(t), u_i(t)) - f(t, y_i(t), u_i(t))\}.$$

Take $\delta_j \downarrow 0$ and $\bar{t} \in S$. For each $j, t \in [S, T]$, let

$$y^j(t) = \begin{cases} y(t) & \text{if } t \in [\bar{t}, (\bar{t}+\delta_j) \wedge T] \\ y_i(t) & \text{if } t \notin [\bar{t}, (\bar{t}+\delta_j) \wedge T]. \end{cases}$$

Write x^j for the state trajectory corresponding to y^j and u_i, with initial value $x_i(S)$. For each j, consider (7.3.5) with $\delta = \delta_j, \xi = 0, y = y^j$ and $u = u^*$. Making use of (7.3.9), we can show that (7.3.7) and (7.3.8) are satisfied, and $\int k_0(t)|y^j(t) - y_i(t)|dt \to 0$ as $j \to \infty$. It follows that

$$E_2(y^j, x^j, y_i, x_i) \to 0 \text{ as } j \to \infty.$$

Since $\bar{t} \in S$, we can pass to the limit in (7.3.5) as $j \to \infty$, to obtain $\phi(\bar{t}, y(\bar{t})) \geq 0$, where

$$\phi(t, y) := k_0(t) \times \left(i \times (|y - x_i(t)|^2 - |y_i(t) - x_i(t)|^2 + \gamma_i \times |y - y_i(t)| \right)$$
$$- p(t) \cdot (f(t, y, u_i(t)) - f(t, y_i(t), u_i(t))).$$

Since y was chosen arbitrarily and S has full measure, we have shown that

$$\int_{[S,T]} \phi(t, y(t))dt \geq 0$$

for all measurable functions y's satisfying $y(t) \in y_i(t) + (1 + k_0(t))^{-1}\mathbb{B}$ for a.e. $t \in [S, T]$.

Take $\epsilon_k \downarrow 0$ and, for each k, consider the multifunction

$$\mathcal{Y}_k(t) := \{y \in y_i(t)+(1+k_0(t))^{-1}\mathbb{B} : \phi(t, y) \leq \inf_{y' \in y_i(t)+(1+k_0(t))^{-1}\mathbb{B}} \phi(t, y')+\epsilon_k\}.$$

$\mathcal{Y}_k$ is a non-empty multifunction such that Gr$\{\mathcal{Y}_k\}$ is an $\mathcal{L} \times \mathcal{B}$ measurable set. By Aumann's theorem, $\mathcal{Y}_k$ has a measurable selection y'. From the preceding analysis, $\int_{[S,T]} \phi(t, y'(t))dt \geq 0$. But then, for each k,

$$0 \leq \int_S^T \phi(t, y'(t))dt \leq \int_S^T \inf_{y \in y_i(t)+(1+k_0(t))^{-1}\mathbb{B}} \phi(t, y)dt + \epsilon_k \times |T - S|.$$

Since k is arbitrary, $\epsilon_k \downarrow 0$ and $\phi(t, y_i(t)) = 0$ a.e.

$$\int_S^T \left(\inf_{y \in y^*(t)+(1+k_0(t))^{-1}\mathbb{B}} \phi(t, y) - \phi(t, y_i(t)) \right)dt \geq 0.$$

7.3 A Preliminary Maximum Principle, for Dynamic Optimization Problems...

From the fact that the integrand is non-positive, we deduce that

$$\phi(t, y_i(t)) = \min\{\phi(t, y) : y \in y_i(t) + (1 + k_0(t))^{-1}\mathbb{B}\}, \text{ a.e. } t \in [S, T].$$

But then

$$-2ik_0(t)(y_i(t) - x_i(t)) \in \partial_x(-p(t) \cdot f(t, y_i(t), u_i(t))) + \gamma_i k_0(t)\mathbb{B} \text{ a.e. } t \in [S, T].$$

From the defining relations for the p's we deduce

$$\dot{p}(t) \in \partial_x(-p(t) \cdot f(t, y_i(t), u_i(t))) + \gamma_i k_0(t)\mathbb{B}, \text{ a.e } t \in [S, T].$$

This implies relation (b)$'$.

Confirmation of (c)$'$: Take $M_k \uparrow \infty$ and, for each k, define the mutifunction $U^k : [S, T] \rightsquigarrow \mathbb{R}^m$

$$U^k(t) := U(t) \cap \{u \in \mathbb{R}^m : |f(t, y_i(t), u) - f(t, y_i(t), u_i(t))| \leq \frac{M_k}{1 + k_0(t)}\mathbb{B}\}.$$

Fix $k > 0$. Choose an arbitrary selector $\hat{u}$ of U^k and define

$$\mathcal{S} := \{\bar{t} \in (S, T) : \bar{t} \text{ is a Lebesgue point of } t \to (m_i(t, \hat{u}(t)) - m_i(t, u_i(t)) \\ - \mathcal{H}_{\lambda=1}(t, y_i(t), p(t), \hat{u}(t)) + \mathcal{H}_{\lambda=1}(t, y_i(t), p(t), u_i(t))\}.$$

Take $\delta_j \downarrow 0$ and $\bar{t} \in \mathcal{S}$. For each $j, t \in [S, T]$, let

$$u^j(t) := \begin{cases} \hat{u}(t) & \text{if } t \in [\bar{t}, (\bar{t} + \delta_j) \wedge T] \\ u_i(t) & \text{if } t \notin [\bar{t}, (\bar{t} + \delta_j) \wedge T]. \end{cases}$$

Write x^j for the state trajectory corresponding to y_i and u^j, with initial value $x_i(S)$. For each j, consider (7.3.5) with $\delta = \delta_j, \xi = 0, y = y_i$, and $u = u^j$. We can show that (7.3.7) and (7.3.8) are satisfied. It follows that

$$E_2(y_i(t), x^j(t), y_i(t), x_i(t)) \to 0 \text{ as } j \to \infty, \text{ a.e..}$$

Since $\bar{t} \in \mathcal{S}$, we can pass to the limit in (7.3.5) as $j \to \infty$, to obtain

$$\gamma_i(m_i(\bar{t}, \hat{u}(\bar{t})) - m_i(\bar{t}, u_i(\bar{t})) - \mathcal{H}_{\lambda=1}\bar{t}, y_i(\bar{t}), p(\bar{t}), u(\bar{t})) \\ + \mathcal{H}_{\lambda=1}(\bar{t}, y_i(\bar{t}), p(\bar{t}), u_i(\bar{t})) \geq 0.$$

Since $\hat{u}$ was chosen arbitrarily and $\mathcal{S}$ has full measure, we can conclude that

$$\int_{[S,T]} \psi(t, u(t))dt \geq 0 \text{ for every selector } u \text{ of } U^k,$$

where

$$\psi(t, u) := \gamma_i(m_i(t, u) - m_i(t, u_i(t))) - \mathcal{H}_{\lambda=1}(t, y_i(t), p(t), u)$$
$$+ \mathcal{H}_{\lambda=1}(t, y_i(t), p(t), u_i(t)).$$

Take $\epsilon_h \downarrow 0$. Fix an integer h. Arguing as above, we can deduce, with the help of Aumann's theorem, there exists a selector u' of U^k such that

$$\psi(t, u'(t)) \leq \inf_{u \in U^k(t)} \psi(t, u) + \epsilon_h \text{ for a.e. } t \in [S, T].$$

(To ensure existence of this selector, we use the fact that $\inf_{u \in U^k(t)} \psi(t, u) > -\infty$ for each t.)

According to the earlier analysis, $\int_{[S,T]} \psi(t, u'(t))dt \geq 0$. It follows that

$$\int_{[S,T]} \inf_{u \in U^k(t)} \psi(t, u) dt \geq -\epsilon_h |T - S|.$$

Since this is true for any h,

$$\int_{[S,T]} \inf_{u \in U^k(t)} \psi(t, u) dt \geq 0.$$

But $\psi(t, u_i(t)) = 0$. It follows that

$$\int_{[S,T]} \left(\inf_{u \in U^k(t)} \psi(t, u) - \psi(t, u_i(t)) \right) dt \geq 0.$$

Noting that the integrand is non-positive, we deduce that

$$\psi(t, u_i(t)) = \inf_{u \in U^k(t)} \psi(t, u) \text{ a.e..}$$

We have shown that

$$-\gamma_i m_i(t, u_i(t)) + \mathcal{H}_{\lambda=1}(t, y_i(t), p(t), u_i(t))$$
$$= \max_{u \in U^k(t)} \{-\gamma_i m_i(t, u) + \mathcal{H}_{\lambda=1}(t, y^*(t), p(t), u)\}.$$

7.3 A Preliminary Maximum Principle, for Dynamic Optimization Problems...

Since $m_i(t, u_i(t)) = 0$ and $|m_i(t, u)| \leq 1$ for all u, we deduce that

$$\mathcal{H}_{\lambda=1}(t, y_i(t), p(t), u_i(t)) \geq \max_{u \in U^k(t)} \{\mathcal{H}_{\lambda=1}(t, y_i(t), p(t), u)\} - \gamma_i.$$

The preceding relation is valid on a set of full measure and

$$U(t) = \lim_{k \to \infty} U^k(t), \quad \text{for all } t \in [S, T].$$

It follows that

$$\mathcal{H}_{\lambda=1}(t, y_i(t), p(t), u_i(t)) \geq \max_{u \in U(t)} \{\mathcal{H}_{\lambda=1}(t, y_i(t), p(t), u)\} - \gamma_i, \quad \text{a.e. } t \in [S, T].$$

The proof of the Lemma is complete.

□

We observe that Lemma 7.3.3 provides perturbed versions of the desired necessary conditions, in terms of a costate function, which we write p_i to emphasize the i dependence. It remains to construct, in the limit as $i \to \infty$, a costate trajectory p, with respect to which conditions in the special case of the maximum principle are satisfied.

Notice that the relation (7.3.2) ensures that, along a subsequence, y_i converges in $L^1_{k_0}$, therefore in L^1 (since $k_0 \geq 1$), and, consequently, a.e., to $\bar{x}(t)$ on $[S, T]$, x_i converges uniformly to $\bar{x}$ and meas$\{t \in [S, T] : u_i(t) \neq \bar{u}\} \to 0$, as $i \to \infty$.

By (d)', the $p_i(T)$'s are uniformly bounded. From (b)' and (A2), $|\dot{p}_i(t)| \leq k_0(t)|p_i(t)|$ a.e.. But then, by Gronwall's lemma, the p_i's are uniformly bounded in the L^∞ norm and the $\dot{p}_i$'s uniformly integrably bounded. By Theorem 6.3.3 (compactness of trajectories), the p_i's converge strongly in L^∞ to some $p \in W^{1,1}$, the $\dot{p}_i$'s converge weakly in L^1 to $\dot{p}$ and p satisfies

$$-\dot{p}(t) \in \text{co } \partial_x (p(t) \cdot f(t, \bar{x}(t), \bar{u}(t))) \text{ a.e. } t \in [S, T],$$

Since the p_i's converge strongly in L^∞, we deduce from (d)'

$$(p(S), -p(T)) \in \nabla \ell(\bar{x}(S), \bar{x}(T)) + \alpha \mathbb{B} \times \{0\} = \partial g(\bar{x}(S), \bar{x}(T))$$

Taking note of (7.3.2), we can arrange, by extracting a subsequence, that

$$\text{meas } \{t : u_i(t) \neq \bar{u}(t) \text{ for all } i \text{ sufficiently large }\} \to 0, \text{ as } i \to \infty.$$

Using this property, we can deduce from (c)' that

$$\mathcal{H}_{\lambda=1}(t, \bar{x}(t), p(t), \bar{u}(t)) \geq \sup_{u \in U(t)} \mathcal{H}_{\lambda=1}(t, \bar{x}(t), p(t), u) \text{ a.e. } t \in [S, T],$$

We have confirmed that all assertions of the proposition, and specified strengthened hypotheses on the structure and regularity of the end-point cost function g.

Step 2: *Completion of the Proof.*

The previous step validates the assertions of the Proposition, under an additional hypothesis concerning the end-point cost function g, namely:

(S): $g(x_0, x_1) = \ell(x_0, x_1) + \alpha |x_0 - \zeta|$, for some C^1 function ℓ, $\zeta \in \mathbb{R}^n$ and $\alpha \geq 0$.

Now suppose that g is merely Lipschitz continuous (with Lipschitz constant k_g), in accordance with hypothesis (F2)'. The final step is to show that the assertions remain true.

For $i = 1, 2, \ldots$, let g^i be the quadratic inf convolution of g (with parameter $\alpha = i$):

$$g^i(z) := \inf_{y \in \mathbb{R}^n \times \mathbb{R}^n} \{g(y) + i \times |y - z|^2\} \text{ for } z \in \mathbb{R}^n \times \mathbb{R}^n . \qquad (7.3.10)$$

Recall the key 'quadratic inf convolution' properties of g^i. Take any $z \in \mathbb{R}^n \times \mathbb{R}^n$ and let $y \in \mathbb{R}^n \times \mathbb{R}^n$ be any vector achieving the infimum in (7.3.10) (one such vector exists). Let

$$\eta^i(z) := -2i(y - z).$$

(i): g^i is Lipschitz continuous with Lipschitz constant k_g,
(ii): $g(z) \geq g^i(z) \geq g(z) - k_g^2 \times i^{-1}$,
(iii): $g^i(z') - g^i(z) \leq \eta^i(z) \cdot (z' - z) + i \times |z' - z|^2$ for all $z' \in \mathbb{R}^n \times \mathbb{R}^n$,
(iv): $\eta^i(z) \in \partial^P g(y)$,
(v): $|y - z| \leq k_g \times i^{-1}$.

Since $(\bar{x}, \bar{u})$ is an L^∞ local minimizer for (F), $(\bar{x}, \bar{u})$ is a minimizer for

$$(Q): \quad \text{Minimize } \{J(x, u) : (x, u) \in \mathcal{B}_\epsilon\},$$

for some $\epsilon > 0$, where

$$J(x, u) := \int_{[S,T]} L(t, u(t))dt + g(x(S), x(T)),$$

and $\mathcal{B}_\epsilon := \{$admissible processes (x, u) for (F) such that $||x - \bar{x}||_{L^\infty} \leq \epsilon\}$.

For each i, consider the problem

$$(Q^i): \quad \text{Minimize } \{J^i(x, u) : (x, u) \in \mathcal{B}_\epsilon\}$$

in which

$$J^i(x, u) := \int_{[S,T]} L(t, u(t))dt + g^i(x(S), x(T)).$$

7.3 A Preliminary Maximum Principle, for Dynamic Optimization Problems...

Equip $\mathcal{B}_\epsilon$ with the metric

$$d_{\mathcal{E}}((x', u'), (x, u)) := \text{meas } \{t \in [S, T] : u'(t) \neq u(t)\} + |x'(S) - x(S)|. \tag{7.3.11}$$

It can be shown that, w.r.t. this metric, $\mathcal{B}_\epsilon$ is complete and J^i is continuous on $\mathcal{B}_\epsilon$.

Now note that, from property (ii) of the quadratic inf convolution construction, $(\bar{x}, \bar{u})$ is a γ_i^2-minimizer for (Q^i), where $\gamma_i^2 := k_g^2 i^{-1}$. In consequence of Ekeland's theorem, there exists $(x_i, u_i) \in \mathcal{B}_\epsilon$ which is a minimizer for $(\tilde{Q}^i)$:

$$(\tilde{Q}^i) \quad \text{Min } \{J^i(x, u) + \gamma_i d_{\mathcal{E}}((x, u), (x_i, u_i)) : (x, u) \in \mathcal{B}_\epsilon\}.$$

and

$$d_{\mathcal{E}}((x_i, u_i), (\bar{x}, \bar{u})) \leq \gamma_i. \tag{7.3.12}$$

The cost function for $(\tilde{Q}^i)$ can be written

$$\tilde{J}^i(x, u) := \int_{[S,T]} (L(t, u(t)) + \gamma_i m_i(t, u(t))) dt + g^i(x(S), x(T)) + \gamma_i |x(S) - x_i(S)|.$$

($m_i(t, u)$ was defined in (7.3.3)). But by properties (iii), (iv) and (v) of quadratic inf convolutions (see above),

$$\begin{cases} g^i(x_0, x_1) - g^i(x_i(S), x_i(T)) \\ \quad \leq \eta_0^i \cdot (x_0 - x_i(S)) + \eta_1^i \cdot (x_1 - x_i(T)) + i \times (|x_0 - x_i(S)|^2 + |x_1 - x_i(T)|^2) \\ \text{and, trivially,} \\ g^i(x_i(S), x_i(T)) - g^i(x_i(S), x_i(T)) \\ \quad = \eta_0^i \cdot (x_i(S) - x_i(S)) + \eta_1^i \cdot (x_i(T) - x_i(T)) \\ \quad + i \times (|x_i(S) - x_i(S)|^2 + |x_i(T) - x_i(T)|^2), \end{cases} \tag{7.3.13}$$

in which

$$(\eta_0^i, \eta_1^i) \in \partial^P g(y_0, y_1) \text{ for some } (y_0, y_1) \in (x_i(S), x_i(T)) + k_g i^{-1}(\mathbb{B} \times \mathbb{B}). \tag{7.3.14}$$

In consequence of the 'upper envelope' property (7.3.13), (x_i, u_i) remains a minimizer for $(\tilde{Q}^i)$, when we replace the additive term '$g^i(x(S), x(T))$' in the cost function by the quadratic function

$$\eta_0^i \cdot (x(S) - x_i(S)) + \eta_1^i \cdot (x(T) - x_i(T)) + i \times (|x(S) - x_i(S)|^2 + |x(T) - x_i(T)|^2).$$

In other words, (x_i, u_i) is a minimizer for

$$(\tilde{Q}_0^i): \quad \text{Minimize } \{\tilde{J}_0^i(x, u) : (x, u) \in \mathcal{B}_\epsilon\},$$

in which

$$\tilde{J}_0^i(x,u) := \int_{[S,T]} (L(t,u(t)) + \gamma_i m_i(t,u(t)))dt + \eta_0^i \cdot (x(S) - x_i(S))$$
$$+ \eta_1^i \cdot (x(T) - x_i(T)) + i \times (|x(S) - x_i(S)|^2$$
$$+ |x(T) - x_i(T)|^2) + \gamma_i |x(S) - x_i(S)|.$$

The data for Problem ($\tilde{Q}_0^i$) satisfies the hypotheses for the validity of the assertions of the Proposition, in the special case covered by Step 1. We conclude the existence of $p_i : [S,T] \to \mathbb{R}^n$ Note, in particular, that, from (d'), $(p_i(S), -p_i(T)) \in (\eta_0^i, \eta_1^i) + \gamma_i \mathbb{B} \times \{0\} \subset \partial^P g(y_0, y_1) + \gamma_i \mathbb{B} \times \{0\}$. From (7.3.14) then,

$$(p_i(S), -p_i(T)) \in (\eta_0^i, \eta_1^i) + \gamma_i \mathbb{B} \times \{0\}$$
$$\subset \bigcup_{z \in (x_i(S), x_i(T)) + k_g i^{-1} \mathbb{B} \times \mathbb{B}} \partial^P g(z) + \gamma_i \mathbb{B} \times \{0\}, \quad (7.3.15)$$

which is a convenient perturbed version of (d), for limit taking. Next define, for each j, $\mathcal{E}_j := \{t \in [S,T] : u_j(t) \neq \bar{u}(t)\}$. From (7.3.12) we know that $meas\{\mathcal{E}_j\} \to 0$ as $j \to \infty$, where $meas$ denotes Lebesgue measure. By extracting a subsequence, we can arrange that $\sum_{j=1}^\infty meas\{\mathcal{E}_j\} < \infty$. But then

$$\sum_{j=k}^\infty meas\{\mathcal{E}_j\} \to 0, \quad \text{as } k \to \infty. \quad (7.3.16)$$

It follows that $meas \cup_{j=k}^\infty \mathcal{E}_j \to 0$, as $k \to \infty$. Now define $\mathcal{T} := [S,T] \setminus \left(\cap_{k=1}^\infty \cup_{j=k}^\infty \mathcal{E}_j \right)$. Notice that $\mathcal{T}$ has representation

$$\mathcal{T} = \{t \in [S,T] : \text{ there exists } N(t) \text{ s.t. } u_i(t) = \bar{u}(t) \text{ for all } i \geq N(t)\}.$$

In view of (7.3.16), $\mathcal{T}$ is a subset of $[S,T]$ with full Lebesgue measure. It follows that

$$u_i(t) = \bar{u}(t), \quad \text{for all } i \text{ sufficiently large, a.e. } t \in [S,T]. \quad (7.3.17)$$

From (7.3.12) we also know that $x_i \to \bar{x}$, uniformly, as $i \to \infty$. On the other hand, the p_i's are uniformly bounded with uniformly integrably bounded derivatives, in view of conditions (c)' and (d)' of Lemma 7.3.3. We may deduce from Ascoli's theorem that, after a further subsequence extraction, p_i converges uniformly to some $W^{1,1}$ function p as $i \to \infty$, and its derivative $\dot{p}_i$ converges weakly in L^1 to $\dot{p}$. The validity of the co-state inclusion (b) (expressed in terms of the limiting p) now follows from the compactness of trajectories theorem Theorem 6.3.3. Condition (c) is obtained in the limit from (c)′, in view of the convergence properties of the x_i's, p_i's and u_i's. Finally, (d) is obtained in the limit from (7.3.15). □

Notice the important role of the convergence property (7.3.17). It has permitted us to derive the costate inclusion and Weierstrass condition, as a limit of the perturbed versions of these conditions, despite the fact the u- dependence of $f(t, x, u)$ and $L(t, u)$ is possibly discontinuous. This would not have been the case, if we had, for example, chosen the metric, for application of Ekeland's theorem in the final step of the proof, to be $d((x', u'), (x, u)) := ||u' - u||_{L^1} + |x'(S) - x(S)|$ in place of (7.3.11). The use of this alternative metric would result in the replacement of (7.3.17) by

$$u_i \to u \text{ strongly in } L^1 \text{ and } a.e.$$

and would not permit limit taking to give the desired necessary conditions, in the absence of additional continuity hypotheses concerning the u-dependence of the data.

7.4 Proof of Theorem 7.2.1

In this section we prove assertions (a)–(d) of the theorem. Proof of the remaining assertion (e), concerning additional properties of $W^{1,1}$ minimizers, when U and $f(t, x, u)$ do not depend on t, will be given in Chap. 9 (Sect. 9.6), as part of a broader study of necessary conditions when additional regularity hypotheses are imposed on the time dependence of the data and when the end-times are included among the choice variables.

Step 1. (Proof of the Theorem Under Strengthed Hypotheses)

In this step we establish validity of the assertions of the Theorem under the following additional hypothesis:

(A): There exist integrable functions $k_0 : [S, T] \to \mathbb{R}$ and $c_0 : [S, T] \to \mathbb{R}$ such that

(i): $|f(t, x, u) - f(t, x', u)| \le k_0(t)|x - x'|$
(ii): $|f(t, x, u)| \le c_0(t)$

for all $x, x' \in \bar{x}(t) + \bar{\epsilon}\mathbb{B}, u \in U(t)$, a.e. $t \in [S, T]$.

and when $(\bar{x}, \bar{u})$ is an L^∞ local minimizer for (P).

Let $\epsilon \in (0, \bar{\epsilon})$ be such that $(\bar{x}, \bar{u})$ is a minimizer for (P) w.r.t. admissible state trajectories x satisfying $||x - \bar{x}||_{L^\infty} \le \epsilon$.

Take $\gamma_i \downarrow 0$. For $i = 1, 2 \ldots$, consider the problem:

$$(P_1^i) \begin{cases} \text{Minimize } J_1^i(x, u) \text{ subject to} \\ \dot{x}(t) = f(t, x(t), u(t)) \\ u(t) \in U(t) \text{ a.e. } t \in [S, T], \\ ||x - \bar{x}||_{L^\infty} \leq \epsilon, \end{cases}$$

in which $J_1^i(x, u) = \max\{g(x(S), x(T)) - g(\bar{x}(T), \bar{x}(T)) + \gamma_i^2, d_C(x(S), x(T))\}$. Since $J_1^i(\bar{x}, \bar{u}) = \gamma_i^2$, and J_1^i is non-negative valued, $(\bar{x}, \bar{u})$ is a γ_i-minimizer.

Let $\mathcal{B}_\epsilon$ denote the set of admissible processes for (P_1^i). Applying Ekeland's theorem to the function $J_1^i : \mathcal{B}_\epsilon \to \mathbb{R}$ with metric $d_{\mathcal{E}}$ given by (7.3.11), we are assured of the existence of a minimizer (x_i, u_i) for

$$(\tilde{P}_1^i) \begin{cases} \text{Minimize } \tilde{J}_1^i(x, u) \text{ subject to} \\ \dot{x}(t) = f(t, x(t), u(t)) \\ u(t) \in U(t) \text{ a.e. } t \in [S, T], \\ ||x - \bar{x}||_{L^\infty} \leq \epsilon, \end{cases}$$

where

$$\tilde{J}_1^i(x, u) := J_1^i(x, u) + \gamma_i \left(\int_{[S,T]} m_i(t, u(t)) \, dt + |x(S) - x_i(S)| \right).$$

Furthermore

$$d_{\mathcal{E}}((x_i, u_i), (\bar{x}, \bar{u})) \leq \gamma_i. \tag{7.4.1}$$

It can be deduced from (7.4.1) that $||x_i - \bar{x}||_{L^\infty} \to 0$ as $i \to \infty$ and therefore that, for i sufficiently large, (x_i, u_i) is an L^∞ local minimizer for $(\tilde{P}_1^i)$, when the constraint '$||x - \bar{x}||_{L^\infty} \leq \epsilon$' is removed. The data for this last problem satisfies the hypotheses of Proposition 7.3.1. We deduce the existence of a costate trajectory p and with properties listed in the Proposition statement. From properties (b) and (c) we obtain immediately the conditions

(b)': $-\dot{p}(t) \in \text{co } \partial_x p(t) \cdot f(t, x_i(t), u_i(t))$, a.e. $t \in [S, T]$,
(c)': $\mathcal{H}(t, x_i(t), p(t), u_i(t)) \geq \max_{u \in U(t)} \mathcal{H}(t, x_i(t), p(t), u) - \gamma_i$ a.e. $t \in [S, T]$.

Let us examine the implications of the transversality condition (d). In this connection, we take note of the following important strict inequality:

$$\max\{g(x_i(S), x_i(T)) - g(\bar{x}(S), \bar{x}(T)) + \gamma_i^2, d_C(x_i(S), x_i(T))\} > 0,$$

for i sufficiently large. \hfill (7.4.2)

Indeed if this were not the case, we would have $g(x_i(S), x_i(T)) - g(\bar{x}(S), \bar{x}(T)) \leq -\gamma_i^2$ and $d_C(x_i(S), x_i(T)) = 0$. This contradicts the L^∞ local optimality of $(\bar{x}, \bar{u})$ for problem (P).

7.4 Proof of Theorem 7.2.1

In consequence of the max rule for limiting subdifferentials, we know that

$$\partial \max\{g(\bar{x}_i(S), \bar{x}_i(T)) - g(\bar{x}(S), \bar{x}(T)) + \gamma_i^2, d_C(\bar{x}_i(S), \bar{x}_i(T))\}$$
$$\subset \lambda \partial g(\bar{x}_i(S), \bar{x}_i(T)) + (1-\lambda) \partial d_C(\bar{x}_i(S), \bar{x}_i(T)),$$

for some $\lambda \in [0, 1]$. Moreover, in view of (7.4.2),

'$1 - \lambda > 0$' $\Longrightarrow$

'$d_C(\bar{x}_i(S), \bar{x}_i(T)) = \max\{g(\bar{x}_i(S), \bar{x}_i(T)) - g(\bar{x}(S), \bar{x}(T)) + \gamma_i^2,$
$$d_C(\bar{x}_i(S), \bar{x}_i(T))\}'. \tag{7.4.3}$$

We deduce from condition (d) that

(d)': $(p(S), -p(T)) \in \lambda \partial g(x_i(S), x_i(T)) + (1-\lambda)\partial d_C(x_i(S), x_i(T)) + \gamma_i \mathbb{B} \times \{0\}$.

We claim that

$$(1 + k_g)\lambda + |(p(S), p(T))| \geq 1 - \gamma_i. \tag{7.4.4}$$

This is obviously true if $\lambda = 1$. If, on the other hand, $\lambda < 1$, we know from (7.4.2) and (7.4.3) that $d_C(x_i(S), x_i(T)) > 0$. But then $\partial d_C(x_i(S), x_i(T)) \subset \partial \mathbb{B}$. It follows then from (d)' that

$$|(p(S), p(T))| \geq (1 - \lambda) - k_g \lambda - \gamma_i.$$

Then $(1 + k_g)\lambda + |(p(S), p(T))| \geq 1 - \gamma_i$, as asserted.

To emphasize the fact that these relations depend on i, we change the notation for p and λ to p_i and λ_i, respectively. Condition (7.4.1) ensures that, along a subsequence,

$$u_i(t) = \bar{u}(t) \text{ for all } i \text{ sufficiently large, a.e. } t \in [S, T]$$

and also that $x_i \to \bar{x}$ uniformly as $i \to \infty$. The p_i's are a uniformly bounded sequence of absolutely continuous functions with uniformly integrably bounded derivatives. It follows that p_i converges to an absolutely continuous function p, and $\dot{p}_i$ converges to $\dot{p}$ weakly in L^1, following extraction of a further subsequence A similar convergence analysis to that employed in the proof of Proposition 7.3.1 permits us to pass to the limit in relations $(b)' - (d)'$, and thereby arrive at the assertions (b)-(d) of the theorem statement. We deduce from (7.4.4), in the limit, that $(1 + k_g)\lambda + |(p(S), p(T))| \geq 1$. Hence $\lambda + ||p||_{L^\infty} > 0$. This is condition (a).

Step 2: *Completion of the Proof*

We have confirmed the assertions of the theorem, but only when $(\bar{x}, \bar{u})$ is an L^∞ local minimizer and the hypotheses (H1)–(H3) are supplemented by (A). It remains to remove both of these restrictions.

Assume, first, that $(\bar{x}, \bar{u})$ is an L^∞ local minimizer for (P) and that the assertions of the Theorem are known to be valid under hypotheses (H1)–(H3) and (A). We show that the assertions remain valid when (H1)–(H3) alone are satisfied.

For each integer $j = 1, 2, \ldots$, consider the dynamic optimization problem

$$(P_j) \begin{cases} \text{Minimize } g(x(S), x(T)) \\ \text{over } x \in W^{1,1}([S, T]; \mathbb{R}^n) \text{ and meas. functions } u : [S, T] \to \mathbb{R}^m, \\ \text{such that} \\ \dot{x}(t) = f(t, x(t), u(t)) \text{ a.e. } t \in [S, T], \\ u(t) \in U_j(t) \text{ a.e. } t \in [S, T], \\ (x(S), x(T)) \in C, \end{cases}$$

in which

$$U_j(t) := \{u \in U(t) : k(t, u) \le k(t, \bar{u}(t)) + j, |f(t, \bar{x}(t), u)| \le |\dot{\bar{x}}(t)| + j\}.$$

For each j, $(\bar{x}, \bar{u})$ remains an L^∞ local minimizer of (P_j). (H1)–(H3) are satisfied. But, because of the modified control constraint multifunction, the data for problem (P_j) also satisfies the supplementary hypothesis (A), in which the integrable functions $k(t, \bar{u}(t)) + j$ and $|\dot{\bar{x}}(t)| + j + \bar{\epsilon} \times (k(t, \bar{u}(t)) + j)$ take the roles of $k_0(t)$ and $c_0(t)$. For each j then, there exists (p_j, λ_j) such that conditions (a), (b) and (d) of the theorem statement are satisfied (when (λ_j, p_j) replaces (λ, p)), together with the restricted-sense Weierstrass condition:

$$\mathcal{H}(t, \bar{x}(t), p_j(t), \bar{u}(t)) \ge \max_{u \in U_j(t)} \mathcal{H}(t, \bar{x}(t), p_j(t), u) \quad \text{for all } t \in \mathcal{S}, \tag{7.4.5}$$

where $\mathcal{S} \subset [S, T]$ is a subset of full measure, which does not depend on j. For each j, we can normalize the non-zero Lagrange multipliers to satisfy $\lambda_j + |p_j(S)| = 1$. From the costate inclusion we deduce that $|\dot{p}_j(t)| \le k(t, \bar{u}(t))|p_j(t)|$, a.e.. It follows from Gronwall's lemma that the p_j's are uniformly bounded and, consequently, the $\dot{p}_j$'s have a uniform integral bound. Along a subsequence then, $\lambda_j \to \lambda$ for some $\lambda \in [0, 1]$ and $p_j \to p$ strongly in L^∞, for some $p \in W^{1,1}$. Furthermore, $\dot{p}_j \to \dot{p}$ weakly in L^1. Passing to the limit, as $j \to \infty$, we deduce that $\lambda + |p(S)| = 1$. This implies condition (a) of the theorem statement.

We recover also conditions (b) and (d) of the theorem statement (expressed in terms of the Lagrange multipliers (p, λ)), in the limit as $j \to \infty$. Making use of the observation:

for any $t \in \mathcal{S}$ and $u \in U(t)$, $u \in U_j(t)$ for j sufficiently large,

we recover also (c), from (7.4.5).

All the assertions of the theorem have been proved, under the hypotheses (H1)-(H3), in the case when $(\bar{x}, \bar{u})$ is an L^∞ local minimizer. Now suppose that $(\bar{x}, \bar{u})$ is merely a $W^{1,1}$ minimizer. Then there exists $\epsilon > 0$ be such that $(\bar{x}, \bar{u})$ is minimizing with respect to admissible processes (x, u) satisfying $\|x - \bar{x}\|_{W^{1,1}} \leq \epsilon$.

We see that $(\bar{z} \equiv 0, \bar{x}, \bar{u})$ is a minimizer for the dynamic optimization problem with augmented state:

$$(P_{\text{aug}}) \begin{cases} \text{Minimize } g(x(S), x(T)) \\ \text{over } x \in W^{1,1}([S, T]; \mathbb{R}^n), \ z \in W^{1,1}(S, T) \text{ and } u \text{ a selector of } U, \\ \text{such that} \\ (\dot{x}(t), \dot{z}(t)) = (f(t, x(t), u(t)), |f(t, x(t), u(t)) - \dot{\bar{x}}(t)|) \text{ a.e. } t \in [S, T], \\ u(t) \in U(t) \text{ a.e. } t \in [S, T], \\ (x(S), x(T)) \in C, \ z(S) = 0, \ |x(S) - \bar{x}(S)| + |z(T)| \leq \epsilon. \end{cases}$$

The data for problem (P_{aug}) satisfies (H1)–(H3). But then, since $(\bar{z}, \bar{x}, \bar{u})$ is certainly an L^∞ minimizer, we can apply the special case of the theorem already proved. We deduce the existence of a set of Lagrange multipliers (λ, p, r), in which the costate trajectory components p and r are associated with the state variables x and z, satisfying the conditions of the maximum principle, expressed in terms of (λ, p, r) and $(\bar{z}, \bar{x}, \bar{u})$.

Since the constraint '$|x(S) - \bar{x}(S)| + |z(T)| \leq \epsilon$' is inactive at $(\bar{x}(S), z(T))$, we can assume, when analysing the implications of the transversality condition associated with (P_{aug}), that this constraint is absent; according, we deduce that $r(T) = 0$ and $(p(S), -p(T)) \in \partial g(\bar{x}(S), \bar{x}(T)) + N_C(g(\bar{x}(S), \bar{x}(T))$. The latter relation is the transversality condition for (P).

The dynamic constraint in (P_{aug}) takes the form of a controlled differential equation that does not depend on the z variable; hence, $\dot{r} \equiv 0$. Since $r(T) = 0$, we know then that $r \equiv 0$. Setting $r \equiv 0$ in all the other maximum principle relations for (P_{aug}) yields the non-triviality condition, costate inclusion and Weierstrass condition for (P). □

7.5 Exercises

7.1 Consider the autonomous dynamic optimization problem

$$(P7) \begin{cases} \text{Minimize } g(x(S), x(T)) \\ \text{subject to} \\ \dot{x}(t) = f(x(t), u(t)) \text{ and } u \in U, \text{ a.e. } t \in [S, T], \\ (x(S), x(T)) \in C. \end{cases}$$

with data: functions $g : \mathbb{R}^n \times \mathbb{R}^n \to \mathbb{R}$, $f : \mathbb{R}^n \times \mathbb{R}^m \to \mathbb{R}^n$ and sets $U \subset \mathbb{R}^m$ and $C \subset \mathbb{R}^n \times \mathbb{R}^n$. Assume

(a): $f(., u)$ is C^1 for each $u \in U$, f is continuous, U is compact, C is closed, g is locally Lipschitz continuous,
(b): there exist $c_f > 0$ and $k_f > 0$ such that

$$|f(x', u) - f(x, u)| \leq k_f |x' - x| \text{ and } |f(x, u)| \leq c_f,$$
$$\text{for all } x', x \in \mathbb{R}^n, t \in [S, T], u \in U.$$

Let $(\bar{x}, \bar{u})$ be an extremal for $(P7)$, in the sense that there exists a costate trajectory and cost multiplier $(p, \lambda) \neq (0, 0)$ satisfying the costate equation, Weierstrass condition and transversality conditions of Theorem 7.2.1 along $(\bar{x}, \bar{u})$. Confirm the 'constancy of the maximized Hamiltonian' condition: there exists $c \in \mathbb{R}$ such that

$$H(\bar{x}(t), p(t)) = c, \text{ for all } t \in [S, T],$$

where $H(x, p) := \max_{u \in U} p \cdot f(x, u)$.

Remark
The implications of this exercise is that, for smooth data, constancy of the maximized Hamiltonian is a consequence of the standard extremality conditions and therefore not an independent condition.

Hint: Show first that $H(x, p)$ is locally Lipschitz continuous and, consequently, $t \to H(\bar{x}(t), p(t))$ is absolutely continuous, as a composition of a Lipschitz continuous function and an absolutely continuous function. Deduce that the function has zero derivative at appropriate points $t \in (S, T)$ comprising a set of full measure, taking as starting point the relations

$$p(t + \epsilon) \cdot f(\bar{x}(t + \epsilon), \bar{u}(t + \epsilon)) - p(t) \cdot f(\bar{x}(t), \bar{u}(t))$$
$$= H(\bar{x}(t + \epsilon), p(t + \epsilon)) - H(\bar{x}(t), p(t))$$

for all ϵ (positive or negative), sufficiently small in magnitude.

7.2 Let $\bar{x}$ be a minimizer for the calculus of variations problem involving higher order derivatives

$$\begin{cases} \text{Minimize } \int_S^T L(t, x(t), D^{(1)}x(t), \ldots, D^{(N)}x(t))dt \\ \text{over } x \in W^{N,1}([S, T]; \mathbb{R}^n) \text{ s.t.} \\ (x, D^{(1)}x, \ldots, D^{(N-1)}x)(S) = d_0 \text{ and} \\ (x, D^{(1)}x, \ldots, D^{(N-1)}x)(T) = d_1, \end{cases}$$

the data for which comprise an integer $N > 0$, a function $L : \mathbb{R}^{2+N} \to \mathbb{R}$, and N-vectors d_0 and d_1. Here, $D^{(k)}x$ denotes the k'th derivative of x. Assume

(a): L is a C^2 function and the minimizer $\bar{x}$ is C^{N+1},

7.5 Exercises

(b): there exists $k_L \in L^1(S, T)$ such that

$$|\nabla_{x_0,\ldots,x_{N-1}} L(t, x_0, \ldots, x_{N-1}, D^{(N)}\bar{x}(t))|_{(x_0,\ldots,x_{N-1})=(\bar{x}(t), D^{(1)}(\bar{x}(t)),\ldots,D^{(N-1)}\bar{x})(t)}|$$
$$\leq k_L(t), \quad \text{for a.e. } t \in [S, T].$$

Derive the following higher order Euler equation

$$\nabla_{x_0} L(t, \bar{x}(t), \ldots, D^{(N)}\bar{x}(t)) - \frac{d}{dt}\left(\nabla_{x_1} L(t, \bar{x}(t), \ldots, D^{(N)}\bar{x}(t))\right) + $$
$$\ldots (-1)^N \frac{d^N}{dt^N}\left(\nabla_{x_N} L(t, \bar{x}(t), \ldots, D^{(N)}\bar{x}(t))\right) = 0, \text{ a.e. } t \in [S, T].$$

Hint: Reformulate the problem as one with block states $x, D^{(1)}x, \ldots, D^{(N-1)}x$ and in which $D^{(N)}x$ is interpreted as a control variable. Show that this problem is normal (cost multiplier must be non-zero) and apply the maximum principle. This yields absolutely continuous functions $(p_0, \ldots, p_{N-1})$ such that condition.

$$(\dot{p}_0, \dot{p}_1 + p_0, \dot{p}_2 + p_1, \ldots, \dot{p}_{N-2} + p_{N-3}, \dot{p}_{N-1} + p_{N-2})(t)$$
$$= \nabla_{x_0,\ldots,x_{N-1}} L(t, \bar{x}(t), D^{(1)}\bar{x}(t), \ldots, D^{(N)}\bar{x}(t)), \text{ for a.e. } t \in [S, T].$$

Eliminate $p_0, \ldots, p_{N-2}$ by differentiating across these equations. Combine the resulting equation for p_{N-1} with the Weierstrass condition to obtain the higher order Euler equation.

7.3 (A Multiprocess Maximum Principle) Consider a collection of control systems: for $i = 1, \ldots, M$

$$(S_i) \begin{cases} \dot{x}_i(t) = f_i(x_i(t), u_i(t)) \text{ a.e. } t \in [s_i, t_i], \\ u_i(t) \in U_i, \text{ a.e. } t \in [s_i, t_i]. \end{cases}$$

Here, for each i, $f_i : \mathbb{R}^{n_i} \times \mathbb{R}^{m_i} \to \mathbb{R}^{n_i}$ is a given function and $U_i \subset \mathbb{R}^{m_i}$ is a given subset. Now consider the dynamic optimization problem

$$(P7) \begin{cases} \text{Minimize } g(\{(s_i, x_i(s_i), t_i, x_i(t_i))\}_{i=1}^M) \\ \text{subject to} \\ ([s_i, t_i], x_i, u_i) \text{ is a (free end-time) process for } (S_i), i = 1, \ldots, M, \\ \{(s_i, x_i(s_i), t_i, x_i(t_i))\}_{i=1}^M \in C, \end{cases}$$

in which $g : \Pi_{i=1}^M (\mathbb{R} \times \mathbb{R}^{n_i} \times \mathbb{R} \times \mathbb{R}^{n_i}) \to \mathbb{R}$ is a given locally Lipschitz function and $C \subset \Pi_{i=1}^M (\mathbb{R} \times \mathbb{R}^{n_i} \times \mathbb{R} \times \mathbb{R}^{n_i})$ is a given closed set.

Let $\{([\bar{s}_i, \bar{t}_i], \bar{x}_i, \bar{u}_i)\}_{i=1}^M$ be a minimizing family of processes for this problem. Assume that the data for each system (S_i), in relation to $(\bar{x}_i, \bar{u}_i)$ satisfy the hypotheses of the free end-time nonsmooth maximum principle (Theorem 9.6.1). Assume furthermore, $\bar{t}_i > \bar{s}_i$ for each i. Derive the following maximum principle for 'optimal multiple processes': there exist $\lambda \geq 0$ and $\{p_i : [\bar{s}_i, \bar{t}_i] \to \mathbb{R}^{n_i}\}$ such that

(i): $\lambda + \sum_{i=1}^M \|p\|_{L^\infty}(\bar{s}_i, \bar{t}_i) \neq 0$,
(ii): $-\dot{p}_i(t) \in \mathrm{co}\,\partial_x p_i(t) \cdot f_i(\bar{x}_i(t), \bar{u}_i(t))$, a.e. $t \in [\bar{s}_i, \bar{t}_i]$, $i = 1, \ldots, M$,
(iii): $p_i(t) \cdot \dot{\bar{x}}_i(t) = \max_{u \in U_i} p_i(t) \cdot f_i(\bar{x}_i(t), \bar{u}_i(t))$, a.e. $t \in [\bar{s}_i, \bar{t}_i]$,
(iv): $\{(-h_i, p_i(\bar{s}_i), h_i, -p_i(\bar{t}_i))\} \in \lambda \partial g(\{(\bar{s}_i, \bar{x}_i(\bar{s}_i), \bar{t}_i, \bar{x}_i(\bar{t}_i))\})$
$\qquad\qquad\qquad\qquad\qquad\qquad\qquad + N_C(\{(\bar{s}_i, \bar{x}_i(\bar{s}_i), \bar{t}_i, \bar{x}_i(\bar{t}_i))\})$,

in which $h_i \in \mathbb{R}$, for each i, is such that

$$h_i = \sup_{u \in U_i} p_i(t) \cdot f_i(\bar{x}_i(t), u), \quad \text{a.e. } t \in [\bar{s}_i, \bar{t}_i].$$

Hint: Consider the standard dynamic optimization problem in which the state and control variables are $((y_1, \tau_1), \ldots, (y_M, \tau_M))$ and $((v_1, w_1), \ldots, (v_M, w_M))$ respectively, and the time domain is $[0, 1]$:

$$\begin{cases} \text{Minimize } g(\{(\tau_i(0), y_i(0), \tau_i(1), y_i(1))\}_{i=1}^M) \\ \text{subject to:} \\ (\dot{\tau}_i(s), \dot{y}_i(s)) = (\bar{t}_i - \bar{s}_i) w_i(s)(1, f_i(y_i(s), v_i(s))) \text{ a.e. } s \in [0, 1], \text{ for } i = 1, \ldots, M, \\ (v_1, \ldots, v_M)(s) \in U_1 \times \ldots \times U_M, \text{ a.e. } s \in [0, 1], \\ (w_1, \ldots, w_M)(s) \in [0.5, 1.5] \times \ldots \times [0.5, 1.5], \text{ a.e. } s \in [0, 1], \\ \{(\tau_i(0), y_i(0), \tau_i(1), y_i(1))\}_{i=1}^M \in C. \end{cases}$$

By considering changes of variable $s \to \tau_i(0) + (\bar{t}_i - \bar{s}_i) \int_0^s w_i(s') ds'$, $i = 1, \ldots, M$ (cf. Sect. 9.6), show that $(\{(\bar{y}_i, \bar{\tau}_i), (\bar{v}_i), \bar{w}_i\})$ is a minimizer for this problem. Here, for each i,

$$\bar{\tau}_i(s) := \bar{s}_i + (\bar{t}_i - \bar{s}_i)s, \quad \bar{y}_i(s) := \bar{x}_i(\bar{s}_i + (\bar{t}_i - \bar{s}_i)s), \quad \bar{v}_i(s)$$
$$:= \bar{u}_i(\bar{s}_i + (\bar{t}_i - \bar{s}_i)s)) \text{ and } \bar{w}_i(s) \equiv 1.$$

Apply the maximum principle (Theorem 7.2.1) and interpret the resulting relations in terms of the data for the original problem.

7.4 (Clarke's Nonsmooth Maximum Principle for Boundary Points of Reachable Sets)
Consider the control system

$$(S) \begin{cases} \dot{x}(t) = f(t, x(t), u(t)) \text{ a.e. } t \in [S, T], \\ u(t) \in U(t), \text{ a.e. } t \in [S, T]. \end{cases}$$

Let $\psi : \mathbb{R}^n \times \mathbb{R}^n \to \mathbb{R}^k$. Define the endpoints reachable set, relative to ψ, to be

$$\mathcal{R}_\psi := \{\psi(x(S), x(T)) : (x, u) \text{ is a process for } (S)\}.$$

(A process is a pair comprising an absolutely continuous x and a measurable u satisfying the constraints of (S)). Let $(\bar{x}, \bar{u})$ be a process such that

$$\psi(\bar{x}(S), \bar{x}(T)) \subset \partial \mathcal{R}_\psi,$$

where $\partial \mathcal{R}_\psi$ denotes 'boundary of $\mathcal{R}_\psi$'. Assume that ψ is C^1 on a neighbourhood of $(\bar{x}(S), \bar{x}(T))$. Assume also that, for some $k_f \in L^1(S, T)$ and $\epsilon > 0$,

(a): $f(., x, .)$ is $\mathcal{L} \times \mathcal{B}^m$ measurable for each $x \in \mathbb{R}^n$ and $\{(t, u) \in [S, T] \times \mathbb{R}^m : u \in U(t)\}$ is a $\mathcal{L} \times \mathcal{B}^m$ measurable set,
(b): $|f(t, x, u) - f(t, x', u)| \leq k_f(t)|x - x'|$ for all $x, x' \in \bar{x}(t) + \epsilon \mathbb{B}$ and $u \in U(t)$, a.e. $t \in [S, T]$.

Show that there exists $p \in W^{1,1}([S, T]; \mathbb{R}^n)$ such that
(i): $-\dot{p}(t) \in \operatorname{co} \partial_x p(t) \cdot f(t, \bar{x}(t), \bar{u}(t))$, a.e. $t \in [S, T]$,
(ii): $p(t) \cdot f(t, \bar{x}(t), \bar{u}(t)) = \max_{u \in U(t)} \left(p(t) \cdot f(t, \bar{x}(t), u) \right)$ a.e. $t \in [S, T]$,
(iii): $(p(S), -p(T)) = \eta D\psi(\bar{x}(S), \bar{x}(T))$, for some $\eta \in \mathbb{R}^k$ such that $|\eta| = 1$.

Here, $D\psi(x_0, x_1)$ denotes the Jacobian of ψ at (x_0, x_1).
Hint: Since $\psi(\bar{x}(S), \bar{x}(T))$ is a boundary point, there exists a sequence of points $\{z_i\}$ in $\mathbb{R}^k \setminus \mathcal{R}_\phi$ such that $z_i \to \psi(\bar{x}(S), \bar{x}(T))$, as $i \to \infty$. Then $(\bar{x}, \bar{u})$ is an α_i minimizer for

$$\begin{cases} \text{Minimize } |z_i - \psi(x(S), x(T))| \\ \text{over processes } (x, u) \text{ for } (S), \end{cases}$$

for some $\alpha_i \downarrow 0$. Now apply Ekeland's theorem and Clarke's nonsmooth maximum principle, and pass to the limit as $i \to \infty$.

7.6 Notes for Chapter 7

The maximum principle came to prominence through the book [167], co-authored by Pontryagin, Boltyanskii, Gamkrelidze and Mischenko, published in Russian in 1961, and in English translation in 1962. It is also referred to as the Pontryagin maximum principle because of Pontryagin's role as leader of the research group at the Steklov Institute, Moscow, which achieved this advance. The first proof is attributed to Boltyanskii [42], however. There is a voluminous Russian literature on the original maximum principle and extensions, which we make no attempt to summarize here. We refer to Milyutin and Osmolovskii's book [153], which covers the important contributions of Dubovitskii/Milyutin and their collaborators.

Prominent among researchers in the West who first entered this field were Neustadt, Warga and Halkin. (See the monographs [163, 164] and [204] and the paper [121].) These authors had their distinctive points of view—Neustadt's and Halkin's approach to deriving necessary conditions were close in spirit to that of Dubovitskii/Milyutin and Gamkrelidze, while Warga worked more in a Western tradition of variational analysis associated with L. C. Young and McShane. They all aimed, however, to axiomatize the proof of Pontryagin, Boltyanskii, Gamkrelidze and Mischenko [167]. The key idea is to show that if a (suitably chosen) convex approximation to some generalized 'target set' intersects the interior of a convex approximation to the reachable set at some control u^*, then the reachable set itself intersects the interior of the target set. This furnishes a contrapositive proof of the maximum principle because the first property is equivalent to 'the maximum principle conditions are not satisfied at u^*', while the second property implies 'u^* is not a minimizer'. We regard this as a 'dual' approach—the assertion 'the maximum principle conditions are not satisfied' is essentially a statement about the non-existence of a hyperplane separating the convex approximations to the target and reachable sets. It is 'axiomatic' to the extent that attention focuses on abstract conditions on convex approximations of sets, consistent with these relations. The elegant mixed Lagrange multiplier rules of Ioffe and Tihomirov [132], yielding necessary conditions in dynamic optimization as direct corollaries, were also grounded in this idea.

The challenge of deriving versions of the maximum principle covering problems with nonsmooth data was taken up in the 1970s. A landmark advance was Clarke's nonsmooth maximum principle [59], treating fully nonsmooth problems with general endpoint constraints, published in 1976. (An earlier version, dealing with the 'free endpoint' case, earlier appeared in Clarke's 1973 PhD thesis.) The techniques used by Clarke to prove his nonsmooth maximum principle [59] had a decidedly 'primal' flavour and were a marked departure from the methodologies of Dobovitski/Milyutin, Neustadt, Halkin and Warga. Clarke's proof built on the fact that simple, variational arguments can be used to derive necessary conditions of optimality for problems with no right end-point constraints. He dealt with the troublesome right end-point constraint by finding a neighbouring process which is a minimizer for a perturbed optimization problem with free right end-point. Necessary conditions for the perturbed, free right end-point problem are invoked. (The perturbed problems needed to be constructed in a subtle manner, usually with the help of a variational principle, to ensure that the minimizing controls for the perturbed problems converge to the nominal control. Naive penalty methods will not do the trick.) Necessary conditions are then obtained for the original problem by passage to the limit. This general approach, deriving necessary conditions for a 'difficult' dynamic optimization problem via necessary conditions for a simpler perturbed problem and passing to the limit, is a very powerful one and adapts to other contexts; it has been used, for example, to obtain necessary conditions for dynamic optimization problems with impulsive control [181, 197], time delays [41], infinite dimensional state spaces [100, 140].

On the other hand, Warga extended the earlier dual approach to prove, independently, another kind of nonsmooth maximum principle for dynamic optimization problems with data Lipschitz continuous with respect to the state variable x [205, 207]. The role of gradients with respect to x in the co-state equation and transversality condition is here taken by another kind of 'generalized' derivative namely a *derivate container*. Because the dynamic optimization literature makes reference to this construct, we note here the definition:

Definition Take a neighbourhood $\mathcal{O}$ of a point $\bar{x} \in \mathbb{R}^n$ and a Lipschitz continuous mapping $\Lambda : \mathcal{O} \to \mathbb{R}^m$. A compact set $\mathcal{L}$ of linear maps from $\mathbb{R}^n$ to $\mathbb{R}^m$ is said to be a derivate container of Λ at $\bar{x}$ if, for every $\epsilon > 0$, there exists a neighbourhood $\mathcal{O}_\epsilon$ of $\bar{x}$ and a sequence $\{\Lambda_j : \mathcal{O}_\epsilon \to \mathbb{R}^m\}$ of C^1 maps such that

(i): $\nabla \Lambda_j(x) \in \mathcal{L} + \epsilon \mathbb{B}$ for all j and $x \in \mathcal{O}_\epsilon$
(ii): $\Lambda_j(x) \to \Lambda$ as $j \to \infty$, uniformly over $x \in \mathcal{O}_\epsilon$.

The choice of derivate container can be tailored to a particular application. In some cases, this choice can be exercised to give a different, and more precise, transversality condition than that derived in this chapter. However for technical reasons to do with the fact that the convex hull of a derivate container to a Lipschitz continuous function contains no more information than the generalized Jacobian [206], the co-state inclusion is essentially the same. The dual approach also yielded a nonsmooth maximum principle [124], via a nonsmooth generalization of Ioffe and Tihomirov's earlier mixed multiplier rule.

Distinctive features of Clarke's nonsmooth maximum principle are, first, the transparent link to the original optimality condition (if the date is smooth, we replace limiting subdifferentials by Fréchet derivatives and immediately recover the classical maximum principle). This link, while present of course, is less obvious in nonsmooth maximum principles of Warga, where reduction to the smooth case involves choice of derivate container and interpretation of an abstract multiplier rule. Second, the Clarke's optimality condition is exceptional for the unrestrictive nature of the hypotheses that are invoked. This is true particularly in relation to the integrably bounded Lipschitz dependence of velocity function w.r.t. its x dependence. Clarke requires $f(t, ., u)$ to be locally Lipschitz continuous on a neighbourhood of $\bar{x}(t)$ (the nominal state trajectory), but then hypothesizes integrability only of the function $t \to k(t, \bar{u}(t))$ (i.e. the Lipschitz bound *evaluated along the nominal control $\bar{u}(t)$*). This is a very weak Lipschitz continuity hypothesis for derivation of necessary conditions in dynamic optimization indeed and, for that reason, is widely employed to this day.

The dual approach to proving the maximum principle has been revisited by Sussmann [185], using constructs related to Warga's derivate containers. Let $(\bar{x}, \bar{u})$ be an optimal process. Scrutiny of the maximum principle conditions reveals that the co-state equation makes sense when we assume that the right side of the differential equation modelling the dynamics is 'differentiable' with respect to the state in some sense *merely along the optimal control function*. Sussman provides a version of maximum principle incorporating the *Lojasiewicz refinement*, namely a version of the maximum principle in which this weaker differentiability hypothesis replaces the usual ones involving *all* control functions. This development was prefigured in earlier multiplier rules of Halkin [122] and Ioffe [126], who merely invoked differentiability of the data at (rather than near) the minimizer under consideration. Primal methods yield a simple, alternative proof of the Lojasiewicz refinement for free right end-point problems (and refinements), though not for general fixed end-point problems.

The price of achieving the Lojasiewicz refinement for fixed end-point dynamic optimization problems would appear to be a hypothesis requiring the solutions of the state equation to be unique for an arbitrary initial condition and control. The Lojasiewicz refinement tells us then that the maximum principle remains valid when a Lipschitz continuity hypothesis, resembling standard sufficient conditions for uniqueness of solutions to the state equation (for an arbitrary control and initial state), is replaced by the hypothesis of uniqueness of solutions to the state equation itself. This suggests that the uniqueness condition is in some sense more fundamental to the derivation of necessary conditions in dynamic optimization than Lipschitz continuity. On the other hand, examples of dynamic optimization problems can be constructed whose data satisfy the Lipschitz continuity hypotheses for derivation of Clarke's nonsmooth maximum principle (these resemble, but are weaker than standard sufficient conditions for uniqueness), yet for which the uniqueness hypothesis is violated. The question of what hypothesis is the more 'fundamental' does not therefore have a simple answer. A counter-example provided by Bressan [46] points to fundamental limitations to the derivation by dual/set separation methods of versions of the maximum principle derived by primal methods, incorporating transversality conditions (expressed in terms of limiting normal cones to the end-point constraint set) and therefore, also, on the scope for unifying the two approaches [186].

In this chapter, the primal techniques pioneered by Clarke are used, once again, to proof his nonsmooth maximum principle. We first prove the necessary conditions in a special case in which the data are smooth and there are no end-point constraints, using elementary needle variation techniques. We extend these conditions to allow for nonsmooth data (but still no endpoint constraints). This is done by approximating the problem by a smooth problem, using inf convolution to regularize the cost (as in [67]) and a decoupling technique to replace the state variable in the dynamic constraint by a control-like variable, which eliminates the need to deal with nonsmoothness of the right side of the controlled differential equation (a decoupling technique also used in [67] and earlier in [200]); we recover necessary conditions by applying Ekeland's theorem and passage to the limit. Finally we derive optimality conditions for the nonsmooth problem with end-point constraints, approximating it by a non-smooth problem with no end-point constraints, which we can now deal with, and making a second application of Ekeland's theorem. This proof strategy is simpler and more direct than those employed Clarke's paper and the subsequent monographs [65, 68] or [194], all of which arrive at the desired necessary conditions, by first of all proving necessary conditions (in the form of either Clarke's Hamiltonian inclusion or the generalized Euler Lagrange inclusion) for a relaxed

7.6 Notes for Chapter 7

version of original dynamic constraint formulated as a differential inclusion. By contrast, our approach avoids excursions into the theory of necessary conditions for differential inclusion problems altogether. It does not involve relaxation procedures or the rather complicated reduction procedures (including approximating velocity sets by sets containing a finite number of points) earlier employed. The underlying simplicity of the approach is evident from the way it is used to prove the classical maximum principle in the Appendix of Chap. 1.

Chapter 8
The Generalized Euler-Lagrange and Hamiltonian Inclusion Conditions

Abstract The subject matter of this chapter is necessary conditions of optimality for differential inclusion problems, by which we mean dynamic optimization problems in which the dynamic constraint takes the form of a differential inclusion. We provide two sets of conditions. These are the generalized Euler Lagrange condition and a later refinement of a set of conditions known as Clarke's Hamiltonian inclusion. As their names imply, these conditions result from reformulating the differential inclusion problem as a generalized Bolza problem in the calculus of variations, in which indicator functions taking account of the dynamic constraint appear in an extended valued Lagrangian, and deriving analogues of the similarly named classical conditions, valid in this more general setting. The conditions, expressed in terms of limiting subgradients and limiting normal cones, are inherently nonsmooth.

We bring fully up to date the generalized Euler Lagrange condition, which has been significantly improved, since the time it was first derived in the 1980s under a convexity hypothesis on velocity sets. The generalized Euler Lagrange condition covered in this chapter, for problems with non-convex, unbounded velocity sets, builds on Clarke's stratified conditions. It also incorporates a recent improvement in the Weierstrass condition, which we refer to as the Ioffe refinement.

The first necessary condition to be proved for differential inclusion problems with convex velocity sets, was Clarke's Hamiltonian inclusion. This chapter includes a subsequent refinement of this condition, the so-called partially convexified Hamiltonian inclusion. It is shown, by means of a duality theorem linking Euler Lagrange and Hamiltonian inclusions, that the partially convexified Hamiltonian inclusion is not an independent condition, but is in fact a consequence of the generalized Euler Lagrange condition.

Questions regarding validity of the Hamiltonian inclusion, for differential inclusion problems with possibly non-convex velocity sets, was the subject of speculation for many years. Current understanding of these issues is conveyed in this chapter. Clarke's (fully convexified) Hamiltonian inclusion for problems with non-convex velocity sets is now known to hold, at local minimizers w.r.t. certain topologies. But the condition may not be valid for local minimizers w.r.t. other topologies.

8.1 Introduction

The distinguishing feature of dynamic optimization problems, as compared with traditional variational problems, is the presence of constraints on the velocity variable. We have seen in the previous chapter that necessary conditions in the form of a maximum principle can be derived when these constraints are formulated in terms of a controlled differential equation governing the evolution of the state variable:

$$\dot{x}(t) = f(t, x(t), u(t)) \quad \text{a.e.,} \qquad (8.1.1)$$

$$u(t) \in U(t) \quad \text{a.e..} \qquad (8.1.2)$$

Another approach to the derivation of necessary conditions in dynamic optimization is to focus attention on the constraints on the velocity variable implicit in the underlying dynamics of the problem. The choice variables are now taken to be arcs satisfying a differential inclusion

$$\dot{x}(t) \in F(t, x(t)) \quad \text{a.e..}$$

This is a broader framework for studying dynamic optimization problems, since constraints described by a controlled differential equation (8.1.1) and (8.1.2) can be reformulated as a differential inclusion constraint by choosing

$$F(t, x) := \{f(t, x, u) : u \in U(t)\}. \qquad (8.1.3)$$

Our goal in this chapter is to derive general necessary conditions for a dynamic optimization problem in which the dynamic constraint takes the form, no longer of a controlled differential equation, but of a differential inclusion:

$$(P) \quad \begin{cases} \text{Minimize } g(x(S), x(T)) \\ \text{over arcs } x \in W^{1,1}([S, T]; \mathbb{R}^n) \text{ satisfying} \\ \dot{x}(t) \in F(t, x(t)) \quad \text{a.e.,} \\ (x(S), x(T)) \in C. \end{cases}$$

Here, $[S, T]$ is a given interval, $g : \mathbb{R}^n \times \mathbb{R}^n \to \mathbb{R}$ is a given function, $F : [S, T] \times \mathbb{R}^n \rightsquigarrow \mathbb{R}^n$ is a given multifunction and $C \subset \mathbb{R}^n \times \mathbb{R}^n$ is a given set.

Problem (P) can be regarded as a generalization of traditional variational problems (in one independent variable) to allow for nonsmooth data. This is because (P) can be expressed as

$$(P)' \quad \begin{cases} \text{Minimize } g(x(S), x(T)) + \int_S^T L(s, x(s), \dot{x}(s))ds \\ \text{over arcs } x \in W^{1,1}([S, T]; \mathbb{R}^n) \text{ satisfying} \\ (x(S), x(T)) \in C, \end{cases}$$

8.1 Introduction

when we choose the cost integrand L to be

$$L(t, x, v) := \Psi_{\mathrm{Gr}F(t,.)}(x, v). \tag{8.1.4}$$

(Ψ_G denotes, as usual, the indicator function of the set G that takes value 0 or $+\infty$ at a point z in its domain, depending on whether z lies in G or its complement, respectively.)

This is an equivalent formulation, in so far as the values of the cost of the new problem and of the one it replaces are the same for an arbitrary arc $x \in W^{1,1}$ satisfying

$$\dot{x}(t) \in F(t, x(t)) \quad \text{a.e..}$$

If, on the other hand, an arc x is not admissible for the original problem, i.e.

$$\dot{x}(t) \notin F(t, x(t))$$

for all t in a subset of positive measure, then the value of the cost for the new problem is infinite and so the arc x is effectively excluded from consideration as a minimizer.

From this perspective, the maximum principle, which concerns an inherently nonsmooth problem, was a remarkable achievement. Pontryagin et al. circumvented the need to examine generalized derivatives of the nonsmooth Lagrangian or Hamiltonian, a general theory for which did not exist at the time, by structuring their necessary conditions around the smooth 'unmaximized' Hamiltonian function.

By contrast, the necessary conditions now available for dynamic optimization problems with dynamics described by a differential inclusion rely heavily on nonsmooth analysis, regarding not merely the proof techniques employed to derive them, but their very statement. Indeed, since the 1980s, research in dynamic optimization and nonsmooth analysis has proceeded hand in hand.

A revealing perspective on necessary conditions for dynamic optimization problems involving differential inclusions is to regard them as generalizations of conditions from the classical calculus of variations. If, in $(P)'$, L were smooth and suitably regular, g smooth and C expressible in terms of smooth 'nondegenerate' functional inequality constraints, classical variational techniques would supply the following necessary conditions for $\bar{x}$ to be a minimizer: there would exist $p \in W^{1,1}([S, T]; \mathbb{R}^n)$ such that

$$(\dot{p}(t), p(t)) = \nabla_{x,v} L(t, \bar{x}(t), \dot{\bar{x}}(t)) \quad \text{a.e. } t \in [S, T], \tag{8.1.5}$$

$$p(t) \cdot \dot{\bar{x}}(t) - L(t, \bar{x}(t), \dot{\bar{x}}(t))$$
$$\geq p(t) \cdot v - L(t, \bar{x}(t), v) \quad \text{for all } v \in \mathbb{R}^n, \text{ a.e. } t \in [S, T] \tag{8.1.6}$$

and

$$(p(S), -p(T)) \in \nabla g(\bar{x}(S), \bar{x}(T)) + N_C(\bar{x}(S), \bar{x}(T)). \qquad (8.1.7)$$

Of course in the present context L, a characteristic function, is certainly not smooth. Nevertheless, under unrestrictive hypotheses on the data for problem (P), the following analogues of the above conditions can be derived: there exist $\lambda \geq 0$ and $p \in W^{1,1}([S, T]; \mathbb{R}^n)$, not both zero, such that

$$\dot{p}(t) \in \mathrm{co}\{\xi : (\xi, p(t)) \in N_{\mathrm{Gr}\, F(t,.)}(\bar{x}(t), \dot{\bar{x}}(t))\} \quad \text{a.e.}, \qquad (8.1.8)$$

$$p(t) \cdot \dot{\bar{x}}(t) \geq p(t) \cdot v \quad \text{for all } v \in F(t, \bar{x}(t)) \quad \text{a.e.} \qquad (8.1.9)$$

and

$$(p(S), -p(T)) \in \lambda \nabla g(\bar{x}(S), \bar{x}(T)) + N_C(\bar{x}(S), \bar{x}(T)). \qquad (8.1.10)$$

These three conditions will be referred to collectively as the generalized Euler Lagrange condition because of the close affinity of (8.1.8) with the classical Euler Lagrange condition. (Condition (8.1.8) alone will sometimes be called the Euler Lagrange inclusion.) The relation between (8.1.8) and the classical condition is evident from the fact that, if $L(t, ., .)$ is the indicator function of the closed set $\mathrm{Gr}\, F(t, .)$ and $(x, v) \in \mathrm{Gr}\, F(t, .)$, we have

$$\partial_{x,v} L(t, x, v) = N_{\mathrm{Gr}\, F(t,.)}(x, v).$$

Condition (8.1.8) may be interpreted then as a 'partially convexified', nonsmooth version of (8.1.5). Equation (8.1.10) differs from (8.1.7) only by the presence of a cost multiplier $\lambda \geq 0$ (to take account of the possibility of certain kinds of degeneracy for variational problems involving end-point and velocity constraints). Finally (8.1.9) is simply another way of writing (8.1.6) when $L(t, ., .)$ is interpreted as the indicator function of $\mathrm{Gr}\, F(t, .)$.

Another point of departure from classical optimality conditions in the calculus of variations, for deriving necessary conditions of optimality in dynamic optimization when the dynamic constraint takes the form of a differential inclusion, is Hamilton's system of equations: a minimizing state trajectory $\bar{x}$ and associated costate trajectory p satisfy

$$(-\dot{p}(t), \dot{\bar{x}}(t)) = \nabla_{x,p} H(t, \bar{x}(t), p(t)) \quad \text{a.e..}$$

Here H, which we refer to as the 'maximized Hamiltonian', is defined to be

$$H(t, x, p) := p \cdot \chi(t, x, p) - L(t, x, \chi(t, x, p))$$

8.1 Introduction

in which $\chi(t, x, p)$ is a stationary point of $v \to p \cdot v - L(t, x, v)$, i.e.

$$\left. \left(p \cdot v - \nabla_v L(t, x, v) \right) \right|_{v = \chi(t,x,p)} = 0.$$

(It is assumed the stationary point $\chi(t, x, p)$ exists and is unique and $\chi(t, ., .)$ is continuously differentiable.) This construction of $H(t, x, .)$ from $L(t, x, .)$ is known as the Legendre-Fenchel transformation.

Since stationary points of $v \to p \cdot v - L(t, x, v)$ are maximizers, when $L(t, x, .)$ is a convex, H can be equivalently defined as the Legendre-Fenchel transformation of $L(t, x, .)$, namely

$$H(t, x, p) = \sup_{v \in \mathbb{R}^n} \{ p \cdot v - L(t, x, v) \}, \tag{8.1.11}$$

in the case that $L(t, x, .)$ is a convex function. Even if $L(t, x, .)$ fails to be convex, (8.1.11) provides a convenient definition of the maximized Hamiltonian, because it makes sense when the onerous conditions (concerning existence, uniqueness and continuous dependence of stationary points) for construction of the Legendre-Fenchel transformation are dropped and, more significantly, because this definition provides a basis for the derivation of optimality conditions for a broad class of dynamic optimization problems. In the case $L(t, ., .)$ is the indicator function of Gr $F(t, .)$, $H(t, x, p)$ is

$$H(t, x, p) = \sup_{v \in F(t,x)} \left\{ p \cdot v - L(t, x, v) \right\}. \tag{8.1.12}$$

Taking (8.1.12) as definition of H, we might seek necessary conditions of optimality for (P) that include the relation

$$(-\dot{p}(t), \dot{\bar{x}}(t)) \in \text{co} \, \partial_{x,p} H(t, \bar{x}(t), p(t)) \text{ a.e.}, \tag{8.1.13}$$

satisfied by the minimizing state trajectory and associated costate trajectory. Here, we acknowledge that $H(t, ., .)$ may fail to be differentiable and consequently we employ, in place of the derivative, the set valued subdifferential co $\partial_{x,p} H$. We would expect the Clarke generalized gradient (i.e. the convexified limiting subdifferential) to be involved, to ensure robustness of the solution of (8.1.13) under perturbations and limit taking. In the spirit of the generalized Euler Lagrange condition, we might hope to replace (8.1.13) by a refined, 'partially convexified' condition

$$-\dot{p}(t) \in \text{co} \{ \xi \in \mathbb{R}^n : (\xi, p(t)) \in \partial_{x,p} H(t, \bar{x}(t), p(t)) \} \text{ a.e..} \tag{8.1.14}$$

The goal of this chapter is to derive necessary conditions of optimality for problem (P), whose dynamic constraint is formulated as a differential inclusion. We provide two sets of conditions. The first includes the Euler Lagrange inclusion (8.1.13). The second, in which (8.1.14) supplements (8.1.13), will be called the

Hamiltonian inclusion. For dynamic optimization problems arising in engineering design, the dynamic constraint usually takes the form of a differential equation parameterized by a control function (8.1.1) and (8.1.2). We may think of the necessary conditions of this chapter, for problems involving differential inclusions, as 'intrinsic' conditions because they depend only on the set of admissible velocities to which the parameterized differential equation gives rise via (8.1.3). The question therefore needs to be addressed whether the intrinsic necessary conditions have anything to offer over the maximum principle which is, after all, specially tailored to these kinds of dynamic constraints. While it is true that the maximum principle is often more convenient to apply, the intrinsic conditions have the significant advantage that they cover certain dynamic optimization problems formulated in terms of a differential equation parameterized by control functions, in cases when the control constraint set U is *state* and time dependent, provided of course that the multifunction

$$F(t, x) = \{f(t, x, u) : u \in U(t, x)\}$$

satisfies the hypotheses under which the intrinsic conditions are valid. By contrast, the classical maximum principle does not apply to problems with state dependent control constraint sets. It is true that there is an extensive literature on extensions of the maximum principle to allow for 'mixed constraints' (pathwise constraints involving both state and control variables). But, as we shall see in Sect. 10.8 of this book, necessary conditions for differential inclusion problems have provided powerful analytical tools for their derivation, under unrestrictive hypotheses.

Even when the control constraint set is not state dependent, the maximum principle and the intrinsic conditions are distinct sets of necessary conditions which we can apply to gain information about optimal controls. Examples can be constructed where the maximum principle excludes candidates for being a minimizer when the intrinsic conditions fail to do so, and vice versa.

8.2 Pseudo Lipschitz Continuity

In this section we introduce two continuity concepts concerning multifunctions $\Gamma : \mathbb{R}^n \rightsquigarrow \mathbb{R}^m$. These have special relevance to dynamic optimization problems in which the dynamic constraint takes the form of a differential inclusion $\dot{x} \in F(t, x)$, when Γ has the interpretation '$\Gamma = F(t, .)$' (for fixed t) and we aim to formulate appropriate Lipschitz continuity hypotheses under which we can derive necessary conditions of optimality.

Definition 8.2.1 (Pseudo-Lipschitz Continuity) Take a multifunction $\Gamma : \mathbb{R}^n \rightsquigarrow \mathbb{R}^m$ and a point $(\bar{x}, \bar{v}) \in \operatorname{Gr} \Gamma$. Take also numbers $\epsilon > 0$, $R > 0$ and $k \geq 0$. We say that Γ is *pseudo-Lipschitz continuous near $(\bar{x}, \bar{v})$ (with parameters ϵ, R and k)* if

$$\Gamma(x') \cap (\bar{v} + R\mathring{\mathbb{B}}) \subset \Gamma(x) + k|x' - x|\mathbb{B} \text{ for all } x', x \in \bar{x} + \epsilon \mathbb{B}.$$

8.2 Pseudo Lipschitz Continuity

Definition 8.2.2 (Bounded Slope Condition) Take a multifunction $\Gamma : \mathbb{R}^n \rightsquigarrow \mathbb{R}^m$ with closed graph $G := \mathrm{Gr}\,\Gamma$ and a point $(\bar{x}, \bar{v}) \in G$. Take also numbers $\epsilon > 0$, $R > 0$ and $k \geq 0$. We say that Γ *satisfies the bounded slope condition near* $(\bar{x}, \bar{v})$ *(with parameters ϵ, R and k)*, if

$$(w, p) \in N_G^P(x, v) \implies |w| \leq k|p|, \text{ for all } (x, v) \in (\bar{x}, \bar{v}) + \epsilon \mathbb{B} \times R\overset{\circ}{\mathbb{B}}.$$

These definitions are more or less equivalent with the caveat that, passing from one to the other, we must modify the parameters ϵ and R involved. The following proposition, which makes precise this assertion, can be regarded as an analogue for multifunctions of the property of real valued functions, lower semi-continuous functions f on a finite dimensional linear space that are bounded below: such a function is Lipschitz continuous with Lipschitz constant at most k on some open neighbourhood of a base point z, if and only if its limiting subgradients are uniformly bounded by k on some neighbourhood of z.

Proposition 8.2.3 *Take a multifunction $\Gamma : \mathbb{R}^n \rightsquigarrow \mathbb{R}^m$ with closed graph $G := \mathrm{Gr}\,\Gamma$. Take $(\bar{x}, \bar{v}) \in G$ and also $\epsilon > 0$, $R > 0$ and $k \geq 0$.*

(a): 'Γ *is pseudo-Lipschitz continuous near* $(\bar{x}, \bar{v})$ *with parameters ϵ, R and k'* $\implies$ 'Γ *satisfies the bounded slope condition near* $(\bar{x}, \bar{v})$ *with parameters ϵ', R and k (as above)*', *for any $\epsilon' \in (0, \epsilon)$,*

(b): 'Γ *satisfies the bounded slope condition near* $(\bar{x}, \bar{v})$ *with parameters ϵ, R and k'* $\implies$ 'Γ *is pseudo-Lipschitz continuous near* $(\bar{x}, \bar{v})$ *with parameters $\frac{R\eta}{3k} \wedge \epsilon'$, $(1 - \eta)R$ and k'*, *for any $\epsilon' \in (0, \epsilon)$ and $\eta \in (0, 1)$.*

Proof

(a): Suppose Γ is pseudo-Lipschitz continuous near $(\bar{x}, \bar{v})$ with parameters $\epsilon > 0$, $R > 0$ and $k \geq 0$. Choose any $\epsilon' \in (0, \epsilon)$. Take $(x, v) \in G \cap \left((\bar{x} + \epsilon'\mathbb{B}) \times (\bar{v} + R\mathbb{B})\right)$ and $(w, p) \in N_G^P(x, v)$. We must show that $|w| \leq k|p|$. With this goal in mind, take $\delta_i \downarrow 0$ and, for each i, write $x_i = x + \delta_i w$. Then, for i sufficiently large, $x_i \in \bar{x} + \epsilon \mathbb{B}$. Invoking the pseudo-Lipschitz continuity hypothesis, we can find, for each i sufficiently large, a point $v_i \in \Gamma(x_i)$ such that $|v_i - v| \leq k|x_i - x| = k\delta_i|w|$. It follows from the proximal normal inequality that there exists $M > 0$ such that, for i sufficiently large,

$$w \cdot (x_i - x) + p \cdot (v_i - v) \leq M(|x_i - x|^2 + |v_i - v|^2).$$

But then

$$\delta_i|w|^2 \leq \delta_i k|p||w| + M(1 + k^2)|w|^2\delta_i^2.$$

Dividing across by δ_i and passing to the limit yields $|w|^2 \leq k|w||p|$, which implies $|w| \leq k|p|$. Assertion (a) has been verified.

(b): Assume that Γ has the bounded slope property near $(\bar{x}, \bar{v})$ with parameters R, ϵ and k. Take any $\epsilon' \in (0, \epsilon)$ and $\eta \in (0, 1)$, $x_1, x_2 \in \bar{x} + \left(\frac{R\eta}{3k} \wedge \epsilon'\right) \mathbb{B}$ and $v_1 \in \Gamma(x_1) \cap (\bar{v} + (1 - \eta) R\overset{\circ}{\mathbb{B}})$. Assertion (b) will follow immediately if we can verify

Claim: There exists a point $v_2 \in \Gamma(x_2)$ such that $|v_2 - v_1| \leq k|x_1 - x_2|$.

We apply the mean value inequality (Theorem 4.5) to the function $f : \mathbb{R}^n \times \mathbb{R}^n \to \mathbb{R} \cup \{+\infty\}$, set Y and base point (x_1, v_1), when

$$f(x, v) := \Psi_G(x, v) \quad \text{and} \quad Y := \{x_2\} \times (v_1 + k|x_1 - x_2|\mathbb{B}).$$

Here Ψ_G is the indicator function of the set G (taking values 0 on G and $+\infty$ on its complement). Suppose the claim is false. Then $(v_1 + k|x_1 - x_2|\mathbb{B}) \cap \Gamma(x_2) = \emptyset$. This implies that

$$\min_{v \in v_1 + k|x_1 - x_2|\mathbb{B}} f(x_2, v) - f(x_1, v_1) = +\infty.$$

The mean value inequality now tells us that, for any real number r and $\tilde{\epsilon} > 0$,

$$r \leq w \cdot (x_2 - x_1) + p \cdot (v - v_1), \quad \text{for all } v \in v_1 + k|x_1 - x_2|\mathbb{B}. \tag{8.2.1}$$

Here, (w, p) is a point in $\partial^P f(z, e)$ and (z, e) is some point in $\mathbb{R}^n \times \mathbb{R}^n$ whose distance to $\operatorname{co}[\{(x_1, v_1)\} \cup Y]$ is not greater than $\tilde{\epsilon}$. Notice, however, that

$$\operatorname{co}[Y \cup \{(x_1, v_1)\}] \subset (\operatorname{co}\{x_1, x_2\}) \times (v_1 + k|x_1 - x_2|\mathbb{B})$$

$$\subset (\bar{x} + \epsilon'\mathbb{B}) \times (\bar{v} + ((1-\eta) + \frac{\eta}{3}) R\overset{\circ}{\mathbb{B}}).$$

We can choose the positive number $\tilde{\epsilon}$, independently of r, such that

$$\operatorname{co}[Y \cup \{(x_1, v_1)\}] + \tilde{\epsilon}\mathbb{B} \subset (\bar{x} + \epsilon\mathbb{B}) \times (\bar{v} + R\overset{\circ}{\mathbb{B}}).$$

Observe that (z, e) is in the set where the bounded slope condition is satisfied. Since $(w, p) \in \partial^P f(z, e) \subset N_G^P(z, e)$, it follows from our assumptions that $|w| \leq k|p|$. So, from (8.2.1),

$$r \leq (|w| - k|p|)|x_1 - x_2| \leq 0.$$

This is not possible since r is an arbitrary positive number. This contradiction confirms the claim. $\square$

8.3 Unbounded Differential Inclusions

Necessary conditions of optimality in the form of the maximum principle, applicable to dynamic optimization problems where the dynamic constraint takes the form of a controlled differential equation $\dot{x}(t) = f(t, x(t), u(t))$, have been derived under the assumption that $f(t, x, u)$ has certain Lipschitz continuity properties w.r.t. the x variable. Now that we are formulating the dynamic constraint as a differential inclusion $\dot{x}(t) \in F(t, x)$, some kind of hypothesis, of a Lipschitz continuity nature, must be imposed on $F(t, x)$ regarding its x dependence, in order to derive necessary conditions in this new framework. What should this hypothesis be? An obvious choice, indeed one which was adopted in the early literature, was to require integrable Lipschitz continuity w.r.t. the Hausdorff metric, i.e., for a given reference trajectory $\bar{x}$,

There exist $\epsilon > 0$ and $k \in L^1$ such that

$$d_H(F(t, x), F(t, x')) \leq k(t)|x - x'| \text{ for all } x, x' \in \bar{x}(t) + \epsilon \mathbb{B} \text{ a.e..}$$

Recall that the Hausdorff distance $d_H(A, B)$ between two closed non-empty subsets of $\mathbb{R}^n$ A and B is defined by (see (6.6.3))

$$d_H(A, B) := \max \left\{ \sup_{a \in A} d_B(a), \sup_{b \in B} d_A(b) \right\}.$$

An equivalent statement of the condition is:
There exist $\epsilon > 0$ and $k \in L^1$ such that

$$F(t, x') \subset F(t, x) + k(t)|x' - x|\mathbb{B} \text{ for all } x', x \in \bar{x}(t) + \epsilon \mathbb{B}, \text{ a.e..} \quad (8.3.1)$$

For unbounded differential inclusions, i.e. in situations where the values of the multifunction F are, possibly, unbounded sets, condition (8.3.1) is usually overly restrictive. This is because, for two unbounded sets which are 'close' in an intuitive sense, the Hausdorff distance between them can be very large. Consider, for example, the multifunction $F : \mathbb{R}^2 \rightsquigarrow \mathbb{R}^2$ defined by

$$F(x_1, x_2) := \{(v_1, v_2) \in \mathbb{R}^2 : v_2 \leq x_1 v_1\}. \quad (8.3.2)$$

Here, values of F are hypographs of linear functions whose slopes depend smoothly on the value of $x = (x_1, x_2)$. A reasonable requirement of a set of necessary conditions for unbounded differential inclusions is that it should allow cases like this. Hypothesis (8.3.1) fails to meet this requirement, because

$$d_H(F(x'), F(x)) = +\infty \text{ for } x_1' \neq x_1,$$

and we are therefore compelled to seek less restrictive hypotheses. This was the motivation behind the imposition, in the later literature, of the less restrictive pseudo-Lipschitz continuity hypothesis (with integrable parameter k):
There exists $R > 0$, $\epsilon > 0$ and $k \in L^1$ such that

$$F(t, x') \cap (\bar{x} + R \overset{\circ}{\mathbb{B}}) \subset F(t, x) + k(t)|x' - x|\mathbb{B}$$

$$\text{for all } x', x \in \bar{x}(t) + \epsilon \mathbb{B}, \text{ a.e..} \qquad (8.3.3)$$

Unfortunately, it is the case that pseudo-Lipschitz continuity does not permit the derivation of Euler Lagrange-type necessary conditions; either this hypothesis must be modified (replacing the parameter R by a time varying function, that we call a *radius function* that is coordinated, in some sense, with the integrable Lipschitz bound k) or the pseudo-Lipschitz continuity hypothesis must be supplemented by other hypotheses. The supplementary hypotheses, which are referred to as *tempered growth* conditions, basically require the distance of the set velocity $F(t, x)$ from the nominal velocity $\dot{\bar{x}}(t)$ to grow not too rapidly, as x deviates from the nominal velocity $\bar{x}$.

8.4 The Generalized Euler Lagrange Condition

This section provides necessary conditions satisfied by $W^{1,1}$ local minimizers, for the dynamic optimization problem (P) of the introduction, namely:

$$(P) \begin{cases} \text{Minimize } g(x(S), x(T)) \\ \text{over arcs } x \in W^{1,1}([S, T]; \mathbb{R}^n) \text{ satisfying} \\ \dot{x}(t) \in F(t, x(t)) \quad \text{a.e.,} \\ (x(S), x(T)) \in C. \end{cases}$$

Here $[S, T]$ is a given interval, $g : \mathbb{R}^n \times \mathbb{R}^n \to \mathbb{R}$ is a given function, $F : [S, T] \times \mathbb{R}^n \leadsto \mathbb{R}^n$ is a given multifunction, and $C \subset \mathbb{R}^n \times \mathbb{R}^n$ is a given closed set. An admissible F trajectory x is an F trajectory that satisfies the constraints of (P).

The necessary conditions assert properties of an admissible arc $\bar{x}$ that is a 'local minimizer' in a sense made precise by the following definition:

Definition 8.4.1 For a specified multifunction $B : [S, T] \leadsto \mathbb{R}^n$ such that $B(t)$ is open for each t, a F trajectory $\bar{x}$ is said to be a $W^{1,1}$ local minimizer for (P) relative to B if there exists $\beta > 0$ such that

$$g(x(S), x(T)) \geq g(\bar{x}(S), \bar{x}(T)), \qquad (8.4.1)$$

for all admissible trajectories such that $\dot{x}(t) \in \dot{\bar{x}}(t) + B(t)$, a.e. and

$$||x - \bar{x}||_{W^{1,1}} \leq \beta. \qquad (8.4.2)$$

8.4 The Generalized Euler Lagrange Condition

An admissible F trajectory such that, for some $\beta > 0$, (8.4.1) is true for all admissible F trajectories x satisfying (8.4.2) is called a $W^{1,1}$ local minimizer for (P). A $W^{1,1}$ local minimizer can, of course, be interpreted as a $W^{1,1}$ local minimizer relative to B, when $B \equiv \mathbb{R}^n$.

The centrepiece of the following necessary conditions is the Euler Lagrange inclusion. They also incorporate versions of the Weierstrass and transversality conditions. The Weierstrass condition employed here is rather subtle and merits prior discussion. Suppose initially that the multifunction $B(t) \equiv \mathbb{R}^n$. The expected form of the Weierstrass condition is the relation

$$p(t) \cdot \dot{\bar{x}}(t) \geq p(t) \cdot e, \text{ for all } e \in F(t, \bar{x}(t)), \text{ a.e.,} \qquad (8.4.3)$$

in which p is the costate arc. However the pseudo Lipschitz continuity hypothesis (8.3.3), now expressed in terms of a possibly time-varying radius function R, places restrictions only on velocities e in $F(t, x) \cap (\dot{\bar{x}}(t) + R(t)\overset{\circ}{\mathbb{B}})$ and, unsurprisingly, earlier efforts to weaken the restrictive hypothesis (8.3.3) focused on confirming the inequality in (8.4.3), only for velocities in the subset of $F(t, \bar{x}(t))$:

$$F(t, \bar{x}(t)) \cap (\dot{\bar{x}}(t) + R(t)\overset{\circ}{\mathbb{B}}). \qquad (8.4.4)$$

(There are technical reasons, related to the fact that proof techniques for the necessary conditions involve replacing $R(t)\mathbb{B}$ by an inner approximation and passage to the limit, why we need to take the intersection in (8.4.4) with the *open* set $\overset{\circ}{\mathbb{B}}$, not $\mathbb{B}$ itself.) But in fact we can do better than this and validate the Weierstrass condition over a larger set of velocities. For this purpose, we introduce the following definition, in which $\bar{x}$ is a given F trajectory:

Definition 8.4.2 The regular velocity set at time t is

$$\Omega_0(t) := \{e \in F(t, \bar{x}(t)) : F(t, .) \text{ is pseudo-Lipschitz}$$
$$\text{continuous near } (\bar{x}(t), e)\}. \qquad (8.4.5)$$

Now suppose the nominal F trajectory $\bar{x}$ is a $W^{1,1}$ local minimizer relative to some multifunction B such that $R(t)\overset{\circ}{\mathbb{B}} \subset B(t)$ a.e.. The version of the necessary conditions below asserts that the Weierstrass condition is valid for all velocities e in

$$\overline{\text{co}}\,(\Omega_0(t) \cap (\dot{\bar{x}}(t) + B(t)))\,.$$

Notice that a version of the Weierstrass condition employing this set is an improvement over one employing $F(t, \bar{x}(t)) \cap (\dot{\bar{x}}(t) + R(t)\overset{\circ}{\mathbb{B}})$. This is because, under the pseudo-Lipschitz continuity hypothesis (8.3.3), every point $e \in F(t, \bar{x}(t)) \cap (\dot{\bar{x}}(t) + R(t)\overset{\circ}{\mathbb{B}})$ lies in $\Omega_0(t)$. From this fact and since $R(t)\overset{\circ}{\mathbb{B}} \subset B(t)$ we deduce that $F(t, \bar{x}(t)) \cap (\dot{\bar{x}}(t) + R(t)\overset{\circ}{\mathbb{B}}) \subset \Omega_0(t) \cap (\dot{\bar{x}}(t) + B(t))$.

We remark that, for many problems of interest, classes of which we identify below, Ω_0 actually coincides with $F(t, \bar{x}(t))$ and, here, the Weierstrass condition can be expressed in terms of the entire velocity set $F(t, \bar{x}(t))$.

Theorem 8.4.3 (The Generalized Euler Lagrange Condition) *Take a measurable multifunction $B : [S, T] \rightsquigarrow \mathbb{R}^n$ such that $B(t)$ is open for a.e. $t \in [S, T]$. Let $\bar{x}$ be a $W^{1,1}$ local minimizer for (P) relative to B. Assume that the following hypotheses are satisfied:*

(G1):   *g is Lipschitz continuous on a neighbourhood of $(\bar{x}(S), \bar{x}(T))$ and C is a closed set,*

(G2):   *$F(t, x)$ is nonempty for each $(t, x) \in [S, T] \times \mathbb{R}^n$, $\mathrm{Gr}\, F(t, .)$ is closed for each $t \in [S, T]$ and F is $\mathcal{L} \times \mathcal{B}^n$ measurable,*

(G3):   *there exist $\epsilon > 0$ and a measurable function $R : [S, T] \to (0, \infty) \cup \{+\infty\}$ (a 'radius function') such that $R(t)\mathring{\mathbb{B}} \subset B(t)$ a.e. and the following conditions are satisfied:*

(a):   *(Pseudo-Lipschitz Continuity) There exists $k \in L^1(S, T)$ such that*

$$F(t, x') \cap (\dot{\bar{x}}(t) + R(t)\mathring{\mathbb{B}}) \subset F(t, x) + k(t)|x' - x|\mathbb{B},$$
$$\text{for all } x, x' \in \bar{x}(t) + \epsilon \mathbb{B}, \text{ a.e. } t \in [S, T], \quad (8.4.6)$$

(b):   *(Tempered Growth) there exists $r \in L^1(S, T)$, $r_0 > 0$ and $\gamma \in (0, 1)$ such that $r_0 \le r(t)$, $\gamma^{-1} r(t) \le R(t)$ a.e. and*
$F(t, x) \cap (\dot{\bar{x}}(t) + r(t)\mathbb{B}) \neq \emptyset$ for all $x \in \bar{x}(t) + \epsilon \mathbb{B}$, a.e. $t \in [S, T]$.

Then there exist an arc $p \in W^{1,1}([S, T]; \mathbb{R}^n)$ and $\lambda \ge 0$, satisfying the following conditions:

(i): $(p, \lambda) \neq (0, 0)$,

(ii): $\dot{p}(t) \in \mathrm{co}\{\eta : (\eta, p(t)) \in N_{\mathrm{Gr}\, F(t,.)}(\bar{x}(t), \dot{\bar{x}}(t))\}$ a.e. $t \in [S, T]$,

(iii): $(p(S), -p(T)) \in \lambda \partial g(\bar{x}(S), \bar{x}(T)) + N_C(\bar{x}(S), \bar{x}(T))$,

(iv): $p(t) \cdot \dot{\bar{x}}(t) \ge p(t) \cdot v$ *for all* $v \in \overline{\mathrm{co}}\, (\Omega_0(t) \cap (\dot{\bar{x}}(t) + B(t)))$, *a.e.* $t \in [S, T]$.

Now assume, also, that $B(t)$ and $F(t, x)$ do not depend on t. Then, in addition to the above conditions, there exists a constant a such that

(v): $p(t) \cdot \dot{\bar{x}}(t) = a$,   *a.e.* $t \in [S, T]$.

The theorem, a proof of which is given in Sect. 8.6 below, is noteworthy both for the unrestrictive nature of the hypotheses under which it is valid and also the precision of the costate inclusion (condition (ii)). We now elaborate on these points.

Concerning these hypotheses, we have pointed out that these will unavoidably include a Lipschitz continuity assumption concerning the x dependence of the velocity set $F(t, x)$ of some nature and, here, the pseudo-Lipschitz continuity hypothesis (G3)(a) is a natural choice, when the F takes values possibly unbounded sets. But in the statement of Theorem 8.4.3, the tempered growth hypothesis is also invoked. The presence of this supplementary hypothesis is, at first sight, surprising and might be thought to pose unnecessary limitations on the range of application of the theorem. We pause therefore to discuss its role.

8.4 The Generalized Euler Lagrange Condition

Observe at the outset that the question of whether, or not, the tempered growth hypothesis is really required arises only when either the integrable Lipschitz bound k in the pseudo-Lipschitz continuity hypothesis (G3)(a) is an unbounded function or the radius function R is not strictly bounded away from 0. Indeed, in the case when k is essentially bounded above by a number k_0 (we can always arrange, by adding a constant, that $k_0 > 0$) and R is essentially bounded below by some positive number $d > 0$, (G3)(b) follows automatically from (G3)(a). (To verify (G3)(b), reduce the size of ϵ to arrange that $k_0\epsilon < d$, and choose $r(t) \equiv k_0\epsilon$ and $\gamma \in (0, 1)$ such that $\gamma^{-1} k_0 \epsilon \leq d$.)

When however k is not essentially bounded, the tempered growth hypothesis cannot be dropped. This is illustrated in the following example.

Example

$$\begin{cases} \text{Minimize} & -x_2(1) \\ \text{over arcs} & x \in W^{1,1}([0, 1]; \mathbb{R}^2) \text{ satisfying} \\ \dot{x}(t) \in F(t, x(t)) \text{ a.e. } t \in [0, 1] \\ x_2(0) = 0, \end{cases}$$

in which

$$F(t, x) := \{(e_1, e_2) \in \mathbb{R}^2 \mid e_1 = 0, \; e_2 = k(t)x_1\}.$$

Here, k is any positive function in $L^1(0, 1)$ which is not essentially bounded.

Take an arbitrary number $R > 0$. Then the F trajectory $\bar{x} = (\bar{x}_1 = 0, \bar{x}_2 \equiv 0))$ is a $W^{1,1}$ local minimizer relative to the multifunction $B \equiv R \overset{\circ}{\mathbb{B}}$. To see this, suppose to the contrary that there exists an admissible F trajectory $x = (x_1, x_2)$ with lower cost and such that

$$\dot{x}(t) \in (0, 0) + R \overset{\circ}{\mathbb{B}} \quad \text{a.e. } t \in [0, 1]. \tag{8.4.7}$$

We have $x_1 > \bar{x}_1 = 0$, whence $\dot{x}(t) = (0, x_1 k(t))$ is not essentially bounded. This implies that, on a set of positive measure, $\dot{x}(t) \notin R \overset{\circ}{\mathbb{B}}$, in contradiction of (8.4.7). Note however that $\bar{x}$ is *not* a $W^{1,1}$ minimizer.

Hypotheses (G1)–(G3)(a) of Theorem 8.4.3 are satisfied for the above choice of radius function. Observe, however, that $(G3)$(b) 'tempered growth' is violated. To see this, take any $\gamma \in (0, 1)$, positive function r, strictly bounded away from 0 and such that $\gamma^{-1} r(t) \leq R$ a.e., and $(x_1, x_2) \in \mathbb{R}^2$ such that $x_1 \neq 0$. Then for times t in a set of positive measure, we have

$$F(t, x) \, (= \{0, x_1 k(t)\}) \not\subset \dot{\bar{x}}(t) + \gamma^{-1} r \mathbb{B}.$$

Since this relation can be expressed $F(t, x) \cap (\dot{\bar{x}}(t) + \gamma^{-1} r \mathbb{B}) = \emptyset$, it is clear that (G3)(b) is not satisfied.

Let us now examine the Euler Lagrange inclusion and transversality condition. They assert the existence of an arc $p = (p_1, p_2)$ and $\lambda \geq 0$, not both zero, satisfying:

$$\dot{p}_2(t) = 0$$
$$\dot{p}_1(t) + p_2(t)k(t) = 0 \tag{8.4.8}$$

and

$$p_1(0) = 0, \quad p_1(1) = 0 \tag{8.4.9}$$
$$p_2(1) = \lambda .$$

If $\lambda = 0$ then also $p \equiv 0$, a contradiction. If, on the other hand, $\lambda > 0$ then (8.4.8) implies

$$p_1(1) = -\lambda \int_0^1 k(t)\, dt < 0$$

in contradiction of (8.4.9). This example tells us that, concerning those aspects of the theorem which deal with situations in which the nominal admissible F trajectory $\bar{x}$ is merely a $W^{1,1}$ minimizer relative to a given multifunction B, the assertions of the theorem (even those omitting the Weierstrass condition) are not true in general, if the tempered growth hypothesis is removed. But no examples have been identified, apparently, that test the need for the tempered growth hypothesis, when $\bar{x}$ is a $W^{1,1}$ local minimizer.

Consider next the nature of differential inclusion (ii) which governs the costate trajectory p. This is a refined version, in which convexification is carried out with respect to just one variable, of the condition

$$(\dot{p}(t), p(t)) \in \operatorname{co} N_{\operatorname{Gr} F(t,.)}(\bar{x}(t), \dot{\bar{x}}(t)), \tag{8.4.10}$$

in which convexification is carried out with respect to two variables. The Euler Lagrange inclusion (ii) is, of course, to be preferred because it provides more precise information about the costate trajectory. But it is in fact a significant improvement on (8.4.10) for the following reasons. As we shall see in Sect. 8.7, condition (ii) implies (for convex valued F's) an alternative necessary condition, generalizing Hamilton's system of equations. (The same cannot be said of condition (8.4.10).) Furthermore, the generalized Euler Lagrange condition for (P) has an important role as an analytical tool for the derivation of necessary conditions for dynamic optimization problems, of a nonstandard nature (problems involving free end-times, discontinuous state trajectories. etc.). Here it is usual practice to write down the necessary conditions for a suitable auxiliary dynamic optimization problem, which is a special case of (P), and to pass to the limit. Now the Euler Lagrange inclusion (ii), applied to the auxiliary problem, yields a costate trajectory p satisfying

$$|\dot{p}(t)| \le k(t)|p(t)|$$

where $k \in L^1$ is as in hypothesis (G3). This bound can be used to justify the use of weak compactness arguments to obtain a costate trajectory for the nonstandard problem when we pass to the limit. Condition (8.4.10) yields no such bound in general, curtailing its usefulness as an analytical tool.

8.5 Special Cases

In this section we demonstrate the unifying power of the 'pseudo-Lipschitz continuity and tempered growth' hypothesis (G3), in Theorem 8.4.3. We identify several sets of conditions, which have served, in past work, as hypotheses for more restrictive versions of the necessary conditions, that are now revealed as sufficient conditions for the validity of (G3).

Proposition 8.5.1 *Take multifunctions* $F : [S, T] \times \mathbb{R}^n \rightsquigarrow \mathbb{R}^n$ *and* $B : [S, T] \rightsquigarrow \mathbb{R}^n$. *Let (G3) be the hypothesis of Theorem 8.4.3, that is:*

(G3): There exist $\epsilon > 0$ *and a measurable function* $R : [S, T] \to (0, \infty) \cup \{+\infty\}$ *(a 'radius function') such that* $R(t) \overset{\circ}{\mathbb{B}} \subset B(t)$ *a.e. and the following conditions are satisfied,*

(a): (Pseudo-Lipschitz Continuity) There exists $k \in L^1(S, T)$ *such that*

$$F(t, x') \cap (\dot{\bar{x}}(t) + R(t)\overset{\circ}{\mathbb{B}}) \subset F(t, x) + k(t)|x' - x|\mathbb{B},$$

for all $x, x' \in \bar{x}(t) + \epsilon \mathbb{B}$, *a.e.* $t \in [S, T]$,

(b): (Tempered Growth) There exist $r \in L^1(S, T)$, $r_0 > 0$ *and* $\gamma \in (0, 1)$ *such that* $r_0 \le r(t)$, $\gamma^{-1} r(t) \le R(t)$ *a.e. and*
$F(t, x) \cap (\dot{\bar{x}}(t) + r(t)\mathbb{B}) \ne \emptyset$ *for all* $x \in \bar{x}(t) + \epsilon \mathbb{B}$, *a.e.* $t \in [S, T]$.

Then (G3) is satisfied if any of the following hypotheses (G3)*, (G3)** or (G3)*** are satisfied.

(G3)*: There exist $\epsilon > 0$, $k \in L^1$ *and a measurable function* $R : [S, T] \to (0, \infty) \cup \{+\infty\}$ *such that* $R(t) \overset{\circ}{\mathbb{B}} \subset B(t)$ *a.e.,*

$$F(t, x') \cap (\dot{\bar{x}}(t) + R(t)\overset{\circ}{\mathbb{B}}) \subset F(t, x) + k(t)|x' - x|\mathbb{B},$$

for all $x, x' \in \bar{x}(t) + \epsilon \mathbb{B}$, *a.e.* (8.5.1)

and either
(a): k is strictly bounded away from 0 and there exists $\omega_0 > 0$ *such that* $\omega_0 k(t) \le R(t)$ *a.e.* $t \in [S, T]$,

or

(b): R is strictly bounded away form 0 and there exists a modulus of continuity θ such that

$$d_{F(t,x)}(\dot{\bar{x}}(t)) \leq \theta(|x - \bar{x}(t)|) \quad \text{for all } x \in \bar{x}(t) + \epsilon \mathbb{B}, \text{ a.e. } t \in [S, T].$$

(G3)**: There exist $\epsilon > 0$, $k \in L^1$ (strictly bounded away from zero), a measurable function $R : [S, T] \to (0, \infty) \cup \{\infty\}$ and $\omega_0 > 0$ such that $\omega_0 k(t) \leq R(t)$ and $R(t)\overset{\circ}{\mathbb{B}} \subset B(t)$, a.e. $t \in [S, T]$ and

$$'(w, p) \in N^P_{Gr\,F(t,.)}(x, v)' \implies '|w| \leq k(t)|p|',$$

for all $v \in \dot{\bar{x}}(t) + R(t)\overset{\circ}{\mathbb{B}}$ and $x \in \bar{x}(t) + \epsilon \mathbb{B}$, a.e. $t \in [S, T]$.

(G3)***: $B \equiv \mathbb{R}^n$. There exist $\epsilon > 0$, $\alpha > 0$, non-negative measurable functions k and β such that k and $t \to \beta(t) k^\alpha(t)$ are integrable on $[S, T]$ and, for each $N \geq 0$,

$$F(t, x') \cap (\dot{\bar{x}}(t) + N\mathbb{B}) \subset F(t, x) + (k(t) + \beta(t) N^\alpha)|x' - x|\mathbb{B},$$

for all $x', x \in \bar{x}(t) + \epsilon \mathbb{B}$ and a.e. $t \in [S, T]$.

The following corollary of Theorem 8.4.3 is an immediate consequence of this proposition.

Corollary 8.5.2 *The assertions of Theorem 8.4.3 remain valid when hypothesis (G3) of Theorem 8.4.3 is replaced by any of the hypotheses (G3)*, (G3)** or (G3)*** in Proposition 8.5.1.*

Remark

We briefly discuss the more restrictive forms (G1)*, (G2)** and (G3)*** of hypothesis (G3) in turn:

(G3)*: The first part of hypothesis (G3)* merely reproduces the 'pseudo-Lipschitz' continuity condition of (G3). Part (a) of (G3)* is a sufficient condition for satisfaction of the 'tempered growth' condition in (G3) and provides the motivation behind the terminology 'tempered growth'. It requires that the radius function has adequate growth, in a sense made precise, in relation to the integral Lipschitz bound k of the multifunction F.

(G3)**: We know from Proposition 8.2.3 that, if $F(t, .)$ satisfies the bounded slope condition, then it is pseudo-Lipschitz continuous (with modified parameters). (G3)** exploits this relation, to provide a restrictive version of (G3), in which a bounded slope condition replaces pseudo-Lipschitz continuity.

(G3)***: (G3)*** combines within a single condition restrictive forms of conditions (a) and (b) in (G3). It requires $F(t, .)$ to satisfy the pseudo-Lipschitz

8.5 Special Cases

condition relative to an arbitrary constant radius function, $R(t) = N$, and places growth conditions on the associated integrable Lipschitz bounds, as N increases. Observe that, if (G3)*** is in force, then the regular velocity set is $\Omega_0(t) = F(t, \bar{x}(t))$ and then, since in this case $B \equiv \mathbb{R}^n$, the Weierstrass condition becomes

$$p(t) \cdot \dot{\bar{x}}(t) \geq p(t) \cdot v, \quad \text{for all } v \in F(t, \bar{x}(t)), \quad \text{a.e. } t \in [S, T].$$

Proof of Proposition 8.5.1 In stages (I), (II) and (III) of the proof, we show that (G3) is satisfied under hypotheses (G1)*, (G2)** and (G3)***, respectfully.

(I): Suppose that (G3)* is satisfied. Assume, first, condition (a) in hypothesis (G3)*. Take any $\gamma \in (0, 1)$ and adjust the size of the parameter $\epsilon > 0$ so that $\epsilon < \gamma \omega_0$. Define $r(t) := \gamma \omega_0 k(t)$. Then, in consequence of condition (a) in (G3)*, we know that r is strictly bounded away from zero and $\gamma^{-1} r(t) \leq R(t)$.

For a.e. $t \in [S, T]$ and all $x \in \bar{x}(t) + \epsilon \mathbb{B}$, we know from (8.5.1) that there exists a point $v \in F(t, x)$ such that $|v - \dot{\bar{x}}(t)| \leq \epsilon k(t)$. Since $\epsilon k(t) \leq r(t)$, it follows that

$$F(t, x) \cap \left(\dot{\bar{x}}(t) + r(t) \mathbb{B} \right) \neq \emptyset.$$

We have verified (G3)(b) in the case when (G3)*(a) is satisfied.

Now assume (b) in (G3)*. Take any $\gamma \in (0, 1)$. Then there exists $r_0 > 0$ such that $\gamma^{-1} r_0 \leq R(t)$ a.e.. By reducing the size of $\epsilon > 0$, if required, we can arrange that $\theta(\epsilon) \leq r_0$. But then condition (b) in (G3)* tells us that, for an arbitrary point $x \in \bar{x}(t) + \epsilon \mathbb{B}$, there exists $e \in F(t, x)$ such that $|e - \dot{\bar{x}}(t)| \leq \theta(|x - \bar{x}(t)|) \leq r_0$. It follows that $F(t, x) \cap (\dot{\bar{x}}(t) + r_0 \mathbb{B}) \neq \emptyset$. We have, once again, verified (G3)(b) (for the constant r function $r \equiv r_0$).

(II): Suppose that (G3)** is satisfied. Choose any $\epsilon_1 \in (0, \epsilon)$ and $\eta \in (0, 1)$. It follows from Proposition 8.2.3(b) that (G3)(a) is satisfied with parameters , $\epsilon' := \frac{\omega_0 \eta}{3} \wedge \epsilon_1$, $R'(t) := (1 - \eta) \omega_0 k(t)$ and $k(t)$.

We now show that (G3)(b) is also satisfied. Take $\gamma' \in (0, 1)$. By reducing the size of ϵ', if necessary, we can arrange that $\epsilon' \leq \gamma'(1 - \eta) \omega_0$. Define $r'(t) = \gamma' R'(t)$. Note that r' is strictly bounded away from zero. Since, as we have shown, (G3)(a) is satisfied (with modified parameters), we know that, for any $x \in \bar{x}(t) + \epsilon' \mathbb{B}$, there exists a point $v \in F(t, x)$ satisfying $|v - \dot{\bar{x}}(t)| \leq k(t) \epsilon' \leq \gamma'(1 - \eta) \omega_0 k(t) = \gamma' R'(t)$. We have shown that (G3)(b) is also satisfied.

(III): We may arrange, by adding a constant if required, that $k \geq 1$. Reduce the size of ϵ so that $\epsilon \in (0, 1)$. Choose the radius function $R := k$. Set $r(t) = \epsilon k(t)$ and $\gamma = \epsilon$. Then r is bounded away from zero, $\gamma \in (0, 1)$ and $\gamma^{-1} r(t) \leq R(t)$ a.e..

From hypothesis (G3)***, in which we set $N = k(t)$, we get

$$F(t, x') \cap (\dot{\bar{x}}(t) + R(t) \overset{\circ}{\mathbb{B}}) \subset F(t, x) + \tilde{k}(t) |x' - x| \mathbb{B}$$

for all $x', x \in \bar{x}(t) + \epsilon \mathbb{B}$, a.e. $t \in [S, T]$. Here $\tilde{k}$ is the integrable function $\tilde{k}(t) := (k(t) + \beta(t) k(t)^\alpha)$.

On the other hand, setting $N = 0$ in (G3)*** yields the relation

$$F(t, x) \cap (\dot{\bar{x}}(t) + \epsilon k(t)\mathbb{B}) \neq \emptyset \text{ for every } x \in \bar{x}(t) + \epsilon \mathbb{B}, \text{ a.e. } t \in [S, T].$$

Recalling that $r(t) = \epsilon k(t)$, we conclude that

$$F(t, x) \cap (\dot{\bar{x}}(t) + r(t)\mathbb{B}) \neq \emptyset, \text{ for every } x \in \bar{x}(t) + \epsilon \mathbb{B}, \text{ a.e. } t \in [S, T].$$

$\square$

8.6 Proof of Theorem 8.4.3

This section provides a proof of all the assertions of the theorem, with the exception of property (vi), namely 'constancy of $t \to p(t) \cdot \dot{\bar{x}}(t)$ when $B(t)$ and $F(t, x)$ do not depend on t'. The proof of (vi) is given in Chap. 9, as part of a broader investigation of dynamic optimization problems involving more general end-point constraints that those present in problem (P). (See the comment following statement of Theorem 9.2.1.)

Step 1 (Necessary Conditions for a Finite Lagrangian Problem)

As a first step to proving Theorem 8.4.3, we establish necessary conditions of optimality for a variational problem of special structure, which is of interest principally because it provides a stepping stone to the derivation of necessary conditions for problems where the dynamic constraint is formulated as a differential inclusion. The variational problem, which makes reference to some trajectory $x' \in W^{1,1}([S, T]; \mathbb{R}^n)$, is as follows:

$$(FL) \begin{cases} \text{Minimize } J(x) := \left(\ell(x(S), x(T))\right) \vee \left(\int_S^T L(t, x(t), \dot{x}(t))dt\right) \\ \quad + \int_S^T L_0(t, x(t), \dot{x}(t))dt + \delta |x(S) - x'(S)| \\ \text{over } x \in W^{1,1}([S, T]; \mathbb{R}^n) \text{ such that} \\ \dot{x}(t) \in D(t) \text{ a.e.} \end{cases}$$

The data comprise an interval $[S, T]$, a number $\delta > 0$, functions $L : [S, T] \times \mathbb{R}^n \times \mathbb{R}^n \to \mathbb{R}$, $L_0 : [S, T] \times \mathbb{R}^n \times \mathbb{R}^n \to \mathbb{R}$ and $\ell : \mathbb{R}^n \times \mathbb{R}^n \to \mathbb{R}$, and a multifunction $D : [S, T] \rightsquigarrow \mathbb{R}^n$.

Problems involving an integral cost, in which the dynamic constraint imposes a constraint on the velocity variables alone, are traditionally referred to as finite Lagrangian problems. (FL) can be regarded as a generalized finite Lagrangian problem, in which the primary integral cost term enters, not additively, but via a 'max' operation. (The secondary integral cost term $\int L_0 dt$ and the left end-point cost term $\delta |x(S) - x'(S)|$ are included in the problem formulation to aid future analysis, in which they will appear as perturbation terms arising from the application of Ekeland's theorem.) A $W^{1,1}$ local minimizer x' for (FL) is defined by analogy with earlier definitions, except that, now, we impose the additional requirement that $t \to (L, L_0)(t, x'(t), \dot{x}'(t))$ is an integrable function.

8.6 Proof of Theorem 8.4.3

Proposition 8.6.1 *Let x' be a $W^{1,1}$ local minimizer for (FL). Assume that the following hypotheses are satisfied: there exist $\epsilon \in (0, 1)$, $N > 0$ and a non-negative function $\tilde{k} \in L^1(S, T)$ such that*

(FL0): *D has closed values, Gr D is $\mathcal{L} \times \mathcal{B}^n$ measurable and $D(t) \subset \dot{x}'(t) + N\mathbb{B}$ a.e.,*

(FL1): *ℓ is Lipschitz continuous on $\epsilon(\mathbb{B} \times \mathbb{B})$,*

(FL2): *$(L, L_0)(., x, v)$ is $\mathcal{L}$-measurable for each $(x, v) \in \mathbb{R}^n \times \mathbb{R}^n$,*

(FL3): *$|(L, L_0)(t, x, v) - (L, L_0)(t, y, w)| \leq \tilde{k}(t)(|x - y| + |v - w|)$, for all $x, y \in \epsilon\mathbb{B}$ and $v, w \in D(t)$, a.e. $t \in [S, T]$.*

Assume, furthermore, that

$$\|x'\|_{W^{1,1}} < \epsilon.$$

Then there exists an arc $p \in W^{1,1}$ and $\lambda \in [0, 1]$ which satisfy

(i): $\begin{cases} (a) : \dot{p}(t) \in (1 - \lambda)\mathrm{co}\, \partial_x L(t, x'(t), \dot{x}'(t)) \\ \qquad\qquad\qquad +\mathrm{co}\, \partial_x L_0(t, x'(t), \dot{x}'(t)) \text{ a.e. } t \in [S, T] \\ \text{and} \\ (b) : \dot{p}(t) \in \mathrm{co}\{\eta : (\eta, p(t)) \in (1 - \lambda)\partial_{x,v} L(t, x'(t), \dot{x}'(t)) \\ \qquad\qquad +\partial_{x,v} L_0(t, x'(t), \dot{x}'(t))\}, \text{ a.e. } t \in [S, T] \text{ such that } \dot{x}'(t) \in \overset{\circ}{D}(t) \end{cases}$

(ii): $(p(S), -p(T)) \in \lambda\partial\ell(x'(S), x'(T)) + \delta\mathbb{B} \times \{0\}$,

(iii): $\quad p(t) \cdot \dot{x}'(t) - (1 - \lambda)L(t, x'(t), \dot{x}'(t)) - L_0(t, x'(t), \dot{x}'(t)) \geq$
$$p(t) \cdot v - (1 - \lambda)L(t, x'(t), v) - L_0(t, x'(t), v),$$
for all $v \in D(t)$, a.e. $t \in [S, T]$.

Furthermore

$$\int_S^T L(t, x'(t), \dot{x}'(t))dt < \ell(x'(S), x'(T)) \implies \lambda = 1. \tag{8.6.1}$$

Proof of Proposition 8.6.1 Since x' is a $W^{1,1}$ local minimizer, there exists $\beta > 0$ such that x' is minimizing, when we add to (FL) the constraint

$$|x(S) - x'(S)| + \int_S^T |\dot{x}(t) - \dot{x}'(t)|dt \leq \beta. \tag{8.6.2}$$

Bearing in mind that $\|x'\|_{W^{1,1}} < \epsilon$, we can arrange, by choosing β sufficiently small, that $\|x\|_{W^{1,1}} < \epsilon$ (and hence also that $\|x\|_{L^\infty} < \epsilon$), for all x satisfying (8.6.2). Here, ϵ is the constant appearing in (FL3).

We note at the outset that hypotheses (FL1) and (FL3) can been replaced by the stronger, global, hypotheses:

(FL1)′: There exists $c_0 > 0$ and $k_\ell > 0$ such that

$$|\ell(x_0, x_1) - \ell(y_0, y_1)| \le k_\ell |(x_0, x_1) - (y_0, y_1)| \text{ and } |\ell(x_0, x_1)| \le c_0$$

for all $(x_0, x_1), (y_0, y_1) \in \mathbb{R}^n \times \mathbb{R}^n$,

(FL3)′: there exist non-negative $k, c \in L^1$ such that

$$\begin{cases} |(L, L_0)(t, x, v) - (L, L_0)(t, y, w)| \le k(t)(|x - y| + |v - w|) \text{ and} \\ |(L, L_0)(t, x, v)| \le c(t) \end{cases}$$

for all $x, y \in \mathbb{R}^n$ and $v, w \in D(t)$, a.e. $t \in [S, T]$.

This is because, if either (FL1)′ or (FL3)′ were not satisfied, we could replace ℓ and (L, L_0) by $\ell'(x_0, x_1) := \ell(\pi_{\epsilon \mathbb{B}}(x_0), \pi_{\epsilon \mathbb{B}}(x_1))$ and $(L', L_0')(t, x, v) = (L, L_0)(t, \pi_{\epsilon \mathbb{B}}(x), v)$ where $\pi_{\epsilon \mathbb{B}}$ is the projection onto $\epsilon \mathbb{B}$. Then, if k_ℓ is the Lipschitz constant of ℓ on $\epsilon(\mathbb{B} \times \mathbb{B})$, ℓ' is Lipschitz continuous on $\mathbb{R}^n \times \mathbb{R}^n$ with Lipschitz constant k_ℓ, which coincides with ℓ on a neighbourhood of $(x'(S), x'(T))$, and we can take $c_0 := |\ell(x'(S), x'(T))| + 2\sqrt{2}\epsilon k_\ell$. The new pair of integrands satisfies (FL3), with integrable bounds $k(t) = 2\epsilon^{-1}\tilde{k}(t)$ and $c(t) := |(L, L_0)(t, x'(t), \dot{x}'(t))| + (2\epsilon + N)k(t)$. Hypotheses (FL0) and (FL2) continue to be satisfied, following these changes in the data, and x' remains a $W^{1,1}$ local minimizer. If we are able to establish existence of some p and λ such that conditions (i)–(iii) are satisfied for x' and the modified data ℓ' and (L', L_0') at x', conditions (i)–(iii) continue to be satisfied for x' and the original data ℓ and (L, L_0) with the same p and λ, because of the 'local' nature of these conditions (w.r.t. the x variable). So we may indeed assume (FL1)′ and (FL3)′, without loss of generality.

By adding a constant if required, we can arrange k in (FL3)′ satisfies $k(t) \ge 1$, a.e..

Take a sequence $K_i \uparrow 0$. Write L_k^1 for the Banach space of measurable functions w such that $t \to k(t)w(t)$ is integrable on $[S, T]$, equipped with the k-weighted L^1 norm $||w||_{L_k^1} := ||kw||_{L^1}$. Write

$$W := \{(\xi, w, v) \in \mathbb{R}^n \times L_k^1 \times L_k^1 : v(t) \in D(t) \text{ a.e.}, ||x_{\xi, v} - x'||_{W^{1,1}} \le \beta\},$$

in which $x_{\xi, v}(t) = \xi + \int_S^t v(s)ds$. Define

$$||(\xi, w, v)||_k := |\xi| + ||kw||_{L^1} + ||kv||_{L^1}.$$

For each i, set

$$\tilde{J}_i(\xi, w, v) := \left(\ell(x_{\xi,v}(S), x_{\xi,v}(T))\right) \vee \left(\int_S^T L(t, w(t), v(t))dt\right)$$

8.6 Proof of Theorem 8.4.3

$$+ \int_S^T L_0(t, w(t), v(t))dt$$

$$+ \delta |x_{\xi,v}(S) - x'(S)| + K_i \int_S^T k(t)|x_{\xi,v}(t) - w(t)|^2 dt.$$

Notice that this cost function differs from that in (FL), by virtue of the facts that, in the cost integrands, the control-like variable w replaces the state variable x and that the discrepancy between w and x is accommodated by a quadratic penalty term, with penalty parameter K_i. □

Claim: For each i, $(W, ||.||_k)$ is a complete metric space and $\tilde{J}_i$ is lower semi-continuous on $(W, ||.||_k)$. There exists a sequence of non-negative numbers $\alpha_i \downarrow 0$ such that, for each i,

$$\tilde{J}_i(x'(S), x', \dot{x}') \leq \inf_{(\xi,w,v) \in W} \tilde{J}_i(\xi, w, v) + \alpha_i^2.$$

We verify the claim. Fix i. W is a subset of the Banach space $\mathbb{R}^n \times L_k^1 \times L_k^1$ with norm $||.||_k$. It suffices to show that the set is strongly closed and that $\tilde{J}_i$ is lower semi-continuous on W. Take an arbitrary sequence $(\xi_j, w_j, v_j) \to (\xi, w, v)$ in $(W, ||.||_k)$. Write $x_j := x_{\xi_j, v_j}$. Then, since $k \geq 1$ a.e., $x_j \to x_{\xi,v}$ in $W^{1,1}$. Restricting attention to a subsequence, we have $v_j(t) \to v(t)$ a.e.. So $v(t) \in D(t)$ a.e. and $||x_{\xi,v} - x'||_{W^{1,1}} \leq \beta$. The limit point (ξ, w, v) satisfies the conditions confirming membership of W, and so W is strongly closed. This establishes that $(W, ||.||_k)$ is complete.

Take again an arbitrary sequence $(\xi_j, w_j, v_j) \to (\xi, w, v)$ in $(W, ||.||_k)$. Set

$$\sigma_0 := \liminf_{j \to \infty} \tilde{J}_i(\xi_j, w_j, v_j).$$

Observe that, from (FL1)′ and (FL3)′, we have $\sigma_0 > -\infty$. By restricting attention to a subsequence, we can arrange that $\tilde{J}_i(\xi_j, w_j, v_j) \to \sigma_0$ as $j \to \infty$. Writing as before $x_j := x_{\xi_j, v_j}$, we have $x_j \to x_{\xi,v}$ strongly in $W^{1,1}$. It follows that $\ell(x_j(S), x_j(T)) \to \ell(x_{\xi,v}(S), x_{\xi,v}(T))$ and $\int_S^T |v_j - v| dt \to 0$ as $i \to \infty$. Along a further subsequence (we do not relabel), $w_j(t) \to w(t)$ and $v_j(t) \to v(t)$ a.e.. In view of the second condition in (FL3)′, which ensures uniform integrable boundedness of the integrands $(L, L_0)(t, w_j(t), v_j(t))$, $j = 1, 2 \ldots$, we have, by the dominated convergence theorem, that

$$\int_S^T (L, L_0)(t, w_j(t), v_j(t))dt \to \int_S^T (L, L_0)(t, w(t), v(t))dt \text{ as } k \to \infty.$$

Bearing in mind that the integrands involved are non-negative, we deduce from Fatou's lemma that

$$\liminf_{j\to\infty} \int_S^T k(t)|x_j(t) - w_j(t)|^2 dt \geq \int_S^T \liminf_{j\to\infty} k(t)|x_j(t) - w_j(t)|^2 dt$$
$$= \int_S^T k(t)|x_{\xi,v}(t) - w(t)|^2 dt.$$

(Note, we allow the possibility that some of these integrals are infinite.) It follows that

$$\sigma_0 \geq \liminf_{j\to\infty}\left[\ell(x_j(S), x_j(T)) \vee \int_S^T L(t, w_j(t), v_j(t))dt \right.$$
$$\left. + \delta|x_j(S) - x'(S)| + \int_S^T L_0(t, w_j(t), v_j(t))dt\right]$$
$$+ \liminf_{j\to\infty} K_i \int_S^T k(t)|x_j(t) - w_j(t)|^2 dt$$
$$\geq \ell(x(S), x(T)) \vee \int_S^T L(t, w(t), v(t))dt + \delta|x_{\xi,v}(S) - x'(S)|$$
$$+ \int_S^T L_0(t, w(t), v(t))dt + K_i \int_S^T k(t)|x_{\xi,v}(t) - w(t)|^2 dt$$
$$= \tilde{J}_i(\xi, w, v).$$

We conclude from this inequality that $\tilde{J}_i$ is lower semi-continuous. The claim is confirmed.

Now define

$$\alpha_i^2 := \tilde{J}_i(x'(S), x', \dot{x}') - \inf_{(\xi,w,v)\in W} \tilde{J}_i(\xi, w, v).$$

(Since $(x'(S), x', \dot{x}') \in W$, the right side is nonnegative.) We show that $\alpha_i \downarrow 0$ as $i \to \infty$.

Choose an arbitrary point $(\xi, w, v) \in W$ such that $\tilde{J}_i(\xi, w, v) < \infty$. Define

$$\tilde{c} := \left(\int_S^T k(t)|w(t) - x_{\xi,v}(t)|^2 dt\right)^{\frac{1}{2}} (<\infty) \text{ and } d := \left(\int_S^T k(t)dt\right)^{\frac{1}{2}}.$$

Then

$$\tilde{J}_i(\xi, w, v) \geq \tilde{J}_i(\xi, x_{\xi,v}, v) - |\int_S^T ((L, L_0)(t, w(t), v(t))$$
$$- (L, L_0)(t, x_{\xi,v}(t), v(t)))dt| + K_i \tilde{c}^2$$
$$\geq \tilde{J}_i(x'(S), x', \dot{x}') - 2\int_S^T k(t)|x_{\xi,v}(t) - w(t)|dt + K_i \tilde{c}^2$$

8.6 Proof of Theorem 8.4.3

(by the minimizing property of $(x'(S), x', \dot{x}')$ and by (FL3)')

$$\geq \tilde{J}_i(x'(S), x', \dot{x}') - \tilde{c}d + K_i\tilde{c}^2$$

(by the Hölder inequality)

$$\geq \tilde{J}_i(x'(S), x', \dot{x}') + \min\{-2c'd + K_i(c')^2 : c' \in \mathbb{R}\}$$
$$= \tilde{J}_i(x'(S), x', \dot{x}') - K_i^{-1}d^2.$$

It follows that

$$0 \leq \alpha_i^2 \left(= \tilde{J}_i(x'(S), x', \dot{x}') - \inf_W \tilde{J}_i(\xi, w, v) \right) \leq K_i^{-1} \times \int_S^T k(t)dt.$$

Since $K_i \uparrow \infty$, we conclude that $\alpha_i \to 0$ as $i \to \infty$. By discarding initial points in the sequence, we can arrange that $\alpha_i \leq 1$, for each i.

Fix i. Since $\tilde{J}_i$ is a lower semi-continuous function and $(x'(S), x', \dot{x}')$ is an α_i^2 minimizer for $\tilde{J}_i$ over W, we know from Ekeland's theorem (Theorem 3.3.1), that there exists $(\xi_i, w_i, v_i) \in W$ which minimizes

$$J_i(\xi, w, v) := \tilde{J}_i(\xi, w, v) + \alpha_i \|(\xi, w, v) - (\xi_i, w_i, v_i)\|_k$$

over W, and

$$\|(\xi_i, w_i, v_i) - (x'(S), x', \dot{x}')\|_k \leq \alpha_i, \text{ for each } i. \tag{8.6.3}$$

Ekeland's theorem also tells us that $\tilde{J}_i(\xi_i, w_i, v_i) \leq J_i(x'(S), x', \dot{x}')(<\infty)$. This implies that $\int_S^T k(t)|x_i(t) - w_i(t)|^2 dt < \infty$. Here, $x_i := x_{\xi_i, v_i}$. Since x_i is bounded, it follows that $t \to k(t)w_i^2(t)$ is integrable.

We can summarize the above discussion in control theoretic terms as follows: $((x_i, y_i, z_i), (v_i, w_i))$ is a minimizer for the dynamic optimization problem

$$(E) \begin{cases} \text{Minimize } \ell(x(S), x(T)) \vee z(T) + \delta |x(S) - x'(S)| \\ \quad + \int_S^T L_0(t, w(t), v(t))dt + \alpha_i \Big(|x(S) - x_i(S)| \\ \quad + \int_S^T (k(t)|v(t) - v_i(t)| + k(t)|w(t) - w_i(t)|)dt \Big) \\ \quad + K_i \int_S^T k(t)|x(t) - w(t)|^2 dt. \\ \text{over arcs } (x, y, z) \in W^{1,1} \times W^{1,1} \times W^{1,1} \text{ satisfying} \\ \dot{x}(t) = v(t), \ \dot{y}(t) = |v(t) - \dot{x}'(t)|, \ \dot{z}(t) = L(t, w(t), v(t)), \text{ a.e.,} \\ w(t) \in \mathbb{R}^n, v(t) \in D(t) \text{ a.e.,} \\ y(S) = z(S) = 0 \text{ and } |x(S) - x'(S)| + y(T) \leq \beta. \end{cases}$$

Here, $y_i(t) := \int_S^t |v_i(s) - \dot{x}'(s)|ds$, and $z_i(t) := \int_S^t L_i(s, w_i(s), v_i(s)))ds$.

(Strictly speaking, the above problem (E) is not equivalent to Min$\{J_i(\xi, w, v) : (\xi, w, v) \in W\}$ because it has domain comprising control functions (v, w) from the larger set $L^1 \times$ {meas. functions $w : [S, T] \to \mathbb{R}^n$}. However the two problems are equivalent, in the sense of having a common set of minimizers since, for any control functions $w, v \notin L_k^1 \times L_k^1$ the cost in (E) is $+\infty$, as is easily shown, and such control functions cannot be candidates for minimizing controls.)

The foregoing dynamic optimization problem is one to which the maximum principle, Theorem 7.2.1, is applicable, following the removal of integral cost terms by state augmentation. The relevant hypotheses are satisfied. (To check hypothesis (H1) in the maximum principle, we make use of the facts that $t \to k(t)w_i^2(t)$ and $t \to k(t)v_i(t)$ are integrable functions.) In consequence of (8.6.3), we know that, for some subsequence,

$$||x_i \to x'||_{L^\infty} \to 0 \text{ and } (v_i, w_i) \to (\dot{x}', x') \text{ in } L^1 \text{ and a.e..}$$

This implies, in particular, that $|x_i(S) - x'(S)| + y_i(T) \to 0$. It follows that the endpoint constraint '$|x_i(S) - x'_i(S)| + y_i(T) \leq \beta$' is inactive at (x_i, y_i), for sufficiently large i.

Because of the decoupled structure of the cost and dynamics, regarding the y and (x, z) state variable components, the costate trajectory component associated with y must be zero; it therefore drops altogether out of the relations. Also, in consequence of the facts that there is no end-point constraint and the end-point cost function is Lischitz continuous near $(x_i(S), x_i(T))$, the maximum principle is valid with cost multiplier 1.

Identifying $p_i \in W^{1,1}$ with the costate trajectory for the x variable, and $-(1-\lambda_i)$ with the (constant) costate trajectory associated with the z variable, we deduce, from the necessary conditions (and the max rule to estimate the limiting subdifferential $\partial(\ell \vee z)$) the following: there exist $p_i \in W^{1,1}$ and $\lambda_i \in [0, 1]$ such that

(A): $-\dot{p}_i(t) = -2K_i k(t)(x_i(t) - w_i(t))$ a.e.,
(B): $(p_i(S), -p_i(T)) \in \lambda_i \partial \ell(x_i(S), x_i(T)) + (\delta + \alpha_i)\mathbb{B} \times \{0\}$,
(C): the function $h_i(w, v) := p_i(t) \cdot v - (1 - \lambda_i)L(t, w, v) - L_0(t, w, v)$

$$- \alpha_i k(t)(|v - v_i(t)| + |w - w_i(t)|) - K_i k(t)|x_i(t) - w|^2$$

achieves its maximum at $(w_i(t), v_i(t))$ over $(w, v) \in \mathbb{R}^n \times D(t)$, a.e..

Notice that, if $\int_S^T L(t, x'(t), \dot{x}'(t))dt < \ell(x'(S), x'(T))$, then $\int_S^T L(t, x_i(t), \dot{x}_i(t))dt < \ell(x_i(S)), x_i(T))$ for i sufficiently large. It follows from the max rule for subdifferentials that $\lambda_i = 1$. But then

$$\int_S^T L(t, x'(t), \dot{x}'(t))dt < \ell(x'(S), x'(T)) \implies \lambda_i = 1, \text{ for } i \text{ sufficiently large}.$$
(8.6.4)

Next, we investigate the consequences of condition (C). Take any $t \in [S, T]$ at which this condition is satisfied. The specified function with constrained maximizer

8.6 Proof of Theorem 8.4.3

$(w_i(t), v_i(t))$ is Lipschitz continuous on a neighbourhood of $(w_i(t), v_i(t))$; write the Lipschitz constant $\rho_i(t)$. It follows from the exact penalization principle that $(w_i(t), v_i(t))$ is also an unconstrained maximizer of the function

$$(w, v) \to p_i(t) \cdot v - (1 - \lambda_i)L(t, w, v) - L_0(t, w, v)$$
$$- \alpha_i k(t)(|v - v_i(t)| + |w - w_i(t)|) - K_i k(t)|x_i(t) - w|^2 - \rho_i(t)d_{D(t)}(v).$$

It follows that

$$(0, p_i(t)) \in \partial_{x,v}((1 - \lambda_i)L + L_0)(t, w_i(t), v_i(t))$$
$$+ \alpha_i k(t)(\mathbb{B} \times \mathbb{B}) + (2K_i k(t)(x_i(t) - w_i(t)), 0) + \{0\} \times \rho_i(t) \partial d_{D(t)}(v).$$

This relation combines with (A) to give:

$$(-\dot{p}_i(t), p_i(t)) \in \partial_{x,v}((1 - \lambda_i)L + L_0)(t, w_i(t), v_i(t))$$
$$+ \alpha_i k(t)(\mathbb{B} \times \mathbb{B}) + \{0\} \times \rho_i(t) \partial d_{D(t)}(v_i(t)). \qquad (8.6.5)$$

Fix $v = v_i(t)$. Then

$$w \to -(1 - \lambda_i)L(t, w, v_i(t)) - L_0(t, w, v_i(t)) - \alpha_i k(t)|w - w_i(t)|$$
$$- K_i k(t)|x_i(t) - w|^2$$

achieves its maximum at $w_i(t)$ over all $w \in \mathbb{R}^n$. In view of (A), this implies

$$\dot{p}_i(t) \in \partial_x((1 - \lambda_i)L + L_0)(t, w_i(t), v_i(t)) + \alpha_i k(t)\mathbb{B}. \qquad (8.6.6)$$

Fix $w = w_i(t)$. Then $v \to p_i(t) \cdot v - (1 - \lambda_i)L(t, w_i(t), v) - L_0(t, w_i(t), v) - \alpha_i k(t)|v - v_i(t)|$ achieves its maximum at $v_i(t)$ over $v \in D(t)$. This implies that, for a.e. $t \in [S, T]$,

$$p_i(t) \cdot v_i(t) - (1 - \lambda_i)L(t, w_i(t), v_i(t)) - L_0(t, w_i(t), v_i(t))$$
$$\geq p_i(t) \cdot v - (1 - \lambda_i)L(t, w_i(t), v) - L_0(t, w_i(t), v)$$
$$- \alpha_i k(t)|v - v_i(t)| \text{ for all } v \in D(t). \qquad (8.6.7)$$

Since $(L, L_0)(t, ., v)$ is Lipschitz continuous with Lipschitz constant $k(t)$ for all $v \in D(t)$, (8.6.6) implies that, for i sufficiently large, $|\dot{p}_i(t)| \leq (2 + \alpha_i)k(t)$. Noting that the $p_i(S)$'s are uniformly bounded (see (B)), we can arrange, by limiting attention to a subsequence, that $p_i \to p$ uniformly and $\dot{p}_i \to \dot{p}$ weakly in L^1 for some $p \in W^{1,1}$. We can also ensure that $\lambda_i \to \lambda$ for some $\lambda \geq 0$. It follows from (8.6.4) that

$$\int_S^T L(t, x'(t), \dot{x}'(t))dt < \ell(x'(S)), x'(T)) \implies \lambda = 1. \qquad (8.6.8)$$

We have confirmed condition (8.6.1). Next, notice that (8.6.7) implies in the limit

$$p(t) \cdot v'(t) - (1-\lambda)L(t, x'(t), v'(t)) - L_0(t, x'(t), v'(t))$$
$$\geq p(t) \cdot v - (1-\lambda)L(t, x'(t), v) - L_0(t, x'(t), v) \text{ for all } v \in D(t) \text{ a.e.}.$$

This is (iii). (B) yields in the limit $(p(S), -p(T)) \in \lambda \partial \ell(x'(S), x'(T)) + \delta \mathbb{B} \times \{0\}$, which is (ii).

We next confirm (i)(b). Fix $\sigma > 0$. Define

$$F^\sigma := \{t \in [S, T] : \dot{x}'(t) + \sigma \mathbb{B} \subset D(t)\}.$$

Define, for $t \in [S, T]$,

$$G^\sigma(t) := \begin{cases} \{q \in \mathbb{R}^n : (q, p(t)) \in \bigcup_{(w,v) \in (x'(t), \dot{x}'(t)) + \sigma \mathbb{B}} \partial_{x,v}((1-\lambda)L + L_0)(t, w, v) \\ \qquad + \sigma k(t)(\mathbb{B} \times \mathbb{B})\} & \text{if } t \in F^\sigma \\ 3k(t)\mathbb{B} & \text{if } t \notin F^\sigma. \end{cases}$$

Define, finally, the monotone sequence of sets E_i^σ, $i = 1, 2, \ldots$,

$$E_i^\sigma := \{t \in [S, T] : |(w_j(t), v_j(t)) - (x'(t), \dot{x}'(t))| \leq \sigma \text{ for all } j \geq i\}.$$

Since $(w_i, v_i) \to (x', \dot{x}')$ a.e.

$$\text{meas}\left([S, T] \setminus E_i^\sigma\right) \to 0 \text{ as } i \to \infty.$$

Taking note also of (8.6.5) and (8.6.6), we deduce that, for all i sufficiently large,

$$\dot{p}_i(t) \in G^\sigma(t), \text{ for a.e. } t \in E_i^\sigma(t).$$

It now follows from the compactness of trajectories theorem (Theorem 6.3.3) that

$$\dot{p}(t) \in \overline{\text{co}}\, G^\sigma(t), \text{ for a.e. } t \in [S, T].$$

This relation is valid for all $\sigma > 0$. Taking account of the upper semicontinuity properties of the limiting subdifferential, we can show that, for a.e. $t \in [S, T]$, $\overline{\text{co}}\, G^\sigma(t)$, $\sigma > 0$, is a monotone family of sets and

$$\lim_{\sigma \downarrow 0} \overline{\text{co}}\, G^\sigma(t) = \text{co}\, \partial_{x,v}((1-\lambda)L + L_0)(t, x'(t), \dot{x}'(t)), \text{ for a.e. } t \text{ s.t. } \dot{x}'(t) \in \text{int}\, D(t).$$

8.6 Proof of Theorem 8.4.3

We have confirmed property (i)(b). A similar analysis, taking as starting point (8.6.6) in place of (8.6.5), yields (i)(a). All the assertions of Proposition 8.6.1 have been verified. The proof is complete.

Step 2 (Hypothesis Reduction)

We now show that, without loss of generality, it suffices to prove the assertions of Theorem 8.4.3 in the special case when some additional hypotheses are imposed on the data and the candidate admissible F trajectory $\bar{x}$. The end result of this reduction step are summarized as Proposition 8.6.2 below.

The following hypothesis is a strengthened form of (G3), in which the function R and r are replaced by constants.

(G3)': There exist $\epsilon > 0$, $R \in (0, \infty)$, $\gamma \in (0, 1)$ and $k \in L^1$ such that $R\overset{\circ}{\mathbb{B}} \subset B(t)$ a.e. and the following conditions are satisfied:

(a): $F(t, x') \cap (\dot{\bar{x}}(t) + R\overset{\circ}{\mathbb{B}}) \subset F(t, x) + k(t)|x' - x|\mathbb{B}$,
for all $x, x' \in \bar{x}(t) + \epsilon \mathbb{B}$, a.e. $t \in [S, T]$.
(b): $F(t, x) \cap (\dot{\bar{x}}(t) + \gamma R\mathbb{B}) \neq \emptyset$ for all $x \in \bar{x}(t) + \epsilon \mathbb{B}$, a.e. $t \in [S, T]$.

Proposition 8.6.2 *Assume the assertions of Theorem 8.4.3 are valid under the hypotheses(G1), (G2) and (G3)' and when $\bar{x} \equiv 0$. Then the assertions of Theorem 8.4.3 are valid under hypotheses (G1)–(G3).*

Proof Suppose that the assertions of the theorem are known to be valid only under the additional assumption that $\bar{x} \equiv 0$. Then, if $\bar{x} \not\equiv 0$, we may consider a modified problem in which $F(t, x)$, $g(x_0, x_1)$ and C are replaced by '$F(t, \bar{x}(t) + x) - \dot{\bar{x}}(t)$', '$g(\bar{x}(S) + x_0, \bar{x}(T) + x_1)$' and '$C - (\bar{x}(S), \bar{x}(T))$', respectively. For the modified problem, $x \equiv 0$ is a $W^{1,1}$ local minimizer relative to the multifunction B. The modified data continue to satisfy (G1)–(G3) and the necessary conditions for the modified problem, which we know to be valid under our assumptions, imply those for the original problem. So we can assume that $\bar{x} \equiv 0$.

Next suppose that the data satisfies (G3) (expressed in terms of $R : [S, T] \to (0, \infty) \cup +\infty$, $r \in L^1$ and $\gamma \in (0, 1)$), with reference to $\bar{x} \equiv 0$ and the multifunction B, but when the assertions of the theorem are known to be valid only in the case when (G3) is replaced by the stronger hypothesis (G3)'. In this situation, we consider the change of independent variable $s = \sigma(t)$ where

$$\sigma(t) := \int_S^t r(t')dt'.$$

σ is a strictly increasing function on its domain $[S, T]$ such that $\sigma(S) = 0$. Write $\tau := \sigma(T)$.

Let us now examine a new problem, which is an example of (P), in which the underlying time interval $[S, T]$ is replaced by $[0, \tau]$ and F is replaced by

$$\tilde{F}(s, x) := \frac{1}{(r \circ \sigma^{-1})(s)} F(\sigma^{-1}(s), x),$$

but, otherwise, the data remains the same. Take $\beta > 0$ to be such that $\bar{x} \equiv 0$ is minimizing with reference to admissible F trajectories x such that $\dot{x} \in B(t)$ a.e. and $||x||_{W^{1,1}} \leq \beta$. Now take $y : [0, \tau] \to \mathbb{R}^n$ to be any admissible $\tilde{F}$ trajectory for the new problem such that $||y||_{W^{1,1}} \leq \beta$ and $\dot{y}(s) \in \tilde{B}(s)$ a.e., where

$$\tilde{B}(s) := \frac{1}{(r \circ \sigma^{-1})(s)} B(\sigma^{-1}(s)).$$

Define $x := y \circ \sigma$. Then x is an F trajectory on $[S, T]$ satisfying $(x(S), x(T)) = (y(0), y(\tau))$ and $\dot{x}(t) \in B(t)$ a.e.. Since y is admissible for the new problem, it follows that x is admissible for the original problem, We have $\dot{x}(t) = (\dot{y} \circ \sigma)(t) r(t)$ a.e.. By Fubini's theorem,

$$\int_S^T |\dot{x}(t)| dt = \int_0^\tau |\dot{y} \circ \sigma|(t) r(t) dt = \int_0^\tau |y(s)| ds \leq \beta.$$

But then, since $x(S) = y(0)$, we know that $||x||_{W^{1,1}} = ||y||_{W^{1,1}} \leq \beta$. Now define $\bar{y} \equiv 0$. Since $\bar{x} \equiv 0$ is admissible for the original problem and $\bar{x} \equiv 0$ maps into $\bar{y} \equiv 0$ under the inverse change of independent variable, we know that $\bar{y} \equiv 0$ is admissible for the new problem. By the optimality properties of $\bar{x} \equiv 0$ and the above relations,

$$g(y(0), y(\tau)) = g(x(S), x(T)) \geq g(\bar{x}(S), \bar{x}(T)) = g(\bar{y}(0), \bar{y}(\tau)).$$

It follows that $\bar{y} \equiv 0$ is a $W^{1,1}$ local minimizer for the new problem relative to $\tilde{B}$.

The data for the new problem satisfies hypotheses (G1) and (G2) of Theorem 8.4.3. We now show that hypothesis (G3)' is also satisfied. Since $\gamma^{-1} r(t) \leq R(t)$ a.e., we know, from (G3), that for all $x, x' \in \epsilon \mathbb{B}$ and a.e. $s \in [0, \tau]$,

$$F(\sigma^{-1}(s), x) \cap (\gamma^{-1} \times r \circ \sigma^{-1})(s) \overset{\circ}{\mathbb{B}} \subset F(\sigma^{-1}(s), x') + (k \circ \sigma^{-1})(s) |x - x'| \mathbb{B}.$$

But then by definition of $\tilde{F}$,

$$\Big((r \circ \sigma^{-1})(s) \tilde{F}(s, x)\Big) \cap (\gamma^{-1} \times r \circ \sigma^{-1})(s) \overset{\circ}{\mathbb{B}}$$
$$\subset (r \circ \sigma^{-1})(s) \tilde{F}(\sigma^{-1}(s), x') + (k \circ \sigma^{-1})(s) |x - x'| \mathbb{B}.$$

It follows that

$$\tilde{F}(s, x) \cap (\gamma^{-1} \overset{\circ}{\mathbb{B}}) \subset \tilde{F}(s, x') + \tilde{k}(s) |x - x'| \mathbb{B} \text{ for all } x, x' \in \epsilon \mathbb{B} \text{ and a.e. } s \in [0, \tau], \quad (8.6.9)$$

8.6 Proof of Theorem 8.4.3

where $\tilde{k}(s) := (k \circ \sigma^{-1})(s)) \frac{1}{r \circ \sigma^{-1}(s)}$. Notice that, in consequence of Fubini's theorem, $\tilde{k}$ is an integrable function. Thus the data for the new problem satisfies condition (G3)'(a), with reference to the nominal $\tilde{F}$ trajectory $\bar{y} \equiv 0$.

We know from (G3)(b) that, for all $x \in \epsilon \mathbb{B}$, $F(t, x) \cap (r(t)\mathbb{B}) \neq \emptyset$ for a.e. $t \in [S, T]$. But then, by definition of $\tilde{F}$,

$$(r \circ \sigma^{-1})(s)\, \tilde{F}(s, x) \cap ((r \circ \sigma^{-1})(s)\mathbb{B}) \neq \emptyset, \quad \text{for all } x \in \epsilon\mathbb{B}, \text{ a.e. } s \in [0, \tau].$$

It follows that

$$\tilde{F}(s, x) \cap \mathbb{B} \neq \emptyset, \quad \text{for all } x \in \epsilon\mathbb{B}, \text{ a.e. } s \in [0, \tau]. \tag{8.6.10}$$

Relations (8.6.9) and (8.6.10) confirm (G3)', with ϵ as before and the other parameters that we now write R' and γ' given by $R' := \gamma^{-1}$ and $\gamma' = \gamma$, in which $\gamma \in (0, 1)$ is the parameter in (G3) (for the data of the original problem).

Since hypotheses (G1), (G2) and (G3)' are satisfied, we are justified, under our assumptions, in applying the necessary conditions of Theorem 8.4.3 to the new problem, with reference to the $W^{1,1}$ minimizer $\bar{y} \equiv 0$ relative to $\tilde{B}$. It is straightforward to deduce the necessary conditions for the original problem, with reference to $\bar{x} \equiv 0$ and B, via the inverse change of independent variable $t = \sigma^{-1}(s)$, in which the Lagrange multipliers $(\tilde{p}, \lambda)$ arising in the new problem are replaced by $(p = \tilde{p} \circ \sigma^{-1}, \lambda)$ in the original one. □

Henceforth in the proof, we may assume that the strengthened hypothesis (G3)' (with parameters $\gamma \in (0, 1)$ and $R > 0$ and integrable Lipschitz bound k) is satisfied since, as has been demonstrated, we may do so without loss of generality.

Step 3 (Construction of an Integral Penalty Integrand)

Our goal in this step is to construct an integral function, with integrand $\rho_S(t, x, v)$, which penalizes violations of the dynamic constraint and restrictions on velocities associated with the multifunction B, namely

$$\dot{x}(t) \in F(t, x(t)) \cap (\dot{\bar{x}}(t) + B(t)),$$

in a suitable manner for the derivation of necessary conditions.

In the earlier literature, which pre-dated consideration of constraints on velocities expressed in terms of the multifunction B, an integral penalty function with integrand constructed from the distance function to $F(t, x)$, namely

$$\rho_F(t, x, v) := \inf\{|v - e| \, : \, e \in F(t, x)\}, \tag{8.6.11}$$

was widely used to reduce problem (P) to an approximating finite Lagrangian problem, as a means to accomplishing this step. But when we introduce an additional constraint on velocities, effectively replacing $F(t, x(t))$ by $F(t, x(t)) \cap (\dot{\bar{x}}(t) + B(t))$, this traditional penalty integrand lacks the crucial regularity properties for the derivation of necessary conditions and an alternative, which we shall refer to as the modified penalty integrand, is required.

Before constructing the modified penalty integrand and exploring its properties, we will find it helpful to have at hand useful information about the standard penalty integrand, assembled in the following lemma.

Lemma 8.6.3 *Take a multifunction* $\Gamma : \mathbb{R}^n \rightsquigarrow \mathbb{R}^n$ *and a point* $(\bar{x}, \bar{v}) \in \mathrm{Gr}\,\Gamma$. *Assume that* Γ *has values non-empty closed sets. Define the function* $\rho_\Gamma : \mathbb{R}^n \times \mathbb{R}^n \to \mathbb{R}$ *to be*

$$\rho_\Gamma(x, v) := \inf\{|v - y| : y \in \Gamma(x)\}.$$

Assume that there exist $\epsilon > 0$, $r > 0$, $k > 0$ *and* $K > 0$ *such that*

$$\Gamma(x') \cap (\bar{v} + r\mathring{\mathbb{B}}) \subset \Gamma(x) + k|x' - x|\mathbb{B}, \text{ for all } x', x \in \bar{x} + \epsilon\mathbb{B}. \qquad (8.6.12)$$

and

$$\Gamma(x) \cap (\bar{v} + K|x - \bar{x}|\mathbb{B}) \neq \emptyset, \text{ for all } x \in \bar{x} + \epsilon\mathbb{B}. \qquad (8.6.13)$$

Then:

(a): *For each* $x \in \mathbb{R}^n$, $\rho_\Gamma(x, .)$ *is Lipschitz continuous with Lipschitz constant* 1. *For each* $v \in \bar{v} + (r/4)\mathbb{B}$, $\rho_\Gamma(., v)$ *is Lipschitz continuous on* $\bar{x} + \min\{\epsilon, r/(4K)\}\mathbb{B}$ *with Lipschitz constant* k.

(b): *Fix elements* $(x, v) \in \mathbb{R}^n \times \mathbb{R}^n$ *and* $(w, p) \in \mathbb{R}^n \times \mathbb{R}^n$ *such that* $|x - \bar{x}| < \min\{\epsilon, r/(3K)\}$ *and* $|v - \bar{v}| < r/3$. *Assume*

$$(w, p) \in \partial \rho_\Gamma(x, v).$$

Then

$$|w| \leq k \text{ and } |p| \leq 1.$$

Furthermore,

$$v \in \Gamma(x) \text{ implies } (w, p) \in N_{\mathrm{Gr}\,\Gamma}(x, v).$$

(c): *Fix elements* $(x, v) \in \mathbb{R}^n \times \mathbb{R}^n$ *and* $(w, p) \in \mathbb{R}^n \times \mathbb{R}^n$ *such that* $|x - \bar{x}| < \epsilon$ *and* $|v - \bar{v}| < r$,

$$\text{`}v \in \Gamma(x) \text{ and } (w, p) \in N_{\mathrm{Gr}\,\Gamma}(x, v).\text{'} \implies \text{`}|w| \leq k|p|\text{'}.$$

(It might be thought that (8.6.13) is a redundant hypothesis, since it is implied by the preceding hypothesis (8.6.12), with K replaced by k. The lemma needs sometimes

8.6 Proof of Theorem 8.4.3

to be applied, however, in situations where $K < k$ and, in this case, the fact that the domain of Lipschitz continuity referred to in part (a) of the lemma is the set $\bar{x} + \min\{\epsilon, r/(4K)\}\mathbb{B}$, not the smaller set $\bar{x} + \min\{\epsilon, r/(4k)\}\mathbb{B}$, can be important.)

Proof
(a): For fixed x, $\rho_\Gamma(x, .)$ is the Euclidean distance function for the set $\Gamma(x)$ and, by the properties of distance functions, is Lipschitz continuous with Lipschitz constant 1. Fix $v \in \bar{v} + (r/3)\mathbb{B}$. Choose any x', $x \in \bar{x} + \min\{\epsilon, r/(3K)\}\mathbb{B}$. Let u' be a closest point to v in $\Gamma(x')$ (and one such point exists), i.e., $|v - u'| = \rho_\Gamma(x', v)$. According to (8.6.13) there exists $e' \in \Gamma(x')$ such that $|e' - \bar{v}| \leq K|x' - \bar{x}| \leq r/4$. Then

$$|u' - \bar{v}| \leq |v - \bar{v}| + |v - u'| \leq |v - \bar{v}| + |v - e'| \leq 2|v - \bar{v}| + |\bar{v} - e'| \leq \frac{2}{3}r + \frac{1}{4}r < r.$$

We have shown that $u' \in \bar{v} + r\overset{\circ}{\mathbb{B}}$. By (8.6.12), there exists $u \in \Gamma(x)$ such that $|u' - u| \leq k|x' - x|$. It follows that

$$\rho_\Gamma(x, v) \leq |v - u| \leq |v - u'| + |u' - u| \leq \rho_\Gamma(x', v) + k|x' - x|.$$

Since the variables x' and x are interchangeable, we conclude that

$$|\rho_\Gamma(x, v) - \rho_\Gamma(x', v)| \leq k|x - x'|.$$

This is the desired Lipschitz continuity property.

(b): We have shown that ρ_Γ is Lipschitz continuous on the closed neighbourhood $\mathcal{N}$ of $(\bar{x}, \bar{v})$

$$\mathcal{N} := (\bar{x}, \bar{v}) + \min\{\epsilon, r/(4K)\}\mathbb{B} \times (r/3)\mathbb{B} \tag{8.6.14}$$

Close scrutiny of the proof of (a) reveals also that

$$(x, v) \in \mathcal{N} \text{ implies } |u - \bar{v}| < r \tag{8.6.15}$$

for any point $u \in \Gamma(x)$ such that $|v - u| = \min\{|v - e| : e \in \Gamma(x)\}$.

Fix (x, v) in the interior of $\mathcal{N}$. Since (as we have shown) ρ_Γ is Lipschitz continuous on a neighbourhood of (x, v), it is meaningful to talk of the limiting subdifferential $\partial \rho_\Gamma(x, v)$. Take any $(w, p) \in \partial \rho_\Gamma(x, v)$. We know that there exist $(x_i, v_i) \to (x, v)$, $(w_i, p_i) \to (w, p)$, $\epsilon_i \downarrow 0$ and a sequence of positive numbers M_i such that

$$w_i \cdot (x' - x_i) + p_i \cdot (v' - v_i) \leq$$
$$\rho_\Gamma(x', v') - \rho_\Gamma(x_i, v_i) + M_i(|x' - x_i|^2 + |v' - v_i|^2) \tag{8.6.16}$$

for all $(x', v') \in (x_i, v_i) + \epsilon_i \mathbb{B}$.

By part (a), $\rho_\Gamma(.,v_i)$ and $\rho_\Gamma(x_i,.)$ are Lipschitz continuous (on appropriate neighbourhoods of (x_i, v_i)) with Lipschitz constants k and 1 respectively, for sufficiently large i. We easily deduce from (8.6.16) that $|w_i| \le k$ and $|p_i| \le 1$. Passing to the limit as $i \to \infty$ gives

$$|w| \le k \text{ and } |p| \le 1.$$

Now suppose that $v \in \Gamma(x)$. For each i, again take u_i to be a closest point to v_i in $\Gamma(x_i)$ (we allow the possibility that $v_i \in \Gamma(x_i)$ for some values of i). It follows from (8.6.12) and (8.6.15) that $u_i \to v$ as $i \to \infty$. Fix i sufficiently large. Choose any (x', u') in Gr Γ close to (x_i, u_i). Now insert $(x', v' = u' + v_i - u_i)$ into (8.6.16). We arrive at

$$w_i \cdot (x' - x_i) + p_i \cdot (u' - u_i) \le$$
$$\rho_\Gamma(x', u' + v_i - u_i) - \rho_\Gamma(x_i, v_i) + M_i(|x' - x_i|^2 + |u' - u_i|^2).$$

Notice however that

$$\rho_\Gamma(x', u' + v_i - u_i) \le |u' + v_i - u_i - u'| = |v_i - u_i| = \rho_\Gamma(x_i, v_i).$$

It follows that

$$w_i \cdot (x' - x_i) + p_i \cdot (u' - u_i) \le M_i(|x' - x_i|^2 + |u' - u_i|^2)$$

for all (x', u') in Gr Γ sufficiently close to (x_i, u_i). This implies $(w_i, p_i) \in N^P_{\text{Gr}\,\Gamma}(x_i, u_i)$. However Gr $\Gamma \cap ((x,v) + \alpha \mathbb{B})$ is a closed set for some $\alpha > 0$. Recalling that $(x_i, u_i) \xrightarrow{\text{Gr}\,\Gamma} (x, v)$, we deduce from the closure properties of the normal cone that $(w, p) \in N_{\text{Gr}\,\Gamma}(x, v)$. This is what we set out to prove.

(c): Take arbitrary pairs of points $(x, v) \in \text{Gr}\,\Gamma(x, v)$ and $(w, p) \in N_{\text{Gr}\,\Gamma}$ with the asserted properties. Assume initially that

$$(w, p) \in N^P_{\text{Gr}\,\Gamma}(x, v).$$

Then, from the proximal normal inequality, there exists $M > 0$ such that

$$(x' - x) \cdot w + (v' - v) \cdot p \le M(|x' - x|^2 + |v' - v|^2)$$

for all $(x', v') \in \text{Gr}\,\Gamma$. Take $\delta_i \downarrow 0$ and, for each i, let $x_i = x + \delta_i w$. Since $|v - \bar{v}| < r$ and in view of (8.6.12) we know that, for each i sufficiently large, there exists $v_i \in \Gamma(x + \delta_i w)$ such that $|v_i - v| \le \delta_i k |w|$. Inserting $x' = x_i$ and $v' = v_i$ into the above relation yields

$$\delta_i |w|^2 \le |v_i - v||p| + \delta_i^2 M(|w|^2 + \delta_i^{-2}|v_i - v|^2) \le \delta_i k|w||p| + \delta_i^2 M(1+k^2)|w|^2.$$

8.6 Proof of Theorem 8.4.3

Suppose $w \neq 0$. Dividing across this relation by $\delta_i |w|$ and passing to the limit as $i \to 0$ yields

$$|w| \leq k|p| \tag{8.6.17}$$

On the other hand, (8.6.17) is obviously true if $w = 0$. We have proved (c) in the special case when (w, p) is a proximal normal to $\operatorname{Gr}\Gamma$ at (x, v). The fact that (8.6.17) continues to be satisfied when (w, p) is merely a limiting normal vector is a direct consequence of the fact that an arbitrary limiting normal vector can be approximated by a proximal normal vector at a neighbouring point. □

Choose $\eta \in (0, 1/2)$ such that (G3)' is satisfied, with $\gamma = (1 - 2\eta)$. Fix $N > R$. Define

$$E^N(t) := \{e \in \Omega_0(t) \cap B(t) : R \leq |e| \leq N\},$$

in which $\Omega_0(t)$ is the regular velocity set relative to $\bar{x} \equiv 0$ and B, defined in (8.4.5), namely

$$\Omega_0(t) := \{e \in F(t, \bar{x}(t)) : F(t, .) \text{ is pseudo-Lipschitz continuous near } (\bar{x}(t), e)\}.$$

Making use of the compactness of the closed ball $N\mathbb{B}$, we can construct a multifunction $E^N_{\text{discrete}} : [S, T] \rightsquigarrow \mathbb{R}^n$ such that

(a): $E^N_{\text{discrete}}(t)$ is an empty or finite set for each $t \in [S, T]$,
(b): E^N_{discrete} is measurable,
(c): $E^N_{\text{discrete}}(t) \subset E^N(t) \subset E^N_{\text{discrete}}(t) + N^{-1}\mathbb{B}$,
  for each $t \in [S, T]$ such that $E^N(t) \neq \emptyset$,
(d): $E^N_{\text{discrete}}(t) \subset E^{N+1}_{\text{discrete}}(t)$ for all integers $N > R$.

Define

$$\theta(t) := \begin{cases} \inf\{|e - e'| : e, e' \in E^N_{\text{discrete}}(t) \text{ and } e \neq e'\} \\ \qquad \wedge \inf\{d_{\partial B(t)}(e) :\} \{e \in E^N_{\text{discrete}}(t)\} \\ \quad \text{if } E^N_{\text{discrete}}(t) \text{ contains at least two elements,} \\ +\infty \quad \text{otherwise}. \end{cases}$$

($\theta(t)$ is the minimum distance between distinct points in $E^N_{\text{discrete}}(t)$, when this set contains two or more points. Since $E^N_{\text{discrete}}(t)$ is a finite set, $\theta(t) > 0$ for each $t \in [S, T]$.)

Take $\delta \in (0, \eta R)$. Define

$$E^\delta_{\text{regular}}(t) := \{e \in E^N_{\text{discrete}}(t) : F(t, .) \text{ is pseudo-Lipschitz continuous}$$

$$\text{near } (\bar{x}(t) = 0, e) \text{ (with parameters}$$

$$\epsilon \geq \delta, \ R \geq \delta \text{ and } k \leq \delta^{-1}) \text{ and } \theta(t) \geq \delta\}.$$

We see that $E^\delta_{\text{regular}}(t) = \emptyset$ if $\theta(t) < \delta$. Observe also that $E^\delta_{\text{regular}}(t) \subset E^N_{\text{discrete}}(t) \subset N\mathbb{B}$. Define

$$D^\delta(t) := (1-\eta)R\mathbb{B} \cup \{e \in e_0 + \tfrac{1}{3}\delta\mathbb{B} : e_0 \in E^\delta_{\text{regular}}(t)\},$$

in which the right side is interpreted as $(1-\eta)R\mathbb{B}$ if $E^\delta_{\text{regular}}(t) = \emptyset$.

Since the distance between distinct points in $E^\delta_{\text{regular}}(t)$ is at least δ and $\delta \le \eta R$, $D^\delta(t)$ is a finite union of disjoint, closed balls. These comprise $(1-\eta)R\mathbb{B}$ and elements from a (possibly empty) collection of disjoint closed balls each with origin lying outside $R\overset{\circ}{\mathbb{B}}$. Observe also that $D^\delta(t) \subset B(t)$ for a.e. $t \in [S,T]$.

Consider next the multifunction $S : [S,T] \times \mathbb{R}^n \rightsquigarrow \mathbb{R}^n$

$$S(t,x) := \{(\chi(|e|)e : e \in F(t,x)\}, \tag{8.6.18}$$

in which $\chi : [0,\infty) \to [1,\infty)$ is the function

$$\chi(d) := 1 + \frac{4(1-\eta)}{\eta R}[d - (1-\eta)R]^+.$$

($\phi^+(d) := \max\{\phi(d),0\}$ is the positive part of the function ϕ.)

Lemma 8.6.4 *The multifunction S takes values closed non-empty sets.*

Proof Fix $(t,x) \in [S,T] \times \mathbb{R}^n$. The set $S(t,x)$ is non-empty because $F(t,x)$ is non-empty. To see that it is closed, take a sequence $\{e_i\}$ in $F(t,x)$ such that the sequence $\{\chi(|e_i|)e_i\}$ converges in $\mathbb{R}^n$. Because $\chi \ge 1$, this implies the $\{e_i\}$ is a bounded sequence. We can arrange then, by extraction of a subseqence, that $e_i \to \bar{e}$, for some $\bar{e}$, which is an element in $F(t,x)$, because the latter set is closed. Since χ is continuous, $\chi(|e_i|)e_i \to \chi(|\bar{e}|)\bar{e}$, which lies in $S(t,x)$. So $S(t,x)$ is closed. □

The function $\rho_S : [S,T] \times \mathbb{R}^n \times \mathbb{R}^n \to [0,+\infty)$, interpreted as a modified penalty integrand, is defined to be

$$\rho_S(t,x,v) := \begin{cases} d_{S(t,x)}(v) & \text{if } |v| \le (1-\eta)R, \\ d_{F(t,x)}(v) & \text{if } |v| > (1-\eta)R. \end{cases}$$

The modified penalty integrand has explicit representation

$$\rho_S(t,x,v) = \begin{cases} \inf\{|v - \chi(|e|)e| : e \in F(t,x)\} & \text{if } |v| \le (1-\eta)R \\ \inf\{|v - e| : e \in F(t,x)\} & \text{if } |v| > (1-\eta)R. \end{cases}$$

We shall make use of the following properties of ρ_S (listed in Lemma 8.6.5). These include the fact that $\rho_S(t, ., .)$ coincides with $\rho_F(t, ., .)$ on a neighbourhood of $(\bar{x}(t), \dot{\bar{x}}(t))$ $(=(0,0))$, for a.e. t. (This is property (v).) Of course $\rho_S \ge 0$; also, according to property (iii), for v confined to a suitable region, $\rho_S(t,x,v) = 0$ only

8.6 Proof of Theorem 8.4.3

if $v \in F(t, x)$. This is a basic requirement of a useful penalty integrand for the velocity constraint. Notice, finally, the crucial Lipschitz continuity property (iv) on a ball about $\bar{x}(t) \equiv 0$ uniformly w.r.t. t, which is, indeed, the *raison d'être* for this particular choice of modified penalty integrand.

Lemma 8.6.5 *Let $\eta \in (0, 1/2)$ be such that (G1)–(G2) and (G3)' are satisfied with parameters $\gamma = (1 - 2\eta)$, R and ϵ and integrable Lipschitz bound k. Take $\delta \in (0, \eta R)$. Then the function ρ_S (whose definition depends on R and η) has the following properties:*

(i): $\rho_S(., x, v)$ is $\mathcal{L}$-measurable for each $(x, v) \in \mathbb{R}^n \times \mathbb{R}^n$,

(ii): *Take any $(t, x) \in [S, T] \times \mathbb{R}^n$ and $v \in D^\delta(t)$. Then $\rho_S(t, x, .)$ is Lipschitz continuous with Lipschitz constant 1 on a relative neighbourhood of v in $D^\delta(t)$. If additionally we assume that $v \in \text{int}\{D^\delta(t)\}$, $\rho_S(t, x, v) > 0$ and $\xi \in \partial_v \rho_S(t, x, v)$, for some $\xi \in \mathbb{R}^n$, then $|\xi| = 1$,*

(iii): *For any $x \in \epsilon \mathbb{B}$ and $v \in D^\delta(t)$ and a.e. $t \in [S, T]$, $\rho_S(t, x, v) = 0$ if and only if $v \in F(t, x)$,*

(iv): *For any $t \in [S, T]$, $v \in D^\delta(t)$, $\rho_S(t, ., v)$ is Lipschitz continuous on $\tilde{\epsilon}\mathbb{B}$ with Lipschitz constant $\tilde{k}(t)$, a.e. $t \in [S, T]$. Here $\tilde{\epsilon}$ and the integrable function $\tilde{k}$ are*

$$\tilde{\epsilon} = \left(\frac{1}{4}\delta^2\right) \wedge \epsilon \quad \text{and} \quad \tilde{k}(t) = \delta^{-1} \vee \left(1 + \frac{12 \times (1 - \eta)}{\eta}\right) k(t), \quad (8.6.19)$$

(v): $\rho_S(t, x, v) = \rho_F(t, x, v)$ *for all $x \in \min\{\epsilon, \frac{1}{8k(t)}(1 - \eta)R\}\mathbb{B}$ and $v \in \frac{1}{8}(1 - \eta)R\mathbb{B}$, a.e. $t \in [S, T]$.*

In view of Lemma 8.6.3 and since $(0, 0) \in \text{Gr}\, F(t, .)$ for a.e. $t \in [S, T]$, (v) implies

(v)(a): $\rho_S(t, ., v)$ *is Lipschitz continuous on $\min\{\tilde{\epsilon}, \frac{1}{8k(t)}(1 - \eta)R\}\mathbb{B}$ with Lipschitz constant $k(t)$, for all $v \in \frac{1}{8}(1 - \eta)R\mathbb{B}$, a.e. $t \in [S, T]$,*

(v)(b): *If $(w, p) \in \partial_{x,v} \rho_S(t, 0, 0)$ then $(w, p) \in N_{\text{Gr}\, F(t,.)}(0, 0)$ and $|w| \leq k(t)|p|$, for a.e. $t \in [S, T]$.*

Proof Take any $(t, x, v) \in [S, T] \times \mathbb{R}^n \times \mathbb{R}^n$. Since $\rho_S(t, x, v)$ is the distance to some non-empty, closed set (either $S(t, x)$ or $F(t, x)$), we know that there exists a point $\bar{e} \in F(t, x)$ achieving the infimum in the definition of $\rho(t, x, v)$. We shall use this fact below.

Property (i) can be deduced from the continuity of χ and the $\mathcal{L} \times \mathcal{B}^n$ measurability of F. Consider (ii). $D^\delta(t)$ is the union on two disjoint sets $D_1 := (1 - \eta)R\mathbb{B}$ and $D_2 := \cap_{j=1}^N (e_j + \frac{1}{3}\delta\mathbb{B})$, in which $\{e_1, \ldots, e_N\}$ is a finite collection of points in $\{e : |e| \geq R\}$. Moreover, $\rho_S(t, x, v) = d_{S(t,x)}(v)$ for $v \in D_1$ and $\rho_S(t, x, v) = d_{F(t,x)}(v)$ for $v \in D_2$. (ii) then follows from basic properties of the distance function to some non-empty, closed set (either $S(t, x)$ or $F(t, x)$).

Consider (iii). Fix $(t, x) \in [S, T] \times \epsilon \mathbb{B}$ and $v \in D^\delta(t)$. Recall that $D^\delta(t)$ is expressible as a union of closed disjoint sets $D_1 = (1 - \eta)R\mathbb{B}$ (on which $\rho_S(t, x, .)$ coincides with $d_{S(t,x)}(.)$) and a closed set D_2 (on which $\rho_S(t, x, .)$ coincides with

$d_{F(t,x)}(.))$. If $|v| > (1-\eta)R\mathbb{B}$ then $\rho_S(t,x,v) = d_{F(t,x)}(v)$ and so $v \in F(t,x)$ if and only if $\rho_S(t,x,v) = 0$. So suppose $|v| \le (1-\eta)R$. If $\rho_S(t,x,v) = 0$ then $v = \chi(|e|)e$, for some $e \in F(t,x)$. Since $\chi \ge 1$ and $|v| \le (1-\eta)R\mathbb{B}$ we know $e \in (1-\eta)R\mathbb{B}$. But then $v \in F(t,x)$. On the other hand, if $v \in F(t,x)$ then $\rho_S(t,x,v) \le |v - \chi(|v|)v| = |v-v| = 0$, since $|v| \le (1-\eta)R$. So $\rho_S(t,x,v) = 0$.

Consider (iv). Take $(t,x) \in [S,T] \times \epsilon\mathbb{B}$ and $v \in D^\delta(t)$. We show that $\rho_S(t,.,v)$ is Lipschitz continuous on $\tilde{\epsilon}\mathbb{B}$, with Lipschitz constant $\tilde{k}(t)$, where $\tilde{\epsilon} > 0$ and $\tilde{k}$ are as in the lemma statement.

Suppose first that $|v| > (1-\eta)R$. In this case, $v \in e_0 + \frac{1}{3}\delta\mathbb{B}$, for some $e_0 \in E^\delta_{\text{regular}}(t)$ and $\rho_S(t,.,v) = d_{F(t,.)}(v)$. Since $F(t,.)$ is pseudo-Lipschitz continuous near $(0, e_0)$ with ϵ, R and k parameters taken to be δ, δ and δ^{-1} respectively, we know from Lemma 8.6.3 (for $K = k = \delta^{-1}$) that $\rho_F(t,.,v)$ is Lipschitz continuous on $(\frac{1}{4}\delta^2 \wedge \epsilon)\mathbb{B}$, with Lipschitz constant δ^{-1}. $\rho_S(t,.,v) = \rho_F(t,.,v)$ is therefore Lipschitz continuous on $\tilde{\epsilon}\mathbb{B}$ with Lipschitz constant $\tilde{k}(t)$.

We may assume then that $v \in (1-\eta)R\mathbb{B}$. Let $\bar{e} \in F(t,x)$ achieve the infimum in the definition of $\rho_S(t,x,v)$. Such a point exists, in consequence of Lemma 8.6.4.
Claim: $|\bar{e}| < R$.
Assume, to the contrary, that $|\bar{e}| \ge R$. Then, from the definition of χ, $\chi(|\bar{e}|) \ge 1 + 4(1-\eta)$. From (G3)', there exists $e_0 \in F(t,x) \cap (1-2\eta)R\mathbb{B}$. Then $\chi(|e_0|) = 1$. We have

$$(1 + 4(1-\eta))R \le |\chi(|\bar{e}|)\bar{e}| \le |v - \chi(|\bar{e}|)\bar{e}| + |v|$$
$$\le |v - \chi(|e_0|)e_0| + (1-\eta)R = |v - e_0| + (1-\eta)R$$
$$\le |v| + |e_0| + (1-\eta)R = (3 - 4\eta)R.$$

Since $R > 0$, this is a contradiction. We have verified the claim.

Take any $t \in [S,T]$, $x, x' \in \epsilon\mathbb{B}$ and $v \in (1-\eta)R\mathbb{B}$. Let $e \in F(t,x)$ and $e' \in F(t,x')$ achieve the minimum in the definition of $\rho_S(t,x,v)$ and $\rho_S(t,x',v)$. We know from the earlier verified 'claim' that $|e| < R$ and $|e'| < R$.

From (G3)'(a), we can find $e_1 \in F(t,x')$ such that

$$|e - e_1| \le k(t)|x - x'|.$$

If $k(t)|x - x'| \ge (2-\eta)R$, then, by (G3)'(b), we can find $e_2 \in F(t,x') \cap (1-2\eta)R\mathbb{B}$. Then

$$|e - e_2| \le |e| + |e_2| < (2-\eta)R \le k(t)|x - x'|.$$

Now choose $\tilde{e} \in F(t,x')$ to be

$$\tilde{e} = \begin{cases} e_1 & \text{if } k(t)|x - x'| < (2-\eta)R \\ e_2 & \text{if } k(t)|x - x'| \ge (2-\eta)R. \end{cases}$$

8.6 Proof of Theorem 8.4.3

From the preceding relations

$$|e - \tilde{e}| \leq (2 - \eta)R \wedge k(t)|x - x'|.$$

Notice that χ is Lipschitz continuous with Lipschitz constant $4(1 - \eta)/(\eta R)$ and

$$|\tilde{e}| \leq |e - \tilde{e}| + |e| \leq (3 - \eta)R.$$

We see also that $\chi(|e|) \leq 1 + 4(1 - \eta)/R$, since $|e| < R$. It follows from these relations that:

$$\begin{aligned}
\rho_S(t, x', v) - \rho_S(t, x, v) &\leq |v - \chi(|\tilde{e}|)\tilde{e}| - |v - \chi(|e|)e| \\
&\leq |\chi(|\tilde{e}|)\tilde{e} - \chi(|e|)e| \\
&\leq |\chi(|\tilde{e}|)\tilde{e} - \chi(|e|)\tilde{e}| + |\chi(|e|)\tilde{e} - \chi(|e|)e| \\
&\leq |\chi(|\tilde{e}|) - \chi(|e|)| |\tilde{e}| + |\chi(|e|)||\tilde{e} - e| \\
&\leq k_1(t)|x - x'|,
\end{aligned}$$

where k_1 is the integrable function $k_1(t) = \left(1 + \frac{12 \times (1-\eta)}{\eta}\right) k(t) \leq \tilde{k}(t)$.

Since the roles of x and x' can be interchanged and x, x' are arbitrary points in $\epsilon \mathbb{B}$, we have shown that $\rho(t, ., v)$ is Lipschitz continuous on $\epsilon \mathbb{B}$, with Lipschitz constant $\tilde{k}(t)$. We have confirmed (iv).

Consider (v). Fix $t \in [S, T]$ such that the conditions in (G3)' is satisfied. Take any $x \in \min\{\epsilon, \frac{1}{8}(1-\eta)Rk^{-1}(t))\}\mathbb{B}$ and $v \in \frac{1}{8}(1-\eta)R\mathbb{B}$. It follows from (G3)'(a) that there exists $e_0 \in F(t, x)$ such that $|e_0| \leq k(t)|x| \leq \frac{1}{8}(1-\eta)R$. In view of the preceding inequality, $\chi(|e_0|) = 1$. But then, because $\rho_S(t, x, .)$ has Lipschitz constant 1 on $(1 - \eta)R\mathbb{B}$,

$$\rho_S(t, x, v) \leq |v| + \rho_S(t, x, 0) \leq \frac{1}{8}(1 - \eta)R + |\chi(|e_0|)e_0| = \frac{1}{4}(1 - \eta)R.$$

Since the infimum in the definition of $\rho_S(t, x, v)$ is achieved, there exists $e^{(1)} \in F(t, x)$ such that

$$|v - \chi(|e^{(1)}|)e^{(1)}| = \rho_S(t, x, v) \leq \frac{1}{4}(1 - \eta)R.$$

But then $|\chi(|e^{(1)}|)e^{(1)}| \leq \frac{5}{8}(1 - \eta)R$. It follows that $\chi(|e^{(1)}|) = 1$. Then

$$\rho_S(t, x, v) = |v - \chi(|e^{(1)}|)e^{(1)}| = |v - e^{(1)}| \geq d_{F(t,x)}(v). \tag{8.6.20}$$

But the infimum in the definition of $d_{F(t,x)}(v)$ is also achieved. So there exists $e^{(2)} \in F(t, x)$ such that

$$d_{F(t,x)}(v) = |v - e^{(2)}| \le |v - e_0| \le |v| + |e_0| \le \frac{1}{4}(1-\eta)R.$$

Then $|e^{(2)}| \le \frac{5}{8}(1-\eta)R$. Consequently, $\chi(|e^{(2)}|) = 1$ and

$$d_{F(t,x)}(v) = |v - e^{(2)}| = |v - \chi(|e^{(2)}|)e^{(2)}| \ge \rho_S(t, x, v). \quad (8.6.21)$$

Equations (8.6.20) and (8.6.21) combine to give $\rho_S(t, x, v) = d_{F(t,x)}(v)$. We have confirmed (v). The assertions (v)(a) and (v)(b) follow directly from (v) in the light of Lemma 8.6.3 and since $(0, 0) \in \mathrm{Gr}\, F(t, .)$. □

Step 4 (Completion of the Proof).

Recall that, under the supplementary hypotheses, the $W^{1,1}$ local minimizer for (P) of interest is $\bar{x} \equiv 0$. Let $\beta \in (0, \epsilon)$ be such that $\bar{x}$ is minimizing w.r.t. all admissible F trajectories x such that $\|x - \bar{x}\|_{W^{1,1}} (= \|x\|_{W^{1,1}}) \le \beta$ and $\dot{x}(t) \in \dot{\bar{x}}(t) + B(t) (= B(t))$ a.e. $t \in [S, T]$.

Take $\eta \in (0, 1/2)$ such that the conditions in (G3)′ are satisfied with $\gamma = (1 - 2\eta)$. For $t \in [S, T]$ and $e \in D^\delta(t)$ define

$$\phi(t, e) := \begin{cases} \frac{1}{2}(|e| - (1-2\eta)R) \vee 0 & \text{if } e \in (1-\eta)R\mathbb{B} \\ \frac{1}{2}(|e - e_0| - \delta/6) \vee 0 & \text{if } e \in e_0 + \frac{1}{3}\delta\mathbb{B} \text{ for some } e_0 \in E^\delta_{\mathrm{regular}}(t). \end{cases}$$

Note the following properties of ϕ, each of which is a simple consequence of the definition of this function: for any $t \in [S, T]$

$$\left.\begin{array}{l} \phi(t, e) = 0 \text{ if } |e| \le (1-2\eta)R \\ \phi(t, .) \text{ is locally Lipschitz continuous on } D^\delta(t) \\ \qquad\qquad \text{with Lipschitz constant } 1/2, \\ \phi(t, e) \ge \frac{\eta R}{2} \wedge \frac{\delta}{12} \text{ if } e \in \partial D^\delta(t). \end{array}\right\} \quad (8.6.22)$$

Take $\alpha_i \downarrow 0$ and, for each i, consider the optimization problem:

$$(P_i) \begin{cases} \text{Minimize } \left(\ell_i(x(S)), x(T)\right) \vee \left(\int_S^T \rho_S(t, x(t), \dot{x}(t))dt\right) + \int_S^T \phi(t, \dot{x}(t))dt \\ \text{over arcs } x \in W^{1,1} \text{ such that} \\ \dot{x}(t) \in D^\delta(t) \text{ a.e. } t \in [S, T], \\ |x(S)| + \int_{[S,T]} |\dot{x}(t)|dt \le \beta. \end{cases}$$

Here,

$$\ell_i(x_0, x_1) := \left(g(x_0, x_1) - g(0, 0) + \alpha_i^2\right) \vee d_C(x_0, x_1).$$

8.6 Proof of Theorem 8.4.3

and $\rho_S(t, x, v)$ is the modified penalty integrand of Step 3, with parameters $\eta \in (0, 1/2)$ and $\delta > 0$ as earlier chosen.

We can formulate this problem as

$$\text{Minimize } \{J_i(x) : x \in \mathcal{S}\}$$

in which

$$\mathcal{S} := \{x \in W^{1,1} : x \text{ satisfies the constraints of } (P_i)\}$$

and

$$J_i(x) := \Big(\ell_i(x(S)), x(T))\Big) \vee \Big(\int_S^T \rho_S(t, x(t), \dot{x}(t))dt\Big) + \int_S^T \phi(t, \dot{x}(t))dt.$$

$\mathcal{S}$ is complete w.r.t. the metric induced by the $\|.\|_{W^{1,1}}$ norm on elements of $\mathcal{S}$. J_i is continuous w.r.t. to this metric. We see that $\bar{x} \equiv 0$ is an α_i^2 minimizer for (P_i). By Ekeland's theorem, we may find $x_i \in \mathcal{S}$ such that x_i is a minimizer for

$$\text{Minimize } \{J_i(x) + \alpha_i\big(|x(S) - x_i(S)| + \int_{[S,T]} |\dot{x}(t) - \dot{x}_i(t)|dt\big) : x \in \mathcal{S}\}$$

and

$$|x_i(S)| + \int_{[S,T]} |\dot{x}_i(t)|dt \leq \alpha_i. \tag{8.6.23}$$

Observe that

$$\Big(\ell_i(x_i(S)), x_i(T))\Big) \vee \Big(\int_S^T \rho_S(t, x_i(t), \dot{x}_i(t))dt\Big) > 0 \tag{8.6.24}$$

for, otherwise, we would have

$$\Big(g(x_i(S), x_i(T)) - g(0, 0) + \alpha_i^2\Big) \vee d_C(x_i(S), x_i(T))$$
$$\vee \Big(\int_S^T \rho_S(t, x_i(t), \dot{x}_i(t))dt\Big) = 0.$$

This implies $\rho_S(t, x_i(t), \dot{x}_i(t)) = 0$ a.e.. But then $\dot{x}(t) \in F(t, x(t))$ a.e., in consequence of Lemma 8.6.5. The relation also tells us $d_C(x_i(S), x_i(T)) = 0$, from which we conclude $(x_i(S), x_i(T)) \in C$, and also, $g(x_i(S), x_i(T)) \leq g(0, 0) - \alpha_i^2$. Since $x_i \in \mathcal{S}$, we know that $\|x_i\|_{W^{1,1}} \leq \beta$. But then x_i is an admissible F trajectory, with $\dot{x}_i(t) \in B(t)$ for a.e. $t \in [S, T]$, that violates the $W^{1,1}$ local optimality of $\bar{x} \equiv 0$. Equation (8.6.24) is confirmed.

For i sufficiently large, $||x_i||_{W^{1,1}} < \beta$. It follows that x_i is a $W^{1,1}$ local minimizer for the problem

$$\begin{cases} \text{Minimize } \ell_i(x(S), x(T)) \vee \left(\int_S^T \rho_S(t, x(t), \dot{x}(t)) dt \right) \\ \quad + \alpha_i \left(|x(S) - x_i(S)| + \int_S^T |\dot{x}(t) - \dot{x}_i(t)| dt \right) + \int_S^T \phi(t, \dot{x}(t)) dt \\ \text{over } x \in W^{1,1} \text{ satisfying} \\ \dot{x}(t) \in D^\delta(t) \text{ a.e. } t \in [S, T]. \end{cases}$$

This problem has the structure of the finite Lagrangian problem introduced in Step 1, a problem for which Proposition 8.6.1 provides necessary conditions of optimality. It is easy to check that the hypotheses for application of the proposition are satisfied, since $\phi(t, .)$ is Lipschitz continuous, uniformly w.r.t. t and in view of the properties of ρ_S established in Lemma 8.6.5.

Using the max rule to estimate $\partial \ell_i$, we deduce that there exist $p_i \in W^{1,1}$, $\bar{\lambda}_i \in [0, 1]$ and $\gamma_i \in [0, 1]$ which satisfy

(A)': $\dot{p}_i(t) \in \text{co}\{\eta : (\eta, p_i(t)) \in (1 - \bar{\lambda}_i) \partial_{x,v} \rho_S(t, x_i(t), \dot{x}_i(t)) + \{0\} \times \alpha_i \mathbb{B}\}$

for a.e. $t \in [S, T]$ such that $\dot{x}_i(t) \in (1 - 2\eta) R \, \overset{\circ}{\mathbb{B}}$

(B)': $(p_i(S), -p_i(T)) \in \bar{\lambda}_i \gamma_i \partial g(x_i(S), x_i(T))$
$\qquad + \bar{\lambda}_i (1 - \gamma_i) \partial d_C(x_i(S), x_i(T)) + \alpha_i \mathbb{B} \times \{0\},$

(C)': $p_i(t) \cdot \dot{x}_i(t) - (1 - \bar{\lambda}_i) \rho_S(t, x_i(t), \dot{x}_i(t)) - \phi(t, \dot{x}_i(t)) \geq$

$$p_i(t) \cdot v - (1 - \bar{\lambda}_i) \rho_S(t, x_i(t), v) - \phi(t, v) - \alpha_i |v - \dot{x}_i(t)|,$$

for all $v \in D^\delta(t)$, a.e. $t \in [S, T]$.

Observe however that

$$\bar{\lambda}_i (1 - \gamma_i) \partial d_C(x_i(S), x_i(T)) = \bar{\lambda}_i (1 - \gamma_i) \Big(\partial d_C(x_i(S), x_i(T)) \cap \partial \mathbb{B} \Big),$$

in which we interpret the right side as $\{0\}$, if $\bar{\lambda}_i (1 - \gamma_i) = 0$. This relation is true when $d_C(x_i(S), x_i(T)) > 0$ since, in this case, $\partial d_C(x_i(S), x_i(T)) \subset \partial \mathbb{B}$, by basic properties of the distance function. On the other hand, it is also true when $d_C(x_i(S), x_i(T)) = 0$ because, then, $\bar{\lambda}_i (1 - \gamma_i) = 0$ in view of (8.6.1), (8.6.24) and the max rule for limiting subdifferentials. But then (B)' can be replaced by the stronger condition:

$$(p_i(S), -p_i(T)) \in \bar{\lambda}_i \gamma_i \partial g(x_i(S), x_i(T)) + \bar{\lambda}_i (1 - \gamma_i)$$
$$\Big(\partial d_C(x_i(S), x_i(T)) \cap \partial \mathbb{B} \Big) + \alpha_i \mathbb{B} \times \{0\}.$$

8.6 Proof of Theorem 8.4.3

Now write $\lambda_i := \bar{\lambda}_i \gamma_i$, $\lambda_i^{(1)} = (1 - \bar{\lambda}_i)$ and $\lambda_i^{(2)} = \bar{\lambda}_i(1 - \gamma_i)$. These non-negative numbers satisfy

$$\lambda_i + \lambda_i^{(1)} + \lambda_i^{(2)} = 1.$$

In view of the preceding observations we can write (A)'-(C)' as

(A): $\dot{p}_i(t) \in \text{co}\{\eta : (\eta, p_i(t)) \in \lambda_i^{(1)} \partial_{x,v} \rho_S(t, x_i(t), \dot{x}_i(t)) + \{0\} \times \alpha_i \mathbb{B}\}$

for a.e. $t \in [S, T]$ such that $\dot{x}_i(t) \in (1 - 2\eta) R \overset{\circ}{\mathbb{B}}$,

(B): $(p_i(S), -p_i(T)) \in \lambda_i \partial g(x_i(S), x_i(T))$
$\qquad + \lambda_i^{(2)} \Big(\partial d_C(x_i(S), x_i(T)) \cap \partial \mathbb{B} \Big) + \alpha_i \mathbb{B} \times \{0\},$

(C): $p_i(t) \cdot \dot{x}_i(t) - \lambda_i^{(1)} \rho_S(t, x_i(t), \dot{x}_i(t)) - \phi(t, \dot{x}_i(t)) \geq$
$\qquad p_i(t) \cdot v - \lambda_i^{(1)} \rho_S(t, x_i(t), v) - \phi(t, v) - \alpha_i |v - \dot{x}_i(t)|,$

for all $v \in D^\delta(t)$, a.e. $t \in [S, T]$.

Proposition 8.6.1 also supplies the information that $\dot{p}_i(t) \in \lambda_i^{(1)} \text{co } \partial_x \rho_S(t, x_i(t), \dot{x}_i(t))$ a.e.. Since $\rho_S(t, ., \dot{x}_i(t))$ is Lipschitz continuous with Lipschitz constant $\tilde{k}(t)$, where $\tilde{k}$ is the integrable Lipschitz constant of Lemma 8.6.5 (iv), we have

$$|\dot{p}_i(t)| \leq \lambda_i^{(1)} \tilde{k}(t) \text{ a.e. } t \in [S, T]. \qquad (8.6.25)$$

From (B) we know that $|p_i(S)| \leq k_g + 1 + \alpha_i$. It follows that the family of costate trajectories p_i, $i = 1, 2, \ldots$ are uniformly bounded and their derivatives are uniformly integrably bounded.

Our aim now is to establish a uniform positive lower bound on the magnitude of (p_i, λ_i). We need to consider separately two possible cases, one of which must occur:

(i): $\rho_S(t, x_i(t), \dot{x}_i(t)) = 0$ a.e.,
(ii): $\rho_S(t, x_i(t), \dot{x}_i(t)) > 0$ on a set of positive $\mathcal{L}$-measure.

Consider case (i). In consequence of (8.6.1), (8.6.24) and the max rule, we know that $\lambda_i + \lambda_i^{(2)} = 1$. It follows then from condition (B) that

$$|(p_i(S), p_i(T))| \geq -\lambda_i k_g + \lambda_i^{(2)} - \alpha_i = 1 - (1 + k_g)\lambda_i - \alpha_i.$$

Hence

$$\sqrt{2} \|p_i\|_{L^\infty} + (1 + k_g)\lambda_i \geq 1 - \alpha_i.$$

Consider case (ii). In this case there is a time $t \in [S, T]$ such that condition (C) is satisfied, $\rho_S(t, x_i(t), \dot{x}_i(t)) > 0$ and $\dot{x}_i(t) \in D^\delta(t)$. We see that $\dot{x}_i(t) \notin F(t, x_i(t))$. This is obviously true if $|\dot{x}_i(t)| > (1 - \eta)R$ because, in this case, $\rho_S(t, x_i(t), \dot{x}_i(t)) = d_{F(t, x_i(t))}(\dot{x}_i(t)))$. So we can assume that $|\dot{x}_i(t)| \le (1 - \eta)R$ and so $\chi(|\dot{x}_i(t)|) = 1$. If, contrary to our assertion, $\dot{x}_i(t) \in F(t, x_i(t))$, it would follow that $\rho_S(t, x_i(t), \dot{x}_i(t)) \le |\dot{x}_i(t) - \chi(|\dot{x}_i(t)|)\dot{x}_i(t)| = |\dot{x}_i(t) - \dot{x}_i(t)| = 0$, which is a contradiction.

To proceed with studying this case, we must distinguish two situations:
(a): $\dot{x}_i(t) \in \partial D^\delta(t)$.
In view of (G3)'(b), there exists $v_0 \in F(t, x_i(t))$ such that $|v_0| \le (1 - 2\eta)R$. Then $\phi(t, v_0) = 0$ and $\rho_S(t, x_i(t), v_0) \le |v_0 - \chi(|v_0|)v_0| = |v_0 - v_0| = 0$. Since $\dot{x}_i(t) \in \partial D^\delta(t)$,
it follows from (8.6.22) that

$$\phi(t, \dot{x}_i(t)) \ge \frac{\eta R}{2} \wedge \frac{\delta}{12}.$$

Using these relations, noting that $\rho_S(t, x_i(t), \dot{x}_i(t)) \ge 0$ and that $|\dot{x}_i(t)| \le N + \delta/3$, and inserting $v = v_0$ in condition (C) yields the inequality

$$|p_i(t)||\dot{x}_i(t) - v_0| \ge \lambda_i^{(1)} \rho_S(t, x_i(t), \dot{x}_i(t)) - 0 + \phi(t, \dot{x}_i(t))$$
$$- 0 - \alpha_i(N + \delta/3 + (1 - 2\eta)R).$$
$$\ge \phi(t, \dot{x}_i(t)) - \alpha_i(N + \delta/3 + (1 - 2\eta)R)$$
$$\ge \left(\frac{\eta R}{2} \wedge \frac{\delta}{12}\right) - \alpha_i(N + \delta/3 + (1 - 2\eta)R).$$

We know $|\dot{x}_i(t)| \le N + \delta/3$ and $|v_0| \le (1 - 2\eta)R$. It follows that

$$\|p_i\|_{L^\infty} \ge (N + \delta/3 + (1 - 2\eta)R)^{-1} \left(\frac{\eta R}{2} \wedge \frac{\delta}{12}\right) - \alpha_i.$$

(b): $\dot{x}_i(t) \in \overset{\circ}{D}^\delta(t)$.
Now, since $\dot{x}_i(t)$ is an unconstrained local minimizer of $v \to -p_i(t) \cdot v + \lambda_i^{(1)} \rho_S(t, x_i(t), v) + \phi(t, v) + \alpha_i |v - \dot{x}_i|$, we have

$$p_i(t) \in \lambda_i^{(1)} \partial_v \rho_S(t, x_i(t), \dot{x}_i(t)) + \partial_v \phi(t, \dot{x}_i(t)) + \alpha_i \mathbb{B}.$$

Since $\rho_S(t, x_i(t), \dot{x}_i(t)) > 0$, we know from Lemma 8.6.5 that elements in $\partial_v \rho_S(t, x_i(t), \dot{x}_i(t))$ have unit length. Taking note also of the fact that $\phi(t, .)$ is locally Lipschitz continuous on $D^\delta(t)$ with Lipschitz constant $1/2$, whence $\partial_v \phi(t, \dot{x}_i(t)) \in (1/2) \mathbb{B}$, we deduce from the preceding relations that

$$\|p_i\|_{L^\infty} \ge \lambda_i^{(1)} - \frac{1}{2} - \alpha_i.$$

8.6 Proof of Theorem 8.4.3

From (B) we know

$$\sqrt{2}\|p_i\|_{L^\infty} \geq |(p_i(S), -p_i(T))| \geq \lambda_i^{(2)} - \lambda_i k_g - \alpha_i = 1 - \lambda_i^{(1)} - (1+k_g)\lambda_i - \alpha_i.$$

Adding the preceding inequalities we obtain

$$(1+\sqrt{2})\|p_i\|_{L^\infty} + (1+k_g)\lambda_i \geq 1 - \frac{1}{2} - 2\alpha_i = \frac{1}{2} - 2\alpha_i.$$

Combining the estimates relating to the cases (i), (ii)(a) and (ii)(b), we arrive at

$$(1+\sqrt{2})\|p_i\|_{L^\infty} + (1+k_g)\lambda_i$$
$$\geq \min\{\frac{1}{2} - 2\alpha_i, (N+\delta/3+(1-2\eta)R)^{-1}\left(\frac{\eta R}{2} \wedge \frac{\delta}{12}\right) - \alpha_i\}. \quad (8.6.26)$$

This is the desired lower bound.

We deduce from (8.6.23) that, along a subsequence, $x_i \to \bar{x} \equiv 0$ uniformly, and $\dot{x}_i \to \dot{\bar{x}} \equiv 0$ in L^1 and a.e.. We have already observed that the $\dot{p}_i$'s are uniformly integrably bounded and the p_i's are uniformly bounded. By further restriction to a subsequence we can then arrange, in consequence of Ascoli's theorem, that $p_i \to p$ uniformly and $\dot{p}_i \to \dot{p}$ weakly in L^1. We can also arrange that $\lambda_i \to \lambda$, $\lambda_i^{(1)} \to \lambda^{(1)}$ and $\lambda_i^{(2)} \to \lambda^{(2)}$, for some $\lambda, \lambda^{(1)}, \lambda^{(2)} \in [0,1]$. A convergence analysis along the lines of the proof of Proposition 8.6.1, permits us to pass to the limit in conditions (A)-(C) and thereby to obtain:

$$(1+\sqrt{2})\|p\|_{L^\infty} + (1+k_g)\lambda$$
$$\geq \min\{\frac{1}{2}, (N+\delta/3+(1-2\eta)R)^{-1}\left(\frac{\eta R}{2} \wedge \frac{\delta}{12}\right)\}, \quad (8.6.27)$$

$$\dot{p}(t) \in \text{co}\{\eta : (\eta, p_i(t)) \in \lambda^{(1)} \partial_{x,v} \rho_S(t, \bar{x}(t), \dot{\bar{x}}(t))\} \text{ for a.e. } t \in [S, T],$$

which, according to Lemma 8.6.5 (v), implies

$$\dot{p}(t) \in \text{co}\{\eta : (\eta, p(t)) \in N_{\text{Gr } F(t,.)}(\bar{x}(t), \dot{\bar{x}}(t))\}, \text{ a.e. } t \in [S, T], \quad (8.6.28)$$

$$(p(S), -p(T)) \in \lambda \partial g(\bar{x}(S), \bar{x}(T)) + N_C(\bar{x}(S), \bar{x}(T)), \quad (8.6.29)$$

and

$$p(t) \cdot \dot{\bar{x}}(t) - \lambda^{(1)} \rho_S(t, \bar{x}(t), \dot{\bar{x}}(t)) - \phi(t, \dot{\bar{x}}(t)) \geq p(t) \cdot v - \lambda^{(1)} \rho_S(t, \bar{x}(t), v) - \phi(t, v),$$

for all $v \in D^\delta(t)$, a.e. $t \in [S, T]$.

Observe that, for all points $v \in (F(t, \bar{x}(t)) \cap (1-2\eta)R\mathbb{B}) \cup E^\delta_{\text{regular}}(t)$, we have $v \in D^\delta(t)$ and $\rho_S(t, \bar{x}(t), v) = \phi(t, v) = 0$. Consequently, the preceding relation implies

$$p(t) \cdot \dot{\bar{x}}(t) \geq p(t) \cdot v, \text{ for all } v \in (F(t, \bar{x}(t)) \cap (1-2\eta)R\mathbb{B}) \cup E^\delta_{\text{regular}}(t). \tag{8.6.30}$$

Lemma 8.6.5 (v)(b) provides the additional information that

$$|\dot{p}(t)| \leq k(t)|p(t)| \text{ for a.e. } t \in [S, T]. \tag{8.6.31}$$

In view of (8.6.27), we can arrange, by scaling the Lagrange multipliers, that

$$\|p\|_{L^\infty} + \lambda = 1. \tag{8.6.32}$$

Relations (8.6.28), (8.6.29), (8.6.30) and (8.6.32) combine to provide a restricted version of the theorem, in which the Weierstrass condition (8.6.30) is affirmed for velocities only in a subset of $\overline{\text{co}}\,(\Omega_0(t) \cap B(t))$.

To derive the full Weierstrass condition, take a sequence $\delta_i \downarrow 0$. Then conditions (8.6.28), (8.6.29), (8.6.30) and (8.6.32) are valid for each i, for some Lagrange multipliers (p, λ) that we now label (p_i, λ_i). From (8.6.31) and (8.6.32) we know the p_i's are uniformly bounded and have uniformly integrable derivatives. We may therefore arrange, by extracting subsequences, that $p_i \to p$ (uniformly), $\dot{p}_i \to \dot{p}$ in the weak L^1 topology and $\lambda_i \to \lambda$, for some $p \in W^{1,1}$ and $\lambda \geq 0$.

Relations (8.6.32), (8.6.29) and (8.6.28) (now expressed in terms of (p_i, λ_i)) are preserved in the limit as $i \to \infty$ (with multipliers (p, λ)). We now attend to the implications of relation (8.6.30). Let $\mathcal{S} \subset [S, T]$ be the subset on which the Weierstrass condition (8.6.30), in which (p_i, λ_i) replaces (p, λ), is satisfied for all values of i. $\mathcal{S} \subset [S, T]$ is a subset of full measure. Take any $t \in \mathcal{S}$. Notice that $E^{\delta_i}_{\text{regular}}(t), i = 1, 2, \ldots$, is an increasing sequence of sets and

$$\lim_i E^{\delta_i}_{\text{regular}}(t) = E^N_{\text{discrete}}(t).$$

We deduce from (8.6.30), in the limit, that $p(t) \cdot \dot{\bar{x}}(t) \geq p(t) \cdot v$, for all $v \in (F(t, \bar{x}(t)) \cap (1 - 2\eta)R\mathbb{B})) \cup E^N_{\text{discrete}}(t)$ for all $t \in \mathcal{S}$.

Now take $N_i \uparrow \infty$ and $\eta_i \downarrow 0$. We know that $E^{N_i}_{\text{discrete}}, i = 1, 2, \ldots$ is an increasing sequence of sets. Then, in consequence of the defining properties of $E^N_{\text{discrete}}(t)$,

$$\Omega_0(t) \cap B(t) \cap \{e \in \mathbb{R}^n : |e| \geq R\} \subset \lim_i E^{N_i}_{\text{discrete}}(t). \tag{8.6.33}$$

We also know that, for each i, $F(t, \bar{x}(t)) \cap (1 - 2\eta_i)R\mathbb{B} = \Omega_0(t) \cap \{e \in \mathbb{R}^n : |e| \leq (1-2\eta_i)R\}$, i=1,2,... (this is because, for every point $e \in F(t, \bar{x}(t))$ such that

$|e| < R$), $F(t, .)$ is pseudo-Lipschitz continuous near $(\bar{x}(t), e)$, in consequence of (G3)$'$. Furthermore, this sequence of sets is increasing and

$$\Omega_0(t) \cap \{e : |e| < R\} \cap B(t) = \Omega_0(t) \cap \{e : |e| < R\}$$
$$\subset \lim_i \left(F(t, \bar{x}(t)) \cap \{e \in \mathbb{R}^n : < (1 - 2\eta_i)R\} \right). \quad (8.6.34)$$

(To obtain the first set identity we have used the fact that $R\overset{\circ}{\mathbb{B}} \subset B(t)$ a.e..) A similar analysis to that undertaken above ensures p can be chosen such that, for a.e. $t \in [S, T]$,

$$p(t) \cdot \dot{\bar{x}}(t) \geq p(t) \cdot v, \quad (8.6.35)$$

for all $v \in \lim_i \left(F(t, \bar{x}(t)) \cap \{e \in \mathbb{R}^n : < (1 - 2\eta_i)R\} \right) \cup \lim_i E_{\text{discrete}}^{N_i}(t)$.

But then, by (8.6.33) and (8.6.34), (8.6.35) is valid for all $v \in \Omega_0(t) \cap B(t)$. Since $v \to p(t) \cdot v$ is a linear function, we may deduce that (8.6.35) is valid for all $v \in \overline{\text{co}}\,(\Omega_0(t) \cap B(t))$. This is the full Weierstrass condition of the theorem statement. The proof is concluded. □

8.7 The Hamiltonian Inclusion for Convex Velocity Sets

We have seen how the Euler Lagrange condition from the classical calculus of variations generalizes to allow for a constraint on the velocity set, in the form of a differential inclusion. We show now that Hamilton's system of equations also generalizes to this broader setting. The new condition will be called the Hamiltonian inclusion. Attention, for the present, is limited to problems in which the velocity sets are convex. For such problems, the Hamiltonian inclusion is implied by the earlier derived Euler Lagrange inclusion, in consequence of the following theorem.

Theorem 8.7.1 (Dualization Theorem) *Take a function $\tilde{L} : \mathbb{R}^n \times \mathbb{R}^m \to \mathbb{R} \cup \{+\infty\}$ and points $(\bar{x}, \bar{v}) \in \text{dom}\,\tilde{L}$ and $\bar{p} \in \mathbb{R}^m$. Let $\tilde{H}(x, .)$ to be the conjugate function of $\tilde{L}(x, .)$, i.e.*

$$\tilde{H}(x, p) := \sup\{p \cdot v - \tilde{L}(x, v) : v \in \mathbb{R}^m\}.$$

Assume that, for some neighbourhoods U, V and W of $\bar{x}$, $\bar{v}$ and $\tilde{L}(\bar{x}, \bar{v})$, the following hypotheses are satisfied:

(H1): $\tilde{L}(x, .)$ is convex for each $x \in U$,
(H2): $\tilde{L}(., .)$ is lower semi-continuous,
(H3): there exists $k > 0$ such that

$$\text{epi}\,\tilde{L}(x', .) \cap (V \times W) \subset \text{epi}\,\tilde{L}(x'', .) + k|x' - x''|\mathbb{B}$$

for all $x', x'' \in U$.

Then

$$\{q : (q, \bar{p}) \in \partial \tilde{L}(\bar{x}, \bar{v})\} \subset \mathrm{co}\{q : (-q, \bar{v}) \in \partial \tilde{H}(\bar{x}, \bar{p})\}. \tag{8.7.1}$$

A proof of the theorem is given in the Appendix at the end of the chapter.

Remarks

1: Under the hypotheses of this theorem, it is not guaranteed that the function $\tilde{H}$ is lower semi-continuous. The limiting subdifferential $\partial \tilde{H}(\bar{z})$ at the point $\bar{z} = (\bar{x}, \bar{p}) \in \mathrm{dom}\, H$ was earlier defined only in the case H is lower semi-continuous. In the present context, $\partial \tilde{H}(\bar{z})$ is taken to be the set of elements ξ having the following properties: there exist sequences of positive numbers $\{\epsilon_i\}$ and $\{\sigma_i\}$ and convergent sequences $z_i \xrightarrow{H} \bar{z}$ and $\xi_i \to \xi$ such that, for each i,

$$\tilde{H}(z) - \tilde{H}(z_i) \geq \xi_i \cdot (z - z_i) - \sigma_i |z - z_i|^2 \text{ for all } z \in z_i + \epsilon_i \mathbb{B}.$$

2: For purposes of dualizing the Euler Lagrange inclusion, only the set inclusion (8.7.1) is required. We shall not use this fact here, but we note Rockafellar [176] has shown that under the hypotheses of Theorem 8.7.1, supplemented by the condition

$\tilde{L}$ is epicontinuous,

the two sets considered do in fact coincide. (Epicontinuity is defined in the Appendix.) That is,

$$\mathrm{co}\{q : (q, \bar{p}) \in \partial \tilde{L}(\bar{x}, \bar{v})\} = \mathrm{co}\{q : (-q, \bar{v}) \in \partial \tilde{H}(\bar{x}, \bar{p})\}.$$

Return now to problem (P), posed in Sect. 8.4. Let $\bar{x}$ be a $W^{1,1}$ local minimizer relative to the multifunction B. We set $\tilde{L}(x, v) := \Psi_{\mathrm{Gr}\, F(t,.,)}(x, v)$, then the function $\tilde{H}$ in the theorem statement coincides with $H(t, ., .)$, where $H : [S, T] \times \mathbb{R}^n \times \mathbb{R}^n \to \mathbb{R}$ is the Hamiltonian function earlier employed in this chapter, namely:

$$H(t, x, p) := \sup_{v \in F(t,x)} p \cdot v.$$

It turns out that, for this identification of $\tilde{L}$ in Theorem 8.4.3, the hypotheses placed on $\tilde{L}$ in the dualization theorem are satisfied, for almost every $t \in [S, T]$, under the hypotheses invoked in Theorem 8.4.3, when we make the additional assumption that F takes values convex sets. The same is true of the earlier stated Corollary 8.5.2 of Theorem 8.4.3. This means that, for problems with convex velocity sets, we can supplement the Euler Lagrange inclusion by the partially convexified Hamiltonian inclusion

$$\dot{p}(t) \in \mathrm{co}\{-q : (q, \dot{\bar{x}}(t)) \in \partial H(t, \bar{x}(t), p(t))\} \text{ a.e..}$$

8.7 The Hamiltonian Inclusion for Convex Velocity Sets

Notice that, when F takes values convex sets, we can also replace the Weierstrass condition in Theorem 8.4.3

$$p(t) \cdot \dot{\bar{x}}(t) \geq p(t) \cdot v, \quad \text{for all } v \in \overline{\text{co}}\,(\Omega_0(t) \cap (\dot{\bar{x}}(t) + B(t))) \text{ a.e. } t \in [S, T] \tag{8.7.2}$$

by the seemingly stronger (but in fact equivalent) *unrestricted* Weierstrass condition

$$p(t) \cdot \dot{\bar{x}}(t) \geq p(t) \cdot v, \quad \text{for all } v \in F(t, \bar{x}(t)), \text{ a.e. } t \in [S, T]. \tag{8.7.3}$$

To see this, suppose that (8.7.3) were not true. Then there would exist a subset of $[S, T]$ of positive measure on which (8.7.2) is satisfied but (8.7.3) is violated. For any time t in this set there exists a point $\bar{v} \in F(t, \bar{x}(t))$ such that $p(t) \cdot \bar{v} > p(t) \cdot \dot{\bar{x}}(t)$. But, under the hypotheses of the Theorem 8.4.3, we know that the set $\Omega_0(t) \cap (\dot{\bar{x}}(t) + B(t))$ contains the neighbourhood of $\dot{\bar{x}}(t)$, $\dot{\bar{x}}(t) + R(t) \overset{\circ}{\mathbb{B}}$. It is therefore possible to choose $\sigma \in (0, 1)$ sufficiently small that $v' := \dot{\bar{x}}(t) + \sigma(\bar{v} - \dot{\bar{x}}(t)) \in \dot{\bar{x}}(t) + R(t) \overset{\circ}{\mathbb{B}}$. Since (by convexity of $F(t, \bar{x}(t))$) $v' \in F(t, \bar{x}(t))$, it follows that $v' \in F(t, \bar{x}(t)) \cap \Omega_0(t) \cap (\dot{\bar{x}}(t) + B(t))$. But then, from (8.7.2), $0 \geq p(t) \cdot v' - p(t) \cdot \dot{\bar{x}}(t) = \sigma(p(t) \cdot \bar{v} - p(t) \cdot \dot{\bar{x}}(t))$, which implies $p(t) \cdot \bar{v} - p(t) \cdot \dot{\bar{x}}(t) \leq 0$ (see Exercise 4.2). This contradiction confirms the unrestricted Weierstrass condition (8.7.3).

The following theorem is a direct consequence of these observations. It tells us that, under the additional hypothesis that F takes values convex sets, the assertions of Theorem 8.4.3 can be supplemented by the partially convexified Hamiltonian inclusion and the information that the Weierstrass condition is satisfied over the entire velocity set $F(t, \bar{x}(t))$.

Theorem 8.7.2 (The Partially Convexified Hamiltonian Inclusion for Convex Velocity Sets) *Take a measurable multifunction $B : [S, T] \rightsquigarrow \mathbb{R}^n$ such that $B(t)$ is open for a.e. $t \in [S, T]$. Let $\bar{x}$ be a $W^{1,1}$ local minimizer for (P) relative to B. Assume that the following hypotheses are satisfied:*

(G1): *g is Lipschitz continuous on a neighbourhood of $(\bar{x}(S), x(\bar{T}))$ and C is closed,*

(G2): *$F(t, x)$ is nonempty for each $(t, x) \in [S, T] \times \mathbb{R}^n$, $\text{Gr}\, F(t, .)$ is closed for each $t \in [S, T]$ and F is $\mathcal{L} \times \mathcal{B}^n$ measurable,*

(G3): *There exist a measurable function $R : [S, T] \to (0, \infty) \cup \{+\infty\}$ and $\epsilon > 0$ such that $R(t) \overset{\circ}{\mathbb{B}} \subset B(t)$ a.e. and the following conditions are satisfied,*

(a): *There exists $k \in L^1$ such that*

$$F(t, x') \cap (\dot{\bar{x}}(t) + R(t) \overset{\circ}{\mathbb{B}}) \subset F(t, x) + k(t)|x' - x|\mathbb{B},$$

for all $x, x' \in \bar{x}(t) + \epsilon \mathbb{B}$, a.e. $t \in [S, T]$,

(b): *There exist $r \in L^1(S, T)$, $r_0 > 0$ and $\gamma \in (0, 1)$ such that $r_0 \leq r(t)$, $\gamma^{-1} r(t) \leq R(t)$ a.e. and*

$$F(t, x) \cap (\dot{\bar{x}}(t) + r(t)\mathbb{B}) \neq \emptyset \text{ for all } x \in \bar{x}(t) + \epsilon \mathbb{B}, \text{ a.e. } t \in [S, T].$$

Assume further that

$$F(t, x) \text{ is convex, for all } x \in \bar{x}(t) + \epsilon \mathbb{B}, \text{ a.e. } t \in [S, T].$$

Then there exist an arc $p \in W^{1,1}([S, T]; \mathbb{R}^n)$ and $\lambda \geq 0$ such that

(i): $(p, \lambda) \neq (0, 0)$,
(ii): $\dot{p}(t) \in \text{co}\{\eta : (\eta, p(t)) \in N_{\text{Gr } F(t,.)}(\bar{x}(t), \dot{\bar{x}}(t))\}$, a.e. $t \in [S, T]$,
(iii): $(p(S), -p(T)) \in \lambda \partial g(\bar{x}(S), \bar{x}(T)) + N_C(\bar{x}(S), \bar{x}(T))$,

(iv): $p(t) \cdot \dot{\bar{x}}(t) \geq p(t) \cdot v$ for all $v \in F(t, \bar{x}(t))$, a.e. $t \in [S, T]$.

Condition (ii) implies

(v): $\dot{p}(t) \in \text{co}\{-\xi : (\xi, \dot{\bar{x}}(t)) \in \partial_{x,p} H(t, \bar{x}(t), p(t))\}$, a.e. $t \in [S, T]$.

Now assume, also, that

$$F(t, x) \text{ does not depend on } t.$$

(In this case we write $F(x)$ in place of $F(t, x)$.) Then, in addition to the above conditions, there exists a constant r such that

(vi): $p(t) \cdot \dot{\bar{x}}(t) \ (= \max_{v \in F(\bar{x}(t))} p(t) \cdot v) = r$ a.e. $t \in [S, T]$.

The following corollary gives alternative hypotheses for the validity of the preceding necessary conditions. The assertions of the corollary are an immediate consequence of Proposition 8.5.1. the Hamiltonian Dualization Theorem (Theorem 8.7.1) and, finally, the preceding observation that the local Weierstrass condition implies the global Weierstrass condition when $F(t, x)$ is convex valued.

Corollary 8.7.3 *The assertions of Theorem 8.7.2 remain valid when hypothesis (G3) of Theorem 8.7.2 is replaced by either (G3)* or (G3)**. If $\bar{x}$ is a $W^{1,1}$ local minimizer (i.e. we can take $B(t) = \mathbb{R}^n$) then (G3) can be replaced by (G3)***. Here,*

(G3)*: There exist a measurable function $R : [S, T] \to (S, T) \cup \{+\infty\}$, strictly bounded away from 0, $k \in L^1$ and $\epsilon > 0$ such that

$$F(t, x') \cap (\dot{\bar{x}}(t) + R(t)\overset{\circ}{\mathbb{B}}) \subset F(t, x) + k(t)|x' - x|\mathbb{B},$$

for all $x, x' \in \bar{x}(t) + \epsilon \mathbb{B}$, a.e.

and either
(a): k is strictly bounded away from zero and there exists $\omega_0 > 0$ such that $\omega_0 k(t)\mathbb{B} \subset R(t)$ a.e. $t \in [S, T]$,

or

(b): There exists a modulus of continuity θ such that

$$d_{F(t,x)}(\dot{\bar{x}}(t)) \leq \theta(|x - \bar{x}(t)|) \quad \text{for all } x \in \bar{x}(t) + \epsilon\mathbb{B}, \text{ a.e. } t \in [S, T].$$

(G3)**: There exist a measurable function $R : [S, T] \to (S, T) \cup \{+\infty\}$, strictly bounded away from 0, $k \in L^1$ and $\epsilon > 0$ such that $R(t)\overset{\circ}{\mathbb{B}} \subset B(t)$, a.e. $t \in [S, T]$ and

$$\text{'}(w, p) \in N_F(t, .)(x, v)\text{'} \implies \text{'}|w| \leq k(t)|p|\text{'}$$

for all $x \in \bar{x}(t) + \epsilon\mathbb{B}$ and $v \in R(t)\overset{\circ}{\mathbb{B}}$ a.e. $t \in [S, T]$,

and either

(a): k is strictly bounded away from zero and there exists $\omega_0 > 0$ such that such that $\omega_0 k(t) \leq R(t)$ a.e. $t \in [S, T]$,

or

(b): There exists a modulus of continuity θ such that

$$d_{F(t,x)}(\dot{\bar{x}}(t)) \leq \theta(|x - \bar{x}(t)|) \quad \text{for all } x \in \bar{x}(t) + \bar{\epsilon}\mathbb{B}, \text{ a.e. } t \in [S, T],$$

(G3)***: There exist $\alpha > 0$, $\epsilon > 0$ and non-negative measurable functions k and β such that k and $t \to \beta(t)k^\alpha(t)$ are integrable and, for each $N \geq 0$,

$$F(t, x') \cap (\dot{\bar{x}}(t) + N\mathbb{B}) \subset F(t, x) + (k(t) + \beta(t)N^\alpha)|x' - x|\mathbb{B},$$

for all $x', x \in \bar{x}(t) + \epsilon\mathbb{B}$, a.e. $t \in [S, T]$.

8.8 The Hamiltonian Inclusion for Non-convex Velocity Sets

We have proved necessary conditions for a $W^{1,1}$ local minimizer $\bar{x}$, relative to a multifunction B, incorporating the partially convexified Hamiltonian inclusion

$$\dot{p}(t) \in \text{co}\{-\xi : (\xi, \dot{\bar{x}}(t)) \in \partial_{x,p} H(t, \bar{x}, p(t))\} \quad \text{a.e. } t \in [S, T],$$

only in the case when the velocity sets $F(t, x)$ are convex. Are they valid, in the absence of convexity? An answer to this question, provided by F.H. Clarke, is a qualified 'yes'; necessary conditions incorporating a Hamiltonian inclusion for problems with possibly non-convex velocity sets can be proved provided, first, we replace the partially convexified Hamiltonian inclusion by the less precise, fully convexified Hamiltonian inclusion and, second, we assume that $\bar{x}$ is an L^∞ local minimizer, not merely a $W^{1,1}$ local minimizer.

Theorem 8.8.1 (Hamiltonian Inclusion for Non-convex F) *Let $\bar{x}$ be an L^∞ local minimizer for (P). Assume*

(G1): g is Lipschitz continuous on a neighbourhood of $(\bar{x}(S), \bar{x}(T))$ and C is closed,
(G2): $F(t, x)$ is nonempty for each $(t, x) \in [S, T] \times \mathbb{R}^n$, $\operatorname{Gr} F(t, .)$ is closed for each $t \in [S, T]$ and F is $\mathcal{L} \times \mathcal{B}^n$ measurable,
(G3)': there exist $\epsilon > 0$ and integrable functions k and c such that

$$F(t, x') \subset F(t, x) + k(t)|x' - x|\mathbb{B} \text{ and } F(t, x) \subset c(t)\mathbb{B}$$

for all $x', x \in \bar{x}(t) + \epsilon \mathbb{B}$, a.e. $t \in [S, T]$.

Then there exist $p \in W^{1,1}([S, T]; \mathbb{R}^n)$ and $\lambda \geq 0$, such that

(i): $(p, \lambda) \neq (0, 0)$,
(ii): $(-\dot{p}(t), \dot{\bar{x}}(t)) \in \operatorname{co} \partial_{x,p} H(t, \bar{x}(t), p(t))$, a.e. $t \in [S, T]$,
(iii): $(p(S), -p(T)) \in \lambda \partial g(\bar{x}(S), \bar{x}(T)) + N_C(\bar{x}(S), \bar{x}(T))$.

Remark
It is unnecessary to add the Weierstrass condition ((iv) of Theorem 8.7.2) to the above conditions of Theorem 8.8.1 because it is implied by the Hamiltonian inclusion (ii) of Theorem 8.8.1 (see Exercise 4.2).

Proof We may assume, without loss of generality that the functions c and k in (G3)' are essentially bounded. (This is because, if not, we can introduce a change of time variable $s = \sigma(t) := \int_S^t (k(t') + c(t'))dt'$, after it has been arranged, by addition of a constant, that $k + c$ are strictly bounded away from 0. This transformation renders the two integrable bounds essentially bounded, in the corresponding hypotheses for the transformed problem; we then prove the stated assertions for the transformed problem, which imply the same assertions for the original problem, as in the proof of Proposition 8.6.2.) By translation of the minimizer $\bar{x}$ to the origin, we can arrange that $\bar{x} \equiv 0$. Reducing the size of $\epsilon > 0$ in (G3)', we can also arrange that the L^∞ local minimizer $\bar{x}$ minimizes $g(x(S), x(T))$ over all admissible F trajectories satisfying $\|x - \bar{x}\|_{L^\infty} \leq \epsilon$, where $\epsilon > 0$ is the constant of hypothesis (G3)'. Finally, by redefining g outside a neighbourhood of $(\bar{x}(S), \bar{x}(T))$, we can arrange that g is globally Lipschitz continuous.

Now take $\alpha_i \downarrow 0$ and, for each $i = 1, 2, \ldots$, define

$$\ell_i(x_0, x_1) := \left(g(x_0, x_1) - g(0, 0) + \alpha_i\right) \vee d_C(x_0, x_1)$$

and consider the optimization problem

$$\text{Minimize } \{J_i(x) : x \in \mathcal{S}\}$$

8.8 The Hamiltonian Inclusion for Non-convex Velocity Sets

in which

$$J_i(x) := \ell_i(x(S), x(T)) + \int_S^T |x(t) - \bar{x}(t)|^2 dt$$

and

$$S := \{x \in W^{1,2} : \dot{x}(t) \in F(t, x(t)) \text{ a.e. and } ||x - \bar{x}||_{L^\infty} \leq \epsilon/2\}.$$

Here $W^{1,2}$ denotes the Hilbert space of absolutely continuous functions with square integrable derivatives on $[S, T]$, with the inner product $\langle x, x' \rangle := x(S) \cdot x'(S) + \int_S^T \dot{x}(t) \cdot \dot{x}'(t) dt$.

Since c is essentially bounded, elements x in S, together with their derivatives $\dot{x}$, are uniformly essentially bounded. We can therefore find constants $k_1 > 0$ and $k_2 > 0$ such that, for any $x \in S$,

$$|g(x(S), x(T))| \leq k_1, \quad \text{and } ||x||_{L^\infty} \leq k_2 ||x||_{W^{1,2}}.$$

(The last inequality is valid because the embedding $W^{1,2}([S, T]; \mathbb{R}^n) \subset L^\infty([S, T]; \mathbb{R}^n)$ is continuous.)

S is nonempty, bounded and strongly closed in $W^{1,2}$, and J_i is continuous on S, w.r.t. the strong $W^{1,2}$ topology and is bounded below on S. We may therefore apply Stegall's theorem (Theorem 3.5.2) to the proper, lower semi-continuous function $\Phi : W^{1,2} \to \mathbb{R} \cup \{+\infty\}$:

$$\Phi(x) := \begin{cases} J_i(x) & \text{if } x \in S \\ +\infty & \text{otherwise}, \end{cases}$$

defined on the Hilbert space $W^{1,2}$. We thereby obtain sequences $\{(d_i, a_i)\}$ in $\mathbb{R}^n \times L^2([S, T]; \mathbb{R}^n)$ and $x_i \in S$, such that $|d_i|^2 + ||a_i||_{L^2}^2 \to 0$ and x_i minimizes

$$x \to J_i(x) + d_i \cdot x(S) + \int_S^T a_i(t) \cdot \dot{x}(t) dt$$

over $x \in S$ for each i. The sequence x_i is uniformly bounded, with uniformly integrably bounded derivatives. It follows that, along a subsequence, $x_i \to x'$ uniformly on $[S, T]$ and $\dot{x}_i \to \dot{x}'$, weakly in L^1, for some $x' \in W^{1,1}([S, T]; \mathbb{R}^n)$. By optimality of x_i, and bearing in mind that $\bar{x} \equiv 0 \in S$, we have

$$\alpha_i = J_i(\bar{x}) \geq \ell_i(x_i(S), x_i(T)) + \int_S^T |x_i(t)|^2 dt + d_i \cdot x_i(S) + \int_S^T a_i(t) \cdot \dot{x}_i(t) dt$$

for each i. Passing to the limit, as $i \to \infty$ on both sides of this relation gives

$$0 \geq \Big(g(x'(S), x'(T)) - g(0,0)\Big) \vee d_C(x'(S), x'(T))$$
$$+ \int_S^T |x'(t)|^2 dt \geq \int_S^T |x'(t)|^2 dt.$$

It follows that $x' \equiv 0$. Notice also that, for each i,

$$\ell_i(x_i(S), x_i(T)) > 0, \qquad (8.8.1)$$

since, otherwise, $g(x_i(S), x_i(T)) \leq g(\bar{x}(S), \bar{x}(T)) - \alpha_i$, $(x_i(S), x_i(T)) \in C$ and $\|x_i - \bar{x}\|_{L^\infty} \leq \epsilon$; these conditions violate the L^∞ optimality of $\bar{x}$.

Let us now investigate the properties of x_i, in the limit as $i \to \infty$. We have established that $x_i \to \bar{x}$ in L^∞ and $(d_i, a_i) \to (0, 0)$ in $\mathbb{R}^n \times L^2$. By extracting subsequences (we do not re-label), we can arrange that $a_i \to 0$ a.e.. Write, for $t \in [S, T]$,

$$z_i(t) := \int_S^t (a_i(s) \cdot \dot{x}_i(s) + |x_i(s)|^2) ds .$$

The minimizing properties of the sequence $\{x_i\}$ can be expressed as follows: for each i sufficiently large, (x_i, z_i) is an L^∞ local minimizer for

$$\begin{cases} \text{Minimize } \ell_i(x(S), x(T)) + z(T) + d_i \cdot x(S) \\ \text{over absolutely continuous functions } (x, z) : [S, T] \to \mathbb{R}^{n+1} \text{ satisfying} \\ (\dot{x}(t), \dot{z}(t)) \in \{(e, a_i(t) \cdot e) : e \in F(t, x(t))\} + \{(0, |x(t) - \bar{x}(t)|^2)\} \text{ a.e. } t \in [S, T], \\ z(S) = 0. \end{cases}$$

Since this is a free right end-point problem, we know from the relaxation theorem (Theorem 6.5.2) that (x_i, z_i) is also an L^∞ local minimizer for the relaxed version of the above problem, in which $F(t, x)$ is replaced by co $F(t, x)$. It is straightforward to check that the hypotheses of Theorem 8.7.2 are satisfied by the data for the relaxed version, with reference to the L^∞ local minimizer (x_i, z_i) and the multifunction $B \equiv \mathbb{R}^n$ (take the parameters $R \equiv 2\|c\|_{L^\infty}$, $r \equiv r_0 := \|c\|_{L^\infty}$, $\gamma = 1/2$, $\tilde{\epsilon} = \epsilon/4$, $\tilde{k}(t) = \sqrt{2}\max\{k(t); |a_i(t)|k(t) + 2\epsilon\}$). In terms of the costate vector block elements, denoted $(p, -r)$, the Hamiltonian is

$$\tilde{H}(t, x, p, r) = \max_{e \in F(t,x)} \{(p - ra_i(t)) \cdot e\} - r|x - \bar{x}(t)|^2$$
$$= H(t, x, p - ra_i(t)) - r|x - \bar{x}(t)|^2 . \qquad (8.8.2)$$

Take any i. Let $(p_i, -r_i)$ be the costate trajectory. Since $\tilde{H}(t, ., ., .)$ does not depend on z, $\dot{r}_i(t) = 0$. But $r_i(T) = 1$, by the transversality condition. We conclude that $r_i \equiv 1$. The Hamiltonian inclusion (condition (v) of Theorem 8.7.2) tells us that

8.8 The Hamiltonian Inclusion for Non-convex Velocity Sets

$$(\dot{p}_i(t), \dot{r}_i(t)) \in \text{co}\{(\xi, \zeta) : ((\xi, \zeta), (x_i(t), z_i(t)))$$
$$\in \partial_{x,z,p,r} \tilde{H}(t, x_i(t), p_i(t), r_i(t))\} \text{ a.e..} \quad (8.8.3)$$

An analysis of limiting subgradients of $\tilde{H}(t, ., ., .)$ at $(x_i(t), p_i(t), r_i(t))$, characterized as limits of neighbouring proximal subgradients, with the help also of Lemma 5.1, permits us to deduce from (8.8.2) and (8.8.3) that

$$(-\dot{p}_i(t), \dot{x}_i(t)) \in \text{co}\,\partial_{x,p} H(t, x_i(t), p_i(t) - a_i(t))$$
$$+ 2|x_i(t) - \bar{x}(t)|\mathbb{B} \times \{0\}; \text{ a.e..} \quad (8.8.4)$$

From the transversality condition and the max rule for subdifferentials we know that there exists $\lambda_i \in [0, 1]$ such that

$$(p_i(S), -p_i(T)) \in \lambda_i \partial g(x_i(S), x_i(T)) + (1 - \lambda_i)\partial d_C(x_i(S), x_i(T)) + d_i \mathbb{B} \times \{0\}.$$

This relation can be strengthened to

$$(p_i(S), -p_i(T)) \in \lambda_i \partial g(x_i(S), x_i(T))$$
$$+ (1 - \lambda_i)(\partial d_C(x_i(S), x_i(T)) \cap \partial \mathbb{B}) + d_i \mathbb{B} \times \{0\}, \quad (8.8.5)$$

in which the set $(1 - \lambda_i)\partial d_C(x_i(S), x_i(T)) \cap \partial \mathbb{B}$ is interpreted as $\{0\}$ if $(1 - \lambda_i) = 0$. This strengthened form of the transversality condition is obviously true in the case $(1 - \lambda_i) = 0$, in view of the above interpretation. If, on the other hand, $(1 - \lambda_i) > 0$, it follows from (8.8.1) and the sum rule that $d_C(x_i(S), x_i(T)) > 0$. But then (8.8.5) is true because, in this case, all elements in ∂d_C have unit Euclidean norm. Since, by (8.8.4) and (8.8.5), p_i and its derivatives $\dot{p}_i$ are uniformly bounded, we can arrange, by subsequence extraction, that

$$p_i \to p \text{ uniformly, and } \dot{p}_i \to \dot{p} \text{ weakly in } L^1,$$

for some absolutely continuous function p. We can also arrange that $\lambda_i \to \lambda$, for some $\lambda \in [0, 1]$.

A similar convergence analysis to that earlier employed, which makes use, now, of the facts that $a_i \to 0$ a.e. and $d_i \to 0$, yields, in the limit,

$$(-\dot{p}(t), \dot{\bar{x}}(t)) \in \text{co}\,\partial_{x,p} H(t, \bar{x}(t), p(t)), \text{ a.e. } t \in [S, T].$$

We have proved the Hamiltonian inclusion. From (8.8.5) we obtain

$$(p(S), -p(T)) \in \lambda \partial g(\bar{x}(S), \bar{x}(T)) + (1 - \lambda)(\partial d_C(\bar{x}(S), \bar{x}(T)) \cap \partial \mathbb{B}).$$

This relation implies

$$(p(S), -p(T)) \in \lambda \partial g(\bar{x}(S), \bar{x}(T)) + N_C(\bar{x}(S), \bar{x}(T)),$$

which is the desired transversality condition. We deduce also from the relation that $|(p(S), -p(T))| \geq (1 - \lambda) - \lambda k_g$, in which k_g is a Lipschitz constant for g. It follows that

$$\sqrt{2}||p||_{L^\infty} + (1 + k_g)\lambda \geq 1.$$

We have verified the Lagrange multiplier non-triviality condition. The proof is complete. □

8.9 Discussion and a Counter-Example

This Section provides a counter-example that gives insights into different forms of the Hamiltonian inclusion condition. We begin however with some historical context. Here, two versions of the optimality condition come into play:

The fully convexified Hamiltonian inclusion:

$$(\dot{p}(t), \dot{\bar{x}}(t)) \in \text{co}\, \partial_{x,p} H(t, \bar{x}(t), p(t)), \text{ a.e.}$$

and

The partially convexified Hamiltonian inclusion:

$$\dot{p}(t) \in \text{co}\{q \,:\, (q, \bar{x}(t)) \in \partial_{x,p} H(t, \bar{x}(t), p(t))\}, \text{ a.e.}.$$

In the above relations, H is, as usual, the Hamiltonian

$$H(t, x, p) := \max_{e \in F(t,x)} p \cdot e.$$

(Note the change of terminology: the formerly named 'Hamiltonian inclusion' of Sect. 8.7 is now referred to as the 'partially convexified Hamiltonian inclusion', to distinguish it from the fully convexified version.) We assume, for simplicity of discussion, that the data (in relation to nominal F trajectory $\bar{x}$ under consideration) satisfies the hypotheses of Theorem 8.8.1.

The first set of necessary conditions for differential inclusion problems, involving the Hamiltonian was proved by F. H. Clarke in 1973. The time-line of the original discovery and subsequent refinements up to 2005 is as follows:

- 1973: Clarke's (fully convexified) Hamiltonian inclusion is valid for L^∞ local minimizers under the convexity hypothesis (C) (of Theorem 8.7.2). (Clarke [56]).
- 1996: The partially convexified Hamiltonian inclusion is valid for $W^{1,1}$ local minimizers under the convexity hypothesis (C). (Loewen/Rockafellar [144]).
- 2005: The fully convexified Hamiltonian inclusion is valid for L^∞ local minimizers in the absence of the convexity hypothesis (Clarke [67]).

8.9 Discussion and a Counter-Example

This sequence of improvements were in part motivated by a trend in refinements of necessary conditions, in the form of the generalized Euler Lagrange condition, which were first proved for L^∞ local minimizers under the convexity hypothesis (C), then for $W^{1,1}$ local minimizers under the convexity hypothesis and, finally, for $W^{1,1}$-minimizers in the absence of (C). While these developments were underway, it seemed reasonable to assume that such a trend could be reproduced also in refinements of necessary conditions involving subgradients of the Hamiltonian.

Looking at the time-line, we see optimality conditions involving the Hamiltonian were first proved only under the convexity hypothesis (C). From the mid-1970s until its resolution by Clarke 25 years later in 2005, the search was on for an answer to the question: is the Hamiltonian inclusion valid, and in what form, in the absence of the convexity hypothesis (C)?

The reasons why a proof of a version of the Hamiltonian inclusion for the nominal trajectory $\bar{x}$, in the absence of the convexity hypothesis (C), was so long coming is fairly easy to understand. The obvious approach is to attempt to bootstrap a proof for problems with non-convex velocities from one requiring the convexity hypothesis is as follows. First, we approximate the dynamic optimization problem by a problem with no right end-point constraint, whose closeness of approximation is calibrated by a parameter α. We then apply Ekeland's theorem to construct a perturbation to the approximate problem that has a minimizer x^α, converging to $\bar{x}$ as $\alpha \downarrow 0$. Because of the absence of a right endpoint constraint and in consequence of the relaxation theorem, x^α remains a minimizer when we replace $F(t, x)$, in the perturbed problem, by its convex hull co $F(t, x)$. We can then appeal to known necessary conditions, applicable when the velocity sets are convex, to write down necessary conditions satisfied by $\bar{x}^\alpha$, and obtain necessary conditions for $\bar{x}$, expressed in terms of the Hamiltonian, in the limit as $\alpha \downarrow 0$. If this proof strategy is to be followed in situations when the nominal trajectory $\bar{x}$ is a $W^{1,1}$ local minimizer, a natural choice of topology on the space of admissible F trajectories for the perturbed problem is the $W^{1,1}$ topology. Because of the presence of the Ekeland perturbation term in the cost of the perturbed problem, associated with this topology, the necessary conditions for the perturbed problem involves the perturbed Hamiltonian

$$H^\alpha(t, x, p) := \max_{e \in F(t,x)} \left(p \cdot e - \alpha |e - \dot{x}^\alpha(t)| \right).$$

The proof strategy requires us to recovery a solution of the unperturbed Hamiltonian inclusion (in fully or partially convexified form) as the limit of solutions of the perturbed Hamiltonian inclusions; this in turn requires the following upper semi-continuity property:

$$\limsup_{(x',p') \to (x,p), \alpha \downarrow 0} \partial_{x,p} H^\alpha(t, x', p') \subset \partial_{x,p} H(t, x, p). \tag{8.9.1}$$

It is precisely at this point that the proof technique fails, because (8.9.1) is not true in general, if F is not convex valued. In order to circumvent this difficulty, Clarke made use of a different variational principle, namely Stegall's theorem, which gives rise to a perturbed Hamiltonian with better stability properties, in place of Ekeland's theorem, as detailed in the proof of Theorem 8.8.1 in Sect. 8.8.

The latest development described in the timeline above was Clarke's proof of the fully convexified Hamiltonian inclusion for L^∞ local minimizers, in the absence of the convexity hypothesis (C). It leaves open the question of whether, in these necessary conditions we can, in general, replace the fully convexified Hamiltonian inclusion by the more precise partially convexified Hamiltonian inclusion and whether such optimality conditions are valid for the larger class of $W^{1,1}$ local minimizers.

The following example sheds light on such queries. It tells us that, in the absence of the convexity hypothesis (C), the fully convexified Hamiltonian inclusion is not valid, in general, when the nominal F trajectory is merely a $W^{1,1}$ local minimizer (not an L^∞ local minimizer). Since the partially convexified Hamiltonian inclusion is a more precise condition than the fully convexified Hamiltonian inclusion, the above statement remains correct, when we substitute 'partially convexified Hamiltonian inclusion' for 'fully convexified Hamiltonian inclusion'.

Example

Let θ be the function $\theta(r) := \min\{r, 0\}$. Consider the optimization problem:

$$(E) \begin{cases} \text{Minimize } \frac{1}{2}x_1(1) - x_2(1) \\ \text{over absolutely continuous functions } x = (x_1, x_2, x_3) \text{ satisfying} \\ \begin{bmatrix} \dot{x}_1(t) \\ \dot{x}_2(t) \\ \dot{x}_3(t) \end{bmatrix} \in \begin{bmatrix} 0 \\ x_1(t) \\ 1 \end{bmatrix} \cup \begin{bmatrix} 0 \\ x_1(t) \\ -1 \end{bmatrix} \cup \begin{bmatrix} 0 \\ \theta(x_1) \\ 0 \end{bmatrix}, \quad \text{a.e. } t \in [0, 1], \\ (x_1(0), x_2(0), x_3(0)) \in [0, \infty) \times \{0\} \times \{0\}, \\ (x_1(1), x_2(1), x_3(1)) \in \mathbb{R} \times \mathbb{R} \times \mathbb{R}. \end{cases}$$

This will be recognized as a special case of (P) in which $F(t, x) = F(x)$ is

$$F(x) = \begin{bmatrix} 0 \\ x_1 \\ 1 \end{bmatrix} \cup \begin{bmatrix} 0 \\ x_1 \\ -1 \end{bmatrix} \cup \begin{bmatrix} 0 \\ \theta(x_1) \\ 0 \end{bmatrix}$$

and

$$g(x^0, x^1) = \frac{1}{2}x_1^1 - x_2^1, \quad C = ([0, \infty) \times \{0\} \times \{0\}) \times \mathbb{R}^3.$$

Salient features of this example are summarized in the following proposition.

8.9 Discussion and a Counter-Example

Proposition 8.9.1 *The data of problem (E) satisfy hypotheses (G1)–(G3) of Theorem 8.8.1, with reference to the admissible F trajectory ($\bar{x} \equiv (0, 0, 0)$), and*

(a): ($\bar{x} \equiv (0, 0, 0)$) *is a* $W^{1,1}$ *local minimizer,*
(b): ($\bar{x} \equiv (0, 0, 0)$) *is not an* L^∞ *local minimizer,*
(c): *There does not exist non-zero* (λ, p) *such that conditions (i)-(iii) in the statement of Theorem 8.8.1 are all satisfied, with reference to* $\bar{x}$.

Proof
(a): Take any $\gamma \in (0, 1/2)$ and any admissible F trajectory (x_1, x_2, x_3) such that

$$\|x\|_{W^{1,1}} \leq \gamma.$$

Then x_1 is constant (write the constant value x_1) and, in view of the left end-point constraint, $x_1 \geq 0$. Since $x_2(0) = 0$, we deduce from $\dot{x}(t) \in F(t, x(t))$ that

$$x_2(1) = \int_I x_1 \, dt + \int_{[0,1]\setminus I} 0 \, dt = x_1 \times \text{meas}\{I\}$$

and

$$|\dot{x}_3(t)| = \begin{cases} 1 & \text{a.e. } t \in I \\ 0 & \text{a.e. } t \in [0, 1]\setminus I. \end{cases}$$

Here, $I := \{t \in [0, 1] : \dot{x}_3(t) \neq 0\}$.
It follows that

$$\|\dot{x}_3\|_{L^1} = \text{meas}\{I\}.$$

Since $x_3(0) = 0$, we know $\|\dot{x}_3\|_{L^1} = \|x_3\|_{W^{1,1}}$. But then

$$\text{meas}\{I\} = \|x_3\|_{W^{1,1}} \leq \|x\|_{W^{1,1}} \leq \gamma,$$

and so $x_2(1) \leq \gamma x_1$. We conclude that

$$g(x(0), x(1)) = \frac{1}{2}x_1 - x_2(1) \geq x_1(1/2 - \gamma) \geq 0.$$

We have shown that x has cost not lower than that of $\bar{x}$.

(b): Take any $\gamma > 0$ and write $\alpha := \gamma/\sqrt{3}$. Let sw be the switching function

$$sw(t) := \begin{cases} +1 & \text{for } t \in [2k - 2, 2k - 1) \\ -1 & \text{for } t \in [2k - 1, 2k), \end{cases} \quad \text{for } k = 1, 2, \ldots$$

Now let $x = (x_1, x_2, x_3)$ be the admissible F trajectory defined by the relations

$$\begin{cases} (\dot{x}_1(t), \dot{x}_2(t), \dot{x}_3(t)) = (0, \alpha, sw(M_\gamma t)) & \text{for } t \in [0, 1], \\ (x_1(0), x_2(0), x_3(0)) = (\alpha, 0, 0), \end{cases}$$

in which M_γ is an even positive integer such that $M_\gamma \geq \sqrt{3}/\gamma$.

We see immediately that $(x_1(1), x_2(1)) = (\alpha, \alpha)$. A simple calculation ('integrate the switching function') yields

$$|x_3(t)| \leq \gamma/\sqrt{3} \text{ for all } t.$$

Since $|x_1(t)|$ and $|x_2(t)|$ are bounded by $\alpha = \gamma/\sqrt{3}$, we have

$$\|x\|_{L^\infty} \leq (3 \times \left(\gamma/\sqrt{3}\right)^2)^{\frac{1}{2}} = \gamma.$$

Also,

$$g(x(0), x(1)) = \frac{1}{2}x_1(1) - x_2(1) = (\frac{1}{2} - 1)\alpha < 0.$$

On the other hand, the F trajectory $\bar{x}$, which is admissible, has cost $g(\bar{x}(0), \bar{x}(1)) = 0$. It follows that there exists an admissible F trajectory arbitrarily close to $\bar{x} \equiv 0$ w.r.t. the L^∞ norm, with lower cost, i.e. $\bar{x}$ is not an L^∞ local minimizer. The F trajectory x will be recognized as a 'chattering' approximation to the admissible co-F trajectory with constant velocity $(0, \alpha, 0)$.

(c): Suppose that, contrary to the assertion, there exists (p, λ) satisfying conditions (i)-(iii) in the statement of Theorem 8.8.1, with reference to $\bar{x} \equiv 0$. In view of hypotheses (G1)–(G3), which are satisfied by the data for (E), and since the right end-points of admissible F trajectories are free, we must have $\lambda > 0$. By scaling (λ, p), we can arrange that $\lambda = 1$. The transversality conditions give:

$$p_1(0) \leq 0, \quad p_1(1) = -\frac{1}{2}, \quad p_2(1) = 1 \quad \text{and} \quad p_3(1) = 0.$$

The Hamiltonian is

$$H(t, x, p) = \max\{(p_2 x_1 + |p_3|), p_2 \times \theta(x_1)\}.$$

The Hamiltonian inclusion yields $\dot{p}_2 \equiv 0$, whence $p_2 \equiv 1$. But then, for each $t \in [0, 1]$, we have on a neighbourhood of $(\bar{x}(t), p(t))$,

$$H(t, x, p) = \max\{(p_2 x_1 + |p_3|), (p_2 \times \min\{x_1, 0\})\} = p_2 x_1 + |p_3|.$$

It follows that

$$\text{co } \partial_{x,p} H(t, \bar{x}(t), p(t)) = \{((p_2(t), 0, 0), (0, 0, \beta) \in \mathbb{R}^6 : \beta \in [-1, +1]\}.$$

Using this information, we deduce from the Hamiltonian inclusion that $\dot{p}_1 = -p_2 \equiv -1$. Since $p_1(1) = -\frac{1}{2}$, it follows that $p_1(0) = +\frac{1}{2}$. But this contradicts $p_1(0) \leq 0$. We have confirmed that the Hamiltonian inclusion conditions are not satisfied. □

Remark
Because $\bar{x}$ is a $W^{1,1}$ local minimizer, it must satisfy the Euler Lagrange inclusion (ii), transversality condition (iii) and Weierstrass condition (iv) in Theorem 8.4.3. With reference to the Euler Lagrange inclusion, the costate trajectory components (p_1, p_2, p_3) satisfy similar conditions to those associated with the Hamiltonian inclusion, with this crucial difference: Now $\dot{p}_1(t) \in [-p_2(t), 0]$, yielding merely the information that $p_1(0) \geq -\frac{1}{2}$. This condition is too weak to establish a contradiction with the transversality condition.

8.10 Appendix: Dualization of the Euler Lagrange Inclusion

Take a function $\tilde{L} : \mathbb{R}^n \times \mathbb{R}^m \to \mathbb{R} \cup \{+\infty\}$ and points $(\bar{x}, \bar{v}) \in \text{dom } \tilde{L}$ and $\bar{p} \in \mathbb{R}^m$. Define $\tilde{H} : \mathbb{R}^n \times \mathbb{R}^m \to \mathbb{R} \cup \{-\infty\} \cup \{+\infty\}$ to be the conjugate functional of $\tilde{L}(x, v)$ with respect to v:

$$\tilde{H}(x, p) := \sup_{v \in \mathbb{R}^m} \{p \cdot v - \tilde{L}(x, v)\}.$$

Our goal is to verify the following inclusion

$$\{q : (q, \bar{p}) \in \partial \tilde{L}(\bar{x}, \bar{v})\} \subset \text{co}\{q : (-q, \bar{v}) \in \partial \tilde{H}(\bar{x}, \bar{p})\}, \quad (8.10.1)$$

under unrestrictive hypotheses on the function $\tilde{L}$. Notice that, since the set on the right is convex, (8.10.1) immediately implies

$$\text{co}\{q : (q, \bar{p}) \in \partial \tilde{L}(\bar{x}, \bar{v})\} \subset \text{co}\{q : (-q, \bar{v}) \in \partial \tilde{H}(\bar{x}, \bar{p})\}. \quad (8.10.2)$$

The validity of (8.10.1) will be proved under a range of hypotheses. Our ultimate objective is to prove the relation under the hypotheses of dualization theorem, Theorem 8.7.1, and thereby to justify incorporating the Hamiltonian inclusion into the necessary conditions of Theorem 8.7.2.

To prepare the ground, it is necessary to introduce some continuity concepts for the function $\tilde{L}$ that focus on the properties of the multifunction $x \rightsquigarrow \text{epi } \tilde{L}(x, .)$.

Definition 8.10.1 Take a function $\tilde{L} : \mathbb{R}^n \times \mathbb{R}^m \to \mathbb{R} \cup \{+\infty\}$. We say that $\tilde{L}$ is *epicontinuous* if, for each $x \in \mathbb{R}^n$ and each $x_i \to x$, we have

$$\lim_{i \to \infty} \text{epi } \tilde{L}(x_i, .) = \text{epi } \tilde{L}(x, .).$$

The defining properties for epicontinuity can be expressed directly in terms of $\tilde{L}$. The proof of the following characterization is straightforward and is therefore omitted:

Proposition 8.10.2 *Take a function* $\tilde{L} : \mathbb{R}^n \times \mathbb{R}^m \to \mathbb{R} \cup \{+\infty\}$. *Then* $\tilde{L}$ *is epicontinuous if and only if*

(a) *Given any point* (x, v) *and any sequence* $(x_i, v_i) \to (x, v)$, *we have*

$$\tilde{L}(x, v) \leq \liminf_{i \to \infty} \tilde{L}(x_i, v_i),$$

and

(b) *Given any point* (x, v) *and any sequence* $x_i \to x$, *there exists a sequence* $v_i \to v$ *such that*

$$\tilde{L}(x, v) \geq \limsup_{i \to \infty} \tilde{L}(x_i, v_i).$$

Definition 8.10.3 Take a function $\tilde{L} : \mathbb{R}^n \times \mathbb{R}^m \to \mathbb{R} \cup \{+\infty\}$ and a point $(\bar{x}, \bar{v}) \in$ dom $\tilde{L}$. We say that $\tilde{L}$ is *epicontinuous near* $(\bar{x}, \bar{v})$ if there exist neighbourhoods U, V and W of $\bar{x}$, $\bar{v}$ and $L(\bar{x}, \bar{v})$ with the property:

(a) given any point $((x, v), \tilde{L}(u, v)) \in U \times V \times W$ and any sequence $(x_i, v_i) \to (x, v)$, we have

$$\tilde{L}(x, v) \leq \liminf_{i \to \infty} \tilde{L}(x_i, v_i),$$

and

(b) given any $(x, v, \tilde{L}(x, v)) \in U \times V \times W$ and any sequence $x_i \to x$, a sequence $v_i \to v$ can be found such that

$$\tilde{L}(x, v) \geq \limsup_{i \to \infty} \tilde{L}(x_i, v_i).$$

Definition 8.10.4 Given a function $\tilde{L} : \mathbb{R}^n \times \mathbb{R}^m \to \mathbb{R} \cup \{+\infty\}$ and a point $(\bar{x}, \bar{v}) \in$ dom $\tilde{L}$, we say that $\tilde{L}$ is *locally epi-Lipschitz* near $(\bar{x}, \bar{v})$ if there exist neighbourhoods U, V and W of $(\bar{x}, \bar{v})$ and $L(\bar{x}, \bar{v})$ respectively and $k > 0$ such that

$$\text{epi}\,\tilde{L}(x', .) \cap (V \times W) \subset \text{epi}\,\tilde{L}(x'', .) + k|x' - x''|B$$

for all $x', x'' \in U$.

8.10 Appendix: Dualization of the Euler Lagrange Inclusion

Observe that $\tilde{L}$ is locally epi-Lipschitz near $(\bar{x}, \bar{v})$ if and only if the multifunction $x \rightsquigarrow \tilde{L}(x,.)$ is pseudo-Lipschitz near $\big(\bar{x}, (\bar{v}, \tilde{L}(\bar{x}, \bar{v}))\big)$.

In the first version of the dualization theorem provided here, a Lipschitz continuity hypothesis on the data is imposed via the conjugate functional $\tilde{H}(x, p)$:

Theorem 8.10.5 *Take a function* $\tilde{L} : \mathbb{R}^n \times \mathbb{R}^m \to \mathbb{R} \cup \{+\infty\}$ *and points* $(\bar{x}, \bar{v}) \in$ dom L *and* $\bar{p} \in \mathbb{R}^m$. *Assume that, for some neighbourhoods* U *and* P *of* $\bar{x}$ *and* $\bar{p}$ *respectively, the following hypotheses are satisfied:*

(H1): $\tilde{L}(x,.)$ *is convex for each* $x \in U$,
(H2): $\tilde{L}$ *is lower semi-continuous and epicontinuous near* $(\bar{x}, \bar{v})$,
(H3): $\tilde{H}(., p)$ *is Lipschitz continuous on* U, *uniformly with respect to all* p*'s in* P.

Then

$$\{q : (q, \bar{p}) \in \partial \tilde{L}(\bar{x}, \bar{v})\} \subset \operatorname{co}\{q : (-q, \bar{v}) \in \partial \tilde{H}(\bar{x}, \bar{p})\}.$$

Proof In view of (H3) we can arrange, by reducing the size of U and P if required, that $\tilde{H}$ is finite-valued and continuous on $U \times P$. From (H3) it can also be deduced that the set

$$\{q \in \mathbb{R}^n : (q, v') \in \partial \tilde{H}(x', p')\}$$

is uniformly bounded as (x', p', v') ranges over $U \times P \times \mathbb{R}^m$. In view of (H2), we can arrange by shrinking the neighbourhood U if necessary, and choosing suitable neighbourhoods V and W of $\bar{v}$ and $\tilde{L}(\bar{x}, \bar{v})$ respectively, that the following assertions are valid: given any $(x, v) \in U \times V$ such that $\tilde{L}(x, v) \in W$ and any sequence $x_i \to x$, there exists a sequence $v_i \to v$ such that

$$\limsup_i \tilde{L}(x_i, v_i) \leq \tilde{L}(x, v).$$

These facts will be used presently. Take a point $\bar{q} \in \mathbb{R}^n$ which satisfies

$$(\bar{q}, \bar{p}) \in \partial \tilde{L}(\bar{x}, \bar{v}).$$

We must show

$$-\bar{q} \in \operatorname{co}\{q : (q, \bar{v}) \in \partial \tilde{H}(\bar{x}, \bar{p})\}.$$

We note at the outset, however, that it suffices to treat only the case when $(\bar{q}, \bar{p})$ is a proximal normal:

$$(\bar{q}, \bar{p}) \in \partial^P \tilde{L}(\bar{x}, \bar{v}). \tag{8.10.3}$$

This is because if merely $(\bar{q}, \bar{p}) \in \partial \tilde{L}(\bar{x}, \bar{v})$, there exist sequences $(x_i, v_i) \xrightarrow{\tilde{L}} (\bar{x}, \bar{v})$ and $(q_i, p_i) \to (\bar{q}, \bar{p})$ such that $(q_i, p_i) \in \partial^P \tilde{L}(x_i, v_i)$ for each i. Applying the special case of the theorem gives

$$-q_i \in \mathrm{co}\{q : (q, v_i) \in \partial \tilde{H}(x_i, p_i)\}$$

for each i sufficiently large.

Since $\tilde{H}$ is continuous on $U \times P$, we have $(x_i, p_i) \xrightarrow{H} (\bar{x}, \bar{p})$. We then deduce from the uniform boundedness of the sets $\{q : (q, v_i) \in \partial \tilde{H}(x_i, p_i)\}$ and Carathéodory's theorem that

$$-\bar{q} \in \mathrm{co}\{q : (q, \bar{v}) \in \partial \tilde{H}(\bar{x}, \bar{p})\}.$$

This confirms that, without loss of generality, we can assume (8.10.3).

By modifying the neighbourhoods U and V if required, we can arrange that V is compact and, for some constant $\sigma > 0$,

$$M(x, v) \geq 0 \text{ for all } (x, v) \in (\bar{x}, \bar{v}) + U \times V$$

where

$$M(x, v) := \tilde{L}(x, v) - \tilde{L}(\bar{x}, \bar{v}) - \bar{q} \cdot (x - \bar{x}) - \bar{p} \cdot (v - \bar{v})$$
$$+ \sigma |x - \bar{x}|^2 + \sigma |v - \bar{v}|^2. \qquad (8.10.4)$$

Since $M(x, .)$ is lower semi-continuous and strictly convex, for each $x \in U$ there exists a unique minimizer v_x over the compact set V. We can deduce from (H2) that

$$\limsup_{x \to \bar{x}} M(x, v_x) \leq 0.$$

By the lower semicontinuity of M, and since $\bar{v}$ is the unique minimizer for $M(\bar{x}, .)$ over V, we have

$$\lim_{x \to \bar{x}} v_x = \bar{v}. \qquad (8.10.5)$$

For all x sufficiently close to $\bar{x}$ then, v_x is interior to V and so, since $M(x, .)$ is convex, v_x is the unique global minimizer. For all such x

$$0 \leq M(x, v_x) = \min_{v \in \mathbb{R}^m} M(x, v). \qquad (8.10.6)$$

From (8.10.4), $\bar{v}$ minimizes

$$v \to \tilde{L}(\bar{x}, v) + \sigma |v - \bar{v}|^2 - \bar{p} \cdot (v - \bar{v}).$$

8.10 Appendix: Dualization of the Euler Lagrange Inclusion

Since $\tilde{L}(\bar{x}, .)$ is convex, this implies that $\bar{p}$ is a subgradient of $\tilde{L}(\bar{x}, .)$, in the sense of convex analysis. But then $p \cdot \bar{v} - \tilde{L}(\bar{x}, \bar{v}) \geq \max\{p \cdot v - \tilde{L}(\bar{x}, v) : v \in \mathbb{R}^m\}$. Taking note of the definition of H, we conclude that

$$\tilde{L}(\bar{x}, \bar{v}) = \bar{p} \cdot \bar{v} - \tilde{H}(\bar{x}, \bar{p}). \tag{8.10.7}$$

Representing $\tilde{L}(x, .)$ at the conjugate function of $\tilde{H}(x, .)$, we obtain

$$\min_{v \in \mathbb{R}^m} M(x, v)$$

$$= \min_{v \in \mathbb{R}^m} \Big\{ \sup_{p \in \mathbb{R}^n} [p \cdot v - \tilde{H}(x, p)] - \tilde{L}(\bar{x}, \bar{v}) - \bar{q} \cdot (x - \bar{x})$$

$$- \bar{p} \cdot (v - \bar{v}) + \sigma |x - \bar{x}|^2 + \sigma |v - \bar{v}|^2 \Big\}$$

$$= \min_{v \in \mathbb{R}^m} \sup_{p \in \mathbb{R}^m} K_x(v, p) + \tilde{H}(\bar{x}, \bar{p}) - \bar{q} \cdot (x - \bar{x}) + \sigma |x - \bar{x}|^2$$

by (8.10.7). Here

$$K_x(v, p) := (p - \bar{p}) \cdot v + \sigma |v - \bar{v}|^2 - \tilde{H}(x, p).$$

It follows from Proposition 3.4.7 (a version of the mini-max theorem for non-compact domains) that

$$\min_{v} \sup_{p} K_x(v, p) = \min_{v} K_x(v, p_x)$$

$$= (p_x - \bar{p}) \cdot \bar{v} - (4\sigma)^{-1} |p_x - \bar{p}|^2 - \tilde{H}(x, p_x),$$

in which

$$p_x := \bar{p} - 2\sigma(v_x - \bar{v}). \tag{8.10.8}$$

We also have that

$$\min_{v} \sup_{p'} K_x(v, p') \geq \min_{v} K_x(v, p) = (p - \bar{p}) \cdot \bar{v} - (4\sigma)^{-1} |p - \bar{p}|^2 - \tilde{H}(x, p),$$

for any p. From (8.10.6) we deduce that

$$0 \leq M(x, v_x) = -(4\sigma)^{-1} |p_x - \bar{p}|^2 - \tilde{H}(x, p_x)$$
$$- \bar{q} \cdot (x - \bar{x}) + \sigma |x - \bar{x}|^2 + (p_x - \bar{p}) \cdot \bar{v} + \tilde{H}(\bar{x}, \bar{p})$$

and

$$0 = M(\bar{x}, \bar{v}) \geq -(4\sigma)^{-1} |p - \bar{p}|^2 - \tilde{H}(\bar{x}, p) + (p - \bar{p}) \cdot \bar{v} + \tilde{H}(\bar{x}, \bar{p})$$

for all p. It follows from these last two inequalities that

$$\tilde{H}(\bar{x}, p) - \tilde{H}(x, p_x) + (4\sigma)^{-1}(|p - \bar{p}|^2 - |p_x - \bar{p}|^2)$$
$$+ (p_x - \bar{p}) \cdot \bar{v} - \bar{q} \cdot (x - \bar{x}) + \sigma |x - \bar{x}|^2 \geq 0, \quad (8.10.9)$$

for all x near $\bar{x}$ and all p.

For fixed x close to $\bar{x}$, define the function $\phi_x(z)$, whose argument z is partitioned as $z = (y, p)$, to be

$$\phi_x(z) := \tilde{H}(y, p) - \tilde{H}(x, p_x) + (4\sigma)^{-1}(|p - \bar{p}|^2 - |p_x - \bar{p}|^2)$$
$$\bar{v} \cdot (p_x - p) - \bar{q} \cdot (x - y) + \sigma |x - y|^2.$$

In consequence of the definition of ϕ_x, and also by (8.10.9),

$$\phi_x(\bar{x}, p) \geq 0 \text{ for all } p \text{ and } \phi(x, p_x) = 0. \quad (8.10.10)$$

By (8.10.5) and (8.10.8) we know that $p_x \to \bar{p}$ as $x \to \bar{x}$. We can choose sequences $\epsilon_i \downarrow 0$ and $\delta_i \downarrow 0$, therefore, which satisfy $\epsilon_i \delta_i^{-1} \to 0$ as $i \to \infty$ and

$$|x - \bar{x}| \leq \epsilon_i \text{ implies } |p_x - \bar{p}| \leq \delta_i/2.$$

Fix i. For each $x \in \bar{x} + \epsilon_i \mathbb{B}$, $x \neq \bar{x}$, apply the generalized mean value inequality theorem (Theorem 4.5.1) to ϕ_x with

$$z_0 := (x, p_x) \text{ and } Z = \{(\bar{x}, p) : p \in \bar{p} + \delta_i \mathbb{B}\}.$$

This supplies $(\zeta'(x), \eta'(x)) \in \partial^P \tilde{H}(y'(x), p'(x))$ for some $(y'(x), p'(x)) \in \mathbb{R}^n \times \mathbb{R}^m$ such that

$$|y'(x) - x| \leq 2\epsilon_i, \quad |p'(x) - \bar{p}| \leq \epsilon_i + \delta_i$$

and (in view of (8.10.10))

$$-\epsilon_i |x - \bar{x}| \leq -\epsilon_i |x - \bar{x}| + \inf_{z \in Z} \phi_x(z) - \phi_x(x, p_x)$$
$$\leq [\zeta'(x) + \bar{q} + 2\sigma(y'(x) - x)] \cdot (\bar{x} - x)$$
$$+ [\eta'(x) - \bar{v} + (2\sigma)^{-1}(p'(x) - \bar{p})] \cdot (p - p_x)$$

for all $p \in \bar{p} + \delta_i \mathbb{B}$.

Since $|p_x - \bar{p}| < \delta_i/2$, taking the minimum over p in the final term on the right, we arrive at

$$-\epsilon_i |x - \bar{x}| \leq [\zeta'(x) + \bar{q} + 2\sigma(y'(x) - x)] \cdot (\bar{x} - x)$$
$$- (\delta_i/2)|\eta'(x) - \bar{v} + (2\sigma)^{-1}(p'(x) - \bar{p})|. \quad (8.10.11)$$

This last inequality is valid for all $x \in \bar{x} + \epsilon_i \mathbb{B}$ such that $x \neq \bar{x}$.

8.10 Appendix: Dualization of the Euler Lagrange Inclusion

Now, $|\zeta'(x)|$ is bounded on $\bar{x} + \epsilon_i \mathbb{B}$ by a constant independent of i and of our choice of $y'(x)$, $p'(x)$ (see the remarks at the beginning of the proof). It follows that there exists K, independent of i, such that, for all $x \in \bar{x} + \epsilon_i \mathbb{B}$, $x \neq \bar{x}$,

$$-\epsilon_i^2 \leq (K + |\bar{q}| + 4\sigma\epsilon_i)\epsilon_i - (\delta_i/2)|\eta'(x) - \bar{v}| + (\delta_i/2)(2\sigma)^{-1}(\epsilon_i + \delta_i).$$

Since $\epsilon_i \delta_i^{-1} \to 0$ as $i \to \infty$ there exists $\gamma_i \downarrow 0$ such that

$$\sup_{x \in \bar{x} + \epsilon_i \mathbb{B}} |\eta'(x) - \bar{v}| < \gamma_i. \quad \text{for each } i. \tag{8.10.12}$$

Inequality (8.10.11) also tells us that

$$-\epsilon_i |x - \bar{x}| \leq [\zeta'(x) + \bar{q} + 2\sigma(y'(x) - x)] \cdot (\bar{x} - x)$$
$$\leq (\zeta'(x) + \bar{q}) \cdot (\bar{x} - x) + 4\sigma\epsilon_i |x - \bar{x}|$$

for all $x \in \bar{x} + \epsilon_i \mathbb{B}$. This means that

$$0 \leq (\zeta'(x) + \bar{q}) \cdot (\bar{x} - x) + \max_{e \in \epsilon_i(1+4\sigma)\mathbb{B}} e \cdot (\bar{x} - x) \tag{8.10.13}$$

for all $x \in \bar{x} + \epsilon_i \mathbb{B}$. (8.10.12) and (8.10.13) yield

$$\sup_{\zeta \in \bar{q} + S_i} \zeta \cdot (\bar{x} - x) \geq 0 \tag{8.10.14}$$

for all $x \in \bar{x} + \epsilon_i \mathbb{B}$. Here

$$S_i := \{\zeta' : (\zeta', v') \in \partial^P \tilde{H}(x', p') + (\epsilon_i(1 + 4\sigma)\mathbb{B}) \times (\gamma_i \mathbb{B})$$
$$|x' - x| \leq 2\epsilon_i, |p' - \bar{p}| \leq \epsilon_i + \delta_i\}.$$

We conclude from (8.10.14) that

$$-\bar{q} \in \overline{\text{co}} S_i.$$

But in view of the remarks at the beginning of the proof, the S_i's are bounded sets and $\tilde{H}$ is continuous on a neighbourhood of $(\bar{x}, \bar{p})$. From Carathéodory's theorem and the closure properties of $\partial \tilde{H}$ we deduce that

$$-\bar{q} \in \text{co}\{\zeta : (\zeta, \bar{v}) : (\zeta, \bar{v}) \in \partial \tilde{H}(\bar{x}, \bar{p})\}.$$

The proof is complete. □

The hypotheses in the above version of the dualization theorem include the requirement that the dual function $\tilde{H}(., p)$ is Lipschitz continuous near $\bar{x}$, in some uniform sense. We wish to replace it by the condition that $\tilde{L}$ is locally epi-Lipschitz near $(\bar{x}, \bar{v})$. Unfortunately, the local epi-Lipschitz condition for $\tilde{L}$ concerns the behaviour of $\tilde{L}$ on a neighbourhood of $(\bar{x}, \bar{v}) \in \text{dom}\,\tilde{L}$, yet the values of $\tilde{H}$ are affected by the *global* properties of $\tilde{L}$. We cannot therefore expect, in general, to guarantee regularity properties of $\tilde{H}$ by hypothesizing $\tilde{L}$ is locally epi-Lipschitz.

The local epi-Lipschitz property does however ensure that, for some $\epsilon > 0$, the related function H_ϵ has useful Lipschitz continuity properties, where

$$\tilde{H}_\epsilon(x, p) := \sup\{p \cdot v - \tilde{L}(x, v) : v \in \bar{v} + \epsilon \mathbb{B}\}. \tag{8.10.15}$$

Here ϵ is some parameter. H_ϵ will be recognized as the conjugate function, with respect to the second variable, of the 'localization' of L:

$$\tilde{L}_\epsilon(x, v) := \begin{cases} \tilde{L}(x, v) & \text{if } v \in \bar{v} + \epsilon \mathbb{B} \\ +\infty & \text{otherwise}. \end{cases}$$

Relevant properties of $\tilde{H}_\epsilon$ are listed in following proposition.

Proposition 8.10.6 *Take a function* $\tilde{L} : \mathbb{R}^n \times \mathbb{R}^m \to \mathbb{R} \cup \{+\infty\}$ *and a point* $(\bar{x}, \bar{v}) \in \text{dom}\,\tilde{L}$. *Assume that*

(i): $\tilde{L}(x, .)$ *is convex for each x in a neighbourhood of $\bar{x}$,*
(ii): $\tilde{L}$ *is lower semi-continuous,*
(iii): $\tilde{L}$ *is locally epi-Lipschitz near $(\bar{x}, \bar{v})$.*

Then for each $c > 0$, there exists $\epsilon > 0$ and $\beta > 0$ such that $H_\epsilon(., p)$ is Lipschitz continuous on $\bar{x} + \beta \mathbb{B}$, uniformly with respect to all p's in $c\mathbb{B}$.

Proof Hypothesis (iii) can be expressed: there exist neighbourhoods U, V and W of $\bar{x}$, $\bar{v}$ and $\tilde{L}(\bar{x}, \bar{v})$ respectively and $k > 0$ such that

$$\text{epi}\,\tilde{L}(x', .) \cap (V \times W) \subset \text{epi}\,\tilde{L}(x'', .) + k|x' - x''|\mathbb{B} \tag{8.10.16}$$

for all $x', x'' \in U$. By increasing k if necessary, we can arrange that $k > 1$. Choose $\alpha > 0$ such that $\tilde{L}(\bar{x}, \bar{v}) + \alpha \mathbb{B} \subset W$.

Assume that the assertions of the proposition are false. We will show that this leads to a contradiction. Under this assumption, there exists $c > 0$ such that for all $\beta > 0$ and $\epsilon > 0$, the function $H_\epsilon(., p)$ is not Lipschitz continuous on $\bar{x} + \beta \mathbb{B}$, uniformly with respect to all p's in $c\mathbb{B}$. We can therefore find a sequence $\beta_i \downarrow 0$ and a positive constant ϵ such that, for all i, $H_\epsilon(., p)$ is not Lipschitz continuous on $\bar{x} + \beta \mathbb{B}$ uniformly with respect to $p \in c\mathbb{B}$ and

(a) the set $\mathcal{K} := (\bar{x} + \beta_i B) \times (\bar{v} + \epsilon \mathbb{B})$ is included in $U \times V$,
(b) $k\beta_i < \epsilon$,
(c) if (x, v) is in $\mathcal{K}$, then $\tilde{L}(x, v) > \tilde{L}(\bar{x}, \bar{v}) - \alpha$,
(d) $2c\epsilon < \alpha/2$.

(We have invoked hypothesis (ii) to ensure (c).)

8.10 Appendix: Dualization of the Euler Lagrange Inclusion

Fix i. Since $\tilde{L}$ is lower semi-continuous on $U \times \mathbb{R}^m$, finite at $(\bar{x}, \bar{v})$ and nowhere takes the value $-\infty$, $\tilde{L}$ is bounded below on the compact subset $\mathcal{K}$. By adding a constant to $\tilde{L}$ (this does not affect the assertions of the proposition), we can arrange that $\tilde{L}(x, v) \geq 0$ for all (x, v) in $\mathcal{K}$. Now for each x in $\bar{x} + \beta_i \mathbb{B}$, we have by (8.10.16) that

$$(\bar{v}, \tilde{L}(\bar{x}, \bar{v})) \in \mathrm{epi}\, \tilde{L}(\bar{x}, \cdot) \cap (V \times W) \subset \mathrm{epi}\, \tilde{L}(x, \cdot) + k|\bar{x} - x|\mathbb{B}.$$

Hence, in particular, there exists v in $\mathbb{R}^m$ such that $|v - \bar{v}| \leq k|x - \bar{x}| < k\beta_i < \epsilon$ and $\tilde{L}(x, v) \leq \tilde{L}(\bar{x}, \bar{v}) + k|\bar{x} - x| < +\infty$. This fact, combined with the non-negativity of $\tilde{L}$ on $\mathcal{K}$, establishes that $\tilde{H}_\epsilon(x, p)$ is finite for all p in $\mathbb{R}^m$ and for all x in $\bar{x} + \beta_i \mathbb{B}$.

By the contradiction hypothesis, we can find points x_i and y_i in $\bar{x} + \beta_i \mathbb{B}$ (with $x_i \neq y_i$) and p_i in $c\mathbb{B}$ such that

$$\tilde{H}_\epsilon(x_i, p_i) - \tilde{H}_\epsilon(y_i, p_i) < -i|x_i - y_i|. \tag{8.10.17}$$

But a lower semi-continuous function which is locally epi-Lipschitz near a point is locally epi-continuous near the point. Since $\{x_i\}$ converges to $\bar{x}$, it follows that there exists a sequence $\{v_i\}$ converging to $\bar{v}$ such that $\limsup_{i \to \infty} \tilde{L}(x_i, v_i) \leq \tilde{L}(\bar{x}, \bar{v})$. We can assume i sufficiently large that v_i is in $\bar{v} + \epsilon \mathbb{B}$ and such that $\tilde{L}(x_i, v_i) < \tilde{L}(\bar{x}, \bar{v}) + \alpha/2$ (by definition of lim sup).

Now since $v \to \tilde{L}(y_i, v)$ is lower semi-continuous on $\mathbb{R}^m$ (by (ii)) and since $\bar{v} + \epsilon \mathbb{B}$ is a compact set, we have by definition of the real number $\tilde{H}_\epsilon(y_i, p_i)$ that there exists w_i in $\bar{v} + \epsilon \mathbb{B}$ such that $\tilde{H}_\epsilon(y_i, p_i) = p_i \cdot w_i - \tilde{L}(y_i, w_i)$. By (8.10.17), this implies that for all v in $\bar{v} + \epsilon \mathbb{B}$,

$$p_i \cdot v - \tilde{L}(x_i, v) - p_i \cdot w_i + \tilde{L}(y_i, w_i) < -i|x_i - y_i|. \tag{8.10.18}$$

Of course $(w_i, \tilde{L}(y_i, w_i))$ belongs to $\mathrm{epi}\, \tilde{L}(y_i, \cdot)$. Let us show that it also belongs to $V \times W$. Certainly, w_i belongs to V. Now since v_i is in $\bar{v} + \epsilon \mathbb{B}$, using inequality (8.10.18), we obtain

$$p_i \cdot v_i - \tilde{L}(x_i, v_i) - p_i \cdot w_i + \tilde{L}(y_i, w_i) < -i|x_i - y_i| \leq 0,$$

which implies that

$$\tilde{L}(y_i, w_i) \leq \tilde{L}(x_i, v_i) - p_i \cdot (v_i - \bar{v}) + p_i \cdot (w_i - \bar{v}) \leq \tilde{L}(\bar{x}, \bar{v}) + \frac{\alpha}{2} + 2c\epsilon.$$

Since (y_i, w_i) is in $\mathcal{K}$, we deduce from (c) and (d) that $\tilde{L}(y_i, w_i)$ is in W. We can now use (8.10.16) (with $x' = y_i$ and $x'' = x_i$) to find $\tilde{v}_i$ in $\mathbb{R}^m$ such that

$$\begin{cases} |\tilde{v}_i - w_i| \leq k|x_i - y_i| \\ \tilde{L}(x_i, \tilde{v}_i) - \tilde{L}(y_i, w_i) \leq k|x_i - y_i|. \end{cases} \tag{8.10.19}$$

There are two cases to consider:

Case A: $|\tilde{v}_i - \bar{v}| \leq \epsilon \mathbb{B}$ for an infinite number of index values.

In this case, by extracting a subsequence if necessary, we can arrange that this assertion is valid for all i. Inserting $v = \tilde{v}_i$ into (8.10.18) gives

$$p_i \cdot \tilde{v}_i - \tilde{L}(x_i, \tilde{v}_i) - p_i \cdot w_i + \tilde{L}(y_i, w_i) < -i|x_i - y_i|.$$

Hence by (8.10.19), we have

$$-(|p_i| + 1)k|x_i - y_i| < -i|x_i - y_i|,$$

which is impossible for large i, because $|p_i| < c$ and $x_i \neq y_i$.

Case B: $|\tilde{v}_i - \bar{v}| > \epsilon$ for all i sufficiently large.

In this case, we deduce from (8.10.16), in which we set $x' = \bar{x}$ and $x'' = x_i$, that, for i sufficiently large, $u_i \in \mathbb{R}^m$ can be found such that

$$\begin{cases} |u_i - \bar{v}| \leq k|x_i - \bar{x}| < k\beta_i < \epsilon, \\ \tilde{L}(x_i, u_i) - \tilde{L}(\bar{x}, \bar{v}) \leq k|x_i - \bar{x}|. \end{cases} \quad (8.10.20)$$

Define v'_i in $\bar{v} + \epsilon \mathbb{B}$ according to

$$v'_i := \mu_i u_i + (1 - \mu_i)\tilde{v}_i,$$

where $\mu_i \in [0, 1]$ is chosen such that $|v'_i - \bar{v}| = \epsilon$ (this is possible since $|u_i - \bar{v}| < \epsilon$ and $|\tilde{v}_i - \bar{v}| > \epsilon$). We see that, for i large enough,

$$\begin{aligned} \epsilon = |v'_i - \bar{v}| &= |\mu_i(u_i - \bar{v}) + (1 - \mu_i)(\tilde{v}_i - \bar{v})| \\ &\leq \mu_i|u_i - \bar{v}| + (1 - \mu_i)(|\tilde{v}_i - w_i| + |w_i - \bar{v}|) \\ &\leq \mu_i k|x_i - \bar{x}| + (1 - \mu_i)(k|x_i - y_i| + \epsilon). \end{aligned}$$

It follows that, for i sufficiently large,

$$\mu_i(\epsilon + k|x_i - y_i| - k|x_i - \bar{x}|) \leq k|x_i - y_i|.$$

But $x_i \to \bar{x}$ and $|x_i - y_i| \to 0$. Noting that, for i sufficiently large, $\epsilon - k|x_i - \bar{x}| > 0$ (this follows from (b)), we have, for i sufficiently large, that

$$\mu_i \leq \frac{2k}{\epsilon}|x_i - y_i|. \quad (8.10.21)$$

Since v'_i is in $\bar{v} + \epsilon \mathbb{B}$, we may insert it into (8.10.18), to get

$$p_i \cdot (\mu_i u_i + (1 - \mu_i)\tilde{v}_i - w_i) - \tilde{L}(x_i, v'_i) + \tilde{L}(y_i, w_i) < -i|x_i - y_i|, \quad (8.10.22)$$

for all i sufficiently large. By convexity, we obtain

$$\tilde{L}(x_i, v'_i) \leq \mu_i \tilde{L}(x_i, u_i) + (1 - \mu_i)\tilde{L}(x_i, \tilde{v}_i).$$

8.10 Appendix: Dualization of the Euler Lagrange Inclusion

Hence, we have, for i sufficiently large,

$$\begin{aligned}
-\tilde{L}(x_i, v_i') &\geq -\mu_i \tilde{L}(x_i, u_i) - (1 - \mu_i)\tilde{L}(x_i, \tilde{v}_i) \\
&\geq -\mu_i(\tilde{L}(\bar{x}, \bar{v}) + k|x_i - \bar{x}|) - (1 - \mu_i)\tilde{L}(x_i, \tilde{v}_i) \\
&\geq -\mu_i(\tilde{L}(\bar{x}, \bar{v}) + k|x_i - \bar{x}|) - (1 - \mu_i)(\tilde{L}(y_i, w_i) + k|x_i - y_i|) \\
&\geq -\mu_i(\tilde{L}(\bar{x}, \bar{v}) + k|x_i - \bar{x}|) - \tilde{L}(y_i, w_i) - k|x_i - y_i|.
\end{aligned}$$

To derive these relations, we have used (8.10.19) and (8.10.20), and noted that, since (y_i, w_i) is in $\mathcal{K}$, $\mu_i(\tilde{L}(y_i, w_i) + k|x_i - y_i|) \geq 0$. This inequality combines with (8.10.22) to give

$$\mu_i p_i \cdot (u_i - \tilde{v}_i) + p_i \cdot (\tilde{v}_i - w_i) - \mu_i(\tilde{L}(\bar{x}, \bar{v}) + k|x_i - \bar{x}|)$$
$$- k|x_i - y_i| < -i|x_i - y_i|.$$

We deduce from (8.10.19) that, for i sufficiently large,

$$-\mu_i c|u_i - \tilde{v}_i| - ck|x_i - y_i| - \mu_i(\tilde{L}(\bar{x}, \bar{v}) + k|x_i - \bar{x}|)$$
$$- k|x_i - y_i| < -i|x_i - y_i|.$$

It follows from (8.10.21)

$$-\frac{2k}{\epsilon}[c|u_i - \tilde{v}_i| + \tilde{L}(\bar{x}, \bar{v}) + k|x_i - \bar{x}|] \times |x_i - y_i|$$
$$- (c + 1)k|x_i - y_i| < -i|x_i - y_i|.$$

Since $\{u_i\}$ and $\{\tilde{v}_i\}$ are bounded sequences and $\{x_i\}$ converges to $\bar{x}$, this is impossible.

In both cases (A) and (B), we have arrive at a contradiction. The assertions of the Proposition must therefore be true. □

The link that Proposition 8.10.6 provides between epi-Lipschitz hypotheses on $\tilde{L}$ and Lipschitz continuity properties of $\tilde{H}_\epsilon$ are now used to prove the following generalization of Theorem 8.10.5.

Theorem 8.10.7 *Take a function $\tilde{L} : \mathbb{R}^n \times \mathbb{R}^m \to \mathbb{R} \cup \{+\infty\}$ and points $(\bar{x}, \bar{v}) \in$ dom $\tilde{L}$ and $\bar{p} \in \mathbb{R}^m$. Assume that, for some neighbourhoods U and P of $\bar{x}$ and $\bar{p}$ respectively, the following hypotheses are satisfied:*

(H1): $\tilde{L}(x, .)$ *is convex for each x in a neighbourhood of $\bar{x}$,*
(H2): $\tilde{L}$ *is lower semi-continuous and epicontinuous near $(\bar{x}, \bar{v})$,*
(H3): *For some $\epsilon > 0$, $\tilde{H}_\epsilon(., p)$ (defined in (8.10.15)) is Lipschitz continuous on U, uniformly with respect to all p's in P.*

Then

$$\{q : (q, \bar{p}) \in \partial \tilde{L}(\bar{x}, \bar{v})\} \subset \mathrm{co}\{q : (-q, \bar{v}) \in \partial \tilde{H}(\bar{x}, \bar{p})\}.$$

Proof Take a point $\bar{q}$ such that $(\bar{q}, \bar{p}) \in \partial \tilde{L}(\bar{x}, \bar{v})$. We must show that

$$-\bar{q} \in \mathrm{co}\{q \in \mathbb{R}^n : (q, \bar{v}) \in \partial \tilde{H}(\bar{x}, \bar{p})\}.$$

We observe that

$$(\bar{q}, \bar{p}) \in \partial \tilde{L}_\epsilon(\bar{x}, \bar{v}),$$

since $\tilde{L}$ and $\tilde{L}_\epsilon$ coincide on a neighbourhood of $(\bar{x}, \bar{v})$.

But the hypotheses of the preceding theorem (Theorem 8.10.5) are satisfied when $\tilde{L}_\epsilon$ replaces $\tilde{L}$. We deduce that

$$\bar{q} \in \mathrm{co}\{q : (-q, \bar{v}) \in \partial \tilde{H}_\epsilon(\bar{x}, \bar{p})\}.$$

It follows from Carathéodory's theorem and the closure properties of the proximal subdifferential that $\bar{q}$ is a convex combination of $(n + 1)$ points of the form

$$q = \lim_i q_i$$

for some sequences $v_i \to \bar{v}$ and $(x_i, p_i) \to (\bar{x}, \bar{p})$ such that

$$(-q_i, v_i) \in \partial^P \tilde{H}_\epsilon(x_i, p_i) \text{ for all } i. \tag{8.10.23}$$

Take such a point q. Let $\{v_i\}, \{(x_i, p_i)\}$ be sequences associated with q as above.

Claim:

$$\tilde{H}_\epsilon(\bar{x}, \bar{p}) = \tilde{H}(\bar{x}, \bar{p}) \tag{8.10.24}$$

and

$$\tilde{H}_\epsilon(x_i, p_i) = \tilde{H}(x_i, p_i) \text{ for all } i \text{ sufficiently large}. \tag{8.10.25}$$

Let us verify these assertions. (8.10.23) implies that, for each i, $\sigma_i > 0$ can be found such that

$$\tilde{H}_\epsilon(x, p) - \tilde{H}_\epsilon(x_i, p_i) \geq -q_i \cdot (x - x_i) + v_i \cdot (p - p_i) - \sigma_i |(x, p) - (x_i, p_i)|^2 \tag{8.10.26}$$

for all (x, p) in some neighbourhood of (x_i, p_i). It follows that, for each i,

$$\tilde{H}_\epsilon(x_i, p) - \tilde{H}_\epsilon(x_i, p_i) \geq v_i \cdot (p - p_i) - \sigma_i |p - p_i|^2,$$

for all p in some neighbourhood of p_i.

8.10 Appendix: Dualization of the Euler Lagrange Inclusion

Since $\tilde{H}_\epsilon(x,.)$ is convex, we deduce that

$$v_i \in \partial \tilde{H}_\epsilon(x_i,.)(p_i),$$

in which $\partial \tilde{H}_\epsilon$ denotes the subdifferential of convex analysis. But this implies that (from Fenchel's inequality)

$$\tilde{H}_\epsilon(x_i, p_i) = p_i \cdot v_i - \tilde{L}_\epsilon(x_i, v_i) = p_i \cdot v_i - \tilde{L}(x_i, v_i), \qquad (8.10.27)$$

for i sufficiently large.

For i sufficiently large, the maximum of the concave function $v \to p_i \cdot v - \tilde{L}(x_i, v)$ over $\bar{v} + \epsilon \mathbb{B}$ is achieved at the point $v_i \in \bar{v} + \epsilon \overset{\circ}{\mathbb{B}}$. We conclude that v_i achieves the maximum of this function over all $v \in \mathbb{R}^n$. It follows that (8.10.25) is true.

Since $\tilde{H}_\epsilon$ is continuous on a neighbourhood of $(\bar{x}, \bar{p})$ and $(x_i, v_i) \to (\bar{x}, \bar{v})$, it follows from (8.10.27) and the lower semicontinuity of $\tilde{L}$ that

$$\tilde{H}_\epsilon(\bar{x}, \bar{p}) = \bar{p} \cdot \bar{v} - \lim_i \tilde{L}(x_i, v_i) \leq \bar{p} \cdot \bar{v} - \tilde{L}(\bar{x}, \bar{v}) \leq \tilde{H}_\epsilon(\bar{x}, \bar{p}).$$

We conclude that $v \to \bar{p} \cdot v - \tilde{L}(\bar{x}, v)$ has a local maximum at $v = \bar{v}$. Since $-\tilde{L}(\bar{x},.)$ is concave, this maximum is in fact a *global* maximum. We deduce (8.10.24). The claim is verified.

Since $\tilde{H}_\epsilon(x, p) \leq \tilde{H}(x, p)$ for all (x, p), we deduce from (8.10.26) and (8.10.25) that, for each i sufficiently large, there exists $\sigma_i > 0$ such that

$$\tilde{H}(x, p) - \tilde{H}(x_i, p_i) \geq -q_i \cdot (x - x_i) + v_i \cdot (p - p_i) - \sigma_i |(x, p) - (x_i, p_i)|^2$$

for all (x, p) in some neighbourhood of (x_i, p_i). This implies

$$(-q_i, v_i) \in \partial^P \tilde{H}(x_i, p_i)$$

for all i sufficiently large. Since $\tilde{H}_\epsilon$ is continuous on a neighbourhood of $(\bar{x}, \bar{p})$, (8.10.24) and (8.10.25) imply

$$\lim_i \tilde{H}(x_i, p_i) = \lim_i \tilde{H}_\epsilon(x_i, p_i) = \tilde{H}_\epsilon(\bar{x}, \bar{p}) = \tilde{H}(\bar{x}, \bar{p}).$$

We deduce from these relations that q satisfies

$$(-q, \bar{v}) \in \partial \tilde{H}(\bar{x}, \bar{p}).$$

But $\bar{q}$ is expressible as a convex combination of such q's. It follows that

$$\bar{q} \in \text{co}\{q : (-q, \bar{v}) \in \partial \tilde{H}(\bar{x}, \bar{p})\}.$$

This is what we set out to prove. $\square$

We conclude from Proposition 8.10.6 and Theorem 8.10.7 the validity of the following assertions:

Take a function $\tilde{L} : \mathbb{R}^n \times \mathbb{R}^m \to \mathbb{R} \cup \{+\infty\}$ and points $(\bar{x}, \bar{v}) \in \text{dom } \tilde{L}$ and $\bar{p} \in \mathbb{R}^m$. Assume that, for some neighbourhood U of $\bar{x}$, the following hypotheses are satisfied:

(H1): $\tilde{L}(x, .)$ is convex for each $x \in U$,
(H2): $\tilde{L}$ is lower semi-continuous,
(H3): $\tilde{L}$ is locally epi-Lipschitz near $(\bar{x}, \bar{v})$.

Then

$$\{q : (q, \bar{p}) \in \partial \tilde{L}(\bar{x}, \bar{v})\} \subset \text{co}\{q : (-q, \bar{v}) \in \partial \tilde{H}(\bar{x}, \bar{p})\}. \quad (8.10.28)$$

The properties listed above, together, will be recognised as a statement of the dualization theorem (Theorem 8.7.1). The proof is complete. □

8.11 Exercises

8.1 (Kaskosz Lojasiewicz-Type Necessary Conditions [133]) Take an interval $[S, T]$, a multifunction $F : [S, T] \times \mathbb{R}^n \rightsquigarrow \mathbb{R}^n$ and $\bar{x} \in W^{1,1}([S, T]; \mathbb{R}^n)$. Assume that F is $\mathcal{L} \times \mathcal{B}^n$ measurable, takes values closed, convex sets and, furthermore, F has the representation

$$F(t, x) := \{f(t, x) : f \in \mathcal{F}\}, \quad \text{for all } (t, x) \in [S, T] \times \mathbb{R}^n.$$

for some family of functions $\mathcal{F}$ with the following properties:

$$\dot{\bar{x}}(t) = \bar{f}(t, \bar{x}(t)), \text{ a.e., for some } \bar{f} \in \mathcal{F}$$

and there exist $\epsilon > 0$, $k_F > 0$ and $c_F > 0$ such that, for each $f \in \mathcal{F}$,

(a): f is $\mathcal{L} \times \mathcal{B}^n$ measurable ,
(b): $|f(t, x)| \le c_F$ and $|f(t, x') - f(t, x)| \le k_F |x' - x|$, for all $x', x \in \bar{x}(t) + \epsilon \mathbb{B}$ and $t \in [S, T]$.

Let $\bar{x}$ be an L^∞ local minimizer for

$$\begin{cases} \text{Minimize } g(x(S), x(T)) \\ \text{over } x \in W^{1,1}([S, T]; \mathbb{R}^n) \text{ such that} \\ \dot{x}(t) \in F(t, x(t)), \text{ for a.e. } t \in [S, T] \\ (x(S), x(T)) \in C. \end{cases}$$

in which $g : \mathbb{R}^n \times \mathbb{R}^n \to \mathbb{R}$ is a given function and $C \subset \mathbb{R}^n \times \mathbb{R}^n$ is a given closed set. Assume that g is Lipschitz continuous on $(\bar{x}(S), \bar{x}(T)) + \epsilon \mathbb{B}$.

8.11 Exercises

Show that there exist $p \in W^{1,1}([S,T]; \mathbb{R}^n)$ and $\lambda \geq 0$, not both zero, such that

(i): $-\dot{p}(t) \in \operatorname{co} \partial_x p(t) \cdot f(t, \bar{x}(t))$, a.e.,
(ii): $p(t) \cdot \dot{\bar{x}}(t) = \max_{e \in F(t, \bar{x}(t))} p(t) \cdot e$ a.e.,
(iii): $(p(S), -p(T)) \in \lambda g(\bar{x}(S), \bar{x}(T)) + N_C(\bar{x}(S), \bar{x}(T))$.

Hint: For each $\alpha \in (0, 1]$ show that $\bar{x}$ remains an L^∞ local minimizer when, in the problem formulation $F(t, x)$ is replaced by $(\bar{f}(t, x) - \alpha[\bar{f}(t, x) - F(t, x)])$. Apply Clarke's Hamiltonian inclusion conditions to the modified problem, to obtain perturbed version of the state necessary conditions. Pass to the limit as $\alpha \downarrow 0$. (See [145].)

8.2 Let $\bar{x}$ be a $W^{1,s}$ local minimizer (see Exercise 6.1 for the definition), for some $s \in [1, \infty)$, for the dynamic optimization problem

$$(P) \begin{cases} \text{Minimize } g(x(S), x(T)) \\ \text{over arcs } x \in W^{1,1}([S,T]; \mathbb{R}^n) \text{ satisfying} \\ \dot{x}(t) \in F(t, x(t)) \text{ a.e. } t \in [S, T], \\ (x(S), x(T)) \in C. \end{cases}$$

Assume that

(H1): g is Lipschitz continuous on a neighbourhood of $(\bar{x}(S), \bar{x}(T))$ and C is closed,
(H2): $F(t, x)$ is nonempty for each $(t, x) \in [S, T] \times \mathbb{R}^n$, $\operatorname{Gr} F(t, .)$ is closed for each $t \in [S, T]$ and F is $\mathcal{L} \times \mathcal{B}^n$ measurable,
(H3): there exist integrable functions c, $k \in L^1(S, T)$ such that

$$F(t, x) \subset c(t)\mathbb{B} \text{ for all } x \in \mathbb{R}^n, \text{ a.e. } t \in [S, T] \text{ and}$$
$$F(t, x') \subset F(t, x) + k(t)|x' - x|\mathbb{B} \text{ for all } x', x \in \mathbb{R}^n, \text{ a.e. } t \in [S, T].$$

Show that there exist an arc $p \in W^{1,1}([S,T]; \mathbb{R}^n)$ and $\lambda \geq 0$ such that

(i) $(p, \lambda) \neq (0, 0)$,
(ii) $\dot{p}(t) \in \operatorname{co}\{\eta : (\eta, p(t)) \in N_{\operatorname{Gr} F(t,.)}(\bar{x}(t), \dot{\bar{x}}(t))\}$, a.e. $t \in [S, T]$,
(iii) $(p(S), -p(T)) \in \lambda \partial g(\bar{x}(S), \bar{x}(T)) + N_C(\bar{x}(S), \bar{x}(T))$,
(iv) $p(t) \cdot \dot{\bar{x}}(t) \geq p(t) \cdot v$ for all $v \in F(t, \bar{x}(t))$, a.e. $t \in [S, T]$.

If, in addition, $F(t, x)$ is convex for each $(t, x) \in [S, T] \times \mathbb{R}^n$, then condition (ii) implies

(v) $\dot{p}(t) \in \operatorname{co}\{-\xi : (\xi, \dot{\bar{x}}(t)) \in \partial_{x,p} H(t, \bar{x}, p(t))\}$, a.e. $t \in [S, T]$.

Hint: Consider the auxiliary dynamic optimization problem

$$(\widetilde{P})\quad \begin{cases} \text{Minimize } g(x(S), x(T)) \\ \text{over arcs } (x, y) \in W^{1,1}([S, T]; \mathbb{R}^{n+1}) \text{ satisfying} \\ (\dot{x}(t), \dot{y}(t)) \in \widetilde{F}(t, x(t)) \quad \text{a.e. } t \in [S, T], \\ (x(S), x(T)) \in C \quad \text{and} \quad (y(S), y(T)) \in \{0\} \times \epsilon\mathbb{B}, \end{cases}$$

where $\epsilon > 0$ is a suitable number and $\widetilde{F} = \{(v, |\dot{x}(t) - v|^s) : v \in F(t, x)\}$. Observe that $(\bar{x}, \bar{y} \equiv 0)$ is an L^∞ local minimizer for problem $(\widetilde{P})$ and apply Theorem 8.4.3.

8.12 Notes for Chapter 8

Rockafellar's optimality conditions for fully convex problems of Bolza type revealed how the classical Euler Lagrange condition and Hamilton's system of equations can be interpreted in terms of subdifferentials (in the sense of convex analysis) of extended valued Lagrangians and Hamiltonian functions [172, 173]. Despite the limitations of the 'fully convex' setting (which, in effect, restricts attention to dynamic optimization problems with affine dynamics and cost integrands jointly convex with respect to state and velocity variables), Rockafellar's conditions provided the template for subsequent necessary conditions of the kind covered in this chapter.

The breakthrough into necessary conditions for broad classes of dynamic optimization problems formulated in terms of a differential inclusion was Clarke's Hamiltonian inclusion of 1976 [58]:

$$(-\dot{p}, \dot{\bar{x}}) \in \mathrm{co}\, \partial H(t, \bar{x}, p). \tag{8.12.1}$$

A feature of Clarke's derivation, distinguishing it from early necessary conditions for 'differential inclusion' problems, was the natural and intrinsic nature of the hypotheses imposed on the velocity set $F(t, x)$. $F(t, x)$ was required to be merely compact, convex valued, integrably bounded, measurable in time and integrably Lipschitz continuous in the state variable, with respect to the Hausdorff metric.

Clarke also derived, under the above hypotheses, an Euler Lagrange type condition for problems reducible to free right end-point problems [61]:

$$(\dot{p}, p) \in \mathrm{co} N_{\mathrm{Gr}\, F(t,.)}(\bar{x}, \dot{\bar{x}}). \tag{8.12.2}$$

Subsequently, Loewen and Rockafellar [143] established the validity of the generalized Euler Lagrange condition

$$\dot{p} \in \mathrm{co}\{q : (q, p) \in N_{\mathrm{Gr}\, F(t,.)}(\bar{x}, \dot{\bar{x}})\}. \tag{8.12.3}$$

8.12 Notes for Chapter 8

This condition, prefigured by related, weaker, conditions of Mordukhovich [155] (later elaborated in [158]) and Smirnov [183], improves on (8.12.2), because it involves convexification only w.r.t. to the first coordinate, not both coordinates. Loewen and Rockafellar also provided necessary conditions involving a sharper version of Clarke's Hamiltonian inclusion, namely

$$\dot{p} \in \text{co}\{q : (-q, \dot{\bar{x}}) \in \partial H(t, \bar{x}, p)\}, \tag{8.12.4}$$

which, like (8.12.3), involves convexification only w.r.t. the first coordinate.

Rockafellar [176] gave conditions under which the Euler Lagrange inclusion (8.12.3) (for convex valued velocity sets $F(t, x)$) and the Hamiltonian inclusion (8.12.4) are in fact equivalent. Conditions for the one way implication '(8.12.3) $\implies$ (8.12.4)', under weaker hypotheses (suitably matched to the derivation of necessary conditions of optimality, under unrestrictive hypotheses), were later provided by Bessis, Ledyaev and Vinter [24]. Thus, in the case of convex velocity sets, the second condition is subsumed in the first.

Subsequently, Loewen and Rockafellar highlighted the unsatisfactory nature of traditional 'Hausdorff metric' Lipschitz continuity hypotheses, for applications involving unbounded $F(t, x)$'s. They introduced in [143] a more appropriate set of hypotheses for such multifunctions, which included pseudo-Lipschitz continuity.

The foregoing discussion relates to dynamic optimization problems with convex valued $F(t, x)$'s. Of course the maximum principle, which predates all these developments, applies to problems with possibly nonconvex velocity sets, in the case that the dynamic constraint can be parameterized as a control-dependent differential equation. Necessary conditions for problems with nonconvex $F(t, x)$'s have long been available under hypotheses ('calmness', 'controllability', or assumptions about the nature of the end-point constraints), the implications of which are that minimizers are minimizers also for related problems in which $F(t, x)$ is replaced by its convex hull. In such cases, necessary conditions can be derived from known necessary conditions for the case $F(t, x)$ is convex, or by studying local approximations of the mapping of the initial state into the set of admissible state trajectories as in [106]. But the derivation of necessary conditions for problems with nonconvex $F(t, x)$'s, under unrestrictive hypotheses, required some new ideas and was a later chapter in the story.

We observe that, for convex $F(t, x)$'s, the generalized Weierstrass condition

$$p \cdot \dot{\bar{x}} := \max_{v \in F(t, \bar{x})} p \cdot v$$

is merely a consequence of the generalized Euler Lagrange condition (8.12.3). For nonconvex $F(t, x)$'s however, it is a distinct condition which can have an important role in the elimination of putative minimizers.

Mordukhovich, using discrete approximation techniques, established the validity of the Euler Lagrange inclusion, for non-convex valued, bounded $F(t, x)$'s, a.e. continuous in t and Lipschitz continuous in x [158], but only under the hypothesis that the nominal state trajectory is a local minimizer of the associated 'relaxed' problem. (Furthermore, almost everywhere continuity of the data w.r.t. time was required in this paper and the condition was not accompanied by the Weierstrass condition, since this condition is not manifested in discrete time approximations). Mordukhovich constructions were subsequently elaborated in [159, Part 2].

The validity of the Euler Lagrange inclusion for measurably time dependent, possibly nonconvex valued $F(t, x)$'s satisfying epi-Lipschitz continuity hypothesis, accompanied by the Weierstrass condition was proved by Ioffe [127], building on earlier joint work with Rockafellar [131], treating the finite Lagrangian problem. Vinter and Zheng [199, 200] provided a simple, independent derivation of the condition for non-convex $F(t, x)$, based on application of the classical 'smooth' maximum principle to an approximating optimization problem with state free dynamics and passage to the limit.

As we have observed, it was a significant advance to replace Lipschitz continuity hypotheses by a less restrictive pseudo-Lipschitz continuity hypothesis, in the derivation of the generalized Euler Lagrange condition for problems having unbounded velocity sets $F(t, x)$. However this was not the end of the story, because the precise form of the pseudo-Lipschitz condition chosen involves setting parameters that govern how fast a certain Lipschitz bound grows as you move away from the nominal velocity. The appropriate choice of these parameters, to convey as much information as possible, depends on the special case under consideration [127, 143, 194]. In 2005, Clarke published a version of (8.12.3), Clarke's stratified Euler Lagrange condition, in which he hypothesized the pseudo-Lipschitz condition supplemented by an extra condition called the 'tempered growth' condition [67]. This not only subsumed earlier special cases: the conditions placed on the pseudo-Lipschitz continuity parameters correspond to choosing different stratification procedures. It also clarified the nature of supplementary conditions required.

A feature of Clarke's stratified Euler Lagrange condition is that it incorporates only a partial form of the Weierstrass condition. To be specific, it asserts maximization of the Hamiltonian not over the entire velocity set at the nominal state $F(t, \bar{x}(t))$, but over only a subset. Subsequently, employing techniques formalized in [128], Ioffe identifies additional points in $F(t, \bar{x}(t))$ with respect to which the Weierstrass condition is valid (the 'Ioffe refinement') [129].

In our discussion of necessary conditions for 'nonconvex' differential inclusion problems we have, so far, dwelt exclusively on the generalized Euler Lagrange condition. What about the Hamiltonian inclusion (8.12.4)? This later condition is not automatically satisfied because the implication '(8.12.3) $\implies$ (8.12.4)' has been proved only for convex velocity sets $F(t, x)$. It has been a a longstanding open question, posed by Clarke in the 1970's, whether Clarke's Hamiltonian inclusion is valid when $F(t, x)$ is not convex. The question was resolved by Clarke himself in 2005, when he established this optimality condition was indeed valid when $F(t, x)$ is not convex, but only for L^∞ local minimizers and under Lipschitz, not pseudo-Lipschitz, hypotheses on the data [67]. A counter-example showing that Clarke's Hamiltonian inclusion (and therefore also the stronger partially convexified

8.12 Notes for Chapter 8

Hamiltonian inclusion) is not always valid for $W^{1,1}$ local minimizers is due to Vinter [195]. Ioffe [130] has recently proved Clarke's Hamiltonian inclusion under less restrictive hypotheses.

This chapter brings the generalized Euler Lagrange condition fully up to date. The centrepiece is Clarke's 'stratified' form of the condition proved under pseudo-Lipschitz and tempered growth hypotheses, with the Ioffe refinement. The derivation involves first proving necessary conditions for a finite Lagrangian problem as in [67] but our chosen Lagrangian is a constructed as a scaled distance to the velocity set, in place of the distance to an embedding of the velocity set in a higher dimensional space (Clarke's 'lifting' technique). This facilitates proof of the Ioffe refinement. The proof that the generalized Euler Lagrange conditions subsumes the generalized Hamiltonian inclusion for convex $F(t, x)$'s follows that in [24]. Further discussion concerning the validity of the Clarke's Hamiltonian inclusion for nonconvex $F(t, x)$'s can be found in [195].

Chapter 9
Free End-Time Problems

Abstract This chapter provides necessary conditions of optimality for free end-time dynamic optimization problems, that is problems in which the left and right end-times are included among the choice variables. Minimum time problems, in which the aim is to drive the state from an initial state to a target set in the state space in minimum time, are important examples of such problems. We shall see that the earlier derived necessary conditions for fixed end-point problems (Clarke's nonsmooth maximum principle when the dynamic constraint is a controlled differential equation and either the generalized Euler Lagrange condition or a condition expressed in terms of the Hamiltonian for differential inclusion problems) can be supplemented by extra conditions to take account of the enlargement of the set of choice variables to include the end-times. It turns out that these extra conditions are boundary conditions on the Hamiltonian evaluated along the minimizing state and co-state trajectories. But here we have a problem because, for problems with measurable time dependence, the Hamiltonian is only almost everywhere defined and conditions involving point evaluation of the Hamiltonian at the optimal end-times require interpretation.

The necessary conditions featuring in this chapter fall into two groups. For problems in the first group, the data is assumed to be Lipschitz continuous w.r.t. time; in this situation the Hamiltonian (evaluated along the minimizing state and co-state trajectories) is Lipschitz continuous. So the boundary conditions on this function can be interpreted in the obvious way. For this group of problems, necessary conditions (including the classically interpreted extra boundary conditions on the Hamiltonian), can be derived by using a family of re-parameterizations of the time variable to reduce the free end-time problem to a fix end-time problem, to which we apply fixed end-time necessary conditions from earlier chapters.

For problems in the second group we allow the data to be merely measurable w.r.t. time. Here it is still possible to derive free end-time necessary conditions. But now the boundary conditions are interpreted as conditions involving the 'essential values' (strictly speaking, the super and sub essential values) of the Hamiltonian. Essential values associated with a given function are set-valued functions that reduce to a point-valued function coinciding with the original function, when the original function is continuous. They provide a means of interpreting boundary conditions on the Hamiltonian, because essential values are invariant under changes to the original function on a nullset. To prove the necessary conditions, for this group of problems, we build up the proofs in stages, providing proofs of the necessary conditions under progressively less stringent hypotheses. A simple variation of the end-times in the first state establishes the required boundary condition (expressed in terms of essential values). Robustness properties of essential values then ensure that these boundary conditions are retained as we progress through all the stages. In this chapter we make use of new, refined concepts of 'essential value' that allow for more precise forms of the necessary conditions than appears in the earlier literature.

9.1 Introduction

Our investigation of the properties of optimal strategies has, up till now, been confined to dynamic optimization problems for which the underlying time interval $[S, T]$ has been fixed. In this chapter, we address a broader class of problems in which the end-times S and T are included among the choice variables. We refer to such problems as free end-time problems. We shall show that the earlier derived necessary conditions for fixed end-point problems can be supplemented to take account of the enlargement of the set of choice variables to include the end-times. We do so in situations where the dynamic constraint takes the form either of either a differential inclusion or a controlled differential equation. The initial discussion takes as starting point the differential inclusion formulation. Consider:

$$(FT) \begin{cases} \text{Minimize } g(S, x(S), T, x(T)) \\ \text{over intervals } [S, T] \text{ and arcs } x \in W^{1,1}([S, T]; \mathbb{R}^n) \\ \text{satisfying} \\ \dot{x}(t) \in F(t, x(t)) \text{ a.e. } t \in [S, T], \\ (S, x(S), T, x(T)) \in C. \end{cases}$$

Here $g : \mathbb{R} \times \mathbb{R}^n \times \mathbb{R} \times \mathbb{R}^n \to \mathbb{R}$ is a given function, $F : \mathbb{R} \times \mathbb{R}^n \rightsquigarrow \mathbb{R}^n$ is a given multifunction, and $C \subset \mathbb{R} \times \mathbb{R}^n \times \mathbb{R} \times \mathbb{R}^n$ is a given set.

Now, an arc $x \in W^{1,1}([S, T]; \mathbb{R}^n)$ will usually be denoted $([S, T], x)$, to emphasize the underlying time-interval $[S, T]$. In the present context, an arc $([S, T], x)$, where x is an F trajectory on $[S, T]$ that satisfies the end-point constraint $(S, x(S), T, x(T)) \in C$, is referred to as an *admissible F trajectory*.

Important special cases of (FT) are minimum time problems, in which the time duration of a manoeuvre is the quantity we aim to minimize. Many practical dynamic optimization problems are of this nature. For example, a strategy for the attitude control of an orbiting satellite is 'bang-bang' control, based on elimination of the deviation of the satellite's orientation from its nominal value in minimum time. Dynamic optimization problems associated with evasion or pursuit are often minimum time problems—escape from, or catch, your adversary as quickly as possible.

9.1 Introduction

Take a minimizer $([\bar S, \bar T], \bar x)$ for (FT). Then $\bar x$ is a minimizer for the related dynamic optimization problem, in which the end-times are frozen at $\bar S$ and $\bar T$. So the minimizer certainly satisfies fixed end-time necessary conditions, such as those derived in Chap. 8. But additional information about the multipliers $\lambda \geq 0$ and $p \in W([\bar S, \bar T]; \mathbb{R}^n)$, featuring in the fixed end-time necessary conditions, is required to take account of the extra degrees of freedom which have been introduced into the optimization problem. The addition information comes in the form of boundary conditions on the Hamiltonian

$$H(t, x, p) := \max_{v \in F(t,x)} p \cdot v \tag{9.1.1}$$

(evaluated along $t \to (\bar x(t), p(t))$. These boundary conditions are implicit in a generalized transversality condition for free end-time dynamic optimization problems, to be derived in this chapter:

$$(-\xi_0, p(\bar S), \xi_1, -p(\bar T)) \in \lambda \partial g(\bar S, \bar x(\bar S), \bar T, \bar x(\bar T)) + N_C(\bar S, \bar x(\bar S), \bar T, \bar x(\bar T)) \tag{9.1.2}$$

in which

$$\xi_0 = H(t, \bar x(t), p(t))|_{t=\bar S} \text{ and } \xi_1 = H(t, \bar x(t), p(t))|_{t=\bar T}.$$

In one extreme case (fixed end-times), C and g are of the form

$$C = \{(\bar S, x_0, \bar T, x_1) : (x_0, x_1) \in \tilde C\} \text{ and } g(S, x_0, T, x_1) = \tilde g(x_0, x_1),$$

for some set $\tilde C \subset \mathbb{R}^n \times \mathbb{R}^n$, fixed interval $[\bar S, \bar T]$ and function $\tilde g : \mathbb{R}^n \times \mathbb{R}^n \to \mathbb{R}$. Here, (9.1.2) reduces to the familiar 'fixed end-time' transversality condition

$$(p(\bar S), -p(\bar T)) \in \lambda \partial \tilde g(\bar x(\bar S), \bar x(\bar T)) + N_{\tilde C}(\bar x(\bar S), \bar x(\bar T))$$

and, appropriately, conveys no information about the optimal end-times.

In another extreme case (unconstrained end-times), C and g are expressible as

$$C = \{(\alpha, x_0, \beta, x_1) : (\alpha, \beta) \in \mathbb{R}^2, (x_0, x_1) \in \tilde C\} \text{ and } g(S, x_0, T, x_1) = \tilde g(x_0, x_1),$$

for some $\tilde C \subset \mathbb{R}^n \times \mathbb{R}^n$ and some $\tilde g : \mathbb{R}^n \times \mathbb{R}^n \to \mathbb{R}$. In this case the boundary conditions on the Hamiltonian are decoupled from the generalized transversality condition and are simply

$$H(t, \bar x(t), p(t))|_{t=\bar S} = 0 \text{ and } H(t, \bar x(t), p(t))|_{t=\bar T} = 0.$$

The interpretation of these boundary conditions on the Hamiltonian is not entirely straightforward however, for the following reason. The set of state trajectories associated with the differential inclusion $\dot x(t) \in F(t, x(t))$ is unaffected by arbitrary modifications on a null-set of the 'point to multifunction' mapping $t \to F(t, .)$. We can therefore regard the dynamic optimization problem as an optimization problem over state trajectories corresponding to an equivalence class of $F(t, .)$'s which differ only on a nullset. The boundary conditions on the Hamiltonian are meaningless in this context, because the values of $t \to H(t, \bar{x}(t), p(t))$ at $t = \bar{S}$ and $t = \bar{T}$ are not the same across F's in an equivalence class.

When the time dependence of $F(t, x)$ is in some sense Lipschitz continuous, the dilemma of how to define boundary conditions on the Hamiltonian is resolved by showing that the equivalence class of functions almost everywhere equal to $t \to H(t, \bar{x}(t), p(t))$ has a continuous representative r. We can then express the boundary condition in terms of r, i.e. condition (9.1.2) is satisfied with

$$(\xi_0, \xi_1) = (r(\bar{S}), r(\bar{T})).$$

The analysis supplies additional information about the Hamiltonian evaluated along $t \to (\bar{x}(t), p(t))$ (or, more precisely, about r): this reduces to the well-known condition that the Hamiltonian is a.e. equal to a constant in the case when F is independent of t.

For measurably time dependent data, there is no convenient representative of the equivalence class, in terms of which the boundary conditions on the Hamiltonian can conveniently be expressed. In this more general setting, we take a different approach. This involves replacing point evaluation of the Hamiltonian by another operation, namely calculating the 'essential values' of the Hamiltonian. Taking essential values or, to be more precise sub and super essential values, is a generalization of point evaluation which, significantly, is unaffected by modifications on a null-set.

We derive conditions that are satisfied not merely by minimizers for (FT) but by $W^{1,1}$ local minimizers. In the present circumstances, the notion of a '$W^{1,1}$ local minimizer' must be modified, to take account of the fact that the underlying time interval $[S, T]$ is a choice variable.

Given a multifunction $B : (-\infty, +\infty) \rightsquigarrow \mathbb{R}^n$, we say that an admissible F trajectory $([\bar{S}, \bar{T}], x)$ is a $W^{1,1}$ *local minimizer for* (FT) *relative to* B if there exists some $\beta > 0$ such that

$$g(S, x(S), T, x(T)) \geq g(\bar{S}, \bar{x}(\bar{S}), \bar{T}, \bar{x}(\bar{T}))$$

for all admissible F trajectories $([S, T], x)$ satisfying the following properties: $\dot{x}(t) \in \dot{\bar{x}}(t) + B(t)$ for a.e. $t \in [\bar{S}, \bar{T}] \cap [S, T]$ and

$$d(([S, T], x), ([\bar{S}, \bar{T}], \bar{x})) \leq \beta. \qquad (9.1.3)$$

Here, d is the following distance function on the space of absolutely continuous, $\mathbb{R}^n$ valued functions x on a finite time interval: given two such functions $([S, T], x)$ and $([S', T'], x')$

$$d(([S, T], x), ([S', T'], x')) := |S - S'| + |T - T'|$$
$$+ |x(S) - x'(S')| + \int_{S \wedge S'}^{T \vee T'} |\dot{x}_e(s) - \dot{x}'_e(s)| ds, \quad (9.1.4)$$

in which, as usual, $S \wedge S' := \min\{S, S'\}$ and $T \vee T' := \max\{T, T'\}$. Here x_e denotes the extension to $(-\infty, +\infty)$ of the function $x : [S, T] \to \mathbb{R}^n$ on the finite interval $[S, T]$, defined as follows:

$$x_e(t) := \begin{cases} x(S) & \text{if } t < S \\ x(t) & \text{if } t \in [S, T] \\ x(T) & \text{if } t > T. \end{cases}$$

Notice that, according to this definition, $\dot{x}_e(t) = 0$, for a.e. $t \notin [S, T]$.

The subscript e indicating 'extension to $(-\infty, +\infty)$' is often omitted, for brevity, for instance in integrals where the domains of functions defining the integrand are smaller than the interval of integration. In all such cases, it is understood that the functions have been replaced by their extensions, as defined above, prior to evaluation of the integral; for example, the integral in (9.1.4) is written $\int_{S \wedge S'}^{T \vee T'} |\dot{x}(t) - \dot{x}'(t)| dt$. The distance function (9.1.4) is a metric on the space of absolutely continuous, $\mathbb{R}^n$ valued functions on finite intervals, which we refer to as the $W^{1,1}$ metric on this space. Obviously, it coincides with the standard $W^{1,1}$ metric when evaluated at two functions that have the same domain.

9.2 Lipschitz Time Dependence

In this section we derive necessary conditions for the free end-time problem (FT), when a Lipschitz continuity hypothesis is invoked, regarding the time dependence of $F(t, x)$. The key feature of the analysis is a transformation of the independent variable, which generates information about the Hamiltonian evaluated along the minimizing state trajectory and associated costate trajectory. This we now briefly review.

Suppose that $([\bar{S}, \bar{T}], \bar{x})$ is a $W^{1,1}$ local minimizer (FT). Consider a new *fixed end-time* problem (we shall call it the 'transformed' problem):

$$(C)' \begin{cases} \text{Minimize } g(\tau(\bar{S}), y(\bar{S}), \tau(\bar{T}), y(\bar{T})) \\ \text{over } (\tau, y) \in W^{1,1}([\bar{S}, \bar{T}]; \mathbb{R}^{1+n}) \text{ satisfying} \\ (\dot{\tau}(t), \dot{y}(t)) \in \{(w, wv) : w \in [1-\rho, 1+\rho], v \in F(\tau(t), y(t))\}, \text{ a.e. } t \in [\bar{S}, \bar{T}], \\ (\tau(\bar{S}), y(\bar{S}), \tau(\bar{T}), y(\bar{T})) \in C. \end{cases}$$

Here, $\rho \in (0, 1/2)$ is a given number. The significance of $(C)'$ is that there exists a transformation $\mathcal{G}$ which carries absolutely continuous arcs (τ, y) satisfying the constraints of problem (C)' into admissible F trajectories for the original problem (FT). Furthermore this transformation preserves the value of the cost. To be precise, the transformation is

$$\mathcal{G}(\tau, y) = ([\tau(\bar{S}), \tau(\bar{T})], y \circ \psi^{-1}),$$

in which $\psi : [\bar{S}, \bar{T}] \to [S, T)]$ is the strictly increasing function

$$\psi(s) = \tau(\bar{S}) + \int_{\bar{S}}^{s} w(\sigma) d\sigma, \text{ for } s \in [\bar{S}, \bar{T}],$$

$S = \tau(\bar{S})$ and $T = \psi(\bar{T})$. Here, $w : [\bar{S}, \bar{T}] \to [1 - \rho, 1 + \rho]$ is some measurable function such that

$$(\dot{\tau}(s), \dot{y}(s)) \in w(s)(\{1\} \times F(\tau(s), y(s))) \quad \text{a.e. } s \in [\bar{S}, \bar{T}].$$

It follows that, since $(\bar{\tau}(s) \equiv s, \bar{x})$ transforms into $([\bar{S}, \bar{T}], \bar{x})$, $(\bar{\tau}, \bar{x})$ is a minimizer for fixed end-time problem (C)'.

Under appropriate hypotheses on the data for (FT), the data for $(C)'$ satisfy the hypotheses for application of the necessary conditions of Chap. 8. These supply a cost multiplier $\lambda \geq 0$ and adjoint arc components $-r$ and p, corresponding to the state components τ and y. Of various conditions satisfied by r and p, particular interest attaches to the Weierstrass and transversality conditions:

$$-r(t) + p(t) \cdot \dot{\bar{x}}(t) = \qquad (9.2.1)$$
$$\max\{w(-r(t) + p(t) \cdot v) : v \in F(t, \bar{x}(t)), w \in [1 - \rho, 1 + \rho]\} \text{ a.e. } t$$

and

$$(-r(\bar{S}), p(\bar{S}), r(\bar{T}), -p(\bar{T})) \in$$
$$\lambda \partial g(\bar{S}, \bar{x}(\bar{S}), \bar{T}, \bar{x}(\bar{T})) + N_C(\bar{S}, \bar{x}(\bar{S}), \bar{T}, \bar{x}(\bar{T})). \qquad (9.2.2)$$

Fixing $w = 1$, we recover from (9.2.1) the Weierstrass Condition

$$p(t) \cdot \dot{\bar{x}}(t) = \max\{p(t) \cdot v : v \in F(t, \bar{x}(t))\}$$

for (FT). However, further information is now obtained if we fix $v = \dot{\bar{x}}(t)$, namely

$$p(t) \cdot \dot{\bar{x}}(t) (= H(t, \bar{x}(t), p(t))) = r(t) \quad \text{a.e.},$$

(where $H(t, x, p)$ is the Hamiltonian (9.1.1)). We see from this relation and (9.2.2) that the transversality condition provides information about the 'boundary values' of the Hamiltonian evaluated along $(\bar{x}, p)$. The boundary values are interpreted as

9.2 Lipschitz Time Dependence

the end-points of some absolutely continuous function which coincides with the Hamiltonian almost everywhere.

Of course this analysis is justified only when the data for $(C)'$ satisfy the hypotheses for application of suitable necessary conditions of optimality. Since the independent variable t in (FT) becomes a state variable τ in $(C)'$ and since currently available necessary conditions require the data to be, in some generalized sense, 'differentiable' with respect to the state variable, the analysis will effectively be limited to problems for which the data for (FT) is regular regarding also its time dependence.

The following theorem provides necessary conditions of optimality for free end-time problems. The derivation is based on the preceding ideas, but modified to allow for F's which are unbounded. This complicating feature necessitates constraining w to lie in some *time dependent* set $w(t) \in [1 - \rho(t), 1 + \rho(t)])$. The use of proof techniques that treat time as a state-like variable affects the nature of the regular velocity set, in terms of which the Weierstrass condition is expressed. The appropriate definition of the regular velocity set, written $\tilde{\Omega}_0(t)$ and called the (t, x)-regular velocity set, now becomes

$$\tilde{\Omega}_0(t) := \{e \in F(t, \bar{x}(t)) : F \text{ is pseudo} - \text{Lipschitz continuous near}((t, \bar{x}(t)), e)\}.$$
(9.2.3)

(Notice that, in this definition, points in the regular velocity set are identified via the pseudo-Lipschitz continuity property of F jointly w.r.t. the t and x variables, not the x variable alone.)

Theorem 9.2.1 (Free-Time Necessary Conditions: Lipschitz Time Dependence) *Let $([\bar{S}, \bar{T}], \bar{x})$ be a $W^{1,1}$ local minimizer for (FT) (relative to $B \equiv \mathbb{R}^n$) such that $\bar{T} - \bar{S} > 0$. Assume that the following hypotheses are satisfied:*

(G1): g is Lipschitz continuous on a neighbourhood of $(\bar{S}, \bar{x}(\bar{S}), \bar{T}, \bar{x}(\bar{T}))$ and C is a closed set,

(G2): $F(t, x)$ is non-empty for all $(t, x) \in \mathbb{R} \times \mathbb{R}^n$ and $\text{Gr } F$ is closed,

(G3): There exist $\epsilon > 0$ and a measurable function $R : [\bar{S}, \bar{T}] \to (0, \infty) \cup \{+\infty\}$ (a 'radius function') such that the following conditions are satisfied:

  (a): (Pseudo-Lipschitz Continuity) There exists $k \in L^1(\bar{S}, \bar{T})$ such that

$$F(t', x') \cap (\dot{\bar{x}}(t) + R(t)\overset{\circ}{\mathbb{B}}) \subset F(t'', x'') + k_F(t) |(t', x') - (t'', x'')|\mathbb{B},$$

  for all $(t', x'), (t'', x'') \in (t, \bar{x}(t)) + \epsilon\mathbb{B}$, a.e. $t \in [\bar{S}, \bar{T}]$,

  (b): (Tempered Growth) There exist $r \in L^1(\bar{S}, \bar{T}), r_0 > 0$ and $\gamma \in (0, 1)$ such that $r_0 \leq r(t), \gamma^{-1}r(t) \leq R(t)$ a.e. $t \in [\bar{S}, \bar{T}]$ and

$$F(t', x') \cap (\dot{\bar{x}}(t) + r(t)\mathbb{B}) \neq \emptyset, \text{ for all } (t', x') \in (t, \bar{x}(t)) + \epsilon\mathbb{B},$$
a.e. $t \in [\bar{S}, \bar{T}]$.

Then there exist absolutely continuous arcs $p \in W^{1,1}([\bar{S}, \bar{T}]; \mathbb{R}^n)$ and $a \in W^{1,1}([\bar{S}, \bar{T}]; \mathbb{R})$ and a number $\lambda \geq 0$ such that

(i): $(p, \lambda) \neq (0, 0)$,
(ii): $(-\dot{a}(t), \dot{p}(t)) \in \text{co}\{(\zeta, \eta) \, : \, ((\zeta, \eta), p(t)) \in N_{\text{Gr}\,F}((t, \bar{x}(t)), \dot{\bar{x}}(t))\}$,
   a.e. $t \in [\bar{S}, \bar{T}]$,
(iii): $(-a(\bar{S}), p(\bar{S}), a(\bar{T}), -p(\bar{T})) \in \lambda \partial g(\bar{S}, \bar{x}(\bar{S}), \bar{T}, \bar{x}(\bar{T})) + N_C(\bar{S}, \bar{x}(\bar{S}), \bar{T}, \bar{x}(\bar{T}))$,
(iv): $p(t) \cdot \dot{\bar{x}}(t) \geq p(t) \cdot v$ for all $v \in \overline{\text{co}}\,\tilde{\Omega}_0(t)$, a.e. $t \in [\bar{S}, \bar{T}]$,
(v): $a(t) = p(t) \cdot \dot{\bar{x}}(t)$ a.e. $t \in [\bar{S}, \bar{T}]$.

If the initial time $\bar{S}$ is fixed, i.e. $C = \{(\bar{S}, x_0, T, x_1) \, : \, (x_0, T, x_1) \in \tilde{C}\}$, for some closed set $\tilde{C} \subset \mathbb{R}^n \times \mathbb{R} \times \mathbb{R}^n$, and $g(\bar{S}, x_0, T, x_1) = \tilde{g}(x_0, T, x_1)$ for some $\tilde{g} : \mathbb{R}^n \times \mathbb{R} \times \mathbb{R}^n \to \mathbb{R}$, then the above conditions are satisfied when (iii) is replaced by

(iii)': $(p(\bar{S}), a(\bar{T}), -p(\bar{T})) \in \lambda \partial g(\bar{x}(\bar{S}), \bar{T}, \bar{x}(\bar{T})) + N_{\tilde{C}}(\bar{x}(\bar{S}), \bar{T}, \bar{x}(\bar{T}))$.

The assertions of the theorem can be similarly modified when the final time $\bar{T}$ is fixed.

Remark

(The Autonomous Case) An important aspect of this theorem is its implications for autonomous problems (problems for which $F(t, x)$ is independent of t; write $F(x)$ in place of $F(t, x)$). In this case, condition (ii) of the theorem implies that the adjoint arc p satisfies

$$\dot{p}(t) \in \text{co}\{\eta \, : \, (\eta, p(t)) \in N_{\text{Gr}\,F}(\bar{x}(t), \dot{\bar{x}}(t))\} \quad \text{a.e.}$$

(the Euler Lagrange Inclusion) and $a(t) = c$ a.e. for some $c \in \mathbb{R}$. Then, from (v),

$$p(t) \cdot \dot{\bar{x}}(t) = c \quad \text{a.e.} \ t \in [\bar{S}, \bar{T}].$$

Information about the constant c is provided by the transversality condition (iii).

Recall how, in Chap. 8, Theorem 8.4.3 (the Euler Lagrange inclusion) served as a starting point for deriving necessary conditions of optimality, under other hypotheses, listed in Proposition 8.5.1, that replace the 'pseudo Lipschitz continuity and tempered growth' hypothesis in the theorem statement. Analogues of each these hypotheses can be employed, in place of hypothesis (G3) in Theorem 9.2.1. In the following corollary, which is an immediate consequence of Proposition 8.5.1 of Chap. 8, we focus merely on the version of Theorem 9.2.1 that features the third alternative hypothesis, namely (G3)***, of Proposition 8.5.1.

Corollary 9.2.2 *The assertions of Theorem 9.2.1 remain valid when hypothesis (G3) is replaced by:*

(G3)***:   There exist numbers $\alpha > 0$ and $\epsilon > 0$, and non-negative measurable functions k and β such that k and $t \to \beta(t)k^\alpha(t)$ are integrable and, for each $N \geq 0$,

9.2 Lipschitz Time Dependence

$$F(t'', x'') \cap (\dot{\bar{x}}(t) + N\mathbb{B}) \subset F(t', x') + k_N(t)|(t'', x'') - (t', x')|\mathbb{B},$$

$$\text{for all } (t'', x''), (t', x') \in (t, \bar{x}(t)) + \epsilon\mathbb{B}, \text{ a.e. } t \in [\bar{S}, \bar{T}]$$

where $k_N(t) := k(t) + \beta(t)N^\alpha$.

In this case, the Weierstrass condition (iv) becomes

$$p(t) \cdot \dot{\bar{x}}(t) \geq p(t) \cdot v, \quad \text{for all } v \in F(t, \bar{x}(t)), \quad \text{a.e. } t \in [\bar{S}, \bar{T}].$$

Proof of Theorem 9.2.1 Assume that $([\bar{S}, \bar{T}], \bar{x})$ is a $W^{1,1}$ local minimizer for (FT) (relative to $B \equiv \mathbb{R}^n$) and (G1)–(G3) are satisfied. We know then that there exists $\beta > 0$ such that $([\bar{S}, \bar{T}], \bar{x})$ is a minimizer with respect to all F trajectories $([S, T], x)$ satisfying the constraints of (FT) and

$$d(([S, T], x), ([\bar{S}, \bar{T}], \bar{x})) \; (= |S - \bar{S}| + |T - \bar{T}| + |x(S) - \bar{x}(\bar{S})|$$

$$+ \int_{S \wedge \bar{S}}^{T \vee \bar{T}} |\dot{x}(t) - \dot{\bar{x}}(t)| dt) \leq \beta. \tag{9.2.4}$$

Take $\alpha > 0$ such that

$$\alpha < \left(\frac{1}{2}\right) \wedge \left(\frac{1}{2r_0}\right) \text{ and } \frac{1+\alpha}{1-\alpha r_0} \leq \frac{2}{1+\gamma}. \tag{9.2.5}$$

Here, $\gamma > 0$ and $r_0 > 0$ are the parameters of hypothesis (G3)(b). Define

$$\rho_\alpha(t) := \frac{\alpha r_0}{1 + |\dot{\bar{x}}(t)|}, \quad t \in [\bar{S}, \bar{T}].$$

Note that $\rho_\alpha(t) \in (0, 1/2)$. Now consider the following fixed end-time dynamic optimization problem (the 'transformed' problem):

$$(\tilde{T}) \begin{cases} \text{Minimize } g(\tau(\bar{S}), y(\bar{S}), \tau(\bar{T}), y(\bar{T})) \\ \text{over } (\tau, y) \in W^{1,1}([\bar{S}, \bar{T}]; \mathbb{R}^{1+n}) \text{ satisfying} \\ (\dot{\tau}(s), \dot{y}(s)) \in \tilde{F}(s, \tau(s), y(s)) \quad \text{a.e., } s \in [\bar{S}, \bar{T}] \\ (\tau(\bar{S}), y(\bar{S}), \tau(\bar{T}), y(\bar{T})) \in C. \end{cases}$$

Here $\tilde{F} : \mathbb{R} \times \mathbb{R} \times \mathbb{R}^n \rightsquigarrow \mathbb{R} \times \mathbb{R}^n$ is the multifunction

$$\tilde{F}(s, \tau, y) := \{(w, wv) : w \in [1 - \rho_\alpha(s), 1 + \rho_\alpha(s)], \; v \in F(\tau, y)\}.$$

Claim: $(\bar{\tau}(s) := s, \bar{x})$ is a $W^{1,1}$ local minimizer for the fixed end-time problem (T).

We confirm the claim. Take $\beta' > 0$, and let (τ, y) be any admissible $\tilde{F}$ trajectory such that

$$|(\tau(\bar{S}), y(\bar{S})) - (\bar{\tau}(\bar{S}), \bar{x}(\bar{S}))| + \int_{\bar{S}}^{\bar{T}} |(\dot{\tau}(s), \dot{y}(s)) - (1, \dot{\bar{x}}(s))| ds \leq \beta'. \qquad (9.2.6)$$

In consequence of the generalized Filippov selection theorem (Theorem 2.3.14), there exist measurable functions w and v such that

$$w(s) \in [1 - \rho_\alpha(s), 1 + \rho_\alpha(s)] \text{ and } v(s) \in F(\tau(s), y(s)) \text{ a.e.,}$$

$$\dot{\tau}(s) = w(s), \quad \dot{y}(s) = w(s)v(s) \text{ a.e..}$$

Consider now the transformation $\psi : [\bar{S}, \bar{T}] \to [S, T]$,

$$\psi(s) := \tau(\bar{S}) + \int_{\bar{S}}^{s} w(\sigma) d\sigma, \qquad (9.2.7)$$

$S := \tau(\bar{S})$ and $T := \tau(\bar{S}) + \int_{\bar{S}}^{\bar{T}} w(\sigma) d\sigma$. The function ψ is strictly increasing and invertible, and ψ and ψ^{-1} are absolutely continuous functions with essentially bounded derivatives. It follows that $x(t) := (y \circ \psi^{-1})(t)$ is an absolutely continuous function. We deduce from Lemma 9.7.1 of the Appendix that $\beta' > 0$ in (9.2.6) and $\alpha > 0$ in (9.2.5) can be chosen independent of our selection of (τ, y) satisfying (9.2.6), such that

$$d(([S, T], x), ([\bar{S}, \bar{T}], \bar{x})) \leq \beta, \qquad (9.2.8)$$

where $\beta > 0$ is as in (9.2.4). Take any $t \in [S, T]$ and let $s := \psi^{-1}(t)$. We deduce from Fubini's theorem and the chain rule that, for each $t \in [S, T]$,

$$\int_S^t (v \circ \psi^{-1})(t') dt' = \int_{\bar{S}}^s v(s') w(s') ds' = \int_{\bar{S}}^s \dot{y}(s') ds'$$
$$= \int_{\bar{S}}^s \frac{d}{ds}(x \circ \psi)(s') ds' = \int_{\bar{S}}^s (\dot{x} \circ \psi)(s') \frac{d}{ds} \psi(s') ds'$$
$$= \int_{\bar{S}}^s (\dot{x} \circ \psi)(s') w(s') ds' = \int_S^t \dot{x}(t') dt'.$$

It follows that $\dot{x}(t) = v \circ \psi^{-1}(t)$ a.e.. But then, since ψ and ψ^{-1} map Lebesgue null-sets into null-sets,

$$\dot{x}(t) = (v \circ \psi^{-1})(t) \in F((\tau \circ \psi^{-1})(t), (y \circ \psi^{-1})(t)) = F(t, x(t)) \text{ a.e..}$$

From the fact that $(\tau(s), y(s)) = (\psi(s), (x \circ \psi)(s))$ for all $s \in [\bar{S}, \bar{T}]$ we deduce that $S = \tau(\bar{S})$, $T = \tau(\bar{T})$, $x(S) = y(\tau(\bar{S}))$ and $x(T) = y(\tau(\bar{T}))$. We have shown that $([S, T], x)$ is an admissible F trajectory for problem (FT). Since

$$(S, x(S), T, x(T)) = (\tau(\bar{S}), y(\bar{S}), \tau(\bar{T}), y(\bar{T})),$$

9.2 Lipschitz Time Dependence

we can deduce from the $W^{1,1}$ local optimality of $([\bar{S}, \bar{T}], \bar{x})$ that

$$g(\tau(\bar{S}), y(\bar{S}), \tau(\bar{T}), y(\bar{T})) = g(S, x(S), T, x(T))$$
$$\geq g(\bar{S}, \bar{x}(S), \bar{T}, \bar{x}(T)) = g(\bar{\tau}(\bar{S}), \bar{x}(\bar{S}), \bar{\tau}(\bar{T}), \bar{x}(\bar{T})).$$

Since (τ, y) was an arbitrary admissible $\tilde{F}$ trajectory satisfying (9.2.6), we conclude that $(\bar{\tau}, \bar{x})$ is a $W^{1,1}$ local minimizer for (T). The claim is confirmed. Simple changes to the preceding analysis (in which the change of independent variable is now taken to be $t := \bar{S} + \int_{\bar{S}}^{s} w(\sigma) d\sigma$) validate the claim, also in the case when the left end-time $\bar{S}$ is fixed, described at the end of the theorem statement. The case when the right end-time is fixed is similarly treated.

We aim next to applying the necessary conditions of Chap. 8 to $(\tau(s) \equiv s, \bar{x})$, regarded as a $W^{1,1}$ local minimizer. First, let us check that the data for (T) satisfies the hypotheses (G1), (G2) and (G3) of Theorem 8.4.3 of Chap. 8.

The data for problem (T) satisfies hypotheses (G1), (G2) of Theorem 8.4.3, in simple consequence of our assumption that the data of problem (FT) satisfies hypotheses (G1) and (G2) of Theorem 9.2.1. In the analysis below, we will refer to the integrable Lipschitz bound k, the radius function R and the parameters r, $r_0, \gamma \in (0, 1)$ appearing in (G3) of Theorem 9.2.1. We now show that (G3) is also satisfied when, in this hypothesis, the function r and parameters r_0 and $\epsilon > 0$ remain the same, but R, γ and k are replaced by $\tilde{R}$, $\tilde{\gamma}$ and $\tilde{k}$, where

$$\tilde{R}(t) = \left(\frac{1+\gamma}{2}\right) R(t), \quad \tilde{\gamma} := \frac{2\gamma}{1+\gamma} \quad \text{and} \quad \tilde{k}(t) = \sqrt{2}(1 + \alpha r_0) k(t).$$

Notice that $\tilde{\gamma} \in (0, 1)$ and $\tilde{k} \in L^1$; furthermore

$$r_0 \leq r(t) \leq \gamma R(t) \leq \gamma \left(\frac{2}{1+\gamma}\right) \tilde{R}(t) = \tilde{\gamma} \tilde{R}(t).$$

Consider (G3)(a). For a.e. $t \in [\bar{S}, \bar{T}]$ take $(\tau, x), (\tau', x') \in (t, \bar{x}(t)) + \epsilon \mathbb{B}$ and $(w, wv) \in \tilde{F}(t, \tau, x) \cap ((1, \dot{\bar{x}})(t)) + \tilde{R}(t) \overset{\circ}{\mathbb{B}})$. Then $|(w, wv) - (1, \dot{\bar{x}})(t)| < \tilde{R}(t)$. It follows that $|v - w^{-1} \dot{\bar{x}}(t)| < w^{-1} \tilde{R}(t)$. Hence

$$|v - \dot{\bar{x}}(t)| \leq |w^{-1} - 1| |\dot{\bar{x}}(t)| + w^{-1} \tilde{R}(t).$$

But $|w^{-1} - 1| |\dot{\bar{x}}(t)| \leq \frac{1}{w} \left(\frac{\alpha r_0}{1+|\dot{\bar{x}}(t)|}\right) \times |\dot{\bar{x}}(t)| \leq \frac{\alpha r_0}{w}$ and $|w^{-1}| \leq (1 - \alpha r_0)^{-1}$. Noting also that $r_0 \leq \tilde{R}(t)$, we see that

$$|v - \dot{\bar{x}}(t)| < \frac{1+\alpha}{1-\alpha r_0} \tilde{R}(t) \leq \frac{2}{1+\gamma} \tilde{R}(t) = R(t).$$

Since $v \in F(\tau, x) \cap (\dot{\bar{x}}(t) + R(t)\overset{\circ}{\mathbb{B}})$, we know from (G3)(a) that there exists $v' \in F(\tau', x')$ such that $|v' - v| \le k(t)(|x - x'| + |\tau - \tau'|)$. But then, since $w \in [1 - \rho_\alpha(t), 1 + \rho_\alpha(t)]$, (w, wv') is an element in $\tilde{F}(t, \tau', x')$ and

$$|(w, wv') - (w, wv)| = w|v' - v| \le |w|k(t)(|x - x'| + |\tau - \tau'|)$$
$$\le (1 + \alpha r_0)k(t)(|x - x'| + |\tau - \tau'|) \le \tilde{k}(t)|(\tau, x) - (\tau', x')|.$$

The existence of a point (w, wv') satisfying this relation implies

$$\tilde{F}(t, \tau, x) \cap ((1, \dot{\bar{x}}(t)) + \tilde{R}(t)\overset{\circ}{\mathbb{B}}) \subset \tilde{F}(t, \tau', x') + \tilde{k}(t)|(\tau, x) - (\tau', x')|\mathbb{B}.$$

We have shown that (G3)(a) is satisfied with the integrable Lipschitz bound $\tilde{k}(t)$.

Take any $(\tau, x) \in (t, \bar{x}(t)) + \epsilon\mathbb{B}$. Then, by (G3)(b), we know that there exists $v \in F(\tau, x)$ such that $|v - \dot{\bar{x}}(t)| \le r(t)$. But then the element $(1, v) \in \tilde{F}(t, \tau, x)$ satisfies $|(1, v) - (1, \dot{\bar{x}}(t))| \le r(t)$. We have shown

$$\tilde{F}(t, \tau, x) \cap (\dot{\bar{x}}(t) + r(t)\mathbb{B}) \ne \emptyset.$$

Since, as we have already observed, $r_0 \le r(t) \le \tilde{\gamma}\tilde{R}(t)$, we have confirmed that (G3)(b) is also satisfied.

The preceding analysis justifies applying Theorem 8.4.3 to (T), with reference to the $W^{1,1}$ local minimizer $(\bar{\tau} \equiv 1, \bar{x})$. We deduce that there exist $\lambda \ge 0$, and costate trajectory components $-a$ and p such that

$$(\lambda, (a, p)) \ne (0, (0, 0)), \tag{9.2.9}$$

$$(-\dot{a}(s), \dot{p}(s)) \in \text{co}\{(\eta_1, \eta_2) : ((\eta_1, \eta_2), (-a(s), p(s)) \tag{9.2.10}$$
$$\in N_{\text{Gr}\tilde{F}(s,.,.)}((s, \bar{x}(s)), (1, \dot{\bar{x}}(s)))\}, \text{ a.e. } s \in [\bar{S}, \bar{T}]$$

and

$$w[-a(s) + p(s) \cdot v] \le -a(s) + p(s) \cdot \dot{\bar{x}}(s)$$
$$\text{for all } (w, v) \in \Omega_0(t), \text{ a.e. } t \in [\bar{S}, \bar{T}]. \tag{9.2.11}$$

Here, $\Omega_0 : [\bar{S}, \bar{T}] \rightsquigarrow \mathbb{R}^{1+n}$ (the regular velocity set for $\tilde{F}$ relative to $(\bar{\tau} \equiv t, \bar{x})$) is the multifunction

$$\Omega_0(t) := \{(e_0, e_1) \in \tilde{F}(t, t, \bar{x}(t)) : \tilde{F}(t, ., .) \text{ is pseudo Lipschitz}$$
$$\text{continuous near } ((t, \bar{x}(t)), (e_0, e_1))\} \text{ for } t \in [\bar{S}, \bar{T}].$$

9.2 Lipschitz Time Dependence

We deduce from the tranversality condition that

$$(-a(\bar{S}), p(\bar{S}), +a(\bar{T}), -p(\bar{T})) \in N_C(\bar{S}, \bar{x}(\bar{S}), \bar{T}, \bar{x}(\bar{T}))$$
$$+ \lambda \partial g(\bar{S}, \bar{x}(\bar{S}), \bar{T}, \bar{x}(\bar{T})). \tag{9.2.12}$$

A careful analysis of the Euler Lagrange inclusion (9.2.10) and applying basic properties of proximal normals, yields the relation

$$(-\dot{a}(t), \dot{p}(t)) \in \mathrm{co}\{(\eta_1, \eta_2) : ((\eta_1, \eta_2), p(t))$$
$$\in N_{\mathrm{Gr}F}((t, \bar{x}(t)), \dot{\bar{x}}(t))\} \text{ a.e..} \tag{9.2.13}$$

Let us now explore the implications of the Weierstrass condition (9.2.11). It is a straightforward exercise to deduce from hypothesis (G3) that

$$\{(w, w\dot{\bar{x}}(t)) \mid |w - 1| \le \rho_\alpha(t)\} \subset \Omega_0(t), \text{ for a.e. } t \in [\bar{S}, \bar{T}] \tag{9.2.14}$$

and

$$\{(1, v) : v \in \tilde{\Omega}_0(t)\} \subset \Omega_0(t), \text{ for a.e. } t \in [\bar{S}, \bar{T}], \tag{9.2.15}$$

in which $\tilde{\Omega}_0(t)$ is (t, x)-regular velocity set defined by (9.2.3).

It follows from (9.2.11) and (9.2.15) and the linearity of $v \to p(s) \cdot v$ that, for a.e. $t \in [\bar{S}, \bar{T}]$,

$$p(t) \cdot \dot{\bar{x}}(t) = \max\{p(t) \cdot v : v \in \overline{\mathrm{co}}\,\tilde{\Omega}_0(t)\}. \tag{9.2.16}$$

On the other hand, it follows from (9.2.11) and (9.2.14) that $w \to w(-a(t) + p(t) \cdot \dot{\bar{x}}(t))$ is maximized over $w \in [1 - \rho_\alpha(t), 1 + \rho_\alpha(t)]$ at $w = 1$. We conclude that

$$a(t) = p(t) \cdot \dot{\bar{x}}(t), \text{ a.e. } t \in [\bar{S}, \bar{T}]. \tag{9.2.17}$$

This last condition implies that $a \equiv 0$ if $p \equiv 0$. Since $(\lambda, a, p) \ne (0, 0, 0)$, it follows that

$$(\lambda, p) \ne (0, 0). \tag{9.2.18}$$

Reviewing relations (9.2.12), (9.2.13), (9.2.16) and (9.2.18), we see that the assertions of the theorem have been validated, in the case that $([\bar{S}, \bar{T}], \bar{x})$ is a $W^{1,1}$ local minimizer, under hypotheses (G1)–(G3).

The final assertions of the theorem, concerning problems in which the left end-time is fixed are proved by a simple modification of the preceding analysis, in which the transformation $\psi(s) = \bar{S} + \int_{\bar{S}}^{s} w(s')ds'$ replaces (9.2.7) and $(\tau(\bar{S}), y(\bar{S}), \tau(\bar{T}), y(\bar{T})) \in \{\bar{S}\} \times \tilde{C}$ replaces the end-point constraint in (T). □

9.3 Essential Values

The operation of taking the 'essential value' (or, to be more specific, the sub or super essential value) of a given real valued function on the real line is a generalization of point evaluation of a continuous function. Its most important property is that the essential values of a function are unaltered if the function is adjusted on a set of Lebesgue measure zero. This will be exploited to interpret necessary conditions for free end-time dynamic optimization problems with measurably time dependent data, specifically to make sense of boundary conditions on the Hamiltonian.

Definition 9.3.1 Take an closed interval $[S, T] \subset \mathbb{R}$, an integrable function $a : [S, T] \to \mathbb{R}$ and a point $t \in (S, T)$.

(a): The sub essential value of f at $\bar{t}$ is the set

$$\text{sub-ess}_{t \to \bar{t}} a(t) :=$$

$$\left\{ \zeta \in \mathbb{R} : \exists t_i \to \bar{t} \text{ and } \zeta_i \to \zeta \text{ s.t.} \right.$$

$$\left. \limsup_{\epsilon \downarrow 0} \epsilon^{-1} \int_{t_i-\epsilon}^{t_i} a(s)ds \leq \zeta_i \leq \liminf_{\epsilon \downarrow 0} \int_{t_i}^{t_i+\epsilon} a(s)ds, \text{ for each } i \right\}.$$

(b): The super essential value of a at t is the set

$$\text{super-ess}_{t \to \bar{t}} a(t) :=$$

$$\left\{ \zeta \in \mathbb{R} : \exists t_i \to \bar{t} \text{ and } \zeta_i \to \zeta \text{ s.t.} \right.$$

$$\left. \limsup_{\epsilon \downarrow 0} \epsilon^{-1} \int_{t_i}^{t_i+\epsilon} a(s)ds \leq \zeta_i \leq \liminf_{\epsilon \downarrow 0} \int_{t_i-\epsilon}^{t_i} a(s)ds, \text{ for each } i \right\}.$$

If $a(t, x)$ is a function of two variables, $\text{sub-ess}_{t \to \bar{t}} a(t, x)$ denotes the sub essential value of $t \to a(t, x)$, for fixed x, etc.

The operations of taking sub and super essential values generate multifunctions

$$t \to \text{sub-ess}_{t \to \bar{t}} a(t) \quad \text{and} \quad t \to \text{super-ess}_{t \to \bar{t}} a(t)$$

taking values closed, possibly unbounded, sets. These multifunctions are of interest in dynamic optimization primarily because of the role they play in describing the sub- and super subdifferentials of an indefinite integral function.

Proposition 9.3.2 Take an interval $[S, T] \subset \mathbb{R}$, an integrable function $a : [S, T] \to \mathbb{R}$ and a point $\bar{t} \in (S, T)$. Fix any $S' \in [S, T]$ and define the function $\psi : [S, T] \to \mathbb{R}$ to be

9.3 Essential Values

$$\psi(t) := \int_{S'}^{t} a(s)ds.$$

Then

$$\partial \psi(\bar{t}) = \underset{t \to \bar{t}}{\text{sub-ess}}\, a(t) \quad \text{and} \quad -\partial(-\psi)(\bar{t}) = \underset{t \to \bar{t}}{\text{super-ess}}\, a(t).$$

In words, 'the limiting subdifferential of ψ at $\bar{t}$ is the sub essential value of a at $\bar{t}$' and 'the limiting super differential of ψ at $\bar{t}$ is the super essential value of a at $\bar{t}$'. Observe that, since the sub essential value of a and the super essential value of a at a point $\bar{t} \in (S, T)$ do not depend on $S' \in [S, T]$, the sub- and super differentials of the indefinite integral function ψ at $\bar{t}$ do not depend on the choice of the point $S' \in [S, T]$ used to define ψ, either.

Proof We provide the proof of only the first assertion, since that of the second is the same (following a reversal of inequalities).

It can be deduced from the definition of strict subdifferential, Proposition 4.4.3 and Theorem 4.6.3 that $\zeta \in \partial \psi(\bar{t})$ if and only if there exists sequences $t_i \to \bar{t}$ and $\zeta_i \to \zeta$ such that, for each i,

$$\liminf_{\epsilon \downarrow 0} \epsilon^{-1}(\psi(t_i + \epsilon) - \psi(t_i)) \geq \zeta_i \quad \text{and} \quad \liminf_{\epsilon \downarrow 0} \epsilon^{-1}(\psi(t_i - \epsilon) - \psi(t_i)) \geq -\zeta_i.$$

It now follows from the definition of ψ that this last condition is equivalent to

$$\limsup_{\epsilon \downarrow 0} \epsilon^{-1} \int_{t_i - \epsilon}^{t_i} a(s)ds \leq \zeta_i \leq \liminf_{\epsilon \downarrow 0} \epsilon^{-1} \int_{t_i}^{t_i + \epsilon} a(s)ds.$$

We have shown that $\zeta \in \partial \psi(\bar{t})$ if and only if $\zeta \in \underset{t \to \bar{t}}{\text{sub-ess}}\, a(t)$.

□

The following proposition provides further information about sub and super essential values of a. It tells us, in particular, that these sets are unaffected by changing the values of a on a (Lebesgue) nullset. It also provides estimates of sub and super essential values in terms of a multifunction which, for some applications, is simpler to calculate while supplying an adequate approximation.

Proposition 9.3.3 *Take an interval $[S, T] \subset \mathbb{R}$, and two integrable functions $a, b : [S, T] \to \mathbb{R}$ and $\bar{t} \in (S, T)$. Then*

(a): If $a(t) = b(t)$ a.e., then

$$\underset{t \to \bar{t}}{\text{sub-ess}}\, a(t) = \underset{t \to \bar{t}}{\text{sub-ess}}\, b(t) \quad \text{and} \quad \underset{t \to \bar{t}}{\text{super-ess}}\, a(t) = \underset{t \to \bar{t}}{\text{super-ess}}\, b(t),$$

(b): $$\text{sub-ess } a(t) \subset [\inf_{\epsilon \downarrow 0}\Big(\operatorname*{ess\ inf}_{t\in[\bar{t}-\epsilon,\bar{t}+\epsilon]} a(t)\Big), \sup_{\epsilon \downarrow 0}\Big(\operatorname*{ess\ sup}_{t\in[\bar{t}-\epsilon,\bar{t}+\epsilon]} a(t)\Big)]$$
$$t\to\bar{t}$$

and

$$\text{super-ess } a(t) \subset [\inf_{\epsilon \downarrow 0}\Big(\operatorname*{ess\ inf}_{t\in[t-\epsilon,t+\epsilon]} a(t)\Big), \sup_{\epsilon \downarrow 0}\Big(\operatorname*{ess\ sup}_{t\in[t-\epsilon,t+\epsilon]} a(t)\Big)],$$
$$t\to\bar{t}$$

(c): *If a is essentially bounded on a neighbourhood of $\bar{t}$ then*

$$\operatorname*{co\ sub-ess}_{t\to\bar{t}} a(t) = \operatorname*{co\ super-ess}_{t\to\bar{t}} a(t)$$

$$= [\inf_{\epsilon \downarrow 0}\Big(\operatorname*{ess\ inf}_{t\in[\bar{t}-\epsilon,\bar{t}+\epsilon]} a(t)\Big), \sup_{\epsilon \downarrow 0}\Big(\operatorname*{ess\ sup}_{t\in[\bar{t}-\epsilon,\bar{t}+\epsilon]} a(t)\Big)].$$

Proof Properties (a) and (b) are straightforward consequences of the definitions. We attend only to the proof of (c). Write $\zeta^+ := \lim_{\epsilon \downarrow 0}\Big(\operatorname*{ess\ sup}_{s\in[\bar{t}-\epsilon,\bar{t}+\epsilon]} a(s)\Big)$ and $\zeta^- := \lim_{\epsilon \downarrow 0}\Big(\operatorname*{ess\ inf}_{s\in[\bar{t}-\epsilon,\bar{t}+\epsilon]} a(s)\Big)$. ($\zeta^+$ and ζ^- are finite numbers because of the 'essential boundedness' hypothesis.) Denote by S the Lebesgue points of $a : [S, T] \to \mathbb{R}$. Since S has full Lebsgue measure, there exists a sequence of points $t_i \to \bar{t}$ such that $t_i \in S$ for each i and $a(t_i) \to \zeta^+$ as $i \to \infty$.

For each i, by the Lebesgue point property, $a(t_i) = \lim_{\epsilon \downarrow 0} \epsilon^{-1} \int_{t_i}^{t_i+\epsilon} a(s)ds = \lim_{\epsilon \downarrow 0} \epsilon^{-1} \int_{t_i-\epsilon}^{t_i} a(s)ds$. But then, for each i,

$$\limsup_{\epsilon \downarrow 0} \epsilon^{-1} \int_{t_i-\epsilon}^{t_i} a(s)ds \le a(t_i) \le \liminf_{\epsilon \downarrow 0} \epsilon^{-1} \int_{t_i}^{t_i+\epsilon} a(s)ds$$

It follows that $\zeta^+ = \lim_{i\to\infty} a(t_i) \in \operatorname*{sub-ess}_{t\to\bar{t}} a(t)$. Likewise we show that $\zeta^- \in \operatorname*{sub-ess}_{t\to\bar{t}} a(t)$.

But then $[\zeta^-, \zeta^+] \subset \operatorname{co}\big(\operatorname*{sub-ess}_{t\to\bar{t}} a(t)\big)$. Combining this inclusion with property (b), we conclude that $[\zeta^-, \zeta^+] = \operatorname{co}\big(\operatorname*{sub-ess}_{t\to\bar{t}} a(t)\big)$. The proof that $[\zeta^-, \zeta^+] = \operatorname{co}\big(\operatorname*{super-ess}_{t\to\bar{t}} a(t)\big)$ is similar.

□

The following closure properties of the sub and super essential values follow directly from the above representation of the limiting sub- and super essential values and the analogous closure properties of sub- and super-differentials.

Proposition 9.3.4 *Take an interval $[S, T] \subset \mathbb{R}$, a set $A \subset \mathbb{R}^n$ and a function $a : [S, T] \times A \to \mathbb{R}$. Assume that*

(a): $t \to a(t, x)$ is integrable on $[S, T]$ for each $x \in A$,
(b): $x \to a(t, x)$ is continuous on A, uniformly over $t \in [S, T]$.

Then, for each $(\bar{t}, x) \in (S, T) \times A$,

(i): $\operatorname*{sub-ess}_{t \to \bar{t}} a(t, x)$ is closed,

(ii): For any sequences $t_i \to \bar{t}$, $x_i \xrightarrow{A} x$ and $\zeta_i \to \zeta$ such that $\zeta_i \in \operatorname*{sub-ess}_{t \to t_i} a(t_i, x_i)$ for each i, we have

$$\zeta \in \operatorname*{sub-ess}_{t \to \bar{t}} a(t, x),$$

(iii): For any sequences $t_i \to \bar{t}$, $x_i \xrightarrow{A} x$ and $\zeta_i \to \zeta$ such that $\zeta_i \in \operatorname*{super-ess}_{t \to t_i} a(t_i, x_i)$ for each i, we have

$$\zeta \in \operatorname*{super-ess}_{t \to \bar{t}} a(t, x).$$

9.4 Measurable Time Dependence

In this section, we derive necessary conditions for (FT), under hypotheses which require the differential inclusion to have right side $F(t, x)$ which is merely measurable with regard to time. The motivation for treating the measurable time dependence case is partly to unify the theory of necessary conditions for fixed and free end-time dynamic optimization problems. A framework which requires the dynamic constraint to be merely measurable with respect to time is widely adopted for fixed end-time problems. Why should extra regularity be required for free end-time problems?

But there are also practical reasons for developing a theory of free end-time problems, which allows the 'dynamic constraint' to be discontinuous with respect to time. Dynamic optimization problems arising in resource economics, for example, typically require us to find harvesting/investment strategies to minimize a cost which involves an integral cost of the form

$$-\int_0^T c(t, x(t), u(t))dt.$$

Here, c is a given function representing the rate of return on harvesting effort and investment. It is natural in this context to consider problems in which T is a choice variable; the company can choose the harvesting period. Special cases are of interest, in which c is discontinuous with respect to time, to take account, for example, of abrupt changes in interest rates (reflecting, perhaps, penalties incurred for late completion). When the integral cost term is absorbed into the dynamics by means of 'state augmentation' the resulting dynamics will be discontinuous with respect to time. Furthermore, since we can expect an optimal strategy to terminate at the instant when there is an abrupt, unfavourable change in the rate of return, it is not satisfactory to develop a theory in which it is assumed that the optimal end-time occurs at a point of continuity of the data; rather we should allow for the possibility that discontinuities and optimal end-times interact.

The necessary conditions in the following theorem bring together versions of the Euler Lagrange inclusion, the Weierstrass condition and the transversality condition for free end-time problems, where the dynamic constraint takes the form of a differential inclusion. Here, the formulation of the Weierstrass condition involves, for a.e. $t \in [\bar{S}, \bar{T}]$, the regular velocity set $\Omega_0(t)$ for the differential inclusion, relative to the nominal F trajectory $\bar{x}$:

$$\Omega_0(t) := \{e \in F(t, \bar{x}(t)) \,:\, F(t, .) \text{ is pseudo Lipschitz}$$
$$\text{continuous near } (\bar{x}(t), e)\}, \text{ for } t \in [\bar{S}, \bar{T}].$$

and the transversality condition is expressed in terms of essential values of the Hamiltonian at the optimal end-times. The Hamiltonian $H(t, x, p)$ is, as usual, taken to be:

$$H(t, x, p) = \sup_{v \in F(t,x)} p \cdot v.$$

Theorem 9.4.1 (Free End-Time Generalized Euler Lagrange Condition: Measurable Time Dependence) *Take a measurable multifunction $B : \mathbb{R} \rightsquigarrow \mathbb{R}^n$ such that $B(t)$ is open for a.e. $t \in \mathbb{R}$. Let $\bar{x}$ be a $W^{1,1}$ local minimizer for (FT) relative to B. Assume that, for some $\epsilon > 0$ and $\sigma \in (0, (\bar{T} - \bar{S})/2)$, the following hypotheses are satisfied:*

(G1): g is Lipschitz continuous on a neighbourhood of $(\bar{S}, \bar{x}(\bar{S}), \bar{T}, \bar{x}(\bar{T}))$ and C is a closed set,

(G2): $F(t, x)$ is a non-empty for all $(t, x) \in \mathbb{R} \times \mathbb{R}^n$, $\operatorname{Gr} F(t, .)$ is closed for each $t \in \mathbb{R}$ and F is $\mathcal{L} \times \mathcal{B}^n$ measurable,

(G3): There exists a measurable function $R : [\bar{S} - \sigma, \bar{T} + \sigma] \to (0, \infty) \cup \{+\infty\}$ (a 'radius function'), such that $R(t)\mathring{\mathbb{B}} \subset B(t)$ a.e. $t \in [\bar{S} - \sigma, \bar{T} + \sigma]$ and the following conditions are satisfied,

(a): (Pseudo-Lipschitz Continuity) There exists $k \in L^1(\bar{S} - \sigma, \bar{T} + \sigma)$ such that

$$F(t, x') \cap (\dot{\bar{x}}(t) + R(t)\mathring{\mathbb{B}}) \subset F(t, x) + k(t)|x' - x|\mathbb{B},$$
$$\text{for all } x, x' \in \bar{x}(t) + \epsilon\mathbb{B}, \text{ a.e. } t \in [\bar{S} - \sigma, \bar{T} + \sigma],$$

9.4 Measurable Time Dependence

(b): *(Tempered Growth) There exist* $r \in L^1(\bar{S} - \sigma, \bar{T} + \sigma), r_0 > 0$ *and* $\gamma \in (0, 1)$ *such that* $r_0 \leq r(t), \gamma^{-1}r(t) \leq R(t),$ *a.e.* $t \in [\bar{S} - \sigma, \bar{T} + \sigma]$, *and*

$$F(t, x) \cap (\dot{\bar{x}}(t) + r(t)\mathbb{B}) \neq \emptyset \text{ for all } x \in \bar{x}(t) + \epsilon\mathbb{B}, \text{ a.e. } t \in [\bar{S} - \sigma, \bar{T} + \sigma],$$

(G4): *There exists* $c_F \geq 0$ *such that, for a.e.* $t \in [\bar{S} - \sigma, \bar{S} + \sigma] \cup [\bar{T} - \sigma, \bar{T} + \sigma]$,

$$F(t, x) \subset c_F \mathbb{B}, \quad \text{for all } x \in \bar{x}(t) + \epsilon\mathbb{B}. \tag{9.4.1}$$

Then there exist $p \in W^{1,1}([\bar{S}, \bar{T}]; \mathbb{R}^n)$ *and* $\lambda \geq 0$ *such that*

(i): $(p, \lambda) \neq (0, 0)$,
(ii): $\dot{p}(t) \in \text{co}\{\zeta : (\zeta, p(t)) \in N_{\text{Gr}F(t,\cdot)}(\bar{x}(t), \dot{\bar{x}}(t))\}, \quad$ *a.e.* $t \in [\bar{S}, \bar{T}]$,
(iii): *there exist* $\xi_0 \in \text{sub-ess}_{t \to \bar{S}} H(t, \bar{x}(\bar{S}), p(\bar{S}))$ *and*
$\xi_1 \in \text{super-ess}_{t \to \bar{T}} H(t, \bar{x}(\bar{T}), p(\bar{T}))$ *such that*

$$(-\xi_0, p(\bar{S}), \xi_1, -p(\bar{T})) \in \lambda \partial g(\bar{S}, \bar{x}(\bar{S}), \bar{T}, \bar{x}(\bar{T})) + N_C(\bar{S}, \bar{x}(\bar{S}), \bar{T}, \bar{x}(\bar{T})),$$

(iv): $p(t) \cdot \dot{\bar{x}}(t) \geq p(t) \cdot v, \quad$ *for all* $v \in \overline{\text{co}}\,(\Omega_0(t) \cap (\dot{\bar{x}}(t) + B(t)))$,
$\quad$ *a.e.* $t \in [\bar{S}, \bar{T}]$.

If the initial time $\bar{S}$ *is fixed, i.e.* $C = \{(\bar{S}, x_0, T, x_1) : (x_0, T, x_1) \in \tilde{C}\}$, *for some closed set* $\tilde{C} \subset \mathbb{R}^n \times \mathbb{R} \times \mathbb{R}^n$, *and* $g(S, x_0, T, x_1) = \tilde{g}(x_0, T, x_1)$ *for some* $\tilde{g} : \mathbb{R}^n \times \mathbb{R} \times \mathbb{R}^n \to \mathbb{R}$, *then the above conditions are satisfied when (iii) is replaced by: there exists* $\xi_1 \in \text{super-ess}_{t \to \bar{T}} H(t, \bar{x}(\bar{T}), p(\bar{T}))$ *such that*

(iii)': $(p(\bar{S}), \xi_1, -p(\bar{T})) \in \lambda \partial \tilde{g}(\bar{x}(\bar{S}), \bar{T}, \bar{x}(\bar{T})) + N_{\tilde{C}}(\bar{x}(\bar{S}), \bar{T}, \bar{x}(\bar{T}))$,

and when condition (9.4.1) in (G4) is satisfied only for a.e. $t \in [\bar{T} - \sigma, \bar{T} + \sigma]$. *The assertions of the theorem can be similarly modified when the final time* $\bar{T}$ *is fixed.*

We can use Proposition 8.5.1 of Chap. 8 immediately to obtain alternative versions of Theorem 9.4.1 in which the 'pseudo Lipschitz continuity' with 'tempered growth' hypothesis (G3) is replaced by alternative, stronger hypotheses. We record here only one such corollary of Theorem 9.4.1, based on the third hypothesis in Proposition 8.5.1, where (G3) is replaced by the requirement that F is pseudo Lipschitz continuous for arbitrary constant radius functions R and that the associated integrable Lipschitz bounds have polynomial growth w.r.t. R.

Corollary 9.4.2 *The assertions of Theorem 9.4.1 remain valid when hypothesis (G3) is replaced by:*

(G3)***: *There exist numbers* $\alpha > 0$ *and* $\epsilon > 0$, *and non-negative measurable functions* k *and* β *such that* k *and* $t \to \beta(t)k^\alpha(t)$ *are integrable and, for each* $N \geq 0$

$$F(t, x'') \cap (\dot{\bar{x}}(t) + N\mathbb{B}) \subset F(t, x') + k_N(t)|x'' - x'|\mathbb{B},$$

for all $x', x'' \in \bar{x}(t) + \epsilon\mathbb{B}$ *and for a.e.* $t \in [\bar{S} - \sigma, \bar{T} + \sigma]$,
where $k_N(t) := k(t) + \beta(t)N^\alpha$.

In this case, the Weierstrass condition (iv) becomes

$$p(t) \cdot \dot{\bar{x}}(t) \geq p(t) \cdot v, \quad \text{for all } v \in F(t, \bar{x}(t)), \quad \text{a.e. } t \in [\bar{S}, \bar{T}].$$

9.5 Proof of Theorem 9.4.1

Preliminaries
We show, by means of a straightforward adaptation of the proof of Proposition 8.6.2 on Chap. 8, that, without loss of generality, (G1) and (G3) can be replaced by the stronger hypotheses:

(G1)′: g is Lipschitz continuous on $\mathbb{R} \times \mathbb{R}^n \times \mathbb{R} \times \mathbb{R}^n$ and C is a closed set,
(G3)′: There exist $\epsilon > 0$, $R \in (0, \infty)$, $\gamma \in (0, 1)$ and $k \in L^1(\bar{S} - \sigma, \bar{T} + \sigma)$ such that $R\overset{\circ}{\mathbb{B}} \subset B(t)$ a.e. $t \in [\bar{S} - \sigma, \bar{T} + \sigma]$ and the following conditions are satisfied,

(a): $F(t, x') \cap (\dot{\bar{x}}(t) + R\overset{\circ}{\mathbb{B}}) \subset F(t, x) + k(t)|x - x'|\mathbb{B}$, for all $x, x' \in \bar{x}(t) + \epsilon\mathbb{B}$, a.e. $t \in [\bar{S} - \sigma, \bar{T} + \sigma]$,
(b): $F(t, x) \cap (\dot{\bar{x}}(t) + \gamma R\mathbb{B}) \neq \emptyset$ for all $x \in \bar{x}(t) + \epsilon\mathbb{B}$, a.e. $t \in [\bar{S} - \sigma, \bar{T} + \sigma]$.

We can arrange, by translation to the origin, that the $W^{1,1}$ local minimizer of interest is $([\bar{S}, \bar{T}], \bar{x} \equiv 0)$. We can also ensure, by redefining g on a neighbourhood of the optimal end-points, that g is (globally) Lipschitz continuous.

To prove Theorem 9.4.1, we will follow a by now familiar pattern of analysis, a key step in which is to apply known necessary conditions to a minimizer $([S', T'], x')$ for a preliminary dynamic optimization problem, with a perturbation term in the cost that results from a choice of metric topology on the space of admissible F trajectories and application of Ekeland's theorem. For the free end-time problems here under consideration, we employ the metric d (see (9.1.4)), which gives rise to a cost perturbation term $\alpha\, d(([S, T], x), ([S', T'], x'))$, where $\alpha > 0$ and

$$d(([S, T], x), ([S', T'], x')) = |S - S'| + |T - T'| + |x(S) - x'(S')|$$
$$+ \int_{S \wedge S'}^{T \vee T'} |\dot{x}_e(s) - \dot{x}'_e(s)|ds.$$

(We emphasize through our use of the subscript e notation that function extensions are used, to interpret the integral on the right side.) Here we encounter a difficulty: the perturbed cost is not of the right form for the application of earlier-derived necessary conditions, because it involves the integral $\int_{S \wedge S'}^{T \vee T'}$, not $\int_S^T$. The following lemma, whose simple proof we omit, overcomes this difficulty:

9.5 Proof of Theorem 9.4.1

Lemma 9.5.1 *Take an interval $[\bar{S}, \bar{T}]$ and a number $\sigma \in (0, (\bar{T} - \bar{S})/2)$. Now take intervals $[S', T']$, $[S, T] \subset [\bar{S} - \sigma, \bar{T} + \sigma]$ and absolutely continuous functions $x' : [S', T'] \to \mathbb{R}^n$ and $x : [S, T] \to \mathbb{R}^n$. Assume that, for some $\bar{c} > 0$,*

$$|\dot{x}_e(t)| \leq \bar{c} \text{ and } |\dot{\bar{x}}_e(t)| \leq \bar{c} \text{ for a.e. } t \in [\bar{S} - \sigma, \bar{S} + \sigma] \cup [\bar{T} - \sigma, \bar{T} + \sigma].$$

Then

$$\int_{S' \wedge S}^{T' \vee T} |\dot{x}'_e(s) - \dot{x}_e(s)| ds \leq \int_{S \vee S'}^{T \wedge T'} |\dot{x}'(s) - \dot{x}(s)| ds + 2\bar{c}\Big(|S - S'| + |T - T'|\Big).$$

Since $[S \vee S', T \wedge T'] \subset [S, T]$, the lemma implies that, for absolutely continuous functions x, x' with the specified domains and satisfying the specified conditions,

$$d(([S, T], x), ([S', T'], x')) \leq \int_{S}^{T} |\dot{x}(s) - \dot{x}'_e(s)| ds + (1 + 2\bar{c})\Big(|S - S'| + |T - T'|\Big).$$

Since this relation holds with equality when $([S, T], x) = ([S', T'], x')$, we conclude that $([S', T'], x')$ remains a minimizer when the perturbation term in the cost is replaced by a scaled version of

$$\int_{S}^{T} |\dot{x}'(s) - \dot{x}(s)| ds + 2\bar{c}\Big(|S' - S| + |T' - T|\Big).$$

Step 1 (A Free End-Point Problem) Our goal is to extend the Euler Lagrange-type conditions of Chap. 8 to cover problems with free end-times, now expressed in terms of essential values of the Hamiltonian at the optimal end-times. This is accomplished in several steps. First we derive necessary conditions for a problem in which there are no constraints on the end-times and end-states, and the differential inclusion $F(t, .)$ is 'globally' Lipschitz continuous, with integrable Lipschitz constant, in the sense that the pseudo-Lipschitz condition is satisfied with radius function $R \equiv +\infty$.

Take a function $g : \mathbb{R} \times \mathbb{R}^n \times \mathbb{R} \times \mathbb{R}^n \to \mathbb{R}$ and a multifunction $F : \mathbb{R} \times \mathbb{R}^n \rightsquigarrow \mathbb{R}^n$. Consider the following free end-point dynamic optimization problem:

$$(FEP) \begin{cases} \text{Minimize } g(S, x(S), T, x(T)) \\ \text{over } [S, T] \subset \mathbb{R} \text{ and } x \in W^{1,1}([S, T]; \mathbb{R}^n) \text{ satisfying} \\ \dot{x}(t) \in F(t, x(t)) \text{ a.e. } t \in [S, T]. \end{cases}$$

Proposition 9.5.2 *Let $([S', T'], x')$ be a $W^{1,1}$ local minimizer for (FEP). Assume that, for some $\epsilon' > 0$ and $\sigma' \in (0, (T' - S')/2)$,*

(FEP1)': g is Lipschitz continuous,
(FEP2)': $F(t, x)$ is a non-empty, closed set for all $(t, x) \in \mathbb{R} \times \mathbb{R}^n$, $\text{Gr } F(t, .)$ is closed for each $t \in \mathbb{R}$ and F is $\mathcal{L} \times \mathcal{B}^n$ measurable,

(FEP3)': There exists a non-negative function $k \in L^1(S' - \sigma', T' + \sigma')$ such that, for a.e. $t \in [S' - \sigma', T' + \sigma']$,

$$F(t, x) \subset F(t, y) + k(t)|x - y|\mathbb{B}, \quad \text{for all } x, y \in x'(t) + \epsilon'\mathbb{B},$$

(FEP4)': There exists $c_F \geq 0$ such that, for a.e. $t \in [S' - \sigma', S' + \sigma'] \cup [T' - \sigma', T' + \sigma']$,

$$F(t, x) \subset c_F \mathbb{B}, \quad \text{for all } x \in x'(t) + \epsilon'\mathbb{B}.$$

Then there exists $p \in W^{1,1}([S', T']; \mathbb{R}^n)$ such that

(A): $\dot{p}(t) \in \text{co}\{\zeta : (\zeta, p(t)) \in N_{Gr\, F(t,\cdot)}(x'(t), \dot{x}'(t))\}$, a.e. $t \in [S', T']$,
(B): $p(t) \cdot \dot{x}'(t) \geq p(t) \cdot v$ for all $v \in F(t, x'(t))$, a.e. $t \in [S', T']$,
(C): there exist $\xi_0 \in \text{sub-ess}_{t \to S'} H(t, x'(S'), p(S'))$ and
$\xi_1 \in \text{super-ess}_{t \to T'} (t, x'(T'), p(T'))$ such that

$$(-\xi_0, p(S'), \xi_1, -p(T')) \in \partial g(S', x'(S'), T', x'(T')).$$

Proof Assume that $\beta > 0$ is a parameter such that the $W^{1,1}$ local minimizer $([S', T'], x')$ is minimizing w.r.t. admissible F trajectories satisfying

$$d(([S, T], x), ([S', T'], x')) \leq \beta.$$

(Here d is the distance function (9.1.4)).

Hypothesis Reduction We now show that, without loss of generality, we can restrict attention to the special case when (FEP3)' is replaced by the following stronger hypothesis, in which the non-negative function k is replaced by a constant:

(FEP3)*: There exists a constant $k_F > 0$ such that, for a.e. $t \in [S' - \sigma', T' + \sigma']$,

$$F(t, x) \subset F(t, y) + k_F |x - y|\mathbb{B},$$

for all $x, y \in x'(t) + \epsilon'\mathbb{B}$.

Indeed, it is not restrictive to assume that $k(t) \geq 1$ for a.e. $t \in [S' - \sigma', T' + \sigma']$ (otherwise we can always replace k by $k \vee 1$). Consider the change of the independent variable $s = \hat{\sigma}(t)$, where $\hat{\sigma} : \mathbb{R} \to \mathbb{R}$ is defined by

$$\hat{\sigma}(t) := \begin{cases} S' - \sigma' + \frac{T' - S' + 2\sigma'}{\|k\|_{L^1}} \int_{S' - \sigma'}^t k(\tau) d\tau & \text{for all } t \in [S' - \sigma', T' + \sigma'] \\ t & \text{otherwise.} \end{cases}$$

(9.5.1)

9.5 Proof of Theorem 9.4.1

(Here, $\|k\|_{L^1} = \|k\|_{L^1(S'-\sigma', T'+\sigma')}$.) Notice that $\hat{\sigma}$ is a strictly increasing function such that $\hat{\sigma}(S' - \sigma') = S' - \sigma'$ and $\hat{\sigma}(T' + \sigma') = T' + \sigma'$. We can then consider a new problem (FEP) in which F is replaced by

$$\hat{F}(s, y) := \frac{\|k\|_{L^1}}{(T' - S' + 2\sigma') \times (k \circ \hat{\sigma}^{-1})(s)} F(\hat{\sigma}^{-1}(s), y), \quad \text{for all } (s, y) \in \mathbb{R} \times \mathbb{R}^n,$$

and for which ($[\hat{S} := \hat{\sigma}^{-1}(S'), \hat{T} := \hat{\sigma}^{-1}(T')], \hat{x} := x' \circ \hat{\sigma}^{-1}$) is a $W^{1,1}$ local minimizer. Observe also that $\hat{F}$ satisfies (FEP3)* with $k_F = \frac{\|k\|_{L^1}}{T'-S'+2\sigma'}$. It is straightforward to deduce the necessary conditions for the original problem, with reference to the minimizer ($[S', T'], x'$), via the inverse change of independent variable $s = \hat{\sigma}(t)$, in which the Lagrange multiplier $\hat{p} \in W^{1,1}([\hat{S}, \hat{T}]; \mathbb{R}^n)$ arising in the new problem is replaced by $p = \hat{p} \circ \hat{\sigma}$ in the original one.

Consider first the special case in which g has the structure

$$g(S, x_0, T, x_1) = \tilde{g}(S, x_0, T, x_1) + e(S, x_0, T, x_1),$$

for some twice differentiable function $\tilde{g}$ and some Lipschitz continuous function e, with Lipschitz constant k_e. We shall prove a coarser version of the necessary conditions in the proposition statement, in which condition (C) is replaced by

$$(-\xi_0, p(S'), \xi_1, -p(T')) \in \nabla \tilde{g}(S', x'(S'), T', x'(T')) + 2\sqrt{2} k_e (1 + c_F^2)^{\frac{1}{2}} \mathbb{B}.$$

Note, first of all, that x' is a $W^{1,1}$ local minimizer for a version of problem (FEP) in which the underlying time interval is fixed at $[S', T']$. The hypotheses of Theorem 8.4.3 of Chap. 8 are satisfied. We deduce existence of $p \in W^{1,1}([S', T']; \mathbb{R}^n)$ satisfying conditions (A) and (B) and the following transversality condition:

$$(p(S'), -p(T')) \in \nabla_{x_0, x_1} \tilde{g}(S', x'(S'), T', x'(T')) + k_e \mathbb{B}. \tag{9.5.2}$$

We immediately see that $d(([S', T' - s], x'), ([S', T'], x')) = s$. But then, for $s \in [0, \beta]$, by $W^{1,1}$ local optimality of $([S', T'], x')$, $([S', T' - s], x')$ must have cost not less than $([S', T'], x')$. It follows that

$$\tilde{g}(S', x'(S'), T' - s, x'(T' - s)) + e(S', x'(S'), T' - s, x'(T' - s))$$
$$\geq \tilde{g}(S', x'(S'), T', x'(T')) + e(S', x'(S'), T', x'(T')).$$

Since $\tilde{g}$ is a C^2 function, there exists $r_1 > 0$ such that for all s sufficiently small

$$0 \geq \tilde{g}(S', x'(S'), T', x'(T')) - \tilde{g}(S', x'(S'), T' - s, x'(T' - s)) - k_e(1 + c_F^2)^{\frac{1}{2}} s$$
$$= \int_{T'-s}^{T'} \left(\nabla_T \tilde{g}(S', x'(S'), t, x'(t)) + \nabla_{x_1} \tilde{g}(S', x'(S'), t, x'(t)) \cdot \dot{x}'(t) \right) dt$$
$$- k_e(1 + c_F^2)^{\frac{1}{2}} s$$
$$\geq \nabla_T \tilde{g}(S', x'(S'), T', x'(T')) s - \int_{T'-s}^{T'} H(t, x'(T'), p(T')) dt$$
$$- r_1 s^2 - k_e(c_F + (1 + c_F^2)^{\frac{1}{2}}) s.$$

To derive the final inequality in the above relation, we have used the following facts: $-p(T') \in \nabla_{x_1} \tilde{g}(S', x(S'), T', x(T')) + k_e \mathbb{B}$; moreover, by (FEP3)* and (FEP4)', we have, for a.e. $t \in [T' - s, T']$,

$$\dot{x}'(t) \in F(t, x'(T')) + k_F |x'(t) - x'(T')|\mathbb{B} \subset F(t, x'(T')) + k_F c_{FS} \mathbb{B},$$

and so, for a.e. $t \in [T' - s, T']$,

$$p(T') \cdot \dot{x}'(t) \leq \max\{p(T') \cdot v : v \in F(t, x'(T'))\} + k_F c_F |p(T')|s$$
$$= H(t, x'(T'), p(T')) + k_F c_F |p(T')|s.$$

We have shown that

$$\begin{cases} T' \text{ is a local minimum over } [S', T'] \text{ of} \\ T \to -\nabla_T \tilde{g}(S', x'(S'), T', x'(T'))(T' - T) \\ \quad - \int_T^{T'} (-H)(t, x'(T'), p(T'))\, dt \\ \quad + r_1 (T - T')^2 + k_e (c_F + (1 + c_F^2)^{\frac{1}{2}})|T - T'|. \end{cases} \quad (9.5.3)$$

Next, select a measurable function $\xi : [T', T' + \sigma'] \to \mathbb{R}^n$ such that $\xi(t) \in F(t, x'(T'))$ a.e. and

$$p(T') \cdot \xi(t) = \max_{v \in F(t, x'(T'))} p(T') \cdot v, \quad \text{a.e. } t \in [T', T' + s'].$$

We deduce, from Filippov's existence theorem (Theorem 6.2.3) that there exist $s' \in (0, \sigma')$ and an F trajectory y on $[T', T' + s']$ such that $y(T') = x'(T')$ and

$$\int_{T'}^{T'+s} |\dot{y}(t) - \xi(t)|dt \leq (1/2) e^{k_F s} k_F c_F s^2, \quad \text{for all } s \in [0, s']. \quad (9.5.4)$$

We now construct an F trajectory $([S', T' + s'], x)$ by concatenating $([S', T'], x')$ and $([T', T' + s'], y)$. For $s' \in (0, \sigma')$ sufficiently small, $d(([S', T' + s], x), ([S', T'], x')) \leq \beta$ for all $s \in [0, s']$, by Lemma 9.5.1. In view of the $W^{1,1}$ local optimality of $([S', T'], x')$, for all $s \in [0, s']$ the cost of $([S', T' + s], x)$ exceeds that of $([S', T'], x')$. It follows that

$$0 \leq (\tilde{g} + e)(S', x'(S'), T' + s, y(T' + s)) - (\tilde{g} + e)(S', x'(S'), T', y(T'))$$
$$\leq \int_{T'}^{T'+s} \Big(\nabla_T \tilde{g}(S', x'(S'), t, x(t)) + \nabla_{x_1} \tilde{g}(S', x'(S'), t, x(t)) \cdot \dot{x}(t) \Big) dt$$
$$+ k_e (1 + c_F^2)^{\frac{1}{2}} s, \quad \text{for all } s \in [0, s'].$$

9.5 Proof of Theorem 9.4.1

Arguing as in the analysis preceding (9.5.3), we can show that, for some constant $r_2 \geq r_1$ and for all $s \in [0, s']$,

$$\int_{T'}^{T'+s} \Big(\nabla_T \tilde{g}(S', x'(S'), t, x(t)) + \nabla_{x_1} \tilde{g}(S', x'(S'), t, x(t)) \cdot \dot{x}(t)\Big) dt$$

$$\leq (\nabla_T \tilde{g}(S', x'(S'), T', x(T')))s + \int_{T'}^{T'+s} (-H)(t, x'(T'), p(T')) \, dt$$

$$+ (1 + (1/2)e^{k_F s'} k_F c_F) r_2 s^2 + k_e c_F s.$$

We have shown, for some $s' \in (0, \sigma')$,

$$\begin{cases} T' \text{ is a local minimum over } [T', T' + s'] \text{ of} \\ T \rightarrow \nabla_T \tilde{g}(S', x'(S'), T', x'(T'))(T - T') + \int_{T'}^{T}(-H)(t, x'(T'), p(T')) \, dt \\ + (1 + (1/2)e^{k_F s'} k_F c_F) r_2 |T - T'|^2 + k_e (c_F + (1 + c_F^2)^{\frac{1}{2}})|T - T'|. \end{cases} \quad (9.5.5)$$

Interpreting, as usual, the integral $\int_{T'}^{T}(-H)(t, x'(T'), p(T'))dt$ as $-\int_{T}^{T'}(-H)(t, x'(T'), p(T'))dt$ for $T < T'$, we deduce from (9.5.3) and (9.5.5) that

$$\begin{cases} T' \text{ is a local minimum over } [S', T' + s'] \text{ of} \\ T \rightarrow \nabla_T \tilde{g}(S', x'(S'), T', x'(T'))(T - T') + \int_{T'}^{T}(-H)(t, x'(T'), p(T')) \, dt \\ + (1 + (1/2)e^{k_F s'} k_F c_F) r_2 (T - T')^2 + k_e (c_F + (1 + c_F^2)^{\frac{1}{2}})|T - T'|. \end{cases}$$

Applying Proposition 9.3.2, in which we take the indefinite integral function ψ to be $\psi(t) := \int_{T'}^{t}(-H)(s, x'(T'), p(T'))ds, t \in [S', T' + s']$, noting that 0 is a limiting subgradient of a function at a minimizing point and also applying the sum rule for limiting subdifferentials of Lipschitz functions, we deduce that

$$-\nabla_T \tilde{g}(S', x'(S'), T', x'(T')) \in \underset{t \to T'}{\text{sub-ess}}\, (-H)(t, \bar{x}(T'), p(T'))$$

$$+ k_e(c_F + (1 + c_F^2)^{\frac{1}{2}}) \mathbb{B}.$$

A similar analysis, in which we take the left end-time, yields

$$-\nabla_S \tilde{g}(S', x'(S'), T', x'(T')) \in \text{sub-ess}_{t \to S'} H(t, x'(S'), p(S'))$$

$$+ k_e(c_F + (1 + c_F^2)^{\frac{1}{2}}) \mathbb{B}.$$

In view of (9.5.2) and Proposition 9.3.2, we have confirmed that there exist elements $\xi_0 \in \text{sub-ess}_{t \to S'} H(t, x'(S'), p(S'))$ and

$$\xi_1 \in \text{super-ess}_{t \to T'} H(t, x'(T'), p(T')) = -\text{sub-ess}_{t \to T'}(-H)(t, x'(T'), p(T'))$$

such that

$$(-\xi_0, p(S'), \xi_1, -p(T')) \in \nabla \tilde{g}(S', x'(S'), T', x'(T')) + 2\sqrt{2}k_e(1+c_F^2)^{\frac{1}{2}}\mathbb{B}.$$

Recall that, up to this point in the proof, we have assumed the cost function g has a special structure. We now drop this assumption; henceforth we assume that g is an arbitrary Lipschitz continuous function.

For $i = 1, 2, \ldots$ let g_i be the quadratic inf convolution of g (with parameter $\alpha = i$). Since $([S', T'], x')$ is a $W^{1,1}$ local minimizer for (FEP), there exists $\beta > 0$ such that $([S', T'], x')$ is a minimizer for (FEP), when we append the constraint $d(([S, T], x), ([S', T'], x')) \le \beta$. Consider, for each i, the following variant on (FEP), in which we replace g by g_i and add a distance constraint:

$$(FEP)_i \begin{cases} \text{Minimize } g_i(S, x(S), T, x(T)) \\ \text{over } [S, T] \subset \mathbb{R} \text{ and } x \in W^{1,1}([S, T]; \mathbb{R}^n) \text{ satisfying} \\ \dot{x}(t) \in F(t, x(t)) \text{ a.e.} \\ d(([S, T], x), ([S', T'], x')) \le \beta. \end{cases}$$

Write $\alpha_i := k_g/\sqrt{i}$, where k_g is a Lipschitz constant for g. According to the properties of quadratic inf convolutions, $([S', T'], x')$ is an α_i^2 minimizer for this problem.

We can express the preceding problem as

$$\text{Minimize } \{J_i([S, T], x) : ([S, T], x) \in \mathcal{M}\}$$

where $\mathcal{M}$ denotes the space of admissible arcs $([S, T], x)$ for problem $(FEP)_i$ and $J_i([S, T], x) := g_i(S, x(S), T, x(S))$. J_i is continuous on the closed set $\mathcal{M}$, equipped with the d metric topology. It follows then from Ekeland's theorem that there exists an element $([S_i, T_i], x_i)$ in $\mathcal{M}$ that minimizes $J_i([S, T], x) + \alpha_i d([S, T], x), ([S_i, T_i], x_i)$ over $\mathcal{M}$ and

$$d(([S_i, T_i], x_i), ([S', T'], x')) \le \alpha_i. \tag{9.5.6}$$

For i sufficiently large, $\alpha_i < \beta/2$. Then, by the triangle inequality applied to the metric d, $([S_i, T_i], x_i)$ remains a minimizer for $(FEP)_i$, when the distance constraint in the definition of $\mathcal{M}$ is replaced by $d(([S, T], x), ([S_i, T_i], x_i)) \le \beta/2$. It now follows from Lemma 9.5.1 that $([S_i, T_i], x_i)$ is a minimizer for

$$\begin{cases} \text{Minimize } g_i(S, x(S), T, x(T)) \\ \qquad + \alpha_i(\int_S^T |\dot{x}(t) - \dot{x}_i(t)|dt + |x(S) - x_i(S_i)| \\ \qquad\qquad + (1 + 2c_F)(|S - S_i| + |T - T_i|)), \\ \text{over } [S, T] \subset \mathbb{R} \text{ and } x \in W^{1,1}([S, T]; \mathbb{R}^n) \text{ satisfying} \\ \dot{x}(t) \in F(t, x(t)), \text{ a.e. } t \in [S, T], \\ d(([S, T], x), ([S_i, T_i], x_i)) \le \beta/2, . \end{cases}$$

9.5 Proof of Theorem 9.4.1

We have shown that $([S_i, T_i], x_i, y_i \equiv 0)$ is a $W^{1,1}$ local minimizer for

$$(FEP)'_i \begin{cases} \text{Minimize } g_i(S, x(S), T, x(T))) \\ \quad + \alpha_i(y(T) - y(S) + |x(S) - x_i(S_i)| \\ \quad + (1 + 2c_F)(|S - S_i| + |T - T_i|)) \\ \text{over } [S, T] \subset \mathbb{R} \text{ and } (x, y) \in W^{1,1}([S, T]; \mathbb{R}^{n+1}) \text{ satisfying} \\ (\dot{x}(t), \dot{y}(t)) \in \tilde{F}(t, x(t)) \text{ a.e.,} \end{cases}$$

in which $\tilde{F} : \mathbb{R} \times \mathbb{R}^n \rightsquigarrow \mathbb{R}^{n+1}$ is the multifunction

$$\tilde{F}(t, x) := \{(v, |v - \dot{x}_i(t)|) : v \in F(t, x)\} \text{ for } (t, x) \in \mathbb{R} \times \mathbb{R}^n.$$

By properties of quadratic inf-convolutions, there exists a quadratic function $\tilde{g}_i$ such that

$$\nabla \tilde{g}_i(S_i, x_i(S_i), T_i, x_i(T_i)) \in \partial^P g(z_i),$$

for some $z_i \in (S_i, x_i(S_i), T_i, x_i(T_i)) + (k_g/i)\mathbb{B}$, and

$$\begin{cases} \tilde{g}_i(z) \geq g_i(z) \text{ for all } z \in \mathbb{R} \times \mathbb{R}^n \times \mathbb{R} \times \mathbb{R}^n \\ \tilde{g}_i(z) = g_i(z) \text{ when } z = (S', x'(S'), T', x'(T')). \end{cases}$$

It follows that $([S_i, T_i], x_i, y_i \equiv 0)$ is also a $W^{1,1}$ local minimizer for a variant of $(FEP)'_i$, in which the quadratic function $\tilde{g}_i$ replaces g_i, namely

$$\begin{cases} \text{Minimize } \tilde{g}_i(S, x(S), T, x(T))) \\ \quad + \alpha_i(y(T) - y(S) + |x(S) - x_i(S_i)| + (1 + 2c_F)(|S - S_i| + |T - T_i|)) \\ \text{over } [S, T] \subset \mathbb{R} \text{ and } (x, y) \in W^{1,1}([S, T]; \mathbb{R}^{n+1}) \text{ satisfying} \\ (\dot{x}(t), \dot{y}(t)) \in \tilde{F}(t, x(t)) \text{ a.e..} \end{cases}$$

This problem is an example of the special case of (FEP), for which the previous analysis of Step 1 provides necessary conditions of optimality. The relevant hypotheses are satisfied. To interpret these conditions, we note that, given a proximal normal vector $(\xi, (\eta, \eta^0)) \in N^P_{\text{Gr } \tilde{F}(t,.)}(\tilde{x}, (\tilde{v}, |\tilde{v} - \dot{x}_i(t)|))$, there exists $M > 0$ such that $(\tilde{x}, \tilde{v})$ is a local minimizer of the function

$$(x, v) \to -\xi \cdot x - \eta \cdot v - \eta^0 |v - \dot{x}_i(t)| + M(|x - \tilde{x}|^2 + |v - \tilde{v}|^2 + (|v - \dot{x}_i(t)| - |\tilde{v} - \dot{x}_i(t)|)^2)$$

over $(x, v) \in \text{Gr } F(t, .)$. But then, by the exact penalization theorem (Theorem 3.2.1), there exists a constant $K > 0$ such that $(\tilde{x}, \tilde{v})$ is a local minimizer of the function

$$(x, v) \to -\xi \cdot x - \eta \cdot v - \eta^0 |v - \dot{x}_i(t)| + K d_{\text{Gr } F(t,.)}(x, v)$$
$$+ M(|x - \tilde{x}|^2 + |v - \tilde{v}|^2 + (|v - \dot{x}_i(t)| - |\tilde{v} - \dot{x}_i(t)|)^2).$$

Since $N_{\operatorname{Gr} F(t,.)} = \{\lambda \partial d_{\operatorname{Gr} F(t,.)} : \lambda \geq 0\}$, it follows that

$$(\xi, \eta) \in N_{\operatorname{Gr} F(t,.)}(\tilde{x}, \tilde{v}) + |\eta_0|\mathbb{B}. \tag{9.5.7}$$

Consideration of limits of proximal normal vectors to $\operatorname{Gr} \tilde{F}$ yields the information that

$$(\xi, \eta, \eta_0) \in N_{\operatorname{Gr} \tilde{F}(t,.)}(x, v, |v - \dot{x}_i(t)|)$$
$$\implies (\xi, \eta) \in N_{\operatorname{Gr} F(t,.)}(x, v) + |\eta_0|\mathbb{B}. \tag{9.5.8}$$

Now apply the earlier derived necessary conditions of Proposition 9.5.2. Notice that, in consequence of the Euler Lagrange inclusion and the transversality condition, the costate arc component associated with the y state component is a constant and takes value $-\alpha_i$. We deduce that there exists $p_i \in W^{1,1}([S', T']; \mathbb{R}^n)$ such that

(A)': $\dot{p}_i(t) \in \operatorname{co}\{\zeta : (\zeta, p_i(t)) \in N_{\operatorname{Gr} F(t,.)}(x_i(t), \dot{x}_i(t))\} + \alpha_i \mathbb{B}$, a.e. $t \in [S_i, T_i]$,
(B)': $p_i(t) \cdot \dot{x}_i(t) + \alpha_i |v - \dot{x}_i(t)| \geq p_i(t) \cdot v$ for all $v \in F(t, x_i(t))$, a.e. $\in [S_i, T_i]$,
(C)': there exist $\xi_0^i \in \text{sub-ess}_{t \to S_i} H(t, x_i(S_i), p(S_i))$ and
$\xi_1^i \in \text{super-ess}_{t \to T_i} H(t, x_i(T_i), p(T_i))$ such that

$$(-\xi_0^i, p_i(S_i), \xi_1^i, -p_i(T_i)) \in \partial g(z_i) + 2\sqrt{2}\alpha_i(1 + 2c_F)^2 \mathbb{B}.$$

We use our 'constant extrapolation' convention to extend the domains of the p_i's to the entire real line. We deduce from (A)' and (C)' that the p_i's are uniformly bounded with uniformly integrably bounded derivatives. We can arrange then, by subsequence extraction, that $p_i \to p$ uniformly and $\dot{p}_i \to \dot{p}$ weakly in L^1 for some $p \in W^{1,1}$. The sequences $\{\xi_0^i\}$ and $\{\xi_1^i\}$ are bounded. A further subsequence extraction ensures that they have limits ξ_0 and ξ_1 respectively. We know however that $([S_i, T_i], x_i) \to ([S', T'], x')$ w.r.t. the d metric and $z_i \to (S', x'(S'), T', x'(T'))$ as $i \to \infty$. A by now familiar convergence analysis permits us to pass to the limit as $i \to \infty$ in conditions (A)'–(C)' and arrive at the relations asserted in the Proposition statement. The novel feature of the analysis, as compared with treatment of fixed time problems is that we now make use of the stability, under limit taking, of relations expressed in terms of the sub and super essential values. We note specifically that, Proposition 9.3.4, the conditions $\xi_0^i \in \text{sub-ess}_{t \to S_i} H(t, x_i(S_i), p(S_i))$ and $\xi_1^i \in \text{super-ess}_{t \to T_i} H(t, x_i(T_i), p(T_i))$, for each i, imply that $\xi_0 \in \text{sub-ess}_{t \to S'} H(t, x'(S'), p(S'))$ and $\xi_1 \in \text{super-ess}_{t \to T'} H(t, x'(T'), p(T'))$. □

Step 2 (Completion of the Proof) Let $\beta > 0$ be such that the $W^{1,1}$ local minimizer $([\bar{S}, \bar{T}], \bar{x})$ relative to B is minimizing w.r.t. admissible F trajectories $([S, T], x)$ that satisfy $d(([S, T], x), ([\bar{S}, \bar{T}], \bar{x})) \leq \beta$ and $\dot{x}(t) \in \dot{\bar{x}}(t) + B(t)$ for a.e. $t \in [\bar{S}, \bar{T}] \cap [S, T]$. Since we may impose (G3)', it may be assumed that $\bar{x} \equiv 0$.

9.5 Proof of Theorem 9.4.1

We shall make use, once again, of the modified integral penalty function $\rho_S(t, x, v)$ earlier encountered to prove the fixed end-time necessary conditions of Chap. 8.

Choose $\eta \in (0, 1/2)$ such that (G3)' is satisfied with $\gamma = (1 - 2\eta)$. Fix $N > R$. For each $t \in [\bar{S} - \sigma, \bar{T} + \sigma]$ define

$$E^N(t) := \{e \in \Omega_0(t) \cap B(t) : R \leq |e| \leq N\}.$$

Here, $\Omega_0(t)$ is the regular velocity set relative to $([\bar{S}, \bar{T}], \bar{x})$, namely

$$\Omega_0(t) := \{e \in F(t, \bar{x}(t)) : F(t, .) \text{ is pseudo-Lipschitz near } (\bar{x}(t), e)\}.$$

We can construct a multifunction $E^N_{\text{discrete}} : [\bar{S} - \sigma, \bar{T} + \sigma] \rightsquigarrow \mathbb{R}^n$ such that

(a): E^N_{discrete} is measurable,

and, for each $t \in [\bar{S} - \sigma, \bar{T} + \sigma]$,

(b): $E^N_{\text{discrete}}(t)$ is an empty or finite set,
(c): $E^N_{\text{discrete}}(t) \subset E^N(t) \subset E^N_{\text{discrete}}(t) + N^{-1}\mathbb{B}$, for each $t \in [\bar{S} - \sigma, \bar{T} + \sigma]$ such that $E^N(t) \neq \emptyset$,
(d): $E^N_{\text{discrete}}(t) \subset E^{N+1}_{\text{discrete}}(t)$ for all integers $N > R$.

Define

$$\theta(t) := \begin{cases} \min\{|e - e'| : e, e' \in E^N_{\text{discrete}}(t) \text{ and } e \neq e'\} \wedge \inf\{d_{\partial B(t)}(e) : e \in E^N_{\text{discrete}}(t)\} \\ \quad \text{if } E^N_{\text{discrete}}(t) \text{ contains at least two elements,} \\ +\infty \quad \text{otherwise}. \end{cases}$$

Take $\delta \in (0, \eta R)$. Define

$$E^\delta_{\text{regular}}(t) := \{e \in E^N_{\text{discrete}}(t) : F(t, .) \text{ is pseudo-Lipschitz near } (e, 0)$$

$$\text{(with parameters } R \geq \delta, \epsilon \geq \delta \text{ and } k \leq \delta^{-1}) \text{ and } \theta(t) \geq \delta\}.$$

By construction, $E^\delta_{\text{regular}}(t) \subset E^N_{\text{discrete}}(t) \subset N\mathbb{B}$. Define $D^\delta : [\bar{S} - \sigma, \bar{T} + \sigma] \rightsquigarrow \mathbb{R}^n$ as

$$D^\delta(t) := (1 - \eta)R\mathbb{B} \cup \{e \in e_0 + \frac{1}{3}\delta\mathbb{B} : e_0 \in E^\delta_{\text{regular}}(t)\}$$

in which the right side is interpreted as $(1 - \eta)R\mathbb{B}$ if $E^\delta_{\text{regular}}(t) = \emptyset$.

$D^\delta(t)$ is a finite union of disjoint, closed balls. These comprise $(1-\eta)R\mathbb{B}$ and elements from a (possibly empty) collection of disjoint closed balls each with origin outside $R\overset{\circ}{\mathbb{B}}$. We also have that $D^\delta(t) \subset B(t)$ for a.e. $t \in [\bar{S} - \sigma, \bar{T} + \sigma]$.

Now consider the multifunction $S : [\bar{S} - \sigma, \bar{T} + \sigma] \times \mathbb{R}^n \rightsquigarrow \mathbb{R}^n$

$$S(t, x) := \{(\chi(|e|)e : e \in F(t, x)\},$$

in which $\chi : [0, \infty) \to [1, \infty)$ is the function

$$\chi(d) := 1 + \frac{4(1-\eta)}{\eta R}[d - (1-\eta)R]^+.$$

Define $\rho_S : [\bar{S} - \sigma, \bar{T} + \sigma] \times \mathbb{R}^n \times \mathbb{R}^n \to [0, \infty)$ to be

$$\rho_S(t, x, v) := \begin{cases} d_{S(t,x)}(v) & \text{if } |v| \leq (1-\eta)R, \\ d_{F(t,x)}(v) & \text{if } |v| > (1-\eta)R. \end{cases}$$

We refer to Lemma 8.6.5 of Chap. 8 for relevant properties of $\rho_S(t, x, v)$.

For $t \in [\bar{S} - \sigma, \bar{T} + \sigma]$ and $e \in D^\delta(t)$ define

$$\phi(t, e) := \begin{cases} \frac{1}{2}(|e| - (1 - 2\eta)R) \vee 0 & \text{if } e \in (1-\eta)R\mathbb{B} \\ \frac{1}{2}(|e - e_0| - \delta/6) \vee 0 & \text{if } e \in e_0 + \frac{1}{3}\delta\mathbb{B} \text{ for some } e_0 \in E^\delta_{\text{regular}}(t). \end{cases}$$

Note the following properties of ϕ, each of which is a simple consequence of the definition of this function: for any $t \in [\bar{S}, \bar{T}]$

$$\left. \begin{array}{l} \phi(t, e) = 0 \text{ if } |e| \leq (1-2\eta)R \\ \\ \phi(t, .) \text{ is locally Lipschitz continuous on } D^\delta(t) \\ \quad \text{with Lipschitz constant } 1/2, \\ \\ \phi(t, e) \geq \frac{\eta R}{2} \wedge \frac{\delta}{12} \text{ if } e \in \partial D^\delta(t). \end{array} \right\} \quad (9.5.9)$$

Take $\alpha_i \downarrow 0$ and, for each i, consider the optimization problem:

$$(P_i) \begin{cases} \text{Minimize } \left(\ell_i(S, x(S), T, x(T))\right) \vee \left(\int_{\bar{S}+\sigma}^{\bar{T}-\sigma} \rho_S(t, x(t), \dot{x}(t))dt\right) \\ \qquad\qquad\qquad\qquad\qquad\qquad + \int_{\bar{S}+\sigma}^{\bar{T}-\sigma} \phi(t, \dot{x}(t))dt \\ \text{over arcs } x \in W^{1,1} \text{ such that} \\ \dot{x}(t) \in D^\delta(t) \\ d((([S, T], x), ([\bar{S}, \bar{T}], \bar{x})) \leq \beta. \end{cases}$$

9.5 Proof of Theorem 9.4.1

Here,

$$\ell_i(S, x_0, T, x_1) := \left(g(S, x_0, T, x_1) - g(\bar{S}, 0, \bar{T}, 0) + \alpha_i^2\right) \vee d_C(S, x_0, T, x_1)$$

and $\rho_S(t, x, v)$ is the modified penalty integrand, with parameters $\eta \in (0, 1/2)$, $\delta > 0$ and N, as earlier chosen.

We can formulate this problem as

$$\text{Minimize } \{J_i(([S, T], x)) : ([S, T], x) \in \mathcal{S}\},$$

in which $\mathcal{S}$ is the class of admissible arcs $([S, T], x)$ for (P_i) and

$$J_i(([S, T], x)) := \left(\ell_i(S, x(S)), T, x(T))\right) \vee \left(\int_{\bar{S}+\sigma}^{\bar{T}-\sigma} \rho_S(t, x(t), \dot{x}(t)) dt\right)$$

$$+ \int_{\bar{S}+\sigma}^{\bar{T}-\sigma} \phi(t, \dot{x}(t)) dt.$$

$\mathcal{S}$ is complete w.r.t. the metric d. We see that $([\bar{S}, \bar{T}], \bar{x} \equiv 0)$ is an α_i^2 minimizer for (P_i). By Ekeland's theorem, we can find $([S_i, T_i], x_i) \in \mathcal{S}$ such that $([S_i, T_i], x_i)$ is a minimizer for

$$\text{Minimize } \{J_i(([S, T], x)) + \alpha_i d(([S, T], x), ([S_i, T_i], x_i))$$

$$\text{s.t. } ([S, T], x) \in \mathcal{S}\} \quad (9.5.10)$$

and

$$d(([S_i, T_i], x_i), ([\bar{S}, \bar{T}], \bar{x})) \leq \alpha_i. \quad (9.5.11)$$

Note that

$$\left(\ell_i(S_i, x_i(S_i), T_i, x_i(T_i))\right) \vee \left(\int_{\bar{S}+\sigma}^{\bar{T}-\sigma} \rho_S(t, x_i(t), \dot{x}_i(t)) dt\right) > 0, \quad (9.5.12)$$

for, otherwise, we would have

$$\left(g(S_i, x_i(S_i), T_i, x_i(T_i)) - g(\bar{S}, \bar{x}(\bar{S}) = 0, \bar{T}, \bar{x}(\bar{T}) = 0) + \alpha_i^2\right)$$

$$\vee d_C(S_i, x_i(S_i), T_i, x_i(T_i)) \vee \left(\int_{\bar{S}+\sigma}^{\bar{T}-\sigma} \rho_S(t, x_i(t), \dot{x}_i(t)) dt\right) = 0.$$

This implies $\rho_S(t, x_i(t), \dot{x}_i(t)) = 0$ a.e. $t \in [\bar{S} + \sigma, \bar{T} - \sigma]$. But then $\dot{x}_i(t) \in F(t, x_i(t))$ a.e., by properties of the modified distance function. The relation also tells us $d_C(S_i, x_i(S_i), T_i, x_i(T_i)) = 0$, from which we conclude $(S_i, x_i(S_i), T_i, x_i(T_i)) \in C$ and, also, $g(S_i, x_i(S_i), T_i, x_i(T_i))) \le g(\bar{S}, 0, \bar{T}, 0) - \alpha_i^2$. Since $([S_i, T_i], x_i) \in \mathcal{S}$, we know that $d(([S_i, T_i], x_i), ([\bar{S}, \bar{T}], \bar{x})) \le \beta$. But then $([S_i, T_i], x_i)$ is an admissible F trajectory, satisfying $\dot{x}(t) \in B(t)$ for a.e. $t \in [\bar{S}, \bar{T}] \cap [S_i, T_i]$, that violates the $W^{1,1}$ local optimality of $([\bar{S}, \bar{T}], \bar{x} \equiv 0)$. (9.5.12) is confirmed.

For i sufficiently large we have $\alpha_i \le \sigma \wedge (\beta/2)$. Then $[S_i, T_i] \subset [\bar{S}-\sigma, \bar{T}+\sigma]$ and $([S_i, T_i], x_i)$ remains a minimizer for problem (9.5.10), when the distance constraint is replaced by $d(([S, T], x), ([S_i, T_i], x_i)) \le \beta/2$. In view of Lemma 9.5.1, $([S_i, T_i], x_i)$ continues to be a minimizer when the cost $J_i([S, T], x)$ in (9.5.10) is replaced by

$$\tilde{J}_i([S, T], x) := J_i([S, T], x) + \Big(\ell_i(S, x(S), T, x(T)) \Big)$$

$$\vee \Big(\int_{\bar{S}+\sigma}^{\bar{T}-\sigma} \rho_S(t, x(t), \dot{x}(t)) dt \Big)$$

$$+ \int_{\bar{S}+\sigma}^{\bar{T}-\sigma} \phi(t, \dot{x}(t)) dt + \alpha_i \Big(\int_S^T |\dot{x}(t) - \dot{x}_i(t)| dt$$

$$+ |x(S) - x_i(S_i)| + (1 + 2c_F)(|S - S_i| + |T - T_i|) \Big).$$

We know then that $([S_i, T_i], (x_i, y_i(t) := \int_{S_i}^t \rho_S(s, x_i(s), \dot{x}_i(s)) ds, z_i(t) := \int_{S_i}^t \phi(s, \dot{x}_i(s)) ds))$ is a $W^{1,1}$ local minimizer for the problem

$$\begin{cases} \text{Minimize } \Big(g(S, x(S), T, x(T)) - g(\bar{S}, 0, \bar{T}, 0) + \alpha_i^2\Big) \vee d_C(S, x(S), T, x(T)) \\ \qquad\qquad \vee (y(T) - y(S)) \\ +\alpha_i |x(S) - x_i(S_i)| + \alpha_i(1 + 2c_F)(|S - S_i| + |T - T_i|) \\ \qquad\qquad + (z(T) - z(S)) \\ \text{over } ([S, T], (x, y, z)) \in W^{1,1}([S, T]; \mathbb{R}^n) \text{ satisfying} \\ (\dot{x}(t), \dot{y}(t), \dot{z}(t)) \in \tilde{F}(t, x(t)), \end{cases}$$

in which $\tilde{F} : \mathbb{R} \times \mathbb{R}^n \rightsquigarrow \mathbb{R}^n \times \mathbb{R} \times \mathbb{R}$ is the multifunction:

$$\tilde{F}(t, x) =$$
$$\begin{cases} \{(v, \rho_S(t, x, v), \alpha_i |v - \dot{x}_i(t)| + \phi(t, v) : v \in D^\delta(t))\} & \text{if } t \in [\bar{S} - \delta, \bar{T} + \delta] \\ \{(0, 0, 0)\} & \text{otherwise}. \end{cases}$$

We have arrived at a problem in which the endpoints are not constrained and to which the necessary conditions of Step 1 (Proposition 9.5.2) are applicable. It is a straightforward exercise to show that the relevant hypotheses are satisfied, for i sufficiently large, when we identify $([S', T'], x')$ with $([S_i, T_i], x_i)$. Note that

9.5 Proof of Theorem 9.4.1

hypothesis (FEP4)' is satisfied, in consequence of (G4) in Theorem 9.4.1, for suitably small parameters $\delta' > 0$ and $\epsilon' > 0$, since (by (9.5.11)) $S_i \to \bar{S}$, $T_i \to \bar{T}$ and $||x_i - \bar{x}||_{L^\infty} \to 0$, as $i \to \infty$.

Using the sum rule to estimate the subdifferential of the end-point cost term, and by examining the properties of proximal normal vectors to $\text{Gr}\,\tilde{F}(t,.)$ at neighbouring points to $(x_i(t), \dot{x}_i(t))$, we deduce the following information, in which the costate trajectory component associated with x_i is denoted p_i. (Note that the costate components associated with y and z are both constant). There exist non-negative numbers $\lambda_i, \lambda_i^{(1)}$ and $\lambda_i^{(2)}$ satisfying $\lambda_i + \lambda_i^{(1)} + \lambda_i^{(2)} = 1$ such that

(a): $\dot{p}_i(t) \in \text{co}\{\eta : (\eta, p_i(t)) \in \lambda_i^{(1)} \partial_{x,v} \rho_S(t, x_i(t), \dot{x}_i(t)) + \left(\{0\} \times \alpha_i \mathbb{B}\right)$
$$+ \{0\} \times \partial_v \phi(t, \dot{x}_i(t))\}$$
for a.e. $t \in [S_i, T_i]$ such that $\dot{x}_i(t) \in D^\delta(t)$,

(b)': there exist $\xi_0^i \in \text{sub-ess}_{t \to S_i}\, H(t, x_i(S_i), p_i(S_i))$ and
$$\xi_1^i \in \text{super-ess}_{t \to T_i}\, H(t, x_i(T_i), p_i(T_i)) \text{ such that}$$

$$(-\xi_0^i, p_i(S_i), \xi_1^i, -p_i(T_i)) \in \lambda_i \partial g(S_i, x_i(S_i), T_i, x_i(T_i))$$
$$+ \lambda_i^{(2)} \partial d_C(S_i, x_i(S), T_i, x_i(T_i)) + \alpha_i (1 + 2(1 + c_F^2)^2)^{\frac{1}{2}} \mathbb{B} \times \{0\},$$

(c): $p_i(t) \cdot \dot{x}_i(t) - \lambda_i^{(1)} \rho_S(t, x_i(t), \dot{x}_i(t)) - \phi(t, \dot{x}_i(t)) \geq$
$$p_i(t) \cdot v - \lambda_i^{(1)} \rho_S(t, x_i(t), v) - \phi(t, v) - \alpha_i |v - \dot{x}_i(t)|,$$
for all $v \in D^\delta(t)$, a.e. $t \in [S_i, T_i]$.

Observe however that

$$\lambda_i^{(2)} \partial d_C(\xi_0^i, x_i(S_i), \xi_1^i, x_i(T_i)) = \lambda_i^{(2)} \left(\partial d_C(x_i(S), x_i(T)) \cap \partial \mathbb{B}\right),$$

in which we interpret the right side as $\{0\}$, if $\lambda_i^{(2)} = 0$. This relation is true when $d_C(S_i, x_i(S_i), T_i, x_i(T_i)) > 0$ since, according to properties of the subdifferential of the distance function, $\partial d_C(S_i, x_i(S_i), T_i, x_i(T_i)) \subset \partial \mathbb{B}$ in this case. It is also true when $d_C(S_i, x_i(S_i), T_i, x_i(T_i)) = 0$ because in this case we have also $\lambda_i^{(2)} = 0$ in view of (9.5.12) and the max rule for subdifferentials. But then (b)' can be replaced by the stronger condition:

(b): $(-\xi_0^i, p_i(S_i), \xi_1^i, -p_i(T_i)) \in \lambda_i \partial g(S_i, x_i(S), T_i, x_i(T_i))$
$$+ \lambda_i^{(2)} \left(\partial d_C(S_i, x_i(S), T_i, x_i(T_i)) \cap \partial \mathbb{B}\right) + \alpha_i \sqrt{1 + 2(1 + 2c_F)^2}\, \mathbb{B} \times \{0\}.$$

We can conclude from condition (a) above and the Lipschitz regularity of the modified penalty integrand $\rho_S(t,.,.)$ (see properties (ii) and (iv) established in Lemma 8.6.5 of Chap. 8) that p_i also satisfies

$$|\dot{p}_i(t)| \leq \tilde{k}(t)(|p_i(t)| + 2), \text{ a.e. } t \in [S_i, T_i]. \tag{9.5.13}$$

Here, $\tilde{k}$ is the integrable function of property (iv) of Lemma 8.6.5 applied to the modified penalty integrand ρ_S.

From (b) we know that $|p_i(S)| \le k_g + 1 + \alpha_i$. It follows then from (9.5.13) that the family of costate trajectories components p_i, $i = 1, 2, \ldots$ are uniformly bounded and their derivatives are uniformly integrably bounded.

To proceed, we need to establish a uniform positive lower bound on the magnitude of (p_i, λ_i). Let us examine the two possible cases that can arise:

(i): $\rho_S(t, x_i(t), \dot{x}_i(t)) = 0$ a.e. $t \in [S_i, T_i]$,
(ii): $\rho_S(t, x_i(t), \dot{x}_i(t)) > 0$ on a subset of $[S_i, T_i]$, which has positive $\mathcal{L}$-measure.

Consider (i). In this case, $z_i(T_i) - z_i(S_i) = 0$ and so, in view of (9.5.12),

$$z_i(T_i) - z_i(S_i) < \left(g(S_i, x_i(S_i), T_i, x(T_i)) - g(\bar{S}, 0, \bar{T}, 0) + \alpha_i^2 \right)$$

$$\vee d_C(S_i, x_i(S_i), T_i, x(T_i)).$$

But then, from the sum rule, $\lambda_i^{(1)} = 0$ and therefore $\lambda_i^{(2)} = 1 - \lambda_i$. It follows then from (b) that

$$|(-\xi_0^i, p_i(S_i), \xi_1^i, -p_i(T_i))| \ge 1 - (1 + k_g)\lambda_i - \alpha_i \sqrt{1 + 2(1 + 2c_F)^2}.$$

Since $|\xi_0^i| \le c_F|p_i(S_i)|$ and $|\xi_1^i| \le c_F|p_i(T_i)|$, we deduce

$$\sqrt{2}(1 + c_F^2)^{\frac{1}{2}} \|p_i\|_{L^\infty} + (1 + k_g)\lambda_i \ge 1 - \alpha_i\sqrt{2 + (1 + 2c_F)^2}.$$

Consider case (ii). In this case there is a time $t \in [S_i, T_i]$ such that the first relation in condition (c) is satisfied, $\rho_S(t, x_i(t), \dot{x}_i(t)) > 0$ and $\dot{x}_i(t) \in D^\delta(t)$. We claim that $\dot{x}_i(t) \notin F(t, x_i(t))$. This is certainly true if $|\dot{x}_i(t)| \ge (1 - \eta)R$ because, in this case, $\rho_S(t, x_i(t), \dot{x}_i(t)) = d_{F(t, x_i(t))}(\dot{x}_i(t)))$. So we may assume that $|\dot{x}_i(t)| < (1 - \eta)R$. Then $\chi(|\dot{x}_i(t)|) = 1$. This means that, if $\dot{x}_i(t) \in F(t, x_i(t))$, then $\rho_S(t, x_i(t), \dot{x}_i(t)) \le |\dot{x}_i(t) - \chi(\dot{x}_i(t)|)\dot{x}_i(t)| = |\dot{x}_i(t) - \dot{x}_i(t)| = 0$. From this contradiction it follows that $\dot{x}_i(t) \notin F(t, x_i(t))$.

To proceed further, we must now distinguish two situations:

First case: $\dot{x}_i(t) \in \partial D^\delta(t)$.

In view of (G3)', there exists $v_0 \in F(t, x_i(t))$ such that $|v_0| \le (1 - 2\eta)R$. Then $\phi(t, v_0) = 0$ and $\rho_S(t, x_i(t), v_0) \le |v_0 - \chi(|v_0|)v_0| = |v_0 - v_0|) = 0$. Since $\dot{x}_i(t) \in \partial D^\delta(t)$, it follows from property (9.5.9) of the function ϕ that

$$\phi(t, \dot{x}_i(t)) \ge \frac{\eta R}{2} \wedge \frac{\delta}{12}.$$

Using these relations, noting that $\rho_S(t, x_i(t), \dot{x}_i(t)) \ge 0$ and $|\dot{x}_i(t)| \le N + \delta/3$, and inserting $v = v_0$ in the first relation in (c), yields the inequality

$$|p_i(t)||(\dot{x}_i(t) - v_0)| \geq \lambda_i^{(1)} \rho_S(t, x_i(t), \dot{x}_i(t)) - 0 + \phi(t, \dot{x}_i(t))$$
$$- 0 - \alpha_i(N + \delta/3 + (1 - 2\eta)R).$$
$$\geq \phi(t, \dot{x}_i(t)) - \alpha_i(N + \delta/3 + (1 - 2\eta)R)$$
$$\geq \left(\frac{\eta R}{2} \wedge \frac{\delta}{12}\right) - \alpha_i(N + (1 - 2\eta)R).$$

Since $|v_0| \leq (1 - 2\eta)R$, it follows that

$$\|p_i\|_{L^\infty} \geq (N + \delta/3 + (1 - 2\eta)R)^{-1}\left(\frac{\eta R}{2} \wedge \frac{\delta}{12}\right) - \alpha_i.$$

Second Case: $\dot{x}_i(t) \in \overset{\circ}{D}^\delta(t)$.

Now, since $\dot{x}_i(t)$ is an unconstrained local minimizer of $v \to -p_i(t) \cdot v + \lambda_i^{(1)} \rho_S(t, x_i(t), v) + \phi(t, v) + \alpha_i |v - \dot{x}_i|$, we have

$$p_i(t) \in \lambda_i^{(1)} \partial_v \rho_S(t, x_i(t), \dot{x}_i(t)) + \partial_v \phi(t, \dot{x}_i(t)) + \alpha_i \mathbb{B}.$$

But $\rho_S(t, x_i(t), \dot{x}_i(t)) > 0$. It follows from property (ii) of Lemma 8.6.5 applied on the modified penalty integrand that elements in $\partial_v \rho_S(t, x_i(t), \dot{x}_i(t))$ have unit length. Taking note also of the fact that $\phi(t, .)$ is Lipschitz continuous with Lipschitz constant $1/2$, whence $\partial_v \phi(t, \dot{x}_i(t)) \in (1/2)\mathbb{B}$, we deduce from the preceding relations that

$$\|p_i\|_{L^\infty} \geq \lambda_i^{(1)} - \frac{1}{2} - \alpha_i.$$

From (b) we know

$$\sqrt{2}(1 + c_F^2)^{\frac{1}{2}} \|p_i\|_{L^\infty} \geq |(-\xi_i^0, p_i(S_i), \xi_i^1, -p_i(T_i))|$$
$$\geq \lambda_i^{(2)} - \lambda_i k_g - \alpha_i \sqrt{2 + (1 + 2c_F)^2}$$
$$= 1 - \lambda_i^{(1)} - (1 + k_g)\lambda_i - \alpha_i \sqrt{1 + 2(1 + 2c_F)^2}.$$

The preceding inequalities yield

$$(1 + \sqrt{2}(1 + c_F^2)^{\frac{1}{2}})\|p_i\|_{L^\infty} + (1 + k_g)\lambda_i \geq \frac{1}{2} - \alpha_i(1 + \sqrt{2 + (1 + 2c_F)^2}).$$

Combining the estimates relating to both cases (i) and (ii), we arrive at

$$(1 + \sqrt{2}(1 + c_F^2)^{\frac{1}{2}})\|p_i\|_{L^\infty} + (1 + k_g)\lambda_i$$
$$\geq \min\{\frac{1}{2}, (N + \frac{\delta}{3} + (1 - 2\eta)R)^{-1}\left(\frac{\eta R}{2} \wedge \frac{\delta}{12}\right)$$
$$- \alpha_i(1 + \sqrt{1 + 2(1 + 2c_F)^2})\}.$$

This is the desired lower bound.

We deduce from (9.5.11) that, along a subsequence, $x_i \to \bar{x} \equiv 0$ uniformly, and $\dot{x}_i \to \dot{\bar{x}} \equiv 0$ in L^1 and a.e.. We have already observed that the $\dot{p}_i$'s are uniformly integrably bounded and the p_i's are uniformly bounded. By further restriction to a subsequence we can then arrange, in consequence of Ascoli's Theorem, that $p_i \to p$ uniformly and $\dot{p}_i \to \dot{p}$ weakly in L^1. We can also arrange that $\lambda_i \to \lambda$, $\lambda_i^{(1)} \to \lambda^{(1)}$ and $\lambda_i^{(2)} \to \lambda^{(2)}$, for some $\lambda, \lambda^{(1)}, \lambda^{(2)} \in [0, 1]$ and $\xi_0^i \to \xi_0$ and $\xi_1^i \to \xi_1$ for such that (owing to Proposition 9.3.4)

$$\xi_0 \in \text{sub-ess}_{t \to \bar{S}} H(t, \bar{x}(\bar{S}), p(\bar{S})) \text{ and } \xi_1 \in \text{super-ess}_{t \to \bar{T}} H(t, \bar{x}(\bar{T}), p(\bar{T}))$$

for some $\xi_0, \xi_1 \in \mathbb{R}$. A convergence analysis along the lines of the proof of Proposition 8.6.1 of Chap. 8 permits us to pass to the limit in conditions (b)–(c) above and thereby to obtain:

$$(1 + \sqrt{2}(1 + c_F^2)^{\frac{1}{2}}) \|p\|_{L^\infty} + (1 + k_g)\lambda$$
$$\geq \min\{\frac{1}{2}, (N + \frac{\delta}{3} + (1 - 2\eta)R)^{-1}(\frac{\eta R}{2} \wedge \frac{\delta}{12})\}, \quad (9.5.14)$$

$$\dot{p}(t) \in \text{co}\{\eta : (\eta, p(t)) \in N_{\text{Gr } F(t,.)}(\bar{x}(t), \dot{\bar{x}}(t))\}, \text{ a.e. } t \in [\bar{S}, \bar{T}], \quad (9.5.15)$$

$$(-\xi_0, p(\bar{S}), \xi_1, -p(\bar{T})) \in \lambda_i \partial g(\bar{S}, \bar{x}(\bar{S}), \bar{T}, \bar{x}(\bar{T}))$$
$$+ N_C(\bar{S}, \bar{x}(S), \bar{T}, \bar{x}(\bar{T})), \quad (9.5.16)$$

$$p(t) \cdot \dot{\bar{x}}(t) - \lambda^{(1)} \rho_S(t, \bar{x}(t), \dot{\bar{x}}(t)) - \phi(t, \dot{\bar{x}}(t)) \geq p(t) \cdot v$$
$$- \lambda^{(1)} \rho_S(t, \bar{x}(t), v) - \phi(t, v),$$

for all $v \in D^\delta(t)$, a.e. $t \in [\bar{S}, \bar{T}]$.

Observe that, for all points $v \in (F(t, \bar{x}(t)) \cap (1 - 2\eta)R\mathbb{B}) \cup E^\delta_{\text{regular}}(t)$, we have $v \in D^\delta(t)$, $v \in F(t, \bar{x}(t))$ and $\rho_S(t, \bar{x}(t), v) = \phi(t, v) = 0$. So the preceding relation implies

$$p(t) \cdot \dot{\bar{x}}(t) \geq p(t) \cdot v \quad (9.5.17)$$

for all $v \in (F(t, \bar{x}(t)) \cap (1 - 2\eta)R\mathbb{B}) \cup E^\delta_{\text{regular}}(t)$, a.e. $t \in [\bar{S}, \bar{T}]$.

Notice that we have made use of property (v)(b) in Lemma 8.6.5 applied to the modified penalty integrand to express relation (9.5.15) in terms of the normal cone to the graph $N_{\text{Gr } F(t,.)}$ in place of the modified distance function $\rho_S(t, ., .)$. We also know that

$$|\dot{p}(t)| \leq k(t)|p(t)|, \text{ a.e. } t \in [\bar{S}, \bar{T}], \quad (9.5.18)$$

9.5 Proof of Theorem 9.4.1

This follows from property (v)(b) of $\rho_S(t, ., .)$. In view of (9.5.14), we can arrange, by scaling the Lagrange multipliers, that

$$\|p\|_{L^\infty} + \lambda = 1. \tag{9.5.19}$$

Relations (9.5.15), (9.5.16), (9.5.17) and (9.5.19) combine to provide a restricted version of the theorem (under all the hypotheses including (G4)), in which the Weierstrass condition (iv)$'$ is affirmed for velocities only in the subset $(F(t, \bar{x}(t)) \cap (1 - 2\eta)R\mathbb{B}) \cup E^\delta_{\text{regular}}(t)$ of $F(t, \bar{x}(t))$. To derive the full Weierstrass condition, we first take a sequence $\delta_i \downarrow 0$. Relations (9.5.15), (9.5.16), (9.5.17) and (9.5.19) are valid for each i, for some Lagrange multipliers (p, λ) that we now label (p_i, λ_i). From (9.5.16) and (9.5.18) we know the p_i's are uniformly bounded and have uniformly integrable derivatives. We may therefore arrange, by extracting subsequences, that $p_i \to p$ (uniformly), $\dot{p}_i \to \dot{p}$ in the weak L^1 topology and $\lambda_i \to \lambda$, for some $p \in W^{1,1}$ and $\lambda \geq 0$.

Relations (9.5.19), (9.5.15) and (9.5.16) (now expressed in terms of (p_i, λ_i)) are preserved in the limit as $i \to \infty$ (with multipliers (p, λ)). Using the fact that

$$\lim_i E^{\delta_i}_{\text{regular}}(t) = E^N_{\text{discrete}}(t), \quad \text{a.e. } t \in [\bar{S}, \bar{T}],$$

we can show that, in the limit, (9.5.17) yields $p(t) \cdot \dot{\bar{x}}(t) \geq p(t) \cdot v$, for all $v \in (F(t, \bar{x}(t)) \cap (1 - 2\eta)R\mathbb{B}) \cup E^N_{\text{discrete}}(t)$, a.e. $t \in [\bar{S}, \bar{T}]$.

Now take $N_i \uparrow \infty$ and $\eta_i \downarrow 0$. Using the facts that

$$\Omega_0(t) \cap B(t) \subset \lim_i \left(F(t, \bar{x}(t)) \cap \{e \in \mathbb{R}^n : < (1 - 2\eta_i)R\} \right) \cup \lim_i E^{N_i}_{\text{discrete}}(t)$$

and noting the linearity of the mapping $v \to p(t) \cdot v$, we deduce from (9.5.17), in the limit, that

$$p(t) \cdot \dot{\bar{x}}(t) \geq p(t) \cdot v, \quad \text{for all } v \in \overline{co}\,(\Omega_0(t) \cap B(t)), \text{ a.e. } t \in [\bar{S}, \bar{T}]. \tag{9.5.20}$$

This is the full Weierstrass condition of the theorem statement.

It remains only to prove the additional assertions that appear at the end of the theorem, concerning the relaxation of hypothesis (G4) when, say, the left end-time is fixed. In this case a simplified version of Step 1 of the proof, in which only the right end-time is varied, yields the transversality condition

$$\begin{cases} (p(\bar{S}), \xi_1, -p(T')) \in \tilde{g}(x'(\bar{S}), T', x'(T')), \\ \xi_1 \in \text{super-ess}_{t \to T'} H(t, x'(T'), p(T')), \end{cases}$$

for a version of the free end-point problem (FEP), in which, now, the end-point vector $(x(\bar{S}), T, x(T))$ is 'free'. (The part of hypothesis (G4) requiring that the velocity set $F(t, .)$ is bounded, uniformly over times in a neighbourhood of $\bar{S}$, is only required when we analyse perturbations to the left endpoint and can therefore be dropped.) Then, in Step 2 where we apply the necessary conditions of Step 1 to a sequence of ancillary dynamic optimization problems and pass to the limit, the ancillary problems are now taken to have a fixed left end-time. Limit taking, which no longer requires aspects of (G4) relating to the left end-time, yields the transversality condition (iii)$'$ in the theorem statement.

□

9.6 A Free End-Time Maximum Principle

We have derived necessary conditions for free end-time dynamic optimization problems in which the dynamic constraint is expressed as a differential inclusion. We now show that the analysis can be adapted to provide necessary conditions for free end-time problems, in the form of a maximum principle, when the dynamics are modelled instead by a control dependent differential equation:

$(FT)'$
$$\begin{cases} \text{Minimize } g(S, x(S), T, x(T)) \\ \text{over intervals } [S, T], \text{ arcs } x \in W^{1,1}([S, T]; \mathbb{R}^n) \\ \quad \text{and measurable functions } u : [S, T] \to \mathbb{R}^m \text{ satisfying} \\ \dot{x}(t) = f(t, x(t), u(t)) \text{ a.e. } t \in [S, T], \\ u(t) \in U(t) \text{ a.e. } t \in [S, T], \\ \text{and} \\ (S, x(S), T, x(T)) \in C. \end{cases}$$

The data for this problem comprise functions $g : \mathbb{R}^{1+n+1+n} \to \mathbb{R}$ and $f : \mathbb{R} \times \mathbb{R}^n \times \mathbb{R}^m \to \mathbb{R}^n$, a non-empty multifunction $U : \mathbb{R} \rightsquigarrow \mathbb{R}^m$ and a closed set $C \subset \mathbb{R}^{1+n+1+n}$.

Earlier terminology is modified to emphasize that the end-points of the underlying time interval $[S, T]$ are now choice variables. A *process* is taken to be a triple $([S, T], x, u)$ in which $[S, T]$ is an interval, x is an element in $W^{1,1}([S, T]; \mathbb{R}^n)$ (the *state trajectory*) and u (the *control function*) is a measurable $\mathbb{R}^m$ valued function on $[S, T]$ satisfying, for a.e. $t \in [S, T]$,

$$\begin{cases} \dot{x}(t) = f(t, x(t), u(t)) \\ u(t) \in U(t). \end{cases}$$

A process $([\bar{S}, \bar{T}], \bar{x}, \bar{u})$, which satisfies the constraints of $(FT)'$ is said to be a $W^{1,1}$ local minimizer if there exists $\delta' > 0$ such that

$$g(S, x(S), T, x(T)) \geq g(\bar{S}, \bar{x}(\bar{S}), \bar{T}, \bar{x}(\bar{T}))$$

for every process $([S, T], x, u)$ which satisfies the constraints of $(FT)'$ and also

9.6 A Free End-Time Maximum Principle

$$d(([S, T], x), ([\bar{S}, \bar{T}], \bar{x})) \leq \delta'.$$

Here, d is the metric (9.1.4).

Let $\mathcal{H}$ denote the un-maximized Hamiltonian function for $(FT)'$:

$$\mathcal{H}(t, x, p, u) := p \cdot f(t, x, u).$$

Theorem 9.6.1 (Free End-Time Maximum Principle: Lipschitz Time Dependence) *Let $([\bar{S}, \bar{T}], \bar{x}, \bar{u})$ be a $W^{1,1}$ local minimizer for $(FT)'$. Assume that*

(H1): g is Lipschitz continuous on a neighbourhood of $(\bar{S}, \bar{x}(\bar{S}), \bar{T}, \bar{x}(\bar{T}))$ and C is a closed set,

(H2): For fixed x, $f(., x, .)$ is $\mathcal{L} \times \mathcal{B}^m$ measurable. There exist $\epsilon > 0$ and a function $k_f : [\bar{S}, \bar{T}] \times \mathbb{R}^m \to \mathbb{R}$ such that $t \to k(t, \bar{u}(t))$ is integrable and, for a.e. $t \in [\bar{S}, \bar{T}]$,

$$|f(t'', x'', u) - f(t', x', u)| \leq k_f(t, u) |(t'', x'') - (t', x')|$$

for all $(t'', x''), (t', x') \in (t, \bar{x}(t)) + \epsilon \mathbb{B}$, a.e. $t \in [\bar{S}, \bar{T}]$,

(H3): $U(t) = U$ for all $t \in \mathbb{R}$, for some Borel set $U \subset \mathbb{R}^m$.

Then there exist $p \in W^{1,1}([\bar{S}, \bar{T}]; \mathbb{R}^n)$, $a \in W^{1,1}([\bar{S}, \bar{T}]; \mathbb{R})$ and $\lambda \geq 0$ such that

(i): $(p, \lambda) \neq (0, 0)$,
(ii): $(\dot{a}(t), -\dot{p}(t)) \in \text{co } \partial_{t,x} \mathcal{H}(t, \bar{x}(t), p(t), \bar{u}(t))$ a.e. $t \in [\bar{S}, \bar{T}]$,
(iii): $(-a(\bar{S}), p(\bar{S}), +a(\bar{T}), -p(\bar{T})) \in \lambda \partial g(\bar{S}, \bar{x}(\bar{S}), \bar{T}, \bar{x}(\bar{T}))$
$$+ N_C(\bar{S}, \bar{x}(\bar{S}), \bar{T}, \bar{x}(\bar{T})),$$
(iv): $\mathcal{H}(t, \bar{x}(t), p(t), \bar{u}(t)) = \max_{u \in U} \mathcal{H}(t, \bar{x}(t), p(t), u)$ a.e. $t \in [\bar{S}, \bar{T}]$,
(v): $\mathcal{H}(t, \bar{x}(t), p(t), \bar{u}(t)) = a(t)$ a.e. $t \in [\bar{S}, \bar{T}]$.

If the initial time $\bar{S}$ is fixed, i.e. $C = \{(\bar{S}, x_0, T, x_1) : (x_0, T, x_1) \in \tilde{C}\}$, for some closed set $\tilde{C} \subset \mathbb{R}^n \times \mathbb{R} \times \mathbb{R}^n$, and $g(S, x_0, T, x_1) = \tilde{g}(x_0, T, x_1)$ for some $\tilde{g} : \mathbb{R}^n \times \mathbb{R} \times \mathbb{R}^n \to \mathbb{R}$, then the above conditions are satisfied when (iii) is replaced by:

(iii)': $(p(\bar{S}), a(\bar{T}), -p(\bar{T})) \in \lambda \partial \tilde{g}(\bar{x}(\bar{S}), \bar{T}, \bar{x}(\bar{T})) + N_{\tilde{C}}(\bar{x}(\bar{S}), \bar{T}, \bar{x}(\bar{T}))$.

The assertions of the theorem can be similarly modified when the final time $\bar{T}$ is fixed.

Proof Choose δ such that $([\bar{S}, \bar{T}], \bar{x}, \bar{u})$ is a minimizer for $(FT)'$, with respect to processes $([S, T], x, u)$ satisfying the constraints of $(FT)'$ and also

$$d(([S, T], x), ([\bar{S}, \bar{T}], \bar{x})) < \delta.$$

Take $\alpha \in (0, 1/2)$ and consider the dynamic optimization problem on the fixed time interval $[\bar{S}, \bar{T}]$:

$$(R) \begin{cases} \text{Minimize } g(\tau(\bar{S}), y(\bar{S}), \tau(\bar{T}), y(\bar{T})) \\ \text{over processes } ((\tau, y), (v, w)) \text{ on } [\bar{S}, \bar{T}] \text{ satisfying} \\ (\dot{\tau}(s), \dot{y}(s)) = \tilde{f}(\tau(s), y(s), v(s), w(s)) \quad \text{a.e. } s \in [\bar{S}, \bar{T}], \\ v(s) \in U \quad \text{a.e.,} \\ w(s) \in [1-\alpha, 1+\alpha] \quad \text{a.e.,} \\ (\tau(\bar{S}), y(\bar{S}), \tau(\bar{T}), y(\bar{T})) \in C, \end{cases}$$

where

$$\tilde{f}(\tau, y, v, w) := (w, wf(\tau, y, v)).$$

We claim that $((\bar{\tau}(s) \equiv s, \bar{x}), (\bar{u}, \bar{w} \equiv 1))$ is a $W^{1,1}$ local minimizer for (R). To confirm this, take $\beta' > 0$ and let $((\tau, y), (v, w))$ be any process (on $[\bar{S}, \bar{T}]$) that is admissible for problem (R) such that

$$|(\tau(\bar{S}), y(\bar{S})) - (\bar{\tau}(\bar{S}) \equiv \bar{S}, \bar{x}(\bar{S}))| + \int_{\bar{S}}^{\bar{T}} |(\dot{\tau}(s), \dot{y}(s)) - (1, \dot{\bar{x}}(s))| ds \le \beta' \tag{9.6.1}$$

Write $S := \tau(\bar{S})$ and $T := \tau(\bar{T})$ and consider the transformation $\psi : [\bar{S}, \bar{T}] \to [S, T]$:

$$\psi(s) := \tau(\bar{S}) + \int_{\bar{S}}^{s} w(s') ds' \quad (= \tau(s)).$$

We find that ψ is a strictly increasing, Lipschitz continuous function, with Lipschitz continuous inverse. It can be deduced that $x : [S, T] \to \mathbb{R}^n$ and $u : [S, T] \to \mathbb{R}^m$ defined by

$$x := y \circ \psi^{-1} \quad \text{and} \quad u := v \circ \psi^{-1}$$

are absolutely continuous and measurable functions respectively which satisfy

$$(S, x(S), T, x(T)) = (\tau(\bar{S}), x(\bar{S}), \tau(\bar{T}), x(\bar{T})),$$

$$\dot{x}(t) = f(t, x(t), u(t)) \quad \text{a.e. } t \in [S, T],$$

$$u(t) \in U \quad \text{a.e. } t \in [S, T].$$

It follows from the foregoing relationships that $([S, T], x, u)$ is an admissible process for the original dynamic optimization problem. Further, according to Lemma 9.7.1 and (9.6.1), we can arrange, by choosing $\beta' > 0$ and α sufficiently small, that

$$d(([S, T], x), ([\bar{S}, \bar{T}], \bar{x})) \le \delta.$$

9.6 A Free End-Time Maximum Principle

In view of the minimizing properties of $([\bar{S}, \bar{T}], \bar{x}, \bar{u})$, we have

$$g(\tau(\bar{S}), y(\bar{S}), \tau(\bar{T}), y(\bar{T})) = g(S, x(S), T, x(T))$$
$$\geq g(\bar{S}, \bar{x}(\bar{S}), \bar{T}, \bar{x}(\bar{T})) = g(\bar{\tau}(\bar{S}), \bar{y}(\bar{S}), \tau(\bar{T}), \bar{y}(\bar{T})).$$

It follows that the process $((\bar{\tau}, \bar{x}), (\bar{u}, \bar{w}(s) \equiv 1))$ is a minimizer for (R), as claimed. The hypotheses are satisfied under which the necessary conditions of Theorem 7.2.1 can be applied to (R), with reference to the minimizer $((\bar{\tau}(s) \equiv s, \bar{x}), (\bar{u}, \bar{w} \equiv 1))$. We deduce that there exist $p \in W^{1,1}([\bar{S}, \bar{T}]; \mathbb{R}^n)$, $a \in W^{1,1}([\bar{S}, \bar{T}]; \mathbb{R})$ and $\lambda \geq 0$, not all zero, such that

$$(\dot{a}(s), -\dot{p}(s)) \in \mathrm{co}\, \partial_{t,x} \mathcal{H}(s, \bar{x}(s), p(s), \bar{u}(s)) \quad \text{a.e.,}$$

$$(-a(\bar{S}), p(\bar{S}), a(\bar{T}), -p(\bar{T})) \in \lambda \partial g(\bar{S}, \bar{x}(\bar{S}), \bar{T}, \bar{x}(\bar{T})) + N_C(\bar{S}, \bar{x}(\bar{S}), \bar{T}, \bar{x}(\bar{T}))$$

and

$$p(s) \cdot f(s, \bar{x}(s), \bar{u}(s)) - a(s) \geq (p(s) \cdot f(s, \bar{x}(s), u) - a(s))w$$

for all $u \in U$ and $w \in [1 - \alpha, 1 + \alpha]$ a.e..

The last relationship implies

$$\mathcal{H}(s, \bar{x}(s), p(s), \bar{u}(s)) = \max_{u \in U} \mathcal{H}(s, \bar{x}(s), p(s), u) \quad \text{a.e.,}$$

$$\mathcal{H}(s, \bar{x}(s), p(s), \bar{u}(s)) = a(s) \quad \text{a.e..}$$

All the assertions of the theorem have been confirmed, in the case that neither of the end-times are fixed. The refinements referred to at the end of the theorem statement, when either end-point is fixed, is achieved by obvious changes to the analysis.

□

We denote by $H : \mathbb{R} \times \mathbb{R}^n \times \mathbb{R}^n \to \mathbb{R}$ the (maximized) Hamiltonian function for $(FT)'$:

$$H(t, x, p) := \sup_{u \in U(t)} p \cdot f(t, x, u).$$

Theorem 9.6.2 (The Free End-Time Maximum Principle: Measurable Time Dependence) *Let $([\bar{S}, \bar{T}], \bar{x}, \bar{u})$ be a $W^{1,1}$ local minimizer for $(FT)'$ such that $\bar{T} - \bar{S} > 0$. Assume that there exist $\epsilon > 0$ and $\sigma \in (0, (\bar{T} - \bar{S})/2)$ such that*

(H1): g is Lipschitz continuous on a neighbourhood of $(\bar{S}, \bar{x}(\bar{S}), \bar{T}, \bar{x}(\bar{T}))$ and C is a closed set,

(H2): Gr U is a $\mathcal{L} \times \mathcal{B}^m$ measurable set,

(H3): For each $x \in \mathbb{R}^n$, $f(., x, .)$ is $\mathcal{L} \times \mathcal{B}^m$ measurable. There exists a function $k_f : \mathbb{R} \times \mathbb{R}^m \to \mathbb{R}$ such that $t \to k_f(t, \bar{u}(t))$ is integrable on $[\bar{T} - \sigma, \bar{T} + \sigma]$ and

$$|f(t, x', u) - f(t', x'', u)| \leq k_f(t, u)|x' - x''|$$

for all $x', x'' \in \bar{x}_e(t) + \epsilon \mathbb{B}$ and $u \in U(t)$, a.e. $t \in [\bar{S} - \sigma, \bar{T} + \sigma]$,

(H4): There exists $\bar{c} \geq 0$ such that, for a.e. $t \in [\bar{S} - \sigma, \bar{S} + \sigma] \cup [\bar{T} - \sigma, \bar{T} + \sigma]$,

$$|f(t, x, u)| \leq \bar{c}$$

for all $x \in \bar{x}_e(t) + \epsilon \mathbb{B}$ and $u \in U(t)$.

Then there exist $p \in W^{1,1}([\bar{S}, \bar{T}]; \mathbb{R}^n)$ and a real number $\lambda \geq 0$ such that

(i): $(p, \lambda) \neq (0, 0)$,
(ii): $-\dot{p}(t) \in \text{co}\partial_x \mathcal{H}(t, \bar{x}(t), p(t), \bar{u}(t))$ a.e. $t \in [\bar{S}, \bar{T}]$,
(iii): there exist $\xi_0 \in \text{sub-ess}_{t \to \bar{S}} H(t, \bar{x}(\bar{S}), p(\bar{S}))$, and
$$\xi_1 \in \text{super-ess}_{t \to \bar{T}} H(t, \bar{x}(\bar{T}), p(\bar{T})) \text{ such that}$$
$(-\xi_0, p(\bar{S}), \xi_1, -p(\bar{T})) \in \lambda \partial g(\bar{S}, \bar{x}(\bar{S}), \bar{T}, \bar{x}(\bar{T})) + N_C(\bar{S}, \bar{x}(\bar{S}), \bar{T}, \bar{x}(\bar{T}))$,
(iv): $\mathcal{H}(t, \bar{x}(t), p(t), \bar{u}(t)) = \max_{u \in U(t)} \mathcal{H}(t, \bar{x}(t), p(t), u)$ a.e. $t \in [\bar{S}, \bar{T}]$.

If the initial time $\bar{S}$ is fixed, i.e. $C = \{(\bar{S}, x_0, T, x_1) : (x_0, T, x_1) \in \tilde{C}\}$, for some closed set $\tilde{C} \subset \mathbb{R}^n \times \mathbb{R} \times \mathbb{R}^n$, and $g(S, x_0, T, x_1) = \tilde{g}(x_0, T, x_1)$ for some $\tilde{g} : \mathbb{R}^n \times \mathbb{R} \times \mathbb{R}^n \to \mathbb{R}$, then the above conditions are satisfied when (iii) is replaced by: there exists $\xi_1 \in \text{super-ess}_{t \to \bar{T}} H(t, \bar{x}(\bar{T}), p(\bar{T}))$ such that

(iii)': $(p(\bar{S}), \xi_1, -p(\bar{T})) \in \lambda \partial \tilde{g}(\bar{x}(\bar{S}), \bar{T}, \bar{x}(\bar{T})) + N_{\tilde{C}}(\bar{x}(\bar{S}), \bar{T}, \bar{x}(\bar{T}))$,

and when, in hypothesis (H4), the stated conditions are required to hold only for $t \in [\bar{T} - \sigma, \bar{T} + \sigma]$. The assertions of the theorem can be similarly modified when the final time $\bar{T}$ is fixed.

Proof

Step 1 (A Free End-Point Problem):

$$(FET)' \begin{cases} \text{Minimize } g(S, x(S), T, x(T)) \\ \text{over intervals } [S, T] \subset \mathbb{R}, \text{ and arcs } x \in W^{1,1}([S, T]; \mathbb{R}^n) \\ \quad \text{and measurable functions } u : [S, T] \to \mathbb{R}^m \text{ satisfying} \\ \dot{x}(t) = f(t, x(t), u(t)) \text{ a.e. } t \in [S, T], \\ u(t) \in U(t) \text{ a.e. } t \in [S, T], \\ |T - T'|, |S - S'| \leq \sigma'. \end{cases}$$

The following proposition provides necessary conditions for problem (FET)', in which $[S', T']$ is a given interval and $\sigma' > 0$ is a given constant such that $\sigma' < (T' - S')/2$.

9.6 A Free End-Time Maximum Principle

Proposition 9.6.3 *Let $([S', T'], x', u')$ be a global minimizer for (FET)'. Assume that, for some $\epsilon' > 0$ and $\sigma' \in (0, (T' - S')/2)$,*

(FET1)': g *is Lipschitz continuous,*
(FET2)': $\mathrm{Gr}\, U$ *is a $\mathcal{L} \times \mathcal{B}^m$ measurable set,*
(FET3)': For each $x \in \mathbb{R}^n$, $f(., x, .)$ is $\mathcal{L} \times \mathcal{B}^m$ measurable. There exists $k_f \in L^1(S' - \sigma', T' + \sigma')$ such that

$$|f(t, x, u) - f(t, x_1, u)| \le k_f(t)|x - x_1|$$

for all $x, x_1 \in x'_e(t) + \epsilon'\mathbb{B}$ and $u \in U(t)$, a.e. $t \in [S' - \sigma', T' + \sigma']$,
(FET4)': There exists $\bar{c} \ge 0$ such that, for a.e. $t \in [S' - \sigma', S' + \sigma'] \cup [T' - \sigma', T' + \sigma']$,

$$|f(t, x, u)| \le \bar{c}$$

for all $x \in x'_e(t) + \epsilon'\mathbb{B}$ and $u \in U(t)$.

Then there exists $p \in W^{1,1}([S', T']; \mathbb{R}^n)$ such that

(i): $-\dot{p}(t) \in \mathrm{co}\,\partial_x \mathcal{H}(t, x'(t), p(t), u'(t))$ *a.e. $t \in [S', T']$,*
(ii): there exist $\xi_0 \in \text{sub-ess}_{t \to S'} H(t, x'(S'), p(S'))$ and
$\xi_1 \in \text{super-ess}_{t \to T'} H(t, x'(T'), p(T'))$ *such that*

$$(-\xi_0, p(S'), \xi_1, -p(T')) \in \partial g(S', x'(S'), T', x'(T')),$$

(iii): $\mathcal{H}(t, x'(t), p(t), u'(t)) = \max_{u \in U(t)} \mathcal{H}(t, x'(t), p(t), u)$ *a.e. $t \in [S', T']$.*

The proof is based on perturbational methods, in which we employ the following distance function on admissible processes for (FET)': given admissible processes $([S, T], x, u)$ and $([S', T'], x', u')$ for (FET)' we define

$$d_{\text{control}}(([S, T], x, u), ([S', T'], x', u')) := |x(S) - x'(S')| + \qquad (9.6.2)$$
$$\mathrm{meas}\,\{t \in [S, T] \cap [S', T'] : u(t) \ne u'(t)\} + |S - S'| + |T - T'|.$$

The required properties of this distance function are summarized in the following lemma, whose straightforward proof we omit.

Lemma 9.6.4 *Assume that, for some interval $[S', T']$ and $\sigma' \in (0, (T' - S')/2)$, hypotheses (FET2)' - (FET4)' of the proposition are satisfied. Let*

$$\mathcal{M} = \{\text{admissible processes } ([S, T], x, u) \text{ for } (FET)'\}.$$

Then $(\mathcal{M}, d_{\text{control}})$ is a complete metric space.

Proof of Proposition 9.6.3 Observe that, without loss of generality, we can reduce attention to the case when the following strengthen form of hypothesis (FET3)$'$, in which the non-negative function k_f is replaced by a constant:

(FET3)*: There exists a constant $\bar{k} > 0$ such that, for a.e. $t \in [S' - \sigma', T' + \sigma']$,

$$|f(t, x, u) - f(t, x_1, u)| \leq \bar{k}|x - x_1|$$

for all $x, x_1 \in x'_e(t) + \epsilon'\mathbb{B}$.

Indeed, we can consider the change of the independent variable $s = \hat{\sigma}(t)$, where $\hat{\sigma} : \mathbb{R} \to \mathbb{R}$ is defined as in the proof of Proposition 9.5.2, replacing in (9.5.1) the non-negative function k by $\hat{k}_f$ (which can be supposed to be greater than 1 for all $t \in [S' - \sigma', T' + \sigma']$). We obtain a new problem (FET)$'$ in which the function f and the multifunction U are respectively replaced by

$$\hat{f}(s, x, u) := \frac{\|k_f\|_{L^1}}{(T' - S' + 2\sigma') \times (k_f \circ \hat{\sigma}^{-1})(s)} f(\hat{\sigma}^{-1}(s), x, u),$$

for all $(s, x, u) \in \mathbb{R} \times \mathbb{R}^n \times \mathbb{R}^m$,

and

$$\hat{U}(s) := U(\hat{\sigma}^{-1}(s)), \quad \text{for all } s \in \mathbb{R}.$$

Observe that the process $([\hat{S} := \hat{\sigma}^{-1}(S'), \hat{T} := \hat{\sigma}^{-1}(T')], \hat{x} := x' \circ \sigma^{-1}, \hat{u} := u' \circ \sigma^{-1})$ is a global minimizer for this new problem, and that $\hat{f}$ satisfies (FEP3)* with $\bar{k} = \frac{\|k_f\|_{L^1}}{T'-S'+2\sigma'}$. Notice also that, since $\hat{\sigma}$ is an absolutely continuous strictly continuous function on $[S' - \sigma', T' + \sigma']$, Gr $\hat{U}$ remains a $\mathcal{L} \times \mathcal{B}^m$ measurable set (cf. Exercise 2.3). It is then easy to derive the necessary conditions for the original problem, with reference to the minimizer $([S', T'], x', u')$, via the inverse change of independent variable $s = \hat{\sigma}(t)$, where the Lagrange multiplier $\hat{p}$ obtained in the new problem is replaced by $p = \hat{p} \circ \hat{\sigma}$ in the original one.

Consider first the special case of (FET)$'$, in which g has the structure

$$g(S, x_0, T, x_1) = \tilde{g}(S, x_0, T, x_1) + e(S, x_0, T, x_1),$$

for some twice differentiable function $\tilde{g}$ and some Lipschitz continuous function e. We shall prove a less precise version of the necessary conditions in the proposition statement, in which (iii) is replaced by

$$(-\xi_0, p(S'), \xi_1, -p(T')) \in \partial g(S', x'(S'), T', x'(T')) + 2\sqrt{2}k_e(1 + \bar{c}^2)^{\frac{1}{2}}\mathbb{B},$$

where k_e is a Lipschitz constant for e.

9.6 A Free End-Time Maximum Principle

(x', u') is a minimizer for a version of problem (FET)$'$, in which the underlying time interval is fixed at $[S', T']$. The hypotheses of Theorem 7.2.1 of Chap. 7 are satisfied. In consequence of the fact that the problem has no state endpoint constraints, the cost multiplier can be set to 1. It follows that there exists $p \in W^{1,1}([S', T']; \mathbb{R}^n)$ satisfying conditions (i), (iii) and the following tranversality condition:

$$(p(S'), -p(T')) \in \nabla_{x_0, x_1} \tilde{g}(S', x'(S'), T', x'(T')) + k_e \mathbb{B}. \tag{9.6.3}$$

For $s \in (0, \sigma']$, $([S', T' - s], x', u')$ is admissible and

$$\tilde{g}(S', x'(S'), T' - s, x'(T' - s)) + e(S', x'(S'), T' - s, x'(T' - s))$$
$$\geq \tilde{g}(S', x'(S'), T', x'(T')) + e(S', x'(S'), T', x'(T')).$$

Since $\tilde{g}$ is a C^2 function, there exists $r_1 > 0$, that does not depend on s, such that for all $s \in (0, \sigma']$

$$0 \geq \tilde{g}(S', x'(S'), T', x'(T')) - \tilde{g}(S', x'(S'), T' - s, x'(T' - s)) - k_e(1 + \bar{c}^2)^{\frac{1}{2}} s$$
$$= \int_{T'-s}^{T'} \left(\nabla_T \tilde{g}(S', x'(S'), t, x'(t)) + \nabla_{x_1} \tilde{g}(S', x'(S'), t, x'(t)) \cdot \dot{x}'(t) \right) dt$$
$$- k_e(1 + \bar{c}^2)^{\frac{1}{2}} s$$
$$\geq \nabla_T \tilde{g}(S', x'(S'), T', x'(T')) s - \int_{T'-s}^{T'} H(t, x'(T'), p(T')) dt - r_1 s^2$$
$$- k_e(\bar{c} + (1 + \bar{c}^2)^{\frac{1}{2}}) s.$$

To derive the final inequality in the above relation, we have used the facts that $-p(T') \in \nabla_{x_1} \tilde{g}(S', x(S'), T', x(T')) + k_e \mathbb{B}$ and that (from (FET3)* and (FET4)$'$) we have

$$p(T') \cdot \dot{x}'(t) \leq H(t, x'(T'), p(T')) + |p(T')| \bar{k} \bar{c} s, \quad \text{a.e. } t \in [T' - s, T'].$$

We have shown that

$$\begin{cases} T' \text{ is a local minimum over } [S', T'] \text{ of} \\ T \to -\nabla_T \tilde{g}(S', x'(S'), T', x'(T'))(T' - T) \\ \quad - \int_T^{T'} (-H)(t, x'(T'), p(T')) dt \\ \quad + r_1(T - T')^2 + k_e(\bar{c} + (1 + \bar{c}^2)^{\frac{1}{2}})|T - T'|. \end{cases} \tag{9.6.4}$$

Next, for each integer $k \geq 1$ we can select a measurable function $u_k : [T', T' + \sigma'] \to \mathbb{R}^m$ such that $u_k(t) \in U(t)$ for a.e. $t \in [T', T' + \sigma']$ and

$$p(T') \cdot f(t, x'(T'), u_k(t)) \geq \sup_{u \in U(t)} p(T') \cdot f(t, x'(T'), u) - \frac{1}{k}, \quad \text{a.e. } t \in [T', T' + \sigma'].$$

By Corollary 6.2.5 of Filippov's existence theorem (Theorem 6.2.3), for each k, there exists a process (y_k, u_k) on $[T', T' + \sigma']$ such that $y_k(T') = x'(T')$; from (FET3)* and (FET4)' we also have

$$\int_{T'}^{T'+s} |f(t, y_k(t), u_k(t)) - f(t, x'(T'), u_k(t))| dt \leq \bar{k}\bar{c}s^2, \quad \text{for all } s \in [0, \sigma']. \tag{9.6.5}$$

We now construct a sequence of admissible processes $([S', T' + \sigma'], x'_k, u'_k)$ by concatenating $([S', T'], x', u')$ and $([T', T' + \sigma'], y_k, u_k)$. By optimality,

$$0 \leq (\tilde{g} + e)(S', x'_k(S'), T' + s, x'_k(T' + s)) - (\tilde{g} + e)(S', x'(S'), T', x'(T'))$$

$$\leq \tilde{g}(S', x'(S'), T' + s, y_k(T' + s)) - \tilde{g}(S', x'(S'), T', x'(T')) + k_e(1 + \bar{c}^2)^{\frac{1}{2}}s$$

$$= \int_{T'}^{T'+s} \Big(\nabla_T \tilde{g}(S', x'(S'), t, y_k(t)) + \nabla_{x_1}\tilde{g}(S', x'(S'), t, y_k(t)) \cdot \dot{y}_k(t)\Big) dt$$

$$+ k_e(1 + \bar{c}^2)^{\frac{1}{2}}s,$$

for all $s \in [0, \sigma']$. Since $\tilde{g}$ is twice continuously differentiable, there exists $r_1 > 0$ (which can be assumed to be same constant appearing in (9.6.4)) independent of s, such that

$$0 \leq \int_{T'}^{T'+s} \Big(\nabla_T \tilde{g}(S', x'(S'), t, y_k(t)) + \nabla_{x_1}\tilde{g}(S', x'(S'), t, y_k(t)) \cdot \dot{y}_k(t)\Big) dt$$

$$+ k_e(1 + \bar{c}^2)^{\frac{1}{2}}s$$

$$\leq \nabla_T \tilde{g}(S', x'(S'), T', x'(T'))s + \int_{T'}^{T'+s} \nabla_{x_1}\tilde{g}(S', x'(S'), T', x'(T'))$$

$$\cdot f(t, x'(T'), u_k(t)) \, dt + r_1 s^2 + k_e(1 + \bar{c}^2)^{\frac{1}{2}}s.$$

Recalling that $-p(T') \in \nabla_{x_1}\tilde{g}(S', x(S'), T', x(T')) + k_e \mathbb{B}$, from (9.6.5) we deduce that for all k

$$\nabla_{x_1}\tilde{g}(S', x'(S'), T', x'(T')) \cdot f(t, x'(T'), u_k(t))$$

$$\leq (-H)(t, x'(T'), p(T')) + \frac{1}{k} + k_e \bar{c} \quad \text{a.e. } t \in [T', T' + s].$$

Therefore, combining the preceding two relations and taking the limit as $k \to +\infty$, we arrive at

$$0 \leq \nabla_T \tilde{g}(S', x'(S'), T', x'(T'))s$$

$$+ \int_{T'}^{T'+s} (-H)(t, x'(T'), p(T')) \, dt + r_1 s^2 + k_e(\bar{c} + (1 + \bar{c}^2)^{\frac{1}{2}})s.$$

9.6 A Free End-Time Maximum Principle

We have shown that

$$\begin{cases} T' \text{ is a local minimum over } [T', T' + \sigma'] \text{ of} \\ T \to \nabla_T \tilde{g}(S', x'(S'), T', x'(T'))(T - T') \\ + \int_{T'}^{T}(-H)(t, x'(T'), p(T'))\,dt \\ +r_1|T - T'|^2 + k_e(\bar{c} + (1+\bar{c}^2)^{\frac{1}{2}})|T - T'|. \end{cases} \quad (9.6.6)$$

Interpreting, as usual, the integral $\int_{T'}^{T}(-H)(t, x'(T'), p(T'))dt$ as $-\int_{T}^{T'}(-H)(t, x'(T'), p(T'))dt$ for $T < T'$, we deduce from (9.6.4) and (9.6.6) that

$$\begin{cases} T' \text{ is a local minimum over } [S', T' + \sigma'] \text{ of} \\ T \to \nabla_T \tilde{g}(S', x'(S'), T', x'(T'))(T - T') + \int_{T'}^{T}(-H)(t, x'(T'), p(T'))\,dt \\ +r_1(T - T')^2 + k_e(\bar{c} + (1+\bar{c}^2)^{\frac{1}{2}})|T - T'|. \end{cases}$$

Applying Proposition 9.3.2, in which we take the indefinite integral function ψ to be $\psi(t) := \int_{T'}^{t}(-H)(s, x'(T'), p(T'))ds$, $t \in [S', T' + \sigma']$, noting that 0 is a limiting subgradient of a function at a minimizing point and also applying the sum rule for limiting subdifferentials of Lipschitz functions, we deduce that

$$-\nabla_T \tilde{g}(S', x'(S'), T', x'(T')) \in \underset{t \to T'}{\text{sub-ess}}\,(-H)(t, x'(T'), p(T')) + k_e(\bar{c} + (1+\bar{c}^2)^{\frac{1}{2}})\mathbb{B}.$$

Then, using Proposition 9.3.2, we obtain that

$$\nabla_T \tilde{g}(S', x'(S'), T', x'(T')) \in \underset{t \to T'}{\text{super-ess}}\, H(t, x'(T'), p(T')) + k_e(\bar{c} + (1+\bar{c}^2)^{\frac{1}{2}})\mathbb{B}.$$

A similar analysis, in which we vary the left end-time, yields

$$-\nabla_S \tilde{g}(S', x'(S'), T', x'(T')) \in \text{sub-ess}_{t \to S'} H(t, x'(S'), p(S')) + k_e(\bar{c} + (1+\bar{c}^2)^{\frac{1}{2}})\mathbb{B}.$$

In view of (9.6.3), we have confirmed that there exist elements $\xi_0 \in \text{sub-ess}_{t \to S'} H(t, x'(S'), p(S'))$ and $\xi_1 \in \text{super-ess}_{t \to T'} H(t, x'(T'), p(T'))$ such that

$$(-\xi_0, p(S'), \xi_1, -p(T')) \in \nabla \tilde{g}(S', x'(S'), T', x'(T')) + 2\sqrt{2}k_e(1+\bar{c}^2)^{\frac{1}{2}}\mathbb{B}.$$

We have proved a special case of the proposition (with weakened transversality condition), when g has a special structure. We now assume that g is an arbitrary Lipschitz continuous function.

For $i = 1, 2, \ldots$ let g_i be the quadratic inf convolution of g (with parameter $\alpha = i$). Consider, for each i, the following variant of (FET)$'$, in which we replace g by g_i:

$(FET)_i'$ $\begin{cases} \text{Minimize } g_i(S, x(S), T, x(T))) \\ \text{over intervals } [S, T], \text{ arcs } x \in W^{1,1}([S, T]; \mathbb{R}^n) \\ \quad \text{and measurable functions } u : [S, T] \to \mathbb{R}^m \text{ satisfying} \\ \dot{x}(t) = f(t, x(t), u(t)) \text{ a.e. } t \in [S, T], \\ u(t) \in U(t) \text{ a.e. } t \in [S, T], \\ |T - T'|, |S - S'| \le \sigma'. \end{cases}$

Write $\alpha_i := k_g/\sqrt{i}$, where k_g is a Lipschitz constant for g. According to the properties of quadratic inf convolutions, $([S', T'], x', u')$ is an α_i^2 minimizer for this problem.

We can express the preceding problem as

Minimize $\{J_i([S, T], x, u) := g_i(S, x(S), T, x(T))) : ([S, T], x, u) \in \mathcal{M}\}$.

($\mathcal{M}$ was defined in the statement of Lemma 9.6.4.) According to this lemma, $(\mathcal{M}, d_{\text{control}})$ is a complete metric space. J_i is continuous on $(\mathcal{M}, d_{\text{control}})$. It follows then from Ekeland's theorem that, for each i, there exists an element $([S_i, T_i], x_i, u_i) \in \mathcal{M}$ that minimizes $J_i([S, T], x, u) + \alpha_i d_{\text{control}}(([S, T], x, u), ([S_i, T_i], x_i, u_i))$ over $\mathcal{M}$ and

$$d_{\text{control}}(([S_i, T_i], x_i, u_i), ([S', T'], x', u')) \le \alpha_i.$$

This implies that $x_{ie} \to x'_e$ uniformly, $S_i \to S'$ and $T_i \to T'$. Fix a measurable selection $\tilde{u} : [S' - \sigma', T' + \sigma'] \to \mathbb{R}^m$ such that $\tilde{u}(t) \in U(t)$ for a.e. $t \in [S' - \sigma', T' + \sigma']$. We write u'_i the measurable extension of the control u_i defined on $[S_i, T_i] \subset [S' - \sigma', T' + \sigma']$ to the interval $[S' - \sigma', T' + \sigma']$ obtained concatenating u_i with $\tilde{u}$: $u'_i(t) := \begin{cases} u_i(t) \text{ if } t \in [S_i, T_i] \\ \tilde{u}(t) \text{ if } t \in [S' - \sigma', T' + \sigma'] \setminus [S_i, T_i] \end{cases}$. Noting that, for each interval $[S, T] \subset [S' - \sigma', T' + \sigma']$,

$$\text{meas } \{t \in [S, T] \cap [S_i, T_i] : u(t) \ne u_i(t)\}$$
$$\le \text{meas } \{t \in [S, T] : u(t) \ne u'_i(t)\} \quad (9.6.7)$$

(with equality if $[S, T] = [S_i, T_i]$), we see that $([S_i, T_i], x_i, u_i)$ is a minimizer for

$(FET)_i$ $\begin{cases} \text{Minimize } g_i(S, x(S), T, x(T))) + \alpha_i(\int_S^T m_i(t, u(t))dt \\ \qquad\qquad + |S - S_i| + |T - T_i| + |x(S) - x_i(S_i)|) \\ \text{over } [S, T] \subset \mathbb{R} \text{ and } x \in W^{1,1}([S, T]; \mathbb{R}^n) \\ \qquad \text{and measurable functions } u : [S, T] \to \mathbb{R}^m \text{ satisfying} \\ \dot{x}(t) \in f(t, x(t), u(t)) \text{ and } u(t) \in U(t), \text{ a.e. } t \in [S, T], \\ |T - T'|, |S - S'| \le \sigma', \end{cases}$

in which $m_i(t, u) := \begin{cases} 0 \text{ if } u = u'_i(t) \\ 1 \text{ if } u \ne u'_i(t) \end{cases}$.

9.6 A Free End-Time Maximum Principle

By properties of the quadratic inf convolution, there exists a quadratic function $\tilde{g}_i$ such that

$$\nabla \tilde{g}_i(S_i, x_i(S_i), T_i, x_i(T_i)) \in \partial g(e_i),$$

for some $e_i \in (S_i, x_i(S_i), T_i, x_i(T_i)) + (k_g/i)\mathbb{B}$, and

$$\begin{cases} \tilde{g}_i(e) \geq g_i(e) \text{ for all } e \in \mathbb{R} \times \mathbb{R}^n \times \mathbb{R} \times \mathbb{R}^n \\ \tilde{g}_i(e) = g_i(e) \text{ when } e = (S', x'(S'), T', x'(T')). \end{cases}$$

It follows that $([S_i, T_i], x_i, u_i)$ is a minimizer for a variant of (FET)$_i$, in which the quadratic function $\tilde{g}_i$ replaces g_i, namely

$$\begin{cases} \text{Minimize } \tilde{g}_i(S, x(S), T, x(T))) + \alpha_i (\int_S^T m_i(t, u(t))dt \\ \qquad + |x(S) - x_i(S_i)| + |S - S_i| + |T - T_i|) \\ \text{over } [S, T] \subset \mathbb{R} \text{ and } x \in W^{1,1}([S, T]; \mathbb{R}^n) \\ \qquad \text{and measurable functions } u : [S, T] \to \mathbb{R}^m \text{ satisfying} \\ \dot{x}(t) \in f(t, x(t), u(t)) \text{ and } u(t) \in U(t), \text{ a.e. } t \in [S, T], \\ |T - T'|, |S - S'| \leq \sigma'. \end{cases}$$

Since $S_i \to S'$ and $T_i \to T'$, we can arrange (by eliminating initial sequence terms) that $|S_i - S'|, |T_i - T'| \leq \sigma'/2$. But then $([S_i, T_i], x_i, u_i)$ remains a minimizer, when we replace the end-time constraint by $|T - T_i|, |S - S_i| \leq \sigma'/2$. Absorbing the integral term in the cost into the dynamic constraint via state augmentation, we arrive at an example of the special case of (FET)$'$, for which analysis employed in the first part of the proof provides necessary conditions of optimality. The relevant hypotheses are satisfied, when $\sigma'/2$ replaces the parameter σ'. It follows that there exists $p_i \in W^{1,1}([S_i, T_i]; \mathbb{R}^n)$ such that

(A)$'$: $-\dot{p}_i(t) \in \text{co} \, \partial_x [p_i(t) \cdot f(t, x_i(t), u_i(t))]$ a.e. $t \in [S_i, T_i]$,
(B)$'$: $p_i(t) \cdot f(t, x_i(t), u_i(t)) \geq p_i(t) \cdot f(t, x_i(t), u) - \alpha_i$ for all $u \in U(t)$
$\qquad\qquad\qquad\qquad\qquad\qquad\qquad\qquad\qquad$ a.e. $t \in [S_i, T_i]$,
(C)$'$: there exist $\xi_0^i \in \text{sub-ess}_{t \to S_i} H(t, x_i(S_i), p(S_i))$ and
$\qquad\qquad\qquad \xi_1^i \in \text{super-ess}_{t \to T_i} H(t, x_i(T_i), p(T_i))$ such that
$$(-\xi_0^i, p_i(S_i), \xi_1^i, -p_i(T_i)) \in \partial g(e_i) + 2\sqrt{2}\alpha_i(1 + \bar{c}^2)^{\frac{1}{2}}\mathbb{B}.$$

We now use our 'constant extrapolation' convention to extend the domains of the p_i's to the entire real line. We deduce from (A)$'$ and (C)$'$ that the p_i's are uniformly bounded with uniformly integrably bounded derivatives. We can arrange then, by subsequence extraction, that $p_i \to p$ uniformly and $\dot{p}_i \to \dot{p}$ weakly in L^1 for some $p \in W^{1,1}$. The sequences $\{\xi_0^i\}$ and $\{\xi_1^i\}$ are bounded. A further subsequence extraction ensures that they have limits ξ_0 and ξ_1 respectively. We know however that $([S_i, T_i], x_i, u_i) \to ([S', T'], x', u')$ w.r.t. the d_{control} metric and $e_i \to (S', x'(S'), T', x'(T'))$. A by now familiar convergence analysis permits us to pass to the limit as $i \to \infty$ in conditions (A)' - (C)' and arrive at the relations asserted in the Proposition statement. As earlier, we make use of the stability, under limit taking, properties of the limiting process, expressed in terms of the sub- and super essential values.

□

Step 2 (Completion of the Proof) By imitating the analysis in Chap. 7, it is a straightforward task to show that the assertions of the theorem are true in general, if we can demonstrate their validity under the additional hypothesis:

(A): There exist integrable functions $k_0 : [\bar{S} - \sigma, \bar{T} + \sigma] \to \mathbb{R}$, $c_0 : [\bar{S} - \sigma, \bar{T} + \sigma] \to \mathbb{R}$, and a constant $\bar{c} \geq 0$ such that

(i): $|f(t, x, u) - f(t, x', u)| \leq k_0(t)|x - x'|$, for all $x, x' \in \bar{x}(t) + \epsilon \mathbb{B}$, $u \in U(t)$, a.e. $t \in [\bar{S} - \sigma, \bar{T} + \sigma]$,

(ii): $|f(t, x, u)| \leq c_0(t)$, for all $x \in \bar{x}(t) + \epsilon \mathbb{B}$, $u \in U(t)$, a.e. $t \in [\bar{S} - \sigma, \bar{T} + \sigma]$,

(iii): $|f(t, x, u)| \leq \bar{c}$ for all $x \in \bar{x}(t) + \epsilon \mathbb{B}$, $u \in U(t)$, a.e. $t \in [\bar{S} - \sigma, \bar{S} + \sigma] \cup [\bar{T} - \sigma, \bar{T} + \sigma]$.

Observe that when we apply the 'reduction technique' which allows us to restrict attention to (A), the constant $\bar{c}$ appearing in (A) (iii) is the constant of hypothesis (H4) of the theorem statement. Therefore, in the limit taking procedures we can make use of the robustness properties of sub and super essential values.

We can also assume, without loss of generality, that $([\bar{S}, \bar{T}], \bar{x}, \bar{u})$ is a (global) minimizer for (FT)'. To see this, assume that $([\bar{S}, \bar{T}], \bar{x}, \bar{u})$ is minimizer merely w.r.t. candidate admissible processes $([S, T], x, u)$ satisfying $d(([S, T], x), ([\bar{S}, \bar{T}], \bar{x})) \leq \beta$, for some $\beta > 0$. Then $([\bar{S}, \bar{T}], \bar{x}, \bar{u})$ is a (global) minimizer for (FT)' when we add the constraint:

$$\int_S^T |f(t, x(t), u(t)) - f(t, \bar{x}(t), \bar{u}(t))| dt$$
$$+ |x(S) - \bar{x}(\bar{S})| + (1 + 2\bar{c})(|T - \bar{T}| + |S - \bar{S}|) \leq \beta.$$

By applying the special case of the theorem, in which $([\bar{S}, \bar{T}], \bar{x}, \bar{u})$ is a minimizer to this modified problem, following absorption of the integral term in the added constraint into the dynamics via state augmentation, and noting the added constraint is inactive, we validate the assertions of the theorem for the original problem, when $([\bar{S}, \bar{T}], \bar{x}, \bar{u})$ is merely a $W^{1,1}$ local minimizer.

Take $\alpha_i \downarrow 0$ and, for each i, define

$$\ell_i(S, x_0, T, x_i) := \max\{g(S, x(S), T, x(T)) - g(\bar{S}, \bar{x}(S), \bar{T}, \bar{x}(T)) + \alpha_i^2,$$
$$d_C(S, x(S), T, x(T))\}.$$

9.6 A Free End-Time Maximum Principle

Now consider the problem:

$$(P_1^i) \begin{cases} \text{Minimize } J_1^i([S, T], x, u) \text{ subject to} \\ \dot{x}(t) = f(t, x(t), u(t)), \text{ a.e. } t \in [S, T], \\ u(t) \in U(t), \text{ a.e. } t \in [S, T], \\ |S - \bar{S}|, |T - \bar{T}| \le \sigma \end{cases}$$

in which $J_1^i([S, T], x, u) := \ell_i(S, x(S), T, x(T))$. Since $J_1^i(([\bar{S}, \bar{T}], \bar{x}, \bar{u}) = \alpha_i^2$, and J_1^i is non-negative valued, $([\bar{S}, \bar{T}], \bar{x}, \bar{u})$ is an α_i^2-minimizer.

Let $\mathcal{M}$ denote the set of admissible processes $([S, T], x, u)$ for (P_1^i). Observe that the data characterizing the admissible processes for (P_1^i) satisfy the assumptions of the data which characterize the admissible processes for $(FET)'$, and, so, from Lemma 9.6.4 we know that $(\mathcal{M}, d_{\text{control}})$ is a complete metric space. Notice also that the function $J_1^i : \mathcal{M} \to \mathbb{R}$ is continuous w.r.t. the metric d_{control}. By Ekeland's theorem, we are assured of the existence of a minimizer $([S_i, T_i], x_i, u_i)$ for a modification of the above optimization problem, in which $J_1^i([S, T], x, u)$ is replaced by $J_1^i([S, T], x, u) + d_{\text{control}}((([S, T], x, u), ([S_i, T_i], x_i, u_i))$. Furthermore

$$d_{\text{control}}(([S_i, T_i], x_i, u_i), ([\bar{S}, \bar{T}], \bar{x}, \bar{u})) \le \alpha_i. \tag{9.6.8}$$

Similarly as before, we fix a measurable selection $\tilde{u} : [\bar{S} - \sigma, \bar{T} + \sigma] \to \mathbb{R}^m$ such that $\tilde{u}(t) \in U(t)$ for a.e. $t \in [\bar{S} - \sigma, \bar{T} + \sigma]$, and we write $\tilde{u}_i$ the measurable extension to $[\bar{S} - \sigma, \bar{T} + \sigma]$ of the control u_i (defined on $[S_i, T_i] \subset [\bar{S} - \sigma, \bar{T} + \sigma]$), which is obtained as follows:

$$\tilde{u}_i(t) := \begin{cases} u_i(t) & \text{if } t \in [S_i, T_i] \\ \tilde{u}(t) & \text{if } t \in [\bar{S} - \sigma, \bar{T} + \sigma] \setminus [S_i, T_i]. \end{cases}$$

Taking account of inequality (9.6.7), we conclude that $([S_i, T_i], x_i, u_i)$ is a minimizer for

$$(\tilde{P}_1^i) \begin{cases} \text{Minimize } \ell_i(S, x(S), T, x(T)) + \alpha_i (\int_S^T m_i(t, u(t))dt \\ \qquad\qquad + |x(S) - x_i(S_i)| + |S - S_i| + |T - T_i|) \\ \text{over } [S, T] \subset \mathbb{R} \text{ and } x \in W^{1,1}([S, T]; \mathbb{R}^n) \\ \text{and measurable functions } u : [S, T] \to \mathbb{R}^m \text{ satisfying} \\ \dot{x}(t) \in f(t, x(t), u(t)) \text{ and } u(t) \in U(t), \text{ a.e. } t \in [S, T], \\ |S - \bar{S}|, |T - \bar{T}| \le \sigma, \end{cases}$$

in which $m_i(t, u) := \begin{cases} 0 & \text{if } u = \tilde{u}_i(t) \\ 1 & \text{if } u \ne \tilde{u}_i(t) \end{cases}$.

It can be deduced from (9.6.8) that $\|x_i - \bar{x}\|_{L^\infty} \to 0$ as $i \to \infty$ (remember our 'extension' convention for interpreting such relations) and $|T_i - \bar{T}|, |T_i - \bar{T}| \to 0$. It follows that, for i sufficiently large, (x_i, u_i) is minimizer for $(\tilde{P}_1^i)$, when the end-time constraint is replaced by '$|T_i-\bar{T}|, |T_i-\bar{T}| \leq \sigma/2$'. Now absorb the integral cost term into the dynamics by state augmentation. The data for the problem we thereby obtain satisfies the hypotheses of Proposition 9.6.3. We deduce the existence of a costate trajectory p_i such that

(b)' : $-\dot{p}_i(t) \in \text{co}\, \partial_x\, p_i(t) \cdot f(t, x_i(t), u_i(t))$, a.e. $t \in [S_i, T_i]$,
(c)': $p_i(t) \cdot f(t, x_i(t), u_i(t)) \geq p_i(t) \cdot f(t, x_i(t), u) - \alpha_i$ for all $u \in U(t)$
a.e. $t \in [S_i, T_i]$,
(d)'': there exist $\xi_0^i \in \text{sub-ess}_{t \to S_i} H(t, x_i(S_i), p(S_i))$ and
$\xi_1^i \in \text{super-ess}_{t \to T_i} H(t, x_i(T_i), p(T_i))$ such that
$(-\xi_0^i, p_i(S_i), \xi_1^i, -p_i(T_i)) \in \partial \ell_i(S_i, x_i(S_i), T_i, x_i(T_i)) + \sqrt{3}\alpha_i\, \mathbb{B}$.

Let us examine the implications of the transversality condition (d)''. We observe that, for i sufficiently large:

$$\max\{g(S_i, x_i(S_i), T_i, x_i(T_i)) - g(\bar{S}, \bar{x}(\bar{S}), \bar{T}, \bar{x}(\bar{T}))$$
$$+ \alpha_i^2, d_C(S_i, x_i(S_i), T_i, x_i(T_i))\} > 0. \qquad (9.6.9)$$

Indeed if this were not the case, we would have $g(S_i, x_i(S_i), T_i, x_i(T_i)) - g(\bar{S}, \bar{x}(\bar{S}), \bar{T}, \bar{x}(\bar{T})) \leq -\alpha_i^2$ and $d_C(S_i, x_i(S_i), T_i, x_i(T_i)) = 0$. This contradicts the L^∞ local optimality of $([\bar{S}, \bar{T}], \bar{x}, \bar{u})$ for problem $(FT)'$. In consequence of the max rule for limiting subdifferentials, there exists $\lambda_i \in [0, 1]$ such that

$$\partial \ell_i(S_i, x_i(S), T_i, x_i(T)) \subset \lambda_i \partial g(S_i, x_i(S), T_i, x_i(T))$$
$$+ (1 - \lambda_i)\partial d_C(S_i, x_i(S), T_i, x_i(T)).$$

A familiar argument, based on the max rule for limiting subdifferentials and limiting subgradient properties of the distance function, allows us to deduce from (9.6.9) that, in the preceding relation, ∂d_C can be replaced by $\partial d_C \cap \partial \mathbb{B}$, thus:

(d)': $(-\xi_0^i, p_i(S_i), \xi_1^i, -p_i(T_i)) \in \lambda_i \partial g(S_i, x_i(S), T_i, x_i(T))$
$+ (1 - \lambda_i)\left(\partial d_C(S_i, x_i(S), T_i, x_i(T)) \cap \partial \mathbb{B}\right) + \sqrt{3}\alpha_i\, \mathbb{B}$.

Since $|\xi_0^i|, |\xi_1^i| \leq \bar{c}\|p_i\|_{L^\infty}$ we deduce from the preceding relation that $2(1 + \bar{c})\|p_i\|_{L^\infty} \geq (1 - \lambda_i) - \lambda_i k_g - \sqrt{3}\alpha_i$. Here k_g is a Lipschitz constant for g and $\bar{c}$ is the constant of hypothesis (H4) in the theorem statement. It follows that

(a)': $2(1 + \bar{c})\|p_i\|_{L^\infty} + (1 + k_g)\lambda_i \geq 1 - \sqrt{3}\alpha_i$.

Under the strengthened hypotheses on $f(t, x, u)$ introduced at the beginning of this step of the proof ('uniform Lipschitz continuity w.r.t. x with integrable Lipschitz bound'), we can deduce from (b)' and (d)', with the help of Gronwall's lemma, that the p_i's (extended to all of $(-\infty, \infty)$ by constant extrapolation) are uniformly bounded, with uniformly integrably bounded derivatives. But then, along a subsequence, $p_i \to p$ uniformly and $\dot{p}_i \to p_i$, weakly in L^1, for some absolutely continuous function p. We can also arrange, by further subsequence extraction, that $\lambda_i \to \lambda$, $\xi_0^i \to \xi_0$ and $\xi_1^i \to \xi_1$.

We will recognize in conditions (a)′–(d)′, perturbed versions of the desired conditions. A standard convergence analysis permits us to pass to the limit as $i \to \infty$. We thereby recover the conditions asserted in the theorem statement. □

9.7 Appendix: Metrics on the Space of Free End-Time Trajectories

At several points of this chapter, we use the technique of time re-parameterization to reduce free end-time, Lipschitz time-dependent, dynamic optimization problems, to fixed end-time dynamic optimization problems, to which we apply earlier derived necessary conditions for $W^{1,1}$ local optimality. (Here, $W^{1,1}$ local optimality is interpreted in terms of the d metric, defined in (9.1.4).) The application of the re-parameterization technique for the study of $W^{1,1}$ local minimizers is possible then, only if we can show that there exists a $W^{1,1}$ neighbourhood of the nominal arc $\bar{x}$ in the space of re-parameterized F trajectories that maps into a specified $W^{1,1}$ neighbourhood of $\bar{x}$ in the space of original F trajectories, under the inverse parameterization. The following lemma provides this step.

Lemma 9.7.1 *Take an interval $[\bar{S}, \bar{T}]$ and $\bar{x} \in W^{1,1}([\bar{S}, \bar{T}]; \mathbb{R}^n)$. Choose any $\beta > 0$. Then there exist $\beta' > 0$ and $\alpha' \in (0, 1/2)$ with the following properties: For any $S \in \mathbb{R}$ and $w : [\bar{S}, \bar{T}] \to \mathbb{R}$, such that $w(t) \in [1 - \alpha', 1 + \alpha']$ a.e., let $\psi : [\bar{S}, \bar{T}] \to [S, T]$ be the monotone, onto, Lipschitz continuous mapping, with Lipschitz continuous inverse*

$$\psi(s) = S + \int_{\bar{S}}^{s} w(\tau) d\tau, \quad \text{for } s \in [\bar{S}, \bar{T}],$$

where $T = \psi(\bar{T})$. Take any $y \in W^{1,1}([\bar{S}, \bar{T}]; \mathbb{R}^n)$ such that

$$\int_{\bar{S}}^{\bar{T}} |\dot{y}(s) - \dot{\bar{x}}(s)| ds + |y(\bar{S}) - \bar{x}(\bar{S})| + |S - \bar{S}| \leq \beta'. \tag{9.7.1}$$

Now define $x \in W^{1,1}([S, T]; \mathbb{R}^n)$ to be the function $x = y \circ \psi^{-1}$. Then

$$\int_{S \wedge \bar{S}}^{T \vee \bar{T}} |\dot{x}(t) - \dot{\bar{x}}(t)| dt + |x(S) - \bar{x}(\bar{S})| + |S - \bar{S}| + |T - \bar{T}| \leq \beta. \tag{9.7.2}$$

(Recall that, according to our convention, $\dot{\bar{x}}(s) = 0$ for $s \notin [\bar{S}, \bar{T}]$, etc.)

Proof Take $\alpha' \in (0, 1/2)$, $\gamma > 0$ and $\beta' > 0$. Take $S \in \mathbb{R}$ and any measurable function $w : [\bar{S}, \bar{T}] \to \mathbb{R}$ such that $w(t) \in [1 - \alpha', 1 + \alpha']$ a.e.. Let $\psi : [\bar{S}, \bar{T}] \to$

$[S, T]$ be as defined in the Lemma statement. Take any $y \in W^{1,1}([\bar{S}, \bar{T}]; \mathbb{R}^n)$ such that (9.7.1) is satisfied and write $x = y \circ \psi^{-1}$.

It is a simple exercise to show that

$$|\psi(s) - s| \le |S - \bar{S}| + \alpha' \times |\bar{T} - \bar{S}|, \quad \text{for all } s \in [\bar{S}, \bar{T}]. \tag{9.7.3}$$

Let $a : \mathbb{R} \to \mathbb{R}^n$ be any twice continuously differentiable function, with uniformly bounded second derivative, such that $a(\bar{S}) = \bar{x}(\bar{S})$ and

$$\int_{(-\infty, +\infty)} |\dot{a}(t) - \dot{\bar{x}}(t)| dt \le \gamma. \tag{9.7.4}$$

(We use our extension convention to interpret the integral.) Define the non-negative function $\bar{\omega} : [0, \infty) \to [0, \infty)$:

$$\bar{\omega}(r) := \sup\{ \int_E |\dot{\bar{x}}(t)| dt \mid E \subset \mathbb{R} \text{ is measurable and } meas\{E\} \le r \}.$$

From well-known properties of integrable functions, we have $\bar{\omega}(r) \to 0$, as $r \downarrow 0$.

Define

$$e_1 := \|\dot{a}\|_{L^1} \alpha'$$
$$e_2 := (1 + \alpha')\|\ddot{a}\|_{L^\infty}(|S - \bar{S}| + \alpha'|\bar{T} - \bar{S}|)|\bar{T} - \bar{S}|$$
$$e_3 := \bar{\omega}(|S - \bar{S}| + |T - \bar{T}|).$$

Since $x \circ \psi(s) = y(s)$ we know that $(\dot{x} \circ \psi)(s) w(s) = \dot{y}(s)$. Since $|w(t) - 1| \le \alpha'$, it follows from (9.7.4) that

$$\int_{\bar{S}}^{\bar{T}} |\dot{y}(s) - \dot{\bar{x}}(s)| ds \ge \int_{\bar{S}}^{\bar{T}} |(\dot{x} \circ \psi)(s) w(s) - \dot{a}(s)| ds - \gamma$$

$$\ge \int_{\bar{S}}^{\bar{T}} w(s) |(\dot{x} \circ \psi)(s) - \dot{a}(s)| ds - \gamma - e_1$$

$$\ge \int_{\bar{S}}^{\bar{T}} w(s) |(\dot{x} \circ \psi)(s) - (\dot{a} \circ \psi)(s)| ds - \gamma - e_1 - e_2$$

$$= \int_{S}^{T} |\dot{x}(t) - \dot{a}(t)| dt - \gamma - e_1 - e_2$$

(we have made a change of independent variable)

$$\ge \int_{S}^{T} |\dot{x}(t) - \dot{\bar{x}}(t)| dt - 2\gamma - e_1 - e_2$$

$$\ge \int_{S \wedge \bar{S}}^{T \vee \bar{T}} |\dot{x}(t) - \dot{\bar{x}}(t)| dt - 2\gamma - e_1 - e_2 - e_3.$$

9.7 Appendix: Metrics on the Space of Free End-Time Trajectories

We conclude that

$$\int_{S \wedge \bar{S}}^{T \vee \bar{T}} |\dot{x}(t) - \dot{\bar{x}}(t)| dt - ((1+\alpha')\|\ddot{a}\|_{L^\infty} |\bar{T} - \bar{S}|) |S - \bar{S}|$$
$$\leq \int_{\bar{S}}^{\bar{T}} |\dot{y}(s) - \dot{\bar{x}}(s)| ds + 2\gamma + (\|\dot{a}\|_{L^1} + (1+\alpha')\|\ddot{a}\|_{L^\infty} |\bar{T} - \bar{S}|^2)\alpha'$$
$$+ \bar{\omega}(|S - \bar{S}| + |T - \bar{T}|).$$

Since $x(S) = y(\bar{S})$ and, according to (9.7.3), $|T - \bar{T}| \leq |S - \bar{S}| + \alpha'|\bar{T} - \bar{S}|$, it follows from the preceding inequalities that

$$\int_{S \wedge \bar{S}}^{T \vee \bar{T}} |\dot{x}(t) - \dot{\bar{x}}(t)| dt + |x(S) - \bar{x}(\bar{S})| + |S - \bar{S}| + |T - \bar{T}|$$
$$\leq \int_{\bar{S}}^{\bar{T}} |\dot{y}(s) - \dot{\bar{x}}(s)| ds + |y(S) - \bar{x}(\bar{S})|$$
$$+ (1 + (1+\alpha')\|\ddot{a}\|_{L^\infty} |\bar{T} - \bar{S}|) |S - \bar{S}| 2\gamma$$
$$+ (\|\dot{a}\|_{L^1} + (1+\alpha')\|\ddot{a}\|_{L^\infty} |\bar{T} - \bar{S}|^2)\alpha'$$
$$+ \bar{\omega}(|S - \bar{S}| + |T - \bar{T}|) + |T - \bar{T}|$$
$$\leq \int_{\bar{S}}^{\bar{T}} |\dot{y}(s) - \dot{\bar{x}}(s)| ds + |y(S) - \bar{x}(\bar{S})|$$
$$+ (2 + (1+\alpha')\|\ddot{a}\|_{L^\infty} |\bar{T} - \bar{S}|) |S - \bar{S}| + 2\gamma$$
$$+ (\|\dot{a}\|_{L^1} + (1+\alpha')\|\ddot{a}\|_{L^\infty} |\bar{T} - \bar{S}|^2 + |\bar{T} - \bar{S}|)\alpha'$$
$$+ \bar{\omega}(2|S - \bar{S}| + \alpha'|\bar{T} - \bar{S}|).$$

Now choose β', γ and α' such that

$$(2 + (1+\alpha')\|\ddot{a}\|_{L^\infty} |\bar{T} - \bar{S}|)\beta' + 2\gamma$$
$$+(\|\dot{a}\|_{L^1} + (1+\alpha')\|\ddot{a}\|_{L^\infty} |\bar{T}-\bar{S}|^2 + |\bar{T} - \bar{S}|)\alpha' + \bar{\omega}(2\beta' + \alpha'|\bar{T} - \bar{S}|) \leq \beta.$$

Then

$$\int_{S \wedge \bar{S}}^{T \vee \bar{T}} |\dot{x}(t) - \dot{\bar{x}}(t)| dt + |x(S) - \bar{x}(\bar{S})| + |S - \bar{S}| + |T - \bar{T}| \leq \beta.$$

This is the desired inequality.

□

9.8 Exercises

9.1 (Alternative Definitions for Essential Values) Take an integrable function $a : [S, T] \to \mathbb{R}$ and $\bar{t} \in (S, T)$. Fix any $S' \in [S, T]$. Define the absolutely continuous function $\psi : [S, T] \to \mathbb{R}$ to be

$$\psi(t) := \int_{S'}^{t} a(s)ds.$$

We know from Prop 9.3.2 that the limiting sub an super differentials of ψ at $\bar{t}$ can be represented in terms of the sub and super essential values of the integrand a at $\bar{t}$ (and do not depend on the choice of $S' \in [S, T]$ to define the indefinite integral function), thus

$$\partial \psi(\bar{t}) = \underset{t \to \bar{t}}{\text{sub-ess}}\, a(t) \quad \text{and} \quad -\partial(-\psi)(\bar{t}) = \underset{t \to \bar{t}}{\text{super-ess}}\, a(t),$$

We recall the definitions of sub and super essential values:

$$\underset{t \to \bar{t}}{\text{sub-ess}}\, a(t) :=$$

$$\left\{ \zeta \in \mathbb{R} : \exists t_i \to \bar{t} \text{ and } \zeta_i \to \zeta \text{ s.t.} \right.$$

$$\left. \limsup_{\epsilon \downarrow 0} \epsilon^{-1} \int_{t_i - \epsilon}^{t_i} a(s)ds \leq \zeta_i \leq \liminf_{\epsilon \downarrow 0} \epsilon^{-1} \int_{t_i}^{t_i + \epsilon} a(s)ds, \text{ for each } i \right\};$$

$$\underset{t \to \bar{t}}{\text{super-ess}}\, a(t) :=$$

$$\left\{ \zeta \in \mathbb{R} : \exists t_i \to \bar{t} \text{ and } \zeta_i \to \zeta \text{ s.t.} \right.$$

$$\left. \limsup_{\epsilon \downarrow 0} \epsilon^{-1} \int_{t_i}^{t_i + \epsilon} a(s)ds \leq \zeta_i \leq \liminf_{\epsilon \downarrow 0} \int_{t_i - \epsilon}^{t_i} a(s)ds, \text{ for each } i \right\}.$$

These relations permit us to derive first order necessary conditions of optimality, expressed in terms of sub and super essential values of the Hamiltonian evaluated along the optimal state and costate trajectories. However sub and super essential values were predated, regarding their role in necessary conditions, by related concepts:

Essential Value

$$\underset{t \to \bar{t}}{\text{ess}}\, a(t) := \left[\lim_{\epsilon \downarrow 0} \underset{\bar{t} - \epsilon \leq \bar{t} \leq t + \epsilon}{\text{ess inf}}\, a(t), \lim_{\epsilon \downarrow 0} \underset{\bar{t} - \epsilon \leq \bar{t} \leq t + \epsilon}{\text{ess sup}}\, a(t) \right],$$

9.8 Exercises

Lebesgue Essential Value

$$\mathcal{L}\text{-ess}_{t\to \bar{t}} a(t) := \Big\{\zeta \in \mathbb{R} :$$

$$\mathcal{L}\text{-meas}\{s \in [S, T] : |a(s)-\zeta| \leq \epsilon, |s-\bar{t}| \leq \epsilon\} > 0 \text{ for all } \epsilon > 0\Big\}.$$

(A): Show that $\partial \psi(\bar{t}) \subset \underset{t\to \bar{t}}{\text{ess}} \, a(t)$.

(B): Show that, if a is essentially bounded on a neighbourhood of $\bar{t}$, then $\partial \psi(\bar{t}) \subset \text{co}\, \mathcal{L}\text{-ess}_{t\to \bar{t}} a(t) = \underset{t\to \bar{t}}{\text{ess}} \, a(t)$.

(C): Consider the essentially bounded functions

$$a_1(t) := \begin{cases} -1 \text{ if } t \in [0, \tfrac{1}{2}] \\ +1 \text{ if } t \in (\tfrac{1}{2}, 1] \end{cases} \text{ and } a_2 := \begin{cases} +1 \text{ if } t \in [0, \tfrac{1}{2}] \\ -1 \text{ if } t \in (\tfrac{1}{2}, 1] \end{cases},$$

and write ψ_1 and ψ_2 for the corresponding indefinite integral functions. Set $\bar{t} = 1/2$. Show that

$$\mathcal{L}\text{-ess}_{t\to \bar{t}} a_1(t) \overset{\text{strict}}{\subset} \partial\psi_1(\bar{t}) \quad \text{and} \quad \partial\psi_2(\bar{t}) \overset{\text{strict}}{\subset} \underset{t\to \bar{t}}{\text{ess}} \, a_2(t).$$

Remark
This exercise shows that the concepts 'essential value', 'Lebesgue essential value' and its convex hull all fail precisely to capture precisely the elements in the limiting subdifferential of indefinite integral functionals.

9.2 (Strengthened Boundary Condition on Maximized Hamiltonian) Consider

$$\begin{cases} \text{Minimize } g(x(T)) \\ \text{over } T > 0, x \in W^{1,1}([0, T]; \mathbb{R}) \text{ and meas } u : [0, T] \to \mathbb{R} \\ \text{s.t.} \\ \dot{x}(t) \in F(t) \text{ a.e. } t \in [0, T], \\ x(0) = 0, \end{cases}$$

in which

$$g(x) := -x \text{ and } F(t) := \begin{cases} [-2, -1] \text{ if } t < 1 \\ [0, 1] \quad\;\; \text{ if } t \geq 1. \end{cases}$$

Write $H(t, p) := \max\{p \cdot v : v \in F(t)\}$. Take the admissible F trajectory $([0, \bar{T} = 1], \bar{x}(t) \equiv -t)$.

The free-time related component of the necessary conditions of Theorem 9.4.1 is the boundary condition

$$0 \in \underset{t \to \bar{T}}{\text{super-ess}}\, H(t, p(\bar{T})) \tag{9.8.1}$$

in which p is the co-state trajectory. This condition replaces previous boundary conditions in, say, [194]

$$0 \in \underset{t \to \bar{T}}{\text{ess}}\, H(t, p(\bar{T})). \tag{9.8.2}$$

Show that $([0, \bar{T} = 1], \bar{x}(t) \equiv -t)$

(i): is not an extremal with reference to the 'super essential value' boundary condition (9.8.1),
(ii): is an extremal with reference to the standard 'essential value' boundary condition (9.8.2).

i.e. the refined free time necessary condition Theorem 9.4.1 excludes $([0, \bar{T} = 1], \bar{x}(t) \equiv -t)$ as a candidate for minimizer, but previous free time necessary conditions fail to do so.

9.9 Notes for Chapter 9

In this chapter we have derived two kinds of necessary conditions for free end-time dynamic optimization problems or, to be more precise, problems for which the end-times are included among the choice variables.

Free end-time necessary conditions of the first kind give information about the nature of the Hamiltonian on the interior and at the end-points of the optimal time interval under consideration, in particular 'constancy of the Hamiltonian' for autonomous dynamics. They are restricted to problems with dynamic constraints which are Lipschitz continuous with respect to time. The approach followed here, to introduce a change of independent variable which replaces the free end-time problem by a fixed end-time problem and to apply the fixed end-time necessary conditions to the transformed problem, is substantially that employed by Clarke [65]. The underlying idea is implicit in earlier literature, however. (See, e.g., [96].)

The fact that necessary conditions of the first kind (including constancy of the Hamiltonian) are valid for arcs which are merely $W^{1,1}$ local minimizers, has been a matter of some past speculation. It is proved here apparently for the first time.

Free time necessary conditions of the second kind give information about the Hamiltonian merely at the optimal end-times, but cover problems for which the time dependence of the dynamic constraint is no longer assumed Lipschitz continuous. They are typically derived by reworking the proofs of the related fixed end-time necessary conditions to allow explicitly for variations of end-times as well as state trajectories.

9.9 Notes for Chapter 9

In the case of continuous time dependence, free time necessary conditions including boundary conditions on the Hamiltonian are evident in the early Russian literature. They are also to be found in Berkovitz' book [21].

Free time necessary conditions for problems with measurably time dependent data were introduced to the Western literature by Clarke and Vinter [79, 80], based on the concepts of 'essential value' and 'Lebesgue essential value' (see Exercise 9.1) of an almost everywhere defined function, to make sense of the boundary condition on the Hamiltonian in this case. (These ideas had precursors in the Russian literature [96].) The Lebesgue essential value is a classical concept (cf. [212]), aimed at identifying 'significant' points in the range of a function defined on a measure space.

Extension to problems with state constraints were carried out by Clarke, Loewen and Vinter [83] and by Rowland and Vinter [179], who employed a more refined analysis to allow for active state constraints at optimal end-times. The improved necessary conditions of this chapter, involving refinements of essential value concepts, namely sub and super essential values, is a very recent development in the theory [29].

Chapter 10
The Maximum Principle for Problems with Pathwise Constraints

Abstract This chapter provides necessary conditions of optimality for dynamic optimization problems involving pathwise constraints. Attention is directed at problems in which the dynamic constraint takes the form of a controlled differential equation. Two kinds of problems are considered. In the first kind, the pathwise constraint is imposed on the state trajectories (the 'pure state constraint' problem) and, in the second kind, it is imposed on both state trajectories and control functions (the 'mixed constraint' problem). For each kind, the necessary conditions resemble the maximum principle, but modified to take account of the pathwise constraint via additional Lagrange multipliers.

Concerning pure state constraint problems, we consider pathwise scalar functional inequality constraints. It might seem restrictive to limit attention to a scalar inequality constraint. But the constraint function, which is assumed merely to be upper semi-continuous and Lipschitz continuous w.r.t. to the state variable, can be constructed to embrace vector inequality constraints, set inclusion constraints and other constraints of interest. Simple examples illustrate that it is not possible, in general, to accommodate a pure state constraint by means of a simple integrable Lagrange multiplier. Instead we must do so, using a measure multiplier. The presence of a measure multiplier gives rise to a measure driven differential inclusion (differential equation in the smooth case) for the co-state trajectory, with discontinuous solutions. The discontinuous nature of the co-state trajectory is somewhat disguised in the standard formulations of the necessary conditions such as those given in this chapter, because these conditions are expressed in terms of a modified, absolutely continuous co-state trajectory, obtained by subtracting off the singular component from the 'true' co-state trajectory. Dealing with measure multipliers (ensuring stability of key relations under perturbations, for example) required additional analytic tools. These are provided in a separated section near the beginning of the chapter. The derivation of necessary conditions is based on introducing an integral penalty term relating to the pathwise constraint. With the help of Ekeland's theorem, we find a minimizer to a perturbed version of the penalized problem, close to the original minimizer. The necessary conditions for the perturbed problem are then interpreted as a perturbation version of the desired conditions. Passage to the limit, as the parameters controlling the perturbations vanish, completes the derivation.

We also address the degeneracy issue associated with necessary conditions for pure state constraint problems. This is connected with the fact that, for certain problems of interest, in which the pure state constraint is active at either of the end-times, the standard necessary conditions are trivial, in the sense that they are satisfied by all admissible state trajectory/control pairs. Following on from work of Arutyunov and Aseev, we show, under additional hypotheses, that this form of degeneracy can be eliminated.

We adopt a general formulation of the mixed state constraint problem which captures, as special cases, other formulations appearing in the literature, including formulations involving combined functional inequality/equality constraints and set inclusion relations. We seek necessary conditions in which, typically, the pathwise constraint is accommodated by an absolutely continuous Lagrange multiplier. The hypotheses we must impose to justify employing multipliers of this nature, which require some kind of stability of the set of controls satisfying the mixed constraint as the state varies, are violated for pure state constraint problems. Thus necessary conditions for mixed constrained problems do not subsume those for pure state constraints. Following Clarke and de Pinho, we derive conditions for mixed constraint problems by introducing a differential inclusion which incorporates the control constraint, the mixed constraint and the dynamic constraint. The desired necessary conditions are obtained by applying the generalized Euler Lagrange conditions of Chap. 8 to the resulting differential inclusion problem.

We consider also formulations of both pure state and mixed constraint problems, in which the end-times are included among the choice variables ('free time' problems). In each case, we provide the extra necessary conditions associated with the free end-times.

10.1 Introduction

We return to the framework of Chap. 7, in which the dynamic constraint takes the form of a differential equation parameterized by control functions. The goal is to extend the earlier derived necessary conditions of optimality, in the form of a maximum principle, to allow for additional pathwise constraints on state and control variables. The material in this chapter follows two well-established research traditions.

The first tradition concerns necessary conditions for problems with a pathwise constraint on the state variable x alone, which are referred to as problems with *pure state constraints*. A common formulation, and one which we adopt in this chapter, is a functional inequality

$$h(t, x(t)) \leq 0 \text{ for all } t \in [S, T],$$

10.1 Introduction

for a specified scalar valued function h. This appears somewhat restrictive but, as we show, many kinds of pure state constraints are covered by the formulation, including multiple functional inequality constraints, implicit constraints and constraints imposed on a closed subset of the underlying time interval. It was early apparent, even for the simplest of examples, that a straightforward extension of the standard maximum principle, expressed in terms of an absolutely continuous costate trajectory, in which the pathwise constraint is accommodated by an absolutely continuous state constraint Lagrange multiplier, cannot be achieved in a setting of significant generality. The extension can be successfully carried however, if we look for the state constraint Lagrange multiplier in the larger class of continuous linear functionals on $C(S, T)$, represented by a Borel measure. We derive necessary conditions with measure multipliers of this nature, for problems with non-smooth data.

The second tradition concerns pathwise constraints on both the state variable x and the control variable u. Such constraints are called *mixed constraints*. A simple formulation is

$$\phi(t, x(t), u(t)) \leq 0 \text{ and } \psi(t, x(t), u(t)) = 0, \text{ a.e. } t \in [S, T],$$

for specified functions ϕ and ψ ('functional inequality and equality' mixed constraints). We can exploit the fact that the constraint functionals also involve the control variable, now to derive broadly applicable necessary conditions of optimality, in which absolutely continuous multipliers are associated with the pathwise constraints. Hypotheses on the data that are applied to make this possible typically require that projections of the (x, u) dependent constraint sets on the u variable are stable in some sense, under perturbations of the x variable. Such hypotheses are typically not satisfied by pure state constraints. So we see that the theory for mixed constraint problems (when we require the Lagrange multipliers to be absolutely continuous) is not subsumed in the pure state constraints theory.

Until recently, the literature on necessary conditions for mixed constraint problems was a panoply of special cases, treating different ways of formulating the pathwise constraints differing over the precise nature of the necessary conditions. A significant unification was achieved by Clarke and de Pinho, following the approach: construct an equivalent dynamic optimization problem with dynamic constraint a differential inclusion whose right side is obtained by combining the dynamic and pathwise constraints, apply the generalized Euler Lagrange conditions and express these conditions directly in terms of the data for the original problem. We follow their approach to derive necessary conditions for a general formulation of the mixed constraints and then examine their implications for numerous special cases of interest.

10.2 Problems with Pure State Constraints: Preliminary Discussion

We consider the problem

$$(P) \begin{cases} \text{Minimize } g(x(S), x(T)) \\ \text{over } x \in W^{1,1}([S,T]; \mathbb{R}^n) \text{ and measurable functions } u \text{ satisfying} \\ \dot{x}(t) = f(t, x(t), u(t)) \quad \text{a.e.,} \\ u(t) \in U(t) \quad \text{a.e.,} \\ h(t, x(t)) \le 0 \text{ for all } t \in [S, T], \\ (x(S), x(T)) \in C, \end{cases}$$

the data for which comprise: an interval $[S, T]$, functions $g : \mathbb{R}^n \times \mathbb{R}^n \to \mathbb{R}$, $f : [S, T] \times \mathbb{R}^n \times \mathbb{R}^m \to \mathbb{R}^n$ and $h : [S, T] \times \mathbb{R}^n \to \mathbb{R}$, a multifunction $U : [S, T] \rightsquigarrow \mathbb{R}^m$ and a closed set $C \subset \mathbb{R}^n \times \mathbb{R}^n$. The new ingredient in problem (P) is the pathwise state constraint:

$$h(t, x(t)) \le 0 \text{ for all } t \in [S, T]. \tag{10.2.1}$$

We have chosen to formulate the state constraint in problem (P) as a scalar functional inequality constraint, partly because this is a kind of state constraint frequently encountered in engineering applications and partly because it is a convenient starting point for deriving necessary conditions for other types of state constraints (multiple state constraints, implicit state constraints, etc.) of interest.

As before, we refer to a measurable function $u : [S, T] \to \mathbb{R}^m$ which satisfies $u(t) \in U(t)$ a.e., as a *control function*. A *process* (x, u) comprises a control function u and an arc $x \in W^{1,1}([S, T]; \mathbb{R}^n)$ which is a solution to the differential equation $\dot{x}(t) = f(t, x(t), u(t))$ a.e.. A *state trajectory* x is the first component of some process (x, u). A process (x, u) is said to be *admissible* if the state trajectory x satisfies the end-point constraint $(x(S), x(T)) \in C$ and the state constraint $h(t, x(t)) \le 0$, for all $t \in [S, T]$.

Consistent with earlier terminology, a process $(\bar{x}, \bar{u})$ is said to be a $W^{1,1}$ *local minimizer* if there exists $\delta > 0$ such that the process $(\bar{x}, \bar{u})$ minimizes $g(x(S), x(T))$ over all admissible processes (x, u) satisfying

$$\|x - \bar{x}\|_{W^{1,1}} \le \delta.$$

If the $W^{1,1}$ norm is replaced by the L^∞ norm, $(\bar{x}, \bar{u})$ is called an L^∞ *local minimizer*.

What effect does the state constraint have on necessary conditions of optimality? We might expect that it can be accommodated by a Lagrange multiplier term

$$\int_S^T h(s, x(s)) m(s) ds$$

10.2 Problems with Pure State Constraints: Preliminary Discussion

added to the cost. That is to say, if $(\bar{x}, \bar{u})$ is a minimizer for (P), then for some appropriately chosen, non-negative valued function m satisfying the *complementary slackness condition* :

$$m(t) = 0 \text{ for } t \in \{s : h(s, \bar{x}(s)) < 0\}, \qquad (10.2.2)$$

$(\bar{x}, \bar{u})$ satisfies first order necessary conditions of optimality also for the state-constraint free problem

$$\begin{cases} \text{Minimize } g(x(S), x(T)) + \int_S^T h(t, x(t))m(t)dt \\ \text{over } x \in W^{1,1}([S, T]; \mathbb{R}^n) \text{ and measurable functions } u \text{ satisfying} \\ \dot{x} = f \text{ a.e., } u(t) \in U(t) \text{ a.e. and } (x(S), x(T)) \in C. \end{cases}$$

Reducing this problem to one with no integral cost term by state augmentation and applying the state constraint-free maximum principle we arrive at the following set of conditions: there exist $q \in W^{1,1}([S, T]; \mathbb{R}^n)$ and $\lambda \geq 0$ such that

$$(\lambda, q, m) \neq (0, 0, 0), \qquad (10.2.3)$$

$$-\dot{q}(t) = q(t)f_x(t, \bar{x}(t), \bar{u}(t)) - \lambda h_x(t, \bar{x}(t))m(t) \quad \text{a.e.}, \qquad (10.2.4)$$

$$(q(S), -q(T)) \in N_C(\bar{x}(S), \bar{x}(T)) + \lambda \nabla g(\bar{x}(S), \bar{x}(T)), \qquad (10.2.5)$$

$$q(t) \cdot f(t, \bar{x}(t), \bar{u}(t)) = \max_{u \in U(t)} q(t) \cdot f(t, \bar{x}(t), u) \quad \text{a.e..} \qquad (10.2.6)$$

These relations capture the essential character of necessary conditions for problems with pure state constraints, though one modification is required: in order to derive broadly applicable necessary conditions we must allow the Lagrange multiplier m (the precise nature of which has not been specified) to be the 'derivative of a function of bounded variation'. That is to say, the conditions assert the existence of a non-decreasing function of bounded variation $\nu : [S, T] \to \mathbb{R}$ such that the preceding relations apply with (10.2.2), (10.2.3) and (10.2.4) replaced by

$$\nu \text{ is constant on any subinterval of } \{t : h(t, x(t)) < 0\}, \qquad (10.2.7)$$

$$(\lambda, q, \nu) \neq (0, 0, 0), \qquad (10.2.8)$$

and, for all $t \in (S, T]$,

$$-q(t) = -q(S) + \int_S^t q(s)f_x(s, \bar{x}(s), \bar{u}(s))ds - \int_{[S,t]} h_x(s, \bar{x}(s))d\nu(s). \qquad (10.2.9)$$

(We have absorbed λ into the state constraint multiplier: $d\nu = \lambda m(t)dt$.) This is not surprising since, if h is a continuous function, problem (P) can be set up as an optimization problem over pairs of elements (x, u) satisfying (among other conditions) the constraint

$$G(x) \in P^-,$$

where $G : W^{1,1}([S, T]; \mathbb{R}^n) \to C([S, T]; \mathbb{R})$ is the function

$$G(x)(t) := h(t, x(t))$$

and P^- is the cone of non-positive valued continuous functions on $[S, T]$. One expects for such problems a multiplier rule to apply, involving a multiplier ξ in the negative polar cone of P^-, regarded as a subset of the topological dual space $C([S, T]; \mathbb{R})$, which satisfies

$$\langle \xi, G(\bar{x}) \rangle = 0.$$

Such a multiplier is represented by a non-decreasing function ν of bounded variation according to

$$\langle \xi, y \rangle = \int_S^T y(s) d\nu(s) \text{ for all } y \in C([S, T]; \mathbb{R}),$$

in which the integral is interpreted in the Stieltjes sense. The anticipated relations then are (10.2.5)–(10.2.9), in which $\lambda m(t)$ has been replaced by $d\nu(t)$, for some non-decreasing function of bounded variation ν.

Of course the costate arc q, which satisfies the integral equation (10.2.9), is a function of bounded variation itself. It is customary to aim for necessary conditions which differ from those just outlined in one small respect. A further modification to the necessary conditions, which is really just a matter of redefinition, is motivated by the desire to supply relations involving an absolutely continuous costate trajectory. The new costate trajectory, p, is obtained simply by subtracting the 'troublesome' bit off q:

$$p(t) := \begin{cases} q(S) & \text{if } t = S \\ q(t) - \int_{[S,t]} h_x(s, \bar{x}(s)) d\nu(s) & \text{if } t \in (S, T] . \end{cases}$$

Conversely, q can be expressed in terms of p as

$$q(t) := \begin{cases} p(S) & \text{if } t = S \\ p(t) + \int_{[S,t]} h_x(s, \bar{x}(s)) d\nu(s) & \text{if } t \in (S, T] . \end{cases} \quad (10.2.10)$$

The conditions now become: there exist $p \in W^{1,1}([S, T]; \mathbb{R}^n)$, $\lambda \geq 0$ and a non-decreasing function of bounded variation ν such that

$$(NC) \begin{cases} (\lambda, p, \nu) \neq 0, \\ -\dot{p}(t) = q(t) \cdot f_x(t, \bar{x}(t), \bar{u}(t)) \text{ a.e.,} \\ (p(S), -q(T)) \in \lambda \nabla g(\bar{x}(S), \bar{x}(T)) + N_C(\bar{x}(S), \bar{x}(T)), \\ q(t) \cdot f(t, \bar{x}(t), \bar{u}(t)) = \max_{u \in U(t)} q(t) \cdot f(t, \bar{x}(t), u) \text{ a.e.,} \\ \nu \text{ is constant on any subinterval of } \{t : h(t, \bar{x}(t)) < 0\}. \end{cases}$$

In these conditions, the function q is determined from p and ν according to (10.2.10).

The final touch is to express the conditions in terms of the regular Borel measure μ on $[S, T]$ associated with the function of bounded variation ν: μ is the unique regular Borel measure such that $\mu(I) = \int_I d\nu(t)$ for all closed sub-intervals $I \subset [S, T]$. The change then is to replace $\int h_x d\nu(s)$ by $\int h_x d\mu(s)$.

In this chapter, we shall validate a generalization of the necessary condition of optimality (NC) above, to allow for non-smooth data. Now, the functions $f(t, ., u)$ and $h(t, .)$ will be assumed to be merely Lipschitz continuous. The limiting subdifferential $\partial_x f(t, x, u)$ appears in place of the classical derivative $f_x(t, x, u)$, etc. and we shall replace $h_x(s, \bar{x}(s))$ by a selector γ of some suitably defined subdifferential of $x \to h(t, x)$. For convenience, we shall refer to $(p, \lambda, \mu, \gamma)$ as 'Lagrange multipliers' for (P). (Strictly speaking, the Lagrange multipliers are (q, λ, μ)) – those associated with the dynamic constraint $\dot{x} = f(t, x, u)$, the cost and the state constraint respectively.) The statement and proof of the nonsmooth maximum principle with pure state constraints is preceded by one section providing information about measure convergence that is required in the analysis.

10.3 Convergence of Measures

As earlier discussed, the presence of state constraints requires us to consider 'Lagrange multipliers' that are elements in the topological dual $C^*([S, T]; \mathbb{R}^k)$ of the space of continuous functions $C([S, T]; \mathbb{R}^k)$ with supremum norm. (Here $[S, T]$ is a given interval.) The norm on $C^*([S, T]; \mathbb{R}^k)$, written $\|\mu\|_{TV}$, is the induced norm. The set of elements in $C^*([S, T]; \mathbb{R})$ taking non-negative values on non-negative valued functions in $C([S, T]; \mathbb{R})$ is denoted $C^\oplus(S, T)$.

As is well-known, elements $\mu \in C^*([S, T]; \mathbb{R}^k)$ can be identified with the set of finite regular vector-valued measures on the Borel subsets of $[S, T]$. We loosely refer then to elements $\mu \in C^*([S, T]; \mathbb{R}^k)$ as 'measures'. Notice that, for $\mu \in C^\oplus(S, T)$, $\|\mu\|_{TV}$, as defined above, coincides with the total variation of μ, $\int_{[S,T]} d\mu(s)$, as the notation would suggest.

The support of a measure $\mu \in C^*([S, T]; \mathbb{R}^k)$, written supp$\{\mu\}$, is the smallest closed subset $A \subset [S, T]$ with the property that for all relatively open subsets $B \subset [S, T] \setminus A$ we have $\mu(B) = 0$.

Given $\mu \in C^{\oplus}(S, T)$, a μ-continuity set is a Borel subset $B \subset [S, T]$ for which $\mu(\text{bdy } B) = 0$. Take $\mu \in C^{\oplus}(S, T)$. Then there is a countable set $\mathcal{S} \subset (S, T)$, such that all sets of the form $[a, b], [a, b), (a, b]$ with $a, b \in ([S, T] \setminus \mathcal{S})$ are μ-continuity sets.

Given a weak* convergent sequence $\mu_i \to \mu$ in $C^*([S, T]; \mathbb{R}^k)$, there exists a countable subset $\mathcal{S} \subset (S, T)$ such that

$$\int_{[S,t]} d\mu_i(s) \to \int_{[S,t]} d\mu(s)$$

for all $t \in ([S, T] \setminus \mathcal{S})$.

Take a weak* convergent sequence $\mu_i \to \mu$ in $C^{\oplus}(S, T)$. Then

$$\int_B d\mu(t) \leq \liminf_{i \to \infty} \int_B d\mu_i(t)$$

for any relatively open subset $B \subset [S, T]$. Also,

$$\int_B h(t) d\mu(t) = \lim_{i \to \infty} \int_B h(t) d\mu_i(t)$$

for any $h \in C([S, T]; \mathbb{R}^k)$ and any μ-continuity set B.

Take closed subsets A and $A_i, i = 1, 2, \ldots$ of $[S, T] \times \mathbb{R}^n$. We denote by $A(.) : [S, T] \rightsquigarrow \mathbb{R}^n$ the multifunction

$$A(t) := \{a : (t, a) \in A\}.$$

The multifunctions $A_i(.)$ are likewise defined. The following proposition will have an important role in justifying limit taking in 'measure' relations of the kind

$$\eta_i = \gamma_i \mu_i, \quad i = 1, 2, \ldots,$$

in which the sequence of Borel measurable functions $\{\gamma_i\}$ satisfies

$$\gamma_i(t) \in A_i(t) \quad \mu_i\text{- a.e..}$$

Conditions will be given under which we can conclude

$$\eta_0 = \gamma_0 \mu_0$$

where η_0 and μ_0 are weak* limits of $\{\eta_i\}$ and $\{\mu_i\}$ respectively and γ_0 is a Borel measurable function satisfying

$$\gamma_0(t) \in A(t) \quad \mu_0\text{- a.e..} \tag{10.3.1}$$

10.3 Convergence of Measures

Proposition 10.3.1 *Take a weak* convergent sequence $\{\mu_i\}$ in $C^{\oplus}(S, T)$, a sequence of Borel measurable functions $\{\gamma_i : [S, T] \to \mathbb{R}^n\}$ and a sequence of closed sets $\{A_i\}$ in $[S, T] \times \mathbb{R}^n$. Take also a closed set A in $[S, T] \times \mathbb{R}^n$ and a measure $\mu_0 \in C^{\oplus}(S, T)$.*

Assume that $A(t)$ is convex for each $t \in \text{dom } A$ and that the sets A and $A_1, A_2, \ldots$ are uniformly bounded. Assume further that

$$\limsup_{i \to \infty} A_i \subset A,$$

$$\gamma_i(t) \in A_i(t) \quad \mu_i\text{-a.e. for } i = 1, 2, \ldots$$

and

$$\mu_i \to \mu_0 \quad \text{weakly}^*.$$

Define $\eta_i \in C^*([S, T]; \mathbb{R}^k)$

$$\eta_i := \gamma_i \mu_i.$$

Then, along a subsequence,

$$\eta_i \to \eta_0 \quad \text{weakly}^*,$$

for some $\eta_0 \in C^*([S, T]; \mathbb{R}^k)$ such that

$$\eta_0 = \gamma_0 \mu_0,$$

in which γ_0 is a Borel measurable function which satisfies

$$\gamma_0(t) \in A(t) \quad \mu_0\text{-a.e.}.$$

Proof Since the sets A_i, $i = 1, 2, \ldots$, are uniformly bounded there exists some constant K such that

$$|\gamma_i(t)| \le K \quad \mu_i\text{-a.e. } t \in [S, T].$$

But $\{\mu_i\}$ is a weak* convergent sequence. It follows that $\{\|\eta_i\|_{\text{TV}}\}$ is a bounded sequence. Along a subsequence then, $\eta_i \to \eta_0$ (weakly*) for some $\eta_0 \in C^*([S, T]; \mathbb{R}^k)$. Given any $\phi \in C([S, T]; \mathbb{R}^n)$

$$\left| \int_{[S,T]} \phi(t) d\eta_0(t) \right| = \lim_i \left| \int_{[S,T]} \phi(t) \gamma_i(t) d\mu_i(t) \right|$$

$$\leq K \lim_i \int_{[S,T]} |\phi(t)| d\mu_i(t)$$

$$= K \int_{[S,T]} |\phi(t)| d\mu_0(t),$$

by weak* convergence. This inequality, which holds for every continuous ϕ, implies that η_0 is absolutely continuous with respect to μ_0. By the Radon-Nikodym theorem then, there exists an $\mathbb{R}^n$ valued Borel measurable and a μ_0-integrable function $\gamma_0 : [S,T] \to \mathbb{R}^n$ such that

$$\int_E d\eta_0(t) = \int_E \gamma_0(t) d\mu_0(t)$$

for all Borel subsets $E \subset [S,T]$.

For each i we express the n-vector valued measure η_i in terms of its components $\eta_i = (\eta_{i1}, \ldots, \eta_{in})$. Let $\eta_{ij} = \eta_{ij}^+ - \eta_{ij}^-$, $j = 1, \ldots, n$, be the Jordan decomposition of η_{ij}. For each j, $\{\eta_{ij}^+\}_{i=1}^\infty$ and $\{\eta_{ij}^-\}_{i=1}^\infty$ are bounded in total variation, since the η_i's are bounded in total variation. By limiting attention to a further subsequence then, we can arrange that $\lim_i \eta_{ij}^+ = \eta_j^+$, $\lim_i \eta_{ij}^- = \eta_j^-$, for some $\eta_j^+, \eta_j^- \in C^\oplus(S,T)$ and $j = 1, \ldots, n$. (Here limits are interpreted as weak* limits.) Since, for each $\phi = (\phi_1, \ldots, \phi_n) \in C([S,T]; \mathbb{R}^n)$,

$$\int \phi(t) \cdot d\eta_0(t) = \lim_i \int \phi(t) \cdot d\eta_i(t)$$

$$= \lim_i \int \sum_j \phi_j(t) [d\eta_{ij}^+ - d\eta_{ij}^-](t)$$

$$= \int \sum_j \phi_j(t) [d\eta_j^+ - d\eta_j^-](t),$$

we see that $\eta_0 = ((\eta_1^+ - \eta_1^-), \ldots, (\eta_n^+ - \eta_n^-))$. Let $\mathcal{C}$ denote the class of Borel sets which are continuity sets of $\eta_1^+, \ldots, \eta_n^+, \eta_1^-, \ldots, \eta_n^-$ and μ_0. Then for any $E \subset \mathcal{C}$ and $h \in C([S,T]; \mathbb{R}^n)$

$$\int_E d\eta_i(t) \to \int_E d\eta_0(t) \quad \text{and} \quad \int_E h(t) d\mu_i(t) \to \int_E h(t) d\mu_0(t). \quad (10.3.2)$$

Fix a positive integer j. Define the set $A^j \subset [S,T] \times \mathbb{R}^n$ to be

$$A^j := (A + j^{-1}B) \cap ([S,T] \times \mathbb{R}^n).$$

Then, since the sets A_i, $i = 1, 2, \ldots$, are uniformly bounded and $\limsup_{i \to \infty} A_i \subset A$, we have $A_i \subset A^j$, for all i sufficiently large.

10.3 Convergence of Measures

Take any relatively open set $E \subset [S, T] \setminus \text{dom } A^j(.)$. Then, since $\eta_i = \gamma_i \mu_i$ and $\text{supp}\{\eta_i\} \subset \text{dom } A^j(.)$, for i sufficiently large, we have

$$0 \geq \liminf_{i \to \infty} \int_E d\eta_{ij}^+(t) \geq \int_E d\eta_j^+(t) \geq 0.$$

It follows that $\eta_j^+(E) = 0$, $j = 1, \ldots, n$. Likewise, we show that $\eta_j^-(E) = 0$, $j = 1, \ldots, n$.

We conclude that $\eta_0(E) = 0$ for all open sets $E \subset [S, T] \setminus \text{dom } A^j(.)$. This means that

$$\text{supp }\{\eta_0\} \subset \text{dom } A^j(.).$$

Fix $q \in \mathbb{R}^n$. Then, for $r > 0$ sufficiently large, the function

$$s_q(t) := \begin{cases} \max\{q \cdot d : d \in A^j(t)\} & \text{if } A^j(t) \neq \emptyset \\ r & \text{otherwise} \end{cases}$$

is upper semi-continuous and bounded on $[S, T]$. By a well-known property of upper semi-continuous functions, there exists a sequence of continuous functions $\{c_q^k : [S, T] \to \mathbb{R}\}_{k=1}^\infty$ such that

$$s_q(t) \leq c_q^k(t) \quad \text{for all } t \in [S, T], \; k = 1, 2, \ldots$$

and

$$s_q(t) = \lim_{k \to \infty} c_q^k(t) \quad \text{for all } t \in [S, T]. \tag{10.3.3}$$

Choose any $E \subset C$. For each i sufficiently large,

$$q \cdot \int_E d\eta_i(t) = \int_E q \cdot \gamma_i(t) d\mu_i(t)$$

$$= \int_{E \cap \text{dom } A^j(\cdot)} q \cdot \gamma_i(t) d\mu_i(t)$$

$$\leq \int_E c_q^k(t) d\mu_i(t).$$

By (10.3.2) and since c_q^k is continuous, we obtain in the limit as $i \to \infty$

$$\int_E q \cdot \gamma_0(t) d\mu_0(t) \leq \int_E c_q^k(t) d\mu_0(t).$$

Since $\mathcal{C}$ generates the Borel sets, we readily deduce that this inequality is valid for all Borel sets E. It follows that,

$$q \cdot \gamma_0(t) \leq c_q^k(t) \quad \mu_0\text{- a.e.}.$$

From (10.3.3) then,

$$q \cdot \gamma_0(t) \leq s_q(t) \, (= \max\{q \cdot d : d \in A^j(t)\}) \quad \mu_0\text{- a.e.}.$$

By σ-additivity, this last relation holds for all q belonging to some countable, dense subset of $\mathbb{R}^n$. Since A^j is a bounded set, the mapping $q \to c_q^k(t)$ is continuous for each $t \in \text{dom } A^j$. We conclude that it is true for all $q \in \mathbb{R}^m$. But then, since $A^j(t)$ is closed and convex for each $t \in \text{dom } A^j$, we have

$$\gamma_0(t) \in A^j(t) \quad \mu_0\text{- a.e.}.$$

Finally we observe that, since A is closed set,

$$\gamma_0(t) \in \cap_j A^j(t) = A(t) \quad \mu_0\text{- a.e.}.$$

We have shown that (along some subsequence) $\eta_i \to \eta_0$ (weakly*) for some η_0 which can be expressed $\eta_0 = \gamma_0 \mu_0$, where γ_0 is a Borel measurable function satisfying

$$\gamma_0(t) \in A(t) \quad \mu_0\text{- a.e.}.$$

This is what we set out to prove. □

10.4 The Maximum Principle (Pure State Constraints)

In this section we state necessary conditions in the form of a maximum principle for an dynamic optimization problem with pure state constraints (P), posed in the introduction. By $NBV([S, T]; \mathbb{R}^n)$ we denote the space of $\mathbb{R}^n$-valued functions of bounded variation defined on $[S, T]$ which are '*normalized*' in the sense that they are right-continuous on (S, T). Define, the un-maximized Hamiltonian $\mathcal{H} : [S, T] \times \mathbb{R}^n \times \mathbb{R}^n \times \mathbb{R}^m \to \mathbb{R}$

$$\mathcal{H}(t, x, p, u) := p \cdot f(t, x, u). \qquad (10.4.1)$$

Theorem 10.4.1 (Maximum Principle for Problems with Pure State Constraints) *Let $(\bar{x}, \bar{u})$ be a $W^{1,1}$ local minimizer for (P). Assume that, for some $\bar{\epsilon} > 0$, the following hypotheses are satisfied:*

10.4 The Maximum Principle (Pure State Constraints)

(H1) For fixed x, $f(., x, .)$ is $\mathcal{L} \times \mathcal{B}^m$ measurable. There exists an $\mathcal{L} \times \mathcal{B}^m$ measurable function $k : [S, T] \times \mathbb{R}^m \to [0, \infty)$ such that $t \to k(t, \bar{u}(t))$ is integrable and, for a.e. $t \in [S, T]$,

$$|f(t, x, u) - f(t, x', u)| \leq k(t, u)|x - x'|$$

for all $x, x' \in \bar{x}(t) + \bar{\epsilon}\mathbb{B}$ and $u \in U(t)$,

(H2): The set $\text{Gr}\, U$ is $\mathcal{L} \times \mathcal{B}^m$ measurable,

(H3): g is Lipschitz continuous on $(\bar{x}(S), \bar{x}(T)) + \bar{\epsilon}\mathbb{B}$ and C is a closed subset of $\mathbb{R}^{n \times n}$,

(H4): $h(., x)$ is upper semi-continuous for each $x \in \mathbb{R}^n$ and there exists $k_h > 0$ such that

$$|h(t, x) - h(t, x')| \leq k_h|x - x'| \text{ for all } x, x' \in \bar{x}(t) + \bar{\epsilon}\mathbb{B}, t \in [S, T].$$

Then there exist $p \in W^{1,1}([S, T]; \mathbb{R}^n)$, $\mu \in C^{\oplus}(S, T)$, a bounded Borel measurable function $\gamma : [S, T] \to \mathbb{R}^n$ and $\lambda \geq 0$, satisfying the following conditions, in which $q \in NBV([S, T]; \mathbb{R}^n)$ is the function

$$q(t) := \begin{cases} p(S) & \text{if } t = S \\ p(t) + \int_{[S,t]} \gamma(s) d\mu(s) & \text{if } t \in (S, T]. \end{cases}$$

(a): $(p, \mu, \lambda) \neq (0, 0, 0)$,

(b): $-\dot{p}(t) \in \text{co}\, \partial_x \mathcal{H}(t, \bar{x}(t), q(t), \bar{u}(t))$ a.e. $t \in [S, T]$,

(c): $\mathcal{H}(t, \bar{x}(t), q(t), \bar{u}(t)) = \max_{u \in U(t)} \mathcal{H}(t, \bar{x}(t), q(t), u)$ a.e. $t \in [S, T]$,

(d): $(q(S), -q(T)) \in \lambda \partial g(\bar{x}(S), \bar{x}(T)) + N_C(\bar{x}(S), \bar{x}(T))$,

(e): $\text{supp}\{\mu\} \subset \{t \in [S, T] : h(t, \bar{x}(t)) = 0\}$ and $\gamma(t) \in \partial_x^> h(t, \bar{x}(t))$

$$\text{for } \mu\text{-a.e. } t \in [S, T],$$

where

$$\partial_x^> h(t, \bar{x}(t)) := \text{co lim sup}\, \{\partial_x h(t_i, x_i) :$$
$$t_i \to t, x_i \to \bar{x}(t) \text{ and } h(t_i, x_i) > 0 \text{ for each } i\}.$$

Now assume, also, that

$$f(t, x, u), h(t, x) \text{ and } U(t) \text{ are independent of } t.$$

Then, in addition to the above conditions, there exists a constant r such that

(f): $\mathcal{H}(t, \bar{x}(t), q(t), \bar{u}(t)) = r \quad$ a.e..

Remarks

The state constraint formulation '$h(t, x(t)) \leq 0$', in (P) encompasses a number of special cases of interest. Key features of this formulation, contributing to its broad applicability, are that $h(t, x)$ is permitted to be merely Lipschitz continuous w.r.t. the x variable and merely upper semi-continuous w.r.t. the t variable. Consider the following examples.

(i): Multiple state constraints $h_k(t, x(t)) \leq 0$ for $t \in [S, T]$, $k = 1, \ldots, M$, in which the $h_k(t, x)$'s are Lipschitz continuous w.r.t. x, can be accommodated by setting $h(t, x) := \max_k \{h_k(t, x)\}$. (The 'composite' state constraint function $h(t, x)$ retains Lipschitz continuity w.r.t. x.) The max rule of nonsmooth calculus can then be used to express the necessary conditions directly in terms of limiting subdifferentials of the constituent $h_k(t, x)$'s.

(ii): Consider an implicit state constraint $x(t) \in A$, for $t \in [S, T]$, in which $A \subset \mathbb{R}^n$ is a given closed set. Let us assume the following regularity condition on the boundary of the A is satisfied:

$$\text{int } \bar{T}_A(x) \neq \emptyset \text{ for all } x \in \partial A. \quad (10.4.2)$$

Here the necessary conditions are valid in a modified form where, in condition (e), the Borel measurable function γ is now required to satisfy

$$\gamma(t) \in \text{co}\,(N_A(\bar{x}(t)) \cap \{\xi \in \mathbb{R}^n : |\xi| = 1\}).$$

These modified conditions can be derived from Theorem 10.4.1 by setting $h(t, x) = d_A(x)$, where d_A is the distance function to the set A, with the help of Proposition 4.8.2 and Lemma 4.8.3. Observe that, by the polarity relation $\bar{T}_A(x) = (N_C(x))^*$ (cf. Theorem 4.10.7), condition (10.4.2) implies that co $N_A(\bar{x}(t))$ is pointed. Recall that a cone $K \subset \mathbb{R}^n$ is said to be 'pointed' when the relation $d_1 + \cdots + d_k = 0$ is never satisfied with $d_i \in K$ unless $d_i = 0$ for all $i = 1, \ldots, k$.

(iii): The requirement that $h(t, x)$ is merely upper semi-continuous w.r.t. the t variable adds useful flexibility to the formulation of the state constraint, because it permits us to cover cases when a state constraint is imposed only at times lying in a closed subset $I \subset [S, T]$. In such cases we replace '$h(t, x(t)) \leq 0$ for all $t \in I$', by '$\tilde{h}(t, x(t)) \leq 0$ for all $t \in [S, T]$', where

$$\tilde{h}(t, x) = \begin{cases} h(t, x) & \text{if } t \in I \\ -R & \text{if } t \notin I, \end{cases}$$

in which R is a suitably large positive number.

10.5 Proof of Theorem 10.4.1

We omit the proof of the constancy of the Hamiltonian condition (f) for autonomous problems, since it follows from Theorem 10.6.1.

We first confirm the assertions of the theorem, in the special case when the following supplementary hypotheses are satisfied. ($\bar{\epsilon}$ is as in ($H1$) and ($\bar{x}, \bar{u}$) is the admissible process of interest.)

(A1): ($\bar{x}, \bar{u}$) is an L^∞ local minimizer (not merely a $W^{1,1}$ local minimizer),

(A2): There exist integrable functions $c_0 : [S, T] \to \mathbb{R}$ and $k_0 : [S, T] \to \mathbb{R}$ such that

    (i): $|f(t, x, u) - f(t, x', u)| \le k_0(t)|x - x'|$,
    (ii): $|f(t, x, u)| \le c_0(t)$,
    for all $x, x' \in \bar{x}(t) + \bar{\epsilon}\mathbb{B}$, $u \in U(t)$, a.e. $t \in [S, T]$.

(A3): $h(., x)$ is piecewise constant in the following sense: there exists a uniform partition of $\{t_0 = S, t_1, \ldots, t_N = T\}$ and Lipschitz continuous functions $h_k : \mathbb{R}^n \to \mathbb{R}^n$, for $k = 0, \ldots, N - 1$, such that, for each $x \in \mathbb{R}^n$, we have

$$h(t, x) = \begin{cases} h_0(x) & \text{if } t \in [S, t_1), \\ h_k(x) & \text{if } t \in (t_k, t_{k+1}) \text{ for some } k \in \{1, \ldots, N-2\}, \\ h_{k-1}(x) \vee h_k(x) & \text{if } t = t_k \text{ for some } k \in \{1, \ldots, N-1\}, \\ h_{N-1}(x) & \text{if } t \in (t_{N-1}, T]. \end{cases}$$

The final step of the proof will be to show that the assertions of the theorem remain valid, when the supplementary hypotheses are removed.

Assume then that the $W^{1,1}$ local minimizer ($\bar{x}, \bar{u}$) for (P) is actually an L^∞ local minimizer (this is a consequence of supplementary hypothesis (A1)) and that, for some $\bar{\epsilon} > 0$, (H1)–(H4) and (A2)–(A3) are satisfied. Since ($\bar{x}, \bar{u}$) is an L^∞ local minimizer, we know that there exists $\epsilon \in (0, \bar{\epsilon}/2)$ such that ($\bar{x}, \bar{u}$) is a minimizer for (P), w.r.t. processes (x, u) satisfying $||x - \bar{x}||_{L^\infty} \le \epsilon$.

Take $\epsilon_i \downarrow 0$ and, for each i, consider the problem

$$(P_i) \begin{cases} \text{Minimize } J_i(x, u) := g_i(x(S), x(T), \max_{t \in [S, T]} h(t, x(t))) \\ \text{subject to} \\ \dot{x}(t) = f(t, x(t), u(t)), \text{ a.e. } t \in [S, T], \\ u(t) \in U(t) \text{ a.e. } t \in [S, T], \\ ||x - \bar{x}||_{L^\infty} \le \epsilon/2. \end{cases}$$

Here

$$g_i(x_0, x_1, z) := \left(g(x_0, x_1) - g(\bar{x}(S), \bar{x}(T)) + \epsilon_i^2\right) \vee d_C(x_0, x_1) \vee z.$$

For each i, J_i is a continuous function on the complete metric space $\mathcal{M}$

$$\mathcal{M} := \{\text{admissible processes}(x, u) \text{ for } (P_i) \text{ satisfying } ||x - \bar{x}||_{L^\infty} \le \epsilon/2\},$$

with metric

$$d_{\mathcal{E}}((x, u), (x', u')) := |x(S) - x'(S)| + \mathcal{L}\text{-meas}\{t \in [S, T] : u(t) \ne u'(t)\}. \tag{10.5.1}$$

Fix i. Notice that, because the cost is non-negative, and the process $(\bar{x}, \bar{u})$ has cost ϵ_i^2, $(\bar{x}, \bar{u})$ is an ϵ_i^2 minimizer of J_i over $\mathcal{M}$. We deduce from Ekeland's theorem that there exists a process (x_i, u_i) which is a minimizer for

$$(P_i') \begin{cases} \text{Minimize } J_i'(x, u) \\ \text{subject to} \\ \dot{x}(t) = f(t, x(t), u(t)), \text{ a.e. } t \in [S, T], \\ u(t) \in U(t) \text{ a.e. } t \in [S, T], \\ ||x - \bar{x}||_{L^\infty} \le \epsilon/2, \end{cases}$$

where

$$J_i'(x, u) := g_i(x(S), x(T), \max_{t \in [S,T]} h(t, x(t)))$$
$$+ \epsilon_i \left(|x(S) - x_i(S)| + \mathcal{L}\text{-meas}\{t \in [S, T] : u(t) \ne u_i(t)\} \right).$$

Furthermore

$$d_{\mathcal{E}}((x_i, u_i), (\bar{x}, \bar{u})) \le \epsilon_i. \tag{10.5.2}$$

Taking note of hypothesis (A2), we deduce from Filippov's existence theorem that

$$||x_i - \bar{x}||_{L^\infty} \to 0 \text{ and } \mathcal{L}\text{-meas}\{t : u_i(t) \ne \bar{u}(t)\} \to 0 \text{ as } i \to \infty. \tag{10.5.3}$$

It is obvious that $g_i(x(S), x(T), \max_{t \in [S,T]} h(t, x(t))) \ge 0$, for all state trajectories x, since the distance function is non-negative valued. Note, also, that

$$g_i(x_i(S), x_i(T), \max_{t \in [S,T]} h(t, x_i(t))) > 0, \text{ for all } i \text{ sufficiently large.} \tag{10.5.4}$$

This reason is that, if '=' replaces '>' in the above relation, then $g(x_i(S), x_i(T)) \le g(\bar{x}(S), \bar{x}(T)) - \epsilon_i^2$, $(x_i(S), x_i(T)) \in C$ and $h(t, x_i(t)) \le 0$, for all $t \in [S, T]$. Furthermore, $||x_i - \bar{x}||_{L^\infty} \le \epsilon/2$. This contradicts the L^∞ local optimality of $(\bar{x}, \bar{u})$.

There are two cases that can arise:

Case 1: $\max_{t \in [S,T]} h(t, x_i(t)) > 0$, for at most a finite number of index values i,

10.5 Proof of Theorem 10.4.1

Case 2: $\max_{t\in[S,T]} h(t, x_i(t)) > 0$, for an infinite number of index values i.

Consider first Case 1. We can arrange, by excluding initial terms in the sequence, that

$$\max_{t\in[S,T]} h(t, x_i(t)) \leq 0 \text{ for all } i \, .$$

Taking note of (10.5.4), we see that (x_i, u_i) is an L^∞ local minimizer for

$$(Q_i) \begin{cases} \text{Minimize } (g(x(S), x(T)) - g(\bar{x}(S), \bar{x}(T)) + \epsilon_i^2) \vee d_C(x(S), x(T)) \\ \qquad\qquad + \epsilon_i\big(|x(S) - x_i(S)| + \int_{[S,T]} m_i(t, u(t))dt\big) \\ \text{subject to} \\ \dot{x}(t) = f(t, x(t), u(t)), \quad \text{a.e. } t \in [S, T], \\ u(t) \in U(t) \text{ a.e. } t \in [S, T], \end{cases}$$

in which $m_i(t, u) := \begin{cases} 0 \text{ if } u = u_i(t) \\ 1 \text{ otherwise} \, . \end{cases}$

In view of (10.5.4) and since $h(t, x_i(t)) \leq 0$ for $t \in [S, T]$,

$$(g(x_i(S), x_i(T)) - g(\bar{x}(S), \bar{x}(T)) + \epsilon_i^2) \vee d_C(x_i(S), x_i(T)) > 0 \, . \tag{10.5.5}$$

Consider now the dynamic optimization problem obtained by removing the state constraint '$h(t, x(t)) \leq 0$' from the original problem (P). Problem (Q_i) above can be interpreted as a perturbation of this 'state-constraint free' problem. It was precisely this perturbation which, together with (10.5.3) and (10.5.5), was used to prove the state-constraint free maximum principle in Chap. 7, in the presence of the supplementary hypotheses (A1)–(A3). Our previous analysis (that is step 1 of Chap. 7, Sect. 7.4) can then be reproduced to prove all the assertions of the maximum principle, Theorem 10.4.1, with state constraint multiplier μ set to be zero. This deals with Case 1.

We now turn to Case 2. We can arrange, by selecting a subsequence, that

$$\max_{t\in[S,T]} h(t, x_i(t)) > 0, \text{ for all } i \, . \tag{10.5.6}$$

Fix i. For arbitrary $K > 0$, consider the optimization problem

$$(P_i^K) \begin{cases} \text{Minimize } J_i^K(x, u, z) \\ \text{over } x \in W^{1,1}([S, T]; \mathbb{R}^n), \text{ measurable functions } u \text{ and } z \in \mathbb{R} \text{ satisfying} \\ \dot{x}(t) = f(t, x(t), u(t)), \text{ a.e. } t \in [S, T], \\ u(t) \in U(t), \text{ a.e. } t \in [S, T], \\ ||x - \bar{x}||_{L^\infty} \leq \epsilon/2, \end{cases}$$

in which

$$J_i^K(x, u, z) := g_i(x(S), x(T), z) + \epsilon_i d_{\mathcal{E}}((x, u), (x_i, u_i))$$
$$+ K \int_S^T (h(t, x(t)) - z) \vee 0)^2 dt \,.$$

Notice that, for each K and any (x, u), we have $J_i^K(x, u, z = \max_{t \in [S,T]} h(t, x(t))) = J_i'(x, u)$. This implies that, for each i, $\inf(P_i^K) \leq \inf(P_i')$. According to the following lemma, the two infima coincide in the limit as $K \to \infty$, however.

Lemma 10.5.1 *For $i = 1, 2, \ldots$,*

$$\lim_{K \to \infty} \inf(P_i^K) = \inf(P_i') \,.$$

Proof Fix i. Assume that assertion of the lemma is false. Since the mapping $K \to \inf(P_i^K)$ is non-decreasing, we may deduce the existence of $\bar{\rho} > 0$ and sequences $K_j \uparrow \infty$ and $\{(x^j, u^j, z^j)\}$ in $\mathcal{M} \times \mathbb{R}$ such that, for each j,

$$g_i(x^j(S), x^j(T), z^j) + \epsilon_i d_{\mathcal{E}}((x^j, u^j), (x_i, u_i))$$
$$+ K_j \int_{[S,T]} ((h(t, x^j(t)) - z^j) \vee 0))^2 dt$$
$$< g_i(x_i(S), x_i(T), \max_{t \in [S,T]} h(t, x_i(t))) - \bar{\rho} \,, \text{ for } j = 1, 2, \ldots \quad (10.5.7)$$

Notice, at the outset, that we can assume

$$\max_{t \in [S,T]} h(t, x^j(t)) - z^j \geq 0 \text{ for all } j \,. \quad (10.5.8)$$

To see this, suppose that, for some j, this condition is violated, i.e. $\max\{h(t, x^j(t)) : t \in [S, T]\} < z^j$. Now replace z^j by the lower number $\max\{h(t, x^j(t)) : t \in [S, T]\}$. We see that, following this change, (10.5.8) is satisfied. Furthermore, the strict inequality (10.5.7) is preserved; this is because the right side of this relation is unaffected. On the other hand, the value of the integral term on the left is reduced to zero while the first term on the left cannot increase, owing to the monotonicity of $z \to g_i(x_0, x_1, z)$. We have confirmed that, without loss of generality, we may assume (10.5.8). Since the x^j's are uniformly bounded in L^∞, the z^j's are uniformly bounded above. It is also clear from the non-negativity of the first two terms and the structure of the integrand of the integral on the left side of (10.5.7) that the z^j's are also bounded below. It follows that $\{z^j\}$ is a bounded sequence. Again, from the non-negativity of the first two terms and since $K_j \to \infty$ as $j \to \infty$, we deduce that

$$\int_{[S,T]} (h(t, x^j(t)) - z^j) \vee 0)^2 dt \to 0, \text{ as } j \to \infty \,. \quad (10.5.9)$$

10.5 Proof of Theorem 10.4.1

Fix j. Let $\bar{t}$ be a point at which $(h(t, x^j(t)) - z^j)$ achieves its maximum over $[S, T]$. Using the facts that the x^j's are uniformly bounded w.r.t. the L^∞ norm and the z^j's are bounded and invoking hypothesis (H4) and (A3), we see that there exists $\bar{\alpha} \in (0, \frac{|T-S|}{2N})$ such that

$$((h(t, x^j(t)) - z^j) \vee 0)^2 \geq \frac{1}{2}((h(\bar{t}, x^j(\bar{t})) - z^j) \vee 0)^2,$$

for all $t \in [S, T]$ s.t. either $t \in [\bar{t} - \bar{\alpha}, \bar{t}]$ or $t \in [\bar{t}, \bar{t} + \bar{\alpha}]$.

Here $1/N$ is the 'mesh-size' parameter of piecewise continuous function $h(., x)$. (See supplementary hypothesis (A3).) Note the role of (A3) in establishing this crucial inequality. But then

$$\max_{t \in [S,T]} h(t, x^j(t)) - z^j$$

$$= h(\bar{t}, x^j(\bar{t})) - z^j \leq (2/\bar{\alpha})^{\frac{1}{2}} \left(\int_{[S,T]} ((h(t, x^j(t)) - z^j) \vee 0)^2 dt \right)^{\frac{1}{2}}.$$

It follows now from (10.5.8) and (10.5.9) that

$$\max_{t \in [S,T]} h(t, x^j(t)) - z^j \to 0, \text{ as } j \to \infty.$$

Since $g_i(x_0, x_1, .)$ is Lipschitz continuous (with Lipschitz constant 1) we deduce from (10.5.7) that, for j sufficiently large,

$$g_i(x^j(S), x^j(T), \max_{t \in [S,T]} h(t, x^j(t))) + \epsilon_i d_\mathcal{E}((x^j, u^j), (x_i, u_i))$$

$$< g_i(x_i(S), x_i(T), \max_{t \in [S,T]} h(t, x_i(t))) - \bar{\rho}/2, \text{ for } j = 1, 2, \ldots \quad (10.5.10)$$

On the other hand, by the optimality of (x_i, u_i) we have

$$g_i(x^j(S), x^j(T), \max_{t \in [S,T]} h(t, x^j(t))) + \epsilon_i d_\mathcal{E}((x^j, u^j), (x_i, u_i))$$

$$\geq g_i(x_i(S), x_i(T), \max_{t \in [S,T]} h(t, x_i(t))).$$

This contradicts the optimality of (x_i, u_i). The lemma is proved. □

In view of (10.5.6), and the continuity of the embedding $(x, u) \to x$ from $\mathcal{M}$ to L^∞, we can find a sequence $\rho_i \downarrow 0$ such that, for each i,

$$\left.\begin{array}{l} (x, u, z) \in \mathcal{M} \times \mathbb{R} \\ d_\mathcal{E}((x, u), (x_i, u_i)) \leq \rho_i \\ |z - \max_{t \in [S,T]} h(t, x_i(t))| \leq \rho_i \end{array}\right\} \implies g_i(x(S), x(T), z) > 0 \text{ and } z > 0.$$

$$(10.5.11)$$

According to the lemma, we can choose $K_i \uparrow \infty$ such that, for each i, $(x_i, u_i, z_i := \max_{t \in [S,T]} h(t, x_i(t)))$ is a ρ_i^2-minimizer for $(P_i^{K_i})$. Ekeland's theorem then tells us that there exists $(x_i', u_i', z_i' \equiv z_i') \in \mathcal{M} \times \mathbb{R}$ that is an L^∞ local minimizer for $(\tilde{Q}_i)$:

$$(\tilde{Q}_i) \begin{cases} \text{Minimize } g_i(x(S), x(T), z(T)) + \epsilon_i \Big(|x(S) - x_i(S)| + \int_{[S,T]} m_i(t, u(t)) dt \Big) \\ + \rho_i \Big(|x(S) - x_i'(S)| + |z(S) - z_i'(S)| + \int_{[S,T]} (m_i'(t, u(t)) dt \Big) \\ + K_i \int_{[S,T]} ((h(t, x(t)) - z(t)) \vee 0)^2 dt \\ \text{over } (z, x) \in W^{1,1}([S,T]; \mathbb{R}^{1+n}), \text{ meas. } u \text{ satisfying} \\ (\dot{z}(t), \dot{x}(t)) = (0, f(t, x(t), u(t))), \text{ a.e. } t \in [S, T] \\ u(t) \in U(t) \text{ a.e. } t \in [S, T]. \end{cases}$$

Here, $m_i'(t, u) := \begin{cases} 0 \text{ if } u = u_i'(t) \\ 1 \text{ otherwise} \end{cases}$. Furthermore

$$d_{\mathcal{E}}((x_i', u_i'), (x_i, u_i)) + |z_i' - \max_{t \in [S,T]} h(t, x_i(t))| \le \rho_i. \tag{10.5.12}$$

This relation combines with (10.5.2) to tell us that, as $i \to \infty$,

$$\|x_i' - \bar{x}\|_{L^\infty} \to 0, \ z_i' \to \max_{t \in [S,T]} h(t, \bar{x}(t))$$

and $\mathcal{L} - \text{meas}\{t \in [S, T] : u_i'(t) \neq \bar{u}(t)\} \to 0$.

It follows from (10.5.11) that

$$g_i(x_i'(S), x_i'(T), z_i') > 0 \text{ and } z_i' > 0, \text{ for each } i. \tag{10.5.13}$$

Now, for each i, apply the earlier derived state-constraint free version of Clarke's nonsmooth maximum principle of Chap. 7 to the problem $(\tilde{Q}_i)$, with reference to the L^∞ local minimizer (z_i', x_i', u_i'), noting that the relevant hypotheses are satisfied. Since $(\tilde{Q}_i)$ has no endpoint constraints, we can set the cost multiplier to 1. We are assured then of the existence of elements $-r_i \in W^{1,1}$ and $q_i \in W^{1,1}$ (interpreted as costate components associated with the state variable components z and x variables respectively) and a Borel measurable function $\gamma_i : [S, T] \to \mathbb{R}^n$ such that

$$-\dot{q}_i(t) \in \text{co } \partial_x q_i(t) \cdot f(t, x_i'(t), u_i'(t)) - \gamma_i(t)(d\mu_i/dt)(t) \text{ a.e. } t \in [S, T], \tag{10.5.14}$$

and

$$\gamma_i(t) \in \partial_x h(t, x_i'(t)) \ \mu_i\text{-a.e. } t \in [S, T], \tag{10.5.15}$$

$$q_i(t) \cdot f(t, x_i'(t), u_i'(t)) \ge \sup_{u \in U(t)} q_i(t) \cdot f(t, x_i'(t), u) - 2(\epsilon_i + \rho_i), \text{ a.e. } t \in [S, T], \tag{10.5.16}$$

10.5 Proof of Theorem 10.4.1

and

$$(q_i(S), -q_i(T), r_i(T)) \in \partial g_i(x_i'(S), x_i'(T), z_i'(T)) + (\epsilon_i + \rho_i)\mathbb{B} \times \{0\} \times \{0\}, \tag{10.5.17}$$

Furthermore, $\dot{r}_i = (d\mu_i/dt)$ a.e. and $r_i(S) \in \rho_i \mathbb{B}$. This implies that

$$r_i(T) \in \int_{[S,T]} (d\mu_i/ds)(s)ds + \rho_i \mathbb{B}. \tag{10.5.18}$$

Here μ_i is a (positive) measure associated with distribution:

$$\mu_i([S, t]) = 2K_i \int_{[S,t]} ((h(s, x_i'(s)) - z_i') \vee 0)ds. \tag{10.5.19}$$

It follows from (10.5.19) that

$$\text{supp}\{\mu_i\} \subset \{t \in [S, T] : h(t, x_i'(t)) \geq z_i'\}. \tag{10.5.20}$$

Recalling the definition of g_i, namely

$$g_i(x_0, x_1, z) := \big(g(x_0, x_1) - g_i(x_0, x_1) + \epsilon_i\big) \vee d_C(x_0, x_1) \vee z,$$

and taking note of max rule for subdifferentials, we deduce from (10.5.17) that

$$(q_i(S), -q_i(T), r_i(T)) \in \lambda_i \partial g(x_i'(S), x_i'(T)) \times \{0\}$$
$$+ \lambda_i^{(1)} \partial d_C(x_i'(S), x_i'(T)) \times \{0\} +$$
$$\lambda_i^{(2)}\{(0, 0, 1)\} + (\epsilon_i + \rho_i)\mathbb{B} \times \{0\} \times \{0\},$$

in which $\lambda_i, \lambda_i^{(1)}, \lambda_i^{(2)}$ are non-negative numbers such that $\lambda_i + \lambda_i^{(1)} + \lambda_i^{(2)} = 1$. The relation can be replaced by the following more precise condition:

$$(q_i(S), -q_i(T), r_i(T)) \in \lambda_i \partial g(x_i'(S), x_i'(T)) \times \{0\}$$
$$+ \lambda_i^{(1)}\big(\partial d_C(x_i'(S), x_i'(T)) \cap \partial \mathbb{B}\big) \times \{0\}$$
$$+\lambda_i^{(2)}\{(0, 0, 1)\} + (\epsilon_i + \rho_i)\mathbb{B} \times \{0\} \times \{0\}, \tag{10.5.21}$$

in which, we recall, $\partial \mathbb{B}$ denotes the boundary of the unit ball in Euclidean space.

We check this assertion. The two relations are equivalent when $\lambda_i^{(1)} = 0$. So we may assume that $\lambda_i^{(1)} > 0$. But then, according to the max rule, the '$d_C((x_i'(S), x_i'(T)))$' term must achieve the maximum in the definition of $g_i(x_i'(S), x_i'(T), z_i')$. Since, however, $g_i(x_i'(S), x_i'(T), z_i') > 0$ (see (10.5.13)), we conclude that $d_C((x_i'(S), x_i'(T))) > 0$. We know from Lemma 4.8.3 that

$\xi \in \partial d_C(z)$ and $z \notin C \implies \xi \in \partial d_C(z) \cap \partial \mathbb{B}$.

(10.5.21) follows from the preceding relation. From (10.5.21),

$$(q_i(S), -q_i(T)) \in \lambda_i \partial g(x_i'(S), x_i'(T)) + N_C(x_i'(S), x_i'(T)) \cap \mathbb{B} + (\epsilon_i + \rho_i) \mathbb{B} \times \{0\}. \quad (10.5.22)$$

From (10.5.18) and (10.5.21), and since $\lambda_i^{(2)} \leq 1$, we see that

$$|(q_i(S), -q_i(T))| \leq 1 + k_g + (\epsilon_i + \rho_i), \quad \|\mu_i\|_{TV} \leq 1 + \rho_i \quad (10.5.23)$$

and $\lambda_i^{(2)} \leq \|\mu_i\|_{TV} + \rho_i$. (10.5.21) also tells us that $\lambda_i^{(1)} \leq |(q_i(S), -q_i(T))| + k_g \lambda_i + (\rho_i + \epsilon_i)$. Since $\lambda_i + \lambda_i^{(1)} + \lambda_i^{(2)} = 1$, it follows from these relations that

$$(1 + k_g)\lambda_i + |(q_i(S), -q_i(T))| + \|\mu_i\|_{TV} \geq 1 - (\epsilon_i + 2\rho_i) \quad (10.5.24)$$

Now define the function $p_i \in W^{1,1}([S, T]; \mathbb{R}^n)$ to be

$$p_i(t) := q_i(t) - \int_{[S,t]} \gamma_i(s)(d\mu_i(s)/ds)(s)ds \quad \text{for } t \in [S, T].$$

Condition (10.5.14) can be expressed in terms of p_i:

$$-\dot{p}_i(t) \in \text{co}\, \partial_x q_i(t) \cdot f(t, x_i'(t), u_i'(t)) \text{ a.e. } t \in [S, T]. \quad (10.5.25)$$

We summarize our findings so far: for $i = 1, 2, \ldots$, there exist $p_i \in W^{1,1}$, $\mu_i \in C^{\oplus}(S, T)$, a Borel measurable function γ_i and $\lambda_i \geq 0$ satisfying conditions (10.5.24), (10.5.25), (10.5.22), (10.5.16), (10.5.15) and (10.5.20) which imply:

(a'): $(1 + k_g)\lambda_i + |(q_i(S), -q_i(T))| + \|\mu_i\|_{TV} \geq 1 - (\epsilon_i + 2\rho_i)$,
(b'): $-\dot{p}_i(t) \in \text{co}\, \partial_x q_i(s) \cdot f(t, x_i'(t), u_i'(t))$ a.e.,
(c'): $q_i(t) \cdot f(t, x_i'(t), u_i'(t)) \geq \sup_{u \in U(t)} q_i(t) \cdot f(t, x_i'(t), u) - 2(\epsilon_i + \rho_i)$, a.e.,
(d'): $(q_i(S), -q_i(T)) \in \lambda_i \partial g(x_i'(S), x_i'(T)) +$
$\qquad \left(N_C(x_i'(S), x_i'(T))\right) \cap \mathbb{B} + (\epsilon_i + \rho_i)\mathbb{B} \times \{0\}$,
(e'): $\gamma_i(t) \in \partial_x h(t, x_i'(t))$, μ_i-a.e. and
$\qquad \text{supp}\{\mu_i\} \subset \{t \in [S, T] : h(t, x_i'(t)) \geq z_i'\}$.

These conditions will be recognized as perturbed versions of conditions (a)–(e) in the maximum principle Theorem 10.4.1. The next step is to capture (a)–(e) from (a')–(e'), in the limit as $i \to \infty$.

In view of (10.5.2) and (10.5.12), we have $\|x_i' - \bar{x}\|_{L^\infty} \to 0$ as $i \to \infty$ and we can arrange, by subsequence extraction, that

$$\mathcal{L}\text{-meas}\{t : u_i'(t) = \bar{u}(t) \text{ for all } i \text{ sufficiently large}\} = |T - S|. \quad (10.5.26)$$

10.5 Proof of Theorem 10.4.1

From (10.5.23) we deduce that $\{\|\mu_i\|_{TV}\}$, $\{q_i(S)\}$ and $\{p_i(S)\}$ are bounded sequences. By (b′) then, and in consequence of Filippov's existence theorem, the q_i's are uniformly bounded, the p_i's are uniformly bounded and the $\dot{p}_i$'s are uniformly integrably bounded. Invoking Proposition 10.3.1, we deduce that, following a subsequence extraction,

$$\|p_i - p\|_{L^\infty} \to 0, \text{ and } \mu_i \to \mu, \ \gamma_i \mu_i \to \gamma\mu \text{ weakly}^*$$

as $i \to \infty$. Furthermore on some subset $\mathcal{S}_1 \subset [S, T]$ of full $\mathcal{L}$-measure we have

$$q_i(t) \to q(t) \text{ for all } t \in \mathcal{S}_1 \cup \{S\} \cup \{T\} \tag{10.5.27}$$

as $i \to \infty$, where

$$q(t) := p(t) + \int_{[S,t]} \gamma(s)(d\mu(s)/ds)(s)ds \text{ for } t \in (S, T],$$

for some $p \in W^{1,1}([S, T]; \mathbb{R}^n)$, $q \in NBV([S, T]; \mathbb{R}^n)$ and $\mu \in C^\oplus(S, T)$, and some Borel measurable function γ.

Bearing in mind that $z'_i > 0$ for each i and $x'_i \to \bar{x}$ uniformly, we deduce from (e′) that, for some $\delta_i \downarrow 0$,

$$\gamma_i(t) \in \cup\{\partial_x h(t, y) : y \in \bar{x}(t) + \delta_i \mathbb{B} \text{ and } h(t, y) > 0\}, \ \mu_i\text{-a.e. } t \in [S, T].$$

We can deduce from this relation, with the help of Proposition 10.3.1 that

$$\gamma(t) \in \partial_x^> h(t, \bar{x}(t)), \ \mu\text{-a.e. } t \in [S, T].$$

Taking note of (10.5.26) and (10.5.27), we deduce from (c′) and (d′) that

$$q(t) \cdot f(t, \bar{x}(t), \bar{u}(t)) \geq \sup_{u \in U(t)} q(t) \cdot f(t, \bar{x}(t), u), \text{ a.e.,}$$

and

$$(q(S), -q(T)) \in \lambda \partial g(\bar{x}(S), \bar{x}(T)) + N_C(\bar{x}(S), \bar{x}(T)).$$

From (a′) we conclude that: $(1 + k_g)\lambda + |(q(S), q(T))| + \|\mu\|_{TV} \geq 1$. This implies that

$$(\lambda, p, \mu) \neq (0, 0, 0),$$

since, when $\mu = 0$, $p = q$. All the assertions of the state constrained maximum principle have been confirmed.

Up to this point in our analysis, we have assumed that $(\bar{x}, \bar{u})$ is an L^∞ local minimizer and when supplementary hypothesis (A1), (A2) and (A3) are imposed. It remains to lift these restrictions.

We deal first with (A3), the hypothesis that imposed additional structure on the state constraint function h. Assume then (A1) and (A2) are satisfied, but not (A3). Take sequence of integers $N_j \uparrow \infty$. For each j, let $\{t_0^j = S, t_1^j, \ldots, t_{N_j-1}^j, t_{N_j}^j = T\}$ be a uniform partition of $[S, T]$. Now define

$$h_j(t, x) = \begin{cases} \max\{h(t, x) : t \in [S, t_1^j]\} & \text{if } t \in [S, t_0^j), \\ \max\{h(t, x) : t \in [t_k^j, t_{k+1}^j]\} & \text{if } t \in (t_k^j, t_{k+1}^j) \text{ for some } \\ & k \in \{1, \ldots, N_j - 2\}, \\ \max\{h(t, x) : t \in [t_{k-1}^j, t_{k+1}^j]\} & \text{if } t = t_k^j \text{ for some } \\ & k \in \{1, \ldots, N_j - 1\}, \\ \max\{h(t, x) : t \in [t_{N_j-1}^j, t_{N_j}^j]\} & \text{if } t \in (t_{N_j-1}^j, t_{N_j}^j]. \end{cases}$$

Taking note of hypotheses (H4) and (A2), we can show

$$\sup_{t \in [S,T]} h(t, x(t)) \leq \sup_{t \in [S,T]} h_j(t, x(t)) \leq \sup_{t \in [S,T]} h(t, x(t)) + \delta_j, \quad (10.5.28)$$

for all state trajectories x satisfying $\|x - \bar{x}\|_{L^\infty} \leq \epsilon/2$, where $\delta_j := k_h \times \theta(2|T - S|/N_j)$ and

$$\theta(\alpha) := \sup\{\int_I c_0(t)dt : \text{Lebesgue sets } I \subset [S, T] \text{ s.t. } \mathcal{L}\text{-meas}\{I\} \leq \alpha\}.$$

Here k_h and c_0 are as in (H4) and (A2) respectively. Notice that $\delta_j \downarrow 0$ as $i \to \infty$.

We now go back to an earlier stage of the proof. Recall, we specified a sequence $\epsilon_i \downarrow 0$ and invoked Ekeland's theorem to obtain a process (x_i, u_i), for each i, that was a minimizer for (P_i') which we reproduce:

$$(P_i') \begin{cases} \text{Minimize } J_i'(x, u) \\ \text{subject to} \\ \dot{x}(t) = f(t, x(t), u(t)), \text{ a.e. } t \in [S, T], \\ u(t) \in U(t) \text{ a.e. } t \in [S, T], \\ \|x - \bar{x}\|_{L^\infty} \leq \epsilon/2, \end{cases}$$

where

$$J_i'(x, u) := g_i(x(S), x(T), \max_{t \in [S,T]} h(t, x(t)))$$
$$+ \epsilon_i \left(|x(S) - x_i(S)| + \mathcal{L}\text{-meas}\{t \in [S, T] : u(t) \neq u_i(t)\} \right)$$

10.5 Proof of Theorem 10.4.1

and

$$d_{\mathcal{E}}((x_i, u_i), (\bar{x}, \bar{u})) \leq \epsilon_i.$$

We observed, furthermore, that

$$||x_i - \bar{x}||_{L^\infty} \to 0 \text{ and } \mathcal{L}\text{-meas}\{t : u_i(t) \neq \bar{u}(t)\} \to 0 \text{ as } i \to \infty$$

and, for some $\rho_i \downarrow 0$ and $\alpha_i \downarrow 0$,

$$\left.\begin{array}{l}(x, u) \in \mathcal{M} \\ d_{\mathcal{E}}((x, u), (x_i, u_i)) \leq \rho_i\end{array}\right\} \implies g_i(x(S), x(T), \max_{t \in [S,T]} h(t, x(t))) > \alpha_i. \tag{10.5.29}$$

(This was relation (10.5.11).) We now take a new direction in the analysis. According to (10.5.28), we can choose, for each i, an integer $j(i)$ such that

$$\sup_{t \in [S,T]} h(t, x(t)) \leq \sup_{t \in [S,T]} h_{j(i)}(t, x(t)) \leq \sup_{t \in [S,T]} h(t, x(t)) + \rho_i^2, \tag{10.5.30}$$

for all state trajectories x satisfying $||x - \bar{x}||_{L^\infty} \leq \epsilon/2$.

Consider a variant on (P_i'), in which $h_{j(i)}$ replaces h:

$$(Q_i') \begin{cases} \text{Minimize } g_i(x(S), x(T), \max_{t \in [S,T]} h_{j(i)}(t, x(t))) \\ \qquad + \epsilon_i\left(|x(S) - x_i(S)| + \int_{[S,T]} m_i(t, u(t))dt\right) \\ \text{over } x \in W^{1,1} \text{ and meas. functions } u \text{ satisfying} \\ \dot{x}(t) = f(t, x(t), u(t)), \text{ a.e. } t \in [S, T] \\ u(t) \in U(t) \text{ a.e. } t \in [S, T], \\ ||x - \bar{x}||_{L^\infty} \leq \epsilon/2 \end{cases}$$

where $m_i(t, u) := \begin{cases} 0 \text{ if } u = u_i(t) \\ 1 \text{ otherwise} \end{cases}$. It will be clear from (10.5.30) that (x_i, u_i) is a ρ_i^2 minimizer for (Q_i'). Invoking Ekeland's theorem, we can find, for i sufficiently large, an L^∞ local minimizer $(x_i', z_i' \equiv \max_{t \in [S,T]} h_{j(i)}(t, x_i'(t)), u_i'(t))$ for the problem

$$\begin{cases} \text{Minimize } g_i(x(S), x(T), z(T)) + \epsilon_i\left(|x(S) - x_i(S)| + \int_{[S,T]} m_i(t, u(t))dt\right) \\ \qquad + \rho_i\left(|x(S) - x_i'(S)| + \int_{[S,T]} m_i'(t, u(t))dt\right) \\ \text{over } x, z \in W^{1,1} \text{ and meas. functions } u \text{ satisfying} \\ (\dot{x}(t), \dot{z}(t)) = (f(t, x(t), u(t)), 0), \text{ a.e. } t \in [S, T] \\ u(t) \in U(t) \text{ a.e. } t \in [S, T], \\ h_{j(i)}(t, x(t)) - z \leq 0 \text{ for all } t \in [S, T] \end{cases}$$

in which, once again, $m'_i(t, u) := \begin{cases} 0 \text{ if } u = u'_i(t), \\ 1 \text{ otherwise} \end{cases}$ and

$$d_{\mathcal{E}}((x'_i, u'_i), (x_i, u_i)) \leq \rho_i.$$

But then, by (10.5.29) and since $h_{i(j)} \geq h$ and $z \to g_i(x_0, x_1, z)$ is an increasing function, we have

$$g_i(x'_i(S), x'_i(T), z'_i = \max_{t \in [S,T]} h_{i(j)}(t, x'_i(t))) > \alpha_i. \tag{10.5.31}$$

Fix i. Taking note of the fact that the state constraint function $h_{j(i)}$ satisfies the supplementary hypothesis (A3), we may apply the special case of the state constrained maximum principle already proved. This asserts existence of multipliers $(p_i, -r_i) \in W^{1,1}$ (associated with the states x and z respectively), $\mu_i \in C^{\oplus}(S, T)$, a bounded Borel measurable function $\gamma_i : [S, T] \to \mathbb{R}^n$ and $\bar{\lambda}_i \geq 0$, satisfying the following conditions:

(A): $(\bar{\lambda}_i, p_i, r_i, \mu_i) \neq (0, 0, 0, 0)$,
(B): $-\dot{p}_i(t) \in \text{co}\, \partial_x(q_i(t) \cdot f(t, x'_i(t), u'_i(t)))$ and $\dot{r}_i(t) \equiv 0$ a.e. $t \in [S, T]$,
(C): $q_i(t) \cdot f(t, x'_i(t), u'_i(t)) \geq \max_{u \in U(t)} q_i(t) \cdot f(t, x'_i(t), u) - \bar{\lambda}_i \times 2(\epsilon_i + \rho_i)$
 a.e. $t \in [S, T]$,
(D): $\left(q_i(S), -q_i(T), r_i(T) + \int_{[S,T]} d\mu_i(t) \right) \in \bar{\lambda}_i \partial g_i(x'_i(S), x'_i(T), z'_i)$, $r(S) = 0$,
(E): $\text{supp}\{\mu_i\} \subset \{t \in [S, T] : h_{j(i)}(t, x'_i(t)) = z'_i\}$ and
 $\gamma_i(t) \in \partial_x^> h_{j(i)}(t, x'_i(t))$ for μ_i-a.e. $t \in [S, T]$,

in which $q_i(t) := \begin{cases} p_i(S) & \text{if } t = S \\ p_i(t) + \int_{[S,t]} \gamma_i(s) d\mu_i(s) & \text{if } t \in (S, T]. \end{cases}$

(B) and (D) imply that $r_i \equiv 0$. We see also from these conditions that $\bar{\lambda}_i \neq 0$ for, otherwise, $(p_i, r_i, \mu_i) = (0, 0, 0)$, in violation of (A). Scaling the multipliers, we can arrange then that $\bar{\lambda}_i = 1$. We can therefore replace (B), (C) and (D) by

(B'): $-\dot{p}_i(t) \in \text{co}\, \partial_x(q_i(t) \cdot f(t, x'_i(t), u'_i(t)))$ a.e. $t \in [S, T]$,
(C'): $q_i(t) \cdot f(t, x'_i(t), q(t), u'_i(t)) \geq \max_{u \in U(t)} q_i(t) \cdot f(t, x'_i(t), u) - 2(\epsilon_i + \rho_i)$
 a.e. $t \in [S, T]$

and

$$\left(q_i(S), -q_i(T), \int_{[S,T]} d\mu_i(t) \right) \in \partial g_i(x'_i(S), x'_i(T), z'_i).$$

Following the pattern on our earlier analysis, we can deduce from the preceding relation and (10.5.31) that there exist $\lambda_i \geq 0$, $\lambda_i^{(1)} \geq 0$ and $\lambda_i^{(2)} \geq 0$ such that $\lambda_i + \lambda_i^{(1)} + \lambda_i^{(2)} = 1$ and

10.5 Proof of Theorem 10.4.1

$$(q_i(S), -q_i(T), \int_{[S,T]} d\mu_i(t)) \in \lambda_i \partial g(x_i'(S), x_i'(T)) \times \{0\}$$
$$+ \lambda_i^{(1)} (\partial d_C(x_i'(S), x_i'(T)) \cap \partial \mathbb{B})) \times \{0\}$$
$$+ \lambda_i^{(2)} \{(0, 0, 1)\} + (\epsilon_i + \rho_i) \mathbb{B} \times \{0\} \times \{0\}. \qquad (10.5.32)$$

This implies

(D'): $(q_i(S), -q_i(T) \in \lambda_i \partial g(x_i'(S), x_i'(T)) \times \{0\}$
$\quad + N_C(x_i'(S), x_i'(T)) \cap \mathbb{B} \times \{0\} + (\epsilon_i + \rho_i) \mathbb{B} \times \{0\}.$

We see from (10.5.32) that $\lambda_i^{(2)} = \int_{[S,T]} d\mu_i(s) = \|\mu_i\|_{\mathrm{TV}}$. In view of (10.5.31),

$$\sup_{t \in [S,T]} h_{j(i)}(t, x_i'(t)) < \alpha_i \implies \mu_i = 0. \qquad (10.5.33)$$

We can also deduce from the relation and the definition of q_i that

$$\sqrt{2}(\|p_i\|_{L^\infty} + \|\mu_i\|_{\mathrm{TV}}) \geq |(q_i(S), q_i(T))| \geq \lambda_i^{(1)} - k_g \lambda_i - \epsilon_i - \rho_i.$$

Then, since $\lambda_i^{(1)} = 1 - \|\mu_i\|_{\mathrm{TV}} - \lambda_i$,

$$\sqrt{2}(\|p_i\|_{L^\infty} + \|\mu_i\|_{\mathrm{TV}}) \geq |(q_i(S), q_i(T))| \geq 1 - \|\mu_i\|_{\mathrm{TV}} - (1 + k_g)\lambda_i - \epsilon_i - \rho_i.$$

It follows that

$$\sqrt{2}\|p_i\|_{L^\infty} + (1 + \sqrt{2})\|\mu_i\|_{\mathrm{TV}} + (1 + k_g)\lambda_i \geq 1 - \epsilon_i - \rho_i. \qquad (10.5.34)$$

We know that $\lambda_i \leq 1$. Equation (10.5.32) implies that $\{p_i(S)\}$ and $\{\|\mu_i\|_{\mathrm{TV}}\}$ are bounded sequences. In consequence of (A2), $\{p_i\}$ is a bounded sequence in L^∞, with integrably bounded derivatives. We may arrange then, by extracting subsequences that $p_i \to p$ and $\dot{p}_i \to \dot{p}$, $\mu_i \to \mu$ (in the appropriate topologies), for some $p \in W^{1,1}([S, T]; \mathbb{R}^n)$, $\mu \in C^\oplus(S, T)$ and $\lambda \geq 0$.

Now take any point t in the support of μ_i. It follows from (10.5.33) and the max rule that $\lambda_i^{(2)} = 0$ and so

$$\sup_{t \in [S,T]} h_{j(i)}(t, x_i'(t)) \geq \alpha_i.$$

Write $\delta_i := |T - S|/N_{j(i)}$. We deduce from the max rule (Theorem 5.7.1) that

$$\partial_x h_{j(i)}(t, x_i'(t)) \subset \cap_{\epsilon' > 0} \mathrm{co}\, \{\xi \in \partial_x h(s, y): \qquad (10.5.35)$$
$$s \in [t - 2\delta_i, t + 2\delta_i] \cap [S, T],\ y \in x_i'(t) + \epsilon' \mathbb{B},\ h(s, y) \geq \alpha_i - \epsilon'\}.$$

(Confirm this relation first for proximal subgradients of $x \to h(t, x)$ using the proximal subgradient inequality. It extends to limiting subgradients by the density properties of proximal subgradients.) This implies

$$\operatorname{supp}\{\mu_i\} \subset \{t \in [S, T] : \max_{s \in [t-2\delta_i, t+2\delta_i]} h(s, x'_i(t)) > 0\}. \tag{10.5.36}$$

But $x'_i \to \bar{x}$ uniformly. It therefore follows from Proposition 10.3.1, together with condition (E), (10.5.36) and condition (10.5.35), which is valid for all $t \in \operatorname{supp}\{\mu_i\}$, that

$$\int_{[0,t]} \gamma_i(s) d\mu_i(s) \to \int_{[0,t]} \gamma(s) d\mu(s) \text{ for all } t \in [S, T]$$

for some Borel measurable function γ such that

$$\gamma(t) \in \cap_{\epsilon' > 0} \overline{\operatorname{co}} \{\xi \in \partial_x h(s, y) :$$
$$s \in [t - \epsilon', t + \epsilon'] \cap [S, T], y \in \bar{x}(t) + \epsilon' \mathbb{B}, h(s, y) > 0\}$$
$$= \partial_x^> h(t, \bar{x}(t)), \mu\text{-a.e.}$$

and $\operatorname{supp}\{\mu\} \subset \{t \in [S, T] : h(t, \bar{x}(t)) = 0\}$. Condition (e) of the theorem statement is confirmed.

Passing to the limit in (10.5.34), as $i \to \infty$, yields $(p, \mu, \lambda) \neq (0, 0, 0)$. This is condition (a) of the theorem statement. The remaining conditions (b) – (d) are obtained from (B$'$) – (D$'$), in the limit as $j \to \infty$.

Finally let us attend to the removal of the supplementary hypotheses (A1) and (A2). This is accomplished as in the final step of the proof of the state-constraint free maximum principle of Chap. 7. To be more precise, a 'state augmentation' argument is used to verify that the assertions of the theorem remain valid when $(\bar{x}, \bar{u})$ is merely a $W^{1,1}$, not an L^∞, local minimizer. (This disposes of (A1).) If (A2) is violated we consider a variant of problem (P) of the introduction in which $U(t)$ is replaced by the 'inner approximation'

$$U_j(t) := \{u \in U(t) : k(t, u) \leq k(t, \bar{u}(t)) + j, |f(t, \bar{x}(t), u)| \leq |\dot{\bar{x}}| + j\}.$$

Here, j is a positive integer. For each j, the special case of the theorem, already proved, can be applied. Validity of the assertions, in the absence of (A2), is now proved by passage to the limit. □

10.6 Maximum Principles for Free End-Time Problems with State Constraints

We provide in this section extensions of the state constrained maximum principle, to allow for free end-times. As in our earlier treatment of free end-time free problems with free end-times in Chap. 9 (where state constraints were absent), the extra

10.6 Maximum Principles for Free End-Time Problems with State Constraints

relations corresponding to the free end-time take the form of boundary conditions on the maximized Hamiltonian.

Consider the following state constrained problem with free end-times:

$$(FT) \begin{cases} \text{Minimize } g(S, x(S), T, x(T)) \\ \text{over } [S, T] \subset \mathbb{R}, \, x \in W^{1,1}([S, T]; \mathbb{R}^n) \\ \text{and measurable functions } u : [S, T] \to \mathbb{R}^m \text{ satisfying} \\ \dot{x}(t) = f(t, x(t), u(t)) \quad \text{a.e. } t \in [S, T], \\ u(t) \in U(t) \quad \text{a.e. } t \in [S, T], \\ h(t, x(t)) \leq 0 \text{ for all } t \in [S, T], \\ (S, x(S), T, x(T)) \in C, \end{cases}$$

the data for which comprise: functions $g : \mathbb{R} \times \mathbb{R}^n \times \mathbb{R} \times \mathbb{R}^n \to \mathbb{R}$, $f : \mathbb{R} \times \mathbb{R}^n \times \mathbb{R}^m \to \mathbb{R}^n$ and $h : \mathbb{R} \times \mathbb{R}^n \to \mathbb{R}$, a multifunction $U : \mathbb{R} \leadsto \mathbb{R}^m$ and a set $C \subset \mathbb{R} \times \mathbb{R}^n \times \mathbb{R} \times \mathbb{R}^n$.

A process is a triple $([S, T], x, u)$ comprising an interval $[S, T]$, a $W^{1,1}([S, T]; \mathbb{R}^n)$ function x and a Lebesgue measurable function $u : [S, T] \to \mathbb{R}^m$ such that $\dot{x}(t) = f(t, x(t), u(t))$ and $u(t) \in U(t)$, a.e. $t \in [S, T]$. The process is said to be admissible if additionally the elements satisfy the constraints of problem (FT). An admissible process $([\bar{S}, \bar{T}], \bar{x}, \bar{u})$ is called a $W^{1,1}$ minimizer relative if there exists $\beta > 0$ such that

$$g(S, x(S), T, x(T)) \geq g(\bar{S}, \bar{x}(\bar{S}), \bar{T}, \bar{x}(\bar{T}))$$

for all admissible processes $([S, T], x, u)$ such that

$$d(([S, T], x), ([\bar{S}, \bar{T}], \bar{x})) \leq \beta.$$

Here, d is the distance function employed in Chap. 9 (cf. (9.1.4)), which can be written also as follows:

$$d(([S, T], x), ([S', T'], x')) := |S - S'| + |T - T'| + |x(S) - x'(S)|$$
$$+ \int_{-\infty}^{+\infty} |\dot{x}_e(t) - \dot{x}'_e(t)| dt.$$

(Here, and elsewhere, the subscript e attached to 'state' like variables such as x indicates extension by constant extrapolation and, attached to 'velocity' related variables such as $\dot{x}$ or u denotes the extension by zero.)

Following the pattern of Chap. 9, we treat the Lipschitz time dependence and the measurable time dependence cases separately. Define the maximized Hamiltonian

$$H(t, x, p) = \sup_{u \in U(t)} p \cdot f(t, x, u).$$

Theorem 10.6.1 (State Constrained, Free End-Time Problems with Lipschitz Time Dependence) *Let* $([\bar{S}, \bar{T}], \bar{x}, \bar{u})$ *be a* $W^{1,1}$ *local minimizer for (FT). Assume that, for some* $\bar{\epsilon} > 0$, *the following hypotheses are satisfied:*

(H1): g is Lipschitz continuous on a neighbourhood of $(\bar{S}, \bar{x}(\bar{S}), \bar{T}, \bar{x}(\bar{T}))$ *and C is a closed set,*

(H2): $U(t) = U$ *for all* $t \in \mathbb{R}$, *for some Borel set* $U \subset \mathbb{R}^m$,

(H3): For fixed x, $f(., x, .)$ *is* $\mathcal{L} \times \mathcal{B}^m$ *measurable. There exists a function* $k_f : [\bar{S}, \bar{T}] \times \mathbb{R}^m \to \mathbb{R}$ *such that* $t \to k(t, \bar{u}(t))$ *is integrable on* $[\bar{S}, \bar{T}]$ *and*

$$|f(t'', x'', u) - f(t', x', u)| \leq k_f(t, u) |(t'', x'') - (t', x')|$$

for all $(t'', x''), (t', x') \in (t, \bar{x}(t)) + \bar{\epsilon}\mathbb{B}$ *and* $u \in U$, *a.e.* $t \in [\bar{S}, \bar{T}]$,

(H4): h is upper semi-continuous and there exists a constant k_h *such that, for all* $t \in [\bar{S}, \bar{T}]$,

$$|h(t', x') - h(t'', x'')| \leq k_h |(t', x') - (t'', x'')|$$

for all $(t', x'), (t'', x'') \in (t, \bar{x}(t)) + \bar{\epsilon}\mathbb{B}$.

Then there exist absolutely continuous arcs $e \in W^{1,1}([\bar{S}, \bar{T}]; \mathbb{R})$ *and* $p \in W^{1,1}([\bar{S}, \bar{T}]; \mathbb{R}^n)$, $\lambda \geq 0$, *a measure* $\mu \in C^{\oplus}(\bar{S}, \bar{T})$ *and a Borel measurable function* $\gamma = (\gamma_0, \gamma_1) : [\bar{S}, \bar{T}] \to \mathbb{R}^{1+n}$, *such that the following conditions, in which* $(r, q) \in NBV([\bar{S}, \bar{T}]; \mathbb{R}^{1+n})$ *is the function*

$$(-r(t), q(t)) := \begin{cases} (-e(\bar{S}), p(\bar{S})) & \text{if } t = \bar{S} \\ (-e(t), p(t)) + \int_{[\bar{S},t]} \gamma(s) d\mu(s) & \text{if } t \in (\bar{S}, \bar{T}], \end{cases} \quad (10.6.1)$$

are satisfied:

(i) $(p, \mu, \lambda) \neq (0, 0, 0)$,
(ii) $(\dot{e}(t), -\dot{p}(t)) \in \mathrm{co}\, \partial_{t,x} q(t) \cdot f(t, \bar{x}(t), \bar{u}(t))$ *a.e.* $t \in [\bar{S}, \bar{T}]$,
(iii) $q(t) \cdot f(t, \bar{x}(t), \bar{u}(t)) \geq q(t) \cdot f(t, \bar{x}(t), u)$, *for all* $u \in U$, *a.e.* $t \in [\bar{S}, \bar{T}]$,
(iv) $(-r(\bar{S}), q(\bar{S}), r(\bar{T}), -q(\bar{T})) \in \lambda \partial g(\bar{S}, \bar{x}(\bar{S}), \bar{T}, \bar{x}(\bar{T})) + N_{\tilde{C}}(\bar{S}, \bar{x}(\bar{S}), \bar{T}, \bar{x}(\bar{T}))$,
(v) $\gamma(t) \in \partial^{>} h(t, \bar{x}(t))$ μ-*a.e.* $t \in [\bar{S}, \bar{T}]$,
(vi) $r(t) = H(t, \bar{x}(t), q(t))$ *for a.e.* $t \in [\bar{S}, \bar{T}]$,

in which

$\partial^{>} h(t, \bar{x}(t))$
$:= \mathrm{co\, lim\, sup}\, \{\partial h(t_i, x_i) : t_i \to t, x_i \to \bar{x}(t)$ *and* $h(t_i, x_i) > 0$ *for each i*$\}$.

The proof, which we omit, involves reformulating (FT) as a fixed time problem by means of a change of independent variable (as in the proof of Theorem 9.6.1). We then apply known necessary conditions to the fixed time problem (necessary

10.6 Maximum Principles for Free End-Time Problems with State Constraints

conditions for state constrained problems Theorem 10.4.1 in this case) and interpret these conditions in terms of the data for the original problem.

Remark
The Autonomous Case: If $f(t, x, u)$ and $h(t, x)$ do not depend on t, then (10.6.1) and conditions (ii) and (v) in the theorem statement imply that r is a constant function and $\gamma_0 = 0$. Condition (vi) then takes the form: there exists a number c such that

$$H(\bar{x}(t), q(t)) = c, \text{ a.e. } t \in [\bar{S}, \bar{T}].$$

If for admissible processes $([S, T], x, u)$ either the initial time S is unconstrained and g does not depend on S, or the final time T is unconstrained and g does not depend on T, then the constant $c = 0$.

Necessary conditions can also be derived for problems with measurably time dependent data. For such problems, the boundary condition on the maximized Hamiltonian is interpreted in the 'sub and super essential value' sense, introduced in Chap. 9. Recall that, given an integrable function $a : [S, T] \to \mathbb{R}$ and a point $t \in (S, T)$, we have (see Definition 9.3.1):

(a): the sub essential value of f at $\bar{t}$ is the set

$$\text{sub-ess}\, a(t) := \left\{ \zeta \in \mathbb{R} : \exists t_i \to \bar{t} \text{ and } \zeta_i \to \zeta \text{ s.t.} \right.$$
$$\left. \limsup_{\epsilon \downarrow 0} \epsilon^{-1} \int_{t_i - \epsilon}^{t_i} a(s)ds \le \zeta_i \le \liminf_{\epsilon \downarrow 0} \int_{t_i}^{t_i+\epsilon} a(s)ds, \text{ for each } i \right\}.$$

(b): the super essential value of a at t is the set

$$\text{super-ess}\, a(t) := \left\{ \zeta \in \mathbb{R} : \exists t_i \to \bar{t} \text{ and } \zeta_i \to \zeta \text{ s.t.} \right.$$
$$\left. \limsup_{\epsilon \downarrow 0} \epsilon^{-1} \int_{t_i}^{t_i+\epsilon} a(s)ds \le \zeta_i \le \liminf_{\epsilon \downarrow 0} \int_{t_i-\epsilon}^{t_i} a(s)ds, \text{ for each } i \right\}.$$

Theorem 10.6.2 (The Free End-Time Maximum Principle: Measurable Time Dependence) Let $([\bar{S}, \bar{T}], \bar{x}, \bar{u})$ be a $W^{1,1}$ local minimizer for (FT) such that $\bar{T} - \bar{S} > 0$. Assume that there exist $\epsilon > 0$ and $\sigma \in (0, (\bar{T} - \bar{S})/2)$ such that

(H1): g is Lipschitz continuous on a neighbourhood of $(\bar{S}, \bar{x}(\bar{S}), \bar{T}, \bar{x}(\bar{T}))$ and C is a closed set,
(H2): Gr U is a $\mathcal{L} \times \mathcal{B}^m$ measurable set,
(H3): For each $x \in \mathbb{R}^n$, $f(., x, .)$ is $\mathcal{L} \times \mathcal{B}^m$ measurable. There exists a function $k_f : \mathbb{R} \times \mathbb{R}^m \to \mathbb{R}$ such that $t \to k_f(t, \bar{u}(t))$ is integrable on $[\bar{S} - \sigma, \bar{T} + \sigma]$ and

$$|f(t, x', u) - f(t, x, u)| \leq k_f(t, u)|x' - x|$$

for all $x', x \in \bar{x}_e(t) + \epsilon \mathbb{B}$ and $u \in U(t)$, a.e. $t \in [\bar{S} - \sigma, \bar{T} + \sigma]$,
(H4): There exists $\bar{c} \geq 0$ such that, for a.e. $t \in [\bar{S} - \sigma, \bar{S} + \sigma] \cup [\bar{T} - \sigma, \bar{T} + \sigma]$,

$$|f(t, x, u)| \leq \bar{c}$$

for all $x \in \bar{x}_e(t) + \epsilon \mathbb{B}$ and $u \in U(t)$,
(H5): h is upper semi-continuous and there exists a constant k_h such that,

$$|h(t, x) - h(t, x')| \leq k_h |x - x'|$$

for all $x, x' \in \bar{x}_e(t) + \epsilon \mathbb{B}$, for all $t \in [\bar{S} - \sigma, \bar{T} + \sigma]$. Furthermore

$$h(\bar{S}, \bar{x}(\bar{S})) < 0 \quad \text{and} \quad h(\bar{T}, \bar{x}(\bar{T})) < 0. \quad (10.6.2)$$

Then there exist $p \in W^{1,1}([\bar{S}, \bar{T}]; \mathbb{R}^n)$, $\lambda \geq 0$, a measure $\mu \in C^{\oplus}(\bar{S}, \bar{T})$ and a Borel measurable function $\gamma : [\bar{S}, \bar{T}] \to \mathbb{R}^n$, such that the following conditions, in which $q \in NBV([\bar{S}, \bar{T}]; \mathbb{R}^n)$ is the function

$$q(t) := \begin{cases} p(\bar{S}) & \text{if } t = \bar{S} \\ p(t) + \int_{[\bar{S},t]} \gamma(s) d\mu(s) & \text{if } t \in (\bar{S}, \bar{T}], \end{cases} \quad (10.6.3)$$

are satisfied:

(i) $(p, \mu, \lambda) \neq (0, 0, 0)$,
(ii) $-\dot{p}(t) \in \mathrm{co}\, \partial_x q(t) \cdot f(t, \bar{x}(t), \bar{u}(t))$ a.e. $t \in [\bar{S}, \bar{T}]$,
(iii) $(-\xi_0, q(\bar{S}), \xi_1, -q(\bar{T})) \in \lambda \partial g(\bar{S}, \bar{x}(\bar{S}), \bar{T}, \bar{x}(\bar{T})) + N_C(\bar{S}, \bar{x}(\bar{S}), \bar{T}, \bar{x}(\bar{T}))$,

for some $\xi_0 \in \text{sub-ess}_{t \to \bar{S}} H(t, \bar{x}(\bar{S}), p(\bar{S}))$

and $\xi_1 \in \text{super-ess}_{t \to \bar{T}} H(t, \bar{x}(\bar{T}), p(\bar{T}))$,
(iv) $q(t) \cdot f(t, \bar{x}(t), \bar{u}(t)) \geq q(t) \cdot f(t, \bar{x}(t), u)$,
for all $u \in U(t)$, a.e. $t \in [\bar{S}, \bar{T}]$,
(v) $\gamma(t) \in \partial_x^{>} h(t, \bar{x}(t))$ μ-a.e. $t \in [\bar{S}, \bar{T}]$.

If the initial time $\bar{S}$ is fixed, i.e. $C = \{(\bar{S}, x_0, T, x_1) : (x_0, T, x_1) \in \tilde{C}\}$, for some closed set $\tilde{C} \subset \mathbb{R}^n \times \mathbb{R} \times \mathbb{R}^n$, and $g(S, x_0, T, x_1) = \tilde{g}(x_0, T, x_1)$ for some $\tilde{g} : \mathbb{R}^n \times \mathbb{R} \times \mathbb{R}^n \to \mathbb{R}$, then the above conditions are satisfied when (iii) is replaced by: there exists $\xi_1 \in \text{super-ess}_{t \to \bar{T}} H(t, \bar{x}(\bar{T}), p(\bar{T}))$ such that

(iii)': $(p(\bar{S}), \xi_1, -p(\bar{T})) \in \lambda \partial \tilde{g}(\bar{x}(\bar{S}), \bar{T}, \bar{x}(\bar{T})) + N_{\tilde{C}}(\bar{x}(\bar{S}), \bar{T}, \bar{x}(\bar{T}))$,

and when, in hypothesis (H4), the stated conditions are required to hold only for $t \in [\bar{T} - \sigma, \bar{T} + \sigma]$, and (10.6.2) in (H5) is replaced by '$h(\bar{T}, \bar{x}(\bar{T})) < 0$'. The assertions of the theorem can be similarly modified when the final time $\bar{T}$ is fixed.

The proof of this theorem, which we omit, follows the pattern of its fixed end-times counterpart Theorem 10.4.1. The proof of this earlier stated theorem was based on an application of Clarke's nonsmooth maximum principle (for fixed end-times) to a series of state constraint-free problems approximating the original problem, in which the state constrained is replaced by an integral penalty term, and passage to the limit. The difference now is that, at the start of the proof, we apply the free end-time necessary conditions of Theorem 9.6.2. The extra information provided by this theorem, expressed in terms of sub and super essential values of the maximized Hamiltonian, is preserved under limit taking and manifests itself as free end-time transversality condition (v) above.

10.7 Non-degenerate Conditions

We have seen how necessary conditions for state constraint-free problems can be adapted to a state constrained setting, by incorporating extra 'measure multiplier' terms associated with the state constraint.

These necessary conditions provide useful information about minimizers in many cases. But there are cases of interest where they are unsatisfactory. In this section we elaborate on this point, and show how the necessary conditions can be supplemented to broaden their applicability.

Consider, once again, the free end-time dynamic optimization problem with pathwise state constraints:

$$(FTP) \begin{cases} \text{Minimize } g(S, x(S), T, x(T)) \\ \text{over } [S, T] \subset \mathbb{R}, x \in W^{1,1}([S, T]; \mathbb{R}^n) \\ \text{and measurable } u : [S, T] \to \mathbb{R}^m \text{ satisfying} \\ \dot{x}(t) = f(t, x(t), u(t)) \quad \text{a.e. } t \in [S, T], \\ u(t) \in U \quad \text{a.e. } t \in [S, T], \\ h(t, x(t)) \leq 0 \text{ for all } t \in [S, T], \\ (S, x(S), T, x(T)) \in C, \end{cases}$$

the data for which comprise: an interval $[S, T]$, functions $g : \mathbb{R} \times \mathbb{R}^n \times \mathbb{R} \times \mathbb{R}^n \to \mathbb{R}$, $f : \mathbb{R} \times \mathbb{R}^n \times \mathbb{R}^m \to \mathbb{R}^n$ and $h : \mathbb{R} \times \mathbb{R}^n \to \mathbb{R}$, and sets $U \subset \mathbb{R}^m$ and $C \subset \mathbb{R} \times \mathbb{R}^n \times \mathbb{R} \times \mathbb{R}^n$.

Subsequent analysis will address problem (FTP) in greater generality. But for clarity of exposition, we temporarily restrict attention to a special case in which $f(t, x, u)$ and $h(t, x)$ are independent of t. (Accordingly, we suppress t in the notation, writing $f(x, u)$ for $f(t, x, u)$ and $h(x)$ for $h(t, x)$). We also assume that $f(., u)$ (for all $u \in \mathbb{R}^m$), g and h are continuously differentiable functions, f is continuous, the underlying time interval and the left end-point are fixed, i.e.

$$C = \{\bar{S}\} \times \{x_0\} \times \{\bar{T}\} \times C_1,$$

for some $[\bar{S}, \bar{T}] \subset \mathbb{R}$, $x_0 \in \mathbb{R}^n$ ($\bar{T} > \bar{S}$) and $C_1 \subset \mathbb{R}^n$, and g depends only on the right end-point of the state trajectory.

Take a minimizer $\bar{x}$ for (FTP). Theorem 10.4.1 provides necessary conditions of optimality. These assert the existence of an absolutely continuous arc $p \in W^{1,1}([\bar{S}, \bar{T}]; \mathbb{R}^n)$, $\lambda \geq 0$, a measure $\mu \in C^{\oplus}(\bar{S}, \bar{T})$ and $r \in \mathbb{R}$ such that the following conditions are satisfied:

(i) $(\lambda, p, \mu) \neq (0, 0, 0)$,
(ii) $-\dot{p}(t) = (p(t) + \int_{[\bar{S},t]} h_x(\bar{x}(s)) d\mu(s)) \cdot f_x(\bar{x}(t), \bar{u}(t))$ a.e. $t \in [\bar{S}, \bar{T}]$,
(iii) $u \to (p(t) + \int_{[\bar{S},t]} h_x(\bar{x}(s)) d\mu(s)) \cdot f(\bar{x}(t), u)$ is maximized
over $u \in U$ at $\bar{u}(t)$, a.e. $t \in [\bar{S}, \bar{T}]$,
(iv) $-(p(\bar{T}) + \int_{[\bar{S},\bar{T}]} h_x(\bar{x}(s)) d\mu(s)) \in \lambda g_x(\bar{x}(\bar{T})) + N_{C_1}(\bar{x}(\bar{T}))$,
(v) $\operatorname{supp} \mu \subset \{t : h(\bar{x}(t)) = 0\}$,
(vi) $(p(t) + \int_{[\bar{S},t]} h_x(\bar{x}(s)) d\mu(s)) \cdot f(\bar{x}(t), \bar{u}(t)) = r$, for a.e. $t \in (\bar{S}, \bar{T})$.

Notice that (vi) is an 'almost everywhere' condition. A case when the deficiencies of these necessary conditions are particularly in evidence is that when the initial state is fixed and lies in the boundary of the state constraint region. To explore this phenomenon, we suppose that

$$h(x_0) = 0. \tag{10.7.1}$$

Here, we find that conditions (i)–(vi) are satisfied (for some p, λ and μ) when $\bar{x}$ is *any* arc satisfying the constraints of (FTP). A possible choice of multipliers is

$$(p \equiv -h_x(\bar{S}), \ \mu = \delta_{\{\bar{S}\}}, \ \lambda = 0) \tag{10.7.2}$$

($\delta_{\{\bar{S}\}}$ denotes the unit measure concentrated at $\{\bar{S}\}$.) Provided $h_x(\bar{S}) \neq 0$, these multipliers are non-zero. Condition (v) is satisfied, by (10.7.2). The remaining conditions (i)–(iv) and (vi) are satisfied since

$$\int_{(\bar{S},t]} h_x(\bar{x}(s)) d\mu(s) = 0 \quad \text{for } t \in (\bar{S}, \bar{T}).$$

The fact that the necessary conditions (i)–(vi) are automatically satisfied by all admissible arcs renders them useless as necessary conditions of optimality in the case (10.7.1).

How should we deal with the degeneracy phenomenon? Extra necessary conditions are clearly required, which (in the case (10.7.1)) eliminate the uninteresting multipliers (10.7.2).

There are now a number of ways to do this. We focus on an extra condition to eliminate degeneracy, originating in the work of Arutyunov, Aseev and Blagodat-Skikh [5]. In the fixed end-point, autonomous case now under consideration, this asserts that the constancy of the Hamiltonian condition (v) on the open interval $(\bar{S}, \bar{T})$ extends to the end-points, i.e. condition (vi) can be supplemented by

10.7 Non-degenerate Conditions

$$\sup_{u \in U} p(\bar{S}) \cdot f(\bar{x}(\bar{S}), u) = r, \qquad (10.7.3)$$

$$\sup_{u \in U} (p(\bar{T}) + \int_{[\bar{S}, \bar{T}]} h_x(\bar{x}(s)) d\mu(s)) \cdot f(\bar{x}(\bar{T}), u)) = r,$$

where r is the constant of condition (vi).

The significance of these conditions becomes evident when we impose the constraint qualification

$$h_x(x_0) \cdot f(x_0, u) < 0 \text{ for some } u \in U. \qquad (10.7.4)$$

This is a condition on the data, requiring the existence of a control value driving the state into the interior of the state constraint set at x_0.

The constancy of the Hamiltonian condition, strengthened in this way, eliminates the degenerate multipliers (10.7.2). Indeed, for this choice of multipliers, $p(t) + \int_{[\bar{S}, t]} h_x(x(s)) d\mu(s) \equiv 0$ for $t \in (\bar{S}, \bar{T}]$, so $r = 0$ by (vi). But then, in view of (10.7.3) and (10.7.4), we arrive at the contradiction:

$$0 = \sup_{u \in U} p(\bar{S}) \cdot f(\bar{x}(\bar{S}), u) = - \inf_{u \in U} h_x(x_0) \cdot f(x_0, u) > 0.$$

The following theorem is the end result of elaborating these ideas, to allow for time dependent data and free end-times.

Consider the problem (FTP). A process $([S, T], x, u)$ comprises an interval $[S, T] \subset \mathbb{R} (T > S)$, $x \in W^{1,1}([S, T]; \mathbb{R}^n)$ and a measurable function $u : [S, T] \to \mathbb{R}^m$ such that $\dot{x}(t) \in f(t, x(t), u(t))$ and $u(t) \in U$, a.e. $t \in [S, T]$. The process is said to be admissible if $(S, x(S), T, x(S)) \in C$ and $h(t, x(t)) \leq 0$, for all $t \in [S, T]$.

An admissible process $([\bar{S}, \bar{T}], \bar{x}, \bar{u})$ is an L^∞ local minimizer for (FTP) if there exists some $\beta > 0$ such that

$$g(S, x(S), T, x(T)) \geq g(\bar{S}, \bar{x}(\bar{S}), \bar{T}, \bar{x}(\bar{T}))$$

for all admissible processes $([S, T], x, u)$ such that

$$|S - \bar{S}| + |T - \bar{T}| + ||x_e - \bar{x}_e||_{L^\infty} \leq \beta.$$

Here, x_e denotes the extension of $x : [S, T] \to \mathbb{R}^n$, by constant extrapolation to the left and right.

Write

$$H(t, x, p) := \sup_{u \in U} p \cdot f(t, x, u) \qquad (10.7.5)$$

and

$$\tilde{C} := \{(t_0, x_0, t_1, x_i) \in C : h(t_0, x_0) \leq 0 \text{ and } h(t_1, x_1) \leq 0\}.$$

Theorem 10.7.1 Let $([\bar{S}, \bar{T}], \bar{x}, \bar{u})$ be an L^∞ local minimizer for (FTP). Assume that, for some $\epsilon > 0$,

(H1) g is Lipschitz continuous on a neighbourhood of $(\bar{S}, \bar{x}(\bar{S}), \bar{T}, \bar{x}(\bar{T}))$ and C is a closed set,

(H2) $U \subset \mathbb{R}^m$ is a Borel measurable set, $(t, u) \to f(t, x, u)$ is $\mathcal{L} \times \mathcal{B}^m$ measurable for fixed x, and there exist $k_f > 0$, $c_f > 0$ such that
$$\begin{cases} |f(t'', x'', u) - f(t', x', u)| \le k_f(|t'' - t'| + |x'' - x'|), \\ |f(t', x', u)| \le c_f, \end{cases}$$
for all $u \in U$, $(t'', x''), (t', x') \in (t, \bar{x}(t)) + \epsilon \mathbb{B}$ and $t \in [\bar{S}, \bar{T}]$,

(H3) there exists $k_h > 0$ such that
$$|h(t'', x'') - h(t', x')| \le k_h(|(t'', x'') - (t', x')|),$$
for all $(t'', x''), (t', x') \in (t, \bar{x}(t)) + \epsilon \mathbb{B}$ and for all $t \in [\bar{S}, \bar{T}]$.

Then there exist absolutely continuous arcs $e \in W^{1,1}([\bar{S}, \bar{T}]; \mathbb{R})$ and $p \in W^{1,1}([\bar{S}, \bar{T}]; \mathbb{R}^n)$, $\lambda \ge 0$, a measure $\mu \in C^\oplus(\bar{S}, \bar{T})$ and a Borel measurable function $\gamma = (\gamma_0, \gamma_1) : [\bar{S}, \bar{T}] \to \mathbb{R}^{1+n}$, such that the following conditions, in which $(r, q) \in NBV([S, T]; \mathbb{R} \times \mathbb{R}^n)$ is the function

$$(-r(t), q(t)) := \begin{cases} (-e(\bar{S}), p(\bar{S})) & \text{if } t = \bar{S} \\ (-e(t), p(t)) + \int_{[\bar{S}, t]} \gamma(s) d\mu(s) & \text{if } t \in (\bar{S}, \bar{T}], \end{cases} \quad (10.7.6)$$

are satisfied:

(i) $(p, \mu, \lambda) \ne (0, 0, 0)$,
(ii) $(\dot{e}(t), -\dot{p}(t)) \in \text{co}\, \partial_{t,x} q(t) \cdot f(t, \bar{x}(t), \bar{u}(t))$ a.e. $t \in [\bar{S}, \bar{T}]$,
(iii) $q(t) \cdot f(t, \bar{x}(t), \bar{u}(t)) \ge q(t) \cdot f(t, \bar{x}(t), u)$, for all $u \in U$, a.e. $t \in [\bar{S}, \bar{T}]$,
(iv) $(-r(\bar{S}), q(\bar{S}), r(\bar{T}), -q(\bar{T})) \in \lambda \partial g(\bar{S}, \bar{x}(\bar{S}), \bar{T}, \bar{x}(\bar{T}))$
$\qquad\qquad\qquad\qquad\qquad + N_{\tilde{C}}(\bar{S}, \bar{x}(\bar{S}), \bar{T}, \bar{x}(\bar{T}))$,
(v) $\gamma(t) \in \partial^> h(t, \bar{x}(t))$ μ-a.e. $t \in [\bar{S}, \bar{T}]$,
(vi) $r(\bar{S}) = H(\bar{S}, \bar{x}(\bar{S}), q(\bar{S}))$,
(vii) $r(t) = H(t, \bar{x}(t), q(t))$ for a.e. $t \in (\bar{S}, \bar{T})$,
(viii) $r(\bar{T}) = H(\bar{T}, \bar{x}(\bar{T}), q(\bar{T}))$.

Remarks

(a): The 'value added' of this theorem, compared with earlier necessary conditions for problems with pure state constraints provided in Sect. 10.6 is, as anticipated in the introductory comments, the additional boundary conditions (vi) and (viii) on the maximized Hamiltonian.

(b): The transversality condition (iv) is perhaps not quite what one might expect, because it is expressed in terms of a modified endpoint constraint set $\tilde{C}$, obtained by augmenting the original endpoint constraints by the pathwise state constraints at the end-times. This refinement is essential to the proof. Of course

10.7 Non-degenerate Conditions

the transversality conditions involving $\tilde{C}$ and C coincide when either we are considering fixed end-point problems ($C = \{(S_0, x_0, T_0, x_1)\}$) or when the pathwise constraint is inactive at the optimal end-times.

In the event we introduce a constraint qualification, regarding the interaction of the dynamics and the state constraint functional at the optimal end-times, the additional boundary conditions on the maximized Hamiltonian serve to strengthen the non-triviality condition (i) and thereby to eliminate the trivial set of Lagrange multipliers alluded to above.

Corollary 10.7.2 *Let* $([\bar{S}, \bar{T}], \bar{x}, \bar{u})$ *be an* L^∞ *local minimizer for (FTP). Assume hypotheses (H1)–(H3) of Theorem 10.7.1. Assume also that the following constraint qualification is satisfied:*

$$\gamma_0' + \inf_{u \in U} \gamma_1' \cdot f(\bar{S}, \bar{x}(\bar{S}), u) < 0 \quad \text{and} \quad \gamma_0'' + \sup_{u \in U} \gamma_1'' \cdot f(\bar{T}, \bar{x}(\bar{T}), u) > 0 \tag{10.7.7}$$

for all $(\gamma_0', \gamma_1') \in \partial^> h(\bar{S}, \bar{x}(\bar{S}))$ *and all* $(\gamma_0'', \gamma_1'') \in \partial^> h(\bar{T}, \bar{x}(\bar{T}))$.

Then there exist absolutely continuous arcs $e \in W^{1,1}([\bar{S}, \bar{T}]; \mathbb{R})$ *and* $p \in W^{1,1}([\bar{S}, \bar{T}]; \mathbb{R}^n)$, $\lambda \geq 0$, *a measure* $\mu \in C^\oplus(\bar{S}, \bar{T})$ *and a Borel measurable function* $\gamma = (\gamma_0, \gamma_1) : [\bar{S}, \bar{T}] \to \mathbb{R}^{1+n}$ *such that conditions (ii)–(viii) of Theorem 10.7.1 are satisfied and also the following strengthened version of the non-triviality condition (i):*

(i)' : $\lambda + \|q\|_{L^\infty} + \int_{(\bar{S}, \bar{T})} d\mu(s) \neq 0$.

Notice that condition (i)' of the corollary is not satisfied when (λ, p, μ) is chosen according to (10.7.2). Thus, the corollary provides hypotheses under which there exists another, 'non-degenerate', family of Lagrange multipliers, for any problem in which either of the terminal states is fixed and is located in the boundary of the state constraint set.

Proof of Corollary 10.7.2 By Theorem 10.7.1 there exist multipliers (λ, p, μ), not all zero, and associated functions q and r, satisfying conditions (ii)–(viii). We must show that, under the constraint qualification (10.7.7), condition (i)' is also satisfied. Suppose, to the contrary, that (i)' fails to hold. Then $\lambda = 0$, $q(s) = 0$ for a.e. $s \in (\bar{S}, \bar{T})$ and there exist non-negative numbers α, β, not both zero, such that $\mu = \alpha \delta_{\{\bar{S}\}} + \beta \delta_{\{\bar{T}\}}$. We shall assume $\alpha > 0$. (The case '$\beta > 0$' is treated analogously.)

Since $q(t) = 0$ a.e., we know that $r(t) = H(t, \bar{x}(t), q(t)) = 0$ a.e.. Then, from (v) and (10.7.6), there exists $(\gamma_0, \gamma_1) \in \partial^> h(\bar{S}, \bar{x}(\bar{S}))$ such that

$$r(\bar{S}) = +\alpha \gamma_0 \text{ and } q(\bar{S}) = -\alpha \gamma_1.$$

It follows then from condition (vi) that

$$\alpha \gamma_0 = r(\bar{S}) = H(\bar{S}, \bar{x}(\bar{S}), q(\bar{S})) = -\alpha \inf_{u \in U} \gamma_1 \cdot f(\bar{S}, \bar{x}(\bar{S}), u).$$

Consequently,

$$\alpha(\gamma_0 + \inf_{u \in U} \gamma_1 \cdot f(\bar{S}, \bar{x}(\bar{S}), u)) = 0.$$

But this relation contradicts the constraint qualification. The corollary is confirmed. □

Proof of Theorem 10.7.1 Let $\epsilon' > 0$ be such that $([\bar{S}, \bar{T}], \bar{x}, \bar{u})$ is a minimizer for (FTP), with respect to arcs x satisfying

$$|S - \bar{S}| + |T - \bar{T}| + \|x_e - \bar{x}_e\|_{L^\infty} \leq \epsilon'.$$

Fix $\rho \in (0, 1/2)$. Consider the 're-parameterized' dynamic optimization problem

$$(R) \begin{cases} \text{Minimize } g((\tau, y)(s_0), (\tau, y)(s_1)) \\ \text{subject to } ([s_0, s_1], (\tau, y), (v, w)) \text{ for the control system} \\ \quad \begin{cases} \dot{y}(s) = (1 + w(s)) f(\tau(s), y(s), v(s)) & \text{a.e. } s \in [s_0, s_1], \\ \dot{\tau}(s) = 1 + w(s) & \text{a.e. } s \in [s_0, s_1], \\ (v(s), w(s)) \in U \times [-\rho, \rho] & \text{a.e. } s \in [s_0, s_1], \end{cases} \\ \text{subject to} \\ h(\tau(s), y(s)) \leq 0 \text{ for all } s \in [s_0, s_1], \\ ((\tau, y)(s_0), (\tau, y)(s_1)) \in \tilde{C}, \\ |s_0 - \bar{S}| + |s_1 - \bar{T}| + \|y_e - \bar{x}_e\|_{L^\infty} \leq \epsilon'', \end{cases}$$

where $\epsilon'' \in (0, \epsilon'/2)$. Notice that the last constraint in problem (R) is inactive, with reference to any process $([s_0', s_1'], y', u')$ such that $|s_0' - \bar{S}| + |s_1' - \bar{T}| + \|y_e' - \bar{x}_e\|_{L^\infty} < \epsilon''$ and so the end-times s_0 and s_1 of the independent variable s are locally unconstrained. Subsequent analysis will exploit this crucial fact. Proof of the following lemma is along similar lines to that of Theorem 9.6.1 (and Lemma 9.7.1) and is therefore omitted. It is based on the change of independent variable $\psi(s) = S + \int_{[s_0, s]} (1 + w(\sigma)) d\sigma$, $s \in [s_0, s_1]$ (for given s_0, s_1 and $w : [s_0, s_1] \to [-\rho, \rho]$).

Lemma 10.7.3 *We can choose $\rho \in (0, 1/2)$ and $\epsilon'' \in (0, \epsilon'/2)$ sufficiently small such that $([\bar{S}, \bar{T}], (\bar{\tau}(s) \equiv s, \bar{y}), (\bar{u}, \bar{w} \equiv 0))$ is an L^∞ (global) minimizer for problem (R).*

Take $\rho_i \downarrow 0$. For each i define

$$g_i(\tau_0, y_0, \tau_1, y_1) := (g(\tau_0, y_0, \tau_1, y_1) - g(\bar{S}, \bar{x}(\bar{S}), \bar{T}, \bar{x}(\bar{T})) + \rho_i^2) \vee 0.$$

Consider the problem

10.7 Non-degenerate Conditions

$$(R_i) \begin{cases} \text{Minimize } J_i(([s_0, s_1], (\tau, y), (v, w))) := g_i(\tau(s_0), y(s_0), \tau(s_1), y(s_1)) \\ \qquad\qquad \vee \max_{s \in [s_0, s_1]} h(\tau(s), y(s))) \\ \text{over } ([s_0, s_1], (\tau, y), (v, w)) \text{ subject to} \\ \dot{y}(s) = (1 + w(s)) f(\tau(s), y(s), v(s)) \quad \text{a.e. } s \in [s_0, s_1], \\ \dot{\tau}(s) = 1 + w(s) \quad \text{a.e. } s \in [s_0, s_1], \\ (v(s), w(s)) \in U \times [-\rho, \rho] \quad \text{a.e. } s \in [s_0, s_1], \\ (\tau(s_0), y(s_0), \tau(s_1), y(s_1)) \in \tilde{C}, \\ |s_0 - \bar{S}| + |s_1 - \bar{T}| + \|y_e - \bar{x}_e\|_{L^\infty} \le \epsilon'', \end{cases}$$

where $\rho \in (0, 1/2)$ and $\epsilon'' \in (0, \epsilon'/2)$ are taken according to Lemma 10.7.3. Define

$$\mathcal{M} := \{\text{processes } ([s_0, s_1], (\tau, y), (v, w)) \text{ for } (R_i)\}.$$

J_i is a continuous function on the complete metric space $(\mathcal{M}, d_\mathcal{M})$, when we define

$$d_\mathcal{M}(([s_0', s_1'], (\tau', y'), (v', w')), ([s_0, s_1], (\tau, y), (v, w)))$$
$$:= |\tau'(s_0') - \tau(s_0)| + |y'(s_0') - y(s_0)|$$
$$+ |s_0' - s_0| + |s_1' - s_1| + \mathcal{L} - \text{meas}\{s \in \mathbb{R} : v_e'(s) \ne v_e(s)\} + \|w_e' - w_e\|_{L^1}.$$

Here, $v_e(s) := \begin{cases} v(s) & \text{if } s \in [s_0, s_1] \\ 0 & \text{otherwise} \end{cases}$. $v_e'(s), w_e(s), w_e'(s)$ are likewise defined.

(Notice that a different extension convention is employed with the controls (u, w) ('zero extrapolation') compared with the state (τ, y) 'constant extrapolation'.)

Clearly, $([\bar{S}, \bar{T}], (\bar{\tau}(s) \equiv s, \bar{x}), (\bar{u}, \bar{w}(s) \equiv 0))$ is a ρ_i^2-minimizer for problem (R_i). We can therefore conclude from Ekeland's theorem that there exists $([s_{0i}, s_{1i}], (\tau_i, y_i), (v_i, w_i)) \in \mathcal{M}$, which is a minimizer for

$$\text{Minimize } \{J_i'([s_0, s_1], (\tau, y), (v, w)) : ([s_0, s_1], (\tau, y), (v, w)) \in \mathcal{M}\}. \quad (10.7.8)$$

Here

$$J_i'([s_0, s_1], (\tau, y), (v, w)) := J_i([s_0, s_1], (\tau, y), (v, w))$$
$$+ \rho_i d_\mathcal{M}(([s_0, s_1], (\tau, y), (v, w)), ([s_{0i}, s_{1i}], (\tau_i, y_i), (v_i, w_i))).$$

Furthermore,

$$d_\mathcal{M}(([s_{0i}, s_{1i}], (\tau_i, y_i), (v_i, w_i)), ([\bar{S}, \bar{T}], (\bar{x}, \bar{\tau}(s) \equiv s), (\bar{v}, \bar{w} \equiv 0))) \le \rho_i \tag{10.7.9}$$

and

$$J_i'([s_{0i}, s_{1i}], (\tau_i, y_i), (v_i, w_i)) \le J_i'([\bar{S}, \bar{T}], (\bar{x}, \bar{\tau}(s) \equiv s), (\bar{v}, \bar{w} \equiv 0)).$$

Relation (10.7.9) implies

$$|\tau_i(s_{0i}) - \bar{S}| + |y_i(s_{0i}) - \bar{x}(\bar{S})| + |s_{0i} - \bar{S}| + |s_{1i} - \bar{T}|$$
$$+ \mathcal{L} - \text{meas}\{s \in \mathbb{R} : v_{ie}(s) \neq \bar{u}_e(s)\} + ||w_i||_{L^1} \to 0 \quad \text{as } i \to \infty. \tag{10.7.10}$$

We can deduce, with the help of Gronwall's lemma, that

$$||(\tau_{ie}, y_{ie}) - (\bar{\tau}_e, \bar{x}_e)||_{L^\infty} \to 0 \quad \text{as } i \to \infty.$$

Note also that

$$g_i(\tau_i(s_{0i}), y_i(s_{0i}), \tau_i(s_{1i}), y_i(s_{1i})) \vee \max_{t \in [s_{0i}, s_{1i}]} h(\tau_i(t), y_i(t))) > 0,$$

$$\text{for all } i \text{ sufficiently large.} \tag{10.7.11}$$

This reason is that, if '=' replaces '>' in the above relation, then

$$g(\tau_i(s_{0i}), y_i(s_{0i}), \tau_i(s_{1i}), y_i(s_{1i})) \leq g(\bar{S}, \bar{x}(\bar{S}), \bar{T}, \bar{x}(\bar{T})) - \rho_i^2,$$

$(\tau_i(s_{0i}), y_i(s_{0i}), \tau_i(s_{1i}), y_i(s_{1i})) \in \tilde{C}$ and $h(\tau_i(t), y_i(t)) \leq 0$, for all $t \in [s_{0i}, s_{1i}]$. Furthermore, $|s_{0i} - \bar{S}| + |s_{1i} - \bar{T}| + ||y_{ie} - \bar{x}_e||_{L^\infty} \leq \epsilon''$. This contradicts the optimality of $([\bar{S}, \bar{T}], \bar{x}, \bar{u})$ for problem (R).

Observe next that $([s_{0i}, s_{1i}], (\tau_i, y_i), (v_i, w_i))$ is a global minimizer for

$$(R_i'') \begin{cases} \text{Minimize } J_i''(([s_0, s_1], (\tau, y), (v, w))) \\ \text{subject to} \\ \dot{y}(s) = (1 + w(s))f(\tau(s), y(s), v(s)) \quad \text{a.e. } s \in [s_0, s_1], \\ \dot{\tau}(s) = 1 + w(s) \quad \text{a.e. } s \in [s_0, s_1], \\ (v(s), w(s)) \in U \times [-\rho, \rho] \quad \text{a.e. } s \in [s_0, s_1], \\ (\tau(s_0), y(s_0), \tau(s_1), y(s_1)) \in \tilde{C}, \\ |s_0 - \bar{S}| + |s_1 - \bar{T}| + ||y_e - \bar{x}_e||_{L^\infty} \leq \epsilon'', \end{cases}$$

in which

$$J_i''(([s_0, s_1], (\tau, y), (v, w))) := g_i(\tau(s_0), y(s_0), \tau(s_1), y(s_1)) \vee \max_{s \in [s_0, s_1]} h(\tau(s), y(s))$$
$$+ \rho_i \Big(|\tau(s_0) - \tau_i(s_{0i})| + |y(s_0) - y_i(s_{0i})| + 4|s_0 - s_{0i}| + 4|s_1 - s_{1i}| $$
$$+ \int_{s_0}^{s_1} m_i(s, v(s))ds + \int_{s_0}^{s_1} |w(s) - w_{ie}(s)|ds \Big).$$

Here, $m_i(s, v) := \begin{cases} 0 & \text{if } v = v_{ie}(s) \\ 1 & \text{otherwise} \end{cases}$. Use of the cost function appearing in this problem formulation, in place of J_i' in problem (10.7.8), is justified by the following relations:

10.7 Non-degenerate Conditions

$$\|w_e - w_{ie}\|_{L^1} \leq \int_{s_0}^{s_1} |w(s) - w_{ie}(s)|ds + 2(|s_0 - s_{0i}| + |s_1 - s_{1i}|) \quad (10.7.12)$$

and

$$\mathcal{L} - \text{meas}\{s \in \mathbb{R} : v_e(s) \neq v_{ie}(s)\} \leq \int_{s_0}^{s_1} m_i(s, v(s))ds + |s_0 - s_{0i}| + |s_1 - s_{1i}|. \quad (10.7.13)$$

There are two cases to consider.

Case 1: $\max\limits_{s \in [s_{0i}, s_{1i}]} h(\tau_i(s), y_i(s)) > 0$ for at most a finite number of index values i,

Case 2: $\max\limits_{s \in [s_{0i}, s_{1i}]} h(\tau_i(s), y_i(s)) > 0$, for an infinite number of index values i.

Consider first Case 1. We can arrange, by excluding initial terms in the sequence, that

$$\max_{s \in [s_{0i}, s_{1i}]} h(\tau_i(s), y_i(s)) \leq 0 \text{ for all } i.$$

Taking note of (10.7.11), we see that $([s_{0i}, s_{1i}], (\tau_i, y_i), (v_i, w_i))$ is an L^∞ minimizer for

$$(\widetilde{R}'_i) \begin{cases} \text{Minimize } \tilde{J}_i(([s_0, s_1], (\tau, y), (v, w))) \\ \text{subject to} \\ \dot{y}(s) = (1 + w(s))f(\tau(s), y(s), v(s)) \quad \text{a.e. } s \in [s_0, s_1], \\ \dot{\tau}(s) = 1 + w(s) \quad \text{a.e. } s \in [s_0, s_1], \\ (v(s), w(s)) \in U \times [-\rho, \rho] \quad \text{a.e. } s \in [s_0, s_1], \\ (\tau(s_0), y(s_0), \tau(s_1), y(s_1)) \in \tilde{C}, \\ |s_0 - \bar{S}| + |s_1 - \bar{T}| + \|y_e - \bar{x}_e\|_{L^\infty} \leq \epsilon'', \end{cases}$$

in which

$$\tilde{J}_i([s_0, s_1], (\tau, y), (v, w)) := g_i((\tau(s_0), y(s_0), \tau(s_1), y(s_1)))$$

$$+\rho_i \Big(|\tau(s_0) - \tau_i(s_{0i})| + |y(s_0) - y_i(s_{0i})| + 4|s_0 - s_{0i}| + 4|s_1 - s_{1i}| +$$

$$\int_{s_0}^{s_1} m_i(s, v(s))ds + \int_{s_0}^{s_1} |w(s) - w_{ie}(s)|ds \Big).$$

From (10.7.11) and since $h(\tau_i(t), y_i(t)) \leq 0$ for $t \in [s_i, t_i]$,

$$g_i(\tau_i(s_{0i}), y_i(s_{0i}), \tau_i(s_{1i}), y_i(s_{1i}))$$
$$= g(\tau_i(s_{0i}), y_i(s_{0i}), \tau_i(s_{1i}), y_i(s_{1i})) - g(\bar{S}, \bar{x}(\bar{S}), \bar{T}, \bar{x}(\bar{T})) + \rho_i^2). \quad (10.7.14)$$

In view of (10.7.10), we can find a Lebesgue set $\mathcal{T} \subset (\bar{S}, \bar{T})$ with the following properties. Following a further extraction of subsequences we have, for each $s \in \mathcal{T}$,

$$v_i(s) = \bar{u}(s), \quad -\rho < w_i(s) < \rho \text{ for all } i \text{ suff. large and } w_i(s) \to 0, \text{ as } i \to \infty. \tag{10.7.15}$$

For i sufficiently large, we may apply the free end-time maximum principle Theorem 9.6.2 to this problem. We may thereby conclude that there exist $\lambda_i \geq 0$, $r_i \in W^{1,1}([s_{0i}, s_{1i}]; \mathbb{R})$ and $q_i \in W^{1,1}([s_{0i}, s_{1i}]; \mathbb{R}^n)$ such that

$$\lambda_i + \|q_i\|_{L^\infty} = 1, \tag{10.7.16}$$

$$(\dot{r}_i(s), -\dot{q}_i(s)) \in \operatorname{co} \partial_{t,x}(q_i(s) \cdot f(\tau_i(s), y_i(s), v_i(s)))(1 + w_i(s)),$$
$$\text{a.e. } s \in [s_{0i}, s_{1i}], \tag{10.7.17}$$

$$(v, w) \to (1 + w)(q_i(s) \cdot f(\tau_i(s), y_i(s), v) - r_i(s))$$
$$- \rho_i \lambda_i (m_i(s, v) + |w - w_{ie}(s)|)$$

is maximized over $U \times [-\rho, \rho]$ at $(v_i(s), w_i(s))$,

$$\text{a.e. } s \in [s_{0i}, s_{1i}], \tag{10.7.18}$$

$$(-r_i(s_{0i}), q_i(s_{0i}), r_i(s_{1i}), -q_i(s_{1i})) \in \lambda_i \partial g_i(\tau_i(s_{0i}), y_i(s_{0i}), \tau_i(s_{1i}), y_i(s_{1i}))$$
$$+ N_{\tilde{C}}(\tau_i(s_{0i}), y_i(s_{0i}), \tau_i(s_{1i}), y_i(s_{1i})) + \sqrt{2}\lambda_i \rho_i \mathbb{B} \times \{(0, 0)\} \tag{10.7.19}$$

and

$$\sup_{(v,w) \in U \times [-\rho, \rho]} (q_i(s) \cdot f(\tau_i(s), y_i(s), v) - r_i(s))(1 + w)$$
$$\in \rho_i \lambda_i (5 + 2\rho)[-1, +1], \text{ for } s = s_{0i} \text{ and } s_{1i}. \tag{10.7.20}$$

Condition (10.7.20) results from the free end-time transversality condition for the optimal endtimes, when we take account of the fact the end-times s_0 and s_1 in problem $(\widetilde{R}'_i)$ are locally unconstrained, with reference to the minimizer $([s_{0i}, s_{1i}], (\tau_i, y_i), (v_i, w_i))$, for i sufficiently large. Notice that the estimation of the essential values of the maximized Hamiltonian at the end-times takes account of the error terms involving $m_i(s, v)$ and $|w - w_{ie}|$.

(10.7.20) implies

$$\sup_{v \in U} q_i(s) \cdot f(\tau_i(s), y_i(s), v) - r_i(s) \in \rho_i \lambda_i (5 + 2\rho)(1 - \rho)^{-1}[-1, +1],$$

$$\text{for } s = s_{0i} \text{ and } s_{1i}. \tag{10.7.21}$$

By removing a null set of points from $\mathcal{T}$, we can arrange that conditions (10.7.17) and (10.7.18) are satisfied all points $s \in \mathcal{T}, i = 1, 2, \ldots$

10.7 Non-degenerate Conditions

Take any $s \in \mathcal{T}$. Considering first minimization w.r.t. the w variable and then the v variable, we deduce from (10.7.15) and (10.7.18) that, for i sufficiently large,

$$q_i(s) \cdot f(\tau_i(s), y_i(s), v_i(s)) - r_i(s) \in \rho_i \lambda_i \mathbb{B} \qquad (10.7.22)$$

and

$$q_i(s) \cdot f(\tau_i(s), y_i(s), v_i(s)) \geq \sup_{v \in U} q_i(s) \cdot f(\tau_i(s), y_i(s), v) - (1-\rho)^{-1} \rho_i \lambda_i .$$
$$\qquad (10.7.23)$$

But then

$$\sup_{v \in U} (q_i(s) \cdot f(\tau_i(s), y_i(s), v) - r_i(s)) \in (2-\rho)(1-\rho)^{-1} \rho_i \lambda_i \mathbb{B}. \qquad (10.7.24)$$

We deduce from (10.7.14), (10.7.19), properties of the distance function and the max rule for limiting subdifferentials that there exists $\lambda_i \in [0, 1]$ such that

$$(-r_i(s_{0i}), q_i(s_{0i}), r_i(s_{1i}), -q_i(s_{1i})) \in \lambda_i \partial g(\tau_i(s_{0i}), y_i(s_{0i}), \tau_i(s_{1i}), y_i(s_{1i}))$$
$$+ N_{\tilde{C}}(\tau_i(s_{0i}), y_i(s_{0i}), \tau_i(s_{1i}), y_i(s_{1i}))) + \sqrt{2} \rho_i \lambda_i \mathbb{B} \times \{(0,0)\}.$$
$$\qquad (10.7.25)$$

Relations (10.7.16), (10.7.17), (10.7.21), (10.7.22), (10.7.24) and (10.7.25) will be recognized as perturbed versions of the desired necessary conditions, in which $\mu_i = 0$ and therefore $(e_i, p_i) = (r_i, q_i)$. After extracting a further subsequence, we are assured of the existence of $q \in W^1$, $r \in W^{1,1}$ and $\lambda \in [0, 1]$ such that $(r_{ie}, q_{ie}) \to (r_e, q_e)$ uniformly, $(\dot{r}_{ie}, \dot{q}_{ie}) \to (\dot{r}_e, \dot{q}_e)$ weakly in L^1 and $\lambda_i \to \lambda$. We can now employ a standard convergence analysis to show that the asserted relations of the theorem are correct, with λ, $q(=p)$ and $r(=e)$ as above.

We now turn to Case 2. We can arrange, by selecting a subsequence, that

$$\max_{s \in [s_{0i}, s_{1i}]} h(\tau_i(s), y_i(s)) > 0, \text{ for all } i. \qquad (10.7.26)$$

Fix i. For arbitrary $K > 0$, consider the problem

$$(R_i^K) \begin{cases} \text{Minimize } J_i^K([s_0, s_1], (\tau, y), (v, w), z) \\ \text{over } [s_0, s_1] \subset \mathbb{R}, \ (\tau, y) \in W^{1,1}, \text{ measurable } (v, w) \text{ and } z \in \mathbb{R} \\ \text{satisfying} \\ \dot{y}(s) = (1 + w(s)) f(\tau(s), y(s), v(s)) \quad \text{a.e. } s \in [s_0, s_1], \\ \dot{\tau}(s) = 1 + w(s) \quad \text{a.e. } s \in [s_0, s_1], \\ (v(s), w(s)) \in U \times [-\rho, \rho] \quad \text{a.e. } s \in [s_0, s_1], \\ (\tau(s_0), y(s_0), \tau(s_1), y(s_1)) \in \tilde{C}, \\ |s_0 - \bar{S}| + |s_1 - \bar{T}| + \|y_e - \bar{x}_e\|_{L^\infty} \leq \epsilon'', \end{cases}$$

in which

$$J_i^K([s_0, s_1], (\tau, y), (v, w), z) := g_i(\tau(s_0), y(s_0), \tau(s_1), y(s_1)) \vee z$$
$$+ \rho_i\Big(|\tau(s_0) - \tau_i(s_{0i})| + |y(s_0) - y_i(s_{0i})| + 4|s_0 - s_{0i}| + 4|s_1 - s_{1i}|$$
$$+ \int_{s_0}^{s_1} m_i(s, v(s))ds + \int_{s_0}^{s_1} |w(s) - w_{ie}(s)|ds\Big)$$
$$+ K \int_{s_0}^{s_1} (h(\tau(s), y(s)) - z) \vee 0)^2 ds.$$

Notice that, for each K and any $([s_0, s_1], (\tau, y), (v, w))$, we have

$$J_i^K([s_0, s_1], (\tau, y), (v, w), z = \max_{s \in [s_0, s_1]} h(\tau(s), y(s)))$$
$$= J_i''([s_0, s_1], (\tau, y), (v, w)).$$

This implies that, for each i, $\inf(R_i^K) \leq \inf(R_i')$. According to the following lemma, the two infima coincide in the limit as $K \to \infty$, however.

Lemma 10.7.4 *For $i = 1, 2, \ldots$,*

$$\lim_{K \to \infty} \inf(R_i^K) = \inf(R_i').$$

Proof Fix i. Assume that the lemma is false. Since the function $K \to \inf(R_i^K)$ is non-decreasing, we may deduce the existence of $\bar{\rho} > 0$ and sequences $K_j \uparrow \infty$ and $\{([s_0^j, s_1^j], (\tau^j, y^j), (v^j, w^j)), z^j\}$ in $\mathcal{M} \times \mathbb{R}$ such that, for each j,

$$g_i(\tau^j(s_0^j), y^j(s_0^j), \tau^j(s_1^j), y^j(s_1^j)) \vee z^j$$
$$+ \rho_i\Big(|\tau^j(s_0^j) - \tau_i(s_{0i})| + |y^j(s_0^j) - y_i(s_{0i})| + 4|s_0^j - s_{0i}| + 4|s_1^j - s_{1i}|$$
$$+ \int_{s_0^j}^{s_1^j} m_i(s, v^j(s))ds + \int_{s_0^j}^{s_1^j} |w^j(s) - w_{ie}(s)|ds\Big)$$
$$+ K_j \int_{s_0^j}^{s_1^j} ((h(\tau^j(s), y^j(s)) - z^j) \vee 0)^2 ds$$
$$< g_i(\tau_i(s_{0i}), y_i(s_{0i}), \tau_i(s_{1i}), y_i(s_{1i})) \vee \max_{s \in [s_{0i}, s_{1i}]} h(\tau_i(s), y_i(s)) - \bar{\rho},$$
$$\text{for } j = 1, 2, \ldots \quad (10.7.27)$$

10.7 Non-degenerate Conditions

Notice, at the outset, that we can assume

$$\max_{s \in [s_0^j, s_1^j]} h(\tau^j(s), y^j(s)) - z^j \geq 0 \text{ for all } j. \tag{10.7.28}$$

To see this, suppose that, for some j, this condition is violated, i.e. $\max\{h(\tau^j(s), y^j(s)) : s \in [s_0^j, s_1^j]\} < z^j$. Now replace z^j by the lower number $\max\{h(\tau^j(s), y^j(s)) : s \in [s_0^j, s_1^j]\}$. We see that, following this change, (10.7.28) is satisfied. Furthermore, the strict inequality (10.7.27) is preserved. This is because the right side of this relation is unaffected, while the left side can only be reduced by this change. We have confirmed that, without loss of generality, we may assume (10.7.28). Since the (τ^j, y^j)'s are uniformly bounded in L^∞, the z^j's are uniformly bounded above. It is also clear from the non-negativity of the first two terms and the structure of the integrand of the integral on the left side of (10.7.27) that the z^j's are also bounded below. It follows that $\{z^j\}$ is a bounded sequence. Again, from the non-negativity of the first two terms and since $K_j \to \infty$ as $j \to \infty$, we deduce that

$$\int_{[s_0^j, s_1^j]} ((h(\tau^j(s), y^j(s)) - z^j) \vee 0)^2 ds \to 0, \text{ as } j \to \infty. \tag{10.7.29}$$

Fix j. Let $\bar{s}$ be a point at which $(h(\tau^j(s), y^j(s)) - z^j)$ achieves its maximum over $[s_0^j, s_1^j]$. Using the facts that the functions $s \to ((h(\tau^j(s), y^j(s)) - z^j) \vee 0)^2$ are uniformly Lipschitz continuous we can find $\bar{\alpha} > 0$ independent of j such that

$$((h(\tau^j(s), y^j(s)) - z^j) \vee 0)^2 \geq \frac{1}{2}((h(\tau^j(\bar{s}), y^j(\bar{s})) - z^j) \vee 0)^2,$$

$$\text{for all } s \in [s_0^j, s_1^j] \cap [\bar{s} - \bar{\alpha}, \bar{s} + \bar{\alpha}].$$

But then

$$\max_{s \in [s_0^j, s_1^j]} (h(\tau^j(s), y^j(s)) - z^j) \vee 0 = (h(\tau^j(\bar{s}), y^j(\bar{s})) - z^j) \vee 0$$

$$\leq (2/\bar{\alpha})^{\frac{1}{2}} \left(\int_{[s_0^j, s_1^j]} (h(\tau^j(t), y^j(t)) - z^j) \vee 0)^2 dt \right)^{\frac{1}{2}}.$$

It follows now from (10.7.28) and (10.7.29) that

$$\max_{s \in [s_0^j, s_1^j]} h(\tau^j(s), y^j(s)) - z^j \to 0, \text{ as } j \to \infty.$$

Since $z \to g_i(s_0, x_0, s_1, x_1) \vee z$ is Lipschitz continuous (with Lipschitz constant 1) we deduce from this relation that, for j sufficiently large,

$$g_i(\tau^j(s_0^j), y^j(s_0^j), \tau^j(s_1^j), y^j(s_1^j)) \vee \max_{s \in [s_0^j, s_1^j]} h(\tau^j(s), y^j(s))$$

$$+ \rho_i \Big(|\tau^j(s_0^j) - \tau^j(s_0^j)| + |y^j(s_0^j) - y^j(s_0^j)| + 4|s_0^j - s_{0i}| + 4|s_1^j - s_{1i}|$$

$$+ \int_{s_0^j}^{s_1^j} m_i(s, v^j(s)) ds + \int_{s_0^j}^{s_1^j} |w^j(s) - w_{ie}(s)| ds \Big)$$

$$< g_i(\tau_i(s_{0i}), y_i(s_{0i}), \tau_i(s_{1i}), y_i(s_{1i})) \vee \max_{s \in [s_{0i}, s_{1i}]} h(\tau_i(s), y_i(s)) - \bar{\rho}/2.$$
(10.7.30)

This contradicts the optimality of $([s_{0i}, s_{1i}], (\tau_i, y_i), (v_i, w_i))$ for problem (R_i'). The lemma is proved. □

In view of (10.7.26), and the continuity of the map $([s_0, s_1], (\tau, y), (v, w)) \to (\tau, y)$ from $\mathcal{M}$ to L^∞, we can find a sequence $\rho_i' \downarrow 0$ such that, for each i, $\rho_i' \leq 1/2$ and

$$\left. \begin{array}{l} ([s_0, s_1], (\tau, y), (v, w)) \in \mathcal{M}, z \in \mathbb{R} \\ d_\mathcal{M}(([s_0, s_1], (\tau, y), (v, w)), ([s_{0i}, s_{1i}], (\tau_i, y_i), (v_i, w_i))) \leq \rho_i' \\ |z - \max_{s \in [s_{0i}, s_{1i}]} h(\tau_i(s), y_i(s))| \leq \rho_i' \end{array} \right\} \quad (10.7.31)$$

$$\Longrightarrow g_i(\tau(s_0), y(s_0), \tau(s_1), y(s_1)) \vee z > 0 \text{ and } z > 0.$$

According to the lemma, we can choose $K_i \uparrow \infty$ such that, for each i, $([s_{0i}, s_{1i}], (\tau_i, y_i), (v_i, w_i), z_i := \max_{s \in [s_{0i}, s_{1i}]} h(\tau_i(s), y_i(s)))$ is a $\rho_i'^2$-minimizer for $(P_i^{K_i})$. Ekeland's theorem, together with relations (10.7.12) and (10.7.13), then tell us that there exists $(([s_{0i}', s_{1i}'], (\tau_i', y_i'), (v_i', w_i')), z_i') \in \mathcal{M} \times \mathbb{R}$ that is an L^∞ minimizer for:

$$(Q_i) \begin{cases} \text{Minimize } g_i(\tau(s_0), y(s_0), \tau(s_1), y(s_1)) \vee z(s_1) + \rho_i' |z(s_0) - z_i'| \\ \qquad + K_i \int_{s_0}^{s_1} ((h(\tau(s), y(s)) - z(s)) \vee 0)^2 ds \\ + \rho_i \Big(|\tau(s_0) - \tau_i(s_{0i})| + |y(s_0) - y_i(s_{0i})| + 4|s_0 - s_{0i}| \\ \qquad + 4|s_1 - s_{1i}| + \int_{s_0}^{s_1} m_i(s, v(s)) ds + \int_{s_0}^{s_1} |w(s) - w_{ie}(s)| ds \Big) \\ + \rho_i' \Big(|\tau(s_0) - \tau_i'(s_{0i}')| + |y(s_0) - y_i'(s_{0i}')| + 4|s_0 - s_{0i}'| \\ \qquad + 4|s_1 - s_{1i}'| + \int_{s_0}^{s_1} m_i'(s, v(s)) ds + \int_{s_0}^{s_1} |w(s) - w_{ie}'(s)| ds \Big) \\ \text{over } [s_0, s_1] \subset \mathbb{R}, (\tau, y, z) \in W^{1,1}(s_0, s_1), \text{ meas. } (v, w) \text{ satisfying} \\ \dot{y}(s) = (1 + w(s)) f(\tau(s), y(s), v(s)) \quad \text{a.e. } s \in [s_0, s_1], \\ \dot{\tau}(s) = 1 + w(s) \quad \text{a.e. } s \in [s_0, s_1], \\ \dot{z}(s) = 0 \quad \text{a.e. } s \in [s_0, s_1], \\ (v(s), w(s)) \in U \times [-\rho, \rho] \quad \text{a.e. } s \in [s_0, s_1], \\ (\tau(s_0), y(s_0), \tau(s_1), y(s_1)) \in \tilde{C}, \\ |s_0 - \bar{S}| + |s_1 - \bar{T}| + \|y_e - \bar{x}_e\|_{L^\infty} \leq \epsilon''. \end{cases}$$

10.7 Non-degenerate Conditions

Here, $m'_i(t, v) := \begin{cases} 0 \text{ if } v = v'_{ie}(t) \\ 1 \text{ otherwise} \end{cases}$. Furthermore

$$d_\mathcal{M}(([s'_{0i}, s'_{1i}], (\tau'_i, y'_i), (v'_i, w'_i)), ([s_{0i}, s_{1i}], (\tau_i, y_i), (v_i, w_i)))$$
$$+ |z'_i - \max_{s \in [s_{0i}, s_{1i}]} h(\tau_i(s), y_i(s))| \le \rho'_i. \tag{10.7.32}$$

and so we have

$$\|(\tau'_{ie}, y'_{ie}) - (\tau_e, y_{ie})\|_{L^\infty} \to 0, \; |z'_i - \max_{s \in [s_{0i}, s_{1i}]} h(\tau_i(s), y_i(s))| \to 0,$$

$$|s'_{0i} - s_{0i}| \to 0, \; |s'_{1i} - s_{1i}| \to 0$$

and $\mathcal{L} - \text{meas}\{s \in \mathbb{R} : v'_{ie}(s) \ne v_{ie}(s)\} \to 0,$

$$\|w'_{ie} - w_{ie}\|_{L^1} \to 0. \tag{10.7.33}$$

This relation combines with (10.7.9) to tell us that, as $i \to \infty$,

$$\|(\tau'_{ie}, y'_{ie}) - (\bar{\tau}_e, \bar{y}_{ie})\|_{L^\infty} \to 0, \; z'_i \to \max_{s \in [\bar{S}, \bar{T}]} h(s, \bar{x}(s)), \; s'_{0i} \to \bar{S}, \; s'_{1i} \to \bar{T}$$

and $\mathcal{L} - \text{meas}\{s \in \mathbb{R} : v'_{ie}(s) \ne \bar{u}_e(s)\} \to 0, \; \|w'_i\|_{L^1} \to 0.$

It follows that, for a subsequence, $w'_{ie}(s) \to 0$ a.e.. We deduce from (10.7.31) that

$$g_i(\tau'_i(s'_{0i}), y'_i(s'_{0i}), \tau'_i(s'_{1i}), y'_i(s'_{1i})) \vee z'_i > 0 \text{ and } z'_i > 0, \text{ for each } i. \tag{10.7.34}$$

Ekeland's theorem also tells us that

$$K_i \int_{s'_{0i}}^{s'_{1i}} ((h(\tau'_i(s), y'_i(s)) - z'_i) \vee 0)^2 ds \; (+ \text{ non-negative terms})$$
$$\le g_i(\tau_i(s_{0i}), y_i(s_{0i}), \tau_i(s_{1i}), y_i(s_{1i})) \vee \max_{s \in [s_{0i}, s_{1i}]} h(\tau_i(s), y_i(s))$$
$$+ \rho'_i |\max_{s \in [s_{0i}, s_{1i}]} h(\tau_i(s), y_i(s)) - z'_i|$$
$$+ \rho'_i \Big(|\tau_i(s_{0i}) - \tau'_i(s'_{0i})| + |y_i(s_{0i}) - y'_i(s'_{0i})| + 4|s_{0i} - s'_{0i}|$$
$$+ 4|s_{1i} - s'_{1i}| + \int_{s_{0i}}^{s_{1i}} m'_i(s, v_i(s)) ds + \int_{s_{0i}}^{s_{1i}} |w_i(s) - w'_{ie}(s)| ds \Big).$$

The number on the right side of this relation has value 0, in the limit as $i \to \infty$, by (10.7.33). It follows that

$$K_i \int_{s'_{0i}}^{s'_{1i}} ((h(\tau'_i(s), y'_i(s)) - z'_i) \vee 0)^2 ds \to 0, \text{ as } i \to \infty.$$

Now, for each i sufficiently large, apply the free end-time version of the maximum principle (Theorem 9.6.2) to the problem (Q_i), with reference to the L^∞ minimizer $(([s'_{0i}, s'_{1i}], (\tau'_i, y'_i), (v'_i, w'_i)), z'_i)$, noting that the relevant hypotheses are satisfied. We are assured then of the existence of $\tilde{\lambda}_i \in [0, 1]$ and also of elements $-d_i \in W^{1,1}$, $-r_i \in W^{1,1}$ and $q_i \in W^{1,1}$ (interpreted as costate components associated with the state variable components z and τ and y variables respectively) and a Borel measurable function $\gamma_i = (\gamma_{0i}, \gamma_{1i}) : [s'_{0i}, s'_{1i}] \to \mathbb{R}^{1+n}$ such that

$$(\tilde{\lambda}_i, q_i) \neq (0, 0), \tag{10.7.35}$$

$$(\dot{r}_i(s), -\dot{q}_i(s)) \in \mathrm{co}\, \partial_{t,x}(q_i(s) \cdot f(\tau'_i(s), y'_i(s), v'_i(s)))(1 + w'_i(s))$$
$$-\gamma_i(s)(d\mu_i/ds)(s), \text{ a.e. } s \in [s'_{0i}, s'_{1i}], \tag{10.7.36}$$

$$\gamma_i(s) \in \partial h(\tau'_i(s), y'_i(s)), \quad \mu_i - \text{ a.e. } s \in [s'_{0i}, s'_{1i}],$$

$$d_i(s) = (d\mu_i/ds)(s), \text{ a.e. and } d_i(s'_{0i}) \in \rho'_i \tilde{\lambda}_i \mathbb{B}, \tag{10.7.37}$$

$$(v, w) \to (1 + w)(q_i(s) \cdot f(\tau'_i(s), y'_i(s), v) - r_i(s))$$
$$-\rho_i \tilde{\lambda}_i (m_i(s, v) + |w - w_{ie}(s)|) - \rho'_i \tilde{\lambda}_i (m'_i(s, v) + |w - w'_{ie}(s)|)$$

is maximized over $U \times [-\rho, \rho]$ at $(v'_i(s), w'_i(s))$, a.e. $s \in [s'_{0i}, s'_{1i}]$, (10.7.38)

$$((-r_i(s'_{0i}), q_i(s'_{0i}), r_i(s'_{1i}), -q_i(s'_{1i})), d_i(s'_{1i})) \in$$
$$\tilde{\lambda}_i \partial \Big(g_i(\tau'_i(s'_{0i}), y'_i(s'_{0i}), \tau'_i(s'_{1i}), y_i(s'_{1i})) \vee z'_i(s'_{1i})) \Big)$$
$$+ (\xi_i, 0) + \sqrt{2}(\rho_i + \rho'_i) \tilde{\lambda}_i \mathbb{B} \times \{0\},$$
for some $\xi_i \in N_{\tilde{C}}(\tau'_i(s'_{0i}), y'_i(s'_{0i}), \tau'_i(s'_{1i}), y'_i(s'_{1i}))$, (10.7.39)

$$\sup_{(v,w) \in U \times [-\rho, \rho]} (q_i(s) \cdot f(\tau'_i(s), y'_i(s), v)) - r_i(s))(1 + w)$$
$$-\tilde{\lambda}_i K_i (h(\tau'_i(s), y'_i(s)) - z'_i(s)) \vee 0)^2$$
$$\in (\rho_i + \rho'_i) \tilde{\lambda}_i (5 + 2\rho)[-1, +1], \text{ for } s = s'_{0i} \text{ and } s'_{1i}. \tag{10.7.40}$$

Here μ_i is the (positive) measure on the Borel sets of the real line, with support in $[s'_{0i}, s'_{1i}]$ and distribution:

$$\mu_i([s'_{0i}, s]) = 2K_i \tilde{\lambda}_i \int_{[s'_{0i}, s]} (h(\tau'_i(\sigma), y'_i(\sigma)) - z'_i(\sigma)) \vee 0) d\sigma. \tag{10.7.41}$$

Clearly,

$$\mathrm{supp}\{\mu_i\} \subset \{s \in [s'_{0i}, s'_{1i}] : h(\tau'_i(s), y'_i(s))) \geq z'_i(s)\}. \tag{10.7.42}$$

10.7 Non-degenerate Conditions

Relation (10.7.40) results from the free end-time transversality condition for the optimal end-times s'_{0i} and s'_{1i} in problem (Q_i), when we take account of the fact the end-times are locally unconstrained, with reference to the minimizer $(([s'_{0i}, s'_{1i}], (\tau'_i, y'_i), (v'_i, w'_i)), z'_i)$, for i sufficiently large. But $K_i((h(\tau'_i(s), y'_i(s)) - z'_i(s)) \vee 0)^2 = 0$ for $s = s'_{0i}, s'_{1i}$, since $z'_i(s)(\equiv z'_i) > 0$ and $h(\tau'_i(s), y'_i(s)) \le 0$ for $s = s'_{0i}$ and $s = s'_{1i}$. (We use here the fact that, for elements $(\tau_0, y_0, \tau_1, y_1) \in \tilde{C}$, we know $h(\tau_0, y_0) \le 0$ and $h(\tau_1, y_1) \le 0$.) (10.7.40) therefore implies that, for $s = s'_{0i}$ and s'_{1i},

$$\sup_{v \in U} q_i(s) \cdot f(\tau'_i(s), y'_i(s), v)) - r_i(s) \in (\rho_i + \rho'_i)\tilde{\lambda}_i \left(\frac{5 + 2\rho}{1 - \rho}\right)[-1, +1]. \tag{10.7.43}$$

We deduce from (10.7.31) and (10.7.39), with the help of the sum rule, that there exists $v_i \in [0, 1]$ such that

$$(-r_i(s'_{0i}), q_i(s'_{0i}), r_i(s'_{1i}), -q_i(s'_{1i})) \in$$
$$\tilde{\lambda}_i v_i \partial g(\tau'_i(s'_{0i}), y'_i(s'_{0i}), \tau'_i(s'_{1i}), y'_i(s'_{1i})) + \xi_i + \sqrt{2}(\rho_i + \rho'_i)\tilde{\lambda}_i \mathbb{B},$$

$d_i(s'_{1i}) = \tilde{\lambda}_i(1 - v_i)$ and

$$\|\mu_i\|_{TV} = \int_{[s'_{0i}, s'_{1i}]} (d\mu/ds)ds \in \tilde{\lambda}_i(1 - v_i) + \rho'_i \tilde{\lambda}_i \mathbb{B}. \tag{10.7.44}$$

We can find a Lebesgue set $\mathcal{S} \subset (\bar{S}, \bar{T})$ of full measure, with the following properties. Following a further extraction of subsequences we have: for each $s \in \mathcal{S}$, (10.7.36) and (10.7.38) are valid for i sufficiently large and

$$v_i(s) = \bar{u}(s), \quad -\rho < w_i(s) < \rho \text{ for all } i \text{ suff. large and } w_i(s) \to 0, \text{ as } i \to \infty. \tag{10.7.45}$$

Take any $s \in \mathcal{S}$. Considering first minimization w.r.t. the w variable and then the v variable, we deduce from (10.7.38) that, for i sufficiently large,

$$q_i(s) \cdot f(\tau'_i(s), y'_i(s), v'_i(s)) - r_i(s) \in (\rho_i + \rho'_i)\tilde{\lambda}_i \mathbb{B}$$

and

$$q_i(s) \cdot f(\tau'_i(s), y'_i(s), v'_i(s)) \ge \sup_{v \in U} q_i(s) \cdot f(\tau'_i(s), y'_i(s), v))$$
$$- (1 - \rho)^{-1}(\rho_i + \rho'_i)\tilde{\lambda}_i. \tag{10.7.46}$$

But then

$$\sup_{u \in U} q_i(s) \cdot f(\tau'_i(s), y'_i(s), v) - r_i(s) \in (2 - \rho)(1 - \rho)^{-1}(\rho_i + \rho'_i)\tilde{\lambda}_i \mathbb{B}. \tag{10.7.47}$$

Now define the function $(e_i, p_i) \in W^{1,1}([s'_{0i}, s'_{1i}]; \mathbb{R}^{1+n})$ according to

$$(-e_i, p_i)(s) := (-r_i, q_i)(s) - \int_{[s'_{0i}, s]} \gamma_i(\sigma)(d\mu_i(\sigma)/d\sigma)(\sigma) d\sigma \text{ for } s \in [s'_{0i}, s'_{1i}].$$

Condition (10.7.36) can be expressed in terms of (e_i, p_i):

$$(e_i, -\dot{p}_i)(s) \in \operatorname{co} \partial_{t,x} ((-r, q_i)(s) \cdot f(\tau'_i(s), y'_i(s), v'_i(s)))(1 + w'_i(s))$$
$$\text{a.e. } s \in [s'_{0i}, s'_{1i}]. \quad (10.7.48)$$

Notice next that $(\tilde{\lambda}_i, \xi_i) \neq (0, 0)$. This is because, otherwise, $q = 0$, in consequence of (10.7.39), (10.7.36) and Gronwall's lemma (Lemma 6.2.4): this contradicts (10.7.35). Now set $\lambda_i := \tilde{\lambda}_i \nu_i$. We see from (10.7.44) that $\|\mu_i\|_{TV} + \lambda_i \geq (1 - \rho'_i)\tilde{\lambda}_i$. It follows that $(\lambda_i, \xi_i, \mu_i) \neq (0, 0, 0)$. Now normalize $q_i, p_i, \xi_i, \lambda_i$ and $\tilde{\lambda}_i$, dividing each by the positive number $\|\mu_i\|_{TV} + \lambda_i + |\xi_i|$. (We do not re-label.) Then, automatically, $\|\mu_i\|_{TV} + \lambda_i + |\xi_i| = 1$. Observe also that, following relabelling, $\tilde{\lambda}_i \leq (1 - \rho'_i)^{-1} \leq 2$, since $\rho'_i \leq 1/2$ for each i.

Assembling all these relations and recalling the definition of the maximized Hamiltonian, we arrive at the following perturbed versions of the desired conditions: for each i,

$$\|\mu_i\|_{TV} + \lambda_i + |\xi_i| = 1, \quad (10.7.49)$$

$$(e_i, -\dot{p}_i)(s) \in \operatorname{co} \partial_{t,x} ((-r, q_i)(s) \cdot f(\tau'_i(s), y'_i(s), v'_i(s)))(1 + w'_i(s)),$$
$$\text{a.e. } s \in [s'_{0i}, s'_{1i}], \quad (10.7.50)$$

$$(-r_i(s'_{0i}), q_i(s'_{0i}), r_i(s'_{1i}), -q_i(s'_{1i})) \in$$
$$\lambda_i \partial g(\tau'_i(s'_{0i}), y'_i(s'_{0i}), \tau'_i(s'_{1i}), y'_i(s'_{1i})) + \xi_i + \sqrt{2}(\rho_i + \rho'_i)\mathbb{B}, \quad (10.7.51)$$

$$\xi_i \in N_{\tilde{C}}(\tau'_i(s'_{0i}), y'_i(s'_{0i}), \tau'_i(s'_{1i}), y'_i(s'_{1i})), \quad (10.7.52)$$

$$H(\tau'_i(s), y'_i(s), q_i(s)) - r_i(s) \in (\rho_i + \rho'_i)\left(\frac{5 + 2\rho}{1 - \rho}\right)[-1, +1],$$
$$\text{for } s = s_{0i} \text{ and } s_{1i},$$

$$(-r_i, q_i)(s) = (-e_i, p_i)(s) + \int_{[s'_{0i}, s]} \gamma_i(\sigma)(d\mu_i(\sigma)/d\sigma)(\sigma) d\sigma$$
$$\text{for } s \in [s'_{0i}, s'_{1i}]. \quad (10.7.53)$$

Also, for each $s \in \mathcal{S}$ and i sufficiently large:

$$q_i(s) \cdot f(\tau'_i(s), y'_i(s), v'_i(s)) \geq \sup_{v \in U} q_i(s) \cdot f(\tau'_i(s), y'_i(s), v))$$
$$-2(1 - \rho)^{-1}(\rho_i + \rho'_i), \quad (10.7.54)$$

$$H(\tau_i'(s), y_i'(s), q_i(s)) - r_i(s) \in (4-2\rho)(1-\rho)^{-1}(\rho_i + \rho_i')\mathbb{B}. \quad (10.7.55)$$

Relations (10.7.49)–(10.7.52) ensure that the sequences with elements μ_i, (e_i, p_i), γ_i, $(\dot{e}_i, \dot{p}_i)$, λ_i and ξ_i, are uniformly bounded (w.r.t. the total variation norm, the uniform norm, the L^1 norm and Euclidean norms, respectively). Observe that $[s_{0i}', s_{1i}'] \subset [\bar{S} - \epsilon'', \bar{T} + \epsilon'']$ for all i. Therefore, we can consider the extension of elements (e_i, p_i) and $(\dot{e}_i, \dot{p}_i)$ on the compact interval $[\bar{S} - \epsilon'', \bar{T} + \epsilon'']$ according to the preceding convention (by 'constant extrapolation' for the state arcs (e_i, p_i)'s and by 'zero extrapolation' for their velocities $(\dot{e}_i, \dot{p}_i)$'s). For each i (sufficiently large) we shall write μ_{ie} the Borel measure on $[\bar{S} - \epsilon'', \bar{T} + \epsilon'']$ defined by

$$\mu_{ie}(E) := \mu_i(E \cap [\bar{S} - \epsilon'', \bar{T} + \epsilon'']),$$

for all Borel subsets $E \subset [\bar{S} - \epsilon'', \bar{T} + \epsilon'']$. The Borel measurable functions γ_i's are extended by 'constant extrapolation' on $[\bar{S} - \epsilon'', \bar{T} + \epsilon'']$. Consequently their extensions satisfy the property

$$\gamma_{ie}(s) \in \partial h(\tau_{ie}'(s), y_{ie}'(s)), \quad \mu_{ie} - \text{a.e. } s \in [\bar{S} - \epsilon'', \bar{T} + \epsilon''].$$

It follows that, after subsequences have been extracted, there exist limit points μ, e, $\dot{e}$, p, $\dot{p}$, γ, λ and ξ with respect to the appropriate topologies (the limit measure and functions are eventually considered restricted to $[\bar{S}, \bar{T}]$). But then there exists a function of bounded variation (r, q) such that $(r_{ie}, q_{ie})(s) \to (r_e, p_e)(s)$ on a subset of $[\bar{S}, \bar{T}]$ of full Lebesgue measure, including the endpoints $\bar{S}$ and $\bar{T}$. (The subscript 'e' here means 'extension by constant extrapolation'.) We can also arrange that $\xi_i \to \xi$, for some $\xi \in \tilde{N}_C(\bar{S}, \bar{x}(\bar{S}), \bar{T}, \bar{x}(\bar{T}))$. We deduce finally, with the help of a standard convergence analysis, the validity of all the assertions in the theorem statement, by passing to the limit in (10.7.49)–(10.7.55) as $i \to \infty$. Notice, in particular that $\|\mu\|_{TV} + \lambda + |\xi| = 1$. This implies the non-triviality condition $(\lambda, p, \mu) \neq (0, 0, 0)$ since '$(\lambda, p, \mu) = (0, 0, 0)$' implies $\xi \neq 0$ and hence $p \neq 0$, which is not possible. To arrive at condition (v) of the theorem statement we have used Proposition 10.3.1 identifying the sets A_i and A of this proposition as follows

$$A_i = \{(t, (\zeta_0, \zeta_1)) \; : \; (\zeta_0, \zeta_1) \in \partial h(\tau_{ie}'(t), y_{ie}'(t)), \; t \in [\bar{S} - \epsilon'', \bar{T} + \epsilon'']\}$$

and

$$A = \{(t, (\zeta_0, \zeta_1)) \; : \; (\zeta_0, \zeta_1) \in \partial^> h(t, \bar{x}_e(t)), \; t \in [\bar{S} - \epsilon'', \bar{T} + \epsilon'']\}. \qquad \square$$

10.8 Mixed Constraints

The necessary conditions of optimality derived in Chap. 7 cover problems involving a pathwise constraint on control functions. In this chapter we have extended these necessary conditions to allow for an addition pathwise constraint on state trajectories, '$h(t, x(t)) \leq 0$'. But dynamic optimization problems of interest arise, with pathwise constraints on control and state variables, which do not separate out into distinct pathwise constraints for each of the variables. Such problems are called 'mixed constraint' problems. An important example is the case when the control constraint set is state dependent; here the control constraint $u(t) \in U(t)$ of the dynamic optimization problem treated in Chap. 7 is replaced by $u(t) \in U(t, x(t))$ where, now, the control constraint set at time t, $U(t, x)$, is x dependent. Such constraints arise in a number of application areas, including aeronautical control, where the thrust may be a control variable and the upper limit on the thrust will depend on altitude, a state variable component. In this section we provide necessary conditions of optimality for problems subject to a rather general form of mixed constraint and then explore the special form these conditions take in a classical setting, where the pathwise constraints are a collection of inequality and equality constraints on the mixed variables.

Our approach to deriving necessary conditions for dynamic optimization problems with mixed constraints will be to reformulate the problem as one in which the dynamic constraint is a differential inclusion with velocity set $F(t, x)$, obtained by combining the controlled differential equation with the mixed constraint. Necessary conditions for the mixed constraint problem can then be obtained by applying the Euler Lagrange-type conditions of Chap. 8 to the reformulated problem. This approach requires that $F(t, x)$ has the requisite regularity properties to make possible this application of the Euler Lagrange-type conditions. What hypotheses must be imposed, when the mixed constraint arises from a state-dependent control constraint set, $u \in U(t, x)$? The Euler Lagrange-type conditions are valid under hypotheses that include '$F(t, .)$ has bounded slope' or, equivalently, '$F(t, .)$ is pseudo-Lipschitz continuous' (see Proposition 8.2.3). We can arrange that the multifunction F in the reformulation of the mixed constraint problem satisfies this hypothesis by imposing conditions on the data of the mixed constraint problem that include:

$U(t, .)$ has the bounded slope property .

In what follows, we will consider different formulations of the mixed constraint and the precise nature of the bounded slope hypothesis imposed will vary according to our choice. This is because we prefer to impose hypotheses directly on the data in terms of which the mixed constraint is formulated rather than on the corresponding state dependent control constraint set, which is only implicitly defined.

Notice that the 'bounded slope' hypotheses will not be satisfied by the data of the problem posed in Sect. 10.2, involving a separated 'pure' state constraint $h(t, x) \leq 0$ and control constraint $u \in U(t)$. This is because the corresponding state dependent control constraint set is

$$U(t, x) := \begin{cases} U(t) & \text{if } h(t, x)) \leq 0 \\ \emptyset & \text{if } h(t, x)) > 0 . \end{cases}$$

10.8 Mixed Constraints

The multifunction $U(t, .)$ will be discontinuous at points in the boundary of the state constraint set $A(t) := \{x \in \mathbb{R}^n : h(t, x) \le 0\}$, so it will certainly not satisfy the pseudo Lipschitz condition, or the equivalent bounded slope condition. The essential difference between problems involving a pure state constraint and those with mixed constraints treated in this chapter is reflected in the nature of the necessary conditions: for pure constraints these involve 'true' costate trajectories q that are discontinuous functions of bounded variation, whereas costate trajectories for mixed constraint problems, derived below, are absolutely continuous.

Consider the following problem, in which the mixed constraint has the general formulation:

$$\phi(t, x(t), u(t)) \in \Phi(t) \text{ and } u(t) \in U(t) \text{ a.e. } t \in [S, T], \qquad (10.8.1)$$

for a given function ϕ and a given multifunction Φ.

$$(G) \begin{cases} \text{Minimize } g(x(S), x(T)) \\ \text{over } x \in W^{1,1}([S, T]; \mathbb{R}^n) \text{ and measurable functions } u : [S, T] \to \mathbb{R}^m \\ \text{such that} \\ \dot{x}(t) = f(t, x(t), u(t)) \text{ a.e. } t \in [S, T], \\ \phi(t, x(t), u(t)) \in \Phi(t) \text{ and } u(t) \in U(t) \text{ a.e. } t \in [S, T] \\ (x(S), x(T)) \in C, \end{cases}$$

the data for which comprise an interval $[S, T]$, integers $n > 0$, $m > 0$, functions $g : \mathbb{R}^n \times \mathbb{R}^n \to \mathbb{R}$, $f : [S, T] \times \mathbb{R}^n \times \mathbb{R}^m \to \mathbb{R}^n$ and $\phi : [S, T] \times \mathbb{R}^n \times \mathbb{R}^m \to \mathbb{R}^k$, a set $C \subset \mathbb{R}^n \times \mathbb{R}^n$ and multifunctions $U : [S, T] \rightsquigarrow \mathbb{R}^m$ and $\Phi : [S, T] \rightsquigarrow \mathbb{R}^k$.

An *admissible process* for (G) is a pair of functions (x, u) satisfying the specified constraints. We say that $(\bar{x}, \bar{u})$ is a $W^{1,1}$ local minimizer, when there exists $\beta > 0$ such that $(\bar{x}, \bar{u})$ is minimizing over all admissible processes (x, u) such that $||x - \bar{x}||_{W^{1,1}} \le \beta$.

For each time t, we define the set of *admissible* (x, u) *pairs*, $S(t) \subset \mathbb{R}^n \times \mathbb{R}^m$ to be

$$S(t) := \{(x, u) \in \mathbb{R}^n \times \mathbb{R}^m : \phi(t, x, u) \in \Phi(t) \text{ and } u \in U(t)\}. \qquad (10.8.2)$$

We also define $S^{\epsilon, R}(t)$, the localization of $S(t)$ about $(\bar{x}(t), \bar{u}(t))$, in which $(\bar{x}, \bar{u})$ is a nominal admissible process and $\epsilon > 0$ and $R > 0$ are specified parameters:

$$S^{\epsilon, R}(t) := \{(x, u) \in S(t) : |x - \bar{x}(t)| \le \epsilon \text{ and } |u - \bar{u}(t)| \le R\}.$$

The set of *admissible controls at state* $\bar{x}(t)$, $\Omega(t)$, is the set

$$\Omega(t) = \{u \in \mathbb{R}^m : (\bar{x}(t), u) \in S(t)\}.$$

For purposes of formulating the appropriate 'bounded slope' hypo thesis to derive necessary conditions for problem (G), we introduce the following 'bounded slope' condition:

$$(BS)_{t,x,u}^{K} : \left. \begin{array}{l} \lambda \in N_{\Phi(t)}(\phi(t,x,u)) \text{ and} \\ (\alpha, \beta) \in \partial(\lambda \cdot \phi(t,.,.))(x,u) + \{0\} \times N_{U(t)}(u) \end{array} \right\} \implies |\lambda| \leq K|\beta|,$$

for given $K > 0$ and $(t, x, u) \in [S, T] \times \mathbb{R}^n \times \mathbb{R}^m$ such that $(x, u) \in S(t)$.

The necessary conditions will include an unrestrictive form of the Weierstrass condition, expressed in terms of a subset $\Omega_0(t) \subset \Omega(t)$. $\Omega_0(t)$, the set of *regular admissible control values at* $\bar{x}(t)$, is defined to be

$$\Omega_0(t) := \{u \in \mathbb{R}^m \; : \; (\bar{x}(t), u) \in S(t) \text{ and there exists } \rho > 0 \text{ such that}$$

$(BS)_{t,x',u'}^{K=\rho^{-1}}$ is satisfied for all points (x', u') in a neighbourhood of

$$S(t) \cap \left((\bar{x}(t), u) + \rho \mathbb{B} \times \rho \mathbb{B} \right) \text{ relative to } S(t)\}. \tag{10.8.3}$$

Theorem 10.8.1 (Necessary Conditions for a General Problem with Mixed Constraints) *Let $(\bar{x}, \bar{u})$ be a $W^{1,1}$ local minimizer for (G). Assume that, for some $\epsilon > 0$, $r_0 > 0$ and measurable function $R : [S, T] \to \mathbb{R} \cup \{+\infty\}$, such that $R(t) > r_0$ for a.e. $t \in [S, T]$, the following hypotheses are satisfied*

(H1): g *is Lipschitz continuous on* $(\bar{x}(S), \bar{x}(T)) + \epsilon(\mathbb{B} \times \mathbb{B})$ *and C is closed,*

(H2): *for each $x \in \mathbb{R}^n$, $f(., x, .)$ is $\mathcal{L} \times \mathcal{B}^m$ measurable; U is a measurable multifunction that takes values non-empty, closed sets, for a.e. $t \in [S, T]$, $f(t, ., .)$ is locally Lipschitz continuous and there exist measurable functions k_x^f and k_u^f such that, for a.e. $t \in [S, T]$,*

$$|f(t, x_1, u_1) - f(t, x_2, u_2)| \leq k_x^f(t)|x_1 - x_2| + k_u^f(t)|u_1 - u_2|$$

for all (x_1, u_1) and (x_2, u_2) in a neighbourhood of $S^{\epsilon, R(t)}(t)$,

(H3): Φ *is a Lebesgue measurable multifunction taking values closed, non-empty subsets of $\mathbb{R}^\kappa$; for each $x \in \mathbb{R}^n$, $\phi(., x, .)$ is $\mathcal{L} \times \mathcal{B}^m$ measurable; for a.e. $t \in [S, T]$, $\phi(t, ., .)$ is locally Lipschitz continuous and there exist measurable functions k_x^ϕ and k_u^ϕ such that, for a.e. $t \in [S, T]$,*

$$|\phi(t, x_1, u_1) - \phi(t, x_2, u_2)| \leq k_x^\phi(t)|x_1 - x_2| + k_u^\phi(t)|u_1 - u_2|$$

for all (x_1, u_1) and (x_2, u_2) in a neighbourhood of $S^{\epsilon, R(t)}(t)$,

(BS): *There exists a measurable function $K : [S, T] \to \mathbb{R}$ such that, for a.e. $t \in [S, T]$, condition $(BS)_{t,x,u}^{K(t)}$ is satisfied for all (x, u) in a neighbourhood of $S^{\epsilon, R(t)}(t)$ relative to $S(t)$.*

Assume also the compatibility condition: for some $\gamma > 0$,

10.8 Mixed Constraints

(C): k_x^f and $t \to K(t)k_x^\phi(t)k_u^f(t)$ are integrable, and $K(t)k_x^\phi(t) > 0$ and
$$R(t) \geq K(t)k_x^\phi(t)\gamma, \text{ a.e..}$$

Then there exist $p \in W^{1,1}([S,T];\mathbb{R}^n)$ *and* $\lambda^0 \geq 0$ *such that*

(a): $(p, \lambda^0) \neq (0, 0)$,
(b): $(-\dot{p}(t), 0) \in co\, \partial_{x,u}(p(t) \cdot f(t, \bar{x}(t), \bar{u}(t)))$
$$-co\{\xi \in \partial_{x,u}(\lambda \cdot \phi(t, \bar{x}(t), \bar{u}(t))) : \lambda \in N_{\Phi(t)}(\phi(t, \bar{x}(t), \bar{u}(t)))\} -$$
$\{0\} \times co\, N_{U(t)}(\bar{u}(t))$ a.e. $t \in [S, T]$,
(c): $(p(S), -p(T)) \in \lambda^0 \partial g(\bar{x}(S), \bar{x}(T)) + N_C(\bar{x}(S), \bar{x}(T))$,
(d): $p(t) \cdot f(t, \bar{x}(t), \bar{u}(t)) \geq p(t) \cdot f(t, \bar{x}(t), u)$ for all $u \in \Omega_0(t)$, a.e. $t \in [S, T]$.

If in addition $\phi(t, ., .)$ *is continuously differentiable near* $(\bar{x}(t), \bar{u}(t))$ *a.e.* $t \in [S, T]$, *then condition (b) can be replaced by*

(b)': $(-\dot{p}(t), 0) \in co\, \partial\{p(t) \cdot f(t, ., .) - \lambda(t) \cdot \phi(t, ., .)\}(\bar{x}(t), \bar{u}(t))$
$$-\{0\} \times co\, N_{U(t)}(\bar{u}(t)) \text{ a.e.,}$$

in which $\lambda : [S, T] \to \mathbb{R}^k$ *is an integrable function that satisfies*

$$\lambda(t) \in co\, N_{\Phi(t)}(\phi(t, \bar{x}(t), \bar{u}(t))) \text{ a.e..} \quad (10.8.4)$$

Furthermore, for K as in (BS),

$$|\lambda(t)| \leq K(t)k_x^\phi(t)k_u^f(t)|p(t)| \text{ a.e..}$$

A proof of Theorem 10.8.1 is given below in this section.

The version of the mixed constraint that has received most attention is a constraint that is a combination of inequality and equality functional constraints imposed on values of processes (x, u), together with a general set constraint on the values of the control function u:

$$h^{(1)}(t, x(t), u(t)) \leq 0 \text{ and } h^{(2)}(t, x(t), u(t)) = 0 \text{ and } u(t) \in U(t), \text{ a.e. } t \in [S, T],$$

for given functions $h^{(1)}$ and $h^{(2)}$ and a given multifunction U. The problem now of interest is:

$$(M) \begin{cases} \text{Minimize } g(x(S), x(T)) \\ \quad \text{over } x \in W^{1,1}([S, T]; \mathbb{R}^n) \text{ and measurable functions } u : [S, T] \to \mathbb{R}^m \\ \quad \text{such that} \\ \dot{x}(t) = f(t, x(t), u(t)), \text{ a.e. } t \in [S, T], \\ h^{(1)}(t, x(t), u(t)) \leq 0 \text{ and } h^{(2)}(t, x(t), u(t)) = 0 \text{ and } u(t) \in U(t), \\ \text{a.e. } t \in [S, T], \\ (x(S), x(T)) \in C, \end{cases}$$

the data for which comprise an interval $[S, T]$, integers $n > 0$, $m > 0$, functions $g : \mathbb{R}^n \times \mathbb{R}^n \to \mathbb{R}$, $f : [S, T] \times \mathbb{R}^n \times \mathbb{R}^m \to \mathbb{R}^n$, $h^{(1)} : [S, T] \times \mathbb{R}^n \times \mathbb{R}^m \to \mathbb{R}^{K_1}$

and $h^{(2)} : [S, T] \times \mathbb{R}^n \times \mathbb{R}^m \to \mathbb{R}^{\kappa_2}$, a set $C \subset \mathbb{R}^n \times \mathbb{R}^n$ and a multifunction $U : [S, T] \rightsquigarrow \mathbb{R}^m$ taking values non-empty, closed sets.

The next theorem provides necessary conditions for problem (M). The assertions of this theorem can be deduced directly from Theorem 10.8.1, by making the following identifications

$$\Phi(t) = \overbrace{(-\infty, 0] \times \ldots \times (-\infty, 0]}^{\kappa_1} \times \overbrace{\{0\} \times \ldots \times \{0\}}^{\kappa_2} \text{ and } \phi = (h^{(1)}, h^{(2)})$$
(10.8.5)

and noting that, with these choices for Φ and ϕ,

$$N_{\Phi(t)}((h^{(1)}, h^{(2)})(t, x, u)) = \{(\lambda_1, \lambda_2) \in (\mathbb{R}_+)^{\kappa_1} \times \mathbb{R}^{\kappa_2} : \lambda_1 \cdot h^{(1)}(t, x, u) = 0\},$$

for any $(t, x, u) \in [S, T] \times \mathbb{R}^n \times \mathbb{R}^m$ such that $h^{(1)}(t, x, u) \leq 0$ and $h^{(2)}(t, x, u) = 0$.

As we pass from problem (G) to problem (M), we will continue to employ the notation $S(t)$, $S^{\epsilon, R}(t)$, $\Omega(t)$ and its subset $\Omega_0(t)$, when we identify $\Phi(t)$ and $\phi(t)$ according to (10.8.5). For problem (M) then,

$$S(t) = \{(x, u) : h^{(1)}(t, x, u) \leq 0, h^{(2)}(t, x, u) = 0, u \in U(t)\},$$
$$S^{\epsilon, R}(t) = \{(x, u) : h^{(1)}(t, x, u) \leq 0, h^{(2)}(t, x, u) = 0, u \in U(t),$$
$$|x - \bar{x}(t)| \leq \epsilon, |u - \bar{u}(t)| \leq R\},$$
$$\Omega(t) = \{u \in \mathbb{R}^m : (\bar{x}(t), u) \in S(t)\}.$$

The condition $(BS)^K_{t,x,u}$, earlier employed to formulate the 'bounded slope' hypothesis associated with the necessary conditions for the general problem (G), is now replaced by:
$(BS^*)^K_{t,x,u}$:

$$\left.\begin{array}{l}\lambda_1 \in (\mathbb{R}_+)^{\kappa_1}, \lambda_2 \in \mathbb{R}^{\kappa_1}, \\ \lambda_1 \cdot h^{(1)}(t, x, u) = 0, \eta \in N_{U(t)}(u)\end{array}\right\}$$
$$\implies |(\lambda_1, \lambda_2)| \leq K|\nabla_u(\lambda_1 \cdot h^{(1)} + \lambda_2 \cdot h^{(2)})(t, x, u) + \eta|,$$

in which (t, x, u) is a specified point in $[S, T] \times \mathbb{R}^n \times \mathbb{R}^m$ such that $(x, u) \in S(t)$.

For each $t \in [S, T]$, the regular control constraint set $\Omega_0(t)$ at $\bar{x}(t)$, corresponding to this bounded slope condition is:

$$\Omega_0^*(t) := \{u \in \mathbb{R}^m : (\bar{x}(t), u) \in S(t) \text{ and there exists } \rho > 0 \text{ such that}$$

$(BS^*)^{K=\rho^{-1}}_{t,x',u'}$ is satisfied for all points (x', u') in a neighbourhood of

$$S(t) \cap \left((\bar{x}(t), u) + \rho \mathbb{B} \times \rho \mathbb{B}\right) \text{ relative to } S(t)\}.$$

Theorem 10.8.2 (Mixed Functional Constraints: I) *Let $(\bar{x}, \bar{u})$ be a $W^{1,1}$ local minimizer for (M). Assume, for some $\epsilon > 0$ and $r_0 > 0$ and measurable function*

10.8 Mixed Constraints

$R : [S, T] \to \mathbb{R}^m \cup \{+\infty\}$, such that $R(t) > r_0$ for all $t \in [S, T]$, the following hypotheses are satisfied.

(H1): g is Lipschitz continuous on $(\bar{x}(S), \bar{x}(T)) + \epsilon(\mathbb{B} \times \mathbb{B})$ and C is closed,

(H2): for each $x \in \mathbb{R}^n$, $f(., x, .)$ is $\mathcal{L} \times \mathcal{B}^m$ measurable; U is a measurable multifunction that takes values closed, non-empty sets; for a.e. $t \in [S, T]$, $f(t, ., .)$ is locally Lipschitz continuous and there exist measurable functions k_x^f and k_u^f such that, for a.e. $t \in [S, T]$,

$$|f(t, x_1, u_1) - f(t, x_2, u_2)| \le k_x^f(t)|x_1 - x_2| + k_u^f(t)|u_1 - u_2|$$

for all (x_1, u_1) and (x_2, u_2) in a neighbourhood of $S^{\epsilon, R(t)}(t)$,

(H3): for each $x \in \mathbb{R}^n$, $h^{(1)}(., x, .)$ and $h^{(2)}(., x, .)$ are $\mathcal{L} \times \mathcal{B}^m$ measurable; for a.e. $t \in [S, T]$, $h^{(1)}(t, ., .)$ and $h^{(2)}(t, ., .)$ are continuously differentiable and there exist measurable functions k_x^h and k_u^h such that, for a.e. $t \in [S, T]$,

$$|(h^{(1)}, h^2)(t, x_1, u_1) - (h^{(1)}, h^2)(t, x_2, u_2)|$$
$$\le k_x^h(t)|x_1 - x_2| + k_u^h(t)|u_1 - u_2|$$

for all (x_1, u_1) and (x_2, u_2) in a neighbourhood of $S^{\epsilon, R(t)}(t)$,

(BS^*): there exists a measurable function $K : [S, T] \to \mathbb{R}$ such that, for a.e. $t \in [S, T]$,

$(BS^*)_{t,x,u}^{K(t)}$ is satisfied for all (x, u) in a neighbourhood of $S^{\epsilon, R(t)}(t)$.

Assume also that, for some $\gamma > 0$,

(C): k_x^f and $t \to K(t) k_x^h(t) k_u^f(t)$ are integrable, and $K(t) k_x^h(t) > 0$ and $R(t) \ge K(t) k_x^h(t) \gamma$, a.e..

Then there exist $p \in W^{1,1}([S, T]; \mathbb{R}^n)$, $\lambda^0 \ge 0$ and integrable functions $\lambda^1 : [S, T] \to (\mathbb{R}_+)^{\kappa_1}$ and $\lambda^2 : [S, T] \to \mathbb{R}^{\kappa_2}$ such that

$$\lambda^1(t) \cdot h^{(1)}(t, \bar{x}(t), \bar{u}(t)) = 0, \text{ a.e.}$$

and

(a): $(p, \lambda^0) \ne (0, 0)$,
(b): $(-\dot{p}(t), 0) \in co\, \partial_{x,u}(p(t) \cdot f(t, \bar{x}(t), \bar{u}(t))) - \{0\} \times co\, N_{U(t)}(\bar{u}(t))$
$\qquad -\lambda^1(t) \cdot \nabla_{x,u} h^{(1)}(t, \bar{x}(t), \bar{u}(t)) - \lambda^2(t) \cdot \nabla_{x,u} h^{(2)}(t, \bar{x}(t), \bar{u}(t))$
a.e. $t \in [S, T]$,
(c): $(p(S), -p(T)) \in \lambda^0 \partial g(\bar{x}(S), \bar{x}(T)) + N_C(\bar{x}(S), \bar{x}(T))$,
(d): $p(t) \cdot f(t, \bar{x}, \bar{u}(t)) \ge p(t) \cdot f(t, \bar{x}, u)$ for all $u \in \Omega_0^*(t)$, a.e. $t \in [S, T]$.

Furthermore, for K as in (BS^*),

$$|(\lambda_1(t), \lambda_2(t))| \le K(t) k_x^h(t) k_u^f(t) |p(t)| \text{ a.e..}$$

Remarks

(i): Notice that, when the data is continuously differentiable w.r.t x, condition (b) above reduces to

$$-\dot{p}(t) = \nabla_x H^{\lambda^1,\lambda^2}(t, \bar{x}(t), \bar{u}(t), p(t)) \text{ a.e}$$

$$0 \in \nabla_u H^{\lambda^1,\lambda^2}(t, \bar{x}(t), \bar{u}(t), p(t)) - \text{co } N_{U(t)}(\bar{u}(t)) \text{ a.e.,}$$

in which

$$H^{\lambda^1,\lambda^2}(t, x, u, p) := p \cdot f(t, x, u) - \lambda^1(t) \cdot h^{(1)}(t, \bar{x}(t)\bar{u}(t))$$
$$- \lambda^2(t) \cdot h^{(2)}(t, \bar{x}(t)\bar{u}(t)).$$

These relations will be recognized as the standard costate equation for the costate function coupled with a weak (local) version of the Weierstrass condition, expressed in terms of an augmented Hamiltonian with Lagrange multiplier terms to take account of the pathwise inequality and equality constraints. The functions λ^1 and λ^2 have the interpretation of Lagrange multipliers for the two constraints. Further information of Weierstrass condition type is supplied by condition (d) in the corollary statement.

(ii): Of course, if hypotheses (H2), (H3) and (BS^*) are strengthened to require that all the conditions of (BS^*) are satisfied for every (t, x, u) such that $(x, u) \in S(t)$, $|x - \bar{x}(t)| \le \epsilon$ (that is, we remove the condition '$|u - \bar{u}(t)| \le R(t)$'), then the set $\Omega_0(t)$ coincides with $\Omega(t)$ and the Weierstrass condition (e) is valid with $\Omega(t)$ replacing $\Omega_0(t)$.

This theorem can be used as a starting point for deriving necessary conditions of optimality for problems with mixed pathwise constraints, under different hypotheses. We provide just one example of a variant on Theorem 10.8.2, in which we replace the bounded slope condition $(BS^*)^K_{t,x,u}$ of Theorem 10.8.2 by the simpler, and less restrictive condition:

$(BS^{**})_{t,x,u}$:

$$\left.\begin{array}{l}(\lambda_1, \lambda_2) \in (\mathbb{R}_+)^{\kappa_1} \times \mathbb{R}^{\kappa_2}, \lambda_1 \cdot h^{(1)}(t, x, u) = 0, v \in N_U(u) \\ \nabla_u(\lambda_1 \cdot h^{(1)} + \lambda_2 \cdot h^{(2)})(t, x, u) + v = 0\end{array}\right\} \implies (\lambda_1, \lambda_2) = (0, 0),$$

while imposing additional hypotheses on the regularity of the mixed constraint functions, the control constraint set and the nominal control function. The condition in $(BS^{**})_{t,x,u}$ is frequently encountered in the field of nonlinear programming (NLP) in connection the constraint set

$$Q(t, x) := \{u \in \mathbb{R}^m : h^{(1)}(t, x, u) \le 0 \text{ and } h^{(2)}(t, x, u) = 0 \text{ and } u \in U(t),$$

10.8 Mixed Constraints

for fixed (t, x), where it is referred to as the *Magasarian Fromowitz condition*; in the (NLP) context, it has a role as a sufficient condition ensuring regularity properties of the multifunction $(t, x) \to Q(t, x)$ and normality of extremals of optimization problems with $u \in Q(t, x)$ as constraint.

The set of regular controls $\Omega_0(t)$ at $\bar{x}(t)$, based on this new bounded slope condition, is

$$\Omega_0^{**}(t) := \{u \in \mathbb{R}^m : (\bar{x}(t), u) \in S(t) \text{ and there exists } \rho > 0 \text{ such that}$$

$(BS^{**})_{t,x',u'}$ is satisfied for all points (x', u') in a neighbourhood of

$$S(t) \cap \Big((\bar{x}(t), u) + \rho \mathbb{B} \times \rho \mathbb{B}\Big) \text{ relative to } S(t)\}.$$

Theorem 10.8.3 (Mixed Functional Constraints: II) *Let $(\bar{x}, \bar{u})$ be a $W^{1,1}$ local minimizer for (M). Assume that, for some $\epsilon > 0$,*

(H1): g is Lipschitz continuous on a neighbourhood of $(\bar{x}(S), \bar{x}(T))$. $t \to U(t)$ is a constant multifunction (write its value U),

(H2): for each $x \in \mathbb{R}^n$, $f(., x, .)$ is $\mathcal{L} \times \mathcal{B}^m$ measurable, and there exist integrable functions k_x^f and k_u^f such that, for a.e. $t \in [S, T]$,

$$|f(t, x_1, u_1) - f(t, x_2, u_2)| \leq k_x^f(t)|x_1 - x_2| + k_u^f(t)|u_1 - u_2|$$

for all (x_1, u_1) and (x_2, u_2) in a neighbourhood of $S(t)$ such that $|x' - \bar{x}(t)| \leq \epsilon$, $|u' - \bar{u}| \leq \epsilon$,

(H3): for each $x \in \mathbb{R}^n$, $h^{(1)}(., x, .)$ and $h^{(2)}(., x, .)$ are $\mathcal{L} \times \mathcal{B}^m$ measurable; for a.e. $t \in [S, T]$, $h^{(1)}(t, ., .)$ and $h^{(2)}(t, ., .)$ are continuously differentiable and there exists a constant k^h such that, for a.e. $t \in [S, T]$,

$$|(h^{(1)}, h^{(2)})(t, x_1, u_1) - (h^{(1)}, h^{(2)})(t, x_2, u_2)|$$
$$\leq k^h(|x_1 - x_2| + |u_1 - u_2|)$$

for all (x_1, u_1) and (x_2, u_2) in a neighbourhood of $S^{\epsilon, \epsilon}(t)$,

(BS^{**}): $(BS)^{**}_{t,x,u}$ is satisfied, for every point (t, x, u) in

$$\text{closure } \{(t, x, u) \in [S, T] \times \mathbb{R}^n \times \mathbb{R}^m : (x, u) \in S^{\epsilon, \epsilon}(t)\}.$$

(Here $\bar{u}(t)$ is interpreted as some version of the equivalence class of bounded, a.e. equal functions.)

(B): $\bar{u}$ is essentially bounded.

Then there exist $p \in W^{1,1}([S, T]; \mathbb{R}^n)$, $\lambda^0 \geq 0$ and integrable functions $\lambda^1 : [S, T] \to (\mathbb{R}_+)^{\kappa_1}$ and $\lambda^2 : [S, T] \to \mathbb{R}^{\kappa_2}$ such that

$$\lambda^1(t) \cdot h^{(1)}(t, \bar{x}(t), \bar{u}(t)) = 0, \text{ a.e.}$$

and

(a): $(p, \lambda^0) \neq (0, 0)$,
(b): $(-\dot{p}(t), 0) \in co\, \partial\{p(t) \cdot f(t, ., .)\}(\bar{x}(t), \bar{u}(t)) - \{0\} \times co\, N_U(\bar{u}(t))$
$\qquad -\lambda^{(1)}(t) \cdot \nabla_{x,u} h^{(1)}(t, \bar{x}(t), \bar{u}(t)) - \lambda^{(2)}(t) \cdot \nabla_{x,u} h^{(2)}(t, \bar{x}(t), \bar{u}(t))$ a.e.,
(c): $(p(S), -p(T)) \in \lambda^0 \partial g(\bar{x}(S), \bar{x}(T)) + N_C(\bar{x}(S), \bar{x}(T))$,
(d): $p(t) \cdot f(t, \bar{x}, \bar{u}(t)) \geq p(t) \cdot f(t, \bar{x}, u)$ for all $u \in \Omega_0^{**}(t)$, a.e. $t \in [S, T]$.

Proof of Theorem 10.8.3 We shall verify:

Claim: Under hypotheses of Theorem 10.8.3 (Functional Mixed Constraints: II), hypothesis (BS^*) of Theorem 10.8.2 (Functional Mixed Constraints: I), in which $R \equiv \epsilon$, is valid for some constant K function. Furthermore, $\Omega_0^{**}(t) \subset \Omega_0^*(t)$, where $\Omega_0^{**}(t)$ and $\Omega_0^*(t)$ are sets of regular admissible control values at $\bar{x}(t)$ corresponding to the bounded slope conditions $(BS^{**})_{t,x,u}$ and $(BS^*)_{t,x,u}^K$, respectively.

The assertions of Theorem 10.8.3 will follow directly from Theorem 10.8.2 applied to the special case when, for the given parameter $\epsilon > 0$, the function R is taken to be $R \equiv \epsilon$, since the verification of the claim will ensure that (BS^*) is satisfied, while all the other hypotheses in Theorem 10.8.3 are the same as, or stronger than, their counterparts in Theorem 10.8.2. (Note that, since K and the Lipschitz bound k^h are constant functions, the compatibility hypothesis (C) in Theorem 10.8.2 is automatically satisfied.) The assertions of Theorem 10.8.3 merely reproduce those of Theorem 10.8.2, except that, in condition (d) of Theorem 10.8.3, Ω_0^* has been replaced by $\Omega_0^{**}(t)$. This substitution is justified since, according to the claim above, $\Omega_0^{**}(t) \subset \Omega_0^*(t)$.

To verify the first claim suppose, to the contrary, it is false. We shall show this leads to a contradiction. Write

$$\Lambda(t, x, u) := \{(\lambda^1, \lambda^2) \in (\mathbb{R}_+)^{\kappa_1} \times \mathbb{R}^{\kappa_1} : \lambda^1 \cdot h^{(1)}(t, x, u) = 0\}.$$

Under our assumption, there exists $K_i \uparrow \infty$ such that, for each i, we can find $(t, x_i, u_i) \in S^{\epsilon, R=\epsilon}(t)$ and $(\lambda_i^1, \lambda_i^2) \in \Lambda(t, x_i, u_i)$ and $v_i \in N_U(u_i)$ such that

$$K_i |(\lambda_i^1, \lambda_i^2) \cdot (\nabla_u(h^{(1)}, h^{(2)})(t, x_i, u_i)) + v_i| < |(\lambda_i^1, \lambda_i^2)|. \qquad (10.8.6)$$

Since $\bar{u}$ (or, more precisely, the relevant selection from the equivalence class) is bounded, the sequence $\{(t, x_i, u_i)\}$ is bounded. Notice that, from (10.8.6), $(\lambda_i^1, \lambda_i^2) \neq 0$. By dividing across (10.8.6) by $(|(\lambda_i^1, \lambda_i^2)| + |v_i|)$ and relabelling $(\lambda_i^1, \lambda_i^2)/(|(\lambda_i^1, \lambda_i^2)| + |v_i|)$ as $(\lambda_i^1, \lambda_i^2)$ and relabelling $v_i/(|(\lambda_i^1, \lambda_i^2)| + |v_i|)$ as v_i (normalization), we can arrange that the variables in relation (10.8.6) satisfy

$$|(\lambda_i^1, \lambda_i^2)| + |v_i| = 1. \qquad (10.8.7)$$

Notice that $\{(\lambda_i^1, \lambda_i^2)\}$ is a bounded sequence and, for each i, v_i, after the preceding normalization step, continues to satisfy $v_i \in N_U(u_i)$. By extracting subsequences

10.8 Mixed Constraints

then, we can arrange that $(t, x_i, u_i) \to (t, x, u)$, $(\lambda_i^1, \lambda_i^2) \to (\lambda^1, \lambda^2)$ and $v_i \to v$, for some $(t, x, u) \in [S, T] \times \mathbb{R}^n \times \mathbb{R}^m$, $(\lambda^1, \lambda^2) \in \Lambda(t, x, u)$. It is easy to deduce from the assumed regularity properties of $h^{(1)}$ and $h^{(2)}$ that $(x, u) \in S(t)$ and $v \in N_U(u)$. Because the left side of (10.8.6) is bounded and $K_i \uparrow \infty$, we can conclude that $|(\lambda_i^1, \lambda_i^2) \cdot (\nabla_u (h^{(1)}, h^{(2)})(t, x_i, u_i)) + v_i| \to 0$ and so

$$|(\lambda^1, \lambda^2) \cdot (\nabla_u (h^{(1)}, h^{(2)})(t, x, u)) + v| = 0. \tag{10.8.8}$$

But then, from $(BS^{**})_{t,x,u}$, we deduce that $(\lambda^1, \lambda^2) = (0, 0)$ and, therefore, by (10.8.8) we obtain also $|v| = 0$. This contradicts condition (10.8.7). The proof of the first assertion is complete.

A similar sequential compactness argument can be used to show that $\Omega_0^{**} \subset \Omega_0^*$. Since the reverse inclusion is obviously true, we have $\Omega_0^{**} = \Omega_0^*$. □

Proof of Theorem 10.8.1 Let $\varepsilon > 0$ be such that $(\bar{x}, \bar{u})$ is a minimizer for (G) with respect to admissible processes (x, u) such that

$$\|x - \bar{x}\|_{W^{1,1}} \leq \varepsilon.$$

Take S to be the multifunction of (10.8.2). Noting that, from the compatibility hypothesis (C), $k_x^\phi(t)K(t) > 0$ a.e., we can define:

$$c(t) := (k_x^f(t) + k_x^\phi(t)K(t)k_u^f(t))/(k_x^\phi(t)K(t)). \tag{10.8.9}$$

Now define the multifunction $F : [S, T] \times \mathbb{R}^n \rightsquigarrow \mathbb{R}^{n+1+m}$ to be

$$F(t, x) := \{(f(t, x, u), \Delta f(t, x, u), \theta(t, u))$$
$$: \text{there exists } u \in \mathbb{R}^m \text{ s.t. } (x, u) \in S(t)\}$$

in which $\Delta f(t, x, u) := |f(t, x, u) - f(t, \bar{x}(t), \bar{u}(t))|$ and

$$\theta(t, u) := c(t)(u - \bar{u}(t)).$$

Now consider the optimization problem

$$(G)' \begin{cases} \text{Minimize } g(x(S), x(T)) \text{ over } (x, y, z) \in W^{1,1} \\ \text{such that} \\ (\dot{x}(t), \dot{y}(t), \dot{z}(t)) \in F(t, x(t)), \text{ a.e.} \\ \text{and} \\ (x(S), x(T)) \in C, \ y(S) = |x(S) - \bar{x}(S)|, \ z(S) = 0 \text{ and } |y(T)| \leq \varepsilon. \end{cases}$$

We see that $(\bar{x}, 0, 0)$ is a minimizer for this problem. Indeed if (x, y, z) is any admissible F trajectory, we can, in consequence of the constraints of problem $(G)'$, select a measurable function u such that $(x(t), u(t)) \in S(t)$ for a.e. $t \in [S, T]$

and $\|x - \bar{x}\|_{W^{1,1}} \leq \varepsilon$. But then (x, u) is an admissible process for (G) such $\|x - \bar{x}\|_{W^{1,1}} \leq \varepsilon$. Since $(\bar{x}, \bar{u})$ is a $W^{1,1}$ local minimizer for (G), $(\bar{x}, 0, 0)$ is a minimizer for $(G)'$.

Our intention is to apply the necessary conditions of Chap. 8 to problem $(G)'$. This necessitates some preliminary subgradient calculations, the results of which are summarized in the following Lemma.

Lemma 10.8.4 *Take any point $t \in [S, T]$ in the set of full measure such that the conditions in hypotheses (H2), (H3) and (BS) are satisfied. Take an arbitrary point $u \in \Omega_0(t)$. Assume that for some set of positive parameters (ϵ', r', K') such that*

$$\left.\begin{array}{l}\lambda \in N_{\Phi(t)}(\phi(t, x', u')) \text{ and} \\ (\psi, \zeta) \in \partial(\lambda \cdot \phi(t, ., .))(x', u') + \{0\} \times N_{U(t)}(u')\end{array}\right\} \implies |\lambda| \leq K'|\zeta| \quad (10.8.10)$$

for all (x', u') in a neighbourhood of $\{(x'', u'') \in S(t) : |x'' - \bar{x}(t)| \leq \epsilon', |u'' - u| \leq r'\}$ relative to $S(t)$.

Write $k_x^{f'}$, $k_u^{f'}$, $k_x^{\phi'}$ and $k_u^{\phi'}$ for constants such that

$$|f(t, x_1, u_1) - f(t, x_2, u_2)| \leq k_x^{f'}|x_1 - x_2| + k_u^{f'}|u_1 - u_2|$$
$$|\phi(t, x_1, u_1) - \phi(t, x_2, u_2)| \leq k_x^{\phi'}|x_1 - x_2| + k_u^{\phi'}|u_1 - u_2|$$

for all points $(x_1, u_1), (x_2, u_2)$ in a neighbourhood of $\{(x', u') \in S(t) : |x' - \bar{x}(t)| \leq \epsilon', |u' - u| \leq r'\}$ relative to $S(t)$.

Take $(x, u) \in S(t)$ such that $(x, u) \in (\bar{x}(t), \bar{u}(t)) + \frac{\epsilon'}{2}\mathbb{B} \times \frac{r'}{2}\mathbb{B}$ and

$$(\alpha, \beta, \sigma, \tau) \in N_{\mathrm{Gr}\,F(t,.)}(x, f(t, x, u), \Delta f(t, x, u), \theta(t, u)). \quad (10.8.11)$$

Then

(i): *there exist $\eta \in N_{\Phi(t)}(\phi(t, x, u))$ and $\nu \in N_{U(t)}(u)$ such that*

$$(\alpha, -\nu) \in \partial_{x,u}(-\beta \cdot f(t, x, u) - \sigma \Delta f(t, x, u)$$
$$- \tau \cdot \theta(t, u)) + \partial_{x,u}(\eta \cdot \phi(t, x, u)), \quad (10.8.12)$$

furthermore,

$$|\alpha| \leq (k_x^{f'} + K' k_x^{\phi'} k_u^{f'})(|\beta| + |\sigma|) + K' k_x^{\phi'} c(t)|\tau|$$
$$\text{and } |\eta| \leq K'(k_u^{f'}(|\beta| + |\sigma|) + c(t)|\tau|). \quad (10.8.13)$$

(ii): *Now take $u = \bar{u}(t)$ and $(\epsilon', r', K') = (\epsilon, R(t), K(t))$. Then (10.8.13) can be equivalently stated*

$$\sqrt{3} \times |\alpha| \leq K_F(|\beta| + |\sigma| + |\tau|) \text{ and } |\eta| \leq K(t)(k_u^f(t)(|\beta| + |\sigma|) + c(t)|\tau|),$$

10.8 Mixed Constraints

in which

$$K_F(t) := \sqrt{3} \times (k_x^f(t) + K(t)k_x^\phi(t)k_u^f(t)). \tag{10.8.14}$$

Proof In the ensuing analysis the t component of points (t, x, u) in the domains of f, Δf, ϕ and θ, is suppressed, to simplify the notation. Consider the assertions of part (i) of the lemma. We begin by dealing with the special case, in which $(\alpha, \beta, \sigma, \tau)$ is a proximal normal vector to $\operatorname{Gr} F(t,.)$ at $(x, f(x, u), \Delta f(x, u), \theta(u))$. We deduce from the proximal normal inequality that, for some constant $C > 0$, $(x, u, y := \phi(t, x, y))$ is a local minimizer for the optimization problem

$$\begin{cases} \text{Minimize } -\alpha \cdot x' - \beta \cdot f(x', u') - \sigma \Delta f(x', u') - \tau \cdot \theta(u') + C \, |(x', u') - (x, u)|^2 \\ \text{over } (x', u', y') \in \mathbb{R}^m \times \mathbb{R}^m \times \mathbb{R}^\kappa \text{ such that} \\ y' = \phi(x', u'), \; y' \in \Phi(t) \text{ and } u' \in U(t). \end{cases}$$

(Notice that we have exploited the local Lipschitz continuity of $f(t, ., .)$, $\Delta f(t, ., .)$ and $\theta(t, .)$ to justify the inclusion of a quadratic term in the cost only involving the $|(x', u') - (x, u)|^2$ quadratic error term.) Now apply the Lagrange multiplier rule (Theorem 5.6.2). This gives $\lambda_0 \in \{0, 1\}$ and vectors $\nu \in N_{U(t)}(u)$ and $\eta \in \mathbb{R}^\kappa$ such that $(\eta, \lambda_0) \neq (0, 0)$ and

$$(0, 0, 0) \in \lambda_0(-\alpha, 0, 0) + \lambda_0 \partial_{x,u}(-\beta \cdot f(x, u) - \sigma \Delta f(x, u) - \tau \cdot \theta(u)) \times \{0\}$$
$$+ (0, 0, -\eta) + \partial_{x,u}(\eta \cdot \phi(x, u)) + (0, \nu, 0) + \{(0, 0)\} \times N_{\Phi(t)}(\phi(x, u)).$$

Consideration of the third coordinate reveals that $\eta \in N_{\Phi(t)}(\phi(x, u))$. We deduce that

$$(\lambda_0 \alpha, -\nu) \in \lambda_0 \partial_{x,u}(-\beta \cdot f(x, u) - \sigma \Delta f(x, u) - \tau \cdot \theta(u)) + \partial_{x,u}(\eta \cdot \phi(x, u)).$$

If $\lambda_0 = 0$, then $(0, -\nu) \in \partial_{x,u}(\eta \cdot \phi(x, u))$. This relation violates (10.8.10) since, when $\lambda_0 = 0$, we must have $\eta \neq 0$. So we can assume that $\lambda_0 = 1$. It follows that

$$(\alpha, -\nu) \in \partial_{x,u}(-\beta \cdot f(x, u) - \sigma \Delta f(x, u) - \tau \cdot \theta(u)) + \partial_{x,u}(\eta \cdot \phi(x, u)). \tag{10.8.15}$$

We see that there exists $(\psi, \zeta) \in \partial_{x,u}(-\beta \cdot f(x, u) - \sigma \Delta f(x, u) - \tau \cdot \theta(u))$ such that

$$(\alpha - \psi, -\nu - \zeta) \in \partial_{x,u}(\eta \cdot \phi(x, u)).$$

But then, by (10.8.10), $|\eta| \leq K'|\zeta|$. Estimating the magnitude of ζ with the help of the Lipschitz continuity properties of $f(t, ., .)$ etc., we arrive at

$$|\eta| \leq K' \times \left(k_u^{f'}(|\beta| + |\sigma|) + c(t)|\tau|\right). \tag{10.8.16}$$

We deduce from (10.8.15) that

$$|\alpha| \le k_x^{f'}(|\beta|+|\sigma|) + K'k_x^{\phi'}\Big(k_u^{f'}(|\beta|+|\sigma|) + c(t)|\tau|\Big), \qquad (10.8.17)$$

$$|v| \le k_u^{f'}(|\beta|+|\sigma|) + c(t)|\tau| + K'k_u^{\phi'}\Big(k_u^{f'}(|\beta|+|\sigma|) + c(t)|\tau|\Big). \qquad (10.8.18)$$

(10.8.15), (10.8.16) and (10.8.17) are the required relations. We have confirmed the assertions in part (i), when $(\alpha, \beta, \sigma, \tau)$ is a proximal normal vector to $N_{\text{Gr}\,F(t,.)}$. It remains to consider the case when $(\alpha, \beta, \sigma, \tau)$ is merely a (limiting) normal vector. Now we can find a sequences of points $(x_i, u_i) \to (x, u)$ and $(\alpha_i, \beta_i, \sigma_i, \tau_i) \to (\alpha, \beta, \sigma, \tau)$. such that, for each i,

$$(\alpha_i, \beta_i, \sigma_i, \tau_i) \in N_{\text{Gr}\,F(t,.)}^P(x_i, f(x_i, u_i), \Delta f(x_i, u_i), \theta(u_i)).$$

For each i sufficiently large, (x_i, u_i) lies in the relevant neighbourhood of the base point (x, u) and, arguing as above, we can show versions of (10.8.12) and (10.8.13) are valid, in which (x_i, u_i), and $(\alpha_i, \beta_i, \sigma_i, \tau_i)$ replace (x, u) and $(\alpha, \beta, \sigma, \tau)$ respectively, for some $v_i \in N_{U(t)}(u_i)$ and $\eta_i \in N_{\Phi(t)}(\phi(x_i, u_i))$. We deduce from the relation (10.8.13) (strictly speaking, its version at the perturbed base point) that $\{\eta_i\}$ is a bounded sequence. It then follows from (10.8.18), that $\{v_i\}$ is also a bounded sequence. Extracting a subsequence, we can arrange that $\eta_i \to \eta$ and $v_i \to v$, for some vectors η and v. We deduce from the closure properties of the limiting normal cone and the continuity of $\phi(t,.,.)$ that $\eta \in N_{\Phi(t)}(\phi(x, u))$ and $v \in N_{U(t)}(u)$. We recover relations (10.8.11), (10.8.12) and (10.8.13), in the limit as $i \to \infty$.

The assertions of part (ii) of the lemma, concerning the case when $u = \bar{u}(t)$, $(\epsilon', r', K') = (\epsilon, R(t), K(t))$, follow immediately from the substitution of the formula for $c(t)$, provided by (10.8.9), into (10.8.13). Since, now, $(k_x^{f'}, k_u^{f'}, k_x^{\phi'}, k_u^{\phi'}) = (k_x^f(t), k_u^f(t), k_x^\phi(t), k_u^\phi(t))$, this gives

$$|\alpha| \le k_x^f(t)(|\beta|+|\sigma|) + K(t)k_x^\phi(t)\Big(k_u^f(t)(|\beta|+|\sigma|) + c(t)|\tau|\Big).$$
$$= (k_x^f(t) + K(t)k_x^\phi(t)k_u^f(t))(|\beta|+|\sigma|+|\tau|) = (\sqrt{3})^{-1}K_F(t)(|\beta|+|\sigma|+|\tau|),$$

which is the desired relation. $\square$

We are now ready to apply the necessary conditions of Theorem 8.4.3 to problem $(G)'$ with reference to the minimizer $(\bar{x}, 0, 0)$. We do so when the necessary conditions take the modified form of Corollary 8.5.2, in which $(G1)$ and $(G2)$ (of Theorem 8.4.3) are imposed, but when the 'pseudo Lipschitz continuity and tempered growth hypothesis' $(G3)$ is replaced by the bounded slope (and associated compatibility condition) hypothesis $(G3)^{**}$.

$(G1)$ is obviously satisfied. Consider $(G2)$. Fix $t \in [S, T]$ and take any convergent sequence $(x_i, e_i) \to (x, e)$ in $\text{Gr}\,F(t,.)$. It follows that, for each i, there exists $u_i \in U(t)$ such that $e_i = (f(t, x_i, u_i), \Delta f(t, x_i, u_i), c(t)(u_i - \bar{u}(t)))$

10.8 Mixed Constraints

and $\phi(t, x_i, u_i) \in \Phi(t)$. Since $c(t) > 0$, $\{u_i\}$ is a bounded sequence. But then, for a subsequence, $u_i \to u$ for some $u \in \mathbb{R}^m$. By the continuity properties of $f(t, ., .)$ and $\phi(t, ., .)$ and since $U(t)$ and $\Phi(t)$ are closed sets, we know that $e = (f(t, x, u), \Delta f(t, x, u), c(t)(u - \bar{u}(t)))$, $u \in U(t)$ and $\phi(t, x, u) \in \Phi(t)$. We have shown that $(x, e) \in F(t, x)$. This confirms that Gr $F(t, .)$ is closed. It can be shown that $F(., x)$ is measurable.

To check $(G3)^{**}$, we set $R_F(t) := c(t) R(t)$ and we take $B \equiv \mathbb{R}^n$. Fix $t \in [S, T]$. Take any $x \in \bar{x}(t) + \epsilon \mathbb{B}$ and $e \in F(t, x)$ such that $|e - (f(t, \bar{x}(t), \bar{u}(t)), 0, 0)| < R_F(t)$. Then there exists $u \in U(t)$ such that $|u - \bar{u}(t)| < R(t)$, $e = (f(t, x, u), \Delta f(t, x, u), c(t)(u - \bar{u}))$ and $\phi(t, x, u) \in \Phi(t)$. Now take any $(\alpha, \beta, \sigma, \tau) \in N_{\text{Gr} F(t,.)}(e)$. Then, according to Lemma 10.8.4,

$$|\alpha| \le (\sqrt{3})^{-1} K_F(t)(|\beta| + |\sigma| + |\tau|) \le K_F(t)|(\beta, \sigma, \tau)|. \quad (10.8.19)$$

Under the hypotheses of Theorem 10.8.1, $k_x^f(t) + K(t)k_x^\phi(t)k_u^f(t)$ is integrable and there exists $\gamma > 0$ such that $R(t) \ge K(t)k_x^\phi(t)\gamma$. It follows that $K_F(t) = (\sqrt{3})(k_x^f(t) + K(t)k_x^\phi(t)k_u^f(t))$ is integrable and

$$R_F(t) = c(t)R(t) \ge (k_x^f(t) + k_x^\phi(t)K(t)k_u^f(t))\gamma = K_F(t)\tilde{\gamma}, \quad (10.8.20)$$

in which $\tilde{\gamma} := \gamma/\sqrt{3}$. Relations (10.8.19) and (10.8.20) confirm that hypothesis $(G3)^{**}$ is satisfied.

Since the relevant hypotheses are satisfied, we deduce from Cor. 8.5.2 that there exist $(p, r, s) \in W^{1,1}$ (the costate trajectories corresponding to (x, y, z)) and λ^0, not all zero, such that

(A): $(\dot{p}, \dot{r}, \dot{s})(t) \in \text{co}\{\xi : (\xi, (p(t), r(t), s(t))) \in N_{\text{Gr} F(t,.)}(\bar{x}(t), (\dot{\bar{x}}(t), 0, 0))\}$ a.e. $t \in [S, T]$,

(B): $(p(S), -p(T)) \in \lambda^0 \partial g(\bar{x}(S), \bar{x}(T)) + N_C(\bar{x}(S), \bar{x}(T))$,

(C): $p(t) \cdot \dot{\bar{x}}(t) \ge (p(t), r(t), s(t)) \cdot e$ for all $e \in \overline{\text{co}}\,\tilde{\Omega}_0(t)$ a.e. $t \in [S, T]$

in which

$$\tilde{\Omega}_0(t) := \{e \in F(t, \bar{x}(t)) : F(t, .)$$

is pseudo-Lipschitz continuous near $((\bar{x}(t), 0, 0), e)\}.$

Since F does not depend on the (y, z) variables, we deduce from Corollary 8.5.2 that r and s are constant functions. In view of the fact that the endpoint constraint $(y(T), z(T)) \in \varepsilon \mathbb{B} \times \mathbb{R}^n$ on these variables is inactive at $(y, z) = (\bar{y}, \bar{z})$, Corollary 8.5.2 also tells us that $(r, s) \equiv 0$. (B) is the desired transversality condition. We deduce from condition (A) and Lemma 10.8.4 that, for a.e. $t \in [S, T]$, $\dot{p}(t)$ lies in the convex hull of the set of vectors ξ such that

$$(\xi, 0) \in \partial_{x,u}\left(-p(t) \cdot f(t, \bar{x}(t), \bar{u}(t))\right) + \partial_{x,u}\left(\eta \cdot \phi(t, \bar{x}(t), \bar{u}(t))\right) + \{0\} \times N_{U(t)}(\bar{u}(t))$$

for some $\eta \in N_{\Phi(t)}(\phi(t, \bar{x}(t), \bar{u}(t)))$ such that $|\eta| \leq K(t)k_u^f(t)|p(t)|$. From the Carathéodory representation theorem, there exists a simplex $(\gamma_0, \ldots, \gamma_n)$ in $\mathbb{R}^n$, and for $j = 0, \ldots, n$, points $\xi_j \in \mathbb{R}^n$, $\eta_j \in N_{\Phi(t)}(\phi(t, \bar{x}(t), \bar{u}(t)))$ and $v_j \in N_{U(t)}(\bar{u}(t))$ such that $\dot{p}(t) = \sum_j \gamma_j \xi_j$ and, for each j,

$$(\xi_j, 0) \in \partial_{x,u}\Big(-p(t) \cdot f(t, \bar{x}(t), \bar{u}(t))\Big) + \partial_{x,u}\Big(\eta_j \cdot \phi(t, \bar{x}(t), \bar{u}(t))\Big) + \{0\} \times \{v_j\}.$$

and $|\eta_j| \leq K(t)k_u^f(t)|p(t)|$. Scaling by the simplex parameters and summing yields the relation

$$(\dot{p}(t), 0) = (\sum_j \gamma_j \xi_j, 0)$$

$$\in \mathrm{co}\, \partial_{x,u}\Big(-p(t) \cdot f(t, \bar{x}(t), \bar{u}(t))\Big)$$

$$+ \{\xi \in \mathrm{co}\, \partial_{x,u}\, \eta \cdot \phi(t, \bar{x}(t), \bar{u}(t)) : \eta \in N_{\Phi(t)}(\phi(t, \bar{x}(t), \bar{u}(t)))$$

$$\text{and } |\eta| \leq K(t)k_u^f(t)|p(t)|\}$$

$$+ \{0\} \times \mathrm{co}\, N_{U(t)}(\bar{u}(t)).$$

Using the sum rule for subdifferentials and noting that, for a locally Lipschitz continuous function d, $\mathrm{co}\, \partial(-d) = -\mathrm{co}\, \partial d$, we deduce that

$$(-\dot{p}(t), 0) \in \mathrm{co}\, \partial_{x,u}\, p \cdot f(t, \bar{x}(t), \bar{u}(t)) - \{0\} \times \mathrm{co}\, N_{U(t)}(\bar{u}(t))$$

$$- \{\xi \in \mathrm{co}\, \partial_{x,u}\, \eta \cdot \phi(t, \bar{x}(t), \bar{u}(t)) : \eta \in N_{\Phi(t)}(\phi(t, \bar{x}(t), \bar{u}(t)))$$

$$\text{and } |\eta| \leq K(t)k_u^f(t)|p(t)|\}.$$

We have confirmed the costate inclusion. Consider finally the consequences of the Weierstrass condition (C). Fix t in a suitable subset of $[S, T]$ of full measure and take any point $u \in \Omega_0(t)$. (Recall the definition of $\Omega_0(t)$ provided by (10.8.3).) Then there exists $\rho > 0$ such that

$$\left.\begin{array}{l} \lambda \in N_{\Phi(t)}(\phi(t, x', u')) \text{ and} \\ (\alpha, \beta) \in \partial(\lambda \cdot \phi(t, ., .))(x', u') + \{0\} \times N_{U(t)}(u') \end{array}\right\} \implies |\lambda| \leq \rho^{-1}|\beta| \quad (10.8.21)$$

for all points (x', u') in a neighbourhood of $\{(x'', u'') \in S(t) : |x'' - \bar{x}(t)| \leq \rho, |u'' - u| \leq \rho\}$ relative to $S(t)$.

Set $e := (f(t, \bar{x}(t), u), \Delta f(t, \bar{x}(t), u), \theta(t, u))$. Write $k_x^{f'}$, $k_u^{f'}$, $k_x^{\phi'}$ and $k_u^{\phi'}$ for constants such that

$$|f(t, x_1, u_1) - f(t, x_2, u_2)| \leq k_x^{f'}|x_1 - x_2| + k_u^{f'}|u_1 - u_2|,$$

$$|\phi(t, x_1, u_1) - \phi(t, x_2, u_2)| \leq k_x^{\phi'}|x_1 - x_2| + k_u^{\phi'}|u_1 - u_2|$$

for all points $(x_1, u_1), (x_2, u_2)$ in a neighbourhood of $\{(x', u') \in S(t) : |x' - \bar{x}(t)| \leq \rho, |u' - u| \leq \rho\}$. Now let $\delta \in (0, \rho]$ be such that

$$(\sqrt{3} \times \delta)^{-1} \geq (k_x^{f'} + \rho^{-1}(k_x^{\phi'} k_u^{f'} + k_x^{\phi'} c(t))) \text{ and } \delta \leq c(t)\rho.$$

Take any $x' \in \bar{x}(t) + \delta \mathbb{B}$, $e' \in F(t, x) \cap (e + \delta \mathbb{B})$ and $(\alpha, \beta, \sigma, \tau)$ such that

$$(\alpha, \beta, \sigma, \tau) \in N_{\operatorname{Gr} F(t,.)}(x', e').$$

Then $e' = (f(t, x', u'), \Delta f(t, x', u'), \theta(t, u'))$ for some u' such that $(x', u') \in S(t)$. We have $c(t)|u' - u| \leq \delta \leq c(t)\rho$ and hence $|u' - u| \leq \rho$. From Lemma 10.8.4

$$|\alpha| \leq (k_x^{f'} + \rho^{-1}(k_x^{\phi'} k_u^{f'} + k_x^{\phi'} c(t)))(|\beta| + |\sigma| + |\tau|)$$
$$\leq (\sqrt{3} \times \delta)^{-1}(|\beta| + |\sigma| + |\tau|) \leq \delta^{-1}|(\beta, \sigma, \tau)|.$$

We known that $F(t, .)$ satisfies the bounded slope condition near $(\bar{x}(t), 0, 0)$, $(f(t, \bar{x}(t), u), \Delta f(t, \bar{x}(t), u), \theta(t, u))$. It follows from Proposition 8.2.3 that $F(t, .)$ is pseudo-Lipschitz near $(\bar{x}(t), 0, 0)$, $(f(t, \bar{x}(t), u), \Delta f(t, \bar{x}(t), u), \theta(t, u))$. We have shown that $(f(t, \bar{x}(t), u), \Delta f(t, \bar{x}(t), u), \theta(t, u)) \in \tilde{\Omega}_0(t)$. But then, by condition (C),

$$p(t) \cdot (f(t, \bar{x}, u)) \geq p(t) \cdot f(t, \bar{x}(t), \bar{u}(t)).$$

We have confirmed the Weierstrass condition. The proof is complete.

□

10.9 Exercises

10.1 (State Constrained Maximum Principle in Gamkrelidze Form) Let $(\bar{x}, \bar{u})$ be a minimizer for the state constrained problem

$$\begin{cases} \text{Minimize } g(x(S), x(T)) \\ \text{subject to } \dot{x}(t) = f(x(t), u(t)), u(t) \in U \text{ a.e.} \\ h(x(t)) \leq 0 \text{ for all } t \in [S, T] \\ (x(S), x(T)) \in C. \end{cases}$$

with data functions $f : \mathbb{R}^n \times \mathbb{R}^m \to \mathbb{R}^n$, $g : \mathbb{R}^n \times \mathbb{R}^n \to \mathbb{R}$, $h : \mathbb{R}^n \to \mathbb{R}$ and sets $U \subset \mathbb{R}^m$ and $C \subset \mathbb{R}^n \times \mathbb{R}^n$.

Assume that hypotheses (H1)–(H3) of Theorem 10.4.1 are satisfied. Assume further that g is C^1, $f(., \bar{u}(t))$ is C^1 a.e. and h is C^2. Show that there exist $p \in W^{1,1}([S, T]; \mathbb{R}^n)$, $\mu \in C^{\oplus}(S, T)$ and $\lambda \geq 0$ such that

(i): $(p, \mu, \lambda) \neq (0, 0, 0)$,
(ii): $-\dot{p}(t) = \left(p(t) + \int_{[S,t]} d\mu(s) h_x(\bar{x}(t))\right) \cdot f_x(\bar{x}(t), \bar{u}(t))$
$\qquad\qquad + \int_{[S,t]} d\mu(s) h_{xx}(\bar{x}(t)) \cdot f(\bar{x}(t), \bar{u}(t))$,
(iii): $u \to (p(t) + \int_{[S,t]} d\mu(s) h_x(\bar{x}(t))) \cdot f(\bar{x}(t), u)$ is maximized over U at $u = \bar{u}(t)$. a.e.,
(iv): supp $\{\mu\} \subset \{t : h(\bar{x}(t)) = 0\}$,
(v): $(p(S), -(p(T) + \int_{[S,T]} d\mu(t) h_x(\bar{x}(T)))) \in \lambda \nabla g(\bar{x}(S), \bar{x}(T)) + N_C(\bar{x}(S), \bar{x}(T))$.

Hint: Show that $(\bar{x}, \bar{z}(t) := h(\bar{x}(t)), \bar{u}(t))$ is a minimizer for

$$\begin{cases} \text{Minimize } g(x(S), x(T)) \\ \text{subject to } (\dot{x}(t), \dot{z}(t)) = (f(x(t), u(t)), h_x(x(t)) \cdot f(x(t), u(t))), u \in U \text{ a.e.} \\ z(t) \leq 0 \text{ for all } t \in [S, T] \\ (x(S), x(T), z(T)) \in \{(x_0, x_1, z_1) : (x_0, x_1) \in C \text{ and } z_1 \geq h(x_1)\}. \end{cases}$$

Now apply the standard state constrained maximum principle (Theorem 10.4.1) to this problem.

10.2 Show that if the outward pointing condition

$$\max_{u \in U} h_x(\bar{x}(t)) \cdot f(\bar{x}(t), u) > 0 \text{ for all } t \in [S, T]$$

is satisfied and $h(\bar{x}(T)) < 0$, then the non-trivially condition (i) in Exercise 10.1 can be replaced by the stronger condition

(i)$'$: $(p, \lambda) \neq (0, 0)$,

Remark
This example illustrates how the Gamkrelidze form of the necessary conditions can be used to strengthen non-trivially conditions, concerning the Lagrange multipliers, under additional hypotheses.

10.3 Consider the problem

$$\begin{cases} \text{Minimize } \int_S^T L(x(t), \dot{x}(t)) dt \\ \text{over } x \in W^{1,1}([S, T]; \mathbb{R}^n) \text{ s.t.} \\ \dot{x}(t) \in U \text{ for a.e. } t \in [S, T], \\ h(x(t)) \leq 0 \text{ for all } t \in I, \\ x(S) = x_0, \end{cases}$$

for which the data comprise an interval $[S, T]$, functions $L : \mathbb{R}^n \times \mathbb{R}^n \to \mathbb{R}$ and $h : \mathbb{R}^n \to \mathbb{R}$, a closed set $I \subset [S, T]$ and a point $x_0 \in \mathbb{R}^n$. Let $\bar{x}$ be a minimizer. Assume that

(A): L and h are continuously differentiable,
(B): U is a compact, convex set and $L(x, .)$ is convex for each $x \in \mathbb{R}^n$.

10.9 Exercises

Define
$$I(\bar{x}) := \{t \in I : h(\bar{x}(t)) = 0\}.$$

(a): Show that

$$\left(\int_S^T [L_x(\bar{x}(t), \dot{\bar{x}}(t)) \cdot y^v(t) + (L(\bar{x}(t), v(t)) - L(\bar{x}(t), \dot{\bar{x}}(t)))] \, dt\right)$$

$$\vee \left(\max_{t \in I(\bar{x})} h_x(\bar{x}(t)) \cdot y^v(t)\right) \geq 0,$$

for all measurable functions v such that $v(t) \in U$, a.e. $t \in [S, T]$, in which $y^v(t) := \int_S^t (v(s) - \dot{\bar{x}}(t)) ds$.

(b): Deduce from the Aubin one-sided minimax theorem (Theorem 3.6.5) that there exist $\lambda \geq 0$ and $\mu \in C^{\oplus}(S, T)$ such that $\text{supp}\,\mu \subset I(\bar{x})$, $\lambda + ||\mu||_{\text{TV}} = 1$ and

$$\int_S^T [\lambda L_x(\bar{x}(t), \dot{\bar{x}}(t)) \cdot y^v(t) + \lambda (L(\bar{x}(t), v(t)) - L(\bar{x}(t), \dot{\bar{x}}(t)))] \, dt$$

$$+ \int_{[S,T]} y^v(t) \cdot h_x(\bar{x}(t)) d\mu(t) \geq 0,$$

for all measurable functions v such that $v(t) \in U$, a.e. $t \in [S, T]$.

(c): Let $p \in W^{1,1}$ be defined by

$$\begin{cases} \dot{p}(t) = \lambda L_x(\bar{x}(t), \dot{\bar{x}}(t)), \text{ a.e. } t \in [S, T], \\ -p(T) = \int_{[S,T]} h_x(\bar{x}(t)) d\mu(t). \end{cases}$$

Show that

$$\left(p(t) + \int_{[S,t]} h_x(\bar{x}(s)) d\mu(s)\right) \cdot \dot{\bar{x}}(t) - \lambda L(\bar{x}(t), \dot{\bar{x}}(t)) =$$

$$\max_{v \in U} \left(p(t) + \int_{[S,t]} h_x(\bar{x}(s)) d\mu(s)\right) \cdot v + \lambda L(\bar{x}(t), \dot{\bar{x}}(t)), \text{ a.e. } t \in [S, T].$$

Remark
The Aubin one-sided minimax theorem has been used to derive necessary conditions for state constrained dynamic optimization problems. (See e.g. [65] and [194].) This exercise illustrates the analysis involved, for a simple velocity constrained problem in the calculus of variations; it is easily generalized to allow for end-point costs and end-point constraints and a different dynamic constraint. A notable feature of this approach is that it can simply accommodate a state constraint imposed only on an

arbitrary closed subset I of the underlying time interval $[S, T]$, including cases in which I is an isolated point $\bar{t}$. The penalty approach employed in this chapter, in which we take account of the state constraint by adding the penalty term

$$+ K \int_I (h(x(t))) \vee 0)^2 dt$$

and passage to the limit as $K \uparrow \infty$, while offering some advantages in other respects (avoiding the need for convexification, for example), cannot be used directly in this 'isolated point case', because when I has Lebesgue measure zero, the penalty term automatically vanishes. It is for this reason that, in applying the penalty approach to problems in which the state constraint is imposed on a subset of I (or, more generally, in which h(t,x) is t dependent and possibly discontinuous), we need to introduce an extra step, in which h(t,x) is approximated by a function which is piecewise constant w.r.t. t.

10.4 (Free End-Time Mixed Constrained Maximum Principle: Measurable Time Dependence) Let $([\bar{S}, \bar{T}], \bar{x}, \bar{u})$ be a $W^{1,1}$ local minimizer (in the sense expressed in Sect. 9.6 of Chap. 9, and w.r.t. some parameter $\delta > 0$) for the free end-time mixed constrained problem

$$(FTM) \begin{cases} \text{Minimize } g(S, x(S), T, x(T)) \\ \text{over intervals } [S, T], \text{ arcs } x \in W^{1,1}([S, T]; \mathbb{R}^n) \\ \qquad \text{and measurable functions } u : [S, T] \to \mathbb{R}^m \text{ satisfying} \\ \dot{x}(t) = f(t, x(t), u(t)) \text{ a.e. } t \in [S, T], \\ \phi(t, x(t), u(t)) \in \Phi(t) \text{ and } u(t) \in U(t) \text{ a.e. } t \in [S, T] \\ u(t) \in U(t) \text{ a.e. } t \in [S, T], \\ \text{and} \\ (S, x(S), T, x(T)) \in C. \end{cases}$$

The data for this problem comprise functions $g : \mathbb{R}^{1+n+1+n} \to \mathbb{R}$, $f : \mathbb{R} \times \mathbb{R}^n \times \mathbb{R}^m \to \mathbb{R}^n$ and $\phi : \mathbb{R} \times \mathbb{R}^n \times \mathbb{R}^m \to \mathbb{R}^\kappa$, non-empty multifunctions $U : \mathbb{R} \rightsquigarrow \mathbb{R}^m$ and $\Phi : \mathbb{R} \rightsquigarrow \mathbb{R}^\kappa$, and a closed set $C \subset \mathbb{R}^{1+n+1+n}$.

Assume that, for some $\epsilon > 0$, $\sigma \in (0, (\bar{T} - \bar{S})/2)$, $r_0 > 0$ and measurable function $R : [\bar{S} - \sigma, \bar{T} + \sigma] \to \mathbb{R} \cup \{+\infty\}$, such that $R(t) > r_0$ for a.e. $t \in [\bar{S} - \sigma, \bar{T} + \sigma]$, the following hypotheses are satisfied:

(H1): g is Lipschitz continuous on a neighbourhood of $(\bar{S}, \bar{x}(\bar{S}), \bar{T}, \bar{x}(\bar{T}))$ and C is a closed set,

(H2): U is a measurable multifunction that takes values non-empty, closed subsets of $\mathbb{R}^m$; Φ is a Lebesgue measurable multifunction taking values closed, non-empty subsets of $\mathbb{R}^\kappa$,

10.9 Exercises

(H3): for each $x \in \mathbb{R}^n$, $f(., x, .)$ is $\mathcal{L} \times \mathcal{B}^m$ measurable; there exist measurable functions on $\mathbb{R}$ k_x^f and k_u^f such that, for a.e. $t \in [\bar{S} - \sigma, \bar{T} + \sigma]$,

$$|f(t, x_1, u_1) - f(t, x_2, u_2)| \leq k_x^f(t)|x_1 - x_2| + k_u^f(t)|u_1 - u_2|$$

for all (x_1, u_1) and (x_2, u_2) in a neighbourhood of $S_e^{\epsilon, R(t)}(t)$,

(H4): for each $x \in \mathbb{R}^n$, $\phi(., x, .)$ is $\mathcal{L} \times \mathcal{B}^m$ measurable; there exist measurable functions on $\mathbb{R}$ k_x^ϕ and k_u^ϕ such that, for a.e. $t \in [\bar{S} - \sigma, \bar{T} + \sigma]$,

$$|\phi(t, x_1, u_1) - \phi(t, x_2, u_2)| \leq k_x^\phi(t)|x_1 - x_2| + k_u^\phi(t)|u_1 - u_2|$$

for all (x_1, u_1) and (x_2, u_2) in a neighbourhood of $S_e^{\epsilon, R(t)}(t)$,

(BS): there exists a measurable function $K : \mathbb{R} \to \mathbb{R}$ such that, for a.e. $t \in [\bar{S} - \sigma, \bar{T} + \sigma]$, condition $(BS)_{t,x,u}^{K(t)}$ is satisfied for all (x, u) in a neighbourhood of $S_e^{\epsilon, R(t)}(t)$ relative to $S(t)$,

(H5): there exists $\bar{c} \geq 0$ such that, for a.e. $t \in [\bar{S} - \sigma, \bar{S} + \sigma] \cup [\bar{T} - \sigma, \bar{T} + \sigma]$,

$$|f(t, x, u)| \leq \bar{c}, \quad |u| \leq \bar{c}, \quad |(k_x^f(t) + k_x^\phi(t)K(t)k_u^f(t))/(k_x^\phi(t)K(t))| \leq \bar{c},$$

for all $x \in \bar{x}_e(t) + \epsilon \mathbb{B}$ and $u \in U(t)$ such that $\phi(t, x, u) \in \Phi(t)$.

Assume also the compatibility condition: for some $\gamma > 0$,

(C): k_x^f and $t \to K(t)k_x^\phi(t)k_u^f(t)$ are integrable on $[\bar{S}-\sigma, \bar{T}+\sigma]$, and $K(t)k_x^\phi(t) > 0$ and $R(t) \geq K(t)k_x^\phi(t)\gamma$, a.e. on $[\bar{S} - \sigma, \bar{T} + \sigma]$.

Show that there exist $p \in W^{1,1}([\bar{S}, \bar{T}]; \mathbb{R}^n)$ and a real number $\lambda^0 \geq 0$ such that

(i): $(p, \lambda^0) \neq (0, 0)$,
(ii): $(-\dot{p}(t), 0) \in \text{co}\, \partial_{x,u}(p(t) \cdot f(t, \bar{x}(t), \bar{u}(t)))$
$-\text{co}\,\{\xi \in \partial_{x,u}(\lambda \cdot \phi(t, \bar{x}(t), \bar{u}(t))) : \lambda \in N_{\Phi(t)}(\phi(t, \bar{x}(t), \bar{u}(t)))\}$
$- \{0\} \times \text{co}\, N_{U(t)}(\bar{u}(t))$ a.e. $t \in [\bar{S}, \bar{T}]$,
(iii): there exist $\xi_0 \in \text{sub-ess}_{t \to \bar{S}} H(t, \bar{x}(\bar{S}), p(\bar{S}))$,
and $\xi_1 \in \text{super-ess}_{t \to \bar{T}} H(t, \bar{x}(\bar{T}), p(\bar{T}))$ such that

$$(-\xi_0, p(\bar{S}), \xi_1, -p(\bar{T})) \in \lambda^0 \partial g(\bar{S}, \bar{x}(\bar{S}), \bar{T}, \bar{x}(\bar{T})) + N_C(\bar{S}, \bar{x}(\bar{S}), \bar{T}, \bar{x}(\bar{T})),$$

(iv): $p(t) \cdot f(t, \bar{x}(t), \bar{u}(t)) \geq p(t) \cdot f(t, \bar{x}(t), u)$, for all $u \in \Omega_0(t)$, a.e. $t \in [\bar{S}, \bar{T}]$.

Here,

$$S_e^{\epsilon, R}(t) := \{(x, u) \in S(t) : |x - \bar{x}_e(t)| \leq \epsilon \text{ and } |u - \bar{u}(t)| \leq R\}.$$

Hint: Consider the free end-time optimization problem

$$(FTM)' \begin{cases} \text{Minimize } g(S, x(S), T, x(T)) \\ \text{over } [S, T] \subset \mathbb{R} \text{ and } (x, y, z) \in W^{1,1}([S, T]; \mathbb{R}^{n+1+m}) \text{ s.t.} \\ (\dot{x}(t), \dot{y}(t), \dot{z}(t)) \in \tilde{F}(t, x(t)), \text{ a.e. } t \in [S, T] \\ \text{and} \\ (S, x(S), T, x(T)) \in C, \ y(S) = |x(S) - \bar{x}(S)|, \ z(S) = 0 \\ \text{and } |y(T)| \leq \delta, \end{cases}$$

where $\tilde{F}$ is a suitable extended (on $\mathbb{R}$) version of the multifunction F defined in the proof of Theorem 10.8.1. Observe that $([\bar{S}, \bar{T}], \bar{x}, 0, 0)$ is a local minimizer for this problem and apply Theorem 9.4.1.

10.5 (Free End-Time Mixed Constrained Maximum Principle: Lipschitz Time Dependence) Let $([\bar{S}, \bar{T}], \bar{x}, \bar{u})$ be a $W^{1,1}$ local minimizer (in the sense expressed in Sect. 9.6 of Chap. 9, and w.r.t. some parameter $\delta > 0$) for the free end-time mixed constrained problem (FTM) of Exercise 10.4. Suppose that condition (H1) of Exercise 10.4 is satisfied. Assume also that, for some constants $\epsilon > 0$, $r_0 > 0$ and integrable function $R : [\bar{S}, \bar{T}] \to \mathbb{R}_+$, such that $R(t) > r_0$ for a.e. $t \in [\bar{S}, \bar{T}]$, the following hypotheses are satisfied:

(H2)': $U(t) \equiv U$ for all $t \in \mathbb{R}$, for some non-empty, closed set $U \subset \mathbb{R}^m$, $\Phi(t) \equiv \Phi$ for all $t \in \mathbb{R}$, for some non-empty, closed set $\Phi \subset \mathbb{R}^\kappa$,

(H3)': for each $x \in \mathbb{R}^n$, $f(., x, .)$ is $\mathcal{L} \times \mathcal{B}^m$ measurable; there exists a constant $k^f > 0$ such that, for a.e. $t \in [\bar{S}, \bar{T}]$,

$$|f(t_1, x_1, u_1) - f(t_2, x_2, u_2)| \leq k^f (|t_1 - t_2| + |x_1 - x_2| + |u_1 - u_2|)$$

for all $t_1, t_2 \in [t - \epsilon, t + \epsilon]$, (x_1, u_1) and (x_2, u_2) in a neighbourhood of $S^{\epsilon, R(t)}(t)$,

(H4)': for each $x \in \mathbb{R}^n$, $\phi(., x, .)$ is $\mathcal{L} \times \mathcal{B}^m$ measurable; there exists a constant $k^\phi > 0$ such that, for a.e. $t \in [\bar{S}, \bar{T}]$,

$$|\phi(s_1, x_1, u_1) - \phi(s_2, x_2, u_2)| \leq k^\phi (|s_1 - s_2| + |x_1 - x_2| + |u_1 - u_2|)$$

for all $s_1, s_2 \in [t - \epsilon, t + \epsilon]$, (x_1, u_1) and (x_2, u_2) in a neighbourhood of $\widetilde{S}^{\epsilon, R(t)}(t)$,

(BS)': there exists a measurable function $K : [\bar{S}, \bar{T}] \to \mathbb{R}_+$ such that, for a.e. $t \in [\bar{S}, \bar{T}]$,

$$\left. \begin{array}{l} \lambda \in N_\Phi(\phi(s, x, u)) \text{ and} \\ (\alpha_t, \alpha_x, \beta) \in \partial(\lambda \cdot \phi)(s, x, u) + \{0, 0\} \times N_U(u) \end{array} \right\} \implies |\lambda| \leq K(t)|\beta|,$$

for all (s, x, u) in a neighbourhood of $\widetilde{S}^{\epsilon, R(t)}(t)$ relative to $\widetilde{S}$.

Assume also the compatibility condition: for some $\gamma > 0$,

(C)': $t \to K(t) \times (|\dot{\bar{x}}(t)| + R(t))$ is integrable on $[\bar{S}, \bar{T}]$, and $K(t)k_x^\phi(t) > 0$ and $R(t) \geq K(t)k_x^\phi(t)\gamma$, a.e. on $[\bar{S}, \bar{T}]$.

Show that there exist $p \in W^{1,1}([\bar{S}, \bar{T}]; \mathbb{R}^n)$, $a \in W^{1,1}([\bar{S}, \bar{T}]; \mathbb{R})$ and a real number $\lambda^0 \geq 0$ such that

(i): $((p, a), \lambda^0) \neq (0, 0, 0)$,
(ii): $((\dot{a}(s), -\dot{p}(s)), 0) \in \mathrm{co}\, \partial_{t,x,u}(p(s) \cdot f(s, \bar{x}(s), \bar{u}(s)))$
$$-\mathrm{co}\,\{\xi \in \partial_{t,x,u}(\lambda \cdot \phi(s, \bar{x}(s), \bar{u}(s))) : \lambda \in N_\Phi(\phi(s, \bar{x}(s), \bar{u}(s)))\}$$
$$- \{0\} \times \mathrm{co}\, N_U(\bar{u}(t)) \text{ a.e. } s \in [\bar{S}, \bar{T}],$$
(iii): $(-a(\bar{S}), p(\bar{S}), a(\bar{T}), -p(\bar{T})) \in \lambda^0 \partial g(\bar{S}, \bar{x}(\bar{S}), \bar{T}, \bar{x}(\bar{T}))$
$$+ N_C(\bar{S}, \bar{x}(\bar{S}), \bar{T}, \bar{x}(\bar{T})),$$
(iv): $p(t) \cdot f(t, \bar{x}, \bar{u}(t)) = \max_{u \in U} p(t) \cdot f(t, \bar{x}(t), u)$ a.e. $t \in [\bar{S}, \bar{T}]$,
(v): $p(t) \cdot f(t, \bar{x}, \bar{u}(t)) = a(t)$ a.e. $t \in [\bar{S}, \bar{T}]$.

Here,

$$\widetilde{S}^{\epsilon, R}(t) := \{(s, x, u) \in \widetilde{S} : (s, x) \in -(t, \bar{x}(t)) + \epsilon \mathbb{B} \text{ and } u \in \bar{u}(t) + R(t)\mathbb{B}\},$$

$$\widetilde{S} := \{(s, x, u) \in \mathbb{R} \times \mathbb{R}^n \times \mathbb{R}^m : \phi(s, x, u) \in \Phi \text{ and } u \in U\}.$$

Hint: Apply the technique used in the proof of Theorem 9.6.1 to construct an auxiliary fixed time interval mixed constraints problem, and observe that Theorem 10.8.1 is applicable to this auxiliary problem.

10.6 Show that the assertions of the special case of Theorem 10.8.1 remain valid when the condition '$\phi(t, ., .)$' is continuously differentiable near $(\bar{x}(t), \bar{u}(t))$ a.e. $t \in [S, T]$' is replaced by the less restrictive hypothesis '$\phi(t, ., .)$' is strictly differentiable at $(\bar{x}(t), \bar{u}(t))$ a.e. $t \in [S, T]$'. Here, a function $\varphi : \mathbb{R}^n \to \mathbb{R}^k$ is said to be *strictly differentiable* at a point $\bar{z}$ if there exists a (continuous) linear map $D_s\varphi(\bar{z}) : \mathbb{R}^n \to \mathbb{R}^k$ such that for each $v \in \mathbb{R}^n$ we have

$$\lim_{h \downarrow 0,\, z \to \bar{z}} h^{-1}[\varphi(z + hv) - \varphi(x)] = D_s\varphi(\bar{z})(v),$$

and provided the convergence is uniform for v in compact sets (see [65]).

10.10 Notes for Chapter 10

Proposition 10.3.1, whose role is to ensure the preservation of necessary conditions involving measure multipliers under limit taking, for pure state constrained problems, is taken from [196]. Additional material on convergence properties of Borel measures can be found in [39].

Necessary conditions for state constrained dynamic optimization problems with optimal state trajectories lying completely in the boundary of the state constraint region go back to the earliest days of dynamic optimization and appear in the monograph of Pontryagin, Boltyanskii, Gamkrelidze and Mischenko [167]. An early version of the maximum principle for dynamic optimization problems with state constraints, which made allowance for a finite number of boundary and interior arcs was obtained by Gamkrelidze [117], under strong regularity hypotheses on the optimal control. Dubovitskii and Milyutin and their research collaborators have been prominent contributors to the theory of necessary conditions for problems with both pure state and mixed pathwise constraints, starting in the 1960s [96]. Their later work is covered in the monograph [153] and review article [91]. See also [147] and [90].

Nonsmooth maximum principles for pure state constrained problems were proved by Warga [207], Vinter and Pappas [196], Clarke [65] and Ioffe [124]. The maximum principle for problems involving a functional inequality constraint, proved in this chapter, is a refinement of Clarke's necessary conditions for problems with state constraints in [65]. Proof techniques employed in this book involve replacing the pure state constraint by an integral penalty term, an idea which was introduced in [196]. Necessary conditions for free end-time problems were given in [201].

The degeneracy phenomenon, namely the fact that standard necessary conditions of optimality convey no useful information for certain state constrained dynamic optimization problems of interest, was first investigated by Arutyunov, Aseev and Blagodatskikh, see [5]. It was the subject of numerous subsequent publications in the Russian, and more recently, the Western literature [194]. We refer to Arutyunov's book [2], and also to [3] and [4], for overviews and further references. The key idea here was that, for autonomous problems, the costate function can be chosen such that the constancy of the Hamiltonian condition, previously known to hold at interior times can be extended to the end-times: this property coupled with an inward pointing hypothesis can be used to derive non-degenerate forms of the maximum principle. The non-degenerate necessary conditions in this chapter (Theorem 10.7.1) are new; they improve on earlier conditions in, say, [2], because they allow general endpoint constraints and data that is nonsmooth w.r.t. the state variable.

The conditions referred to above are restricted to problems whose data is Lipschitz continuous w.r.t. time. The non-degenerate conditions of Ferreira and Vinter [101] allow measurable time dependence but place restrictions on the nature of the optimal state trajectories near the endtimes, which are difficult to test in practice. Non-degenerate necessary conditions, in the form of the generalized Euler Lagrange inclusion, derived by Rampazzo and Vinter [170], do not pre-suppose any structural properties of minimizers and allow measurable time dependence, nonconvex velocity sets and general end-point constraints.

The approach followed in this chapter to deriving necessary conditions of optimality for mixed problems is that pioneered by Clarke and de Pinho [69]: apply the generalized Euler Lagrange condition to a 'differential inclusion' problem, in which the differential inclusion results from merging the dynamic and mixed

10.10 Notes for Chapter 10

constraints and express these conditions directly in terms of the data for the original problem. The necessary conditions in this chapter include an improved Weierstrass condition, obtained by applying the generalized Euler Lagrange condition with the Ioffe refinement to the differential inclusion problem. Following [69], we show how these conditions unify earlier necessary conditions for mixed constraint problems in the literature. Clarke, Ledyaev and de Pinho [87] showed how the approach can be extended to provide necessary conditions for mixed constraint problems not covered by their original theory.

In this chapter, we have treated pure and mixed constraints separately. But the two kinds of constraints can be treated simultaneously [40], once again, using the Clarke/de Pinho approach; now we apply the pure state constrained version of the generalized Euler Lagrange condition, a derivation of which is given in Chap. 11, to the related differential inclusion problem. These conditions feature a measure multiplier for the pure state constraint and an absolutely continuous multiplier for the mixed constraint. Dmitruk [90] has also derived (smooth) necessary conditions of this nature. We mention that, in the Russian literature, versions of the maximum principle have been derived for problems involving irregular mixed constraints, that is where the pathwise constraint set cannot be expressed as the intersection of a pure state constraint set and a mixed constraint set satisfying, say, the hypotheses of [92].

Chapter 11
The Euler-Lagrange and Hamiltonian Inclusion Conditions in the Presence of State Constraints

Abstract This chapter concerns necessary conditions of optimality for dynamic optimization problems with pathwise state constraints, when the dynamic constraint takes the form of a differential inclusion. Here, the pathwise constraint is expressed as a time-dependent scalar functional inequality constraint. Through redefinition of the state constraint function, which is merely required to be upper semicontinuous and Lipschitz continuous w.r.t. the state variable, we can subsume within this framework other formulations of the pathwise state constraint (vector inequality constraints, combined equality/inequality constraints, set inclusion constraints, etc.)

We have seen in Chap. 10, which concerned necessary conditions for path-constrained dynamic optimization problems involving controlled differential inclusions, how it is not possible, in general, to accommodate the state constraint by means of an absolutely continuous Lagrange multiplier; instead we must introduce a measure multiplier. The same is true when, as in this chapter, we substitute a differential inclusion for a controlled differential equation in the problem formulation. Once again, the necessary conditions involve a costate trajectory that is discontinuous, though we hide this fact, expressing the conditions in terms of a modified, absolutely continuous, co-state trajectory, obtained by subtracting off the singular component from the 'true' co-state trajectory.

The necessary conditions for state constrained differential inclusion problems appearing in this chapter resemble the generalized Euler Lagrange condition of Chap. 8, but now including a measure multiplier associated with the state constraint. They are derived under unrestrictive conditions that allow the velocity sets to be non-convex and unbounded. Under an additional hypothesis that the velocity sets are convex, we can prove, with the help of the duality theorem relating Euler Lagrange and Hamiltonian conditions, also a Hamiltonian version of the condition. The generalized Euler Lagrange condition for state constrained problems of this chapter incorporates the stratified conditions of Clarke and the Ioffe refinement. The proof technique mimics that used in Chap. 8, where the full necessary conditions were built up in stages, starting with an application of the maximum principle to a simple version of the problem. The difference is that, now, we use the state-constrained maximum principle in the first stage. We also address the degeneracy issue associated with necessary conditions for pure state constraint problems.

We showed in Chap. 10 that, for state constrained problems involving controlled differential equations, standard necessary conditions are sometimes degenerate when the state constraint is active at either end-time. The same is true when the dynamic constraint takes the form of a differential inclusion. This chapter also provides non-degenerate necessary conditions, under extra hypotheses, covering these situations.

Formulations of the state constrained differential inclusion problem, in which the end-times are included among the choice variables, are also considered. The extra necessary conditions associated with the free end-times take, as usual, the form of boundary conditions on the Hamiltonian evaluated along the minimizing state and costate trajectories.

11.1 Introduction

In this chapter, we broaden our earlier investigation, in Chap. 8, of necessary conditions for dynamic optimization problems, in which the dynamic constraint is formulated as a differential inclusion. Now, we introduce a pathwise state constraint

$$h(t, x(t)) \leq 0 \quad \text{for all } t \in [S, T].$$

Attention focuses then on:

$$(P) \begin{cases} \text{Minimize } g(x(S), x(T)) \\ \text{over absolutely continuous arcs } x : [S, T] \to \mathbb{R}^n \text{ satisfying} \\ \dot{x}(t) \in F(t, x(t)) \text{ a.e.,} \\ (x(S), x(T)) \in C, \\ h(t, x(t)) \leq 0 \quad \text{for all } t \in [S, T]. \end{cases}$$

Here $[S, T] \subset \mathbb{R}$ is a given interval, $g : \mathbb{R}^n \times \mathbb{R}^n \to \mathbb{R}$ and $h : [S, T] \times \mathbb{R}^n \to \mathbb{R}$ are given functions, $F : [S, T] \times \mathbb{R}^n \rightsquigarrow \mathbb{R}^n$ is a given multifunction, and $C \subset \mathbb{R}^n \times \mathbb{R}^n$ is a given set. We shall also study a free end-time version of this problem.

Admissible F trajectories for (P) are $W^{1,1}$ arcs $x : [S, T] \to \mathbb{R}^n$ that satisfy the constraints of problem (P), which now include the state constraint. Given a multifunction $B : [S, T] \rightsquigarrow \mathbb{R}^n$, we say that an admissible F trajectory $\bar{x}$ is a $W^{1,1}$ local minimizer relative to B if there exists $\beta > 0$ such that

$$g(x(S), x(T)) \geq g(\bar{x}(S), \bar{x}(T))$$

for all admissible F trajectories x such that $||x - \bar{x}||_{W^{1,1}} \leq \beta$ and $\dot{x}(t) \in \dot{\bar{x}}(t) + B(t)$, a.e..

As we shall see, variational techniques employed in Chap. 8 can be adapted to take account of the pathwise state constraint. These techniques can be used, together with the methods of Chap. 9, also to derive necessary conditions of optimality for the free end-time problem.

11.2 The Euler Lagrange Inclusion

In Chap. 10, concerning state constrained problems with dynamic constraint formulated as a controlled differential equation, we drew attention to the fact that, for certain problems of interest (including problems in which the end-states are fixed and located in the state constraint boundary) the standard necessary conditions are automatically satisfied by any admissible process. The remedy to this 'degeneracy' phenomenon was to derive supplementary boundary conditions on the maximized Hamiltonian along the optimal state trajectory. We shall show, also for state constrained problems with dynamic constraint formulated as a differential inclusion, degeneracy can be eliminated by introducing additional hypotheses that include boundary conditions on the maximized Hamiltonian.

11.2 The Euler Lagrange Inclusion

This section provides necessary conditions for the state constrained dynamic optimization problem (P) formulated above. These take the form of the generalized Euler Lagrange inclusion and accompanying conditions.

As in Chap. 8, the following subset of the velocity set F in problem (P), evaluated along a nominal F trajectory $\bar{x}$, is used to formulate the Weierstrass condition: the regular velocity set relative to $\bar{x}$ is the multifunction $\Omega_0 : [S, T] \rightsquigarrow \mathbb{R}^n$:

$$\Omega_0(t) := \{e \in F(t, \bar{x}(t)) \,:\, F(t, .) \text{ is pseudo-Lipschitz continuous near } (\bar{x}(t), e)\}. \tag{11.2.1}$$

(The 'pseudo-Lipschitz continuity near $(\bar{x}(t), e)$' condition was defined in Chap. 8, Definition 8.2.1: it requires that there exists $\epsilon' > 0$, $R' > 0$, $k' > 0$ such that $F(t, x) \cap (e + R'\mathbb{B}) \subset F(t, x') + k'|x - x'|\mathbb{B}$ for all $x, x' \in \bar{x}(t) + \epsilon'\mathbb{B}$.) Consistent with earlier notation, we write $C^{\oplus}(S, T)$ for the subset of the topological dual of $C(S, T)$ comprising elements that take non-negative values on the space of non-negative functions in $C(S, T)$. By $NBV([S, T]; \mathbb{R}^n)$ we denote the space of $\mathbb{R}^n$-valued functions of bounded variation defined on $[S, T]$ which are right-continuous on (S, T). (See Chap. 10, Sect. 10.2 for related discussion.)

Theorem 11.2.1 (The Euler Lagrange Inclusion with State Constraints) *Take a measurable multifunction* $B : [S, T] \rightsquigarrow \mathbb{R}^n$ *such that* $B(t)$ *is open for a.e.* $t \in [S, T]$. *Let* $\bar{x}$ *be a* $W^{1,1}$ *local minimizer for* (P) *relative to B. Assume that, for some* $\epsilon > 0$, *the following hypotheses are satisfied:*

(G1): g is Lipschitz continuous on a neighbourhood of $(\bar{x}(S), \bar{x}(T))$. $h(., x)$ is upper semi-continuous for each $x \in \mathbb{R}^n$ and there exists $k_h > 0$ such that

$$|h(t, x) - h(t, x')| \leq k_h |x - x'| \text{ for all } x, x' \in \bar{x}(t) + \epsilon \mathbb{B}, \text{ for all } t \in [S, T],$$

(G2): $F(t, x)$ is nonempty for each $(t, x) \in [S, T] \times \mathbb{R}^n$, $Gr\, F(t, .)$ is closed for each $t \in [S, T]$ and F is $\mathcal{L} \times \mathcal{B}^n$ measurable,

(G3): There exists a measurable function $R : [S, T] \to (0, \infty) \cup \{+\infty\}$ (a 'radius function') such that $R(t)\overset{\circ}{\mathbb{B}} \subset B(t)$ a.e. and the following conditions are satisfied,

(a): (Pseudo-Lipschitz Continuity) There exists $k \in L^1$ such that

$$F(t, x') \cap (\dot{\bar{x}}(t) + R(t)\overset{\circ}{\mathbb{B}}) \subset F(t, x) + k(t)|x' - x|\mathbb{B}, \quad (11.2.2)$$
$$\text{for all } x', x \in \bar{x}(t) + \epsilon\mathbb{B}, \text{ a.e. } t \in [S, T],$$

(b): (Tempered Growth) There exist $r \in L^1$, $r_0 > 0$ and $\gamma \in (0, 1)$ such that $r_0 \leq r(t)$, $\gamma^{-1} r(t) \leq R(t)$ a.e. and

$$F(t, x) \cap (\dot{\bar{x}}(t) + r(t)\mathbb{B}) \neq \emptyset \text{ for all } x \in \bar{x}(t) + \epsilon\mathbb{B}, \text{ a.e. } t \in [S, T].$$

Then there exist an arc $p \in W^{1,1}([S, T]; \mathbb{R}^n)$, $\mu \in C^{\oplus}(S, T)$, a bounded Borel measurable function $\gamma : [S, T] \to \mathbb{R}^n$ and $\lambda \geq 0$, satisfying the following conditions, in which $q \in NBV([S, T]; \mathbb{R}^n)$ is the function

$$q(t) = \begin{cases} p(S) & \text{if } t = S \\ p(t) + \int_{[S,t]} \gamma(s) d\mu(s) & \text{if } t \in (S, T], \end{cases} \quad (11.2.3)$$

(i) $(\lambda, p, \mu) \neq (0, 0, 0)$,
(ii) $\dot{p}(t) \in co\{\eta : (\eta, q(t)) \in N_{Gr\, F(t,.)}(\bar{x}(t), \dot{\bar{x}}(t))\}$, a.e. $t \in [S, T]$,
(iii) $(q(S), -q(T)) \in \lambda \partial g(\bar{x}(S), \bar{x}(T)) + N_C(\bar{x}(S), \bar{x}(T))$,
(iv) $q(t) \cdot \dot{\bar{x}}(t) \geq q(t) \cdot v$ for all $v \in \overline{co}(\Omega_0(t) \cap (\dot{\bar{x}}(t) + B(t)))$, a.e. $t \in [S, T]$,
(v) $supp\{\mu\} \subset \{t \in [S, T] : h(t, \bar{x}(t)) = 0\}$
and $\gamma(t) \in \partial_x^{>} h(t, \bar{x}(t))$, μ-a.e. $t \in [S, T]$.

Here,

$$\partial_x^{>} h(t, \bar{x}(t)) := co \lim\sup \{\partial_x h(t_i, x_i) : \quad (11.2.4)$$
$$t_i \to t, x_i \to \bar{x}(t) \text{ and } h(t_i, x_i) > 0 \text{ for each } i\}.$$

Remark

(i): The 'complementary slackness' condition '$supp\{\mu\} \subset \{t \in [S, T] : h(t, \bar{x}(t)) = 0\}$' in (v) is included for emphasis. It is in fact implied by the relation $\gamma(t) \in \partial_x^{>} h(t, \bar{x}(t))$, μ-a.e. $t \in [S, T]$ characterizing the function γ, since this relation only makes sense if $\partial_x^{>} h(t, \bar{x}(t))$ is non-empty μ a.e. and since $\partial_x^{>} h(t, \bar{x}(t))$ is empty at all times t at which $h(t, \bar{x}(t)) < 0$.

(ii): In consequence of the unrestrictive nature of hypotheses placed on the state constraint function, the necessary conditions cover a range of state constrained

11.2 The Euler Lagrange Inclusion

dynamic optimization problems, in which the state constraint is formulated in different ways. Examples include multiple functional inequality constraints, functional inequality constraints imposed on a fixed closed subset of $[S, T]$ or implicit constraints $x(t) \in A(t)$ for all $t \in [S, T]$, for some given multifunction $A : [S, T] \rightsquigarrow \mathbb{R}^n$. We refer to the remarks following the statement of Theorem 10.4.1, for information about how these other types of state constraints are subsumed in the framework of problem (P), and how they affect the statement of the necessary conditions.

(iii): *(Convex Velocity Sets)* Assume that

$$F(t, x) \text{ is convex, for all } x \in \bar{x}(t) + \epsilon \mathbb{B}, \text{ a.e. } t \in [S, T].$$

Then, in consequence of the dualization theorem (Theorem 8.7.1), the generalized Euler Lagrange condition (ii) in the theorem statement implies the Hamiltonian inclusion (see Theorem 8.7.2), namely

$$\dot{p}(t) \in \mathrm{co}\{-\xi : (\xi, \dot{\bar{x}}(t)) \in \partial_{x,p} H(t, \bar{x}(t), q(t))\}, \text{ a.e. } t \in [S, T].$$

Furthermore, the Weierstrass condition (iv) is valid in a stronger form, in which $F(t, \bar{x}(t))$ replaces the set $\overline{\mathrm{co}}(\Omega_0(t) \cap (\dot{\bar{x}}(t) + B(t)))$. This last assertion is confirmed in the discussion that precedes the statement of Theorem 8.7.2 in Chap. 8. (See also Example 4.2.)

As in Chap. 8, we can use the 'pseudo Lipschitz continuity and tempered growth' hypothesis (G3) in Theorem 11.2.1 as a springboard for proving other versions of the theorem in which (G3) is replaced by alternative hypotheses which are, perhaps, more easily verified in particular applications. In consequence of Proposition 8.5.1, any of the sets of conditions (G3)*–(G3)*** listed in that proposition can be inserted in place of (G3), to give different versions of Theorem 11.2.1. One example of necessary conditions obtained in this way is:

Corollary 11.2.2 *The assertions of Theorem 11.2.1 remain valid when $\bar{x}$ is a $W^{1,1}$ local minimizer (i.e. $B \equiv \mathbb{R}^n$) and hypothesis (G3) is replaced by:*

(G3)***: *There exist $\alpha > 0$ and non-negative measurable functions k and β such that k and $t \to \beta(t) k^\alpha(t)$ are integrable on $[S, T]$ and, for each $N \geq 0$,*

$$F(t, x') \cap (\dot{\bar{x}}(t) + N\mathbb{B}) \subset F(t, x) + (k(t) + \beta(t) N^\alpha) |x' - x| \mathbb{B},$$

for all $x', x \in \bar{x}(t) + \epsilon \mathbb{B}$ and a.e. $t \in [S, T]$.

Furthermore, the regular velocity set $\Omega_0(t) = F(t, \bar{x}(t))$ and, in consequence, the Weierstrass condition (iv) takes the form

$$q(t) \cdot \dot{\bar{x}}(t) \geq q(t) \cdot v \text{ for all } v \in F(t, \bar{x}(t)) \text{ a.e. } t \in [S, T].$$

11.3 Proof of Theorem 11.2.1

Step 1 (Necessary Conditions for a Finite Lagrangian Problem)
We begin by deriving necessary conditions for a $W^{1,1}$ local minimizer x' for the following problem:

$$(FL) \begin{cases} \text{Minimize } J(x) := \\ \quad \left(\ell(x(S), x(T)) \right) \vee \left(\int_S^T L(t, x(t), \dot{x}(t))dt \right) \vee \left(\max_{t \in [S,T]} h(t, x(t)) \right) \\ \quad + \int_S^T L_0(t, x(t), \dot{x}(t))dt + \delta |x(S) - x'(S)| \\ \text{over } x \in W^{1,1}([S, T]; \mathbb{R}^n) \text{ such that} \\ \dot{x}(t) \in D(t) \text{ a.e. } t \in [S, T]. \end{cases}$$

The data comprise an interval $[S, T]$, a number $\delta > 0$, functions $L : [S, T] \times \mathbb{R}^n \times \mathbb{R}^n \to \mathbb{R}$, $L_0 : [S, T] \times \mathbb{R}^n \times \mathbb{R}^n \to \mathbb{R}$, $\ell : \mathbb{R}^n \times \mathbb{R}^n \to \mathbb{R}$ and $h : [S, T] \times \mathbb{R}^n \to \mathbb{R}$, and a multifunction $D : [S, T] \rightsquigarrow \mathbb{R}^n$.

A $W^{1,1}$ local minimizer x' for (FL) is defined by analogy with our earlier definition for (P), except that, now, we impose the additional requirement that $t \to (L, L_0)(t, x'(t), \dot{x}'(t))$ is integrable.

Proposition 11.3.1 *Let x' be a $W^{1,1}$ local minimizer for (FL). Assume that the following hypotheses are satisfied: there exist $\epsilon > 0$, $N > 0$, $k_h > 0$ and a non-negative function $\tilde{k} \in L^1(S, T)$ such that*

(FL0): D has closed values, $\text{Gr } D$ is $\mathcal{L} \times \mathcal{B}^n$ measurable and $D(t) \subset \dot{x}'(t) + N\mathbb{B}$ a.e.,

(FL1): ℓ is Lipschitz continuous on a neighbourhood of $(0, 0)$. $h(., x)$ is upper semi-continuous for each $x \in \mathbb{R}^n$ such that

$$|h(t, x) - h(t, y)| \leq k_h |x - y| \text{ for all } x, y \in \bar{x}(t) + \epsilon \mathbb{B}, \ t \in [S, T],$$

(FL2): $(L, L_0)(., x, v)$ is $\mathcal{L}$-measurable for each $(x, v) \in \mathbb{R}^n \times \mathbb{R}^n$,
(FL3): $|(L, L_0)(t, y, w) - (L, L_0)(t, x, v)| \leq \tilde{k}(t)(|y - x| + |w - v|)$, for all $y, x \in \epsilon \mathbb{B}$ and $w, v \in D(t)$, a.e. $t \in [S, T]$.

Assume, furthermore, that

$$\|x'\|_{W^{1,1}} < \epsilon.$$

Then there exist an arc $p \in W^{1,1}([S, T]; \mathbb{R}^n)$, $\mu \in C^{\oplus}(S, T)$, a bounded Borel measurable function $\gamma : [S, T] \to \mathbb{R}^n$, $\lambda \geq 0$ and $\lambda^{(1)} \geq 0$, satisfying the following conditions, in which $q \in NBV([S, T]; \mathbb{R}^n)$ is the function

$$q(t) = \begin{cases} p(S) & \text{if } t = S \\ p(t) + \int_{[S,t]} \gamma(s) d\mu(s) & \text{if } t \in (S, T]. \end{cases}$$

11.3 Proof of Theorem 11.2.1

(i): $\lambda + \lambda^{(1)} + ||\mu||_{TV} = 1$,

(ii):
$$\begin{cases} (a): \dot{p}(t) \in \lambda^{(1)} co\, \partial_x L(t, x'(t), \dot{x}'(t)) + co\, \partial_x L_0(t, x'(t), \dot{x}'(t)) \text{ a.e. } t\in [S, T] \\ \text{and} \\ (b): \dot{p}(t) \in co\{\eta : (\eta, q(t)) \in \lambda^{(1)} \partial_{x,v} L(t, x'(t), \dot{x}'(t)) + \partial_{x,v} L_0(t, x'(t), \dot{x}'(t))\}, \\ \qquad\qquad\qquad\qquad\qquad\qquad \text{for a.e. such that } \dot{x}'(t) \in \overset{\circ}{D}(t) \end{cases}$$

(iii) $(q(S), -q(T)) \in \lambda \partial \ell(x'(S), x'(T)) + \delta \mathbb{B} \times \{0\}$,

(iv) $q(t) \cdot \dot{x}'(t) - \lambda^{(1)} L(t, x'(t), \dot{x}'(t)) - L_0(t, x'(t), \dot{x}'(t)) \geq$
$$q(t) \cdot v - \lambda^{(1)} L(t, x'(t), v) - L_0(t, x'(t), v),$$
$$\text{for all } v \in D(t), \text{ a.e. } t \in [S, T],$$

(v) $supp\{\mu\} \subset \{t \in [S, T] : h(t, x'(t)) = \max_{t \in [S,T]} h(t, x'(t))\}$
and $\gamma(t) \in \partial_x^> h(t, x'(t))$ for μ-a.e. $t \in [S, T]$.

Furthermore

$$\int_S^T L(t, x'(t), \dot{x}'(t))dt < \ell(x'(S), x'(T)) \vee \max_{t \in [S,T]} h(t, x'(t)) \implies \lambda^{(1)} = 0, \tag{11.3.1}$$

$$\max_{t \in [S,T]} h(t, x'(t)) < \ell(x'(S), x'(T)) \vee \int_S^T L(t, x'(t), \dot{x}'(t))dt \implies \mu = 0 \tag{11.3.2}$$

and

$$\ell(x'(S), x'(T)) < \max_{t \in [S,T]} h(t, x'(t)) \vee \int_S^T L(t, x'(t), \dot{x}'(t))dt \implies \lambda = 0. \tag{11.3.3}$$

Proof of Proposition 11.3.1 Since x' is a $W^{1,1}$ local minimizer, there exists $\beta > 0$ such that x' is minimizing, when we add to (FL) the constraint

$$|x(S) - x'(S)| + \int_S^T |\dot{x}(t) - \dot{x}'(t)|dt \leq \beta. \tag{11.3.4}$$

Bearing in mind that $||x'||_{W^{1,1}} < \epsilon$, we can arrange, by choosing β sufficiently small, that $||x||_{W^{1,1}} < \epsilon$ (and hence also that $||x||_{L^\infty} < \epsilon$), for all x satisfying (11.3.4). Here, ϵ is the constant appearing in proposition statement.

As justified in the proof of Proposition 8.6.1 in Chap. 8 (the analysis is unaffected by the presence of a pathwise state constraint), we may assume that ℓ is Lipschitz continuous, $h(t, .)$ is Lipschitz continuous with Lipschitz constant k_h for each $t \in [S, T]$ and (FL3) has been replaced by the hypothesis:

(FL3)': there exist non-negative functions $k, c \in L^1$ such that $k(t) \geq 1$ a.e. and
$$\begin{cases} |(L, L_0)(t, y, w) - (L, L_0)(t, x, v)| \leq k(t)(|y - x| + |w - v|) \text{ and} \\ |(L, L_0)(t, x, v)| \leq c(t) \end{cases}$$
for all $y, x \in \mathbb{R}^n$ and $w, v \in D(t)$, a.e. $t \in [S, T]$.

Take a sequence $K_i \uparrow \infty$. Write L_k^1 for the Banach space of measurable functions w such that $t \to k(t)w(t)$ is integrable on $[S, T]$, equipped with the k-weighted L^1 norm $||w||_{L_k^1} := ||kw||_{L^1}$. Write

$$W := \{(\xi, w, v) \in \mathbb{R}^n \times L_k^1 \times L_k^1 : v(t) \in D(t) \text{ a.e.}, ||x_{\xi,v} - x'||_{W^{1,1}} \le \beta\},$$

in which $x_{\xi,v}(t) = \xi + \int_S^t v(s)ds$. Define

$$||(\xi, w, v)||_k := |\xi| + ||kw||_{L^1} + ||kv||_{L^1}.$$

For each i, set

$$\tilde{J}_i(\xi, w, v) :=$$
$$\left(\ell(x_{\xi,v}(S), x_{\xi,v}(T))\right) \vee \left(\int L(t, w(t), v(t))dt\right) \vee \left(\max_{t \in [S,T]} h(t, x(t))\right)$$
$$+ \int L_0(t, w(t), v(t))dt + \delta |x_{\xi,v}(S) - x'(S)|$$
$$+ K_i \int k(t) |x_{\xi,v}(t) - w(t)|^2 dt.$$

Taking note of the fact that the mapping $(\xi, v) \to \max_{t \in [S,T]} h(t, x_{\xi,v}(t))$ is a continuous function on $(W, ||.||_k)$, we can verify, by a simple adaptation of the 'state constraint-free' analysis in Chap. 8:

Claim: For each i, $(W, ||.||_k)$ is a complete metric space and $\tilde{J}_i$ is lower semi-continuous on $(W, ||.||_k)$. There exists a sequence of non-negative numbers $\alpha_i \downarrow 0$ such that, for each i,

$$\tilde{J}_i(x'(S), x', \dot{x}') \le \inf_{(\xi,w,v) \in W} \tilde{J}_i(\xi, w, v) + \alpha_i^2.$$

By Ekeland's theorem (Theorem 3.3.1), there exists $(\xi_i, w_i, v_i) \in W$ which minimizes

$$J_i(\xi, w, v) := \tilde{J}_i(\xi, w, v) + \alpha_i ||(\xi, w, v) - (\xi_i, w_i, v_i)||_k$$

over W. Furthermore

$$||(\xi_i, w_i, v_i) - (x'(S), x', \dot{x}')||_k \le \alpha_i, \text{ for each } i. \tag{11.3.5}$$

11.3 Proof of Theorem 11.2.1

We also know from Ekeland's theorem that $\tilde{J}_i(\xi_i, w_i, v_i) \leq J_i(x'(S), x', \dot{x}')(< \infty)$. This implies that $\int k(t)|x_i(t) - w_i(t)|^2 dt < \infty$, where $x_i := x_{\xi_i, v_i}$. Since x_i is bounded, it follows that $t \to k(t)w_i^2(t)$ is integrable.

Taking note of the fact that the mapping $z \to \ell(x_0, x_1) \vee z$ is monotone, we may deduce from the preceding analysis that $((x_i, y_i, r_i, z_i), (v_i, w_i))$ is a minimizer for the dynamic optimization problem

$$(E_i) \begin{cases} \text{Minimize } \ell(x(S), x(T)) \vee r(T) \vee z(T) + \delta|x(S) - x'(S)| \\ \quad + \int L_0(t, w(t), v(t))dt + \alpha_i \Big(|x(S) - x_i(S)| + \int k(t)(|v(t) - v_i(t)| \\ \quad + |w(t) - w_i(t)|)dt \Big) + K_i \int k(t)|x(t) - w(t)|^2 dt. \\ \text{over arcs } (x, y, r, z) \in W^{1,1} \times W^{1,1} \times W^{1,1} \times W^{1,1} \text{ satisfying} \\ \dot{x}(t) = v(t), \; \dot{y}(t) = |v(t) - \dot{x}'(t)|, \; \dot{r}(t) = L(t, w(t), v(t)), \dot{z}(t) = 0 \text{ a.e.,} \\ w(t) \in \mathbb{R}^n, v(t) \in D(t) \text{ a.e.,} \\ h(t, x(t)) - z(t) \leq 0 \text{ for all } t \in [S, T], \\ y(S) = r(S) = 0 \text{ and } |x(S) - x'(S)| + y(T) \leq \beta. \end{cases}$$

Here
$$y_i(t) := \int_S^t |v_i(s) - \dot{x}_i(s)|ds, \; z_i \equiv \max_{t \in [S,T]} h(t, x_i(t)))$$
and $r_i(t) := \int_S^t L_i(s, w_i(s), v_i(s)))ds$.

(Notice that the problem (E_i) is not strictly equivalent to $\text{Min}\{J_i(\xi, w, v) : (\xi, w, v) \in W\}$ because it has domain comprising control functions (v, w) from the larger set $L^1 \times \{\text{meas. functions } w : [S, T] \to \mathbb{R}^n\}$. However the two problems are equivalent, in the sense of having a common set of minimizers since, for any control functions $(w, v) \notin L_k^1 \times L_k^1$ the cost in (E_i) is $+\infty$, as is easily shown, and such control functions cannot be candidates for minimizing controls.)

In consequence of (11.3.5), we know that, for some subsequence,
$$\|x_i \to x'\|_{L^\infty} \to 0 \text{ and } (v_i, w_i) \to (\dot{x}', x') \text{ in } L^1 \text{ and a.e.}$$

and $y_i(.) = \int_{[S,.]} |v_i(s) - \dot{x}'(s)|ds \to 0$ uniformly. It follows that, for i sufficiently large, $|x_i(S) - x'(S)| + \int_S^T |\dot{x}_i(t) - \dot{x}'(t)|dt < \beta$. This tells as that the final constraint in problem (E_i) is not active.

Fix i. The foregoing dynamic optimization problem (E_i) is one to which the state constrained maximum principle Theorem 10.4.1 is applicable, following absorption of the integral cost term into the dynamic constraint by state augmentation. The relevant hypotheses are satisfied. (To check hypothesis (H1) in the maximum principle, we make use of the facts that $t \to k(t)w_i^2(t)$ and $t \to k(t)v_i(t)$ are integrable functions.)

Because the inequality constraint is inactive, the costate trajectory component associated with y must be zero. Let us identify p_i, $p_i^{(1)}$ and $p_i^{(2)}$ as the costate trajectories associated with the remaining states x, r and z. ($p_i^{(1)}$ and $p_i^{(2)}$ are constant arcs.) Write $\lambda^{(c)}$ for the cost multiplier. A preliminary analysis of the costate inclusion and the transversality condition of the state constrained maximum principle based on the max rule, tells us that, for some $\mu_i \in C^{\oplus}(S, T)$, bounded Borel measurable function $\gamma_i : [S, T] \to \mathbb{R}^n$ and $\lambda^{(c)} \geq 0$, $\lambda_i \geq 0$ $\lambda_i^{(1)} \geq 0$ and $\lambda_i^{(2)} \geq 0$ such that $\lambda_i + \lambda_i^{(1)} + \lambda_i^{(2)} = 1$ the following relations hold:

$$(p_i(S), -p_i(T) - \int_{[S,T]} \gamma_i(s) d\mu_i(s))$$
$$= \lambda_i \lambda^{(c)} \partial \ell(x_i(S), x_i(T)) + (\delta + \alpha_i) \lambda^{(c)} \mathbb{B} \times \{0\},$$
$$-\dot{p}_i(t) = -2\lambda^{(c)} K_i k(t)(x_i(t) - w_i(t)), \text{ a.e. } t \in [S, T],$$
$$-p^{(1)} = \lambda_i^{(1)} \lambda^{(c)}, \ p^{(2)} = 0 \text{ and } \int_{[S,T]} d\mu_i(s) = \lambda_i^{(2)} \lambda^{(c)}.$$

Furthermore, $(\lambda^{(c)}, p_i, p_i^{(1)}, p_i^{(2)}, \mu_i) \neq (0, 0, 0, 0, 0)$. We now observe that $\lambda^{(c)} \neq 0$. Indeed, if this were not true, it would follow from the above relations and Gronwall's lemma that $(p_i, p_i^{(1)}, p_i^{(2)}, \mu_i) = (0, 0, 0, 0)$, which contradicts the fact that costate arcs cannot all vanish. Since $\lambda^{(c)} \neq 0$, we can scale the Lagrange multipliers to ensure that $\lambda^{(c)} = 1$. Notice then that $\|\mu_i\|_{TV} = \int_{[S,T]} d\mu_i(s) = \lambda_i^{(2)}$. We have, after making these changes,

(A): $\lambda_i + \lambda_i^{(1)} + \|\mu_i\|_{TV} = 1$,
(B): $-\dot{p}_i(t) = -2K_i k(t)(x_i(t) - w_i(t))$, a.e. $t \in [S, T]$,
(C): $(q_i(S), -q_i(T)) \in \lambda_i \partial \ell(x_i(S), x_i(T)) + (\delta + \alpha_i) \mathbb{B} \times \{0\}$,
(D): $(w, v) \to q_i(t) \cdot v - \lambda_i^{(1)} L(t, w, v) - L_0(t, w, v)$
$$-\alpha_i k(t)(|v - v_i(t)| + |w - w_i(t)|) - K_i k(t)|x_i(t) - w|^2$$
achieves its maximum at $(w_i(t), v_i(t))$ over $(w, v) \in \mathbb{R}^n \times D(t)$,
a.e. $t \in [S, T]$,
(E): $\text{supp}\{\mu_i\} \subset \{t \in [S, T] : h(t, x_i(t)) = \max_{t \in [S,T]} h(t, x_i(t))\}$ and
$$\gamma(t) \in \partial_x^{>} h(t, x_i(t)) \text{ for } \mu_i\text{-a.e. } t \in [S, T].$$

Here, $q_i(t) = \begin{cases} p_i(S) & \text{if } t = S \\ p_i(t) + \int_{[S,t]} \gamma_i(s) d\mu_i(s) & \text{if } t \in (S, T]. \end{cases}$

Next we investigate the consequences of condition (D). Take any $t \in [S, T]$ at which this condition is satisfied. The specified function with constrained maximizer $(w_i(t), v_i(t))$ is Lipschitz continuous on a neighbourhood of $(w_i(t), v_i(t))$; write the Lipschitz constant $\rho_i(t)$. According to the exact penalization principle then, $(w_i(t), v_i(t))$ is also an unconstrained maximizer of the function:

$$(w, v) \to q_i(t) \cdot v - \lambda_i^{(1)} L(t, w, v)$$
$$-L_0(t, w, v) - \alpha_i k(t)(|v - v_i(t)| + |w - w_i(t)|)$$
$$-K_i k(t)|x_i(t) - w|^2 - \rho_i(t) d_{D(t)}(v).$$

11.3 Proof of Theorem 11.2.1

It follows that

$$(0, q_i(t)) \in \lambda_i^{(1)} \partial_{x,v} L(t, w_i(t), v_i(t)) + \partial_{x,v} L_0(t, w_i(t), v_i(t))$$
$$+ \alpha_i k(t)(\mathbb{B} \times \mathbb{B}) + (2K_i k(t)(x_i(t) - w_i(t)), 0) + \{0\} \times \rho_i(t) \partial d_{D(t)}(v).$$

This relation combines with (B) to give:

$$(\dot{p}_i(t), q_i(t)) \in \lambda_i^{(1)} \partial_{x,v} L(t, w_i(t), v_i(t)) + \partial_{x,v} L_0(t, w_i(t), v_i(t))$$
$$+ \alpha_i k(t)(\mathbb{B} \times \mathbb{B}) + \{0\} \times \rho_i(t) \partial d_{D(t)}(v_i(t)). \quad (11.3.6)$$

Fix $v = v_i(t)$. Then

$$w \to -\lambda_i^{(1)} L(t, w, v_i(t)) - L_0(t, w, v_i(t)) - \alpha_i k(t)|w - w_i(t)| - K_i k(t)|x_i(t) - w|^2$$

achieves its maximum at $w_i(t)$ over all $w \in \mathbb{R}^n$. In view of (B), this implies

$$\dot{p}_i(t) \in \lambda_i^{(1)} \partial_x L(t, w_i(t), v_i(t)) + \partial_x L_0(t, w_i(t), v_i(t)) + \alpha_i k(t) \mathbb{B}. \quad (11.3.7)$$

Fix $w = w_i(t)$. Then
$$v \to q_i(t) \cdot v - \lambda_i^{(1)} L(t, w_i(t), v) - L_0(t, w_i(t), v) - \alpha_i k(t)|v - v_i(t)|$$

achieves its maximum at $v_i(t)$ over $v \in D(t)$. We see that, for a.e. $t \in [S, T]$,

$$q_i(t) \cdot v_i(t) - \lambda_i^{(1)} L(t, w_i(t), v_i(t)) - L_0(t, w_i(t), v_i(t))$$
$$\geq q_i(t) \cdot v - \lambda_i^{(1)} L(t, w_i(t), v) - L_0(t, w_i(t), v) - \alpha_i k(t)|v - v_i(t)| \text{ for all } v \in D(t). \quad (11.3.8)$$

Since $(L, L_0)(t, ., v)$ is Lipschitz continuous with Lipschitz constant $k(t)$ for all $v \in D(t)$, (11.3.7) implies that, for i sufficiently large, $|\dot{p}_i(t)| \leq (2 + \alpha_i)k(t)$. Noting that the $p_i(S)$'s, λ_i's, $\lambda_i^{(1)}$'s and $||\mu_i||_{TV}$'s are uniformly bounded (see (A) and (C)), we can arrange, by subsequence extraction, that $p_i \to p$ uniformly, $\dot{p}_i \to \dot{p}$ weakly in L^1, $\mu_i \to \mu$ weakly*, $dq_i \to dq$ weakly*, $\lambda_i \to \lambda$, $\lambda_i^{(1)} \to \lambda^{(1)}$, some $p \in W^{1,1}$, $\mu \in C^{\oplus}(S, T)$, $\lambda \geq 0$, $\lambda^{(1)} \geq 0$ and $q \in NBV([S, T]; \mathbb{R}^n)$ satisfying

$$q(t) = \begin{cases} p(S) & \text{if } t = S \\ p(t) + \int_{[S,t]} \gamma(s) d\mu(s) & \text{if } t \in (S, T]. \end{cases}$$

From (A)

$$\lambda + \lambda^{(1)} + ||\mu||_{TV} = 1.$$

This is condition (i) of the proposition statement.
Notice next that, if $\int_S^T L(t, x'(t), \dot{x}'(t)) dt < \left(\ell(x'(S), x'(T)) \right) \vee \left(\max_{t \in [S,T]} h(t, x'(t)) \right)$,
we know that

$$\int_S^T L(t, x_i(t), \dot{x}_i(t))dt < \left(\ell(x_i(S), x_i(T))\right) \vee \left(\max_{t \in [S,T]} h(t, x_i(t))\right)$$

for all i sufficiently large.

But then, by the sum rule, $\lambda_i^{(1)} = 0$ for all i sufficiently large. It follows that $\lambda^{(1)} = 0$. This confirms (11.3.1). Likewise we show, with the help of the sum rule, that

$$\max_{t \in [S,T]} h(t, x_i(t)) < \ell(x'(S), x'(T)) \vee \int_S^T L(t, x'(t), \dot{x}'(t))dt \implies \mu = 0,$$

$$\ell(x'(S), x'(T)) < \max_{t \in [S,T]} h(t, x'(t)) \vee \int_S^T L(t, x'(t), \dot{x}'(t))dt \implies \lambda = 0.$$

We have confirmed (11.3.1), (11.3.2) and (11.3.3). Now observe that, in the limit, (11.3.8) implies

$$q(t) \cdot v'(t) - \lambda^{(1)} L(t, x'(t), v'(t)) - L_0(t, x'(t), v'(t))$$
$$\geq q(t) \cdot v - \lambda^{(1)} L(t, x'(t), v) - L_0(t, x'(t), v) \text{ for all } v \in D(t) \text{ a.e.}.$$

This is (iv). (C) yields, in the limit, $(q(S), -q(T)) \in \lambda \partial \ell(x'(S), x'(T)) + \delta \mathbb{B} \times \{0\}$, which is (iii). With the help of Proposition 10.3.1 of Chap. 10, we deduce from (E) the existence of a Borel measurable function γ such that

$$\text{supp}\{\mu\} \subset \{t \in [S, T] : h(t, x'(t)) = \max_{t \in [S,T]} h(t, x'(t))\} \text{ and}$$

$$\gamma(t) \in \partial_x^> h(t, x'(t)) \text{ for } \mu\text{-a.e. } t \in [S, T].$$

We have confirmed (v). Notice next that, since $dq_i \to dq$ in the weak* topology and $q_i(S) \to q(S)$, we know that $q_i(t) \to q(t)$ for all times $t \in [S, T]$ in some set of full measure that contains S and T. Using this fact, we can carry out an almost identical analysis to that appearing in the proof Proposition 8.6.1 (necessary conditions for a generalized finite Lagrangian problem without a pathwise state constraint) to pass to the limit as $i \to \infty$ in (11.3.6) and obtain the two relations

$$\dot{p}(t) \in \text{co}\{\eta : (\eta, q(t)) \in \lambda^{(1)} \partial_{x,v} L(t, x'(t), \dot{x}'(t)) + \partial_{x,v} L_0(t, x'(t), \dot{x}'(t)),$$

a.e. $t \in [S, T]$ such that $\dot{x}'(t) \in \overset{\circ}{D}(t)$,

$$\dot{p}(t) \in \lambda^{(1)} \text{co } \partial_x L(t, x'(t), \dot{x}'(t)) + \text{co } \partial_x L_0(t, x'(t), \dot{x}'(t)) \text{ a.e. } t \in [S, T].$$

This is condition (ii). All the assertions of Proposition 11.3.1 have been verified. The proof is complete. $\square$

11.3 Proof of Theorem 11.2.1

Step 2 (Hypothesis Reduction)
Analogous arguments to those used in Chap. 8 to prove the Euler Lagrange inclusion and related conditions, in the case that there are no state constraints, permit us to impose, without loss of generality, some additional hypotheses, as is asserted in the following proposition.

Proposition 11.3.2 *Assume the assertions of Theorem 11.2.1 are valid under the hypotheses: (G1), (G2) and (G3)', and when $\bar{x} \equiv 0$. Then the assertions of Theorem 11.2.1 are valid under hypotheses (G1)–(G3) alone.*

Here, (G3)', for the given parameter $\epsilon > 0$, is the hypothesis

(G3)': there exist $R > 0$, $k \in L^1$ and $\gamma' \in (0, 1)$ such that $R\overset{\circ}{\mathbb{B}} \subset B(t)$ for a.e. $t \in [S, T]$ and the following conditions are satisfied:

(i): $F(t, x') \cap (\dot{\bar{x}}(t) + R\overset{\circ}{\mathbb{B}}) \subset F(t, x) + k(t)|x' - x|\mathbb{B}$ for all $x', x \in \epsilon\mathbb{B}$,
  a.e. $t \in [S, T]$,

(ii): $F(t, x) \cap (\dot{\bar{x}}(t) + \gamma' R\mathbb{B}) \neq \emptyset$ for all $x \in \epsilon\mathbb{B}$, a.e. $t \in [S, T]$.

Step 3 (An Integral Penalty Function)
We shall make use of the modified integral penalty function, earlier used to derive the Euler Lagrange-type conditions in the absence of state constraints, to take account of the dynamic constraint also in a state constrained setting.

Choose $\eta \in (0, 1/2)$ such that $(G3)'$ is satisfied with $\gamma' = (1 - 2\eta)$. Fix $N > R$ and define

$$E^N(t) := \{e \in \Omega_0(t) \cap B(t) \,:\, R \leq |e| \leq N\},$$

in which $\Omega_0(t)$ is the regular velocity set (11.2.1), namely

$$\Omega_0(t) := \{e \in F(t, \bar{x}(t)) \,:\, F(t, .) \text{ is pseudo-Lipschitz near } (\bar{x}(t), e)\}.$$

We can construct a multifunction $E^N_{\text{discrete}} : [S, T] \rightsquigarrow \mathbb{R}^n$ such that

(a): $E^N_{\text{discrete}}(t)$ is an empty or finite set for each $t \in [S, T]$,
(b): $E^N_{\text{discrete}}(t)$ is measurable,
(c): $E^N_{\text{discrete}}(t) \subset E^N(t) \subset E^N_{\text{discrete}}(t) + N^{-1}\mathbb{B}$, for each $t \in [S, T]$ such that $E^N(t) \neq \emptyset$,
(d): $E^N_{\text{discrete}}(t) \subset E^{N+1}_{\text{discrete}}(t)$ for all integers $N > R$.

Define

$$\theta(t) := \begin{cases} \min\{|e - e'| \,:\, e, e' \in E^N_{\text{discrete}}(t) \text{ and } e \neq e'\} \\ \quad \wedge \inf\{d_{\partial B(t)}(e) \,:\, e \in E^N_{\text{discrete}}(t)\} \\ \quad \text{if } E^N_{\text{discrete}}(t) \text{ contains at least two elements,} \\ +\infty \quad \text{otherwise}. \end{cases}$$

For $\delta \in (0, \eta R)$, define

$$E^\delta_{\text{regular}}(t) := \{e \in E^N_{\text{discrete}}(t) : F(t,.) \text{ is pseudo-Lipschitz near } (e, 0)$$

(with parameters $R \geq \delta, \epsilon \geq \delta$ and $k \leq \delta^{-1}$) and $\theta(t) \geq \delta\}$.

By construction, $E^\delta_{\text{regular}}(t) \subset E^N_{\text{discrete}}(t) \subset N\mathbb{B}$. Define

$$D^\delta(t) := (1 - \eta)R\mathbb{B} \cup \{e \in e_0 + \frac{1}{3}\delta\mathbb{B} : e_0 \in E^\delta_{\text{regular}}(t)\}$$

in which the right side is interpreted as $(1 - \eta)R\mathbb{B}$ if $E^\delta_{\text{regular}}(t) = \emptyset$.
$D^\delta(t)$ is a finite union of disjoint, closed balls. These comprise $(1 - \eta)R\mathbb{B}$ and elements from a (possibly empty) collection of disjoint closed balls each with origin outside $R\overset{\circ}{\mathbb{B}}$. Moreover, $D^\delta(t) \subset B(t)$ for a.e. $t \in [S, T]$.
Now consider the multifunction $S : [S, T] \times \mathbb{R}^n \rightsquigarrow \mathbb{R}^n$

$$S(t, x) := \{(\chi(|e|)e : e \in F(t, x)\}, \tag{11.3.9}$$

in which $\chi : [0, \infty) \to [1, \infty)$ is the function

$$\chi(d) := 1 + \frac{4(1 - \eta)}{\eta R}[d - (1 - \eta)R]^+.$$

Define $\rho_S : [S, T] \times \mathbb{R}^n \times \mathbb{R}^n \to [0, \infty)$ to be

$$\rho_S(t, x, v) := \begin{cases} d_{S(t,x)}(v) & \text{if } |v| \leq (1 - \eta)R, \\ d_{F(t,x)}(v) & \text{if } |v| > (1 - \eta)R. \end{cases}$$

Relevant properties of ρ_S are assembled in Lemma 8.6.5 of Chap. 8.

Step 4 (Completion of the Proof)
Under the supplementary hypothesis (G3)$'$, the $W^{1,1}$ local minimizer for (P) of interest is $\bar{x} \equiv 0$. Let $\beta \in (0, \epsilon)$ be such that $\bar{x}$ is minimizing w.r.t. all admissible F trajectories x such that $\|x - \bar{x}\|_{W^{1,1}} (= \|x\|_{W^{1,1}}) \leq \beta$. Take $\eta > 0$ and $\delta > 0$ as in Step 3.

For $t \in [S, T]$ and $e \in D^\delta(t)$ define

$$\phi(t, e) := \begin{cases} \frac{1}{2}(|e| - (1 - 2\eta)R) \vee 0 & \text{if } e \in (1 - \eta)R\mathbb{B} \\ \frac{1}{2}(|e - e_0| - \delta/6) \vee 0 & \text{if } e \in e_0 + \frac{1}{3}\delta\mathbb{B} \text{ for some } e_0 \in E^\delta_{\text{regular}}(t). \end{cases}$$

Note the following properties of ϕ, each of which is a simple consequence of the definition of this function: for any $t \in [S, T]$

11.3 Proof of Theorem 11.2.1

$$\left.\begin{array}{l} \phi(t, e) = 0 \text{ if } |e| \leq (1 - 2\eta)R \\[6pt] \phi(t, .) \text{ is locally Lipschitz continuous on } D^\delta(t) \\ \qquad\qquad \text{with Lipschitz constant } 1/2 \\[6pt] \phi(t, e) \geq \frac{\eta R}{2} \wedge \frac{\delta}{12} \text{ if } e \in \partial D^\delta(t). \end{array}\right\} \qquad (11.3.10)$$

Take $\alpha_i \downarrow 0$ and, for each i, consider the optimization problem:

$$(P_i) \begin{cases} \text{Minimize} \\ \quad \left(\ell_i(x(S)), x(T))\right) \vee \left(\int_S^T \rho_S(t, x(t), \dot{x}(t))dt\right) \vee \left(\max_{t \in h(t, x(t))} h(t, x(t))\right) \\ \qquad\qquad\qquad\qquad\qquad\qquad\qquad\qquad + \int_S^T \phi(t, \dot{x}(t))dt \\ \text{over arcs } x \in W^{1,1} \text{ such that} \\ \dot{x}(t) \in D^\delta(t) \text{ a.e. } t \in [S, T], \\ |x(S)| + \int_{[S,T]} |\dot{x}(t)|dt \leq \beta. \end{cases}$$

Here,

$$\ell_i(x_0, x_1) := \left(g(x_0, x_1) - g(0, 0) + \alpha_i^2\right) \vee d_C(x_0, x_1)$$

and $\rho_S(t, x, v)$ is the modified penalty integrand of Step 3, with parameters $\eta \in (0, 1/2)$ and $\delta > 0$ as earlier chosen. This problem can be expressed as

$$\text{Minimize } \{J_i(x) : x \in \mathcal{S}\}$$

in which

$$\mathcal{S} := \{x \in W^{1,1} : x \text{ satisfies the constraints of } (P_i)\}$$

and

$$J_i(x) := \left(\ell_i(x(S)), x(T))\right) \vee \left(\int_S^T \rho_S(t, x(t), \dot{x}(t))dt\right) \vee \left(\max_{t \in [S,T]} h(t, x(t))\right) \\ + \int_S^T \phi(t, \dot{x}(t))dt.$$

$\mathcal{S}$ is complete w.r.t. the metric induced by the $||.||_{W^{1,1}}$ norm on elements of $\mathcal{S}$. J_i is continuous w.r.t. to this metric. Since $\bar{x} \equiv 0$ is an α_i^2 minimizer for (P_i) we can, by Ekeland's theorem, find $x_i \in \mathcal{S}$ such that x_i is a minimizer for

$$\text{Minimize } \{J_i(x) + \alpha_i\left(|x(S) - x_i(S)| + \int_{[S,T]} |\dot{x}(t) - \dot{x}_i(t)|dt\right) : x \in \mathcal{S}\}$$

and

$$|x_i(S)| + \int_{[S,T]} |\dot{x}_i(t)| dt \leq \alpha_i .\qquad (11.3.11)$$

We note that

$$\left(\ell_i(x_i(S)), x_i(T))\right) \vee \left(\int_S^T \rho_S(t, x_i(t), \dot{x}_i(t)) dt\right) \vee \left(\max_{t \in [S,T]} h(t, x(t))\right) > 0 \qquad (11.3.12)$$

for, otherwise, we would have

$$\left(g(x_i(S), x_i(T)) - g(0, 0) + \epsilon_i^2\right) \vee d_C(x_i(S), x_i(T))$$
$$\vee \left(\int_S^T \rho_S(t, x_i(t), \dot{x}_i(t)) dt\right) \vee \left(\max_{t \in [S,T]} h(t, x(t))\right) = 0 .$$

This implies $\rho_S(t, x_i(t), \dot{x}_i(t)) = 0$ a.e.. But then $\dot{x}_i(t) \in F(t, x_i(t))$ a.e., in consequence of Lemma 8.6.5 of Chap. 8. By definition of $D^\delta(t)$, $\dot{x}_i(t) \in B(t)$. The preceding relations also tell us that $d_C(x_i(S), x_i(T)) = 0$, which implies $(x_i(S), x_i(T)) \in C$, $\max_{t \in [S,T]} h(t, x(t)) dt \leq 0$ and $g(x_i(S), x_i(T)) \leq g(0, 0) - \alpha_i^2$. Since $x_i \in \mathcal{S}$, we know that $\|x_i\|_{W^{1,1}} \leq \beta$. But then x_i is an admissible F trajectory, with $\dot{x}_i(t) \in B(t)$ for a.e. $t \in [S, T]$, that violates the $W^{1,1}$ local optimality of $\bar{x} \equiv 0$. (11.3.12) is confirmed.

We have shown that x_i is a $W^{1,1}$ local minimizer for the problem

$$\begin{cases} \text{Minimize } \ell_i(x(S), x(T)) \vee \left(\int_S^T \rho_S(t, x(t), \dot{x}(t)) dt\right) \vee \left(\max_{t \in [S,T]} h(t, x(t))\right) \\ \quad + \alpha_i \left(|x(S) - x_i(S)| + \int_S^T |\dot{x}(t) - \dot{x}_i(t)| dt\right) + \int_S^T \phi(t, \dot{x}(t)) dt \\ \text{over } x \in W^{1,1} \text{ satisfying} \\ \dot{x}(t) \in D^\delta(t) \text{ a.e. } t \in [S, T]. \end{cases}$$

This problem is an example of the finite Lagrangian problem of Step 1. The hypotheses are satisfied for the application of Proposition 11.3.1.

Using the max rule to estimate $\partial \ell_i$, we deduce that there exist $p_i \in W^{1,1}$, $\mu_i \in C^\oplus(S, T)$, a bounded Borel measurable function $\gamma_i : [S, T] \to \mathbb{R}^n$ $\bar{\lambda}_i \in [0, 1]$, $\lambda_i^{(1)} \in [0, 1]$ and $\sigma_i \in [0, 1]$ such that the following conditions are satisfied, in which $q_i \in NBV([S, T]; \mathbb{R}^n)$ is the function

$$q_i(t) = \begin{cases} p_i(S) & \text{if } t = S \\ p_i(t) + \int_{[S,t]} \gamma_i(s) d\mu_i(s) & \text{if } t \in (S, T], \end{cases}$$

(A)': $\bar{\lambda}_i + \lambda_i^{(1)} + \|\mu_i\|_{TV} = 1$,

11.3 Proof of Theorem 11.2.1

(B)': $\dot{p}_i(t) \in \text{co}\{\eta : (\eta, q_i(t)) \in \lambda_i^{(1)} \partial_{x,v} \rho_S(t, x_i(t), \dot{x}_i(t)) + \alpha_i \mathbb{B} \times \{0\}\}$,

for a.e. $t \in [S, T]$ such that either $\dot{x}_i(t) \in \dot{\bar{x}}(t) + (1 - 2\eta) R \, \overset{\circ}{\mathbb{B}}$,

(C)': $(q_i(S), -q_i(T)) \in \bar{\lambda}_i \sigma_i \partial g(x_i(S), x_i(T)) + \bar{\lambda}_i(1 - \sigma_i) \partial d_C(x_i(S), x_i(T))$
$\qquad + \alpha_i \mathbb{B} \times \{0\}$,

(D)': $q_i(t) \cdot \dot{x}_i(t) - \lambda_i^{(1)} \rho_S(t, x_i(t), \dot{x}_i(t)) - \phi(t, \dot{x}_i(t)) \geq$

$$q_i(t) \cdot v - \lambda_i^{(1)} \rho_S(t, x_i(t), v) - \phi(t, v) - \alpha_i |v - \dot{x}_i(t)|,$$

for all $v \in D^\delta(t)$, a.e. $t \in [S, T]$,

(E)': $\text{supp}\{\mu_i\} \subset \{t \in [S, T] : h(t, \bar{x}_i(t)) = \max\limits_{t \in [S,T]} h(t, x_i(t))\}$ and

$$\gamma_i(t) \in \partial_x^> h(t, x_i(t)) \text{ for } \mu_i \text{ - a.e. } t \in [S, T].$$

Note that

$$\bar{\lambda}_i(1 - \sigma_i) \partial d_C(x_i(S), x_i(T)) = \bar{\lambda}_i(1 - \sigma_i) \Big(\partial d_C(x_i(S), x_i(T)) \cap \partial \mathbb{B} \Big), \quad (11.3.13)$$

in which we interpret the right side of (11.3.13) as $\{0\}$, if $\bar{\lambda}_i(1 - \sigma_i) = 0$. This is because, if $(x_i(S), x_i(T)) \notin C$, then $\partial d_C(x_i(S), x_i(T)) \subset \partial \mathbb{B}$. If, on the other hand, $(x_i(S), x_i(T)) \in C$ then $d_C(x_i(S), x_i(T)) = 0$. In view of (11.3.12), the max rule and condition (11.3.3) of Proposition 11.3.1, either $\bar{\lambda}_i = 0$ or $1 - \sigma_i = 0$; in both these cases $\bar{\lambda}_i(1 - \sigma_i) = 0$, then, (11.3.13) is true.

Notice also that, from (11.3.12) and condition (11.3.2) of Proposition 11.3.1, $\max_{t \in [S,T]} h(t, x_i(t)) \leq 0 \implies \mu_i = 0$. It follows that $\partial_x h^>(t, x_i(t)) = \text{co} \, \partial_x h(t, x_i(t))$ at all points t in the support of μ_i and

$$\text{supp}\{\mu_i\} \subset \{t \in [S, T] : h(t, \bar{x}_i(t)) > 0\}.$$

Now write $\lambda_i := \sigma_i \bar{\lambda}_i$ and $\lambda^{(2)} := (1 - \sigma_i) \bar{\lambda}_i$. Then $\lambda_i \in [0, 1]$, $\lambda^{(2)} \in [0, 1]$ and

$$\lambda_i + \lambda_i^{(1)} + \lambda_i^{(2)} + ||\mu_i||_{TV} = \bar{\lambda}_i + \lambda^{(1)} + ||\mu_i||_{TV} = 1.$$

Furthermore, we can deduce from (11.3.12), and condition (11.3.1) in Proposition 11.3.1

$$\text{`} \int_{[S,T]} \rho_S(t, x_i(t), \dot{x}_i(t)) dt = 0 \text{'} \implies \text{`} \lambda_i^{(1)} = 0 \text{'}. \quad (11.3.14)$$

With the benefit of the preceding analysis, we can replace (A)'–(E)' by

(A): $\lambda_i + \lambda_i^{(1)} + \lambda_i^{(2)} + ||\mu_i||_{TV} = 1$ and $\lambda_i^{(1)} = 0$ if $\int_{[S,T]} \rho_S(t, x_i(t), \dot{x}_i(t)) dt = 0$,

(B): $\dot{p}_i(t) \in \text{co}\{\eta : (\eta, q_i(t)) \in \lambda_i^{(1)} \partial_{x,v} \rho_S(t, x_i(t), \dot{x}_i(t)) + \{0\} \times \alpha_i \mathbb{B}\}$

for a.e. $t \in [S, T]$ such that $\dot{x}_i(t) \in \dot{\bar{x}}(t) + (1 - 2\eta) R \, \overset{\circ}{\mathbb{B}}$,

(C): $(q_i(S), -q_i(T)) \in \lambda_i \partial g(x_i(S), x_i(T)) + \lambda_i^{(2)}(\partial d_C(x_i(S), x_i(T)) \cap \partial \mathbb{B})$
$\quad + \alpha_i \mathbb{B} \times \{0\}$,

(D): $q_i(t) \cdot \dot{x}_i(t) - \lambda_i^{(1)} \rho_S(t, x_i(t), \dot{x}_i(t)) - \phi(t, \dot{x}_i(t)) \geq$

$\qquad q_i(t) \cdot v - \lambda_i^{(1)} \rho_S(t, x_i(t), v) - \phi(t, v) - \alpha_i |v - \dot{x}_i(t)|$, for all $v \in D^\delta(t)$,
a.e. $t \in [S, T]$,

(E): $\operatorname{supp}\{\mu_i\} \subset \{t \in [S, T] : h(t, x_i(t)) > 0\}$ and $\gamma_i(t) \in \operatorname{co} \partial_x^> h(t, x_i(t))$ for μ_i - a.e.. $t \in [S, T]$.

We also know from condition (ii) (b) of Proposition 11.3.1 that

$$\dot{p}_i(t) \in \lambda_i^{(1)} \partial_x \rho_S(t, x_i(t), \dot{x}_i(t)) t \in [S, T]. \text{ a.e.} \qquad (11.3.15)$$

Since $\rho_S(t, ., \dot{x}_i(t))$ is Lipschitz continuous with Lipschitz constant $\tilde{k}(t)$, where $\tilde{k}$ is the integrable function of Lemma 8.6.5 (iv), we have

$$|\dot{p}_i(t)| \leq \lambda_i^{(1)} \tilde{k}(t) \text{ a.e. } t \in [S, T].$$

From (C) we have $|p_i(S)| \leq k_g + 1 + \alpha_i$. It follows from Gronwall's lemma and the preceding two relations that the family of costate trajectories p_i, $i = 1, 2, \ldots$ are uniformly bounded and their derivatives are uniformly integrably bounded.

We next seek to establish a uniform positive lower bound on the magnitude of (q_i, λ_i, μ_i). We need to consider separately two possible cases, one of which must occur:

(i): $\rho_S(t, x_i(t), \dot{x}_i(t)) = 0$ a.e.
(ii): $\rho_S(t, x_i(t), \dot{x}_i(t)) > 0$ on a set of positive $\mathcal{L}$-measure.

Consider case (i). We know, from (11.3.14), that $\lambda_i^{(1)} = 0$. But then, from conditions (A) and (C), and since $\sqrt{2}\|q_i\|_{L^\infty} \geq |(q_i(S), q_i(T))|$,

$$\sqrt{2}\|q_i\|_{L^\infty} + (1 + k_g)\lambda_i + \|\mu\|_{TV} \geq 1 - \alpha_i.$$

Consider case (ii). In this case there is a time $t \in [S, T]$ such that condition (C) is satisfied, $\rho_S(t, x_i(t), \dot{x}_i(t)) > 0$ and $\dot{x}_i(t) \in D^\delta(t)\mathbb{B}$. We see that $\dot{x}_i(t) \notin F(t, x_i(t))$. This is obviously true if $|\dot{x}_i(t)| > (1 - \eta)R$ because, in this case, $\rho_S(t, x_i(t), \dot{x}_i(t)) = d_{F(t,x_i(t))}(\dot{x}_i(t))$. So we can assume that $|\dot{x}_i(t)| \leq (1 - \eta)R$. Suppose, contrary to our assertion, $\dot{x}_i(t) \in F(t, x_i(t))$. Then $\chi(|\dot{x}_i(t)|) = 1$. It follows that $\rho_S(t, x_i(t), \dot{x}_i(t)) \leq |\dot{x}_i(t) - \chi(|\dot{x}_i(t)|)\dot{x}_i(t)| = |\dot{x}_i(t) - \dot{x}_i(t)| = 0$, which is a contradiction.

We now distinguish the two possible situations:

(a): $\dot{x}_i(t) \in \partial D^\delta(t)$.

In view of (G3)', there exists $v_0 \in F(t, x_i(t))$ such that $|v_0| \leq (1 - 2\eta)R$. Then $\phi(t, v_0) = 0$ and $\rho_S(t, x_i(t), v_0) \leq |v_0 - \chi(|v_0|)v_0| = |v_0 - v_0|) = 0$.

11.3 Proof of Theorem 11.2.1

Since $\dot{x}_i(t) \in \partial D^\delta(t)$, it follows from (11.3.10) that

$$\phi(t, \dot{x}_i(t)) \geq (\eta R/2) \wedge (\delta/12).$$

Using these relations, noting that $\rho_S(t, x_i(t), \dot{x}_i(t)) \geq 0$ and that $|\dot{x}_i(t)| \leq N + \delta/3$, and inserting $v = v_0$ in condition (C) yields the inequality

$$|q_i(t)||\dot{x}_i(t) - v_0| \geq \lambda_i^{(1)} \rho_S(t, x_i(t), \dot{x}_i(t)) - 0 + \phi(t, \dot{x}_i(t))$$
$$- 0 - \alpha_i(N + \delta/3(1 - 2\eta)R).$$
$$\geq \phi(t, \dot{x}_i(t)) - \alpha_i(N + \delta/3 + (1 - 2\eta)R)$$
$$\geq (\eta R/2) \wedge (\delta/12) - \alpha_i(N + \delta/3 + (1 - 2\eta)R).$$

Since $|\dot{x}_i(t)| \leq N + \delta/3$ and $|v_0| \leq (1 - 2\eta)R$, it follows that

$$\|q_i\|_{L^\infty} \geq (N + \delta/3 + (1 - 2\eta)R)^{-1}\Big((\eta R/2) \wedge (\delta/12)\Big) - \alpha_i.$$

(b): $\dot{x}_i(t) \in \overset{\circ}{D}{}^\delta(t)$.

Now, since $\dot{x}_i(t)$ is an unconstrained local minimizer of $v \to -q_i(t) \cdot v + \lambda_i^{(1)} \rho(t, x_i(t), v) + \phi(t, v) + \alpha_i|v - \dot{x}_i|$, we have

$$q_i(t) \in \lambda_i^{(1)} \partial_v \rho_S(t, x_i(t), \dot{x}_i(t)) + \partial_v \phi(t, \dot{x}_i(t)) + \alpha_i \mathbb{B}.$$

Because we have assumed $\rho_S(t, x_i(t), \dot{x}_i(t)) > 0$, we know that elements in $\partial_v \rho_S(t, x_i(t), \dot{x}_i(t))$ have unit length. Taking note also of the fact that $\phi(t, .)$ is Lipschitz continuous with Lipschitz constant $1/2$, whence $\partial_v \phi(t, \dot{x}_i(t)) \in (1/2)\mathbb{B}$, we deduce from the preceding relations that

$$\|q_i\|_{L^\infty} \geq \lambda_i^{(1)} - \frac{1}{2} - \alpha_i.$$

However, from (A) and (C),

$$\sqrt{2}\|q_i\|_{L^\infty} \geq \lambda^{(2)} - k_g \lambda_i - \alpha_i = -(1 + k_g)\lambda_i - \lambda_i^{(1)} - \|\mu_i\|_{TV} + 1 - \alpha_i.$$

Adding these inequalities yields

$$(1 + \sqrt{2})\|q_i\|_{L^\infty} + (1 + k_g)\lambda_i + \|\mu_i\|_{TV} \geq 1 - 1/2 - \alpha_i = 1/2 - 2\alpha_i.$$

Combining the estimates relating to the cases (i), (ii)(a) and (ii)(b), we obtain

$$(1+\sqrt{2})\|q_i\|_{L^\infty} + (1+k_g)\lambda_i + \|\mu_i\|_{TV}$$
$$\geq \min\{\frac{1}{2} - 2\alpha_i \;;\; (N+\delta/3+(1-2\eta)R)^{-1}\left((\eta R/2) \wedge (\delta/12)\right) - \alpha_i\}. \tag{11.3.16}$$

This is the desired lower bound.

We deduce from (11.3.11) that, along a subsequence, $x_i \to \bar{x} \equiv 0$ uniformly, and $\dot{x}_i \to \dot{\bar{x}} \equiv 0$ in L^1 and a.e.. We have already observed that the $\dot{p}_i$'s are uniformly integrably bounded and the p_i's are uniformly bounded. By further restriction to a subsequence we can then arrange, in consequence of Ascoli's theorem, that $p_i \to p$ uniformly and $\dot{p}_i \to \dot{p}$ weakly in L^1. We can also arrange that $\lambda_i \to \lambda$, $\lambda_i^{(1)} \to \lambda^{(1)}$ and $\lambda_i^{(2)} \to \lambda^{(2)}$, for some $\lambda \in [0,1]$, $\lambda^{(1)} \in [0,1]$ and $\lambda^{(2)} \in [0,1]$, $\mu_i \to \mu$ weakly* for some $\mu \in C^\oplus(S,T)$ and $q_i(t) \to q(t)$, for all t in some set of full Lebesgue measure (including the times S and T), for some $q \in NBV([S,T]; \mathbb{R}^n)$. q is related to p and μ according to (11.2.3), for some Borel measurable function γ satisfying $\gamma(t) \in \partial_x^> h(t, \bar{x}(t))$ for μ - a.e..

A convergence analysis along the lines of the proof of Proposition 11.3.1, permits us to pass to the limit in conditions (B)–(E). and thereby to obtain:

$$\dot{p}(t) \in \text{co}\{\eta : (\eta, q(t)) \in \lambda^{(1)} \partial_{x,v} \rho_S(t, \bar{x}(t), \dot{\bar{x}}(t))\}, \text{ a.e. } t \in [S,T], \tag{11.3.17}$$

and also

(C)'': $(q(S), -q(T)) \in \lambda \partial g(\bar{x}(S), \bar{x}(T)) + N_C(\bar{x}(S), \bar{x}(T))$,

(E)'': $\text{supp}\{\mu\} \subset \{t \in [S,T] : h(t, \bar{x}(t)) = 0\}$,
$\gamma(t) \in \partial_x^> h(t, \bar{x}(t))$ for μ - a.e. $t \in [S,T]$

and

$$q(t) \cdot \dot{\bar{x}}(t) - \lambda^{(1)} \rho_S(t, \bar{x}(t), \dot{\bar{x}}(t)) - \phi(t, \dot{\bar{x}}(t))$$
$$\geq p(t) \cdot v - \lambda_i^{(1)} \rho_S(t, \bar{x}(t), v) - \phi(t, v), \text{ for all } v \in D^\delta(t), \text{ a.e. } t \in [S,T].$$

Observe that, for all points $v \in (F(t, \bar{x}(t)) \cap (1-2\eta)R\mathbb{B})) \cup E^\delta_{\text{regular}}(t)$, we have $v \in D^\delta(t)$ and $\rho_S(t, \bar{x}(t), v) = \phi(t, v) = 0$. Consequently, the preceding relation implies

(D)'': $q(t) \cdot \dot{\bar{x}}(t) \geq q(t) \cdot v$,
for all $v \in (F(t, \bar{x}(t)) \cap (1-2\eta)R\mathbb{B})) \cup E^\delta_{\text{regular}}(t)$ a.e. $t \in [S,T]$.

From Lemma 8.6.5 of Chap. 8 and from (B) it follows that

(B)'': $\dot{p}(t) \in \text{co}\{\eta : (\eta, q(t)) \in N_{\text{Gr} F(t,\cdot)}(\bar{x}(t), \dot{\bar{x}}(t))\}$, a.e. $t \in [S,T]$

and

$$|\dot{p}(t)| \leq k(t)|q(t)| \text{ for a.e. } t \in [S,T]. \tag{11.3.18}$$

From (11.3.16),

$$(1+\sqrt{2})\|q\|_{L^\infty} + (1+k_g)\lambda + \|\mu\|_{TV}$$
$$\geq \min\{\frac{1}{2}, (N+\delta/3+(1-2R\eta))^{-1}((\eta R/2) \wedge (\delta/12))\} > 0.$$

Since $q = p$ if $\mu = 0$, we deduce that $(p, \lambda, \mu) \neq (0, 0, 0)$. We can therefore arrange, by scaling the Lagrange multipliers, that

(A)''': $\|p\|_{L^\infty} + \lambda + \|\mu\|_{TV} = 1$.

Conditions (A)''–(E)'' above provide a restricted form of theorem, in which the Weierstrass condition (D)'' is affirmed only for velocities in the subset (recalling that $\bar{x} \equiv 0$)

$$(F(t, \bar{x}(t)) \cap (1-2\eta)R\mathbb{B})) \cup E^N_{\text{regular}}(t) \subset \overline{\text{co}}\,(\Omega_0(t) \cap B(t)).$$

To recover the full Weierstrass condition, we employ an analogous proof technique to that one of Chap. 8. The first step is to validate the condition for all $v \in (F(t, \bar{x}(t)) \cap (1-2\eta)R\mathbb{B})) \cup E^N_{\text{discrete}}(t)$, by taking a sequence $\delta_i \downarrow 0$, carrying out the above constructions for each i and passing to the limit, using the fact that, for a.e. $t \in [S, T]$,

$$\lim_i E^{\delta_i}_{\text{regular}}(t) = E^N_{\text{discrete}}(t).$$

The second step is to validate it for all $v \in \Omega_0(t) \cap B(t)$, by taking sequences $N_i \uparrow \infty$ and $\eta_i \downarrow 0$ and using the fact that, for a.e. $t \in [S, T]$,

$$\Omega_0(t) \cap B(t) \subset \lim_{i \to \infty}\left((F(t, \bar{x}(t)) \cap (1-2\eta_i)R\mathbb{B})) \cup E^{N_i}_{\text{discrete}}(t)\right).$$

Finally, we can replace $\Omega_0(t) \cap B(t)$ by $\overline{\text{co}}\,(\Omega_0(t) \cap B(t))$, using the linearity of the mapping $e \to q(t) \cdot e$. The proof is complete. □

11.4 Free End-Time Problems with State Constraints

This section supplies necessary conditions that are 'state constrained' analogues of the free end-time necessary conditions of Chap. 9. As before, the extra conditions to take account of the free end-times are boundary conditions on the Hamiltonian. Consider the following state constrained problem with free end-times:

$$(FT) \begin{cases} \text{Minimize } g(S, x(S), T, x(T)) \\ \text{over intervals } [S, T] \text{ and arcs } x \in W^{1,1}([S, T]; \mathbb{R}^n) \text{ satisfying} \\ \dot{x}(t) \in F(t, x(t)) \quad \text{a.e. } t \in [S, T], \\ (S, x(S), T, x(T)) \in C, \\ h(t, x(t)) \le 0 \quad \text{for all } t \in [S, T]. \end{cases}$$

Here $g : \mathbb{R} \times \mathbb{R}^n \times \mathbb{R} \times \mathbb{R}^n \to \mathbb{R}$ and $h : \mathbb{R} \times \mathbb{R}^n \to \mathbb{R}$ are given functions, $F : \mathbb{R} \times \mathbb{R}^n \rightsquigarrow \mathbb{R}^n$ is a given multifunction, and $C \subset \mathbb{R} \times \mathbb{R}^n \times \mathbb{R} \times \mathbb{R}^n$ is a given closed set. F trajectories for (FT) are now written $([S, T], x)$ to emphasize the underlying time interval $[S, T]$ of the arc x. F trajectories that satisfy the constraints of (FT) are said to be admissible.

Take a multifunction $B : (-\infty, +\infty) \to \mathbb{R}^n$. An admissible F trajectory $([\bar{S}, \bar{T}], \bar{x})$ is called a $W^{1,1}$ minimizer relative to B if there exists $\beta > 0$ such that

$$g(S, x(S), T, x(T)) \ge g(\bar{S}, \bar{x}(\bar{S}), \bar{T}, \bar{x}(\bar{T}))$$

for all admissible F trajectories such that $\dot{x}(t) \in \dot{\bar{x}}(t) + B(t)$ a.e. $t \in [\bar{S}, \bar{T}] \cap [S, T]$ and

$$d(([S, T], x), ([\bar{S}, \bar{T}], \bar{x})) \le \beta.$$

Here, d is the distance function employed in Chap. 9:

$$d(([S, T], x), ([S', T'], x')) :=$$
$$|x(S) - x'(S)| + \int_{-\infty}^{+\infty} |\dot{x}_e(t) - \dot{x}'_e(t)| dt + |S - S'| + |T - T'|.$$

(Recall that x_e denotes the extension, by constant extrapolation to left and right, of $x : [S, T] \to \mathbb{R}^n$, etc. Often, for brevity, the subscript 'e' is omitted; any integral in which the domain of an integrand is not compatible with the limits of integration is interpreted in terms of this extension convention.)

As in Chap. 9, we treat the Lipschitz time dependence and the measurable time dependence cases separately. Define, as usual,

$$H(t, x, p) = \sup_{v \in F(t,x)} p \cdot v .$$

Theorem 11.4.1 (State Constrained, Free End-Time Problems with Lipschitz Time Dependence) *Let $([\bar{S}, \bar{T}], \bar{x})$ be a $W^{1,1}$ local minimizer for (FT) such that $\bar{T} - \bar{S} > 0$. Assume that, for some $\epsilon > 0$, the following hypotheses are satisfied:*

(G1): g is Lipschitz continuous on a neighbourhood of $(\bar{S}, \bar{x}(\bar{S}), \bar{T}, \bar{x}(\bar{T}))$ and C is a closed set,

11.4 Free End-Time Problems with State Constraints

(G2): $F(t, x)$ is non-empty for all $(t, x) \in \mathbb{R} \times \mathbb{R}^n$ and Gr F is closed,
(G3): There exist $\epsilon > 0$ and a measurable function $R : [\bar{S}, \bar{T}] \to (0, \infty) \cup \{+\infty\}$
(a 'radius function') such that the following conditions are satisfied:
(a): (Pseudo-Lipschitz Continuity) There exists $k \in L^1(\bar{S}, \bar{T})$ such that

$$F(t', x') \cap (\dot{\bar{x}}(t) + R(t)\overset{\circ}{\mathbb{B}}) \subset F(t'', x'') + k_F(t) |(t', x') - (t'', x'')|\mathbb{B},$$

for all $(t', x'), (t'', x'') \in (t, \bar{x}(t)) + \epsilon \mathbb{B}$, a.e. $t \in [\bar{S}, \bar{T}]$,

(b): (Tempered Growth) There exists $r \in L^1(\bar{S}, \bar{T})$, $r_0 > 0$ and $\gamma \in (0, 1)$ such that $r_0 \leq r(t)$, $\gamma^{-1} r(t) \leq R(t)$ a.e. $t \in [\bar{S}, \bar{T}]$ and

$$F(t', x') \cap (\dot{\bar{x}}(t) + r(t)\mathbb{B}) \neq \emptyset, \text{ for all } (t', x') \in (t, \bar{x}(t)) + \epsilon \mathbb{B}, \text{ a.e. } t \in [\bar{S}, \bar{T}],$$

(G4): h is upper semi-continuous and there exists a constant k_h such that, for all $t \in [\bar{S}, \bar{T}]$,

$$|h(t', x') - h(t'', x'')| \leq k_h |(t', x') - (t'', x'')|$$

for all $(t', x'), (t'', x'') \in (t, \bar{x}(t)) + \epsilon \mathbb{B}.$

Then there exist absolutely continuous arcs

$$a \in W^{1,1}([\bar{S}, \bar{T}]; \mathbb{R}) \quad \text{and} \quad p \in W^{1,1}([\bar{S}, \bar{T}]; \mathbb{R}^n),$$

a non-negative number λ, a measure $\mu \in C^\oplus(\bar{S}, \bar{T})$ and a Borel measurable function $\gamma = (\gamma_0, \gamma_1) : [\bar{S}, \bar{T}] \to \mathbb{R}^{1+n}$ such that the following conditions are satisfied, in which

$$q(t) := \begin{cases} p(\bar{S}) & \text{if } t = \bar{S} \\ p(t) + \int_{[\bar{S}, t]} \gamma(s) d\mu(s) & \text{if } t \in (\bar{S}, \bar{T}]. \end{cases}$$

(i) $\lambda + \|p\|_{L^\infty} + \|\mu\|_{TV} = 1$,
(ii) $(-\dot{a}(t), \dot{p}(t)) \in \text{co}\{(\alpha, \beta) : (\alpha, \beta, q(t)) \in N_{\text{Gr } F}(t, \bar{x}(t), \dot{\bar{x}}(t))\}$,
a.e. $t \in [\bar{S}, \bar{T}]$,
(iii) $q(t) \cdot \dot{\bar{x}}(t) \geq q(t) \cdot v$ for all $v \in \overline{\text{co}} \tilde{\Omega}_0(t)$, a.e. $t \in [\bar{S}, \bar{T}]$,
(iv) $(-a(\bar{S}), q(\bar{S}), a(\bar{T}) - \int_{[\bar{S}, \bar{T}]} \gamma_0(s) d\mu(s), -q(\bar{T}))$
$\in \lambda \partial g(\bar{S}, \bar{x}(\bar{S}), \bar{T}, \bar{x}(\bar{T})) + N_C(\bar{S}, \bar{x}(\bar{S}), \bar{T}, \bar{x}(\bar{T}))$,
(v) $a(t) = \int_{[\bar{S}, t)} \gamma_0(s) d\mu(s) + H(t, \bar{x}(t), q(t))$, a.e. $t \in [\bar{S}, \bar{T}]$,
(vi) $\text{supp } \mu \subset \{t : h(t, \bar{x}(t)) = 0\}$ and $\gamma(t) \in \partial^> h(t, \bar{x}(t))$, μ-a.e. $t \in [\bar{S}, \bar{T}]$.

If the initial time $\bar{S}$ is fixed, i.e. $C = \{(\bar{S}, x_0, T, x_1) : (x_0, T, x_1) \in \tilde{C}\}$, for some closed set $\tilde{C} \subset \mathbb{R}^n \times \mathbb{R} \times \mathbb{R}^n$, and $g(\bar{S}, x_0, T, x_1) = \tilde{g}(x_0, T, x_1)$ for some $\tilde{g} : \mathbb{R}^n \times \mathbb{R} \times \mathbb{R}^n \to \mathbb{R}$, then the above conditions are satisfied when (iii) is replaced by

(iii)': $(q(\bar{S}), a(\bar{T}), -q(\bar{T})) \in \lambda \partial \tilde{g}(\bar{x}(\bar{S}), \bar{T}, \bar{x}(\bar{T})) + N_C(\bar{S}, \bar{x}(\bar{S}), \bar{T}, \bar{x}(\bar{T}))$.

The assertions of the theorem can be similarly modified when the final time $\bar{T}$ is fixed. Here,

$$\partial^> h(t, \bar{x}(t)) := \text{co lim sup } \{\partial h(t_i, x_i) : t_i \to t, x_i \to \bar{x}(t) \text{ and } h(t_i, x_i) > 0 \text{ for each } i\},$$

and

$$\tilde{\Omega}_0(t) := \{e \in F(t, \bar{x}(t)) : F \text{ is pseudo} - \text{Lipschitz continuous near}((t, \bar{x}(t)), e)\}.$$

The proof closely patterns that of Theorem 9.2.1 and is therefore omitted. The essential idea is to reformulate (FT) as a fixed time problem by means of a change of independent variable, apply known necessary conditions to the fixed time problem (necessary conditions for state constrained problems in this case) and to interpret these conditions in terms of the original problem.

Remark
The Autonomous Case: If $F(t, x)$ and $h(t, x)$ do not depend on t, conditions (ii) and (v) in the theorem statement imply that $a(.)$ is a constant and $\gamma_0 = 0$. Condition (v) then takes the form: there exists a number c such that

$$H(t, \bar{x}(t), p(t) + \int_{[\bar{S},t]} \gamma_1(s)d\mu(s)) = c, \text{ a.e. } t \in [\bar{S}, \bar{T}].$$

If for admissible F trajectories $([S, T], x)$ either the initial time S is unconstrained and g does not depend on S, or the final time T is unconstrained and g does not depend on T, then the constant $c = 0$.

We can use Proposition 8.5.1 of Chap. 8 to prove other versions of Theorem 11.4.1, in which hypothesis (G3) is replaced by alternative, stronger, hypotheses. For example, choosing the third set of conditions in Proposition 8.5.1 serving this purpose, we arrive at:

Corollary 11.4.2 *The assertions of Theorem 11.4.1 remain valid when $\bar{x}$ is a $W^{1,1}$ local minimizer (i.e. $B \equiv \mathbb{R}^n$) and hypothesis is (G3) is replaced by:*

$(G3)^{***}$: *There exist $\epsilon > 0$, $\alpha > 0$ and non-negative measurable functions k and β such that k and $t \to \beta(t)k^\alpha(t)$ are integrable on $[\bar{S}, \bar{T}]$ and, for each $N \geq 0$,*

$$F(t', x') \cap (\dot{\bar{x}}(t) + N\mathbb{B}) \subset F(t'', x'') + (k(t) + \beta(t)N^\alpha)|x' - x''|\mathbb{B},$$

for all (t', x'), $(t'', x'') \in (t, \bar{x}(t)) + \epsilon\mathbb{B}$ and a.e. $t \in [\bar{S}, \bar{T}]$.

In this case, the Weierstrass condition (iii) of Theorem 11.4.1 becomes

$$p(t) \cdot \dot{\bar{x}}(t) \geq p(t) \cdot v, \quad \text{for all } v \in F(t, \bar{x}(t)), \quad \text{a.e. } t \in [\bar{S}, \bar{T}].$$

11.4 Free End-Time Problems with State Constraints

Necessary conditions can also be derived for problems with measurably time dependent data. For such problems, the boundary condition on the Hamiltonian is interpreted in the 'sub and super essential values' sense, introduced in Chap. 9. Recall that, given an integrable function $a : [S, T] \to \mathbb{R}$ and a point $t \in (S, T)$, we have (see Definition 9.3.1):

(a): the sub essential value of f at $\bar{t}$ is the set

$$\underset{t \to \bar{t}}{\text{sub-ess}\,} a(t) := \bigg\{ \zeta \in \mathbb{R} \,:\, \exists t_i \to \bar{t} \text{ and } \zeta_i \to \zeta \text{ s.t.}$$

$$\limsup_{\epsilon \downarrow 0} \epsilon^{-1} \int_{t_i - \epsilon}^{t_i} a(s)ds \le \zeta_i \le \liminf_{\epsilon \downarrow 0} \int_{t_i}^{t_i + \epsilon} a(s)ds, \text{ for each } i \bigg\}.$$

(b): the super essential value of a at t is the set

$$\underset{t \to \bar{t}}{\text{super-ess}\,} a(t) := \bigg\{ \zeta \in \mathbb{R} \,:\, \exists t_i \to \bar{t} \text{ and } \zeta_i \to \zeta \text{ s.t.}$$

$$\limsup_{\epsilon \downarrow 0} \epsilon^{-1} \int_{t_i}^{t_i + \epsilon} a(s)ds \le \zeta_i \le \liminf_{\epsilon \downarrow 0} \int_{t_i - \epsilon}^{t_i} a(s)ds, \text{ for each } i \bigg\}.$$

Theorem 11.4.3 (State Constrained, Free End-Time Problems with Measurable Time Dependence) *Take a measurable multifunction $B : \mathbb{R} \rightsquigarrow \mathbb{R}^n$ such that $B(t)$ is open for a.e. $t \in \mathbb{R}$. Let $([\bar{S}, \bar{T}], \bar{x})$ be a $W^{1,1}$ local minimizer for (FT) relative to B such that $\bar{T} - \bar{S} > 0$. Assume that, for some $\epsilon > 0$ and $\sigma > 0$, the data satisfy the following hypotheses:*

(G1): g *is Lipschitz continuous on a neighbourhood of* $(\bar{S}, \bar{x}(\bar{S}), \bar{T}, \bar{x}(\bar{T}))$ *and C is a closed set,*

(G2): $F(t, x)$ *is nonempty for each* $(t, x) \in \mathbb{R} \times \mathbb{R}^n$, $\text{Gr}\, F(t, .)$ *is closed for each $t \in \mathbb{R}$ and F is $\mathcal{L} \times \mathcal{B}^n$ measurable,*

(G3): *There exists a measurable function* $R : [\bar{S} - \sigma, \bar{T} + \sigma] \to (0, \infty) \cup \{+\infty\}$ *(a 'radius function'), such that $R(t) \mathring{\mathbb{B}} \subset B(t)$ a.e. $t \in [\bar{S} - \sigma, \bar{T} + \sigma]$ and the following conditions are satisfied:*

(a): *(Pseudo-Lipschitz Continuity) There exists $k \in L^1(\bar{S} - \sigma, \bar{T} + \sigma)$ such that*

$$F(t, x') \cap (\dot{\bar{x}}(t) + R(t)\mathring{\mathbb{B}}) \subset F(t, x) + k(t)|x' - x|\mathbb{B}, \tag{11.4.1}$$

for all $x, x' \in \bar{x}(t) + \epsilon \mathbb{B}$, *a.e.* $t \in [\bar{S} - \sigma, \bar{T} + \sigma]$,

(b): *(Tempered Growth) There exist $r \in L^1(\bar{S} - \sigma, \bar{T} + \sigma)$, $r_0 > 0$ and $\alpha \in (0, 1)$ such that $r_0 \le r(t)$, $\alpha^{-1} r(t) \le R(t)$, a.e. $t \in [\bar{S} - \sigma, \bar{T} + \sigma]$, and*

$$F(t, x) \cap (\dot{\bar{x}}(t) + r(t)\mathbb{B}) \neq \emptyset \text{ for all } x \in \bar{x}(t) + \epsilon \mathbb{B}, \text{ a.e. } t \in [\bar{S} - \sigma, \bar{T} + \sigma],$$

(G4): *There exists $c_F \ge 0$ such that, for a.e. $t \in [\bar{S} - \sigma, \bar{S} + \sigma] \cup [\bar{T} - \sigma, \bar{T} + \sigma]$,*

$$F(t, x) \subset c_F \mathbb{B}, \quad \text{for all } x \in \bar{x}(t) + \epsilon \mathbb{B}. \tag{11.4.2}$$

(G5): *h is upper semi-continuous and there exists a constant k_h such that, for a.e. $t \in [\bar{S} - \sigma, \bar{T} + \sigma]$*

$$|h(t, x) - h(t, x')| \leq k_h |x - x'|$$

for all $x, x' \in \bar{x}_e(t) + \epsilon \mathbb{B}$.
Furthermore,

$$h(\bar{S}, \bar{x}(\bar{S})) < 0 \quad \text{and} \quad h(\bar{T}, \bar{x}(\bar{T})) < 0. \tag{11.4.3}$$

Then there exist an absolutely continuous arc $p \in W^{1,1}([\bar{S}, \bar{T}]; \mathbb{R}^n)$, real numbers $\lambda \geq 0$, ξ_0, ξ_1, a measure $\mu \in C^{\oplus}(\bar{S}, \bar{T})$ and a μ-integrable function $\gamma : [\bar{S}, \bar{T}] \to \mathbb{R}^n$ such that the following conditions are satisfied, in which

$$q(t) = \begin{cases} p(\bar{S}) & \text{if } t = \bar{S} \\ p(t) + \int_{[\bar{S},t]} \gamma(s) d\mu(s) & \text{if } t \in (\bar{S}, \bar{T}]. \end{cases}$$

(i) $\lambda + \|p\|_{L^\infty} + \|\mu\|_{TV} = 1$,
(ii) $\dot{p}(t) \in \text{co}\{\zeta : (\zeta, q(t)) \in N_{\text{Gr}F(t,\cdot)}(\bar{x}(t), \dot{\bar{x}}(t))\}$, a.e. $t \in [\bar{S}, \bar{T}]$,
(iii) $(-\xi_0, q(\bar{S}), \xi_1, -q(\bar{T})) \in \lambda \partial g(\bar{S}, \bar{x}(\bar{S}), \bar{T}, \bar{x}(\bar{T})) + N_C(\bar{S}, \bar{x}(\bar{S}), \bar{T}, \bar{x}(\bar{T}))$,
(iv) $q(t) \cdot \dot{\bar{x}}(t) \geq q(t) \cdot v$ for all $v \in \overline{co}(\Omega_0(t) \cap (\dot{\bar{x}}(t) + B(t)))$ a.e. $t \in [\bar{S}, \bar{T}]$,
(v) $\gamma(t) \in \partial_x^> h(t, \bar{x}(t))$ μ-a.e. $t \in [\bar{S}, \bar{T}]$,
(vi) $\xi_0 \in \text{sub-ess}_{t \to \bar{S}} H(t, \bar{x}(\bar{S}), q(\bar{S}))$,
(vii) $\xi_1 \in \text{super-ess}_{t \to \bar{T}} H(t, \bar{x}(\bar{T}), q(\bar{T}))$.

(In condition (i), $\Omega_0(t)$ is as defined by (11.2.1).)
If the initial time is fixed (i.e. $C = \{(\bar{S}, x_0, T, x_1) : (x_0, T, x_1) \in \tilde{C}\}$ for some set $\tilde{C}$) and $g(S, x_0, T, x_1) = \tilde{g}(x_0, T, x_1)$, for some function $\tilde{g} : \mathbb{R}^n \times \mathbb{R} \times \mathbb{R}^n \to \mathbb{R}$, then the above assertions (except (vi)) remain true when condition (11.4.2) in (G4) is satisfied only for a.e. $t \in [\bar{T} - \sigma, \bar{T} + \sigma]$ and (11.4.3) in (G5) is replaced by '$h(\bar{T}, \bar{x}(\bar{T})) < 0$'; in this case, (iii) can be written

$$(q(\bar{S}), \xi_1, -q(\bar{T})) \in \lambda \partial \tilde{g}(\bar{x}(\bar{S}), \bar{T}, \bar{x}(\bar{T})) + N_{\tilde{C}}(\bar{x}(\bar{S}), \bar{T}, \bar{x}(\bar{T})).$$

Hypotheses (G4)–(G5) and the assertions of the theorem are modified similarly, when the right end-time is fixed.

Remark
(Convex Velocity Sets): If

$$F(t, x) \text{ is convex, for all } x \in \bar{x}(t) + \epsilon \mathbb{B}, \text{ a.e. } t \in [S, T],$$

then, according to the dualization theorem (Theorem 8.7.1), the generalized Euler Lagrange condition (ii) implies the Hamiltonian inclusion, namely

$$\dot{p}(t) \in \text{co}\{-\xi : (\xi, \dot{\bar{x}}(t)) \in \partial_{x,p} H(t, \bar{x}, q(t))\}, \text{ a.e. } t \in [S, T];$$

also, the Weierstrass condition (iii) is valid in a stronger form, in which $F(t, \bar{x}(t))$ replaces the set $\overline{\text{co}}(\Omega_0(t) \cap (\dot{\bar{x}}(t) + B(t)))$.

The proof of Theorem 11.4.3 (providing necessary conditions for state constrained, free end-time problems and measurably time depend data) is along similar lines to that its fixed end-times counterpart Theorem 11.2.1 and, for this reason, we merely describe here its main features. The proof comprises four steps. In Step 1 we prove necessary conditions for a free time version of the finite Lagrangian problem. These necessary conditions, which incorporate a transversality condition expressed in terms of sub and super essential values of the Hamiltonian, are proved by the techniques used in the proof of the free end-time necessary condition Theorem 9.4.1 of Chap. 9. (This involves, first, proving the necessary conditions when the endpoint cost is continuously differentiable, in which case the transversality conditions are deduced from fixed time necessary conditions by applying variations to the endtimes; the analysis is then extended to cover Lipschitz continuous end-point cost functions, by means of 'inf convolution' arguments and an application of Ekeland's theorem). Step 2 (hypothesis reduction) and Step 3 (construction of an integral penalty function) are exactly the same as in the fixed end-time setting. In Step 4 (completion of the proof) we apply the finite Lagrangian necessary conditions of Step 1 to a sequence of intermediate dynamic optimization problems involving the integral penalty function of Step 3. The intermediate problems are the same as those employed in Step 4 of the proof the fixed end-time necessary conditions (Theorem 11.2.1) except that, now, the end-times are free. The necessary conditions of Theorem 11.4.3 (in a restricted sense) are obtained by passing to the limit in the necessary for the intermediate problem. The full necessary conditions are finally obtained by another passage to the limit, now involving the parameters used to define the integral penalty function.

Other versions of Theorem 11.4.3 can be proved with the help of Proposition 8.5.1 of Chap. 8, in which hypothesis (G3) is replaced by other hypotheses. For example

Corollary 11.4.4 *The assertions of Theorem 11.4.3 remain valid when $\bar{x}$ is a $W^{1,1}$ local minimizer (i.e. $B \equiv \mathbb{R}^n$) and hypothesis (G3) is replaced by:*

(G3)***: *There exist $\alpha > 0$, $\epsilon > 0$ and non-negative measurable functions k and β such that k and $t \to \beta(t)k^\alpha(t)$ are integrable and, for each $N \geq 0$,*

$$F(t, x') \cap (\dot{\bar{x}}(t) + N\mathbb{B}) \subset F(t, x) + (k(t) + \beta(t)N^\alpha)|x' - x|\mathbb{B},$$

for all $x', x \in \bar{x}(t) + \epsilon\mathbb{B}$ and a.e. $t \in [\bar{S} - \sigma, \bar{T} + \sigma]$.

In this case, the Weierstrass condition (iii) of Theorem 11.4.4 becomes

$$p(t) \cdot \dot{\bar{x}}(t) \geq p(t) \cdot v, \quad \text{for all } v \in F(t, \bar{x}(t)), \quad \text{a.e. } t \in [\bar{S}, \bar{T}].$$

11.5 Non-degenerate Necessary Conditions

The preceding Euler Lagrange-type necessary conditions, for state constrained dynamic optimization problems, provide useful information about minimizers in many cases. But there are cases when they are degenerate, notably when the initial or terminal state is fixed and either of these fixed end-states are located in the state constraint set boundary. Suppose, indeed, that we restrict attention (for clarity of exposition) to a special case in which F and h are independent of t, h is a continuously differentiable function, g depends only on the right endpoint of the state trajectory, the underlying time interval and the left endpoint are fixed, i.e.

$$C = \{\bar{S}\} \times \{x_0\} \times \{\bar{T}\} \times C_1,$$

for some $[\bar{S}, \bar{T}] \subset \mathbb{R}$, $x_0 \in \mathbb{R}^n$ and $C_1 \subset \mathbb{R}^n$, and

$$h(x_0) = 0.$$

For such problems, conditions (i)–(vi) of Theorem 11.4.1 are satisfied by *any* admissible F trajectory $\bar{x}$, with Lagrange multipliers taken to be

$$(p \equiv -\nabla h(\bar{x}(\bar{S})), \; \mu = \delta'_{\{\bar{S}\}}, \; \lambda = 0). \tag{11.5.1}$$

Provided $\nabla h(\bar{x}(\bar{S})) = \nabla h(x_0) \neq 0$, these multipliers are non-zero. In these cases, the Euler Lagrange type necessary conditions fail to give useful information about minimizers.

We have already explored this 'degeneracy' phenomenon in Chap. 10, in connection with the state-constrained maximum principle. The earlier discussion is equally relevant to Euler Lagrange-type necessary conditions for state constrained problems in which the dynamic constraint is a differential inclusion. The remedy will be the same as before. It will be to supplement the necessary conditions with extra conditions, asserting that the regularity properties of the maximized Hamiltonian on the interior of the optimal time interval $[\bar{S}, \bar{T}]$ extend to the end-times. In the presence of an appropriate constraint qualification, these extra conditions exclude the degenerate Lagrange multipliers (11.5.1) and tell us that there must exist other Lagrange multiplier choices with possibly non-trivial information content.

Consider, once again, the free end-time dynamic optimization problem with pathwise state constraints:

$$(FT) \begin{cases} \text{Minimize } g(S, x(S), T, x(T)) \\ \text{over } [S, T] \subset \mathbb{R}, x \in W^{1,1}([S, T]; \mathbb{R}^n) \\ \text{satisfying} \\ \dot{x}(t) \in F(t, x(t)) \quad \text{a.e. } t \in [S, T], \\ h(t, x(t)) \leq 0 \text{ for all } t \in [S, T], \\ (S, x(S), T, x(T)) \in C, \end{cases}$$

11.5 Non-degenerate Necessary Conditions

the data for which comprise: functions $g : \mathbb{R} \times \mathbb{R}^n \times \mathbb{R} \times \mathbb{R}^n \to \mathbb{R}$ and $h : \mathbb{R} \times \mathbb{R}^n \to \mathbb{R}$, a multifunction $F : \mathbb{R} \times \mathbb{R}^n \rightsquigarrow \mathbb{R}^n$ and a set $C \subset \mathbb{R} \times \mathbb{R}^n \times \mathbb{R} \times \mathbb{R}^n$.

Consistent with earlier terminology (cf. Chap. 9), an F trajectory $([S, T], x)$ is a pair comprising an interval $[S, T] \subset \mathbb{R}$ $(T > S)$ and an arc $x \in W^{1,1}([S, T]; \mathbb{R}^n)$ such that $\dot{x}(t) \in F(t, x(t))$ a.e. $t \in [S, T]$. The F trajectory is said to be admissible if $(S, x(S), T, x(S)) \in C$ and $h(t, x(t)) \leq 0$, for all $t \in [S, T]$. An admissible F trajectory $([\bar{S}, \bar{T}], \bar{x})$ is an L^∞ local minimizer for (FT) if there exists some $\beta > 0$ such that $g(S, x(S), T, x(T)) \geq g(\bar{S}, \bar{x}(\bar{S}), \bar{T}, \bar{x}(\bar{T}))$ for all admissible F trajectories $([S, T], x)$ such that $|S - \bar{S}| + |T - \bar{T}| + ||x_e - \bar{x}_e||_{L^\infty} \leq \beta$. Here, x_e denotes the extension of $x : [S, T] \to \mathbb{R}^n$, by constant extrapolation to the left and right.

Write

$$H(t, x, p) := \sup_{v \in F(t,x)} p \cdot v \qquad (11.5.2)$$

and

$$\tilde{C} := \{(t_0, x_0, t_1, x_i) \in C : h(t_0, x_0) \leq 0 \text{ and } h(t_1, x_1) \leq 0\}.$$

Theorem 11.5.1 *Let* $([\bar{S}, \bar{T}], \bar{x})$ *be an* L^∞ *local minimizer for* (FT). *Assume that, for some* $\epsilon > 0$,

(H1) g *is Lipschitz continuous on a neighbourhood of* $(\bar{S}, \bar{x}(\bar{S}), \bar{T}, \bar{x}(\bar{T}))$ *and* C *is a closed set,*

(H2) F *takes values non-empty, compact sets, and there exist* $K_f > 0$, $c_f > 0$ *such that*
$$\begin{cases} F(t', x') \subset F(t'', x'') + k_F(|t' - t''| + |x' - x''|)\mathbb{B}, \\ F(t', x') \subset c_F \mathbb{B}, \end{cases}$$
for all $(t', x'), (t'', x'') \in (t, \bar{x}(t)) + \epsilon \mathbb{B}$, *for all* $t \in [\bar{S}, \bar{T}]$,

(H3) h *is Lipschitz continuous on a neighbourhood of* $\operatorname{Gr} \bar{x}$.

Then there exist absolutely continuous arcs $e \in W^{1,1}([\bar{S}, \bar{T}]; \mathbb{R})$ *and* $p \in W^{1,1}([\bar{S}, \bar{T}]; \mathbb{R}^n)$, $\lambda \geq 0$, *a measure* $\mu \in C^\oplus(\bar{S}, \bar{T})$ *and a Borel measurable function* $\gamma = (\gamma_0, \gamma_1) : [\bar{S}, \bar{T}] \to \mathbb{R}^{1+n}$ *such that the following relations, in which* $(r, q) \in NBV([\bar{S}, \bar{T}]; \mathbb{R}^{1+n})$ *is the function*

$$(-r(t), q(t)) := \begin{cases} (-e(\bar{S}), p(\bar{S})) & \text{if } t = \bar{S} \\ (-e(t), p(t)) + \int_{[\bar{S},t]} \gamma(s) d\mu(s) & \text{if } t \in (\bar{S}, \bar{T}], \end{cases} \qquad (11.5.3)$$

are satisfied:

(i) $(\lambda, p, \mu) \neq (0, 0, 0)$,
(ii) $\gamma(t) \in \partial^> h(t, \bar{x}(t))$ μ- *a.e.* $t \in [\bar{S}, \bar{T}]$,
(iii) $(-\dot{e}(t), \dot{p}(t)) \in \operatorname{co}\{(\zeta, \eta) : ((\zeta, \eta), q(t)) \in N_{\operatorname{Gr} F}((t, \bar{x}(t)), \dot{\bar{x}}(t))\}$,
$$\text{a.e. } t \in [\bar{S}, \bar{T}],$$
(iv) $q(t) \cdot \dot{\bar{x}}(t) = \sup_{v \in F(t, \bar{x}(t))} q(t) \cdot v$, *a.e.* $t \in [\bar{S}, \bar{T}]$,

(v) $(-r(\bar{S}), q(\bar{S}), r(\bar{T}), -q(\bar{T})) \in \lambda \partial g(\bar{S}, \bar{x}(\bar{S}), \bar{T}, \bar{x}(\bar{T}))$
$\qquad + N_{\tilde{C}}(\bar{S}, \bar{x}(\bar{S}), \bar{T}, \bar{x}(\bar{T}))$,
(vi) $r(\bar{S}) = H(\bar{S}, \bar{x}(\bar{S}), q(\bar{S}))$,
(vii) $r(t) = H(t, \bar{x}(t), q(t))$, for all $t \in (\bar{S}, \bar{T})$,
(viii) $r(\bar{T}) = H(\bar{T}, \bar{x}(\bar{T}), q(\bar{T}))$.

Remarks

(a): The extra information provided by this theorem is of course the pair of conditions (vi) and (viii), asserting that condition (vii) extends to the end-times. Notice that, in contrast to the non-degenerate state constrained maximum principle of Chap. 10, condition (vii) is an 'everywhere', not an 'almost everywhere' condition on the open interval $(\bar{S}, \bar{T})$. The 'almost everywhere' condition can be strengthened to the 'everywhere' version, in consequence of the facts that H is a continuous function and q is a bounded variation function with everywhere left and right limits.

(b): Notice that, as with the non-degenerate state-constrained maximum principle, the transversality condition (v) is expressed in terms of the modified end-point constraint set $\tilde{C}$, in place of C.

Corollary 11.5.2 *Let $([\bar{S}, \bar{T}], \bar{x})$ be an L^∞ local minimizer for (FT). Assume hypotheses (H1)–(H3) of Theorem 11.5.1. Assume also that the following constraint qualification is satisfied:*

$$\gamma_0' + \min_{v \in F(\bar{S}, \bar{x}(\bar{S}))} \gamma_1' \cdot v < 0 \text{ and } \gamma_0'' + \max_{v \in F(\bar{T}, \bar{x}(\bar{T}))} \gamma_1'' \cdot v > 0$$

for all $(\gamma_0', \gamma_1') \in \partial^> h(\bar{S}, \bar{x}(\bar{S}))$ and all $(\gamma_0'', \gamma_1'') \in \partial^> h(\bar{T}, \bar{x}(\bar{T}))$.

Then there exist absolutely continuous arcs $e \in W^{1,1}([\bar{S}, \bar{T}]; \mathbb{R})$ and $p \in W^{1,1}([\bar{S}, \bar{T}]; \mathbb{R}^n)$, $\lambda \geq 0$, a measure $\mu \in C^\oplus(\bar{S}, \bar{T})$ and a Borel measurable function $\gamma = (\gamma_0, \gamma_1) : [\bar{S}, \bar{T}] \to \mathbb{R}^{1+n}$ such that conditions (ii)–(viii) of Theorem 11.5.1 are satisfied and also the following strengthened version of the non-triviality condition (i):

(i)': $\lambda + \|q\|_{L^\infty} + \int_{(\bar{S}, \bar{T})} d\mu(s) ds \neq 0$.

Notice that, for problems with a fixed left endpoint discussed above, the degenerate choice of multipliers (11.5.1), in which $q(t) (= p(t) + \int_{[\bar{S},t]} h_x d\mu(s) ds) = 0$ for $t \in (\bar{S}, \bar{T})$, violates the strengthened non-triviality condition (i)' in the corollary.

Proof of Corollary 11.5.2 The proof is analogous to that of Corollary 10.7.2 of Chap. 10.

Proof of Theorem 11.5.1 Let $\epsilon' > 0$ be such that $([\bar{S}, \bar{T}], \bar{x})$ is a minimizer for (FT), with respect to arcs $([S, T], x)$ satisfying

$$|S - \bar{S}| + |T - \bar{T}| + \|x_e - \bar{x}_e\|_{L^\infty} \leq \epsilon'.$$

11.5 Non-degenerate Necessary Conditions

Fix $\rho \in (0, 1/2)$. Consider the 're-parameterized' dynamic optimization problem

$$(R) \begin{cases} \text{Minimize } g((\tau, y)(s_0), (\tau, y)(s_1)) \\ \text{over intervals } [s_0, s_1] \subset \mathbb{R} \text{ and } (\tau, y) \in W^{1,1}([s_0, s_1]; \mathbb{R}^{1+n}) \\ \text{satisfying} \\ (\dot{\tau}(s), \dot{y}(s)) \in \{(w, wv) : w \in [1 - \rho, 1 + \rho], v \in F(\tau(s), y(s))\}, \\ \qquad\qquad\qquad\qquad\qquad\qquad\qquad\qquad\qquad \text{a.e. } s \in [s_0, s_1], \\ \text{subject to} \\ h(\tau(s), y(s)) \leq 0 \text{ for all } s \in [s_0, s_1], \\ ((\tau, y)(s_0), (\tau, y)(s_1)) \in \tilde{C}, \\ |s_0 - \bar{S}| + |s_1 - \bar{T}| + \|y_e - \bar{x}_e\|_{L^\infty} \leq \epsilon'', \end{cases}$$

in which $\epsilon'' \in (0, \epsilon'/2)$. Set $\tilde{F}(\tau, y) := \{(w, wv) : w \in [1-\rho, 1+\rho], v \in F(\tau, y)\}$. As in the proof of Theorem 10.7.1 the last constraint in problem (R) is inactive, with reference to any $\tilde{F}$ trajectory $([s_0, s_1], (\tau, y))$ for (R) such that $|s_0 - \bar{S}| + |s_1 - \bar{T}| + \|y_e - \bar{x}_e\|_{L^\infty} < \epsilon''$ and so the end-times s_0 and s_1 of the independent variable s are locally unconstrained. This fact will play a crucial role in subsequent analysis.

A change of independent variable analysis, similar to that employed in the proof of Theorem 9.6.1 (and Lemma 9.7.1), permits us to conclude:

Lemma 11.5.3 *We can choose $\rho \in (0, 1/2)$ and $\epsilon'' \in (0, \epsilon'/2)$ sufficiently small such that $([\bar{S}, \bar{T}], (\bar{\tau}(s) \equiv s, \bar{y}))$ is an L^∞ local minimizer for problem (R).*

Take $\rho_i \downarrow 0$. For each i define

$$g_i(\tau_0, y_0, \tau_1, y_1) := (g(\tau_0, y_0, \tau_1, y_1) - g(\bar{S}, \bar{x}(\bar{S}), \bar{T}, \bar{x}(\bar{T})) + \rho_i^2) \vee 0.$$

Consider the problem

$$(R_i) \begin{cases} \text{Minimize } J_i(([s_0, s_1], (\tau, y))) := g_i(\tau(s_0), y(s_0), \tau(s_1), y(s_1)) \\ \qquad\qquad\qquad\qquad\qquad \vee \max_{s \in [s_0, s_1]} h(\tau(s), y(s))) \\ \text{subject to} \\ (\dot{\tau}(s), \dot{y}(s)) \in \{(w, wv) : w \in [1 - \rho, 1 + \rho], v \in F(\tau(s), y(s))\}, \\ \qquad\qquad\qquad\qquad\qquad\qquad\qquad\qquad\qquad \text{a.e. } s \in [s_0, s_1], \\ (\tau(s_0), y(s_0), \tau(s_1), y(s_1)) \in \tilde{C}, \\ |s_0 - \bar{S}| + |s_1 - \bar{T}| + \|y_e - \bar{x}_e\|_{L^\infty} \leq \epsilon''. \end{cases}$$

Define

$$\mathcal{M} := \{\tilde{F} \text{ trajectories } ([s_0, s_1], (\tau, y)) \text{ for } (R_i)\}.$$

J_i is a continuous function on the complete metric space $(\mathcal{M}, d_\mathcal{M})$, when we define

$$d_\mathcal{M}(([s_0', s_1'], (\tau', y')), ([s_0, s_1], (\tau, y))) := |\tau'(s_0') - \tau(s_0)| + |y'(s_0') - y(s_0)|$$
$$+ |s_0' - s_0| + |s_1' - s_1| + \|\dot{\tau}_e' - \dot{\tau}_e\|_{L^1} + \|\dot{y}_e' - \dot{y}_e\|_{L^1}.$$

Clearly, $([\bar{S}, \bar{T}], (\bar{\tau}(s) \equiv s, \bar{x}))$ is a ρ_i^2-minimizer for problem (R_i). We can therefore conclude from Ekeland's theorem that there exists $([s_{0i}, s_{1i}], (\tau_i, y_i)) \in \mathcal{M}$, which is a minimizer for

$$\text{Minimize } \{J_i'(([s_0, s_1], (\tau, y))) : ([s_0, s_1], (\tau, y)) \in \mathcal{M}\}. \quad (11.5.4)$$

Here

$$J_i'(([s_0, s_1], (\tau, y))) := J_i(([s_0, s_1], (\tau, y))) +$$
$$\rho_i d_{\mathcal{M}}(([s_0, s_1], (\tau, y)), ([s_{0i}, s_{1i}], (\tau_i, y_i))).$$

Furthermore,

$$d_{\mathcal{M}}(([s_{0i}, s_{1i}], (\tau_i, y_i)), ([\bar{S}, \bar{T}], (\bar{\tau}(s) \equiv s, \bar{x}))) \leq \rho_i \quad (11.5.5)$$

and

$$J_i'(([s_{0i}, s_{1i}], (\tau_i, y_i))) \leq J_i'(([\bar{S}, \bar{T}], (\bar{\tau}(s) \equiv s, \bar{x}))).$$

We can find Lebesgue measurable functions $w_i : [s_{0i}, s_{1i}] \to \mathbb{R}$ and $v_i : [s_{0i}, s_{1i}] \to \mathbb{R}^n$ such that

$$w_i(s) \in [1 - \rho, 1 + \rho], \ v_i(s) \in F(\tau_i(s), y_i(s))$$
$$\text{and } (\dot{\tau}_i(s), \dot{y}_i(s)) = (w_i(s), w_i(s) v_i(s)), \text{ a.e..}$$

Relation (11.5.5) implies that, for a subsequence,

$$|\tau_i(s_{0i}) - \bar{S}|, |y_i(s_{0i}) - \bar{x}(\bar{S})|, |s_{0i} - \bar{S}|, |s_{1i} - \bar{T}| \to 0 \text{ and}$$
$$\|(w_{ie} - \chi_{[\bar{S}, \bar{T}]}(=\dot{\bar{\tau}}_e), v_{ie} - \dot{\bar{x}}_e)\|_{L^1} \to 0 \text{ as } i \to \infty. \quad (11.5.6)$$

(The subscripts for the velocity related variables w_{ie}, v_{ie}, etc., indicate extension by 0, not by constant extrapolation as for the state-related variables. $\chi_{[\bar{S}, \bar{T}]}$ is the characteristic function of the set $[\bar{S}, \bar{T}]$.) Furthermore $(w_{ie}(s) - \chi_{[\bar{S}, \bar{T}]}(s), v_{ie}(s) - \dot{\bar{x}}_e(s)) \to (0, 0)$ a.e.. We can deduce, with the help of Gronwall's lemma that

$$\|(\tau_{ie}, y_{ie}) - (\bar{\tau}_e, \bar{x}_e)\|_{L^\infty} \to 0 \text{ as } i \to \infty.$$

Note also that

$$g_i(\tau_i(s_{0i}), y_i(s_{0i}), \tau_i(s_{1i}), y_i(s_{1i})) \vee \max_{t \in [s_{0i}, s_{1i}]} h(\tau_i(t), y_i(t))) > 0,$$

for all i sufficiently large. \hfill (11.5.7)

11.5 Non-degenerate Necessary Conditions

This reason is that, if '=' replaces '>' in the above relation, then

$$g(\tau_i(s_{0i}), y_i(s_{0i}), \tau_i(s_{1i}), y_i(s_{1i})) \le g(\bar{S}, \bar{x}(\bar{S}), \bar{T}, \bar{x}(\bar{T})) - \rho_i^2,$$

$(\tau_i(s_{0i}), y_i(s_{0i}), \tau_i(s_{1i}), y_i(s_{1i})) \in \tilde{C}$ and $h(\tau_i(t), y_i(t)) \le 0$, for all $t \in [s_{0i}, s_{1i}]$. Furthermore, $\|y_{ie} - \bar{x}_e\|_{L^\infty} \le \epsilon'$. This contradicts the L^∞ local optimality of $([\bar{S}, \bar{T}], \bar{x})$ for (FT).

Observe next that $([s_{0i}, s_{1i}], (\tau_i, y_i))$ is a global minimizer for

$$(R_i') \begin{cases} \text{Minimize } J_i''(([s_0, s_1], (\tau, y))) \\ \text{subject to} \\ (\dot{\tau}(s), \dot{y}(s)) \in \{(w, wv) : w \in [1 - \rho, 1 + \rho], v \in F(\tau(s), y(s))\}, \\ \qquad\qquad \text{a.e. } s \in [s_0, s_1], \\ (\tau(s_0), y(s_0), \tau(s_1), y(s_1)) \in \tilde{C}, \\ |s_0 - \bar{S}| + |s_1 - \bar{T}| + \|y_e - \bar{x}_e\|_{L^\infty} \le \epsilon'', \end{cases}$$

in which

$$J_i''(([s_0, s_1], (\tau, y))) := g_i(\tau(s_0), y(s_0), \tau(s_1), y(s_1)) \vee \max_{s \in [s_0, s_1]} h(\tau(s), y(s))$$
$$+ \rho_i \Big(|\tau(s_0) - \tau_i(s_{0i})| + |y(s_0) - y_i(s_{0i})| + c(|s_0 - s_{0i}| + |s_1 - s_{1i}|) +$$
$$\int_{s_0}^{s_1} |\dot{\tau}_e(s) - \dot{\tau}_{ie}(s)| ds + \int_{s_0}^{s_1} |\dot{y}_e(s) - \dot{y}_{ie}(s)| ds \Big).$$

Here, $c := 1 + \max\{2\rho, 4\rho c_F\}$.

Use of the cost function appearing in this problem formulation, in place of J_i' in problem (11.5.4), is justified by the following relations:

$$\|\dot{y}_e - \dot{y}_{ie}\|_{L^1} \le \int_{s_0}^{s_1} |\dot{y}_e(s) - \dot{y}_{ie}(s)| ds + 4\rho c_F(|s_0 - s_{01}| + |s_1 - s_{i1}|),$$
$$\|\dot{\tau}_e - \dot{\tau}_{ie}\|_{L^1} \le \int_{s_0}^{s_1} |\dot{\tau}_e(s) - \dot{\tau}_{ie}(s)| ds + 2\rho(|s_0 - s_{01}| + |s_1 - s_{i1}|). \tag{11.5.8}$$

There are two cases to consider.

Case 1: $\max_{s \in [s_{0i}, s_{1i}]} h(\tau_i(s), y_i(s)) > 0$ for at most a finite number of index values i,

Case 2: $\max_{s \in [s_{0i}, s_{1i}]} h(\tau_i(s), y_i(s)) > 0$, for an infinite number of index values i.

Consider first Case 1. We can arrange, by excluding initial terms in the sequence, that

$$\max_{s \in [s_{0i}, s_{1i}]} h(\tau_i(s), y_i(s)) \le 0 \text{ for all } i.$$

Taking note of (11.5.7), we see that $([s_{0i}, s_{1i}], (\tau_i, y_i))$ is an L^∞ local minimizer for

$$(\widetilde{R}_i) \begin{cases} \text{Minimize } \widetilde{J}_i(([s_0, s_1], (\tau, y))) \\ \text{subject to } (\dot{\tau}(s), \dot{y}(s)) \in \{(w, wv) : w \in [1-\rho, 1+\rho], v \in F(\tau(s), y(s))\}, \\ \qquad \qquad \text{a.e. } s \in [s_0, s_1], \\ (\tau(s_0), y(s_0), \tau(s_1), y(s_1)) \in \widetilde{C}, \\ |s_0 - \bar{S}| + |s_1 - \bar{T}| + \|y_e - \bar{x}_e\|_{L^\infty} \le \epsilon'', \end{cases}$$

in which

$$\widetilde{J}_i(([s_0, s_1], (\tau, y), (v, w))) := g_i((\tau(s_0), y(s_0), \tau(s_1), y(s_1))$$
$$+ \rho_i \Big(|\tau(s_0) - \tau_i(s_{0i})| + |y(s_0) - y_i(s_{0i})| + c(|s_0 - s_{0i}| + |s_1 - s_{1i}|)$$
$$+ \int_{s_0}^{s_1} |\dot{\tau}(s) - \dot{\tau}_{ie}(s)| ds + \int_{s_0}^{s_1} |\dot{y}(s) - \dot{y}_{ie}(s)| ds \Big).$$

From (11.5.7) and since $h(\tau_i(t), y_i(t)) \le 0$ for $t \in [s_i, t_i]$,

$$g_i(\tau_i(s_{0i}), y_i(s_{0i}), \tau_i(s_{1i}), y_i(s_{1i})) = g(\tau_i(s_{0i}), y_i(s_{0i}), \tau_i(s_{1i}), y_i(s_{1i}))$$
$$- g(\bar{S}, \bar{x}(\bar{S}), \bar{T}, \bar{x}(\bar{T})) + \rho_i^2. \qquad (11.5.9)$$

For i sufficiently large, we may apply the free end-time Euler Lagrange necessary conditions Theorem 9.4.1 of Chap. 9 to this problem. (Since these optimality conditions apply to problems with no running cost it is required, as a first step, to reformulate the above problem as a running cost-free problem by state augmentation.) Taking account of (11.5.9), we thereby conclude that there exist $\lambda_i \ge 0, r_i \in W^{1,1}([s_{0i}, s_{1i}]; \mathbb{R})$ and $q_i \in W^{1,1}([s_{0i}, s_{1i}]; \mathbb{R}^n)$ such that

$$\lambda_i + \|q_i\|_{L^\infty} = 1, \qquad (11.5.10)$$

$$(-\dot{r}_i(s), \dot{q}_i(s)) \in \text{co}\{(\zeta, \eta) : ((\zeta, \eta), q_i(s)) \in N_{\text{Gr } F}((\tau_i(s), y_i(s)), v_i(s))$$
$$+ \{0\} \times (|w_i(s) - 1| |q_i(s)| + \lambda_i \rho_i (1+\rho)) \mathbb{B}\}, \text{ a.e. } s \in [s_{0i}, s_{1i}],$$

$$(v, w) \to w(q_i(s) \cdot v - r_i(s)) - \rho_i \lambda_i (|w - w_i(s)| + |wv - w_i(s)v_i(s)|)$$

is maximized over $\qquad (11.5.11)$

$(v, w) \in F(\tau_i(s), y_i(s)) \times [1-\rho, 1+\rho]$ at $(w_i(s), v_i(s))$,

a.e. $s \in [s_{0i}, s_{1i}], \qquad (11.5.12)$

$$(-r_i(s_{0i}), q_i(s_{0i}), r_i(s_{1i}), -q_i(s_{1i})) \in \lambda_i \partial g(\tau_i(s_{0i}), y_i(s_{0i}), \tau_i(s_{1i}), y_i(s_{1i}))$$
$$+ N_{\widetilde{C}}(\tau_i(s_{0i}), y_i(s_{0i}), \tau_i(s_{1i}), y_i(s_{1i})) + \sqrt{2} \lambda_i \rho_i \mathbb{B} \times \{(0, 0)\} \qquad (11.5.13)$$

and, for $s = s_{0i}$ and s_{1i},

$$\sup_{(v,w) \in F(\tau_i(s), y_i(s)) \times [1-\rho, 1+\rho]} (q_i(s) \cdot v - r_i(s)) w \in \rho_i \lambda_i c_1 [-1, +1]. \qquad (11.5.14)$$

Here, $c_1 = 2((1+\rho)(1+c_F) + c/2)$.

11.5 Non-degenerate Necessary Conditions

Equation (11.5.14) results from the free-time transversality condition for the optimal end-times, when we take account of the fact the end-times s_0 and s_1 in problem $(\widetilde{R}_i)$ are locally unconstrained, with reference to the minimizer $([s_{0i}, s_{1i}], (\tau_i, y_i))$, for i sufficiently large. Notice that the estimation of the sub and super essential values of the maximized Hamiltonian at the end-times takes account of the integral error term $\rho_i \int (|\dot{\tau} - \dot{\tau}_{ie}| + |\dot{y} - \dot{y}_{ie}|) ds$ in the cost.

(11.5.14) implies

$$\sup_{v \in F(\tau_i(s), y_i(s))} q_i(s) \cdot v - r_i(s) \in \rho_i \lambda_i (1-\rho)^{-1} c_1 [-1, +1],$$

for $s = s_{0i}$ and s_{1i}. (11.5.15)

In view of (11.5.6), we can find a Lebesgue set $\mathcal{T} \subset (\bar{S}, \bar{T})$ with the following properties. Following a further extraction of subsequences we have, for each $s \in \mathcal{T}$,

$$1 - \rho < w_i(s) < 1 + \rho \text{ for all } i \text{ suff. large and} \tag{11.5.16}$$

$$(w_i(s) - 1, w_i(s) v_i(s) - \dot{\bar{x}}(s)) \to (0, 0), \text{ as } i \to \infty.$$

By removing a null set of points from $\mathcal{T}$, we can arrange that conditions (11.5.11) and (11.5.12) are satisfied all points $s \in \mathcal{T}, i = 1, 2, \ldots$

Take any $s \in \mathcal{T}$. Considering first minimization w.r.t. the w variable and then the v variable, we deduce from (11.5.16) and (11.5.12) that, for i sufficiently large,

$$q_i(s) \cdot v_i(s) - r_i(s) \in \rho_i \lambda_i (1 + c_F) \mathbb{B} \tag{11.5.17}$$

and

$$q_i(s) \cdot v_i(s) \geq \sup_{v \in F(\tau_i(s), y_i(s))} q_i(s) \cdot v - 2c_F (1-\rho)^{-1} \rho_i \lambda_i. \tag{11.5.18}$$

But then

$$\sup_{v \in F(\tau_i(s), y_i(s))} q_i(s) \cdot v - r_i(s) \in \rho_i \lambda_i c_2 \mathbb{B}. \tag{11.5.19}$$

Here, $c_2 := (1 + c_F) + 2c_F (1-\rho)^{-1}$.

Relations (11.5.10), (11.5.11), (11.5.13), (11.5.15), (11.5.17) and (11.5.19) will be recognized as perturbed versions of the desired necessary conditions, in which $\mu_i = 0$ and therefore $(e_i, p_i) = (r_i, q_i)$. After extracting a further subsequence, we are assured of the existence of $q \in W^{1,1}$, $r \in W^{1,1}$ and $\lambda \in [0, 1]$ such that $(r_{ie}, q_{ie}) \to (r_e, q_e)$ uniformly, $(\dot{r}_{ie}, \dot{q}_{ie}) \to (\dot{r}_e, \dot{q}_e)$ weakly in L^1 and $\lambda_i \to \lambda$. We can now employ a standard convergence analysis to show that the asserted relations of the theorem are correct, with λ, $q(= p)$ and $r(= e)$ as above.

We now turn to Case 2. We can arrange, by selecting a subsequence, that

$$\max_{s \in [s_{0i}, s_{1i}]} h(\tau_i(s), y_i(s)) > 0, \text{ for all } i. \quad (11.5.20)$$

Fix i. For arbitrary $K > 0$, consider the problem

$$(R_i^K) \begin{cases} \text{Minimize } J_i^K(([s_0, s_1], (\tau, y), z)) \\ \text{over } [s_0, s_1] \subset \mathbb{R}, \ (\tau, y) \in W^{1,1} \text{ and } z \in \mathbb{R} \text{ satisfying} \\ (\dot{\tau}(s), \dot{y}(s)) \in \{(w, wv) : w \in [1-\rho, 1+\rho], v \in F(\tau(s), y(s))\}, \\ \qquad\qquad\qquad\qquad\qquad\qquad\qquad \text{a.e. } s \in [s_0, s_1], \\ (\tau(s_0), y(s_0), \tau(s_1), y(s_1)) \in \tilde{C}, \\ |s_0 - \tilde{S}| + |s_1 - \tilde{T}| + \|y_e - \tilde{x}_e\|_{L^\infty} \le \epsilon'', \end{cases}$$

in which

$$J_i^K(([s_0, s_1], (\tau, y), z)) := g_i(\tau(s_0), y(s_0), \tau(s_1), y(s_1)) \vee z$$
$$+ \rho_i \Big(|\tau(s_0) - \tau_i(s_{0i})| + |y_i(s_0) - y(s_{0i})| + c(|s_0 - s_{0i}| + |s_1 - s_{1i}|)$$
$$+ \int_{s_0}^{s_1} |\dot{\tau}(s) - \dot{\tau}_{ie}(s)| ds + \int_{s_0}^{s_1} |\dot{y}(s) - \dot{y}_{ie}(s)| ds \Big)$$
$$+ K \int_{s_0}^{s_1} ((h(\tau(s), y(s)) - z) \vee 0)^2 ds.$$

Notice that, for each K and any $([s_0, s_1], (\tau, y))$, we have

$$J_i^K \big(([s_0, s_1], (\tau, y), z = \max_{s \in [s_0, s_1]} h(\tau(s), y(s)))\big) = J_i''(([s_0, s_1], (\tau, y))).$$

This implies that, for each i, $\inf(R_i^K) \le \inf(R_i')$. An analysis similar to that employed in the proof of Lemma 10.7.4, Chap. 10, establishes that the two infima coincide in the limit as $K \to \infty$. Thus:

Lemma 11.5.4 *For $i = 1, 2, \ldots$,*

$$\lim_{K \to \infty} \inf(R_i^K) = \inf(R_i').$$

In view of (11.5.20), and the continuity of the map $([s_0, s_1], (\tau, y)) \to (\tau, y)$ from $\mathcal{M}$ to L^∞, we can find a sequence $\rho_i' \downarrow 0$ such that, for each i, $\rho_i' \le 1/2$ and

$$\left. \begin{array}{l} ([s_0, s_1], (\tau, y)) \in \mathcal{M}, z \in \mathbb{R} \\ d_{\mathcal{M}}(([s_0, s_1], (\tau, y)), ([s_{i0}, s_{i1}], (\tau_i, y_i))) \le \rho_i' \\ |z - \max_{s \in [s_{0i}, s_{1i}]} h(\tau_i(s), y_i(s))| \le \rho_i' \end{array} \right\} \quad (11.5.21)$$
$$\implies g_i(\tau(s_0), y(s_0), \tau(s_1), y(s_1)) \vee z > 0 \text{ and } z > 0.$$

11.5 Non-degenerate Necessary Conditions

According to the lemma, we can choose $K_i \uparrow \infty$ such that, for each i, $([s_{i0}, s_{i1}], (\tau_i, y_i), z_i := \max_{s \in [s_{0i}, s_{1i}]} h(\tau_i(s), y_i(s)))$ is a $\rho_i'^2$-minimizer for $(P_i^{K_i})$. Ekeland's theorem, together with relations (11.5.8), then tell us that there exists $([s_{0i}', s_{1i}'], (\tau_i', y_i'), z_i') \in \mathcal{M} \times \mathbb{R}$ that is an L^∞ minimizer for:

$$(Q_i) \begin{cases} \text{Minimize } g_i(\tau(s_0), y(s_0), \tau(s_1), y(s_1)) \vee z(s_1) + \rho_i' |z(s_0) - z_i'| \\ \quad + K_i \int_{s_0}^{s_1} ((h(\tau(s), y(s)) - z(s)) \vee 0)^2 ds \\ \quad + \rho_i \Big(|\tau(s_0) - \tau_i(s_{0i})| + |y(s_0) - y_i(s_{0i})| + c(|s_0 - s_{0i}| + |s_1 - s_{1i}|) \\ \qquad + \int_{s_0}^{s_1} |\dot{\tau}(s) - \dot{\tau}_{ie}(s)| ds + \int_{s_0}^{s_1} |\dot{y}(s) - \dot{y}_{ie}(s)| ds \Big) \\ \quad + \rho_i' \Big(|\tau(s_0) - \tau_i'(s_{0i}')| + |y(s_0) - y_i'(s_{0i}')| + c(|s_0 - s_{0i}'| + |s_1 - s_{1i}'|) \\ \qquad + \int_{s_0}^{s_1} |\dot{\tau}(s) - \dot{\tau}_{ie}'(s)| ds + \int_{s_0}^{s_1} |\dot{y}(s) - \dot{y}_{ie}'(s)| ds \Big) \\ \text{over } [s_0, s_1] \subset \mathbb{R}, (\tau, y, z) \in W^{1,1}([s_0, s_1]; \mathbb{R}^{1+n+1}) \text{ satisfying} \\ (\dot{\tau}(s), \dot{y}(s), \dot{z}(s)) \in \{(w, wv, 0) : w \in [1-\rho, 1+\rho], v \in F(\tau(s), y(s))\}, \\ \hspace{8cm} \text{a.e. } s \in [s_0, s_1], \\ (\tau(s_0), y(s_0), \tau(s_1), y(s_1)) \in \tilde{C}, \\ |s_0 - \bar{S}| + |s_1 - \bar{T}| + \|y_e - \bar{x}_e\|_{L^\infty} \leq \epsilon''. \end{cases}$$

Furthermore

$$d_\mathcal{M}(([s_{0i}', s_{1i}'], (\tau_i', y_i')), ([s_{0i}, s_{1i}], (\tau_i, y_i)))$$
$$+ |z_i' - \max_{s \in [s_{0i}, s_{1i}]} h(\tau_i(s), y_i(s))| \leq \rho_i'. \quad (11.5.22)$$

We can find Lebesgue measurable functions $w_i' : [s_{0i}', s_{1i}'] \to \mathbb{R}$ and $v_i' : [s_{0i}', s_{1i}'] \to \mathbb{R}^n$ such that

$$w_i'(s) \in [1-\rho, 1+\rho], \; v_i'(s) \in F(\tau_i'(s), y_i'(s))$$
$$\text{and } (\dot{\tau}_i'(s), \dot{y}_i'(s)) = (w_i'(s), w_i'(s)v_i'(s)), \text{ a.e.}.$$

From (11.5.22) it follows that

$$|\tau_i'(s_{0i}') - \tau_i(s_{0i})|, |y_i'(s_{0i}') - y_i(s_{0i})|, |s_{0i}' - s_{0i}|, |s_{1i}' - s_{1i}| \to 0 \text{ and}$$
$$\|(w_{ie}' - w_{ie}, v_{ie}' - v_{ie})\|_{L^1} \to 0 \text{ as } i \to \infty. \quad (11.5.23)$$

We can deduce, with the help of Gronwall's lemma, that

$$\|(\tau_{ie}', y_{ie}') - (\tau_{ie}, y_{ie})\|_{L^\infty} \to 0. \quad (11.5.24)$$

Relations (11.5.5) and (11.5.23) imply that

$$|\tau_i'(s_{0i}') - \bar{S}|, |y_i'(s_{0i}') - \bar{x}(\bar{S})|, |s_{0i}' - \bar{S}|, |s_{1i}' - \bar{T}| \to 0 \text{ and}$$
$$\|(w_{ie}' - \chi_{[\bar{S}, \bar{T}]}(= \dot{\bar{\tau}}_e), v_{ie}' - \dot{\bar{x}}_e)\|_{L^1} \to 0 \text{ as } i \to \infty.$$

Furthermore, for a subsequence, $(w'_{ie}(s) - \chi_{[\bar{s},\bar{T}]}(s), v'_{ie}(s) - \dot{\bar{x}}_e(s)) \to (0,0)$ a.e..
We deduce from (11.5.21) and (11.5.22) that

$$g_i(\tau'_i(s'_{0i}), y'_i(s'_{0i}), \tau'_i(s'_{1i}), y'_i(s'_{1i})) \vee z'_i > 0 \text{ and } z'_i > 0, \text{ for each } i. \qquad (11.5.25)$$

Ekeland's theorem also tells us that

$$K_i \int_{s'_{0i}}^{s'_{1i}} ((h(\tau'_i(s), y'_i(s)) - z'_i) \vee 0)^2 ds \ (+ \text{ non-negative terms})$$

$$\leq g_i(\tau_i(s_{0i}), y_i(s_{0i}), \tau_i(s_{1i}), y_i(s_{1i})) \vee \max_{s \in [s_{0i}, s_{1i}]} h(\tau_i(s), y_i(s))$$

$$+\rho'_i | \max_{s \in [s_{0i}, s_{1i}]} h(\tau_i(s), y_i(s)) - z'_i|$$

$$+\rho'_i \Big(|\tau_i(s_{0i}) - \tau'_i(s'_{0i})| + |y_i(s_{0i}) - y'_i(s'_{0i})| + c(|s_{0i} - s'_{0i}| + |s_{1i} - s'_{1i}|)$$

$$+ \int_{s_{0i}}^{s_{1i}} |\dot{\tau}_i(s) - \dot{\tau}'_{ie}(s)|ds + \int_{s_{0i}}^{s_{1i}} |\dot{y}_i(s) - \dot{y}'_{ie}(s)|ds \Big).$$

The right side of this relation has value 0, in the limit as $i \to \infty$, by (11.5.23) and (11.5.24). It follows that

$$K_i \int_{s'_{0i}}^{s'_{1i}} ((h(\tau'_i(s), y'_i(s)) - z'_i) \vee 0)^2 ds \to 0, \text{ as } \to \infty.$$

Now, for each i sufficiently large, apply the free end-time Euler Lagrange condition Theorem 9.4.1 (Chap. 9) to the problem (Q_i), with reference to the L^∞ minimizer $(([s'_{0i}, s'_{1i}], (\tau'_i, y'_i)), z'_i)$, noting that (following state augmentation and modification of the differential inclusion constraint to accommodate the integral cost terms) the relevant hypotheses are satisfied. We are assured then of the existence of $\tilde{\lambda}_i \in [0, 1]$ and also of elements $-d_i \in W^{1,1}$, $-r_i \in W^{1,1}$ and $q_i \in W^{1,1}$ (interpreted as costate components associated with the state variable components z and τ and y variables respectively) and a Borel measurable function $\gamma_i = (\gamma_{0i}, \gamma_{1i}) : [s'_{0i}, s'_{1i}] \to \mathbb{R}^{1+n}$ such that

$$(\tilde{\lambda}_i, q_i) \neq (0, 0), \qquad (11.5.26)$$

$$(-\dot{r}_i(s), \dot{q}_i(s)) \in \text{co}\{(\zeta, \eta) : ((\zeta, \eta), q_i(s))$$

$$\in N_{\text{Gr} F}((\tau'_i(s), y'_i(s)), v'_i(s)) \qquad (11.5.27)$$

$$+\{0\} \times (|w'_i(s) - 1| |q_i(s)| + \tilde{\lambda}_i(\rho_i + \rho'_i)(1 + \rho))\mathbb{B}\}$$

$$+\gamma_i(s)(d\mu_i/ds)(s), \qquad \text{a.e. } s \in [s'_{0i}, s'_{1i}]$$

$$\gamma_i(s) \in \partial h(\tau'_i(s), y'_i(s)), \ \mu_i - \text{ a.e. } s \in [s'_{0i}, s'_{1i}],$$

11.5 Non-degenerate Necessary Conditions

$$\dot{d}_i(s) = (d\mu_i/ds)(s), \text{ a.e. and } d_i(s'_{0i}) \in \rho'_i \tilde{\lambda}_i \mathbb{B}, \tag{11.5.28}$$

$$(v, w) \to w(q_i(s) \cdot v - r_i(s)) - \rho_i \tilde{\lambda}_i (|w - w_i(s)|$$
$$+ |wv - w_i(s)v_i(s)|) - \rho'_i \tilde{\lambda}_i (|w - w'_i(s)| + |wv - w'_i(s)v'_i(s)|)$$

is maximized over $F(\tau'_i(s), y'_i(s)) \times [1 - \rho, 1 + \rho]$

$$\text{at } (v'_i(s), w'_i(s)), \qquad \text{a.e. } s \in [s'_{0i}, s'_{1i}], \tag{11.5.29}$$

$$((-r_i(s_{0i}), q_i(s_{0i}), r_i(s_{1i}), -q_i(s_{1i})), d_i(s_{1i})) \in$$

$$\tilde{\lambda}_i \partial \Big(g_i(\tau'_i(s'_{0i}), y'_i(s'_{0i}), \tau'_i(s'_{1i}), y_i(s_{1i})) \vee z'_i(s'_{1i}) \Big) + (\xi_i, 0)$$

$$+ \sqrt{2}\tilde{\lambda}_i(\rho_i + \rho'_i) \mathbb{B} \times \{(0, 0)\},$$

for some $\xi_i \in N_{\tilde{C}}(\tau'_i(s'_{0i}), y'_i(s'_{0i}), \tau'_i(s'_{1i}), y'_i(s'_{1i}))$, \qquad (11.5.30)

$$\sup_{(v,w) \in F(\tau'_i(s), y'_i(s)) \times [1-\rho, 1+\rho]} (q_i(s) \cdot v - r_i(s))w$$

$$- \tilde{\lambda}_i K_i((h(\tau'_i(s), y'_i(s)) - z'_i(s)) \vee 0)^2 \in (\rho_i + \rho'_i)\tilde{\lambda}_i c_1 [-1, +1],$$

$$\text{for } s = s'_{0i} \text{ and } s'_{1i}. \tag{11.5.31}$$

Here, $c_1 := 2((1+\rho)(1+c_F) + c/2)$ (as in Case 1) and μ_i is the (positive) measure on the Borel sets of the real line, with support in $[s'_{0i}, s'_{1i}]$ and distribution:

$$\mu_i([s'_{0i}, s]) = 2K_i \tilde{\lambda}_i \int_{[s'_{0i}, s]} (h(\tau'_i(\sigma), y'_i(\sigma)) - z'_i) \vee 0) d\sigma. \tag{11.5.32}$$

We see that

$$\text{supp}\{\mu_i\} \subset \{s \in [s'_{0i}, s'_{1i}] : h(\tau'_i(s), y'_i(s))) \geq z'_i\}. \tag{11.5.33}$$

Relation (11.5.31) results from the free-time transversality condition for the optimal end-times s'_{0i} and s'_{1i} in problem (Q_i). But $K_i((h(\tau'_i(s), y'_i(s)) - z'_i(s)) \vee 0)^2 = 0$ for $s = s'_{0i}, s'_{1i}$, since $z'_i(s) > 0$ and $h(\tau'_i(s), y'_i(s)) \leq 0$ for $s = s_{0i}$ and s_{1i}. (We use here the fact that, for elements $(\tau_0, y_0, \tau_1, y_1) \in \tilde{C}$, we know $h(\tau_0, y_0) \leq 0$ and $h(\tau_1, y_1) \leq 0$.) (11.5.31) therefore implies that, for $s = s'_{0i}$ and s'_{0i},

$$\sup_{v \in F(\tau'_i(s), y'_i(s))} q_i(s) \cdot v - r_i(s) \in (\rho_i + \rho'_i)\tilde{\lambda}_i \left(\frac{c_1}{1-\rho}\right) [-1, +1]. \tag{11.5.34}$$

We deduce from (11.5.21) and (11.5.30), with the help of the max and sum rules, that there exists $\nu_i \in [0, 1]$ such that

$$(-r_i(s'_{0i}), q_i(s'_{0i}), r_i(s'_{1i}), -q_i(s'_{1i})) \in$$
$$\tilde{\lambda}_i v_i \partial g(\tau'_i(s'_{0i}), y'_i(s'_{0i}), \tau'_i(s'_{1i}), y'_i(s'_{1i})) + \xi_i + \sqrt{2}\tilde{\lambda}_i(\rho_i + \rho'_i)\mathbb{B}$$
$$\text{for some } \xi_i \in N_{\tilde{C}}(\tau'_i(s'_{0i}), y'_i(s'_{0i}), \tau'_i(s'_{1i}), y'_i(s'_{1i})) \tag{11.5.35}$$

and

$$\|\mu_i\|_{TV} = \int_{[s'_{0i},s'_{1i}]} (d\mu/ds)ds \in \tilde{\lambda}_i(1 - v_i) + \rho'_i \tilde{\lambda}_i \mathbb{B}. \tag{11.5.36}$$

We can find a Lebesgue set $\mathcal{S} \subset (\bar{S}, \bar{T})$ of full measure, with the following properties. Following a further extraction of subsequences we have: for each $s \in \mathcal{S}$, (11.5.27) and (11.5.29) are valid,

$$1 - \rho < w_i(s) < 1 + \rho \text{ for all } i \text{ suff. large and} \tag{11.5.37}$$

$$(w'_{ie}(s), v'_{ie}(s)) \to (1, \dot{\bar{x}}(s)), \text{ as } i \to \infty.$$

Take any $s \in \mathcal{S}$. Considering first minimization w.r.t. the w variable and then the v variable, we deduce from (11.5.29) that, for i sufficiently large,

$$q_i(s) \cdot v'_i(s) - r_i(s) \in (\rho_i + \rho'_i)\tilde{\lambda}_i(1 + c_F)\mathbb{B}$$

and

$$q_i(s) \cdot v'_i(s) \geq \sup_{v \in F(\tau'_i(s), \tau'_i(s))} q_i(s) \cdot v - 2(1 - \rho)^{-1}(\rho_i + \rho'_i)\tilde{\lambda}_i(\rho + (1 + \rho)c_F).$$
$$\tag{11.5.38}$$

But then

$$H(\tau'_i(s), y'_i(s), r_i(s), q_i(s))\Big(= \sup_{v \in F(\tau'_i(s), \tau'_i(s))} q_i(s) \cdot v - r_i(s)\Big) \in c_3(\rho_i + \rho'_i)\tilde{\lambda}_i \mathbb{B},$$
$$\tag{11.5.39}$$

in which $c_3 := (1 + c_F) + 2(1 - \rho)^{-1}(\rho + (1 + \rho)c_F)$. Now define the function $(e_i, p_i) \in W^{1,1}([s'_{0i}, s'_{1i}]; \mathbb{R}^{1+n})$ according to

$$(-e_i, p_i)(s) := (-r_i, q_i)(s) - \int_{[s'_{0i},s]} \gamma_i(\sigma)(d\mu_i(\sigma)/d\sigma)(\sigma)d\sigma \text{ for } s \in [s'_{0i}, s'_{1i}].$$
$$\tag{11.5.40}$$

Condition (11.5.27) can be expressed in terms of (e_i, p_i):

$$(-\dot{e}_i(s), \dot{p}_i(s)) \in \text{co}\{(\zeta, \eta) : ((\zeta, \eta), q_i(s)) \in N_{\text{Gr}\,F}((\tau'_i(s), y'_i(s)), v'_i(s))$$
$$+ \{0\} \times (|w'_i(s) - 1| |q_i(s)| + \tilde{\lambda}_i(\rho_i + \rho'_i)(1 + \rho))\mathbb{B}\},$$
$$\text{a.e. } s \in [s'_{0i}, s'_{1i}]. \tag{11.5.41}$$

11.5 Non-degenerate Necessary Conditions

Notice next the $(\tilde{\lambda}_i, \xi_i) \neq (0, 0)$. This is because, otherwise, $q_i = 0$, in consequence of (11.5.28), (11.5.30) and Gronwall's lemma: this contradicts (11.5.26). Now set $\lambda_i := \tilde{\lambda}_i v_i$. We see from (11.5.36) that $||\mu_i||_{TV} + \lambda_i \geq (1 - \rho'_i)\tilde{\lambda}_i$. It follows that $(\lambda_i, \xi_i, \mu_i) \neq (0, 0, 0)$.

Normalize q_i, p_i, ξ_i, λ_i and $\tilde{\lambda}_i$, dividing each by the positive number $||\mu_i||_{TV} + \lambda_i + |\xi_i|$. (We do not re-label.) Then, automatically,

$$||\mu_i||_{TV} + \lambda_i + |\xi_i| = 1. \tag{11.5.42}$$

Observe also that, following relabelling, $\tilde{\lambda}_i \leq (1 - \rho'_i)^{-1} \leq 2$, since $\rho'_i \leq 1/2$ for each i.

Replacing $\tilde{\lambda}_i v_i$ by λ_i in the transversality condition (11.5.35), we recognize relations (11.5.34), (11.5.35), (11.5.39), (11.5.40), (11.5.41) and (11.5.42) as perturbed versions of the desired conditions. The relations imply that the sequences with elements μ_i, (e_i, p_i), γ_i, $(\dot{e}, \dot{p}_i)$, λ_i and ξ_i, are uniformly bounded (w.r.t. the total variation norm, the uniform norm, the L^1 norm and Euclidean norms, respectively). Observe that $[s'_{0i}, s'_{1i}] \subset [\bar{S} - \epsilon'', \bar{T} + \epsilon'']$ for all i. Therefore, we can consider the extension of elements (e_i, p_i) and $(\dot{e}_i, \dot{p}_i)$ on the compact interval $[\bar{S} - \epsilon'', \bar{T} + \epsilon'']$ according to the preceding convention (by 'constant extrapolation' for the state arcs (e_i, p_i)'s and by 'zero extrapolation' for their velocities (e_i, p_i)'s). For each i (sufficiently large) we shall write μ_{ie} the Borel measure on $[\bar{S} - \epsilon'', \bar{T} + \epsilon'']$ defined by

$$\mu_{ie}(E) := \mu_i(E \cap [\bar{S} - \epsilon'', \bar{T} + \epsilon'']),$$

for all Borel subsets $E \subset [\bar{S} - \epsilon'', \bar{T} + \epsilon'']$. The Borel measurable functions γ_i's are extended by 'constant extrapolation' on $[\bar{S} - \epsilon'', \bar{T} + \epsilon'']$. Consequently their extension satisfy the property

$$\gamma_{ie}(s) \in \partial h(\tau'_{ie}(s), y'_{ie}(s)), \quad \mu_{ie} - \text{a.e. } s \in [\bar{S} - \epsilon'', \bar{T} + \epsilon''].$$

It follows that, after subsequences have been extracted, there exist limit points μ, e, $\dot{e}$, p, $\dot{p}$, γ, λ and ξ with respect to the appropriate topologies. But then there exists a function of bounded variation (r, q) such that $(r_{ie}, q_{ie})(s) \to (r_e, p_e)(s)$ on a subset of $[\bar{S}, \bar{T}]$ of full Lebesgue measure, including the endpoints $\bar{S}$ and $\bar{T}$. (The subscript 'e' here means 'extension by constant extrapolation'.) We can also arrange that $\xi_i \to \xi$, for some $\xi \in \tilde{N}_C(\bar{S}, \bar{x}(\bar{S}), \bar{T}, \bar{x}(\bar{T}))$. We deduce finally, with the help of a standard convergence analysis, the validity of all the assertions in the theorem statement, by passing to the limit as $i \to \infty$. Notice, in particular that $||\mu||_{TV} + \lambda + |\xi| = 1$. This implies the non-triviality condition $(\lambda, p, \mu) \neq (0, 0, 0)$ since '$(\lambda, p, \mu) = (0, 0, 0)$' implies $\xi \neq 0$ and hence $p \neq 0$, which is not possible. $\square$

11.6 Exercises

11.1 Theorem 8.8.1 provides necessary conditions in the form of Clarke's Hamiltonian inclusion for problems with dynamic constraints formulated as a differential inclusion. Consider a generalization of the optimization problem considered in Sect. 8.8, now to include the pathwise state constraint $h(t, x(t)) \leq 0$, for all $t \in [S, T]$:

$$(P11) \begin{cases} \text{Minimize } g(x(S), x(T)) \\ \text{over } x \in W^{1,1}([S, T]; \mathbb{R}^n) \text{ such that} \\ \dot{x}(t) \in F(t, x(t)), \text{ for a.e. } t \in [S, T] \\ h(t, x(t)) \leq 0, \text{ for all } t \in [S, T] \\ (x(S), x(T)) \in C. \end{cases}$$

Let $\bar{x}$ be an L^∞ local minimizer for $(P11)$. Assume that, for some $\epsilon > 0$ conditions (G1), (G2) and (G3)′ of Theorem 8.8.1 are satisfied. Suppose also that $h(., x)$ is upper semi-continuous for each $x \in \mathbb{R}^n$ and there exists $k_h > 0$ such that

$$|h(t, x) - h(t, x')| \leq k_h |x - x'| \text{ for all } x, x' \in \bar{x}(t) + \bar{\epsilon}\mathbb{B}, \ t \in [S, T].$$

Then, prove that there exist $p \in W^{1,1}([S, T]; \mathbb{R}^n)$, $\mu \in C^{\oplus}(S, T)$, a bounded Borel measurable function $\gamma : [S, T] \to \mathbb{R}^n$ and $\lambda \geq 0$, satisfying the following conditions, in which $q \in NBV([S, T]; \mathbb{R}^n)$ is the function

$$q(t) := \begin{cases} p(S) & \text{if } t = S \\ p(t) + \int_{[S,t]} \gamma(s) d\mu(s) & \text{if } t \in (S, T]. \end{cases}$$

(i): $(\lambda, p, \mu) \neq (0, 0, 0)$,

(ii): $(-\dot{p}(t), \dot{\bar{x}}(t)) \in \text{co } \partial_{x,p} H(t, \bar{x}(t), q(t))$, a.e. $t \in [S, T]$,

(iii): $(q(S), -q(T)) \in \lambda \partial g(\bar{x}(S), \bar{x}(T)) + N_C(\bar{x}(S), \bar{x}(T))$,

(iv): $q(t) \cdot \dot{\bar{x}}(t) \geq q(t) \cdot v$ for all $v \in F(t, \bar{x}(t))$ a.e. $t \in [S, T]$,

(v): $\text{supp}\{\mu\} \subset \{t \in [S, T] : h(t, \bar{x}(t)) = 0\}$ and $\gamma(t) \in \partial_x^> h(t, \bar{x}(t))$ for μ-a.e. $t \in [S, T]$.

Hint: Pattern the earlier 'state constraint-free' analysis, employed in the proof of Theorem 8.8.1 and based on Stegall's theorem, but now use the necessary conditions of Theorem 11.2.1 (and the remarks after Theorem 11.2.1), which allow for state constraints.

11.7 Notes for Chapter 11

Clarke [65] first proved the Hamiltonian inclusion for dynamic optimization problems with pathwise state constraints and dynamics modelled by a differential inclusion with convex valued right side ('convex velocity sets'). The validity of the generalized Euler Lagrange inclusion and the generalized Hamiltonian inclusion, also for convex velocity sets, was established by Loewen and Rockafellar [143].

The generalized Euler Lagrange inclusion, for state constrained problems with unbounded, possibly nonconvex, velocity sets satisfying a one sided Lipschitz continuity condition, was first derived by Vinter and Zheng [200]. An improved version of this theorem, proved under unrestrictive epi-Lipschitz continuity and tempered growth hypotheses on $F(t, x)$ (as in Clarke's stratified conditions for state-constraint free problems [67]), is due to Bettiol et al. [34]. Recently, Ioffe [129] achieved a further improvement to the optimality condition, by incorporating his refinement.

This chapter provides the latest version of the Euler Lagrange inclusion, including the Ioffe refinement and proved under unrestrictive 'stratified' hypotheses. The proof technique is similar to that used in Chap. 8, to derive necessary conditions for state constraint-free problems. That is, necessary conditions for the original problem are obtained as the limit of necessary conditions for a 'finite Lagrangian' problem, constructed from the scaled distance function. Necessary conditions for the finite Lagrangian problem, in which a state constraint is present, are derived, now with the help of the smooth state constrained maximum principle.

Free end-time necessary conditions for state constrained differential inclusion problems, with data measurable in time, were first derived by Clarke, Loewen and Vinter [83]. Refinements, to allow for state constraints which are active at the optimal end-times, were investigated in [179].

The non-degeneracy phenomenon was earlier investigated in Chap. 10, in the context of state-constrained dynamic optimization problems involving a controlled differential equation. The phenomenon is encountered also for problems involving a differential inclusion. Non-degenerate forms of the Hamiltonian inclusion for such problems were derived by Aseev [6] and Vinter [194] under the assumption that the velocity set is convex. The necessary conditions in this chapter (Theorem 11.5.1) are new [31]; they improve on earlier conditions because they allow the velocity set to be non-convex and they incorporate the (partially convexified) Euler Lagrange inclusion.

Chapter 12
Regularity of Minimizers

Abstract This chapter provides conditions under which minimizers for dynamic optimization problems possess regularity properties of interest, such as Lipschitz continuity or boundedness of higher derivatives. We discuss the significance of regularity analysis, regarding the derivation of necessary conditions of optimality, the selection of numerical schemes for the computation of optimal strategies and the implementation of such strategies.

Tonelli identified conditions (Tonelli existence hypotheses), under which the basic problem in the calculus of variations has a minimizer in the class of absolutely continuous functions that satisfy specified end-point conditions. Tonelli's existence hypotheses do not guarantee, however, that minimizers satisfy the Euler Lagrange condition; this is a set-back to solution techniques based on seeking a minimizer among arcs satisfying standard necessary conditions of optimality. For problems with smooth Lagrangians and in which the arcs are scalar valued, Tonelli showed however that, under his existence hypotheses, the velocity of the minimizing arcs is locally essentially bounded at all times in an open set of full Lebesgue measure. This regularity property can be exploited, to show that minimizers satisfy the Euler Lagrange condition under additional hypotheses. In sections of this chapter dealing with Tonelli regularity, we cover subsequent extensions of Tonelli's original discoveries: the velocities of minimizers are essentially locally bounded on a relatively open subset of full measure of the reference time interval, under Tonelli's existence hypotheses, even when we allow vector valued arcs and nonsmooth Lagrangians. We show that this extended Tonelli regularity property leads to improved criteria for validity of the Euler Lagrange condition. The proof of these extensions, which is based on the construction of auxiliary Lagrangians, is a showcase for application of nonsmooth methods, which introduce a new flexibility into the use of Tonelli's methods, by allowing us to construct nonsmooth auxiliary Lagrangians. These aspects of Tonelli regularity theory relate to calculus of variations problems. We briefly consider extensions of the theory gives conditions for the essential boundedness of the minimizing control, for dynamic optimization problems associated with a linear control system.

The chapter also includes material on another approach to establishing regularity of minimizers, in circumstances when standard necessary conditions are not valid and therefore cannot be (directly) used in the regularity analysis. This is based on reparameterization of the minimizer by a change of independent variable. Applications of this approach, when first introduced, were restricted to autonomous problems satisfying the superlinear growth conditions in the Tonelli existence hypotheses. We take account of recent extensions of reparameterization methods, which can be used to establish regularity properties of minimizers, specifically Lipchitz continuity, in some situations where the problem concerned is non-autonomous and the superlinear growth hypothesis is replaced by a less restrictive, slow growth hypothesis.

In the final section of this chapter, we restrict attention to certain dynamic optimization problems with dynamic constraint a controlled differential equation, where minimizers are known to satisfy the maximum principle. Here we show how these necessary conditions can be applied directly to establish Lipschitz continuity of state trajectories.

12.1 Introduction

In this chapter we seek information about regularity of minimizers. When do minimizing arcs have essentially bounded derivatives, higher order derivatives or other qualitative properties of interest in applications?

Minimizer regularity has a number of important implications. In the computation of minimizers for example, the choice of efficient discretization schemes and numerical procedures depends on minimizer differentiability. Again, in control engineering applications, where an dynamic optimization strategy is implemented digitally, the quality of the approximation of an ideal 'continuous time' strategy depends on minimizer regularity. Minimizer regularity can also help us to predict natural phenomena, governed by variational principles. In particular, studying the regularity of minimizers for variational problems arising in nonlinear elasticity, gives insights into mechanisms for material failure.

We begin however by focusing on the fundamental role of regularity theory, discussed in Chap. 1, to provide a solid foundation for Tonelli's direct method. Recall that this involves the following steps:

Step 1: Establish that the problem has a minimizer.

Step 2: Derive necessary conditions for an arc to be a minimizer.

Step 3: Search among 'extremals' (arcs satisfying the necessary conditions) for an arc with minimum cost; this will be a minimizer.

The direct method fails when it is applied to any class of dynamic optimization problems for which minimizers exist, but when necessary conditions of optimality are available only under hypotheses which the minimizers may possibly violate. An example of this phenomenon is given in Chap. 1. In this example, finding an arc which satisfies the necessary conditions yields a unique extremal. However this extremal is not a minimizer, because the minimizers are not located among the extremals.

12.1 Introduction

Regularity analysis enters the picture because it helps us to identify classes of problems, for which all minimizers satisfy known necessary conditions of optimality. It thereby justifies searching for minimizers among extremals.

At the same time, we anticipate difficulties in investigating minimizer regularity, with a view to justifying the direct method. An obvious approach is to draw conclusions about minimizer regularity from the necessary conditions. Yet, to justify the direct method, we need to establish regularity of precisely those minimizers for which satisfaction of the necessary conditions is not guaranteed a priori.

Much of this chapter is devoted to deriving regularity properties of solutions to the basic problem in the calculus of variations, concerning the minimization of an integral functional (with finite Lagrangian) over absolutely continuous vector valued arcs with fixed end-points:

$$(BP) \begin{cases} \text{Minimize } \int_S^T L(t, x(t), \dot{x}(t)) dt \\ \text{over } x \in W^{1,1}([S, T]; \mathbb{R}^n) \text{ satisfying} \\ (x(S), x(T)) = (x_0, x_1). \end{cases}$$

Here, $[S, T]$ is a fixed interval with $T > S$, $L : [S, T] \times \mathbb{R}^n \times \mathbb{R}^n \to \mathbb{R}$ is a given function and x_0, x_1 are given n-vectors.

No apology is made for limiting attention to fixed end-point problems. If an arc $\bar{x}$ is a minimizer for any other kind of end-point condition, then $\bar{x}$ is a minimizer also for the related fixed end-point problem with $x_0 := \bar{x}(S)$, $x_1 := \bar{x}(T)$. So regularity properties for fixed end-point problems imply the same regularity for all other kinds of boundary conditions.

On the other hand, the fact that the formulation (BP) covers only dynamic optimization problems for which the velocity variable in unconstrained (traditional variational problems) is a genuine shortcoming. We remedy it, to some extent, by studying some generalizations in the closing sections of the chapter, to allow for dynamic constraints.

Our starting point is the following well-known set of conditions (HE1)–(HE3) for the existence of minimizers to (BP). Because of their close affinity to conditions considered by Tonelli in his pioneering work on existence of minimizers, we refer to them as the *Tonelli existence hypotheses*.

Theorem 12.1.1 (Tonelli Existence Theorem) *Assume that the data for (BP) satisfy the following hypotheses*

(HE1) (Convexity, etc.) L *is bounded on bounded sets, $L(., x, v)$ is measurable for each (x, v) and $L(t, x, .)$ is convex for each (t, x),*

(HE2) (Uniform Local Lipschitz Continuity) *For each $N > 0$, there exist $k_N > 0$ such that*

$$|L(t, x, u) - L(t, x', u')| \le k_N |(x, v) - (x', v')|$$

for all $(x, v), (x', v') \in N\mathbb{B}$, a.e. $t \in [S, T]$,

(HE3) (Coercivity) *There exist $\alpha \geq 0$ and a convex function $\theta : [0, \infty) \to [0, \infty)$ satisfying $\lim_{r \to \infty} \theta(r)/r = +\infty$ such that*

$$L(t, x, v) \geq \theta(|v|) - \alpha|x| \quad \text{for all } (x, v) \in \mathbb{R}^n \times \mathbb{R}^n, \text{ a.e. } t \in [S, T].$$

Then (BP) has a minimizer.

This is a special case of the existence theorem for the generalized Bolza problem proved in Chap. 2. (See Theorem 2.4.1.)

Chapter 8 provides the following necessary conditions of optimality for (BP):

Proposition 12.1.2 *Let $\bar{x}$ be an L^∞ local minimizer for (BP). Assume that, in addition to (HE1)–(HE2), there exists $k \in L^1$ and $\epsilon > 0$ such that*

$$|L(t, x, v) - L(t, x', u')| \leq k(t)|(x, v) - (x', v')| \tag{12.1.1}$$

for all $(x, v), (x', v') \in (\bar{x}(t), \dot{\bar{x}}(t)) + \epsilon \mathbb{B}$, a.e..

Then, there exists an arc $p \in W^{1,1}([S, T]; \mathbb{R}^n)$ such that

$$\dot{p}(t) \in \text{co}\{q : (q, p(t)) \in \partial L(t, \bar{x}(t), \dot{\bar{x}}(t))\} \quad \text{a.e.} \tag{12.1.2}$$

(where ∂L denotes the limiting subdifferential of $L(t, ., .))$ and

$$p(t) \cdot \dot{\bar{x}}(t) - L(t, \bar{x}(t), \dot{\bar{x}}(t)) = \max_{v \in \mathbb{R}^n} \{p(t) \cdot v - L(t, \bar{x}(t), v)\} \quad \text{a.e..} \tag{12.1.3}$$

Furthermore, if L is independent of t (write $L(x, v)$ in place of $L(t, x, v)$), then there exists a constant a such that

$$a = p(t) \cdot \dot{\bar{x}}(t) - L(\bar{x}(t), \dot{\bar{x}}(t)) \quad \text{a.e..}$$

Note that, since $L(t, x, .)$ is a convex function and the subdifferential for convex finite valued functions on $\mathbb{R}^n$ coincides with the limiting subdifferential, (12.1.3) implies

$$p(t) \in \partial_v L(t, \bar{x}(t), \dot{\bar{x}}(t)) \quad \text{a.e..}$$

Proof The arc $\left(\bar{x}, \bar{z}(t) = \int_S^t L(s, \bar{x}(s), \dot{\bar{x}}(s)) ds\right)$ is an L^∞ local minimizer for the optimal dynamic optimization problem

$$\begin{cases} \text{Minimize } z(T) \\ \text{over } (x, z) \in W^{1,1}([S, T]; \mathbb{R}^{n+1}) \text{ satisfying} \\ (\dot{x}(t), \dot{z}(t)) \in F(t, x(t), z(t)), \\ (x(S), z(S)) = (x_0, 0), \\ (x(T), z(T)) \in \{x_1\} \times \mathbb{R}. \end{cases}$$

12.1 Introduction

Here,
$$F(t, x, z) := \{(v, L(t, x, v)) : v \in \dot{\bar{x}}(t) + \mathbb{B}\}$$

The hypotheses hold under which the generalized Euler Lagrange condition (Theorem 8.4.3) is valid (observe that condition (G3)*** in Proposition 8.5.1 is satisfied and so Corollary 8.5.2 guarantees the validity of the necessary conditions of Theorem 8.4.3). We deduce the existence of an arc $(p, r) \in W^{1,1}([S, T]; \mathbb{R}^{n+1})$ and a constant $\lambda \geq 0$ such that $((p, r), \lambda) \neq (0, 0)$, $r(T) = \lambda$ and

$$(\dot{p}(t), \dot{r}(t)) \in \text{co}\{(q, w) : (q, w, p(t), r(t))$$
$$\in N_{\text{Gr } F(t,\cdot)}(\bar{x}(t), \bar{z}(t), \dot{\bar{x}}(t), \dot{\bar{z}}(t))\} \quad \text{a.e. } t \in [S, T].$$

By considering limits of proximal normal vectors we readily deduce that

'$(q, w, p, -r) \in N_{\text{Gr } F(t,\cdot)}(\bar{x}, \bar{z}, \dot{\bar{x}}, \dot{\bar{z}})$' implies '$w = 0$'.

and therefore $r \equiv \lambda \ (\geq 0)$. Clearly, $\lambda = 0$ would imply that $(p \equiv 0, r \equiv 0)$. It follows that necessarily we have $r \equiv \lambda > 0$. We can therefore arrange by scaling the multipliers that $r = \lambda = 1$. The analysis of limits of proximal normals vectors now provides the following implication

'$(q, 0, p, -1) \in N_{\text{Gr } F(t,\cdot)}(\bar{x}, \bar{z}, \dot{\bar{x}}, \dot{\bar{z}})$' implies '$(q, p)$
$\in \partial L(t, \bar{x}, \dot{\bar{x}})$ and $p \in \partial_v L(t, \bar{x}, \dot{\bar{x}})$'.

So we deduce the validity of (12.1.1), (12.1.2) and, since $v \to L(t, x, v)$ is convex, $\partial_v L(\bar{x}(t), \dot{\bar{x}}(t))$ coincides with the subdifferential of $v \to L(t, \bar{x}(t), v)$ (at $v = \dot{\bar{x}}(t)$) in the sense of the convex analysis. It follows that

$$p(t) \cdot (v - \dot{\bar{x}}(t)) \leq L(t, \bar{x}(t), v) - L(t, \bar{x}(t), \dot{\bar{x}}(t)), \text{ for all } v \in \mathbb{R}^n, \text{ a.e. } t \in [S, T],$$

confirming (12.1.3). If L is independent of t, Corollary 8.5.2 (cf. condition (v) of Theorem 8.4.3) also tells us that p can be chosen to satisfy

$$a = p(t) \cdot \dot{\bar{x}}(t) - \lambda L(\bar{x}(t), \dot{\bar{x}}(t)) \quad \text{a.e. } t \in [S, T],$$

for some constant a. Then, the assertions of the proposition are all confirmed. □

Proposition 12.1.2 falls short of requirements, from the point of view of identifying a large class of problems, of type (BP), for which there exist minimizers and for which all minimizers satisfy known necessary conditions of optimality. This is because of the 'extra' hypothesis (12.1.1) therein invoked, to justify the necessary conditions. (Note that hypothesis (HE2) does not imply (12.1.1) when $\dot{\bar{x}}$ fails to be essentially bounded.)

At the outset, we need to ask whether the presence of the troublesome hypothesis (12.1.1) in Proposition 12.1.2 merely reflects a weakness in our analysis and can be dispensed with in situations in which the Tonelli existence hypotheses are satisfied, or whether it points to genuine restrictions on classes of problems for which the optimality conditions of Proposition 12.1.2 can be confirmed.

This question was resolved by an example of Ball and Mizel, exposing the gap between the Tonelli existence Hypotheses and hypotheses needed for the derivation of Euler Lagrange-type necessary conditions.

Example (Ball Mizel)

$$\begin{cases} \text{Minimize } \int_0^1 \{r\dot{x}^2(t) + (x^3(t) - t^2)^2 \dot{x}^{14}(t)\}\, dt \\ \text{over } x \in W^{1,1}([0, 1]; \mathbb{R}) \text{ satisfying} \\ x(0) = 0, \quad x(1) = k. \end{cases}$$

Here, $r > 0$ and $k > 0$ are constants, linked by the relation

$$r = (2k/3)^{12}(1 - k^3)(13k^3 - 7).$$

It can be shown that there exists $\epsilon > 0$ such that, for all $k \in (1 - \epsilon, 1)$, the arc

$$\bar{x}(t) := kt^{2/3} \tag{12.1.4}$$

is the unique minimizer for this problem. (This is by no means a straightforward undertaking!)

It is a simple matter to check by direct substitution that $\bar{x}$ satisfies, a.e., a pointwise version of the Euler condition:

$$d/dt\, L_v(t, \bar{x}(t), \dot{\bar{x}}(t)) = L_x(t, \bar{x}(t), \dot{\bar{x}}(t)),$$

in which

$$L(t, x, v) = rv^2 + (x^3 - t^2)^2 v^{14}.$$

However this cannot be expressed as an Euler Lagrange type condition in terms of an absolutely continuous adjoint arc p,

$$(\dot{p}(t), p(t)) = \nabla L((t, \bar{x}(t), \dot{\bar{x}}(t))),$$

because the only candidate for adjoint arc p, namely

$$p(t) = -\alpha t^{-1/3}$$

(for some $\alpha \neq 0$ depending on k), is not an absolutely continuous function.

12.1 Introduction

The train of thought behind this example, incidentally, is that the Lagrangian $(x^3 - t^2)^{2m}|\dot{x}|^{2k}$, for any integers $m > 0$ and $k > 0$, has a minimizer $x(t) = t^{2/3}$ with unbounded derivative. It fails however to exhibit 'bad' behaviour under the Tonelli existence hypotheses, because the coercivity condition (HE3) is violated. To ensure satisfaction also of (HE3) we now add a small quadratic coercive term '$\epsilon |\dot{x}|^2$ to the Lagrangian. It then turns out that the values of m and k can be adjusted to ensure that $x(t) = kt^{3/2}$ is a pointwise solution to the Euler equation, for some $k \in (0, 1)$.

The Ball Mizel example is of interest not only for confirming that the Tonelli existence hypotheses alone fail to ensure validity of standard necessary conditions such as the Euler Lagrange condition. It also helps us to predict the nature of minimizers under these hypotheses.

A salient feature of the minimizing arc (12.1.4) is that the minimizer is in some sense badly behaved only on some small subset of the underlying time interval $[0, 1]$, namely $\{0\}$.

It is a remarkable fact that *all* minimizers to (BP) share this property. The key result of the chapter, the generalized Tonelli regularity theorem, makes precise this assertion: it says that, merely under the Tonelli existence hypotheses, the bad behaviour of each minimizer $\bar{x}$ is confined to a closed set of zero measure. In this context, $\bar{t}$ is defined to be a point of bad behaviour for $\bar{x}$ if $\bar{x}$ fails to be Lipschitz continuous on any neighbourhood of $\bar{t}$ in $[S, T]$.

We commented earlier on an inherent difficulty in establishing regularity properties of minimizers, namely that we would like to exploit regularity implications of standard necessary conditions, but cannot do this in interesting cases when the hypotheses, under which these necessary conditions can been derived, are violated. This brings us to one of the most valuable aspects of the generalized Tonelli regularity theorem. It permits us to derive a weakened, 'local' version of the generalized Euler Lagrange condition merely under the Tonelli existence hypotheses. We can combine information about minimizers inherent in this optimality condition with additional hypotheses on the data for (BP), to establish refined regularity properties for special classes of problems.

The role of the generalized Tonelli regularity theorem to generate refined regularity theorems for special classes of problems is investigated at length in Sect. 12.4. A particularly striking result, first proved via the generalized Tonelli regularity theorem, is Proposition 12.4.2 below:

Suppose that $L(t, x, v)$ is independent of t. Then, under the Tonelli existence hypotheses, all minimizers for (BP) are Lipschitz continuous.

To paraphrase, existence of points of bad behaviour is a phenomenon associated solely with non-autonomous problems.

A related topic to regularity analysis is investigation of the so-called *Lavrentiev phenomenon*. This is the surprising fact that, for certain variational problems posed over spaces of arcs $x : [S, T] \to \mathbb{R}^n$, the infimum cost over the space of Lipschitz continuous functions is strictly less that the infimum cost over the space of absolutely continuous functions; surprising because the Lipschitz functions comprise a dense subset of the space of absolutely continuous functions, with respect to the strong $W^{1,1}$ topology. Typical applications of the generalized Tonelli regularity theorem identify classes of problems for which minimizers over the class of absolutely continuous functions are in fact Lipschitz continuous. The link is of course that all such applications inform us about special cases of (BP) for which the Lavrentiev phenomenon cannot occur.

The concluding sections concern regularity of minimizers for optimization problems with dynamic and pathwise constraints. We indicate how the analysis underlying the generalized Tonelli regularity theorem can, in some cases, be extended to cover dynamic optimization problems involving linear dynamic constraints. We also examine some situations in which the application of 'standard' necessary conditions for dynamic optimization problems with dynamic constraints and pathwise state constraints leads directly to regularity information.

The choice of topics in this chapter on minimizer regularity is motivated in part by a desire to illustrate the benefits of applying nonsmooth analysis, in smooth as well as nonsmooth settings. To the authors' knowledge, the only available proof of the generalized Tonelli regularity theorem, for vector valued arcs, is via nonsmooth analysis, even when the data is smooth.

The proof techniques involve the construction of an auxiliary Lagrangian and the comparison of minimizers for the origin and auxiliary Lagrangians on suitable subintervals. Since we are unable to apply the standard necessary conditions to minimizers for the original problem, we establish regularity properties indirectly, by applying them to the problem with the auxiliary Lagrangian, which does satisfy the relevant hypotheses. So everything hinges on the construction of the auxiliary Lagrangian. This is most simply carried out by taking convex hulls of certain epigraph sets, operations which generate nonsmooth functions. Nonsmooth necessary conditions are then required, to analyse the solutions to the auxiliary problem, even if the original Lagrangian is smooth.

In the $n = 1$ case (scalar valued arcs) and for strictly convex Lagrangians, Tonelli showed how to construct a smooth auxiliary Lagrangian with suitable properties. But the task of carrying out such an exercise for vector valued arcs would appear to be a formidable one and, since nonsmooth necessary conditions are now available, unnecessary.

In this chapter we also consider another approach to establishing regularity of minimizers, in situations when standard necessary conditions are not available and therefore cannot be (directly) exploited for regularity analysis. The approach involves introducing a family of reparameterizations of the minimizer by a means of transformations of independent variable. A variational analysis, in which 'variations' are generated by these transformations, can be used to extract regularity information about the minimizer. Applications of this approach were restricted to autonomous problems satisfying the superlinear growth conditions in the Tonelli existence hypotheses. This chapter supplies recent extensions of reparameterization methods, which can be used to establish Lipschitz continuity of minimizers, in some situations where the problem concerned is non-autonomous and the superlinear growth hypothesis is replaced by a less restrictive, slow growth hypothesis.

In the final section of this chapter, we restrict attention to certain dynamic optimization problems with dynamic constraint a controlled differential equation, where minimizers are known to satisfy the maximum principle. Here we how show these necessary conditions can be applied directly to establish Lipschitz continuity of state trajectories.

12.2 Tonelli Regularity

In a landmark paper published in 1915, Tonelli [189] developed a technique for establishing regularity properties of solutions to (BP), under hypotheses similar to (HE1)–(HE3), based on the construction of auxiliary Lagrangians. The regularity properties of minimizers proved in this chapter, which go beyond Tonelli's results in a number of significant respects, are the fruits of combining Tonelli's technique and modern nonsmooth analytical methods.

A cornerstone of the Tonelli's regularity analysis is the concept of a regular point:

Definition 12.2.1 Take a function $y : [S, T] \to \mathbb{R}^n$ and a point $\tau \in [S, T]$. We say τ is a *regular point* of y if

$$\liminf_{\substack{S \leq s_i \leq \tau \leq t_i \leq T \\ s_i \to \tau, t_i \to \tau, s_i \neq t_i}} \frac{|y(t_i) - y(s_i)|}{|t_i - s_i|} < +\infty. \tag{12.2.1}$$

Otherwise expressed, τ is a regular point of y if there exist $s_i \to \tau$, $t_i \to \tau$ and $K > 0$ such that, for all i

$$s_i \neq t_i \quad \text{and} \quad S \leq s_i < t_i \leq T$$

and

$$|y(t_i) - y(s_i)| \leq K|t_i - s_i|.$$

Because the definition involves limits of difference quotients, it would at first sight appear that assuming y is regular at t comes close to assuming that it is Lipschitz continuous near t. In fact the definition of 'regular point' is much less restrictive than this. For example, $t = 0$ is a regular point of $y(s) = |s|^{\frac{1}{2}}$, $-1 \leq s \leq +1$, because

$$\frac{|y(t_i) - y(s_i)|}{|t_i - s_i|} = 0 \quad \text{for all } i,$$

when we choose $s_i = -i^{-1}$ and $t_i = +i^{-1}$ for $i = 1, 2, \ldots$ Yet this function has unbounded slope near $t = 0$.

Theorem 12.2.2 (Generalized Tonelli Regularity Theorem) *Assume that the Tonelli existence hypotheses (HE1)–(HE3) are satisfied. A solution to (BP) exists. Let $\bar{x}$ be any L^∞ local minimizer for (BP). Take any regular point τ for $\bar{x}$. Then there exists a relatively open subinterval $I \subset [S, T]$ with the properties: $\tau \in I$ and the restriction of $\bar{x}$ to I is Lipschitz continuous.*

We defer proof of the theorem until the next section. Here we examine some of its implications.

Take an arbitrary L^∞ local minimizer $\bar{x}$ for (BP). Let D be the subset of $[S, T]$ comprising points at which $\bar{x}$ is differentiable. Since $\bar{x}$ is an absolutely continuous function, it is differentiable almost everywhere. Consequently D has full measure.

Take any $t \in D$. Then, since $\bar{x}$ is differentiable at t,

$$\lim_i \frac{|\bar{x}(t_i) - \bar{x}(s_i)|}{|t_i - s_i|} < +\infty$$

for some sequences $t_i \to t$ and $s_i \to s$ such that $S \leq s_i \leq t \leq t_i \leq T$ and $t_i > s_i$ for each i. We see that t is a regular point of $\bar{x}$. According to the preceding theorem, we can choose a relatively open interval $I_t \subset [S, T]$, containing t, such that x is Lipschitz continuous on I_t.

Now set

$$\Omega = \cup_{t \in D} I_t.$$

Ω is a relatively open set because it is a union of relatively open sets. It has full measure because the subset D has full measure. We see also that $\bar{x}$ is locally Lipschitz continuous on Ω, in the sense that for any $t \in \Omega$ we can find a neighbourhood of t (I_t serves the purpose) on which $\bar{x}$ is Lipschitz continuous. We have drawn the following important conclusions from Theorem 12.2.2:

Corollary 12.2.3 *Assume the Tonelli existence hypotheses (HE1)–(HE3) are satisfied. Take any L^∞ local minimizer $\bar{x}$ for (BP). Then there exists a relatively open set of full measure, Ω, such that $\bar{x}$ is locally Lipschitz continuous on Ω.*

For a given absolutely continuous function $\bar{x}$, let $\Omega_{\max}(\bar{x})$ be the union of all relatively open sets Ω with the property that $\bar{x}$ is locally Lipschitz continuous on Ω. Then $\Omega_{\max}$ is a relatively open subset of $[S, T]$, of full measure, and $\bar{x}$ is locally Lipschitz continuous on $\Omega_{\max}(\bar{x})$. We define

$$\mathcal{S} := [S, T] \setminus \Omega_{\max}(\bar{x})$$

to be the *Tonelli set* for $\bar{x}$. The Tonelli set, which we think of as the set of times at which $\bar{x}$ exhibits bad behaviour can be alternatively described as the set of points t in $[S, T]$ such that $\bar{x}$ has unbounded slope in the vacinity of t.

12.2 Tonelli Regularity

Summarizing and extending these results, we arrive at

Theorem 12.2.4 *Assume that the data for (BP) satisfy the Tonelli existence hypotheses (HE1)–(HE3). Take any L^∞ local minimizer $\bar{x}$. Then the Tonelli set $\mathcal{S}$ for $\bar{x}$ is a (possibly empty) closed set of zero measure. We have:*

(i) *Given any closed subinterval $I \subset [S, T]\backslash\mathcal{S}$, $\bar{x}$ is Lipschitz continuous on I,*
(ii) *Given any closed interval $I \subset [S, T]\backslash\mathcal{S}$, there exists some $p \in W^{1,1}(I; \mathbb{R}^n)$ such that, for a.e. $t \in I$,*

$$\dot{p}(t) \in \mathrm{co}\{q : (q, p(t)) \in \partial L(t, \bar{x}(t), \dot{\bar{x}}(t))\}$$

and

$$p(t) \cdot \dot{\bar{x}}(t) - L(t, \bar{x}(t), \dot{\bar{x}}(t)) = \max_{v \in \mathbb{R}^n} \{p(t) \cdot v - L(t, \bar{x}(t), v)\},$$

(iii) *Suppose in addition to (HE1)–(HE3) that, for each $t \in [S, T]$ and $w \in \mathbb{R}^n$, $L(., \bar{x}(t), w)$ is continuous at t and $L(t, \bar{x}(t), .)$ is strictly convex. Then $\bar{x}$ is continuously differentiable on $[S, T]\backslash\mathcal{S}$,*
(iv) *Suppose in addition to the hypotheses of (iii) that, for each $t \in [S, T]$, the function L is C^r on a neighbourhood of $(t, \bar{x}(t), \dot{\bar{x}}(t))$ (for some integer $r \geq 2$) and $L_{vv}(t, \bar{x}(t), \dot{\bar{x}}(t)) > 0$. Then $\bar{x}$ is r-times continuously differentiable on $[S, T]\backslash\mathcal{S}$.*

In (iii) and (iv), the assertion '$\bar{x}$ is C^r on $[S, b)$' is taken to mean that $\bar{x}$ is of class C^r on the open set (S, b) and all derivatives of order r or less have limits as $t \downarrow S$. We interpret '$\bar{x}$ is C^r on $(b, T]$, etc., likewise.

Proof We know already that $\mathcal{S}$ is a closed set of zero measure, outside of which $\bar{x}$ is locally Lipschitz continuous.

(i): Take a closed interval $I \subset [S, T]\backslash\mathcal{S}$, Since $\bar{x}$ is locally Lipschitz continuous on the compact interval I, $\bar{x}$ is (globally) Lipschitz continuous on I,
(ii): Take any closed interval $[a, b] \subset [S, T]\backslash\mathcal{S}$. Then $\bar{x}$ restricted to $[a, b]$ is an L^∞ local minimizer for (BP) with end-point constraints $x(a) = \bar{x}(a)$ and $x(b) = \bar{x}(b)$. Because $\bar{x}$ is Lipschitz continuous on $[a, b]$ and the relevant Lipschitz continuity hypothesis (12.1.1) is satisfied, the asserted necessary conditions follow from Proposition 12.1.2,
(iii): Take any point t in the relatively open set $[S, T]\backslash\mathcal{S}$. Then there exists some relatively open interval containing t, with end-points a and b and such that $[a, b] \subset [S, T]\backslash\mathcal{S}$. We know L is Lipschitz continuous on $[a, b]$ and that there exists $p \in W^{1,1}([a, b]; \mathbb{R}^n)$ such that, for a.e. $t \in [a, b]$,

$$p(t) \cdot \dot{\bar{x}}(t) - L(t, \bar{x}(t), \dot{\bar{x}}(t)) \geq p(t) \cdot v - L(t, \bar{x}(t), v) \quad \text{for all } v \in \mathbb{R}^n. \quad (12.2.2)$$

If the bounded function $\dot{\bar{x}}$ is not almost everywhere equal to a continuous function on $[a, b]$ there exists $r \in [a, b]$ and sequences $\{c_i\}$ and $\{d_i\}$ converging to r in $[a, b]$, such that (12.2.2) is satisfied for $t = r$ and $t = c_i, t = d_i, i = 1, 2, \ldots$, the limits

$$\alpha := \lim_{i \to \infty} \dot{\bar{x}}(c_i), \quad \beta = \lim_{i \to \infty} \dot{\bar{x}}(d_i)$$

exist and $\alpha \neq \beta$.

Fix $v \in \mathbb{R}^n$. Under the additional hypotheses we have

$$L(r, \bar{x}(r), v) = \lim_{i \to \infty} L(c_i, \bar{x}(c_i), v),$$

$$L(r, \bar{x}(r), \alpha) = \lim_{i \to \infty} L(c_i, \bar{x}(c_i), \dot{\bar{x}}(c_i)).$$

Setting $t = c_i$ in (12.2.2) and passing to the limit as $i \to \infty$ gives

$$p(r) \cdot v - L(r, \bar{x}(r), v) \leq p(r) \cdot \alpha - L(r, \bar{x}(r), \alpha).$$

Using the same arguments in relation to the sequence $\{d_i\}$, we arrive at

$$p(r) \cdot v - L(r, \bar{x}(r), v) \leq p(r) \cdot \beta - L(r, \bar{x}(r), \beta).$$

These inequalities are satisfied for arbitrary points $v \in \mathbb{R}^n$. We have shown that the function

$$v \to p(r) \cdot v - L(r, \bar{x}(r), v)$$

achieves its maximum at the points $v = \alpha$ and $v = \beta$. Since it is strictly concave, we deduce that $\alpha = \beta$. From this contradiction we conclude that an arbitrary point $t \in [S, T] \setminus \mathcal{S}$ is contained in a relatively open interval on which $\dot{\bar{x}}$ is almost everywhere equal to a continuous function. It follows that $\dot{\bar{x}}$ is continuous on $[S, T] \setminus \mathcal{S}$.

(iv): Take any point $s \in [S, T] \setminus \mathcal{S}$. Then s is contained in some relatively open subinterval of $[S, T]$ with end-points a and b, such that $[a, b] \subset [S, T] \setminus \mathcal{S}$. $\bar{x}$, restricted to $[a, b]$, is an L^∞ local minimizer to (BP) for 'end-point data' $(a, \bar{x}(a), b, \bar{x}(b))$. Since, by (iii), this subarc has essentially bounded derivative, we deduce from Proposition 12.1.2 the existence of a vector d such that

$$L_v(t, \bar{x}(t), \dot{\bar{x}}(t)) = d + \int_a^t L_x(\sigma, \bar{x}(\sigma), \dot{\bar{x}}(\sigma)) d\sigma \tag{12.2.3}$$

for all $t \in [a, b]$. The right side is C^1 and so is $(t, v) \to L_v(t, \bar{x}(t), v)$. Since $L_{vv} > 0$, it follows from the implicit function Theorem that $\dot{\bar{x}}$ is C^1, from which it follows that $\bar{x}$ is C^2. (12.2.3) therefore implies

$$d^2 \bar{x}(t)/dt^2 = [L_{vv}]^{-1} \{L_x - L_{vt} - L_{vx} \cdot \dot{\bar{x}}(t)\}, \tag{12.2.4}$$

on (a, b). (The derivatives are evaluated at $(t, \bar{x}(t), \dot{\bar{x}}(t))$.) It follows that $d^2\bar{x}/dt^2$ is continuous (and has limits at a and b). Now suppose we know that $\bar{x}$ is C^{r-1} on $[a, b]$ and that L is C^r (for some integer $r \geq 2$). The right side of (12.2.4) is C^{r-2}, and therefore $d^2\bar{x}/dt^2$ is C^{r-2}. It follows that $\bar{x}$ restricted to I is C^r (and if $d^{r-1}\bar{x}/dt^{r-1}$ has limits at the end-points of I, then so does $d^r\bar{x}/dt^r$.) The fact that $\bar{x}$ is C^r on a relatively open subinterval containing s now follows by induction. □

Thus far, we have placed no restrictions on the dimension n of the state variable. If $n = 1$, then minimizing arcs have the surprising property that they are differentiable at *every* point in $[S, T]$, even on their Tonelli sets. To justify such assertions however, we must allow derivatives which take values $+\infty$ or $-\infty$.

Theorem 12.2.5 *Assume that, in addition to the Tonelli Existence Hypotheses (HE1)–(HE3), for each $t \in [S, T]$ and $w \in \mathbb{R}^n$ $L(., \bar{x}(t), w)$ is continuous at t and $L(t, \bar{x}(t), .)$ is strictly convex. Suppose further that*

$$n = 1.$$

Let $\bar{x}$ be a minimizer. Then $\bar{x}$ is everywhere differentiable on $[S, T]$, in the sense that the following limit exists (finite or infinite) for each $\tau \in [S, T]$:

$$\lim_{\substack{t \to \tau \\ a \leq t \leq b}} \frac{\bar{x}(t) - \bar{x}(\tau)}{t - \tau}. \tag{12.2.5}$$

Proof If τ is such that the left side of (12.2.1) is finite then, as we have shown in (iii) of Theorem 12.2.4, $\bar{x}$ is C^1 near τ. In this case, the limit (12.2.5) exists and is finite. So we may assume that the left side of (12.2.1) is infinite.

Suppose that $\tau = S$. The limit (12.2.5) can fail to exist only if

$$\limsup_{t \downarrow S} \frac{\bar{x}(t) - \bar{x}(S)}{t - S} = +\infty \quad \text{and} \quad \liminf_{t \downarrow S} \frac{\bar{x}(t) - \bar{x}(S)}{t - S} = -\infty. \tag{12.2.6}$$

But since $\bar{x}$ is continuous, we deduce from these two relations that there exist points t, arbitrarily close to S, with $\bar{x}(t) = \bar{x}(S)$. It follows that the left side of (12.2.1) is finite, a contradiction. We show similarly that limit (12.2.5) exists also when $\tau = T$.

It remains to consider the case when τ lies in (S, T) and the left side of (12.2.1) is infinite. Reasoning as above, we justify restricting attention to the case when

$$\lim_{t \downarrow \tau} \frac{\bar{x}(t) - \bar{x}(\tau)}{t - \tau} = +\infty \quad \text{and} \quad \lim_{s \uparrow \tau} \frac{\bar{x}(s) - \bar{x}(\tau)}{s - \tau} = -\infty. \tag{12.2.7}$$

(The related case, in which the limits from right and left are $-\infty$ and $+\infty$ respectively, is treated analogously.)

Fix $\epsilon > 0$ such that $[\tau - \epsilon, \tau + \epsilon] \subset [S, T]$. We claim that there exists $\delta > 0$ with the following property: corresponding to any $r \in (0, \delta)$, a point $s \in (\tau - \epsilon, \tau)$ can be found such that $\bar{x}(s) = \bar{x}(\tau) + r$. Indeed, if this were not the case, we could deduce from the continuity of $\bar{x}$ that $\bar{x}(s) \leq \bar{x}(\tau)$ for all $s \in (\tau - \epsilon, \tau)$. This contradicts the second condition in (12.2.7).

Arguing in similar fashion, we can show that (possibly after reducing the size of δ) there exists $t \in (\tau, \tau + \epsilon)$ and $r \in (0, \delta)$ for which $\bar{x}(t) = \bar{x}(\tau) + r$, for otherwise we obtain a contradiction of the first condition in (12.2.7). We know that $\tau \in (s, t)$. Also, $\bar{x}(s) = \bar{x}(t)$ and $|t - s| \leq 2\epsilon$. Since ϵ is an arbitrary positive number, the left side of (12.2.1) is zero, again a contradiction. □

12.3 Proof of The Generalized Tonelli Regularity Theorem

We prove Theorem 12.2.2. Existence of minimizers is assured by Theorem 12.1.1. Take any L^∞ local minimizer $\bar{x}$. Note at the outset that, without loss of generality, we can assume:

(HE4): $\theta : [0, \infty) \to \mathbb{R}$ is a non-negative valued, non-decreasing function.

Indeed, if θ fails to satisfy this condition we can replace it by

$$\tilde{\theta}(r) := \inf\{\theta(r') : r' \geq r\} - \inf\{\theta(r') : r' \geq 0\}.$$

It is a straightforward task to deduce from the convexity and superlinear growth of $L(t, x, .)$ that this new function meets the requirements of hypothesis (HE3) and also satisfies (HE4), for the Lagrangian

$$L(t, x, v) - \inf\{\theta(r') : r' \geq 0\}.$$

Since minimizers are unaffected by the addition of a constant to L, we have confirmed that we can add (HE4) to the hypotheses.

We can also assume

(HE5):   In condition (HE3), the inequality is strict and $\alpha = 0$.

To see this, take $k > 0$ such that $\|\bar{x}\|_{L^\infty} < k$. Consider a new problem, in which L in problem (BP) is replaced by

$$L'(t, x, v) := \max\{L(t, x, v); -\alpha k + \theta(|v|)\} + \alpha k + 1.$$

Hypotheses (HE1)–(HE2) continue to be satisfied (with $\theta = \tilde{\theta}$ and $\alpha = 0$). However the inequality in (HE3) is now strict. We have $L' \geq L + \alpha k + 1$ everywhere and $L'(t, x, v) = L(t, x, v) + \alpha k + 1$ for all $v \in \mathbb{R}^n$ and points (t, x) in some tube about $\bar{x}$. So $\bar{x}$ remains an L^∞ local minimizer and the assertions of the theorem for L' imply those for the original L. This confirms that we can add (HE5) to the hypotheses.

12.3 Proof of The Generalized Tonelli Regularity Theorem

Take a regular point $\tau \in [S, T]$ of $\bar{x}$. Then there exist sequences $s_i \to \tau$, $t_i \to \tau$ such that $S \le s_i \le \tau \le t_i \le T$ for all i, and

$$\lim_{i \to \infty} \frac{|\bar{x}(t_i) - \bar{x}(s_i)|}{|t_i - s_i|} < \infty. \tag{12.3.1}$$

Since $\bar{x}$ is continuous, we can arrange (by decreasing each s_i and increasing each t_i if necessary) that, for each i, $t_i > \tau$ if $\tau < T$ and $s_i < \tau$ if $\tau > S$. This means that, for each i, τ is contained in a relatively open subinterval of $[S, T]$ with end-points s_i and t_i.

For each i let $y : [s_i, t_i] \to \mathbb{R}^n$ be the linear interpolation of the end-points of $\bar{x}$ restricted to $[s_i, t_i]$:

$$y_i(t) := \bar{x}(s_i) + \frac{t - s_i}{|t_i - s_i|}(\bar{x}(t_i) - \bar{x}(s_i)).$$

Lemma 12.3.1 *There exist constants R_0 and M such that*

$$\|\bar{x}\|_{L^\infty} \le M, \quad \|y_i\|_{L^\infty} \le M \quad \text{for all } i$$

and

$$\|\dot{y}_i\|_{L^\infty} < R_0 \quad \text{for all } i.$$

Furthermore, if $\{z_i \in W^{1,1}([s_i, t_i]; \mathbb{R}^n)\}$ is any sequence of arcs such that, for each i,

$$z_i(s_i) = \bar{x}(s_i) \quad \text{and} \quad z_i(t_i) = \bar{x}(t_i)$$

and

$$\frac{1}{2}\int_{s_i}^{t_i} \theta(|\dot{z}_i(t)|)dt \le \int_{s_i}^{t_i} L(t, y_i(t), \dot{y}_i(t))dt,$$

then

$$\|z_i\|_{L^\infty} \le M \quad \text{for all } i.$$

Proof The fact that M and R_0 can be chosen to satisfy all the stated conditions is obvious, with the exception of '$\|z_i\|_{L^\infty} \le M$'. To show that this condition can also be satisfied, we choose $\alpha > 0$ such that $\theta(\alpha') > \alpha'$ whenever $\alpha' \ge \alpha$. Then, for any i and $t \in [s_i, t_i]$,

$$|z_i(t)| \leq |z_i(s_i)| + \int_{s_i}^{t_i} |\dot{z}_i(s)| ds$$

$$\leq |z_i(s_i)| + \alpha|t_i - s_i| + \int_{s_i}^{t_i} \theta(|\dot{z}_i(s)|) ds$$

$$\leq |z_i(s_i)| + \alpha|t_i - s_i| + 2\int_{s_i}^{t_i} L(t, y_i(s), \dot{y}_i(s)) ds.$$

We have now merely to note that all terms on the right side of this inequality are bounded by a constant which does not depend on i or $t \in [s_i, t_i]$. □

Define

$$c_0 := \max\{|L(t, x, v)| : t \in [S, T], |x| \leq M \text{ and } |v| \leq R_0\},$$

and the function $d : \mathbb{R} \to \mathbb{R}$

$$d(\sigma) := \inf\{|v| : p \in \partial_v L(t, x, v), t \in [S, T], |x| \leq M, |p| \geq \sigma\}.$$

We deduce from (HE2) that

$$\lim_{\sigma \to \infty} d(\sigma) = +\infty. \qquad (12.3.2)$$

Fix $\epsilon > 0$. Choose $R_1 \geq R_0$ to satisfy

$$\theta \circ d\left(\frac{\theta(r')}{2r'} - \frac{c_0}{r'} - 2\epsilon\right) > 2c_0 \qquad (12.3.3)$$

whenever $r' \geq R_1, t \in [S, T]$ and $|x| \leq M$. This is possible by (12.3.2) and since θ has superlinear growth. Define

$$c_1 := \max\{|L(t, x, v)| : t \in [S, T], |x| \leq M, |v| \leq R_1\} \qquad (12.3.4)$$

and

$$\sigma_1 := \max\{|\xi| : \xi \in \partial_v L(t, x, v)|, t \in [S, T], |x| \leq M, |v| \leq R_1\}. \qquad (12.3.5)$$

Choose $R_2 > R_1$ such that

$$\frac{1}{2}\theta(r) \geq \sigma_1(R_1 + r) + c_1 \quad \text{if } r \geq R_2. \qquad (12.3.6)$$

Set

$$\phi(v) := \frac{1}{2}\max\{\theta(|v|); \theta(R_2)\}$$

12.3 Proof of The Generalized Tonelli Regularity Theorem

and, for each $(t, x) \in [S, T] \times \mathbb{R}^n$ and $v \in \mathbb{R}^n$, define

$$\tilde{L}(t, x, v) := \inf \left\{ \alpha : (v, \alpha) \in \text{co} \left[\text{epi}\phi \cup \text{epi}\left\{ L(t, x, .) + \Psi_{R_2 \mathbb{B}} \right\} \right] \right\}.$$

The expression on the right summarizes the following construction:

$\tilde{L}(t, x, .)$ *is the function with epigraph set* E, *where* E *is the convex hull of the unions of the epigraph sets of* ϕ *and of the function* $v \to L(t, x, v)$, *restricted to the ball* $R_2 \mathbb{B}$.

An alternative representation is as follows:

$$\tilde{L}(t, x, v) = \inf \{ \lambda L(t, x, u) + (1 - \lambda) \phi(w) :$$
$$0 \le \lambda \le 1, \ |u| \le R_2 \ \text{and} \ \lambda u + (1 - \lambda) w = v \}.$$

This 'auxiliary Lagrangian' $\tilde{L}$ has the following properties:

Lemma 12.3.2

(a) $\tilde{L}(t, x, v)$ *is locally bounded, measurable in* t *and convex in* v,
(b) $\tilde{L}(t, x, v)$ *is locally Lipschitz continuous in* (x, v) *uniformly in* $t \in [S, T]$,
(c) $\tilde{L}(t, x, v) \ge \theta(|v|)/2$ *for all* (t, x, v),
(d) *for all* $t \in [S, T]$ *and* $x \in M\mathbb{B}$ *we have*

$$\tilde{L}(t, x, v) = L(t, x, v) \quad \text{if} \quad |v| \le R_1,$$
$$\tilde{L}(t, x, v) \le L(t, x, v) \quad \text{if} \quad |v| \le R_2,$$
$$\tilde{L}(t, x, v) < L(t, x, v) \quad \text{if} \quad |v| > R_2,$$

(e) *For* $(t, x) \in [S, T] \times \mathbb{R}^n$

$$\tilde{L}(t, x, v) = \theta(|v|)/2 \quad \text{if} \quad |v| \ge R_2.$$

Proof

(a): $\tilde{L}$ is convex in v, by construction. It is locally bounded since $0 \le \tilde{L} \le \phi$ and ϕ is locally bounded. To see that $\tilde{L}$ is measurable in t we use the fact that, for fixed x and v, $\tilde{L}$ can be expressed as a pointwise infimum of a countable family of measurable functions

$$t \to \lambda L(t, x, v) + (1 - \lambda) \phi(w)$$

obtained by allowing (λ, v, w) to range over a countable dense subset of $[S, T] \times \mathbb{R}^n \times \mathbb{R}^n$. (a) has been proved.

(c): This property follows from the facts that both L and ϕ satisfy the desired inequality and that $v \to \frac{1}{2}\theta(|v|)$ is a convex function.

(e): Take arbitrary points $w' \in \mathbb{R}^n$, $|w'| > R_2$ and $(t, x) \in [S, T] \times \mathbb{R}^n$. Choose $\zeta \in \partial\phi(w')$. The subgradient inequality for convex functions gives

$$\phi(w) - \phi(w') - \zeta \cdot (w - w') \geq 0 \quad \text{for all } w \in \mathbb{R}^n. \tag{12.3.7}$$

Since θ is continuous and strictly increasing on $[R_2, \infty)$, $\phi(w)$ and $\frac{1}{2}\theta(|w|)$ coincide on a neighbourhood of $w = w'$. It follows that ζ is a subgradient also of $w \to \frac{1}{2}\theta(|w|)$ at $w = w'$ and, for all $u \in \mathbb{R}^n$ and $(t, x) \in [S, T] \times \mathbb{R}^n$ we have

$$L(t, x, u) - \phi(w') - \zeta \cdot (u - w') \geq \frac{1}{2}\theta(|u|) - \frac{1}{2}\theta(|w'|) - \zeta \cdot (u - w') \geq 0. \tag{12.3.8}$$

From (12.3.7) and (12.3.8) we deduce that

$$\tilde{L}(t, x, v) = \inf\{\lambda L(t, x, u) + (1 - \lambda)\phi(w) :$$
$$\lambda \in [0, 1], |u| \leq R_2, v = \lambda u + (1 - \lambda)w\}$$
$$\geq \phi(w') - \zeta \cdot (v - w').$$

Setting $v = w'$ we see that $\tilde{L}(t, x, w') \geq \phi(w')$. But then $\tilde{L}(t, x, v) = \frac{1}{2}\theta(|v|)$ in the region $\{v : |v| > R_2\}$, since ϕ and $v \to \frac{1}{2}\theta(|v|)$ coincide here and ϕ majorizes $\tilde{L}$. This remains true in the region $\{v : |v| \geq R_2\}$, by the continuity properties of convex functions.

(b): Take any $k_1 > 0$. Let K be a Lipschitz constant for $x \to L(t, x, v)$ uniformly valid for $t \in [S, T]$, $|v| \leq R_2$ and $|x| \leq k_1$. Take $x_1, x_2 \in \mathbb{R}^n$ such that $|x_1|$ and $|x_2| \leq k_1$. Then for any $\delta > 0$, $t \in [S, T]$ and w we can choose u, w and λ in the definition of $\tilde{L}$ such that

$$\tilde{L}(t, x_1, v) \leq \lambda L(t, x_1, u) + (1 - \lambda)\phi(w)$$
$$\leq \lambda L(t, x_2, u) + K|x_1 - x_2| + (1 - \lambda)\phi(w)$$
$$\leq \tilde{L}(t, x_2, v) + K|x_1 - x_2| + \delta.$$

Since x_1 and x_2 are interchangeable and $\delta > 0$ is arbitrary, it follows that $x \to \tilde{L}(t, x, v)$ has Lipschitz constant at most K on $k_1 \mathbb{B}$, for all $t \in [S, T]$ and $v \in R_2 \mathbb{B}$.

Now take $k_2 \geq R_2$. We shall show that $v \to \tilde{L}(t, x, v)$ is Lipschitz continuous in the region $k_2 \mathbb{B}$, uniformly over $(t, x) \in [S, T] \times \mathbb{R}^n$. It will follow that $\tilde{L}$ is locally Lipschitz continuous jointly in the variables x, v, uniformly in $t \in [S, T]$, since it has this property with respect to these variables individually.

Choose $(t, x) \in [S, T] \times \mathbb{R}^n$, and let $v \to p \cdot v + q$ be an arbitrary, non-constant affine function which is majorized by $v \to \tilde{L}(t, x, v)$. By (e), we must have

12.3 Proof of The Generalized Tonelli Regularity Theorem

$$p \cdot v + q \leq \frac{1}{2}\theta(|v|)$$

for all v such that $|v| \geq k_2$. Setting $v = (k_2 + 1)p/|p|$ we obtain

$$(k_2 + 1)|p| + q \leq \frac{1}{2}\theta(k_2 + 1). \tag{12.3.9}$$

However since $\tilde{L} \geq 0$ we also have

$$\tilde{L}(t, x, v) - p \cdot v - q \geq -|p|k_2 - q \tag{12.3.10}$$

for $v \in k_2 \mathbb{B}$. Equations (12.3.9) and (12.3.10) yield

$$\tilde{L}(t, x, v) - p \cdot v - q \geq +|p| - \frac{1}{2}\theta(k_2 + 1)$$

for $v \in k_2 \mathbb{B}$. Set $K_1 := \frac{1}{2}\theta(k_2 + 1) + 1$. It follows that

$$\tilde{L}(t, x, v) - p \cdot v - q \geq 1 \tag{12.3.11}$$

for $v \in k_2 \mathbb{B}$ and $|p| \geq K_1$.

Now the function $v \to \tilde{L}(t, x, v)$ is expressible as the pointwise supremum of affine functionals majorized by $\tilde{L}$. However inequality (12.3.11) tells us that, to evaluate the pointwise supremum in the region $|v| \leq k_2$, we can restrict attention to affine functions with Lipschitz constant at most K_1. It follows then that $v \to \tilde{L}(t, x, v)$ has Lipschitz constant at most K_1 in this region, uniformly with respect to $(t, x) \in [S, T] \times \mathbb{R}^n$.

(d): The cases $|v| \geq R_2$ and $R_2 > |v| > R_1$ follow from (e) since L majorizes $\tilde{L}$ and L strictly majorizes $u \to \frac{1}{2}\theta(|u|)$. It remains to show that $L(t, x, v) = \tilde{L}(t, x, v)$ for all $(t, x) \in [S, T] \times M\mathbb{B}$ and $v \in R_1 \mathbb{B}$.

Take (t, x) and v as above and choose $\zeta \in \partial_v L(t, x, v)$. By (12.3.6) and the definition of the constants c_1 and σ_1 (see (12.3.4) and (12.3.5)) we have that

$$\phi(w) \geq \frac{1}{2}\theta(|w|) \geq \sigma_1(R_1 + |w|) + c_1 \geq L(t, x, v) + \zeta \cdot (w - v),$$

for all points w which satisfy $|w| \geq R_2$. On the other hand, we also know that

$$\phi(w) \geq \frac{1}{2}\theta(R_2) \geq \sigma_1(R_1 + R_2) + c_1 \geq L(t, x, v) + \zeta \cdot (w - v)$$

for all points w which satisfy $|w| \leq R_2$. By the subgradient inequality however

$$L(t, z, u) - L(t, z, v) - \zeta \cdot (u - v) \geq 0$$

for all $u \in \mathbb{R}^n$. Scaling and adding these inequalities, we arrive at

$$\tilde{L}(t, x, v') = \inf\{\lambda L(t, x, u) + (1 - \lambda)\phi(w) :$$
$$0 \leq \lambda \leq 1, |u| \leq R_2, v' = \lambda u + (1 - \lambda)w\}$$
$$\leq L(t, x, v) + \zeta \cdot (v' - v)$$

for all points $v' \in \mathbb{R}^n$. Setting $v' = v$ yields $\tilde{L}(t, x, v) \geq L(t, x, v)$. Since however L majorizes $\tilde{L}$ we can replace inequality here by equality. This is what we set out to prove. □

Consider the optimization problems (P_i), $i = 1, 2 \ldots$,

$$(P_i) \begin{cases} \text{Minimize } \int_{s_i}^{t_i} \tilde{L}(t, x(t), \dot{x}(t))dt \\ \quad \text{over } x \in W^{1,1}([s_i, t_i]; \mathbb{R}^n) \text{ satisfying} \\ x(s_i) = \bar{x}(s_i), \ x(t_i) = \bar{x}(t_i). \end{cases}$$

In view of properties (a)–(c) of the auxiliary Lagrangian, we deduce from Theorem 12.1.1 that (P_i) has a minimizer, which we denote by x_i, for each i. Notice that, for each i,

$$\frac{1}{2}\int_{s_i}^{t_i} \theta(|\dot{x}_i(s)|)ds \leq \int_{s_i}^{t_i} \tilde{L}(t, x_i(t), \dot{x}_i(t))dt$$
$$\leq \int_{s_i}^{t_i} \tilde{L}(t, y_i(t), \dot{y}_i(t))dt$$
$$\leq \int_{s_i}^{t_i} L(t, y_i(t), \dot{y}_i(t))dt,$$

by properties (c) and (d) of $\tilde{L}$ and since $\|\dot{y}_i\|_{L^\infty} \leq R_1$.

We conclude from Lemma 12.3.1 that

$$\|x_i\|_{L^\infty} \leq M \quad \text{for all } i.$$

Observe next that $(x_i, z_i(t) = \int_{s_i}^t \tilde{L}(s, x_i(s), \dot{x}_i(s))ds)$ is a solution to the dynamic optimization problem:

$$\begin{cases} \text{Minimize } z(t_i) \\ \text{over } (x, z) \in W^{1,1}([s_i, t_i]; \mathbb{R}^{n+1}) \text{ satisfying} \\ (\dot{x}(t), \dot{z}(t)) \in F(t, x(t), z(t)) \quad \text{a.e. } t \in [s_i, t_i], \\ x(s_i) = \bar{x}(s_i), \ z(s_i) = 0, \ x(t_i) = \bar{x}(t_i), \end{cases}$$

where $F(t, x, z) := \text{epi } \tilde{L}(t, x, .) \ (= \{(v, \alpha) \in \mathbb{R}^n \times \mathbb{R} \ : \ \alpha \geq \tilde{L}(t, x, v), \ v \in \mathbb{R}^n\})$. It is straightforward to confirm the hypotheses under which the necessary

12.3 Proof of The Generalized Tonelli Regularity Theorem

conditions of Theorem 8.7.2 (the Hamiltonian inclusion for convex velocity sets) are valid, with reference to the minimizer (x_i, z_i).

Notice the crucial role of property (e) of $\tilde{L}$ which, together with (b), ensures that $\tilde{L}$ satisfies the Lipschitz continuity hypothesis: there exists $\beta > 0$ such that

$$|\tilde{L}(t, x, v) - \tilde{L}(t, x', v)| \leq \beta |x - x'| \qquad (12.3.12)$$

for all $v \in \mathbb{R}^n$ and all $(t, x), (t, x')$ in some tube about x_i.

We deduce existence of an arc $p_i \in W^{1,1}$ such that

$$\dot{p}_i(t) \in \text{co}\{-\xi \;:\; (\xi, \dot{x}_i(t), \dot{z}_i(t)) \in \partial_{x,p,q} \tilde{H}(t, x_i(t), p_i(t), -1)\} \qquad (12.3.13)$$

$$\dot{z}_i(t) = \tilde{L}(t, x_i(t), \dot{x}_i(t)) \quad \text{a.e. } t \in [s_i, t_i],$$

where $\tilde{H}(t, x, p, q) := \sup_{(v,\alpha) \in \text{epi } \tilde{L}(t,x,\cdot)} (p, q) \cdot (v, \alpha)$. Moreover, the Weierstrass condition tells us that

$$p_i(t) \cdot \dot{x}_i(t) - \lambda \tilde{L}(t, x_i(t), \dot{x}_i(t)) \geq \max_{v \in \mathbb{R}^n} \{p_i(t) \cdot v - \tilde{L}(t, x_i(t), v)\}, \qquad (12.3.14)$$

which yields

$$p_i(t) \in \partial_v \tilde{L}(t, x_i(t), \dot{x}_i(t)) \quad \text{a.e. } t \in [s_i, t_i]. \qquad (12.3.15)$$

Observe that from (12.3.12) and (12.3.13) we deduce that $|\dot{p}_i(t)| \leq \beta$ a.e. $t \in [s_i, t_i]$, and so

$$|p_i(t_i) - p_i(s_i)| \leq \beta |t_i - s_i|$$

for all i. Let i_0 be the smallest integer such that

$$\beta |t_j - s_j| \leq \epsilon \quad \text{for all } j \geq i_0.$$

Choose any $i \geq i_0$. Then, by (12.3.15),

$$p_i(s_i) \in \partial_v \tilde{L}(t, x_i(t), \dot{x}_i(t)) + \epsilon \mathbb{B} \quad \text{a.e. } t \in [s_i, t_i]. \qquad (12.3.16)$$

We claim that

$$\|\dot{x}_i\|_{L^\infty} \leq R_1. \qquad (12.3.17)$$

Indeed, assume to the contrary that for all points $\bar{t}$ in some subset of $[s_i, t_i]$ of positive measure we have

$$|\dot{x}_i(\bar{t})| > R_1.$$

According to (12.3.16) then, we can arrange that

$$\tilde{L}(\bar{t}, x_i(\bar{t}), 0) - \tilde{L}(\bar{t}, x_i(\bar{t}), \dot{x}_i(\bar{t})) \geq -p_i(s_i) \cdot \dot{x}_i(\bar{t}) - \epsilon |\dot{x}_i(\bar{t})|.$$

So,

$$|p_i(s_i)| \cdot |\dot{x}_i(\bar{t})| \geq \tilde{L}(\bar{t}, x_i(\bar{t}), \dot{x}_i(\bar{t})) - \tilde{L}(\bar{t}, x_i(\bar{t}), 0) - \epsilon |\dot{x}_i(\bar{t})|.$$

It follows that

$$|p_i(s_i)| \geq \frac{\theta(r)}{2r} - \frac{c_0}{r} - \epsilon,$$

for some $r > R_1$. By (12.3.16), for a.e. $t \in [s_i, t_i]$

$$|\dot{x}_i(t)| \geq d(|p_i(s_i)| - \epsilon) \geq d\left(\frac{\theta(r)}{2r} - \frac{c_0}{r} - 2\epsilon\right).$$

Since θ is a monotone function, recalling also (12.3.3), we obtain

$$\frac{1}{2}\theta(|\dot{x}_i(t)|) \geq \frac{1}{2}\theta \circ d\left(\frac{\theta(r)}{2r} - \frac{c_0}{r} - 2\epsilon\right) > c_0 \quad \text{a.e..}$$

It follows now from properties (c) and (d) of $\tilde{L}$ (see Lemma 12.3.2) that

$$\int_{s_i}^{t_i} \tilde{L}(t, x_i(t), \dot{x}_i(t)) dt \geq \frac{1}{2} \int_{s_i}^{t_i} \theta(|\dot{x}_i(t)|) dt$$

$$> |t_i - s_i| c_0$$

$$\geq \int_{s_i}^{t_i} L(t, y_i(t), \dot{y}_i(t)) dt$$

$$= \int_{s_i}^{t_i} \tilde{L}(t, y_i(t), \dot{y}_i(t)) dt.$$

But this contradicts the optimality of x_i. Condition (12.3.17) is verified.
We claim finally that

$$\|\dot{x}(t)\|_{L^\infty([s_i, t_i]; \mathbb{R}^n)} \leq R_2. \tag{12.3.18}$$

This will imply that $\dot{x}$ is locally essentially bounded on $[s_i, t_i]$.

Suppose, to the contrary, that there exists a subset $\mathcal{D} \subset [s_i, t_i]$ of positive measure such that

$$|\dot{\bar{x}}(t)| > R_2 \quad \text{for all } t \in \mathcal{D}.$$

Since $\|\dot{\bar{x}}\|_{L^\infty} \leq M$, we have from property (d) of L that

$$\tilde{L}(t, \bar{x}(t), \dot{\bar{x}}(t)) < L(t, \bar{x}(t), \dot{\bar{x}}(t)) \quad \text{for } t \in \mathcal{D}. \tag{12.3.19}$$

Then

$$\int_{s_i}^{t_i} L(t, \bar{x}(t), \dot{\bar{x}}(t))dt \leq \int_{s_i}^{t_i} L(t, x_i(t), \dot{x}_i(t))dt$$

(by optimality of $\bar{x}$)

$$= \int_{s_i}^{t_i} \tilde{L}(t, x_i(t), \dot{x}_i(t))dt$$

(since $\|\dot{x}_i\|_{L^\infty([s_i,t_i];\mathbb{R}^n)} \leq R_1$ and by property (d) of $\tilde{L}$)

$$\leq \int_{s_i}^{t_i} \tilde{L}(t, \bar{x}(t), \dot{\bar{x}}(t))dt$$

(by optimality of x_i)

$$< \int_{s_i}^{t_i} L(t, \bar{x}(t), \dot{\bar{x}}(t))dt$$

(by (12.3.19) and since $L \geq \tilde{L}$).

From this contradiction, we deduce that (12.3.18) is true. We have confirmed that $\dot{\bar{x}}$ is essentially bounded on the relatively open subinterval $[s_i, t_i]$ containing τ.

12.4 Lipschitz Continuous Minimizers

Our aim in this section is to explore the implications of the generalized Tonelli regularity theorem for particular classes of problems. The idea is to provide a more detailed, qualitative description of minimizers than that of Sect. 12.2, when the Tonelli existence hypotheses are supplemented by additional hypotheses.

The generalized Tonelli regularity theorem is a very fruitful source of refined regularity theorems, supplying information about the Tonelli set in special cases. It can be used, for example, to show that for a large class of problems with polynomial Lagrangians, the Tonelli set is a countable set with a finite number of accumulation points [77].

We concentrate here, however, on just one application area: identifying hypotheses which, when added to the Tonelli existence hypotheses, assure that all minimizers of (BP) over the class of absolutely continuous functions are, in fact, Lipschitz continuous. An equivalent property is that the Tonelli set is empty.

The significance of establishing that a minimizer for (BP) is Lipschitz continuous is that, on one hand, the hypotheses are then met under which the standard necessary conditions, such as those summarized as Proposition 12.1.2, can be used to investigate minimizers in detail. One the other, it rules out pathological behaviour, which might otherwise give rise to difficulties in the computation of minimizers, associated with the Lavrentiev phenomenon.

We shall make repeated use of the following lemma, which gives sufficient conditions under which an interval on which a minimizer is Lipschitz continuous can be extended to include an end-point of $[S, T]$.

Lemma 12.4.1 *Assume the Tonelli Existence Hypotheses (HE1)–(HE3). Let $\bar{x}$ be a minimizer for (BP) and let S be the Tonelli set of $\bar{x}$. Take $\bar{t} \in [S, T]\setminus S$. Suppose that there exists $k > 0$ with the following property: for any $t' \in (\bar{t}, T)$ such that $[\bar{t}, t'] \subset [S, T]\setminus S$, there exists $p \in W^{1,1}([\bar{t}, t']; \mathbb{R}^n)$ satisfying*

$$\dot{p}(t) \in \mathrm{co}\{q : (q, p(t)) \in \partial L(t, \bar{x}(t), \dot{\bar{x}}(t))\} \quad a.e.,$$

$$|p(t)| \leq k \quad \text{for all } t \in [t, t'].$$

Then $\dot{\bar{x}}$ is Lipschitz continuous on $[\bar{t}, T]$.

Proof Define

$$\tau_{\max} := \sup\{\tau \in (\bar{t}, T] : \dot{\bar{x}} \text{ is essentially bounded on } [\bar{t}, \tau]\}.$$

(The set over which the supremum is taken is non-empty because of Theorem 12.2.2.)

Choose a sequence $\{t_i\}$ in $(\bar{t}, \tau_{\max})$ such that $t_i \uparrow \tau_{\max}$. According to the hypotheses, for each i there exists $p_i \in W^{1,1}([\bar{t}, t_i]; \mathbb{R}^n)$ and $k \geq 0$ such that

$$\dot{p}_i(t) \in \mathrm{co}\{q : (q, p_i(t)) \in \partial L(t, \bar{x}(t), \dot{\bar{x}}(t))\} \quad \text{a.e. } t \in [\bar{t}, t_i] \quad (12.4.1)$$

and

$$|p_i(t)| \leq k \quad \text{for all } t \in [\bar{t}, t_i]. \quad (12.4.2)$$

Since $L(t, x, .)$ is convex, (12.4.1) implies that

$$p_i(t) \in \partial_v L(t, \bar{x}(t), \dot{\bar{x}}(t)) \quad \text{a.e. } t \in [\bar{t}, t_i]. \quad (12.4.3)$$

We claim that $\dot{\bar{x}}$ is essentially bounded on $[\bar{t}, \tau_{\max}]$. If this were not the case, there would exist a point σ_i in $(\bar{t}, t_i)$ for $i = 1, 2, \ldots$, such that $\sigma_i \uparrow \tau_{\max}$, (12.4.3) is

12.4 Lipschitz Continuous Minimizers

satisfied at $t = \sigma_i$ and

$$|\dot{x}_i(\sigma_i)| \to +\infty. \tag{12.4.4}$$

Since $L(\sigma_i, \bar{x}(\sigma_i), .)$ is convex, (12.4.3) implies

$$p_i(\sigma_i) \cdot \dot{\bar{x}}(\sigma_i) - L(\sigma_i, \bar{x}(\sigma_i), \dot{\bar{x}}(\sigma_i)) \geq \min_{|v| \leq 1}\{p_i(\sigma_i) \cdot v - L(\sigma_i, \bar{x}(\sigma_i), v)\}$$
$$\geq -|p_i(\sigma_i)| - \alpha$$

for $i = 1, 2, \ldots$, where

$$\alpha := \sup\{L(\sigma, \bar{x}(\sigma), v) : \sigma \in [S, T], |v| \leq 1\}.$$

It follows that, for $i = 1, 2, \ldots$

$$L(\sigma_i, \bar{x}(\sigma_i), \dot{\bar{x}}(\sigma_i))/|\dot{\bar{x}}(\sigma_i)| \leq (|p_i(\sigma_i)|(1 + |\dot{\bar{x}}(\sigma_i)| + \alpha)/|\dot{\bar{x}}(\sigma_i)|.$$

As $i \to \infty$, the left side of this inequality has limit $+\infty$ by (12.4.4). But the right side is bounded above, in view of (12.4.2). From this contradiction we deduce that $\dot{\bar{x}}$ is essentially bounded on $[\bar{t}, \tau_{\max}]$.

If $\tau_{\max} < T$, then $\tau_{\max}$ is a regular point of $\bar{x}$ and $\dot{\bar{x}}$ is essentially bounded on a relatively open neighbourhood of $\tau_{\max}$ in $[S, T]$. This contradicts the defining property of $\tau_{\max}$. It follows that $\dot{\bar{x}}$ is essentially bounded on all of $[\bar{t}, T]$. □

Our first application of the lemma is to show that, if L does not depend on time (the 'autonomous' case), then all minimizers are Lipschitz continuous.

Proposition 12.4.2 *Assume that, in addition to the Tonelli Existence Hypotheses (HE1)–(HE3), $L(t, x, v)$ is independent of t. Then all L^∞ local minimizers for (P) are Lipschitz continuous.*

Proof Write $L(x, v)$ in place of $L(t, x, v)$. Take an L^∞ local minimizer $\bar{x}$. Choose a regular point $\bar{t} \in (S, T)$ of $\bar{x}$. (This is possible since the regular points have full measure.) We show that $\dot{\bar{x}}$ is essentially bounded on $[\bar{t}, T]$. A similar argument can be used to confirm that $\dot{\bar{x}}$ is essentially bounded on $[S, \bar{t}]$. It will follow that $\bar{x}$ is Lipschitz continuous on $[S, T]$.

Take any $t' \in (\bar{t}, T]$ such that $\dot{\bar{x}}$ is essentially bounded on $[\bar{t}, t']$. Because L is independent of t, we can apply the necessary conditions of Proposition 12.1.2, including the constancy of the Hamiltonian condition, to the optimal subarc $\bar{x}$ restricted to $[\bar{t}, t']$. We know then that there exist an arc $p \in W^{1,1}([\bar{t}, t']; \mathbb{R}^n)$ and a number c such that, for a.e. $t \in [\bar{t}, t']$,

$$c = p(t) \cdot \dot{\bar{x}}(t) - L(\bar{x}(t), \dot{\bar{x}}(t)) \tag{12.4.5}$$

$$p(t) \in \partial_v L(\bar{x}(t), \dot{\bar{x}}(t)). \tag{12.4.6}$$

Fix $\epsilon > 0$ such that $\dot{\bar{x}}$ is essentially bounded on $[\bar{t}, \bar{t} + \epsilon]$. It follows from the local Lipschitz continuity of L and (12.4.6) that there exists a constant $k_1 > 0$, independent of t', such that

$$|p(t)| \leq k_1 \quad \text{a.e. } t \in [\bar{t}, t'] \cap [\bar{t}, \bar{t} + \epsilon].$$

But then (12.4.5) implies that there exists k_2 (independent of t') such that

$$|c| \leq k_2.$$

We deduce from (12.4.5) and (12.4.6) that, for a.e. $t \in [\bar{t}, t']$,

$$\begin{aligned} c &= p(t) \cdot \dot{\bar{x}}(t) - L(\bar{x}(t), \dot{\bar{x}}(t)) \\ &\geq \max_{|v| \leq 1} \{p(t) \cdot v - L(\bar{x}(t), v)\} \\ &\geq |p(t)| - \alpha, \end{aligned}$$

where α, defined by

$$\alpha := \inf\{L(\bar{x}(t), v) : S \leq t \leq T, |v| \leq 1\},$$

does not depend on t'.

Since p is continuous, we have

$$|p(t)| \leq \alpha + c \quad \text{for all } t \in [\bar{t}, t'].$$

We see that p is bounded on $[\bar{t}, t']$ by a constant which does not depend on t'. It follows from Lemma 12.4.1 that $\dot{\bar{x}}$ is essentially bounded on $[\bar{t}, T]$. □

Another case when the Lemma 12.4.1 can be used to establish Lipschitz continuity of minimizers is when $L(t, x, v)$ is convex, jointly in (x, v):

Proposition 12.4.3 *Assume, in addition to the Tonelli Existence Hypotheses (HE1)–(HE3), that $L(t, x, v)$ is convex in (x, v) for each $t \in [S, T]$. Then all L^∞ local minimizers for (BP) are Lipschitz continuous.*

Proof Take an L^∞ local minimizer $\bar{x}$. Choose a regular point $\tau \in (S, T)$ of $\bar{x}$. As in the proof of the previous proposition, we content ourselves with showing that $\dot{\bar{x}}$ is essentially bounded on $[\tau, T]$. The demonstration that the same is true on $[S, \tau]$ is along precisely similar lines.

Take any $t' \in (\bar{t}, T]$ such that $\dot{\bar{x}}$ is essentially bounded on $[\bar{t}, t']$. Observe next that $(\bar{x}, \bar{z}(t) \equiv \int_{\bar{t}}^{t} \tilde{L}(s, \bar{x}(s), \dot{\bar{x}}(s)) ds)$ is a solution to the dynamic optimization problem:

12.4 Lipschitz Continuous Minimizers

$$\begin{cases} \text{Minimize } z(t_i) \\ \text{over } (x, z) \in W^{1,1}([\bar{t}, t']; \mathbb{R}^{n+1}) \text{ satisfying} \\ (\dot{x}(t), \dot{z}(t)) \in F(t, x(t), z(t)) \quad \text{a.e. } t \in [\bar{t}, t'], \\ x(\bar{t}) = \bar{x}(\bar{t}), \ z(\bar{t}) = 0, \ x(t') = \bar{x}(t'), \end{cases}$$

where $F(t, x, z) := \text{epi } L(t, x, .)$. We know from Theorem 8.4.3 that there exists $p \in W^{1,1}([\bar{t}, t']; \mathbb{R}^n)$ such that

$$\dot{p}(t) \in \text{co}\{\eta \,:\, (\eta, 0, p(t), -1) \in N_{\text{Gr } F(t,.,.)}(\bar{x}(t), \bar{z}(t), \dot{\bar{x}}(t), \dot{\bar{z}}(t))\} \text{ a.e. } t \in [\bar{t}, t']. \tag{12.4.7}$$

Since $L(t, ., .)$ is convex, we deduce that $\text{Gr } F(t, .)$ is convex, and so from (12.4.7) it follows that

$$(\dot{p}(t), p(t)) \in \partial L(t, \bar{x}(t), \dot{\bar{x}}(t)) \quad \text{a.e. } t \in [\bar{t}, t'], \tag{12.4.8}$$

where $\partial L(t, ., .)$ is the subdifferential in the usual sense of convex analysis.

Since $\dot{\bar{x}}$ is essentially bounded on a neighbourhood of $\bar{t}$, $L(t, ., .)$ is locally Lipschitz continuous uniformly in t and p is continuous, (12.4.8) implies that there exists $k_1 > 0$ (independent of t') such that

$$|p(\bar{t})| \leq k_1. \tag{12.4.9}$$

Since the 'convex' subdifferential is employed, (12.4.8) implies that, for a.e. $t \in [\bar{t}, t']$,

$$L(t, y, v) - L(t, \bar{x}(t), \dot{\bar{x}}(t))$$
$$\geq (y - \bar{x}(t)) \cdot \dot{p}(t) + (v - \dot{\bar{x}}(t)) \cdot p(t) \quad \text{for all } y \in \mathbb{R}^n, \ v \in \mathbb{R}^n.$$

By examining the implications of this inequality when $y = \bar{x} + u$ and $v = 0$, for an arbitrary unit vector u, we deduce that, for a.e. $t \in [\bar{t}, t']$,

$$|\dot{p}(t)| = \max_{|u| \leq 1} \dot{p} \cdot u$$
$$\leq \max_{|u| \leq 1} L(t, \bar{x}(t) + u, 0) - L(t, \bar{x}(t), \dot{\bar{x}}(t)) + p(t) \cdot \dot{\bar{x}}(t).$$

It follows that, for some integrable functions, γ_1 and γ_2 which do not depend on t',

$$|\dot{p}(t)| \leq \gamma_1(t)|p(t)| + \gamma_2(t) \quad \text{a.e. } t \in [\bar{t}, t'].$$

We deduce from this inequality and (12.4.9), with the help of Gronwall's lemma (Lemma 6.2.4), that there exists $k_2 > 0$ (independent of t') such that

$$|p(t)| \leq k_2 \quad \text{for all } t \in [\bar{t}, t'].$$

It follows now from Lemma 12.4.1 that $\dot{\bar{x}}$ is essentially bounded on $[\bar{t}, T]$. □

Finally, we illustrate how the Tonelli regularity theorem can be used to reduce the hypotheses under which Euler Lagrange type necessary conditions of optimality have traditionally been derived. As in the previous applications, this is done via the intermediary of Lemma 12.4.1.

Proposition 12.4.4 *Take an L^∞ local minimizer $\bar{x}$ for (BP). Assume that, in addition to the Tonelli Existence Hypotheses (HE1)–(HE3), there exist integrable functions c and γ such that, for a.e. $t \in [S, T]$,*

$$\sup_{\xi \in P_x[\mathrm{co}\partial L](t)} |\xi| \leq c(t) \inf_{\eta \in \partial_v L} |\eta| + \gamma(t), \qquad (12.4.10)$$

where $P_x[\mathrm{co}\partial L](t)$ denotes the projection of $\mathrm{co}\partial L$ onto the first coordinate

$$P_x[\mathrm{co}\partial L](t) := \{\xi : (\xi, \eta) \in \mathrm{co}\,\partial L(t, \bar{x}(t), \dot{\bar{x}}(t)) \text{ for some } \eta \in \mathbb{R}^n\}$$

and $\partial_v L$ is evaluated at $(t, \bar{x}(t), \dot{\bar{x}}(t))$.
Then $\bar{x}$ is Lipschitz continuous.

Notice that the supplementary hypothesis (12.4.10) reduces to

$$|L_x(t, \bar{x}(t), \dot{\bar{x}}(t))| \leq c(t)|L_v(t, \bar{x}(t), \dot{\bar{x}}(t))| + \gamma(t) \quad \text{a.e. } t \in [S, T]$$

when L is smooth.

This hypothesis is less restrictive than that invoked in standard necessary conditions, such as those of Proposition 12.1.2, in two respects. Firstly the presence of the non-negative term $c(t)|L_v|$ on the right side reduces the severity of the inequality. Secondly, the inequality is required to hold precisely along the L^∞ local minimizer $\bar{x}$, not over a tube about $\bar{x}$.

Proof Choose a regular point $\tau \in (S, T)$ of $\bar{x}$. As usual, we show merely that $\dot{\bar{x}}$ is essentially bounded on $[\tau, T]$, since a similar analysis can be used to demonstrate the essential boundedness of $\dot{\bar{x}}$ also on $[S, \tau]$.

Take any $t' \in (\bar{t}, T]$ such that $\dot{\bar{x}}$ is essentially bounded on $[\bar{t}, t']$. The necessary conditions of Proposition 12.1.2 supply $p \in W^{1,1}([\bar{t}, t']; \mathbb{R}^n)$ such that

$$\dot{p}(t) \in \mathrm{co}\{q : (q, p(t)) \in \partial L(t, \bar{x}(t), \dot{\bar{x}}(t))\} \quad \text{a.e. } t \in [\bar{t}, t'].$$

This implies that, for a.e. $t \in [\bar{t}, t']$,

$$\dot{p}(t) \in P_x[\mathrm{co}\partial L](t)$$

and

$$p(t) \in \partial_v L(t, \bar{x}(t), \dot{\bar{x}}(t)) \qquad (12.4.11)$$

Since $L(t,.,.)$ is locally Lipschitz continuous uniformly in t, $\bar{x}$ is essentially bounded on a neighbourhood of $\bar{t}$ and p is continuous, we deduce from (12.4.11), that there exists some k_1 (independent of t'), such that

$$|p(\bar{t})| \leq k_1.$$

It follows from supplementary hypothesis (12.4.10) that

$$|\dot{p}(t)| \leq c(t)|p(t)| + \gamma(t) \quad \text{a.e. } t \in [\bar{t}, t'].$$

We now deduce from Gronwall's lemma (Lemma 6.2.4) that there exists $k_2 > 0$ (independent of t') such that

$$|p(t)| \leq k_2 \quad \text{for all } t \in [\bar{t}, t'].$$

It follows from Lemma 12.4.1 that $\dot{\bar{x}}$ is essentially bounded on $[\bar{t}, T]$. □

12.5 Nonautonomous Variational Problems with State Constraints

In the preceding section, we applied the generalized Tonelli regularity theorem Theorem 12.2.2 to show that minimizers of the basic problem (BP) over the space of absolutely continuous functions are Lipschitz continuous in certain cases of interest.

An alternative approach to showing Lipschitz continuity of minimizers is based on an application of the maximum principle and time re-parameterization techniques. This has the merit of simplicity. Besides, the hypotheses which it is necessary to impose are weaker in some instances and the methods extend to cover problems with pathwise state constraints.

Consider the optimization problem

$$(CV) \begin{cases} \text{Minimize } \int_S^T L(t, x(t), \dot{x}(t))dt \\ \text{over arcs } x \in W^{1,1}([S, T]; \mathbb{R}^n) \text{ satisfying} \\ x(S) = x_0, \; x(T) = x_1, \\ \dot{x}(t) \in C \quad \text{a.e. } t \in [S, T], \\ x(t) \in A \quad \text{for all } t \in [S, T], \end{cases}$$

the data for which comprise an interval $[S, T]$, a function $L : [S, T] \times \mathbb{R}^n \times \mathbb{R}^n \to \mathbb{R}$, points $x_0, x_1 \in \mathbb{R}^n$, a cone $C \subset \mathbb{R}^n$, and a set $A \subset \mathbb{R}^n$. We lay stress on the presence of the pathwise state constraint

$$x(t) \in A \quad \text{for all } t \in [S, T].$$

Given a given reference arc $\bar{x} \in W^{1,1}([S, T]; \mathbb{R}^n)$, we shall consider the following conditions, in which $k : [S, T] \times (0, +\infty) \to \mathbb{R}$ is a $\mathcal{L} \times \mathcal{B}^1$ measurable function such that $k(., 1) \in L^1(S, T)$:

(A1): there exists $\varepsilon_* > 0$ such that for a.e. $t \in [S, T]$ and for all $\sigma > 0$ we have

$$|L(s_2, \bar{x}(t), \sigma \dot{\bar{x}}(t)) - L(s_1, \bar{x}(t), \sigma \dot{\bar{x}}(t))| \le k(t, \sigma) |s_2 - s_1| \quad (12.5.1)$$

whenever $s_1, s_2 \in [t - \varepsilon_*, t + \varepsilon_*] \cap [S, T]$,

(A2): for every selection $Q(t, v)$ of $\tilde{\partial}_r L(t, \bar{x}(t), rv)_{r=1}$, we can find $c > \operatorname{ess\,inf}\{|\dot{\bar{x}}(t)| : t \in [S, T]\}$ and a negligible set $\mathcal{N}$ such that,

$$\lim_{R \to +\infty} \sup_{\substack{|v| \ge R, v \in C \\ \tilde{\partial}_r L(t, \bar{x}(t), rv)_{r=1} \ne \emptyset \\ t \in [S, T] \setminus \mathcal{N}}} \{L(t, \bar{x}(t), v) - Q(t, v)\} + \int_S^T k(s, 1)\, ds$$

$$< \inf_{\substack{|v| < c, v \in C \\ \tilde{\partial}_r L(t, \bar{x}(t), rv)_{r=1} \ne \emptyset \\ t \in [S, T] \setminus \mathcal{N}}} \{L(t, \bar{x}(t), v) - Q(t, v)\}. \quad (12.5.2)$$

Here, given a point $(t, x, v) \in [S, T] \times \mathbb{R}^n \times \mathbb{R}^n$ such that $L(t, x, v) < +\infty$, we denote by $\tilde{\partial}_r L(t, x, rv)_{r=1}$ the subdifferential in the sense of convex analysis of the function $(0, +\infty) \ni r \mapsto L(t, x, rv)$ at $r = 1$:

$$\tilde{\partial}_r L(t, x, rv)_{r=1} := \{q \in \mathbb{R} : q \times (r - 1) \le L(t, x, rv)$$
$$- L(t, x, v) \text{ for all } r > 0\}. \quad (12.5.3)$$

We shall interpret condition (12.5.2) to be satisfied whenever the term on the left side of the inequality takes value '$-\infty$' or in situations where there exists $R_0 > 0$ such that $\tilde{\partial}_r L(t, \bar{x}(t), rv)_{r=1} = \emptyset$ for all $|v| \ge R_0$.

Theorem 12.5.1 *Let $\bar{x}$ be a $W^{1,1}$ local minimizer for (CV). Assume that L is Borel measurable and hypotheses (A1) and (A2) are satisfied. Then $\bar{x}$ is Lipschitz continuous.*

A proof of Theorem 12.5.1 is provided at the end of this section.

Corollary 12.5.2 *Let $\bar{x}$ be a $W^{1,1}$ local minimizer for (CV). Assume that*

(i) *L is Borel measurable and satisfies (A1) of Theorem 12.5.1,*
(ii) *L is bounded on bounded subsets of $[S, T] \times A \times C$,*
(iii) *there exists an increasing function $\theta : [0, \infty) \to [0, \infty)$ such that*

$$\lim_{r \to \infty} \theta(r)/r = +\infty$$

12.5 Nonautonomous Variational Problems with State Constraints

and a constant α such that

$$L(t, x, v) > \theta(|v|) - \alpha |v| \quad \text{for all } (x, v) \in \mathbb{R}^n \times \mathbb{R}^n, \text{ a.e. } t \in [S, T].$$

Then $\bar{x}$ is Lipschitz continuous.

Remark
Observe that, if L is independent of t, then (A1) is automatically satisfied (with $k = 0$). Therefore, an immediate consequence of Theorem 12.5.1 is that, in the 'autonomous case', if the Lagrangian is merely Borel measurable and condition (A2) is satisfied (or, in particular, conditions (ii)–(iii) of Corollary 12.5.2 are satisfied), then a $W^{1,1}$ local minimizer for (CV) is always Lipschitz continuous.

Proof of Corollary 12.5.2 Consider any $Q(t, v) \in \tilde{\partial}_r L(t, \bar{x}(t), rv)_{r=1}$ where $0 \neq v \in C$ and $t \in [S, T]$. Then

$$L\left(t, \bar{x}(t), rv\right) - L(t, \bar{x}(t), v) \geq Q(t, v)(r - 1), \quad \text{for all } r > 0. \tag{12.5.4}$$

Taking $r = \dfrac{1}{|v|}$ (> 0) in (12.5.4) we obtain

$$L\left(t, \bar{x}(t), \frac{v}{|v|}\right) \geq L(t, \bar{x}(t), v) + Q(t, v)\left(\frac{1}{|v|} - 1\right)$$

$$\geq \left(L(t, \bar{x}(t), v) - Q(t, v)\right)\left(1 - \frac{1}{|v|}\right) + \frac{1}{|v|} L(t, \bar{x}(t), v).$$

Bearing in mind conditions (ii) and (iii), and the fact that C is a cone (so $\dfrac{v}{|v|}$ remains in C), we deduce, when $|v| \geq 2$, that

$$M - \frac{\theta(|v|)}{|v|} + \alpha \geq \frac{1}{2}\left(L(t, \bar{x}(t), v) - Q(t, v)\right), \quad \text{a.e. } t \in [S, T],$$

for a suitable constant M. Therefore, given a selection $Q(t, v)$ of $\tilde{\partial}_r L(t, \bar{x}(t), rv)_{r=1}$ there exists a subset $\mathcal{N} \subset [S, T]$ of zero measure such that

$$\lim_{R \to +\infty} \sup_{\substack{|v| \geq R, v \in C \\ \tilde{\partial}_r L(t, \bar{x}(t), rv)_{r=1} \neq \emptyset \\ t \in [S, T] \setminus \mathcal{N}}} \{L(t, \bar{x}(t), v) - Q(t, v)\}$$

$$\leq \lim_{R \to +\infty} \sup_{|v| \geq R} 2\left(M - \frac{\theta(|v|)}{|v|} + \alpha\right) = -\infty.$$

It follows that condition (A2) is satisfied (for any selection $Q(t, v)$ and for any $c > \operatorname{ess\,inf}\{|\dot{\bar{x}}(t)| : t \in [S, T]\}$), and so we can apply Theorem 12.5.1 since all the required hypotheses are in force. □

As a preliminary step in the proof of Theorem 12.5.1, we establish a technical lemma and Theorem 12.5.4 below which provides a Weierstrass-type variational inequality: this plays a central role in the proof of Theorem 12.5.1.

Lemma 12.5.3 *Let $\{y_h\}_{h \in \mathbb{N}}$ be a sequence of functions in $W^{1,1}([S, T]; \mathbb{R})$ satisfying the following conditions:*

(i) $y_h(S) = S$ and $y_h(T) = T$, for all $h \in \mathbb{N}$,
(ii) there exists $\omega > 0$ such that, for every $h \in \mathbb{N}$, $\dot{y}_h(t) \geq \omega$ for a.e. $t \in [S, T]$,
(iii) the sequence $\{y_h\}$ converges to $\bar{y}(t) := t$ in $W^{1,1}([S, T]; \mathbb{R})$.

Then, for every $\bar{x} \in W^{1,1}([S, T]; \mathbb{R}^n)$, the sequence $\{\bar{x} \circ y_h^{-1}\}$ admits a subsequence which converges to $\bar{x}$ in $W^{1,1}([S, T]; \mathbb{R}^n)$.

Proof Observe that the sequence of the inverse functions $\{y_h^{-1}\}$ converges to $\bar{y}(t) := t$ in $W^{1,1}([S, T]; \mathbb{R})$. Indeed, recalling that $y_h(S) = S = \bar{y}(S)$, by means of the change of variable $t := y_h(s)$, for each h we obtain:

$$\|y_h^{-1} - \bar{y}\|_{W^{1,1}} = \|(d/dt)(y_h^{-1}) - 1\|_{L^1}$$

$$= \int_S^T \left|\frac{1 - \dot{y}_h(y_h^{-1}(t))}{\dot{y}_h(y_h^{-1}(t))}\right| dt = \int_S^T \left|\frac{1 - \dot{y}_h(s)}{\dot{y}_h(s)}\right| \dot{y}_h(s)\, ds$$

$$= \|\dot{y}_h - 1\|_{L^1} = \|y_h - \bar{y}\|_{W^{1,1}}.$$

Therefore, extracting a subsequence (we do not relabel), we obtain that the sequence $\{(d/dt)(y_h^{-1})\}$ converges to $\dot{\bar{y}} \equiv 1$ almost everywhere on $[S, T]$.

For every $h \in \mathbb{N}$ let ϕ_h be the linear operator $\phi_h : W^{1,1}([S, T]; \mathbb{R}^n) \to W^{1,1}([S, T]; \mathbb{R}^n)$ defined by

$$\phi_h(w) := w \circ y_h^{-1}.$$

The lemma statement is confirmed if we show that $\{\phi_h(\bar{x})\}$ converges to $\bar{x}$ in $W^{1,1}([S, T]; \mathbb{R}^n)$. To see this, notice first of all that, for all $w \in W^{1,1}([S, T]; \mathbb{R}^n)$, we have

$$(d/dt)\phi_h(w)(t) = (\dot{w} \circ y_h^{-1}(t))(d/dt)(y_h^{-1})(t) \quad \text{a.e. } t \in [S, T], \quad (12.5.5)$$

and so

$$\lim_{h \to +\infty} (d/dt)\phi_h(w)(t) = \dot{w}(t) \quad \text{a.e. } t \in [S, T].$$

Moreover, applying the change of variable $s := y_h^{-1}(t)$ we deduce that

12.5 Nonautonomous Variational Problems with State Constraints

$$\|(d/dt)\phi_h(w)\|_{L^1} = \int_S^T |\dot{w}(y_h^{-1}(t))|[(d/dt)(y_h^{-1})(t)]\,dt$$

$$= \int_S^T |\dot{w}(s)|\,ds = \|\dot{w}\|_{L^1}$$

and, consequently,

$$\|\phi_h(w)\|_{W^{1,1}} = |\phi_h(w)(S)| + \|(d/dt)\phi_h(w)\|_{L^1} \leq \|w\|_{W^{1,1}}.$$

It follows that the operators ϕ_h's are equi-bounded.

Now, we claim that, if $w \in C^2([S, T]; \mathbb{R}^n)$, then the sequence $\phi_h(w)$ converges to w in $W^{1,1}([S, T]; \mathbb{R}^n)$. Indeed, since the family of functions $\{y_h^{-1}\}_h$ is equi-Lipschitz continuous (with Lipschitz constant $1/\omega$), from (12.5.5) it follows that

$$|(d/dt)\phi_h(w)(t)| \leq \frac{1}{\omega}\|\dot{w}\|_{L^\infty} \quad \text{a.e. } t \in [S, T].$$

The dominated convergence theorem implies the convergence of $\{(d/dt)\phi_h(u)\}$ to $\dot{w}$ in L^1 as $h \to +\infty$, and, recalling that $\phi_h(w)(S) = w(S)$ for all h, we deduce that $\|\phi_h(w) - w\|_{W^{1,1}} \to 0$ as $h \to +\infty$, confirming the claim above.

Now, consider a sequence $\{w_k\}$ in $C^2([S, T]; \mathbb{R}^n)$ that converges to $\bar{x}$ w.r.t. the $W^{1,1}$ norm. For each $h, k \in \mathbb{N}$ we have:

$$\phi_h(\bar{x}) - \bar{x} = (\phi_h(\bar{x}) - \phi_h(w_k)) + (\phi_h(w_k) - w_k) + (w_k - \bar{x}).$$

The analysis above tells us that:

$$\|\phi_h(\bar{x}) - \bar{x}\|_{W^{1,1}} \leq \|\phi_h(\bar{x}) - \phi_h(w_k)\|_{W^{1,1}} + \|\phi_h(w_k) - w_k\|_{W^{1,1}} + \|w_k - \bar{x}\|_{W^{1,1}}$$

$$\leq 2\|w_k - \bar{x}\|_{W^{1,1}} + \|\phi_h(w_k) - w_k\|_{W^{1,1}},$$

and passing to the limit as $h \to +\infty$ first, and then as $k \to +\infty$, we deduce that $\phi_h(\bar{x}) \to \bar{x}$ as $h \to +\infty$.

This concludes the proof of the lemma. □

Theorem 12.5.4 *Let $\bar{x}$ be a $W^{1,1}$ local minimizer for (CV). Assume that L is Borel measurable and satisfies assumption (A1). Then there exists an arc $p \in W^{1,1}([S, T]; \mathbb{R})$ such that:*

$$L\!\left(t, \bar{x}(t), \frac{\dot{\bar{x}}(t)}{u}\right)u - L(t, \bar{x}(t), \dot{\bar{x}}(t)) \geq p(t)(u - 1)$$

$$\text{for all } u > 0, \text{ a.e. } t \in [S, T], \qquad (12.5.6)$$

$$\dot{p}(t) \in \operatorname{co}\partial_t L(t, \bar{x}(t), \dot{\bar{x}}(t)) \quad \text{a.e. } t \in [S, T] \qquad (12.5.7)$$

and

$$|\dot{p}(t)| \le k(t, 1) \quad \text{a.e. } t \in [S, T]. \tag{12.5.8}$$

Proof In what follows we fix an integer $j \ge 2$ and we take $\bar{y}(t) := t$. We prove first the validity of (12.5.6) for '$u \ge 1/j$' instead of '$u > 0$' for some absolutely continuous arc p_j which satisfies also (12.5.7) and (12.5.8). Then a compactness argument, applied to the sequence of arcs $\{p_j\}$, will allow us to pass to the case '$u > 0$' in a limit-taking procedure.

Step 1. We introduce the following auxiliary Lagrangian: for all $t \in [S, T]$, $y \in \mathbb{R}$ and $u \in \mathbb{R}$ we set

$$\ell(t, y, v) := \begin{cases} \widetilde{L}\left(y, \bar{x}(t), \dfrac{\dot{\bar{x}}(t)}{u}\right) u & \text{if } u \ge 1/j \text{ and } \dot{\bar{x}}(t) \text{ exists,} \\ 0 & \text{otherwise.} \end{cases}$$

where $\widetilde{L}$ is the Borel extension of L to $\mathbb{R} \times \mathbb{R}^n \times \mathbb{R}^n$ defined by

$$\widetilde{L}(t, x, v) := \begin{cases} L(S, x, v) & \text{if } t \le S \\ L(t, x, v) & \text{if } t \in [S, T] \\ L(T, x, v) & \text{if } t \ge T, \end{cases}$$

for all $x, v \in \mathbb{R}^n$. Consider the following auxiliary dynamic optimization problem:

$$(P) \begin{cases} \text{Minimize } \int_S^T \ell(t, y(t), \dot{y}(t)) dt \\ \text{over measurable functions } u : [S, T] \to \mathbb{R} \text{ and arcs} \\ \qquad\qquad y \in W^{1,1}([S, T]; \mathbb{R}) \text{ satisfying} \\ \dot{y}(t) = u \quad \text{a.e. } t \in [S, T], \\ u(t) \in [1/j, \infty) \quad \text{a.e. } t \in [S, T], \\ y(S) = S, \quad y(T) = T, \end{cases}$$

Observe that the function $\ell(t, y, u)$ is $\mathcal{L} \times \mathcal{B}^1 \times \mathcal{B}^1$ measurable, therefore, from Corollary 2.3.3 it follows that the map $t \mapsto \ell(t, y(t), u(t))$ is Lebesgue measurable for every pair (y, v) of Lebesgue measurable functions.

The next lemma establishes that if $\bar{x}$ is a $W^{1,1}$ local minimizer for (CV) then $(\bar{y}(t) := t, \bar{u}(t) \equiv 1)$ is a $W^{1,1}$ local minimizer for (P).

Lemma 12.5.5 *If $\bar{x}$ is a $W^{1,1}$ local minimizer for (CV), then the process $(\bar{y}(t) := t, \bar{u}(t) \equiv 1)$ is a $W^{1,1}$ local minimizer for (P).*

12.5 Nonautonomous Variational Problems with State Constraints

Proof Assume that $\bar{x}$ is a $W^{1,1}$ local minimizer for (CV), which means that there exists $\varepsilon > 0$ such that

$$\int_S^T L(t, \bar{x}(t), \dot{\bar{x}}(t))\, dt \leq \int_S^T L(t, x(t), \dot{x}(t))\, dt,$$

for all admissible arcs x for (CV) such that $\|x - \bar{x}\|_{W^{1,1}} \leq \varepsilon$. We claim that there exists $\rho > 0$ for which, for every $y \in W^{1,1}([S, T]; \mathbb{R})$ such that $y(S) = S$, $y(T) = T$, $\dot{y} \geq 1/j$ a.e. on $[S, T]$ and $\|y - \bar{y}\|_{W^{1,1}} \leq \rho$, we have

$$\|\bar{x} \circ y^{-1} - \bar{x}\|_{W^{1,1}} \leq \varepsilon.$$

Indeed, suppose that the claim is false. Then we can find a sequence $\{y_h\}$ of functions in $W^{1,1}([S, T]; \mathbb{R})$ such that, for all h, $\dot{y}_h \geq 1/j$ a.e. on $[S, T]$ and

$$y_h(S) = S,\ y_h(T) = T,\ \|y_h - \bar{y}\|_{W^{1,1}} \leq \frac{1}{h+1}, \quad \text{but} \quad \|\bar{x} \circ y_h^{-1} - \bar{x}\|_{W^{1,1}} > \varepsilon.$$

On the other hand we know from Lemma 12.5.3 (taking $\omega = 1/j$) that a subsequence of $\{\bar{x} \circ y_h^{-1}\}_h$ converges to $\bar{x}$ in $W^{1,1}([S, T]; \mathbb{R})$, arriving at a contradiction of our premise.

Now, let (y, u) be any process satisfying the constraints of the optimal problem (P) such that $\|y - \bar{y}\|_{W^{1,1}} \leq \rho$. Since $\dot{y}(t) = u(t) \geq 1/j$ almost everywhere on $[S, T]$, we deduce that $y : [S, T] \to [S, T]$ is strictly increasing and

$$|y(t_2) - y(t_1)| = \left|\int_{t_1}^{t_2} \dot{y}(t)\, dt\right| \geq \frac{1}{j}|t_2 - t_1| \quad \text{for all } t_1, t_2 \in [S, T].$$

Therefore y is invertible and its inverse y^{-1} is Lipschitz continuous with Lipschitz constant j. We consider the absolutely continuous function

$$x(\tau) := \bar{x}(y^{-1}(\tau)), \quad \text{for all } \tau \in [S, T].$$

Notice that $x(S) = \bar{x}(S)$, $x(T) = \bar{x}(T)$, and $x(\tau) \in \bar{x}([S, T]) \subset A$. Moreover, $\dot{\bar{x}}(y^{-1}(\tau))$ and $(d/d\tau)(y^{-1})(\tau)$ exist for almost every $\tau \in [S, T]$. We deduce that

$$(d/d\tau)(y^{-1})(\tau) = \frac{1}{\dot{y}(y^{-1}(\tau))}, \quad \text{and} \quad \dot{x}(\tau) = \frac{\dot{\bar{x}}(y^{-1}(\tau))}{\dot{y}(y^{-1}(\tau))} \in C, \text{ a.e. } \tau \in [S, T],$$

and, so, x is an admissible arc for problem (CV) such that $\|x - \bar{x}\|_{W^{1,1}} \leq \varepsilon$. (The fact that x is absolutely continuous and the validity of the chain rule for the derivative follows from the fact that y^{-1} is a monotone absolutely continuous function.) It follows that

$$\int_S^T L(\tau, x(\tau), \dot{x}(\tau))\, d\tau \geq \int_S^T L(t, \bar{x}(t), \dot{\bar{x}}(t))\, dt. \tag{12.5.9}$$

Of course we have

$$\int_S^T L(t, \bar{x}(t), \dot{\bar{x}}(t))\, dt = \int_S^T \ell(t, \bar{y}(t) \equiv t, \bar{u}(t) \equiv 1)\, dt.$$

Now, using the change of variables $t = y^{-1}(\tau)$ on the left term of (12.5.9), we deduce that

$$\int_S^T L(\tau, x(\tau), \dot{x}(\tau))\, d\tau = \int_S^T L\!\left(\tau, \bar{x}(y^{-1}(\tau)), \frac{\dot{\bar{x}}(y^{-1}(\tau))}{\dot{y}(y^{-1}(\tau))}\right) d\tau$$

$$= \int_S^T L\!\left(y(t), \bar{x}(t), \frac{\dot{\bar{x}}(t)}{\dot{y}(t)}\right) \dot{y}(t)\, dt$$

$$= \int_S^T \ell(t, y(t), \dot{y}(t))\, dt.$$

This confirms the lemma statement. □

Step 2. Set

$$k^*(t, u) := \begin{cases} k\!\left(t, \dfrac{1}{u}\right) u & \text{if } (t, u) \in [S, T] \times [1/j, +\infty) \\ 0 & \text{if } (t, u) \in [S, T] \times (-\infty, 1/j), \end{cases}$$

where k is the $\mathcal{L} \times \mathcal{B}^1$ measurable function provided by assumption (A1). Observe that k^* is $\mathcal{L} \times \mathcal{B}^1$ measurable and $k^*(., \bar{u} \equiv 1)(= k(., 1)) \in L^1(S, T)$. Moreover, for almost every $t \in [S, T]$, we have

$$|\ell(t, y_2, u) - \ell(t, y_1, u)|$$
$$\leq k^*(t, u)|y_2 - y_1|, \quad \text{for all } y_1, y_2 \in [t - \varepsilon_*, t + \varepsilon_*] \text{ and } u \geq 1/j. \tag{12.5.10}$$

Indeed, let t be such that (12.5.1) holds. For any $y_1, y_2 \in [t - \varepsilon_*, t + \varepsilon_*]$ and $u \geq 1/j$ we have

$$|\ell(t, y_2, u) - \ell(t, y_1, u)| = \left|\tilde{L}\!\left(y_2, \bar{x}(t), \frac{\dot{\bar{x}}(t)}{u}\right) - \tilde{L}\!\left(y_1, \bar{x}(t), \frac{\dot{\bar{x}}(t)}{u}\right)\right| u. \tag{12.5.11}$$

We claim that

$$\left|\tilde{L}\!\left(y_2, \bar{x}(t), \frac{\dot{\bar{x}}(t)}{u}\right) - \tilde{L}\!\left(y_1, \bar{x}(t), \frac{\dot{\bar{x}}(t)}{u}\right)\right| \leq k\!\left(t, \frac{1}{u}\right)|y_2 - y_1|. \tag{12.5.12}$$

12.5 Nonautonomous Variational Problems with State Constraints

Observe that it is not restrictive to assume that $\varepsilon_* < (T-S)/2$ and $y_1 \leq y_2$. Clearly the claim is trivial if $y_1 \leq y_2 \leq S$ or $T \leq y_1 \leq y_2$. So we can restrict attention to the following three cases.

Case 1: $y_1 \leq S \leq y_2$. Then, we have

$$\left| \tilde{L}\left(y_2, \bar{x}(t), \frac{\dot{\bar{x}}(t)}{u}\right) - \tilde{L}\left(y_1, \bar{x}(t), \frac{\dot{\bar{x}}(t)}{u}\right) \right|$$

$$= \left| L\left(y_2, \bar{x}(t), \frac{\dot{\bar{x}}(t)}{u}\right) - L\left(S, \bar{x}(t), \frac{\dot{\bar{x}}(t)}{u}\right) \right|$$

$$\leq k\left(t, \frac{1}{u}\right) |y_2 - S| \leq k\left(t, \frac{1}{u}\right) |y_2 - y_1|.$$

Case 2: $S \leq y_1 \leq y_2 \leq T$. Then $\tilde{L}\left(y_i, \bar{x}(t), \frac{\dot{\bar{x}}(t)}{u}\right) = L\left(y_i, \bar{x}(t), \frac{\dot{\bar{x}}(t)}{u}\right)$, for $i=1,2$, and (12.5.12) follows directly from (12.5.1).

Case 3: $y_1 \leq T \leq y_2$. In this case the analysis can be carried out in a similar way to that one outlined in case 1. The validity of (12.5.10) is a direct consequence of (12.5.11) and (12.5.12).

As a consequence, taking $(\bar{y}(t) = t, \bar{u}(t) \equiv 1)$ as a reference $W^{1,1}$ local minimizer, since all the required hypotheses are satisfied we can apply the maximum principle Theorem 7.2.1 to the above dynamic optimization problem (P), after reducing it to a (right) end-point cost problem by state augmentation.

The (un-maximized) Hamiltonian, which may depend also on the cost multiplier $\lambda \geq 0$, takes the form:

$$\mathcal{H}_\lambda(t, y, p, u) := pu - \lambda \ell(t, y, u).$$

We deduce the existence of an absolutely continuous function $p_j : [S, T] \to \mathbb{R}$ and (normalizing if necessary) a number $\lambda \in \{0, 1\}$ satisfying the nontriviality condition

$$(p_j(t), \lambda) \neq (0, 0) \quad \text{for all } t \in [S, T], \tag{12.5.13}$$

the co-state inclusion

$$-\dot{p}_j(t) \in \text{co}\, \partial_y \mathcal{H}_\lambda(t, \bar{y}(t), p_j(t), \bar{u}(t))$$

$$= -\lambda \text{co}\, \partial_y \ell\left(t, \bar{y}(t), \bar{u}(t)\right) \quad \text{a.e. } t \in [S, T], \tag{12.5.14}$$

and the Weierstrass condition

$$\mathcal{H}_\lambda(t, \bar{y}(t), p_j(t), \bar{u}(t)) = \max_{u \geq 1/j} \mathcal{H}_\lambda(t, \bar{y}(t), p_j(t), u) \quad \text{a.e. } t \in [S, T]. \tag{12.5.15}$$

Taking into account that $\bar{y}(t) = t$ and $\bar{u}(t) \equiv 1$, (12.5.15) yields

$$p_j(t) - \lambda\ell(t,t,1) \geq p_j(t)u - \lambda\ell\left(t,t,u\right) \quad \text{for all } u \geq 1/j, \quad \text{a.e. } t \in [S,T],$$

which, in terms of the reference Lagrangian L, becomes

$$p_j(t) - \lambda L(t,\bar{x}(t),\dot{\bar{x}}(t)) \geq p_j(t)u$$

$$- \lambda L\left(t,\bar{x}(t),\frac{\dot{\bar{x}}(t)}{u}\right)u \quad \text{for all } u \geq 1/j, \quad \text{a.e. } t \in [S,T]. \quad (12.5.16)$$

If $\lambda = 0$, then from (12.5.16) we would obtain that, for almost every $t \in [S,T]$,

$$p_j(t) \geq p_j(t)u \quad \text{for all } u \geq 1/j,$$

which would imply that $p_j(t) = 0$, contradicting (12.5.13). Thus we necessarily have $\lambda = 1$, and from (12.5.14) and (12.5.16) it follows that

$$L\left(t,\bar{x}(t),\frac{\dot{\bar{x}}(t)}{u}\right)u - L(t,\bar{x}(t),\dot{\bar{x}}(t)) \geq p_j(t)(u-1)$$

$$\text{for all } u \geq 1/j, \quad \text{a.e. } t \in [S,T], \quad (12.5.17)$$

$$\dot{p}_j(t) \in \text{co } \partial_t L(t,\bar{x}(t),\dot{\bar{x}}(t)) \quad \text{a.e. } t \in [S,T], \quad (12.5.18)$$

and

$$|\dot{p}_j(t)| \leq k(t,1) \quad \text{a.e. } t \in [S,T]. \quad (12.5.19)$$

Step 3. To conclude the proof of the theorem, we must show that there exists an absolutely continuous function $p \in W^{1,1}([S,T];\mathbb{R})$ such that

$$L\left(t,\bar{x}(t),\frac{\dot{\bar{x}}(t)}{u}\right)u - L(t,\bar{x}(t),\dot{\bar{x}}(t)) \geq p(t)(u-1)$$

$$\text{for all } u > 0, \text{ a.e. } t \in [S,T], \quad (12.5.20)$$

$$\dot{p}(t) \in \text{co } \partial_t L(t,\bar{x}(t),\dot{\bar{x}}(t)) \quad \text{and} \quad |\dot{p}(t)| \leq k(t,1)$$

$$\text{a.e. } t \in [S,T]. \quad (12.5.21)$$

First observe that, employing a standard countable additivity argument, there exists a set of full measure $E \subset [S,T]$ such that (12.5.17) holds for every $t \in E$ and for every $j \geq 2$. Fix $t_0 \in E$. Let $u_1, u_2 \geq 1/2$ such that $u_1 < 1 < u_2$. By applying

12.5 Nonautonomous Variational Problems with State Constraints

(12.5.17) respectively taking $u = u_1$ and $u = u_2$, we obtain the following relations, which are valid for *all* $j \geq 2$:

$$\frac{1}{1-u_1}\left(L(t_0, \bar{x}(t_0), \dot{\bar{x}}(t_0)) - L\left(t_0, \bar{x}(t_0), \frac{\dot{\bar{x}}(t_0)}{u_1}\right)u_1\right) \leq p_j(t_0), \quad (12.5.22)$$

$$p_j(t_0) \leq \frac{1}{u_2-1}\left(L\left(t_0, \bar{x}(t_0), \frac{\dot{\bar{x}}(t_0)}{u_2}\right)u_2 - L(t_0, \bar{x}(t_0), \dot{\bar{x}}(t_0))\right). \quad (12.5.23)$$

Therefore the sequence $\{p_j(t_0)\}_j$ is bounded and, using also (12.5.19), we deduce that $\{p_j(S)\}_j$ is a bounded sequence. Then the compactness of trajectories theorem (Theorem 6.3.3) guarantees that along a subsequence (we do not relabel) p_j converges weakly in $W^{1,1}$ to an arc $p \in W^{1,1}([S, T]; \mathbb{R})$ satisfying (12.5.21). Fix any $t \in E$ and let $u > 0$. Take $N \geq 1$ with $1/N < u$. Since (12.5.17) is valid for every $j \geq 2$, passing to the limit as $j \to +\infty$, we obtain that the inequality in (12.5.20) is satisfied for every $t \in E$. Recalling that $E \subset [S, T]$ is of full measure we can conclude that (12.5.20) is confirmed.

Proof of Theorem 12.5.1 Since we assume that (A1) is satisfied, from Theorem 12.5.4 we know that we can find an absolutely continuous function $p \in W^{1,1}([S, T]; \mathbb{R})$ satisfying properties (12.5.6), (12.5.7) and (12.5.8). Let $t \in [S, T]$ be a point such that the Weierstrass-type condition (12.5.6) is satisfied. The change of variable $r = \frac{1}{u}$ gives

$$L(t, \bar{x}(t), r\dot{\bar{x}}(t))\frac{1}{r} - L(t, \bar{x}(t), \dot{\bar{x}}(t))$$

$$\geq p(t)\left(\frac{1}{r} - 1\right), \quad \text{for all } r > 0, \quad \text{a.e. } t \in [S, T]. \quad (12.5.24)$$

Multiplying both terms of (12.5.24) by r we obtain

$$L(t, \bar{x}(t), r\dot{\bar{x}}(t)) - rL(t, \bar{x}(t), \dot{\bar{x}}(t))$$

$$\geq p(t)(1 - r) \quad \text{for all } r > 0, \quad \text{a.e. } t \in [S, T]. \quad (12.5.25)$$

Adding $(r-1)L(t, \bar{x}(t), \dot{\bar{x}}(t))$ to both terms of (12.5.25) we deduce that

$$L(t, \bar{x}(t), r\dot{\bar{x}}(t)) - L(t, \bar{x}(t), \dot{\bar{x}}(t))$$

$$\geq [L(t, \bar{x}(t), r\dot{\bar{x}}(t)) - p(t)](r-1) \quad \text{for all } r > 0, \quad \text{a.e. } t \in [S, T].$$

This means that $L(t, \bar{x}(t), \dot{\bar{x}}(t)) - q(t) \in \tilde{\partial}_r L(t, \bar{x}(t), r\dot{\bar{x}}(t))_{r=1}$ (see (12.5.3) for the definition of $\tilde{\partial}_r L$). Setting $q(t) := L(t, \bar{x}(t), \dot{\bar{x}}(t)) - p(t)$, it follows that

$$\begin{cases} L(t, \bar{x}(t), \dot{\bar{x}}(t)) - q(t) = p(t), \\ q(t) \in \tilde{\partial}_r L(t, \bar{x}(t), r\dot{\bar{x}}(t))_{r=1} \end{cases} \quad \text{a.e. } t \in [S, T]. \tag{12.5.26}$$

Let $Q(t, v)$ be a selection of $\tilde{\partial}_r L(t, \bar{x}(t), rv)_{r=1}$ (at those points where the latter set is nonempty) such that $Q(t, \dot{\bar{x}}(t)) = q(t)$ a.e. on $[S, T]$.

Recall now that we are assuming also (A2). So, if $\tilde{\partial}_r L(t, \bar{x}(t), rv)_{r=1} = \emptyset$ or is not defined when $|v| > R_0$, for some $R_0 > 0$, then (12.5.26) implies that $\|\dot{\bar{x}}\|_{L^\infty} \leq R_0$. Otherwise, there exist $c \in \mathbb{R}$ and a negligible set $\mathcal{N}$ such that (12.5.2) holds. Then we can find a set $E_0 \subset [S, T]$ of positive measure such that, for each point $t_0 \in E_0$, $|\dot{\bar{x}}(t_0)| < c$. Let p, q be as in (12.5.26). Fixing a point $t_0 \in E_0 \setminus \mathcal{N}$, we have

$$p(t) = L(t, \bar{x}(t), \dot{\bar{x}}(t)) - q(t) = p(t_0) + \int_{t_0}^{t} \dot{p}(s)\,ds \quad \text{a.e. } t \in [S, T].$$

Since $\dot{\bar{x}}(t_0) \in C$ it follows that

$$p(t) \geq \inf_{\substack{|v|<c, v \in C \\ \tilde{\partial}_r L(\tau, \bar{x}(\tau), rv)_{r=1} \neq \emptyset \\ \tau \in [S,T] \setminus \mathcal{N}}} \{L(\tau, \bar{x}(\tau), v) - Q(\tau, v)\} - \int_S^T k(s, 1)\,ds.$$

From condition (A2) we can find $\bar{R} > 0$ such that, for a.e. $t \in [S, T]$,

$$p(t) = L(t, \bar{x}(t), \dot{\bar{x}}(t)) - q(t) > \sup_{\substack{|v| \geq \bar{R}, v \in C \\ \tilde{\partial}_r L(t, \bar{x}(t), rv)_{r=1} \neq \emptyset}} \{L(t, \bar{x}(t), v) - Q(t, v)\}.$$

We conclude that $|\dot{\bar{x}}(t)| \leq \bar{R}$ a.e. $t \in [S, T]$. □

12.6 Bounded Controls

Up to now, we have restricted attention to regularity properties of minimizers for variational problems with no dynamic constraints. Our investigations have centred on conditions under which derivatives are locally essentially bounded on some relatively open subset of full measure and on the implications of such conditions. What regularity properties of minimizers can be established for problems with nonlinear dynamic constraints, under standard hypotheses guaranteeing existence of a minimizer? General results are lacking. Our earlier analysis adapts however to yield conditions for optimal controls to be locally essentially bounded (and hence for optimal state trajectories to have locally essentially bounded derivatives) in the case of a time invariant linear dynamic constraint:

$$\dot{x}(t) = Ax(t) + Bu(t) + d(t).$$

12.6 Bounded Controls

Consider the problem:

$$(L) \begin{cases} \text{Minimize } \int_S^T L(t, x(t), u(t))dt \\ \text{over } x \in W^{1,1}([S, T]; \mathbb{R}^n) \text{ and measurable } u : [S, T] \to \mathbb{R}^m \text{ satisfying} \\ \dot{x}(t) = Ax(t) + Bu(t) + d(t) \quad \text{a.e.,} \\ x(S) = x_0, \; x(T) = x_1, \end{cases}$$

the data for which comprise an interval $[S, T]$, functions $L : [S, T] \times \mathbb{R}^n \times \mathbb{R}^m \to \mathbb{R}$ and $d : [S, T] \to \mathbb{R}^n$, matrices $A \in \mathbb{R}^{n \times n}$ and $B \in \mathbb{R}^{n \times m}$ and points $x_0, x_1 \in \mathbb{R}^n$.

Theorem 12.6.1 *Suppose that the data for (L) satisfy the following hypotheses*

(H1): $L(t, x, u)$ is bounded on bounded sets, measurable in t and convex in u. d is an integrable function,

(H2): For each bounded set $M \subset \mathbb{R}^n \times \mathbb{R}^m$, there exists a constant K such that for all $t \in [S, T]$ and $(x_1, u_1), (x_2, u_2) \in M$

$$|L(t, x_1, u_1) - L(t, x_2, u_2)| \leq K|(x_1 - x_2, u_1 - u_2)|,$$

(H3): There exist a number $c \geq 0$ and a convex function $\theta : [0, \infty) \to [0, \infty)$ such that $\theta(r)/r \to \infty$ as $r \to \infty$ and

$$L(t, x, v) \geq -c|x| + \theta(|v|)$$

for all $(t, x, v) \in [S, T] \times \mathbb{R}^n \times \mathbb{R}^m$.

Suppose that there exists a process that satisfies the end-point constraints for (L).

Take any minimizer $(\bar{x}, \bar{u})$. (Under the hypotheses a minimizer exists.) Then there exists a closed subset $\Omega \subset [S, T]$ of zero measure with the following property: for any $t' \in [S, T] \setminus \Omega$, $\bar{u}$ is essentially bounded on a relative neighbourhood of t'. Furthermore, there exists a measurable function $p : [S, T] \to \mathbb{R}^n$ which is locally Lipschitz continuous on $[S, T] \setminus \Omega$, such that

$$-\dot{p}(t) \in p(t)A - \text{co}\, \partial_x L(t, \bar{x}(t), \bar{u}(t)) \quad \text{a.e.}$$

and

$$\mathcal{H}(t, \bar{x}(t), \bar{u}(t), p(t)) = \max_{u \in \mathbb{R}^m} \mathcal{H}(t, \bar{x}(t), u, p(t)) \quad \text{a.e.}$$

where $\mathcal{H}$ denotes

$$\mathcal{H}(t, x, u, p) := p \cdot (Ax + Bu) - L(t, x, u)$$

and $\partial_x L$ denotes the limiting subgradient with respect to x.

We deduce from Theorem 2.4.1 that (L) has a minimizer. However the earlier derived maximum principle, Theorem 7.2.1, cannot be applied to this problem, because the hypotheses of Theorem 12.6.1, which are tailored to the requirements of existence theory, do not imply those of Theorem 7.2.1. Theorem 12.6.1 asserts validity of a weaker form of maximum principle, in which the adjoint arc is not required to be absolutely continuous, but is required instead to be merely locally Lipschitz continuous on a relatively open subset of $[S, T]$, of full measure.

Theorem 12.6.1 is proved by reducing problem (L) to a variational problem without a dynamic constraint, but in which the cost integrand depends on x and its higher derivatives $Dx, \ldots, D^{\tilde{n}-1}x$, for some $\tilde{n} \leq n$, and then carrying out a similar, but more intricate, analysis to that of Sect. 12.3. Details are given in [82].

Not surprisingly, Theorem 12.6.1 serves as a stepping stone to proving essential boundedness of optimal controls in special cases. One such case we now consider. In applications of dynamic optimization to control system design, it is usually necessary to take account of magnitude constraints on control variables, which reflect actuator limitations, safety considerations or permissible regions of the control variable space for validity of the dynamic model. The presence of constraints can, however, greatly complicate the computation of optimal controls. The standard technique for bypassing these difficulties is to drop the constraint and to add, instead, a term $\epsilon|u|^r$ to the cost integrand which penalizes excessive control action. In quadratic cost control it is known that inclusion of this penalty term ensures that optimal controls are bounded, for any $\epsilon > 0$ and $r = 2$. Furthermore, the larger ϵ, the smaller is the uniform bound on optimal controls. This is shown by direct calculation, an approach which is not possible for (L) in general. We can however use Theorem 12.6.1 to assess how large the exponent r in the penalty term must be, at least to ensure boundedness of controls.

Proposition 12.6.2 *Let $(\bar{x}, \bar{u})$ be a minimizer for (L). Assume that hypotheses (H1)–(H3) of Theorem 12.6.1 are satisfied and that L has the form*

$$L(t, x, u) = L_1(t, x, u) + \epsilon|u|^r$$

in which $L_1 : [S, T] \times \mathbb{R}^n \times \mathbb{R}^m \to \mathbb{R}$ is a given non-negative valued function and $r \geq 1$ and $\epsilon > 0$ are given numbers. Suppose further that:

Given any compact set $D \subset [S, T] \times \mathbb{R}^n$, there exists a number c such that

$$\max\{|a| + |b| : (a, b) \in \text{co}\partial L(t, x, u)\} \leq c(1 + |u|^r)$$

for all $(t, x) \in D$ and $u \in \mathbb{R}^m$.

Then $\bar{u}$ is essentially bounded on $[S, T]$.

A similar analysis to that of Sect. 12.4 can be used to prove the proposition. See [82] for details.

Recall the Ball Mizel example, which can be reformulated as an dynamic optimization problem:

$$\begin{cases} \text{Minimize } \int_0^1 \{(x^3(t) - t^2)^2 |u(t)|^{14} + \epsilon |u(t)|^2\} dt \\ \text{over } x \in W^{1,1}([0,1]; \mathbb{R}) \text{ and measurable } u : [0,1] \to \mathbb{R} \text{ satisfying} \\ \dot{x}(t) = u(t) \quad \text{a.e.,} \\ x(S) = 0, \quad x(1) = k. \end{cases}$$

It follows from the discussion of Sect. 12.1, that there is a unique optimal control for this problem, namely the unbounded function $u(t) = kt^{-1/3}$, for suitable choices of the constants $k > 0$ and $\epsilon > 0$. Proposition 12.6.2 provides an engineering perspective on this problem. It tells us that $r = 2$ is too small a value for the exponent in the 'penalty term' $\epsilon |u(t)|^r$, to ensure boundedness of optimal controls. On the other hand, it is a straightforward matter to check by applying Proposition 12.6.2 that optimal controls are bounded if the penalty term is taken to be

$$\epsilon \int_S^T |u(t)|^r dt$$

for any $\epsilon > 0$ and $r \geq 14$.

12.7 Lipschitz Continuous Controls

In certain cases, necessary conditions of optimality can be used directly to establish regularity properties of minimizers. Indeed, examining implications of known necessary conditions is the traditional approach to regularity analysis. A simple example is Hilbert's proof that arcs satisfying the Euler Lagrange condition for the basic problem in the calculus of variations are automatically of class C^r if the Lagrangian $L(t, x, v)$ is of class C^r ($r \geq 2$) and strictly convex in v. (The main steps are reproduced in the proof above of assertion (iv) of Theorem 12.2.4.) We illustrate the method by giving conditions under which a control function satisfying the state constrained maximum principle of Chap. 10 is Lipschitz continuous.

Consider the dynamic optimization problem

$$(R) \begin{cases} \text{Minimize } l(x(S), x(T)) + \int_S^T [L(t, x(t)) + \frac{1}{2} u^T(t) R u(t)] dt \\ \text{over arcs } x \in W^{1,1}([S, T]; \mathbb{R}^n) \text{ and} \\ \quad \text{measurable functions } u : [S, T] \to \mathbb{R}^m \text{ satisfying} \\ \dot{x}(t) = f(t, x(t)) + G(t, x(t)) u(t) \quad \text{a.e.,} \\ h(x(t)) \leq 0 \quad \text{for all } t \in [S, T], \\ (x(S), x(T)) \in C, \end{cases}$$

with data an interval $[S, T]$, functions $L : [S, T] \times \mathbb{R}^n \to \mathbb{R}$, $l : \mathbb{R}^n \times \mathbb{R}^n \to \mathbb{R}$, $f : [S, T] \times \mathbb{R}^n \to \mathbb{R}^n$, $G : [S, T] \times \mathbb{R}^n \to \mathbb{R}^{n \times m}$, $h : \mathbb{R}^n \to \mathbb{R}$, a closed set $C \subset \mathbb{R}^n \times \mathbb{R}^n$ and a symmetric $m \times m$ matrix R.

It is to be expected that control functions for (R) satisfying the maximum principle (with non-zero cost multiplier) are Lipschitz continuous, when the corresponding state trajectories are interior. In this case, the regularity property follows directly from the generalized Weierstrass condition and strict convexity of the Unmaximized Hamiltonian with respect to the control variable. The fact that optimal controls are Lipschitz continuous also for problems with active state constraints comes, on the other hand, as something of a surprise. It means that, for the class of problems here considered, optimal state trajectories do not instantly change direction when they strike the boundary of the state constraint set.

Write

$$\mathcal{H}(t, x, p, u) := p \cdot [f(t, x) + G(t, x)u] - [L(t, x) + (1/2)u^T R u]$$

and, for $x \in W^{1,1}$,

$$I(x) := \{t \in [S, T] : h(x(t)) = 0\}.$$

We say that a process $(\bar{x}, \bar{u})$, satisfying the constraints of (R) is a *normal extremal* if there exist $p \in W^{1,1}([S, T]; \mathbb{R}^n)$ and $\mu \in C^*(S, T)$ such that

$$-\dot{p}(t) \in \partial_x \mathcal{H}(t, \bar{x}(t), p(t) + \int_{[S,t]} \nabla h(\bar{x}(s)) d\mu(s), \bar{u}(t)) \quad \text{a.e.},$$

$$(p(S), -[p(T) + \int_{[S,T]} \nabla h(\bar{x}(t)) d\mu(t)]) \in \partial l(\bar{x}(S), \bar{x}(T)) + N_C(\bar{x}(S), \bar{x}(T)),$$

$$\operatorname{supp}\{\mu\} \subset I(\bar{x}),$$

$$\mathcal{H}(t, \bar{x}(t), p(t) + \int_{[S,t]} \nabla h(\bar{x}(s)) d\mu(s), \bar{u}(t)) \tag{12.7.1}$$
$$= \max_{u \in \mathbb{R}^n} \mathcal{H}(t, \bar{x}(t), p(t) + \int_{[S,t]} \nabla h(\bar{x}(s)) d\mu(s), u). \quad \text{a.e..}$$

In other words, a normal extremal is an admissible process, for which the maximum principle is satisfied with cost multiplier $\lambda = 1$.

Theorem 12.7.1 (Lipschitz Continuity of Optimal Controls) *Take a normal extremal $(\bar{x}, \bar{u})$ for (R). Assume that*

(H1): L, f, G and l are locally Lipschitz continuous,

12.7 Lipschitz Continuous Controls

(H2): h *is of class* $C^{1,1}$, *i.e.* h *is everywhere differentiable with a derivative which is locally Lipschitz continuous,*

(H3): $\nabla h^T(\bar{x}(t))G(t, \bar{x}(t)) \neq 0$ *for all* $t \in I(\bar{x})$,

(H4): R *is positive definite.*

Then $\bar{u}$ is Lipschitz continuous.

Remarks

(i): The proof to follow can be adapted to allow for multiple state constraints. Extensions are also possible to cover problems in which the cost integrand is strictly convex in the control variable, but possibly non-quadratic.

(ii): The normality hypothesis in Theorem 12.7.1 is of an intrinsic nature. Sufficient conditions, open to direct verification, can be given for normality. These typically require that local approximations to the dynamics are controllable in some sense, by means of controls which maintain strict feasibility of the pathwise state constraint.

(iii): Hager [120] and Malanowski [146] have shown that optimal controls are Lipschitz continuous (and have estimated the Lipschitz constants of optimal controls and Lagrange multipliers) for certain classes of problems involving control constraints as well as state constraints. Theorem 12.7.1 does not cover problems with control constraints but, on the other hand, departs from [120] and [146] by allowing nonsmooth data and also initial states which lie in the state constraint boundary.

Proof In view of the special structure of the Hamiltonian, we deduce from (12.7.1) that

$$\bar{u}(t) = R^{-1}G^T(t, \bar{x}(t))\left[p(t) + \int_{[S,t]} \nabla h(\bar{x}(s))d\mu(s)\right] \quad \text{a.e..} \quad (12.7.2)$$

It is immediately evident from this expression that $\bar{u}$ is essentially bounded. It follows also that $\bar{u}$ can be chosen to have left and right limits on (S, T) and to be continuous at $t = S$ and $t = T$. We deduce from the state and adjoint equation that p and x are both Lipschitz continuous.

Step 1: We show that μ has no atoms in (S, T) and, for any $t \in (S, T) \cap I(\bar{x})$,

$$\nabla h^T(\bar{x}(t))\left[f + GR^{-1}G^T\left(p(t) + \int_{[S,t]} \nabla h(\bar{x}(s))d\mu(s)\right)\right] = 0. \quad (12.7.3)$$

(In this relation, f and G are evaluated at $(t, \bar{x}(t))$.)

Take any point $t \in (S, T) \cap I(\bar{x})$. Then the fact that $\bar{x}(t)$ satisfies the state constraint permits us to conclude that, for all $\delta > 0$ sufficiently small,

$$\delta^{-1}(h(\bar{x}(t + \delta)) - h(\bar{x}(t))) \leq 0,$$

$$\delta^{-1}(h(\bar{x}(t)) - h(\bar{x}(t-\delta))) \geq 0.$$

Passing to the limit as $\delta \downarrow 0$, we deduce that

$$\nabla h^T(\bar{x}(t))\left[f + GR^{-1}G^T\left(p(t) + \int_{[S,t]} \nabla h(\bar{x}(s))d\mu(s)\right)\right] \leq 0,$$

$$\nabla h^T(\bar{x}(t))\left[f + GR^{-1}G^T\left(p(t) + \int_{[S,t)} \nabla h(\bar{x}(s))d\mu(s)\right)\right] \geq 0.$$

Subtracting these inequalities gives

$$\nabla h^T(\bar{x}(t))G(t,\bar{x}(t))R^{-1}G^T(t,\bar{x}(t))\nabla h(\bar{x}(t))\mu(\{t\}) \leq 0.$$

Since $\nabla h^T(\bar{x}(t))G(t,\bar{x}(t))R^{-1}G^T(t,\bar{x}(t))\nabla h(\bar{x}(t)) > 0$, it follows that

$$\mu(\{t\}) = 0.$$

These relations also imply (12.7.3). Since the support of μ is contained in $I(\bar{x})$, we conclude that μ has no atoms in (S, T).

Step 2: We show that $t \to \int_{[S,t]} d\mu(s)$ is Lipschitz continuous on (S, T). As $\bar{u}$ is continuous at $t = S$ and $t = T$, the Lipschitz continuity of $\bar{u}$ on $[S, T]$ then follows directly from (12.7.2).

Assume to the contrary that the function is not Lipschitz continuous on (S, T). Then there exist $K_i \uparrow \infty$ and a sequence of intervals $\{[s_i, t_i]\}$ in (S, T) such that, for each i,

$$s_i \neq t_i \text{ and } \int_{s_i}^{t_i} d\mu(s) = K_i|t_i - s_i|. \tag{12.7.4}$$

Since $\text{supp}\{\mu\} \subset I(\bar{x})$, it follows that $[s_i, t_i] \cap I(\bar{x}) \neq \emptyset$. Furthermore, we can arrange by increasing s_i and decreasing t_i if necessary that

$$s_i, t_i \in (S, T) \cap I(\bar{x}).$$

In view of (12.7.4), we can ensure by subsequence extraction that either

(A): $\int_{s_i}^{\frac{s_i+t_i}{2}} d\mu(s) \geq \frac{1}{2}\int_{s_i}^{t_i} d\mu(s),$ for all i
or
(B): $\int_{\frac{s_i+t_i}{2}}^{t_i} d\mu(s) \geq \frac{1}{2}\int_{s_i}^{t_i} d\mu(s),$ for all i.

Assume first (A). Under the hypotheses and since ∇h is continuous, there exists $\beta > 0$ such that, for each i sufficiently large,

$$\nabla h^T(\bar{x}(t))GR^{-1}G^T\nabla h(\bar{x}(s)) > \beta \quad \text{for all } s, t \in [s_i, t_i].$$

12.7 Lipschitz Continuous Controls

Since $h(\bar{x}(s_i)) = 0$,

$$\begin{aligned} h(\bar{x}(t_i)) &= 0 + \int_{s_i}^{t_i} \tfrac{d}{dt} h(\bar{x}(s)) ds \\ &= \int_{s_i}^{t_i} \nabla h^T(\bar{x}(t)) \left[f + GR^{-1} G^T \left(p(t) + \int_{[S,t]} \nabla h(\bar{x}(s)) d\mu(s) \right) \right] dt \\ &= \int_{s_i}^{t_i} [D_i(t) + E_i(t)] dt, \end{aligned}$$

where

$$D_i(t) := \nabla h^T(\bar{x}(t)) \left[f + GR^{-1} G^T \left(p(t) + \int_{[S,s_i]} \nabla h(\bar{x}(s)) d\mu(s) \right) \right]$$

and

$$E_i(t) := \nabla h^T(\bar{x}(t)) GR^{-1} G^T \int_{(s_i,t]} \nabla h(\bar{x}(s)) d\mu(s).$$

Under the hypotheses, the functions $D_i : [s_i, t_i] \to \mathbb{R}$, $i = 1, \ldots$, are Lipschitz continuous with a common local Lipschitz constant (write it K). Also, by (12.7.3),

$$D_i(s_i) = 0 \quad \text{for all } i.$$

It follows that

$$\int_{s_i}^{t_i} D_i(t) dt = \int_{s_i}^{t_i} (t_i - t) \frac{d}{dt} D_i(t) dt \geq -K \frac{(t_i - s_i)^2}{2}.$$

Also,

$$\begin{aligned} \int_{s_i}^{t_i} E_i(t) dt &= \int_{s_i}^{t_i} \nabla h^T(\bar{x}(t)) GR^{-1} G^T \int_{(s_i,t]} \nabla h(\bar{x}(s)) d\mu(s) dt \\ &\geq \beta \int_{s_i}^{t_i} \int_{s_i}^{t} d\mu(s) dt \\ &= \beta [-\int_{s_i}^{t_i} (t - s_i) d\mu(t) + |t_i - s_i| \int_{s_i}^{t_i} d\mu(t)] \\ &= \beta \int_{s_i}^{t_i} (t_i - t) d\mu(t). \end{aligned}$$

But by (A),

$$\begin{aligned} \int_{s_i}^{t_i} (t_i - t) d\mu(t) &= \int_{s_i}^{\frac{s_i+t_i}{2}} (t_i - t) d\mu(t) + \int_{\frac{s_i+t_i}{2}}^{t_i} (t_i - t) d\mu(t) \\ &\geq \left(t_i - \tfrac{s_i+t_i}{2} \right) \int_{s_i}^{\frac{s_i+t_i}{2}} d\mu(t) + 0 \\ &\geq \tfrac{t_i - s_i}{2} \tfrac{K_i}{2} (t_i - s_i). \end{aligned}$$

Therefore,

$$h(\bar{x}(t_i)) \geq -K \frac{(t_i - s_i)^2}{2} + \beta \frac{K_i}{4} (t_i - s_i)^2, \quad \text{for all } i.$$

Since $K_i \uparrow \infty$ it follows that $h(\bar{x}(t_i)) > 0$, for i sufficiently large. This contradicts the fact that $\bar{x}$ satisfies the state constraint.

Similar reasoning leads to a contradiction in case (B) also. Specifically, we examine the properties of the functions

$$\tilde{D}_i(t) := \nabla h^T(\bar{x}(t))\left[f(t,\bar{x}(t)) + GR^{-1}G^T\left(p(t) + \int_{[S,t_i]} \nabla h(\bar{x}(s))d\mu(s)\right)\right]$$

and

$$\tilde{E}_i(t) := -\nabla h^T(\bar{x}(t))GR^{-1}G^T\int_{(t,t_i]} \nabla h(\bar{x}(s))d\mu(s)$$

in place of D_i and E_i and show that, for i sufficiently large,

$$\int_{s_i}^{t_i} [\tilde{D}_i + \tilde{E}_i]dt < 0.$$

Since $h_j(\bar{x}(s_i)) = 0$ we have

$$h_i(\bar{x}(t_i)) = 0 - \int_{s_i}^{t_i} [\tilde{D}_i + \tilde{E}_i]dt > 0,$$

which is not possible. $\square$

12.8 Exercises

12.1 Let $\bar{x} \in W^{1,1}([S,T];\mathbb{R}^n)$ be a minimizer for

$$\begin{cases} \text{Minimize } \int_S^T L(t,x(t),\dot{x}(t))dt \\ \text{over } x \in W^{1,1} \text{ s.t.} \\ x(S) = x_0 \text{ and } x(T) = x_1 \end{cases}$$

in which L is a given function and x_0 and x_1 are given points in $\mathbb{R}^n$. Assume that

(i): L is a C^k function ($k \geq 2$),

(ii): $\nabla_v^2 L(t,\bar{x}(t),v)$ is positive definite, for all $(t,v) \in [S,T] \times \mathbb{R}^n$.

Suppose it is known that $\bar{x} \in W^{1,\infty}([S,T];\mathbb{R}^n)$. Show that $\bar{x} \in C^k([S,T];\mathbb{R}^n)$. *Hint: Use the Euler Lagrange condition (Theorem 1.2.1) to show that $\bar{x} \in C^{j-1}([S,T];\mathbb{R}^n)$ implies $\bar{x} \in C^j([S,T];\mathbb{R}^n)$, for $j = 2,\ldots,k$.*

12.8 Exercises

12.2 Let $\bar{x}$ be a minimizer for the calculus of variations problem involving a second order derivative:

$$\begin{cases} \text{Minimize } \int_S^T L(x(t), \dot{x}(t), \ddot{x}(t))dt \\ \text{over } x \in W^{2,1}([S, T]; \mathbb{R}^n) \text{ s.t.} \\ (x, \dot{x})(S) = (x_0, v_0) \quad \text{and} \quad (x, \dot{x})(T) = (x_1, v_1), \end{cases}$$

in which $L : \mathbb{R}^n \times \mathbb{R}^n \times \mathbb{R}^n \to \mathbb{R}$ is a given function and (x_0, v_0) and (x_1, v_1) are given points in $\mathbb{R}^n \times \mathbb{R}^n$. Assume

(i): L is locally Lipschitz continuous,
(ii): $w \to L(x, v, w)$ is convex each fixed (x, v),
(iii): there exists $k_L \in L^1$ such that

$$\partial_x L(\bar{x}(t), \dot{\bar{x}}(t), \ddot{\bar{x}})(t)|_{x=\bar{x}(t)} \in k_L(t)\mathbb{B}, \text{ for a.e. } t \in [S, T],$$

(iv): there exists $\theta : (0, \infty) \to (0, \infty)$ such that $\theta(0) = 0$, $\lim_{r \uparrow \infty} \theta(r)/r = \infty$ and

$$L(x, v, w) \geq \theta(|w|) \quad \text{for all } (x, v, w) \in \mathbb{R}^n \times \mathbb{R}^n.$$

Show that $\ddot{\bar{x}}$ is essentially bounded.
Hint: Reformulate the problem as one involving a first order linear dynamic constraint, by state augmentation. Now use a similar time transformation technique to that one employed in Sect. 12.5 to identify $\bar{x}$ is a minimizer for an dynamic optimization problem for which necessary conditions of optimality are known. Then deduce essential boundedness of $\ddot{\bar{x}}$ from these necessary conditions.

12.3 (A Nonsmooth Erdmann – Du Bois-Reymond Condition) Let $\bar{x}$ be a $W^{1,1}$ local minimizer for the optimization problem

$$(CV) \begin{cases} \text{Minimize } \int_S^T L(t, x(t), \dot{x}(t))dt \\ \text{over arcs } x \in W^{1,1}([S, T]; \mathbb{R}^n) \text{ satisfying} \\ x(S) = x_0, \; x(T) = x_1, \\ \dot{x}(t) \in C \quad \text{a.e. } t \in [S, T], \\ x(t) \in A \quad \text{for all } t \in [S, T], \end{cases}$$

the data for which comprise an interval $[S, T]$, a function $L : [S, T] \times \mathbb{R}^n \times \mathbb{R}^n \to \mathbb{R}$, points $x_0, x_1 \in \mathbb{R}^n$, a cone $C \subset \mathbb{R}^n$, and a set $A \subset \mathbb{R}^n$. Assume that the Lagrangian L is Borel measurable and satisfies assumption (A1) of Theorem 12.5.4. Show that there exist an arc $p \in W^{1,1}([S, T]; \mathbb{R})$ and an integrable function $q : [S, T] \to \mathbb{R}$ such that

$$\begin{cases} L(t, \bar{x}(t), \dot{\bar{x}}(t)) - q(t) = p(t), \\ q(t) \in \tilde{\partial}_r L(t, \bar{x}(t), r\dot{\bar{x}}(t))_{r=1} \end{cases} \quad \text{a.e. } t \in [S, T]. \tag{12.8.1}$$

Show that if, in particular, $L = L(x, v)$ is autonomous and just Borel measurable, then condition (12.8.1) is satisfied and p is a constant. (Recall that $\tilde{\partial}_r L(t, x, rv)_{r=1}$ denotes the subdifferential in the sense of convex analysis of the function $(0, +\infty) \ni r \mapsto L(t, x, rv)$ at $r = 1$, see (12.5.3).)

Hint: Use conditions (12.5.6) and (12.5.7) provided by Theorem 12.5.4 and the change of variable $r = 1/u$ ($u > 0$).

12.4 Show that Theorems 12.5.1, 12.5.4, Corollary 12.5.2 and the result stated in Exercise 12.3 can be extended to a $W^{1,s}$ local minimizer $\bar{x}$ for problem (CV) of Exercise 12.3 (or of Sect. 12.5), when $s \in [1, \infty)$. (For the definition of '$W^{1,s}$ local minimizer' see Exercise 6.1.)

Hint: Extend Lemmas 12.5.3 and 12.5.5 to the case in which $\bar{x}$ is a $W^{1,s}$ arc.

12.9 Notes for Chapter 12

Regularity properties of minimizers for variational problems in one independent variable were studied extensively by Tonelli [190] in the early years of the 20th century, for inherent interest, no doubt, but also motivated by a desire to fill the gap between hypotheses for existence of minimizers and hypotheses needed to derive first order necessary conditions of optimality and thereby to validate the direct method. The regularity issue has remained a central one in multidimensional calculus of variations (see [115]). However this aspect of Tonelli's legacy, notably the discovery that under the hypotheses of existence theory bad behaviour can be confined to a closed set of zero measure (the Tonelli set), was unaccountably overlooked when one dimensional calculus of variations evolved into dynamic optimization. Interest in Tonelli regularity was reawakened by Ball and Mizel, who saw potential applications to the field of nonlinear elasticity in which material failure can be associated with the existence of non-empty Tonelli sets. Ball and Mizel studied the structure of Tonelli sets and gave the first examples of problems satisfying the hypotheses of existence theory and yet having non-empty Tonelli sets [16], including the Ball Mizel example of Sect. 11.1. A brief proof by Vinter and Clarke that the Ball Mizel example exhibits the pathological behaviour of interest, based on the construction of a nonsmooth verification function, appears in [74].

In a series of papers Clarke and Vinter brought together Tonelli's proof techniques, based on the application of necessary conditions to suitably regular auxiliary Lagrangians, and methods of nonsmooth analysis, to explore further the properties of minimizers under the hypotheses of existence theory. The scope for constructing nonsmooth Lagrangians adds greatly to the flexibility of the approach. [75] generalizes Tonelli's earlier results [189] on the structure of Tonelli sets, to allow for vector

12.9 Notes for Chapter 12

valued arcs, nonsmooth Lagrangians and to eliminate the need for strict convexity. [75] supplied the first proof that minimizers for autonomous problems satisfying the hypotheses of existence theory are Lipschitz continuous. Other sufficient conditions for Lipschitz continuity are proved, a sample of which are reproduced in this chapter. [76] concerns properties of the Tonelli set for noncoercive problems. In [77] it is established that Tonelli sets for polynomial Lagrangians are countable, with a finite number of accumulation points. Generalizations to variational problems involving higher derivatives and to dynamic optimization problems with affine dynamic constraints are provided in [81] and [82] respectively.

The use of time reparameterization to supply an independent proof that minimizers for autonomous problems satisfying the hypotheses of Tonelli Existence Theory are Lipschitz continuous originates with Ambrosio, Ascenzi and Buttazzo [1]. A streamlined proof of this regularity property for solutions to autonomous problems was devised by Clarke [68]. Theorem 12.5.1 is a recent departure, due to Bettiol and Mariconda, based on this approach but applicable to non-autonomous problems [25, 26]. Hypothesis (A1) in Theorem 12.5.1 is a local Lipschitz condition of the map $s \mapsto L(s, \bar{x}(t), \sigma \dot{\bar{x}}(t))$ for a.e. $t \in [S, T]$ and for all $\sigma > 0$ (for a reference minimizer $\bar{x}$), and represents a nonsmooth extension of the classical Cesari condition (S) [54, §2.7 A]. Assumption (A2) is a variation on a condition which was first introduced by Clarke in [66] (for Lagrangians $L(t, x, v)$ that are convex in the velocity variable v); hypothesis (A2) is satisfied not only when the Lagrangian is coercive (w.r.t. v) but also in many situations in which the Lagrangian has merely a slow growth condition (see papers [25, 26] and [149] for a discussion, generalizations also to the case of extended valued Lagrangians, and related results such as the avoidance of the 'Lavrentiev phenomenon').

The final section of the Chapter concerns the direct application of necessary conditions to establish regularity of optimal controls. A significant early advance was Hager's proof of Lipschitz continuity of optimal controls, for dynamic optimization problems with affine dynamics, a smooth, coercive cost integrand jointly convex with respect to state and control variables, and with unilateral state and control constraints satisfying an independence condition [120]. Extensions to allow for nonlinear dynamics were carried out by Malanowski [146]. A simple, independent proof of Hager's regularity theorem, in the case of linear quadratic problems with affine state constraints, based on discrete approximations, was provided by Dontchev and Hager [94]. A more refined regularity analysis of this class of problems was undertaken by Dontchev and Kolmanovsky [95], who have given conditions for optimal controls to be piecewise analytic. The assertions of Theorem 12.7.1 are those of Malanowski's regularity theorem in the case of a single state constraint, no control constraints and a quadratic control term in the cost. The proof, which allows nonsmooth data and a milder constraint qualification, strong normality, is new.

Chapter 13
Dynamic Programming

Abstract Dynamic programming is an approach to solving dynamic optimization problems, centred on properties of the value function, that is the minimum cost parameterized by the initial time and state. For the dynamic optimization problems considered in this book the value function is a solution to the Hamilton Jacobi equation (HJE). The central question is, how should we define 'solution' in such a manner that the HJE has a unique solution and that this solution coincides with the value function? Other matters of interest opened up by this approach include techniques for verifying the optimality of a putative minimizer, feedback representation of optimal strategies and computational methods, some of which are discussed in this chapter. The goal of representing the value function as the unique solution, appropriately defined, of the HJE has been arrived at along two different paths. The first involves viscosity solutions, as introduced by Crandall and Lions. With this solution concept it is possible to show directly, and without consideration of state trajectories, that the Hamilton Jacobi equation has a unique solution. The second path is system theoretic, in the sense that it is intimately connected with properties of state trajectories; invariance theorems are employed to show that a solution to the Hamilton Jacobi equation provides a lower bound to the cost of an arbitrary state trajectory and this lower bound is achieved by some state trajectory.

This chapter is an up to date treatment of dynamic programming that places emphasis on the system theoretic point of view. We consider dynamic optimization problems for which the value function is a, possibly discontinuous, lower semi-continuous function. Various, equivalent, definitions of 'solution' of HJE are involved, prominent among which is that of proximal solution of Clarke. The chapter begins with a study of system invariance leading, on the one hand, to conditions under which there exists a state trajectory satisfying a given pathwise constraint and, on the other, to conditions under which all state trajectories have this property. The desired characterization of the value function as the unique proximal solution of the HJE results from applying the invariance theorems to an extended control system with pathwise constraint set constructed from an epigraph set, corresponding to the proximal solution of the HJE under consideration. This link between the value function and proximal solutions to HJE is established for various formulations of the dynamic optimization problem. Problems with finite time horizons, discounted-cost problems with an infinite horizon, minimum time problems and problems with pathwise state constraints all make their appearance. In the earlier literature a full characterization of the value function as the unique lower semi-continuous solution of HJE was achieved only for continuously time-dependent dynamics. A notable feature of theory presented in this chapter, the result of recent research, is that we allow discontinuous time dependence of the dynamics.

Other topics are covered in this chapter. These include verification techniques of dynamic programming type and the role of semiconcavity in dynamic programming.

A rounded exposition on the subject of dynamic programming necessarily embraces both viscosity solution and system theoretic methods. The final sections include discussion, comparing and contrasting the methods. Finally a simple proof of comparison theorem relating to an infinite horizon problem is given, to convey the flavour of viscosity techniques.

13.1 Introduction

Consider the dynamic optimization problem:

$$(P) \begin{cases} \text{Minimize } g(x(T)) \\ \text{over arcs } x \in W^{1,1}([S, T]; \mathbb{R}^n) \text{ satisfying} \\ \dot{x}(t) \in F(t, x(t)) \quad \text{a.e.,} \\ x(S) = x_0, \end{cases}$$

the data for which comprise an interval $[S, T] \subset \mathbb{R}$, a function $g : \mathbb{R}^n \to \mathbb{R} \cup \{+\infty\}$, a multifunction $F : [S, T] \times \mathbb{R}^n \rightsquigarrow \mathbb{R}^n$ and a point $x_0 \in \mathbb{R}^n$.

Notice that the cost function g is allowed to take value $+\infty$. Implicit in this formulation then is the end-point constraint:

$$x(T) \in C,$$

where

$$C := \{x \in \mathbb{R}^n : g(x) < +\infty\}.$$

Dynamic programming, as it relates to the above problem, concerns the relation between, on the one hand, minimizers and the infimum cost of (P) and, on the other, solutions to the Hamilton Jacobi equation:

$$\phi_t(t, x) + \min_{v \in F(t,x)} \phi_x(t, x) \cdot v = 0 \quad \text{for all } (t, x) \in D \quad (13.1.1)$$

$$\phi(T, x) = g(x) \quad \text{for all } x \in D_1. \quad (13.1.2)$$

Here, D and D_1 are given subsets of $[S, T] \times \mathbb{R}^n$ and $\mathbb{R}^n$ respectively.

13.1 Introduction

Equation (13.1.1) can alternatively be expressed in terms of the Hamiltonian

$$H(t, x, p) := \sup_{v \in F(t,x)} p \cdot v,$$

thus

$$-\phi_t(t, x) + H(t, x, -\phi_x(t, x)) = 0 \quad \text{for all } (t, x) \in D.$$

The link between the dynamic optimization problem (P) and the Hamilton Jacobi equation is the value function $V : [S, T] \to \mathbb{R} \cup \{+\infty\}$: for each $(t, x) \in [S, T] \times \mathbb{R}^n$, $V(t, x)$ is defined to be the infimum cost for the problem

$$(P_{t,x}) \begin{cases} \text{Minimize } g(y(T)) \\ \text{over arcs } y \in W^{1,1}([t, T]; \mathbb{R}^n) \text{ satisfying} \\ \dot{y}(s) \in F(s, y(s)) \quad \text{a.e.,} \\ y(t) = x, \end{cases}$$

in which (t, x) replaces the initial data (S, x_0) in (P). We write this relation

$$V(t, x) = \inf(P_{t,x}).$$

The elementary theory of dynamic programming (see Chap. 1) tells us that, if V is a C^1 function then, under appropriate hypotheses on the data for (P), V is a solution to (13.1.1) and (13.1.2) when $D = (S, T) \times \mathbb{R}^n$ and $D_1 = \mathbb{R}^n$.

Of course knowledge of the value function V provides the minimum cost for (P): it is simply V evaluated at (S, x_0). But in favourable circumstances, it also supplies information about minimizers. Suppose that V is a C^1 function and also that, for each (t, x),

$$\chi(t, x) := \{v \in F(t, x) : V_x(t, x) \cdot v = \min_{v' \in F(t,x)} V_x(t, x) \cdot v'\} \quad (13.1.3)$$

is non-empty and single valued. Finally suppose that

$$\dot{y}(s) = \chi(s, y(s)), \quad y(t) = x \quad (13.1.4)$$

has a $W^{1,1}$ solution y on $[t, T]$, for any $(t, x) \in [S, T] \times \mathbb{R}^n$. Then y is a minimizer for $(P_{t,x})$. Indeed, the calculation:

$$V(t, x) = V(T, y(T)) - \int_t^T \frac{d}{ds} V(s, y(s)) ds$$

$$= V(T, y(T)) - \int_t^T [V_t(s, y(s)) + V_x(s, y(s)) \cdot \chi(s, y(s))] ds$$

$$= V(T, y(T)) - \int_t^T [V_t(s, y(s)) + \min_{v \in F(t,x)} V_x(s, y(s)) \cdot v] ds$$
$$= V(T, y(T)) - 0 = g(y(T))$$

shows that y has cost $V(t, x)$, i.e. y is a minimizer for $(P_{t,x})$. In particular, selecting $(t, x) = (S, x_0)$ and solving the differential Eq. (13.1.4) yields a minimizer for (P). An advantage to this approach is that it supplies the minimizer in 'feedback' form, favoured in control engineering applications. In a sense then, dynamic programming reduces the dynamic optimization problem to one of solving a partial differential equation.

In more recent dynamic programming research, the role of the Hamilton Jacobi equation in characterizing the value function is emphasized. But the value function as an object of interest in its own right is a relative newcomer to variational analysis. Traditionally, the Hamilton Jacobi equation has had a different role: that of providing sufficient conditions of optimality, to test whether a putative minimizer (arrived at by finding an arc which satisfies some set of necessary conditions, say) is truly a minimizer. Using the Hamilton Jacobi equation in this spirit is called the Carathéodory method. The essential character of the approach is captured by the following sufficient condition of optimality:

Let $\bar{x}$ be an arc which satisfies the constraints of (P). Then $\bar{x}$ is an L^∞ local minimizer if, for some $\epsilon > 0$, a C^1 function[1] $\phi : T(\bar{x}, \epsilon) \to \mathbb{R}$ can be found satisfying the Hamilton Jacobi equation (13.1.1) and (13.1.2) with $D = \text{int } T(\bar{x}, \epsilon)$ and $D_1 = (\bar{x}(T) + \epsilon \mathbb{B}) \cap \text{dom } g$ and if

$$\phi(S, x_0) = g(\bar{x}(T)). \tag{13.1.5}$$

In the above, $T(\bar{x}, \epsilon)$ denotes the ϵ-tube about $\bar{x}$:

$$T(\bar{x}, \epsilon) := \{(t, y) \in [S, T] \times \mathbb{R}^n : y \in \bar{x}(t) + \epsilon \mathbb{B}\}.$$

To justify these assertions, we have merely to note that, for any arc x satisfying the constraints of (P) and also the condition $||x - \bar{x}||_{L^\infty} < \epsilon$, we have

$$\phi(S, x_0) = \phi(T, x(T)) - \int_t^T \frac{d}{ds} \phi(s, x(s)) ds$$
$$= \phi(T, x(T)) - \int_t^T [\phi_t(s, x(s)) + \phi_x(s, x(s)) \cdot \dot{x}(s)] ds$$
$$\leq g(x(T)) + 0.$$

---

[1] Given a closed set $A \subset \mathbb{R}^k$ and a function $\psi : A \to \mathbb{R}^m$, we say that ψ is a C^1 function if it is continuous, if it is of class C^1 on the interior of A and if $\nabla \psi$ extends, as a continuous function, to all of A.

13.1 Introduction

This inequality combines with (13.1.5) to confirm the L^∞ local optimality of $\bar{x}$. Appropriately, functions ϕ used in this way are called *verification functions*. The application of the Carathéodory method is illustrated, for example, in L C Young's book [209], where it is used to solve a number of classical problems in the calculus of variations. For many of these problems, it is comparatively straightforward to determine a candidate for minimizer by solving Euler's equation (a necessary condition for an arc to be a minimizer). Confirming this arc is truly a minimizer can be accomplished in favourable circumstances by constructing a verification function.

An important feature of verification functions is that they do not have to be the value function for (P). In many cases, *finite* verification functions serve to confirm the local minimality of a particular arc, even when the value function is infinite at some points in its domain. This flexibility can simplify the task of finding verification functions in specific applications.

How have these early, elementary ideas evolved? Consider the relation between the Hamilton Jacobi equation and the value function. The first issue to be settled, if we are to regard solving the Hamilton Jacobi equation as a means to generating the value function, is whether the Hamilton Jacobi equation has a unique solution which coincides with the value function. This question is a challenging one because, for many dynamic optimization problems of interest, the value function is not continuously differentiable. A simple example of this phenomenon was discussed in Chap. 1. At the outset then we must come up with a suitable concept of generalized solution to the Hamilton Jacobi equation. Minimal requirements are:

1. The (possibly non-differentiable) value function is a generalized solution,

and

2. There exists a unique generalized solution.

A plausible approach to providing the desired characterization of the value function is based on the following solution concept: a locally Lipschitz continuous function $V : [S, T] \times \mathbb{R}^n$ is an *almost everywhere* solution of (HJE) if

$$-\phi_t(t, x) + H(t, x, -\phi_x(t, x)) = 0$$

at almost all points $(t, x) \in (S, T) \times \mathbb{R}^n$, w.r.t. Lebesgue measure, at which V is Fréchet differentiable, and also

$$\phi(T, x) = g(x), \quad \text{for all } x \in \mathbb{R}^n.$$

Almost everywhere solutions might be expected to furnish a satisfactory theory since, by Rademacher's theorem, a locally Lipschitz continuous function is almost everywhere Frechét differentiable. But unfortunately, this is not the case, as is illustrated by the following example, due to F. H. Clarke [68], in which there are multiple almost everywhere solutions.

Example
Consider

$$\begin{cases} \text{Minimize } g(x(1)) := |x(1)| \\ \text{over arcs } x \in W^{1,1}([0,1]; \mathbb{R}) \text{ s.t.} \\ \dot{x}(t) \in [-1, +1] \quad \text{a.e. } t \in [0,1] \\ x(0) = x_0, \end{cases}$$

where $x_0 \in \mathbb{R}$ is a given state. The value function is

$$V(t, x) = \max\{|x| + t - 1, 0\} \quad \text{for all } (t, x) \in [0, 1] \times \mathbb{R}.$$

It is easy to check that V is a Lipschitz continuous function that is an almost everywhere solution to (HJE). However it is not uniquely so. Another, distinct, Lipschitz continuous function that is also an almost everywhere solution is

$$\tilde{V}(t, x) = |x| + t - 1, \quad \text{for all } (t, x) \in [S, T] \times \mathbb{R}^n.$$

Notice however that

$$\tilde{V}(t, x) \leq V(t, x), \quad \text{for all } (t, x) \in [S, T] \times \mathbb{R}^n.$$

We shall show (Proposition 13.4.3, below) that this inequality is true for *any* locally Lipschitz continuous function that is an almost everywhere solution of (HJE). Our analysis has revealed that, while the almost everywhere solution concept is not useful when it comes to characterizing the value function, this class of solutions do at least provide lower bounds on the value function.

The lesson learned from Example 13.1 is that we must look elsewhere for a suitable definition of 'solution'. Viscosity solutions [88], whose appearance on the scene in the 1980's marked an important advance in the theory of dynamic programming, met all the requirements, in situations were the value function is continuous. (Extensions allowing for discontinuous value functions 'lower semi-continuous viscosity' solutions were subsequently introduced [19].) The key ingredients in the definition are as follows:

A *continuous function* $\phi : [S, T] \times \mathbb{R}^n \to \mathbb{R}$ *is said to be a* viscosity solution *of*

$$-\phi_t(t, x) + H(t, x, -\phi_x(t, x)) = 0 \quad \text{for all } (t, x) \in (S, T) \times \mathbb{R}^n \quad (13.1.6)$$

if it satisfies the following conditions:

(a) *for any point* $(t, x) \in (S, T) \times \mathbb{R}^n$ *and any* C^1 *function* $w : \mathbb{R} \times \mathbb{R}^n \to \mathbb{R}$ *such that* $(t', x') \to \phi(t', x') - w(t', x')$ *has a local minimum at* (t, x) *we have*

$$-\xi^0 + H(t, x, -\xi^1) \geq 0$$

where $(\xi^0, \xi^1) = \nabla w(t, x)$,

13.1 Introduction

(b) *for any point $(t, x) \in (S, T) \times \mathbb{R}^n$ and any C^1 function $w : \mathbb{R} \times \mathbb{R}^n \to \mathbb{R}$ such that $(t', x') \to \phi(t', x') - w(t', x')$ has a local maximum at (t,x) we have*

$$-\xi^0 + H(t, x, -\xi^1) \leq 0$$

where $(\xi^0, \xi^1) = \nabla w(t, x)$.

If a function ϕ satisfies condition (a) is said to be a viscosity supersolution (of (13.1.6)). If it satisfies (b), it is said to be a viscosity subsolution of (13.1.6).

The theory of viscosity solutions employs *directly* the defining relation of super- and sub- viscosity solutions to derive comparison relations of the following kind.

Comparison Principle: Take viscosity super- and sub-solutions W^+ and W^- respectively. Then

$$W^+(t, x) \geq W^-(t, x) \quad \text{for all } (t, x) \in \{T\} \times \mathbb{R}^n$$

$$\implies W^+(t, x) \geq W^-(t, x) \quad \text{for all } (t, x) \in [S, T] \times \mathbb{R}^n$$

Take any viscosity solution W (i.e. W is simultaneously a super- and sub-solution) satisfying the boundary condition $W(T, .) = g$. Then, by the comparison principle,

$$W(t, x) \geq V(t, x) \quad \text{and} \quad V(t, x) \geq W(t, x) \quad \text{for all } (t, x) \in [S, T] \times \mathbb{R}^n$$
(13.1.7)

which implies $W = V$. Thus the comparison principle tells us a little bit more than 'V is the unique solution to the Hamilton Jacobi equation satisfying the boundary condition'.

In parallel with these developments, efforts were made to establish relation (13.1.7), governing the value function V, for an arbitrary 'generalized' solution to the Hamilton Jacobi equation by linking this relation to invariance properties of differential inclusions. This alternative approach has been variously referred to as the *control theoretic* approach, because it exploits properties of the class of state trajectories associated with the underlying control system, the *viability* approach, because it is based on viability/invariance theorems from the theory of differential inclusions and the *generalized solutions* approach, since it involves an interpretation of 'solution' to the Hamilton Jacobi equation in an appropriate 'generalized' sense, that is somewhat different to that employed in the viscosity solutions literature.

When we turn attention to problems with right end-point constraints it is no longer tenable to assume that the value function is continuous. The value function is lower semi-continuous, however, under unrestrictive, verifiable hypotheses on the data. This suggests that we should aim to characterize value functions in terms of lower semi-continuous solutions to the Hamilton Jacobi equation, appropriately defined.

The centerpiece of this chapter is a control theoretic framework for studying lower semi-continuous 'solutions' to the Hamilton Jacobi equation which relates them to the corresponding value function. Fundamental to our approach is the following concept of generalized solution to the Hamilton Jacobi equation, introduced by Clarke et al. [84].

A lower semi-continuous function $\phi : [S, T] \times \mathbb{R}^n \to \mathbb{R} \cup \{+\infty\}$ *is said to be a proximal solution to the Hamilton-Jacobi equation if*

(a)' *at each point* $(t, x) \in ((S, T) \times \mathbb{R}^n) \cap \mathrm{dom}\,\phi$ *such that* $\partial^P \phi(t, x)$ *is non-empty,*

$$-\xi^0 + H(t, x, -\xi^1) = 0 \quad \text{for all } (\xi^0, \xi^1) \in \partial^P \phi(t, x).$$

Notice that condition (a) in the earlier definition of viscosity solution implies

$$-\xi^0 + H(t, x, -\xi^1) \geq 0 \quad \text{for all } (\xi^0, \xi^1) \in \partial^P \phi(t, x)$$

at every point $(t, x) \in (S, T) \times \mathbb{R}^n$ such that $\partial^P \phi(t, x)$ is non-empty, because $(\xi^0, \xi^1) \in \partial^P \phi(t, x)$ means that (t, x) is a local minimizer of $\phi - w$, where w is the quadratic function

$$w(t', x') = (t', x') \cdot (\xi^0, \xi^1) - M(|(t', x') - (t, x)|^2),$$

for some $M > 0$.

We see that the definition of proximal solution (the solution concept employed according to the generalized solutions approach) is arrived by discarding condition (b) in the definition of viscosity solutions, relaxing condition (a) somewhat (the 'generalized solutions' approach only requires consideration of quadratic test functions w) and replacing inequality by equality.

If we wish to allow for $F(t, x)$'s that are possibly discontinuous w.r.t. the t variable, it is natural to seek a characterization of the value function in terms of the unique generalized solution to the Hamilton-Jacobi equation in an 'almost everywhere w.r.t. time' sense, i.e., when we replace conditions (a)' above by: there exists a subset $\mathcal{T} \subset [S, T]$ of zero Lebesgue measure such that

(a)'' at each point $(t, x) \in (((S, T) \setminus \mathcal{T}) \times \mathbb{R}^n) \cap \mathrm{dom}\,\phi$ such that $\partial^P \phi(t, x)$ is non-empty,

$$-\xi^0 + H(t, x, -\xi^1) = 0 \quad \text{for all } (\xi^0, \xi^1) \in \partial^P \phi(t, x).$$

This is not possible[2] within a theory aiming to show the value function is the unique lower semi-continuous function that is a generalized solution of the Hamilton Jacobi

---

[2] Frankowska, Plaskacz and Rzezuchowski [113] have shown, however, that, for problems in which F is measurably time dependent, the value function is the unique lower Dini solution of the Hamilton Jacobi equation, in the class of lower semi-continuous functions that are also epicontinuous.

13.1 Introduction

equation, in the above 'almost everywhere sense'. The following example confirms this assertion.

Example
Consider

$$\begin{cases} \text{Minimize } g(x(1)) := x(1) \\ \text{over arcs } x \in W^{1,1}([0,1]; \mathbb{R}) \text{ s.t.} \\ \dot{x}(t) = 0 \quad \text{a.e. } t \in [0,1] \\ x(0) = x_0, \end{cases}$$

where $x_0 \in \mathbb{R}$ is a given state. The value function is $V(t, x) = x$ for all $(t, x) \in [0, 1] \times \mathbb{R}$. However

$$W(t, x) := \begin{cases} x - 1 & \text{if } t \leq \frac{1}{2} \\ x & \text{if } t > \frac{1}{2} \end{cases}$$

is also a lower semi-continuous function that satisfies the 'almost everywhere' condition (a)'. (We choose $\mathcal{T} = \{\frac{1}{2}\}$, to exclude consideration of the troublesome point $\frac{1}{2}$ at which $W(t, x)$ fails to satisfy the original conditions (a).) This confirms that the value function is not the unique lower semi-continuous function satisfying the almost everywhere conditions (a)'.

Our analysis will cover situations in which the velocity set is discontinuous w.r.t. time. We restrict the nature of discontinuities considered, however, by requiring that the multifunction $t \to F(t, x)$ has everywhere left and right limits and is continuous on the complement of a set of measure zero. A special case covered by this hypothesis is when $t \to F(t, x)$ has bounded variation w.r.t. time (many time discontinuities encountered in applied dynamic optimization are of this nature).

In this chapter the link between lower semi-continuous solutions to the Hamilton Jacobi equation and value functions is achieved, as a by-product of theorems on the invariance properties of solutions to differential inclusions. This control theoretic approach is consistent with analytical techniques employed elsewhere in this book, and illustrates further areas of application of generalized subdifferentials.

We mention that the invariance theorems on which the analysis is based are of great independent interest and have been applied in many areas of dynamic systems theory and in nonlinear analysis. (The systematic use of invariance theorems, such as those proved in this chapter, is called 'viability theory').

Of course showing that the value function is a generalized solution of the Hamilton Jacobi equation falls somewhat short of obtaining detailed information about minimizers for (P). What is required is an analysis of the feedback map χ of (13.1.3), or some nonsmooth analogue of it, and also means of interpreting solutions to the 'closed loop equation' $\dot{x} \in \chi(t, x)$ (we can expect the right side to be multivalued) which generates minimizers for (P). This important area of study,

Optimal Synthesis as it is called, requires rather different analytical techniques to those used elsewhere in this book and will not be entered into here.

Another direction of research into Carathéodory's method concerns the inverse problem: what concept of verification function should we adopt in order that, corresponding to every local minimizer, there exists a verification function? This area of research aims at simplifying the task of finding a verification function to confirm the optimality of a specified putative minimizer: the smaller the class of verification functions, one of which can be guaranteed to confirm the optimality of a given minimizer, the narrower will be the search and therefore the 'better' the inverse theorem. The main result in this area, proved in this chapter, is that, under a mild non-degeneracy hypothesis, there always exists a Lipschitz continuous verification function, even in situations when there are right end-point constraints and when the value function (which is after all the obvious choice for verification function) is not even continuous!

Another topic covered in this chapter is the interpretation of costate arcs in terms of generalized gradients of value functions. Connections are thereby made between the necessary conditions of earlier chapters and dynamic programming.

The relevant relations were sketched in Chap. 1, where the costate arcs were associated with the maximum principle. We review them, in the context of the dynamic optimization problem (P). Now, gradients of the value function are related to the costate arc appearing in the generalized Euler Lagrange condition.

Let $\bar{x}$ be a minimizer. To simplify the analysis, let us suppose that V is a C^2 function, that g is a C^1 function and that $F(t, .)$ has closed graph. We deduce from the fact that g coincides with $V(T, .)$ that

$$V_x(T, \bar{x}(T)) = g_x(\bar{x}(T)).$$

Since $\bar{x}$ is a minimizer, we have $V(S, x_0) = g(\bar{x}(T)) = V(T, \bar{x}(T))$. It then follows from (13.1.1) and the identity

$$V(S, x_0) = V(T, \bar{x}(T)) - \int_S^T \{V_t(t, \bar{x}(t)) + V_x(t, \bar{x}(t)) \cdot \dot{\bar{x}}(t)\} dt$$

that

$$V_t(t, \bar{x}(t)) + V_x(t, \bar{x}(t)) \cdot \dot{\bar{x}}(t) = 0 =$$
$$\min[V_t(t, x) + V_x(t, x) \cdot v : x \in \mathbb{R}^n, \, v \in F(t, x)] \quad \text{a.e..}$$

Clearly,

$$-V_x(t, \bar{x}(t)) \cdot \dot{\bar{x}}(t) = \max_{v \in F(t, \bar{x}(t))} (-V_x(t, \bar{x}(t))) \cdot v \quad \text{a.e..}$$

Furthermore, from the multiplier rule (Theorem 5.6.1), we have that

13.1 Introduction

$$-(V_{tx}(t,\bar{x}(t)) + V_{xx}(t,\bar{x}(t)) \cdot \dot{\bar{x}}(t), V_x(t,\bar{x}(t))) \in N_{\text{Gr } F(t,\cdot)}(\bar{x}(t), \dot{\bar{x}}(t)) \quad \text{a.e..}$$

Now define

$$p(t) := -V_x(t,\bar{x}(t)),$$
$$h(t) := \max_{v \in F(t,\bar{x}(t))} p(t) \cdot v \ (= p(t) \cdot \dot{\bar{x}}(t)).$$

Since $\dot{p}(t) = -V_{tx}(t,\bar{x}(t)) - V_{xx}(t,\bar{x}(t)) \cdot \dot{\bar{x}}(t)$ we deduce from these relations that

$$(\dot{p}(t), p(t)) \in N_{\text{Gr } F(t,\cdot)}(\bar{x}(t), \dot{\bar{x}}(t)) \quad \text{a.e.,}$$
$$p(t) \cdot \dot{\bar{x}}(t) = \max_{v \in F(t,\bar{x}(t))} p(t) \cdot v \quad \text{a.e.,}$$
$$-p(T) = g_x(\bar{x}(T)),$$
$$(h(t), -p(t)) = \nabla V(t, \bar{x}(t)) \quad \text{a.e..} \tag{13.1.8}$$

We have arrived at set of necessary conditions including the Euler Lagrange inclusion, in which the Hamiltonian, evaluated along the optimal trajectory, and the costate arc are interpreted in terms of gradients of the value function. In general, V will not be a C^2 function and the above arguments cannot be justified. We will however derive, by means of a more sophisticated analysis, necessary conditions of optimality involving an costate arc p which satisfies a nonsmooth version of (13.1.8), namely p satisfies (at the same time) the following 'partial' and 'full' sensitivity relations:

$$-p(t) \in \text{co } \partial_x V(t, \bar{x}(t)) \quad \text{a.e.} \tag{13.1.9}$$

$$(h(t), -p(t)) \in \text{co } \partial V(t, \bar{x}(t)) \quad \text{a.e..} \tag{13.1.10}$$

We conclude this introduction with two elementary relations satisfied by the value function for (P), ones to which frequent reference will be made. Take points s_1, $s_2 \in [S, T]$ with $s_1 \leq s_2$, a minimizer $\bar{x} : [s_1, T] \to \mathbb{R}^n$ (for problem $(P_{s_1, \bar{x}(s_1)})$) and also an F trajectory $x : [s_1, T] \to \mathbb{R}^n$. Then

$$V(s_1, x(s_1)) \leq V(s_2, x(s_2)) \tag{13.1.11}$$

$$V(s_1, \bar{x}(s_1)) = V(s_2, \bar{x}(s_2)). \tag{13.1.12}$$

These relations are collectively referred to as the *principle of optimality*, which can be paraphrased as

The value function for (P) is non-decreasing along an arbitrary F trajectory, and is constant along a minimizing F trajectory.

We validate this principle. Suppose that (13.1.11) is false. Then there exists $\epsilon > 0$ such that

$$V(s_1, x(s_1)) > V(s_2, x(s_2)) + \epsilon.$$

By definition of V, there exists an F trajectory $z : [s_2, T] \to \mathbb{R}^n$ such that $z(s_2) = x(s_2)$ and

$$V(s_2, x(s_2)) > g(z(T)) - \epsilon/2.$$

But then the F trajectory $y : [S, T] \to \mathbb{R}^n$, defined as

$$y(t) = \begin{cases} x(t) & s_1 \leq t < s_2 \\ z(t) & s_2 \leq t \leq T \end{cases}$$

satisfies

$$V(s_1, x(s_1)) \leq g(y(T)) = g(z(T)) < V(s_2, x(s_2)) + \epsilon/2$$
$$< V(s_1, x(s_1)) - \epsilon/2.$$

From this contradiction we deduce that (13.1.11) is true.

As for (13.1.12), we note that, again by definition of V,

$$V(s_2, \bar{x}(s_2)) \leq g(\bar{x}(T)) = V(s_1, \bar{x}(s_1)).$$

This combines with condition (13.1.11) to give condition (13.1.12).

13.2 Invariance Theorems

Invariance theorems concern solutions to a differential inclusion, which satisfy a specified constraint. Theorems giving conditions for existence of at least one solution satisfying the constraint are called *weak invariance theorems* or *viability theorems*. Those asserting that *all* solutions satisfy the constraint are called *strong invariance theorems*.

These theorems have far reaching implications, in stability theory, dynamic programming, robust controller design and differential games (to name but a few applications areas!). Here, however, we concentrate on versions of the theorems suitable for characterizing value functions in dynamic optimzation.

The starting point for all the important results in this section is a theorem providing conditions on an 'autonomous' multifunction $F : \mathbb{R}^n \rightsquigarrow \mathbb{R}^n$ and a closed set $D \subset \mathbb{R}^n$ such that, for a given initial state x_0 in D, the differential inclusion and accompanying constraint:

13.2 Invariance Theorems

$$\begin{cases} \dot{x} \in F(x) & \text{a.e. } t \in [S, \infty) \\ x(S) = x_0 \\ x(t) \in D & \text{for all } t \in [S, \infty) \end{cases}$$

have a locally absolutely continuous solution. Typically, the hypotheses invoked in theorems of this nature include the requirement that, corresponding to any point $x \in D$, $F(x)$ contains a tangent vector to D, i.e. there is an admissible velocity pointing into the set D. In the following theorem, this 'inward pointing' condition has a dual formulation in terms of normal vectors, rather than tangent vectors: for every point $x \in D$ such that $N_D^P(x)$ is non-empty,

$$\min_{v \in F(t,x)} \xi \cdot v \leq 0, \quad \text{for all } \xi \in N_D^P(x).$$

The condition imposes restrictions on (F, D) only at points x in $\mathcal{D}$:

$$\mathcal{D} := \{x \in D : N_D^P(x) \neq \emptyset\}.$$

$\mathcal{D}$ can be a rather sparse subset of D; it is surprising that such an economical condition suffices to guarantee existence of an F trajectory satisfying the constraint.

Theorem 13.2.1 (*Weak Invariance Theorem for Autonomous Systems*) *Take a multifunction $F : \mathbb{R}^n \leadsto \mathbb{R}^n$ and a closed set $D \subset \mathbb{R}^n$. Assume:*

(i) $F(x)$ *is a non-empty, convex set for each $x \in D$,*
(ii) $\operatorname{Gr} F$ *is closed at every $x \in D$,*

$$x_i \to x, \quad v_i \to v, \quad x_i \in D, \quad v_i \in F(x_i) \quad \Rightarrow \quad v \in F(x),$$

(iii) *There exists $c > 0$ such that*

$$F(x) \subset c(1 + |x|)\mathbb{B} \quad \text{for all } x \in D.$$

Assume further that, for every $x \in D$ such that $N_D^P(x) \neq \{0\}$,

$$\min_{v \in F(x)} \zeta \cdot v \leq 0 \quad \text{for all } \zeta \in N_D^P(x). \tag{13.2.1}$$

Then, given any $x_0 \in D$ and $S \in \mathbb{R}$, there exists a locally absolutely continuous function $x : [S, \infty) \to \mathbb{R}^n$ satisfying

$$\begin{cases} \dot{x}(t) \in F(x(t)) & \text{a.e. } t \in [S, +\infty), \\ x(S) = x_0, \\ x(t) \in D & \text{for all } t \in [S, +\infty). \end{cases}$$

Proof It suffices to prove existence of an arc x satisfying the stated conditions on an arbitrary finite interval $[S, T]$, since we can then generate an arc on $[0, +\infty)$ by concatenating a countable number of such arcs.

Observe that only the points on the boundary of D are really involved in condition (13.2.1), since, when $x \in \text{int} D$, we have $N_D^P(x) = \{0\}$ and, in this case, (13.2.1) is automatically satisfied. Moreover, the hypothesis (iii) in the theorem statement can be replaced by the stronger hypothesis:

(H′) There exists $K > 0$ such that $F(x) \subset K\mathbb{B}$ for all $x \in \mathbb{R}^n$,

without loss of generality. To see this, choose any r satisfying

$$r > \left[e^{c|T-S|} (|x_0| + c|T - S|) \right].$$

Define

$$\tilde{F}(x) := \begin{cases} F(x) & \text{if } |x| \leq r \\ \text{co}\,((F\,(rx/|x|) \cup \{0\}) & \text{if } |x| > r. \end{cases}$$

$\tilde{F}$ and D satisfy the hypotheses of the theorem statement, and also (H′), in which $K = c(1+r)$. If the assertions of the theorem are valid under the stronger hypotheses then, there exists an $\tilde{F}$ trajectory z such that $z(S) = x_0$ and $z(t) \in D$ for all $t \in [S, T]$. But

$$|\dot{z}| \leq c(1 + |z|),$$

so by Gronwall's lemma (Lemma 6.2.4)

$$|z(t)| \leq e^{c|T-S|} (|x_0| + c|T - S|) < r.$$

Since F and $\tilde{F}$ coincide on $D \cap r\mathbb{B}$, z is also an F trajectory. It follows that the assertions of the 'finite interval' version of the theorem are true with $x = z$. So we can assume (H′).

Fix an integer $m > 0$. Let $\{t_0 = S, t_1, \ldots, t_m = T\}$ be a uniform partition of $[S, T]$. Set $h_m := |T - S|/m$.

Define sequences $\{x_0^m, \ldots, x_m^m\}$, $\{v_0^m, \ldots, v_{m-1}^m\}$ and $\{y_0^m, \ldots, y_{m-1}^m\}$ recursively as follows: take $y_0^m = x_0^m = x_0$. If, for some $i \in \{0, \ldots, m-1\}$, x_i^m is given then choose $y_i^m \in D$ to satisfy

$$|x_i^m - y_i^m| = \inf\{|x_i^m - y| \mid y \in D\}.$$

We now make use of the fact that, since y_i^m is a closest point to x_i^m in D, and $x_i^m - y_i^m \in N_D^P(y_i^m)$ whenever $x_i^m \notin D$. Under the hypotheses then, $v_i^m \in F(y_i^m)$ can be chosen to satisfy

13.2 Invariance Theorems

$$v_i^m \cdot (x_i^m - y_i^m) \leq 0, \tag{13.2.2}$$

indeed, in this choice, when $x_i^m \in D$, we can take any $v_i^m \in F(x_i^m)$, otherwise we use condition (13.2.1). Finally, set

$$x_{i+1}^m := x_i^m + h_m v_i^m.$$

Now we define $z^m : [S, T] \to \mathbb{R}^n$ to be the polygonal arc whose graph is the linear interpolant between the node points $\{(t_0^m, x_0^m), \ldots, (t_m^m, x_m^m)\}$. We have, for $i = 0, 1, \ldots, m-1$,

$$z^m(t) = x_i^m + (t - t_i^m) v_i^m \quad \text{for all } t \in [t_i^m, t_{i+1}^m].$$

Since $|v_0^m| \leq K$ and $x_0^m \in D$,

$$d_D^2(x_1^m) \leq |x_1^m - x_0^m|^2 = h_m^2 |v_0|^2 \leq K^2 (h_m)^2. \tag{13.2.3}$$

Fix $i \in \{1, \ldots, m\}$. Since $y_i^m \in D$, we have

$$d_D^2(x_i^m) \leq |x_i^m - y_{i-1}^m|^2$$
$$= |x_{i-1}^m - y_{i-1}^m|^2 + |x_i^m - x_{i-1}^m|^2 + 2 \left(x_{i-1}^m - y_{i-1}^m \right) \cdot \left(x_i^m - x_{i-1}^m \right)$$
$$\leq d_D^2(x_{i-1}^m) + K^2 (h_m)^2 + 0 \tag{13.2.4}$$

(by (13.2.2)).

Relations (13.2.3) and (13.2.4), which are valid for for arbitrary $i \in \{1, \ldots, m\}$, combine to tell us that

$$d_D^2(x_i^m) \leq m K^2 (h_m)^2 = |T - S|^2 K^2 m^{-1}, \tag{13.2.5}$$

for $i = 1, \ldots, m$.

The functions $\{z^m\}_{m=1}^\infty$ are uniformly bounded and have common Lipschitz constant K. By the compactness of trajectories theorem (Theorem 6.3.3), there exists a Lipschitz continuous arc z such that, following the extraction of a suitable subsequence, we have

$$z^m \to z \quad \text{uniformly as } m \to \infty \tag{13.2.6}$$

and

$$\dot{z}^m \to \dot{z} \quad \text{weakly in } L^1.$$

In view of (13.2.5) and the continuity of the distance function, we have

$$d_D(z(t)) = 0 \quad \text{for all } t \in [S, T].$$

Since D is closed, $z(t) \in D$ for all $t \in [S, T]$.

It remains to check that z is an F trajectory. Suppose to the contrary that $\dot{z}(t) \notin F(z(t))$ on a set of positive measure. We deduce from the upper semi-continuous of F that there exists $\delta > 0$ such that

$$\dot{z}(t) \notin F_\delta(t)$$

on a set of positive measure. Here

$$F_\delta(t) := \text{co } \{v : v \in F(\xi) \text{ for some } \xi \in z(t) + \delta \mathbb{B}\} + \delta \mathbb{B}.$$

Arguing as in the proof of the compactness of trajectories theorem (Theorem 6.3.3), employing well-known properties on the support functions, we can find some $p \in \mathbb{R}^n$ and a subset $A \subset [S, T]$ of positive measure such that

$$p \cdot \dot{z}(t) > h(t) \quad \text{for all } t \in A. \tag{13.2.7}$$

Here, h is the bounded, measurable function

$$h(t) = \max_{v \in F_\delta(t)} p \cdot v.$$

By (13.2.6) however,

$$\dot{z}^m(t) \in F_\delta(t) \quad \text{a.e.}$$

for all m sufficiently large. This implies that

$$p \cdot \dot{z}^m(t) \leq h(t) \quad \text{a.e.}$$

for all m sufficiently large. By weak convergence of $\{\dot{z}^m\}$ however,

$$\int_A h(t)dt \geq \lim_{m \to \infty} \int_A p \cdot \dot{z}^m(t)dt = \int_A p \cdot \dot{z}(t)dt.$$

This contradicts (13.2.7). It follows that z is an F trajectory. □

Theorem 13.2.2 (Weak Invariance Theorem for Time Varying Systems) *Take multifunctions $F : [S, T] \times \mathbb{R}^n \rightsquigarrow \mathbb{R}^n$, $P : [S, T] \rightsquigarrow \mathbb{R}^n$. Take also $\bar{S} \in [S, T)$ and a point $x_0 \in P(\bar{S})$. Assume that*

(i) *Gr F is closed and F takes values non-empty convex sets,*

13.2 Invariance Theorems

(ii) There exist $c > 0$ such that
$$F(t, x) \subset c(1 + |x|) \mathbb{B} \quad \text{for all } (t, x) \in \operatorname{Gr} P,$$

(iii) $\operatorname{Gr} P$ is closed and, if $\bar{S} = S$,
$$x_0 \in \limsup\nolimits_{t \downarrow \bar{S}} P(t). \tag{13.2.8}$$

(iv) for every $(t, x) \in \operatorname{Gr} P \cap ((S, T) \times \mathbb{R}^n)$ such that $N^P_{\operatorname{Gr} P}(t, x) \neq \{0\}$,
$$\zeta^0 + \min_{e \in F(t,x)} \zeta^1 \cdot e \leq 0 \quad \text{for all } (\zeta^0, \zeta^1) \in N^P_{\operatorname{Gr} P}(t, x).$$

Then there exists an absolutely continuous function x satisfying
$$\begin{cases} \dot{x}(t) \in F(t, x(t)) & \text{a.e. } t \in [\bar{S}, T], \\ x(S) = x_0, \\ x(t) \in P(t) & \text{for all } t \in [\bar{S}, T]. \end{cases}$$

Proof Define $\tilde{F} : \mathbb{R} \times \mathbb{R}^n \rightsquigarrow \mathbb{R}^{1+n}$, $\tilde{P} : \mathbb{R} \rightsquigarrow \mathbb{R}^n$ and $\tilde{D} \subset \mathbb{R}^{1+n}$ to be

$$\tilde{F}(t, x) := \begin{cases} \{1\} \times F(t, x) & \text{for } S < t < T \\ \operatorname{co}(\{(0, 0)\} \cup (\{1\} \times F(S, x))) & \text{for } t \leq S \\ \operatorname{co}(\{(0, 0)\} \cup (\{1\} \times F(T, x))) & \text{for } t \geq T, \end{cases}$$

$$\tilde{P}(t) := \begin{cases} P(t) & \text{for } S < t < T \\ P(S) & \text{for } t \leq S \\ P(T) & \text{for } t \geq T \end{cases}$$

and
$$\tilde{D} := \operatorname{Gr} \tilde{P}.$$

If $\bar{S} = S$, according to (13.2.8), we can choose sequences $\{t_0^i\} \subset (S, T)$ and $\{x_0^i\} \subset \mathbb{R}^n$ such that $t_0^i \to \bar{S} = S$, $x_0^i \to x_0$ and $x_0^i \in P(t_0^i)$ for all i. In the case '$\bar{S} \in (S, T)$' we choose $t_0^i = \bar{S}$ and $x_0^i = x_0$ for all i.

Fix i and apply the autonomous weak invariance theorem (Theorem 13.2.1) to $(\tilde{F}, \tilde{D})$, with initial time t_0^i and initial state (t_0^i, x_0^i). (It is a straightforward matter to check that the relevant hypotheses are satisfied.) This supplies a locally absolutely continuous arc $(\tau^i, x^i) : [t_0^i, T] \to \mathbb{R} \times \mathbb{R}^n$ such that

$$\begin{cases} (\dot{\tau}^i(t), \dot{x}^i(t)) \in \tilde{F}(\tau^i(t), x^i(t)) & \text{a.e. } t \in [t_0^i, +\infty) \\ (\tau^i, x^i)(t_0^i) = (t_0^i, x_0^i) \\ (\tau^i(t), x^i(t)) \in \tilde{D} & \text{for all } t \in [t_0^i, +\infty). \end{cases}$$

Clearly,
$$0 \leq \dot{\tau}^i(t) \leq 1.$$

It follows that
$$S < t_0^i \leq \tau^i(t) < T \quad \text{a.e. } t \in [t_0^i, T).$$

By definition of $\tilde{F}$ then,
$$\dot{\tau}^i(t) = 1 \quad \text{a.e. } t \in [t_0^i, T].$$

We conclude that $\tau^i(t) = t$ for all $t \in [t_0^i, T]$, whence x^i satisfies
$$\dot{x}^i(t) \in F(t, x^i(t)) \quad \text{a.e. } t \in [t_0^i, T]$$
$$x^i(t) \in P(t) \quad \text{for all } t \in [t_0^i, T].$$

If $\bar{S} \in (S, T)$, then the proof is completed by taking the F trajectory $x = x^i$. So we continue considering the case $\bar{S} = S$. Now extend the domain of x^i to $[\bar{S}, +\infty)$ by constant extrapolation from the right and restrict the resulting function to $[\bar{S}, T]$. This last function we henceforth denote x_i.

Consider the sequence of arcs $x_i : [\bar{S}, T] \to \mathbb{R}^n$. The x_i's are uniformly bounded and the $\dot{x}_i$'s are uniformly integrably bounded. According to the compactness of trajectories theorem (Theorem 6.3.3), there exists an absolutely continuous arc x, which is an F trajectory on $[S, T]$ and
$$x_i(t) \to x(t) \quad \text{for all } t \in [S, T].$$

Because the x_i's are equicontinuous, $x_i(t_0^i) = x_0^i$ for each i and P has closed graph, it follows that the F trajectory x satisfies the conditions $x(\bar{S}) = x_0$ and $x(t) \in P(t)$ for all $t \in [\bar{S}, T]$. This is what we set out to prove. □

We now consider strong invariance properties of differential inclusions. In contrast to weak invariance theorems, in which the existence of some F trajectory is asserted under regularity hypotheses on F which merely require F to have closed graph, strong invariance theorems typically invoke stronger, 'Lipschitz' regularity hypotheses concerning F.

The first strong invariance theorem we state covers autonomous differential inclusions and constraints. The proof, due to Clarke and Ledyaev, is remarkable for its simplicity and for the fact that it covers cases in which F is possibly non-convex valued and fails even to have closed values.

Theorem 13.2.3 *(Strong Invariance Theorem for Autonomous Problems)* Take a multifunction $F : \mathbb{R}^n \rightsquigarrow \mathbb{R}^n$ and a closed set $D \subset \mathbb{R}^n$. Assume that

13.2 Invariance Theorems

(i) F takes values non-empty sets,
(ii) there exists $K > 0$ such that

$$F(x') \subset F(x'') + K|x' - x''|\mathbb{B} \quad \text{for all } x', x'' \in \mathbb{R}^n,$$

(iii) for each $x \in D$ such that $N_D^P(x)$ is non-empty

$$\sup_{v \in F(x)} \zeta \cdot v \leq 0 \quad \text{for all } \zeta \in N_D^P(x). \tag{13.2.9}$$

Then for any Lipschitz continuous function $x : [S, T] \to \mathbb{R}^n$ satisfying

$$\begin{cases} \dot{x}(t) \in F(x(t)) & \text{a.e. } t \in [S, T] \\ x(S) = x_0 \end{cases}$$

we have

$$x(t) \in D \quad \text{for all } t \in [S, T].$$

Proof Since x is Lipschitz continuous and d_D^2 is locally Lipschitz continuous, the function $s \to d_D^2(x(s))$ is Lipschitz continuous. Choose a point $t \in (S, T)$ such that x and $s \to d_D^2(x(s))$ are differentiable at t and also $\dot{x}(t) \in F(x(t))$. The set of such points has full measure.

Let z be a closest point to $x(t)$ in D. Then $z \in D$ and $x(t) - z \in N_D^P(z)$. Under the hypotheses, there exists $v \in F(z)$ such that

$$|v - \dot{x}(t)| \leq K|x(t) - z|.$$

In view of (13.2.9),

$$(x(t) - z) \cdot v \leq 0.$$

Since

$$(x(t) - z) \cdot \dot{x}(t) = (x(t) - z) \cdot v + (x(t) - z) \cdot (\dot{x}(t) - v)$$
$$\leq 0 + |x(t) - z| \cdot |\dot{x}(t) - v| \leq K|x(t) - z|^2,$$

it follows that

$$(x(t) - z) \cdot \dot{x}(t) \leq K d_D^2(x(t)). \tag{13.2.10}$$

Now

$$d_D^2(x(t)) = |x(t) - z|^2$$

and, for any $\delta \in (0, T - t)$,

$$d_D^2(x(t+\delta)) = \inf_{y \in D}|x(t+\delta) - y|^2 \leq |x(t+\delta) - z|^2.$$

We conclude that

$$\begin{aligned}(d/dt)d_D^2(x(t)) &= \lim_{\delta \downarrow 0}\delta^{-1}\left[d_D^2(x(t+\delta)) - d_D^2(x(t))\right]\\ &\leq \lim_{\delta \downarrow 0}\delta^{-1}[\psi(x(t+\delta)) - \psi(x(t))]\\ &= \nabla\psi \cdot \dot{x}(t) = 2(x(t) - z(t)) \cdot \dot{x}(t)\end{aligned}$$

in which $\psi : \mathbb{R}^n \to \mathbb{R}$ is the analytic function

$$\psi(x) := |x - z|^2.$$

We deduce from (13.2.10) that

$$(d/dt)d_D^2(x(t)) \leq 2Kd_D^2(x(t)).$$

This inequality holds for almost every $t \in [S, T]$. Since $d_D^2(x(S)) = 0$, we conclude from Gronwall's lemma (Lemma 6.2.4) that

$$d_D^2(x(t)) = 0 \quad \text{for all } t \in [S, T].$$

Since D is closed, this implies that $x(t) \in D$ for all $t \in [S, T]$. □

The autonomous strong invariance Theorem 13.2.3 can be extended to cover time dependent multifunctions F, by means of state augmentation. This is possible however only if F is Lipschitz continuous with respect to the time variable (since state augmentation accords the time variable the status of a state variable component, and therefore requires that F satisfies the same regularity hypotheses with respect to the time variable as it does with respect to the original state variable). The following strong invariance theorem does not require F to be Lipschitz continuous with respect to the time variable (though it invokes additional hypotheses on F in other respects).

Theorem 13.2.4 (Strong Invariance Theorem for Time Varying Systems) *Take multifunctions $F : [S, T] \times \mathbb{R}^n \rightsquigarrow \mathbb{R}^n$ and $P : [S, T] \rightsquigarrow \mathbb{R}^n$. Take also $\bar{S} \in [S, T)$ and an F trajectory $\bar{x} : [\bar{S}, T] \to \mathbb{R}^n$ such that*

$$\bar{x}(\bar{S}) \in P(\bar{S}).$$

13.2 Invariance Theorems

Assume that, for some constant $\delta > 0$,

(i) *F takes values non-empty, closed, convex sets, and $F(.,x)$ is $\mathcal{L}(S,T)$-measurable for all $x \in \mathbb{R}^n$,*

(ii) *There exists $c > 0$ such that*

$$F(t,x) \subset c\mathbb{B} \quad \text{for all } x \in \bar{x}(t) + \delta\mathbb{B}, \text{ a.e. } t \in [\bar{S}, T],$$

(iii) *there exists a non-negative function $k_F \in L^1$ such that $t \to 1/(1+k_F(t))$ is a.e. equal to a continuous function and*

$$F(t,x') \subset F(t,x) + k_F(t)|x-x'|\mathbb{B} \quad \text{for all } x, x' \in \bar{x}(t) + \delta\mathbb{B},$$
$$\text{for a.e. } t \in [\bar{S}, T],$$

(iv) *for each $s \in [\bar{S}, T)$ and $t \in (\bar{S}, T]$ the following one-sided set-valued limits exist and are non-empty:*

$$F(s^+, x) := \lim_{s' \downarrow s} F(s', x) \quad \text{for all } x \in \bar{x}(s) + \delta\mathbb{B} \quad \text{and}$$

$$F(t^-, x) := \lim_{t' \uparrow t} F(t', x) \quad \text{for all } x \in \bar{x}(t) + \delta\mathbb{B},$$

and there exists a set $\mathcal{S} \subset (\bar{S}, T)$ of full Lebesgue measure such that, for every $t \in \mathcal{S}$,

$$s \to F(s,x) \text{ is continuous at } t \text{ for each } x \in \bar{x}(t) + \delta\mathbb{B},$$

(v) *Gr P is closed and, if $\bar{S} = S$,*

$$\bar{x}(\bar{S}) \in \limsup\nolimits_{t \downarrow \bar{S}} P(t),$$

(vi) *for every $(t,x) \in \text{Gr } P \cap T_\delta(\bar{x}) \cap ((S,T) \times \mathbb{R}^n)$ at which $N^P_{\text{Gr } P}(t,x) \neq \emptyset$,*

$$\zeta^0 + \left(\max_{v \in F(t^-, x)} \zeta^1 \cdot v\right) \wedge \left(\max_{v \in F(t^+, x)} \zeta^1 \cdot v\right) \leq 0 \quad \text{for all } (\zeta^0, \zeta^1) \in N^P_{\text{Gr } P}(t,x),$$

in which $T_\delta(\bar{x}) := \{(t,x) \in [\bar{S}, T] : |x - \bar{x}(t)| \leq \delta\}$.

Then

$$x(t) \in P(t) \quad \text{for all } t \in [\bar{S}, T].$$

Proof We reserve for future use the following facts: by reducing the size of δ we can arrange that

$$\left.\begin{aligned} F(t^-, x') &\subset F(t^-, x) + k_F(t)|x - x'| \mathbb{B} \\ &\quad \text{for all } x, x' \in \bar{x}(t) + \delta \mathbb{B} \text{ and } t \in (S, T] \\ F(t^+, x') &\subset F(t^+, x) + k_F(t)|x - x'| \mathbb{B} \\ &\quad \text{for all } x, x' \in \bar{x}(t) + \delta \mathbb{B} \text{ and } t \in [S, T) \end{aligned}\right\} \qquad (13.2.11)$$

and

(These relations are simple consequences of hypotheses (iii) and (iv).)

We simplify the analysis by making, at the outset, a hypothesis reduction: we may assume, without loss of generality,

(HS): k_F is a constant function,

To justify imposing the first extra hypothesis (HS), suppose that k_F is not a constant function. Introduce a change of independent variable $s = \sigma(t)$, where

$$\sigma(t) = \int_{\bar{S}}^{t} (1 + k_F(t'))dt'.$$

σ is a monotone increasing function that maps $[S, T]$ onto $[0, \sigma_1]$, where $\sigma_1 = \int_{\bar{S}}^{T} (1 + k_F(t'))dt'$. Define $y(s) := (\bar{x} \circ \sigma^{-1})(s)$ for $s \in [0, \sigma_1]$. y is a G trajectory on $[\sigma_0, \sigma_1]$, where $G : [0, \sigma_1] \times \mathbb{R}^n \rightsquigarrow \mathbb{R}^n$ is the multifunction

$$G(s, y) := (r \circ \sigma^{-1})(s) \, F(\sigma^{-1}(s), y) \, .$$

Here, r is the continuous selector of the equivalence class of almost everywhere equal functions $t \to 1/(1 + k_F(t))$. (See hypothesis (iii).) Now define the multifunction $Q : [0, \sigma_1] \rightsquigarrow \mathbb{R}^n$ to be $Q(s) := P(\sigma^{-1}(s))$. We find that the pair of multifunctions (G, Q) satisfies the hypotheses of the theorem statement, with time variable $s \in [0, \sigma_1]$, in which the parameters c and δ remain the same, k_F is replaced by the measurable function $s \to \frac{(k_F \circ \sigma^{-1})(s)}{1+(k_F \circ \sigma^{-1})(s)}$ (which is dominated by the constant function taking value 1) and the subset $\mathcal{S} \subset [\bar{S}, T]$ of full Lebesgue measure is replaced by the subset $\sigma(\mathcal{S}) \subset [0, \bar{\sigma}]$ of full Lebesgue measure. Since the transformed data satisfies the hypotheses of the theorem with constant Lipschitz bound (in fact, we can take the constant to be 1), we deduce from the special case of the theorem that $y(s) \in Q(s)$ for all $s \in [0, \sigma_1]$. But then $x(t) = (y \circ \sigma)(t) \in Q(\sigma(t)) = P(t)$, for all $t \in [\bar{S}, T]$, as required.

The key step in the proof is to confirm the following:

Claim: For any subset $[t_0, t_f] \subset [S, T]$ such that $|t_f - t_0| \leq \delta/(8c)$,

'$\bar{x}(t_0) \in P(t_0)$' implies '$\bar{x}(t) \in P(t)$ for all $t \in [t_0, t_f]$'.

13.2 Invariance Theorems

Notice that, if this claim is justified, then the assertions of the theorem are confirmed. To see this, we take a sequence $\alpha_i \downarrow 0$ and, for each i, consider a partition $\mathcal{P}_i = \{t_0 = \bar{S}, \ldots, t_N = t_f\}$ of $[S, T - \alpha_i]$ such that diam $\mathcal{P}_i \leq \delta/(8c)$. Invoking the claim successively on the intervals $[t_0, t_1], \ldots, [t_{N-1}, t_N]$, we deduce that $x(t) \in P(t)$ for all $t \in [t_0, \bar{T} - \alpha_i]$. But then, since $\bar{x}$ is continuous and P has closed graph, we can conclude that $x(t) \in P(t)$ for all $t \in [\bar{S}, T]$, as required.

It remains then to confirm the claim, when the hypotheses are supplemented by (HS). Take a subinterval $[t_0, t_f] \subset [S, T]$ such that $|t_f - t_0| < \delta/(8c)$. Then, by hypothesis (v), we can find sequences $\{s_i\}$ and $\{\xi_i\}$ in $[t_0, t_f)$ and $\mathbb{R}^n$ respectively such that

(a): $s_i \downarrow t_0$, $\xi_i \to \bar{x}(t_0)$ and $x_i \in P(s_i)$ for all i, if $t_0 = S$,
(b): $s_i = t_0$ and $\xi_i = \bar{x}(t_0)$ for all i, if $t_0 > S$.

Take $\epsilon_i \downarrow 0$ such that $\epsilon_i \in (0, \delta/4)$, for each i. Since $\bar{x}$ is continuous we can arrange, by eliminating terms in the sequence, that $|\xi_i - \bar{x}(s_i)| < \epsilon_i$.

Fix i. Now let $y_i : [s_i, t_f] \to \mathbb{R}^n$ be a continuously differentiable function such that $y_i(s_i) = \xi_i$ and

$$|\xi_i - \bar{x}(s_i)| + \|\dot{y}_i - \dot{\bar{x}}\|_{L^1(s_i, t_f)} \leq \epsilon_i . \tag{13.2.12}$$

Notice that $\|y_i - \bar{x}\|_{L^\infty(s_i, t_f)} \leq \epsilon_i (\leq \delta/4)$. Define the multifunctions

$$\tilde{F}_i^-(t, x) := \{v \in F(t^-, x) : |v - \dot{y}_i(t)| \leq k_F |x - y_i(t)| \\ + d_{F(t^-, y_i(t))}(\dot{y}_i(t)) \vee d_{F(t^+, y_i(t))}(\dot{y}_i(t))\},$$

$$\tilde{F}_i^+(t, x) := \{v \in F(t^+, x) : |v - \dot{y}_i(t)| \leq k_F |x - y_i(t)| \\ + d_{F(t^-, y_i(t))}(\dot{y}_i(t)) \vee d_{F(t^+, y_i(t))}(\dot{y}_i(t))\} .$$

$$\tilde{F}_i(t, x) := \tilde{F}_i^-(t, x) \cup \tilde{F}_i^+(t, x)$$

Now define $\Gamma_i : [s_i, t_f] \rightsquigarrow \mathbb{R}^{1+n}$:

$$\Gamma_i(t, x) := \begin{cases} \{1\} \times \operatorname{co} \tilde{F}_i(t, x)\} & \text{if } t \in (s_i, t_f) \text{ and } |x - y_i(t)| \leq \delta/2 \\ \operatorname{co}\{(0, 0)\} \cup (\{1\} \times c\mathbb{B}) & \text{otherwise .} \end{cases}$$

and $D := \operatorname{Gr} P$.

We next apply the autonomous weak invariance Theorem 13.2.1 to (Γ_i, D), when the underlying time interval is taken to be $[s_i, t_f]$ and the initial state is (s_i, x_i). To justify this, we must check that the hypotheses are satisfied.

We show first that $\tilde{F}_i^+(t, x)$ and $\tilde{F}_i^+(t, x)$ are both non-empty sets. In view of the definition, we need to check this property only when $|x - y_i(t)| \leq \delta/2$ and t is any point in $[s_i, t_f]$. Take $e \in F(t^-, y_i(t))$ such that $|\dot{y}_i(t) - e| = d_{F(t^-, y_i(t))}(\dot{y}_i(t))$.

Then, by (13.2.11), there exists $v \in F(t^-, x)$ such that $|v - e| \le k_F |x - y_i(t)|$. It follows that

$$|v - \dot{y}_i(t)| \le |v - e| + |e - \dot{y}_i(t)| \le k_F |x - y_i(t)| + d_{F(t^-, y_i(t))}(\dot{y}_i(t)).$$

Clearly, v satisfies conditions for membership of the set $\tilde{F}_i^-(t, x)$; this set then is non-empty. A similar analysis establishes non-emptiness of $\tilde{F}_i(t^+, x)$. Using the fact that the function $t \to d_{F(t^-, y_i(t))}(\dot{y}_i(t)) \vee d_{F(t^+, y_i(t))}(\dot{y}_i(t))$ is upper semi-continuous, it is easy to show that the multifunction $\tilde{F}_i$ has closed graph. The multifunction Γ_i then, too, takes values non-empty convex sets and has closed graph.

It remains to check the inward pointing hypothesis. This will be satisfied if we can show

$$\zeta^0 + \min_{v \in \operatorname{co} \tilde{F}_i(t,x)} \zeta^1 \cdot v \le 0 \quad \text{for all } x \in y_i(t) + (\delta/2)\mathbb{B}, t \in [s_i, t_f].$$

Let v^- and v^+ be minimizers of $v \to \zeta^1 \cdot v$ over the non-empty, compact sets $\tilde{F}_i^-(t, x)$ and $\tilde{F}_i^+(t, x)$, respectively. Then, since $v^- \in F(t^-, x)$ and $v^+ \in F(t^+, x)$,

$$\min_{v \in \tilde{F}_i(t,x)} \zeta^1 \cdot v = \left(\zeta^1 \cdot v^-\right) \wedge \left(\zeta^1 \cdot v^+\right) \le (\max_{v \in F(t^-, x)} \zeta^1 \cdot v) \wedge (\max_{v \in F(t^+, x)} \zeta^1 \cdot v)$$

So, in consequence of hypothesis (vi) in the theorem statement,

$$\zeta^0 + \min_{v \in \operatorname{co} \tilde{F}(t,x)} \zeta^1 \cdot v = \zeta^0 + \min_{v \in \tilde{F}_i(t,x)} \zeta^1 \cdot v$$

$$\le \zeta^0 + \left(\max_{v \in F(t^-, x)} \zeta^1 \cdot v\right) \wedge \left(\max_{v \in F(t^+, x)} \zeta^1 \cdot v\right) \le 0.$$

We have arrived at the desired inequality.

The autonomous weak invariance Theorem 13.2.1 now tells us that there exists a Γ_i trajectory (τ_i, x_i) on $[s_i, t_f]$ such that $\tau_i(s_i) = s_i$, $x_i(s_i) = \bar{x}(s_i)$ and $(\tau_i(t), x_i(t)) \in \Gamma_i(t)$ for all $t \in [s_i, T]$. From the properties of F and the definition of Γ_i we know that $\dot{\tau}_i = 1$ and $|\dot{x}_i(t)| \le c$, a.e.. But then

$$x_i(t) \in P(t) \quad \text{for all } t \in [s_i, t_f]. \tag{13.2.13}$$

In view of (13.2.12), since $|t_f - s_i| < \delta/(8c)$ and since, both x_i and $\bar{x}$ are F trajectories (and consequently $|\dot{x}_i(t)|$ and $|\dot{\bar{x}}(t)|$ have magnitudes bounded by c), we have, for all $t \in [s_i, t_f]$,

$$|x_i(t) - y_i(t)| \le |x_i(t) - \bar{x}(t)| + |\bar{x}(t) - y_i(t)|$$

$$\le \int_{s_i}^{t_f} |\dot{x}_i(t) - \dot{\bar{x}}(t)| dt + |\xi_i - \bar{x}(s_i)| + |\bar{x}(t) - y_i(t)|$$

$$< \delta/4 + \delta/4 + \delta/4 = 3\delta/4.$$

It now follows from the definition of Γ_i that $\dot{x}_i(t) \in F(t, x_i(t))$, a.e. $t \in [s_i, t_f]$ and, recalling also hypothesis (iv),

$$|\dot{x}_i(t) - \dot{y}_i(t)| \leq k_F |x_i(t) - y_i(t)| + d_{F(t, y_i(t))}(\dot{y}_i(t)), \quad \text{a.e. } t \in [s_i, t_f].$$

Since $d_{F(t,x)}(v)$ is Lipschitz continuous w.r.t. to x and v, with Lipschitz constants k_F and 1 respectively, we know however that

$$d_{F(t, y_i(t))}(\dot{y}_i(t)) \leq d_{F(t, \bar{x}(t))}(\dot{\bar{x}}(t)) + k_F |y_i(t) - \bar{x}(t)| + |\dot{y}_i(t) - \dot{\bar{x}}(t)|$$
$$\leq 0 + k_F |y_i(t) - \bar{x}(t)| + |\dot{y}_i(t) - \dot{\bar{x}}(t)|$$

It now follows from (13.2.12) that

$$\int_{s_i}^{t_f} d_{F(t, y_i(t))}(\dot{y}_i(t)) dt \leq (1 + k_F |T - S|) \times \epsilon_i.$$

But then, by Gronwall's lemma,

$$|x_i(t) - \bar{x}(t)| \leq e^{k_F |T-S|} |\xi_i - \bar{x}(s_i)| + e^{k_F |T-S|}(1 + k_F |T - S|) \times \epsilon_i$$

for all $t \in [s_i, t_f]$. \hfill (13.2.14)

Let x_i^e be the extension of x_i to $[t_0, t_f]$, by constant extrapolation to the left. By the compactness of trajectories theorem, x_i^e converges, uniformly, to an F trajectory x^* on $[t_0, t_f]$. It follows from (13.2.14) that $x_i^e \to x^*$, uniformly on $[t_0, t_f]$, and $x^* = \bar{x}$. Since P has closed graph, we conclude from (13.2.13) that $\bar{x}(t) \in P(t)$ for all $t \in [t_0, t_f]$. The claim has been confirmed and the proof is complete. □

13.3 The Value Function and Generalized Solutions of the Hamilton Jacobi Equation

With the apparatus of invariance now at hand, we are ready to return to the dynamic optimization problem:

$$(P) \begin{cases} \text{Minimize } g(x(T)) \\ \text{over arcs } x \in W^{1,1}([S, T]; \mathbb{R}^n) \text{ satisfying} \\ \dot{x}(t) \in F(t, x(t)) \quad \text{a.e.,} \\ x(S) = x_0, \end{cases}$$

in which $g : \mathbb{R}^n \to \mathbb{R} \cup \{+\infty\}$ is a given function, $[S, T]$ is a given interval, $F : [S, T] \times \mathbb{R}^n \rightsquigarrow \mathbb{R}^n$ is a given multifunction and $x_0 \in \mathbb{R}^n$ a given point.

Recall that the value function for (P), $V : [S, T] \times \mathbb{R}^n \to \mathbb{R} \cup \{+\infty\}$, is the function

$$V(t, x) := \inf (P_{t,x}) \quad \text{for all } (t, x) \in [S, T] \times \mathbb{R}^n,$$

in which $\inf (P_{t,x})$ denotes the infimum cost of the optimization problem:

$$(\text{P}_{t,x}) \begin{cases} \text{Minimize } g(y(T)) \\ \text{over arcs } y \in W^{1,1}([t, T]; \mathbb{R}^n) \text{ satisfying} \\ \dot{y}(s) \in F(t, y(s)) \quad \text{a.e.,} \\ x(t) = x, \end{cases}$$

thus

$$V(t, x) := \inf(P_{t,x}), \quad \text{for all } (t, x) \in [S, T] \times \mathbb{R}^n.$$

Our goal is to characterize V as a generalized solution, appropriately defined, to the Hamilton Jacobi equation:

$$\phi_t(t, x) + \min_{v \in F(t,x)} \phi_x(t, x) \cdot v = 0 \quad \text{for all } (t, x) \in (S, T) \times \mathbb{R}^n \qquad (13.3.1)$$

$$\phi(T, x) = g(x) \text{ for all } x \in \mathbb{R}^n, \qquad (13.3.2)$$

in various senses.

To achieve this, we use the properties of two new constructs from nonsmooth analysis.

Definition 13.3.1 Consider a set $D \subset \mathbb{R}^k$, a function $\varphi : D \to \mathbb{R} \cup \{+\infty\}$, a point $x \in \text{dom } \varphi$ and a vector $d \in \mathbb{R}^k$. The *lower Dini (directional) derivative* of φ at x in the direction $d \in \mathbb{R}^k$ is defined to be:

$$D_\uparrow \varphi(x; d) := \liminf_{h \downarrow 0,\ e \to d} h^{-1} [\varphi(x + he) - \varphi(x)] .$$

(For purposes of evaluating the limits on the right side, $\varphi(x + he)$ is assigned the value $+\infty$ when $x + he \notin D$.)

The following lemma, which provides alternative characterizations of the lower Dini derivative in the case D=$[S, T] \times \mathbb{R}^n$ and also the case when φ is locally Lipschitz continuous, is a straightforward consequence of the definitions and is stated without proof. (The differences are in the choice of sequences used to evaluate the 'lim inf'.)

Lemma 13.3.2 *Take a function* $\varphi : [S, T] \times \mathbb{R}^n \to \mathbb{R} \cup \{+\infty\}$, *a point* $(t, x) \in ([S, T] \times \mathbb{R}^n) \cap \text{dom } \varphi$ *and a vector* $v \in \mathbb{R}^n$. *We have*

(i) $D_\uparrow \varphi((t, x); (1, v))$

$$= \liminf \{h^{-1}(\varphi(t + h, x + hw) - \varphi(t, x) : h \downarrow 0,\ w \to v\},$$

13.3 The Value Function and Generalized Solutions of the Hamilton Jacobi...

(ii) if φ is Lipschitz continuous on a neighbourhood of (t, x), then

$$D_\uparrow \varphi((t, x); (1, v)) = \liminf\{h^{-1}(\varphi(t + h, x + hv) - \varphi(t, x)) : h \downarrow 0\},$$

(iii) if φ is Fréchet differentiable at $(t, x) \in (S, T) \times \mathbb{R}^n$, then

$$D_\uparrow \varphi((t, x); d) = \nabla \varphi(t, x) \cdot d, \quad \text{for all } d \in \mathbb{R}^{n+1}.$$

We now make precise two notions of 'generalized solution' to the Hamilton Jacobi equation (13.3.1)–(13.3.2), when the multifunction F of the underlying differential inclusion is possibly discontinuous with respect to time, but is required to have left and right limits: for each $s \in [S, T)$, $t \in (S, T]$ and $x \in \mathbb{R}^n$, the following one-sided set-valued limits (in the Kuratowski sense) are non-empty:

$$F(s^+, x) := \lim_{s' \downarrow s} F(s', x) \quad \text{and} \quad F(t^-, x) := \lim_{t' \uparrow t} F(t', x). \tag{13.3.3}$$

Definition 13.3.3 A function $\phi : [S, T] \times \mathbb{R}^n \to \mathbb{R} \cup \{+\infty\}$ is a *lower Dini solution* to (13.3.1)–(13.3.2) if

(i) $\inf_{v \in F(t^+, x)} D_\uparrow \phi((t, x); (1, v)) \leq 0$
    for all $(t, x) \in ([S, T) \times \mathbb{R}^n) \cap \operatorname{dom} V$,

(ii) $\sup_{v \in F(t^-, x)} D_\uparrow \phi((t, x); (-1, -v)) \leq 0$
    for all $(t, x) \in ((S, T] \times \mathbb{R}^n) \cap \operatorname{dom} V$,

(iii) $\phi(T, x) = g(x)$ for all $x \in \mathbb{R}^n$.

Definition 13.3.4 A lower semi-continuous function $\phi : [S, T] \times \mathbb{R}^n \to \mathbb{R} \cup \{+\infty\}$ is a *proximal solution* to (13.3.1)–(13.3.2) if

(i) for all $(t, x) \in ((S, T) \times \mathbb{R}^n) \cap \operatorname{dom} \phi$, $(\xi^0, \xi^1) \in \partial^P \phi(t, x)$

$$\xi^0 + \inf_{v \in F(t^+, x)} \xi^1 \cdot v \leq 0,$$

(ii) for all $(t, x) \in ((S, T) \times \mathbb{R}^n) \cap \operatorname{dom} \phi$, $(\xi^0, \xi^1) \in \partial^P \phi(t, x)$

$$\xi^0 + \inf_{v \in F(t^-, x)} \xi^1 \cdot v \geq 0,$$

(iii) for all $x \in \mathbb{R}^n$,

$$\liminf_{\{(t', x') \to (S, x) : t' > S\}} \phi(t', x') = \phi(S, x)$$

and

$$\liminf_{\{(t', x') \to (T, x) : t' < T\}} \phi(t', x') = \phi(T, x) = g(x).$$

These definitions are consistent with the classical notion of 'solution' to (13.3.1), in the case that $t \to F(t, x)$ is continuous, since any C^1 function $\phi : [S, T] \times \mathbb{R}^n \to \mathbb{R}$ satisfying

$$\phi_t(t, x) + \min_{v \in F(t,x)} \phi_x(t, x) \cdot v = 0 \quad \text{for all } (t, x) \in (S, T) \times \mathbb{R}^n,$$

is automatically a lower Dini solution and also a proximal solution in the sense just defined.

The following proposition relates these two solution concepts for the Hamilton Jacobi equation (13.3.1)–(13.3.2).

Proposition 13.3.5 *Take a lower semi-continuous function* $\phi : [S, T] \times \mathbb{R}^n \to \mathbb{R} \cup \{+\infty\}$. *Concerning the data for problem (P), we assume that the (Kuratowski) limits (13.3.3) exist and are non-empty sets, for all $x \in \mathbb{R}^n$, $s \in (S, T]$ and $t \in [S, T)$ and $F(t^+, x)$ is a compact set, for all $(t, x) \in (S, T) \times \mathbb{R}^n$. Then*

'ϕ *is a lower Dini solution to (13.3.1)–(13.3.2)' implies 'ϕ is proximal solution to (13.3.1)–(13.3.2)'*.

Proof Let ϕ be a lower semi-continuous function which is also a lower Dini solution to (13.3.1)–(13.3.2). Take any $x \in \mathbb{R}^n$. Then the condition

$$\phi(S, x) \geq \liminf_{t' \downarrow S, x' \to x} \phi(t', x') \tag{13.3.4}$$

is certainly satisfied if $(S, x) \notin \operatorname{dom} \phi$. Suppose that $(S, x) \in \operatorname{dom} \phi$. By assumption, $F(S^+, x)$ is non-empty. It follows from the definition of lower Dini solution that there exist a vector $v \in F(S^+, x)$ and sequences $h_i \downarrow 0$ and $v_i \to v$, $v_i \in \mathbb{R}^n$ for all i, such that

$$\lim_i h_i^{-1}(\phi(S + h_i, x + h_i v_i) - \phi(S, x)) \leq 0.$$

But then

$$\phi(S, x) \geq \lim_i \phi(t_i, x_i),$$

where, for each i, $t_i = S + h_i$ and $x_i = x + h_i v_i$. Thus (13.3.4) has been confirmed in this case too. Since ϕ is lower semi-continuous, the reverse inequality also holds. We conclude that

$$\phi(S, x) = \liminf_{t' \downarrow S, x' \to x} \phi(t', x').$$

A similar analysis, in which we take as starting point a vector v in the non-empty set $F(T^-, x)$ and use the second defining relation of lower Dini solutions, yields

13.3 The Value Function and Generalized Solutions of the Hamilton Jacobi...

$$\liminf_{t'\uparrow T, x'\to x} \phi(t', x') = \phi(T, x) = g(x).$$

Now suppose that $(t, x) \in ((S, T) \times \mathbb{R}^n) \cap \operatorname{dom} \phi$ and $(\xi^0, \xi^1) \in \partial^P \phi(t, x)$. This means that there exist $M > 0$ and $\epsilon > 0$ such that

$$\xi^0(t' - t) + \xi^1 \cdot (x' - x) \leq \phi(t', x') - \phi(t, x) + M(|t' - t|^2 + |x' - x|^2)$$

for all $(t', x') \in (t, x) + \epsilon \mathbb{B}$.

Since ϕ is a lower Dini solution of (13.3.1)–(13.3.2) and $F(t^+, x)$ is compact, there exist sequences $h_i \downarrow 0$ and $\{v_i\}$ in $\mathbb{R}^n$ such that $v_i \to \bar{v}$ for some $\bar{v} \in F(t^+, x)$ and

$$\lim_i h_i^{-1}(\phi(t + h_i, x + h_i v_i) - \phi(t, x)) \leq 0.$$

Setting $(t', x') = (t + h_i, x + h_i v_i)$, we see that, for i sufficiently large,

$$\xi^0 + \xi^1 \cdot v_i \leq h_i^{-1}(\phi(t + h_i, x + h_i v_i) - \phi(t, x)) + Mh_i(1 + |v_i|^2).$$

Since $v_i \to \bar{v}$ and $\bar{v} \in F(t^+, x)$, we obtain, in the limit as $i \to \infty$, that

$$\xi^0 + \inf_{v \in F(t^+, x)} \xi^1 \cdot v \leq \xi^0 + \xi^1 \cdot \bar{v} = \xi^0 + \lim_i \xi^1 \cdot v_i \leq 0. \qquad (13.3.5)$$

Now for any given $v \in F(t^-, x)$, since ϕ a lower Dini solution of (13.3.1)–(13.3.2), there exist sequences $h_i \downarrow 0$ and $\{v_i\}$ in $\mathbb{R}^n$, such that $v_i \to v$ and

$$\limsup_{i \to \infty} h_i^{-1}(\phi(t - h_i, x - h_i v_i) - \phi(t, x)) \leq 0.$$

According the definition of $\partial^P \phi(t, x)$, setting $(t', x') = (t - h_i, x - h_i v_i)$, we obtain

$$-(\xi^0 + \xi^1 \cdot v_i) \leq h_i^{-1}(\phi(t - h_i, x - h_i v_i) - \phi(t, x)) + Mh_i(1 + |v_i|^2),$$

for i sufficiently large. Since $v_i \to v$, we deduce from these relations that

$$\xi^0 + \xi^1 \cdot v \geq 0.$$

Since v was an arbitrary element in $F(t^-, x)$, we have shown that

$$\inf_{v \in F(t^-, x)} (\xi^0 + \xi^1 \cdot v) \geq 0,$$

The proof is concluded. □

The following proposition assembles some useful regularity properties about the value function:

Proposition 13.3.6 *Assume that the data for (P) satisfies the following hypotheses*

(i): $F : [S, T] \times \mathbb{R}^n \rightsquigarrow \mathbb{R}^n$ *takes closed non-empty values, F is $\mathcal{L} \times \mathcal{B}^n$-measurable, the graph of $F(t, .)$ is closed for each $t \in [S, T]$,*

(ii): *there exists $c \in L^1(S, T)$ such that*

$$F(t, x) \subset c(t)(1 + |x|) \, \mathbb{B} \quad \text{for all } x \in \mathbb{R}^n \quad \text{and for a.e. } t \in [S, T] \,,$$

(iii): $g : \mathbb{R}^n \to \mathbb{R} \cup \{+\infty\}$ *is lower semi-continuous.*

Then

(a): *Then $V(t, x) > -\infty$ for all $(t, x) \in [S, T] \times \mathbb{R}^n$,*

(b): *If in addition F takes convex values, then V is lower semi-continuous,*

(c): *If in addition, for every $R_0 > 0$, there exists $k_F \in L^1(S, T)$ such that*

$$F(t, x') \subset F(t, x) + k_F(t)|x - x'|\mathbb{B} \quad \text{for all } x, x' \in R_0 \mathbb{B} \text{ and a.e. } t \in [S, T]$$

and g is continuous, then V is continuous,

(d): *If in addition, for every $R_0 > 0$, there exists $k_F \in L^1(S, T)$ and $c_0 > 0$ such that*

$$F(t, x') \subset F(t, x) + k_F(t)|x - x'|\mathbb{B} \quad \text{for all } x, x' \in R_0 \mathbb{B} \text{ and a.e. } t \in [S, T]$$

$$F(t, x) \subset c_0 \, \mathbb{B} \quad \text{for all } (t, x) \in [S, T] \times R_0 \mathbb{B} \,,$$

and g is locally Lipschitz continuous, then V is locally Lipschitz continuous.

Proof

(a): Take any $(t, x) \in [S, T] \times \mathbb{R}^n$. It follows from hypothesis (ii) that, for any state trajectory on y on $[t, T]$ such that $y(t) = x$, there exists a number $r > 0$, independent of y, such that $|y(T)| \leq r$. But $V(t, x) \geq \inf_{x' \in r\mathbb{B}} g(x')$. Since $g : \mathbb{R}^n \to \mathbb{R} \cup \{+\infty\}$ is lower semi-continuous function, its infimum value on the compact set $r\mathbb{B}$ cannot take value $-\infty$. It follows that $V(t, x) > -\infty$.

(b): Fix $(t, x) \in [S, T] \times \mathbb{R}^n$. Then, by Proposition 6.4.1, we know that $(P_{t,x})$ has a minimizer y. (We allow the possibly that $(P_{t,x})$ has infinite cost). It follows that

$$V(t, x) \, (= \, g(y(T)) \,) \, > \, -\infty.$$

Now take any sequence $\{(t_i, x_i)\}$ in $[S, T] \times \mathbb{R}^n$, such that $(t_i, x_i) \to (t, x)$ and $\lim_i V(t_i, x_i)$ exists. To prove that V is lower semi-continuous we must show that

$$V(t, x) \, \leq \, \lim_i V(t_i, x_i).$$

13.3 The Value Function and Generalized Solutions of the Hamilton Jacobi...

For each i, (P_{t_i, x_i}) has a minimizer, which we write $y_i : [t_i, T] \to \mathbb{R}^n$. Of course $y_i(t_i) = x_i$. Extend y_i to all $[S, T]$ by constant extrapolation from the right on $[S, t_i]$. From hypothesis (ii) and Gronwall's lemma (Lemma 6.2.4), all F trajectories having initial data in $[S, T] \times (|x| + 1)\mathbb{B}$ evolve in $[S, T] \times R_0 \mathbb{B}$, where $R_0 := e^{\int_S^T c(t)\, dt}(|x| + 2)$. Observe that it is not restrictive to assume that $x_i \in (|x| + 1)\mathbb{B}$ for all i. Then, we deduce that the y_i's are uniformly integrably bounded and their derivatives are uniformly integrably bounded. The compactness of trajectories theorem (Theorem 6.3.3) ensures that, following extraction of a subsequence, we have $y_i \to y$, uniformly, for some absolutely continuous arc whose restriction to $[t, T]$ is an F trajectory. Since the $\dot{y}_i$'s are uniformly integrably bounded,

$$y(t) = \lim_i y_i(t_i) = \lim_i x_i = x.$$

But then,

$$V(t, x) \leq g(y(T)) \leq \lim_i g(y_i(T)) = \lim_i V(t_i, x_i).$$

It follows that V is lower semi-continuous.

(c): Fix any $r_0 > 0$. Notice that, from hypothesis (ii) and Gronwall's lemma (Lemma 6.2.4), for any $\tau \in [S, T]$ and any F trajectory x on $[\tau, T]$ such that $x(\tau) \in r_0 \mathbb{B}$ we have: $x(t) \in R_0 \mathbb{B}$ for all $t \in [t, T]$, where $R_0 := \exp\{\int_S^T c(t)\, dt\}(r_0 + 1)$. Take two points $(t_1, x_1), (t_2, x_2) \in [S, T] \times \mathbb{R}^n$ such that $x_1, x_2 \in r_0 \mathbb{B}$. It is not restrictive to assume that $t_1 \leq t_2$. Fix any $\varepsilon > 0$. Then there exists an F trajectory y_ε such that $y_\varepsilon(t_1) = x_1$ and

$$V(t_1, x_1) \geq g(y_\varepsilon(T)) - \varepsilon. \tag{13.3.6}$$

We know that $|x_2 - y_\varepsilon(t_2)| \leq |x_1 - x_2| + (1 + R_0) \int_{t_1}^{t_2} c(s)\, ds$. Consider the point $(t_2, y_\varepsilon(t_2))$. From the Filippov's existence theorem (Theorem 6.2.3) there exists an F trajectory y_2 on $[t_2, T]$ such that $y_2(t_2) = x_2$ and

$$\|y_2 - y_\varepsilon\|_{L^\infty(t_2, T)} \leq K |x_2 - y_\varepsilon(t_2)| \leq K \left(|x_1 - x_2| + (1 + R_0) \int_{t_1}^{t_2} c(s)\, ds\right), \tag{13.3.7}$$

where $K := e^{\int_S^T k_F(s)\, ds}$. Write ω_g the modulus of continuity of g. Therefore, from (13.3.6) and (13.3.7) we have

$$V(t_2, x_2) - V(t_1, x_1) \leq g(y_2(T)) - g(x_\varepsilon(T)) + \varepsilon$$
$$\leq \omega_g(|y_2(T) - x_\varepsilon(T)|) + \varepsilon$$
$$\leq \omega_g\left(K\left(|x_1 - x_2| + (1 + R_0) \int_{t_1}^{t_2} c(s)\, ds\right)\right) + \varepsilon.$$

We can also find an F trajectory z_ε on $[t_2, T]$ such that $z_\varepsilon(t_2) = x_2$ and

$$V(t_2, x_2) \geq g(z_\varepsilon(T)) - \varepsilon. \tag{13.3.8}$$

Consider now an F trajectory z_1 on $[t_1, t_2]$ (if $t_1 < t_2$) and extend it by means of Filippov's existence theorem to an F trajectory on $[t_1, T]$ still written z_1 such that

$$\|z_1 - z_\varepsilon\|_{L^\infty(t_2,T)} \leq K |x_2 - z_1(t_2)| \leq K (|x_2 - x_1| + (1 + R_0) \int_{t_1}^{t_2} c(s)\, ds). \tag{13.3.9}$$

So, arguing as above, we obtain the reverse inequality

$$V(t_1, x_1) - V(t_2, x_2) \leq \omega_g(K (|x_1 - x_2| + (1 + R_0) \int_{t_1}^{t_2} c(s)\, ds)) + \varepsilon.$$

and, taking the limit as ε tends to zero, it follows that

$$|V(t_1, x_1) - V(t_2, x_2)| \leq \omega_g(K (|x_1 - x_2| + (1 + R_0) \int_{t_1}^{t_2} c(s)\, ds)).$$

Clearly this inequality is valid whenever we fix any $x_1 \in r_0 \text{int}\, \mathbb{B}$ and take an arbitrary $x_2 \in r_0 \text{int}\, \mathbb{B}$ in a neighbourhood of x_1, implying the continuity of V at (t_1, x_1). This confirms assertion (b) of the proposition.

(d): We employ the same argument used in (c), taking into account that g is local Lipschitz continuous and that $F(t, x) \subset c_0 \mathbb{B}$ for all $(t, x) \in [S, T] \times R_0 \mathbb{B}$. Write k_g the Lipschitz constant of g on $R_0 \mathbb{B}$. Then, we have

$$|V(t_1, x_1) - V(t_2, x_2)| \leq k_g \sqrt{2} c_0 (1 + R_0) K |(t_1, x_1) - (t_2, x_2)|.$$

$\square$

We are now ready to identify the value function as the unique lower semi-continuous function that is a solution to the Hamilton Jacobi equation (13.3.1)–(13.3.2), in either the lower Dini or proximal sense.

Theorem 13.3.7 (Characterization of Lower semi-continuous Value Functions)
Assume that the data for (P) satisfies the following hypotheses:

(H1): $F : [S, T] \times \mathbb{R}^n \rightsquigarrow \mathbb{R}^n$ takes closed, convex, non-empty values, $F(., x)$ is $\mathcal{L}(S, T)$-measurable for all $x \in \mathbb{R}^n$,

(H2): (i) there exists $c \in L^1(S, T)$ such that

$F(t, x) \subset c(t)(1 + |x|)\, \mathbb{B}$ *for all* $x \in \mathbb{R}^n$ *and for a.e.* $t \in [S, T]$,

and

13.3 The Value Function and Generalized Solutions of the Hamilton Jacobi... 675

(ii) for every $R_0 > 0$, there exists $c_0 > 0$ such that

$$F(t,x) \subset c_0 \mathbb{B} \quad \text{for all } (t,x) \in [S,T] \times R_0\mathbb{B},$$

(H3): for every $R_0 > 0$, there exist $k_F \in L^1(S,T)$ and a modulus of continuity $\omega_F : \mathbb{R}_+ \to \mathbb{R}_+$ such that $t \to 1/(1+k_F(t))$ is almost everywhere equal to a continuous function,

(i) $\quad F(t,x') \subset F(t,x) + k_F(t)|x-x'|\,\mathbb{B},$
$\quad\quad$ for all $x, x' \in R_0\mathbb{B}$ and a.e. $t \in [S,T]$,

and

(ii) $\quad d_H(F(t,x'), F(t,x)) \leq \omega_F(|x-x'|)$
$\quad\quad$ for all $x, x' \in R_0\mathbb{B}$ for all $t \in [S,T]$,

(H4): (i) for each $s \in [S,T)$, $t \in (S,T]$ and $x \in \mathbb{R}^n$ the following one-sided set-valued limits exist and are non-empty:

$$F(s^+,x) := \lim_{s' \downarrow s,\, x' \to x} F(s',x') \quad \text{and} \quad F(t^-,x) := \lim_{t' \uparrow t,\, x' \to x} F(t',x'),$$

(ii) and there exists a subset $\mathcal{S} \subset (S,T)$ of full Lebesgue measure such that,

$$s \to F(s,x) \text{ is continuous at } s=t, \text{ for all } t \in \mathcal{S} \text{ and } x \in \mathbb{R}^n.$$

(H5): $g : \mathbb{R}^n \to \mathbb{R} \cup \{+\infty\}$ is lower semi-continuous.

Take a function $V : [S,T] \times \mathbb{R}^n \to \mathbb{R} \cup \{+\infty\}$. Then, assertions (a)–(c) below are equivalent:

(a) V is the value function for (P),
(b) V is lower semi-continuous on $[S,T] \times \mathbb{R}^n$ and

(i) for all $(t,x) \in ([S,T) \times \mathbb{R}^n) \cap \operatorname{dom} V$

$$\inf_{v \in F(t^+,x)} D_\uparrow V((t,x); (1,v)) \leq 0,$$

(ii) for all $(t,x) \in ((S,T] \times \mathbb{R}^n) \cap \operatorname{dom} V$

$$\sup_{v \in F(t^-,x)} D_\uparrow V((t,x); (-1,-v)) \leq 0,$$

(iii) for all $x \in \mathbb{R}^n$

$$V(T, x) = g(x).$$

(c) *V is lower semi-continuous on* $[S, T] \times \mathbb{R}^n$ *and*

(i) for all $(t, x) \in ((S, T) \times \mathbb{R}^n) \cap \operatorname{dom} V$, $(\xi^0, \xi^1) \in \partial^P V(t, x)$

$$\xi^0 + \inf_{v \in F(t^+, x)} \xi^1 \cdot v \leq 0,$$

(ii) for all $(t, x) \in ((S, T) \times \mathbb{R}^n) \cap \operatorname{dom} V$, $(\xi^0, \xi^1) \in \partial^P V(t, x)$

$$\xi^0 + \inf_{v \in F(t^-, x)} \xi^1 \cdot v \geq 0,$$

(iii) for all $x \in \mathbb{R}^n$,

$$\liminf_{\{(t', x') \to (S, x) : t' > S\}} V(t', x') = V(S, x)$$

and

$$\liminf_{\{(t', x') \to (T, x) : t' < T\}} V(t', x') = V(T, x) = g(x).$$

Proof The proof structure will be to demonstrate '(a) $\Rightarrow$ (b)', '(b) $\Rightarrow$ (c)' and '(c) $\Rightarrow$ (a)'. In Proposition 13.3.5 it has already been shown that '(b) $\Rightarrow$ (c)'. It remains to supply the missing links: '(a) $\Rightarrow$ (b)' and '(c) $\Rightarrow$ (a)'. The key step of the proof is '(c) $\Rightarrow$ (a)'. This involves showing that, for an arbitrary point (t, x) in the domain of a function V satisfying condition (c), (A): $V(t, x)$ is the cost of some state trajectory originating from (t, x) and (B): $V(t, x)$ is a lower bound on the cost of an arbitrary state trajectory. To demonstrate these properties of V we use a weak invariance theorem to construct the state trajectory x with property (A) and a strong invariance theorem to confirm the lower bound property (B) of V.

'(a) $\Rightarrow$ (b)'
The lower semicontinuity of V follows from Proposition 13.3.6. Take any point $(t, x) \in \operatorname{dom} V \cap ([S, T) \times \mathbb{R}^n)$. Choose $r_0 > 0$ such that $r_0 > |x| + 1$. Consider a minimizer y for $(P_{t,x})$, which exists under the stated hypotheses. Observe also that by hypothesis (H2) $y(s) \in R_0 \mathbb{B}$, for all $s \in [t, T]$, where $R_0 := \exp\{\int_S^T c(s) ds\}(r_0 + 1)$. Invoking the principle of optimality for the value function, we have:

$$V(t + \delta, y(t + \delta)) - V(t, x) = 0 \qquad (13.3.10)$$

13.3 The Value Function and Generalized Solutions of the Hamilton Jacobi...

for all $\delta \in (0, T - t]$. Using the fact that y is an F trajectory, we also have

$$\delta^{-1}(y(t + \delta) - y(t)) = \delta^{-1} \int_t^{t+\delta} \dot{y}(s) ds$$

for all $\delta \in (0, T - t]$, and

$$\dot{y}(s) \in F(s, y(s)) \quad \text{a.e. } s \in [t, T].$$

In view of (H4), this implies

$$\dot{y}(s) \in F(s^+, y(s)) \quad \text{a.e. } s \in [t, T].$$

Now take an arbitrary sequence $\delta_i \downarrow 0$. For each i we write

$$v_i := \delta_i^{-1} \int_t^{t+\delta_i} \dot{y}(s) ds.$$

Then, condition (H2)(ii) implies that y is Lipschitz continuous, and therefore $\{v_i\}$ is a bounded sequence. Extracting a subsequence (we do not relabel), we can arrange that

$$v_i \to \bar{v} \quad \text{as } i \to \infty$$

for some $\bar{v} \in \mathbb{R}^n$. Fix any $p \in \mathbb{R}^n$, and observe that

$$p \cdot v_i = \delta_i^{-1} \int_t^{t+\delta_i} p \cdot \dot{y}(s) ds \leq \delta_i^{-1} \int_t^{t+\delta_i} \max_{v \in F(s^+, y(s))} p \cdot v \, ds.$$

Noting that $s \to \max_{v \in F(s^+, y(s))} p \cdot v$ is a right-continuous function (at $s = t$), we can pass to the limit as $i \to \infty$ in this relation to deduce:

$$p \cdot \bar{v} \leq \max_{v \in F(t^+, x)} p \cdot v. \tag{13.3.11}$$

But (13.3.11) is valid for all $p \in \mathbb{R}^n$ and $F(t^+, x)$ is convex. It follows that $\bar{v} \in F(t^+, x)$.

For each i, we have $y(t + \delta_i) = x + \delta_i v_i$. Then, from (13.3.10), we deduce that

$$\delta_i^{-1}[V(t + \delta_i, x + \delta_i v_i) - V(t, x)] = 0.$$

Since $v_i \to \bar{v}$ and $\bar{v} \in F(t^+, x)$, we obtain that

$$\inf_{v \in F(t^+, x)} D_\uparrow V((t, x), (1, v)) \leq \lim_i \delta_i^{-1}[V(t + \delta_i, x + \delta_i v_i) - V(t, x)] = 0,$$

confirming condition (b)(i).

Now, fix any $(t, x) \in \text{dom } V \cap (S, T] \times \mathbb{R}^n$. Choose any $\tilde{v} \in F(t^-, x)$. Take a sequence $\delta_i \downarrow 0$ such that $\delta_i < |t - S|$ for all i. Set $y(s) := x + (s - t)\tilde{v}$ for all $s \in [S, t]$. In consequence of the generalized Filippov existence theorem (Theorem 6.2.3), there exists, for each i sufficiently large, an F trajectory z_i on $[t - \delta_i, t]$ such that $z_i(t) = x$ and

$$|z_i(t - \delta_i) - [x - \delta_i \tilde{v}]| \le K \int_{t-\delta_i}^{t} d_{F(s, y(s))}(\tilde{v})\, ds \quad \text{for all } i,$$

in which $K := \exp\{\int_S^T k_F(s)\, ds\}$ is a number that does not depend on i. Then,

$$|\tilde{v} - v_i| \le K \delta_i^{-1} \int_{t-\delta_i}^{t} d_{F(s, y(s))}(\tilde{v})\, ds,$$

where

$$v_i := \delta_i^{-1}(x - z_i(t - \delta_i)) \quad \text{for each } i.$$

Since $F(t, .)$ is continuous (from (H3)) and $\tilde{v} \in F(t^-, x)$, it follows that

$$\lim_{s \uparrow t} d_{F(s, y(s))}(\tilde{v}) = 0.$$

Then, we deduce that $v_i \to \tilde{v}$ as $i \to \infty$. Applying the principle of optimality,

$$V(t - \delta_i, x - \delta_i v_i) \le V(t, z_i(t)) \quad \text{for all } i,$$

an so

$$\delta_i^{-1}(V(t - \delta_i, x - \delta_i v_i) - V(t, x)) \le 0 \quad \text{for all } i.$$

This condition yields

$$D_\uparrow V((t, x), (-1, -\tilde{v})) (\le \limsup_{i \to \infty} \delta_i^{-1}(V(t - \delta_i, x - \delta_i v_i) - V(t, x))) \le 0.$$

Since $\tilde{v}$ was an arbitrary vector in $F(t^-, x)$, we deduce that

$$\sup_{v \in F(t^-, x)} D_\uparrow V((t, x), (-1, -v)) \le 0,$$

confirming (b)(ii).

'(c) $\Rightarrow$ (a)'

13.3 The Value Function and Generalized Solutions of the Hamilton Jacobi...

Suppose that V is a lower semi-continuous function that satisfies conditions c(i)–(iii) of the theorem statement. We shall show that V is the value function. We shall do so initially under the assumption that (H2) and (H3) have been replaced by the more restrictive hypotheses

(H2)': there exists $c_0 > 0$ such that $F(t, x) \subset c_0 \mathbb{B}$ for all $(t, x) \in [S, T] \times \mathbb{R}^n$,
(H3)': there exist $k_F \in L^1(S, T)$ such that $t \to 1/(1 + k_F(t))$ is almost everywhere equal to a continuous function and

$$F(t, x') \subset F(t, x) + k_F(t)|x - x'| \mathbb{B}$$

for all $x, x' \in \mathbb{R}^n$ and a.e. $t \in [S, T]$.

We shall show that it suffices to assume (H2) and (H3) at a later stage of the proof.

Take an arbitrary function $V : [S, T] \times \mathbb{R}^n \to \mathbb{R} \cup \{+\infty\}$ satisfying condition (c) and an arbitrary point $(\bar{t}, \bar{x}) \in [S, T) \times \mathbb{R}^n$. Our aim is to prove that

$$V(\bar{t}, \bar{x}) = \inf(P_{\bar{t}, \bar{x}}).$$

Step 1 We show that

$$V(\bar{t}, \bar{x}) \geq \inf(P_{\bar{t}, \bar{x}}).$$

To prove this inequality, it suffices to find an F trajectory $\bar{x}$ on $[\bar{t}, T]$ which satisfies $x(\bar{t}) = \bar{x}$, and

$$V(\bar{t}, \bar{x}) \geq g(\bar{x}(T)).$$

We can assume that $V(\bar{t}, \bar{x}) < +\infty$, since otherwise the inequality above is automatically satisfied. Define the multifunction $\Gamma : [S, T] \times \mathbb{R}^{n+1} \rightsquigarrow \mathbb{R}^{n+2}$ to be

$$\Gamma(\tau, x, a) := \begin{cases} \text{co}\,(\{(0, 0, 0)\} \cup (\{1\} \times F(S^+, x) \times \{0\})) & \text{if } \tau = S \\ \{1\} \times \text{co}\,\{F(\tau^-, x) \cup F(\tau^+, x)\} \times \{0\} & \text{if } \tau \in (S, T) \\ \text{co}\,(\{(0, 0, 0)\} \cup (\{1\} \times F(T^-, x) \times \{0\})) & \text{if } \tau = T. \end{cases}$$

Consider the following differential inclusion and state constraint:

$$(S) \begin{cases} (\dot{\tau}, \dot{x}, \dot{a}) \in \Gamma(\tau, x, a) & \text{a.e. } t \in [\bar{t}, T] \\ (\tau(t), x(t), a(t)) \in \text{epi}\, V & \text{for all } t \in [\bar{t}, T] \\ (\tau(\bar{t}), x(\bar{t}), a(\bar{t})) = (\bar{t}, \bar{x}, V(\bar{t}, \bar{x})). \end{cases}$$

Our intention is to apply the autonomous weak invariance theorem (Theorem 13.2.1) to this system. But first we must check that the relevant hypotheses are satisfied. Assumptions (H1), (H2)', (H3)' and (H4) guarantee that Gr Γ is closed at every point of $D :=$ epi V, Γ takes values non-empty convex sets and satisfies

$$\Gamma(\tau, x, a) \subset (c_0 + 1)\,\mathbb{B} \quad \text{for all } (\tau, x, a) \in [S, T] \times \mathbb{R}^n \times \mathbb{R}\,.$$

(Here, c_0 is as in (H2)$'$.) It remains to show that the 'inward pointing condition' is satisfied. Take any $(\tau, x, \alpha) \in$ epi V and any $\xi \in N^P_{\text{epi } V}(\tau, x, \alpha)$. By the nature of proximal normal vectors to epigraph sets, $\xi = (\xi^0, \xi^1, -\lambda)$, for some $\lambda \geq 0$. We must show

$$\min_{w \in \Gamma(\tau, x, \alpha)} (\xi^0, \xi^1, -\lambda) \cdot w \leq 0\,. \tag{13.3.12}$$

Observe that, if $\tau = S$ or $\tau = T$, (13.3.12) is automatically true, since

$$\min_{w \in \Gamma(\tau, x, \alpha)} \xi \cdot w \leq \xi \cdot (0, 0, 0) = 0\,.$$

So we can assume that $S < \tau < T$. Note that we need to check (13.3.12) only when $\alpha = V(\tau, x)$. This is because, in the case $\alpha > V(\tau, x)$, '$(\xi^0, \xi^1, -\lambda) \in N^P_{\text{epi } V}(\tau, x, \alpha)$' implies $\lambda = 0$ and $(\xi^0, \xi^1, 0) \in N^P_{\text{epi } V}(\tau, x, V(\tau, x))$. (These facts are simple consequences of definition of the proximal normal cone and of the structure of epigraph sets.) We shall show:

$$\xi^0 + \min_{v \in F(t^-, x) \cup F(\tau^+, x)} \xi^1 \cdot v \leq 0\,. \tag{13.3.13}$$

This implies

$$\min_{w \in \Gamma(\tau, x, \alpha)} (\xi^0, \xi^1, -\lambda) \cdot w \leq \xi^0 + \min_{v \in \text{co}\,\{F(\tau^-, x) \cup F(\tau^+, x)\}} \xi^1 \cdot v \leq 0\,.$$

which is the inward pointing condition (13.3.12). To check (13.3.13) we need to consider two cases.

Case 1: $\lambda > 0$. In this case $((1/\lambda)\xi^0, (1/\lambda)\xi^1, -1) \in N^P_{\text{epi } V}((\tau, x), V(\tau, x))$. It follows that

$$((1/\lambda)\xi^0, (1/\lambda)\xi^1) \in \partial^P V(\tau, x)\,.$$

But then, by (c)(i), $(1/\lambda)\left(\xi^0 + \min_{v \in F(\tau^+, x)} \xi^1 \cdot v\right) \leq 0$. This implies (13.3.13).

Case 2: $\lambda = 0$. In this case, $(\xi^0, \xi^1) \in \partial^\infty_P V(t, x)$ and we know from Theorem 4.6.2 that there exist $(\xi^0_i, \xi^1_i) \to (\xi^0, \xi^1)$, $\lambda_i \downarrow 0$ and $(t_i, x_i) \to (t, x)$ such that, for each i,

$$(\lambda_i^{-1}\xi^0_i, \lambda_i^{-1}\xi^1_i) \in \partial^P V(t_i, x_i)\,.$$

13.3 The Value Function and Generalized Solutions of the Hamilton Jacobi...

Then, by condition (c)(i), for each i there exists $v_i \in F(t_i^+, x_i)$ such that

$$\xi_i^0 + \xi_i^1 \cdot v_i \leq 0.$$

But $\{v_i\}$ is a bounded sequence. We can therefore arrange, by extracting a subsequence we do not relabel), that $v_i \to v$, for some $v \in \mathbb{R}^n$. Since $(t, x) \rightsquigarrow F(t^-, x) \cup F(t^+, x)$ is an upper semi-continuous multifunction, it follows that $v \in F(t^-, x) \cup F(t^+, x)$. So, in the limit as $i \to \infty$, we obtain

$$0 \geq \xi^0 + \xi^1 \cdot v \geq \xi^0 + \inf_{v \in F(t^-,x) \cup F(t^+,x)} \xi^1 \cdot v.$$

We have confirmed (13.3.13) in this case too. We are justified then in applying the autonomous weak invariance Theorem 13.2.1 to system (S).

Suppose first that $\bar{t} > S$. We deduce the existence of a Γ trajectory (τ, x, a) on $[\bar{t}, T]$ such that $(\tau(t), x(t), a(t)) \in$ epi V, for all $t \in [\bar{t}, T]$. Taking note of the fact $\bar{t} > S$ (this condition excludes the possibility that $\tau \equiv S$) and also (H4)(ii), we deduce from the structure of the multifunction Γ that x is a F trajectory, $\tau(t) \equiv t$ and a is a constant (we write it also a) such that and $a \geq V(t, x(t))$, for all $t \in [\bar{t}, T]$. Then, since $a(\bar{t}) = V(\bar{t}, \bar{x})$, we have

$$g(\bar{x}(T)) = V(T, \bar{x}(T)) \leq \bar{a}(T) = \bar{a}(\bar{t}) = V(\bar{t}, \bar{x}).$$

If, on the other hand, $\bar{t} = S$, we can, in view of (c)(iii), take a sequence $(S_i, x_i) \to (S, \bar{x})$ such that $S_i \downarrow S$ and $\lim_{i \to \infty} V(S_i, x_i) = V(S, \bar{x})$. Repeating the preceding analysis, we show, for each $i \geq 0$, that there exists an F trajectory y_i such that

$$V(S_i, x_i) \geq g(y_i(T)). \tag{13.3.14}$$

Extending y_i to all $[S, T]$ by constant extrapolation from the right on $[S, S_i]$, with the help of the compactness of trajectories theorem (Theorem 6.3.3) we can arrange, by extracting a suitable subsequence, that $y_i \to \bar{y}$ uniformly, for some F trajectory $\bar{y}$ such that $\bar{y}(\bar{t}) = \bar{x}$. Since g is lower semi-continuous, we may pass to the limit as $i \to \infty$ in the preceding relation and deduce that

$$V(S, \bar{x}) = \liminf_{i \to \infty} V(S_i, x_i) \geq \liminf_{i \to \infty} g(y_i(T)) \geq g(\bar{y}(T)).$$

This completes step 1.

Step 2 We now prove that

$$V(\bar{t}, \bar{x}) \leq \inf(P_{\bar{t}, \bar{x}}).$$

Take an arbitrary F trajectory y on $[\bar{t}, T]$ such that $y(\bar{t}) = \bar{x}$. We shall establish that

$$V(\bar{t}, \bar{x}) \leq g(y(T)).$$

We can assume, without loss of generality, that $g(y(T)) < +\infty$ for, otherwise, the inequality above holds automatically.

Define

$$\tilde{y}(s) := y(T-s), \quad \tilde{F}(s,x) := -F(T-s,x) \quad \text{and} \quad \tilde{V}(s,x) := V(T-s,x)$$

for $(s,x) \in [0, T-\bar{t}] \times \mathbb{R}^n$. It is a straightforward matter to check that $\tilde{y}$ satisfies

$$\begin{cases} \dot{\tilde{y}}(s) \in \tilde{F}(s, \tilde{y}(s)) & \text{a.e. } t \in [0, T-\bar{t}] \\ \tilde{y}(0) = y(T). \end{cases}$$

We readily deduce from the fact that V satisfies condition (c)(ii) that

$$\xi^0 + \inf_{v \in F(t^-, x)} \xi^1 \cdot v \geq 0, \tag{13.3.15}$$

for all $(t,x) \in ((S,T) \times \mathbb{R}^n) \cap \text{dom } V$, and all $(\xi^0, \xi^1) \in \partial^P V(t,x)$. Condition (c)(iii) implies that

$$\tilde{V}(0, \tilde{y}(0)) = \liminf_{s' \downarrow 0, x' \to \tilde{y}(0)} \tilde{V}(s', x'). \tag{13.3.16}$$

Now define the multifunctions $\tilde{\Gamma} : [0, T-\bar{t}] \times \mathbb{R}^{n+1} \rightsquigarrow \mathbb{R}^{n+1}$ and $\tilde{P} : [0, T-\bar{t}] \rightsquigarrow \mathbb{R}^{n+1}$ to be

$$\tilde{\Gamma}(s, (z,a)) := \tilde{F}(s,z) \times \{0\},$$

$$\tilde{P}(s) := \{(x, \alpha) : \tilde{V}(s,x) \leq \alpha\}$$

respectively. We shall apply the strong invariance theorem (Theorem 13.2.4) to the differential inclusion and accompanying constraint

$$\begin{cases} (\dot{z}, \dot{a}) \in \tilde{\Gamma}(s, (z,a)) & \text{a.e } s \in [0, T-\bar{t}], \\ z(0) = y(T), \; a(0) = \tilde{V}(0, \tilde{y}(0)), \\ (z(s), a(s)) \in \tilde{P}(s) & \text{for all } s \in [0, T-\bar{t}]. \end{cases}$$

(The boundary condition on a makes sense, since $\tilde{V}(0, \tilde{y}(0))\ (= g(y(T))) < +\infty$.) To do so, we need first to check that the relevant hypotheses are satisfied. According to (13.3.16), we can find $s_i \downarrow 0$, $x_i \to y(T)$ such that $\tilde{V}(s_i, x_i) \to \tilde{V}(0, y(T))$. Writing $a_i := \tilde{V}(s_i, x_i)$, we see that $(x_i, a_i) \to (y(T), \tilde{V}(0, y(T)))$ and $(s_i, (x_i, a_i)) \in \tilde{P}(s_i)$ for all i. This confirms that

$$(z(0), a(0)) \in \limsup_{s' \downarrow 0} \tilde{P}(s').$$

$\tilde{\Gamma}$ and $\tilde{P}$ both clearly possess the required closure and regularity properties for application of the strong invariance theorem (Theorem 13.2.4).

13.3 The Value Function and Generalized Solutions of the Hamilton Jacobi... 683

It remains to check the 'inward pointing' hypothesis. With this objective in mind, we consider any point $(s, (z, \alpha)) \in \text{Gr } \tilde{P}$ such that $0 < s < T - S$, and any vector

$$(\zeta^0, \zeta^1, -\lambda) \in N^P_{\text{Gr } \tilde{P}}(s, (z, \alpha)). \tag{13.3.17}$$

We must show that

$$\zeta^0 + \left(\max_{w \in \tilde{\Gamma}(s^-, (z,\alpha))} \zeta^1 \cdot v\right) \wedge \left(\max_{w \in \tilde{\Gamma}(s^+, (z,\alpha))} \zeta^1 \cdot w\right) \leq 0$$

for all $(\zeta^0, \zeta^1, -\lambda) \in N^P_{\text{Gr } \tilde{P}}(s, (z, \alpha))$.

Taking note of the structure of $\tilde{\Gamma}$, this last condition can be expressed as

$$\zeta^0 + \left(\max_{\tilde{v} \in \tilde{F}(s^-, z)} \zeta^1 \cdot \tilde{v}\right) \wedge \left(\max_{\tilde{v} \in \tilde{F}(s^+, z)} \zeta^1 \cdot \tilde{v}\right) \leq 0. \tag{13.3.18}$$

Since $\text{Gr } \tilde{P}$ is an epigraph set, $\lambda \geq 0$. We deduce from (13.3.17) that

$$(\zeta^0, \zeta^1, -\lambda) \in N^P_{\text{epi } \tilde{V}}(s, z, \alpha).$$

As noted at the similar stage of the analysis in Step 1, we need to check (13.3.18) only when $\alpha = \tilde{V}(s, z)$. We observe also that $(\zeta^0, \zeta^1, -\lambda) \in N^P_{\text{epi } \tilde{V}}(s, z, a)$ if and only if $(-\zeta^0, \zeta^1, -\lambda) \in N^P_{\text{epi } V}(T - s, z, a)$. It follows that (13.3.18) is equivalent to the following condition: for every $(t, z) \in (\bar{t}, T) \times \mathbb{R}^n$ and $(\xi^0, \xi^1, -\lambda) \in N^P_{\text{epi } V}(t, z, V(t, z))$ (with $\lambda \geq 0$) we have:

$$\left(\max_{\tilde{v} \in -F(t^-, z)} (-\xi^0 + \xi^1 \cdot \tilde{v})\right) \wedge \left(\max_{\tilde{v} \in -F(t^+, z)} (-\xi^0 + \xi^1 \cdot \tilde{v})\right) \leq 0.$$

which is equivalent to

$$\left(\min_{v \in F(t^-, z)} (\xi^0 + \xi^1 \cdot v)\right) \vee \left(\min_{v \in F(t^+, z)} (\xi^0 + \xi^1 \cdot v)\right) \geq 0. \tag{13.3.19}$$

Consider first the case $\lambda > 0$. Then $\lambda^{-1}(\xi^0, \xi^1) \in \partial^P V(t, z)$ and so, from condition c(ii),

$$\xi^0 + \min_{v \in F(t^-, z)} \xi^1 \cdot v \geq 0,$$

which implies (13.3.18). If $\lambda = 0$, there exist $(\xi^0_i, \xi^1_i) \to (\xi^0, \xi^1)$, $\lambda_i \downarrow 0$ and $(t_i, z_i) \to (t, z)$ such that, for each i,

$$(\lambda_i^{-1} \xi^0_i, \lambda_i^{-1} \xi^1_i) \in \partial^P V(t_i, z_i).$$

But then, by condition (c)(ii),

$$\xi_i^0 + \min_{v \in F(t_i^-, z_i)} \xi_i \cdot v \geq 0.$$

By extracting a subsequence we can arrange that, either $t_i \leq t$ for all i, or $t_i > t$ for all i. Since $(t_i, z_i) \to (t, z)$ and in consequence of Hypothesis (H4)(i), we can pass to the limit as $i \to \infty$ in the preceding relation to obtain

$$\xi^0 + \min_{v \in F(t^-, z)} \xi^1 \cdot v \geq 0$$

if $t_i \leq t$ for all i. Other other hand, if $t_i > t$ for all t, passage to the limit gives

$$\xi^0 + \min_{v \in F(t^+, z)} \xi^1 \cdot v \geq 0.$$

In either case then, (13.3.19) is verified.

The strong invariance theorem (Theorem 13.2.4) tells us that the $\tilde{\Gamma}$ trajectory $(\tilde{y}, a \equiv \tilde{V}(0, \tilde{y}(0)))$ satisfies

$$(\tilde{y}(s), a(s)) \in \tilde{P}(s) \quad \text{for all } s \in [0, T - \bar{t}].$$

Setting $s = T - \bar{t}$, we deduce that

$$\tilde{V}(0, \tilde{0}) = a(0) = a(T - \bar{t}) \geq \tilde{V}(T - \bar{t}, \tilde{y}(T - \bar{t})).$$

Since $V(T, y(T)) = \tilde{V}(0, \tilde{y}(0))$ and $V(\bar{t}, y(\bar{t})) = \tilde{V}(T - \bar{t}, \tilde{y}(T - \bar{t}))$, it follows that

$$g(y(T)) = V(T, y(T)) \geq V(\bar{t}, y(\bar{t})) = V(\bar{t}, \bar{x}).$$

This is the desired inequality and completes Step 2.

Conclusion of Proof Steps 1 and 2 above combine to tell us that

$$V(\bar{t}, \bar{x}) = \inf(P_{\bar{t}, \bar{x}}) \tag{13.3.20}$$

at the arbitrary point $(\bar{t}, \bar{x}) \in [S, T] \times \mathbb{R}^n$. This confirms '(c) $\Rightarrow$ (a)', under the added hypotheses (H2)' and (H3)'.

To validate (13.3.20) when either (H2)' or (H3)' are not satisfied, we choose $r_0 > |\bar{x}|$ and consider the multifunction

$$\tilde{F}(s, y) := \begin{cases} F(s, y) & \text{if } |y| \leq R_0 \\ F(s, R_0 y/|y|) & \text{if } |y| > R_0, \end{cases}$$

in which $R_0 = \exp\{\int_S^T c(t)dt\}(1+r_0)$. The multifunction $\tilde{F}$ satisfies hypotheses (H1), (H2)', (H3)' and (H4). The preceding analysis tell us that $V(t,x)$ is the infimum cost for the modified dynamic optimization problem, in which $\tilde{F}$ replaces F, and for initial data (t,x), where V is any lower semi-continuous function that satisfies conditions *(c)(i)–(c)(iii)* (in relation to the multifunction $\tilde{F}$), restricted to $[S,T] \times R_0 \mathbb{B}$. But, by standard *a priori* estimates, the F trajectories emanating from (t,x) also evolve in $[S,T] \times R_0 \mathbb{B}$. It follows that the families of state trajectories are the same for the two multifunctions F and $\tilde{F}$. From this we conclude that, for arbitrary initial data $(t,x) \in [S,T] \times r_0 \mathbb{B}$, the infimum costs are the same for the two multifunctions. Also the classes of lower semi-continuous functions satisfying hypotheses *(c)(i)–(c)(iii)* on $[S,T] \times R_0 \mathbb{B}$ are the same, because F and $\tilde{F}$ coincide on this set. It follows that $V(t,x)$ is the minimum cost for the original dynamic optimization problem (for an initial data point $(t,x) \in [S,T] \times r_0 \mathbb{B}$), where V is any lower semi-continuous function satisfying conditions *(c)(i)–(c)(iii)* (for the original multifunction F). It follows that (13.3.20) is valid, since $(\bar{t}, \bar{x}) \in [S,T] \times r_0 \mathbb{B}$. □

13.4 Local Verification Theorems

A traditional role of the Hamilton Jacobi equation in variational analysis has been to provide sufficient conditions of optimality. Consider the dynamic optimization problem

$$(Q) \begin{cases} \text{Minimize } g(x(T)) \\ \text{over arcs } x \in W^{1,1}([S,T]; \mathbb{R}^n) \text{ satisfying} \\ \dot{x}(t) \in F(t, x(t)) \quad \text{a.e.,} \\ x(S) = x_0, \\ x(T) \in C, \end{cases}$$

the data for which comprise: a function $g : \mathbb{R}^n \to \mathbb{R}$, a subinterval $[S,T] \subset \mathbb{R}$, a multifunction $F : [S,T] \times \mathbb{R}^n \rightsquigarrow \mathbb{R}^n$, a point $x_0 \in \mathbb{R}^n$ and a closed set $C \subset \mathbb{R}^n$.

It is convenient at this stage to adopt a formulation in which the terminal cost function g is assumed to be finite valued and constraints on terminal values of state trajectories are expressed as a set inclusion $x(T) \in C$.

Take an arc $x \in W^{1,1}$ satisfying the constraints of problem (Q). A smooth local verification function for $\bar{x}$ is a C^1 function $\phi : [S,T] \times \mathbb{R}^n \to \mathbb{R}$ with the properties: there exists $\delta > 0$ such that

$$\phi_t(t,x) + \min_{v \in F(t,x)} \phi_x(t,x) \cdot v \geq 0 \quad \text{for all } (t,x) \in \text{int } T(\bar{x}, \delta)$$

$$\phi(T,x) \leq g(x) \quad \text{for all } x \in C \cap (\bar{x}(T) + \delta \mathbb{B})$$

$$\phi(0, x_0) = g(\bar{x}(T)).$$

Here $T(\bar{x}, \delta)$ is the δ tube about $\bar{x}$:

$$T(\bar{x}, \delta) := \{(t, x) \in [S, T] \times \mathbb{R}^n : |x - \bar{x}(t)| \leq \delta\}.$$

Existence of a smooth verification function for $\bar{x}$ is a sufficient condition for $\bar{x}$ to be an L^∞ local minimizer for (Q). It is of interest to know how broad the class of dynamic optimization problems is, for which the local optimality of local minimizers can be confirmed by some verification function. Inverse verification theorems, which provide conditions for the existence of verification functions associated with a minimizer address this issue.

Unfortunately continuously differentiable verification theorems provide a rather restrictive framework for inverse verification theorems. For dynamic optimization problems whose solutions exhibit 'bang bang' behaviour indeed, non-existence of continuously differentiable verification functions is to be expected.

Clearly, we need a less restrictive concept of smooth local verification functions, which can still be used in sufficient conditions of optimality, but whose existence can be guaranteed under unrestrictive hypotheses.

These considerations lead to the following definition, which makes reference to the data for problem (Q) above and an arc $\bar{x} \in W^{1,1}([S, T] \times \mathbb{R}^n)$.

Definition 13.4.1 A function $\phi : T(\bar{x}, \delta) \to \mathbb{R} \cup \{+\infty\}$ is called a lower semi-continuous local verification function for $\bar{x}$ (with parameter $\delta > 0$) if ϕ is lower semi-continuous and the following conditions are satisfied:

(i) For every $(t, x) \in (\text{int } T(\bar{x}, \delta)) \cap \text{dom } \phi$ such that $\partial^P \phi(t, x)$ is non-empty,

$$\xi^0 + \inf_{v \in F(t^-, x)} \xi^1 \cdot v \geq 0 \quad \text{for all } (\xi^0, \xi^1) \in \partial^P \phi(t, x),$$

(ii) $\phi(T, x) \leq g(x)$ for all $x \in C$,
(iii) $\liminf_{t' \uparrow T, x' \to x} \phi(t', x') = \phi(T, x)$ for all $x \in C \cap (\bar{x}(T) + \delta \mathbb{B})$,
(iv) $\phi(S, x_0) = g(\bar{x}(T))$.

The next proposition tells us that sufficient conditions for local minimality can be formulated in terms of lower semi-continuous local verification functions. Furthermore, the existence of a lower semi-continuous verification function confirming optimality of a given local minimizer is guaranteed, under unrestrictive hypotheses on the data.

Proposition 13.4.2 *Let the arc $\bar{x} \in W^{1,1}([S, T]; \mathbb{R}^n)$ satisfy the constraints of problem (Q). Assume that, for some $\epsilon > 0$, we have*

(i) *F takes values non-empty, closed, convex sets, and $F(., x)$ is $\mathcal{L}(S, T)$-measurable for all $x \in \mathbb{R}^n$,*
(ii) *There exists $c > 0$ such that*

$$F(t, x) \subset c\mathbb{B} \quad \text{for all } x \in \bar{x}(t) + \delta \mathbb{B}, \text{ a.e. } t \in [S, T],$$

13.4 Local Verification Theorems

(iii) there exist $k_F \in L^1(S, T)$ and a modulus of continuity $\omega_F : \mathbb{R}_+ \to \mathbb{R}_+$ such that $t \to 1/(1 + k_F(t))$ is almost everywhere equal to a continuous function,

$$F(t, x') \subset F(t, x) + k_F(t)|x - x'| \mathbb{B}$$
$$\text{for all } x, x' \in \bar{x}(t) + \delta \mathbb{B}, \text{ for a.e. } t \in [S, T],$$

and

$$d_H(F(t, x'), F(t, x)) \leq \omega_F(|x - x'|)$$
$$\text{for all } x, x' \in \bar{x}(t) + \delta \mathbb{B}, \text{ and for all } t \in [S, T],$$

(iv) for each $s \in [S, T)$ and $t \in (S, T]$ the following one-sided set-valued limits exist and are non-empty:

$$F(s^+, x) := \lim_{s' \downarrow s, \, x' \to x} F(s', x') \quad \text{for all } x \in \bar{x}(s) + \delta \mathbb{B} \quad \text{and}$$

$$F(t^-, x) := \lim_{t' \uparrow t, \, x' \to x} F(t', x') \quad \text{for all } x \in \bar{x}(t) + \delta \mathbb{B},$$

and there exists a subset $\mathcal{S} \subset (S, T)$ of full Lebesgue measure such that

$$s \to F(s, x) \text{ is continuous at } s = t, \quad \text{for all } t \in \mathcal{S} \quad \text{and} \quad \bar{x}(t) + \delta \mathbb{B},$$

(v) g is lower semi-continuous on $\bar{x}(T) + \epsilon \mathbb{B}$.

We have

(a) Suppose that there exists a lower semi-continuous local verification function for $\bar{x}$ (with parameter $\delta > 0$). Then $\bar{x}$ is an L^∞ local minimizer for (Q).

Conversely,

(b) Suppose that $\bar{x}$ is an L^∞ local minimizer for (Q) and that g is bounded on $\bar{x}(T) + \epsilon \mathbb{B}$. Then there exists a lower semi-continuous local verification function for $\bar{x}$.

Proof
(a): Suppose that there exists a lower semi-continuous local verification function for $\bar{x}$ (with parameter δ). Reduce the size of δ if necessary, to ensure that $\delta < \epsilon$.

Let x be any F trajectory satisfying the constraints for (Q) and such that $\|x - \bar{x}\|_{L^\infty} \leq \delta/2$. It is required to show that

$$g(x(T)) \geq g(\bar{x}(T)).$$

Apply the strong invariance theorem to the differential inclusion

$$\begin{cases} (\dot{z}, \dot{a}) \in -F(T-s, z) \times \{0\} & \text{a.e. } s \in [0, T-S] \\ (z(0), a(0)) = (x(T), \phi(T, x(T))) \end{cases}$$

and the accompanying constraint

$$(z(s), a(s)) \in \{(z, \alpha) \in \mathbb{R}^n \times \mathbb{R} : \phi(T-s, z) \leq \alpha \text{ and } |z - \bar{x}(T-s)| \leq \delta\}.$$

This gives

$$\phi(S, x_0) \leq \phi(T, x(T)).$$

The arguments involved are almost identical to those employed in the proof of Theorem 13.3.7 (see step 2 of '(c) $\Rightarrow$ (a)'). But then

$$\phi(S, x_0) \; (\leq \phi(T, x(T)) \;) \; \leq \; g(x(T)).$$

Since $\phi(S, x_0) = g(\bar{x}(T))$, $\bar{x}$ is an L^∞ local minimizer as claimed.

(b): Now suppose that $\bar{x}$ is an L^∞ local minimizer. Choose $\epsilon_1 \in (0, \epsilon)$ such that $\bar{x}$ is a minimizer with respect to arcs x satisfying the constraints of (Q) together with $\|x - \bar{x}\|_{L^\infty} \leq \epsilon_1$. Choose $\epsilon' \in (0, \epsilon_1)$. Now consider the dynamic optimization problem $(\tilde{P})$ (this involves a constant k, whose value will be set presently):

$$(\tilde{P}) \begin{cases} \text{Minimize } \tilde{g}(x(T)) + ky(T) \\ \text{over arcs } (x, y) \in W^{1,1}([S, T]; \mathbb{R}^{n+1}) \text{ satisfying} \\ (\dot{x}(t), \dot{y}(t)) \in \tilde{F}(t, x(t)) \times \{\tilde{L}(t, x(t))\} \quad \text{a.e.,} \\ (x(S), y(S)) = (x_0, 0), \\ (x(T), y(T)) \in C \times \mathbb{R}. \end{cases}$$

Here $\tilde{g}$ and $\tilde{F}$ are 'localized' versions of g and F, namely

$$\tilde{g}(x) := g(\bar{x}(T) + \text{tr}_{\epsilon_1}(x - \bar{x}(T))), \quad \tilde{F}(t, x) := F(t, \bar{x}(t) + \text{tr}_{\epsilon_1}(x - \bar{x}(t)))$$

and

$$\tilde{L}(t, x) := \max\{|x - \bar{x}(t)| - \epsilon', 0\}$$

in which

$$\text{tr}_\alpha(z) := \begin{cases} z & \text{if } |z| \leq \alpha \\ \alpha z/|z| & \text{if } |z| > \alpha. \end{cases}$$

Choose constants $\kappa > 0$ and $k > 0$ which satisfy

13.4 Local Verification Theorems

$$\kappa > \max\{|g(x)| : x \in \bar{x}(T) + \epsilon_1 \mathbb{B}\},$$

$$k > |\epsilon_1 - \epsilon'|^{-1} |\epsilon_1 + \epsilon'|^{-1} 16c\kappa .$$

Claim: $\bar{x}$ is a global minimizer for $(\tilde{P})$.
To verify this, take any other arc satisfying the constraints of $(\tilde{P})$. It must be shown that:

$$\tilde{g}(x(T)) + k \int_S^T \tilde{L}(t, x(t))dt \geq \tilde{g}(\bar{x}(T)) + k \int_S^T \tilde{L}(t, \bar{x}(t))dt \ (= g(\bar{x}(T))).$$

There are two cases to consider:
(A): $\|x - \bar{x}\|_{L^\infty} < \epsilon_1$. In this case, $\tilde{F}(t, x(t)) = F(t, x(t))$ for all $t \in [S, T]$ and $\tilde{g}(x(T)) = g(x(T))$. So x is actually an F trajectory. But then, by local optimality of $\bar{x}$, we have

$$\tilde{g}(x(T)) + k \int_S^T \tilde{L}(t, x(t))dt$$

$$\geq g(x(T)) \geq g(\bar{x}(T)) = \tilde{g}(\bar{x}(T)) + k \int_S^T \tilde{L}(t, \bar{x}(t))dt,$$

as required.
(B): There exists $\bar{t} \in [S, T]$ such that

$$|x(\bar{t}) - \bar{x}(\bar{t})| = \epsilon_1.$$

In this case, since $t \to |x(t) - \bar{x}(t)|$ is Lipschitz continuous with Lipschitz constant $2c$ and $|x(S) - \bar{x}(S)| = 0$, we have

$$\text{meas}\{t \mid |x(t) - \bar{x}(t)| \geq (\epsilon' + \epsilon_1)/2\} \geq (\epsilon_1 - \epsilon')/(4c).$$

But then

$$\tilde{g}(x(T)) + k \int_S^T \tilde{L}(t, x(t))dt$$

$$= \tilde{g}(x(T)) + k \int_S^T \max\{|x(t) - \bar{x}(t)| - \epsilon', 0\}dt$$

$$\geq -\kappa + (k/8c)(\epsilon_1 + \epsilon')(\epsilon_1 - \epsilon')$$

$$> \kappa \geq g(\bar{x}(T)).$$

The inequality holds in this case too then; the claim is confirmed.

Now let $\tilde{V}(t, x, y)$ be the value function for $(\tilde{P})$. Since $(\tilde{P})$ is, in effect, a reformulation of an dynamic optimization problem with end-point and integral cost terms as an dynamic optimization problem with end-point cost term alone, $\tilde{V}(t, x, y)$ can be expressed

$$\tilde{V}(t, x, y) = \phi(t, x) + y$$

for some function $\phi : [S, T] \times \mathbb{R}^n \to \mathbb{R} \cup \{+\infty\}$.

Notice that $F \equiv \tilde{F}$ and $\tilde{L} \equiv 0$ on $T(\bar{x}, \epsilon')$ and $\tilde{g} \equiv g$ on $\bar{x}(T) + \epsilon'\mathbb{B}$. Now we use information about value functions supplied by Theorem 13.3.7, in which the terminal constraint functional is taken to be the extended valued function

$$\tilde{g}(x(T)) + ky(T) + \Psi_C(x(T)).$$

We deduce that ϕ is lower semi-continuous and, for every $(t, x) \in (\text{int } T(\bar{x}, \epsilon')) \cap \text{dom } \phi$, such that $\partial^P \phi(t, x)$ is non-empty,

$$\xi^0 + \inf_{v \in F(t^-, x)} \xi^1 \cdot v \geq 0 \quad \text{for all } (\xi^0, \xi^1) \in \partial^P V(t, x)$$

$$\phi(T, x) = g(x) \quad \text{for all } x \in (\bar{x}(T) + \epsilon'\mathbb{B}) \cap C,$$

$$\liminf_{t' \uparrow T, x' \to x} \phi(t', x') = \phi(T, x) \quad \text{for all } x \in C \cap (\bar{x}(T) + \epsilon'\mathbb{B}).$$

Finally, since $\bar{x}$ is a minimizer for $(\tilde{P})$,

$$\phi(S, x_0) = g(\bar{x}(T)).$$

We have confirmed that there exists some lower semi-continuous, local verification function for $\bar{x}$, namely ϕ. □

In many cases, when a candidate local verification function ϕ is constructed from a field of extremals, on some subset E of $[S, T] \times \mathbb{R}^n$, ϕ is Lipschitz continuous and continuously differentiable on a open subset of E of full Lebesgue measure. The following proposition serves as a convenient aid to verify that ϕ is indeed a local verification function: we have merely to check that ϕ is a classical solution of the Hamilton Jacobi equation at each point on a neighbourhood of which it is continuously differentiable.

Proposition 13.4.3 *Consider problem (Q) involving the data g, F, x_0 and C. Take an admissible F trajectory $\bar{x}$ and a multifunction $E : [S, T] \rightsquigarrow \mathbb{R}^n$ such that $\bar{x}(t) \in E(t)$, for all $t \in [S, T]$. Assume that*

(i) there exists $c > 0$ such that

$$F(t, x) \subset c\mathbb{B} \quad \text{for all } (t, x) \in \text{Gr } E,$$

(ii) there exists a function $\psi : [S, T] \times \mathbb{R}^n \to \mathbb{R}$ that is Lipschitz continuous on bounded sets and, for some $\delta > 0$, satisfies

13.4 Local Verification Theorems

$$\begin{cases} \nabla_t \psi(t,x) + \min_{v \in F(t,x)} \{\nabla_x \psi(t,x) \cdot v\} \geq 0 \\ \text{for a.e. point } (t,x) \in [S,T] \times \mathbb{R}^n \text{ at which} \\ \psi \text{ is differentiable and } x \in E(t) + \delta \mathbb{B}, \\ \psi(T,x) = g(x) \text{ for all } x \in C, \end{cases} \quad (13.4.1)$$

(iii) $\psi(S, x_0) = g(\bar{x}(T))$.

Then, for any admissible F trajectory x on $[S, T]$ such that $x(S) = x_0$ and $x(t) \in E(t)$ for all $t \in [S, T]$, we have

$$g(x(T)) \geq g(\bar{x}(T)). \quad (13.4.2)$$

Proof Take any F trajectory x on $[S, T]$ such that $x(S) = x_0$ and $x(t) \in E(t)$ for all $t \in [S, T]$. From (i) we know that the F trajectory x is Lipschitz continuous on $[S, T]$ and, therefore, remains confined in a ball $R_0 \mathbb{B}$ (for some radius $R_0 > 0$). Owing to (ii) ψ is Lipschitz continuous on the bounded set $[S, T] \times R_0 \mathbb{B}$. It follows that $t \to \psi(t, x(t))$ is a Lipschitz continuous function on $[S, T]$, and, so, it is a.e. differentiable and

$$\psi(S, x(S)) = \psi(T, x(T)) - \int_S^T (d/dt)\psi(t, x(t))dt.$$

Now define $\mathcal{N} \subset (S, T)$ to be the subset comprising points t in (S, T) such at $d/dt\, \psi(t, x(t))$ exists, $dx(t)/dt$ exists, $\dot{x}(t) \in F(t, x(t))$ and t is a Lebesgue point of $\dot{x}$. The subset $\mathcal{N}$ has full Lebesgue measure.

Take any $t \in \mathcal{N}$ and $h_i \downarrow 0$. Then, for each i sufficiently large, there exists, by Lebourg's two-sided mean value theorem (Theorem 4.5.3), $\lambda_i \in [0, 1]$ and

$$(\zeta_i^0, \zeta_i^1) \in \mathrm{co}\, \partial \psi(t + \lambda_i h_i, x(t) + \lambda_i(x(t + h_i) - x(t)))$$

such that

$$h_i^{-1}[\psi(t + h_i, x(t + h_i)) - \psi(t, x(t))] = (\zeta_i^0, \zeta_i^1) \cdot (1, h_i^{-1} \int_t^{t+h_i} \dot{x}(s)ds). \quad (13.4.3)$$

The sequence $\{(\zeta_i^0, \zeta_i^1)\}$ is bounded, since ψ is Lipschitz continuous on bounded sets. We can therefore arrange, by extracting a subsequence, that, for some (ζ^0, ζ^1),

$$(\zeta_i^0, \zeta_i^1) \to (\zeta^0, \zeta^1), \text{ as } i \to \infty.$$

By the upper semicontinuity properties of the Clarke subdifferential co $\partial \psi$, we know that $(\zeta^0, \zeta^1) \in \mathrm{co}\, \partial \psi(t, x(t))$.

Since the left side of (13.4.3) converges to $(d/dt)\psi(t, x(t))$ and t is a Lebesgue point of $\dot{x}$, we deduce from (13.4.3) that

$$(d/dt)\psi(t, x(t)) = \lim_{h \downarrow 0} h^{-1}[\psi(t+h, x(t+h)) - \psi(t, x(t))] = (\zeta^0, \zeta^1) \cdot (1, \dot{x}(t)).$$
(13.4.4)

In consequence of Theorem 4.7.7, there exist, for $k = 0, 1, \ldots, n+1$, $\alpha_k \in [0, 1]$ and a sequence of points $(t_j^k, x_j^k) \to (t, x(t))$, at each point along which ψ is differentiable, such that $\sum_{k=0}^{n+1} \alpha_k = 1$ and, as $j \to \infty$,

$$\left(\sum_{k=0}^{n+1} (\nabla_t \psi(t_j^k, x_j^k), \nabla_x \psi(t_j^k, x_j^k))\right) \cdot (1, \dot{x}(t)) \to (\zeta^0, \zeta^1) \cdot (1, \dot{x}(t)).$$

Since $\dot{x}(t) \in F(t, x(t))$, it follows from our assumptions concerning ψ that

$$\nabla_t \psi(t_j^k, x_j^k) + \nabla_x \psi(t_j^k, x_j^k) \cdot \dot{x}(t) \geq 0 \text{ for each } j.$$

The preceding two relations yield, in the limit as $j \to \infty$, the information

$$(\zeta^0, \zeta^1) \cdot (1, \dot{x}(t)) \geq 0.$$

But then, from (13.4.4), $(d/dt)\psi(t, x(t)) \geq 0$. Since this relation is true for all t's in a set of full measure, it follows that

$$\psi(S, x(S)) = \psi(T, x(T)) - \int_S^T d/dt \, \psi(t, x(t)) dt \leq g(x(T)) + 0.$$

Bearing in mind that this relation is valid for any admissible F trajectory and noting that, by assumption (iii) $\psi(S, x(S)) = g(\bar{x}(T))$, we deduce that

$$g(\bar{x}(T)) \leq g(x(T)).$$

This concludes the proof. □

Proposition 13.4.2 tells us that an L^∞ local minimizer can, in principle, be confirmed as such by some lower semi-continuous, local verification function. There follows an inverse verification theorem which aims to simplifying the task of finding a local verification function. It gives conditions under which the search can be confined to the class of functions which are *Lipschitz continuous* on a tube about the putative minimizer $\bar{x}$. An important point is that we can establish existence of Lipschitz continuous local verification functions, even in the presence of an endpoint constraint

$$x(T) \in C.$$

13.4 Local Verification Theorems

In such cases, Lipschitz continuous functions are typically *distinct* from the value function V for the original problem. (If, for example an L^∞ local minimizer $\bar{x}$ is such that $\bar{x}(T) \in \partial C$, then the value function will take value $+\infty$ at some points in any tube about $\bar{x}$ and so certainly cannot serve as a Lipschitz continuous local verification function.)

The conditions under which there exists a Lipschitz continuous local verification function include a constraint qualification, requiring that necessary conditions of optimality, in the form of the Hamiltonian inclusion, are 'normal' at the locally minimizing arc $\bar{x}$ under examination.

(CQ) For any $p \in W^{1,1}([S,T]; \mathbb{R}^n)$ and $\lambda \geq 0$ such that

$$\begin{cases} \|p\|_{L^\infty} + \lambda \neq 0, \\ \dot{p}(t) \in \operatorname{co}\{-q \,:\, (q, \dot{\bar{x}}(t)) \in \partial H(t, \bar{x}(t), p(t))\} \quad \text{a.e.,} \\ -p(T) \in \lambda \partial g(\bar{x}(T)) + N_C(\bar{x}(T)), \end{cases}$$

we have $\lambda \neq 0$.

Here, as usual, $H(t, x, p)$ is the Hamiltonian

$$H(t, x, p) = \max_{v \in F(t,x)} p \cdot v.$$

Theorem 13.4.4 (Existence of Lipschitz Continuous Verification Functions) *Let the arc $\bar{x} \in W^{1,1}([S,T]; \mathbb{R}^n)$ satisfy the constraints of problem (Q). Assume that, for some $\epsilon > 0$,*

(i) *F takes values non-empty, closed, convex sets, and $F(., x)$ is $\mathcal{L}(S, T)$-measurable for all $x \in \mathbb{R}^n$,*

(ii) *There exists $c > 0$ such that*

$$F(t, x) \subset c\mathbb{B} \text{ for all } x \in \bar{x}(t) + \delta\mathbb{B}, \text{ a.e. } t \in [S, T],$$

(iii) *there exist $k_F \in L^1(S, T)$ and a modulus of continuity $\omega_F : \mathbb{R}_+ \to \mathbb{R}_+$ such that $t \to 1/(1 + k_F(t))$ is almost everywhere equal to a continuous function,*

$$F(t, x') \subset F(t, x) + k_F(t)|x - x'|\mathbb{B}$$

for all $x, x' \in \bar{x}(t) + \delta\mathbb{B}$, for a.e. $t \in [S, T]$,

and

$$d_H(F(t, x'), F(t, x)) \leq \omega_F(|x - x'|)$$

for all $x, x' \in \bar{x}(t) + \delta\mathbb{B}$, and for all $t \in [S, T]$,

(iv) *for each* $s \in [S, T)$ *and* $t \in (S, T]$ *the following one-sided set-valued limits exist and are non-empty:*

$$F(s^+, x) := \lim_{s' \downarrow s, x' \to x} F(s', x') \quad \textit{for all } x \in \bar{x}(s) + \delta \mathbb{B} \quad \textit{and}$$

$$F(t^-, x) := \lim_{t' \uparrow t, x' \to x} F(t', x') \quad \textit{for all } x \in \bar{x}(t) + \delta \mathbb{B},$$

and there exists a subset $\mathcal{S} \subset (S, T)$ *of full Lebesgue measure such that*

$$s \to F(s, x) \textit{ is continuous at } s = t, \textit{ for all } t \in \mathcal{S} \textit{ and } \bar{x}(t) + \delta \mathbb{B},$$

(v) *g is Lipschitz continuous on* $\bar{x}(T) + \epsilon \mathbb{B}$.

We have

(a) *If there exists a Lipschitz continuous local verification function for* $\bar{x}$, *then* $\bar{x}$ *is an* L^∞ *local minimizer for* (Q),
(b) *If* $\bar{x}$ *is an* L^∞ *local minimizer and* (CQ) *is satisfied, then there exists a Lipschitz continuous local verification function for* $\bar{x}$.

Proof

(a): This is, of course, just a special case of Proposition 13.4.2, already proved.

(b): Let $\bar{x}$ be an L^∞ local minimizer for (Q). We shall construct a Lipschitz continuous local verification function for $\bar{x}$.

Choose a constant $\bar{\epsilon} \in (0, \epsilon)$, such that $\bar{x}$ is a minimizer with respect to all arcs satisfying the constraints of (Q) and also $||x - \bar{x}||_{L^\infty} \leq \bar{\epsilon}$. Let κ be a constant such that $|g(x)| \leq \kappa$ for $x \in \bar{x}(T) + \bar{\epsilon}\mathbb{B}$. Take any sequence $\epsilon_i \downarrow 0$ such that $\epsilon_i \in (0, \bar{\epsilon})$ for all i. Take also sequences $k_i \uparrow +\infty$, $K_i \uparrow +\infty$ such that

$$k_i > 64 c \kappa \epsilon_i^{-2} \text{ for each } i \quad \text{and} \quad K_i/k_i \to \infty \text{ as } i \to \infty.$$

For each i, define

$$L_i(t, x) := \max\{|x - \bar{x}(t)| - \epsilon_i/2, 0\}.$$

Define also

$$\tilde{g}(x) := g(\bar{x}(T)) + \mathrm{tr}_{\bar{\epsilon}}(x - \bar{x}(T)), \quad \tilde{F}(t, x) := F(t, \bar{x}(t)) + \mathrm{tr}_{\bar{\epsilon}}(x - \bar{x}(t))$$

For each i, consider the dynamic optimization problem $(\tilde{P}_i)$:

13.4 Local Verification Theorems

$$(\tilde{P}_i) \begin{cases} \text{Minimize } \tilde{g}(x(T)) + k_i z(T) + K_i d_C(x(T)) \\ \text{over arcs } (x, z) \in W^{1,1}([S, T]; \mathbb{R}^{n+1}) \text{ satisfying} \\ (\dot{x}(t), \dot{z}(t)) \in \tilde{F}(t, x(t)) \times \{L_i(t, x(t))\} \quad \text{a.e.} \\ (x(S), z(S)) = (x_0, 0). \end{cases}$$

We shall show presently that,

$$(\bar{x}, z(t) \equiv 0) \text{ is a minimizer for } (\tilde{P}_{i_0}) \quad (13.4.5)$$

for some index value i_0.

Completion of the proof is then a straightforward matter. Indeed, let $V(t, x, z)$ be the value function for $(\tilde{P}_{i_0})$. Clearly

$$V(t, x, z) = \phi(t, x) + z$$

for some function $\phi : [S, T] \times \mathbb{R}^n \to \mathbb{R} \cup \{+\infty\}$. We also note that

$$\tilde{F}(t, x) = F(t, x) \quad \text{and} \quad L(t, x) = 0 \text{ for } (t, x) \in T(\bar{x}, \epsilon_{i_0}/2)$$

and

$$\tilde{g}(x) + K_{i_0} d_C(x) = g(x) \quad \text{for } x \in C \cap (\bar{x}(T) + \epsilon_{i_0}/2 \mathbb{B}).$$

The information about value functions supplied by Theorem 13.3.7, applied to $V(t, x, z)$, tells us that ϕ is a local verification function for (P_{i_0}). (P_{i_0}) has a Lipschitz continuous terminal cost function, however, and there is no right endpoint constraint. It follows from Proposition 13.3.6 that V (and therefore ϕ also) is locally Lipschitz continuous. This is what we set out to prove.

It remains then to confirm (13.4.5). Here we make use of a contradiction argument. Suppose that, for each i, $\bar{x}$ is not a minimizer for $(\tilde{P}_i)$.

Under the hypotheses it is known that, for each i, $(\tilde{P}_i)$ has a minimizer: write it

$$(x_i, t \to k_i \int_S^t L_i(s, x_i(s))ds).$$

Because it is assumed that $\bar{x}$ is not a minimizer for $(\tilde{P}_i)$, we have

$$\tilde{g}(x_i(T)) + k_i \int_S^T L_i(t, x_i(t))dt + K_i d_C(x_i(T)) < \tilde{g}(\bar{x}(T)) + 0 + 0 = g(\bar{x}(T)). \quad (13.4.6)$$

We now note two important properties of x_i:

$$||x_i - \bar{x}||_{L^\infty} \leq \epsilon_i \quad (13.4.7)$$

$$x_i(T) \notin C. \quad (13.4.8)$$

Indeed, if (13.4.7) is not true, then there exists $\bar{t} \in [S, T]$ such that $|x_i(\bar{t}) - \bar{x}(\bar{t})| = \epsilon_i$. Since $t \to |x_i(t) - \bar{x}(t)|$ is Lipschitz continuous with Lipschitz constant $2c$, it follows that

$$\text{meas}\{t \in [S, T] : |x_i(t) - \bar{x}(t)| > (3/4)\epsilon_i\} \geq \epsilon_i/(8c).$$

But then

$$\tilde{g}(x_i(T)) + k_i \int_S^T L_i(t, x_i(t))dt + K_i d_C(x_i(T))$$
$$> -\kappa + k_i \epsilon_i^2/(32c) > \kappa \geq g(\bar{x}(T)),$$

in contradiction of (13.4.6). So (13.4.7) is true.

Suppose (13.4.8) is not true. Then x_i satisfies the constraints of (Q) and also $\|x_i - \bar{x}\| \leq \bar{\epsilon}$. So, by local optimality of $\bar{x}$,

$$g(x_i(T)) \geq g(\bar{x}(T)).$$

But then, in view of (13.4.6),

$$\tilde{g}(x_i(T)) + k_i \int_S^T L_i(t, x_i(t))dt + K_i d_C(x_i(T))$$
$$> \tilde{g}(x_i(t)) + 0 + 0 = g(x_i(T)) \geq g(\bar{x}(T)).$$

This contradicts (13.4.6). So (13.4.8) is true.

Now apply the necessary conditions of Theorem 8.7.2 (the Hamiltonian inclusion for convex velocity sets) to problem $(\tilde{P}_i)$, with reference to the L^∞ local minimizer x_i. These provide a costate arc $p_i \in W^{1,1}([S, T]; \mathbb{R}^n)$ and $\lambda_i \geq 0$ satisfying:

$$\|p_i\|_{L^\infty} + \lambda_i = 1,$$
$$\dot{p}(t) \in \text{co}\{-q \mid (q, \dot{x}_i(t)) \in \partial H(t, x_i(t), p_i(t))$$
$$+ \lambda_i k_i/K_i \text{co}\, \partial L(t, x_i(t))\} \quad \text{a.e.},$$
$$-p_i(T) \in \lambda_i(1/K_i)\partial \tilde{g}(x_i(T)) + \lambda_i \partial d_C(x_i(T)).$$

Here, $H(t, x, p)$ is the Hamiltonian for the original problem. Notice that, before applying the necessary conditions, we have scaled the cost by $(1/K_i)$.

Since $x_i(T) \notin C$,

$$\partial d_C(x_i(t)) \in \{\xi \in \mathbb{R}^n : |\xi| = 1\}.$$

We deduce that

$$|p_i(T)| \geq \lambda_i(1 - k_g/K_i).$$

(k_g is a Lipschitz constant for g on $\bar{x}(T) + \bar{\epsilon}\mathbb{B}$.)

A by now familiar convergence analysis can be used to justify passing to the limit in the above relations: along a subsequence, $p_i \to p$ uniformly and $\lambda_i \to \alpha$, for some $p \in W^{1,1}([S, T]; \mathbb{R}^n)$ and $\alpha \geq 0$, satisfying

$$\|p\|_{L^\infty} + \alpha = 1, \qquad (13.4.9)$$

$$\dot{p}(t) \in \text{co}\{-q \mid (q, \dot{\bar{x}}(t)) \in \partial H(t, \bar{x}(t), p(t))\} \quad \text{a.e.},$$

$$-p(T) \in \alpha \partial d_C(\bar{x}(T)) \subset N_C(\bar{x}(T)),$$

$$\|p\|_{L^\infty} \geq \alpha.$$

Notice that $p \neq 0$ since, if $p = 0$, then $\alpha = 0$ which contradicts (13.4.9). We have exhibited a costate trajectory violating the constraint qualification (CQ). It follows that the original claim, namely that $\bar{x}$ is minimizer for (P_i) for some i, is true. The proof is complete. □

13.5 Costate Trajectories and Gradients of the Value Function

In this section we again consider the dynamic optimization problem of Sect. 13.1:

$$(P) \begin{cases} \text{Minimize } g(x(T)) \\ \text{over arcs } x \in W^{1,1}([S, T]; \mathbb{R}^n) \text{ satisfying} \\ \dot{x}(t) \in F(t, x(t)) \quad \text{a.e.}, \\ x(S) = x_0, \end{cases}$$

the data for which comprise a function $g : \mathbb{R}^n \to \mathbb{R}$, an interval $[S, T] \subset \mathbb{R}$, a multifunction $F : [S, T] \times \mathbb{R}^n \rightsquigarrow \mathbb{R}^n$ and a vector $x_0 \in \mathbb{R}^n$.

Note, however, that g is taken to be finite valued. End-point constraints are therefore excluded.

Let $\bar{x}$ be a minimizer. Then, under appropriate hypotheses on the data, the necessary conditions of Theorem 8.4.3 assert the existence of a costate trajectory $p \in W^{1,1}([S, T]; \mathbb{R}^n)$ such that

$$\dot{p}(t) \in \text{co}\{q : (q, p(t)) \in N_{\text{Gr } F(t,.)}(\bar{x}(t), \dot{\bar{x}}(t))\} \quad \text{a.e.},$$

$$p(t) \cdot \dot{\bar{x}}(t) = h(t) \quad \text{a.e.},$$

$$-p(T) \in \partial g(\bar{x}(T)),$$

where

$$h(t) := \max_{v \in F(t, \bar{x}(t))} p(t) \cdot v .$$

(Since there are no right end-point constraints, we are justified in assuming that the cost multiplier is $\lambda = 1$.)

Let V be the value function for (P). We shall invoke hypotheses under which V is locally Lipschitz continuous. Earlier discussion leads us to expect that the costate trajectory p (and h) will be related to V according to

$$(h(t), -p(t)) \in \text{co}\, \partial V(t, \bar{x}(t)) \quad \text{a.e. } t \in (S, T). \tag{13.5.1}$$

(To be more precise, we would expect this inclusion to hold for some kind of subdifferential of V; with hindsight we adopt the convexified limiting subdifferential.)

The aim of this section is to confirm the 'sensitivity relation' (13.5.1). Before entering into the details of the arguments involved, we briefly examine two different approaches which hold out hope of simpler proof techniques, if only to dismiss them. The first is to take an arbitrary selector of the multifunction $t \to \text{co}\, \partial V(t, \bar{x}(t))$, partitioned as $(h, -p)$, such that p is absolutely continuous, and attempt to show that it is a costate trajectory. The other is to take an arbitrary costate trajectory and Hamiltonian (evaluated along this costate trajectory and state trajectory $\bar{x}$) and attempt to show that they satisfy the sensitivity relation (13.5.1). Their appeal is that they involve simply checking the properties of a plausible candidate for a costate trajectory. The inadequacies of these approaches, at least for problems with nonsmooth data, are made evident by the following example, which reveals that, on the one hand, a pair of functions $(h, -p)$ chosen from the set of selectors for $t \to \text{co}\, \partial V(t, \bar{x}(t))$, in which p is absolutely continuous, may fail to generate costate trajectories and, on the other, costate trajectories may fail to generate a pair of functions that satisfy the sensitivity relation.

Example

$$\begin{cases} \text{Minimize } g(x(1)) \text{ over } x \in W^{1,1}([0, 1]; \mathbb{R}) \text{ satisfying} \\ \dot{x}(t) \in \{x(t)u : u \in [0, 1]\}, \\ x(0) = 0, \end{cases}$$

where

$$g(\xi) := \begin{cases} -\xi & \text{if } \xi > 0 \\ -e^{0.5}\xi & \text{if } \xi \leq 0 . \end{cases}$$

The only admissible arc for this problem is $\bar{x}(t) \equiv 0$. This then is the minimizer. The value function V is easily calculated:

13.5 Costate Trajectories and Gradients of the Value Function

$$V(t, \xi) = \begin{cases} -e^{(1-t)}\xi & \text{if } \xi > 0 \\ -e^{0.5}\xi & \text{if } \xi \leq 0. \end{cases}$$

Evidently

$$\text{co}\, \partial V(t, 0) = \{0\} \times \text{co}\{-e^{0.5}, -e^{1-t}\}.$$

The Hamiltonian is

$$H(t, x, p) = \max\{px, 0\}.$$

For this problem a costate trajectory is any absolutely continuous function p which satisfies

$$-\dot{p}(t) = \alpha(t)p(t) \quad \text{a.e. } t \in [0, 1] \tag{13.5.2}$$

and

$$p(1) \in [1, e^{0.5}], \tag{13.5.3}$$

for some measurable function $\alpha : [0, 1] \to [0, 1]$.

Notice that $p \equiv e^{0.5}$ is a costate trajectory (it corresponds to the choice $\alpha \equiv 0$) which satisfies

$$(H(t, \bar{x}(t), p(t)), -p(t)) \in \text{co}\, \partial V(t, \bar{x}(t)) \quad \text{for all } t \in [0, 1].$$

So, for this example there is a costate trajectory satisfying the anticipated relation.

However, $p_1 \equiv 1$ is also a costate trajectory (it too corresponds to the choice $\alpha \equiv 0$) with the property that

$$(H(t, \bar{x}(t), p_1(t)), -p_1(t)) \notin \text{co}\, \partial V(t, \bar{x}(t)) \quad \text{for all } t \in [0, 1).$$

This means that a costate trajectory exists which, on a set of full measure, fails to satisfy the sensitivity relation.

On the other hand, for any number $\omega > 0$, p_2 given by

$$p_2(t) = e^{0.5} + (e^{1-t} - e^{0.5})\sin(\omega t)$$

is a continuously differentiable function which satisfies

$$(H(t, \bar{x}(t), p_2(t)), -p_2(t)) \in \text{co}\, \partial V(t, \bar{x}(t)) \quad \text{for all } t \in [0, 1].$$

We note however that

$$dp_2/dt(0) = \omega e^{0.5}(e^{0.5} - 1).$$

By Gronwall's lemma (Lemma 6.2.4), if p_2 is also a solution to (13.5.2) and (13.5.3) then $p_2(0) \leq e^{1.5}$. Hence $\dot{p}_2(0) \leq e^{1.5}$. It follows that p_2 cannot be a costate trajectory if $\omega > e(e^{0.5} - 1)^{-1}$.

We have shown that there also exists an absolutely continuous function which, combined with associated function h, satisfies the sensitivity relation but which is not a costate trajectory.

In order to pick out a special costate trajectory which, with its associated function h, satisfies the sensitivity relation, even in situations such as that described in the above example, we make use of quite different techniques, the flavour of which we now attempt to convey. The underlying ideas become more transparent when we switch to problems having a traditional controlled differential equation formulation and when we assume that the value function V is a C^1 solution to the Hamilton Jacobi equation. Consider for the time being then:

$$\begin{cases} \text{Minimize } g(x(T)) \text{ over } x \in W^{1,1}([S, T]; \mathbb{R}^n) \\ \qquad \text{and measurable functions } u : [S, T] \to \mathbb{R}^n \text{ satisfying} \\ \dot{x} = f(t, x(t), u(t)) \quad \text{a.e. } [S, T], \\ u(t) \in U \quad \text{a.e. } [S, T], \\ x(S) = x_0, \end{cases}$$

where $[S, T]$ is a given time interval, $g : \mathbb{R}^n \to \mathbb{R}$ and $f : [S, T] \times \mathbb{R}^n \times \mathbb{R}^m \to \mathbb{R}^n$ are given functions, $U \subset \mathbb{R}^m$ is a given set and $x_0 \in \mathbb{R}^n$ is a given point.

Let $(\bar{x}, \bar{u})$ be a minimizer. Under appropriate hypotheses, the value function satisfies

$$V_t(t, \xi) + V_x(t, \xi) \cdot f(t, \xi, u) \geq 0 \qquad (13.5.4)$$

$$\text{for all } (t, \xi) \in (S, T) \times \mathbb{R}^n \quad \text{and} \quad u \in U,$$

and $V(T, x) = g(x)$ for all $x \in \mathbb{R}^n$. Also,

$$V_t(t, \bar{x}(t)) + V_x(t, \bar{x}(t)) \cdot f(t, \bar{x}(t), \bar{u}(t)) = 0 \quad \text{a.e. } t \in [S, T]. \qquad (13.5.5)$$

Let $u : [S, T] \to \mathbb{R}^m$, $v : [S, T] \to \mathbb{R}^n$ and $w : [S, T] \to \mathbb{R}$ be arbitrary measurable functions and x be an absolutely continuous arc such that

$$\dot{x}(t) = (1 + w(t))(f(t, x(t), u(t)) + v(t)) \quad \text{a.e.}$$

$$(u(t), w(t), v(t)) \in U \times \mathbb{B} \times [-1, +1] \quad \text{a.e..}$$

Inserting $(\xi, u) = (x(t), u(t))$ into (13.5.4), multiplying across the inequality by the nonnegative number $(1 + w(t))$ and adding and subtracting terms, we obtain the inequality

13.5 Costate Trajectories and Gradients of the Value Function

$$V_t(t, x(t)) + V_x(t, x(t)) \cdot (f(t, x(t), u(t)) + v(t))(1 + w(t))$$
$$+ \eta(t, x(t), v(t), w(t)) \geq 0 \quad \text{a.e..}$$

Here, η is the function

$$\eta(t, x, v, w) := -(1 + w)V_x(t, x) \cdot v + wV_t.$$

Noting that the first two terms on the left of this inequality can be written $dV(t, x(t))/dt$ and also the boundary condition $V(T, .) = g$, we deduce that

$$J(x, u, v, w) \geq 0,$$

in which J is defined to be the functional

$$J(x, u, v, w) :=$$
$$g(x(T)) - V(S, x(S)) + \int_S^T \eta(t, x(t), v(t), w(t))dt.$$

Similar reasoning applied to (13.5.5) gives

$$J(\bar{x}, \bar{u}, \bar{v} \equiv 0, \bar{w} \equiv 0) = 0.$$

We deduce that $(\bar{x}, \bar{u}, \bar{v} \equiv 0, \bar{w} \equiv 0)$ is an L^∞ local minimizer for the optimization problem:

$$\begin{cases} \text{Minimize } g(x(T)) - V(S, x(S)) + \int_S^T \eta(t, x(t), v(t), w(t))dt \\ \text{over } x \in W^{1,1} \text{ and measurable functions } (u, v, w) : [S, T] \to \mathbb{R}^{m+n+1} \\ \text{satisfying} \\ \dot{x}(t) = (1 + w(t))(f(t, x(t), u(t)) + v(t)) \quad \text{a.e.,} \\ (u(t), v(t), w(t)) \in U \times \mathbb{B} \times [-1, +1] \quad \text{a.e..} \end{cases}$$

This is an dynamic optimization problem to which the maximum principle is applicable, with reference to the L^∞ local minimizer $(\bar{x}, (\bar{u}, \bar{v} \equiv 0, \bar{w} \equiv 0))$. It turns out that the a costate trajectory p for this derived problem associated with the cost multiplier $\lambda = 1$ (we can make such a choice of cost multiplier since the dynamic optimization problem has free right end-point) is also a costate trajectory for the original dynamic optimization problem. But the Weierstrass condition for the derived problem gives us the extra information that, for almost every t,

$$d(t, \bar{v}(t) = 0, \bar{w}(t) = 0) = \max\{d(t, v, w) : |v| \leq 1, |w| \leq 1\},$$

where

$$d(t, v, w) := (1+w)[p(t) \cdot f(t, \bar{x}(t), \bar{u}(t)) - V_t t(t, \bar{x}(t))]$$
$$+ (1+w)[p(t) + V_x(t, \bar{x}(t))] \cdot v.$$

In particular, $d(t, 0, 0) \le d(t, 0, w)$ for every $w \in [-1, 1]$ and $d(t, 0, 0) \le d(t, v, 0)$ for every $v \in \mathbb{B}$, from which we conclude that

$$(p(t) \cdot f(t, \bar{x}(t), \bar{u}(t)), -p(t)) = (V_t(t, \bar{x}(t)), V_x(t, \bar{x}(t))) \quad \text{a.e..}$$

This is precisely the sensitivity relation.

What we have done is to modify the problem so that $(\bar{x}, \bar{u})$ becomes a minimizer with respect to a richer class of controls. It is reasonable to expect that necessary conditions of optimality for the derived problem will convey more information about $(\bar{x}, \bar{u})$ than those for the original problem; the extra information is the sensitivity relation.

The reader will be justifiably sceptical at this stage as to whether these elementary arguments can be modified to handle situations where V is possibly nonsmooth, since then the cost integrand in the derived problem, which involves derivatives of V, may be discontinuous with respect to the state variable and altogether unsuitable for application of the standard first order necessary conditions. We shall get round this difficulty by replacing the offending function $\eta(t, x, v, w)$ by an approximation which does not depend on the state variable at all.

We are ready to prove

Theorem 13.5.1 *(Sensitivity Relations) Let $\bar{x}$ be an L^∞ local minimizer for (P). Assume that the following hypotheses are satisfied:*

(a): *F takes values non-empty, closed sets, F is $\mathcal{L} \times \mathcal{B}^n$-measurable, and $F(., x)$ is continuous from the right on $[S, T)$, for each $x \in \mathbb{R}^n$,*
(b): *there exists $c > 0$ such that*

$$F(t, x) \subset c(1 + |x|)\mathbb{B} \quad \text{for all } (t, x) \in [S, T] \times \mathbb{R}^n,$$

(c): *there exists $k_F \in L^1(S, T)$ such that*

$$F(t, x) \subset F(t, x') + k_F(t)|x - x'|\mathbb{B}, \quad \text{for all } x, x' \in \mathbb{R}^n, \text{ a.e. } t \in [S, T],$$

(d): *g is locally Lipschitz continuous.*

Let V be the value function for problem (P). Then there exists $p \in W^{1,1}([S, T]; \mathbb{R}^n)$ such that

(i): $\dot{p}(t) \in \text{co}\{\xi : (\xi, p(t)) \in N_{\text{Gr } F(t,.)}(\bar{x}(t), \dot{\bar{x}}(t))\}$ *a.e. $t \in [S, T]$,*
(ii): $p(t) \cdot \dot{\bar{x}}(t) = \max_{v \in F(t, \bar{x}(t))} p(t) \cdot v$ *a.e. $t \in [S, T]$,*
(iii): $-p(T) \in \partial g(\bar{x}(T))$,
(iv): $p(S) \in \partial_x(-V)(S, \bar{x}(S))$.

13.5 Costate Trajectories and Gradients of the Value Function

Furthermore,

$$(H(t, \bar{x}(t), p(t)), -p(t)) \in co\, \partial V(t, \bar{x}(t)) \quad a.e.\ t \in [S, T]. \tag{13.5.6}$$

Remark

(a): Notice that the sensitivity relation (13.5.6) implies

$$-p(t) \in \mathrm{Proj}_x\, co\, \partial V(t, \bar{x}(t)) \quad a.e.\ t \in [S, T], \tag{13.5.7}$$

where Proj_x denotes 'projection onto the x block coordinate'. It is possible to show [35], though we do not do so here, that condition (13.5.6) can be supplemented by the (partial) sensitivity relation:

$$-p(t) \in co\, \partial_x V(t, \bar{x}(t)) \quad a.e.\ t \in [S, T]. \tag{13.5.8}$$

Conditions (13.5.6) and (13.5.8) are distinct necessary conditions since, in some situations,

$$\mathrm{Proj}_x\{\partial V(t, x)\} \not\subset \partial_x^0 V(t, x).$$

For instance for $f : \mathbb{R}_+ \times \mathbb{R} \to \mathbb{R}$ given by $f(t, x) = |x + 1 - t| - |x|$ we have $f(1, x) = 0$ for all x and therefore $\partial_x f(1, 0) = \{0\}$. On the other hand, $\partial^0 f(1, 0)$ is equal to the convex hull of the set $\{\pm(1, 0), \pm(1, -2)\}$. Thus, $\mathrm{Proj}_x\{\partial f(1, 0)\} \not\subset \partial_x f(1, 0)$.

(b): The assertions of Theorem 13.5.1 remain valid (with possibly different p) if we replace the Euler Lagrange inclusion (i) with the partially convexified Hamiltonian inclusion:

(i)′ $\dot{p}(t) \in co\,\{\xi : (-\xi, \dot{\bar{x}}(t)) \in \partial_{x,p} H(t, \bar{x}(t), p(t))\}$ for a.e. $t \in [S, T]$.

To show this, we note that if F is replaced by its convex hull co F the hypotheses invoked in Theorem 13.5.1 are still valid. Furthermore, $\bar{x}$ remains a minimizer and the value function is unaffected by the change. Now apply Theorem 13.5.1 with co F in place of F. This yields the assertions of the theorem (possibly with a different p), except that, now, co F appears in place of F in condition (i). But this modified condition implies the partially convexified Hamiltonian inclusion, in consequence of the dualization Theorem 8.7.1 of Chap. 8.

Proof According to Proposition 13.3.6, V is a locally Lipschitz continuous function. For $\epsilon \in (0, 1)$ we define $G_\epsilon : [S, T] \rightsquigarrow \mathbb{R}^{1+n}$ and $\sigma_\epsilon : \mathbb{R} \times \mathbb{R}^n \times \mathbb{R} \to \mathbb{R}$ to be

$$G_\epsilon(t) := \{(\alpha, \beta) : \mathbb{R}^{1+n} : (\alpha, \beta) \in co\, \partial V(s, y)$$
$$\text{for some } (s, y) \in ((t, \bar{x}(t)) + \epsilon \mathbb{B}) \cap ((S, T) \times \mathbb{R}^n)\}$$

and

$$\sigma_\epsilon(t, v, w) := \sup_{(\alpha, \beta) \in G_\epsilon(t)} (\alpha, \beta) \cdot (w, -(1 + w)v). \qquad (13.5.9)$$

It is a straightforward matter to check that, for fixed $\epsilon \in (0, 1)$, σ_ϵ is bounded on bounded sets, $\sigma_\epsilon(., v, w)$ is measurable for all $(v, w) \in \mathbb{R}^n \times \mathbb{R}$, and (by known properties on limiting subdifferentials and support functions) $\sigma_\epsilon(t, ., .)$ is locally Lipschitz continuous for each $t \in [S, T]$.

Now consider the following dynamic optimization problem:

$$(P_\epsilon) \begin{cases} \text{Minimize } J(x(S), x(T), y(T)) \\ \text{over arcs } (x, y, v, w) \in W^{1,1}([S, T]; \mathbb{R}^{n+1+n+1}) \text{ satisfying} \\ (\dot{x}(t), \dot{y}(t), \dot{v}(t), \dot{w}(t)) \in \tilde{F}_\epsilon(t, x(t)), \quad \text{a.e. } t \in [S, T] \\ (x(S), y(S), v(S), w(S)) \in \mathbb{R}^n \times \{0\} \times \{0\} \times \{0\} \end{cases}$$

in which

$$\tilde{F}_\epsilon(t, x) := \{((e + v)(1 + w), \sigma_\epsilon(t, v, w), a, b) : e \in F(t, x), a \in \epsilon\mathbb{B}, b \in \epsilon\mathbb{B}\}$$

and

$$J(x_0, x_1, y_1) := g(x_1) - V(S, x_0) + y_1.$$

We verify:

Claim: For any $\epsilon \in (0, 1)$, $(\bar{x}, \bar{y} \equiv 0, \bar{v} \equiv 0, \bar{w} \equiv 0)$ is an L^∞ local minimizer for (P_ϵ).

To confirm the claim, take any arc (x, y, v, w) satisfying the constraints of (P_ϵ) and such that $\|(x, y, v, w) - (\bar{x}, \bar{y}, \bar{v}, \bar{w})\|_{L^\infty} \leq \epsilon/2$. Notice to begin with that at every point $t \in (S, T)$ which is a Lebesgue point of $s \to \dot{x}(s)$ and which is a differentiability point of the Lipschitz continuous function $s \to V(s, x(s))$ (such points comprise a set of full measure) we have

$$\begin{aligned} d/dt V(t, x(t)) &= \lim_{h \downarrow 0} h^{-1}[V(t + h, x + \int_t^{t+h} \dot{x}(s)ds) - V(t, x)], \\ &= \lim_{h \downarrow 0} h^{-1}[V(t + h, x + \dot{x}(t)h) - V(t, x)] \\ &\geq D_\uparrow V((t, x(t)); (1, \dot{x}(t))). \qquad (13.5.10) \end{aligned}$$

Since (x, y, v, w) satisfies the constraints of (P_ϵ), it can be deduced from the measurable selection theorem that there exist measurable functions e, a and b such that, for all points $t \in (S, T)$ in a set of full measure, we have

13.5 Costate Trajectories and Gradients of the Value Function

$$\dot{y}(t) = \sigma_\epsilon(t, v(t), w(t)),$$

$$\dot{x}(t) = (e(t) + v(t))(1 + w(t)),$$

$$e(t) \in F(t, x(t)),$$

$$|v(t)|, |w(t)| \leq \epsilon|T - S|.$$

Since V is locally Lipschitz continuous, we deduce from Lemma 13.3.2(ii), the continuity of $F(., x)$ from the right, hypothesis (c) and the principle of optimality that, for a.e. $t \in [S, T]$,

$$D_\uparrow V((t, x(t)); (1, e(t))) = \liminf_{h \downarrow 0}[V(t+h, x+he(t)) - V(t, x)]h^{-1} \geq 0.$$

Then, for all points $t \in (S, T)$ in a set of full measure, the following relations, in which we write $x(t)$, $e(t)$, $v(t)$ and $w(t)$ briefly as x, e, v and w, are valid:

$$0 \leq \liminf_{h \downarrow 0}[V(t+h, x+he) - V(t, x)]h^{-1}(1+w)$$

(we have used the fact that $(1+w) > 0$)

$$\leq \liminf_{h \downarrow 0}[V(t+h(1+w), x+h(1+w)e) - V(t, x)]h^{-1}$$

(by positive homogeneity)

$$\leq \liminf_{h \downarrow 0}[V(t+h, x+h(e+v)(1+w)) - V(t, x)]h^{-1}$$

$$+ \limsup_{h \downarrow 0}[V(t_h + hw, x_h - hv(1+w)) - V(t_h, x_h)]h^{-1}$$

(in which $t_h := t+h$ and $x_h := x+h(e+v)(1+w)$)

$$\leq D_\uparrow V((t, x); (1, (e+v)(1+w))) + D^0 V((t, x); (w, -v(1+w))).$$

In the final expression $D^0 V$ denotes the generalized directional derivative.

We deduce from the fact that the generalized directional derivative $D^0 V$ is the support function of $\mathrm{co}\, \partial V$ (observe that, since V is locally Lipschitz continuous, we can invoke Prop. 4.7.4) that

$$D^0 V(t, x); (w, -v(1+w))$$

$$\leq \sup_{(\alpha, \beta) \in G_\epsilon(t)} (\alpha, \beta) \cdot (w, -v(1+w)) = \sigma_\epsilon(t, v, w).$$

It follows

$$D_\uparrow V((t, x(t)); (1, (e(t) + v(t))(1 + w(t))))$$

$$+ \sigma_\epsilon(t, v(t), w(t)) \geq 0 \quad \text{a.e. } t \in [S, T].$$

But then, since $V(T, .) = g$ and in view of (13.5.10),

$$J(x(S), x(T), y(T)) = g(x(T)) - V(S, x(S)) + \int_S^T \sigma_\epsilon(t, v(t), w(t))dt$$

$$= V(T, x(T)) - V(S, x(S)) + \int_S^T \sigma_\epsilon(t, v(t), w(t))dt$$

$$= \int_S^T [D_\uparrow V((t, x(t)); (1, \dot{x}(t))) + \sigma_\epsilon(t, v(t), w(t))]dt$$

$$\geq 0 = J(\bar{x}(S), \bar{x}(T), \bar{y}(T)).$$

The claim is verified.

We now apply the necessary conditions of Theorem 8.4.3 to (P_ϵ), with reference to the minimizer $(\bar{x}, \bar{y} \equiv 0, \bar{v} \equiv 0, \bar{w} \equiv 0)$. Since the problem has free right endpoint, we may set the cost multiplier $\lambda = 1$. It follows that there exists a costate trajectory $\tilde{p} = (p_x, p_y, p_v, p_w)$ (associated with the state arc $(\bar{x}, \bar{y} \equiv 0, \bar{v} \equiv 0, \bar{w} \equiv 0)$) such that

$$\dot{\tilde{p}}(t) \in \mathrm{co}\{\eta : (\eta, \tilde{p}(t)) \in N_{\mathrm{Gr}\, \tilde{F}_\epsilon(t,.)}((\bar{x}(t), 0, 0, 0), (\dot{\bar{x}}(t), 0, 0, 0)) \text{ a.e.},$$
(13.5.11)

$$p_x(t) \cdot \dot{\bar{x}}(t) = \max\{p_x(t) \cdot e : e \in F(t, \bar{x}(t))\} \quad \text{a.e.,} \quad (13.5.12)$$

$$-\tilde{p}(T) \in \partial g(\bar{x}(T)) \times \{1\} \times \{0\} \times \{0\} \quad (13.5.13)$$

$$p_x(S) \in \partial_x(-V)(S, \bar{x}(S)).$$

An analysis of proximal normals and strict normals approximating vectors in the limiting normal cone $N_{\mathrm{Gr}\, \tilde{F}_\epsilon}$ permits us to conclude from (13.5.11) that $\eta_y = 0$, $p_v = 0$, $p_w = 0$,

$$(\eta_x, p_x) \in N_{\mathrm{Gr}\, F(t,.)}(\bar{x}(t), \dot{\bar{x}}(t)),$$

and

$$(\eta_v + p_x, \eta_w + p_x \cdot \dot{\bar{x}}(t)) \in -p_y \mathrm{co}\partial_{v,w}\sigma_\epsilon(t, \bar{v}(t) = 0, \bar{w}(t) = 0). \quad (13.5.14)$$

Then from (13.5.11)–(13.5.13) we deduce also that $\dot{p}_v \equiv 0$, $\dot{p}_w \equiv 0$, $p_y \equiv -1$, and, writing $p = p_x$, we obtain

$$\dot{p}(t) \in \mathrm{co}\{q : (q, p(t)) \in N_{\mathrm{Gr}\, F(t,.)}(\bar{x}(t), \dot{\bar{x}}(t))\}$$

and

13.5 Costate Trajectories and Gradients of the Value Function

$$-p(T) \in \partial g(\bar{x}(T)), \quad p(S) \in \partial_x(-V)(S, \bar{x}(S)).$$

Denote by H the Hamiltonian for the original problem, namely

$$H(t, x, p) := \max_{v \in F(t,x)} p \cdot v.$$

Now, recalling the definition of σ_ϵ, from (13.5.14) it follows that

$$(H(t, \bar{x}(t), p(t)), -p(t)) \in \operatorname{co} G_\epsilon(t). \tag{13.5.15}$$

Thus far, $\epsilon \in (0, 1)$ has been treated as a constant. Now choose a sequence $\epsilon_i \downarrow 0$. For each i, write p_i in place of p and ϵ_i in place of ϵ in the above relations. It can be deduced from (13.5.15) and the local Lipschitz continuity of V that $\{p_i\}_{i \geq 1}$ is bounded in $L^\infty([S, T]; \mathbb{R}^n)$. We can also obtain from Lemma 7.2.3 that

$$|\dot{p}_i(t)| \leq k_F(t)|p_i(t)| \quad \text{for a.e. } t \in [S, T].$$

Consequently, the p_i's are uniformly bounded and $\dot{p}_i$'s are uniformly integrably bounded. So, by extracting a subsequence if necessary (without relabelling), we can arrange that $p_i \to p$ uniformly, for some absolutely continuous function p, and $\dot{p}_i$ converges to $\dot{p}$ weakly in $L^1([S, T]; \mathbb{R}^n)$.

A by now familiar convergence analysis can be used to show that the following conditions are satisfied

$$\dot{p}(t) \in \operatorname{co}\{q : (q, p(t)) \in N_{\operatorname{Gr} F(t,\cdot)}(\bar{x}(t), \dot{\bar{x}}(t))\}, \quad \text{a.e.} \tag{13.5.16}$$

$$p(t) \cdot \dot{\bar{x}}(t) = \max_{v \in F(t, \bar{x}(t))} p(t) \cdot v \quad \text{a.e.} \tag{13.5.17}$$

$$-p(T) \in \partial g(\bar{x}(T)),$$

$$-p(S) \in \partial_x(-V)(S, \bar{x}(S)).$$

Write $H(t, \bar{x}(t), p(t))$ briefly as $h(t)$. From (13.5.15) we obtain in the limit the information that, for some subinterval $\mathcal{S} \subset [S, T]$ of full measure,

$$(h(t), -p(t)) \in \cap_{\epsilon' > 0} G_{\epsilon'}(t) \quad \text{for all } t \in \mathcal{S}. \tag{13.5.18}$$

Finally, we use these relations to show that

$$(h(t), -p(t)) \in \operatorname{co} \partial V(t, \bar{x}(t)) \quad \text{for all } t \in \mathcal{S}. \tag{13.5.19}$$

This will complete the proof.

Suppose to the contrary that (13.5.19) is false at some $t \in \mathcal{S}$. Then we can strictly separate the point $(h(t), -p(t))$ and the closed convex set $\operatorname{co} \partial V(t, \bar{x}(t))$. In other words, there exist $\alpha \in \mathbb{R}$, $\beta \in \mathbb{R}^n$ and $\gamma > 0$ such that

$$\alpha h(t) - p(t) \cdot \beta - \gamma > \max\{\alpha\tau + \xi \cdot \beta : (\tau, \xi) \in \mathrm{co}\, \partial V(t, \bar{x}(t))\}$$
$$= D^0 V((t, \bar{x}(t)); (\alpha, \beta)).$$

Again we have used here the fact that $D^0 V$ is the support function of $\mathrm{co}\,\partial V$.

However $D^0 V$ is upper semi-continuous with respect to all its arguments. It follows that, for some $\epsilon_1 > 0$,

$$\alpha h(t) - p(t) \cdot \beta - \gamma/2 > D^0 V((s, y); (\alpha, \beta))$$

for all points $(s, y) \in ((t, \bar{x}(t)) + \delta_1 \mathbb{B}) \cap ([S, T] \times \mathbb{R}^n)$ We conclude that

$$\alpha h(t) - p(t) \cdot \beta - \gamma/2 > \sup_{(\tau,\xi) \in G_{\epsilon_1}(t)} \alpha\tau + \xi \cdot \beta.$$

But then

$$(h(t), -p(t)) \notin \overline{\mathrm{co}}\, G_{\epsilon_1}(t),$$

contradicting (13.5.18). It follows that (13.5.19) is true. □

13.6 State Constrained Problems

This section provides characterizations of the value function in terms of solutions to the Hamilton Jacobi equation, when the formulation of the underlying dynamic optimization problem is broadened to include a pathwise state constraint. We also show the earlier established relations between costate trajectories and gradients of the value function continue to apply in this wider context. Consider then:

$$(SC) \begin{cases} \text{Minimize } g(x(T)) \\ \text{over arcs } x \in W^{1,1}([S, T]; \mathbb{R}^n) \text{ satisfying} \\ \dot{x}(t) \in F(t, x(t)) \quad \text{a.e.,} \\ x(t) \in A \quad \text{for all } t \in [S, T], \\ x(S) = x_0, \end{cases}$$

with data an interval $[S, T] \subset \mathbb{R}$, a function $g : \mathbb{R}^n \to \mathbb{R} \cup \{+\infty\}$, a multifunction $F : [S, T] \times \mathbb{R}^n \rightsquigarrow \mathbb{R}^n$, a closed set $A \subset \mathbb{R}^n$ (the 'state constraint set') and a point $x_0 \in \mathbb{R}^n$.

Write $(SC_{t,x})$ for problem (SC) when the initial data (t, x) replaces (S, x_0). Denote by $V : [S, T] \times A \to \mathbb{R} \cup \{+\infty\}$ the value function for (SC). That is, for each $(t, x) \in [S, T] \times A$, $V(t, x)$ is the infimum cost of $(SC_{t,x})$. The cost of any state trajectory y violating the state constraint, $y(s) \in A$ for all $s \in [t, T]$, is

13.6 State Constrained Problems

taken to be $+\infty$. Notice that, consistent with this interpretation, V takes value $+\infty$ at points (t, x) such that $x \notin A$.

It is to be expected that the value function V for (S) will be the unique solution (appropriately defined) to the Hamilton Jacobi equation

$$V_t + \min_{v \in F(t,x)} V_x \cdot v = 0 \quad \text{for } (t, x) \in (S, T) \times A,$$

accompanied by a boundary condition on $\{T\} \times A$ and also a boundary condition on $(S, T) \times \partial A$.

Under the hypotheses here imposed, it turns out that the appropriate boundary condition on $(S, T) \times \partial A$ is a boundary inequality

$$V_t + \min_{v \in F(t,x)} V_x \cdot v \leq 0 \quad \text{for } (t, x) \in (S, T) \times \partial A,$$

suitably interpreted.

The following theorems tell us that a characterization of the value function for state constrained problems along the above lines is possible, when certain compatibility hypotheses are satisfied, concerning the interaction of the state constraint set A and the velocity set F. These compatibility hypotheses, which allow alternative, 'inward-pointing' or 'outward-pointing' versions, require that there exist velocities driving the state strictly away from the boundary of the state constraint set.

Earlier sections, in which state constraints are not present, have covered problems in which the velocity set $F(t, x)$ is possibly discontinuous w.r.t. t. Specifically, we have hypothesized '$t \to F(t, x)$ has left and right limits and is almost everywhere continuous'. Passing to problems with state constraints, we shall now require that $F(t, x)$, as a function of t, has 'bounded variation' (BV); the (BV) hypothesis is a strengthening of the earlier t-dependence hypothesis that, nonetheless, allows t discontinuity.

(BV): For each $R_0 > 0$, $F(., x)$ has bounded variation uniformly over $x \in R_0 \mathbb{B}$, in the following sense: there exists a bounded variation function $\eta : [S, T] \to \mathbb{R}$ such that, for every $[s, t] \subset [S, T]$ and $x \in R_0 \mathbb{B}$,

$$d_H(F(s, x), F(t, x)) \leq \eta(t) - \eta(s) .$$

As before, the left and right limit sets of $F(., x)$, namely $F(t^+, x)$ and $F(t^-, x)$, have the following meaning: for all $s \in [S, T)$ and $t \in (S, T]$, and for all $x \in \mathbb{R}^n$

$$F(s^+, x) := \lim_{s' \downarrow s, x' \to x} F(s', x') \quad \text{and} \quad F(t^-, x) := \lim_{t' \uparrow t, x' \to x} F(t', x') .$$

Theorem 13.6.1 (Characterization of Value Functions for State Constrained Problems (I): Outward-Pointing Condition) *Consider* (SC_{S,x_0}). *Assume (H1), (H2)(i), (H3), (H4)(i) and (H5) of Theorem 13.3.7 and (BV). Suppose in addition that*

$(CQ)_{outward}$: *for each $s \in [S, T)$, $t \in (S, T]$ and $x \in \partial A$,*

$$F(t^-, x) \cap \left(-\operatorname{int} T_A(x)\right) \neq \emptyset \quad \text{and} \quad F(s^+, x) \cap \left(-\operatorname{int} T_A(x)\right) \neq \emptyset.$$

Take a function $V : [S, T] \times \mathbb{R}^n \to \mathbb{R} \cup \{+\infty\}$. Then assertions (a)–(c) below are equivalent:

(a) *V is the value function for (SC_{S,x_0}),*
(b) *V is lower semi-continuous on $[S, T] \times \mathbb{R}^n$, $V(t, x) = +\infty$ if $x \notin A$, and*

   (i) *for all $(t, x) \in ([S, T) \times A) \cap \operatorname{dom} V$*

$$\inf_{v \in F(t^+, x)} D_\uparrow V((t, x); (1, v)) \leq 0,$$

   (ii) *for all $(t, x) \in ((S, T] \times \operatorname{int} A) \cap \operatorname{dom} V$*

$$\sup_{v \in F(t^-, x)} D_\uparrow V((t, x); (-1, -v)) \leq 0,$$

   (iii) *for all $x \in A$*

$$\liminf_{\{(t', x') \to (T, x) : t' < T, x' \in \operatorname{int} A\}} V(t', x') = V(T, x) = g(x).$$

(c) *V is lower semi-continuous on $[S, T] \times \mathbb{R}^n$, $V(t, x) = +\infty$ if $x \notin A$, and*

   (i) *for all $(t, x) \in ((S, T) \times A) \cap \operatorname{dom} V$, $(\xi^0, \xi^1) \in \partial^P V(t, x)$*

$$\xi^0 + \inf_{v \in F(t^+, x)} \xi^1 \cdot v \leq 0,$$

   (ii) *for all $(t, x) \in ((S, T) \times \operatorname{int} A) \cap \operatorname{dom} V$, $(\xi^0, \xi^1) \in \partial^P V(t, x)$*

$$\xi^0 + \inf_{v \in F(t^-, x)} \xi^1 \cdot v \geq 0,$$

   (iii) *for all $x \in A$,*

$$\liminf_{\{(t', x') \to (S, x) : t' > S\}} V(t', x') = V(S, x)$$

   and

$$\liminf_{\{(t', x') \to (T, x) : t' < T,\ x' \in \operatorname{int} A\}} V(t', x') = V(T, x) = g(x).$$

Theorem 13.6.2 *(Characterization of Value Functions for State Constrained Problems (II): Inward-Pointing Condition)* Consider (SC_{S,x_0}). Assume (H1), (H2)(i), (H3), (H4)(i) and (H5) of Theorem 13.3.7 and (BV). Suppose in addition g is continuous on A and
$(CQ)_{inward}$: for each $s \in [S, T)$, $t \in (S, T]$ and $x \in \partial A$,

$$F(t^-, x) \cap \text{int } T_A(x) \neq \emptyset \quad \text{and} \quad F(s^+, x) \cap \text{int } T_A(x) \neq \emptyset.$$

Take a function $V : [S, T] \times \mathbb{R}^n \to \mathbb{R} \cup \{+\infty\}$. Then the assertions (a)–(c) of Theorem 13.6.1 remain equivalent.

Remarks

1. Theorems 13.6.1 and 13.6.2 differ principally according to whether an 'inward pointing' and 'outward pointing' constraint qualification is invoked. Note however that Theorem 13.6.2 is more restrictive, to the extent that it requires 'g is continuous on A', a hypothesis that automatically excludes problems with endpoint constraints.
2. It can been shown that, under the hypotheses of Theorem 13.6.2, including the 'inward pointing condition', the value function is continuous on $[S, T] \times A$. Notice however that both of the characterizations, (b) and (c), are in terms of lower semi-continuous functions. Thus the theorem not only identifies the value function as the unique generalized solution to the Hamilton-Jacobi equation, but gives the extra information that there are no 'hidden' generalized solutions in the larger class of lower semi-continuous functions.

13.7 Proofs of Theorems 13.6.1 and 13.6.2

We shall make repeatedly use of the 'L^∞ distance estimate theorem' Theorem 6.6.2 for state constrained differential inclusions. We observe that among the hypotheses invoked in Theorems 13.6.1 and 13.6.2 we assume that the multifunction F takes convex values and satisfies condition (H4)(i) (of Theorem 13.3.7); in consequence, the inward pointing condition (IPC)' of Theorem 6.6.2 can now be stated in the simpler form:

(CQ)': for each $s \in [S, T)$, $t \in (S, T]$ and $x \in R\mathbb{B} \cap \partial A$,

$$F(t^-, x) \cap \text{int } T_A(x) \neq \emptyset \quad \text{and} \quad F(s^+, x) \cap \text{int } T_A(x) \neq \emptyset.$$

The following proposition assembles some useful regularity properties of the value function. Notice in particular that the value function inherits the Lipschitz regularity of the final cost function, a property that follows from the preceding L^∞ distance estimate.

Proposition 13.7.1 *Let V be the value function for (SC). Assume that the following hypotheses are satisfied*

(i): $F : [S, T] \times \mathbb{R}^n \rightsquigarrow \mathbb{R}^n$ *takes closed non-empty values, F is $\mathcal{L} \times \mathcal{B}^n$-measurable, the graph of $F(t, .)$ is closed for each $t \in [S, T]$,*
(ii) there exists $c \in L^1(S, T)$ such that

$$F(t, x) \subset c(t)(1 + |x|) \mathbb{B} \quad \text{for all } x \in \mathbb{R}^n \quad \text{and for a.e. } t \in [S, T],$$

(iii) $g : \mathbb{R}^n \to \mathbb{R} \cup \{+\infty\}$ *is lower semi-continuous.*

(a) Then $V(t, x) > -\infty$ for all $(t, x) \in [S, T] \times \mathbb{R}^n$,
(b) If in addition F takes convex values, then V is lower semi-continuous and $V(t, x) > -\infty$ for all $(t, x) \in [S, T] \times \mathbb{R}^n$,
(c) If, in addition, the hypotheses (H1), (H2), (H3)(i) of Theorem 13.3.7 together with (BV) and (CQ)' are satisfied and g is locally Lipschitz continuous on A (resp. continuous on A), then V is locally Lipschitz continuous on $[S, T] \times A$ (resp. continuous on $[S, T] \times A$).

Proof

(a) and (b) The proof is exactly as in the proof of Proposition 13.3.6. We highlight only the fact that, even though the extensions to $[S, T]$ of the minimizers y_i, $i = 1, 2, \ldots$ might not to be admissible, the limiting F trajectory y is admissible.

(c) We show here that V is locally Lipschitz continuous on $[S, T] \times A$ when g is locally Lipschitz continuous on A, since the case continuous on $[S, T] \times A$ can be similarly proved (see Proposition 13.3.6). Observe that all the hypotheses of the distance estimate theorem (Theorem 6.6.2) are satisfied. Fix $r_0 > 0$. Notice that for any $\tau \in [S, T]$ and any F trajectory x on $[\tau, T]$ such that $x(\tau) \in r_0 \mathbb{B}$ we have: $x(t) \in R_0 \mathbb{B}$ for all $t \in [t, T]$, where $R_0 := \exp\{\int_S^T c(t) \, dt\}(r_0 + 1)$. Fix two points (t_1, x_1), $(t_2, x_2) \in [S, T] \times A$ such that $x_1, x_2 \in A \cap r_0 \mathbb{B}$. It is not restrictive to assume that $t_1 \leq t_2$. Fix any $\varepsilon > 0$. Then there exists an admissible F trajectory x_ε such that $x_\varepsilon(t_1) = x_1$ and

$$V(t_1, x_1) \geq g(x_\varepsilon(T)) - \varepsilon. \tag{13.7.1}$$

Consider the point $(t_2, x_\varepsilon(t_2))$. From the Filippov's existence theorem (Theorem 6.2.3) there exists an F trajectory (not necessarily admissible) $\hat{x}$ on $[t_2, T]$ such that $\hat{x}(t_2) = x_2$ and

$$\|\hat{x} - x_\varepsilon\|_{L^\infty(t_2, T)} \leq e^{\int_S^T k_F(s) \, ds} |x_2 - x_\varepsilon(t_2)|. \tag{13.7.2}$$

Then, by Theorem 6.6.2, there exists an admissible F trajectory x such that $x(t_2) = x_2$ satisfying the estimate

$$\|x - \hat{x}\|_{L^\infty(t_2, T)} \leq \widetilde{K} \max\{d_A(\hat{x}(t)) : t \in [t_2, T]\}. \tag{13.7.3}$$

13.7 Proofs of Theorems 13.6.1 and 13.6.2

But $\max\{d_A(\hat{x}(t)) : t \in [t_2, T]\} \leq \|\hat{x} - x_\varepsilon\|_{L^\infty(t_2, T)}$ and $|x_2 - x_\varepsilon(t_2)| \leq |x_1 - x_2| + c_0|t_1 - t_2|$, where $c_0 > 0$ is a constant provided by (H2)(ii). From (13.7.2) and (13.7.3), it follows that

$$\|x - x_\varepsilon\|_{L^\infty(t_2, T)} \leq \|\hat{x} - x_\varepsilon\|_{L^\infty(t_2, T)} + \|x - \hat{x}\|_{L^\infty(t_2, T)}$$
$$\leq (\tilde{K} + 1) \|\hat{x} - x_\varepsilon\|_{L^\infty(t_2, T)}$$
$$\leq \sqrt{2}c_0(\tilde{K} + 1)e^{\int_S^T k_F(s)\,ds} |z_1 - z_2|.$$

Write k_g the Lipschitz constant of g on $R_0\mathbb{B}$. Therefore we have

$$V(t_2, x_2) - V(t_1, x_1) \leq g(x(T)) - g(x_\varepsilon(T)) + \varepsilon$$
$$\leq k_g \|x - x_\varepsilon\|_{L^\infty(t_2, T)} + \varepsilon$$
$$\leq K_0 |(t_1, x_1) - (t_2, x_2)| + \varepsilon,$$

where the constant $K_0 := k_g \sqrt{2}c_0(\tilde{K} + 1)e^{\int_S^T k_F(s)\,ds}$ is a constant which does not depend on ε, (t_1, x_1) or (t_2, x_2). The reverse inequality

$$V(t_1, x_1) - V(t_2, x_2) \leq K_0 |(t_1, x_1) - (t_2, x_2)| + \varepsilon$$

is easily obtained by means of similar arguments. The assertion (c) of the proposition now follows, by taking the limit as ε tends to zero.

□

Proof of Theorem 13.6.1

'(a) $\Rightarrow$ (b)'. Let V be the value function. Using the facts that whenever $\bar{x}$ is a minimizer for $(P_{t,x})$, the function $s \to V(s, \bar{x}(s))$ is constant on $[t, T]$ and, if x is an admissible F trajectory satisfying the initial condition $x(t) = x$, then $s \to V(s, x(s))$ is non-decreasing on $[t, T]$, we can reproduce the analysis in the state constraints-free case to confirm (b)(i) and (b)(ii). (Observe that the analysis employed in the proof of Theorem 13.3.7 can also be used to prove (b)(ii), since in this case $(t, x) \in ((S, T] \times \text{int } A)$ and, therefore, there exists $\bar{\delta} > 0$ such that any F trajectory emanating from x is admissible on $[t, t + \bar{\delta}]$.)

It remains to prove (b)(iii). Take any $x \in A$. Since V is lower semi-continuous, we have that

$$\liminf_{\{(t', x') \to (T, x) : t' < T, x' \in \text{int } A\}} V(t', x') \geq V(T, x).$$

To complete the verification of condition (b)(iii), consider the differential inclusion

$$\begin{cases} \dot{y}(s) \in \tilde{F}(s, y(s)) \text{ for a.e. } s \in [0, T - S] \\ y(0) = x, \end{cases}$$

where $\widetilde{F}(s, y) := -F(T - s, y)$. There exists an admissible $\widetilde{F}$ trajectory $\tilde{z}$ on $[0, T - S]$ such that $\tilde{z}(0) = x$ and

$$\tilde{z}(s) \in \text{int } A, \quad \text{for all } s \in (0, T - S].$$

(To justify this assertion we note that, by Filippov's existence theorem, there exists an $\widetilde{F}$ trajectory $\hat{z}$ on $[0, T - S]$ such that $\hat{z}(0) = x$. The existence of the $\widetilde{F}$ trajectory $\tilde{z}$ with the stated properties is then established by applying Theorem 6.6.2 with reference trajectory $\hat{z}$.) Notice next that $y := \tilde{z}(T - \cdot)$ is an admissible F trajectory on $[S, T]$ such that $y(T) = x$ and

$$y(t) \in \text{int } A, \quad \text{for all } t \in [S, T).$$

By the principle of optimality, however, $V(t, x(t)) \leq V(T, x(T)) = V(t, x)$ for all $t \in [S, T]$. We deduce from the preceding relations that

$$V(T, x) \leq \liminf_{\{(t',x') \to (T,x): t' < T, x' \in \text{int } A\}} V(t', x') \leq \liminf_{t \uparrow T} V(t, x(t)) \leq V(T, x) = g(x).$$

This is condition (b)(iii).

'(b) $\Rightarrow$ (c)'. The relations '(b)(i) $\Rightarrow$ (c)(i)' and '(b)(ii) $\Rightarrow$ (c)(ii)' have already been established in the proof of Theorem 13.3.7, the analysis in which is the same whether a state constraint is present or not. To complete the proof we have merely to note that the first relation in (c)(iii) is an immediate consequence of (b)(i).

'(c) $\Rightarrow$ (a)'. In view of our convention that F trajectories that violate the state constraint have cost $+\infty$, it is clear that the validity of this relation will be a consequence of completing the following two steps, in which $V : [S, T] \times \mathbb{R}^n \to \mathbb{R} \cup \{+\infty\}$ in an arbitrary lower semi-continuous function satisfying condition (c) and $(\bar{t}, \bar{x})$ is an arbitrary point on $[S, T] \times \mathbb{R}^n$.

Step 1 We show

$$V(\bar{t}, \bar{x}) \geq g(\bar{x}(T))$$

for some admissible F trajectory $\bar{x}$ on $[\bar{t}, T]$ such that $\bar{x}(\bar{t}) = \bar{x}$.

Step 2 We show

$$V(\bar{t}, \bar{x}) \leq g(x(T)),$$

in which x is an arbitrary admissible F trajectory on $[\bar{t}, T]$ such that $x(\bar{t}) = \bar{x}$.

Step 1 has already been accomplished. Indeed scrutiny of the proof of Theorem 13.3.7, reveals that the analysis leading to the desired conclusions is the same, whether or not state constraints are present. We attend then only to Step 2.

13.7 Proofs of Theorems 13.6.1 and 13.6.2

We can also assume that $g(x(T)) < +\infty$ since, otherwise, the inequality above is automatically satisfied. From the second relation in (c)(iii) we have

$$\liminf_{\{(t',x') \to (T,x(T)): t' < T, x' \in \text{int } A\}} V(t', x') = V(T, x(T)) = g(x(T)).$$

There exists, in consequence, a sequence $\{(t_i, \xi_i)\}$ in $(\bar{t}, T) \times \text{int } A$ such that

$$V(t_i, \xi_i) \to g(x(T)), \quad t_i \uparrow T \quad \text{and} \quad \xi_i \to x(T), \quad \text{as} \quad i \to +\infty. \tag{13.7.4}$$

Notice that, for each i, the arc $s \to x(T - s)$ is an admissible $\widetilde{F}$ trajectory on $[T - t_i, T - \bar{t}]$, where $\widetilde{F}(s, y) := -F(T - s, y)$. Applying the Filippov existence theorem to the differential inclusion

$$\begin{cases} \dot{y}(s) \in \widetilde{F}(s, y(s)) \text{ for a.e. } s \in [T - t_i, T - \bar{t}] \\ y(T - t_i) = \xi_i, \end{cases}$$

we obtain a sequence of $\widetilde{F}$ trajectories $y_i : [\bar{t}, t_i] \to \mathbb{R}^n$, $i = 1, 2, \ldots$, such that, for all i sufficiently large, $y_i(t_i) = \xi_i$ and

$$\|y_i - x(T - \cdot)\|_{L^\infty(\bar{T} - t_i, T - \bar{t})} \leq \exp\{\int_S^T k_F(t) \, dt\} |x(t_i) - \xi_i|. \tag{13.7.5}$$

Since x is an admissible F trajectory, it follows that

$$\max\{d_A(y_i(t)) \mid t \in [\bar{t}, t_i]\} \leq \exp\{\int_S^T k_F(t) \, dt\} |x(t_i) - \xi_i|. \tag{13.7.6}$$

Write $\rho_i := \exp\{\int_S^T k_F(t) \, dt\} |x(t_i) - \xi_i|$ and notice that $\rho_i \downarrow 0$ as $i \to \infty$. Application of Theorem 6.6.2 to the differential inclusion $\dot{y} \in \widetilde{F}$ yields a sequence of admissible F trajectories $z_i : [\bar{t}, t_i] \to \mathbb{R}^n$, $i = 1, 2, \ldots$, such that, for each i, $z_i(T - t_i) = \xi_i$

$$z_i(s) \in \text{int } A, \quad \text{for all } s \in [T - t_i, T - \bar{t}]$$

and

$$\|y_i - z_i\|_{L^\infty(\bar{t}, t_i)} \leq \widetilde{K} \rho_i,$$

in which $\widetilde{K}$ is some number that does not depend on i. This inequality combines with (13.7.6) to give

$$\|z_i - x(T - \cdot)\|_{L^\infty(T - t_i, T - \bar{t})} \leq (\widetilde{K} + 1) \rho_i.$$

With the help of Filippov's existence theorem (in which we take as reference trajectory $s \to x(T-s)$ restricted to $[0, T-t_i]$), we can extend each z_i as an $\tilde{F}$ trajectory to all of $[0, T-\bar{t}]$. (The extension is also written 'z_i'.) It follows from the Filippov existence theorem that the extended $\tilde{F}$ trajectories satisfy:

$$\|z_i - x(T-\cdot)\|_{L^\infty(0, T-\bar{t})} \le (\tilde{K}+1)\rho_i \,. \tag{13.7.7}$$

Furthermore, for some sequence $\delta_i \downarrow 0$,

$$z_i(s) + \delta_i \mathbb{B} \subset \text{int } A \quad \text{for all } s \in [T-t_i, T-\bar{t}], \tag{13.7.8}$$

and

$$V(t_i, z_i(T-t_i)) = V(t_i, \xi_i) \to g(x(T)), \quad \text{as } i \to \infty \,.$$

Fix i. We now employ a similar analysis to that used in Step 2 of the proof of the relation '(c) $\Rightarrow$ (a)' in Theorem 13.3.7, based on the strong invariance theorem (Theorem 13.2.4): we consider the constrained differential inclusion

$$\begin{cases} (\dot{z}, \dot{a}) \in \tilde{\Gamma}(s, (z, a)) & \text{a.e } s \in [T-t_i, T-\bar{t}], \\ z(T-t_i) = \xi_i, \; a(T-t_i) = V(t_i, \xi_i), \\ (z(s), a(s)) \in \tilde{P}(s) & \text{for all } s \in [T-t_i, T-\bar{t}], \end{cases}$$

in which $\tilde{\Gamma} : [T-t_i, T-\bar{t}] \times \mathbb{R}^{n+1} \rightsquigarrow \mathbb{R}^{n+2}$ and $\tilde{P} : [T-t_i, T-\bar{T}] \rightsquigarrow \mathbb{R}^{n+1}$ are

$$\tilde{\Gamma}(s, (z, a)) := -F(T-s, z) \times \{0\}, \quad \text{for all } (s, (z, a)) \in [T-t_i, T-\bar{T}] \times \mathbb{R}^{n+1}$$

$$\tilde{P}(s) := \{(z, \alpha) : V(T-s, z) \le \alpha\}, \quad \text{for all } s \in [T-t_i, T-\bar{T}],$$

respectively. The multifunctions $\tilde{\Gamma}$ and $\tilde{P}$ both have the required closure and regularity for application of the strong invariance theorem (Theorem 13.2.4). Take as reference trajectory $(z_i, a_i \equiv V(t_i, \xi_i))$. Since $T-t_i > 0$ and $(z_i(T-t_i), a_i(T-t_i)) = (\xi_i, V(t_i, \xi_i)) \in \tilde{P}(T-t_i)$ the boundary condition, in which the initial time $\bar{S}$ is interpreted as $\bar{S} = T-t_i$, is also valid. Concerning the 'inward pointing' hypothesis, we deduce from (c)(ii) and (13.7.8), employing the same analysis as used in the proof of Theorem 13.3.7 (step 2 of '(c) $\Rightarrow$ (a)'), that condition (vi) of Theorem 13.2.4 is certainly satisfied on a δ_i-tube around the $\tilde{\Gamma}$ trajectory $(z_i, a_i \equiv V(t_i, \xi_i))$. So, from Theorem 13.3.7,

$$(z_i(s), V(t_i, \xi_i)) \in \tilde{P}(s) \quad \text{for all } s \in [T-t_i, T-\bar{t}].$$

Setting $s = T-\bar{t}$ yields

$$V(\bar{t}, z_i(T-\bar{t})) \le V(T-t_i, \xi_i), \quad \text{for each } i \,. \tag{13.7.9}$$

From (13.7.4), (13.7.7), (13.7.9) and the lower semi-continuous of V, we finally arrive

$$V(\bar{t}, x(\bar{t})) \leq \liminf_{i \to \infty} V(\bar{t}, z_i(T - \bar{t})) \leq \lim_{i \to \infty} V(t_i, \xi_i) = g(x(T)).$$

This concludes Step 2. $\square$

Proof of Theorem 13.6.2 The proof is a modification of that of Theorem 13.6.1, to take account of the fact that we have replaced the outward pointing constraint qualification by an inward pointing one. The analysis will require that g is continuous on A.

The analysis required to show (a) $\Rightarrow$ (b) $\Rightarrow$ (c) for Theorem 13.6.1 is the same as that employed in the proof of Theorem 13.6.2, but simpler because the boundary condition

$$\liminf_{\{(t', x') \to (T, x) : t' < T, x' \in \text{int } A\}} V(t', x') = V(T, x),$$

is an immediate consequence of the continuity of the value function, a property established in Proposition 13.7.1. (The outward pointing constraint qualification was used in this part of the proof of Theorem 13.6.1 only to show that the value function satisfied this condition and is not required here.)

It remains then to show '(c) $\Rightarrow$ (a)'. As before, this involves the two steps: for any given, $(\bar{t}, \bar{x}) \in [S, T] \times \mathbb{R}^n$.

Step 1 We show

$$V(\bar{t}, \bar{x}) \geq g(\bar{x}(T))$$

for some admissible F trajectory $\bar{x}$ on $[\bar{t}, T]$ such that $\bar{x}(\bar{t}) = \bar{x}$.

Step 2 We show

$$V(\bar{t}, \bar{x}) \leq g(x(T)),$$

in which x is an arbitrary admissible F trajectory on $[\bar{t}, T]$ such that $x(\bar{t}) = \bar{x}$.

Step 1 (which does not involve the constraint qualification) is exactly the same as in the proof of Theorem 13.6.1. It remains only to attend to Step 2. Take then any admissible F trajectory x on $[\bar{t}, T]$ emanating from $\bar{x}$. Invoking the inward point constraint qualification and applying Theorem 6.6.2, we can construct a sequence of admissible F trajectories $\{x_i\}$ on $[\bar{t}, T]$ such that $x_i(\bar{t}) = \bar{x}$ and

$$||x_i - x||_{L^\infty(\bar{t}, T)} \to 0 \quad \text{as } i \to \infty, \tag{13.7.10}$$

$$x_i(t) \in \text{int } A, \quad \text{for all } t \in (\bar{t}, T] \quad \text{and} \quad i = 1, 2, \ldots.$$

Take a sequence $t_i \downarrow \bar{t}$ and note that, for each i sufficient large, there exists $\delta_i > 0$ such that

$$x_i(t) + \delta_i \mathbb{B} \subset \text{int } A, \quad \text{for all } t \in [t_i, T]. \tag{13.7.11}$$

Take the multivalued function $\tilde{F}(s, y) := -F(T - s, y)$ (already used in step 2 of the proof of Theorem 13.6.1), and consider the $\tilde{F}$ trajectories z_i on $[0, T - \bar{t}]$, defined as $z_i(s) := x_i(T - s)$. From (13.7.11) we deduce that

$$z_i(s) + \delta_i \mathbb{B} \subset \text{int } A, \quad \text{for all } s \in [0, T - t_i]. \tag{13.7.12}$$

Fix i and consider the constrained differential inclusion

$$\begin{cases} (\dot{z}, \dot{a}) \in \tilde{\Gamma}(s, (z, a)) & \text{a.e } s \in [0, T - t_i], \\ z(0) = x_i(T), \ a(0) = V(T, x_i(T)), \\ (z(s), a(s)) \in \tilde{P}(s) & \text{for all } s \in [0, T - t_i], \end{cases}$$

where $\tilde{\Gamma} : [0, T - \bar{t}] \times \mathbb{R}^{n+1} \rightsquigarrow \mathbb{R}^{n+2}$ and $\tilde{P} : [0, T - \bar{t}] \rightsquigarrow \mathbb{R}^{n+1}$ are defined as follows

$$\tilde{\Gamma}(s, (z, a)) := \tilde{F}(s, z) \times \{0\}, \quad \text{for all } (s, (z, a)) \in [0, T - \bar{t}] \times \mathbb{R}^{n+1}$$

$$\tilde{P}(s) := \{(x, \alpha) : V(T - s, x) \le \alpha\}, \quad \text{for all } s \in [0, T - \bar{t}].$$

Observe that condition (c)(iii) guarantees that $(z(0), a(0)) = (x_i(T), V(T, x_i(T))) \in \limsup_{s \downarrow 0} \tilde{P}(s)$. Note also that $\tilde{\Gamma}$ and $\tilde{P}$ satisfy the hypotheses of the strong invariance theorem (Theorem 13.2.4). Moreover, by means of a similar analysis to that in Step 2 of the proof of Theorem 13.3.7, in which we show '(c) $\Rightarrow$ (a)', we can deduce from (c)(ii) and (13.7.12) that the 'inward pointing' hypothesis (vi) of Theorem 13.2.4 is satisfied on a δ_i-tube around the $\tilde{\Gamma}$ trajectory $(z_i, a \equiv V(t_i, \xi_i))$. The strong invariance theorem is therefore applicable. It follows that

$$(z_i(s), a(s) \equiv V(t_i, \xi_i)) \in \tilde{P}(s) \quad \text{for all } s \in [0, T - t_i].$$

As a consequence, for each i, we have:

$$V(T, x_i(T)) \ge V(t_i, z_i(T - t_i)) \ (= V(t_i, x_i(t_i))). \tag{13.7.13}$$

Noting (13.7.10), (13.7.9) and (c)(iii), and using the facts that V is lower semi-continuous and that g is continuous on A, we conclude:

$$V(\bar{t}, x(\bar{t})) \le \liminf_{i \to \infty} V(t_i, x_i(t_i)) \le \lim_{i \to \infty} V(T, x_i(T))$$

$$= \lim_{i \to \infty} g(x_i(T)) = g(x(T)).$$

This completes Step 2. $\square$

13.8 Costate Trajectories and Gradients of the Value Functions for State-Constrained Problems

In Sect. 13.5, we showed that the costate trajectory p, appearing in the generalized Euler Lagrange condition, with reference to a minimizing state trajectory x, could be chosen to satisfy the 'sensitivity relation'

$$(H(t, \bar{x}(t), p(t)), -p(t)) \in \text{co}\, \partial V(t, \bar{x}(t)) \quad \text{a.e. } t \in [S, T].$$

Is a version of this relation valid, when a state constraint '$x(t) \in A$ for all $t \in [S, T]$' is added to the underlying problem formulation? The answer is qualified 'yes': it is indeed valid, but only if we interpret 'costate arc' appropriately for this context. We provide motivation behind the correct interpretation. For ease of exposition, we assume that the dynamic constraint takes the form of a controlled differential equation.

$$\begin{cases} \dot{x}(t) = f(t, x(t), u(t)) \\ u(t) \in U \end{cases}$$

and the state constraint has the functional representation

$$A := \{x\, :\, h(x) \leq 0\},$$

for some C^1 function $h : \mathbb{R}^n \to \mathbb{R}$.

Recall the discussion at the beginning of Chap. 10, concerning the manner in which necessary conditions of optimality should be modified to accommodate a state constraint $x(t) \in A$. We pointed out that if the necessary conditions for the state constrained problem correspond to necessary conditions for a related problem in which the state constrained is removed and, instead, a term

$$\int_S^T h(\bar{x}(t)) dv(t)$$

(in which the integrator dv is interpreted as the Lagrange multiplier for the state constraint) is added to the cost, then the Weierstrass and transversality conditions are satisfied for a costate trajectory q of bounded variation, that satisfies the costate equation

$$-dq(t) = q(t) f_x(s, \bar{x}(t), u(t)) dt - h_x(\bar{x}(s)) dv(t).$$

The set of optimality conditions can be expressed in terms of modified costate trajectory

$$p(t) := q(t) - \int h_x dv,$$

that satisfies the modified costate equation

$$-\dot{p}(t) = (p(t) + \int_{[S,t]} h_x(\bar{x}(s))dv(t))f_x(s, \bar{x}(t), u(t))dt.$$

Of course, the distinction between q and p disappears if there are no state constraints because, then, $v \equiv 0$.

It can be argued that the costate trajectory q is more fundamental than the modified costate trajectory p and introduction of the modified costate trajectory is merely a cosmetic step to simplify the statement of the necessary conditions. This is because q, and not p, can be interpreted as the sensitivity of the cost to perturbations of the dynamic constraint, a property that was identified Clarke and Loewen [72]. Not surprisingly then, the sensitivity relations for state constrained problems, providing costate interpretations of the gradients of the value function involve the original costate trajectory q and not the modified costate trajectory p.

Theorem 13.8.1 (Sensitivity Relations in the Presence of State Constraints) *Consider problem (SC) (of Sect. 13.6). Denote by $V : [S, T] \times \mathbb{R}^n \to \mathbb{R} \cup \{+\infty\}$ the value function. Assume that*

(a): F takes values in the space of closed, non-empty sets. $F(., x)$ is Lebesgue measurable for each $x \in \mathbb{R}^n$,
(b): there exists $c > 0$ such that

$$F(t, x) \subset c(1 + |x|)\mathbb{B} \quad \text{for all } (t, x) \in [S, T] \times \mathbb{R}^n,$$

(c): there exists $k_F \in L^1(S, T)$ such that

$$F(t, x) \subset F(t, x') + k_F(t)|x - x'|\mathbb{B}, \quad \text{for all } x, x' \in \mathbb{R}^n, \text{ a.e. } t \in [S, T],$$

(d): g is locally Lipschitz continuous,
(e): A is a nonempty closed set,

and also:

(BV): for each $R_0 > 0$, $F(., x)$ has bounded variation uniformly over $x \in R_0\mathbb{B}$, in the following sense: there exists a bounded variation function $\eta : [S, T] \to \mathbb{R}$ such that, for every $[s, t] \subset [S, T]$ and $x \in R_0\mathbb{B}$,

$$d_H(F(s, x), F(t, x)) \le \eta(t) - \eta(s),$$

(IPC): (Inward Pointing Condition) for each $t \in [S, T)$, $s \in (S, T]$ and $x \in \partial A$

13.8 Costate Trajectories and Gradients of the Value Functions for State... 721

$$\left(\liminf_{x' \xrightarrow{A} x} \operatorname{co} F(t^+, x')\right) \cap \operatorname{int} T_A(x) \neq \emptyset,$$

$$\left(\liminf_{x' \xrightarrow{A} x} \operatorname{co} F(s^-, x')\right) \cap \operatorname{int} T_A(x) \neq \emptyset.$$

Let $\bar{x}$ be an L^∞ local minimizer for (SC). Then there exist $p \in W^{1,1}([S, T]; \mathbb{R}^n)$, $\mu \in C^{\oplus}(S, T)$, a bounded Borel measurable function $\gamma : [S, T] \to \mathbb{R}^n$ and $\lambda \geq 0$, satisfying the following conditions, in which $q \in NBV([S, T]; \mathbb{R}^n)$ is the function

$$q(t) := \begin{cases} p(S) & \text{if } t = S \\ p(t) + \int_{[S,t]} \gamma(s) d\mu(s) & \text{if } t \in (S, T]. \end{cases}$$

(i): $\gamma(t) \in (\overline{\operatorname{co}} N_A(\bar{x}(t))) \cap \mathbb{B}$ μ − a.e. $t \in [S, T]$,
(ii): $\dot{p}(t) \in \operatorname{co}\{\xi : (\xi, q(t)) \in N_{\operatorname{Gr}\{F(t,\cdot)\}}(\bar{x}(t), \dot{\bar{x}}(t))\}$ a.e. $t \in [S, T]$,
(iii): $-q(T) \in \partial g(\bar{x}(T))$, $q(S) \in \partial(-V + \Psi_A)(\bar{x}(S))$.

in which $\Psi_A(x) := 0$ if $x \in A$, and takes value $+\infty$ if $x \notin A$;

(iv): $q(t) \cdot \dot{\bar{x}}(t) = \max_{v \in F(t, \bar{x}(t))} q(t) \cdot v$ a.e.$t \in [S, T]$.

Furthermore,

$$(H(t, \bar{x}(t), q(t)), -q(t)) \in \partial^0 V(t, \bar{x}(t)) \text{ a.e. } t \in (S, T], \tag{13.8.1}$$

in which, for $(t, x) \in [S, T] \times A$,

$$\partial^0 V(t, x) := \cap_{\epsilon > 0} \overline{\operatorname{co}} \cup_{\{(t', x') \in ((t,x) + \epsilon \mathbb{B}) \cap [S,T] \times \operatorname{int} A\}} \partial V(t', x').$$

Remarks

(a): The 'full' sensitivity relation (13.8.1) can be supplemented by the 'partial' sensitivity relation

$$-q(t) \in \partial_x^0 V(t, \bar{x}(t)) \text{ for a.e. } t \in (S, T], \tag{13.8.2}$$

in which, for $(t, x) \in [S, T] \times A$,

$$\partial_x^0 V(t, x) := \cap_{\epsilon > 0} \overline{\operatorname{co}} \cup_{\{x' \in (x + \epsilon \mathbb{B}) \cap \operatorname{int} A\}} \partial_x V(t, x'),$$

(13.8.1) and (13.8.2) are distinct conditions; see the discussion following the statement of Theorem 13.5.1.

(b): The two sensitivity relations (13.8.1) and (13.8.2) are more precise sensitivity relations than those expressed in terms of the Clarke subdifferential co $\partial V(t, \bar{x}(t))$ and partial Clarke subdifferential co $\partial_x V(t, \bar{x}(t))$, because the definitions of $\partial^0 V(t,\bar{x}(t))$ and $\partial^0_x V(t,\bar{x}(t))$ involve limit taking from within the interior of the state constraint set. Indeed

$$\partial^0_x V(t,x) = \operatorname{co} \partial_x V(t,x) \quad \text{and} \quad \partial^0 V(t,x) = \operatorname{co} \partial V(t,x)$$

for $(t,x) \in [S,T] \times \operatorname{int} A$ and

$$\partial^0_x V(t,x) \overset{\text{strict}}{\subset} \operatorname{co} \partial_x V(t,x) \quad \text{and} \quad \partial^0 V(t,x) \overset{\text{strict}}{\subset} \operatorname{co} \partial V(t,x)$$

for $(t,x) \in [S,T] \times \partial A$.

(c): The assertions of Theorem 13.8.1 remain valid (with possibly different p, μ and γ) if we replace the Euler Lagrange inclusion (ii) with the partially convexified Hamiltonian inclusion:

(ii)' $\quad \dot{p}(t) \in \operatorname{co}\{\xi : (-\xi, \dot{\bar{x}}(t)) \in \partial_{x,p} H(t, \bar{x}(t), q(t))\} \quad$ for a.e. $t \in [S,T]$.

Proof (Sketch) We employ the penalty function $\sigma^A_\epsilon : [S,T] \times \mathbb{R}^n \times \mathbb{R} \to \mathbb{R}$ defined by

$$\sigma^A_\epsilon(t, v, w) := \sup_{(\alpha,\beta) \in G^A_\epsilon(t)} (\alpha, \beta) \cdot (w, -(1+w)v),$$

in which $G^A_\epsilon : [S,T] \rightsquigarrow \mathbb{R}^{1+n}$ to be

$$G^A_\epsilon(t) := \{(\alpha, \beta) : \mathbb{R}^{1+n} : (\alpha, \beta) \in \operatorname{co} \partial V(s,y)$$

for some $(s,y) \in ((t,\bar{x}(t)) + \epsilon \mathbb{B}) \cap ((S,T) \times \mathbb{R}^n)$ s.t. $y \in \operatorname{int} A\}$

and $\epsilon > 0$ is a (small) parameter. (σ^A_ϵ coincides with the penalty function σ_ϵ appearing in the proof of Theorem 13.5.1 in the no state-constraints case, that is when $A = \mathbb{R}^n$.) Now consider the following dynamic optimization problem:

$$(AP) \begin{cases} \text{Minimize } g(x(T)) - V(S, x(S)) + \int_S^T \sigma^A_\epsilon(t, v(t), w(t)) dt \\ \text{over arcs } x \text{ satisfying} \\ \dot{x}(t) = \{(e + v(t)(1 + w(t)) : e \in F(t, x(t))\} \quad \text{a.e.} \\ \text{for some functions } v : [S,T] \to \epsilon \mathbb{B}, w : [S,T] \to \epsilon \mathbb{B}, \\ (x(S), v(S), w(S)) \in \mathbb{R}^n \times \{0\} \times \{0\} \\ x(t) \in A \quad \text{for all } t \in [S,T]. \end{cases}$$

(At this stage, we are intentionally vague about the function spaces from which the functions v and w must be chosen in the specification of the dynamic constraint.) Let us recall that, in the proof of the state constraint-free sensitivity relation Theorem 13.5.1, we showed that $\bar{x}$ (and perturbation functions $\bar{v} \equiv 0$, $\bar{w} \equiv 0$) was a minimizer for the 'auxiliary problem' (AP). By applying standard necessary conditions to (AP) and passage to the limit as $\epsilon \downarrow 0$, we show that the generalized Euler Lagrange conditions are valid for the original problem, but now augmented by the sensitivity relation. This basic idea is retained, when we take account of the state constraint, but the details of how exactly it is applied change. Once again, it can be shown that $(\bar{x}, \bar{v} \equiv 0, \bar{w} \equiv 0)$ is a minimizer for (AP). This is because the proof that $\bar{x}$ (combined with $\bar{v} \equiv 0$ and $\bar{w} \equiv 0$) is a minimizer for (AP) involves showing that, given an admissible state trajectory for (AP) (associated with some function pair (v, w)) we can find an admissible state trajectory (associated with the same function pair (v, w)) that is interior to A on $(S, T]$. The distance estimate of Theorem 6.6.1 that is used to establish the existence of this interior state trajectory is valid only under the inward pointing hypothesis and when restrictions are placed on the regularity of v and w. (The inward pointing hypothesis is also required to establish that V is locally Lipschitz continuous on A.) It is then possible to show that $\bar{x}$ satisfies the usual generalized Euler Lagrange conditions, but now supplemented by the stated sensitivity relation, by applying the known necessary conditions to (AP), with reference to the minimizer $(\bar{x}, \bar{v} \equiv 0, \bar{w} \equiv 0)$. Full details of the proof of a closely related 'sensitivity' theorem, in which the bounded variation hypothesis (BV) in theorem statement is replaced by a stronger 'absolutely continuous' hypothesis, appears in [35]; the substance of the proof remains the same under the (BV) hypothesis. □

13.9 Semiconcavity and the Value Function

In this section, we introduce an important property of the value function, called semiconcavity, and discuss its significance.

Definition 13.9.1 Take set $K \subset \mathbb{R}^k$ and $\bar{x} \in K$. We say that a function $\phi : \mathbb{R}^k \to \mathbb{R}$ is semiconcave near $\bar{x}$ relative to K, if there exist some $\epsilon > 0$ and a modulus of continuity $\theta : [0, \infty) \to [0, \infty)$ such that

$$\lambda \phi(x) + (1 - \lambda)\phi(x') - \phi(\lambda x + (1 - \lambda)x') \leq \lambda(1 - \lambda)|x - x'| \times \theta(|x - x'|),$$

for all $x, x' \in (\bar{x} + \epsilon \mathbb{B}) \cap K$ and $\lambda \in [0, 1]$.

The statement 'ϕ is semiconcave on K' is interpreted as 'ϕ is semiconcave near all points x in K, relative to K'. If $K = \mathbb{R}^k$, we omit the qualifier 'relative to K'. We say that ϕ is 'locally semiconcave' if it is semiconcave on K for every compact subset $K \subset \mathbb{R}^k$.

Recall the definition of 'ϕ is concave near $\bar{x}$' : for some $\epsilon > 0$,

$$\lambda \phi(x) + (1 - \lambda)\phi(x') - \phi(\lambda x + (1 - \lambda)x') \leq 0,$$

for all $x, x', \in \bar{x} + \epsilon \mathbb{B}$ and $\lambda \in [0, 1]$. Thus, semiconcavity is a weakened version of concavity (on a neighbourhood of a given base point), in which '0' on the right side of the defining inequality is replaced by a scaled, superlinear error term.

The following proposition identifies one situation in which a function is semi-concave near a point in its domain.

Proposition 13.9.2 *Take a point $\bar{x} \in \mathbb{R}^k$ and a function $\phi : \mathbb{R}^k \to \mathbb{R}$ that is continuously differentiable on a neighbourhood of $\bar{x}$. Then ϕ is semiconcave near $\bar{x}$.*

Proof Let $\epsilon > 0$ be such that ϕ is continuously differentiable on an open set containing $\bar{x} + \epsilon \mathbb{B}$. Take any $\lambda \in [0, 1]$ and $x, x' \in \bar{x} + \epsilon \mathbb{B}$. Then, by the mean value theorem, there exists $t, t' \in [0, 1]$ such that

$$\phi(\lambda x + (1-\lambda)x') = \phi(x) + \nabla\phi(x + t(1-\lambda)(x' - x))(1-\lambda)(x' - x)$$

and

$$\phi(\lambda x + (1-\lambda)x') = \phi(x') + \nabla\phi(x' + t'\lambda(x - x'))\lambda(x - x').$$

Multiplying across the first and second equalities by λ and $(1 - \lambda)$ respectively and addition yields

$$\phi(\lambda x + (1-\lambda)x') = \lambda\phi(x) + (1-\lambda)\phi(x')$$
$$+(\nabla\phi(x + t(1-\lambda)(x' - x))$$
$$-\nabla\phi(x' + t'\lambda(x - x')))\lambda(1-\lambda)(x' - x).$$

It follows that

$$\lambda\phi(x) + (1-\lambda)\phi(x') - \phi(\lambda x + (1-\lambda)x') \leq \lambda(1-\lambda)|x - x'| \times \theta(|x - x'|),$$

where θ is a modulus of continuity of $\nabla\phi$ on $\bar{x} + \epsilon\mathbb{B}$. □

If we know that a function is semiconcave, we can draw a number of useful conclusions about subgradients to the function. These are well-documented in the literature (See [52]). Of special significance for dynamic programming is the following property:

Proposition 13.9.3 *Take point $\bar{x} \in \mathbb{R}^k$ and a function $\phi : \mathbb{R}^k \to \mathbb{R}$. Assume that ϕ is semiconcave near $\bar{x}$ and, for some $\xi \in \mathbb{R}^k$,*

$$\xi \in \partial^P \phi(\bar{x}).$$

Then ϕ is Fréchet differentiable at $\bar{x}$ and $\nabla\phi(\bar{x}) = \{\xi\}$.

Proof Take $\xi \in \partial^P \phi(\bar{x})$. By definition of the proximal subgradient, there exists $\epsilon > 0$ and $M > 0$ such that

$$\xi \cdot d \leq \phi(\bar{x} + d) - \phi(\bar{x}) + M|d|^2, \quad \text{for all } d \in \epsilon\mathbb{B}. \tag{13.9.1}$$

13.9 Semiconcavity and the Value Function

Using the fact that ϕ is semiconcave near $\bar{x}$, we can arrange (by reducing the size of ϵ and choosing and appropriate continuity modulus θ) that, for all $d \in \epsilon \mathbb{B}$,

$$\phi(\bar{x}) \, (= \, \phi(\frac{1}{2}(\bar{x}+d) + \frac{1}{2}(\bar{x}-d)) \geq \frac{1}{2}\phi(\bar{x}+d) + \frac{1}{2}\phi(\bar{x}-d) - \frac{1}{2}|d| \times \theta(2|d|) \,,$$

This inequality can be equivalently expressed:

$$\phi(\bar{x}) - \phi(\bar{x}-d) \geq \phi(\bar{x}+d) - \phi(\bar{x}) - |d| \times \theta(2|d|) \,. \tag{13.9.2}$$

We deduce from (13.9.1) and (13.9.2) that

$$\xi \cdot d \geq \phi(\bar{x}) - \phi(\bar{x}-d) - M|d|^2 \geq \phi(\bar{x}+d) - \phi(\bar{x})$$
$$- |d| \times \theta_1(|d|), \quad \text{for all } d \in \epsilon \mathbb{B},$$

in which θ_1 is the continuity modulus $\theta_1(r) := M|r| + \theta(2r)$. Combining this inequality with (13.9.1) yields

$$|\xi \cdot d - (\phi(\bar{x}+d) - \phi(\bar{x}))| \leq |d| \times \theta_2(|d|), \quad \text{for all } d \in \epsilon \mathbb{B},$$

in which θ_2 is the continuity modulus $\theta_2(r) := M|r|^2 + \theta_1(r)$. We have confirmed that η is Fréchet differentiable at $\bar{x}$. □

We now bring the value function into the picture. Recall that the pointwise infimum of a family of concave functions, which have a uniform lower bound, is also concave. As we have shown, smooth functions are semiconcave. We might expect then that, under appropriate hypotheses, the value function, which after all is defined as a pointwise infimum of more regular functions, is semiconcave. This is indeed the case for smooth dynamic optimization problems with no right endpoint constraints, in which the dynamic constraint takes the form of a controlled differential equation.

We restrict attention henceforth to the dynamic optimization problem:

$$(Q) \begin{cases} \text{Minimize } g(x(T)) \\ \text{over arcs } x \in W^{1,1}([S,T];\mathbb{R}^n) \text{ satisfying} \\ \dot{x}(t) \in f(t, x(t), u(t)) \quad \text{a.e.}, \\ u(t) \in U \quad \text{a.e.}, \\ x(S) = x_0, \end{cases}$$

the data for which comprise an interval $[S, T] \subset \mathbb{R}$, a function $g : \mathbb{R}^n \to \mathbb{R}$, a function $f : [S, T] \times \mathbb{R}^n \times \mathbb{R}^m \to \mathbb{R}^n$, a (nonempty) subset $U \subset \mathbb{R}^m$ and a point $x_0 \in \mathbb{R}^n$.

The value function $V : [S, T] \times \mathbb{R}^n \to \mathbb{R}$ for (Q) is defined in the usual way: $V(t, x) := \inf\{Q_{t,x}\}$, where the right side denotes the infimum cost of a modified version of problem (Q), in which the generic initial data point (t, x) replaces (S, x_0).

Proposition 13.9.4 *Let V be the value function for problem (Q). Assume that*

(H1): f *is continuous, the set* $\{f(t, x, u) : u \in U\}$ *is closed for all* $(t, x) \in [S, T] \times \mathbb{R}^n$, *and there exists* $c_f > 0$ *such that* $|f(t, x, u)| \leq c_f(1 + |x|)$ *for all* $(t, x, u) \in [S, T] \times \mathbb{R}^n \times U$,

(H2): *there exists* $k_f > 0$ *such that* $|f(t, x, u) - f(t', x', u)| \leq k_f(|t - t'| + |x - x'|)$ *for all* $t, t' \in [S, T]$, $x, x' \in \mathbb{R}^n$ *and* $u \in U$,

(H3): *for each* $R > 0$ *there exists a continuity modulus* $\theta : [0, \infty) \to [0, \infty)$ *such that*

$$|\lambda f(t, x, u) + (1 - \lambda) f(t, x', u) - f(t, \lambda x + (1 - \lambda) x', u)|$$
$$\leq \lambda(1 - \lambda)|x - x'| \times \theta(|x - x'|)$$

for all $t \in [S, T]$, $x, x' \in \bar{x} + R\mathbb{B}$ *and* $u \in U$,

(H4): *g is locally Lipschitz continuous and locally semiconcave.*

Then V is locally Lipschitz continuous and locally semiconcave on $[S, T] \times \mathbb{R}^n$.

A proof of this proposition, which we omit, is given in [52]. The analysis involved is similar to that used earlier in this chapter to prove local Lipschitz continuity properties of the value function, but adapted to exploit hypotheses (H3) and (H4).

Remark

Hypothesis (H3) is satisfied in the special case when $f(t, ., u)$ is continuously differentiable for each $t \in [S, T]$, $u \in U$ and, for each $R > 0$, there exists a continuity modulus $\theta : [0, \infty) \to [0, \infty)$ such that

$$|\nabla_x f(t, x_1, u) - \nabla_x f(t, x_2, u)| \leq \theta(|x_1 - x_2|)$$

for all $t \in [S, T]$, $x_1, x_2 \in R\mathbb{B}$ and $u \in U$.

We shall say that a locally Lipschitz function $\phi : [S, T] \times \mathbb{R}^n \to \mathbb{R}$ is a classical Lipschitz solution to the Hamiltonian Jacobi equation when

$$(HJE)_Q \begin{cases} \nabla_t \phi(t, x) + \inf_{u \in U} \nabla_x \phi(t, x) \cdot f(t, x, u) = 0 \\ \quad \text{for all points } (t, x) \in (S, T) \times \mathbb{R}^n \\ \quad \text{at which } \phi \text{ is Fréchet differentiable} \\ \phi(T, x) = g(x), \quad \text{for all } x \in \mathbb{R}^n. \end{cases}$$

(Notice that the notion of classical Lipschitz solution is similar to, but slightly different from, that of 'almost everywhere solution' referred to in the introductory discussion to this chapter. The latter requires satisfaction of the Hamilton Jacobi equation *only at almost every point* at which ϕ is Fréchet differentiable. But to qualify as classical Lipschitz solution, it must satisfy the equation at *every* such point.)

13.9 Semiconcavity and the Value Function

It is now a simple step to provide a characterization of the value function in terms of locally Lipschitz continuous, semiconcave functions on $[S, T] \times \mathbb{R}^n$ that are classical Lipschitz solutions to $(HJE)_Q$.

Theorem 13.9.5 *Consider problem (Q). Assume hypotheses (H1)–(H4) of Proposition 13.9.4. Then the value function V is the unique classical Lipschitz solution to $(HJE)_Q$, in the class of functions $\phi : [S, T] \times \mathbb{R}^n \to \mathbb{R}$ that are locally Lipschitz continuous and semiconcave on $[S, T] \times \mathbb{R}^n$.*

Proof Notice to begin with that, in consequence of the relaxation theorem (Theorem 6.5.2), the value function V for problem (Q) coincides with the value function for its relaxed counterpart:

$$(Q)_{\text{relaxed}} \begin{cases} \text{Minimize } g(x(T)) \\ \text{over arcs } x \in W^{1,1}([S, T]; \mathbb{R}^n) \text{ satisfying} \\ \dot{x}(t) \in \operatorname{co} F(t, x(t)) \quad \text{a.e.,} \\ x(S) = x_0, \end{cases}$$

in which $\operatorname{co} F(t, x) := \{\sum_{i=0}^n \lambda_i f(t, x, u_i) : \lambda \in \Lambda^n$ and $u_0, \ldots, u_n \in U\}$. Here $\Lambda^n := \{\lambda_0, \ldots, \lambda_n\} \in (\mathbb{R}_+)^{n+1} : \sum_{i=0}^n = 1\}$. The data for $(Q)_{\text{relaxed}}$ satisfies the hypotheses of Theorem 13.3.7.

We know from Proposition 13.9.4 that the value function V belongs to the class of locally Lipschitz continuous functions that are semiconcave on $[S, T] \times \mathbb{R}^n$. From Theorem 13.3.7, V is a lower Dini solution to $(HJE)_Q$. This implies that, for any point $(t, x) \in (S, T) \times \mathbb{R}^n$ at which V is Fréchet differentiable (cf. Lemma 13.3.2(iii)),

$$-\nabla_t V(t, x) - \nabla_x V(t, x) \cdot f(t, x, u) = D_\uparrow((t, x); (-1, -f(t, x, u))) \le 0$$

for all $u \in U$ and

$$\nabla_t V(t, x) + \sum_{i=0}^n \lambda_i \nabla_x V(t, x) \cdot f(t, x, u_i) = D_\uparrow((t, x); (1, \sum_{i=0}^n \lambda_i f(t, x, u_i))) \le 0$$

for some $u_0, \ldots, u_n \in U$ and $\lambda \in \Lambda^n$. We deduce from the preceding relation, by means of a simple contradiction argument, that

$$\nabla_t V(t, x) + \sum_{i=0}^n \lambda_i \nabla_x V(t, x) \cdot f(t, x, \tilde{u}) \le 0$$

for some $\tilde{u} \in U$. But then $\nabla_t V(t, x) + \min_{u \in U} \nabla_x V(t, x) \cdot f(t, x, u) = 0$. Since, by definition of the value function, $V(T, x) = g(x)$ for all $x \in \mathbb{R}^n$, we have shown that V is a classical Lipschitz solution to the Hamilton Jacobi equation for (Q).

It remains then to show that any locally Lipschitz continuous, semiconcave function ϕ on $[S, T] \times \mathbb{R}^n$ that is a classical Lipschitz solution ϕ to $(HJE)_Q$ must be the value function.

Take such a function ϕ. Let $(t, x) \in (S, T) \times \mathbb{R}^n$ and a vector ξ be such that

$$(\xi_0, \xi_1) \in \partial^P \phi(t, x).$$

Then, according to Proposition 13.9.3, ϕ is Fréchet differentiable at (t, x) and $(\xi_0, \xi_1) = \nabla \phi(t, x)$. Since ϕ is a classical Lipschitz solution to $(HJE)_Q$, it follows that

$$\xi_0 + \inf_{u \in U} \xi_1 \cdot f(t, x, u) = 0.$$

The function ϕ is continuous and therefore automatically satisfies the relevant boundary conditions. It follows that V is a proximal solution to the Hamilton Jacobi equation for (Q). But then, according to Theorem 13.3.7, $\phi = V$. This is what we set out to show. □

Remark

In the introduction, we raised the question of whether the value function V could be identified as the unique locally Lipchitz function that satisfies the Hamilton Jacobi equation at almost all points at which the function is Fréchet differentiable. Example 13.1 of the introductory discussion tells us the answer is 'no'; here we constructed a locally Lipschitz continuous function, different from the value function, which satisfies both the relevant boundary condition and the Hamilton Jacobi equation at all points of Fréchet differentiability. We observe that this example is not in conflict with Theorem 13.9.5, because the extra solution $\tilde{V}$ is not semiconcave on $[S, T] \times \mathbb{R}^n$.

13.10 The Infinite Horizon Problem

In this section we study the following dynamic optimization problem, with discounted running cost over an infinite horizon:

$$(P_\infty)(x_0) \begin{cases} \text{Minimize } \int_0^\infty e^{-\lambda t} L(x(t), u(t)) dt \\ \text{over } x \in W^{1,1}_{\text{loc}}([0, \infty); \mathbb{R}^n) \text{ and} \\ \qquad \text{measurable functions } u : [0, \infty) \to \mathbb{R}^m \text{ satisfying} \\ \dot{y}(t) = f(y(t), u(t)), \text{ a.e. } t \in [0, \infty), \\ u(t) \in U, \text{ a.e. } t \in [0, \infty), \\ y(0) = x_0, \end{cases}$$

13.10 The Infinite Horizon Problem

in which $f : \mathbb{R}^n \times \mathbb{R}^m \to \mathbb{R}^n$ and $L : \mathbb{R}^n \times \mathbb{R}^m \to \mathbb{R}$ are given functions, $U \subset \mathbb{R}^m$ is a given set, $x_0 \in \mathbb{R}^n$ and $\lambda > 0$.

Given $t_0 \in \mathbb{R}$, a pair of functions $(x : [t_0, \infty) \to \mathbb{R}^n, u : [t_0, \infty) \to \mathbb{R}^m)$ is said to be a process on $[t_0, \infty)$ if x is locally absolutely continuous, u is measurable, $\dot{x}(t) = f(x(t), u(t))$ and $u(t) \in U$, for a.e. $t \in [t_0, \infty)$. If $t_0 = 0$ we simply say 'process'.

Our earlier study of finite horizon problems would suggest that dynamic programming for infinite horizon problems should focus on properties of the value function $W : \mathbb{R} \times \mathbb{R}^n \times \mathbb{R} \to \mathbb{R}$, generated by the family of perturbed problems in which the initial value of the time, state and the accumulated cost are arbitrary points $(t_0, x_0, y_0) \in \mathbb{R} \times \mathbb{R}^n \times \mathbb{R}$:

$$(Q_\infty)(t_0, x_0, y_0) \begin{cases} \text{Minimize } y_0 + \int_{t_0}^\infty e^{-\lambda s} L(x(s), u(s)) ds \\ \text{over processes } (x, u) \text{ on } [t_0, \infty) \text{ such that } x(t_0) = x_0. \end{cases}$$

Thus,

$$W(t_0, x_0, y_0) := \inf \{ y_0 + \int_{t_0}^\infty e^{-\lambda s} L(x(s), u(s)) ds :$$
$$(x, u) \text{ is a process on } [t_0, \infty) \text{ s.t. } x(t_0) = x_0 \}.$$

Notice that functions f and L and the set U do not depend on the independent variable t. This means that, if (x, u) is a process on $[0, \infty)$, then $t \to (x(t-h), u(t-h))$ is also a process, now on $[h, \infty)$, for any time shift h. Consequently

$$W(t_0, x_0, y_0) = y_0 + e^{-\lambda t_0} V(x_0) \quad \text{for all } (t_0, x_0, y_0) \in \mathbb{R} \times \mathbb{R}^n \times \mathbb{R},$$

where $V : \mathbb{R}^n \to \mathbb{R}$ is the function

$$V(x_0) := \inf \{ \int_0^\infty e^{-\lambda s} L(y(s), u(s)) ds :$$
$$(y, u) \text{ is a process on } [0, \infty) \text{ s.t. } y(t) = x_0 \}.$$

As we shall see, V is associated with the Hamilton Jacobi equation

$$-\lambda V(x) + \inf_{u \in U} \Big(\nabla V(x) \cdot f(x, u) + L(x, u) \Big) = 0. \tag{13.10.1}$$

The function V provides the same information as W, but more economically, since it is a function of the state variable alone. For this reason, dynamic programming for infinite horizon problems with a discounted integral cost, in which f, L and U are independent of time, is typically centred on the one-parameter value function V and on efforts to identify this function as a unique solution (in some sense) to (13.10.1).

We shall impose the following hypotheses

(H1): $\lambda > 0$, f is continuous, U is compact and

$$\{(v, \alpha) \in \mathbb{R}^n \times \mathbb{R} : v \in f(x, u), \alpha \geq L(x, u), \text{ for some } u \in U\}$$
is convex for each $x \in \mathbb{R}^n$,

(H2): there exist positive constants k_f and c_f such that

$$|f(x, u) - f(y, u)| \leq k_f |x - y| \quad \text{and} \quad |f(x, u)| \leq c_f(1 + |x|)$$
for all $x, y \in \mathbb{R}^n$ and $u \in \mathbb{R}^m$,

(H3): there exist positive constants k_L and c_L such that

$$|L(x, u) - L(y, u)| \leq k_L |x - y| \quad \text{and} \quad |L(x, u)| \leq c_L$$
for all $x, y \in \mathbb{R}^n$ and $u \in \mathbb{R}^m$.

Take $t_0 \in \mathbb{R}$, $x_0 \in \mathbb{R}^n$ and a arbitrary measurable $u : [t_0, \infty) \to \mathbb{R}^m$ such that $u(t) \in U$, a.e.. Then, in consequence of (H2), there exists a unique locally absolutely continuous $x : [t_0, \infty)$ such that (x, u) is a process on $[t_0, \infty)$ satisfying $x(t_0) = x_0$.

In more general treatments of the infinite horizon problem, the definition of the integral in the cost requires special attention, in situations where the integrand is not integrable for some processes. Note that, under our hypotheses, such difficulties do not arise, because, for any control process (x, u) on $[0, \infty)$, the absolute value of integrand in dominated by the integrable function $t \to c_L e^{-\lambda t}$. Note also that, under these hypotheses, the cost admits the uniform bound

$$\left| \int_0^\infty e^{-\lambda t} L(x(t), u(t)) dt \right| \leq c_L \lambda^{-1} \text{ for all processes } (x, u).$$

This means that V is well defined (as an infimum over a non-empty set) and V is bounded.

Proposition 13.10.1 *Assume (H1)–(H3).*

(i): For each $x_0 \in \mathbb{R}^n$ $(P_\infty)(x_0)$ has a minimizing process on $[0, \infty)$,
(ii): For any $x_0 \in \mathbb{R}^n$, $T \geq 0$ and process (x, u) such that $x(0) = x_0$ we have

$$V(x(0)) \leq \int_0^T e^{-\lambda s} L(x(s), u(s)) ds + e^{-\lambda T} V(x(T)),$$

(iii): For any $x_0 \in \mathbb{R}$, $T \geq 0$ and minimizing process $(\bar{x}, \bar{u})$ for $(P_\infty)(x_0)$ we have

$$V(\bar{x}(0)) = \int_0^T e^{-\lambda s} L(\bar{x}(s), \bar{u}(s)) ds + e^{-\lambda T} V(\bar{x}(T)),$$

13.10 The Infinite Horizon Problem

Proof The proofs of (ii) and (iii) (a statement of the principle of optimality for infinite horizon problems) are based on simple contradiction arguments. We therefore attend only to the proof of (i).

(i) Let (x_i, u_i) be a minimizing sequence for $(P_\infty)(x_0)$. Then $t \to (x_i(t), y_i := \int_0^t e^{-\lambda s} L(x_i(s), u_i(s))ds)$ is a solution to the differential inclusion

$$(\dot{x}, \dot{y}) \in \tilde{F}(t, x), \text{ a.e. } t \in [0, \infty), \ (x(0), y(0)) = (x_0, 0),$$

in which

$$\tilde{F}(t, x) := \{(f(x, u), \alpha) : \text{ for some } \alpha \in e^{-\lambda t}[L(x, u), c_L] \text{ and } u \in U\}. \tag{13.10.2}$$

The restrictions of the functions (x_i, y_i) to any interval $[0, T]$, $i = 1, 2, \ldots$ is uniformly bounded with uniformly integrably bounded derivatives. We can use this property to extract subsequences to ensure uniform convergence of the functions and weak L^1 convergence of their derivatives on each interval in a sequence of finite intervals $[0, T_j]$, $T_j \uparrow \infty$, and then extract an appropriate diagonal sequence to ensure that the limiting function does not depend on the index j. In this way, and with the help of Theorem 6.3.3 (notice that, because of (H1), the relevant convexity hypothesis is satisfied), we can find a solution to the differential inclusion $(\bar{x}, \bar{y})$ on $[0, \infty)$ with the following properties. $(\dot{\bar{x}}, \dot{\bar{y}})$ is an integrable function such that $(\dot{\bar{x}}, \dot{\bar{y}}) = (f(\bar{x}(t), \bar{u}(t)), \bar{\alpha}(t))$, $\bar{u}(t) \in U$ a.e. $t \in [0, \infty)$, for measurable functions $\bar{\alpha}$ and $\bar{u}$ such that $\bar{\alpha}(t) \geq e^{-\lambda t} L(\bar{x}(t), \bar{u}(t))$ a.e. $t \in [0, \infty)$, $(\bar{x}, \bar{u})$ is a process satisfying $\bar{x}(0) = x_0$ and

$$\lim_{i \to \infty} \int_0^\infty e^{-\lambda t} L(x_i(t), u_i(t))dt \geq \int_0^\infty e^{-\lambda t} L(\bar{x}(t), \bar{u}(t))dt.$$

The last relation confirms that $(\bar{x}, \bar{u})$ is a minimizer for $(P_\infty)(x_0)$. □

The following proposition provides regularity properties of the value function V and tells us that it is a proximal solution to the Hamilton Jacobi equation (13.10.1).

Proposition 13.10.2 *Assume (H1)–(H3).*

(i): $V : \mathbb{R}^n \to \mathbb{R}$ is a bounded, uniformly continuous function,
(ii): for any $x \in \mathbb{R}^n$ such that $\partial^P V(x) \neq \emptyset$,

$$-\lambda V(x) + \inf_{u \in U}(\eta \cdot f(x, u) + L(x, u)) = 0 \ \text{ for all } \eta \in \partial^P V(x).$$

Proof
(i): We have already observed that V is a bounded function. We show that V is also uniformly continuous.

We may assume, without loss of generality, that $k_f > \lambda$. Take any $\delta > 0$ and $x_0 \in \mathbb{R}^n$. For $\epsilon > 0$, choose any $y_0 \in x_0 + \epsilon \mathbb{B}$. Then $V(x_0) = \int_0^\infty e^{-\lambda t} L(x(t), u(t)) dt$ for some process (x, u) such that $x(0) = x_0$ (a minimizer for $(P_\infty)(x_0)$). It follows from Cor. 6.2.5 of Filippov's existence theorem (Theorem 6.2.3) that there exists $y \in W^{1,1}(0, \infty)$ such that (y, u) is a process satisfying $y(0) = y_0$. Take $T > 0$. We have

$$V(x_0) \geq \int_0^T e^{-\lambda t} L(x(t), u(t)) dt - \lambda^{-1} c_L e^{-\lambda T}$$

$$\geq \int_0^T e^{-\lambda t} L(y(t), u(t)) dt - k_L |x_0 - y_0| \int_0^T e^{(k_f - \lambda)t} dt - \lambda^{-1} c_L e^{-\lambda T}$$

$$= \int_0^T e^{-\lambda t} L(y(t), u(t)) dt - k_L |x_0$$

$$- y_0|(k_f - \lambda)^{-1}(e^{(k_f - \lambda)T} - 1) - \lambda^{-1} c_L e^{-\lambda T}$$

$$= \int_0^\infty e^{-\lambda t} L(y(t), u(t)) dt - k_L |x_0 - y_0|(k_f - \lambda)^{-1}(e^{(k_f - \lambda)T} - 1)$$

$$- 2\lambda^{-1} c_L e^{-\lambda T}$$

$$\geq V(y_0) - k_L |x_0 - y_0|(k_f - \lambda)^{-1}(e^{(k_f - \lambda)T} - 1) - 2\lambda^{-1} c_L e^{-\lambda T}.$$

This means that, if we choose T such that $2\lambda^{-1} c_L e^{-\lambda T} \leq \delta/2$ and then adjust the size of $\epsilon > 0$ such that

$$k_L (k_f - \lambda)^{-1}(e^{(k_f - \lambda)T} - 1)\epsilon \leq \delta/2,$$

then $V(x_0) - V(y_0) \geq -\delta$. Reversing the roles of x_0 and y_0 gives $V(y_0) - V(x_0) \geq -\delta$. So $|V(x_0) - V(y_0)| \leq \delta$. Notice that the choice of $\epsilon > 0$ to achieve this estimate did not depend on x_0. It follows that V is uniformly continuous.

(ii): Take any $x_0 \in \mathbb{R}^n$ such that $\partial^P V(x_0)$ is non-empty. Let $\eta \in \partial^P V(x_0)$.

Fix any $v \in U$. Take $h_0 > 0$ sufficiently small and write $\tilde{y}$ for the solution to the equation $\dot{y}(t) = \tilde{f}(y(t), v)$ a.e. $t \in [0, h_0]$ such that $y(0) = x_0$, where $\tilde{f}(y, v) := -f(y, v)$. Define $x(t) := \tilde{y}(-t)$ for all $t \in [-h_0, 0]$. Observe that $x(-h_0) = \tilde{y}(h_0)$, $x(0) = \tilde{y}(0)$ and $\dot{x}(t) = f(x(t), v)$, a.e. $t \in [-h_0, 0]$. We extend the process (x, v) (defined on $[-h_0, 0]$) to a process (still written (x, v)) on $[-h_0, \infty)$. Using the fact that η is a proximal subgradient and applying (ii) of Proposition 13.10.1 to the process (x, v) with $T = h$, we deduce the existence of $c > 0$ such that, for all $h \in [0, h_0]$,

13.10 The Infinite Horizon Problem

$$\eta \cdot \int_0^{-h} f(x(s), v) ds - ch^2$$

$$\leq V(x_0 + \int_0^{-h} f(x(s), v) ds) - V(x_0)$$

$$= V(x(-h)) - e^{-\lambda h} V(x(0)) - (1 - e^{-\lambda h}) V(x_0)$$

$$\leq e^{\lambda h} \int_0^h e^{-\lambda s} L(x(s-h), v) ds - (1 - e^{-\lambda h}) V(x_0).$$

Dividing across this inequality by h and passing to the limit as $h \downarrow 0$ gives

$$-\lambda V(x_0) + \eta \cdot f(x_0, v) + L(x_0, v) \geq 0. \tag{13.10.3}$$

Now let $(\bar{x}, \bar{u})$ be a minimizing process for $(P_\infty)(x_0)$. Apply (iii) of Proposition 13.10.1 to the process $(\bar{x}, \bar{u})$ with $T = h$. Again using also the fact that η is a proximal subgradient, we deduce the existence of $c > 0$ such that, for $h > 0$ sufficiently small,

$$\eta \cdot \int_0^h f(\bar{x}(s), \bar{u}(s)) ds - ch^2 \leq V(x_0 + \int_0^h f(\bar{x}(s), \bar{u}(s)) ds) - V(x_0)$$

$$= e^{-\lambda h} V(x_0 + \int_0^h f(\bar{x}(s), \bar{u}(s)) ds) - V(x_0) + (1 - e^{-\lambda h}) V(x_0$$

$$+ \int_0^h f(\bar{x}(s), \bar{u}(s)) ds)$$

$$= -\int_0^h e^{-\lambda s} L(\bar{x}(s), \bar{u}(s)) ds + (1 - e^{-\lambda h}) V(x_0 + \int_0^h f(\bar{x}(s), \bar{u}(s)) ds).$$

$$\tag{13.10.4}$$

We deduce from the convexity hypothesis (H1) that there exists $h_i \downarrow 0$ and $\bar{v} \in U$ such that

$$h_i^{-1} \int_0^{h_i} f(\bar{x}(s), \bar{u}(s)) ds \to f(x_0, \bar{v}) \text{ and}$$
$$\lim_{i \to \infty} h_i^{-1} \int_0^{h_i} e^{-\lambda s} L(\bar{x}(s), \bar{u}(s)) ds \geq L(x_0, \bar{v}).$$

Set $h = h_i$ in (13.10.4). Dividing across (13.10.4) by h_i and passing to the limit as $i \to \infty$ gives

$$-\lambda V(x_0) + \eta \cdot f(x_0, \bar{v}) + L(x_0, \bar{v}) \leq 0. \tag{13.10.5}$$

We conclude from (13.10.3), which is satisfied for all $v \in U$, and from (13.10.5), which is valid for some $\bar{v} \in U$, that

$$-\lambda V(x_0) + \inf_{v \in U}\Big(\eta \cdot f(x_0, v) + L(x_0, v)\Big) = 0.$$

Since this last relation is true for any $x_0 \in \mathbb{R}^n$ at which $\partial^P V(x) \neq \emptyset$ and for any $\eta \in \partial^P V(x) \neq \emptyset$, we have confirmed that V is a proximal solution to the Hamilton Jacobi equation. □

Theorem 13.10.3 *Assume (H1)–(H3). Then the value function V is the unique bounded continuous function such that, for each $x \in \mathbb{R}^n$ such that $\partial^P V(x) \neq \emptyset$,*

$$-\lambda V(x) + \inf_{u \in U}\Big(\eta \cdot f(x, u) + L(x, u)\Big) = 0 \quad \text{for all } \eta \in \partial^P V(x).$$

Proof We have shown that the value function V is a continuous, bounded proximal solution to the Hamilton Jacobi equation. It remains then to prove that it is the unique such solution.

Let V' be an arbitrary continuous, bounded proximal solution of the Hamilton Jacobi equation. Take any $x_0 \in \mathbb{R}^n$, $y_0 \in \mathbb{R}$ and $T > 0$. Let $W : (-\infty, \infty) \times \mathbb{R}^n \times \mathbb{R} \to \mathbb{R}$ be the function

$$W(t, x, y) := y + e^{-\lambda t} V'(x).$$

Now consider the problem

$$(R_\infty)(x_0, y_0) \begin{cases} \text{Minimize } y(T) + e^{-\lambda T} V'(x(T)) \\ \text{over } (x, y) \in W^{1,1}_{\text{loc}}([0, \infty); \mathbb{R}^{n+1}) \text{ satisfying} \\ (\dot{x}, \dot{y})(t) \in \{(f(x(t), u), \alpha) : \\ \qquad \alpha \in e^{-\lambda t}[L(x(t), u), c_L] \text{ and } u \in U\}, \\ x(0) = x_0 \text{ and } y(0) = y_0. \end{cases}$$

Take any point $(t, x, y) \in \mathbb{R} \times \mathbb{R}^n \times \mathbb{R}$ at which $\partial^P W(t, x, y)$ is non-empty and $(\xi_0, \xi_1, \xi) \in \partial^P W(t, x, y)$. It follows from the definition of W that $\partial^P V'(x) \neq \emptyset$ and $(\xi_0, \xi_1, \xi_2) = (-\lambda e^{\lambda t} V'(x), e^{\lambda t} \eta, 1)$ for some $\eta \in \partial^P V'(x)$. But V' is a proximal solution to the Hamilton Jacobi equation so

$$-\lambda V'(x) + \inf_{u \in U}\Big(\eta \cdot u + L(x, u)\Big) = 0.$$

It follows that

$$-e^{-\lambda t}\lambda V'(x) + \inf_{u \in U}\{e^{-\lambda t}\eta \cdot v^{(1)} + v^{(2)} : (v^{(1)}, v^{(2)}) \in \tilde{F}(t, y, x)\} = 0.$$

($\tilde{F}$ was defined in (13.10.2)).

But then

$$\xi_0 + \min\{\xi_1 \cdot v^{(1)} + \xi_2 v^{(2)} : (v^{(1)}, v^{(2)}) \in \tilde{F}(t, y, x)\} = 0.$$

We have confirmed that W is a proximal solution of the Hamilton Jacobi equation for problem $(R_\infty)(x_0, y_0)$. Notice also that W also satisfies the right boundary condition

$$W(T, x, y) = y + e^{-\lambda T} V'(x) \quad \text{for all } x, y \in \mathbb{R}^n.$$

But we know from Theorem 13.3.7 that W is then the value function for problem $(R_\infty)(x_0, y_0)$. It follows in particular that

$$W(0, x_0, 0) = \inf\{\int_0^T e^{-\lambda t} L(x(t), u(t)) dt$$
$$+ e^{-\lambda T} V'(x(T)) \; : \; (x, u) \text{ is a process s.t. } x(0) = x_0\}.$$

We can now deduce from the boundedness of V' and L the uniform bound

$$|\int_0^T e^{-\lambda t} L(x(t), u(t)) dt + e^{-\lambda T} V'(x(T))$$
$$- \int_0^\infty e^{-\lambda t} L(x(t), u(t)) dt| \leq (c_0 + c_L \lambda^{-1}) e^{-\lambda T},$$

over all processes (x, u) on $[0, \infty)$ such that $x(0) = x_0$, in which $c_0 := \sup_{x \in \mathbb{R}^n} V'$. Since this estimate is valid for all $T > 0$, we can conclude that

$$V'(x_0) = W(0, x_0, 0)$$
$$= \inf\{\int_0^\infty e^{-\lambda t} L(x(t), u(t)) dt \; : \; (x, u) \text{ is a process on } [0, \infty) \text{ s.t. } x(0) = x_0\}.$$

We have shown that V' is the value function for $(P_\infty)(x_0)$. □

13.11 The Minimum Time Problem

In minimum time control, the goal is to find a control strategy that drives the given initial state of a control system to a specified 'target' set in the state space in minimum time. In this section we treat the following version of the minimum time problem in which the dynamic constraint takes the form of a differential inclusion. For each $x \in \mathbb{R}^n$ consider then

$$(P_x) \begin{cases} \text{Minimize } T \\ \text{over } T \geq 0 \text{ and } x \in W^{1,1}([0,T]; \mathbb{R}^n) \text{ satisfying} \\ \dot{y}(t) \in F(y(t)), \text{ a.e. } t \in [0,T), \\ y(0) = x \text{ and } y(T) \in S. \end{cases}$$

in which $F : \mathbb{R}^n \rightsquigarrow \mathbb{R}^n$ is a given multifunction and $S \subset \mathbb{R}^n$ is a given closed set. The value function $T : \mathbb{R}^n \to \mathbb{R} \cup +\infty$ is

$$T(x) := \inf\{(P_x)\}, \text{ for each } x \in \mathbb{R}^n.$$

In other words, $T(x)$ is the infimum cost of problem (P_x). (We interpret the infimum cost $T(x) = +\infty$ if there is no F trajectory y, starting at x, that reaches S in finite time.)

Throughout this section, we shall refer to the following hypotheses:

(H1): $F(x)$ is a closed, convex, non-empty set for each $x \in \mathbb{R}^n$,

(H2): there exist constants k_F and c_F such that

$$F(x') \subset F(x) + k_F |x - x'| \mathbb{B} \quad \text{and} \quad F(x) \subset c_F \mathbb{B} \quad \text{for all } x', x \in \mathbb{R}^n.$$

Proposition 13.11.1 (Properties of the Value Function) *Let T be the value function. Assume (H1) and (H2). Then*

(a): *(P_x) has a minimizer for each x (we allow the possibility that $\inf\{P_x\} = +\infty$, i.e. there are no F trajectories starting at the specified x and arriving at S in finite time). T is non-negative valued and lower semi-continuous,*

(b): *Take any $x \in \mathbb{R}^n$. Then*

$$1 + \min_{v \in F(x)} \xi \cdot v \geq 0, \quad \text{for all } \xi \in \partial^P T(x),$$

(c): *Take any $x \in \mathbb{R}^n \setminus S$. Then*

$$1 + \min_{v \in F(x)} \xi \cdot v \leq 0, \quad \text{for all } \xi \in \partial^P T(x).$$

Proof

(a): Obviously the value function is non-negative valued. Existence of a minimizers for (P_x) and the lower semicontinuity of the value function are proved by a straightforward adaptation (to allow for processes with variable end-times) of the proof of Proposition 6.5.3 and Proposition 13.3.6.

(b): We can assume $T(x) < +\infty$ since otherwise $\partial^P T(x) = \emptyset$. Take any $v \in F(x)$, $\xi \in \partial^P T(x)$ and $\epsilon > 0$. Then, in consequence of Filippov's existence theorem, there exists an F trajectory $y : [-\epsilon, 0] \to \mathbb{R}^n$ such that $y(0) = x$ and

13.11 The Minimum Time Problem

$$\epsilon^{-1}(y(0) - y(-\epsilon)) \to v, \text{ as } \epsilon \downarrow 0.$$

A simple contradiction argument tells us that $T(y(-\epsilon)) < +\infty$ and

$$T(y(-\epsilon)) - T(x) \le \epsilon.$$

(This inequality can be regarded as part the principle of optimality for the minimum time problem.) Then by the definition of proximal subgradients and the Lipschitz continuity of y, there exists $M > 0$ such that, for $\epsilon > 0$ sufficiently small,

$$\xi \cdot (y(-\epsilon) - y(0)) \le T(y(-\epsilon)) - T(y(0)) + M\epsilon^2 \le \epsilon + M\epsilon^2.$$

Dividing across by ϵ, using the preceding relations and passing to the limit as $\epsilon \downarrow 0$ yields

$$-\xi \cdot v = \lim_{\epsilon \downarrow 0} \xi \cdot \epsilon^{-1}(y(-\epsilon) - y(0)) \le 1$$

Since v was an arbitrary point on $F(x)$, we have shown that $1 + \min_{v \in F(x)} \xi \cdot v \ge 0$.

(c): We can, once again, assume without loss of generality that $T(x) < +\infty$. Since $x \notin S$, it follows that (P_x) has a minimizer $y : [0, T] \to \mathbb{R}^n$ such that $T > 0$. A simple contradiction tells us that there exists $\epsilon \downarrow 0$ such that, for each i

$$T(y(0)) = T(y(\epsilon_i)) + \epsilon_i.$$

(This equality is the other part of the principle of optimality.) Since y is Lipschitz continuous, $\{\epsilon_i^{-1}(y(\epsilon_i) - y(0))\}$ is a bounded sequence. After extraction of a suitable subsequence then

$$\epsilon_i^{-1}(y(\epsilon_i) - y(0)) \to \bar{v}, \text{ as } i \to \infty,$$

for some $\bar{v} \in \mathbb{R}^n$. A straightforward contradiction argument, based on the convexity of $F(x)$, the regularity properties of F and application of the separation theorem, tells us that $v \in F(x)$ (see the proof of Theorem 13.3.7).

Take any $\xi \in \partial^P T(x)$. By definition of proximal subgradients and the Lipschitz continuity of y, there exists $M > 0$ such that, for all i sufficiently large,

$$\xi \cdot (y(\epsilon_i) - y(0)) \le T(y(\epsilon_i)) - T(y(0)) + M\epsilon_i^2 = -\epsilon_i + M\epsilon_i^2.$$

The preceding relations tell us that

$$1 + \xi \cdot \bar{v} = 1 + \lim_{i \to \infty} \epsilon_i^{-1}(y(\epsilon_i)) - y(0)) \le 0.$$

We have shown that, for some $\bar{v} \in F(x)$, $1 + \xi \cdot \bar{v} \le 0$. $\square$

Theorem 13.11.2 *Consider problem* (P_x). *Assume* (H1) *and* (H2). *The value function* T *is the unique function* V *that satisfies the following conditions*

(i): *V is lower semi-continuous and bounded below,*
(ii): *for every* $x \in \mathbb{R}^n \setminus S$,

$$1 + \min_{v \in F(x)} \xi \cdot v = 0, \quad \text{for all } \xi \in \partial^P V(x),$$

(iii): *for every* $x \in S$,

$$1 + \min_{v \in F(x)} \xi \cdot v \geq 0, \quad \text{for all } \xi \in \partial^P V(x),$$

(iv): $V(x) = 0$, *for* $x \in S$.

Remarks

(A): This theorem, due to Wolenski and Zhuang [208] and building on earlier work by Clarke et al. [84], is remarkable for the unrestrictive nature of the hypotheses that are invoked. The target set is required merely to be a closed subset of the state-space. In these circumstances the value function may be discontinuous and take infinite values at points in its domain. To achieve a dynamic programming type characterization of the value function, we might expect the need for additional hypotheses involving the regularity of the boundary of the target set or concerning the way the target set interacts with the dynamic constraint. But any such hypotheses are entirely absent from the theorem statement.

(B): Condition (iii) is automatically satisfied on the interior of S, so it is of interest only on the boundary of this set ∂S. It is helpful to regard (ii) as the statement that the value function satisfies the Hamilton Jacobi equation for the minimum time problem (in the proximal normal sense), with domain the open set $\mathbb{R}^n \setminus S$ and to interpret condition (iii) as boundary condition on the boundary points of $\mathbb{R}^n \setminus S$.

Proof We have already shown that the value function $T : \mathbb{R}^n \to \mathbb{R} \cup \{+\infty\}$ satisfies conditions (i)–(iv). Now take any function V satisfying these conditions. It remains to show that $V(x) = T(x)$ for all $x \in \mathbb{R}^n$. To prove the theorem it suffices to complete the following two steps which, together, tell us that $V(x) = \inf\{(P_x)\} = T(x)$.

Step 1: Take any $x \in \mathbb{R}^n$. We show that

$$V(x) \geq \inf\{(P_x)\}.$$

13.11 The Minimum Time Problem

We can assume that $x \in \mathbb{R}^n \setminus S$ since the relation is obviously satisfied for $x \in S$. We can further assume that $V(x) < +\infty$ for otherwise there is nothing to prove. Define $W(y, \tau) := V(y) + \tau$. To carry out this step, we consider the following control system with pathwise state constraint:

$$\begin{cases} (\dot{y}(t), \dot{\tau}(t), \dot{a}(t)) \in \tilde{F}(y(t)), & \text{a.e. } t \in [0, \infty), \\ (y(t), \tau(t), a(t)) \in \text{epi}\{W\} & \text{for all } t \in [0, \infty), \\ y(0) = x, \tau(0) = 0 \text{ and } a(0) = W(x, \tau). \end{cases}$$

Here,

$$\tilde{F}(y) := \begin{cases} F(y) \times \{1\} \times \{0\} & \text{if } y \in \mathbb{R}^n \setminus S, \\ c_F \mathbb{B} \times [-1, 1] \times [-1, 1] & \text{if } y \in S. \end{cases}$$

$\tilde{F}$ is a bounded, upper semi-continuous multifunction with values convex sets and its graph is closed. We now examine the inward pointing condition for application of a weak invariance theorem. Take any $(y, \tau, a) \in \text{epi } W$ and $(\eta_0, \eta_1, -\lambda) \in N^P_{\text{epi } W}(y, \tau, a)$. Since the constraint set is an epigraph set, $\lambda \geq 0$. We must show that

$$\min_{e \in \tilde{F}(y,\tau)} (\eta_0, \eta_1, -\lambda) \cdot e \leq 0.$$

This is obviously true for $y \in S$ because, in this case, $\tilde{F}(y, \tau)$ contains 0 as an interior point. So assume that $y \in \mathbb{R}^n \setminus S$. Then $\tilde{F}(y) = F(y) \times \{1\} \times \{0\}$. Suppose first that $\lambda > 0$. Then $(\eta_0/\lambda, \eta_1/\lambda, -1) \in N^P_{\text{epi } W}(y, \tau, a)$, which (owing to Lemma 5.1.2) implies that $(\eta_0/\lambda) \in \partial^P V(y)$ and $\eta_1/\lambda = 1$. By condition (ii) in the theorem statement then, $\min_{v \in F(y)} (\eta_0/\lambda) \cdot v + \eta_1/\lambda \leq 0$. It follows

$$\min_{e \in \tilde{F}(y,\tau)} (\eta_0, \eta_1, -\lambda) \cdot e = \lambda \min_{v \in F(y)} \left[(\eta_0/\lambda) \cdot v + \eta_1/\lambda + 0 \right] \leq 0.$$

The other case to be considered is that when $\lambda = 0$. We show the inward pointing condition is true also in this case, by means of the earlier used technique of approximating 'vertical' proximal subgradients to the epigraph of V by non-vertical ones (cf. the proof of Theorem 13.3.7).

All the hypotheses have been verified under which we can apply the weak invariance Theorem 13.2.1 for autonomous state-constrained systems. This tells us that there exists an $\tilde{F}$ trajectory $(y, \tau, a) : [0, \infty) \to \mathbb{R}^n \times \mathbb{R} \times \mathbb{R}$ such $(y, \tau, a)(0) = (x, 0, V(x))$ and $a(t) \geq W(y(t), \tau(t))$ for all $t \geq 0$.

Recall that $x \in \mathbb{R}^n \setminus S$. Define $\bar{T} := \inf\{T' : y(T') \notin \mathbb{R}^n \setminus S\}$. (We allow $\bar{T} = \infty$, i.e. the case when $y(t) \in \mathbb{R}^n \setminus S$ for all $t \geq 0$.) Take any finite $T' < \bar{T}$. Then, for a.e. $t \in [0, T']$, $\dot{y}(t) \in F(y(t))$, $\tau(t) = t$ and $\dot{a}(t) = 0$. Thus y is an F trajectory on $[0, \bar{T}]$ and, since $(y(t), \tau(t), a(t)) \in \text{epi}\{W\}$ for all $t \in [0, T']$

$$V(x) = W(x, 0) = a(0) = a(T') \geq W(y(T'), \tau(T')) = V(y(T')) + T'.$$

Since V is bounded below, the T''s satisfying this relation must be uniformly bounded, This implies that $\bar{T}$ is finite. Then, since y is continuous, $y(\bar{T}) \in S$ and so $V(y(\bar{T})) = 0$. Recalling that V is lower semi-continuous, it follows that

$$V(x) \geq \liminf_{T' \uparrow \bar{T}} V(y(T')) + \bar{T} \geq 0 + \bar{T} \geq \inf\{(P_x)\}.$$

Step 2: Take any $x \in \mathbb{R}^n$. We show that

$$V(x) \leq \inf\{(P_x)\}.$$

We can assume that $\inf\{(P_x)\} < \infty$ for, otherwise, there is nothing to prove. Take an arbitrary F trajectory $y : [0, \bar{T}] \to \mathbb{R}^n$ such that $y(0) = x$ and $y(\bar{T}) \in S$. Define $W(y, \tau) := V(y) + \tau$. Consider the control system:

$$\begin{cases} (\dot{z}(s), \dot{\tau}(s), \dot{a}(s)) \in -F(z(s)) \times \{-1\} \times \{0\}, \text{ a.e. } t \in [0, \bar{T}], \\ (y(s), \tau(s), a(s)) \in \text{epi}\{W\} \quad \text{for all } t \in [0, \bar{T}], \\ z(0) = y(\bar{T}), \tau(0) = \bar{T} \text{ and } a(0) = \bar{T}. \end{cases}$$

We verify the inward pointing condition for application of a strong invariance theorem. Take any $(z, \tau, a) \in \text{epi } W$ and $(\eta_0, \eta_1, -\lambda) \in N^P_{\text{epi } W}(z, \tau, a)$. Then $\lambda \geq 0$. We must show that

$$\sup_{e \in (-F(z)) \times \{-1\} \times \{0\}} (\eta_0, \eta_1, -\lambda) \cdot e \leq 0.$$

We limit attention to the case $\lambda > 0$, since the case '$\lambda = 0$' can be treated by looking at limits of non-vertical proximal gradients. Then $(\eta_0/\lambda, \eta_1/\lambda, -1) \in N^P_{\text{epi } W}(y, \tau, a)$, which implies that $\eta_0/\lambda \in \partial^P V(z)$ and $\eta_1/\lambda = 1$. By conditions (ii) and (iii) in the theorem statement then, $(\eta_0/\lambda) \cdot v' + \eta_1/\lambda \geq 0$ for all $v' \in F(z)$. It follows

$$\sup_{e \in (-F)(y) \times \{-1\} \times \{0\}} (\eta_0, \eta_1, -\lambda) \cdot e = \lambda \sup_{v' \in F(y)} \left[-(\eta_0/\lambda) \cdot v' - \eta_1/\lambda) + 0 \right] \leq 0.$$

The inward pointing condition has been confirmed. Notice also that the closed-valued multifunction $-F \times \{-1\} \times \{0\}$ has the requisite Lipschitz continuity property.

Now apply the strong invariance Theorem 13.2.3 for autonomous state-constrained systems with reference to the $-F \times \{-1\} \times \{0\}$ trajectory $(z(s), \tau(s), a(s)) := (y(\bar{T} - s), \bar{T} - s, \bar{T})$. Notice that $z(0) = y(\bar{T})$, $\tau(0) = \bar{T}$ and $a(0) = \bar{T} = V(y(\bar{T})) + \bar{T} = W(y(\bar{T}), \bar{T})$. So $(z(0), \tau(0), a(0)) \in \text{epi } W$. The theorem tells us that $a(s) \geq W(z(s), \tau(s))$ for all $s \in [0, \bar{T}]$. But then

$$\bar{T} = a(0) = a(\bar{T}) \geq W(z(\bar{T}), \tau(\bar{T})) = V(y(0)) + 0 = V(x).$$

Since $\bar{T} = \inf\{T' : y(T') \notin \mathbb{R}^n \setminus S\}$ is the cost of an arbitrary F trajectory y on $[0, \bar{T}]$ such that $y(0) = x$ and $y(\bar{T}) \in S$, we have $V(x) < \infty$ and

$$V(x) \leq \inf\{(P_x)\}.$$

This completes the proof. □

13.12 Viscosity Solutions of the Hamilton Jacobi Equation

Historically, the characterization of the value function for dynamic optimization problems in terms of viscosity solutions to the Hamilton Jacobi equation preceded those involving generalized solutions (by which we mean either lower Dini solutions or proximal solutions) developed in the previous sections. The viscosity solutions approach aims to derive comparison principles, uniqueness of solutions for appropriate boundary data, and stability of solutions under approximation, for Hamilton Jacobi equations, broadly interpreted. The proof techniques employed are based on a direct analysis of the Hamilton Jacobi equation and, apart from the proof that the value function is a viscosity solution of the Hamilton Jacobi equation that satisfies the specified boundary conditions, makes no use of properties of state trajectories. By contrast, the control theoretic approach pursued in this chapter is based, as we have seen, on a study of the invariance properties of state trajectories. The viscosity solutions approach is more general, in the sense that it can be applied to certain Hamilton Jacobi equations, even higher order partial differential equations, which are not associated with any deterministic dynamic optimization problem. On the other hand, the control theoretic approach has certain advantages, notably in the way it can take account of constraints, some of which we discuss in this section.

The viscosity solutions approach is in widespread use. It is of interest then to examine how the information it supplies about dynamic optimization problems compares with that obtained by the constructive methods of this chapter. The purpose of this section is to make such comparisons.

The Hamilton Jacobi equation (HJE) for problem (P) of the introduction can be written

$$-W_t(t, x) + H(t, x, -W_x(t, x)) = 0. \qquad (13.12.1)$$

in which H is, as usual, the Hamiltonian function

$$H(t, x, p) := \max_{e \in F(t,x)} p \cdot e.$$

In the viscosity solutions literature, interpretation of (possibly discontinuous) solutions to (HJE) in the class locally bounded (possibly discontinuous) functions W, is based on the following definitions, in which W_* and W^* (referred to as the upper semi-continuous envelope and the lower semi-continuous envelope of W, respectively) are the functions:

$$W^*(t, x) := \limsup_{(t',x') \to (t,x)} W(t', x') \text{ and } W_*(t, x)$$

$$:= \liminf_{(t',x') \to (t,x)} W(t', x'), \text{ for all } (t, x) \in (S, T) \times \mathbb{R}^n.$$

Definition 13.12.1

(i): W is a *viscosity supersolution* of (13.12.1) if, for any point $(t, x) \in (S, T) \times \mathbb{R}^n$ and any C^1 function $\psi : \mathbb{R} \times \mathbb{R}^n \to \mathbb{R}$ such that $(t', x') \to W_*(t', x') - \psi(t', x')$ has a local minimum at (t, x), we have

$$-\psi_t(t, x) + H(t, x, -\psi_x(t, x)) \geq 0$$

(ii): W is a *viscosity subsolution* of (13.12.1) if, for any point $(t, x) \in (S, T) \times \mathbb{R}^n$ and any C^1 function $\psi : \mathbb{R} \times \mathbb{R}^n \to \mathbb{R}$ such that $(t', x') \to W^*(t', x') - w(t', x')$ has a local maximum at (t, x), we have

$$-\psi_t(t, x) + H(t, x, -\psi_x(t, x)) \leq 0$$

(iii): W is a *viscosity solution* of (13.12.1) if it is both a viscosity supersolution and a viscosity subsolution. If W is a lower semi-continuous function satisfying conditions (i) and (ii) in the above definition, we say that W is a lower semi-continuous viscosity solution. (Notice, we can replace W_* by W in condition (ii) when W is lower semi-continuous, since, in this case, the two functions coincide.)

Under suitable hypotheses on the data for problem (P) including a continuity requirement on $F(t, x)$ regarding its t dependence, it is known that the value function for (P) is indeed the unique lower semi-continuous viscosity solution to (13.12.1) satisfying appropriate boundary conditions.

The following theorem provides a characterization of this nature. While, as we have remarked, such characterizations are associated with the viscosity solutions literature, we offer an independent proof, based on constructive methods. We allow the multifunction $F(t, x)$ to be discontinuous w.r.t. the t variable; supersolutions must therefore be interpreted in terms of the limit set $F(t^+, x)$ (in place of $F(t, x)$). We write H^+ for the modified Hamiltonian functions, in which $F(t, x)$ is replaced by $F(t^+, x)$:

13.12 Viscosity Solutions of the Hamilton Jacobi Equation

$$H^+(t, x, p) := \max_{e \in F(t^+, x)} p \cdot e.$$

Theorem 13.12.2 (*Viscosity Solution Characterization of Lower semi-continuous Value Functions*) Assume (H1)–(H5) of Theorem 13.3.7. Suppose, in addition, that

(HS) *g is locally bounded and satisfies $(g^*)_* = g$.*

Take a lower semi-continuous, locally bounded function $V : [S, T] \to \mathbb{R}$.

Then V is the value function for (P) if and only if V is a lower semi-continuous viscosity solution of (13.12.1) in the following sense: V is lower semi-continuous and

(i) (V is a viscosity supersolution) for any point $(t, x) \in (S, T) \times \mathbb{R}^n$ and any C^1 function $\psi : \mathbb{R} \times \mathbb{R}^n \to \mathbb{R}$ such that

$$(t', x') \to V(t', x') - \psi(t', x')$$

has a local minimum at (t, x) we have

$$-\psi_t(t, x) + H^+(t, x, -\psi_x(t, x)) \geq 0,$$

(ii) (V is a viscosity subsolution) for any point $(t, x) \in (S, T) \times \mathbb{R}^n$ and any C^1 function $\psi : \mathbb{R} \times \mathbb{R}^n \to \mathbb{R}$ such that

$$(t', x') \to V^*(t', x') - \psi(t', x')$$

has a local maximum at (t, x) we have

$$-\psi_t(t, x) + H^+(t, x, -\psi_x(t, x)) \leq 0,$$

(iii) (boundary conditions) For all $x \in \mathbb{R}^n$,

$$\liminf_{\{(t', x') \to (S, x) : t' > S\}} V(t', x') = V(S, x),$$

$$V(T, x) = g(x) \quad \text{and} \quad V^*(T, x) = g^*(x).$$

Remarks

(a): We take this opportunity to compare the 'generalized solutions' and 'viscosity solutions' characterizations of the value function, provided by Theorems 13.3.7 and 13.12.2. Dynamic optimization problems with discontinuous value functions are of interest primarily, as we have already remarked, because they include problems with a right endpoint constraint $x(T) \in C$, in which C is a given closed, strict subset of $\mathbb{R}^n$. In the generalized solutions characterization of Theorem 13.3.7, we accommodate such a constraint by replacing the original cost g by an extended valued cost

$$g^{(1)}(x) := \begin{cases} g(x) \text{ if } x \in C \\ +\infty \text{ otherwise} \end{cases}$$

In the framework of Theorem 13.12.2, we can still deal with a right endpoint constraint, even though we now require the end-point cost function to be bounded on bounded sets; in place of $g^{(1)}$ we employ

$$g^{(2)}(x) := \begin{cases} g(x) \text{ if } x \in C \\ +K \text{ otherwise} \end{cases}$$

(If the velocity set F is bounded on bounded sets, we can arrange that the value function of the problem with cost function $g^{(2)}$ will coincide with that of the original problem on an arbitrary bounded domain, by choosing the constant K sufficiently large.) A more serious issue is that, to apply Theorem 13.12.2, we need to assume that

$$(g^*)_* = g. \tag{13.12.2}$$

Notice that the function $g^{(2)}$ fails to satisfy this hypothesis if the set C has empty interior or, indeed, if $\overline{\text{int } C} \neq C$. This means that Theorem 13.12.2 cannot be applied to dynamic optimization problems with a fixed endpoint ($C = \{x_1\}$, for some x_1). This difficulty does not arise, however, in the application of Theorem 13.3.7, which does not require imposition of hypothesis (13.12.2).

The following example shows that hypothesis (13.12.2) cannot be removed from the hypotheses of Theorem 13.12.2.

$$\begin{cases} \text{Minimize } g(x(1)) \\ \text{subject to} \\ \dot{x}(t) \in [-1, +1] \text{ a.e. } t \in [0, 1] \end{cases}$$

in which

$$g(x) := \begin{cases} 0 \text{ if } x \in C \\ 1 \text{ otherwise} \end{cases}$$

and $C := \{0\}$. The value function is

$$V(t, x) = \begin{cases} 0 \text{ if } |x| \leq 1 - t \\ 1 \text{ otherwise.} \end{cases}$$

The data for this problem satisfy the hypotheses of Theorem 13.3.7 and, accordingly, V, as given above, is the unique lower semi-continuous function satisfying conditions of this theorem. Notice however that the function g in the example fails to satisfy (13.12.2) and the function V is not the unique function that satisfies conditions (i)–(iii) of Theorem 13.12.2; the function

$$\tilde{V}(t, x) = \begin{cases} 0 \text{ if } x = 0 \\ 1 \text{ otherwise} \end{cases}$$

also satisfies the conditions.

(b): In the case when the end-point cost function satisfies hypothesis (HS), Theorems 13.12.2 and 13.3.7 provide two different descriptions of the value function, in terms of viscosity solutions and proximal solutions to the Hamilton Jacobi equation. It follows that, under this additional hypothesis, a lower semi-continuous function is a viscosity solution to the Hamilton Jacobi equation if and only if it is a proximal solution. The equivalence of these solution concepts was partly anticipated by Barron and Jensen [19] (see also [18]), who proposed a definition of generalized solution to the Hamilton Jacobi equation for lower semi-continuous functions, similar to that of 'proximal' solution, in which the strict subdifferential replaces the proximal subdifferential; they showed directly that continuous functions which were solutions to the Hamilton Jacobi equation in this sense are also viscosity solutions, in the sense of Def. 13.12.1.

(c): The role of the right limits (see the definition of H^+) in the value function viscosity solution characterization of Theorem 13.12.2 is crucial: this characterization would become in general false if we try to substitute an Hamiltonian defined in terms of the left limit set $F(t^-, x)$ in place of H^+ (cf. Exercise 13.3).

Proof The proof has the following structure. Assume (H1)–(H5). In Step 1 we show that, if g is locally bounded (in this step we do not require $(g^*)_* = g$), then the value function satisfies conditions (i), (ii) and (iii) of the theorem statement. Steps 2 and 3 are devoted to showing that a lower semi-continuous function satisfying conditions (i), (ii) and (iii) is the value function. More precisely, in Step 2 we prove that if V is a lower semi-continuous function, which satisfies conditions (i) and (iii), then 'inf($P_{t,x}$) $\leq V(t, x)$'. In Step 3 we show that if V is a lower semi-continuous locally bounded function satisfying conditions (ii) and (iii) and, in addition, (HS), then 'inf($P_{t,x}$) $\geq V(t, x)$'.

Step 1 Let V be the value function. Assume that, in addition to (H1)–(H5), g is local bounded. We deduce from Proposition 13.3.6 that V is lower semi-continuous and locally bounded and so, in particular, dom $V = [S, T] \times \mathbb{R}^n$.

Fix any point $(t, x) \in (S, T) \times \mathbb{R}^n$. Take a test function ψ such that $V - \psi$ achieves a local minimum at (t, x). Then, since ψ is continuously differentiable, for all (t', x') in a neighbourhood of (t, x), we have

$$\nabla_t \psi(t, x)(t' - t) + \nabla_x \psi(t, x) \cdot (x' - x) \leq V(t', x') \\ - V(t, x) + o(|(t' - t, x' - x)|) \quad (13.12.3)$$

in which $o : \mathbb{R}_+ \to \mathbb{R}_+$ is a function satisfying $o(\epsilon)/\epsilon \to 0$ as $\epsilon \downarrow 0$. Arguing as in the proof of Theorem 13.3.7, we can find sequences $h_i \downarrow 0$ and $v_i \to \bar{v}$, for some $\bar{v} \in F(t^+, x)$, such that

$$\lim_{i \to +\infty} h_i^{-1}(V(t + h_i, x + h_i v_i) - V(t, x)) \leq 0.$$

Taking $(t', x') = (t + h_i, x + h_i v_i)$ in (13.12.3), and dividing across by h_i, we obtain, for i sufficiently large,

$$\nabla_t \psi(t, x) + \nabla_x \psi(t, x) \cdot v_i \leq h_i^{-1}(V(t+h_i, x+h_i v_i) - V(t, x)) + h_i^{-1} o(h_i \sqrt{1 + |v_i|^2}).$$

Therefore, in the limit as $i \to \infty$, it follows that

$$\nabla_t \psi(t, x) + \inf_{v \in F(t^+, x)} \nabla_x \psi(t, x) \cdot v \leq \nabla_t \psi(t, x) + \nabla_x \psi(t, x) \cdot \bar{v} \leq 0.$$

We have confirmed that the value function satisfies condition (i).

Fix $(t, x) \in (S, T) \times \mathbb{R}^n$. There exists a sequence of points $(t_i, x_i) \in (S, T) \times \mathbb{R}^n \setminus \{(t, x)\}$ such that $(t_i, x_i) \to (t, x)$ as $i \to \infty$ and

$$\lim_{i \to \infty} V(t_i, x_i) = V^*(t, x). \tag{13.12.4}$$

We claim that we can restrict attention to the case when $t_i > t$ for all i. Indeed, either we can extract a subsequence (we do not relabel) such that $t_i > t$ for all i, or there exists i_0 such that $t_i \leq t$ for all $i \geq i_0$. In this latter case, we take a strictly decreasing sequence of elements $\tau_i \in (t, T]$ such that $\tau_i \downarrow t$. Fix any $i \geq i_0$, and consider any F trajectory $x_i \in W^{1,1}([t_i, T]; \mathbb{R}^n)$ such that $x_i(t_i) = x_i$. By the principle of optimality,

$$V(t_i, x_i) \leq V(\tau_i, x_i(\tau_i)), \quad \text{for all } i.$$

Since V^* is upper semi-continuous, we deduce:

$$V^*(t, x) = \limsup_{i \to \infty} V(t_i, x_i) \leq \limsup_{i \to \infty} V(\tau_i, x_i(\tau_i))$$
$$\leq \limsup_{i \to \infty} V^*(\tau_i, x_i(\tau_i)) \leq V^*(t, x).$$

Hence, $\limsup_{i \to \infty} V(\tau_i, x_i(\tau_i)) = V^*(t, x)$. But then, along a subsequence (again, we do not relabel),

$$\lim_{i \to \infty} V(\tau_i, x_i(\tau_i)) = V^*(t, x).$$

13.12 Viscosity Solutions of the Hamilton Jacobi Equation

We have shown that, in the latter case, we can replace the original sequence $\{(t_i, x_i)\}$ by a new sequence $\{(\tau_i, x_i(\tau_i))\}$ such that $\tau_i > t$ for all i and (13.12.4) is satisfied. The claim is confirmed.

Now take any $\tilde{v} \in F(t^+, x)$. Then there exists a sequence of vectors v_i such that $v_i \in F(t_i^+, x_i)$ and $\lim_{i \to \infty} v_i = \tilde{v}$. For every $i \in \mathbb{N}$, define the arc

$$y_i(s) := x_i + (s - t_i)v_i, \text{ for all } s \in [t_i, T].$$

It follows from the Filippov existence theorem that, for each i, there exists an F trajectory z_i such that $z_i(t_i) = x_i$ and, for any $h \in (0, T - t_i]$,

$$\|z_i - y_i\|_{L^\infty([t_i, t_i+h], \mathbb{R}^n)} \leq K \left(\int_{t_i}^{t_i+h} d_{F(s, y_i(s))}(v_i) ds \right),$$

where $K = \exp\left(\int_S^T k_F(s) ds \right)$. In consequence of hypothesis (H2)(i), we can choose $R_0 > 0$ such that, for each $i \in \mathbb{N}$, $|y_i(s)| \leq R_0$ for all $s \in [t_i, T]$. Observe also that, from (H2)(ii), $|\dot{y}_i(s)| \leq c_0$ for a.e. $s \in [t_i, T]$, for each i.

Define, for each i, $\delta_i = \max\{|V(t_i, x_i) - V^*(t, x)|, |x_i - x|, |t_i - t|\}$. Take a strictly decreasing sequence $h_i \downarrow 0$ such that, for each i, $h_i \geq \sqrt{\delta_i}$.

Fix i, and let $w_i := \frac{1}{h_i} \int_{t_i}^{t_i+h_i} \dot{z}_i(s) ds$. Note that

$$|v_i - w_i| \leq K \left(\int_{t_i}^{t_i+h_i} d_{F(s, y_i(s))}(v_i) ds \right) \leq K \theta_i(h_i),$$

where

$$\theta_i(h) := \begin{cases} \sup\{d_H(F(s, y), F(t_i^+, x_i)) : & \text{if } h \neq 0, \\ \quad 0 < s - t_i \leq h \text{ and } |x_i - y| \leq c_0 h\} \\ 0 & \text{otherwise.} \end{cases}$$

There exists $\tau \in [t_i, t_i + h_i]$ and $z \in \mathbb{R}^n$ such that $|x_i - z| \leq c_0 h_i$ and

$$\theta_i(h_i) \leq d_H(F(\tau, z), F(t_i^+, x_i)) + \frac{1}{i+1}.$$

It follows that

$$\theta_i(h_i) \leq d_H(F(\tau, z), F(t^+, x)) + d_H(F(t^+, x), F(t_i^+, x_i)) + \frac{1}{i+1}.$$

Notice that:

$$\tau - t \leq (\tau - t_i) + (t_i - t) \leq h_i + h_i^2$$

and $|x - z| \leq |z - x_i| + |x_i - x| \leq h_i c_0 + h_i^2.$

We deduce that

$$\theta_i(h_i) \le \sup\left\{d_H(F(s,y), F(t^+, x)) : 0 < s - t \le h_i + h_i^2\right.$$
$$\left. \text{and } |y - x| \le c_0 h_i + h_i^2\right\}$$
$$+ \sup\left\{d_H(F(t^+, x), F(s^+, y)) : 0 < s - t \le h_i^2\right.$$
$$\left. \text{and } |y - x| \le h_i^2\right\}$$
$$+ (1+i)^{-1}.$$

This implies that $\theta_i(h_i) \to 0$, as $i \to \infty$. Since, for each i, $|\tilde{v} - w_i| \le \theta_i(h_i) + |v_i - \tilde{v}|$, it follows that

$$w_i \to \tilde{v}, \text{ as } i \to \infty.$$

Define, for each i, $e_i := 1 - \frac{t - t_i}{h_i}$ and $\tilde{w}_i := w_i - \frac{x - x_i}{h_i}$. Since $(e_i, \tilde{w}_i) \to (1, \tilde{v})$, as $i \to \infty$, we have

$$\limsup_{i \to \infty} h_i^{-1} \left[V^*(t + h_i e_i, x + h_i \tilde{w}_i) - V^*(t, x)\right]$$
$$= \limsup_{i \to \infty} h_i^{-1} \left[V^*(t_i + h_i, x_i + h_i w_i) - V^*(t, x)\right]$$
$$\ge \limsup_{i \to \infty} h_i^{-1} \left[V(t_i + h_i, x_i + h_i w_i) - V^*(t, x)\right].$$

Fix i. Then

$$V(t_i + h_i, x_i + h_i w_i) - V^*(t, x) \ge V(t_i + h_i, x_i + h_i w_i) - V(t_i, x_i) - \delta_i. \quad (13.12.5)$$

From the principle of optimality,

$$V(t_i + h_i, x_i + h_i w_i) - V(t_i, x_i) \ge 0.$$

Dividing across inequality (13.12.5) by h_i and passing to the limit, while recalling $\frac{\delta_i}{h_i} \le \sqrt{\delta_i}$ for each i, yields:

$$\limsup_{i \to \infty} h_i^{-1} \left[V^*(t + h_i e_i, x + h_i \tilde{w}_i) - V^*(t, x)\right] \ge 0. \quad (13.12.6)$$

Now, take a test function ψ such that $V^* - \psi$ achieves a local maximum at (t, x). Then, for all (t', x') in a neighbourhood of (t, x), we obtain

$$V^*(t', x') - V^*(t, x) - \nabla_t \psi(t, x)(t' - t) - \nabla_x \psi(t, x) \cdot (x' - x) \le o(|y - x|) \quad (13.12.7)$$

13.12 Viscosity Solutions of the Hamilton Jacobi Equation

where $o : \mathbb{R}_+ \to \mathbb{R}_+$ is a function such that $o(\epsilon)/\epsilon \to 0$ as $\epsilon \downarrow 0$. Setting $(t', x') = (t + h_i e_i, x + h_i \tilde{w}_i)$ in (13.12.7) and dividing across by h_i yields

$$h_i^{-1}(V^*(t + h_i e_i, x + h_i \tilde{w}_i) - V^*(t, x)) - \nabla_t \psi(t, x) e_i - \nabla_x \psi(t, x) \cdot \tilde{w}_i$$
$$\leq h_i^{-1} o(h_i |(e_i, \tilde{w}_i)|).$$

It follows from these relations, in the limit as $i \to \infty$, that

$$- \nabla_t \psi(t, x) - \nabla_x \psi(t, x) \cdot \tilde{v} \leq 0.$$

This inequality is valid for all points $\tilde{v} \in F(t^+, x)$. We have confirmed condition (ii).

Consider, finally, (iii). The first condition can be verified by the same argument as that employed in the proof Theorem 13.3.7. It remains to prove that $V^*(T, .) = g^*$. It follows immediately from the relation $V(T, .) = g$ that $V^*(T, x) \geq g^*(x)$ for every $x \in \mathbb{R}^n$. We show that the converse inequality is also true.

Fix $x \in \mathbb{R}^n$. There exists a sequence $\{(t_i, x_i)\}$ in $[S, T] \times \mathbb{R}^n \setminus \{(T, x)\}$ converging to (T, x) such that:

$$\lim_{i \to \infty} V(t_i, x_i) = \limsup_{(t,y) \to (T,x)} V(t, y) = V^*(T, x).$$

For every i, there exists an F trajectory $x_i \in W^{1,1}([t_i, T]; \mathbb{R}^n)$ such that $x_i(t_i) = x_i$. By the principle of optimality,

$$V(t_i, x_i) \leq V(T, x_i(T)) = g(x_i(T)).$$

Since $x_i(T) \to x$ as $i \to \infty$, we deduce that

$$V^*(T, x) \leq \limsup_{i \to \infty} g(x_i(T)) \leq \limsup_{y \to x} g(y) = g^*(x).$$

We have shown that $V^*(T, x) \leq g^*(x)$. This relation combines with the earlier inequality to yield $V^*(T, x) = g^*(x)$.

Step 2 Assume that V is a lower semi-continuous locally bounded function satisfying conditions (i) and (iii). Take any point $(t, x) \in (S, T) \times \mathbb{R}^n$ and $(\xi^0, \xi^1) \in \partial^P V(t, x)$. Then there exists $M > 0$ such that the test function $\psi(s, y) := \xi^0(s - t) + \xi^1(y - x)M|(s, y) - (t, x)|^2$ is a test function such that $V(s, y) - \psi(s, y)$ is minimized at (t, x). It follows from condition (i) that

$$\xi^0 + \min_{v \in F(t^+, x)} \xi^1 \cdot v \leq 0, \text{ for all } (\xi^0, \xi^1) \in \partial^P V(t, x).$$

Thus V satisfies (c)(i) of Theorem 13.3.7. Since V also satisfies condition (iii), we can use the same arguments employed in the proof of Theorem 13.3.7 to show that

$$\inf(P_{\bar{t},\bar{x}}) \leq V(\bar{t},\bar{x}), \text{ for every } (\bar{t},\bar{x}) \in [S,T] \times \mathbb{R}^n.$$

Step 3 Assume that V is a lower semi-continuous, locally bounded function satisfying conditions (ii) and (iii). Suppose also that $(g^*)_* = g$. Fix $(\bar{t},\bar{x}) \in [S,T] \times \mathbb{R}^n$. Let $x \in W^{1,1}([\bar{t},T];\mathbb{R}^n)$ be an F trajectory such that $x(\bar{t}) = \bar{x}$. We want to prove prove that:

$$V(\bar{t},\bar{x}) \leq g(x(T)).$$

First observe that we can find a sequence of points $\xi_j \in \mathbb{R}^n$ such that $\xi_j \to x(T)$ as $j \to \infty$ and

$$\lim_{j \to +\infty} g^*(\xi_j) = (g^*)_*(x(T)).$$

In consequence of the Filippov existence theorem, there exists, for each j, an F trajectory x_j on $[\bar{t},T]$ such that $x_j(T) = \xi_j$ and

$$\|x_j - x\|_{L^\infty} \leq K|\xi_j - x(T)| \quad \text{for all } j,$$

in which $K := \exp\{\int_S^T k_F(s)\,ds\}$. Write $\bar{x}_j := x_j(\bar{t})$.

Fix j. Consider the system

$$\begin{cases} (\dot{x},\dot{b}) \in \Gamma(t,(x,b)) & \text{a.e. } t \in [\bar{t},T] \\ (x(t),b(t)) \in P(t) & \text{for all } t \in [\bar{t},T], \\ (x(\bar{t}),b(\bar{t})) = (\bar{x}_j, V^*(\bar{t},\bar{x}_j)), \end{cases} \quad (13.12.8)$$

where the multifunctions $\Gamma : [\bar{t},T] \times \mathbb{R}^{n+1} \leadsto \mathbb{R}^{n+1}$ and $P : [\bar{t},T] \leadsto \mathbb{R}^{n+1}$ are defined as follows

$$\Gamma(t,(x,b)) := F(t,x) \times \{0\},$$

$$P(t) := \{(x,\beta) : V^*(t,x) \geq \beta\}$$

respectively. We shall apply the strong invariance theorem (Theorem 13.2.4) to the constrained differential inclusion (13.12.8). The multifunctions Γ and P clearly possess the necessary closure and regularity properties for this application of the theorem. Let us check the hypothesis

$$(\bar{x}_j, V^*(S,\bar{x}_j)) \in \limsup_{s' \downarrow S} P(s'). \quad (13.12.9)$$

According to the definition of the upper semi-continuous envelope function, there exists a sequence of points $\{(s_i,x_i)\}$ in $[S,T]$ such that

13.12 Viscosity Solutions of the Hamilton Jacobi Equation

$$(s_i, x_i) \to (S, \bar{x}_j) \quad \text{and} \quad V(s_i, x_i) \to V^*(S, \bar{x}_j), \text{ as } i \to \infty. \qquad (13.12.10)$$

Notice that we can assume $s_i > S$ for each i. Indeed, if $s_i = S$ for some i, then, invoking the first condition of (iii), we can find a sequence of points $\{(s_i^k, x_i^k)\}_{k=1}^\infty$ in $(S, T]$ such that $(s_i^k, x_i^k) \to (s_i, x_i)$ and $V(s_i^j, x_i^k) \to V(s_i, x_i)$, as $j \to \infty$. We can then replace the element (s_i, x_i) by some element from the sequence $\{(s_i^k, x_i^k)\}_{j=1}^\infty$, such that (13.12.10) continues to be satisfied but, now, $s_i > S$, for each i.

We know that $(x_i, V(s_i, x_i)) \in P(s_i)$ for each i, since $V^* \geq V$. This fact combines with (13.12.10) to yield the information that $(\bar{x}_j, V^*(S, \bar{x}_j)) \in \limsup_{i \to \infty} P(s_i)$. Since $s_i > S$ for each i, we deduce (13.12.9).

It remains to check the 'inward pointing' hypothesis. Take any point $(s, (x, \beta)) \in \operatorname{Gr} P$, satisfying $S < s < T$, and any vector $(\zeta^0, \zeta^1, \lambda) \in N_{\operatorname{Gr} P}^P(s, (x, \beta))$. It follows from the proximal normal inequality that

$$(\zeta^0, \zeta^1, \lambda) \in N_{\text{hyp } V^*}^P(s, x, \beta).$$

We must show that

$$\zeta^0 + \min\left\{ \max_{w \in \Gamma(s^-, (x, \beta))} \zeta^1 \cdot v, \max_{w \in \Gamma(s^+, (x, \beta))} \zeta^1 \cdot w \right\} \leq 0.$$

In view of the definition of Γ, this relation can be equivalently expressed:

$$\zeta^0 + \min\left\{ \max_{v \in F(s^-, x)} \zeta^1 \cdot v; \max_{v \in F(s^+, x)} \zeta^1 \cdot v \right\} \leq 0. \qquad (13.12.11)$$

Since $\operatorname{Gr} P$ is the hypograph of an upper semi-continuous function, the proximal normal vector $(\zeta^0, \zeta^1, \lambda)$ must be such that $\lambda \geq 0$. Notice also that (s, x, β) is an interior point to the hypograph, if $\beta < V^*(s, x)$; in this case $(\zeta^0, \zeta^1) = (0, 0)$ and (13.12.11) is automatically satisfied. So we can assume that $\beta = V^*(s, x)$. We need consider two distinct cases.

$\lambda > 0$: in this case, $\lambda^{-1}(\zeta^0, \zeta^1) \in \partial^P(-V^*)(s, x)$. Then, by the proximal normal inequality for lower semi-continuous functions, there exist $M > 0$ with the following properties: the test function $\psi(s', x') := -\lambda^{-1}\zeta^0(s' - s) - \lambda^{-1}\zeta^1 \cdot (x' - x) + M|(s', x') - (s, x)|^2$ is such that $V^* - \psi$ has a local maximum at (s, x) and $(\nabla_t \psi, \nabla_x \psi)(s, x) = -\lambda^{-1}(\zeta^0, \zeta^1)$. It follows from condition (ii) that

$$\lambda^{-1}\zeta^0 + \sup_{v \in F(s^+, x)} \lambda^{-1}\zeta^1 \cdot v = -\nabla_t \psi(s, x) - \min_{v \in F(s^+, x)} \nabla_x \psi(s, x) \cdot v \leq 0.$$

This implies (13.12.11).

$\lambda = 0$: in this case, there exist sequences $(\zeta_i^0, \zeta_i^1) \to (\zeta^0, \zeta^1)$, $\lambda_i \downarrow 0$ and $(s_i, x_i) \to (s, x)$ such that, for each $i \geq 0$,

$$(\lambda_i^{-1}\zeta_i^0, \lambda_i^{-1}\zeta_i^1) \in \partial^P(-V^*)(s_i, x_i).$$

Considering the test function $\psi_i(s', x') := -\lambda_i^{-1}\zeta_i^0(s' - s) - \lambda^{-1}\zeta_i^1 \cdot (x' - x) + M_i|(s', x') - (s, x)|^2$, for suitably chosen $M_i > 0$, we can show, as in the preceding case, that

$$\zeta_i^0 + \sup_{v \in F(s_i^+, x_i)} \zeta_i^1 \cdot v \leq 0.$$

By extracting suitable subsequences and arguing as in the proof of Theorem 13.3.7 we deduce that (13.12.11) is verified also in this case.

We have verified that the hypotheses are satisfied for application of the strong invariance theorem (Theorem 13.2.4) is applicable to the differential inclusion and constraint (13.12.8) and therefore may conclude that the Γ trajectory $(x_j, b_j \equiv V(\bar{t}, \bar{x}_j))$ satisfies

$$(x_j(t), b_j(t)) \in P(t) \quad \text{for all } t \in [\bar{t}, T].$$

Taking $t = T$, we see that

$$V^*(\bar{t}, \bar{x}_j) \leq V^*(T, x_j(T)) = g^*(x_j(T)) \, (= g^*(\xi_j)), \quad \text{for every } j. \tag{13.12.12}$$

Since $\|x_j - x\|_{L^\infty} \to 0$ as $j \to \infty$, $V \leq V^*$, and V is lower semi-continuous, we may pass to the limit in (13.12.12) to obtain

$$V(\bar{t}, \bar{x}) \leq \liminf_{j \to \infty} g^*(\xi_j).$$

Since $\lim_{j \to +\infty} g^*(\xi_j) = (g^*)_*(x(T))$ and $(g^*)_* = g$, it follows that

$$V(\bar{t}, \bar{x}) \leq g(x(T)).$$

Recalling that x was an arbitrary F trajectory such that $x(\bar{t}) = \bar{x}$, we deduce that

$$V(\bar{t}, \bar{x}) \leq \inf(P_{\bar{t},\bar{x}}).$$

Step 3 is complete.

□

Theorem 13.12.3 (Viscosity Solution Characterization of Value Functions for State Constrained Problems: Inward/Outward-Pointing Condition) *Consider (SC). Assume (H1), (H2)(i), (H3), (H4)(i) and (H5) of Theorem 13.3.7, (BV), and (CQ)$_{outward}$ and (CQ)$_{inward}$ of Theorems 13.6.1 and 13.6.2. Suppose, in addition, that $g_{|A}$ is locally bounded and satisfies $((g_{|A})^*)_* = g_{|A}$. Take a lower semi-continuous, locally bounded function $V : [S, T] \times \mathbb{R}^n \to \mathbb{R}$ such that $V(t, x) = +\infty$ when $x \notin A$.*

13.12 Viscosity Solutions of the Hamilton Jacobi Equation

Then V is the value function for (SC) if and only if V is a locally bounded function on $[S, T] \times A$, lower semi-continuous constrained viscosity solution of (13.12.1) in the following sense:

(i) (V is a viscosity supersolution) for any point $(t, x) \in (S, T) \times A$ and any C^1 function $\psi : \mathbb{R} \times \mathbb{R}^n \to \mathbb{R}$ such that

$$(t', x') \to V(t', x') - \psi(t', x')$$

has a local minimum at (t, x) (relative to $[S, T] \times A$) we have

$$-\psi_t(t, x) + H^+(t, x, -\psi_x(t, x)) \geq 0,$$

(ii) (V is a viscosity subsolution) for any point $(t, x) \in (S, T) \times \text{int } A$ and any C^1 function $\psi : \mathbb{R} \times \mathbb{R}^n \to \mathbb{R}$ such that

$$(t', x') \to V^*(t', x') - \psi(t', x')$$

has a local maximum at (t, x) (relative to $[S, T] \times A$) we have

$$-\psi_t(t, x) + H^+(t, x, -\psi_x(t, x)) \leq 0,$$

(iii) for all $x \in A$

$$\liminf_{\{(t',x') \to (S,x) | t' > S\}} V(t', x') = V(S, x),$$

$$(V_{|[S,T] \times A})^*(T, x) = (g_{|A})^*(x) \quad \text{and} \quad V(T, x) = g(x).$$

Proof (Sketch) The proof is based on the following two relations: 'the value function is a constrained viscosity solution (i.e. satisfies (i), (ii) and (iii) of (13.12.1)' and 'if V is a locally bounded function on $[S, T] \times A$ which is also a lower semi-continuous constrained viscosity solution of (13.12.1) then V is the value function'.

The first relation is actually valid even if the outward pointing constraint qualification and condition '$((g_{|A})^*)_* = g_{|A}$' are not in force. Indeed, making use also of the distance estimate of Theorem 6.6.2, one can prove the following proposition. □

Proposition 13.12.4 *Assume (H1), (H2)(i), (H3), (H4)(i) and (H5) of Theorem 13.3.7, (BV), and* $(CQ)_{inward}$*. Suppose, in addition, that $g_{|A}$ is locally bounded. Then V satisfies (i), (ii) and (iii) of Theorem 13.12.3.*

To complete the proof of Theorem 13.12.3 one must show the second relation. With this objective in mind we assume that (H1), (H2)(i), (H3) and (H5) of Theorem 13.3.7 and (BV) are satisfied, and consider a lower semi-continuous function $V : [S, T] \times \mathbb{R}^n \to \mathbb{R} \cup \{+\infty\}$, locally bounded function on

$[S, T] \times A$, such that $V(t, x) = +\infty$ whenever $x \notin A$. One can prove the following properties:

(A) if V satisfies (d)(i) and $V(T, x) = g(x)$ for all $x \in A$, then $\inf(P_{t,x}) \leq V(t, x)$ for any $(t, x) \in [S, T] \times A$,

(B) if, in addition, (CQ)$_{\text{outward}}$ holds, g is a locally bounded on A such that $((g_{|A})^*)_* = g_{|A}$, and V satisfies (d)(ii) and (d)(iii), then $\inf(P_{t,x}) \geq V(t, x)$ for any $(t, x) \in [S, T] \times A$.

It is not difficult to see that the proof of (A) can be reduced to the analysis employed in the proof of Thm 13.6.1. On the other hand, the proof of (B) requires a different construction of several sequences of arcs, taking into account the condition '$((g_{|A})^*)_* = g_{|A}$', and the notions of lower/upper semi-continuous envelopes.

13.13 A Comparison Theorem for Viscosity Solutions

A key feature of viscosity solutions methods is to provide *direct* proofs of comparison theorems, which in turn lead to uniqueness of viscosity solutions to the Hamilton Jacobi equation of dynamic programming. This contrasts with approaches to proving uniqueness employed earlier in this chapter, based on invariance properties of state trajectories, where the proof that the Hamilton Jacobi equation has a unique generalized solution *and* that it coincides the value function comes as a combined package. Since the viscosity solutions methods are used to prove uniqueness alone, it is unsurprising then that they are capable of establishing this property, under hypotheses that are in some respects weaker than those required by the system theoretic approach. Indeed, viscosity solutions methods can be used to prove uniqueness of viscosity solutions, even when the Hamilton Jacobi equation does not arise from a dynamic optimization problem at all.

Our aim in this section is to give the flavour of viscosity solutions methods, by proving a comparison theorem and examining its implications. The comparison theorem concerns the Hamilton Jacobi equation connected with the infinite horizon problem, investigated earlier in the chapter by means of system theoretic techniques. Comparison theorems can be proved also for finite horizon problems, but the viscosity solutions techniques involved are more complicated.

For given $x \in \mathbb{R}^n$, consider then

$$(P_\infty)(x) \begin{cases} \text{Minimize } \int_0^\infty e^{-\lambda t} L(x(t), u(t)) dt \\ \text{over } x \in W^{1,1}_{\text{loc}}([0, \infty); \mathbb{R}^n) \text{ and} \\ \qquad \text{measurable functions } u : [0, \infty) \to \mathbb{R}^m \text{ satisfying} \\ \dot{y}(t) = f(y(t), u(t)), \text{ a.e. } t \in [0, \infty), \\ u(t) \in U, \text{ a.e. } t \in [0, \infty), \\ y(0) = x, \end{cases}$$

13.13 A Comparison Theorem for Viscosity Solutions

in which $f : \mathbb{R}^n \times \mathbb{R}^m \to \mathbb{R}^n$ and $L : \mathbb{R}^n \times \mathbb{R}^m \to \mathbb{R}$ are given functions, $U \subset \mathbb{R}^m$ is a given set and $\lambda > 0$.

The value function $V : \mathbb{R}^n \to \mathbb{R}$ associated with this problem is

$$V(x) := \inf \int_0^\infty e^{-\lambda t} L(y(t), u(t)) dt,$$

in which the infimum is taken over infinite horizon processes (y, u) for $(P_\infty)(x)$ (as earlier defined) satisfying the specified constraints and such that $t \to L(y(t), u(t))$ is integrable.

Depending on the hypotheses imposed on the data for $(P)_\infty$, the value function V a 'solution' (in the classical, viscosity or proximal sense) to the Hamilton Jacobi equation

$$\lambda v(x) + H(x, v_x(x)) = 0, \quad \text{for all } x \in \mathbb{R}^n, \tag{13.13.1}$$

in which the function $H : \mathbb{R}^n \times \mathbb{R}^n \to \mathbb{R}$ is the function

$$H(x, p) := \sup_{u \in U} \left(- p \cdot f(x, u) - L(x, u) \right). \tag{13.13.2}$$

In this section we deal with value functions which, under the standing assumptions, turn out to be continuous. Accordingly, we introduce the notion of viscosity solution of the Hamilton Jacobi equation (13.13.1) in a continuous context. We shall say that a continuous function $v : \mathbb{R}^n \to \mathbb{R}$ is a *viscosity subsolution* of (13.13.1) if, for any C^1 function $\phi : \mathbb{R}^n \to \mathbb{R}$, we have

$$\lambda v(x) + H(x, \nabla \phi(x)) \leq 0, \quad \text{for all } x \in \mathbb{R}^n$$

at any local maximum point $\bar{x}$ of $v - \phi$. A continuous function $v : \mathbb{R}^n \to \mathbb{R}$ is a *viscosity supersolution* of (13.13.1) if, for any C^1 function $\phi : \mathbb{R}^n \to \mathbb{R}$, we have

$$\lambda v(x) + H(x, \nabla \phi(x)) \geq 0, \quad \text{for all } x \in \mathbb{R}^n$$

at any local minimum point $\bar{x}$ of $v - \phi$. v is a *viscosity solution* if it is simultaneously a sub- and supersolution.

We refer to the following hypotheses. Note that they are expressed directly in terms of the function H appearing in the Hamilton Jacobi equation, consistent with the goal of proving properties of viscosity solutions without reference to the underlying control system.

(A1): There exists a modulus of continuity $\omega_1 : \mathbb{R}_+ \to \mathbb{R}_+$ such that

$$|H(x, p) - H(y, p)| \leq \omega_1(|x-y|(1+|p|)), \quad \text{for all } x, y \in \mathbb{R}^n \text{ and } p \in \mathbb{R}^n.$$

(A2): For all $R > 0$ there exists a modulus of continuity $\omega_2 : \mathbb{R}_+ \to \mathbb{R}_+$ such that

$$|H(x, p) - H(x, q)| \leq \omega_2(|p - q|), \quad \text{for all } x \in R\mathbb{B} \text{ and } p, q \in \mathbb{R}^n.$$

To achieve a characterization of the value function for $(P_\infty)(x)$ based on Theorem 13.13.1 below, we need first to show that the value function is a bounded and continuous viscosity solution of (13.13.1). This is accomplished by the following proposition.

Proposition 13.13.1 *Assume (H1)–(H3) (of Sect. 13.10). Then, the value function $V : \mathbb{R}^n \to \mathbb{R}$ is a bounded, uniformly continuous function that is a viscosity solution of (13.13.1).*

Proof From Proposition 13.10.2 we already know that V is a bounded, uniformly continuous function. So, here, we only show that V is a viscosity solution of (13.13.1).

Consider a C^1 function $\phi : \mathbb{R}^n \to \mathbb{R}$ and a point $x_0 \in \mathbb{R}^n$ which is a local maximum point for $V - \phi$. Therefore, there exists $\delta > 0$ such that $\phi(x_0) - \phi(x) \leq V(x_0) - V(x)$ for all $x \in x_0 + \delta\mathbb{B}$. Take any $v \in U$. Let $(x, u \equiv v)$ be the process on $[0, \infty)$ satisfying $x(0) = x_0$. For $h_0 > 0$ sufficiently small $x(t) \in x_0 + \delta\mathbb{B}$ for all $t \in [0, h_0]$. It follows that

$$\phi(x_0) - \phi(x(t)) \leq V(x_0) - V(x(t)), \quad \text{for all } t \in [0, h_0].$$

Applying (ii) of Proposition 13.10.1 to the process $(x, u \equiv v)$, we deduce that, for all $t \in [0, h_0]$,

$$\phi(x_0) - \phi(x(t)) \leq \int_0^t e^{-\lambda s} L(x(s), v) ds + (e^{-\lambda t} - 1) V(x(t)).$$

Dividing across this inequality by $t \in (0, h_0]$ and passing to the limit as $t \downarrow 0$ we obtain

$$-\nabla \phi(x_0) \cdot f(x_0, v) \leq L(x_0, v) - \lambda V(x_0).$$

Taking into account that $v \in U$ is arbitrary, we conclude that

$$\lambda V(x_0) + \sup_{v \in U} \left(-\nabla \phi(x_0) \cdot f(x_0, v) - L(x_0, v) \right) \leq 0.$$

This confirms that V is a viscosity subsolution.

Now, we prove that V is a viscosity supersolution. Take any a C^1 function $\phi : \mathbb{R}^n \to \mathbb{R}$ and a point $x_0 \in \mathbb{R}^n$ that is a local minimum for $V - \phi$: we can find $\delta > 0$ such that $V(x_0) - \phi(x_0) \leq V(x) - \phi(x)$ for all $x \in x_0 + \delta\mathbb{B}$. Let $(\bar{x}, \bar{u})$ be a minimizing process for $(P_\infty)(x_0)$. There exists $h_0 > 0$ sufficiently small $\bar{x}(h) \in x_0 + \delta\mathbb{B}$ for all $h \in [0, h_0]$. Apply (iii) of Proposition 13.10.1 to the process $(\bar{x}, \bar{u})$ with $T = h$. We deduce that, for all $h \in [0, h_0]$,

13.13 A Comparison Theorem for Viscosity Solutions

$$0 \leq \phi(x_0) - \phi(\bar{x}(h)) + V(\bar{x}(h)) - V(x_0)$$

$$= \phi(x_0) - \phi(\bar{x}(h)) - \int_0^h e^{-\lambda s} L(\bar{x}(s), \bar{u}(s)) ds$$

$$+ (1 - e^{-\lambda h}) V(\bar{x}(h)). \quad (13.13.3)$$

By means of arguments similar to those used in the proof of (ii) of Proposition 13.10.2, we deduce that there exists $h_i \downarrow 0$ and $\bar{v} \in U$ such that

$$h_i^{-1} \int_0^{h_i} \nabla \phi(\bar{x}(s)) \cdot f(\bar{x}(s), \bar{u}(s)) ds \to \nabla \phi(x_0) \cdot f(x_0, \bar{v}) \text{ and}$$
$$\lim_{i \to \infty} h_i^{-1} \int_0^{h_i} e^{-\lambda s} L(\bar{x}(s), \bar{u}(s)) ds \geq L(x_0, \bar{v}).$$

Setting $h = h_i$ in (13.13.3), dividing across (13.13.3) by h_i and passing to the limit as $i \to \infty$ yields

$$0 \leq -\nabla \phi(x_0) \cdot f(x_0, \bar{v}) - L(x_0, \bar{v}) + \lambda V(x_0).$$

We conclude that V is also a viscosity supersolution. We have confirmed that V is a viscosity solution to the Hamilton Jacobi equation (13.13.1) and the proof is complete. □

Theorem 13.13.2 (Comparison Theorem) *Assume (A1) and (A2). Take bounded continuous functions $v_1, v_2 : \mathbb{R}^n \to \mathbb{R}$ such that v_1 is a viscosity subsolution and v_2 is a viscosity supersolution of (13.13.1). Then*

$$v_1(x) \leq v_2(x) \quad \text{for all } x \in \mathbb{R}^n.$$

In the case that v_1, v_2 are viscosity solutions, v_1, v_2 are viscosity super- and subsolutions respectively. Reversing the roles of v_1, v_2 we deduce from the theorem also that $v_1(x) - v_2(x) \geq 0$ for all x and so $v_1 = v_2$. We have shown:

Corollary 13.13.3 *Assume (A1) and (A2). The there is at most one viscosity solution to (13.13.1) in the class of bounded continuous functions.*

Remarks

(a): Notice that (A1) and (A2) are always satisfied when (H2) and (H3) (of Sect. 13.10) are in force,
(b): The function H derived from the data for the dynamic optimization problem according to (13.13.2) automatically satisfies the convexity condition

$$H(x, .) \text{ is convex for all } x \in \mathbb{R}^n,$$

in consequence of the fact that it is defined via a supremum operation. The fact that this convexity condition is absent from the hypotheses of Theorem 13.13.2 is highly significant because it means that Theorem 13.13.2 can be applied to zero-sum two player differential games, with dynamic constraint $\dot{x} = f(x, u, v)$, $(u, v) \in U \times V$ and running cost $L(x, u, v)$ involving the control variables u and v of the two players. Here the value V, parameterized by the initial state x is, under appropriate hypotheses, the unique viscosity solution of the Hamilton Jacobi equation (13.13.2) when, now, H is defined to be

$$H(x, p) := \inf_{v \in V} \sup_{u \in U} \Big(- p \cdot f(x, u, v) - L(x, u, v) \Big).$$

Proof of Theorem 13.13.2 Suppose the assertions of the theorem are false. Then there exists a point $\tilde{x} \in \mathbb{R}^n$ and $\delta > 0$ such that $v_1(\tilde{x}) - v_2(\tilde{x}) > \delta$.

Define the function $g : \mathbb{R}^n \to \mathbb{R}$ to be $g(x) = \frac{1}{2} \log_e(1 + |x|^2)$. Let $\beta > 0$ be a number that satisfies

$$\lambda^{-1} \omega_2(2\beta) \leq \delta/4 \quad \text{and} \quad \beta g(\tilde{x}) \leq \delta/4.$$

For given $\epsilon > 0$ define the function $\Phi : \mathbb{R}^n \times \mathbb{R}^n \to \mathbb{R}$ to be

$$\Phi(x, y) := v_1(x) - v_2(y) - \frac{|x - y|^2}{2\epsilon} - \beta(g(x) + g(y)).$$

The function Φ is continuous. In consequence of the facts that $v_1 - v_2$ is a bounded function and g is a non-negative function such that $\lim_{|x| \to \infty} g(x) = \infty$, we know that Φ has compact level sets. It follows that Φ has a maximizer over $\mathbb{R}^n \times \mathbb{R}^n$, which we write (x_ϵ, y_ϵ). We have

$$v_1(x_\epsilon) - v_2(y_\epsilon) \geq \Phi(x_\epsilon, y_\epsilon) \geq \Phi(\tilde{x}, \tilde{x})$$
$$= v_1(\tilde{x}) - v_2(\tilde{x}) - 2\beta g(\tilde{x}) \geq \delta - \delta/2 = \delta/2. \qquad (13.13.4)$$

Since $\Phi(x_\epsilon, y_\epsilon) \geq 0$, we have

$$\beta(g(x_\epsilon) + g(y_\epsilon)) + \frac{|x_\epsilon - y_\epsilon|^2}{2\epsilon} \leq v_1(x_\epsilon) - v_2(y_\epsilon) \leq C,$$

in which $C := \sup\{|v_1(x) - v_2(x)| : x \in \mathbb{R}^n\} < \infty$.

It follows from this relation that $|x_\epsilon - y_\epsilon| \to 0$ as $\epsilon \downarrow 0$.

Since g is a coercive, non-negative function, we can also deduce from this relation that there exists $R > 0$, independent of ϵ, such that $|(x_\epsilon, y_\epsilon)| \leq R$.

Since (x_ϵ, y_ϵ) is a maximizer for Φ, $2\Phi(x_\epsilon, y_\epsilon) \geq \Phi(x_\epsilon, x_\epsilon) + \Phi(y_\epsilon, y_\epsilon)$. Expanding this inequality and cancelling terms gives

$$\frac{|x_\epsilon - y_\epsilon|^2}{\epsilon} \leq v_1(x_\epsilon) - v_1(y_\epsilon) + v_2(x_\epsilon) - v_2(y_\epsilon).$$

Now the continuous functions v_1, v_2 are uniformly continuous on the compact set $R\mathbb{B}$. But then, in consequence of the preceding relation and the fact $|x_\epsilon - y_\epsilon| \to 0$,

$$\frac{|x_\epsilon - y_\epsilon|^2}{\epsilon} \to 0 \text{ as } \epsilon \downarrow 0. \qquad (13.13.5)$$

Now define the C^1 test functions

$$\phi(x) := \frac{|x - y_\epsilon|^2}{2\epsilon} + \beta g(x) \quad \text{and} \quad \psi(y) := -\frac{|y - x_\epsilon|^2}{2\epsilon} - \beta g(y).$$

From the facts that $\Phi(x, y_\epsilon)$ is maximized at $x = x_\epsilon$ and $-\Phi(x_\epsilon, y)$ is minimized at y_ϵ we deduce that $v_1 - \phi$ is maximized at x_ϵ and $v_2 - \psi$ is minimized at y_ϵ. Since v_1 and v_2 are viscosity sub- and supersolutions of (13.13.1) respectively, we know then that

$$\lambda v_1(x_\epsilon) + H(x_\epsilon, \epsilon^{-1}(x_\epsilon - y_\epsilon) + \beta \nabla g(x_\epsilon) \leq 0$$

and

$$\lambda v_2(y_\epsilon) + H(y_\epsilon, \epsilon^{-1}(x_\epsilon - y_\epsilon) - \beta \nabla g(y_\epsilon) \geq 0.$$

Subtracting these inequalities, dividing across by λ, taking note of hypotheses (A1) and (A2), and observing that $|\nabla g(x)| \leq 1$ for all x, we see that

$$v_1(x_\epsilon) - v_2(y_\epsilon) \leq \lambda^{-1}\omega_1(|x_\epsilon - y_\epsilon|(\epsilon^{-1}|x_\epsilon - y_\epsilon| + \beta) + \lambda^{-1}\omega_2(2\beta).$$

(Here we use that fact that the modulus of continuity ω_2 is taken with respect to the radius R and that $|(x_\epsilon, y_\epsilon)| \leq R$.) In view of (13.13.5) we can arrange, by choosing ϵ sufficiently small, that

$$\lambda^{-1}\omega_1(|x_\epsilon - y_\epsilon|(\epsilon^{-1}|x_\epsilon - y_\epsilon| + \beta) < \delta/4.$$

Since $\lambda^{-1}\omega_2(2\beta) \leq \delta/4$, we can conclude that

$$v_1(x_\epsilon) - v_2(y_\epsilon) < \delta/2.$$

We have arrived at a contradiction of (13.13.4) and the proof is complete. □

13.14 Exercises

13.1 *(Problems with Running Costs.)* Consider the dynamic optimization problem with running cost, parameterized by the initial state $x_0 \in \mathbb{R}^n$:

$$(P)(x_0) \begin{cases} \text{Minimize } g(x(T)) + \int_S^T L(t, x(t), \dot{x}(t))dt \\ \text{over } x \in W^{1,1}([S, T]; \mathbb{R}^n) \text{ satisfying} \\ \dot{x}(t) \in F(t, x(t)), \text{ a.e. } t \in [S, T], \\ x(S) = x_0, \end{cases}$$

in which $[S, T]$ is a given interval, $F : [S, T] \times \mathbb{R}^n \rightsquigarrow \mathbb{R}^n$ is a given multifunction and $g : \mathbb{R}^n \to \mathbb{R} \cup \{+\infty\}$ and $L : [S, T] \times \mathbb{R}^n \times \mathbb{R}^n \to \mathbb{R}$ are given functions. Assume

(H1): $g : \mathbb{R}^n \to \mathbb{R} \cup \{+\infty\}$ is lower semi-continuous,
(H2): F is continuous and takes closed, non-empty values. L is continuous and

$$\{(v, \alpha) \in \mathbb{R}^n \times \mathbb{R} : v \in F(t, x) \text{ and } \alpha \geq L(t, x, v)\}$$

is convex for all $(t, x) \in [S, T] \times \mathbb{R}^n$,

(H3): there exists $c > 0$ such that

$$|v| \leq c \text{ and } |L(t, x, v)| \leq c \text{ for all } (t, x, v) \in \text{Gr } F,$$

(H4): there exists $k > 0$ such that
$$F(t, x') \subset F(t, x) + k|x - x'| \mathbb{B}, \text{ for all } x, x' \in \mathbb{R}^n \text{ and } t \in [S, T],$$
and
$$|L(t, x', v) - L(t, x, v)| \leq k|x' - x|,$$
$$\text{for all } x', x \in \mathbb{R}^n, t \in [S, T] \text{ and } v \in \mathbb{R}^n.$$

Show that a lower semi-continuous function $V : [S, T] \times \mathbb{R}^n \to \mathbb{R} \cup \{+\infty\}$ is the value function if and only if

(i) for all $(t, x) \in ((S, T) \times \mathbb{R}^n) \cap \text{dom } V$, $(\xi^0, \xi^1) \in \partial^P V(t, x)$

$$\xi^0 + \inf_{v \in F(t,x)} \{\xi^1 \cdot v + L(t, x, v)\} = 0,$$

(ii) for all $x \in \mathbb{R}^n$,

$$\liminf_{\{(t',x') \to (S,x) : t' > S\}} V(t', x') = V(S, x)$$

and

$$\liminf_{\{(t',x') \to (T,x) : t' < T\}} V(t', x') = V(T, x) = g(x).$$

Hint: Reformulate the problem as one without running cost, by means of state augmentation.

13.14 Exercises

$$(P)(x_0, z_0) \begin{cases} \text{Minimize } g(x(T)) + z(T) \\ \text{over } (x, z) \in W^{1,1} \text{ satisfying} \\ (\dot{x}, \dot{z})(t) \in \{(v, \alpha) : v \in F(t, x) \text{ and } \alpha \in [L(t, x, v), c]\} \\ \text{a.e. } t \in [S, T], \\ x(S) = x_0, z(S) = z_0. \end{cases}$$

Characterize the value function $W(t, x, z)$ for the state augmented problem, using the theory of Sect. 13.3, noting that it has structure $W(t, x, z) = V(t, x) + z$, where V is the value function of the original problem.

13.2 *(Decrease Properties of Control Systems.)* Consider the control system

$$\begin{cases} \dot{x}(t) \in F(x(t)), \text{ a.e. } t \in [0, +\infty), \\ x(0) = x_0, \end{cases}$$

in which $F : \mathbb{R}^n \rightsquigarrow \mathbb{R}^n$ is a given multifunction and $x_0 \in \mathbb{R}^n$. Assume

(H1): F is continuous and takes closed convex non-empty values,
(H2): there exist constants $c > 0$ and $k > 0$ such that
$$F(x) \subset c\mathbb{B} \quad \text{and} \quad F(x') \subset F(x) + k|x - x'|\,\mathbb{B}, \text{ for all } x, x' \in \mathbb{R}^n.$$

Take any continuous function $V : \mathbb{R}^n \to \mathbb{R}$.

(i) *(Weak Decrease.)* Assume, for all $x \in \mathbb{R}^n$,

$$\min_{v \in F(x)} v \cdot \xi \leq 0, \quad \text{for all } \xi \in \partial^P V(x).$$

Show that there exists an F trajectory x on $[0, \infty)$ such that $x(0) = x_0$ and

$$V(x(t)) \leq V(x_0) \quad \text{for all } t \in [0, \infty).$$

(ii) *(Strong Decrease.)* Assume

$$\min_{v \in F(x)} v \cdot \xi \geq 0, \quad \text{for all } \xi \in \partial^P V(x), x \in \mathbb{R}^n.$$

Show that for *all* F trajectories x on $[0, \infty)$ such that $x(0) = x_0$ and

$$V(x(t)) \leq V(x_0).$$

Hint: (i) Apply the weak invariance theorem for autonomous systems to the state augmented control system

$$\begin{cases} (\dot x(t)), \dot a(t)) \in F(x(t)) \times \{0\}, \text{ a.e. } t \in [0, \infty) \\ (x(t), a(t)) \in \text{epi } V \quad \text{for all } t \in [0, \infty) \\ x(0) = x_0, \; z(0) = V(x_0). \end{cases}$$

(ii) Take any F trajectory x on $[0, \infty)$ such that $x(0) = x_0$. Fix any $T > 0$. Apply the strong invariance theorem for autonomous systems to the above extended-state control system, restricted to the time interval $[0, T]$, in reverse time.

13.3 Consider the dynamic optimization problem

$$(P_{t_0, x_0}) \begin{cases} \text{Minimize } g(x(1)) + \int_{t_0}^{1} L(t, x(t), \dot x(t)) dt \\ \text{over arcs } x \in W^{1,1}([t_0, 1]; \mathbb{R}) \text{ such that} \\ \dot x(t) \in F(t) \quad \text{for a.e. } t \in [t_0, 1], \\ x(t_0) = x_0, \end{cases}$$

where $t_0 \in [0, 1]$, $x_0 \in \mathbb{R}$,

$$F(t) := \begin{cases} [-1, 1], & \text{if } 0 \le t \le \tfrac{1}{2}, \\ [-\tfrac{1}{2}, \tfrac{1}{2}], & \text{if } \tfrac{1}{2} < t \le 1, \end{cases}$$

$$g(x) := \begin{cases} 1 + x, & \text{if } x > 0, \\ x, & \text{if } x \le 0, \end{cases}$$

and

$$L(t, x, v) := \begin{cases} (v + 1)^2, & \text{if } 0 < t \le \tfrac{1}{2}, \\ (v + \tfrac{1}{2})^2 + 2, & \text{if } \tfrac{1}{2} < t \le 1. \end{cases}$$

(i) Write the explicit expression of the value function $V : [0, 1] \times \mathbb{R} \to \mathbb{R}$.

(ii) Show that V satisfies conditions (b)–(c) of Theorem 13.3.7 and the characterization provided by Theorem 13.12.2.

(iii) Take the point $(t_0, x_0) = (\tfrac{1}{2}, \tfrac{1}{4})$ and show that the characterization provided by Theorem 13.12.2 must involve the right limits in the Hamiltonian (that means $F(t^+, x)$ and $L(t^+, x, v)$), for otherwise, if the limits were taken from the other side, the characterization of Theorem 13.12.2 would not be satisfied.

13.15 Notes for Chapter 13

Dynamic programming, initiated by Richard Bellman [20], is a large field and our focus in this chapter is restricted to some of the more important topics. These include the relation between the value function and generalized solutions of the Hamilton

Jacobi equation for deterministic dynamic optimization problems (over a finite or infinite horizon), extensions to allow for state constraints, minimum time problems, existence of verification functions and the interpretation of costate trajectories as gradients of value functions.

In his 1942 paper, Nagumo studied conditions under which a differential equation has a solution which evolves in a given closed set [162]. Extensions to differential inclusions, supplying on the one hand conditions for the existence of a solution to a differential equation evolving in a given closed set and, on the other, conditions for all solutions to evolve in the set, have been of continuing interest since the 1970's. Following Clarke et al. [84], we refer to them as weak and strong invariance theorems respectively. These are the main tools used in this chapter to characterize value functions as generalized solutions to the Hamilton Jacobi equation.

Weak invariance theorems can be regarded as existence theorems for solutions to differential inclusions on general closed domains—a viewpoint stressed by Deimling [93]. Weak invariance theorems are widely referred to as viability theorems (Aubin's terminology). They are the cornerstone of viability theory, in which broad issues are addressed, relating to existence of invariant trajectories (nature of 'viability domains', etc.).

Nagumo showed that, under an inward pointing hypothesis expressed in terms of the Bouligand tangent cone, there exists an invariant trajectory for a differential equation which can be constructed as a limit of polygonal arcs satisfying the given state constraint at mesh points. An abstract compactness argument assures the existence of node points for the polygonal arcs, with appropriate limiting properties.

Generalizations to differential inclusions were considered by a number of authors. Under an inward pointing hypothesis involving the Clarke tangent cone, existence of invariant trajectories (possessing also a monotonicity property) was established by Clarke [57] for Lipschitz multifunctions and for continuous multifunctions by Aubin-Clarke [12]. Weak invariance theorems for upper semicontinuous, convex-valued multifunctions (and, more generally, for multifunctions 'with memory') were proved by Haddad [119], under an inward pointing hypothesis involving the Bouligand tangent cone.

The Bouligand contingent cone condition is given a central role in Aubin's book [10]. This is because it is in some sense necessary for weak invariance and because it can be shown to be equivalent to various, apparently less restrictive, conditions which arise in applications.

We follow a different approach to building up sets of conditions for weak invariance, mapped out by Clarke et al. in [84]. It is to hypothesize an inward pointing condition, expressed in term of the proximal normal cone of the state constraint set (wherever it is non-empty), to construct a polygonal arc on a uniform mesh by 'proximal aiming' and obtain an invariant arc in the limit as the mesh size tends to zero. The advantage of the proximal aiming approach is simplicity, regarding both the proof of weak invariance theorems and also the investigation of alternative sets of conditions for weak invariance. In earlier proofs of weak invariance, taking inspiration from Nagumo's original ideas, abstract compactness arguments are required to select a special, non-uniform mesh for construction of polygonal arcs, to ensure convergence. The proximal aiming approach, by contrast, is 'robust' regarding mesh selection—we can, for convenience, choose a uniform mesh. The other point is that proximal aiming yields weak invariance theorems under an inward pointing hypothesis formulated in terms of the proximal normal cone, which yield as direct corollaries weak invariance under other commonly encountered formulations of the hypothesis, because the proximal normal formulation is the least restrictive among them.

The proof given here of the strong invariance theorem for time-dependent systems, based on application of the weak invariance theorem to a modified differential inclusion which in some sense penalizes deviations from the arc under consideration, draws on ideas from [84], adapted to allow for discontinuous time dependence. A simple, alternative proof based on 'Lipschitz' parameterizations, as in [107], can also be given.

Gonzalez' paper [116] establishing that locally Lipschitz value functions are almost everywhere solutions of the Hamilton Jacobi equation is typical of the early literature linking nonsmooth value functions and the Hamilton Jacobi equation. Clarke et al. [62] and Offin [165] showed that locally Lipschitz continuous value functions are generalized solutions of the Hamilton Jacobi equation, not in an almost every sense but according to a new definition of generalized solution (based on generalized gradients) that looked ahead to the 'pointwise' concepts of generalized solution which now dominate the field. Information about uniqueness of solutions was lacking at this stage, though a characterization of locally Lipschitz continuous value functions in terms of the Hamilton Jacobi equation was achieved, namely as the upper envelope of almost everywhere solutions of the Hamilton Jacobi inequality [116].

In a key advance [88], Crandall and Lions introduced the concept of viscosity solutions to Hamilton Jacobi equation in the class of uniformly continuous functions together with analytical techniques for establishing uniqueness of such solutions. The Hamilton Jacobi equations considered embraced not merely those arising in dynamic optimization, but also in differential games and the second order equations of stochastic dynamic optimization. An important subsequent development [19] was Barron and Jensen's characterization of lower semi-continuous, finite-valued value functions as unique 'lower semi-continuous viscosity solutions' to the Hamilton Jacobi equation of dynamic optimization. These authors adopted a simple new 'single differential' definition of viscosity solution to replace former definitions involving two partial differential inequalities. Bardi and Caputzo Docetta's monograph [17] provides extensive coverage of viscosity methods in dynamic optimization and differential games and of computational methods. (See also [105]).

The viscosity approach is to prove directly that the relevant Hamilton Jacobi equation has a unique solution (in the viscosity sense). It is then shown, as a separate undertaking, that the value function is such solution.

An alternative approach is to use invariance theorems to show directly that an arbitrary generalized solution simultaneously majorizes and minorizes the value function and, therefore, coincides with it. This approach is 'system theoretic', to the extent that it involves the construction of state trajectories and the analysis of

13.15 Notes for Chapter 13

the monotonicity properties of state trajectories. It has its roots in interpretations of 'generalized' solutions to Hamilton Jacobi equations and Lyapunov inequalities by Clarke (later distilled in [73]) and Aubin [9], predating viscosity solutions, used to establish basic properties of verification functions and Lyapunov functions in a nonsmooth setting, and also in early work (referenced in Subbotin's book [184]) in the games theory literature.

Frankowska, in [107], gave an independent system theoretic proof of Barron and Jensen's characterization of lower semi-continuous value functions (for $F(t, x)$'s continuous w.r.t. time), based on applications of a weak invariance theorem forward in time and a strong invariance theorem backward in time. She also provided alternative Hamilton Jacobi equation solution concepts that can be used in this characterization.

In Sect. 13.3, we follow the system theoretic approach, to achieve a characterization of lower semi-continuous value functions. The analysis is based, as in [107], on application of invariance theorems in forward and backward time. But, because our analysis is based on the invariance theorems of Clarke et al. [84], in which inward pointing conditions are expressed in terms of proximal normals to the constraint set, the characterization is in terms of 'proximal' solutions to the Hamilton Jacobi equation. (The class of lower semi-continuous functions that are proximal solutions is larger that of lower semi-continuous viscosity solutions.)

Initial advances in the application of system theoretic methods to link lower semi-continuous value functions and generalized solutions to the Hamilton Jacobi equation, were based on the assumption that the t dependence of the velocity set $F(t, x)$ was continuous.

The value function characterizations in this chapter make use of recent research by Bernis et al. [23] (building on [28]), in which $t \to F(t, x)$ is assumed to belong to the class of discontinuous multifunctions that have bounded variation. As shown by Frankowska et al. [112, 113], it is possible to establish a characterization of lower semi-continuous value functions, for problems in which $t \to F(t, x)$ is merely measurable. But this characterization is only partial, because it is expressed only in terms of a subset of the class of generalized solutions, namely solutions that also satisfy a regularity condition ('epicontinuity'). See also [171, 198].

Inverse verification theorems give conditions under which the existence of a verification function is guaranteed, confirming the optimality of a putative minimizer. In contrast to investigations of the relation of the value function and solutions to the Hamilton Jacobi equation, where we seek as unrestrictive conditions on value functions as possible for validity of the stated characterizations, improvements to inverse verification theorems are achieved by *restricting* the class of verification functions considered: they narrow the search for a verification function to suit a particular application. The link between normality hypotheses and the existence of Lipschitz continuous verification functions for dynamic optimization problems with end-point constraints was established in [62] and elaborated in [73].

No mention is made in this chapter of a powerful approach to the derivation of inverse verification theorems, based on the application of convex analysis. It allows free end-times and general end-point and pathwise state constraints. The approach, which has roots in L.C. Young's theory of flows for parametric problems in the calculus of variations, has been employed by Fleming, Ioffe, Klötzler, Vinter and others. (See [193].) Here, in effect, a dual problem to the dynamic optimization problem at hand is set up, in which a Hamilton Jacobi inequality features as a constraint. In this context, a verification theorem involves a sequence of smooth functions (a maximizing sequence for the dual problem) satisfying a Hamilton Jacobi inequality.

The sensitivity relation, providing an interpretation of the costate trajectory, the Hamiltonian and gradients of the value function, is implicit in the early heuristic 'dynamic programming' proof of the maximum principle outlined in Chap. 1. A rigorous analysis, confirming the sensitivity relation for some costate trajectory, selected from the set of all possible costate trajectories satisfying nonsmooth necessary conditions, and the Hamiltonian is given in [192]. An earlier proof of the sensitivity relation involving the costate variable alone appears in [78]. Improved versions of these relations are given in [35].

Soner [182] achieved a characterization of the value functions for dynamic optimization problems with state constraints as unique continuous viscosity-type solutions to the 'steady state' Hamilton Jacobi equation. An inward pointing hypothesis is invoked both to assist the uniqueness analysis and to ensure the value function has the required continuity properties.

The system theoretic approach was followed by Frankowska and Vinter [111], to characterize lower semi-continuous as unique proximal solutions to the Hamilton Jacobi equation under an outward pointing condition, in the case $F(t, x)$ is continuous in the time variable. (See also [112].) Bettiol and Vinter [28] provided a similar characterization for problems in which $t \to F(t, x)$ is discontinuous, with bounded variation. This characterization was subsequently refined in [23], avoiding using asymptotic proximal subgradients and covering problems involving a running cost is only continuous w.r.t. the state variable.

The discussion of semi-concavity in Sect. 13.9 focuses on its role as an additional regularity condition to ensure the uniqueness of Lipschitz continuous, almost everywhere solutions to the value function. This however only scratches the surface. Broader implications of value function semi-concavity are addressed in, for example, [52] and [51].

Study of the Hamilton Jacobi equation for dynamic optimization problems with infinite horizon has been undertaken mainly in the context of viscosity solutions. Soner's paper [182], which allow for state constraints, was an important early contribution. (See [17] and the references therein.) In Sect. 13.10 we see how control theoretic techniques developed for finite horizon problems can be simply adapted also to allow for infinite horizons.

Minimum time problems comprise perhaps the most extensively covered subclass of dynamic optimization problems in which the end-time is a choice variable ('free end-time' problems). Studies of the minimum time function (the value function in this case) centred on the Hamilton Jacobi equation feature prominently the viscosity solutions literature (see [17] and the references therein.) Sect. 13.11, in which control theoretic methods are used to provide a Hamilton Jacobi equation

13.15 Notes for Chapter 13

characterization of the minimum time problem in the absence of any controllability hypotheses, makes use of ideas in [208].

In Sect. 13.12, relations between viscosity solutions methods and system theoretic methods are explored (see [22] for an extension to the case in which we have also an integral cost which is merely continuous w.r.t. the state variable). Section 13.13 conveys the flavour of viscosity solution methods by applying them to prove the uniqueness of viscosity solutions to an infinite horizon problem. This section draws on material from [17].

References

1. L. Ambrosio, O. Ascenzi, G. Buttazzo, Lipschitz regularity for minimizers of integral functionals with highly discontinuous integrands. J. Math. Anal. Appl. **142**, 301–316 (1989)
2. A. Arutyunov, *Optimality Conditions: Abnormal and Degenerate Problems*, vol. 526 (Springer, Science and Business Media, New York, 2000)
3. A.V. Arutyunov, S.M. Aseev, Investigation of the degeneracy phenomenon of the maximum principle for optimal control problems with state constraints. SIAM J. Control Optim. **35**, 930–952 (1997)
4. A. Arutyunov, D. Karamzin, A Survey on regularity conditions for state-constrained optimal control problems and the non-degenerate maximum principle. J. Optim. Theory Appl. **184**, 697–723 (2020)
5. A.V. Arutyunov, S.M. Aseev, V.I. Blagodatskikh, First order necessary conditions in the problem of optimal control of a differential inclusion with phase constraints. Russ. Acad. Sci. Sb. Math. **79**, 117–139 (1994)
6. S.M. Aseev, *Method of Smooth Approximations in the Theory of Necessary Conditions for Differential Inclusions*. Mathematics, vol. 61 (Izvestiya, Moscow, 1997)
7. H. Attouch, *Variational Convergence for Functions and Operators*. Applicable Mathematics Series (Pitman, London, 1984)
8. J.-P. Aubin, *Applied Functional Analysis* (Wiley Interscience, New York, 1978)
9. J.-P. Aubin, *Contingent Derivatives of Set-Valued Maps and Existence of Solutions to Nonlinear Inclusions and Differential Inclusions*. Advances in Mathematics, Supplementary Studies, ed. by L. Nachbin (1981), pp. 160–232
10. J.-P. Aubin, *Viability Theory* (Birkhäuser, Boston, 1991)
11. J.-P. Aubin, A. Cellina, *Differential Inclusions* (Springer-Verlag, Berlin, 1984)
12. J.-P. Aubin, F.H. Clarke, Monotone invariant solutions to differential inclusions. J. Lond. Math. Soc. **16**, 357–366 (1977)
13. J.-P. Aubin, I. Ekeland, *Applied Nonlinear Analysis* (Wiley-Interscience, New York, 1984)
14. J.-P. Aubin, H. Frankowska, *Set-Valued Analysis* (Birkhäuser, Boston, 1990)
15. R.J. Aumann, Integrals of set-valued functions. J. Math. Anal. Appl. **12**, 15–26 (1967)
16. J. Ball, V. Mizel, One-dimensional variational problems whose minimizers do not satisfy the Euler-Lagrange equation. Arch. Rat. Mech. Anal. **90**, 325–388 (1985)
17. M. Bardi, I. Capuzzo-Dolcetta, *Optimal Control and Viscosity Solutions of Hamilton-Jacobi Equations* (Birkhäuser, Boston, 1997)
18. G. Barles, *Solutions de Viscosité des Equations de Hamilton-Jacobi*. Mathématiques et Applications, vol. 17 (Springer, Paris, 1994)

19. E.N. Barron, R. Jensen, Semicontinuous viscosity solutions for Hamilton-Jacobi equations with convex Hamiltonians. Commun. Partial Differ. Equ. **15**, 1713–1742 (1990)
20. R. Bellman, *Dynamic Programming* (Princeton University Press, Princeton, 1957)
21. L.D. Berkovitz, *Optimal Control Theory* (Springer-Verlag, New York, 1974)
22. J. Bernis, P. Bettiol, Hamilton-Jacobi equation for state constrained Bolza problems with discontinuous time dependence: a characterization of the value function. J. Convex Anal. **30**(2), 591–614 (2023)
23. J. Bernis, P. Bettiol, R.B. Vinter, Solutions to the Hamilton-Jacobi equation for state constrained Bolza problems with discontinuous time dependence. J. Differ. Equ. **341**, 589–619 (2022)
24. D. Bessis, Y.S. Ledyaev, R.B. Vinter, Dualization of the Euler and Hamiltonian inclusions. Nonlinear Anal. **43**, 861–882 (2001)
25. P. Bettiol, C. Mariconda, A new variational inequality in the calculus of variations and Lipschitz regularity of minimizers. J. Differ. Equ. **268**(5), 2332–2367 (2020)
26. P. Bettiol, C. Mariconda, A Du Bois-Reymond convex inclusion for nonautonomous problems of the calculus of variations and regularity of minimizers. Appl. Math. Optim. **83**, 2083–2107 (2021)
27. P. Bettiol, R.B. Vinter, Trajectories satisfying a smooth state constraint: improved estimates. IEEE Trans. Automat. Control **56**(5), 1090–1096 (2011)
28. P. Bettiol, R. B. Vinter, The Hamilton Jacobi equation for optimal control problems with discontinuous time dependence. SIAM J. Control Optim. **55**, 1199–1225 (2017)
29. P. Bettiol, R.B. Vinter, L^∞ Estimates on trajectories confined to a closed subset, for control systems with bounded time variation. Math. Program. Ser. B Math. Program. **168**(1–2), 201–228 (2018)
30. P. Bettiol, R.B. Vinter, Improved first order necessary conditions for dynamic optimization problems with free end-times (to appear)
31. P. Bettiol, R.B. Vinter, Non-degenerate necessary conditions for non-convex differential inclusions (to appear)
32. P. Bettiol, A. Bressan, R. Vinter, On trajectories satisfying a state constraint: $W^{1,1}$ estimates and counterexamples. SIAM J. Control Optim. Equ. **48**, 2267–2281 (2011)
33. P. Bettiol, H. Frankowska, R.B. Vinter, L^∞ estimates on trajectories confined to a closed subset. J. Differ. Equ. **252**, 1912–1933 (2012)
34. P. Bettiol, A. Boccia, R.B. Vinter, Stratified necessary conditions for differential inclusions with state constraints. SIAM J. Control Optim. **51**(5), 3903–3917 (2013)
35. P. Bettiol, H. Frankowska, R.B. Vinter, Improved sensitivity relations in state constrained optimal control. Appl. Math. Optim. **71**(2), 353–377 (2015)
36. P. Bettiol, N. Khalil, R.B. Vinter, Normality of generalized Euler-Lagrange conditions for state constrained optimal control problems. J. Convex Anal. **23**(1), 291–311 (2016)
37. P. Bettiol, M. Quincampoix, R.B. Vinter, Existence and characterization of the values of two player differential games with state constraints. Appl. Math. Optim. **80**(3), 765–799 (2019)
38. J.T. Betts, *Practical Methods for Optimal Control and Estimation Using Nonlinear Programming*, 2nd edn. (Cambridge University Press, New York, 2009)
39. P. Billingsley, *Convergence of Probability Measures* (John Wiley and Sons, New York, 1968)
40. A. Boccia, M.D.R. de Pinho, R.B. Vinter, Optimal control problems with mixed and pure state constraints. SIAM J. Control Optim. **54**, 3061–3083 (2016)
41. A. Boccia, R.B. Vinter, The maximum principle for optimal control problems with time delays. SIAM J. Control Optim. **55**, 2905–2935 (2017)
42. V.G. Boltyanskii, The maximum principle in the theory of optimal processes (in Russian). Dokl. Akad. Nauk SSSR, **119**, 1070–1073 (1958)
43. J.M. Borwein, D. Preiss, A smooth variational principle with applications to subdifferentiability and to differentiability of convex functions. Trans. Am. Math. Soc. **303**, 517–527 (1987)
44. J.M. Borwein, Q.J. Zhu, A survey of subdifferential calculus with applications. Nonlinear Anal. Theory Methods Appl. **38**(6), 687–773 (1999)

References

45. J.M. Borwein, Q.J. Zhu, *Techniques of Variational Analysis*. CMS Books in Mathematics/Ouvrages de Mathématiques de la SMC, vol. 20 (Springer, New York, 2005)
46. A. Bressan, On the intersection of a Clarke cone with a Boltyanskii cone. SIAM J. Control Optim. **45**, 2054–2064 (2007)
47. A. Bressan, G. Facchi, Trajectories of differential inclusions with state constraints. J. Differ. Equ. **250**(4), 2267–2281 (2011)
48. A.E. Bryson, *Dynamic Optimization* (Addison Wesley, New York, 1999)
49. A.E. Bryson, Y.-C. Ho, *Applied Optimal Control* (Blaisdell, New York, 1969). And (in revised addition) Halstead Press (a division of John Wiley and Sons), New York, 1975
50. R. Bulirsch, F. Montrone, H.J. Pesch, Abort landing in the presence of windshear as a minimax optimal problem, Part 1: necessary conditions and Part 2: multiple shooting and homotopy. J. Opt. Theory Appl. **70**, 1–23, 223–254 (1991)
51. P. Cannarsa, L. Rifford, Semiconcavity results for optimal control problems admitting no singular minimizing controls. Ann. I. H. Poincaré **52**, 773–802 (2008)
52. P. Cannarsa, C. Sinestrari, *Semiconcave Functions, Hamilton-Jacobi Equations, and Optimal Control* (Birkhäuser, New York, 2004)
53. C. Castaing, M. Valadier, *Convex Analysis and Measurable Multifunctions*. Springer Lecture Notes in Mathematics, vol. 580 (Springer-Verlag, New York, 1977)
54. L. Cesari, *Optimization - Theory and Applications: Problems with Ordinary Differential Equations* (Springer Verlag, New York, 1983)
55. C.W. Clark, *Mathematical Bioeconomics: The Optimal Management of Renewable Resources* (Wiley, New York, 1990)
56. F.H. Clarke, *Necessary Conditions for Nonsmooth Problems in Optimal Control and the Calculus of Variations*. Ph.D. Dissertation, University of Seattle, Washington, 1973
57. F.H. Clarke, Generalized gradients and applications. Trans. Am. Math. Soc. **205**, 247–262 (1975)
58. F.H. Clarke, Necessary conditions for a general control problem, in *Calculus of Variations and Control Theory*, ed. by D.L. Russell (Academic Press, New York, 1976), pp. 259–278
59. F.H. Clarke, The maximum principle under minimal hypotheses. SIAM J. Control Optim. **14**, 1078–1091 (1976)
60. F.H. Clarke, A new approach to Lagrange multipliers. Math. Oper. Res. **1**, 165–174 (1976)
61. F.H. Clarke, Optimal solutions to differential inclusions. J. Optim. Theory Appl. **19**, 469–478 (1976)
62. F.H. Clarke, The applicability of the Hamilton-Jacobi verification technique, in *Proceedings of the 10th IFIP Conference on New York, 1981*, ed. by R.F. Drenick, F. Kozin. System Modeling and Optimization Series, vol. 38 (Springer-Verlag, New York, 1982), pp. 88–94
63. F.H. Clarke, Perturbed optimal control problems. IEEE Trans. Automat. Control **31**, 535–542 (1986)
64. F.H. Clarke, Methods of dynamic and nonsmooth optimization, in *CBMS/NSF Regional Conference Series in Applied Mathematics*, vol. 57 (SIAM, Philadelphia, 1989)
65. F.H. Clarke, *Optimization and Nonsmooth Analysis* (Wiley-Interscience, New York, 1983). Reprinted as vol. 5 of Classics in Applied Mathematics, SIAM, Philadelphia, PA, 1990
66. F.H. Clarke, An indirect method in the calculus of variations. Trans. Am. Math. Soc. **336**, 655–673 (1993)
67. F.H. Clarke, Necessary conditions in dynamic optimization. Mem. Am. Math. Soc. **173** (2005)
68. F.H. Clarke, *Functional Analysis, Calculus of Variations and Optimal Control*. Graduate Texts in Mathematics, vol. 264 (Springer Verlag, New York, 2013)
69. F.H. Clarke, M.D.R. de Pinho, Optimal control problems with mixed constraints. SIAM J. Control Optim. **48**(7), 4500–4524 (2010)
70. F.H. Clarke, Y.S. Ledyaev, Mean value inequalities. Proc. Am. Math. Soc. **122**, 1075–1083 (1994)
71. F.H. Clarke, Y.S. Ledyaev, Mean value inequalities in Hilbert space. Trans. Am. Math. Soc. **344**, 307–324 (1994)

72. F.H. Clarke, P.D. Loewen, The value function in optimal control: sensitivity, controllability and time-optimality. SIAM J. Control Optim. **24**, 243–263 (1986)
73. F.H. Clarke, R.B. Vinter, Local optimality conditions and Lipschitzian solutions to the Hamilton-Jacobi equation. SIAM J. Control Optim. **21**, 856–870 (1983)
74. F.H. Clarke, R.B. Vinter, On the conditions under which the Euler equation or the maximum principle hold. Appl. Math. Optim. **12**, 73–79 (1984)
75. F.H. Clarke, R.B. Vinter, Regularity properties of solutions to the basic problem in the calculus of variations. Trans. Am. Math. Soc. **289**, 73–98 (1985)
76. F.H. Clarke, R.B. Vinter, Existence and regularity in the small in the calculus of variations. J. Differ. Equ. **59**, 336–354 (1985)
77. F.H. Clarke, R.B. Vinter, Regularity of solutions to variational problems with polynomial Lagrangians. Bull. Polish Acad. Sci. **34**, 73–81 (1986)
78. F.H. Clarke, R.B. Vinter, The relationship between the maximum principle and dynamic programming. SIAM J. Control Optim. **25**, 1291–1311 (1987)
79. F.H. Clarke, R.B. Vinter, Optimal multiprocesses. SIAM J. Control Optim. **27**, 1072–1091 (1989)
80. F.H. Clarke, R.B. Vinter, Applications of optimal multiprocesses. SIAM J. Control Optim. **27**, 1048–1071 (1989)
81. F.H. Clarke, R.B. Vinter, A regularity theory for problems in the calculus of variations with higher order derivatives. Trans. Am. Math. Soc. **320**, 227–251 (1990)
82. F.H. Clarke, R.B. Vinter, Regularity properties of optimal controls. SIAM J. Control Optim. **28**, 980–997 (1990)
83. F.H. Clarke, P.D. Loewen, R.B. Vinter, Differential inclusions with free time. Ann. l'Inst. Henri Poincaré (An. Nonlin.) **5**, 573–593 (1989)
84. F.H. Clarke, Y.S. Ledyaev, R.J. Stern, P.R. Wolenski, Qualitative properties of trajectories of control systems: a survey. J. Dyn. Control Syst. **1**, 1–47 (1995)
85. F.H. Clarke, Y.S. Ledyaev, R.J. Stern, P.R. Wolenski, *Nonsmooth Analysis and Control Theory*. Graduate Texts in Mathematics, vol. 178 (Springer-Verlag, New York, 1998)
86. F.H. Clarke, L. Rifford, R.J. Stern, Feedback in state constrained optimal control. ESAIM Control Optim. Calc. Var. **7**, 97–133 (2002)
87. F.H. Clarke, Y. Ledyaev, M.D.R. de Pinho, An extension of the Schwarzkopf multiplier rule in optimal control. SIAM J. Control Optim. **49**(2), 599–610 (2011)
88. M.G. Crandall, P.L. Lions, Viscosity solutions of Hamilton-Jacobi equations. Trans. Am. Math. Soc. **277**, 1–42 (1983)
89. G. Debreu. Integration of correspondences, in *Proceedings of the Fifth Berkeley Symposium on Mathematical Statistics and Probability II*, ed. by L. LeCam, J. Neyman, E. Scott (University of California Press, Berkeley, California, 1967), pp. 351–372
90. A.V. Dmitruk, Maximum principle for a general optimal control problem with state and regular mixed constraints. Comput. Math. Model. **4**, 364–377 (1993)
91. A.V. Dmitruk, On the development of Pontryagin's maximum principle in the works of A.Ya. Dubovitskii and A.A. Milyutin. Control Cybern. **38**(4A), 923–957 (2009)
92. A.V. Dmitruk, N.P. Osmolovskii, Local minimum principle for an optimal control problem with a nonregular mixed constraint. SIAM J. Control Optim. **60**(4), 1919–1941 (2022)
93. K. Diemling, *Multivalued Differential Equations* (de Gruyter, Berlin, 1992)
94. A.L. Dontchev, W.W. Hager, A new approach to Lipschitz continuity in state constrained optimal control. Syst. Control Lett. **35**, 137–143 (1998)
95. A.L. Dontchev, I. Kolmanovsky, On regularity of optimal control. Recent developments in optimization, in *Proceedings of the French-German Conference on Optimization*, ed. by R. Durier, C. Michelot. Lecture Notes in Economics and Mathematical Systems, vol. 429 (Springer Verlag, Berlin, 1995), pp. 125–135
96. A.J. Dubovitskii, A.A. Milyutin, Extremal problems in the presence of restrictions. U.S.S.R. Comput. Math. Math. Phys. **5**, 1–80 (1965)
97. N. Dunford, J.T. Schwartz, *Linear Operators. Part I: General Theory* (Interscience, London, 1958). Reissued by Wiley-Interscience (Wiley Classics Library), 1988

98. I. Ekeland, On the variational principle. J. Math. Anal. Appl. **47**, 324–353 (1974)
99. I. Ekeland, Nonconvex minimization problems. Bull. Am. Math. Soc. (N.S.) **1**, 443–474 (1979)
100. H.O. Fattorini, *Infinite-Dimensional Optimization and Control Theory* (Cambridge University Press, Cambridge, 1999)
101. M.M.A. Ferreria, R.B. Vinter, When is the maximum principle for state-constrained problems degenerate? J. Math. Anal. Appl. **187**, 432–467 (1994)
102. A.F. Filippov, On certain questions in the theory of optimal control. J. SIAM Control (A) **1**, 76–84 (1962)
103. A. Filippov, Classical solutions of differential equations with multivalued right-hand side. SIAM J. Control **5**, 609–621 (1967)
104. W.H. Fleming, R.W. Rishel, *Deterministic and Stochastic Optimal Control* (Springer Verlag, New York, 1975)
105. W.H. Fleming, H.M. Soner, *Controlled Markov Processes and Viscosity Solutions* (Springer Verlag, New York, 1993)
106. H. Frankowska, The maximum principle for an optimal solution to a differential inclusion with end point constraints. SIAM J. Control Optim. **25**, 145–157 (1987)
107. H. Frankowska, Lower semicontinuous solutions of the Hamilton-Jacobi equation. SIAM J. Control Optim. **31**, 257–272 (1993)
108. H. Frankowska, M. Mazzola. Discontinuous solutions of Hamilton-Jacobi-Bellman equation under state constraints. Calc. Var. Partial Differ. Equ. **46**, 725–747 (2013)
109. H. Frankowska, M. Mazzola. On relations of the adjoint state to the value function for optimal control problems with state constraints. Nonlinear Differ. Equ. Appl. **20**, 361–383 (2013)
110. H. Frankowska, F. Rampazzo, Filippov's and Filippov-Wazewski's theorems on closed domains. J. Differ. Equ. **161**, 449–478 (2000)
111. H. Frankowska, R.B. Vinter, Existence of neighbouring feasible trajectories: applications to dynamic programming for state constrained optimal control problems. J. Optim. Theory Applic. **104**(1), 21–40 (2000)
112. H. Frankowska, M. Plaskacz, Semicontinuous solutions of Hamilton Jacobi equations with state constraints, in *Differential Inclusions and Optimal Control*. Lecture Notes in Nonlinear Analysis, vol. 2, ed. by J. Andres, L. Gorniewicz, P. Nistri (Juliusz Schauder Center for Nonlinear Studies, Toruá, 1998), pp. 145–161
113. H. Frankowska, M. Plaskacz, T. Rzezuchowski, Measurable viability theorems and the Hamilton-Jacobi-Bellman equation. J. Differ. Equ. **116**, 265–305 (1995)
114. D. Gay, The AMPL modeling language: An aid to formulating and solving optimization problems. Numer. Anal. Optim. **57**, 95–116 (2015)
115. M. Giaquinta, *Multiple Integrals in the Calculus of Variations and Nonlinear Elliptic Systems* (Princeton University Press, Princeton, 1983)
116. M.R. Gonzalez, Sur l'existence d'une solution maximale de l'equation de Hamilton Jacobi. C. R. Acad. Sci. **282**, 1287–1280 (1976)
117. R.V. Gramkrelidze, Optimal control processes with restricted phase coordinates (in Russian). Izv. Akad. Nauk SSSR Ser. Math. **24**, 315–356 (1960)
118. R.V. Gramkrelidze, On sliding optimal regimes. Soviet Math. Dokl. **3**, 559–561 (1962)
119. G. Haddad, Monotone trajectories of differential inclusions and functional differential inclusions with memory. Isr. J. Math. **39**, 83–100 (1981)
120. W.W. Hager, Lipschitz continuity for constrained processes. SIAM J. Control and Optim. **17**, 321–338 (1979)
121. H. Halkin, On the necessary condition for the optimal control of nonlinear systems. J. Analyse Math. **12**, 1–82 (1964)
122. H. Halkin, Implicit functions and optimization problems without continuous differentiability of the data. SIAM J. Control **12**, 239–236 (1974)
123. A.D. Ioffe, Approximate subdifferentials and applications I: the finite dimensional theory. Trans. Am. Math. Soc. **281**, 389–416 (1984)

124. A.D. Ioffe, Necessary conditions in nonsmooth optimization. Math. Oper. Res. **9**, 159–189 (1984)
125. A.D. Ioffe, Proximal analysis and approximate subdifferentials. J. Lond. Math. Soc. **41**, 175–192 (1990)
126. A.D. Ioffe, A Lagrange multiplier rule with small convex-valued subdifferentials for nonsmooth problems of mathematical programming involving equality and non-functional constraints. Math. Prog. **72**, 137–145 (1993)
127. A.D. Ioffe, Euler-Lagrange and Hamiltonian formalisms in dynamic optimization. Trans. Am. Math. Soc. **349**, 2871–2900 (1997)
128. A.D. Ioffe, *Variational Analysis of Regular Mappings* (Springer-Verlag, Berlin, 2017)
129. A.D. Ioffe, On generalized Bolza problem and its application to dynamic optimization. J. Optim. Theory Appl. **182**, 285–309 (2019)
130. A.D. Ioffe, Maximum principles for optimal control problems with differential inclusions. SIAM J. Control Optim. (to appear)
131. A.D. Ioffe, R.T. Rockafellar, The Euler and Weierstrass conditions for nonsmooth variational problems. Calc. Var. Partial Differ. Equ. **4**, 59–87 (1996)
132. A.D. Ioffe, V.M Tihomirov, *Theory of Extremal Problems* (North-Holland, Amsterdam, 1979)
133. B. Kaskosz, S. Lojasiewicz, A maximum principle for generalized control systems Nonlinear Anal. Theory, Methods Appl. **19**, 109–130 (1992)
134. A.Y. Kruger, Properties of generalized differentials. Siberian Math. J. **26**, 822–832 (1985)
135. K. Kuratowski. *Topologie, I and II*. Panstowowe Wyd Nauk, Warsaw. (Academic Press, New York, 1966)
136. G. Lebourg, Valeur moyenne pour gradient généralisé. C. R. Acad. Sci. Paris **281**, 795–797 (1975)
137. U. Ledzewicz, H. Shättler, *Geometric Optimal Control Theory, Methods and Examples*. Series on Interdisciplinary and Applied Mathematics, vol. 38 (Springer, Berlin, 2012)
138. U. Ledzewicz, H. Shättler, *Optimal Control for Mathematical Models of Cancer Therapies*. Series on Interdisciplinary and Applied Mathematics, vol. 42 (Springer, Berlin, 2015)
139. U. Ledzewicz, H. Maurer, H. Schättler, Optimal and suboptimal protocols for a mathematical model for tumor anti-angiogenesis in combination with chemotherapy. Math. Biosci. Eng. **8**(2), 307–328 (2011)
140. X. Li, J. Yong, *Optimal Control Theory for Infinite Dimensional Systems* (Birkhäuser, Boston, 1995)
141. P.D. Loewen, Optimal Control via Nonsmooth Analysis, in *CRM Proceedings of the Lecture Notes*, vol. 2 (American Mathematical Society, Providence, 1993)
142. P.D. Loewen, A mean value theorem for Fréchet subgradients. Nonlinear Anal. Theory Methods Appl. **23**, 1365–1381 (1995)
143. P.D. Loewen, R.T. Rockafellar, Optimal control of unbounded differential inclusions. SIAM J. Control Optim. **32**, 442–470 (1994)
144. P.D. Loewen, R.T. Rockafellar, New necessary conditions for the generalized problem of Bolza. SIAM J. Control Optim. **34**, 1496–1511 (1996)
145. P.D. Loewen, R.B. Vinter, Pontryagin-type necessary conditions for differential inclusion problems. Syst. Control Lett. **9**, 263–265 (1987)
146. K. Malanowski, On the regularity of solutions to optimal control problems for systems linear with respect to control variable. Arch. Auto. i Telemech. **23**, 227–241 (1978)
147. K. Makowski, L.W. Neustadt, Optimal Control Problems with mixed control-phase variable equality and inequality constraints. SIAM J. Control **12**, 184–228 (1974)
148. K. Malanowski, H. Maurer, Sensitivity analysis for state constrained optimal control problems. Discrete Contin. Dyn. Syst. **4**, 241–272 (1998)
149. C. Mariconda, Equi-Lipschitz minimizing trajectories for non coercive, discontinuous, non convex Bolza controlled-linear optimal control problems. Trans. Am. Math. Soc. Ser. B **8**, 899–947 (2021)
150. D.Q. Mayne, E. Polak, An exact penalty function algorithm for control problems with control and terminal equality constraints, Parts I and II. J. Optim. Theory Appl. **32**, 211–246, 345–363 (1980)

151. D.Q. Mayne, R.B. Vinter, First-order necessary conditions in optimal control. J. Optim. Theory Appl. **189**, 716–743 (2021)
152. K. Miao, R.B. Vinter, Optimal control of a growth/consumption model, in *Optimal Control, Applications and Methods*, vol. 42 (2021), pp. 1672–1688
153. A.A. Milyutin, N.P. Osmolovskii, *Calculus of Variations and Optimal Control*. Translations of Mathematical Monographs (American Mathematical Society, Providence, 1998)
154. B.S. Mordukhovich, Maximum principle in the optimal time control problem with non-smooth constraints. Prikl. Math. Mech. **40**, 1004–1023 (1976)
155. B.S. Mordukhovich, Metric approximations and necessary optimality conditions for general classes of nonsmooth extremal problems. Soviet Math. Doklady **22**, 526–530 (1980)
156. B.S. Mordukhovich, Complete characterization of openness, metric regularity, and Lipschitz properties of multifunctions. Trans. Am. Math. Soc. **340**, 1–36 (1993)
157. B.S. Mordukhovich, Generalized differential calculus for nonsmooth and set-valued mappings. J. Math. Anal. Appl. **183**, 250–282 (1994)
158. B.S. Mordukhovich, Discrete approximations and refined Euler-Lagrange conditions for non-convex differential inclusions. SIAM J. Control Optim. **33**, 882–915 (1995)
159. B.S. Mordukhovich, *Variational Analysis and Generalized Differentiation*, vols. I and II (Springer Verlag, Berlin, 2006)
160. B.S. Mordukhovich, *Variational Analysis and Applications*. Springer Monographs in Mathematics (2018)
161. J. Murray, Some optimal control problems in cancer chemotherapy with a toxicity limit. Math. Biosci. **100**(1), 49–67 (1990)
162. M. Nagumo, Uber die lage der integralkurven gewöhnlicher differentialgleichungen. Proc. Phys. Math. Soc. Jpn. **24**, 551–559 (1942)
163. L.W. Neustadt, A general theory of extremals. J. Comput. Sci. **3**, 57–92 (1969)
164. L.W. Neustadt, *Optimization* (Princeton University Press, Princeton, 1976)
165. D. Offin, *A Hamilton-Jacobi approach to the differential inclusion problem*. M.Sc. Thesis, University of British Columbia, Canada, 1979
166. E. Polak, *Optimization: Algorithms and Consistent Approximations* (Springer Verlag, New York, 1997)
167. L.S. Pontryagin, V.G. Boltyanskii, R.V. Gamkrelidze, E.F. Mischenko, *The Mathematical Theory of Optimal Processes* (Transl.), ed. by K.N. Tririgoff, L.W. Neustadt (Wiley, New York, 1962)
168. R. Pytlak, Runge-Kutta based procedure for optimal control of differential-algebraic equations. J. Opt. Theory Appl. **97**, 675–705 (1998)
169. R. Pytlak, R.B. Vinter, A feasible directions algorithm for optimal control problems with state and control constraints: convergence analysis. SIAM J. Control Optim. **36**, 1999–2019 (1998)
170. F. Rampazzo, R.B. Vinter, Nondegenerate necessary conditions for nonconvex optimal control problems with state constraints. SIAM J. Control Optim. **39**(4), 989–1007 (2000)
171. A.E. Rapaport, R.B. Vinter, Invariance properties of time measurable differential inclusions and dynamic programming. J. Dyn. Control Syst. **2**, 423–448 (1996)
172. R.T. Rockafellar, Generalized Hamiltonian equations for convex problems of Lagrange. Pac. J. Math. **33**, 411–428 (1970)
173. R.T. Rockafellar, Existence and duality theorems for convex problems of Bolza. Trans. Am. Math. Soc. **159**, 1–40 (1971)
174. R.T. Rockafellar, Existence theorems for general control problems of Bolza and Lagrange. Adv. Math. **15**, 312–333 (1975)
175. R.T. Rockafellar, Proximal subgradients, marginal values and augmented Lagrangians in nonconvex optimization. Math. Oper. Res. **6**, 424–436 (1982)
176. R.T. Rockafellar, Equivalent subgradient versions of Hamiltonian and Euler-Lagrange equations in variational analysis. SIAM J. Control Optim. **34**, 1300–1314 (1996)
177. R.T. Rockafellar, R.J.-B. Wets, *Variational Analysis*. Grundlehren der Mathematischen Wissenschaften, vol. 317 (Springer-Verlag, New York, 1998)

178. J. Rosenblueth, R.B. Vinter, *Relaxation procedures for time delay systems*. J. Math. Anal. Appl. **162**, 542–563 (1991)
179. J.D.L. Rowland, R.B. Vinter, Dynamic optimization problems with free time and active state constraints. SIAM J. Control Optim. **31**, 677–697 (1993)
180. O. Sharomi, T. Malik, Optimal control in epidemiology. Ann. Oper. Res. **251**, 55–71 (2017)
181. G.N. Silva, R.B. Vinter, Necessary conditions for optimal impulsive control problems. SIAM J. Control Optim. **35**, 1829–1846 (1998)
182. H.M. Soner, Optimal control with state-space constraints. SIAM J. Control Optim. **24**, 552–561 (1986)
183. G.V. Smirnov, Discrete approximations and optimal solutions to differential inclusions. Cybernetics **27**, 101–107 (1991)
184. A.I. Subbotin, *Generalized Solutions of First-Order PDEs* (Birkhäuser, Boston, 1995)
185. H.J. Sussmann, Geometry and optimal control, in *Mathematical Control Theory*, ed. by J. Baillieul, J.C. Willems (Springer Verlag, New York, 1999), pp. 140–194
186. H.J. Sussmann, Set separation, approximating multicones, and the Lipschitz maximum principle. J. Differ. Equ. **243**, 448–488 (2007)
187. H.J. Sussmann, J.C. Willems, 300 years of optimal control: from the brachystochrone to the maximum principle. IEEE Control Syst. Mag. **17**(3), 32–44 (1997)
188. P.L. Tan, H. Maurer, J. Kanesan, J.H. Chuah, Optimal control of cancer chemotherapy with delays and state constraints. J. Optim. Theory Appl. **194**(3), 749–770 (2022)
189. L. Tonelli, Sur une méthode direct du calcul des variations. Rend. Circ. Math. Palermo **39**, 233–264 (1915)
190. L. Tonelli, *Fondamenti di Calcolo delle Variazioni*, vols. 1 and 2 (Zanichelli, Bologna, 1921/1923)
191. J. L. Troutman, *Variational Calculus with Elementary Convexity* (Springer-Verlag, New York, 1983)
192. R.B. Vinter, New results on the relationship between dynamic programming and the maximum principle. Math. Control Signals Syst. **1**, 97–105 (1988)
193. R.B. Vinter, Convex duality and nonlinear optimal control. SIAM J. Control Optim. **31**, 518–538 (1993)
194. R.B. Vinter, *Optimal Control* (Birkhäuser, Boston, 2000)
195. R.B. Vinter, The Hamiltonian inclusion for non-convex velocity sets. SIAM J. Control and Optim. **52**(2), 1237–125 (2014)
196. R.B. Vinter, G. Pappas, A maximum principle for non-smooth optimal control problems with state constraints. J. Math. Anal. Appl. **89**, 212–232 (1982)
197. R.B. Vinter, F.L. Pereira, A maximum principle for optimal processes with discontinuous trajectories. SIAM J. Control Optim. **26**, 205–229 (1988)
198. R.B. Vinter, P. Wolenski, Hamilton Jacobi theory for optimal control problems with data measurable in time. SIAM J. Control Optim. **28**, 1404–1419 (1990)
199. R.B. Vinter, H. Zheng, The extended Euler-Lagrange condition for nonconvex variational problems. SIAM J. Control Optim. **35**, 56–77 (1997)
200. R.B. Vinter, H. Zheng, Necessary conditions for optimal control problems with state constraints. Trans. Am. Math. Soc. **350**, 1181–1204 (1998)
201. R.B. Vinter, H. Zheng, Necessary conditions for free end-time, measurably time dependent optimal control problems with state constraints. J. Set Valued Anal. **8**, 1–19 (2000)
202. A. Wächler, L. Biegler, On the implementation of a primal-dual interior point filter line search algorithm for large-scale nonlinear programming. Math. Program. **106**, 25–57 (2006)
203. D.H. Wagner, Survey of measurable selection theorems. SIAM. J. Control Optim. **15**, 859–903 (1977)
204. J. Warga, *Optimal Control of Differential and Functional Equations* (Academic Press, New York, 1972)
205. J. Warga, Derivate containers, inverse functions and controllability in *Calculus of Variations and Control Theory*, ed. by D.L. Russell (Academic Press, New York, 1976)

206. J. Warga, Fat homeomorphisms and unbounded derivate containers. J. Math. Anal. Appl. **81**, 545–560 (1981)
207. J. Warga, Optimization and controllability without differentiability assumptions. SIAM J. Control Optim. **21**, 239–260 (1983)
208. P.R. Wolenski, Y. Zhuang, Proximal analysis and the minimal time function. SIAM J. Control Optim. **36**(3), 1048–1072 (1998)
209. L.C. Young, *Lectures on the Calculus of Variations and Optimal Control Theory* (Saunders, Philadelphia, 1991)
210. J. Zabzyk, *Mathematical Control Theory: An Introduction* (Birkhäuser, Boston, 1992)
211. D. Zagrodny, Approximate mean value theorem for upper subderivatives. Nonlinear Anal. Theory Methods Appl. **12**, 1413–1428 (1988)
212. J. Zimmer, *Essential Results of Functional Analysis* (University of Chicago Press, Chicago, 1990)

Index

A
Action principle, 13, 120
Adjoint equation, 30
Aumann
    integrals of multifunctions, 266
    measurable selection theorem, 104

B
Bolza problem, generalized, 107
Borwein and Preiss' theorem, 133
Brachistochrone problem, 10

C
Calculus of variations, 9
Carathéodory
    verification technique, 26
Carathéodory function, 97
Caristi's fixed point theorem, 145
Castaing Representation, 99
Chain rule, 51
    nonsmooth, 230
Compactness of trajectories theorem, 259
Complementary slackness condition, 475
Conjugate of a function, 109
Constraint qualification, Mangasarian Fromovitz, 234
Continuity, $-\mu$, 478

D
Decrease Properties of Control Systems, 761
Directional derivative
    generalized, 178
    lower dini, 668
Dirichlet's principle, 12, 120
Distance estimate theorem, 271
Distance function, 184–189
Dualization, 391
Du Bois-Reymond Lemma, 15
Dunford Pettis theorem, 257
Dynamic programming, 36–41, 759

E
Ekeland's theorem, 53, 54, 123–127
Elasticity, nonlinear, 66
Epicontinuous function, 391
    near a point, 392
Epi-Lipschitz function, local, 392
Erdmann
    first condition, 17
    second condition, 17
Erdmann – Du Bois-Reymond Condition
    nonsmooth, 639
Essential values, 424–427, 466
Euler equation, 14
    higher order, 325
Euler Lagrange condition, 17, 21
    generalized, free end-time, 428
Exact penalization, 54, 121
Existence of minimizers, 22–593

F
Fermat's principle, 120
Field of extremals, 29

Filippov
  Generalized Existence Theorem, 247
  Generalized selection theorem, 104
  selection theorem, 105
Finite element methods, 12
Free end-time problems, 412
  state constraints, 498, 567
Fuzzy calculus, 216

G
Galerkin methods, 12
Gronwall's inequality, 254
  Proximal, 293

H
Hamilton condition, 20, 21
Hamiltonian, 18
  constancy condition, 17
  inclusion, 60
  un-maximized, 33
Hamilton Jacobi equation, viii, 25–30, 667
  lower Dini solutions, 669
  proximal solutions, 669
Hausdorff distance function, 273

I
Inf convolution, 128, 207
Infinite Horizon Control, 88
Infinite horizon problems, 728
  minimizers existence, 292
Invariance
  Strong, Autonomous, 660
  Strong, Time Varying, 662
  Weak, Autonomous, 655
  Weak, Time Varying, 658

J
Jacobi condition, 25
Jensen's inequality, 110

L
Lagrange multiplier rule, 232–236
Lavrentiev phenomenon, 597
Lebesgue essential value, 466
Legendre condition, 25
Legendre-Fenchel transformation, 109

Legendre transformation, 19
Lipschitz continuity, criteria, 190, 194
Lower semi-continuous envelope, 742

M
Magasarian Fromowitz condition, 529
Marginal function, 106, 220
Maximum orbit transfer problem, 2
Maximum principle, 30–36, 68, 295
  Clarke nonsmooth, 297
  Clarke's nonsmooth, boundary points of
    reachable sets, 326
  free end-time, 448
  Gamkrelidze Form, 537
  multiprocess, 325
Max rule, 51
  finite family of functions, 231
  infinite family of functions, 236
Mean value inequality, 167–173
  proximal, 168
  two sided, 172
Measures, weak* convergence, 477
Minimax theorem, 134–145
  Aubin one-sided, 137
  Von Neumann, 136
Minimizer
  L^∞ local, 68, 474
  $W^{1,1}$ local, 474
  $W^{1,1}$ local minimizer relative to a
    multifunction B, 342
  $W^{1,1}$ local, free end-time, 414
  $W^{1,1}$ local, free end-time process, 448
  $W^{1,s}$ local, 405
  intermediate, 292
Moreau-Yosida envelope, 128
Multifunction
  lower/upper semi-continuous, 94
  measurable, 95
  measurable selection, 103
Multiplier rule, 52, 55

N
Necessary conditions
  Kaskosz Lojasiewicz-Type, 404
  mixed constraints, 524
  mixed constraints, free end-time, 540
  non-degenerate, 574
Nonsmooth analysis, 45, 59, 150–216
  optimal control, 41

Normal cone, 151–155
  of convex analysis, 155
  limiting, 46, 152, 194
  proximal, 45, 151
  strict, 152, 194

P

Picard iteration, 248
Poisson's equation, 12
Principle of optimality, 653
Product Rule, 232
Proper function, 109
Proximal normal inequality, 151, 162

R

Rademacher's theorem, 49
Regularity, 592–638
  Tonelli, 598, 600
Regular point, tonelli, 599
Relaxation theorem, 267
Running cost problem, 759

S

Saddlepoint, 135
Sensitivity, 40, 698
Set Convergence, 92
Snell's Laws, 11
State augmentation, 299
Stegall's theorem, 133
Strictly Differentiable Function, 543
Subdifferential, 156–161
  asymptotic limiting, 158
  asymptotic proximal, 158
  asymptotic strict, 158
  characterization, 173–177
  Clarke, 180
  of convex analysis, 161

limiting, 47, 157
limiting, indefinite integral function, 425
partial limiting, 224
proximal, 47, 157
strict, 157
Sub essential value, 424
Sum rule, 51, 226
Superdifferential
  limiting, indefinite integral function, 425
Super essential value, 424
Synthesis, optimal, 38, 652

T

Takahashi's minimization theorem, 145
Tangent cone, 194–201
  bouligand, 194
  clarke, 194
  clarke, interior, 201
Tonelli, direct method, 22–25
Tonelli set, 600
Transversality condition, 21
Truncation function, 247
Two point boundary value problem, 34

U

Upper semi-continuous envelope, 742

V

Value function, 28, 645
Verification function, 647
Verification theorem, 685–697
Viscosity solutions, 741
  comparison theorem, 757

W

Weierstrass condition, 18, 21

GPSR Compliance

The European Union's (EU) General Product Safety Regulation (GPSR) is a set of rules that requires consumer products to be safe and our obligations to ensure this.

If you have any concerns about our products, you can contact us on ProductSafety@springernature.com

In case Publisher is established outside the EU, the EU authorized representative is:

Springer Nature Customer Service Center GmbH
Europaplatz 3
69115 Heidelberg, Germany

Batch number: 08237845

Printed by Printforce, the Netherlands